MW00845978

Book companion website:

S^2: ss2.eecs.umich.edu

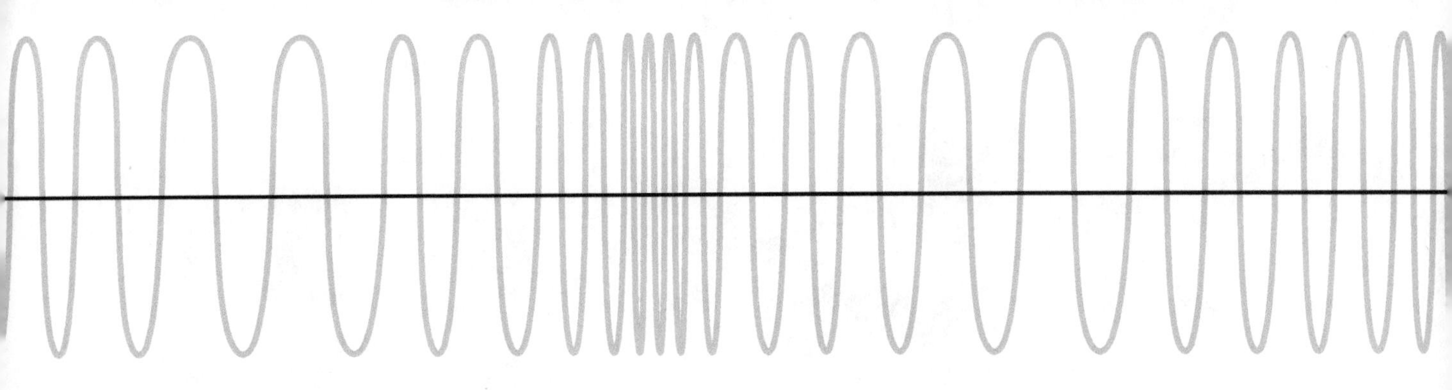

SIGNALS AND SYSTEMS:

Theory and Applications

Fawwaz T. Ulaby
The University of Michigan

Andrew E. Yagle
The University of Michigan

This book is published by Michigan Publishing under an agreement with the authors. It is made available free of charge in electronic form to any student or instructor interested in the subject matter.

Published in the United States of America by
Michigan Publishing
Manufactured in the United States of America

ISBN 978-1-60785-486-9 (hardcover)
ISBN 978-1-60785-487-6 (electronic)

This book is dedicated to the memories of

Mrs. Margaret Cunningham and **Mr. Lyle Cunningham,**

Professor Raymond A. Yagle, and

Mrs. Anne Yagle

Brief Contents

Chapter 1	Signals	1
Chapter 2	Linear Time-Invariant Systems	30
Chapter 3	Laplace Transform	85
Chapter 4	Applications of the Laplace Transform	131
Chapter 5	Fourier Analysis Techniques	192
Chapter 6	Applications of the Fourier Transform	253
Chapter 7	Discrete-Time Signals and Systems	346
Chapter 8	Applications of Discrete-Time Signals and Systems	420
Chapter 9	Filter Design, Multirate, and Correlation	474
Chapter 10	Image Processing, Wavelets, and Compressed Sensing	545
Appendix A	Symbols, Quantities, and Units	619
Appendix B	Review of Complex Numbers	621
Appendix C	Mathematical Formulas	625
Appendix D	MATLAB® and MathScript	628
Appendix E	A Guide to Using LabVIEW Modules	636
Appendix F	Answers to Selected Problems	637
Index		639

List of LabVIEW Modules

2.1 Convolution of Exponential Functions 47
2.2 Automobile Suspension Response 76
4.1 Oven Temperature Response 172
4.2 Inverted Pendulum Response 182
6.1 Notch Filter to Remove Sinusoid from Trumpet Signal 290
6.2 Comb Filter to Separate Two Trumpet Signals 291
6.3 Filtering a Noisy Trumpet Signal 304
8.1 Discrete-Time Frequency Response from Poles and Zeros 426
8.2 Discrete-Time Notch Filter to Eliminate One of Two Sinusoids 432
8.3 Discrete-Time Notch Filter to Eliminate Sinusoid from Trumpet Signal 432
8.4 Discrete-Time Comb Filter to Eliminate Periodic Signal from Sinusoid 438
8.5 Discrete-Time Comb Filter to Separate Two Trumpet Signals 439
8.6 Dereverberation of a Simple Signal 444
8.7 Denoising a Periodic Signal by Thresholding 456
8.8 Separating Two Trumpet Signals Using the DFT 457
8.9 Computing Spectra of Discrete-Time Periodic Signals Using the DTFS 460
8.10 Computing Continuous-Time Fourier Transforms Using the DFT 468
9.1 Discrete-Time Lowpass Filter Design Using Windowing 485
9.2 Spectrogram of Tonal Version of "The Victors" 492
9.3 Spectrogram of a Chirp Signal 492
9.4 Use of Autocorrelation to Estimate Period 533
9.5 Use of Cross-Correlation to Estimate Time Delay 533
10.1 Effect of Lowpass Filtering an Image 553
10.2 Denoising a Noisy Image Using Lowpass Filtering 560
10.3 Deconvolution from a Noisy Blurred Image Using Wiener Filter 569
10.4 Wavelet Transform of Shepp-Logan Phantom Using Haar Wavelets 593
10.5 Compression of Clown Image Using Daubechies Wavelets 595
10.6 Wavelet Denoising of Clown Image Using Daubechies Wavelets 597
10.7 Inpainting of Clown Image Using Daubechies Wavelets and IST Algorithm 608

Contents

Preface

Coming Attractions

Chapter 1 Signals 1

Overview 2
1-1 Types of Signals 3
1-2 Signal Transformations 6
1-3 Waveform Properties 9
1-4 Nonperiodic Waveforms 11
1-5 Signal Power and Energy 21
Summary 24
Problems 25

Chapter 2 Linear Time-Invariant Systems 30

Overview 31
2-1 Linear Time-Invariant Systems 31
2-2 Impulse Response 35
2-3 Convolution 40
2-4 Graphical Convolution 46
2-5 Convolution Properties 50
2-6 Causality and BIBO Stability 58
2-7 LTI Sinusoidal Response 61
2-8 Impulse Response of Second-Order LCCDEs 65
2-9 Car Suspension System 72
Summary 78
Problems 79

Chapter 3 Laplace Transform 85

Overview 86
3-1 Definition of the (Unilateral) Laplace Transform 86
3-2 Poles and Zeros 89
3-3 Properties of the Laplace Transform 90
3-4 Circuit Analysis Example 98
3-5 Partial Fraction Expansion 99
3-6 Transfer Function **H**(**s**) 106
3-7 Poles and System Stability 108
3-8 Invertible Systems 111
3-9 Bilateral Transform for Continuous-Time Sinusoidal Signals 113
3-10 Interrelating Different Descriptions of LTI Systems 114
3-11 LTI System Response Partitions 117
Summary 123
Problems 124

Chapter 4 Applications of the Laplace Transform 131

Overview 132
4-1 **s**-Domain Circuit Element Models 132
4-2 **s**-Domain Circuit Analysis 134
4-3 Electromechanical Analogs 140

4-4 Biomechanical Model of a Person Sitting in a Moving Chair 146
4-5 Op-Amp Circuits 149
4-6 Configurations of Multiple Systems 154
4-7 System Synthesis 157
4-8 Basic Control Theory 160
4-9 Temperature Control System 167
4-10 Amplifier Gain-Bandwidth Product 171
4-11 Step Response of a Motor System 174
4-12 Control of a Simple Inverted Pendulum on a Cart 178
Summary 183
Problems 183

Chapter 5 Fourier Analysis Techniques 192

Overview 193
5-1 Phasor-Domain Technique 193
5-2 Fourier Series Analysis Technique 195
5-3 Fourier Series Representations 197
5-4 Computation of Fourier Series Coefficients 198
5-5 Circuit Analysis with Fourier Series 213
5-6 Parseval's Theorem for Periodic Waveforms 216
5-7 Fourier Transform 218
5-8 Fourier Transform Properties 223
5-9 Parseval's Theorem for Fourier Transforms 230
5-10 Additional Attributes of the Fourier Transform 232
5-11 Phasor vs. Laplace vs. Fourier 235
5-12 Circuit Analysis with Fourier Transform 236
5-13 The Importance of Phase Information 238
Summary 243
Problems 244

Chapter 6 Applications of the Fourier Transform 253

Overview 254
6-1 Filtering a 2-D Image 254
6-2 Types of Filters 256
6-3 Passive Filters 263
6-4 Active Filters 272
6-5 Ideal Brick-Wall Filters 275
6-6 Filter Design by Poles and Zeros 278
6-7 Frequency Rejection Filters 281
6-8 Spectra of Musical Notes 287
6-9 Butterworth Filters 289
6-10 Denoising a Trumpet Signal 298
6-11 Resonator Filter 300
6-12 Modulation 303
6-13 Sampling Theorem 319
Summary 334
Problems 336

Chapter 7 Discrete-Time Signals and Systems 346

Overview 347
7-1 Discrete Signal Notation and Properties 348
7-2 Discrete-Time Signal Functions 351
7-3 Discrete-Time LTI Systems 356
7-4 Properties of Discrete-Time LTI Systems 359
7-5 Discrete-Time Convolution 363
7-6 The $\mathbf{z}$-Transform 366
7-7 Properties of the $\mathbf{z}$-Transform 369
7-8 Inverse $\mathbf{z}$-Transform 374
7-9 Solving Difference Equations with Initial Conditions 378
7-10 System Transfer Function 380
7-11 BIBO Stability of $\mathbf{H}(\mathbf{z})$ 381
7-12 System Frequency Response 384
7-13 Discrete-Time Fourier Series (DTFS) 389
7-14 Discrete-Time Fourier Transform (DTFT) 394
7-15 Discrete Fourier Transform (DFT) 400
7-16 Fast Fourier Transform (FFT) 407
7-17 Cooley–Tukey FFT 411
Summary 414
Problems 415

Chapter 8 Applications of Discrete-Time Signals and Systems 420

Overview 421
8-1 Discrete-Time Filters 421
8-2 Notch Filters 427

8-3 Comb Filters 434
8-4 Deconvolution and Dereverberation 439
8-5 Bilateral **z**-Transforms 445
8-6 Inverse Bilateral **z**-Transforms 447
8-7 ROC, Stability, and Causality 449
8-8 Deconvolution and Filtering Using the DFT 450
8-9 Computing Spectra of Periodic Signals 457
8-10 Computing Spectra of Nonperiodic Signals 462
Summary 469
Problems 469

Chapter 9 Filter Design, Multirate, and Correlation 474

Overview 475
9-1 Data Windows 475
9-2 Spectrograms 485
9-3 Finite Impulse Response (FIR) Filter Design 492
9-4 Infinite Impulse Response (IIR) Filter Design 503
9-5 Multirate Signal Processing 512
9-6 Downsampling 513
9-7 Upsampling 516
9-8 Interpolation 517
9-9 Multirate Signal Processing Examples 518
9-10 Oversampling by Upsampling 520
9-11 Audio Signal Processing 525
9-12 Correlation 527
9-13 Biomedical Applications 534
Summary 538
Problems 539

Chapter 10 Image Processing, Wavelets, and Compressed Sensing 545

Overview 546
10-1 Image Processing Basics 546
10-2 Discrete-Space Fourier Transform 549
10-3 2-D DFT 553
10-4 Downsampling and Upsampling of Images 554
10-5 Image Denoising 555
10-6 Edge Detection 559
10-7 Image Deconvolution 565
10-8 Overview of the Discrete-Time Wavelet Transform 569
10-9 Haar Wavelet Transform 572
10-10 The Family of Wavelet Transforms 577
10-11 Non-Haar Single-Stage Perfect Reconstruction 581
10-12 Daubechies Scaling and Wavelet Functions 584
10-13 2-D Wavelet Transform 590
10-14 Denoising by Thresholding and Shrinking 595
10-15 Compressed Sensing 599
10-16 Computing Solutions to Underdetermined Equations 601
10-17 Landweber Algorithm 604
10-18 Compressed Sensing Examples 606
Summary 614
Problems 614

Appendix A Symbols, Quantities, and Units 619

Appendix B Review of Complex Numbers 621

Appendix C Mathematical Formulas 625

Appendix D MATLAB® and MathScript 628

Appendix E A Guide to Using LabVIEW Modules 636

Appendix F Answers to Selected Problems 637

Index 639

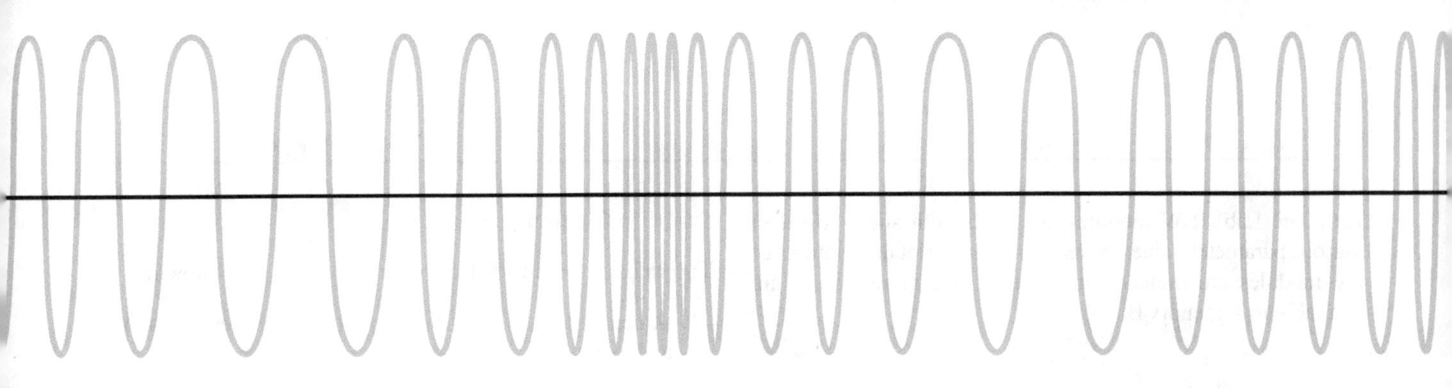

Preface

I hear and I forget. I see and I remember. I do and I understand.—Confucius

This is a signals and systems textbook with a difference: *Engineering applications of signals and systems are integrated into the presentation as equal partners with concepts and mathematical models*, instead of just presenting the concepts and models and leaving the student to wonder how it all relates to engineering.

Book Contents

The first six chapters of this textbook cover the usual basic concepts of continuous-time signals and systems, including the Laplace and Fourier transforms. Chapters 7 and 8 present the discrete-time version of Chapters 1–6, emphasizing the similarities and analogies, and often using continuous-time results to derive discrete-time results. The two chapters serve to introduce the reader to the world of discrete-time signals and systems. Concepts highlighted in Chapters 1–8 include: compensator feedback configuration (Ch. 4); energy spectral density, group delay, expanded coverage of exponential Fourier series (Ch. 5); filtering of images, Hilbert transform, single-sideband (SSB), zero and first-order hold interpolation (Ch. 6); the Cooley-Tukey FFT (Ch. 7); bilateral **z**-transform and use for non-minimum-phase deconvolution (Ch. 8).

Chapter 9 covers the usual concepts of discrete-time signal processing, including data windows, FIR and IIR filter design, multirate signal processing, and auto-correlation and cross-correlation. It also includes some nontraditional concepts, including spectrograms, application of multirate signal processing, and the musical circle of fifths to audio signal processing, and some biomedical applications of auto-correlation and cross-correlation.

Chapter 10 covers image processing, discrete-time wavelets (including the Smith-Barnwell condition and the Haar and Daubechies discrete-time wavelet expansions), and an introduction to compressed sensing. This is the first sophomore-junior level textbook the authors are aware of that allows students to apply compressed sensing concepts. Applications include: image denoising using 2-D filtering; image denoising using thresholding and shrinkage of image wavelet transforms; image deconvolution using Wiener filters; "valid" image deconvolution using ISTA; image inpainting; tomography and the projection-slice theorem, and image reconstruction from partial knowledge of 2-D DFT values. Problems allow students to apply these techniques to actual images and learn by doing, not by only reading.

LabVIEW Modules

An important feature of this book is the 32 LabView modules that support the more complicated examples in the text. The LabVIEW modules use GUIs with slides to select various parameter values. Screen shots and brief descriptions of the modules are included in the text, and more details are available in Appendix E.

Applications

The systems applications presented in this textbook include: spring-mass-damper automobile suspension systems, electromechanical analogues with specific application to a biomechanical model, **s**-domain circuit analysis, oven temperature control, motor system control, and inverted pendulum control.

Signals applications include: implementation of a notch filter to remove an interfering tone from the sound of a trumpet (which inspired the idea for the book cover), implementation of a comb filter to eliminate one of two trumpets playing two different notes simultaneously, and implementation of a resonator filter to remove most of the noise from a noisy trumpet signal. These signals applications are repeated using discrete-time signal processing, along with dereverberation, deconvolution (both real-time and batch), DFT-based noise filtering, and use of the DFT to compute spectra of both periodic (the trumpet signal) and non-periodic signals.

It amazes one of us (AEY) that almost all books on signal processing, even discrete-time signal processing, simply present the mathematical theory of the topic and show various methods for designing filters without ever explaining what they are for, let alone implementing them. Studying signal processing without filtering real-world signals is like studying a cookbook without ever turning on an oven or even a stove. This textbook implements the techniques presented on an actual trumpet signal (due to its simplicity and for unity of presentation) and provides the software programs so that students can experiment with altering parameters and see for themselves how the techniques work.

Website Contents (S²)

The website (see inside front cover) contains the following:

1. A detailed set of solutions to all of the concept questions and exercises.
2. A detailed description of each computer-based example, including a program listing and a sample output plot.
3. Forty .m files, fifteen .mat files, and thirty-two .vi files for examples, problems, and LabVIEW modules.
4. 32 interactive LabView modules.
5. Over 200 "Test Your Understanding" questions.

Textbook Suitability for Courses

This textbook is designed for a two-course sequence, with the first being a sophomore-level or early junior-level introductory course on signals and systems. The continuous-time material of Chapters 1 through 6 can be covered in one semester. Following the first two chapters on basics of signals and LTI systems, the book is structured such that odd-numbered chapters present theory and each even-numbered chapter presents applications of the theory presented in the chapter that preceded it.

The textbook is now suitable for any of the following:

1. A single semester course on continuous-time signals and systems (Chs. 1-6);
2. A two-quarter course on continuous-time signals and systems and their discrete-time counterparts (Chs. 1-8);
3. A two-semester sequence on continuous-time (Chs. 1–6) and then discrete-time (Chs. 7–10) signals and systems. In addition to the usual DSP topics (Chs. 7-9), the latter course can also cover image processing, discrete-time wavelets, and compressed sensing (Ch. 10).

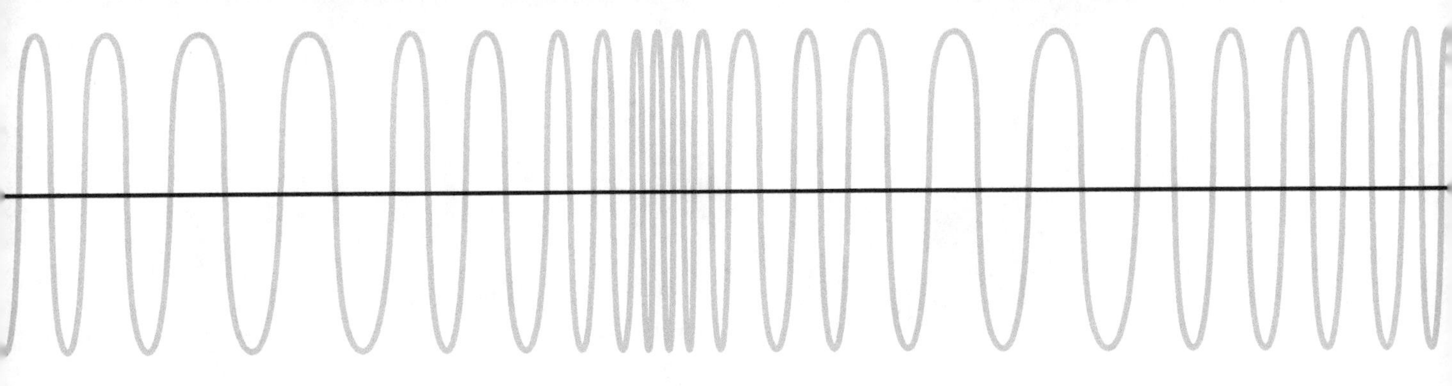

Acknowledgments

Of the many individuals who lent their support to make this book possible, Richard Carnes deserves singular credit and recognition. Richard—who is a professional pianist and music teacher—is the LaTeX wizard responsible for typing the manuscript and drafting most of its figures and charts. His attention to detail is ever-present throughout the book. We are also grateful to Ms. Rose Anderson for the elegant cover design she developed for the book.

The interactive LabVIEW modules used in this book were developed by Dr. Eric Luo. We are most grateful to Dr. Luo for his time, effort, and creative designs.

For their reviews of the manuscript and for offering many constructive criticisms, we are grateful to Professors Stephane Lafortune and Kim Winick of the University of Michigan, Ahmad Safaai-Jazi of Virginia Tech, and Ivan Selesnick of Polytechnic University of New York.

The manuscript was also scrutinized by a highly discerning group of University of Michigan students: Connor Field, Melissa Pajerski, Paul Rigge, Brett Kuprel, Kyle Harman, Nathan Sawicki, Mehrdad Damsaz, Nicholas Fava, and Doug Krussell. The book website was developed by Dr. Leland Pierce and his wife Janice Richards, and Peter Wiley helped with many of the illustrations. We are truly grateful to all of them.

FAWWAZ ULABY AND ANDREW YAGLE

Coming Attractions

The following examples are illustrative of the applications covered in this book.

Automobile Suspension System Model

Section 2-9 presents and analyzes a spring-mass-damper model for a car suspension system (Fig. P1(a)) and applies it to compute the displacement of the car as it travels over a 10 cm deep pothole. The plots in Fig. P1(b) provide a comparison of the car's response for an underdamped system and for a critically damped system. The underdamped response provides a much smoother ride over the pothole.

Biomechanical Model of a Person Sitting in a Moving Chair

Section 4-4 presents a biomechanical model of a person sitting in a moving chair, and analyzes it using electromechanical analogues and **s**-domain circuit analysis (Fig. P2).

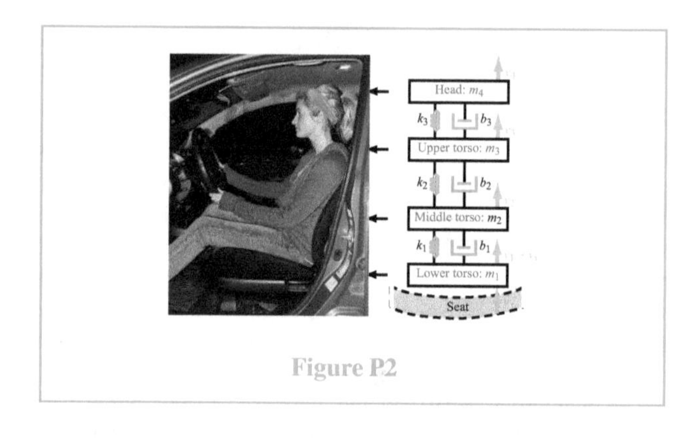

Figure P2

Oven Temperature Control System

Section 4-9 presents and analyzes an oven temperature control system. The system and its response to a sharp change in desired temperature, with and without feedback, are all shown in Fig. P3. Feedback increases the speed of the oven response.

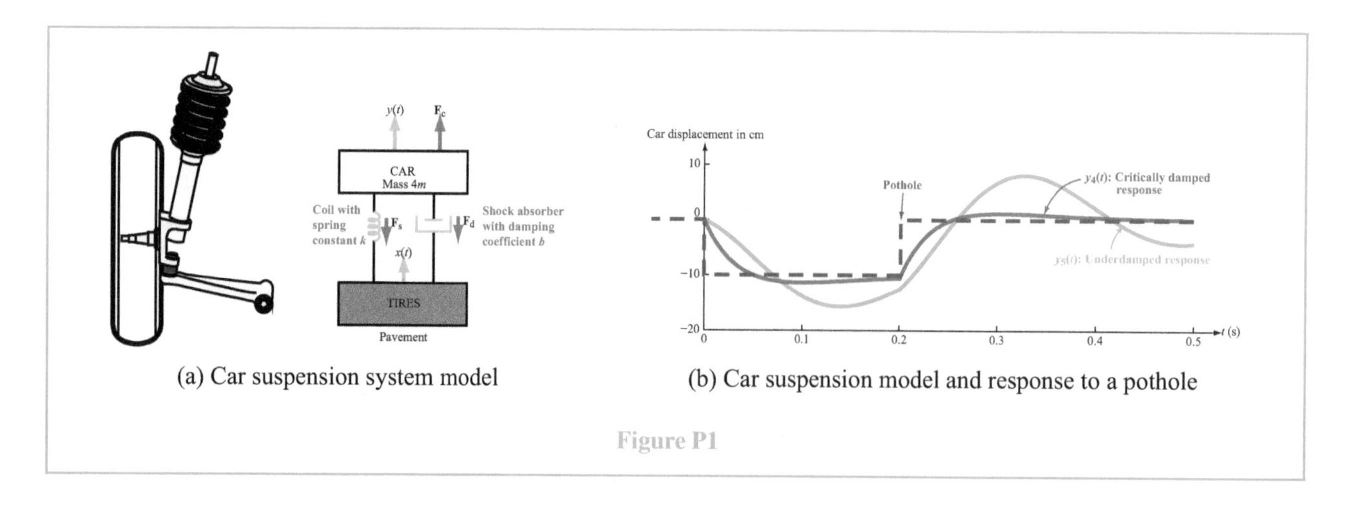

(a) Car suspension system model

(b) Car suspension model and response to a pothole

Figure P1

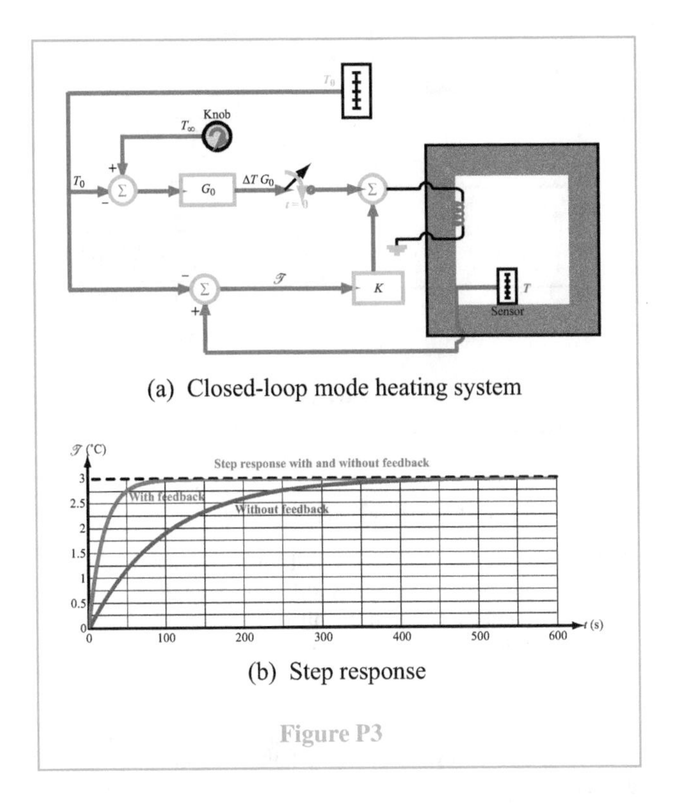

(a) Closed-loop mode heating system

(b) Step response

Figure P3

Step Response of a Motor System

Section 4-9 presents and analyzes a motor, both with and without feedback. The system is shown in Fig. P4. Feedback stabilizes the system.

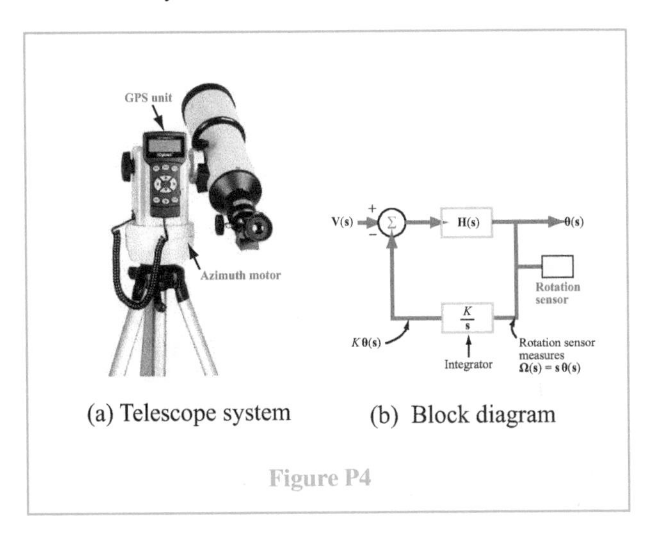

(a) Telescope system

(b) Block diagram

Figure P4

Control of Simple Inverted Pendulum on Cart

Stabilization of an inverted pendulum on a moving cart is a classic control problem that illustrates many aspects of feedback control. Section 4-12 presents and analyzes this problem, shown in Fig. P5(a), and Fig. P5(b) shows how the inverted pendulum responds to a tiny shove, using various types of feedback.

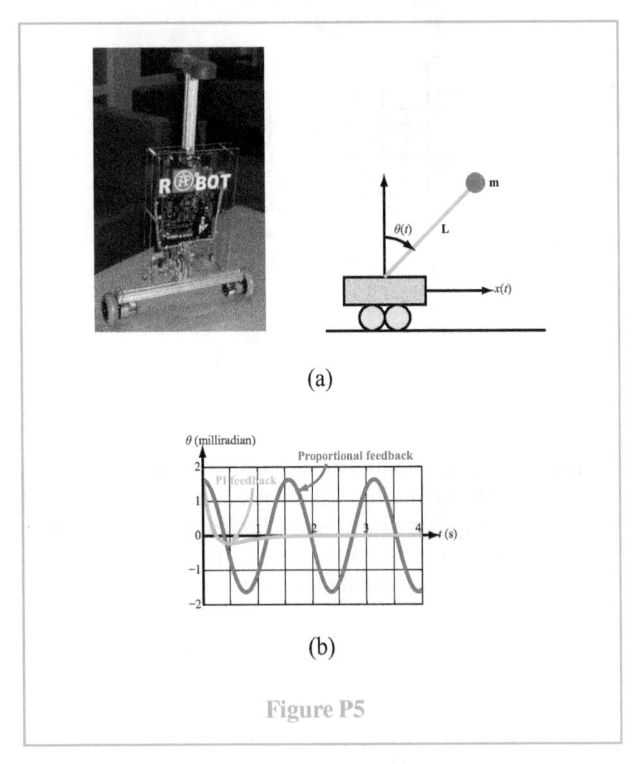

(a)

(b)

Figure P5

Spectra of Musical Notes

It is well known that the sound of a musical instrument playing a single note consists of a fundamental and overtones, which are pure tones (sinusoidal signals) at frequencies that are integer multiples of the reciprocal of the period of the signal. This is a good illustration of a Fourier series expansion. Section 6-8 presents spectra of musical notes. Figure P6 shows the periodic signal of a trumpet sound and its associated spectrum (note the fundamental at 494 Hz and the overtones, called harmonics in signal processing).

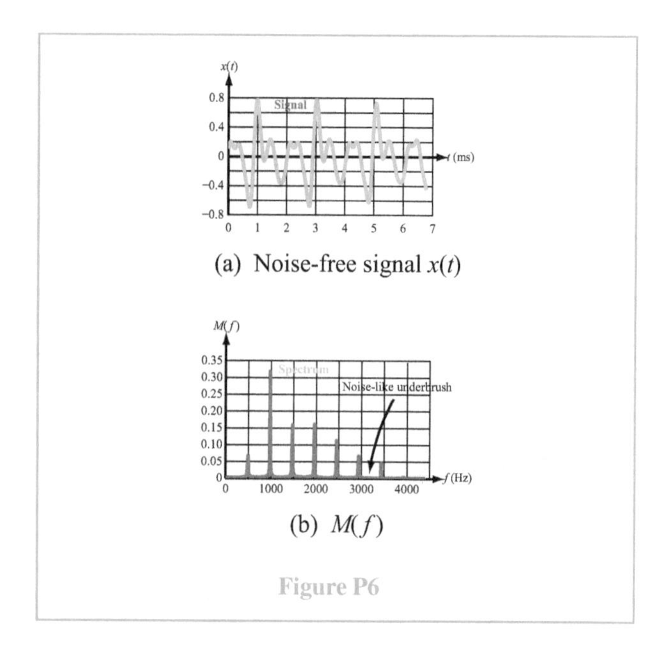

Figure P6

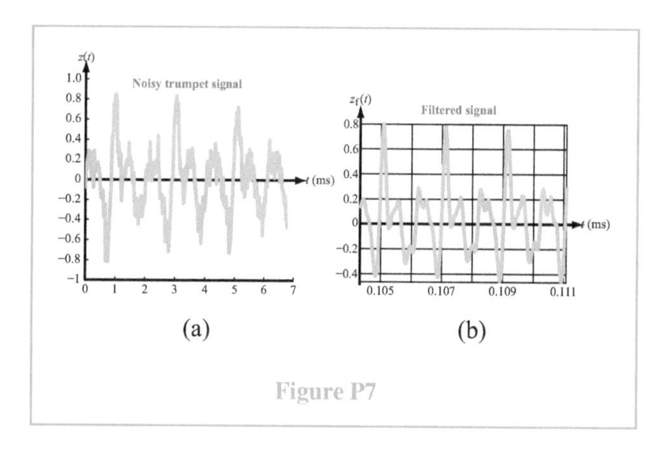

Figure P7

Denoising a Trumpet Signal

Section 6-9 presents an example of how to denoise a noisy trumpet signal. The result of applying a resonator filter, which passes only the harmonics and eliminates all frequencies between them, converts the noisy signal in Fig. P7(a) into the denoised signal in part (b) of the figure.

Frequency Division Multiplexing

Radio and analog cable TV all use frequency division multiplexing to transmit many different signals at different carrier frequencies through the air or a cable. The basic idea is illustrated in Fig. P8.

Sampling Theorem

The sampling theorem, presented in Section 6-13, makes digital signal and image processing possible. Many electronic devices are digital. Sampling a continuous-time signal makes its spectrum periodic in frequency, with a period equal to the sampling rate. A lowpass filter can then recover the original, continuous-time signal. The concept is illustrated in Fig. P9.

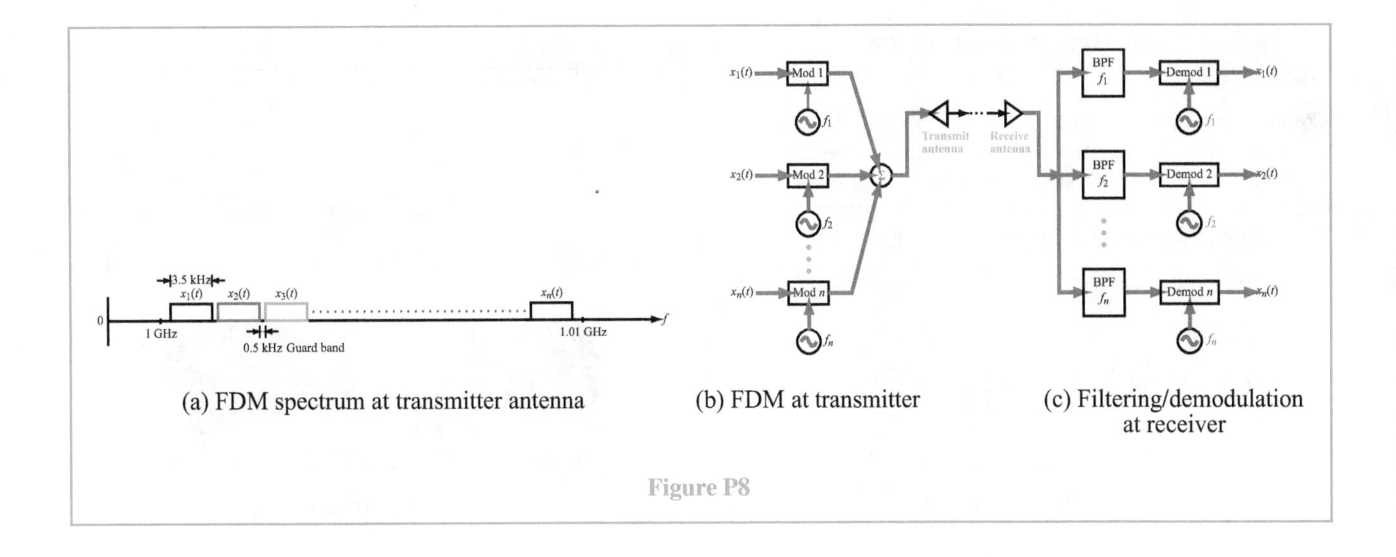

(a) FDM spectrum at transmitter antenna (b) FDM at transmitter (c) Filtering/demodulation at receiver

Figure P8

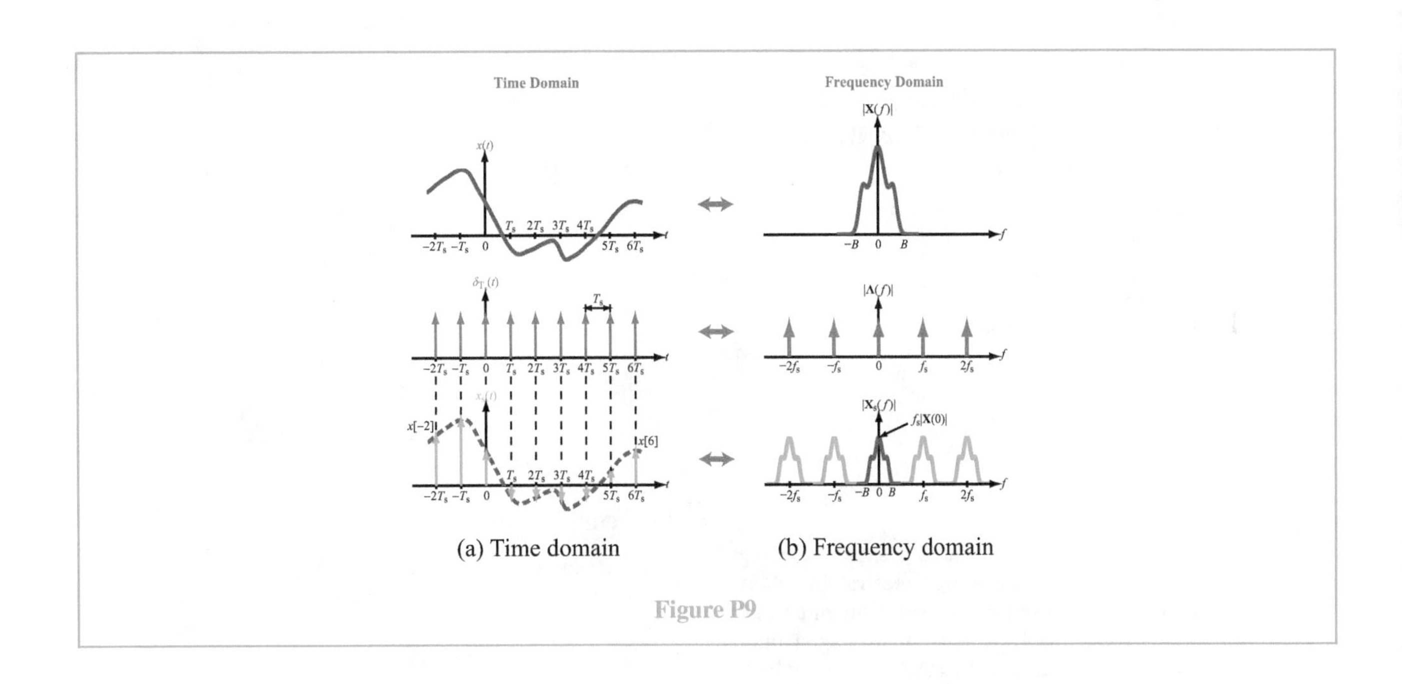

(a) Time domain (b) Frequency domain

Figure P9

Comb Filters for Removal of Periodic Interference

Comb filters, presented in Section 8-3, can be used to separate two simultaneously played trumpet notes because their spectra do not overlap (in most cases). A comb filter can be used to eliminate the harmonics of one trumpet while leaving the other unaffected, as illustrated by Fig. P10.

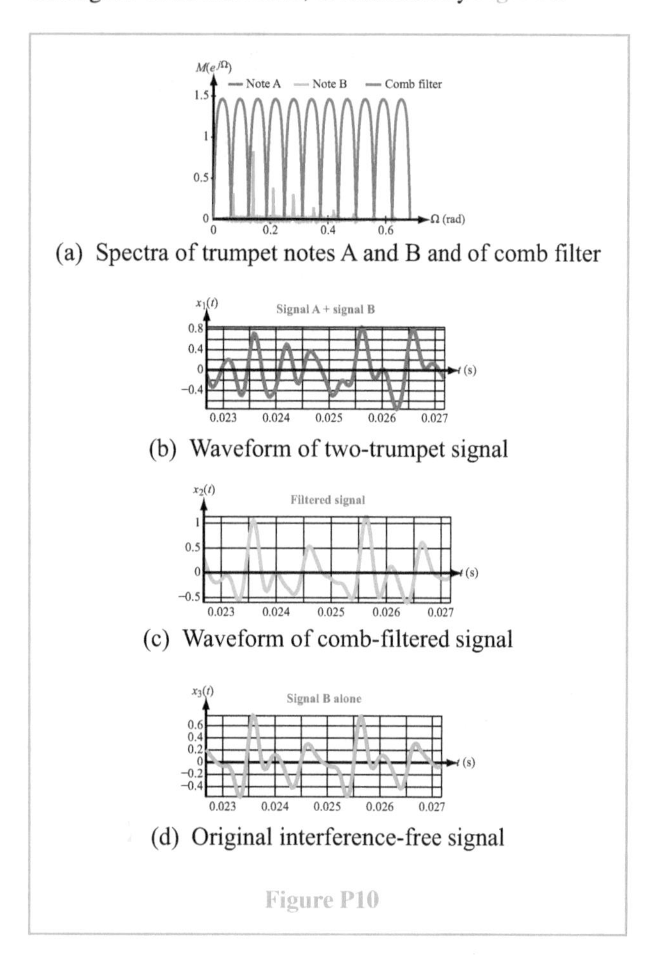

(a) Spectra of trumpet notes A and B and of comb filter

(b) Waveform of two-trumpet signal

(c) Waveform of comb-filtered signal

(d) Original interference-free signal

Figure P10

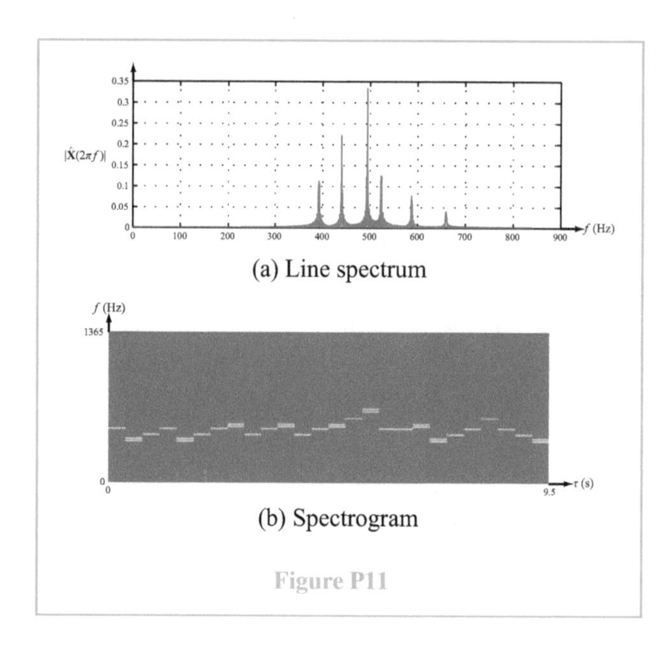

(a) Line spectrum

(b) Spectrogram

Figure P11

Multirate Signal Processing

Given a snippet of a trumpet playing a single note, multirate signal processing and the "Circle of Fifths" of music, shown in Fig. P12, can be used to alter the pitch of the note to any of the other eleven notes. Multirate signal processing is the discrete-time version of a variable-speed tape recorder or record player. It has many other uses as well.

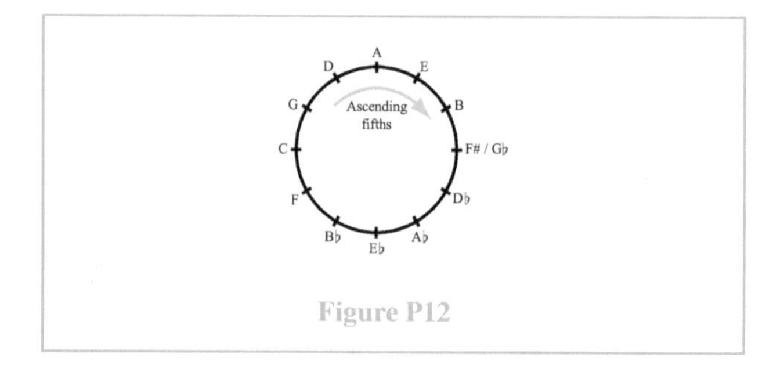

Figure P12

Spectrograms

The spectrogram of a signal is essentially a time-varying spectrum, produced by concatenating spectra of short segments of the signal. The spectrum and spectrum of a tonal version of "The Victors," the University of Michigan fight song, are shown in Fig. P11. The spectrum reveals which specific notes are included, not when they are played. But the spectrogram reveals which notes are played as well as when they are played.

Autocorrelation

Autocorrelation is the sum of products of the values of a signal with a time-delayed version of itself. It can be used to compute the period of a noisy periodic signal, as in the extremely noisy EKG signals in Fig. P13.

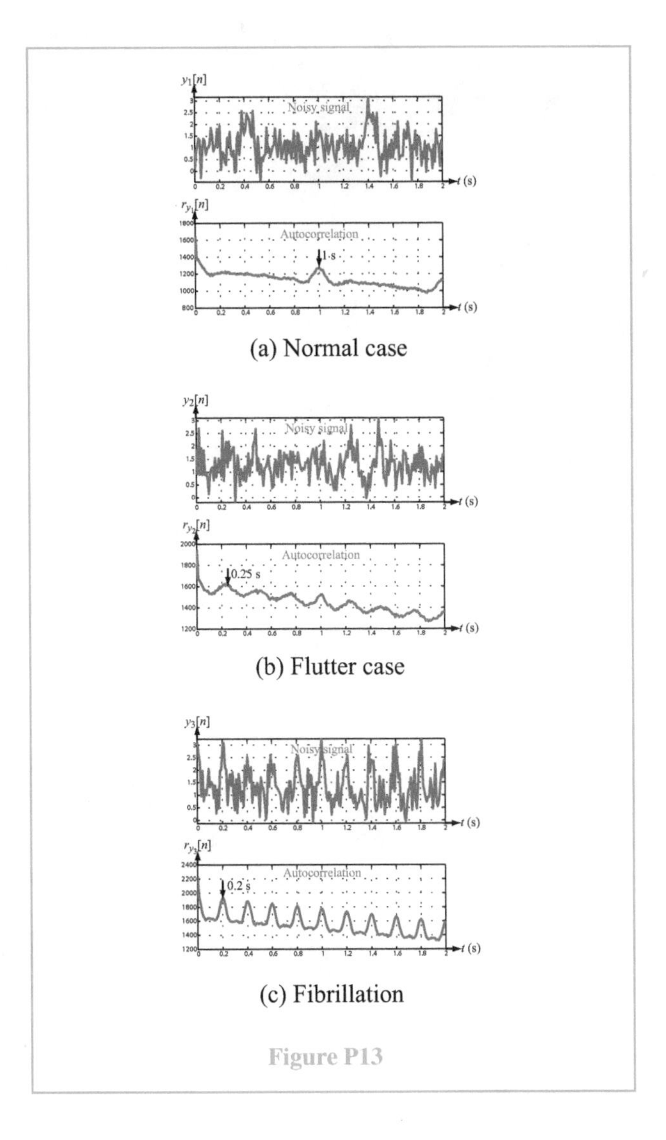

(a) Normal case

(b) Flutter case

(c) Fibrillation

Figure P13

Edge Detection in Images

Edge detection in images is presented in Section 10-6, an example of which is shown in Fig. P14.

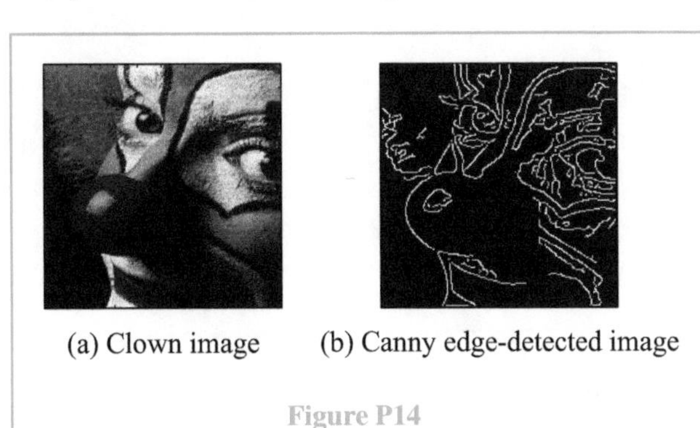

(a) Clown image (b) Canny edge-detected image

Figure P14

Deconvolution (Deblurring) of Images

Deconvolution is performed to undo the distorting effect of a system, such as an out-of-focus lens. A simple example, extracted from Section 10-7, is shown in Fig. P15.

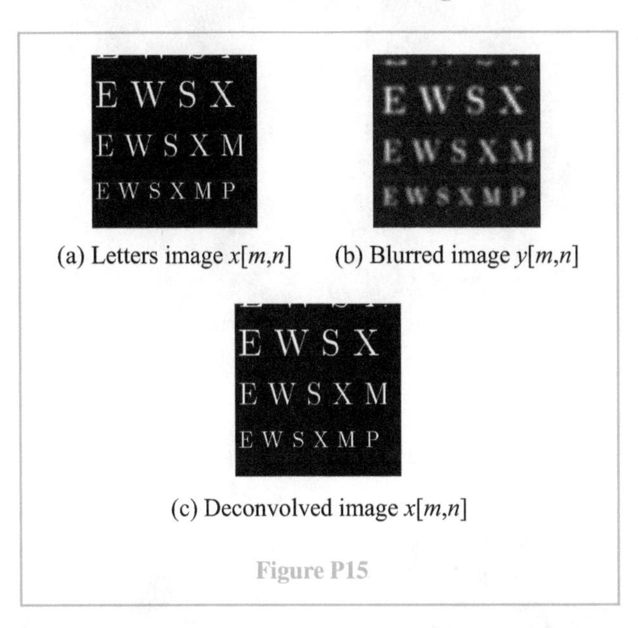

(a) Letters image $x[m,n]$ (b) Blurred image $y[m,n]$

(c) Deconvolved image $x[m,n]$

Figure P15

Wavelet-Based Denoising

The wavelet transform, presented in Sections 10-8 through 10-13, represents signals and images using basis functions that are localized in space or time, and also in frequency. The wavelet transform of a signal or image is sparse (mostly zero-valued), which makes it useful for compression, denoising, and reconstruction from few measurements.

Denoising an image using thresholding and shrinkage of its wavelet transform is illustrated in Fig. P16.

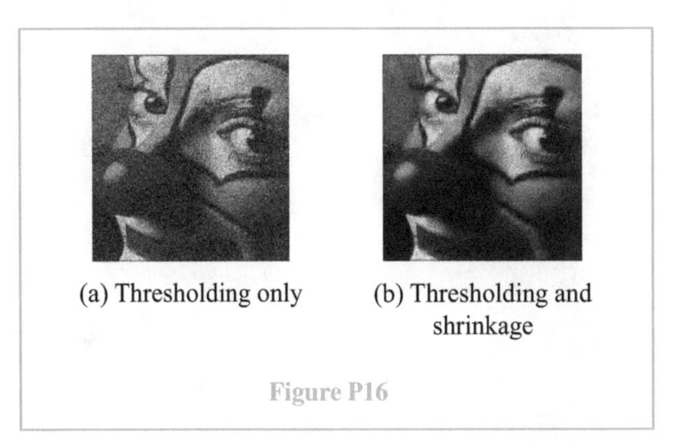

(a) Thresholding only (b) Thresholding and shrinkage

Figure P16

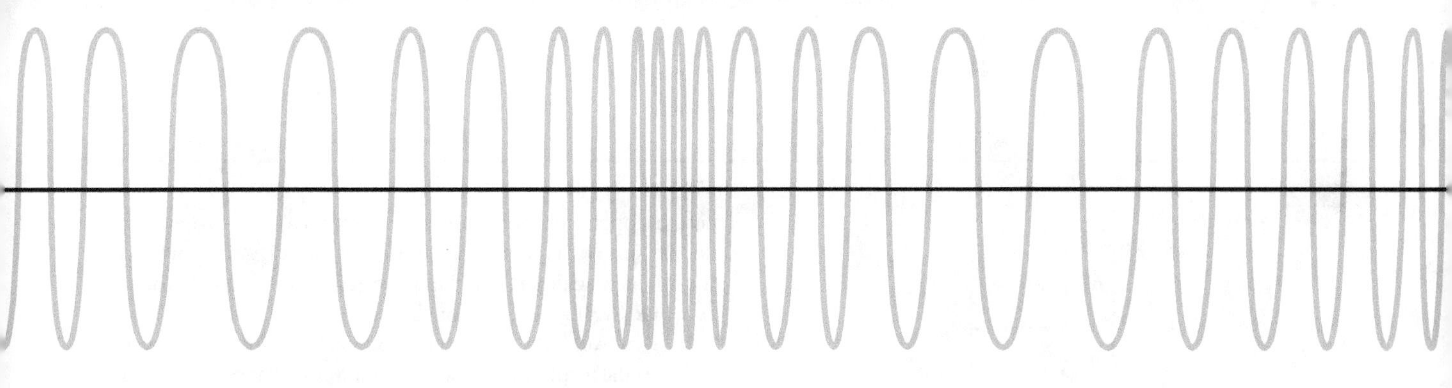

1 Signals

Contents

Overview, 2
1-1 Types of Signals, 3
1-2 Signal Transformations, 6
1-3 Waveform Properties, 9
1-4 Nonperiodic Waveforms, 11
1-5 Signal Power and Energy, 21
Summary, 24
Problems, 25

Objectives

Learn to:

- Perform transformations on signals.
- Use step, ramp, pulse, and exponential waveforms to model simple signals.
- Model impulse functions.
- Calculate power and energy contents of signals.

Signals come in many forms: continuous, discrete, analog, digital, periodic, nonperiodic, with even or odd symmetry or no symmetry at all, and so on. Signals with special *waveforms* include ramps, exponentials, and impulses. This chapter introduces the vocabulary, the properties, and the ***transformations*** commonly associated with signals in preparation for exploring in future chapters ***how signals interact with systems***.

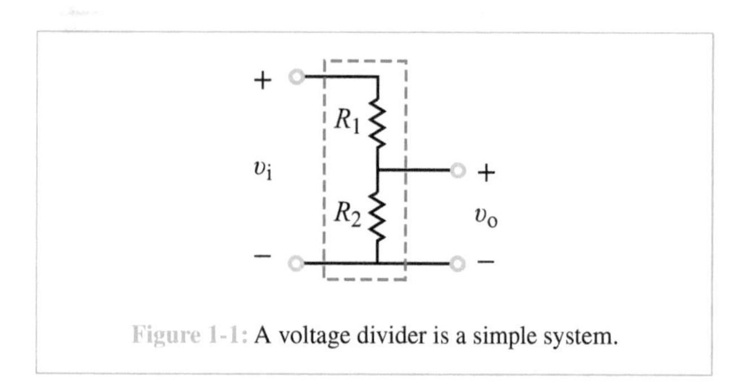

Figure 1-1: A voltage divider is a simple system.

Overview

This book is about how signals interact with systems. More precisely, it is about how a system transforms input signals (***excitations***) into output signals (***responses***) to perform a certain operation (or multiple operations). A system may be as simple as the voltage divider in Fig. 1-1, wherein the divider scales down input voltage υ_i to output voltage $\upsilon_o = [R_2/(R_1 + R_2)]\upsilon_i$, or as complex as a human body (Fig. 1-2). Actually, the human body is a ***system of systems***; it includes the respiratory, blood circulation, and nervous systems, among many others. Each can be modeled as a system with one or more input signals and one or more output signals. When a person's fingertip touches a hot object (Fig. 1-3), a nerve ending in the finger senses the elevated temperature and sends a message (input signal) to the ***central nervous system*** (***CNS***), consisting of the brain and spinal cord. Upon processing the input signal, the CNS (the system) generates several output signals directed to various muscles in the person's arm, ordering them to move the finger away from the hot object.

By modeling signals and systems mathematically, we can use the system model to predict the output resulting from a specified

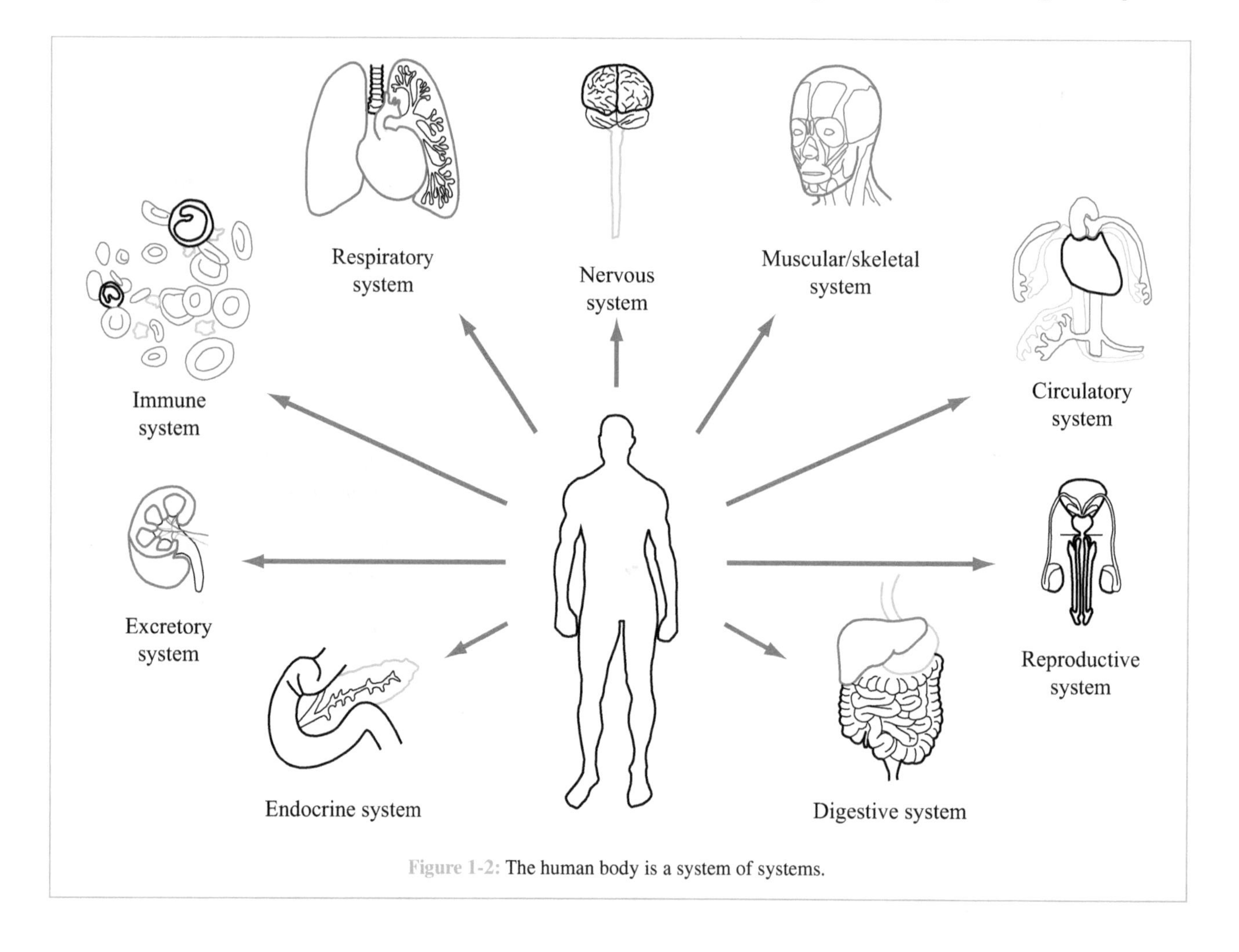

Figure 1-2: The human body is a system of systems.

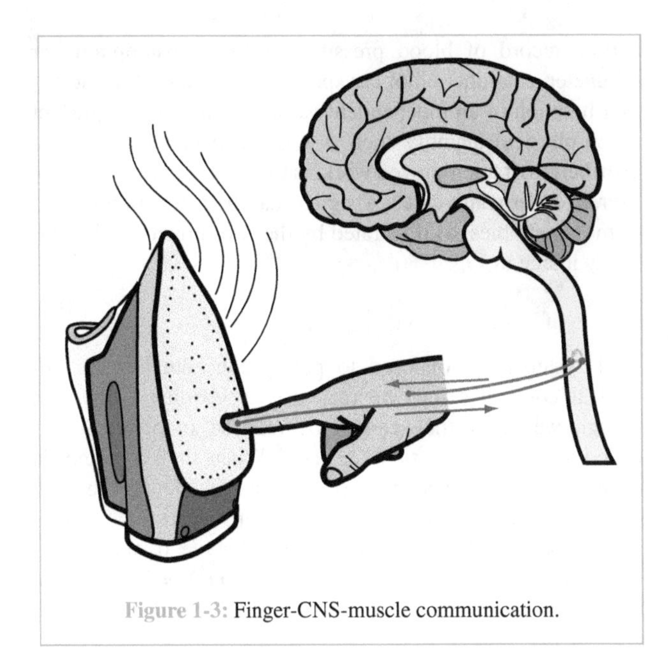

Figure 1-3: Finger-CNS-muscle communication.

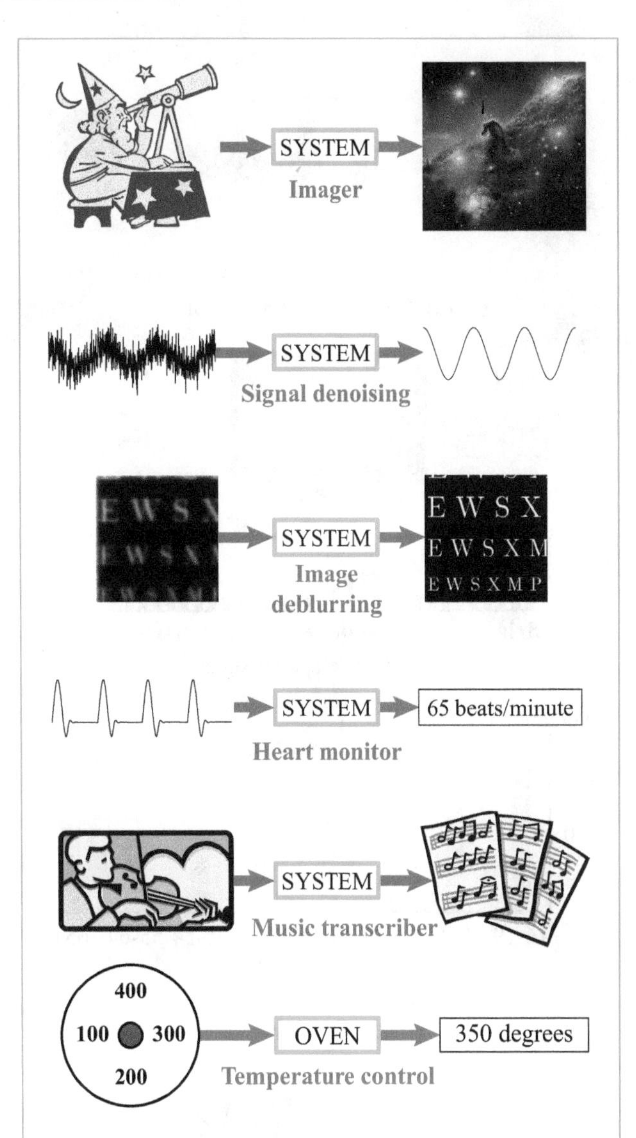

Figure 1-4: A system transforms a continuous input signal $x(t)$ into an output signal $y(t)$ or a discrete input signal $x[n]$ into a discrete output signal $y[n]$. Such system transformations exist not only in the few examples shown here but also in countless electrical, mechanical, biological, acoustic, and financial domains, among many others.

input. We can also design systems to perform operations of interest. A few illustrative examples are depicted in Fig. 1-4. Signals and systems are either continuous or discrete. Both types are treated in this book, along with numerous examples of practical applications.

To set the stage for a serious study of signals and systems and how they interact with one another, we devote the current chapter to an examination of the various mathematical models and attendant properties commonly used to characterize physical signals, and then we follow suit in Chapter 2 with a similar examination for systems.

1-1 Types of Signals

1-1.1 Continuous vs. Discrete

The acoustic pressure waveform depicted in Fig. 1-5(a) is a ***continuous-time signal*** carrying music between a source (the trumpet) and a receiver (the listener's ear). The waveform varies with both spatial location and time. At a given instant in time, the waveform is a plot of acoustic pressure as a function of the spatial dimension x, but to the listener's eardrum, the intercepted waveform is a time-varying function at a fixed value of x.

▶ Traditionally, a signal has been defined as any quantity that exhibits a variation with either time, space, or both. Mathematically, however, variation of any quantity as a function of any independent variable would qualify as a signal as well. ◀

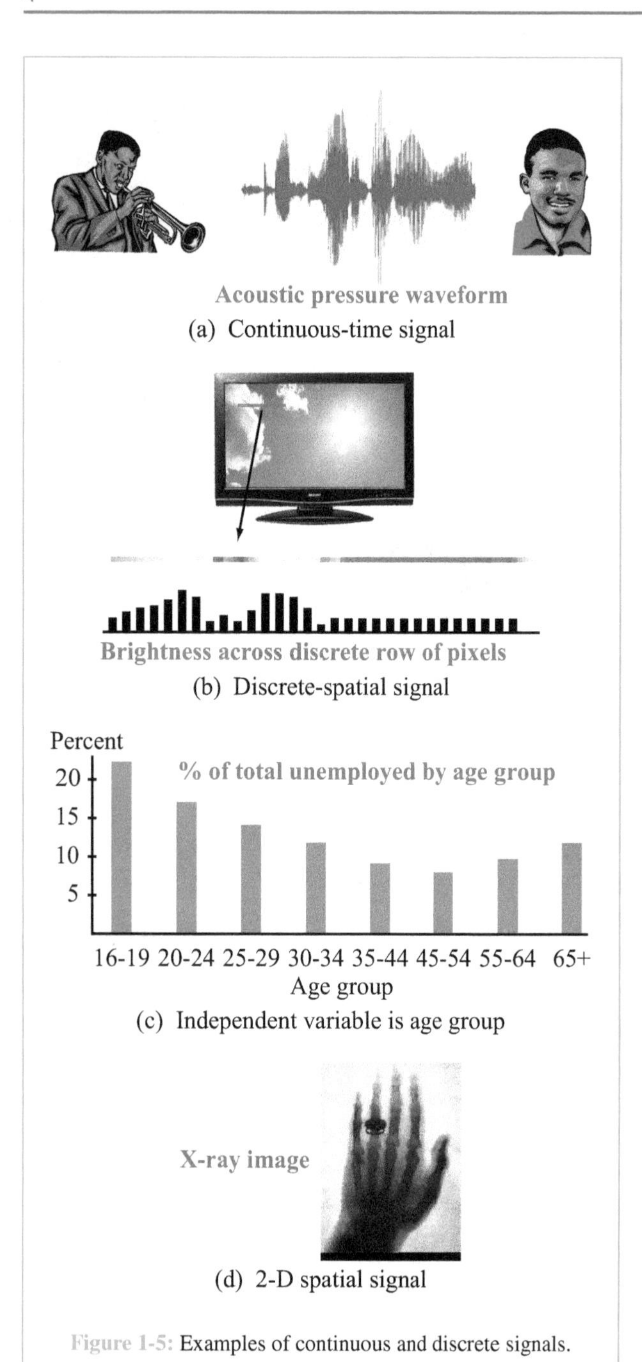

Figure 1-5: Examples of continuous and discrete signals.

In contrast with the continuous-time signal shown in Fig. 1-5(a), the brightness variation across the row of pixels on the computer display of Fig. 1-5(b) constitutes a *discrete-space signal* because the brightness is specified at only a set of discrete locations. In either case, the signal may represent a physical quantity, such as the altitude profile of atmospheric temperature, a time record of blood pressure, or fuel consumption per kilometer as a function of car speed— wherein each is plotted as a function of an independent variable—or it may represent a non-physical quantity such as a stock market index or the distribution of unemployed workers by age group (Fig. 1-5(c)). Moreover, in some cases, a signal may be a function of two or more variables, as illustrated by the two-dimensional (2-D) X-ray image in Fig. 1-5(d).

1-1.2 Causal vs. Noncausal

Real systems—as opposed to purely conceptual or mathematical constructs that we may use as learning tools even though we know they cannot be realized in practice—are called ***physically realizable systems***. When such a system is excited by an input signal $x(t)$, we usually define the time dimension such that $t = 0$ coincides with when the signal is first introduced. Accordingly, $x(t) = 0$ for $t < 0$, and $x(t)$ is called a ***causal signal***. By extension, if $x(t) \neq 0$ for $t < 0$, it is called ***noncausal***, and if $x(t) = 0$ for $t > 0$, it is called ***anticausal***.

► A signal $x(t)$ is:

- **causal** if $x(t) = 0$ for $t < 0$ (starts at or after $t = 0$)
- **noncausal** if $x(t) \neq 0$ for any $t < 0$ (starts before $t = 0$)
- **anticausal** if $x(t) = 0$ for $t > 0$ (ends at or before $t = 0$) ◄

Even though (in practice) our ultimate goal is to evaluate the interaction of causal signals with physically realizable systems, we will occasionally use mathematical techniques that represent a causal signal in terms of artificial constructs composed of sums and differences of causal and anticausal signals.

1-1.3 Analog vs. Digital

Consider an electronic sensor designed such that its output voltage υ is linearly proportional to the air temperature T surrounding its temperature-sensitive thermistor. If the sensor's output is recorded continuously as a function of time (Fig. 1-6(b)), the resulting voltage record $\upsilon(t)$ would be *analogous* to the pattern of the actual air temperature $T(t)$. Hence, $\upsilon(t)$ is regarded as an ***analog signal*** representing $T(t)$. The term analog (short for analogue) conveys the similarity between the measured signal and the physical quantity it represents. It also implies that because both υ and t are continuous variables, the resolution associated with the recorded $\upsilon(t)$ is infinite along both dimensions.

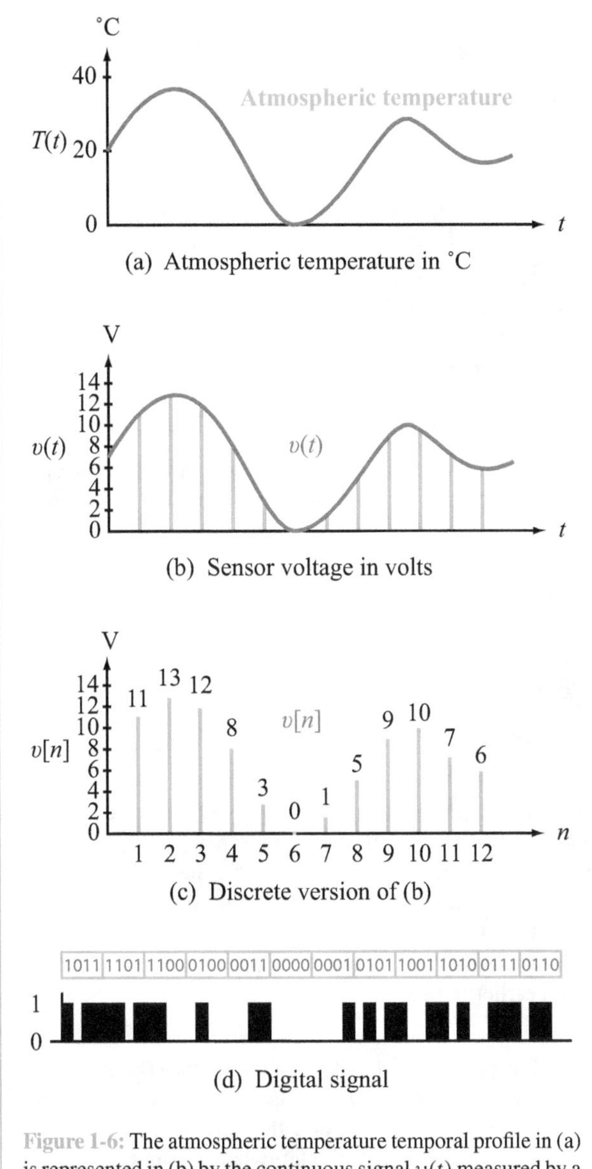

Figure 1-6: The atmospheric temperature temporal profile in (a) is represented in (b) by the continuous signal $\upsilon(t)$ measured by a temperature sensor. The regularly spaced sequence $\upsilon[n]$ in (c) is the discrete version of $\upsilon(t)$. The discrete signal $\upsilon[n]$ is converted into a digital sequence in (d) using a 4-bit encoder.

Had the temperature sensor recorded υ at only a set of equally spaced, discrete values of time, the outcome would have looked like the ***discrete-time signal*** $\upsilon[n]$ displayed in Fig. 1-6(c), in which the dependent variable υ continues to enjoy infinite resolution in terms of its own magnitude but not along the independent variable t.

► To distinguish between a continuous-time signal $\upsilon(t)$ and a discrete-time signal $\upsilon[n]$, the independent variable t in $\upsilon(t)$ is enclosed in curved brackets, whereas for discrete-time signal $\upsilon[n]$, the index n is enclosed in square brackets. ◄

If, in addition to discretizing the signal in time, we were to quantize its amplitudes $\upsilon[n]$ using a 4-bit encoder, for example, we would end up with the ***digital discrete-time signal*** shown in Fig. 1-6(d). By so doing, we have sacrificed resolution along both dimensions, raising the obvious question: *Why is it that the overwhelming majority of today's electronic and mechanical systems—including cell phones and televisions—perform their signal conditioning and display functions in the digital domain?* The most important reason is so that signal processing can be implemented on a digital computer. Computers process finite sequences of numbers, each of which is represented by a finite number of bits. Hence, to process a signal using a digital computer, it must be in discrete-time format, and its amplitude must be encoded into a binary sequence.

Another important reason for using digital signal processing has to do with ***noise***. Superimposed on a signal is (almost always) an unwanted random fluctuation (noise) contributed by electromagnetic fields associated with devices and circuits as well as by natural phenomena (such as lightning). Digital signals are more immune to noise interference than their analog counterparts.

The terms continuous-time, discrete-time, analog, and digital can be summarized as follows:

- A signal $x(t)$ is ***analog and continuous-time*** if both x and t are continuous variables (infinite resolution). Most real-world signals are analog and continuous-time (Chapters 1 through 6).

- A signal $x[n]$ is ***analog and discrete-time*** if the values of x are continuous but time n is discrete (integer-valued). Chapters 7 through 10 deal with analog discrete-time signals.

- A signal $x[n]$ is ***digital and discrete-time*** if the values of x are discrete (i.e., quantized) and time n also is discrete (integer-valued). Computers store and process digital discrete-time signals. This class of signals is outside the scope of this book.

- A signal $x(t)$ is ***digital and continuous-time*** if $x(t)$ can only take on a finite number of values. An example is the class of logic signals, such as the output of a flip-flop, which can only take on values 0 or 1.

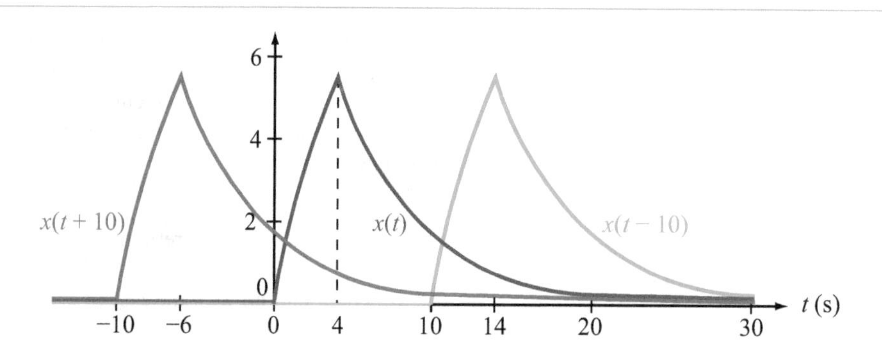

Figure 1-7: Waveforms of $x(t)$, $x(t-10)$, and $x(t+10)$. Note that $x(t-10)$ reaches its peak value 10 s later than $x(t)$, and $x(t+10)$ reaches its peak value 10 s sooner than $x(t)$.

As was stated earlier, a signal's independent variable may not always be time t, and in some cases, the signal may depend on more than one variable (as in 2-D images and 3-D X-ray tomographs). Nevertheless, in the interest of brevity when introducing mathematical techniques, we will use the symbol t as our independent variable exclusively. This does not preclude using other, more appropriate, symbols when applying the techniques to specific applications, nor does it limit expanding the formulation to 2-D or 3-D when necessary.

Concept Question 1-1: What is the difference between a continuous-time signal and a discrete-time signal? Between a discrete-time signal and a digital signal? (See S^2)

Concept Question 1-2: What is the definition of a *causal* signal? *Noncausal* signal? *Anticausal* signal? (See S^2)

1-2 Signal Transformations

▶ A system transforms an input signal into an output signal. The transformation may entail the modification of some attribute of the input signal, the generation of an entirely new signal (based on the input signal), or the extraction of information from the input signal for display or to initiate an action. ◀

For example, the system may delay, compress or stretch out the input signal, or it may filter out the noise accompanying it. If the signal represents the time profile of a car's acceleration $a(t)$, as measured by a microelectromechanical sensor, the system may perform an integration to determine the car's velocity: $\upsilon(t) = \int_0^t a(\tau)\, d\tau$. In yet another example, the system may be an algorithm that generates signals to control the movements of a manufacturing robot, using the information extracted from multiple input signals.

1-2.1 Time-Shift Transformation

If $x(t)$ is a continuous-time signal, a ***time-shifted*** version with delay T is given by

$$y(t) = x(t-T), \tag{1.1}$$

wherein t is replaced with $(t-T)$ everywhere in the expression and/or plot of $x(t)$, as illustrated in Fig. 1-7. If $T > 0$, $y(t)$ is *delayed* by T seconds relative to $x(t)$; the peak value of the waveform of $y(t)$ occurs T seconds later in time than does the peak of $x(t)$. Conversely, if $T < 0$, $y(t)$ is *advanced* by T seconds relative to $x(t)$, in which case the peak value of $y(t)$ occurs earlier in time.

▶ While preserving the shape of the signal $x(t)$, the ***time-shift transformation*** $x(t-T)$ is equivalent to sliding the waveform to the right along the time axis when T is positive and sliding it to the left when T is negative. ◀

1-2.2 Time-Scaling Transformation

Figure 1-8 displays three waveforms that are all similar (but not identical) in shape. Relative to the waveform of $x(t)$, the waveform of $y_1(t)$ is compressed along the time axis, while that of $y_2(t)$ is expanded (stretched out). Waveforms of signals $y_1(t)$ and $y_2(t)$ are *time-scaled* versions of $x(t)$:

$$y_1(t) = x(2t), \tag{1.2a}$$

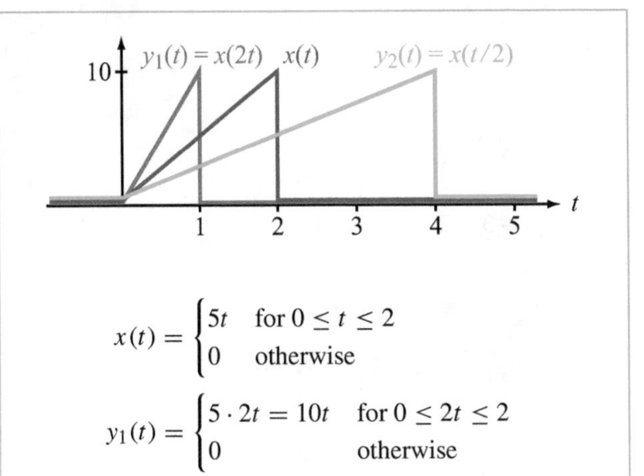

$$x(t) = \begin{cases} 5t & \text{for } 0 \leq t \leq 2 \\ 0 & \text{otherwise} \end{cases}$$

$$y_1(t) = \begin{cases} 5 \cdot 2t = 10t & \text{for } 0 \leq 2t \leq 2 \\ 0 & \text{otherwise} \end{cases}$$

$$y_2(t) = \begin{cases} 5t/2 = 2.5t & \text{for } 0 \leq t/2 \leq 2 \\ 0 & \text{otherwise} \end{cases}$$

Figure 1-8: Waveforms of $x(t)$, a compressed replica given by $y_1(t) = x(2t)$, and an expanded replica given by $y_2(t) = x(t/2)$.

and

$$y_2(t) = x(t/2). \tag{1.2b}$$

Mathematically, the ***time-scaling transformation*** can be expressed as

$$y(t) = x(at), \tag{1.3}$$

where a is a compression or expansion factor depending on whether its absolute value is larger or smaller than 1, respectively. For the time being, we will assume a to be positive. As we will see shortly in the next subsection, a negative value of a causes a time-reversal transformation in addition to the compression/expansion transformation.

▶ Multiplying the independent variable t in $x(t)$ by a constant coefficient a results in a temporally compressed replica of $x(t)$ if $|a| > 1$ and by a temporally expanded replica if $|a| < 1$. ◀

1-2.3 Time-Reversal Transformation

▶ Replacing t with $-t$ in $x(t)$ generates a signal $y(t)$ whose waveform is the mirror image of that of $x(t)$ with respect to the vertical axis. ◀

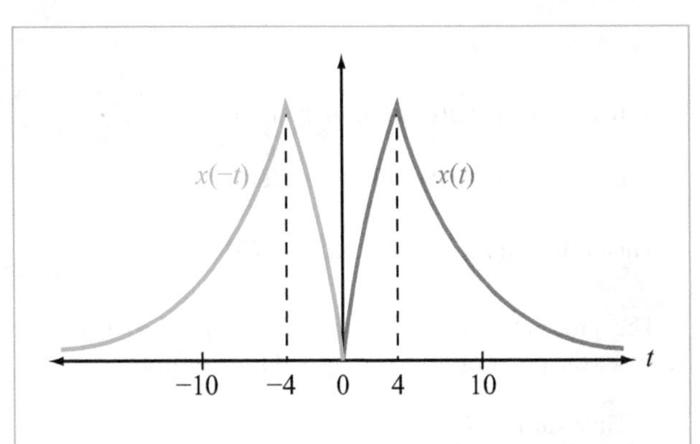

Figure 1-9: Waveforms of $x(t)$ and its time reversal $x(-t)$.

The ***time-reversal transformation*** is expressed as

$$y(t) = x(-t), \tag{1.4}$$

and is illustrated by the waveforms in Fig. 1-9.

1-2.4 Combined Transformation

The three aforementioned transformations can be combined into a generalized transformation:

$$y(t) = x(at - b) = x\left(a\left(t - \frac{b}{a}\right)\right) = x(a(t - T)), \tag{1.5}$$

where $T = b/a$. We recognize T as the time shift and a as the compression/expansion factor. Additionally, the sign of a ($-$ or $+$) denotes whether or not the transformation includes a time-reversal transformation.

The procedure for obtaining $y(t) = x(a(t - T))$ from $x(t)$ is as follows:

(1) Scale time by a:

- If $|a| < 1$, then $x(t)$ is expanded.
- If $|a| > 1$, then $x(t)$ is compressed.
- If $a < 0$, then $x(t)$ is also reflected.

This results in $z(t) = x(at)$.

(2) Time shift by T:

- If $T > 0$, then $z(t)$ shifts to the right.
- If $T < 0$, then $z(t)$ shifts to the left.

This results in $z(t - T) = x(a(t - T)) = y(t)$.

The procedure for obtaining $y(t) = x(at - b)$ from $x(t)$ reverses the order of time scaling and time shifting:

(1) Time shift by b.

(2) Time scale by a.

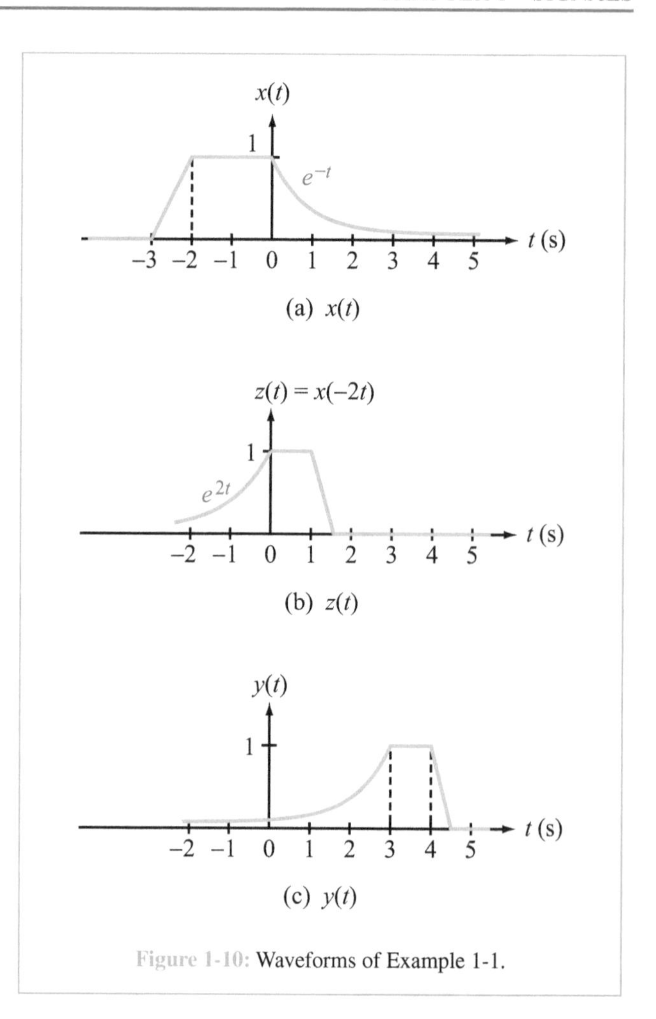

Figure 1-10: Waveforms of Example 1-1.

Example 1-1: Multiple Transformations

For signal $x(t)$ profiled in Fig. 1-10(a), generate the corresponding profile of $y(t) = x(-2t + 6)$.

Solution: We start by recasting the expression for the dependent variable into the standard form given by Eq. (1.5),

$$y(t) = x\left(-2\left(t - \frac{6}{2}\right)\right)$$

$$= x(-2(t - 3)).$$

Reversal Compression factor Time-shift

We need to apply the following transformations:

(1) Scale time by $-2t$: This causes the waveform to reflect around the vertical axis and then compresses time by a factor of 2. These steps can be performed in either order. The result, $z(t) = x(-2t)$, is shown in Fig. 1-10(b).

(2) Delay waveform $z(t)$ by 3 s: This shifts the waveform to the right by 3 s (because the sign of the time shift is negative). The result, $y(t) = z(t - 3) = x(-2(t - 3))$, is displayed in Fig. 1-10(c).

Concept Question 1-3: Is the shape of a waveform altered or preserved upon applying a time-shift transformation? Time-scaling transformation? Time-reversal transformation? (See S^2)

Exercise 1-1: If signal $y(t)$ is obtained from $x(t)$ by applying the transformation $y(t) = x(-4t - 8)$, determine the values of the transformation parameters a and T.

Answer: $a = -4$ and $T = -2$. (See S^2)

Exercise 1-2: If $x(t) = t^3$ and $y(t) = 8t^3$, are $x(t)$ and $y(t)$ related by a transformation?

Answer: Yes, because $y(t) = 8t^3 = (2t)^3 = x(2t)$. (See S^2)

Exercise 1-3: What type of transformations connect $x(t) = 4t$ to $y(t) = 2(t+4)$?

Answer: $y(t) = x(\frac{1}{2}(t+4))$, which includes a time-scaling transformation with a factor of 1/2 and a time-shift transformation with a time advance of 4 s. (See S²)

1-3 Waveform Properties

1-3.1 Even Symmetry

▶ A signal $x(t)$ exhibits ***even symmetry*** if its waveform is symmetrical with respect to the vertical axis. ◀

The shape of the waveform on the left-hand side of the vertical axis is the mirror image of the waveform on the right-hand side. Mathematically, a signal $x(t)$ has *even symmetry* if

$$x(t) = x(-t) \qquad \textbf{(even symmetry).} \tag{1.6}$$

A signal has even symmetry if reflection about the vertical axis leaves its waveform unaltered.

The signal displayed in Fig. 1-11(b) is an example of a waveform that exhibits even symmetry. Other examples include $\cos(\omega t)$ and t^n for even integers n.

1-3.2 Odd Symmetry

In contrast, the waveform in Fig. 1-11(c) has ***odd symmetry***.

▶ A signal exhibits odd symmetry if the shape of its waveform on the left-hand side of the vertical axis is the *inverted* mirror image of the waveform on the right-hand side. ◀

Equivalently,

$$x(t) = -x(-t) \qquad \textbf{(odd symmetry).} \tag{1.7}$$

A signal has odd symmetry if reflection about the vertical axis, followed by reflection about the horizontal axis, leaves its waveform unaltered. Examples of odd signals include $\sin(\omega t)$ and t^n for odd integers n.

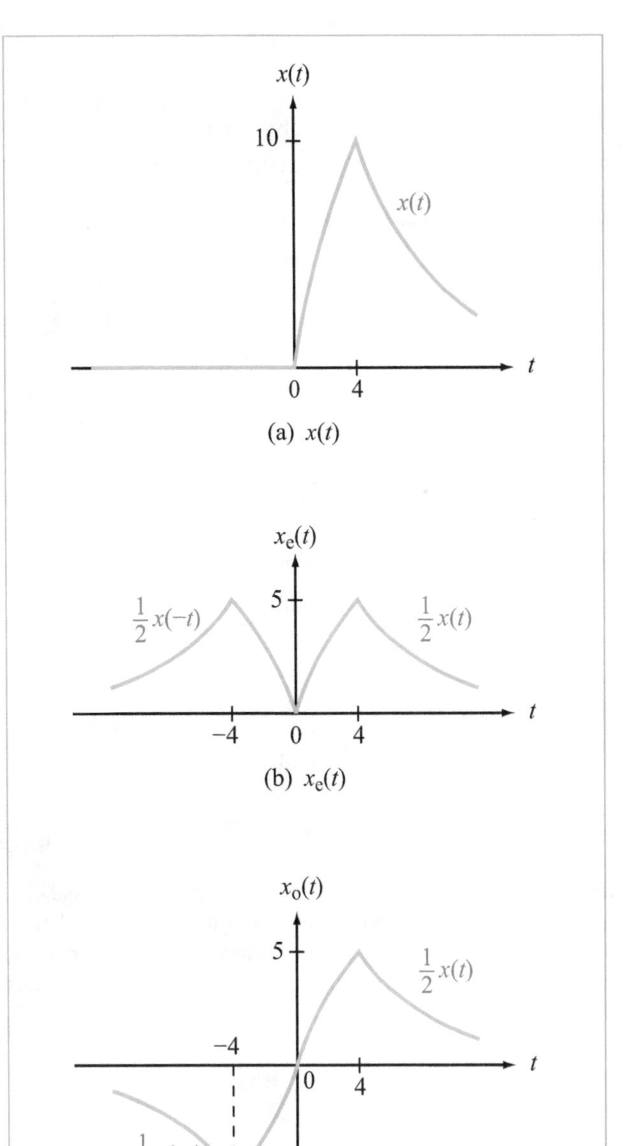

Figure 1-11: Signal $x(t)$ and its even and odd components.

We should note that if signal $y(t)$ is equal to the product of two signals, namely

$$y(t) = x_1(t)\, x_2(t), \tag{1.8}$$

then $y(t)$ will exhibit even symmetry if $x_1(t)$ and $x_2(t)$ both exhibit the same type of symmetry (both even or both odd)

Table 1-1: **Signal transformations.**

Transformation	Expression	Consequence
Time shift	$y(t) = x(t - T)$	Waveform is shifted along $+t$ direction if $T > 0$ and along $-t$ direction if $T < 0$.
Time scaling	$y(t) = x(at)$	Waveform is compressed if $\|a\| > 1$ and expanded if $\|a\| < 1$.
Time reversal	$y(t) = x(-t)$	Waveform is mirror-imaged relative to vertical axis.
Generalized	$y(t) = x(at - b)$	Shift by b, then scale by a, time reversal if $a < 0$.
Even / odd synthesis	$x(t) = x_e(t) + x_o(t)$	$x_e(t) = \frac{1}{2}\{x(t) + x(-t)\}$, $x_o(t) = \frac{1}{2}\{x(t) - x(-t)\}$.

and $y(t)$ will exhibit odd symmetry if $x_1(t)$ and $x_2(t)$ exhibit different forms of symmetry. That is,

$$(\text{even}) \times (\text{even}) = \text{even},$$

$$(\text{even}) \times (\text{odd}) = \text{odd},$$

and

$$(\text{odd}) \times (\text{odd}) = \text{even}.$$

1-3.3 Even / Odd Synthesis

In Chapter 5, we will find it easier to analyze a signal if it possesses even or odd symmetry than if it possesses neither. In that case, it may prove advantageous to synthesize a signal $x(t)$ as the sum of two component signals, one with even symmetry and another with odd symmetry:

$$x(t) = x_e(t) + x_o(t), \tag{1.9}$$

with

$$x_e(t) = \frac{1}{2}[x(t) + x(-t)], \tag{1.10a}$$

$$x_o(t) = \frac{1}{2}[x(t) - x(-t)]. \tag{1.10b}$$

As the graphical example shown in Fig. 1-11 demonstrates, adding a copy of its time reversal, $x(-t)$, to any signal $x(t)$ generates a signal with even symmetry. Conversely, subtracting a copy of its time reversal from $x(t)$ generates a signal with odd symmetry.

Table 1-1 provides a summary of the linear transformations examined thus far.

1-3.4 Periodic vs. Nonperiodic

A signal's waveform may be ***periodic*** or ***nonperiodic*** (also called ***aperiodic***).

► A periodic signal $x(t)$ of period T_0 satisfies the ***periodicity*** property:

$$x(t) = x(t + nT_0) \tag{1.11}$$

for all integer values of n and all times t. ◄

The periodicity property states that the waveform of $x(t)$ repeats itself every T_0 seconds. Examples of periodic signals are displayed in Fig. 1-12.

Note that if a signal is periodic with period T_0, it is also periodic with period $2T_0$, $3T_0$, etc. The ***fundamental period*** of a periodic signal is the smallest value of T_0 such that Eq. (1.11) is satisfied for all integer values of n. In future references, the term "period" shall refer to the fundamental period T_0.

The most important family of periodic signals are sinusoids. A sinusoidal signal $x(t)$ has the form

$$x(t) = A\cos(\omega_0 t + \theta), \qquad -\infty < t < \infty,$$

where

$$A = \text{amplitude}; \quad x_{\max} = A \quad \text{and} \quad x_{\min} = -A,$$

$$\omega_0 = \text{angular frequency in rad/s},$$

$$\theta = \text{phase-angle shift in radians or degrees}.$$

Related quantities include

$$f_0 = \omega_0/2\pi = \text{circular frequency in Hertz},$$

$$T_0 = 1/f_0 = \text{period of } x(t) \text{ in seconds}.$$

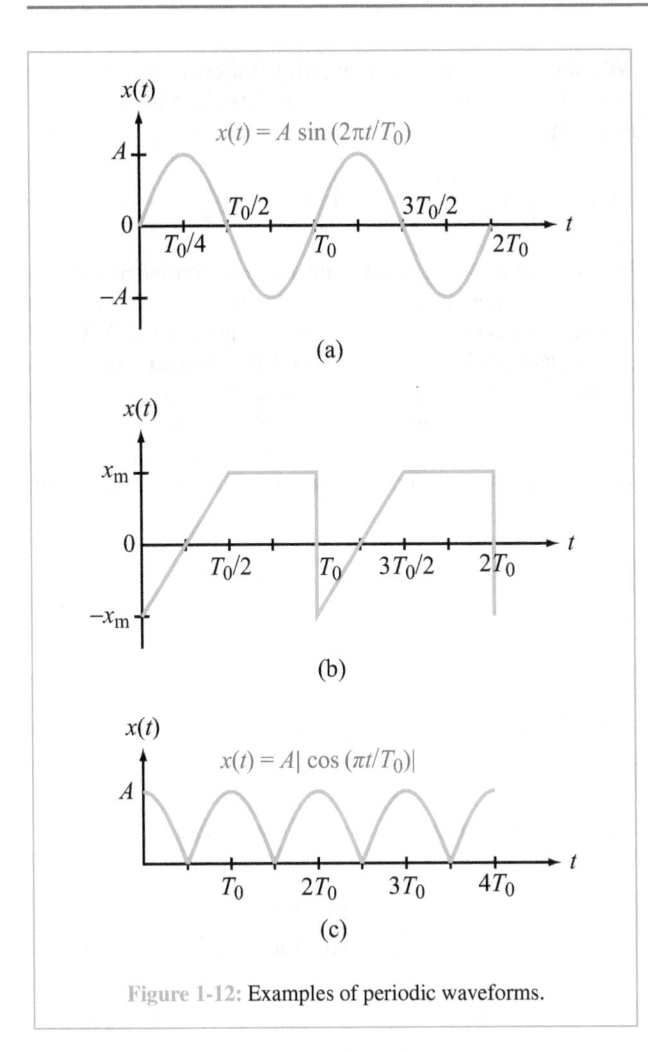

Figure 1-12: Examples of periodic waveforms.

Another important periodic signal is the complex exponential given by

$$x(t) = Ae^{j\omega_0 t} = |A|e^{j(\omega_0 t+\theta)},$$

where, in general, A is a complex amplitude given by

$$A = |A|e^{j\theta}.$$

By Euler's formula,

$$x(t) = |A|e^{j(\omega_0 t+\theta)} = |A|\cos(\omega_0 t + \theta) + j|A|\sin(\omega_0 t + \theta).$$

Hence, the complex exponential is periodic with period $T_0 = 2\pi/\omega_0$.

Concept Question 1-4: Define *even-symmetrical* and *odd-symmetrical* waveforms. (See S²)

Concept Question 1-5: State the periodicity property. (See S²)

Exercise 1-4: Which of the following functions have even-symmetrical waveforms, odd-symmetrical waveforms, or neither? (a) $x_1(t) = 3t^2$, (b) $x_2(t) = \sin(2t)$, (c) $x_3(t) = \sin^2(2t)$, (d) $x_4(t) = 4e^{-t}$, (e) $x_5(t) = |\cos 2t|$.

Answer: (a), (c), and (e) have even symmetry; (b) has odd symmetry; (d) has no symmetry. (See S²)

1-4 Nonperiodic Waveforms

► A nonperiodic signal is any signal that does not satisfy the periodicity property. ◄

Many real-world signals are often modeled in terms of a core set of elementary waveforms which includes the step, ramp, pulse, impulse, and exponential waveforms, and combinations thereof. Accordingly, we will use this section to review their properties and mathematical expressions and to point out the connections between them.

1-4.1 Step-Function Waveform

The waveform of signal $u(t)$ shown in Fig. 1-13(a) is an (ideal) ***unit step function***: It is equal to zero for $t < 0$, at $t = 0$ it makes a *discontinuous* jump to 1, and from there on forward it remains at 1. Mathematically, $u(t)$ is defined as

$$u(t) = \begin{cases} 0 & \text{for } t < 0, \\ 1 & \text{for } t > 0. \end{cases} \quad (1.12)$$

Because $u(t)$ does not have a unique value at $t = 0$, its derivative is infinite at $t = 0$, qualifying $u(t)$ as a singularity function.

► A ***singularity function*** is a function such that either itself or one (or more) of its derivatives is (are) not finite everywhere. ◄

Occasionally, it may prove more convenient to model the unit step function as a ramp over an infinitesimal interval extending

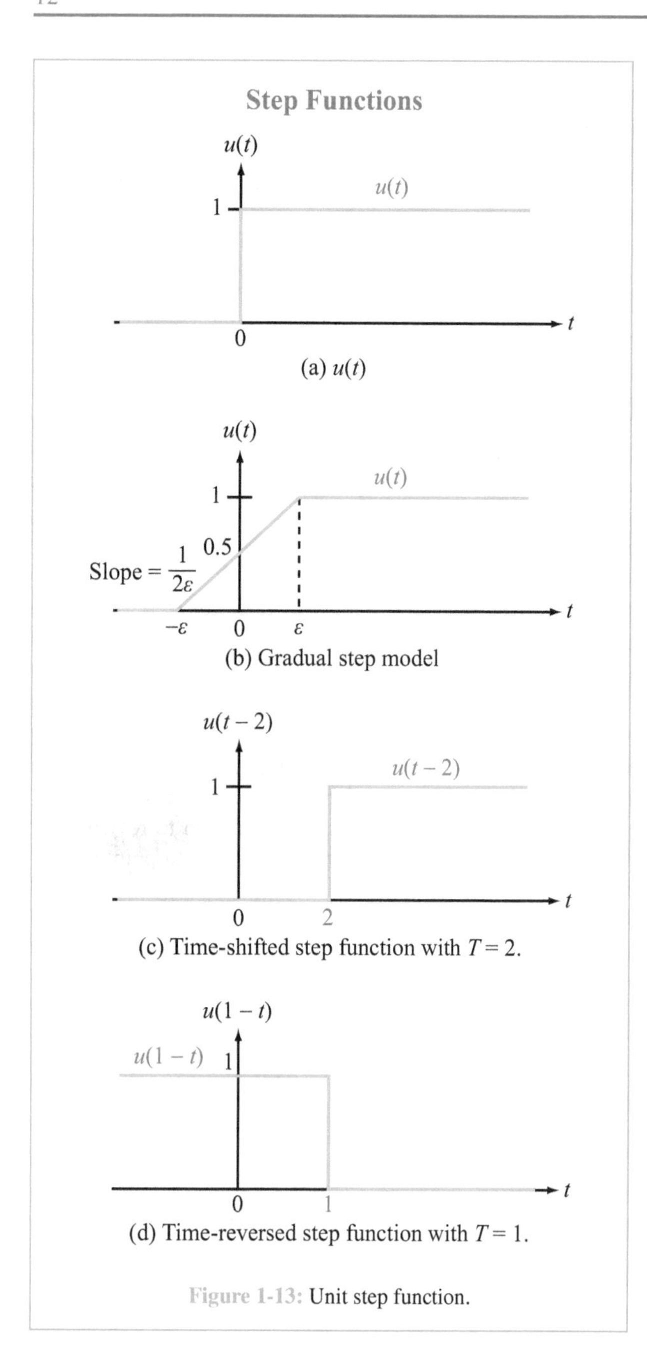

Figure 1-13: Unit step function.

between $-\epsilon$ and $+\epsilon$, as shown in Fig. 1-13(b). Accordingly, $u(t)$ can be defined as

$$u(t) = \lim_{\epsilon \to 0} \begin{cases} 0 & \text{for } t \leq -\epsilon, \\ \left[\frac{1}{2}\left(\frac{t}{\epsilon} + 1\right)\right] & \text{for } -\epsilon \leq t \leq \epsilon, \\ 1 & \text{for } t \geq \epsilon. \end{cases} \tag{1.13}$$

With this alternative definition, $u(t)$ is a continuous function everywhere, but in the limit as $\epsilon \to 0$ its slope in the interval $(-\epsilon, \epsilon)$ is

$$u'(t) = \lim_{\epsilon \to 0} \frac{d}{dt}\left[\frac{1}{2}\left(\frac{t}{\epsilon} + 1\right)\right] = \lim_{\epsilon \to 0}\left(\frac{1}{2\epsilon}\right) \to \infty. \tag{1.14}$$

The slope of $u(t)$ still is not finite at $t = 0$, consistent with the formal definition given by Eq. (1.12), which describes the unit step function as making an *instantaneous* jump at $t = 0$. As will be demonstrated later in Section 1-4.4, the alternative definition for $u(t)$ given by Eq. (1.13) will prove useful in establishing the connection between $u(t)$ and the impulse function $\delta(t)$.

The ***unit time-shifted step function*** $u(t-T)$ is a step function that transitions between its two levels when its argument $(t - T) = 0$:

$$u(t - T) = \begin{cases} 0 & \text{for } t < T, \\ 1 & \text{for } t > T. \end{cases} \tag{1.15}$$

► For any unit step function, its value is zero when its argument is less than zero and one when its argument is greater than zero. ◄

Extending this definition to the ***time-reversed step function***, we have

$$u(T - t) = \begin{cases} 1 & \text{for } t < T, \\ 0 & \text{for } t > T. \end{cases} \tag{1.16}$$

By way of examples, Figs. 1-13(c) and (d) display plots of $u(t - 2)$ and $u(1 - t)$, respectively.

1-4.2 Ramp-Function Waveform

The ***unit ramp function*** $r(t)$ and the ***unit time-shifted ramp function*** $r(t - T)$ are defined as

$$r(t) = \begin{cases} 0 & \text{for } t \leq 0, \\ t & \text{for } t \geq 0, \end{cases} \tag{1.17a}$$

and

$$r(t - T) = \begin{cases} 0 & \text{for } t \leq T, \\ (t - T) & \text{for } t \geq T. \end{cases} \tag{1.17b}$$

Two ramp-function examples are displayed in Fig. 1-14. In each case, the ramp function is zero when its argument $(t - T)$ is

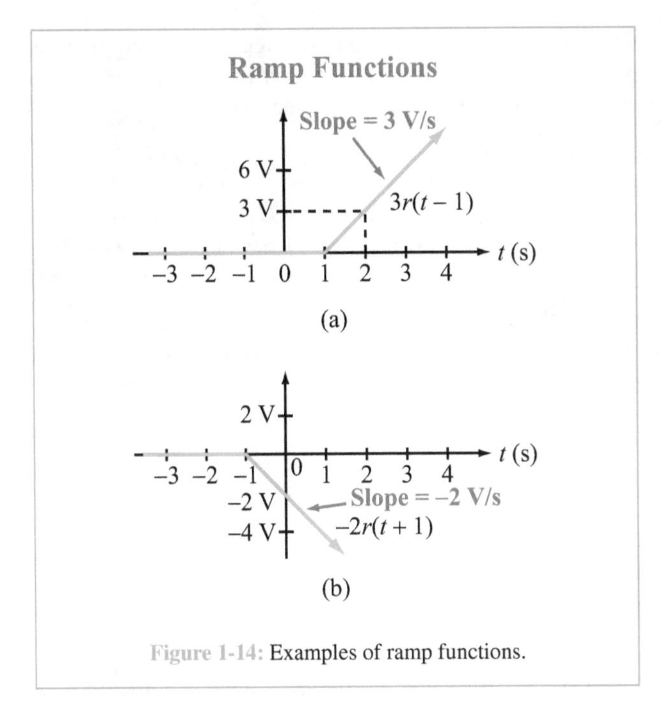

Figure 1-14: Examples of ramp functions.

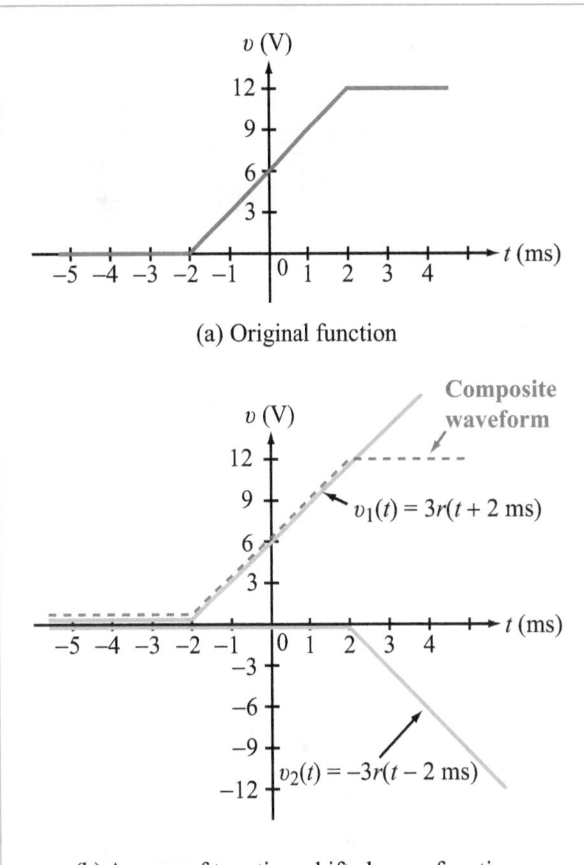

Figure 1-15: Step waveform of Example 1-2.

smaller than zero, and equal to its own argument when the value of t is such that the argument is greater than zero. The slope of a ramp function $x(t) = ar(t - T)$ is specified by the constant coefficient a.

Because the time-derivative of $r(t - T)$—i.e., its slope—is discontinuous at $t = T$, the ramp function qualifies as a ***singularity function***.

The unit ramp function is related to the unit step function by

$$r(t) = \int_{-\infty}^{t} u(\tau)\, d\tau = t\, u(t), \qquad (1.18)$$

and for the time-shifted case,

$$r(t - T) = \int_{-\infty}^{t} u(\tau - T)\, d\tau = (t - T)\, u(t - T). \qquad (1.19)$$

Example 1-2: Synthesizing a Step-Waveform

For the (realistic) step waveform $\upsilon(t)$ displayed in Fig. 1-15, develop expressions in terms of ramp and ideal step functions. Note that $\upsilon(t)$ is in volts (V) and the time scale is in milliseconds.

Solution: The voltage $\upsilon(t)$ can be synthesized as the sum of two time-shifted ramp functions (Fig. 1-15(b)): One starts at $t = -2$ ms and has a positive slope of 3 V/ms and a second starts at $t = 2$ ms but with a slope of -3 V/ms. Thus,

$$\upsilon(t) = \upsilon_1(t) + \upsilon_2(t) = 3r(t + 2 \text{ ms}) - 3r(t - 2 \text{ ms}) \quad \text{V}.$$

In view of Eq. (1.19), $\upsilon(t)$ also can be expressed as

$$\upsilon(t) = 3(t + 2 \text{ ms})\, u(t + 2 \text{ ms}) - 3(t - 2 \text{ ms})\, u(t - 2 \text{ ms}) \quad \text{V}.$$

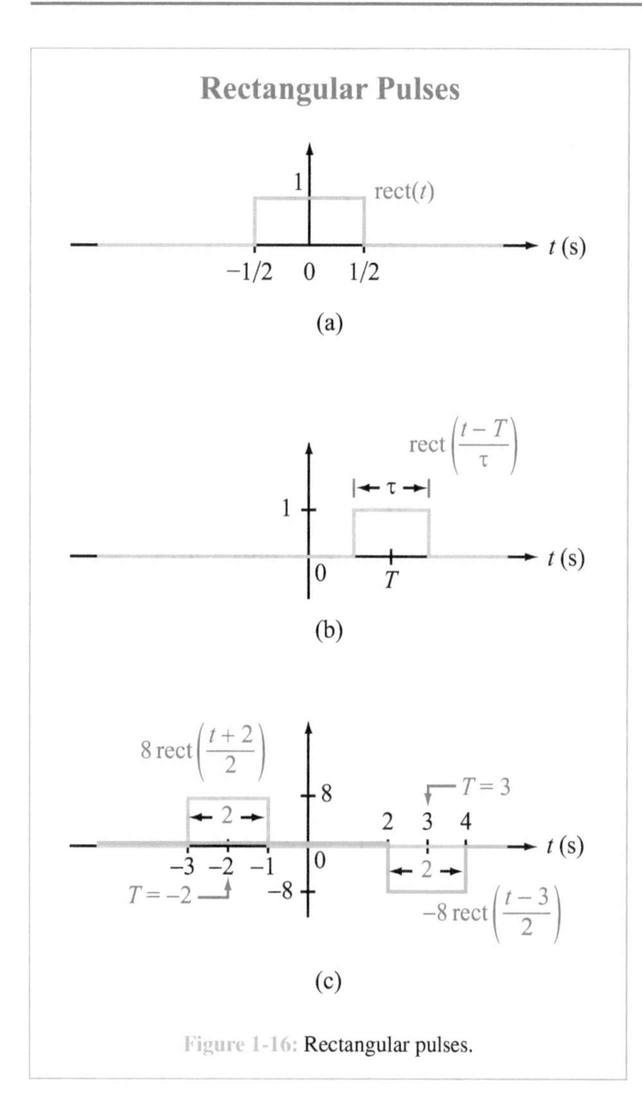

Figure 1-16: Rectangular pulses.

1-4.3 Pulse Waveform

The ***rectangular function*** rect(t) is defined as

$$\text{rect}(t) = \begin{cases} 1 & \text{for } |t| < \frac{1}{2} \\ 0 & \text{for } |t| > \frac{1}{2} \end{cases} = u\left(t + \frac{1}{2}\right) - u\left(t - \frac{1}{2}\right),$$

and its waveform is displayed in Fig. 1-16(a). Note that the rectangle is of width 1 s, height 1 unit, and is centered at $t = 0$.

In general, a rectangular pulse can be described mathematically by the rectangular function rect$[(t - T)/\tau]$. Its two parameters are T, which defines the location of the center of the pulse along the t-axis, and τ, which is the duration of the pulse (Fig. 1-16(b)). Examples are shown in Fig. 1-16(c). The general rectangular function is defined as

$$\text{rect}\left(\frac{t - T}{\tau}\right) = \begin{cases} 0 & \text{for } t < (T - \tau/2), \\ 1 & \text{for } (T - \tau/2) < t < (T + \tau/2), \\ 0 & \text{for } t > (T + \tau/2). \end{cases} \tag{1.20}$$

We note that because the rectangular function is discontinuous at its two edges (namely at $t = T - \tau/2$ and $t = T + \tau/2$), it is a bona fide member of the family of singularity functions.

Example 1-3: Rectangular and Trapezoidal Pulses

Develop expressions in terms of ideal step functions for (a) the rectangular pulse $\upsilon_a(t)$ in Fig. 1-17(a) and (b) the more-realistic trapezoidal pulse $\upsilon_b(t)$ in Fig. 1-17(b).

Solution: (a) The amplitude of the rectangular pulse is 4 V, its duration is 2 s, and its center is at $t = 3$ s. Hence,

$$\upsilon_a(t) = 4\,\text{rect}\left(\frac{t - 3}{2}\right) \quad \text{V}.$$

The sequential addition of two time-shifted step functions, $\upsilon_1(t)$ at $t = 2$ s and $\upsilon_2(t)$ at $t = 4$ s, as demonstrated graphically in Fig. 1-17(c), accomplishes the task of synthesizing the rectangular pulse in terms of two step functions:

$$\upsilon_a(t) = \upsilon_1(t) + \upsilon_2(t) = 4[u(t - 2) - u(t - 4)] \quad \text{V}.$$

Generalizing, a unit rectangular function rect$[(t - T)/\tau]$ always can be expressed as

$$\text{rect}\left(\frac{t - T}{\tau}\right) = u\left[t - \left(T - \frac{\tau}{2}\right)\right] - u\left[t - \left(T + \frac{\tau}{2}\right)\right]. \tag{1.21}$$

(b) The trapezoidal pulse exhibits a change in slope at $t = 0$, $t = 1$ s, $t = 3$ s, and $t = 4$ s, each of which can be accommodated by the introduction of a time-shifted ramp function with the appropriate slope. Building on the procedure used in Example 1-2, $\upsilon_b(t)$ can be synthesized as the sum of

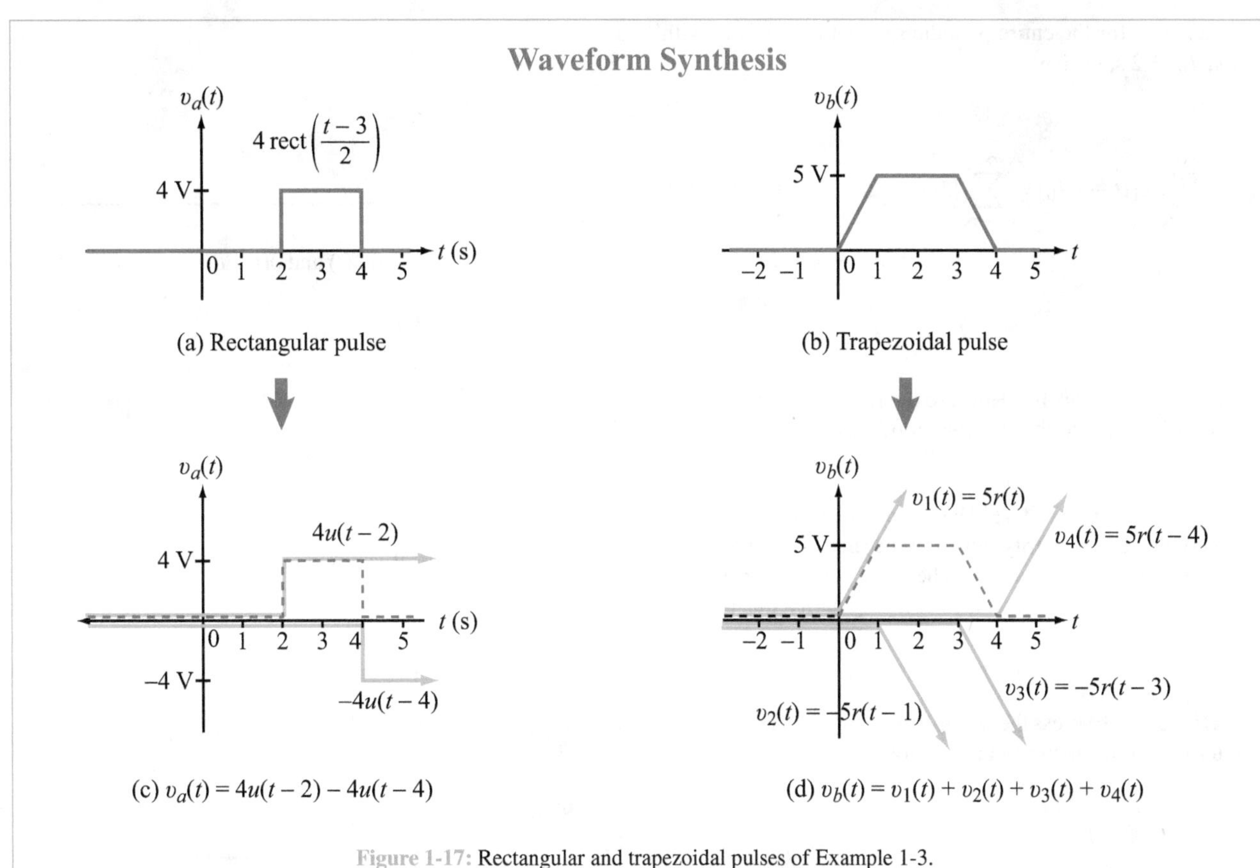

Figure 1-17: Rectangular and trapezoidal pulses of Example 1-3.

the four ramp functions shown in Fig. 1-17(d):

$$\begin{aligned} v_b(t) &= v_1(t) + v_2(t) + v_3(t) + v_4(t) \\ &= 5[r(t) - r(t-1) - r(t-3) + r(t-4)] \\ &= 5[t\,u(t) - (t-1)\,u(t-1) \\ &\quad - (t-3)\,u(t-3) + (t-4)\,u(t-4)] \text{ V}, \end{aligned}$$

where in the last step, we used the relation given by Eq. (1.19).

Example 1-4: Periodic Sawtooth Waveform

Express the periodic sawtooth waveform shown in Fig. 1-18 in terms of step and ramp functions.

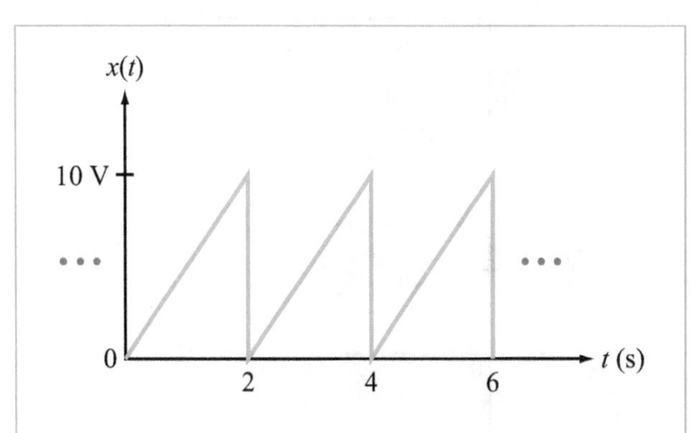

Figure 1-18: Periodic sawtooth waveform of Example 1-4.

Solution: The segment between $t = 0$ and $t = 2$ s is a ramp with a slope of 5 V/s. To effect a sudden drop from 10 V down to zero at $t = 2$ s, we need to (a) add a negative ramp function at $t = 2$ s and (b) add a negative offset of 10 V in the form of a delayed step function. Hence, for this time segment,

$$x_1(t) = [5r(t) - 5r(t-2) - 10u(t-2)] \text{ V}, \quad 0 \le t < 2 \text{ s}.$$

By extension, for the entire periodic sawtooth waveform with period $T_0 = 2$ s, we have

$$x(t) = \sum_{n=-\infty}^{\infty} x_1(t - nT_0) = \sum_{n=-\infty}^{\infty} [5r(t-2n) - 5r(t-2-2n) - 10u(t-2-2n)] \text{ V}.$$

Concept Question 1-6: How are the ramp and rectangle functions related to the step function? (See S²)

Concept Question 1-7: The step function $u(t)$ is considered a singularity function because it makes a discontinuous jump at $t = 0$. The ramp function $r(t)$ is continuous at $t = 0$. Yet it also is a singularity function. Why? (See S²)

Exercise 1-5: Express the waveforms shown in Fig. E1-5 in terms of unit step or ramp functions.

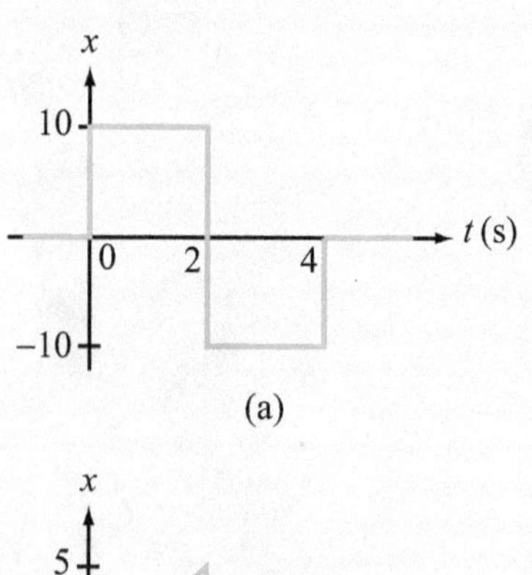

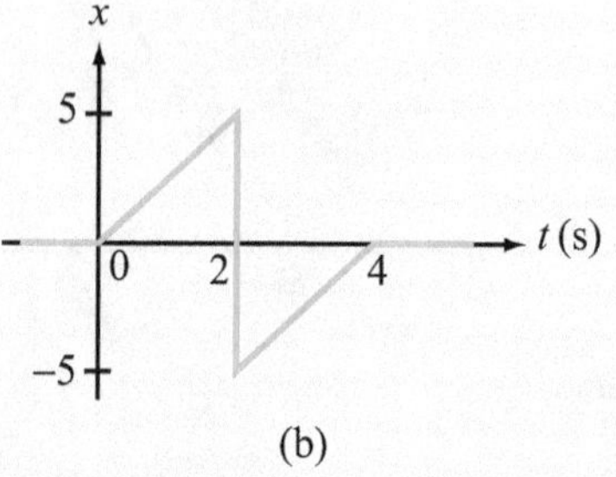

Figure E1-5

Answer: (a) $x(t) = 10\,u(t) - 20\,u(t-2) + 10\,u(t-4)$, (b) $x(t) = 2.5\,r(t) - 10\,u(t-2) - 2.5\,r(t-4)$. (See S²)

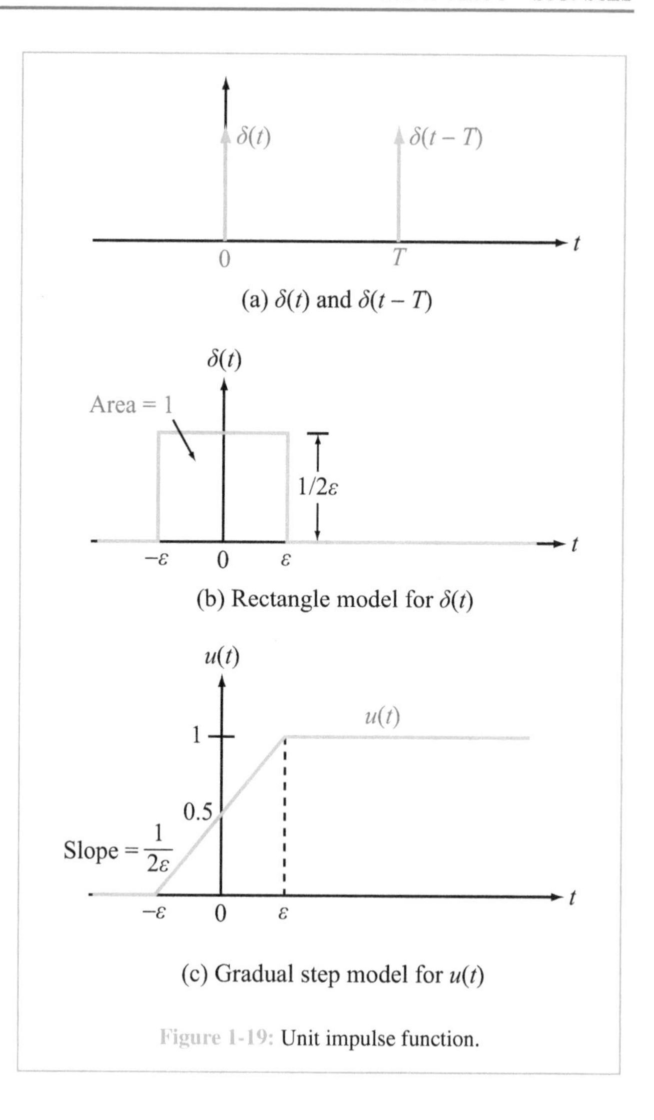

Figure 1-19: Unit impulse function.

Exercise 1-6: How is $u(t)$ related to $u(-t)$?

Answer: They are mirror images of one another (with respect to the y-axis). (See S²)

1-4.4 Impulse Function

Another member of the family of singularity functions is the ***unit impulse function***, which is also known as the ***Dirac delta function*** $\delta(t)$. Graphically, it is represented by a vertical arrow, as shown in Fig. 1-19(a). If its location is time-shifted to $t = T$, it is designated $\delta(t - T)$. For any specific location T, the impulse function is defined through the combination of two properties:

$$\delta(t-T)=0 \qquad \text{for } t \neq T \tag{1.22a}$$

and

$$\int_{-\infty}^{\infty} \delta(t-T)\,dt = 1. \tag{1.22b}$$

▶ The first property states that the impulse function $\delta(t-T)$ is zero everywhere, except at its own location $(t=T)$, but its value is infinite at that location. The second property states that the total area under the unit impulse is equal to 1, regardless of its location. ◀

To appreciate the meaning of the second property, we can represent the impulse function by the rectangle shown in Fig. 1-19(b) with the understanding that $\delta(t)$ is defined in the limit as $\epsilon \to 0$. The rectangle's dimensions are such that its width, $w=2\epsilon$, and height, $h=1/(2\epsilon)$, are reciprocals of one another. Consequently, the area of the rectangle is always unity, even as $\epsilon \to 0$.

According to the rectangle model displayed in Fig. 1-19(b), $\delta(t)=1/(2\epsilon)$ over the narrow range $-\epsilon<t<\epsilon$. For the gradual step model of $u(t)$ shown in Fig. 1-19(c), its slope also is $1/(2\epsilon)$. Hence,

$$\frac{du(t)}{dt}=\delta(t). \tag{1.23}$$

Even though this relationship between the unit impulse and unit step functions was obtained on the basis of specific geometrical models for $\delta(t)$ and $u(t)$, its validity can be demonstrated to be true always. The corresponding expression for $u(t)$ is

$$u(t)=\int_{-\infty}^{t} \delta(\tau)\,d\tau, \tag{1.24}$$

and for the time-shifted case,

$$\frac{d}{dt}\,[u(t-T)]=\delta(t-T), \tag{1.25a}$$

$$u(t-T)=\int_{-\infty}^{t} \delta(\tau-T)\,d\tau. \tag{1.25b}$$

By extension, a scaled impulse $k\,\delta(t)$ has an area k and

$$\int_{-\infty}^{t} k\,\delta(\tau)\,d\tau = k\,u(t). \tag{1.26}$$

Example 1-5: Alternative Models for Impulse Function

Show that models $x_1(t)$ and $x_2(t)$ in Fig. 1-20 qualify as unit impulse functions in the limit as $\epsilon \to 0$.

Solution: To qualify as a unit impulse function, a function must: (1) be zero everywhere except at $t=0$, (2) be infinite at $t=0$, (3) be even, and (4) have a unit area.

(a) Triangle Model $x_1(t)$

(1) As $\epsilon \to 0$, $x_1(t)$ is indeed zero everywhere except at $t=0$.

(2) $\lim_{\epsilon\to 0} x_1(0) = \lim_{\epsilon\to 0} \frac{1}{\epsilon} = \infty$; hence infinite at $t=0$.

(3) $x_1(t)$ is clearly an even function (Fig. 1-20(a)).

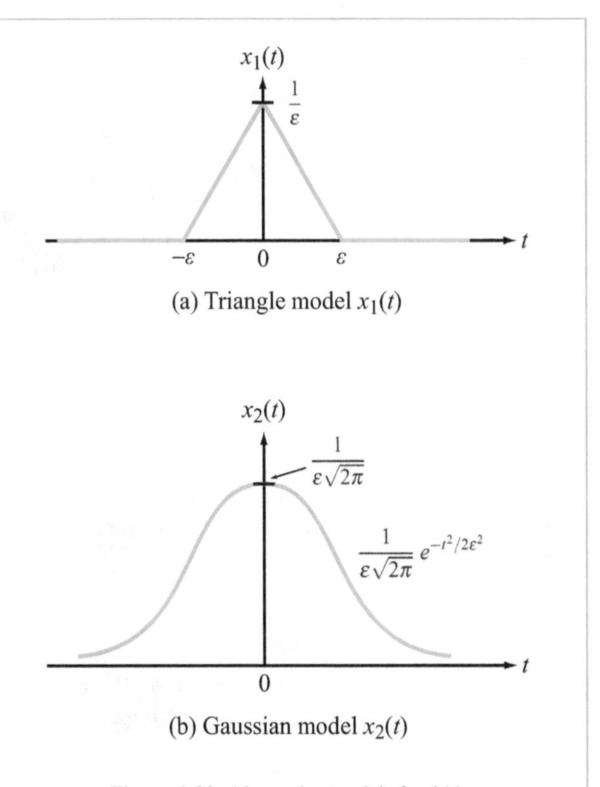

Figure 1-20: Alternative models for $\delta(t)$.

(4) Area of triangle $= \frac{1}{2}\left(2\epsilon \times \frac{1}{\epsilon}\right) = 1$, regardless of the value of ϵ.

Hence, $x_1(t)$ does qualify as a unit impulse function.

(b) Gaussian Model $x_2(t)$

(1) Except at $t = 0$, as $\epsilon \to 0$, the magnitude of the exponential $e^{-t^2/2\epsilon^2}$ always will be smaller than $\epsilon\sqrt{2\pi}$. Hence, $x_2(t) \to 0$ as $\epsilon \to 0$, except at $t = 0$.

(2) At $t = 0$,

$$\lim_{\epsilon\to 0}\left[\frac{1}{\epsilon\sqrt{2\pi}}\, e^{-t^2/2\epsilon^2}\right]_{t=0} = \lim_{\epsilon\to 0}\left[\frac{1}{\epsilon\sqrt{2\pi}}\right] = \infty.$$

(3) $x_2(t)$ is clearly an even function (Fig. 1-20(b)).

(4) The area of the Gaussian model is

$$A = \int_{-\infty}^{\infty} \frac{1}{\epsilon\sqrt{2\pi}}\, e^{-t^2/2\epsilon^2}\, dt.$$

Applying the integral formula

$$\int_{-\infty}^{\infty} e^{-a^2x^2}\, dx = \frac{\sqrt{\pi}}{a}$$

leads to $A = 1$. Hence, $x_2(t)$ qualifies as a unit impulse function.

1-4.5 Sampling Property of $\delta(t)$

As was noted earlier, multiplying an impulse function by a constant k gives a scaled impulse of area k. Now we consider what happens when a continuous-time function $x(t)$ is multiplied by $\delta(t)$. Since $\delta(t)$ is zero everywhere except at $t = 0$, it follows that

$$x(t)\,\delta(t) = x(0)\,\delta(t), \tag{1.27}$$

provided that $x(t)$ is continuous at $t = 0$. By extension, multiplication of $x(t)$ by the time-shifted impulse function $\delta(t - T)$ gives

$$x(t)\,\delta(t - T) = x(T)\,\delta(t - T). \tag{1.28}$$

► Multiplication of a time-continuous function $x(t)$ by an impulse located at $t = T$ generates a scaled impulse of magnitude $x(T)$ at $t = T$, provided $x(t)$ is continuous at $t = T$. ◄

One of the most useful features of the impulse function is its *sampling (or sifting) property*. For any function $x(t)$ known to be continuous at $t = T$:

$$\int_{-\infty}^{\infty} x(t)\,\delta(t - T)\,dt = x(T). \tag{1.29}$$

(sampling property)

Derivation of the sampling property relies on Eqs. (1.22b) and (1.28):

$$\int_{-\infty}^{\infty} x(t)\,\delta(t - T)\,dt = \int_{-\infty}^{\infty} x(T)\,\delta(t - T)\,dt$$
$$= x(T)\int_{-\infty}^{\infty} \delta(t - T)\,dt = x(T).$$

1-4.6 Time-Scaling Transformation of $\delta(t)$

To determine how time scaling affects impulses, let us evaluate the area of $\delta(at)$:

$$\int_{-\infty}^{\infty} \delta(at)\,dt = \int_{-\infty}^{\infty} \delta(\tau)\,\frac{d\tau}{|a|} = \frac{1}{|a|}.$$

Hence, $\delta(at)$ is an impulse of area $1/|a|$. It then follows that

$$\delta(at) = \frac{1}{|a|}\,\delta(t) \qquad \text{for } a \neq 0. \tag{1.30}$$

(time-scaling property)

This result can be visualized for $|a| > 1$ by recalling that scaling time by a compresses the time axis by $|a|$. The area of the uncompressed rectangle in Fig. 1-19(b) is

$$\text{Area of } \delta(t): \qquad \frac{1}{2\epsilon}[\epsilon - (-\epsilon)] = 1.$$

Repeating the calculation for a compressed rectangle gives

$$\text{Area of } \delta(at): \qquad \frac{1}{2\epsilon}\left[\frac{\epsilon}{|a|} - \left(-\frac{\epsilon}{|a|}\right)\right] = \frac{1}{|a|}.$$

Also note that the impulse is an even function because

$$\delta(-t) = \frac{1}{|-1|}\,\delta(t) = \delta(t),$$

$$\delta(at - b) = \frac{1}{|a|}\,\delta\left(t - \frac{b}{a}\right).$$

Example 1-6: Impulse Integral

Evaluate $\int_1^2 t^2\,\delta(2t - 3)\,dt$.

Solution: Using the time-scaling property, the impulse function can be expressed as

$$\delta(2t - 3) = \delta\left(2\left(t - \frac{3}{2}\right)\right)$$
$$= \frac{1}{2}\,\delta\left(t - \frac{3}{2}\right).$$

Hence,

$$\int_1^2 t^2\,\delta(2t - 3)\,dt = \frac{1}{2}\int_1^2 t^2\,\delta\left(t - \frac{3}{2}\right)\,dt$$
$$= \frac{1}{2}\left(\frac{3}{2}\right)^2$$
$$= \frac{9}{8}.$$

We note that $\delta(t - (3/2)) \neq 0$ only at $t = 3/2$, which is included in the interval of integration, $1 \leq t \leq 2$.

Concept Question 1-8: How is $u(t)$ related to $\delta(t)$? (See S²)

Concept Question 1-9: Why is Eq. (1.29) called the *sampling property* of the impulse function? (See S²)

Exercise 1-7: If $x(t)$ is the rectangular pulse shown in Fig. E1-7(a), determine its time derivative $x'(t)$ and plot it.

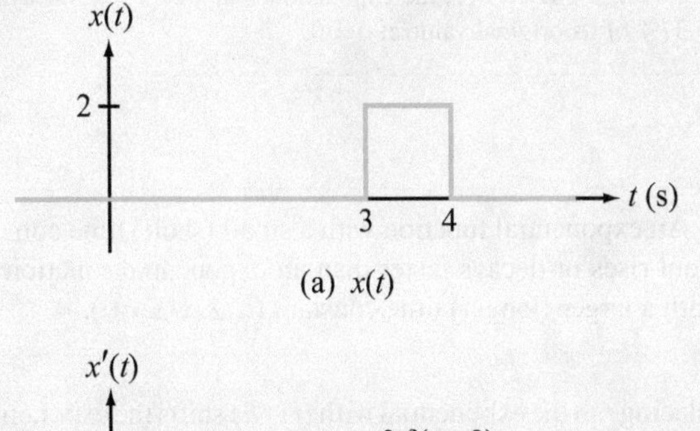

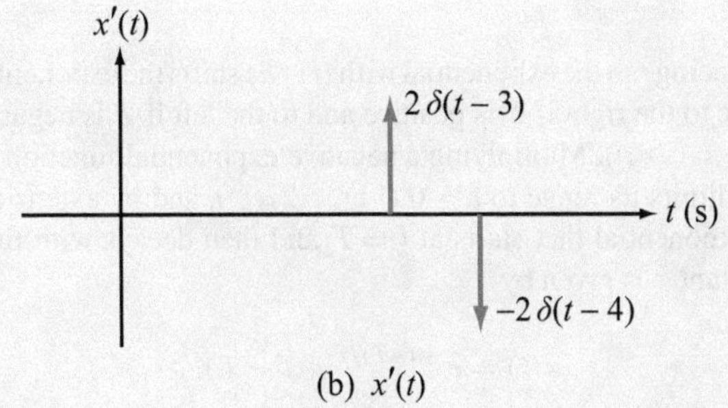

Figure E1-7

Answer: $x'(t) = 2\delta(t - 3) - 2\delta(t - 4)$. (See S²)

1-4.7 Exponential Waveform

The ***exponential function*** is a particularly useful tool for characterizing fast-rising and fast-decaying waveforms. Figure 1-21 displays plots for

$$x_1(t) = e^{t/\tau}$$

and

$$x_2(t) = e^{-t/\tau}$$

for $\tau > 0$. The rates of increase of $x_1(t)$ and decrease of $x_2(t)$ are governed by the magnitude of the ***time constant*** τ.

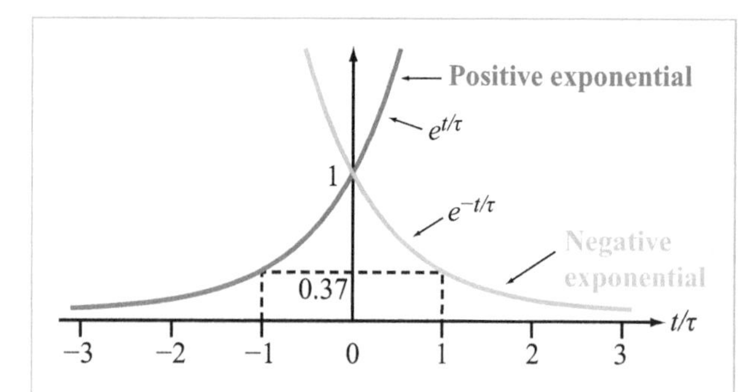

Figure 1-21: By $t = \tau$, the exponential function $e^{-t/\tau}$ decays to 37% of its original value at $t = 0$.

▶ An exponential function with a small (short) time constant rises or decays faster than an exponential function with a larger (longer) time constant (Fig. 1-22(a)). ◀

Replacing t in the exponential with $(t-T)$ shifts the exponential curve to the right if T is positive and to the left if T is negative (Fig. 1-22(b)). Multiplying a negative exponential function by $u(t)$ limits its range to $t > 0$ (Fig. 1-22(c)), and by extension, an exponential that starts at $t = T$ and then decays with time constant τ is given by

$$x(t) = e^{-(t-T)/\tau}\ u(t-T).$$

Its waveform is displayed in Fig. 1-22(d).

Occasionally, we encounter waveforms with the shape shown in Fig. 1-22(e), wherein $x(t)$ starts at zero and builds up as a function of time towards a saturation value. An example is the voltage response of an initially uncharged capacitor,

$$\upsilon(t) = V_0(1 - e^{-t/\tau})\ u(t).$$

Table 1-2 provides a general summary of the shapes and expressions of the five nonperiodic waveforms we reviewed in this section.

Concept Question 1-10: If the time constant of a negative exponential function is doubled in value, will the corresponding waveform decay faster or slower? (See S²)

Concept Question 1-11: What is the approximate shape of the waveform described by the function $(1 - e^{-|t|})$? (See S²)

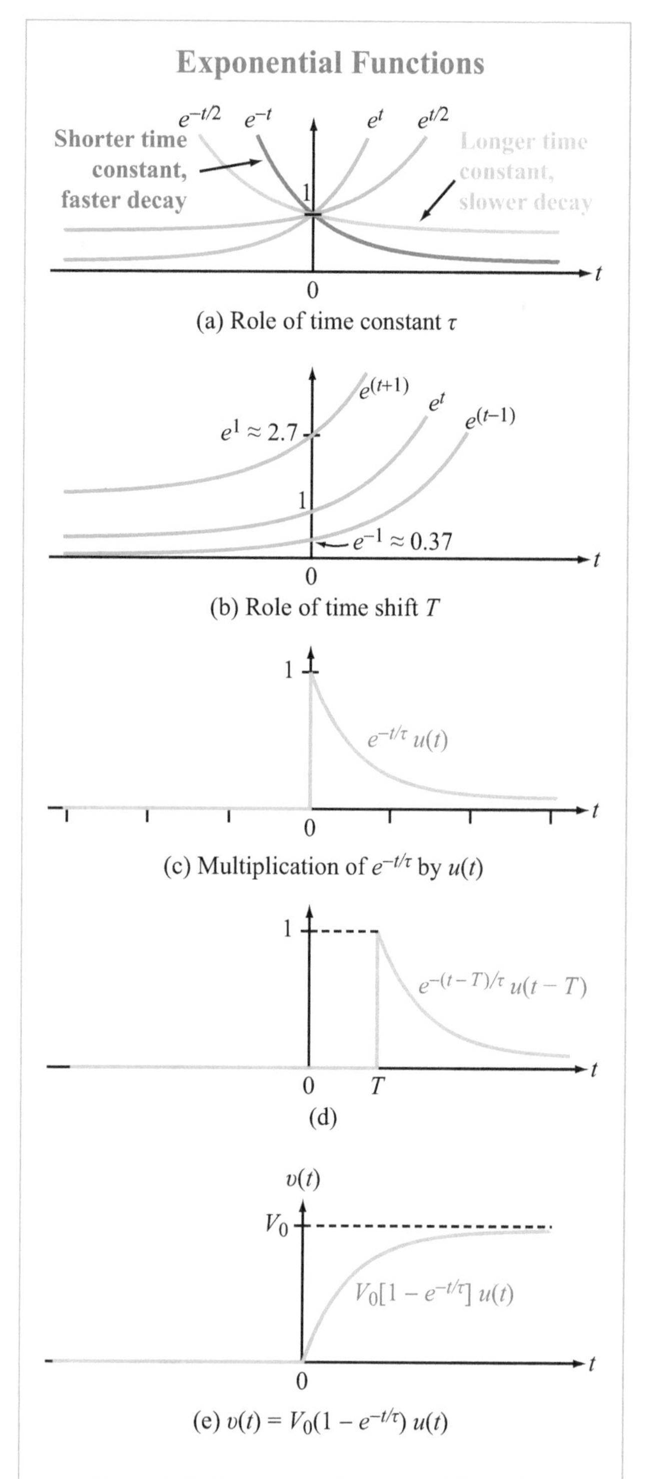

Figure 1-22: Properties of the exponential function.

Table 1-2: **Common nonperiodic functions.**

Function	Expression	General Shape
Step	$u(t-T) = \begin{cases} 0 & \text{for } t < T \\ 1 & \text{for } t > T \end{cases}$	
Ramp	$r(t-T) = (t-T)\,u(t-T)$	
Rectangle	$\text{rect}\left(\frac{t-T}{\tau}\right) = u(t-T_1) - u(t-T_2)$ $T_1 = T - \frac{\tau}{2}; \quad T_2 = T + \frac{\tau}{2}$	
Impulse	$\delta(t-T)$	
Exponential	$\exp[-(t-T)/\tau]\,u(t-T)$	

Exercise 1-8: The radioactive decay equation for a certain material is given by $n(t) = n_0 e^{-t/\tau}$, where n_0 is the initial count at $t = 0$. If $\tau = 2 \times 10^8$ s, how long is its half-life? [Half-life $t_{1/2}$ is the time it takes a material to decay to 50% of its initial value.]

Answer: $t_{1/2} = 1.386 \times 10^8$ s $\approx$ 4 years. (See S²)

Exercise 1-9: If the current $i(t)$ through a resistor R decays exponentially with a time constant τ, what is the ratio of the power dissipated in the resistor at $t = \tau$ to its value at $t = 0$?

Answer: $p(t) = i^2 R = I_0^2 R(e^{-t/\tau})^2 = I_0^2 R e^{-2t/\tau}$, $p(\tau)/p(0) = e^{-2} = 0.135$, or 13.5%. (See S²)

1-5 Signal Power and Energy

The instantaneous power $p(t)$ dissipated in a resistor R due to the flow of current $i(t)$ through it is

$$p(t) = i^2(t)\, R. \tag{1.31}$$

Additionally, the associated energy expended over a time interval $t_1 < t < t_2$ is

$$E = \int_{t_1}^{t_2} p(t)\, dt. \tag{1.32}$$

The expressions for power and energy associated with a resistor can be extended to characterize the ***instantaneous power*** and total ***energy*** of any signal $x(t)$—whether electrical or not and whether real or complex—as

$$p(t) = |x(t)|^2 \tag{1.33a}$$

and

$$E = \lim_{T\to\infty} \int_{-T}^{T} |x(t)|^2 \, dt = \int_{-\infty}^{\infty} |x(t)|^2 \, dt, \qquad (1.33b)$$

where E is defined as the *total energy* over an infinite time interval ($-\infty < t < \infty$).

If $|x(t)|^2$ does not approach zero as $t \to \pm\infty$, the integral in Eq. (1.33b) will not converge. In that case, E is infinite, rendering it unsuitable as a measure of the signal's energy capacity. An alternative measure is the *(time) average power* P_{av}, which is defined as the power $p(t)$ averaged over all time:

$$P_{av} = \lim_{T\to\infty} \frac{1}{T} \int_{-T/2}^{T/2} p(t) \, dt = \lim_{T\to\infty} \frac{1}{T} \int_{-T/2}^{T/2} |x(t)|^2 \, dt. \qquad (1.34)$$

Conversely, if E is finite, P_{av} becomes the unsuitable measure, because in view of Eq. (1.33b), we have

$$P_{av} = \lim_{T\to\infty} \frac{E}{T} = 0 \qquad (E \text{ finite}). \qquad (1.35)$$

▶ P_{av} and E define three classes of signals:

(a) *Power signals*: P_{av} is finite and $E \to \infty$

(b) *Energy signals*: $P_{av} = 0$ and E is finite

(c) *Non-physical signals*: $P_{av} \to \infty$ and $E \to \infty$ ◀

For a periodic signal of period T_0, it is not necessary to evaluate the integral in Eq. (1.34) with $T \to \infty$; integration over a single period is sufficient. That is,

$$P_{av} = \frac{1}{T_0} \int_{-T_0/2}^{T_0/2} |x(t)|^2 \, dt \quad \textbf{(periodic signal)}. \qquad (1.36)$$

Most periodic signals have finite average power; hence, they qualify as power signals.

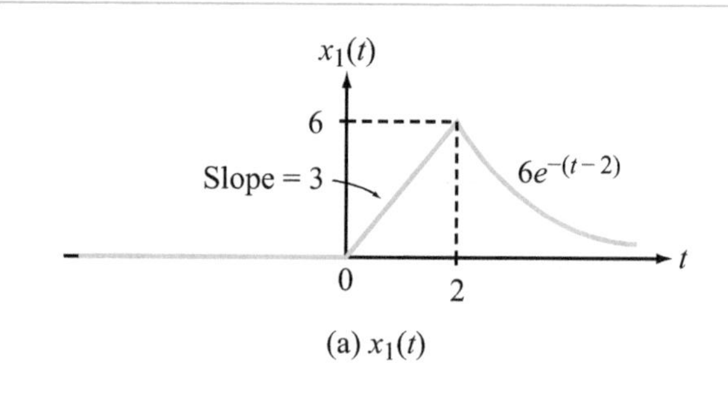

(a) $x_1(t)$

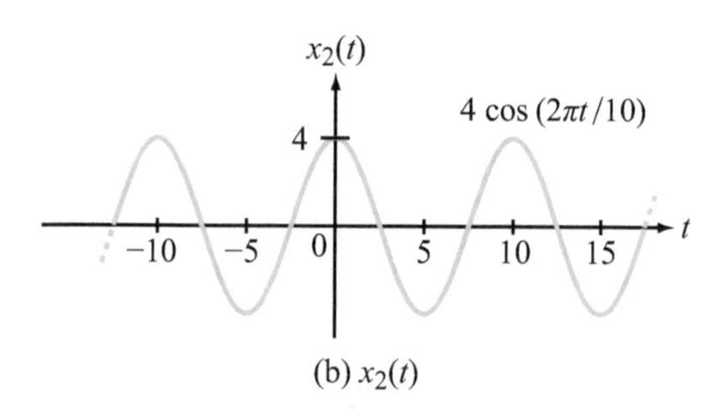

(b) $x_2(t)$

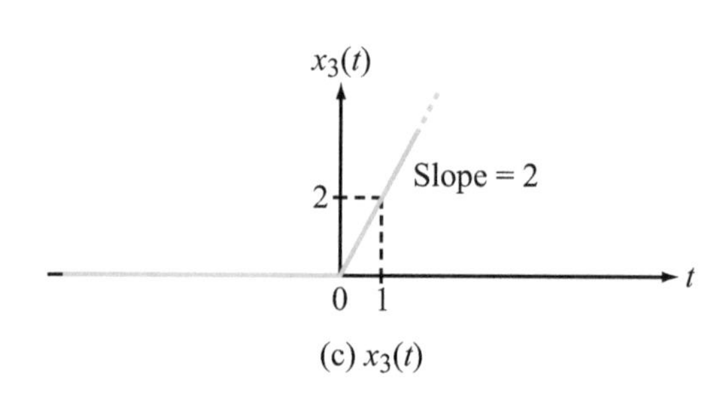

(c) $x_3(t)$

Figure 1-23: Signal waveforms for Example 1-7.

Example 1-7: Power and Energy

Evaluate P_{av} and E for each of the three signals displayed in Fig. 1-23.

Solution:

(a) Signal $x_1(t)$ is given by

$$x_1(t) = \begin{cases} 0 & \text{for } t \leq 0, \\ 3t & \text{for } 0 \leq t \leq 2, \\ 6e^{-(t-2)} & \text{for } t \geq 2. \end{cases}$$

Its total energy E is

$$\begin{aligned} E_1 &= \int_0^2 (3t)^2\,dt + \int_2^\infty [6e^{-(t-2)}]^2\,dt \\ &= \int_0^2 9t^2\,dt + \int_2^\infty 36e^{-2(t-2)}\,dt \\ &= \left.\frac{9t^3}{3}\right|_0^2 + 36e^4 \int_2^\infty e^{-2t}\,dt \\ &= 24 + 36e^4 \left(\left.\frac{-e^{-2t}}{2}\right|_2^\infty \right) \\ &= 42. \end{aligned}$$

Note that the second integral represents the energy of $6e^{-t}\,u(t)$ delayed by 2 s. Since delaying a signal does not alter its energy, an alternative method for evaluating the second integral is

$$\begin{aligned} \int_0^\infty (6e^{-t})^2\,dt &= 36 \int_0^\infty e^{-2t}\,dt \\ &= 18. \end{aligned}$$

Since E_1 is finite, it follows from Eq. (1.35) that $P_{\text{av}_1} = 0$.

(b) Signal $x_2(t)$ is a periodic signal given by

$$x_2(t) = 4\cos\left(\frac{2\pi t}{10}\right).$$

From the argument of $\cos(2\pi t/10)$, the period is 10 s. Hence, application of Eq. (1.36) leads to

$$\begin{aligned} P_{\text{av}_2} &= \frac{1}{10} \int_{-5}^{5} \left[4\cos\left(\frac{2\pi t}{10}\right) \right]^2 dt \\ &= \frac{1}{10} \int_{-5}^{5} 16\cos^2\left(\frac{2\pi t}{10}\right) dt \\ &= 8. \end{aligned}$$

The integration was facilitated by the integral relation

$$\frac{1}{T_0} \int_{-T_0/2}^{T_0/2} \cos^2\left(\frac{2\pi n t}{T_0} + \phi\right) dt = \frac{1}{2}, \tag{1.37}$$

which is valid for any value of ϕ and any integer value of n equal to or greater than 1. In fact, because of Eq. (1.37),

$$P_{\text{av}} = \frac{A^2}{2} \quad \left(\begin{array}{l}\textbf{for any sinusoidal}\\ \textbf{signal of amplitude } \boldsymbol{A}\\ \textbf{and nonzero frequency}\end{array}\right). \tag{1.38}$$

If $\omega_0 = 0$, $P_{\text{av}} = A^2$, not $A^2/2$. Since P_{av_2} is finite, it follows that $E_2 \to \infty$.

(c) Signal $x_3(t)$ is given by

$$x_3(t) = 2r(t) = \begin{cases} 0 & \text{for } t \le 0, \\ 2t & \text{for } t \ge 0. \end{cases}$$

The time-averaged power associated with $x_3(t)$ is

$$\begin{aligned} P_{\text{av}_3} &= \lim_{T\to\infty} \frac{1}{T} \int_0^{T/2} 4t^2\,dt \\ &= \lim_{T\to\infty} \frac{1}{T} \left[\left.\frac{4t^3}{3}\right|_0^{T/2} \right] \\ &= \lim_{T\to\infty} \left[\frac{1}{T} \times \frac{4T^3}{24} \right] \\ &= \lim_{T\to\infty} \left[\frac{T^2}{6} \right] \to \infty. \end{aligned}$$

Moreover, $E_3 \to \infty$ as well.

Concept Question 1-12: Signals are divided into three power/energy classes. What are they? (See S²)

Exercise 1-10: Determine the values of P_{av} and E for a pulse signal given by $x(t) = 5\,\text{rect}\left(\frac{t-3}{4}\right)$.

Answer: $P_{\text{av}} = 0$ and $E = 100$. (See S²)

Summary

Concepts

- A signal may be continuous, discrete, analog, or digital. It may vary with time, space, or some other independent variable and may be single or multidimensional.
- Signals are classified as causal, noncausal, or anticausal, according to when they start and end.
- Signals can undergo time-shift, time-scaling, and time-reversal transformations.
- A signal may exhibit even or odd symmetry. A signal with neither form of symmetry can be synthesized as the sum of two component signals: one with even symmetry and the other with odd symmetry.
- Real-world signal waveforms often are modeled in terms of a set of elementary waveforms, which include the step, ramp, pulse, impulse, and exponential waveforms.
- A signal's energy capacity is characterized by its average power P_{av} and total energy E. These attributes are defined for any signal, whether electrical or not.

Mathematical and Physical Models

Signal Transformations

Time shift	$y(t) = x(t - T)$
Time scaling	$y(t) = x(at)$
Time reversal	$y(t) = x(-t)$

Signal Symmetry

Even	$x(t) = x(-t)$
Odd	$x(t) = -x(-t)$
Even part	$x_{\text{e}}(t) = \frac{1}{2}\{x(t) + x(-t)\}$
Odd part	$x_{\text{o}}(t) = \frac{1}{2}\{x(t) - x(-t)\}$
Sum	$x(t) = x_{\text{e}}(t) + x_{\text{o}}(t)$

Signal Waveforms

See Table 1-2

Signal Power and Energy

$$P_{\text{av}} = \lim_{T \to \infty} \frac{1}{T} \int_{-T/2}^{T/2} |x(t)|^2 \, dt$$

$$E = \lim_{T \to \infty} \int_{-T}^{T} |x(t)|^2 \, dt = \int_{-\infty}^{\infty} |x(t)|^2 \, dt$$

Important Terms

Provide definitions or explain the meaning of the following terms:

analog signal	even symmetry	physically realizable system	time constant
anticausal signal	exponential waveform	pulse waveform	time reversal
causal signal	impulse function	ramp function	time-scaled
continuous signal	noncausal signal	sampling property	time-shifted
digital signal	nonperiodic (aperiodic)	signal power	unit rectangular
Dirac delta function	odd symmetry	signal energy	unit step function
discrete signal	periodic	singularity	

PROBLEMS

Section 1-1: Types of Signals

1.1 Is each of these 1-D signals:

- Analog or digital?
- Continuous-time or discrete-time?

(a) Daily closes of the stock market
(b) Output from phonograph-record pickup
(c) Output from compact-disc pickup

1.2 Is each of these 2-D signals:

- Analog or digital?
- Continuous-space or discrete-space?

(a) Image in a telescope eyepiece
(b) Image displayed on digital TV
(c) Image stored in a digital camera

1.3 The following signals are 2-D in space and 1-D in time, so they are 3-D signals. Is each of these 3-D signals:

- Analog or digital?
- Continuous or discrete?

(a) The world as you see it
*(b) A movie stored on film
(c) A movie stored on a DVD

Section 1-2: Signal Transformations

1.4 Given the waveform of $x_1(t)$ shown in Fig. P1.4(a), generate and plot the waveform of:

(a) $x_1(-2t)$
(b) $x_1[-2(t-1)]$

1.5 Given the waveform of $x_2(t)$ shown in Fig. P1.4(b), generate and plot the waveform of:

(a) $x_2[-(t+2)/2]$
(b) $x_2[-(t-2)/2]$

1.6 Given the waveform of $x_3(t)$ shown in Fig. P1.4(c), generate and plot the waveform of:

*Answer(s) in Appendix F.

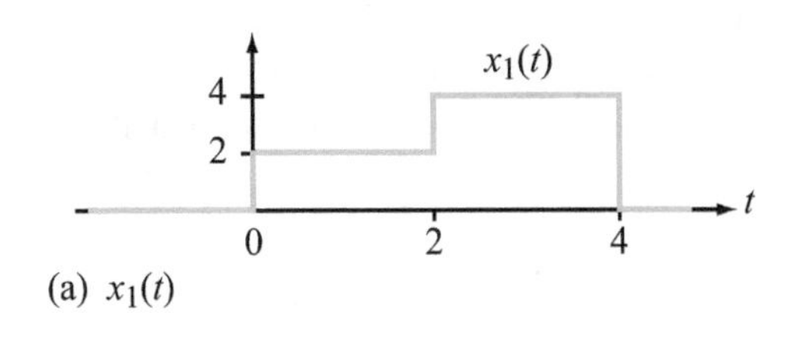

(a) $x_1(t)$

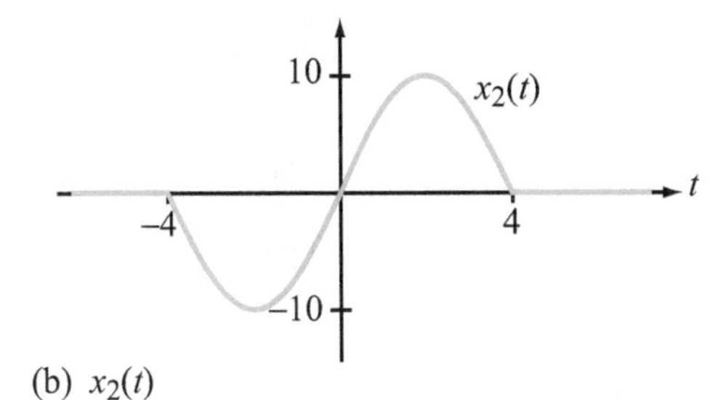

(b) $x_2(t)$

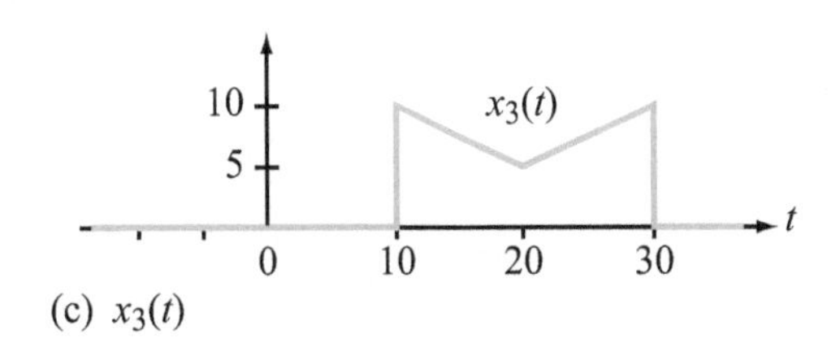

(c) $x_3(t)$

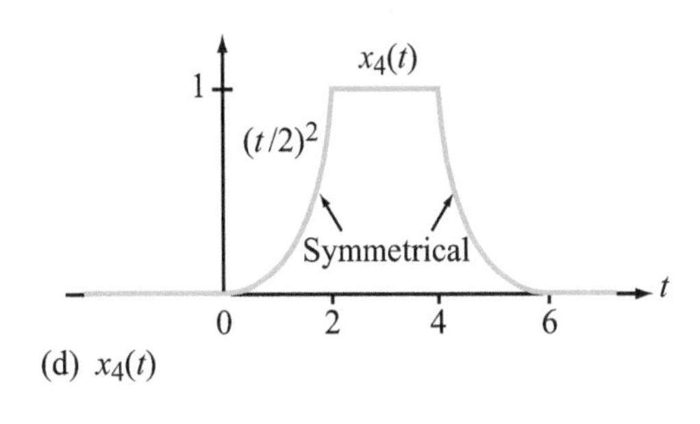

(d) $x_4(t)$

Figure P1.4: Waveforms for Problems 1.4 to 1.7.

*(a) $x_3[-(t+40)]$
(b) $x_3(-2t)$

1.7 The waveform shown in Fig. P1.4(d) is given by

$$x_4(t) = \begin{cases} 0 & \text{for } t \le 0, \\ \left(\frac{t}{2}\right)^2 & \text{for } 0 \le t \le 2 \text{ s}, \\ 1 & \text{for } 2 \le t \le 4 \text{ s}, \\ f(t) & \text{for } 4 \le t \le 6 \text{ s}, \\ 0 & \text{for } t \ge 6 \text{ s}. \end{cases}$$

(a) Obtain an expression for $f(t)$, which is the segment covering the time duration between 4 s and 6 s.

(b) Obtain an expression for $x_4(-(t-4))$ and plot it.

1.8 If

$$x(t) = \begin{cases} 0 & \text{for } t \leq 2 \\ (2t-4) & \text{for } t \geq 2, \end{cases}$$

plot $x(t)$, $x(t+1)$, $x\left(\frac{t+1}{2}\right)$, and $x\left(-\frac{(t+1)}{2}\right)$.

1.9 Given $x(t) = 10(1 - e^{-|t|})$, plot $x(-t+1)$.

1.10 Given $x(t) = 5\sin^2(6\pi t)$, plot $x(t-3)$ and $x(3-t)$.

1.11 Given the waveform of $x(t)$ shown in Fig. P1.11(a), generate and plot the waveform of:

(a) $x(2t+6)$

*(b) $x(-2t+6)$

(c) $x(-2t-6)$

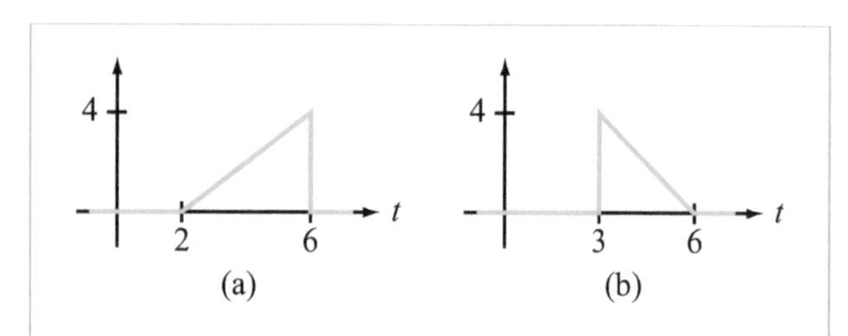

Figure P1.11: Waveforms for Problems 1.11 and 1.12.

1.12 Given the waveform of $x(t)$ shown in Fig. P1.11(b), generate and plot the waveform of:

(a) $x(3t+6)$

(b) $x(-3t+6)$

(c) $x(-3t-6)$

1.13 If $x(t) = 0$ unless $a \leq t \leq b$, and $y(t) = x(ct+d)$ unless $e \leq t \leq f$, compute e and f in terms of a, b, c, and d. Assume $c > 0$ to make things easier for you.

1.14 If $x(t)$ is a musical note signal, what is $y(t) = x(4t)$? Consider sinusoidal $x(t)$.

1.15 Give an example of a non-constant signal that has the property $x(t) = x(at)$ for all $a > 0$.

Sections 1-3 and 1-4: Waveforms

1.16 For each of the following functions, indicate if it exhibits even symmetry, odd symmetry, or neither one.

(a) $x_1(t) = 3t^2 + 4t^4$

*(b) b $x_2(t) = 3t^3$

1.17 For each of the following functions, indicate if it exhibits even symmetry, odd symmetry, or neither one.

(a) $x_1(t) = 4[\sin(3t) + \cos(3t)]$

(b) $x_2(t) = \dfrac{\sin(4t)}{4t}$

1.18 For each of the following functions, indicate if it exhibits even symmetry, odd symmetry, or neither one.

(a) $x_1(t) = 1 - e^{-2t}$

(b) $x_2(t) = 1 - e^{-2t^2}$

1.19 Generate plots for each of the following step-function waveforms over the time span from -5 s to $+5$ s.

(a) $x_1(t) = -6u(t+3)$

(b) $x_2(t) = 10u(t-4)$

(c) $x_3(t) = 4u(t+2) - 4u(t-2)$

1.20 Generate plots for each of the following step-function waveforms over the time span from -5 s to $+5$ s.

(a) $x_1(t) = 8u(t-2) + 2u(t-4)$

*(b) $x_2(t) = 8u(t-2) - 2u(t-4)$

(c) $x_3(t) = -2u(t+2) + 2u(t+4)$

1.21 Provide expressions in terms of step functions for the waveforms displayed in Fig. P1.21.

1.22 Generate plots for each of the following functions over the time span from -4 s to $+4$ s.

(a) $x_1(t) = 5r(t+2) - 5r(t)$

(b) $x_2(t) = 5r(t+2) - 5r(t) - 10u(t)$

*(c) $x_3(t) = 10 - 5r(t+2) + 5r(t)$

(d) $x_4(t) = 10\,\text{rect}\left(\dfrac{t+1}{2}\right) - 10\,\text{rect}\left(\dfrac{t-3}{2}\right)$

(e) $x_5(t) = 5\,\text{rect}\left(\dfrac{t-1}{2}\right) - 5\,\text{rect}\left(\dfrac{t-3}{2}\right)$

1.23 Provide expressions for the waveforms displayed in Fig. P1.23 in terms of ramp and step functions.

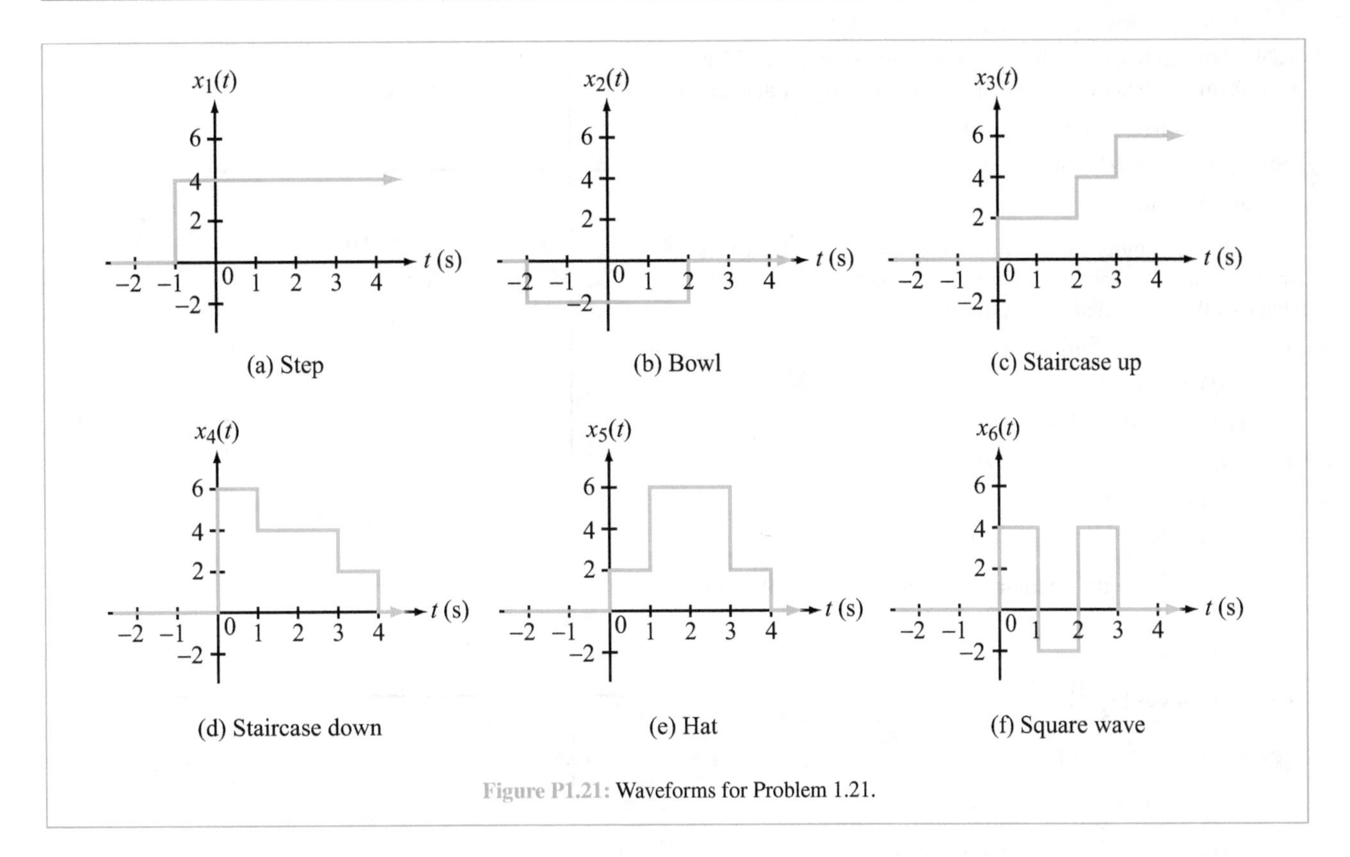

Figure P1.21: Waveforms for Problem 1.21.

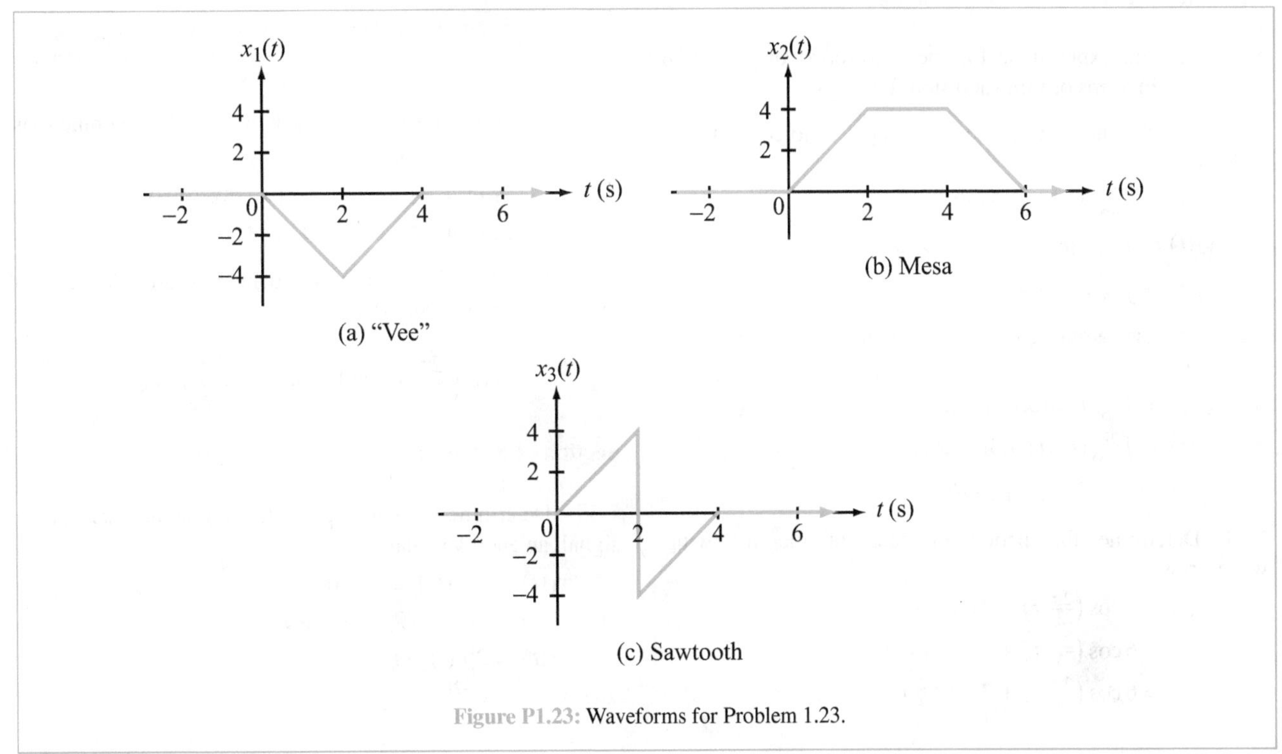

Figure P1.23: Waveforms for Problem 1.23.

1.24 For each of the following functions, indicate if its waveform exhibits even symmetry, odd symmetry, or neither.

(a) $x_1(t) = u(t-3) + u(-t-3)$

(b) $x_2(t) = \sin(2t)\cos(2t)$

(c) $x_3(t) = \sin(t^2)$

1.25 Provide plots for the following functions over a time span and with a time scale that will appropriately display the shape of the associated waveform of:

(a) $x_1(t) = 100e^{-2t}\, u(t)$

(b) $x_2(t) = -10e^{-0.1t}\, u(t)$

(c) $x_3(t) = -10e^{-0.1t}\, u(t-5)$

(d) $x_4(t) = 10(1 - e^{-10^3 t})\, u(t)$

(e) $x_5(t) = 10e^{-0.2(t-4)}\, u(t)$

(f) $x_6(t) = 10e^{-0.2(t-4)}\, u(t-4)$

1.26 Determine the period of each of the following waveforms.

(a) $x_1(t) = \sin 2t$

(b) $x_2(t) = \cos\left(\frac{\pi}{3}\, t\right)$

(c) $x_3(t) = \cos^2\left(\frac{\pi}{3}\, t\right)$

*(d) $x_4(t) = \cos(4\pi t + 60°) - \sin(4\pi t + 60°)$

(e) $x_5(t) = \cos\left(\frac{4}{\pi}\, t + 30°\right) - \sin(4\pi t + 30°)$

1.27 Provide expressions for the waveforms displayed in Fig. P1.27 in terms of ramp and step functions.

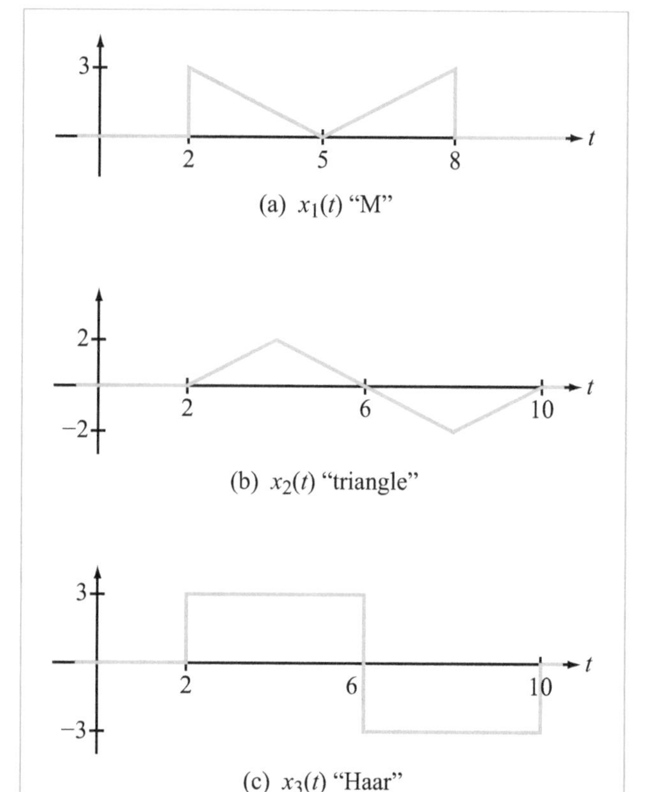

Figure P1.27: Waveforms for Problem 1.27.

1.28 Use the sampling property of impulses to compute the following.

(a) $y_1(t) = \int_{-\infty}^{\infty} t^3\, \delta(t-2)\, dt$

(b) $y_2(t) = \int_{-\infty}^{\infty} \cos(t)\, \delta(t - \pi/3)\, dt$

(c) $y_3(t) = \int_{-3}^{-1} t^5\, \delta(t+2)\, dt$

1.29 Use the sampling property of impulses to compute the following.

(a) $y_1(t) = \int_{-\infty}^{\infty} t^3\, \delta(3t-6)\, dt$

*(b) $y_2(t) = \int_{-\infty}^{\infty} \cos(t)\, \delta(3t - \pi)\, dt$

(c) $y_3(t) = \int_{-3}^{-1} t^5\, \delta(3t+2)\, dt$

1.30 Determine the period of each of the following waveforms.

(a) $x_1(t) = 6\cos\left(\frac{2\pi}{3}\, t\right) + 7\cos\left(\frac{\pi}{2}\, t\right)$

(b) $x_2(t) = 6\cos\left(\frac{2\pi}{3}\, t\right) + 7\cos(\pi\sqrt{2}\, t)$

(c) $x_3(t) = 6\cos\left(\frac{2\pi}{3}\, t\right) + 7\cos\left(\frac{2}{3}\, t\right)$

1.31 Determine the period of each of the following functions.

(a) $x_1(t) = (3 + j2)e^{j\pi t/3}$

(b) $x_2(t) = (1 + j2)e^{j2\pi t/3} + (4 + j5)e^{j2\pi t/6}$

(c) $x_3(t) = (1 + j2)e^{jt/3} + (4 + j5)e^{jt/2}$

1.32 If M and N are both positive integers, provide a general expression for the period of

$$A\cos\left(\frac{2\pi}{M}\, t + \theta\right) + B\cos\left(\frac{2\pi}{N}\, t + \phi\right).$$

Sections 1-5: Power and Energy

1.33 Determine if each of the following signals is a power signal, an energy signal, or neither.

(a) $x_1(t) = 3[u(t+2) - u(t-2)]$

(b) $x_2(t) = 3[u(t-2) - u(t+2)]$

(c) $x_3(t) = 2[r(t) - r(t-2)]$

(d) $x_4(t) = e^{-2t}\, u(t)$

1.34 Determine if each of the following signals is a power signal, an energy signal, or neither.

(a) $x_1(t) = [1 - e^{2t}]\, u(t)$

*(b) $x_2(t) = [t \cos(3t)]\, u(t)$

(c) $x_3(t) = [e^{-2t} \sin(t)]\, u(t)$

1.35 Determine if each of the following signals is a power signal, an energy signal, or neither.

(a) $x_1(t) = [1 - e^{-2t}]\, u(t)$

(b) $x_2(t) = 2\sin(4t)\cos(4t)$

(c) $x_3(t) = 2\sin(3t)\cos(4t)$

1.36 Use the notation for a signal $x(t)$:

$E[x(t)]$: total energy of the signal $x(t)$,
$P_{\text{av}}[x(t)]$: average power of the signal $x(t)$ if $x(t)$ is periodic.

Prove each of the following energy and power properties.

(a)
$$E[x(t+b)] = E[x(t)]$$
and
$$P_{\text{av}}[x(t+b)] = P_{\text{av}}[x(t)]$$
(time shifts do not affect power or energy).

(b)
$$E[ax(t)] = |a|^2\, E[x(t)]$$
and
$$P_{\text{av}}[ax(t)] = |a|^2\, P_{\text{av}}[x(t)]$$
(scaling by a scales energy and power by $|a|^2$).

(c)
$$E[x(at)] = \frac{1}{a}\, E[x(t)]$$
and
$$P_{\text{av}}[x(at)] = P_{\text{av}}[x(t)]$$
if $a > 0$ (time scaling scales energy by $\frac{1}{a}$ but doesn't affect power).

1.37 Use the properties of Problem 1.36 to compute the energy of the three signals in Fig. P1.27.

1.38 Compute the energy of the following signals.

(a) $x_1(t) = e^{-at}\, u(t)$ for $a > 0$

(b) $x_2(t) = e^{-a|t|}$ for $a > 0$

(c) $x_3(t) = (1 - |t|)\,\text{rect}(t/2)$

1.39 Compute the average power of the following signals.

(a) $x_1(t) = e^{jat}$ for real-valued a

(b) $x_2(t) = (3 - j4)e^{j7t}$

*(c) c $x_3(t) = e^{j3}e^{j5t}$

1.40 Prove these energy properties.

(a) If the even-odd decomposition of $x(t)$ is
$$x(t) = x_{\text{e}}(t) + x_{\text{o}}(t),$$
then
$$E[x(t)] = E[x_{\text{e}}(t)] + E[x_{\text{o}}(t)].$$

(b) If the causal-anticausal decomposition of $x(t)$ is $x(t) = x(t)\, u(t) + x(t)\, u(-t)$, then
$$E[x(t)] = E[x(t)\, u(t)] + E[x(t)\, u(-t)].$$

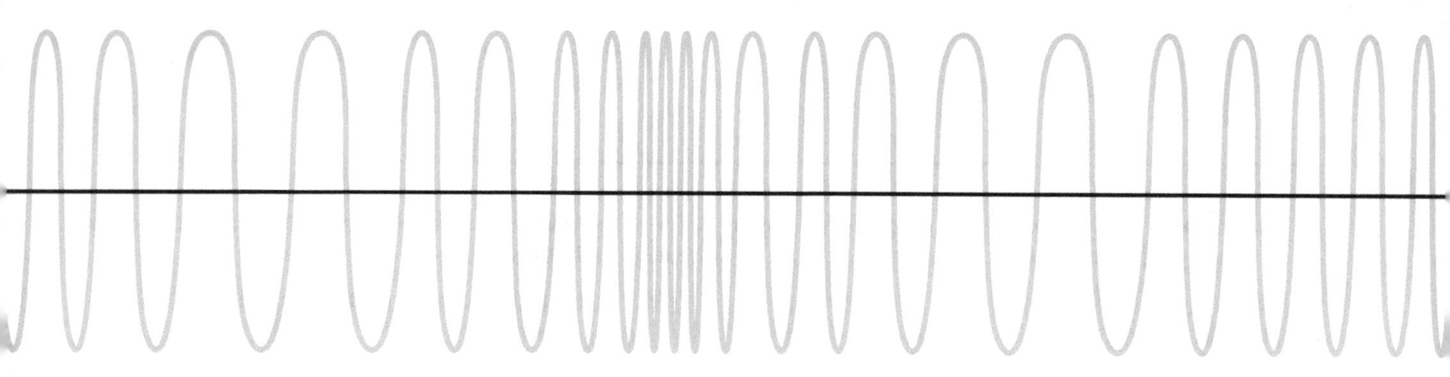

2 Linear Time-Invariant Systems

Contents

Overview, 31
2-1 Linear Time-Invariant Systems, 31
2-2 Impulse Response, 35
2-3 Convolution, 40
2-4 Graphical Convolution, 46
2-5 Convolution Properties, 50
2-6 Causality and BIBO Stability, 58
2-7 LTI Sinusoidal Response, 61
2-8 Impulse Response of Second-Order LCCDEs, 65
2-9 Car Suspension System, 72
Summary, 78
Problems, 79

Objectives

Learn to:

- Describe the properties of LTI systems.
- Determine the impulse and step responses of LTI systems.
- Perform convolution of two functions.
- Determine causality and stability of LTI systems.
- Determine the overdamped, underdamped, and critically damped responses of second-order systems.
- Determine a car's response to various pavement profiles.

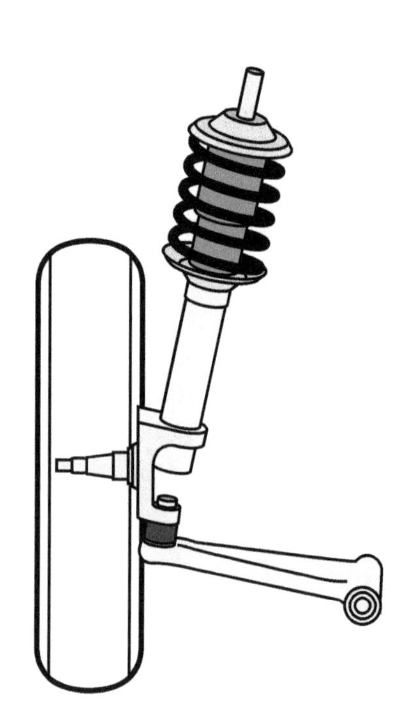

By modeling a car suspension system in terms of a differential equation, we can determine ***the response*** of the car's body to any pavement profile it is made to drive over. The same approach can be used to compute the response of any ***linear system*** to any ***input excitation***. This chapter provides the language, the mathematical models, and the tools for characterizing linear, time-invariant systems.

Overview

Recall from the overview of Chapter 1 that a ***system*** is a device or process that accepts as an input a signal $x(t)$ and produces as an output another signal $y(t)$. It is customary to use the notation $x(t)$ to designate the input and $y(t)$ to designate the output:

Input $x(t)$ ⟶ System ⟶ $y(t)$ Output.

This notation should not be misconstrued to mean that $y(t)$ at time t depends only on $x(t)$ at time t; it may also depend on past or future values of $x(t)$.

Among the various types of systems encountered in science and engineering, ***linear time-invariant*** (*LTI*) systems stand out as the most prominent. We note that:

1. Many physical systems, including circuits and mechanical systems with linear elements, are LTI.
2. The input-output behavior of an LTI system can be completely characterized by determining or observing its response to an impulsive excitation at its input. Once this is known, the system's response to any other input can be computed using ***convolution*** (see Section 2-3).
3. An LTI system can alter the amplitude and phase (but not the frequency) of an input sinusoid or complex exponential signal (see Section 2-7). This feature is fundamental to the design of systems to filter noisy signals and images (see Chapter 6).
4. The ease with which an LTI system can be analyzed facilitates the inverse process, namely that of designing the system to perform a desired task.

This chapter introduces the reader to the properties, characterizations, and applications of LTI systems. Following a brief discussion of the scaling, additivity, and time-invariance properties of LTI systems, we will discuss what the ***impulse response*** of an LTI system means—physically and mathematically—and how to compute it. The next major topic is ***convolution***, which is often regarded by students as a rather difficult concept to understand and apply. Accordingly, we devote much attention to its derivation, properties, and computation. Next, we discuss causality and stability of an LTI system and examine its response to a complex exponential or sinusoid. Finally, we analyze an LTI spring-mass-damper model of an automobile suspension using the techniques developed in this chapter.

Throughout this chapter, we assume that all initial conditions are zero, so the system has no initial stored energy. We will learn how to incorporate non-zero initial conditions in Section 3-11.

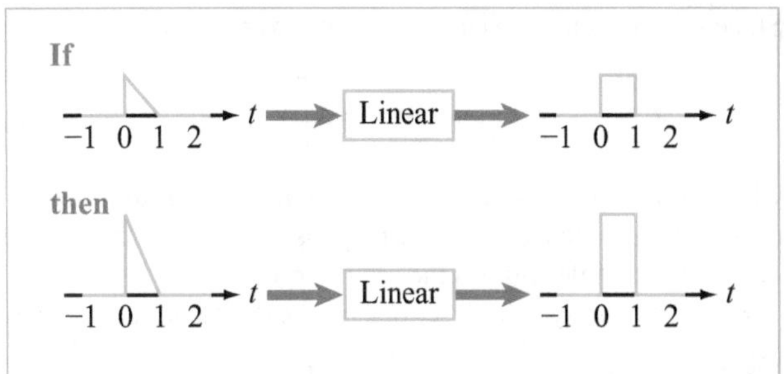

Figure 2-1: In a linear system, scaling the input signal $x(t)$ by a constant multiplier c results in an output $c\ y(t)$.

2-1 Linear Time-Invariant Systems

▶ A system is ***linear*** if it has the ***scaling*** and ***additivity*** properties. ◀

2-1.1 Scaling Property

If the response of a system to input $x(t)$ is output $y(t)$, and if this implies that the response to $c\ x(t)$ is $c\ y(t)$ for any constant c, the system is ***scalable***. The scaling property can be depicted in symbolic form as:

Given: $x(t)$ ⟶ System ⟶ $y(t)$,

then the system is ***scalable*** (has the scaling property) if

$c\ x(t)$ ⟶ System ⟶ $c\ y(t)$.

Thus, scaling the input also scales the output. The ***scaling property*** is also known as the ***homogeneity*** or ***scalability property***.

A graphical example of the scaling property is shown in Fig. 2-1.

Next, consider a system whose output $y(t)$ is linked to its input $x(t)$ by the differential equation

$$\frac{d^2y}{dt^2} + 2\frac{dy}{dt} + 3y = 4\frac{dx}{dt} + 5x. \tag{2.1}$$

Upon replacing $x(t)$ with $c\ x(t)$ and $y(t)$ with $c\ y(t)$ in all terms, we end up with

$$\frac{d^2}{dt^2}(cy) + 2\frac{d}{dt}(cy) + 3(cy) = 4\frac{d}{dt}(cx) + 5(cx).$$

Since c is constant, we can rewrite the expression as

$$c\left[\frac{d^2y}{dt^2}+2\,\frac{dy}{dt}+3y\right]=c\left[4\,\frac{dx}{dt}+5x\right], \tag{2.2}$$

which is identical to the original equation, but multiplied by the constant c. Hence, since the response to $c\,x(t)$ is $c\,y(t)$, the system is scalable and has the scaling property.

However, the system described by the differential equation

$$\frac{d^2y}{dt^2}+2\,\frac{dy}{dt}+3y=4\,\frac{dx}{dt}+5x+6 \tag{2.3}$$

is *not scalable*, because the last term on the right-hand side is independent of both $x(t)$ and $y(t)$, so we cannot factor out c. Therefore the system is not linear.

The scaling property provides a quick mechanism to test if a system is not linear. A good rule of thumb that works often (but not always) is to test whether the system has the scaling property with $c = 2$.

► If doubling the input does not double the output, the system is not linear. But if doubling the input doubles the output, the system is *probably* (but not necessarily) linear. ◄

2-1.2 Additivity Property

If the system responses to N inputs $x_1(t),\ x_2(t),\ldots,x_N(t)$ are respectively $y_1(t),\ y_2(t),\ldots,\ y_N(t)$, then the system is ***additive*** if

$$\sum_{i=1}^{N} x_i(t) \longrightarrow \boxed{\text{System}} \longrightarrow \sum_{i=1}^{N} y_i(t). \tag{2.4}$$

That is, *the response of the sum is the sum of the responses.*

For an additive system, additivity must hold for both finite and infinite sums, as well as for integrals.

► The combination of scalability and additivity is also known as the ***superposition property***, whose application is called the ***superposition principle***. ◄

If we denote $y_1(t)$ as the response to $x_1(t)$ in Eq. (2.1), and similarly $y_2(t)$ as the response to $x_2(t)$, then

$$\frac{d^2y_1}{dt^2}+2\,\frac{dy_1}{dt}+3y_1=4\,\frac{dx_1}{dt}+5x_1, \tag{2.5a}$$

$$\frac{d^2y_2}{dt^2}+2\,\frac{dy_2}{dt}+3y_2=4\,\frac{dx_2}{dt}+5x_2. \tag{2.5b}$$

Next, if we add the two equations and denote $x_3 = x_1 + x_2$ and $y_3 = y_1 + y_2$, we get

$$\frac{d^2y_3}{dt^2}+2\,\frac{dy_3}{dt}+3y_3=4\,\frac{dx_3}{dt}+5x_3. \tag{2.5c}$$

Hence, the system characterized by differential equation (2.1) has the additivity property. Since it was shown earlier also to have the scaling property, the system is indeed linear.

2-1.3 Linear Differential Equations

Many physical systems are described by a ***linear differential equation*** (*LDE*) of the form

$$\sum_{i=0}^{n} a_{n-i}\,\frac{d^iy}{dt^i}=\sum_{i=0}^{m} b_{m-i}\,\frac{d^ix}{dt^i}, \tag{2.6}$$

where coefficients a_0 to a_n and b_0 to b_m may or may not be functions of time t. In either case, Eq. (2.6) represents a linear system, because it has both the scaling and additivity properties. If the coefficients are time-invariant (i.e., constants), then the equation is called a ***linear, constant coefficient, differential equation*** (*LCCDE*), and the system it represents is not only linear but time-invariant as well. Time invariance is discussed in Section 2-1.5.

2-1.4 Significance of the Linearity Property

To appreciate the significance of the linearity property, consider the following scenario. We are given a linear system characterized by a third-order differential equation. The system is excited by a complicated input signal $x(t)$. Our goal is to obtain an analytical solution for the system's output response $y(t)$.

Fundamentally, we can pursue either of the following two approaches:

Option 1: Direct "brute-force" approach: This involves solving the third-order differential equation with the complicated signal $x(t)$. While feasible, this may be mathematically demanding and cumbersome.

Option 2: Indirect, linear-system approach, comprised of the following steps:

(a) Synthesize $x(t)$ as the ***linear combination*** of N relatively simple signals $x_1(t),\ x_2(t),\ \ldots,\ x_N(t)$:

$$x(t) = c_1\,x_1(t) + c_2\,x_2(t) + \cdots + c_N\,x_N(t),$$

with signals $x_1(t)$ to $x_N(t)$ chosen so that they yield straightforward solutions of the third-order differential equation.

(b) Solve the differential equation N times, once for each of the N input signals acting alone, yielding outputs $y_1(t)$ through $y_N(t)$.

(c) Apply the linearity property to obtain $y(t)$ by summing the outputs $y_i(t)$:

$$y(t) = c_1\, y_1(t) + c_2\, y_2(t) + \cdots + c_N\, y_N(t).$$

Even though the second approach entails solving the differential equation N times, the overall solution may prove significantly more tractable than the solution associated with option 1. The procedure will be demonstrated using numerous examples scattered throughout the book.

Systems described by LCCDEs encompass all circuits containing ideal resistors, inductors, capacitors, op-amps, and dependent sources, so long as the circuit contains only a *single* independent current or voltage source acting as the input. The output is any individual voltage or current in the circuit, since (by definition) a system has only one input and one output.

Limiting the circuit to a single independent source (the input) may seem overly restrictive, as many circuits have more than one independent source. Fortunately, the limitation is circumvented by applying the ***superposition principle***. To analyze a circuit with N independent sources and zero initial conditions (i.e., all capacitors are uncharged and all inductors have zero currents flowing through them):

(a) We set all but one source to zero and then analyze the (now linear) circuit using that source as the input. A voltage source set to zero becomes a short circuit, and a current source set to zero becomes an open circuit. We designate $y_1(t)$ as the circuit response to source $x_1(t)$ acting alone.

(b) We then rotate the non-zero choice of source among all of the remaining independent sources, setting all other sources to zero each time. The process generates responses $y_2(t)$ to $y_N(t)$, corresponding to sources $x_2(t)$ to $x_N(t)$, respectively.

(c) We add responses $y_1(t)$ through $y_N(t)$.

We shall see in Section 4-1 that non-zero initial conditions are equivalent to additional independent sources.

Example 2-1: Superposition

Apply the superposition principle to determine the current I through resistor R_2 in the circuit of Fig. 2-2(a).

Solution: (a) The circuit contains two sources, I_0 and V_0. We start by transforming the circuit into the sum of two new circuits: one with I_0 alone and another with V_0 alone, as shown in parts (b) and (c) of Fig. 2-2, respectively. The current through R_2 due to I_0 alone is labeled I_1, and that due to V_0 alone is labeled I_2.

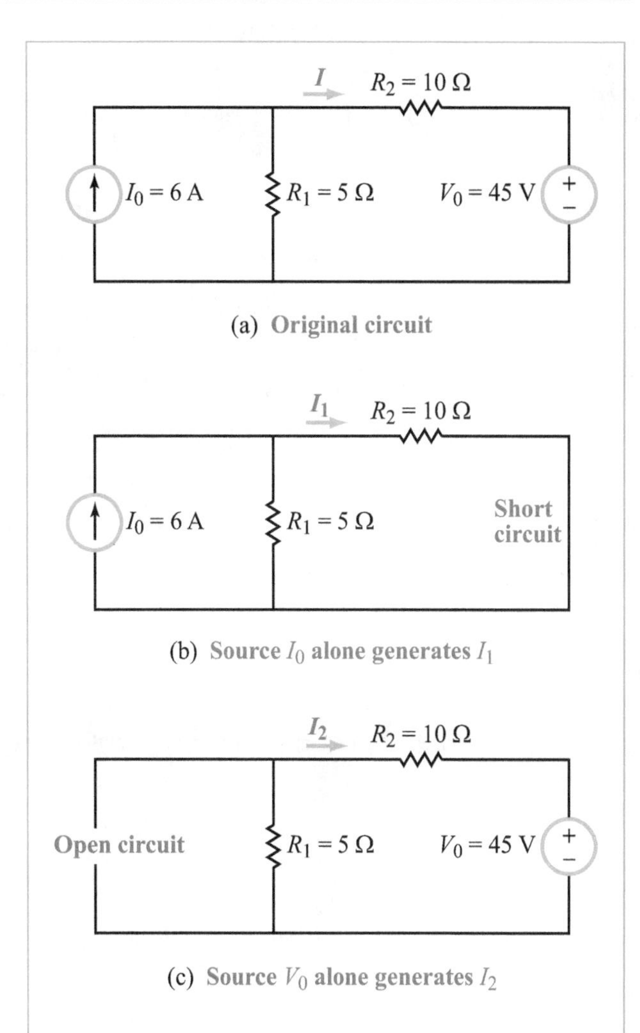

Figure 2-2: Application of the superposition technique to the circuit of Example 2-1.

Application of current division in the circuit of Fig. 2-1(b) gives

$$I_1 = \frac{I_0 R_1}{R_1 + R_2} = \frac{6 \times 5}{5 + 10} = 2\text{ A}.$$

In the circuit of Fig. 2-1(c), Ohm's law gives

$$I_2 = \frac{-V_0}{R_1 + R_2} = \frac{-45}{5 + 10} = -3\text{ A}.$$

Hence,

$$I = I_1 + I_2 = 2 - 3 = -1\text{ A}.$$

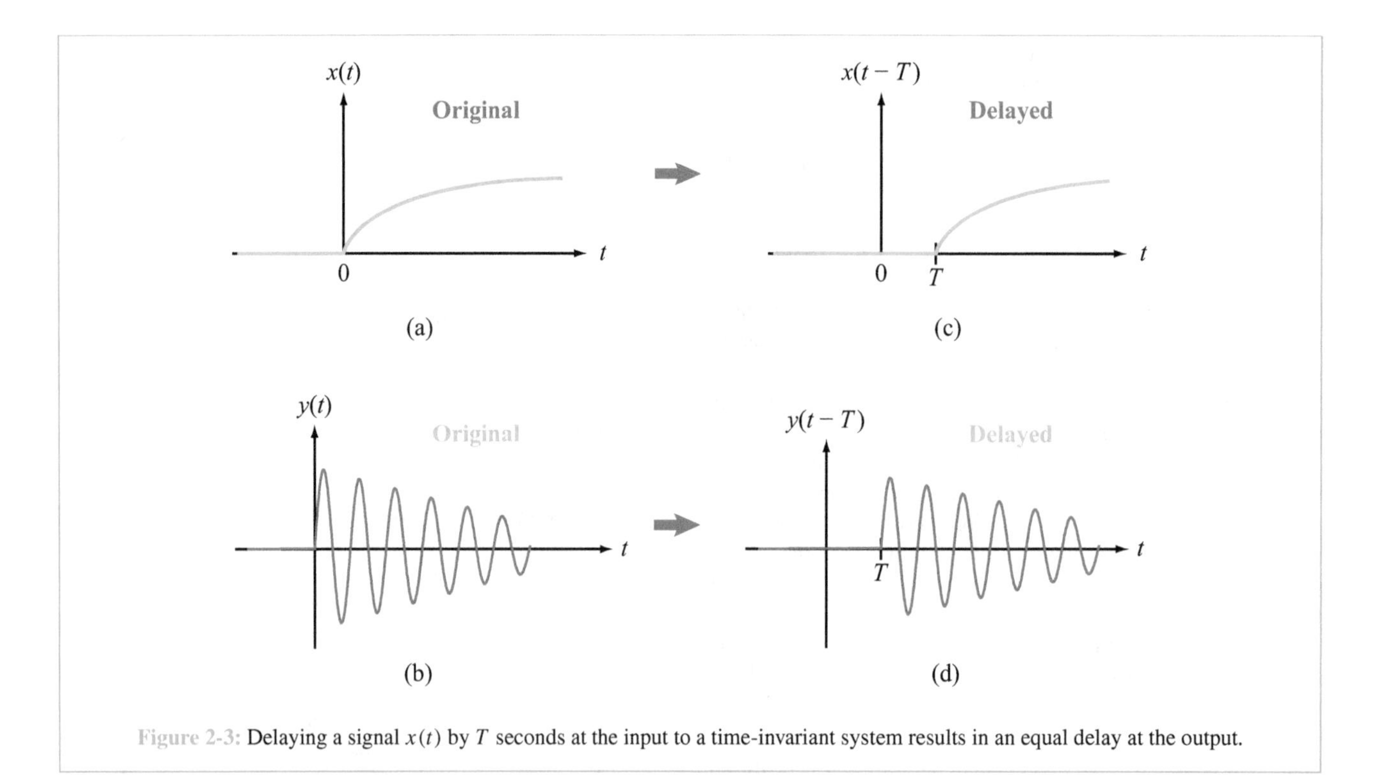

Figure 2-3: Delaying a signal $x(t)$ by T seconds at the input to a time-invariant system results in an equal delay at the output.

2-1.5 Time-Invariant Systems

▶ A system is ***time-invariant*** if delaying the input signal $x(t)$ by any constant T generates the same output $y(t)$, but delayed by exactly T. ◀

Given: $x(t)$ ⟶ System ⟶ $y(t)$,

then the system is ***time-invariant*** if

$x(t-T)$ ⟶ System ⟶ $y(t-T)$.

The process is illustrated graphically by the waveforms shown in Figs. 2-3 and 2-4.

Physically, a system is time-invariant if it has no internal clock. If the input signal is delayed, the system has no way of knowing it, so it accepts the delayed input and delivers a correspondingly delayed output.

How can one tell if a system is time-invariant? A good rule of thumb that *almost* always works is:

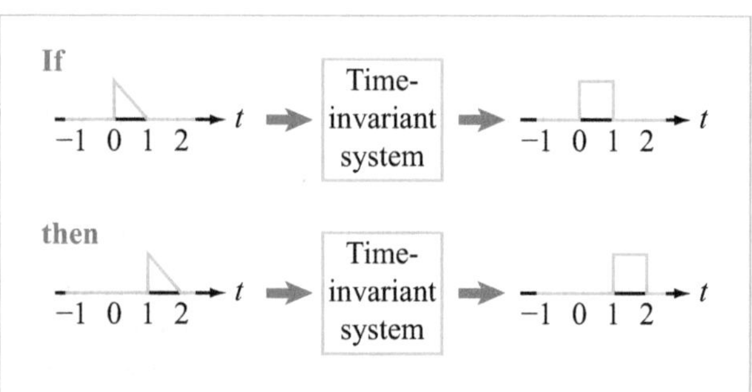

Figure 2-4: If the input of a time-invariant system is delayed by 1 s, the ouput will be delayed by 1 s also.

If t appears explicitly only in the expression for $x(t)$ or in derivatives or limits of integration in the equation describing the system, the system is likely *time-invariant.*

Examples of ***time-invariant systems*** include

(a) $y_1(t) = 3\,\dfrac{d^2x}{dt^2}$,

(b) $y_2(t) = \sin[x(t)]$, and

(c) $y_3(t) = \dfrac{x(t+2)}{x(t-1)}$,

where $x(t+2)$ is $x(t)$ advanced by 2 seconds and $x(t-1)$ is $x(t)$ delayed by 1 second. We note that while all three systems are time-invariant, systems (b) and (c) are not linear.

Examples of ***non-time-invariant systems*** include

(d) $y_4(t) = t\,x(t)$,

(e) $y_5(t) = x(t^2)$, and

(f) $y_6(t) = x(-t)$.

We note that systems (d) to (f) all are linear but not time-invariant, and systems (e) and (f) are not time-invariant even though they do pass the rule-of-thumb test stated earlier—namely that t is part of the expression of $x(t)$ and not in a multiplying function or coefficient. These are the two exceptions to the rule of thumb.

► For the remainder of this book, we will deal with LTI systems exclusively. ◄

Concept Question 2-1: What three properties must an LTI system have? (See S²)

Concept Question 2-2: Does a system described by a linear differential equation qualify as LTI or just as linear? (See S²)

Exercise 2-1: Does the system $y(t) = x^2(t)$ have the scaling property?

Answer: No, because substituting $cx(t)$ for $x(t)$ gives $y(t) = [cx(t)]^2 = c^2x^2(t)$, which is different from $cy(t) = cx^2(t)$. (See S²)

Exercise 2-2: Which of the following systems is linear? (a) $y_1(t) = |\sin(3t)|\,x(t)$, (b) $y_2(t) = a\,\frac{dx}{dt}$ (c) $y_3(t) = |x(t)|$, (d) $y_4(t) = \sin[x(t)]$

Answer: Systems (a) and (b) are linear, but (c) and (d) are not. (See S²)

Exercise 2-3: Which systems are time-invariant?

(a) $y(t) = \frac{dx}{dt} + \sin[x(t-1)]$

(b) $\frac{dy}{dt} = 2\sin[x(t-1)] + 3\cos[x(t-1)]$

Answer: Both are time-invariant. (See S²)

2-2 Impulse Response

The ***impulse response*** $h(t)$ of a system is (logically enough) the response of the system to an impulse $\delta(t)$. Similarly, the ***step response*** $y_{\text{step}}(t)$ is the response of the system to a unit step $u(t)$. In symbolic form:

$$\delta(t) \longrightarrow \boxed{\text{LTI}} \longrightarrow h(t) \tag{2.7a}$$

and

$$u(t) \longrightarrow \boxed{\text{LTI}} \longrightarrow y_{\text{step}}(t). \tag{2.7b}$$

The significance of the impulse response is that, if we know $h(t)$ for an LTI system, we can compute the response to any other input $x(t)$ using the convolution integral derived in Section 2-3. The significance of the step response is that, in many physical situations, we are interested in how well a system "tracks" (follows) a step input. We will use the step response extensively in control-related applications in Chapter 4.

2-2.1 Static and Dynamic Systems

► A system for which the output $y(t)$ at time t depends only on the input $x(t)$ at time t is called a ***static*** or ***memoryless*** system. ◄

An example of such a system is

$$y(t) = \frac{\sin[x(t)]}{x^2(t)+1}.$$

To compute the output of a memoryless system from its input, we just plug into its formula. Parenthetically, we note that this system is not linear.

The only memoryless system that is also LTI is $y(t) = ax(t)$ for any constant a. We remind the reader that the ***affine*** system $y(t) = ax(t) + b$ is not a linear system, although a plot of its formula is a straight line. In the physical world, most systems are ***dynamic*** in that the output $y(t)$ at time t depend on past (or future) as well as present values of the input $x(t)$. Such systems often involve integrals and derivatives of either the input, the output, or both.

A system may ***physically exist***, and its output may be measured in response to physically realizable inputs, such as a gradual step function (Section 1-4.1) or a sinusoid, or it may be described by a ***mathematical model*** (usually in the form of a differential equation constructed from an analysis of the physics, chemistry, or biology governing the behavior of the system). In either case, the system behavior is characterized by its impulse response $h(t)$.

To develop familiarity with what the impulse response means and represents, we will use a simple RC-circuit to derive and demonstrate the physical interpretation of its impulse response $h(t)$.

2-2.2 Computing $h(t)$ and $y_{\text{step}}(t)$ of RC Circuit

Consider the RC circuit shown in Fig. 2-5(a). The input is the voltage source $x(t)$, and the output is the voltage $y(t)$ across the capacitor, which is initially uncharged, so $y(0^-) = 0$. The system in this case is the circuit inside the dashed box. To distinguish between the state of the system (or its output response) before and after the introduction of the input excitation, we denote $t = 0^-$ and $t = 0$ as the times immediately before and immediately after the introduction of $x(t)$, respectively. The conditions prevailing at $t = 0^-$ are called ***initial conditions***. Accordingly, $y(0^-) = 0$ represents the initial condition of the system's output.

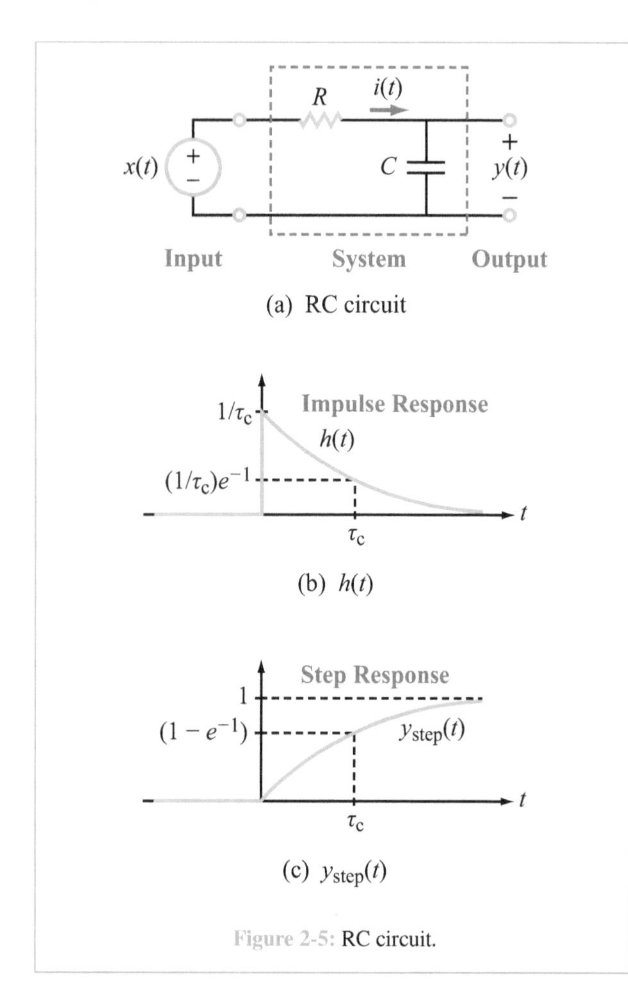

Figure 2-5: RC circuit.

Application of Kirchhoff's voltage law around the loop yields

$$R\,i(t) + y(t) = x(t). \tag{2.8}$$

For the capacitor, the current through it, $i(t)$, is related to the voltage across it, $y(t)$, by

$$i(t) = C\,\frac{dy}{dt}\,. \tag{2.9}$$

Substituting Eq. (2.9) in Eq. (2.8) and then dividing all terms by RC leads to

$$\frac{dy}{dt} + \frac{1}{RC}\,y(t) = \frac{1}{RC}\,x(t). \tag{2.10}$$

To compute the ***impulse response***, we label $x(t) = \delta(t)$ and $y(t) = h(t)$ and obtain

$$\frac{dh}{dt} + \frac{1}{RC}\,h(t) = \frac{1}{RC}\,\delta(t). \tag{2.11}$$

Next, we introduce the ***time constant*** $\tau_c = RC$ and multiply both sides of the differential equation by the ***integrating factor*** e^{t/τ_c}. The result is

$$\frac{dh}{dt}\,e^{t/\tau_c} + \frac{1}{\tau_c}\,e^{t/\tau_c}\,h(t) = \frac{1}{\tau_c}\,e^{t/\tau_c}\,\delta(t). \tag{2.12}$$

The left side of Eq. (2.12) is recognized as

$$\frac{dh}{dt}\,e^{t/\tau_c} + \frac{1}{\tau_c}\,e^{t/\tau_c}\,h(t) = \frac{d}{dt}\,[h(t)\,e^{t/\tau_c}], \tag{2.13a}$$

and the sampling property of the impulse function given by Eq. (1.27) reduces the right-hand side of Eq. (2.12) to

$$\frac{1}{\tau_c}\,e^{t/\tau_c}\,\delta(t) = \frac{1}{\tau_c}\,\delta(t). \tag{2.13b}$$

Incorporating these two modifications in Eq. (2.12) leads to

$$\frac{d}{dt}\,[h(t)\,e^{t/\tau_c}] = \frac{1}{\tau_c}\,\delta(t). \tag{2.14}$$

Integrating both sides from 0^- to t gives

$$\int_{0^-}^{t} \frac{d}{d\tau}\,[h(\tau)\,e^{\tau/\tau_c}]\,d\tau = \frac{1}{\tau_c}\int_{0^-}^{t} \delta(\tau)\,d\tau, \tag{2.15}$$

where τ is a dummy variable of integration and has no connection to the time constant τ_c. We will use τ as a dummy variable throughout this chapter. By Eq. (1.24), we recognize the integral in the right-hand side of Eq. (2.15) as the step function $u(t)$. Hence,

$$h(\tau)\, e^{\tau/\tau_c}\Big|_{0^-}^{t} = \frac{1}{\tau_c}\, u(t),$$

or

$$h(t)\, e^{t/\tau_c} - h(0^-) = \frac{1}{\tau_c}\, u(t). \tag{2.16}$$

The function $h(t)$ represents $y(t)$ for the specific case when the input is $\delta(t)$. We are told that the capacitor was initially uncharged, so $y(0^-) = 0$ for any input. Hence, $h(0^-) = 0$, and Eq. (2.16) reduces to

$$h(t) = \frac{1}{\tau_c}\, e^{-t/\tau_c}\, u(t). \tag{2.17}$$

(impulse response of the RC circuit)

The plot displayed in Fig. 2-5(b) indicates that the capacitor voltage $h(t)$ jumps instantaneously at $t = 0$, which contradicts the natural behavior of a capacitor. This contradiction is a consequence of the fact that a pure $\delta(t)$ is a physically unrealizable singularity function. However, as discussed in the next subsection, we can approximate an impulse by a finite-duration pulse and then use it to measure the impulse response of the circuit.

To compute the ***step response*** of the circuit, we start with the same governing equation given by Eq. (2.10), but this time we label $x(t) = u(t)$ and $y(t) = y_{\text{step}}(t)$. The procedure leads to

$$y_{\text{step}}(t) = [1 - e^{-t/\tau_c}]\, u(t). \tag{2.18}$$

(step response of the RC circuit)

As the initially uncharged capacitor builds up charge, the voltage across it builds up monotonically from zero at $t = 0$ to 1 V as $t \to \infty$ (Fig. 2-5(c)).

▶ Once we know the step response of an LTI system, we can apply the superposition principle to compute the response of the system to any input that can be expressed as a linear combination of scaled and delayed step functions. ◀

2-2.3 Measuring $h(t)$ and $y_{\text{step}}(t)$ of RC Circuits

Suppose we have a physical system for which we do not have an LCCDE model. To physically measure its impulse and step responses, we need to be able to apply impulse and step signals at its input terminals. As ideal impulse and step signals are physically unrealizable, we resort to techniques that offer reasonable approximations. In general, the approximate techniques used for representing $u(t)$ and $\delta(t)$ must be tailor-made to suit the system under consideration. We will illustrate the procedure by *measuring* the step and impulse responses of an RC circuit (Fig. 2-5(a)) with $\tau_c = RC = 2$ s.

Step response

A unit step function at the input terminals can be represented by a 1-V dc voltage source connected to the RC circuit via a switch that is made to close at $t = 0$ (Fig. 2-6(a)). In reality, the process of closing the switch has a finite (but very short) switching time τ_s. That is, it takes the voltage at the input terminals τ_s seconds to transition between zero and 1 volt. For the RC circuit, its time constant τ_c is a measure of its ***reaction time***, so if $\tau_s \ll \tau_c$, the circuit will *sense* the switching action as if it were instantaneous, in which case the measured output

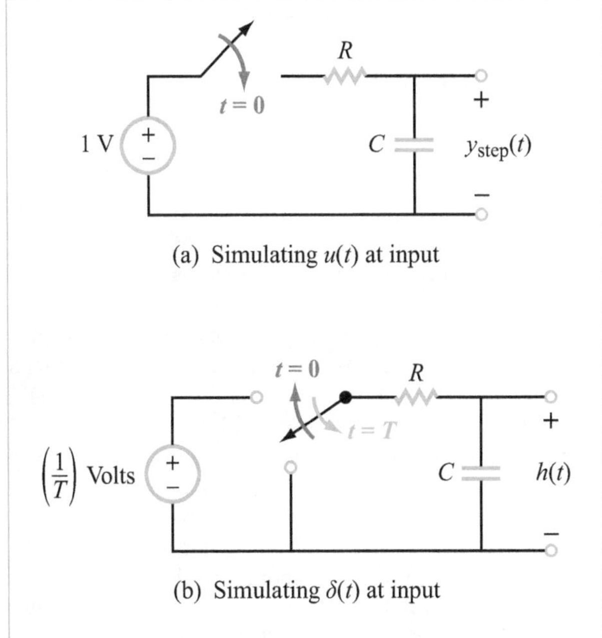

Figure 2-6: Simulating inputs $u(t)$ and $\delta(t)$ through the use of switches.

voltage $y_{\text{step}}(t)$ would closely match the computed expression given by Eq. (2.18),

$$y_{\text{step}}(t) = [1 - e^{-t/2}]\, u(t), \tag{2.19}$$

where we have replaced τ_c with 2 s.

Impulse response

A unit impulse $\delta(t)$ can be simulated by a rectangle model consisting of two step functions:

$$x(t) = \frac{1}{T}\,[u(t) - u(t-T)] \qquad \text{as } T \to 0. \tag{2.20}$$

The rectangular pulse has a duration of T and its amplitude (in volts) is always chosen such that it is numerically equal to $1/T$, thereby maintaining the area of the pulse at 1. Physically, the rectangle excitation is realized by connecting a voltage source of magnitude $(1/T)$ volts to the circuit via a switch that connects the RC circuit to the source at $t = 0$ and then connects it to a short circuit at $t = T$ (Fig. 2-6(b)).

Since the RC circuit is LTI, the superposition principle applies. In view of Eq. (2.19), which is the response to the input excitation $u(t)$, the response $y_{\text{pulse}}(t)$ to the pulse excitation modeled by Eq. (2.20) is

$$\begin{aligned} y_{\text{pulse}}(t) &= \frac{1}{T}[y_{\text{step}}(t) - y_{\text{step}}(t-T)] \\ &= \frac{1}{T}\left\{(1 - e^{-t/2})\, u(t) - (1 - e^{-(t-T)/2})\, u(t-T)\right\}. \end{aligned} \tag{2.21}$$

Our goal is to answer the following pair of questions:

(1) Is it possible to measure the impulse response of the RC circuit by exciting it with a rectangular pulse of amplitude $(1/T)$ and duration T? (2) Are there any constraints on the choice of the value of T?

To explore the answers to these two questions, we display in Fig. 2-7 plots of $y_{\text{pulse}}(t)$—all computed in accordance with Eq. (2.21)—for three different values of T, namely, 5 s, 0.05 s, and 0.01 s. The first case was chosen to illustrate the response of the circuit to a pulse whose duration T is greater than its own time constant τ_c (i.e., $T = 5$ s and $\tau_c = 2$ s). Between $t = 0$ and $t = 5$ s, the capacitor charges up to a maximum voltage of $(1 - e^{-5/2})/5 = 0.184$ V (Fig. 2-7(a)), almost to the level of the input voltage (0.2 V). Upon disconnecting the voltage source and replacing it with a short circuit at $t = T = 5$ s, the capacitor starts to discharge towards zero volts as $t \to \infty$. During both the charge-up and discharge periods, the rates of increase and decay are governed by the time constant $\tau_c = 2$ s.

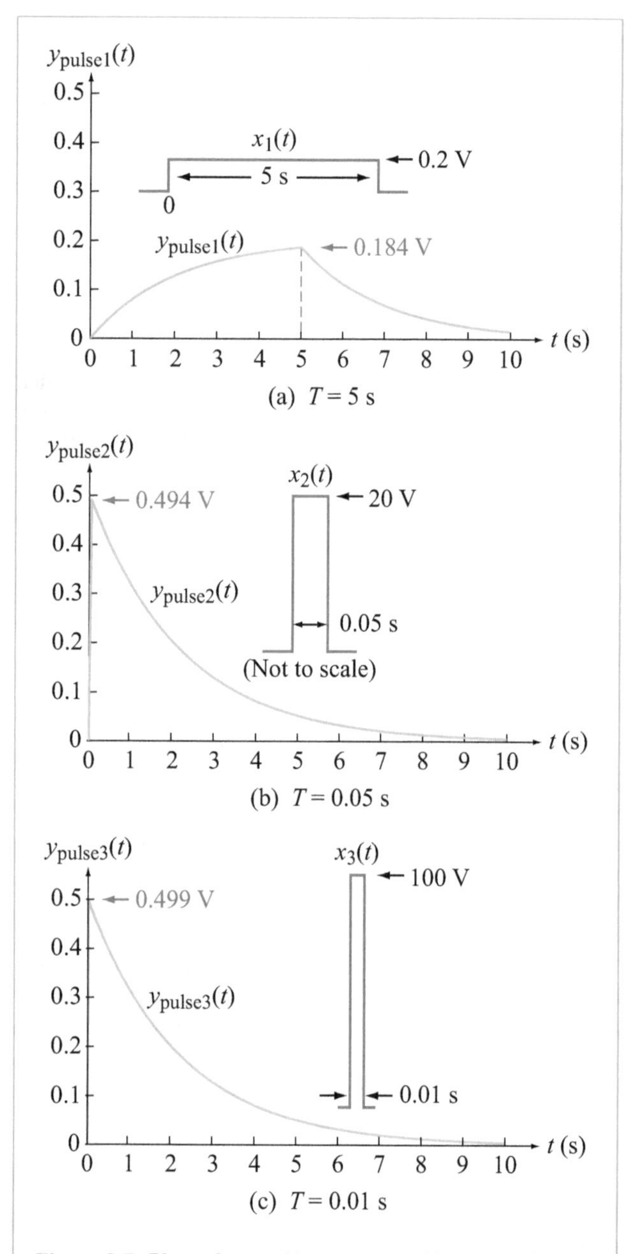

Figure 2-7: Plots of $y_{\text{pulse}}(t)$, as expressed by Eq. (2.21), for three values of T.

Next, we examine the case of $T = 0.05$ s (Fig. 2-7(b)); the pulse duration now is much shorter than the time constant $(T/\tau_c = 0.025)$. Also, the pulse amplitude is $1/T = 20$ V (compared with only 0.2 V for the previous case). The response charges up to $20(1 - e^{-0.05/2}) = 0.494$ V at $t = 0.05$ s and then discharges to zero as $t \to \infty$. The plot in Fig. 2-7(b) exhibits a close resemblance to the theoretical response shown earlier in Fig. 2-5(b). For the theoretical response, the peak value is $1/\tau_c = 1/2 = 0.5$ V.

As the pulse duration is made shorter still, the measured response approaches both the shape and peak value of the theoretical response. The plot in Fig. 2-7(c) (corresponding to $T = 0.01$ s) exhibits a peak value of $100(1 - e^{-0.01/2}) = 0.499$ V, which is very close to the theoretical peak value of 0.5 V.

Hence, the combined answer to the questions posed earlier is *yes*, the impulse response of the RC circuit can indeed be determined experimentally, provided the excitation pulse used at the input is of unit area and its duration is much shorter than the circuit's time constant τ_c. This statement can be generalized:

▶ The response of an RC circuit to a sufficiently fast pulse of any shape is $(A/\tau_c)e^{-t/\tau_c}\, u(t)$, where A is the area under the pulse. "Sufficiently fast" means that the duration of the pulse $T \ll \tau_c$. ◀

2-2.4 Impulse and Ramp Responses from Step Response

We now extend the procedure we used to determine $h(t)$ of the RC circuit to LTI systems in general:

Step 1: Physically *measure* the step response $y_{\text{step}}(t)$.

Step 2: Differentiate it to obtain

$$h(t) = \frac{dy_{\text{step}}}{dt}. \tag{2.22}$$

To demonstrate the validity of Eq. (2.22), we start with the definition of the step response given by Eq. (2.7b). We then use the time-invariance property of LTI systems to delay both the input and output by a constant delay ϵ:

$$u(t-\epsilon) \longrightarrow \boxed{\text{LTI}} \longrightarrow y_{\text{step}}(t-\epsilon). \tag{2.23}$$

Next, using the additivity property of LTI systems, we subtract Eq. (2.23) from Eq. (2.7b), and using the scaling property, we multiply the result by $1/\epsilon$:

$$\frac{u(t) - u(t-\epsilon)}{\epsilon} \longrightarrow \boxed{\text{LTI}} \longrightarrow \frac{y_{\text{step}}(t) - y_{\text{step}}(t-\epsilon)}{\epsilon}. \tag{2.24}$$

The two steps were justified by the additivity and scaling properties of LTI systems. Finally, if we let $\epsilon \to 0$, the left-hand side of Eq. (2.24) becomes du/dt, and the right-hand side becomes dy_{step}/dt. Moreover, from Eq. (1.23), $du/dt = \delta(t)$. Hence,

$$\frac{du}{dt} = \delta(t) \longrightarrow \boxed{\text{LTI}} \longrightarrow h(t) = \frac{dy_{\text{step}}}{dt}. \tag{2.25}$$

Example 2-2: RC Circuit Impulse Response from Step Response

Obtain (a) the impulse response of the RC circuit in Fig. 2-5(a) by applying Eq. (2.22) and (b) the circuit response to a ramp-function input $x(t) = r(t)$.

Solution: (a) Using the expression for $y_{\text{step}}(t)$ given by Eq. (2.18), the impulse response of the RC circuit is

$$\begin{aligned} h(t) &= \frac{d}{dt}\left[(1 - e^{-t/\tau_c})\, u(t)\right] \\ &= \delta(t) - \delta(t)\, e^{-t/\tau_c} + \frac{1}{\tau_c}\, e^{-t/\tau_c}\, u(t) \\ &= \frac{1}{\tau_c}\, e^{-t/\tau_c}\, u(t), \end{aligned} \tag{2.26}$$

where we used $\delta(t) = du/dt$ and $x(t)\,\delta(t) = x(0)\,\delta(t)$ for any $x(t)$. Because the step response $y_{\text{step}}(t)$ was available to us, the process leading to Eq. (2.26) was quite painless and straightforward (in contrast with the much longer differential-equation solution contained in Section 2-2.2).

(b) From Eq. (2.7b), we have

$$u(t) \longrightarrow \boxed{\text{LTI}} \longrightarrow y_{\text{step}}(t). \tag{2.27}$$

The ramp function $r(t)$ is related to the step function $u(t)$ by Eq. (1.18) as

$$r(t) = \int_{-\infty}^{t} u(\tau)\, d\tau.$$

Integration is a linear operation, so its application to both sides of Eq. (2.27) gives

$$r(t) = \int_{-\infty}^{t} u(\tau)\, d\tau \longrightarrow \boxed{\text{LTI}} \longrightarrow y_{\text{ramp}}(t) \tag{2.28}$$

with

$$\begin{aligned} y_{\text{ramp}}(t) &= \int_{-\infty}^{t} y_{\text{step}}(\tau)\, d\tau \\ &= \int_{-\infty}^{t} (1 - e^{-\tau/\tau_c})\, u(\tau)\, d\tau \\ &= \int_{0}^{t} (1 - e^{-\tau/\tau_c})\, d\tau \\ &= [t - \tau_c(1 - e^{-t/\tau_c})]\, u(t). \end{aligned} \tag{2.29}$$

Concept Question 2-3: How can you measure the impulse response of a real system? (See S²)

Concept Question 2-4: How can you compute the impulse response of a system from its step response? (See S²)

Exercise 2-4: Determine the impulse response of a system whose step response is

$$y_{\text{step}}(t) = \begin{cases} 0, & t \le 0 \\ t, & 0 \le t \le 1 \\ 1, & t \ge 1. \end{cases}$$

Answer: $h(t) = u(t) - u(t-1)$. (See S²)

Exercise 2-5: The RC circuit of Fig. 2-5(a) is excited by $x(t) = (1 - 1000t)[u(t) - u(t - 0.001)]$. Compute the capacitor voltage $y(t)$ for $t > 0.001$ s, given that $\tau_c = 1$ s.

Answer: $x(t)$ is a very short wedge-shaped pulse with area $= (0.5)(1)(0.001) = 0.0005$. Therefore, $y(t) = 0.0005e^{-t}\, u(t)$. (See S²)

2-3 Convolution

We now derive a remarkable property of LTI systems.

► The response $y(t)$ of an LTI system with impulse response $h(t)$ to *any* input $x(t)$ can be computed *explicitly* using the ***convolution integral***

$$y(t) = \int_{-\infty}^{\infty} x(\tau)\, h(t - \tau)\, d\tau = h(t) * x(t).$$

(convolution integral) (2.30)

All initial conditions must be zero. ◄

The convolution operation denoted by $*$ in Eq. (2.30) is *not* the same as multiplication. The term *convolution* refers to the convoluted nature of the integral. Note the following:

(a) τ is a dummy variable of integration.

(b) $h(t - \tau)$ is obtained from the impulse response $h(t)$ by (1) replacing t with the variable of integration τ, (2) reversing $h(\tau)$ along the τ axis to obtain $h(-\tau)$, and (3) delaying $h(-\tau)$ by time t to obtain $h(t - \tau) = h(-(\tau - t))$.

(c) As will be demonstrated later in this section, convolution is commutative; we can interchange x and h in Eq. (2.30).

According to the convolution property represented by Eq. (2.30), once $h(t)$ of an LTI system has been determined, the system's response can be readily evaluated for any specified input excitation $x(t)$ by performing the convolution integration. The integration may be carried out analytically, graphically, or numerically—all depending on the formats of $x(t)$ and $h(t)$. Multiple examples are provided in the latter part of this section.

2-3.1 Derivation of Convolution Integral

The derivation of the convolution integral follows directly from the definition of the impulse response and the properties of LTI systems. We need only the five steps outlined in Fig. 2-8.

Step 1: From the definition of the impulse response given by Eq. (2.7a), we have

$$\delta(t) \longrightarrow \boxed{\text{LTI}} \longrightarrow h(t). \tag{2.31}$$

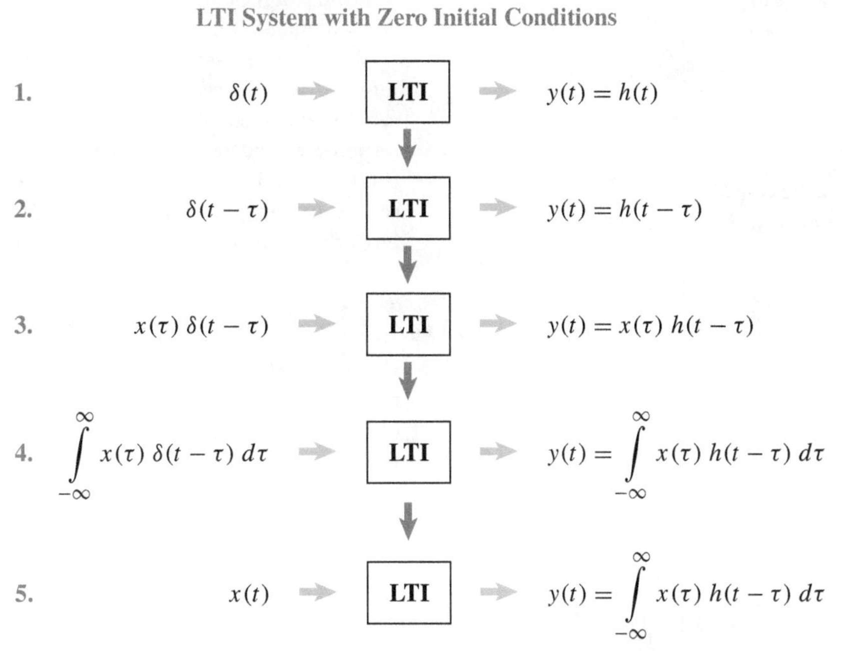

LTI System with Zero Initial Conditions

1. $\delta(t)$ → **LTI** → $y(t) = h(t)$

2. $\delta(t-\tau)$ → **LTI** → $y(t) = h(t-\tau)$

3. $x(\tau)\,\delta(t-\tau)$ → **LTI** → $y(t) = x(\tau)\,h(t-\tau)$

4. $\int_{-\infty}^{\infty} x(\tau)\,\delta(t-\tau)\,d\tau$ → **LTI** → $y(t) = \int_{-\infty}^{\infty} x(\tau)\,h(t-\tau)\,d\tau$

5. $x(t)$ → **LTI** → $y(t) = \int_{-\infty}^{\infty} x(\tau)\,h(t-\tau)\,d\tau$

Figure 2-8: Derivation of the convolution integral for a linear time-invariant system.

Step 2: According to the ***time-invariance property*** of LTI systems, delaying the input $\delta(t)$ by a constant τ will delay the output $h(t)$ by the same constant τ:

$$\delta(t-\tau) \longrightarrow \boxed{\text{LTI}} \longrightarrow h(t-\tau). \tag{2.32}$$

Step 3: From the ***scaling property*** of LTI systems, if the input has an amplitude $x(\tau)$, the output will scale by the same factor:

$$x(\tau)\,\delta(t-\tau) \longrightarrow \boxed{\text{LTI}} \longrightarrow x(\tau)\,h(t-\tau). \tag{2.33}$$

Step 4: According to Eq. (2.33), an impulsive excitation $x(\tau_1)\,\delta(t-\tau_1)$ at a specific value of τ (namely, τ_1) will generate a corresponding output $x(\tau_1)\,h(t-\tau_1)$:

$$x(\tau_1)\,\delta(t-\tau_1) \longrightarrow \boxed{\text{LTI}} \longrightarrow x(\tau_1)\,h(t-\tau_1). \tag{2.34a}$$

Similarly, a second impulsive excitation of area $x(\tau_2)$ at $\tau = \tau_2$ leads to

$$x(\tau_2)\,\delta(t-\tau_2) \longrightarrow \boxed{\text{LTI}} \longrightarrow x(\tau_2)\,h(t-\tau_2). \tag{2.34b}$$

If the LTI system is excited by both inputs simultaneously, the ***additivity property*** assures us that the output will be equal to the sum of the outputs in Eqs. (2.34a and b). For a continuous-time input function $x(\tau)$ extending over $-\infty < \tau < \infty$, the input and output of the LTI system become definite integrals:

$$\int_{-\infty}^{\infty} x(\tau)\,\delta(t-\tau)\,d\tau \;\downarrow\; \boxed{\text{LTI}} \;\downarrow\; \int_{-\infty}^{\infty} x(\tau)\,h(t-\tau)\,d\tau. \tag{2.35}$$

Step 5: From the ***sampling property*** of impulses, we have

$$\int_{-\infty}^{\infty} x(\tau)\,\delta(t-\tau)\,d\tau = x(t). \tag{2.36}$$

Hence, Eq. (2.35) simplifies to

$$x(t) \longrightarrow \boxed{\text{LTI}} \longrightarrow y(t) = \int_{-\infty}^{\infty} x(\tau)\, h(t-\tau)\, d\tau. \tag{2.37}$$

As a shorthand notation, the convolution integral is represented by an asterisk *, so output $y(t)$ is expressed as

$$y(t) = x(t) * h(t) = \int_{-\infty}^{\infty} x(\tau)\, h(t-\tau)\, d\tau. \tag{2.38}$$

By changing variables inside the integral of Eq. (2.38) from τ to $(t-\tau)$, the convolution can be expressed by the equivalent form

$$y(t) = h(t) * x(t) = \int_{-\infty}^{\infty} h(\tau)\, x(t-\tau)\, d\tau. \tag{2.39}$$

Equivalency of the integrals in Eqs. (2.38) and (2.39) implies that the convolution operation is ***commutative***; that is,

$$x(t) * h(t) = h(t) * x(t). \tag{2.40}$$

2-3.2 Causal Signals and Systems

The integral given by Eq. (2.38) implies that the output $y(t)$ at time t depends on all excitations $x(t)$ occurring at the input, including those that will occur at times later than t. Since this cannot be true for a ***physically realizable system***, the upper integration limit should be replaced with t instead of ∞. Also, if we choose our time scale such that no excitation exists before $t = 0$, we can then replace the lower integration limit with zero. Hence, for causal signals and systems, Eq. (2.38) becomes

$$\begin{aligned} y(t) &= u(t) \int_{0}^{t} x(\tau)\, h(t-\tau)\, d\tau \\ &= u(t) \int_{0}^{t} x(t-\tau)\, h(\tau)\, d\tau. \end{aligned} \tag{2.41}$$

(causal signals and systems)

Multiplication of the integral by $u(t)$ is a reminder that the convolution of two causal functions is itself causal.

Note that the response $y(t)$ provided by the convolution integral assumes that all initial conditions of the system are zero. If the system has non-zero initial conditions, its total response will be the sum of two responses: one due to the initial conditions and another due to the input excitation $x(t)$. Through several examples, we will show in Chapter 3 how to account properly for the two contributions of the total response.

2-3.3 Computing Convolution Integrals

Given a system with an impulse response $h(t)$, the output response $y(t)$ can be determined for any specified input excitation $x(t)$ by computing the convolution integral given by Eq. (2.38) or by Eq. (2.41) if $x(t)$ is causal. The process may be carried out:

(a) analytically, by performing the integration to obtain an expression for $y(t)$, and then evaluating $y(t)$ as a function of time t,

(b) graphically, by *simulating* the integration using plots of the waveforms of $x(t)$ and $h(t)$,

(c) analytically with simplification, by taking advantage of the expedient convolution properties outlined in Section 2-5 to simplify the integration algebra, or

(d) numerically on a digital computer.

As we will see shortly through Examples 2-3 and 2-4, the analytical approach is straightforward, but the algebra often is cumbersome and tedious. The graphical approach (Section 2-4) allows the user to "visualize" the convolution process, but it is slow and repetitious. The two approaches offer complementary insight; the choice of one over the other depends on the specifics of $x(t)$ and $h(t)$.

After learning to use approaches (a) and (b), the reader will appreciate the advantages offered by approach (c); invoking the convolution properties simplifies the algebra considerably. Finally, numerical integration always is a useful default option, particularly if $x(t)$ is a measured signal or difficult to model mathematically.

Example 2-3: RC Circuit Response to Triangular Pulse

An RC circuit with $\tau_c = RC = 1$ s is excited by the triangular pulse shown in Fig. 2-9(a). The capacitor is initially uncharged. Determine the output voltage across the capacitor.

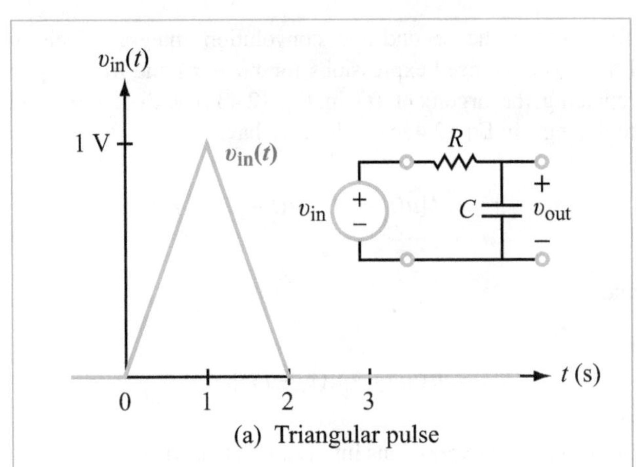

(a) Triangular pulse

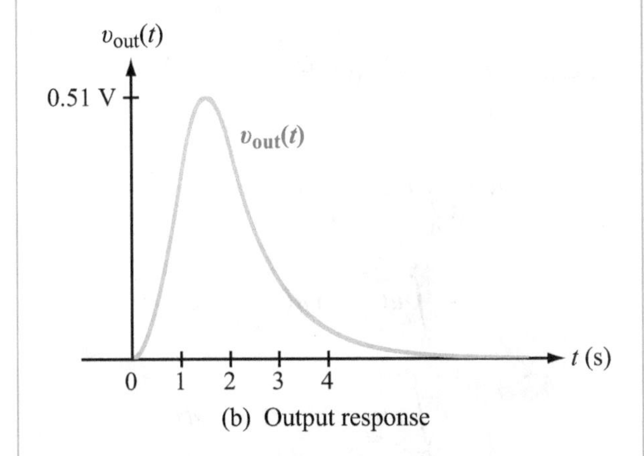

(b) Output response

Figure 2-9: Triangular pulse exciting an RC circuit with $\tau_c = 1$ s (Example 2-3).

Solution: The input signal, measured in volts, is given by

$$v_{in}(t) = \begin{cases} 0 & \text{for } t \le 0, \\ t & \text{for } 0 \le t \le 1 \text{ s}, \\ 2 - t & \text{for } 1 \le t \le 2 \text{ s}, \\ 0 & \text{for } t \ge 2 \text{ s}, \end{cases}$$

and according to Eq. (2.17), the impulse response for $\tau_c = 1$ is

$$h(t) = \frac{1}{\tau_c} e^{-t/\tau_c} u(t) = e^{-t} u(t).$$

According to Eq. (2.41), the output is given by

$$v_{out}(t) = v_{in}(t) * h(t) = \int_0^t v_{in}(\tau)\, h(t-\tau)\, d\tau$$

with

$$h(t-\tau) = e^{-(t-\tau)} u(t-\tau) = \begin{cases} 0 & \text{for } t < \tau, \\ e^{-(t-\tau)} & \text{for } t > \tau. \end{cases} \tag{2.42}$$

The convolution integral will be evaluated for each of the four time segments associated with $v_{in}(t)$ separately.

(1) $t < 0$:

The lowest value that the integration variable τ can assume is zero. Therefore, when $t < 0$, $t < \tau$ and $h(t - \tau) = 0$. Consequently,

$$v_{out}(t) = 0 \qquad \text{for } t < 0.$$

(2) $0 \le t \le 1$ s:

$$h(t-\tau) = e^{-(t-\tau)}, \qquad v_{in}(\tau) = \tau,$$

and

$$v_{out}(t) = \int_0^t \tau e^{-(t-\tau)}\, d\tau = e^{-t} + t - 1, \qquad \text{for } 0 \le t \le 1 \text{ s}.$$

(3) $1 \text{ s} \le t \le 2$ s:

$$v_{in}(\tau) = \begin{cases} \tau & \text{for } 0 \le \tau \le 1 \text{ s}, \\ 2 - \tau & \text{for } 1 \text{ s} \le \tau \le 2 \text{ s}, \end{cases}$$

and

$$\begin{aligned} v_{out}(t) &= \int_0^1 \tau e^{-(t-\tau)}\, d\tau + \int_1^t (2-\tau) e^{-(t-\tau)}\, d\tau \\ &= (1 - 2e)e^{-t} - t + 3, \qquad \text{for } 1 \text{ s} \le t \le 2 \text{ s}. \end{aligned}$$

(4) $t \geq 2$ s:

$$\upsilon_{\text{out}}(t) = \int_0^1 \tau e^{-(t-\tau)}\, d\tau + \int_1^2 (2-\tau) e^{-(t-\tau)}\, d\tau$$
$$= (1 - 2e + e^2)e^{-t} \qquad \text{for } t \geq 2 \text{ s}.$$

The cumulative response covering all four time segments is displayed in Fig. 2-9(b).

Note that the four expressions for $\upsilon_{\text{out}}(t)$ provide identical values at junctions between adjoining time intervals. The convolution of two signals that contain no impulses must be a continuous signal. This is a good check when computing the convolution integral over multiple intervals.

2-3.4 Convolution of Two Rectangular Pulses

Given the rectangular waveforms $x(t)$ and $h(t)$ shown in Fig. 2-10, our goal is to evaluate their convolution. To that end, we start by expressing $x(t)$ and $h(t)$ in terms of step functions:

$$x(t) = A[u(t) - u(t - T_1)] \tag{2.43a}$$

and (for $T_2 > T_1$)

$$h(t) = B[u(t) - u(t - T_2)]. \tag{2.43b}$$

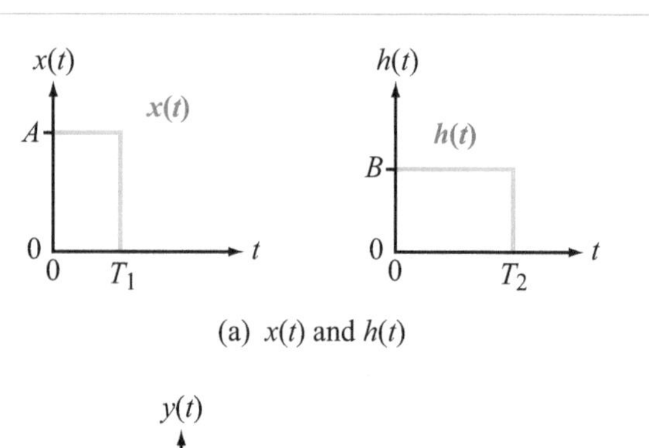

(a) $x(t)$ and $h(t)$

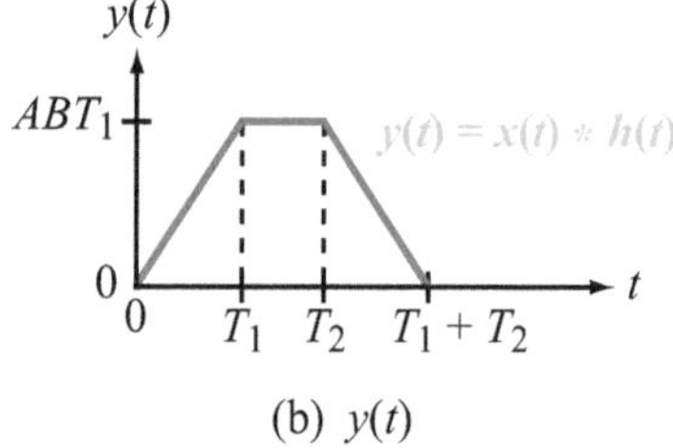

(b) $y(t)$

Figure 2-10: The convolution of the two rectangular waveforms in (a) is the pyramid shown in (b).

To evaluate the second-line convolution integral given in Eq. (2.41), we need expressions for $x(t-\tau)$ and $h(\tau)$. Upon replacing the argument (t) in Eq. (2.43a) with $(t-\tau)$ and replacing t in Eq. (2.43b) with τ, we have

$$x(t-\tau) = A[u(t-\tau) - u(t - T_1 - \tau)] \tag{2.44a}$$

and

$$h(\tau) = B[u(\tau) - u(\tau - T_2)]. \tag{2.44b}$$

Inserting these expressions into Eq. (2.41) leads to

$$\begin{aligned} y(t) &= u(t) \int_0^t x(t-\tau)\, h(\tau)\, d\tau \\ &= AB\, u(t) \left\{ \int_0^t u(t-\tau)\, u(\tau)\, d\tau \right. \\ &\quad - \int_0^t u(t-\tau)\, u(\tau - T_2)\, d\tau \\ &\quad - \int_0^t u(t - T_1 - \tau)\, u(\tau)\, d\tau \\ &\quad \left. + \int_0^t u(t - T_1 - \tau)\, u(\tau - T_2)\, d\tau \right\}. \end{aligned} \tag{2.45}$$

We will now examine each term individually. Keeping in mind that the integration variable is τ, and because the upper integration limit is t, the difference $(t-\tau)$ is never smaller than zero (over the range of integration). Also, T_1 and T_2 are both non-negative numbers.

Term 1: Since over the range of integration, both $u(\tau)$ and $u(t-\tau)$ are equal to 1, the integral in the first term simplifies to

$$u(t) \int_0^t u(t-\tau)\, u(\tau)\, d\tau = \left[\int_0^t d\tau \right] u(t) = t\, u(t). \tag{2.46}$$

Term 2: The unit step function $u(\tau - T_2)$ is equal to zero unless $\tau > T_2$, requiring that the lower integration limit be replaced with T_2 and the outcome be multiplied by $u(t - T_2)$:

$$\int_0^t u(t-\tau)\, u(\tau - T_2)\, d\tau = \left[\int_{T_2}^t d\tau\right] u(t-T_2)$$
$$= (t - T_2)\, u(t - T_2). \quad (2.47)$$

Term 3: The step function $u(t - T_1 - \tau)$ is equal to zero, unless $\tau < t - T_1$, requiring that the upper limit be replaced with $t - T_1$. Additionally, to satisfy the inequality, the smallest value that t can assume at the lower limit ($\tau = 0$) is $t = T_1$. Consequently, the outcome of the integration should be multiplied by $u(t - T_1)$:

$$\int_0^t u(t - T_1 - \tau)\, u(\tau)\, d\tau = \left[\int_0^{t-T_1} d\tau\right] u(t - T_1)$$
$$= (t - T_1)\, u(t - T_1). \quad (2.48)$$

Term 4: To accommodate the product

$$u(t - T_1 - \tau)\, u(\tau - T_2),$$

we need to (a) change the lower limit to T_2, (b) change the upper limit to $(t - T_1)$, and (c) multiply the outcome by $u(t - T_1 - T_2)$:

$$\int_0^t u(t - T_1 - \tau)\, u(\tau - T_2)\, d\tau$$
$$= \left[\int_{T_2}^{t-T_1} d\tau\right] u(t - T_1 - T_2)$$
$$= (t - T_1 - T_2)\, u(t - T_1 - T_2). \quad (2.49)$$

Collecting the results given by Eqs. (2.46) through (2.49) gives

$$y(t) = AB[t\, u(t) - (t - T_2)\, u(t - T_2) - (t - T_1)\, u(t - T_1)$$
$$+ (t - T_1 - T_2)\, u(t - T_1 - T_2)]. \quad (2.50)$$

The waveform of $y(t)$ is displayed in Fig. 2-10(b).

Building on the experience gained from the preceding example, we can generalize the result as follows.

Convolution Integral

For functions $x(t)$ and $h(t)$ given by

$$x(t) = f_1(t)\, u(t - T_1) \quad (2.51a)$$

and

$$h(t) = f_2(t)\, u(t - T_2), \quad (2.51b)$$

where $f_1(t)$ and $f_2(t)$ are any constants or time-dependent signals and T_1 and T_2 are any non-negative numbers, their convolution is

$$y(t) = x(t) * h(t)$$
$$= u(t) \int_0^t x(t - \tau)\, h(\tau)\, d\tau$$
$$= \int_0^t f_1(t - \tau)\, f_2(\tau)\, u(t - T_1 - \tau)\, u(\tau - T_2)\, d\tau$$
$$= \left[\int_{T_2}^{t-T_1} f_1(t - \tau)\, f_2(\tau)\, d\tau\right] u(t - T_1 - T_2). \quad (2.52)$$

The convolution result represented by Eq. (2.52) will prove useful and efficient when analytically evaluating the convolution integral of signals described in terms of step functions.

Example 2-4: RC Circuit Response to Rectangular Pulse

Given the RC circuit shown in Fig. 2-11(a), determine the output response to a 1-s-long rectangular pulse. The pulse amplitude is 1 V.

Solution: The time constant of the RC circuit is $\tau_c = RC = (0.5 \times 10^6) \times 10^{-6} = 0.5$ s. In view of Eq. (2.17), the impulse response of the circuit is

$$h(t) = \frac{1}{\tau_c}\, e^{-t/\tau_c}\, u(t) = 2e^{-2t}\, u(t). \quad (2.53)$$

The input voltage is

$$\upsilon_{in}(t) = [u(t) - u(t - 1)] \text{ V}. \quad (2.54)$$

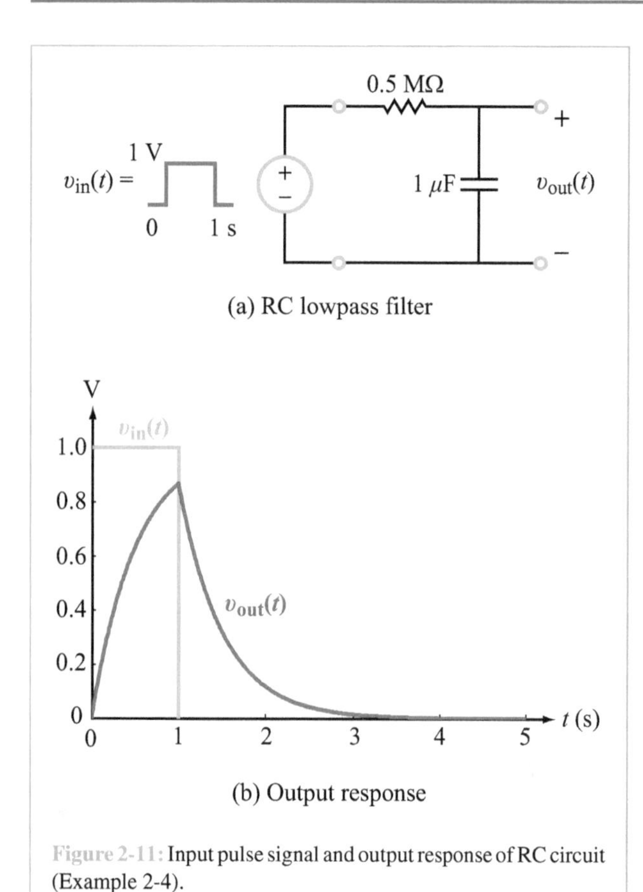

Figure 2-11: Input pulse signal and output response of RC circuit (Example 2-4).

Application of Eq. (2.41) to the expressions for $h(t)$ and $v_{in}(t)$, given respectively by Eqs. (2.53) and (2.54), gives

$$\begin{aligned} v_{out}(t) &= v_{in}(t) * h(t) \\ &= u(t) \int_0^t v_{in}(\tau)\, h(t-\tau)\, d\tau \\ &= u(t) \int_0^t [u(\tau) - u(\tau-1)] \\ &\qquad \times 2e^{-2(t-\tau)}\, u(t-\tau)\, d\tau \\ &= u(t) \int_0^t 2e^{-2(t-\tau)}\, u(\tau)\, u(t-\tau)\, d\tau \\ &\quad - u(t) \int_0^t 2e^{-2(t-\tau)}\, u(\tau-1)\, u(t-\tau)\, d\tau. \end{aligned} \tag{2.55}$$

Upon application of the recipe described by Eq. (2.52), $v_{out}(t)$ becomes

$$\begin{aligned} v_{out}(t) &= \left[\int_0^t 2e^{-2(t-\tau)}\, d\tau\right] u(t) \\ &\quad - \left[\int_1^t 2e^{-2(t-\tau)}\, d\tau\right] u(t-1) \\ &= \frac{2}{2}\, e^{-2(t-\tau)}\Big|_0^t\, u(t) - \frac{2}{2}\, e^{-2(t-\tau)}\Big|_1^t\, u(t-1) \\ &= [1 - e^{-2t}]\, u(t) - [1 - e^{-2(t-1)}]\, u(t-1) \text{ V}, \end{aligned} \tag{2.56}$$

where we reintroduced the unit step functions $u(t)$ and $u(t-1)$ associated with the two integration terms. Figure 2-11(b) displays the temporal waveform of $v_{out}(t)$.

2-4 Graphical Convolution

The convolution integral given by

$$y(t) = \int_0^t x(\tau)\, h(t-\tau)\, d\tau$$

can be evaluated graphically by computing it at successive values of t.

Graphical Convolution Technique

Step 1: On the τ-axis, display $x(\tau)$ and $h(-\tau)$ with the latter being an image of $h(\tau)$ folded about the vertical axis.

Step 2: Shift $h(-\tau)$ to the right by a small increment t to obtain $h(t-\tau) = h(-(\tau - t))$.

Step 3: Determine the product of $x(\tau)$ and $h(t-\tau)$ and integrate it over the τ-domain from $\tau = 0$ to $\tau = t$ to get $y(t)$. The integration is equal to the area overlapped by the two functions.

Step 4: Repeat steps 2 and 3 for each of many successive values of t to generate the complete response $y(t)$.

By way of illustration, let us consider the RC circuit of Example 2-4. Plots of the input excitation $v_{in}(t)$ and the impulse

Module 2.1 **Convolution of Exponential Functions** This module computes the convolution of $e^{-at}\,u(t)$and $e^{-bt}\,u(t)$. The values of exponents a and b are selectable.

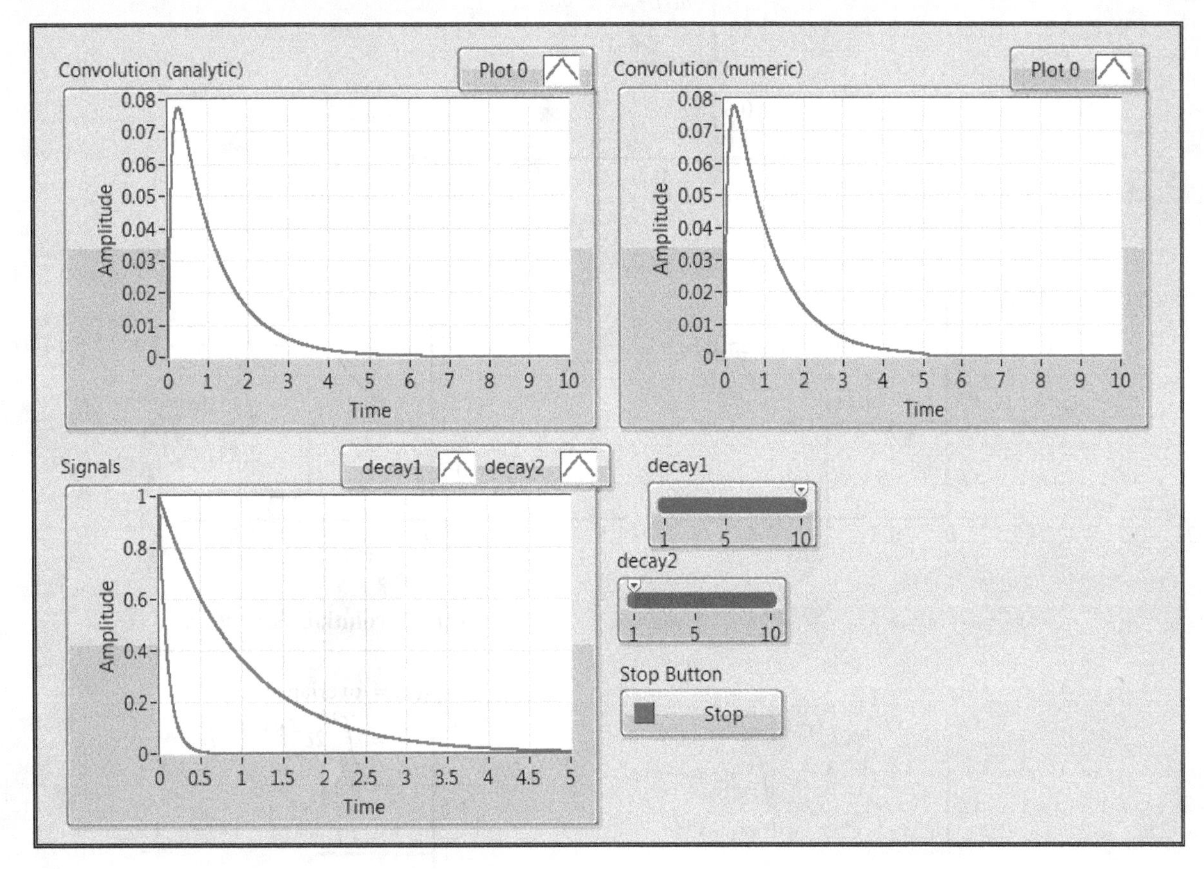

response $h(t)$ are displayed in Fig. 2-12(a). To perform the convolution

$$\upsilon_{\text{out}}(t) = u(t) \int_0^t \upsilon_{\text{in}}(\tau)\; h(t-\tau)\; d\tau, \tag{2.57}$$

we need to plot $\upsilon_{\text{in}}(\tau)$ and $h(t-\tau)$ along the τ axis. Figure 2-12(b) through (e) show these plots at progressive values of t, starting at $t = 0$ and concluding at $t = 2$ s. In all cases, $\upsilon_{\text{in}}(\tau)$ remains unchanged, but $h(t-\tau)$ is obtained by "folding" the original function across the vertical axis to generate $h(-\tau)$ and then shifting it to the right along the τ-axis by an amount t. The output voltage is equal to the integrated product of $\upsilon_{\text{in}}(\tau)$ and $h(t-\tau)$, which is equal to the shaded overlapping areas in the figures.

At $t = 0$ (Fig. 2-12(b)), no overlap exists; hence, $\upsilon_{\text{out}}(0) = 0$. Sliding $h(t-\tau)$ by $t = 0.5$ s, as shown in Fig. 2-12(c), leads to $\upsilon_{\text{out}}(0.5) = 0.63$. Sliding $h(t-\tau)$ further to the right leads to a greater overlap, reaching a maximum at $t = 1$ s (Fig. 2-12(d)). Beyond $t = 1$ s, the overlap is smaller in area, as illustrated by part (e) of Fig. 2-12, which corresponds to a shift $t = 2$ s. If the values of $\upsilon_{\text{out}}(t)$, as determined through this graphical integration method at successive values of t, are plotted as a function of t, we would get the same circuit response curve shown earlier in Fig. 2-11(b).

Example 2-5: Graphical Convolution

Given the waveforms shown in Fig. 2-13(a), apply the graphical convolution technique to determine the response $y(t) = x(t) * h(t)$.

Solution: Figure 2-13(b) shows waveforms $x(\tau)$ and $h(-\tau)$, plotted along the τ-axis. The waveform $h(-\tau)$ is the mirror

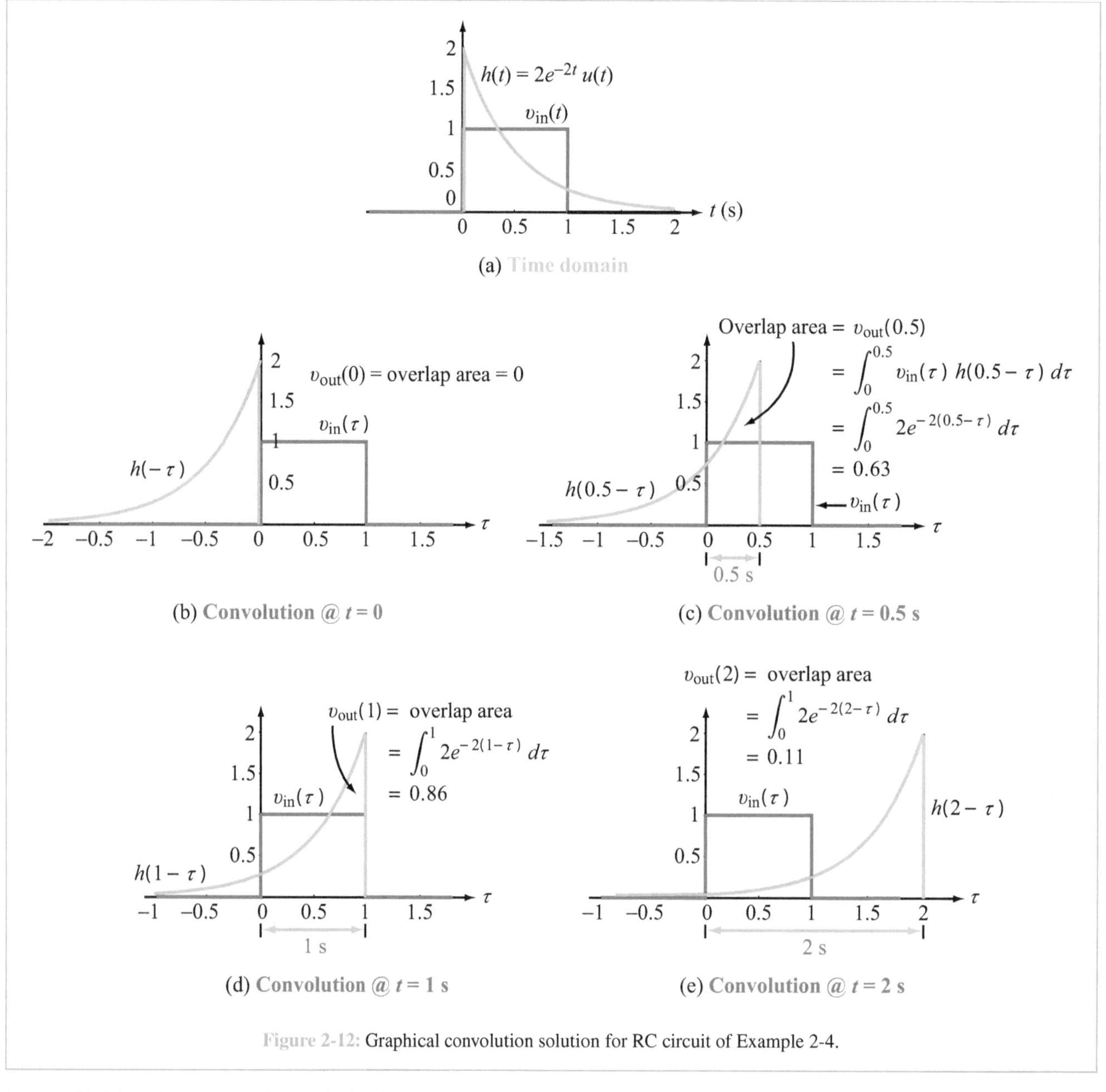

Figure 2-12: Graphical convolution solution for RC circuit of Example 2-4.

image of $h(\tau)$ with respect to the vertical axis. In Fig. 2-13(c) through (e), $h(t-\tau)$ is plotted for $t = 1$ s, 1.5 s, and 2 s, respectively. In each case, the shaded area is equal to $y(t)$. We note that when $t > 1$ s, one of the shaded areas contributes a positive number to $y(t)$ while the other contributes a negative number. The overall resultant response $y(t)$ generated by this process of sliding $h(t-\tau)$ to the right is displayed in Fig. 2-13(f).

We note that when two functions with finite time widths T_1 and T_2 are convolved, the width of the resultant function will be equal to $(T_1 + T_2)$, regardless of the shapes of the two functions.

Concept Question 2-5: Describe the process involved in the application of the graphical convolution technique. (See S^2)

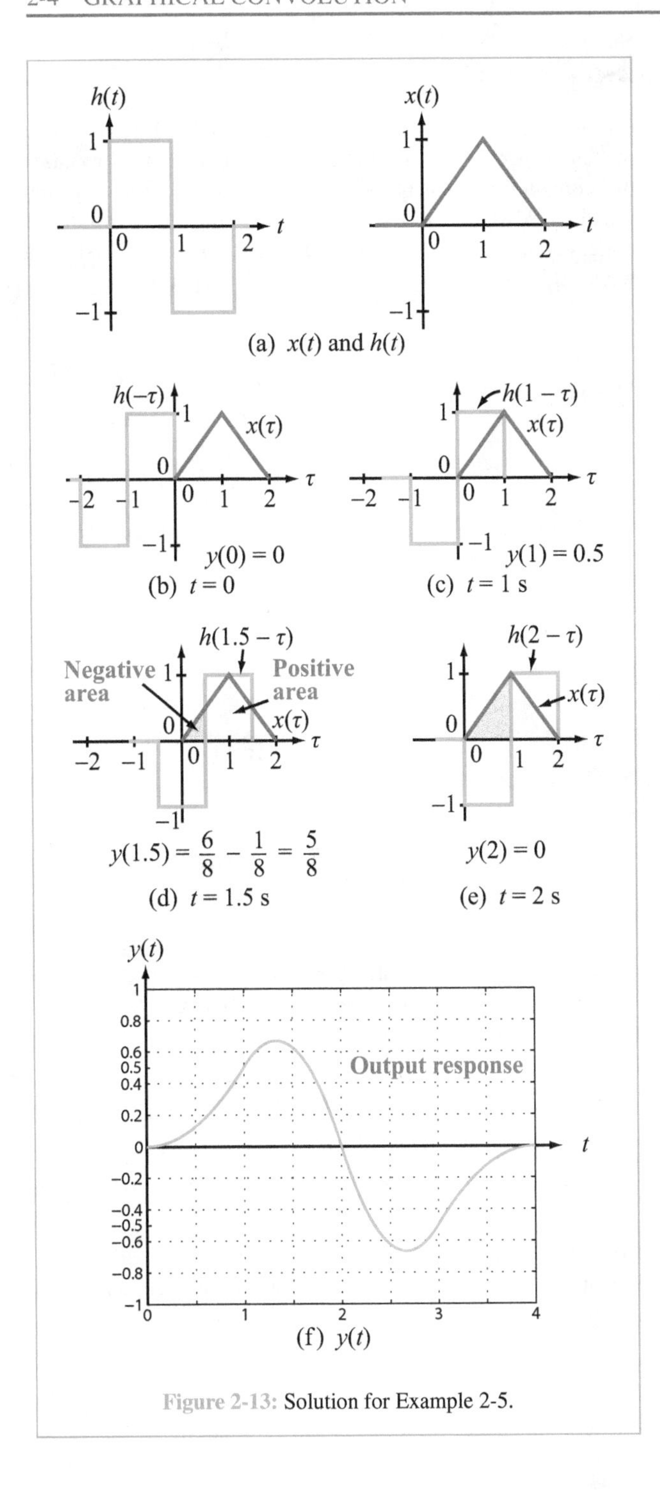

Figure 2-13: Solution for Example 2-5.

Exercise 2-6: Apply graphical convolution to the waveforms of $x(t)$ and $h(t)$ shown in Fig. E2-6 to determine $y(t) = x(t) * h(t)$.

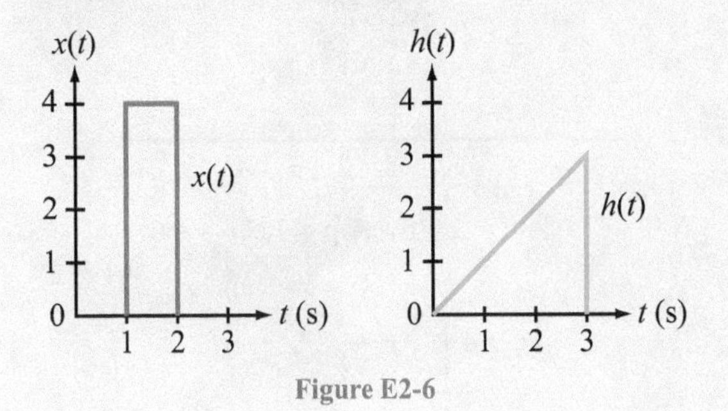

Figure E2-6

Answer:

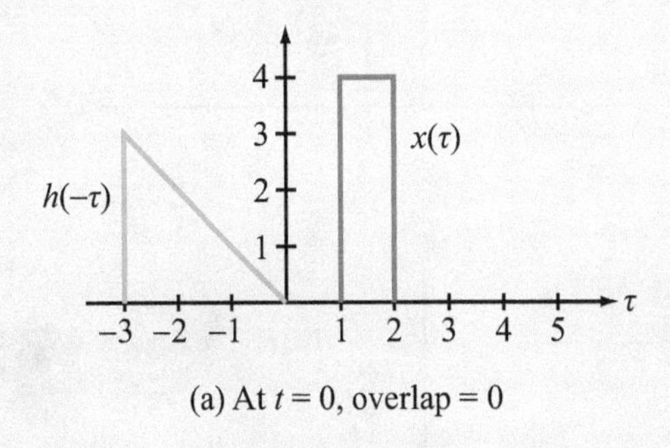

(a) At $t = 0$, overlap = 0

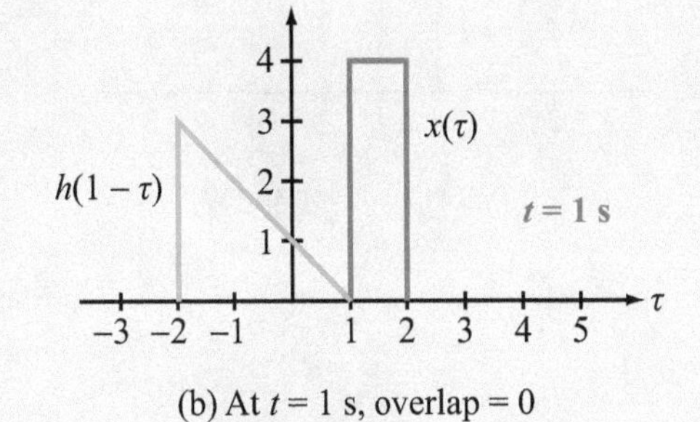

(b) At $t = 1$ s, overlap = 0

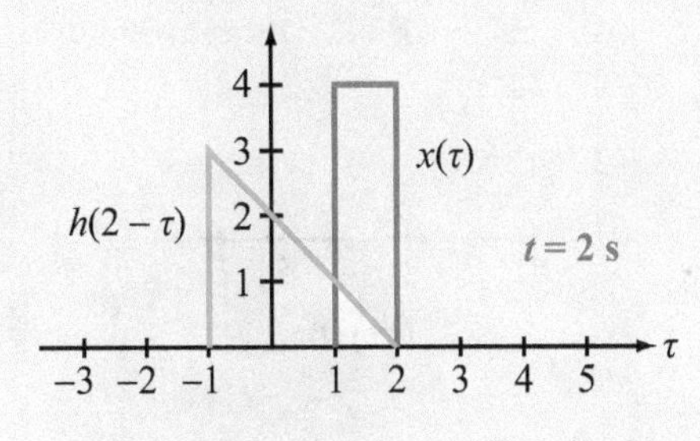

(c) At $t = 2$ s, overlap = $1/2 \times 4 = 2$

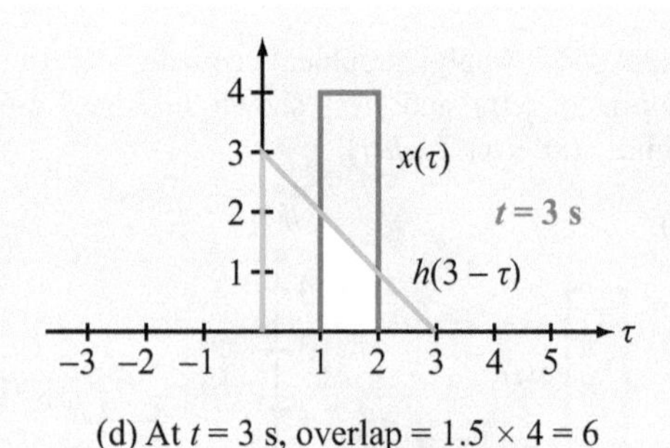

(d) At $t = 3$ s, overlap $= 1.5 \times 4 = 6$

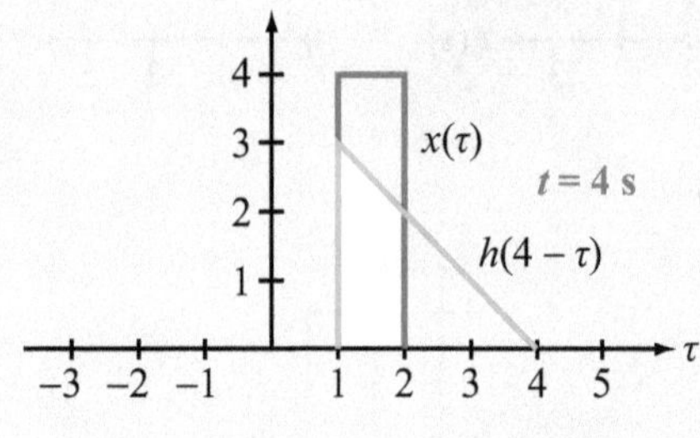

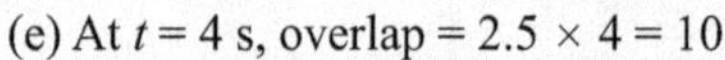

(e) At $t = 4$ s, overlap $= 2.5 \times 4 = 10$

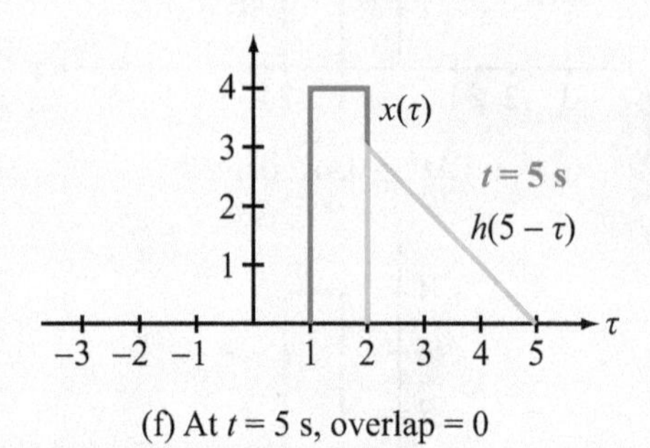

(f) At $t = 5$ s, overlap $= 0$

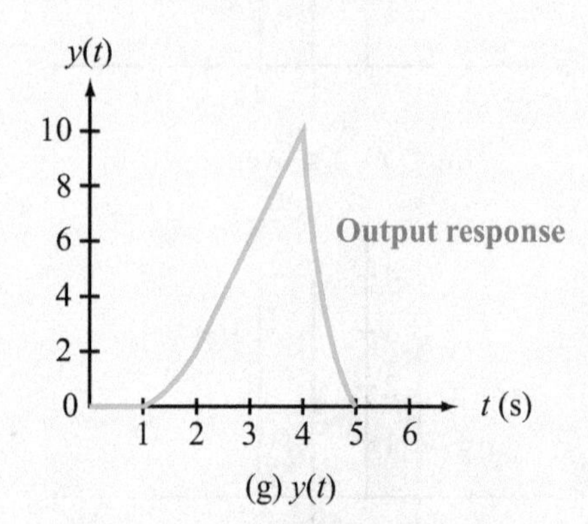

(g) $y(t)$

(See S²)

2-5 Convolution Properties

Computation of convolutions can be greatly simplified by using the ten properties outlined in this section. In fact, in many cases the convolutions can be determined without computing any integrals. Also, to help the user with both the computation and understanding of the convolution operation, we will attach a physical interpretation to each of the ten properties.

2-5.1 Commutative Property

This property states that

$$y(t) = h(t) * x(t) = x(t) * h(t). \tag{2.58}$$

Its validity can be demonstrated by changing variables from τ to $\tau' = t - \tau$ in the convolution integral:

$$\begin{aligned} y(t) = h(t) * x(t) &= \int_{-\infty}^{\infty} h(\tau)\, x(t-\tau)\, d\tau \\ &= \int_{\infty}^{-\infty} h(t-\tau')\, x(\tau')\, (-d\tau') \\ &= \int_{-\infty}^{\infty} x(\tau')\, h(t-\tau')\, d\tau' = x(t) * h(t). \end{aligned} \tag{2.59}$$

Physically, the commutative property means that we can interchange the input and the impulse response of the LTI system without changing its output:

If $x(t)$ ➡ $h(t)$ ➡ $y(t)$,

then $h(t)$ ➡ $x(t)$ ➡ $y(t)$.

Computationally, we can choose to time-shift and reverse whichever of $x(t)$ and $h(t)$ that makes the computation or graphical calculation of the convolution integral easier.

2-5.2 Associative Property

The combined convolution of three functions is the same, regardless of the order in which the convolution is performed:

$$y(t) = [g(t) * h(t)] * x(t) = g(t) * [h(t) * x(t)]. \tag{2.60}$$

Proving this property in the time domain requires much algebra, but proving it after introducing the Laplace transform in Chapter 3 is straightforward. Hence, we will accept it at face value for the time being.

► Physically, the associative property implies that the impulse response of two LTI systems connected in series is the convolution of their individual impulse responses. ◄

To see why, consider two LTI systems with impulse responses $h_1(t)$ and $h_2(t)$ connected ***in series***:

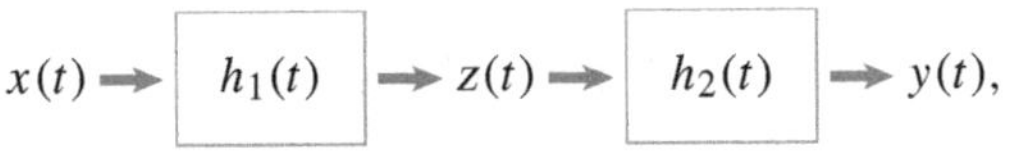

$$x(t) \rightarrow \boxed{h_1(t)} \rightarrow z(t) \rightarrow \boxed{h_2(t)} \rightarrow y(t),$$

where $z(t)$ is an intermediate signal between the two systems. Separately, the two systems are characterized by

$$z(t) = x(t) * h_1(t) \tag{2.61a}$$

and

$$y(t) = z(t) * h_2(t). \tag{2.61b}$$

Combining the two parts of Eq. (2.61) gives

$$y(t) = [x(t) * h_1(t)] * h_2(t). \tag{2.62}$$

The associative property allows us to rewrite Eq. (2.62) as

$$y(t) = x(t) * [h_1(t) * h_2(t)] \tag{2.63}$$

and to represent the process symbolically (Fig. 2-14) as

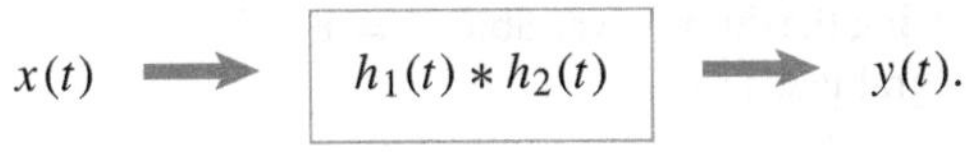

$$x(t) \longrightarrow \boxed{h_1(t) * h_2(t)} \longrightarrow y(t).$$

► The joint impulse response of two series-connected LTI systems is equivalent to the convolution of their individual impulse responses carried out in either order. A series connection of two systems also is known as a ***cascade connection***. ◄

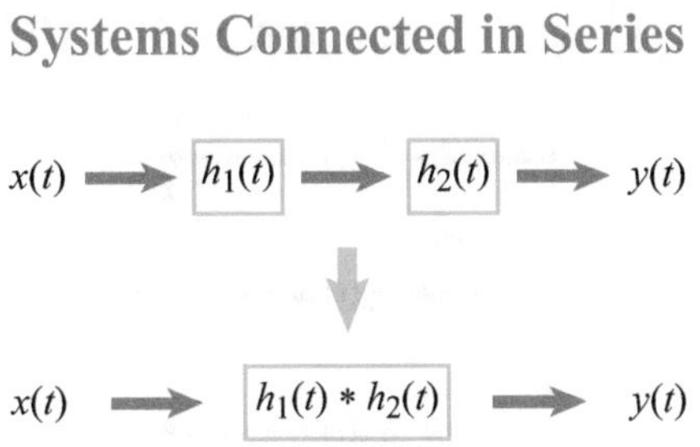

Systems Connected in Series

$$x(t) \rightarrow \boxed{h_1(t)} \rightarrow \boxed{h_2(t)} \rightarrow y(t)$$

$$\Downarrow$$

$$x(t) \rightarrow \boxed{h_1(t) * h_2(t)} \rightarrow y(t)$$

Figure 2-14: The impulse response of two systems connected in series is equivalent to the convolution of their individual impulse responses.

2-5.3 Distributive Property

The distributive property, which follows directly from the definition of convolution, allows us to perform the convolution operation on the sum of multiple input signals as

$$\begin{aligned} &h(t) * [x_1(t) + x_2(t) + \cdots + x_N(t)] = \\ &h(t) * x_1(t) + h(t) * x_2(t) + \cdots + h(t) * x_N(t). \end{aligned} \tag{2.64}$$

Conversely, if it is possible to model a complicated impulse response $h(t)$ as the sum of two or more simpler impulse responses,

$$h(t) = h_1(t) + h_2(t) + \cdots + h_N(t), \tag{2.65}$$

then application of the distributive property serves to simplify computation of the convolution integral by replacing a computation involving the complicated impulse response with multiple, but simpler, computations involving its constituent components. That is,

$$\begin{aligned} x(t) * h(t) &= x(t) * h_1(t) + x(t) * h_2(t) \\ &\quad + \cdots + x(t) * h_N(t). \end{aligned} \tag{2.66}$$

► The impulse response of the parallel connection of two or more LTI systems is the sum of their impulse responses. ◄

The preceding statements are encapsulated by the symbolic diagram in Fig. 2-15.

Systems Connected in Parallel

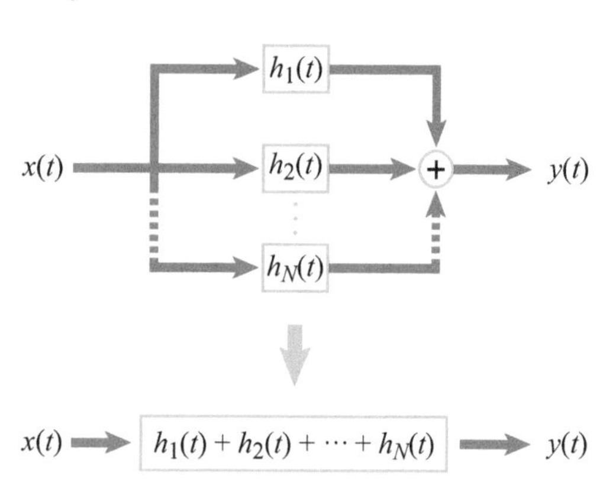

Figure 2-15: The impulse response $h(t)$ of multiple LTI systems connected in parallel is equivalent to the sum of their individual impulse responses.

In combination, the associative and distributive properties state that for LTI systems:

(1) Systems in series: Impulse responses convolved.
(2) Systems in parallel: Impulse responses added.

Example 2-6: Four Interconnected Systems

Determine the overall impulse response $h(t)$ of the LTI system depicted in Fig. 2-16.

Solution: Since $h_1(t)$ and $h_2(t)$ are in series, their joint impulse response is $h_1(t) * h_2(t)$. Similarly, for the lower branch, the joint response is $h_3(t) * h_4(t)$.

The two branches are in parallel, so their combined impulse response is the sum of the impulse responses of the individual branches:

$$h(t) = h_1(t) * h_2(t) + h_3(t) * h_4(t).$$

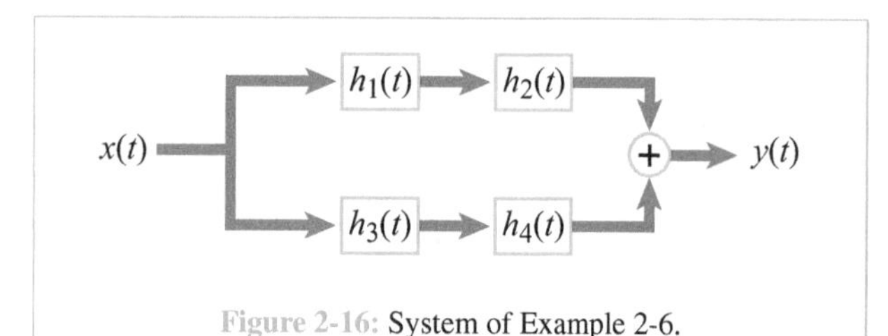

Figure 2-16: System of Example 2-6.

2-5.4 Causal * Causal = Causal

Recall from Chapter 1 that a signal $x(t)$ is causal if $x(t) = 0$ for $t < 0$. Moreover, the output $y(t)$ of a real-world *causal system* can depend on only present and past values of its input signal $x(t)$. Hence, if its input is zero for $t < 0$, its output also must be zero for $t < 0$. This argument led us in Section 2-3.2 to change the limits of the convolution integral to $[0, t]$ and to multiply the integral by $u(t)$. That is,

$$\begin{aligned} y(t) &= u(t) \int_0^t h(\tau)\, x(t-\tau)\, d\tau \\ &= u(t) \int_0^t x(\tau)\, h(t-\tau)\, d\tau. \end{aligned} \tag{2.67}$$

▶ Convolution of a causal signal with another causal signal or with a causal impulse response of an LTI system generates a causal output. Moreover, the impulse response of a causal LTI system is causal. ◀

2-5.5 Time-Shift Property

Given the convolution integral

$$y(t) = h(t) * x(t) = \int_{-\infty}^{\infty} h(\tau)\, x(t-\tau)\, d\tau, \tag{2.68}$$

the convolution of $h(t)$ delayed by T_1 and $x(t)$ delayed by T_2 is

$$h(t-T_1) * x(t-T_2) = \int_{-\infty}^{\infty} h(\tau - T_1)\, x(t-T_2-\tau)\, d\tau. \tag{2.69}$$

Introducing the dummy variable $\tau' = \tau - T_1$ everywhere inside the integral leads to

$$\begin{aligned} h(t-T_1) * x(t-T_2) &= \int_{-\infty}^{\infty} h(\tau')\, x(t-T_1-T_2-\tau')\, d\tau' \\ &= \int_{-\infty}^{\infty} h(\tau')\, x(t'-\tau')\, d\tau', \end{aligned} \tag{2.70}$$

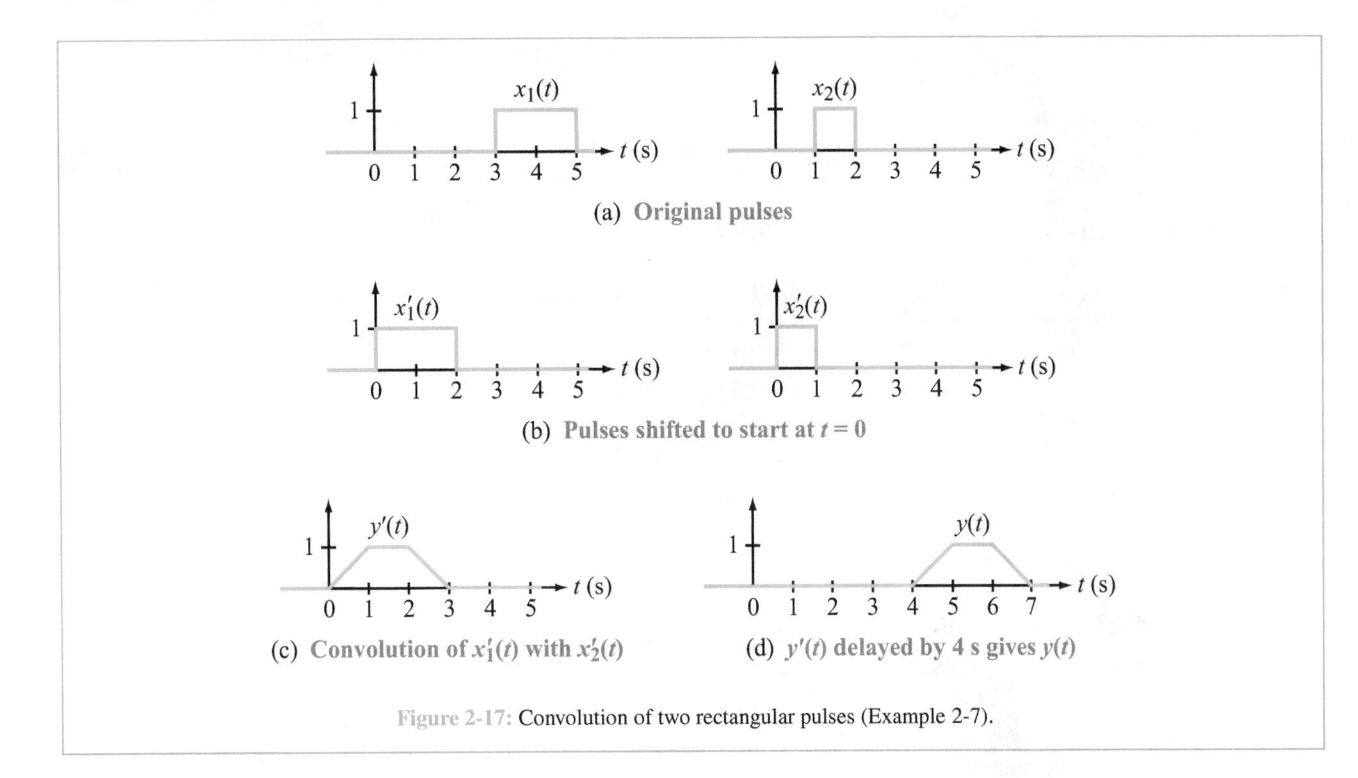

Figure 2-17: Convolution of two rectangular pulses (Example 2-7).

where we introduced the time-shifted variable $t' = t - T_1 - T_2$. The form of the integral in Eq. (2.70) matches the definition of the convolution integral given by Eq. (2.68), except that t has been replaced with t'. Hence, the output $y(t)$ becomes

$$y(t') = y(t - T_1 - T_2). \tag{2.71}$$

In conclusion,

$$h(t - T_1) * x(t - T_2) = y(t - T_1 - T_2). \tag{2.72}$$

(time-shift property)

Physically, the time-shift property is just a statement of time-invariance; delaying the input of a time-invariant system and/or if the system has a built-in delay causes the output to be delayed by the sum of the two delays. Computationally, the shift property allows us to delay or advance signals to take advantage of their symmetry or causality to make the convolution operation simpler.

Example 2-7: Convolution of Two Delayed Rectangular Pulses

Compute the convolution of the rectangular pulses $x_1(t)$ and $x_2(t)$ shown in Fig. 2-17(a).

Solution: Our plan is to take advantage of as many convolution properties as possible. We start by shifting $x_1(t)$ and $x_2(t)$ to the left so that they both start at $t = 0$ (Fig. 2-17(b)). The shifted pulses are

$$x_1'(t) = x_1(t + 3) = u(t) - u(t - 2),$$

and

$$x_2'(t) = x_2(t + 1) = u(t) - u(t - 1),$$

and their convolution is

$$\begin{aligned} x_1'(t) * x_2'(t) &= [u(t) - u(t - 2)] * [u(t) - u(t - 1)] \\ &= u(t) * u(t) - u(t) * u(t - 1) \\ &\quad - u(t) * u(t - 2) + u(t - 1) * u(t - 2). \end{aligned}$$

For the first term, using the causality property, we have

$$u(t) * u(t) = u(t) \int_0^t 1 \, d\tau = t \, u(t) = r(t). \tag{2.73}$$

Applying the time-shift property to the remaining three terms gives

$$u(t) * u(t - 1) = r(t - 1),$$

$$u(t) * u(t - 2) = r(t - 2),$$

and

$$u(t-1) * u(t-2) = r(t-3).$$

Hence,

$$x_1'(t) * x_2'(t) = r(t) - r(t-1) - r(t-2) + r(t-3),$$

which is plotted in Fig. 2-17(c). Finally, we apply the time-shift property again to restore the combined shift of 4 s that was applied earlier to $x_1(t)$ and $x_2(t)$. The result is shown in Fig. 2-17(d).

2-5.6 Convolution with an Impulse

Convolution of a signal $x(t)$ with an impulse function $\delta(t-T)$ simply delays $x(t)$ by T:

$$\begin{aligned} x(t) * \delta(t-T) &= \int_{-\infty}^{\infty} x(\tau)\, \delta(t-T-\tau)\, d\tau \\ &= x(t-T). \end{aligned} \tag{2.74}$$

The result follows from the sampling property of the impulse function given by Eq. (1.29), namely that $\delta(t-T-\tau)$ is zero except at $\tau = t - T$.

Example 2-8: Four Interconnected Systems

Compute the overall impulse response $h(t)$ of the system diagrammed in Fig. 2-18.

Solution: In the top branch, $u(t)$ and $\delta(t+1)$ are in series, so their joint impulse response is

$$h_1(t) = u(t) * \delta(t+1) = u(t+1).$$

Similarly, for the lower branch, we have

$$h_2(t) = -\delta(t) * u(t) = -u(t).$$

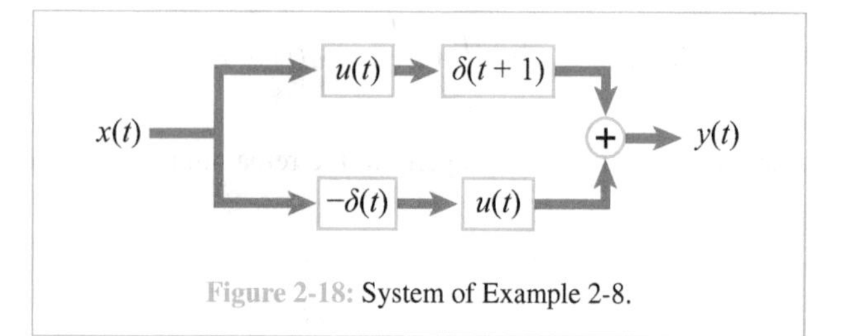

Figure 2-18: System of Example 2-8.

For the combination of the two branches,

$$h(t) = h_1(t) + h_2(t) = u(t+1) - u(t).$$

Note that the output is

$$y(t) = x(t) * h(t) = \int_{-\infty}^{\infty} x(\tau)\, [u(t+1-\tau) - u(t-\tau)]\, d\tau.$$

Using the recipe given by Eq. (2.52) or, equivalently, the knowledge that a step function is zero if its argument is negative, leads to

$$y(t) = \int_{t}^{t+1} x(\tau)\, d\tau.$$

Since $y(t)$ depends on $x(t)$ beyond time t (upper limit is $t+1$), the system is noncausal.

2-5.7 Width Property

▶ When a function of time width T_1 is convolved with another of time width T_2, the width of the resultant function is $(T_1 + T_2)$, regardless of the shapes of the two functions. ◀

This property was noted earlier in connection with the graphical convolution examples.

If $h(t) = 0$ when t is outside the interval $a \le t \le b$ and $x(t) = 0$ when t is outside the interval $c \le t \le d$, then $y(t) = 0$ when t is outside the interval $(a+c) \le t \le (b+d)$. Hence, we have the

$$\text{width of } y(t) = (b+d) - (a+c). \tag{2.75}$$

In Example 2-7, the widths of $x_1(t)$ and $x_2(t)$ are 2 s and 1 s, respectively. The width of their convolution is 3 s (Fig. 2-17(d)).

2-5.8 Area Property

▶ The area under the convolution $y(t)$ of two functions $x(t)$ and $h(t)$ is the product of the areas under the individual functions. ◀

The area under $y(t)$ is obtained by integrating $y(t)$ over $(-\infty, \infty)$. That is,

$$\begin{aligned}
\text{Area of } y(t) &= \int_{-\infty}^{\infty} y(t)\, dt \\
&= \int_{-\infty}^{\infty} \left[\int_{-\infty}^{\infty} h(\tau)\, x(t-\tau)\, d\tau \right] dt \\
&= \int_{-\infty}^{\infty} \int_{-\infty}^{\infty} h(\tau)\, x(t-\tau)\, dt\, d\tau \\
&= \int_{-\infty}^{\infty} h(\tau) \left[\int_{-\infty}^{\infty} x(t-\tau)\, dt \right] d\tau \\
&= \left[\int_{-\infty}^{\infty} h(\tau)\, d\tau \right] \left[\int_{-\infty}^{\infty} x(t-\tau)\, dt \right] \\
&= \text{area of } h(t) \times \text{area of } x(t). \qquad (2.76)
\end{aligned}$$

In step 3 of the derivation, we interchanged the order of dt and $d\tau$ (because they are independent variables), and in step 4 we used the fact that the areas of $x(t)$ and $x(t-\tau)$ are the same when evaluated over $(-\infty, \infty)$.

2-5.9 Convolution with a Step Function

From the definition of convolution, we have

$$\begin{aligned}
x(t) * u(t) &= \int_{-\infty}^{\infty} x(\tau)\, u(t-\tau)\, d\tau \\
&= \int_{-\infty}^{t} x(\tau)\, d\tau, \qquad (2.77)
\end{aligned}$$

(ideal integrator)

where we used the fact that $u(t-\tau) = 0$ for $t - \tau < 0$ or, equivalently, $\tau > t$. Operationally, the right-hand side of Eq. (2.77) constitutes an ideal integrator, and for a causal signal, the lower limit on the integral should be zero.

▶ An LTI system whose impulse response is equivalent to a unit step function performs like an ideal integrator. ◀

2-5.10 Differentiation and Integration Properties

Given the convolution

$$y(t) = \int_{-\infty}^{\infty} x(\tau)\, h(t-\tau)\, d\tau,$$

taking the derivative with respect to time gives

$$\begin{aligned}
\frac{d}{dt}[y(t)] &= \frac{d}{dt} \int_{-\infty}^{\infty} x(\tau)\, h(t-\tau)\, d\tau \\
&= \int_{-\infty}^{\infty} x(\tau)\, \frac{d}{dt}[h(t-\tau)]\, d\tau. \qquad (2.78)
\end{aligned}$$

Hence,

$$\frac{d}{dt}[y(t)] = x(t) * \frac{d}{dt}[h(t)]. \qquad (2.79)$$

Similarly, we can show that

$$\frac{d}{dt}[y(t)] = \frac{d}{dt}[x(t)] * h(t). \qquad (2.80)$$

Combining these two relations and generalizing to higher order derivatives, we get

$$\left(\frac{d^m x}{dt^m}\right) * \left(\frac{d^n h}{dt^n}\right) = \frac{d^{m+n} y}{dt^{m+n}}. \qquad (2.81)$$

In a similar fashion, it is easy to show that

$$\begin{aligned}
\int_{-\infty}^{t} y(\tau)\, d\tau &= x(t) * \left[\int_{-\infty}^{t} h(\tau)\, d\tau \right] \\
&= \left[\int_{-\infty}^{t} x(\tau)\, d\tau \right] * h(t). \qquad (2.82)
\end{aligned}$$

The ten convolution properties are summarized in Table 2-1, followed by a list of commonly encountered convolutions in Table 2-2.

Table 2-1: **Convolution properties.**

Convolution Integral	$y(t) = h(t) * x(t) = \int_{-\infty}^{\infty} h(\tau)\, x(t-\tau)\, d\tau$
• Causal Systems and Signals:	$y(t) = h(t) * x(t) = u(t) \int_{0}^{t} h(\tau)\, x(t-\tau)\, d\tau$
Property	**Description**
1. Commutative	$x(t) * h(t) = h(t) * x(t)$
2. Associative	$[g(t) * h(t)] * x(t) = g(t) * [h(t) * x(t)]$
3. Distributive	$x(t) * [h_1(t) + \cdots + h_N(t)] = x(t) * h_1(t) + \cdots + x(t) * h_N(t)$
4. Causal $*$ Causal = Causal	$y(t) = u(t) \int_{0}^{t} h(\tau)\, x(t-\tau)\, d\tau$
5. Time-shift	$h(t - T_1) * x(t - T_2) = y(t - T_1 - T_2)$
6. Convolution with Impulse	$x(t) * \delta(t - T) = x(t - T)$
7. Width	Width of $y(t)$ = width of $x(t)$ + width of $h(t)$
8. Area	Area of $y(t)$ = area of $x(t)$ × area of $h(t)$
9. Convolution with $u(t)$	$y(t) = x(t) * u(t) = \int_{-\infty}^{t} x(\tau)\, d\tau$ (Ideal integrator)
10a. Differentiation	$\left(\frac{d^m x}{dt^m}\right) * \left(\frac{d^n h}{dt^n}\right) = \frac{d^{m+n} y}{dt^{m+n}}$
10b. Integration	$\int_{-\infty}^{t} y(\tau)\, d\tau = x(t) * \left[\int_{-\infty}^{t} h(\tau)\, d\tau\right] = \left[\int_{-\infty}^{t} x(\tau)\, d\tau\right] * h(t)$

Example 2-9: RC-Circuit with Triangular Pulse—Revisited

In Example 2-3, we computed the convolution of a triangular pulse (shown again in Fig. 2-19) given by

$$x(t) = r(t) - 2r(t-1) + r(t-2),$$

with the impulse response of an RC circuit, namely,

$$h(t) = e^{-t}\, u(t).$$

Recompute the convolution integration by taking advantage of the properties listed in Table 2-1.

Solution: Using the distributive property (#3 in Table 2-1) and the time-shift property (#5), $\upsilon_{\text{out}}(t)$ can be expressed as

$$\begin{aligned} \upsilon_{\text{out}}(t) &= h(t) * \upsilon_{\text{in}}(t) \\ &= h(t) * r(t) - 2h(t) * r(t-1) + h(t) * r(t-2) \\ &= z(t) - 2z(t-1) + z(t-2), \end{aligned} \quad (2.83)$$

Table 2-2: **Commonly encountered convolutions.**

1. $u(t) * u(t) = t\, u(t)$
2. $e^{at}\, u(t) * u(t) = \left(\dfrac{e^{at}-1}{a}\right) u(t)$
3. $e^{at}\, u(t) * e^{bt}\, u(t) = \left[\dfrac{e^{at}-e^{bt}}{a-b}\right] u(t), \qquad a \neq b$
4. $e^{at}\, u(t) * e^{at}\, u(t) = te^{at}\, u(t)$
5. $te^{at}\, u(t) * e^{bt}\, u(t) = \dfrac{e^{bt} - e^{at} + (a-b)te^{at}}{(a-b)^2}\, u(t),$ $a \neq b$
6. $te^{at}\, u(t) * e^{at}\, u(t) = \frac{1}{2}\, t^2 e^{at}\, u(t)$
7. $\delta(t - T_1) * \delta(t - T_2) = \delta(t - T_1 - T_2)$

where $z(t)$ is an intermediary signal given by

$$\begin{aligned} z(t) = h(t) * r(t) &= u(t) \int_0^t h(\tau)\, r(t-\tau)\, d\tau \\ &= u(t) \int_0^t e^{-\tau}(t-\tau)\, d\tau \\ &= (e^{-t} + t - 1)\, u(t). \end{aligned} \tag{2.84}$$

Using Eq. (2.84) and its time-shifted versions in Eq. (2.83) leads to

$$\begin{aligned} \upsilon_{\text{out}}(t) = (e^{-t} + t - 1)\, u(t) &- 2[e^{-(t-1)} + (t-1) - 1] \\ &\times u(t-1) + [e^{-(t-2)} + (t-2) - 1]\, u(t-2). \end{aligned} \tag{2.85}$$

As expected, the plot of $\upsilon_{\text{out}}(t)$ shown in Fig. 2-19(b), which is based on the expression given by Eq. (2.85), is identical with the convolution plot shown earlier in Fig. 2-9(b) in conjunction with Example 2-3.

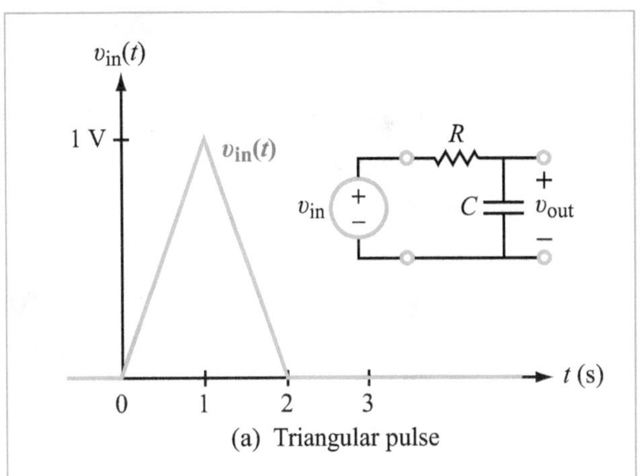

(a) Triangular pulse

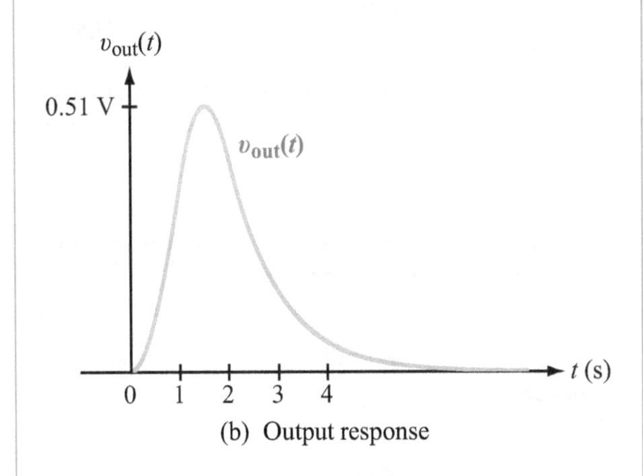

(b) Output response

Figure 2-19: Triangular pulse exciting an RC circuit (Example 2-9). This is the same as Fig. 2-9.

In Section 2-3.4, we demonstrated through lengthy integration steps that if the waveforms of $x(t)$ and $h(t)$ are rectangular pulses given by

$$x(t) = A[u(t) - u(t - T_1)] \tag{2.86a}$$

and

$$h(t) = B[u(t) - u(t - T_2)], \tag{2.86b}$$

their convolution is given by Eq. (2.50) as

$$\begin{aligned} y(t) &= x(t) * h(t) \\ &= AB[t\, u(t) - (t - T_2)\, u(t - T_2) - (t - T_1)\, u(t - T_1) \\ &\quad + (t - T_1 - T_2)\, u(t - T_1 - T_2)]. \end{aligned} \tag{2.87}$$

We will now derive the result given by Eq. (2.87) in a few simple steps, using the distributive property (#3 in Table 2-1), the time-shift property (#5), and Eq. (2.73), namely, $u(t) * u(t) = t\, u(t)$.

Thus,

$$\begin{aligned} y(t) &= x(t) * h(t) \\ &= A[u(t) - u(t - T_1)] * B[u(t) - u(t - T_2)] \\ &= AB[u(t) * u(t) - u(t) * u(t - T_2) - u(t) * u(t - T_1) \\ &\quad + u(t - T_1) * u(t - T_2)] \\ &= AB[t\, u(t) - (t - T_2)\, u(t - T_2) - (t - T_1)\, u(t - T_1) \\ &\quad + (t - T_1 - T_2)\, u(t - T_1 - T_2)], \end{aligned} \quad (2.88)$$

which is identical to Eq. (2.87).

It should be evident by now that using the convolution properties of Table 2-1 can greatly simplify the computation of convolutions!

In practice, convolutions are computed numerically using the *fast Fourier transform* (*FFT*) algorithm discussed in detail in Chapter 7.

Concept Question 2-6: What initial conditions does the convolution operation require? (See S²)

Concept Question 2-7: Describe the time-shift property of convolution. (See S²)

Concept Question 2-8: What is the outcome of convolving a signal with a step function? With an impulse function? (See S²)

Concept Question 2-9: What is the area property of convolution? (See S²)

Exercise 2-7: Evaluate

$$u(t) * \delta(t-3) - u(t-4) * \delta(t+1).$$

Answer: 0. Use convolution properties #5 and #6. (See S²)

Exercise 2-8: Evaluate $\lim_{t \to \infty} [e^{-3t}\, u(t) * u(t)]$.

Answer: $1/3 =$ area under $e^{-3t}\, u(t)$. Use convolution properties #8 and #9 in Table 2-1. (See S²)

2-6 Causality and BIBO Stability

Convolution leads to necessary and sufficient conditions for two important LTI system properties: ***causality*** and ***bounded-input/bounded output*** (***BIBO***) stability. We derive and illustrate these properties in this section.

2-6.1 Causality

We define a ***causal system*** as a system for which the present value of the output $y(t)$ can only depend on present and past values of the input $\{x(\tau),\ \tau \leq t\}$. For a noncausal system, the present output could depend on future inputs. Noncausal systems are also called ***anticipatory*** systems, since they anticipate the future.

A physical system must be causal, because a noncausal system must have the ability to see into the future! For example, the noncausal system $y(t) = x(t+2)$ must know the input two seconds into the future to deliver its output at the present time. This is clearly impossible in the real world.

> ▶ An LTI system is causal ***if and only if*** its impulse response is a causal function: $h(t) = 0$ for $t < 0$. ◀

2-6.2 BIBO Stability: Definition

"BIBO stable" may sound like a hobbit from *The Lord of the Rings*, but it is an extremely important and desirable property of LTI systems. It can be stated succinctly as: "If the input doesn't blow up, the output won't either."

A signal $x(t)$ is ***bounded*** if there is a constant C, so that $|x(t)| \leq C$ for all t. Examples of bounded signals include:

- $\cos(3t)$, $7e^{-2t}\, u(t)$, and $e^{2t}\, u(1-t)$.

Examples of unbounded signals include:

- t^2, $e^{2t}\, u(t)$, e^{-t}, and $1/t$.

A system is ***BIBO*** (***bounded input/bounded output***) stable if every bounded input $x(t)$ results in a bounded output $y(t)$, as depicted by the simple example shown in Fig. 2-20. It does not require that an unbounded input result in a bounded output (this would be unreasonable). Stability is obviously a desirable property for a system.

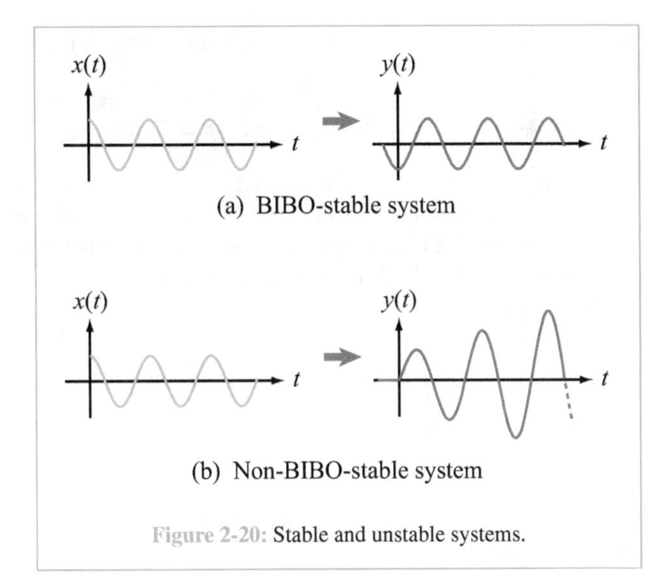

Figure 2-20: Stable and unstable systems.

2-6.3 BIBO Stability: Condition and Proof

How can we ascertain if an LTI system is BIBO stable? The answer is provided by the following theorem:

▶ An LTI system is BIBO stable *if and only if* its impulse response $h(t)$ is ***absolutely integrable*** (i.e., if and only if $\int_{-\infty}^{\infty} |h(t)|\, dt$ is finite). ◀

We will go through the proof of this result in some detail, because it is a useful exercise to learn how to prove mathematical theorems.

A theorem statement that includes the phrase "if and only if" requires a bidirectional proof. If B represents BIBO stability and A represents absolute integrability of $h(t)$, to prove the bidirectional statement $A \leftrightarrow B$ requires us to prove both $A \rightarrow B$ and $B \rightarrow A$. To prove $A \rightarrow B$, we *suppose* A is true and use it to prove B is true. Alternatively, we can use the ***contrapositive*** of $A \rightarrow B$, wherein we suppose B is false and use it to prove A is false. That is,

$$A \rightarrow B \equiv \overline{B} \rightarrow \overline{A}, \tag{2.89}$$

where $\overline{A}$ is shorthand for A being false.

"If" part of proof: $A \rightarrow B$

With

$$A = \text{absolute integrability of } h(t)$$

and

$$B = \text{BIBO stability},$$

we suppose A is true and use it to prove $A \rightarrow B$. Absolute integrability of $h(t)$ implies that there exists a constant L of finite magnitude such that

$$\int_{-\infty}^{\infty} |h(t)|\, dt = L < \infty. \tag{2.90}$$

To prove B, namely, that the system is BIBO stable, we have to suppose that the input $x(t)$ is bounded, which means that there exists a constant M of finite magnitude such that

$$|x(t)| \leq M.$$

The goal now is to prove that the output $y(t)$ is bounded. Having properly formulated the problem, it is now clear how to proceed. Applying the ***triangle inequality*** for integrals to the absolute value of the convolution integral given by Eq. (2.30) leads to

$$\begin{aligned} |y(t)| = \left| \int_{-\infty}^{\infty} h(\tau)\, x(t-\tau)\, d\tau \right| &\leq \int_{-\infty}^{\infty} |h(\tau)\, x(t-\tau)|\, d\tau \\ &\leq \int_{-\infty}^{\infty} |h(\tau)| M\, d\tau = LM, \end{aligned} \tag{2.91}$$

which states that $|y(t)|$ is bounded by the finite constant LM. Hence, if $h(t)$ is absolutely integrable, a bounded input results in a bounded output, and the system is BIBO stable.

"Only if" part of proof: $B \rightarrow A$

We can suppose B (system is BIBO stable) is true and then try to prove A ($h(t)$ is absolutely integrable) is true, but it is easier to prove the contrapositive. We suppose $h(t)$ is *not* absolutely integrable (i.e., A not true), and then identify a bounded input that results in an unbounded output (B not true). In short, we need a counterexample.

Since we are supposing $\int_{-\infty}^{\infty} |h(t)|\, dt \rightarrow \infty$, we seek a bounded input that would lead to an unbounded output. A convenient candidate is the noncausal input $x(t)$ given by

$$x(t) = \begin{cases} \dfrac{h(-t)}{|h(-t)|} = \dfrac{|h(-t)|}{h(-t)} = \pm 1 & \text{if } h(-t) \neq 0, \\ 0 & \text{if } h(-t) = 0. \end{cases}$$

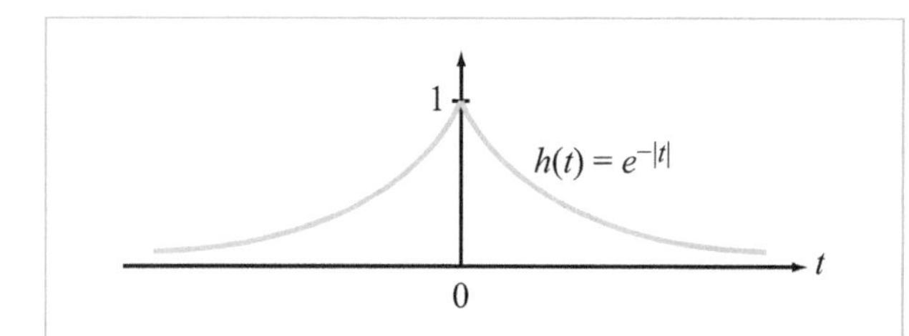

Figure 2-21: Impulse response of system in Example 2-10.

At $t = 0$, the corresponding output $y(t)$ is

$$y(0) = \int_{-\infty}^{\infty} h(\tau)\, x(0-\tau)\, d\tau = \int_{-\infty}^{\infty} h(\tau) \frac{|h(\tau)|}{h(\tau)}\, d\tau = \int_{-\infty}^{\infty} |h(\tau)|\, d\tau \to \infty. \quad (2.92)$$

This result proves that a system whose impulse response is not absolutely integrable is not BIBO stable ($\overline{A} \to \overline{B}$), which is equivalent to $B \to A$.

Example 2-10: Causality and BIBO Stability

A system is characterized by the impulse response $h(t) = e^{-|t|}$. Determine if the system is (a) causal and (b) BIBO stable.

Solution: (a) The specified impulse response is a two-sided exponential function (Fig. 2-21). Since $h(t) \neq 0$ for $t < 0$, the system is noncausal.

(b)

$$\int_{-\infty}^{\infty} |h(t)|\, dt = \int_{-\infty}^{\infty} e^{-|t|}\, dt = 2\int_{0}^{\infty} e^{-t}\, dt = 2.$$

Hence, the system is BIBO stable.

2-6.4 BIBO Stability of System with Decaying Exponentials

Consider a causal system with an impulse response

$$h(t) = Ce^{\gamma t}\, u(t), \quad (2.93)$$

where C is a finite constant and γ is, in general, a finite complex coefficient given by

$$\gamma = \alpha + j\beta, \qquad \alpha = \mathfrak{Re}[\gamma], \quad \text{and} \quad \beta = \mathfrak{Im}[\gamma]. \quad (2.94)$$

Such a system is BIBO stable if and only if $\alpha < 0$ (i.e., $h(t)$ is a one-sided exponential with an exponential coefficient whose real part is negative). To verify the validity of this statement, we test to see if $h(t)$ is absolutely integrable. Since $|e^{j\beta t}| = 1$ and $e^{\alpha t} > 0$,

$$\int_{-\infty}^{\infty} |h(t)|\, dt = \int_{0}^{\infty} |Ce^{\alpha t} e^{j\beta t}|\, dt = |C| \int_{0}^{\infty} e^{\alpha t}\, dt. \quad (2.95)$$

(a) $\alpha < 0$

If $\alpha < 0$, we can rewrite it as $\alpha = -|\alpha|$ in the exponential, which leads to

$$\int_{-\infty}^{\infty} |h(t)|\, dt = |C| \int_{0}^{\infty} e^{-|\alpha| t}\, dt = \frac{|C|}{|\alpha|} < \infty. \quad (2.96)$$

Hence, $h(t)$ is absolutely integrable and the system is BIBO stable.

(b) $\alpha \geq 0$

If $\alpha \geq 0$, Eq. (2.95) becomes

$$\int_{-\infty}^{\infty} |h(t)|\, dt = |C| \int_{0}^{\infty} e^{\alpha t}\, dt \to \infty,$$

thereby proving that the system is not BIBO stable when $\alpha \geq 0$.

► By extension, for any positive integer N, an impulse response composed of a linear combination of N exponential signals

$$h(t) = \sum_{i=1}^{N} C_i\, e^{\gamma_i t}\, u(t) \quad (2.97)$$

is absolutely integrable, and its LTI system is BIBO stable, if and only if all of the exponential coefficients γ_i have negative real parts. This is a fundamental attribute of LTI system theory. ◄

Some books define an LTI system as ***marginally stable*** if one or more of the exponential coefficients γ_i have *zero* real parts and the remaining coefficients have negative real parts. Accordingly, an LTI system with $h(t) = e^{jt}\, u(t)$ should be marginally stable, because $\gamma = 0 + j1$.

However, a marginally stable system is not BIBO stable. We can test the validity of the assertion by evaluating the system's output for a causal input signal $x(t) = e^{jt}\, u(t)$,

$$y(t) = \int_{-\infty}^{\infty} x(\tau)\, h(t-\tau)\, d\tau$$

$$= u(t) \int_{0}^{t} e^{j\tau} e^{j(t-\tau)}\, d\tau = u(t)\, e^{jt} \int_{0}^{t} d\tau = te^{jt}\, u(t).$$

Since t in the final expression can grow to ∞, the system is not BIBO stable. Hence, this marginally stable LTI system is *not* BIBO stable.

Concept Question 2-10: What constitutes a complete proof of a statement of the form "A is true if and only if B is true"? (See S²)

Concept Question 2-11: Given an expression for the impulse response $h(t)$ of an LTI system, how can you determine if the system is (a) causal and (b) BIBO stable? (See S²)

Exercise 2-9: A system's impulse response is $h(t) = u(t-1)/t^2$. Is the system BIBO stable?

Answer: Yes, because $\int_1^\infty |1/(t^2)|\, dt = 1$. (See S²)

Exercise 2-10: A system's impulse response is $h(t) = u(t-1)/t$. Is the system BIBO stable?

Answer: No, because $\int_1^\infty 1/|t|\, dt = \log(|t|)|_1^\infty \to \infty$. (See S²)

Exercise 2-11: A system's impulse response is

$$h(t) = (3 + j4)e^{-(1-j2)t}\, u(t) + (3 - j4)e^{-(1+j2)t}\, u(t).$$

Is the system BIBO stable?

Answer: Yes, because real parts of both exponential coefficients are negative. (See S²)

2-7 LTI Sinusoidal Response

This section introduces a vitally important property of LTI systems.

> ► The response of an LTI system to a complex exponential input signal $x(t) = Ae^{j\omega t}$ is another complex exponential signal $y(t) = \mathbf{H}(\omega)\, Ae^{j\omega t}$, where $\mathbf{H}(\omega)$ is a complex coefficient that depends on ω. A similar statement applies to sinusoidal inputs and outputs. ◄

In Chapter 5, we will show that a complicated periodic signal $x(t)$ of period T_0 and angular frequency $\omega_0 = 2\pi/T_0$ can be expressed as a linear combination of an infinite number of complex exponential signals

$$x(t) = \sum_{n=-\infty}^{\infty} \mathbf{x}_n e^{jn\omega_0 t} \tag{2.98}$$

with constant coefficients $\mathbf{x}_n$.

> ► A similar process applies to nonperiodic signals, wherein $\sum_{n=-\infty}^{\infty}$ becomes an integral $\int_{-\infty}^{\infty}$. ◄

According to the sinusoidal-response property, the corresponding output $y(t)$ will assume the form

$$y(t) = \sum_{n=-\infty}^{\infty} \mathbf{H}_n \mathbf{x}_n e^{jn\omega_0 t} \tag{2.99}$$

with constant coefficients $\mathbf{H}_n$.

This sinusoidal-response property is at the heart of how ***filtering*** of signals is performed by an LTI system to remove noise, attenuate undesirable frequency components, or accentuate contrast or other features in voice signals and images. The relevant steps are covered as follows:

Chapter 2: We will show in later parts of the present section why an LTI system excited by an input signal $x(t) = e^{j\omega t}$ generates an output signal $y(t) = \mathbf{H}(\omega)\, e^{j\omega t}$, and we will learn how to compute $\mathbf{H}(\omega)$ from the system's impulse response $h(t)$.

Chapter 5: We will learn how to perform the decomposition of periodic signals, as described by Eq. (2.98). We will also learn how to similarly express nonperiodic signals in terms of frequency integrals.

Chapter 6: We will design a special LTI system called a ***Butterworth*** filter with a special $\mathbf{H}(\omega)$. Additionally, we will use LTI systems to perform multiple *filtering* examples.

2-7.1 LTI System Response to Complex Exponential Signals

Let us examine the response of an LTI system to a complex exponential $x(t) = Ae^{j\omega t}$. Since the system is LTI, we may (without loss of generality) set $A = 1$. Hence,

$$y(t) = h(t) * x(t) = h(t) * e^{j\omega t} = \int_{-\infty}^{\infty} h(\tau)\, e^{j\omega(t-\tau)}\, d\tau$$

$$= e^{j\omega t} \int_{-\infty}^{\infty} h(\tau)\, e^{-j\omega\tau}\, d\tau$$

$$= \mathbf{H}(\omega)\, e^{j\omega t}, \qquad (2.100)$$

where the ***frequency response function*** $\mathbf{H}(\omega)$ is defined as

$$\mathbf{H}(\omega) = \int_{-\infty}^{\infty} h(\tau)\, e^{-j\omega\tau}\, d\tau. \qquad (2.101)$$

The result given by Eq. (2.100) validates the statement made earlier that the output of an LTI system excited by a complex exponential signal is also a complex exponential signal at the same angular frequency ω.

The frequency response function $\mathbf{H}(\omega)$ bears a one-to-one correspondence to $h(t)$; it is completely specified by the expression for $h(t)$ and the value of ω. As alluded to earlier in connection with Eqs. (2.98) and (2.99), we can express time signals in terms of sums of signals at multiple frequencies. The ability to *transform* signals between the time domain and the ω-domain will prove very useful in the analysis and design of LTI systems. It is in the spirit of such transformations that we describe $\mathbf{H}(\omega)$ as the *frequency-domain equivalent of $h(t)$*.

In symbolic form, Eq. (2.100) is

$$e^{j\omega t} \longrightarrow \boxed{h(t)} \longrightarrow \mathbf{H}(\omega)\, e^{j\omega t}. \qquad (2.102)$$

In addition to working with $\mathbf{H}(\omega)$ to determine a system's response to a specified input, we will occasionally find it convenient to also work with its mirror image $\mathbf{H}(-\omega)$. Repetition of the procedure leading to Eq. (2.101) yields

$$y(t) = \mathbf{H}(-\omega)\, e^{-j\omega t} \qquad (2.103)$$

with

$$\mathbf{H}(-\omega) = \int_{-\infty}^{\infty} h(\tau)\, e^{j\omega\tau}\, d\tau. \qquad (2.104)$$

In symbolic form,

$$e^{-j\omega t} \longrightarrow \boxed{h(t)} \longrightarrow \mathbf{H}(-\omega)\, e^{-j\omega t}. \qquad (2.105)$$

Example 2-11: $\mathbf{H}(\omega)$ of an RC Circuit

Determine the frequency response of an RC circuit characterized by an impulse response $h(t) = e^{-t}\, u(t)$. The capacitor contained no charge prior to $t = 0$.

Solution: Application of Eq. (2.101) leads to

$$\mathbf{H}(\omega) = \int_{-\infty}^{\infty} h(\tau)\, e^{-j\omega\tau}\, d\tau$$

$$= \int_{0}^{\infty} e^{-\tau} e^{-j\omega\tau}\, d\tau = \frac{1}{1 + j\omega}. \qquad (2.106)$$

2-7.2 Properties of $\mathbf{H}(\omega)$

The frequency response function is characterized by the following properties:

(a) $\mathbf{H}(\omega)$ is independent of t.

(b) $\mathbf{H}(\omega)$ is a function of $j\omega$, not just ω.

(c) $\mathbf{H}(\omega)$ is defined for an input signal $e^{j\omega t}$ that exists for all time, $-\infty < t < \infty$; i.e., $x(t) = e^{j\omega t}$, not $e^{j\omega t}\, u(t)$.

If $h(t)$ is real, $\mathbf{H}(\omega)$ is related to $\mathbf{H}(-\omega)$ by

$$\mathbf{H}(-\omega) = \mathbf{H}^*(\omega) \qquad \text{(for } h(t) \text{ real).} \qquad (2.107)$$

The condition defined by Eq. (2.107) is known as ***conjugate symmetry***. It follows from the definition of $\mathbf{H}(\omega)$ given by Eq. (2.101), namely,

$$\mathbf{H}^*(\omega) = \int_{-\infty}^{\infty} h^*(\tau)\, (e^{-j\omega\tau})^*\, d\tau$$

$$= \int_{-\infty}^{\infty} h(\tau)\, e^{j\omega\tau}\, d\tau = \mathbf{H}(-\omega). \qquad (2.108)$$

▶ Since $h(t)$ is real, it follows from Eq. (2.107) that the magnitude $|\mathbf{H}(\omega)|$ is an even function and the phase $\underline{/\mathbf{H}(\omega)}$ is an odd function:

- $$|\mathbf{H}(\omega)| = |\mathbf{H}(-\omega)| \tag{2.109a}$$

and

- $$\underline{/\mathbf{H}(-\omega)} = -\underline{/\mathbf{H}(\omega)}\,. \tag{2.109b}$$

◀

The frequency response function of the RC filter, namely, $\mathbf{H}(\omega) = 1/(1 + j\omega) = (1 - j\omega)/(1 + \omega^2)$, satisfies the aforementioned relationships.

2-7.3 H(ω) of LCCDE

If an LTI system is described by a linear, constant-coefficient, differential equation, we can determine its frequency response function $\mathbf{H}(\omega)$ without having to know its impulse response $h(t)$. Consider, for example, the LCCDE

$$\frac{d^2y}{dt^2} + 2\,\frac{dy}{dt} + 3y = 4\,\frac{dx}{dt} + 5x. \tag{2.110}$$

According to Eq. (2.100), $y(t) = \mathbf{H}(\omega)\, e^{j\omega t}$ when $x(t) = e^{j\omega t}$. Substituting these expressions for $x(t)$ and $y(t)$ in Eq. (2.110) leads to the ***frequency domain equivalent*** of Eq. (2.110):

$$(j\omega)^2\, \mathbf{H}(\omega)\, e^{j\omega t} + 2(j\omega)\, \mathbf{H}(\omega)\, e^{j\omega t} + 3\mathbf{H}(\omega)\, e^{j\omega t} = 4(j\omega)e^{j\omega t} + 5e^{j\omega t}. \tag{2.111}$$

After canceling $e^{j\omega t}$ (which is never zero) from all terms and then solving for $\mathbf{H}(\omega)$, we get

$$\mathbf{H}(\omega) = \frac{5 + j4\omega}{-\omega^2 + 2j\omega + 3} = \frac{5 + j4\omega}{(3 - \omega^2) + j2\omega}\,. \tag{2.112}$$

▶ We note that every d/dt operation in the time domain translates into a multiplication by $j\omega$ in the frequency domain. ◀

More generally, for any LCCDE of the form

$$a_0\,\frac{d^ny}{dt^n} + \cdots + a_{n-2}\,\frac{d^2y}{dt^2} + a_{n-1}\,\frac{dy}{dt} + a_n y = b_0\,\frac{d^mx}{dt^m} + \cdots + b_{m-2}\,\frac{d^2x}{dt^2} + b_{m-1}\,\frac{dx}{dt} + b_m x\,,$$

or equivalently,

$$\sum_{i=0}^{n} a_{n-i}\,\frac{d^iy}{dt^i} = \sum_{i=0}^{m} b_{m-i}\,\frac{d^ix}{dt^i}\,, \tag{2.113}$$

its $\mathbf{H}(\omega)$ can be determined by substituting $x(t) = e^{j\omega t}$ and $y(t) = \mathbf{H}(\omega)\, e^{j\omega t}$ in the LCCDE and using

$$(d^i/dt^i)e^{j\omega t} = (j\omega)^i e^{j\omega t}.$$

The process leads to

$$\mathbf{H}(\omega) = \frac{\displaystyle\sum_{i=0}^{m} b_{m-i}(j\omega)^i}{\displaystyle\sum_{i=0}^{n} a_{n-i}(j\omega)^i}\,. \tag{2.114}$$

In Eqs. (2.113) and (2.114), m and n are the orders of the highest derivatives of $x(t)$ and $y(t)$, respectively. We note that $\mathbf{H}(\omega)$ is a ***rational function*** of $j\omega$ (ratio of two polynomials).

Example 2-12: H(ω) of an LCCDE

Compute $\mathbf{H}(1)$ for the system described by

$$3\,\frac{d^3y}{dt^3} + 6\,\frac{dy}{dt} + 4y(t) = 5\,\frac{dx}{dt}\,.$$

Solution: To determine $\mathbf{H}(\omega)$, we apply Eq. (2.114) with $m = 1$ and $n = 3$:

$$\mathbf{H}(\omega) = \frac{b_1(j\omega)^0 + b_0(j\omega)^1}{a_3(j\omega)^0 + a_2(j\omega)^1 + a_1(j\omega)^2 + a_0(j\omega)^3}\,.$$

From the specified differential equation, $a_0 = 3$, $a_1 = 0$, $a_2 = 6$, $a_3 = 4$, $b_0 = 5$, and $b_1 = 0$. Hence,

$$\mathbf{H}(\omega) = \frac{5(j\omega)}{4 + 6(j\omega) + 3(j\omega)^3}\,.$$

At $\omega = 1$ rad/s,

$$\mathbf{H}(1) = \frac{j5}{4 + j6 - j3} = \frac{j5}{4 + j3} = 1e^{j53^\circ}.$$

2-7.4 LTI Response to Sinusoids

We now examine the response of an LTI system with real-valued $h(t)$ to a pure sinusoid of the form $x(t) = \cos\omega t$. When analyzing signals and systems, operations involving sinusoids are most easily realized by expressing them in terms of complex exponentials. In the present case,

$$x(t) = \cos \omega t = \frac{1}{2}[e^{j\omega t} + e^{-j\omega t}]. \tag{2.115}$$

Using the additivity property of LTI systems, we can determine the system's response to $\cos \omega t$ by adding the responses expressed by Eqs. (2.102) and (2.105):

$$\begin{array}{ccccc} \frac{1}{2}\, e^{j\omega t} & \longrightarrow & \boxed{h(t)} & \longrightarrow & \frac{1}{2}\, \mathbf{H}(\omega)\, e^{j\omega t} \\ + & & & & + \\ \frac{1}{2}\, e^{-j\omega t} & \longrightarrow & \boxed{h(t)} & \longrightarrow & \frac{1}{2}\, \mathbf{H}(-\omega)\, e^{-j\omega t}. \end{array} \tag{2.116}$$

Since $h(t)$ is real, Eq. (2.107) stipulates that

$$\mathbf{H}(-\omega) = \mathbf{H}^*(\omega).$$

If we define

$$\begin{aligned} \mathbf{H}(\omega) &= |\mathbf{H}(\omega)| e^{j\theta}, \\ \mathbf{H}(-\omega) &= |\mathbf{H}(\omega)| e^{-j\theta}, \end{aligned} \tag{2.117}$$

where $\theta = \underline{/\mathbf{H}(\omega)}$, the sum of the two outputs in Eq. (2.116) becomes

$$\begin{aligned} y(t) &= \frac{1}{2}\, [\mathbf{H}(\omega)\, e^{j\omega t} + \mathbf{H}(-\omega)\, e^{-j\omega t}] \\ &= \frac{1}{2}\, |\mathbf{H}(\omega)| [e^{j(\omega t+\theta)} + e^{-j(\omega t+\theta)}] \\ &= |\mathbf{H}(\omega)| \cos(\omega t + \theta). \end{aligned}$$

In summary,

$$\begin{gathered} y(t) = |\mathbf{H}(\omega)| \cos(\omega t + \theta) \\ (\text{for } x(t) = \cos \omega t). \end{gathered} \tag{2.118}$$

In symbolic form,

$$\cos \omega t \longrightarrow \boxed{h(t)} \longrightarrow |\mathbf{H}(\omega)| \cos(\omega t + \theta), \tag{2.119}$$

and by applying the scaling and time-invariance properties of LTI systems, this result can be generalized to any cosine function of amplitude A and phase angle ϕ

$$\begin{array}{c} A \cos(\omega t + \phi) \\ \downarrow \\ \boxed{h(t)} \\ \downarrow \\ A|\mathbf{H}(\omega)| \cos(\omega t + \theta + \phi). \end{array} \tag{2.120}$$

The full implication of this result will become apparent when we express periodic signals as linear combinations of complex exponentials or sinusoids.

Example 2-13: Response of an LCCDE to a Sinusoidal Input

$$5\cos(4t + 30^\circ) \longrightarrow \boxed{\frac{dy}{dt} + 3y = 3\frac{dx}{dt} + 5x} \longrightarrow ?$$

Solution: Inserting $x(t) = e^{j\omega t}$ and $y(t) = \mathbf{H}(\omega)\, e^{j\omega t}$ in the specified LCCDE gives

$$j\omega\, \mathbf{H}(\omega)\, e^{j\omega t} + 3\mathbf{H}(\omega)\, e^{j\omega t} = 3j\omega e^{j\omega t} + 5e^{j\omega t}.$$

Solving for $\mathbf{H}(\omega)$ gives

$$\mathbf{H}(\omega) = \frac{5 + j3\omega}{3 + j\omega}.$$

The angular frequency ω of the input signal is 4 rad/s. Hence,

$$\mathbf{H}(4) = \frac{5 + j12}{3 + j4} = \frac{13e^{j67.38^\circ}}{5e^{j53.13^\circ}} = \frac{13}{5}\, e^{j14.25^\circ}.$$

Application of Eq. (2.120) with $A = 5$ and $\phi = 30^\circ$ gives

$$y(t) = 5 \times \frac{13}{5} \cos(4t + 14.25^\circ + 30^\circ) = 13\cos(4t + 44.25^\circ).$$

Concept Question 2-12: If the input to an LTI system is a sinusoid, which of the following attributes can change between the input and the output: amplitude, frequency, or phase? (See S²)

Concept Question 2-13: What kind of function is $\mathbf{H}(\omega)$ for a system described by an LCCDE? (See S²)

Exercise 2-12:

$\cos(t)$ → $h(t) = e^{-t}\, u(t)$ → ?

Answer: The output is $(1/\sqrt{2})\cos(t - 45°)$. (See S²)

Exercise 2-13:

$2\cos(t)$ → System → $2\cos(2t) + 2$.

Initial conditions are zero. Is this system LTI?

Answer: No. An LTI cannot create a sinusoid at a frequency different from that of its input. (See S²)

Exercise 2-14:

$\cos(2t)$ → System → 0.

Can we say that the system is not LTI?

Answer: No. An LTI can make an amplitude = 0. (See S²)

2-8 Impulse Response of Second-Order LCCDEs

Many physical systems are described by second-order LCCDEs of the form

$$\frac{d^2y}{dt^2} + a_1 \frac{dy}{dt} + a_2\, y(t) = b_1 \frac{dx}{dt} + b_2\, x(t), \qquad (2.121)$$

where a_1, a_2, b_1, and b_2 are constant coefficients. In this section, we examine how to determine the impulse response function $h(t)$ for such a differential equation, and in Section 2-9, we demonstrate how we use that experience to analyze a spring-mass-damper model of an automobile suspension system.

2-8.1 LCCDE with No Input Derivatives

For simplicity, we start by considering a version of Eq. (2.121) without the dx/dt term, and then we use the result to treat the more general case in the next subsection.

For $b_1 = 0$ and $b_2 = 1$, Eq. (2.121) becomes

$$\frac{d^2y}{dt^2} + a_1 \frac{dy}{dt} + a_2\, y(t) = x(t). \qquad (2.122)$$

Step 1: Roots of characteristic equation

Assuming $y(t)$ has a general solution of the form $y(t) = Ae^{st}$, substitution in the homogeneous form of Eq. (2.122)—i.e., with $x(t) = 0$—leads to the *characteristic equation*:

$$s^2 + a_1 s + a_2 = 0. \qquad (2.123)$$

If p_1 and p_2 are the roots of Eq. (2.123), then

$$s^2 + a_1 s + a_2 = (s - p_1)(s - p_2), \qquad (2.124)$$

which leads to

$$p_1 + p_2 = -a_1, \qquad p_1 p_2 = a_2, \qquad (2.125)$$

and

$$p_1 = -\frac{a_1}{2} + \sqrt{\left(\frac{a_1}{2}\right)^2 - a_2}\,,$$
$$p_2 = -\frac{a_1}{2} - \sqrt{\left(\frac{a_1}{2}\right)^2 - a_2}\,. \qquad (2.126)$$

Roots p_1 and p_2 are

(a) real if $a_1^2 > 4a_2$,

(b) complex conjugates if $a_1^2 < 4a_2$, or

(c) identical if $a_1^2 = 4a_2$.

Step 2: Two coupled first-order LCCDEs

The original differential equation given by Eq. (2.122) now can be rewritten as

$$\frac{d^2y}{dt^2} - (p_1 + p_2)\frac{dy}{dt} + (p_1 p_2)\, y(t) = x(t), \qquad (2.127a)$$

which can in turn be cast in the form

$$\left[\frac{d}{dt} - p_1\right]\left[\frac{d}{dt} - p_2\right] y(t) = x(t). \qquad (2.127b)$$

Furthermore, we can split the *second-order differential equation* into *two coupled first-order equations* by introducing an intermediate variable $z(t)$:

$$\frac{dz}{dt} - p_1\, z(t) = x(t) \qquad (2.128a)$$

and

$$\frac{dy}{dt} - p_2\, y(t) = z(t). \qquad (2.128b)$$

These coupled first-order LCCDEs represent a *series (or cascade) connection* of LTI systems, each described by a first-order LCCDE. In symbolic form, we have

$$x(t) \longrightarrow \boxed{h_1(t)} \longrightarrow z(t)$$
$$z(t) \longrightarrow \boxed{h_2(t)} \longrightarrow y(t), \tag{2.129}$$

where $h_1(t)$ and $h_2(t)$ are the impulse responses corresponding to Eqs. (2.128a and b), respectively.

Step 3: Impulse response of cascaded LTI systems

By comparison with Eq. (2.10) and its corresponding impulse response, Eq. (2.17), we conclude that

$$h_1(t) = e^{p_1 t}\, u(t) \tag{2.130a}$$

and

$$h_2(t) = e^{p_2 t}\, u(t). \tag{2.130b}$$

Using convolution property #2 in Table 2-1, the impulse response of the series connection of two LTI systems is the convolution of their impulse responses. Utilizing entry #3 in Table 2-2, the combined impulse response becomes

$$\begin{aligned} h_c(t) &= h_1(t) * h_2(t) = e^{p_1 t}\, u(t) * e^{p_2 t}\, u(t) \\ &= \left[\frac{1}{p_1 - p_2}\right] [e^{p_1 t} - e^{p_2 t}]\, u(t). \end{aligned} \tag{2.131}$$

2-8.2 LCCDE with Input Derivative

We now consider the more general case of a second-order LCCDE that contains a first-order derivative on the input side of the equation

$$\frac{d^2 y}{dt^2} + a_1 \frac{dy}{dt} + a_2\, y(t) = b_1 \frac{dx}{dt} + b_2\, x(t). \tag{2.132}$$

By defining the right-hand side of Eq. (2.132) as an intermediate variable $w(t)$, the system can be represented as

$$x(t) \longrightarrow \boxed{w(t) = b_1 \frac{dx}{dt} + b_2\, x(t)} \longrightarrow w(t)$$
$$w(t) \longrightarrow \boxed{h_c(t)} \longrightarrow y(t), \tag{2.133}$$

where $h_c(t)$ is the impulse response given by Eq. (2.131) for the system with $b_1 = 0$ and $b_2 = 1$.

To determine the impulse response of the overall system, we need to compute the convolution of $h_c(t)$ with the (yet to be determined) impulse response representing the other box in Eq. (2.133). A more expedient route is to use convolution property #1 in Table 2-1. Since convolution is commutative, we can reverse the order of the two LTI systems in Eq. (2.133),

$$x(t) \longrightarrow \boxed{h_c(t)} \longrightarrow \upsilon(t)$$
$$\upsilon(t) \longrightarrow \boxed{y(t) = b_1 \frac{d\upsilon}{dt} + b_2\, \upsilon(t)} \longrightarrow y(t), \tag{2.134}$$

where $\upsilon(t)$ is another intermediate variable created for the sake of convenience. By definition, when $x(t) = \delta(t)$, the output $y(t)$ becomes the impulse response $h(t)$ of the overall system. That is, if we set $x(t) = \delta(t)$, which results in $\upsilon(t) = h_c(t)$ and $y(t) = h(t)$, the system becomes

$$\delta(t) \longrightarrow \boxed{h_c(t)} \longrightarrow h_c(t)$$
$$h_c(t) \longrightarrow \boxed{h(t) = b_1 \frac{dh_c}{dt} + b_2\, h_c(t)} \longrightarrow h(t). \tag{2.135}$$

Finally, the impulse response $h(t)$ of the overall system is

$$\begin{aligned} h(t) &= b_1 \frac{dh_c}{dt} + b_2\, h_c(t) \\ &= \left[b_1 \frac{d}{dt} + b_2\right] \left[\frac{1}{p_1 - p_2}\right] [e^{p_1 t} - e^{p_2 t}]\, u(t) \\ &= \frac{b_1 p_1 + b_2}{p_1 - p_2}\, e^{p_1 t}\, u(t) - \frac{b_1 p_2 + b_2}{p_1 - p_2}\, e^{p_2 t}\, u(t). \end{aligned} \tag{2.136}$$

Having established in the form of Eq. (2.136) an explicit expression for the impulse response of the general LCCDE given by Eq. (2.132), we can now determine the response $y(t)$ to any causal input excitation $x(t)$ by evaluating

$$y(t) = u(t) \int_0^t h(\tau)\, x(t - \tau)\, d\tau. \tag{2.137}$$

2-8.3 Parameters of Second-Order LCCDE

Mathematically, our task is now complete. However, we can gain much physical insight into the nature of the system's response by examining scenarios associated with the three states of roots p_1 and p_2 [as noted earlier in connection with Eq. (2.126)].

(a) p_1 and p_2 are real and distinct (different).

(b) p_1 and p_2 are complex conjugates of one another.

(c) p_1 and p_2 are real and equal.

Recall that p_1 and p_2 are defined in terms of coefficients a_1 and a_2 in the LCCDE, so different systems characterized by LCCDEs with identical forms but different values of a_1 and a_2 may behave quite differently.

Before we examine the three states of p_1 and p_2 individually, it will prove useful to express p_1 and p_2 in terms of physically meaningful parameters. To start with, we reintroduce the expressions for p_1 and p_2 given by Eq. (2.126):

$$p_1 = -\frac{a_1}{2} + \sqrt{\left(\frac{a_1}{2}\right)^2 - a_2}\,, \qquad (2.138a)$$

$$p_2 = -\frac{a_1}{2} - \sqrt{\left(\frac{a_1}{2}\right)^2 - a_2}\,. \qquad (2.138b)$$

Based on the results of Section 2-6.4, in order for the system described by Eq. (2.136) to be BIBO stable, it is necessary that the real parts of both p_1 and p_2 be negative. We now show that this is true if and only if $a_1 > 0$ and $a_2 > 0$. Specifically, we have the following:

(a) If both p_1 and p_2 are real, distinct, and negative, Eq. (2.138) leads to the conclusion that $a_1^2 > 4a_2$, $a_1 > 0$, and $a_2 > 0$.

(b) If p_1 and p_2 are complex conjugates with negative real parts, it follows that $a_1^2 < 4a_2$, $a_1 > 0$, and $a_2 > 0$.

(c) If p_1 and p_2 are real, equal, and negative, then $a_1^2 = 4a_2$, $a_1 > 0$, and $a_2 > 0$.

▶ The LTI system described by the LCCDE Eq. (2.132) is BIBO stable if and only if $a_1 > 0$ and $a_2 > 0$. ◀

We now introduce three new non-negative, physically meaningful parameters:

$$\alpha = \frac{a_1}{2} = \textit{attenuation coefficient} \quad \text{(Np/s)}, \qquad (2.139a)$$

$$\omega_0 = \sqrt{a_2} = \textit{undamped natural frequency} \quad \text{(rad/s)}, \qquad (2.139b)$$

and

$$\xi = \frac{\alpha}{\omega_0} = \frac{a_1}{2\sqrt{a_2}} = \textit{damping coefficient} \quad \text{(unitless)}. \qquad (2.139c)$$

The unit Np is short for nepers, named after the inventor of the logarithmic scale, John Napier. In view of Eq. (2.139), p_1 and p_2 can be written as

$$p_1 = -\alpha + \sqrt{\alpha^2 - \omega_0^2} = \omega_0\left[-\xi + \sqrt{\xi^2 - 1}\right] \qquad (2.140a)$$

and

$$p_2 = -\alpha - \sqrt{\alpha^2 - \omega_0^2} = \omega_0\left[-\xi - \sqrt{\xi^2 - 1}\right] \qquad (2.140b)$$

The damping coefficient ξ plays a critically important role, because its value determines the character of the system's response to any input $x(t)$. The system exhibits markedly different responses depending on whether

(a) $\xi > 1$ ➡ *overdamped* response,

(b) $\xi = 1$ ➡ *critically damped* response, or

(c) $\xi < 1$ ➡ *underdamped* response.

The three names, overdamped, underdamped, and critically damped, refer to the shape of the system's response. Figure 2-22 displays three system step responses, each one of which starts at zero at $t = 0$ and rises to $y = 1$ as $t \to \infty$, but the shapes of their waveforms are quite different. The overdamped response exhibits the slowest path towards $y = 1$; the underdamped response is very fast, but it includes an oscillatory component; and the critically damped response represents the fastest path without oscillations.

2-8.4 Overdamped Case ($\xi > 1$)

For convenience, we rewrite Eq. (2.136) as

$$h(t) = A_1 e^{p_1 t}\, u(t) + A_2 e^{p_2 t}\, u(t) \qquad (2.141)$$

(overdamped impulse response)

with

$$A_1 = \frac{b_1 p_1 + b_2}{p_1 - p_2} \quad \text{and} \quad A_2 = \frac{-(b_1 p_2 + b_2)}{p_1 - p_2}. \qquad (2.142)$$

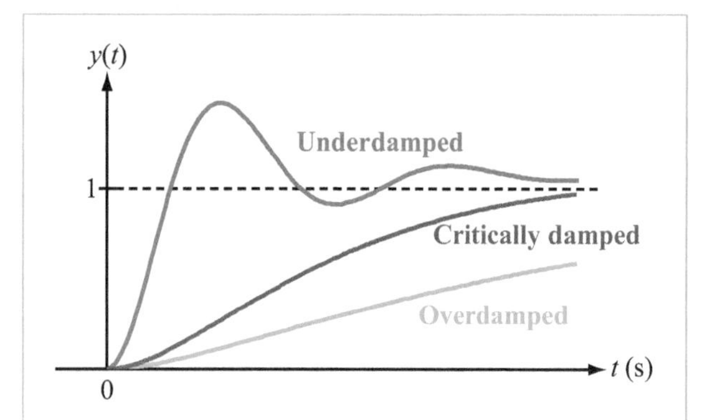

Figure 2-22: Comparison of overdamped, underdamped, and critically damped responses. In each case, the response starts at zero at $t = 0$ and approaches 1 as $t \to \infty$, but the in-between paths are quite different.

If $\xi > 1$—which corresponds to $\alpha^2 > \omega_0^2$, or equivalently, $a_1^2 > 4a_2$—roots p_1 and p_2 are both negative real numbers with $|p_2| > |p_1|$. The step response $y_{\text{step}}(t)$ is obtained by inserting Eq. (2.141) into Eq. (2.137) and setting $x(t-\tau) = u(t-\tau)$ and $y(t) = y_{\text{step}}(t)$:

$$y_{\text{step}}(t) = \int_0^t [A_1 e^{p_1\tau} \, u(\tau) + A_2 e^{p_2\tau} \, u(\tau)] \, u(t-\tau) \, d\tau. \tag{2.143}$$

Over the range of integration $(0, t)$, $u(\tau) = 1$ and $u(t-\tau) = 1$. Hence,

$$y_{\text{step}}(t) = \left[\int_0^t (A_1 e^{p_1\tau} + A_2 e^{p_2\tau}) \, d\tau \right] u(t),$$

which integrates to

$$y_{\text{step}}(t) = \left[\frac{A_1}{p_1} (e^{p_1 t} - 1) + \frac{A_2}{p_2} (e^{p_2 t} - 1) \right] u(t).$$

(overdamped step response) (2.144)

Example 2-14: Overdamped Response

Compute and plot the step response $y_{\text{step}}(t)$ of a system described by the LCCDE

$$\frac{d^2 y}{dt^2} + 12 \frac{dy}{dt} + 9y(t) = 4 \frac{dx}{dt} + 25x(t).$$

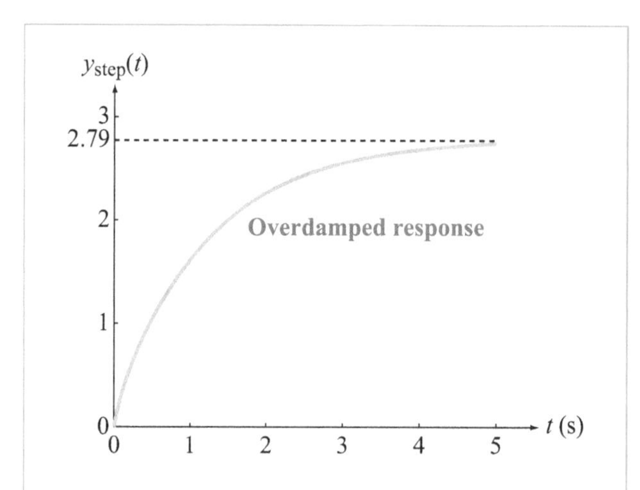

Figure 2-23: Step response of overdamped system in Example 2-14.

Solution: From the LCCDE, $a_1 = 12$, $a_2 = 9$, $b_1 = 4$, and $b_2 = 25$. When used in Eqs. (2.139), (2.140), and (2.142), we obtain the values

$$\alpha = \frac{a_1}{2} = \frac{12}{2} = 6 \text{ Np/s},$$

$$\omega_0 = \sqrt{a_2} = \sqrt{9} = 3 \text{ rad/s}, \qquad \xi = \frac{\alpha}{\omega_0} = \frac{6}{3} = 2,$$

$$p_1 = -0.8 \text{ Np/s}, \qquad p_2 = -11.2 \text{ Np/s},$$

$$\frac{A_1}{p_1} = -2.62, \text{ and } \frac{A_2}{p_2} = -0.17.$$

Since $\xi > 1$, the response is overdamped, in which case Eq. (2.144) applies:

$$y_{\text{step}}(t) = \left[2.62(1 - e^{-0.8t}) + 0.17(1 - e^{-11.2t}) \right] u(t).$$

The plot of $y_{\text{step}}(t)$ is shown in Fig. 2-23.

2-8.5 Underdamped Case ($\xi < 1$)

If $\xi < 1$, or equivalently, $\alpha^2 < \omega_0^2$, the square root in Eq. (2.140) becomes negative, causing roots p_1 and p_2 to become complex numbers. For reasons that will become apparent shortly, this condition leads to an ***underdamped*** step response with a waveform that oscillates at a ***damped natural frequency*** ω_d defined as

$$\omega_d = \sqrt{\omega_0^2 - \alpha^2} = \omega_0 \sqrt{1 - \xi^2} \, . \tag{2.145}$$

In terms of ω_d, roots p_1 and p_2 [Eq. (2.140)] become

$$p_1 = -\alpha + j\omega_d \quad \text{and} \quad p_2 = -\alpha - j\omega_d. \qquad (2.146)$$

Inserting these expressions into the impulse response given by Eq. (2.141) leads to

$$\begin{aligned} h(t) &= [A_1 e^{-\alpha t} e^{j\omega_d t} + A_2 e^{-\alpha t} e^{-j\omega_d t}]\, u(t) \\ &= [A_1(\cos\omega_d t + j\sin\omega_d t) \\ &\quad + A_2(\cos\omega_d t - j\sin\omega_d t)] e^{-\alpha t}\, u(t) \\ &= [(A_1 + A_2)\cos\omega_d t + j(A_1 - A_2)\sin\omega_d t] e^{-\alpha t}\, u(t), \end{aligned}$$

which can be contracted into

$$h(t) = [B_1 \cos\omega_d t + B_2 \sin\omega_d t] e^{-\alpha t}\, u(t)$$
(underdamped impulse response) (2.147)

by introducing two new coefficients, B_1 and B_2, given by

$$B_1 = A_1 + A_2 = \frac{b_1 p_1 + b_2}{p_1 - p_2} - \frac{b_1 p_2 + b_2}{p_1 - p_2} = b_1 \qquad (2.148a)$$

and

$$B_2 = j(A_1 - A_2) = \frac{b_2 - b_1\alpha}{\omega_d}. \qquad (2.148b)$$

The negative exponential $e^{-\alpha t}$ in Eq. (2.147) signifies that $h(t)$ has a damped waveform, and the sine and cosine terms signify that $h(t)$ is oscillatory with an angular frequency ω_d and a corresponding ***time period***

$$T = \frac{2\pi}{\omega_d}. \qquad (2.149)$$

Example 2-15: Underdamped Response

Compute and plot the step response $y_{\text{step}}(t)$ of a system described by the LCCDE

$$\frac{d^2y}{dt^2} + 12\frac{dy}{dt} + 144y(t) = 4\frac{dx}{dt} + 25x(t).$$

Solution: From the LCCDE, $a_1 = 12$, $a_2 = 144$, $b_1 = 4$, and $b_2 = 25$.

The damping coefficient is

$$\xi = \frac{a_1}{2\sqrt{a_2}} = \frac{12}{2\sqrt{144}} = 0.5.$$

Hence, this is an underdamped case and the appropriate impulse response is given by Eq. (2.147). The step response $y_{\text{step}}(t)$ is obtained by convolving $h(t)$ with $x(t) = u(t)$:

$$\begin{aligned} y_{\text{step}}(t) &= h(t) * u(t) \\ &= \left[\int_0^t (B_1\cos\omega_d\tau + B_2\sin\omega_d\tau)e^{-\alpha\tau}\, d\tau\right] u(t). \end{aligned}$$

Performing the integration by parts leads to

$$\begin{aligned} y_{\text{step}}(t) &= \frac{1}{\alpha^2 + \omega_d^2} \\ &\quad \cdot \{[-(B_1\alpha + B_2\omega_d)\cos\omega_d t \\ &\quad + (B_1\omega_d + B_2\alpha)\sin\omega_d t]e^{-\alpha t} \\ &\quad + (B_1\alpha + B_2\omega_d)\}\, u(t). \end{aligned} \qquad (2.150)$$
(underdamped step response)

For the specified constants,

$$\alpha = 6 \text{ Np/s}, \qquad \omega_0 = \sqrt{a_2} = 12 \text{ rad/s}, \qquad \xi = 0.5,$$
$$\omega_d = \omega_0\sqrt{1 - \xi^2} = 10.4 \text{ rad/s},$$
$$B_1 = 4, \qquad B_2 = 9.62 \times 10^{-2}, \text{ and}$$
$$y_{\text{step}}(t) = \left\{[-0.17\cos 10.4t + 0.29\sin 10.4t]\, e^{-6t} + 0.17\right\} u(t).$$

Figure 2-24 displays a plot of $y_{\text{step}}(t)$, which exhibits an oscillatory pattern superimposed on the exponential response. The oscillation period is $T = 2\pi/10.4 = 0.6$ s.

2-8.6 Critically Damped Case ($\xi = 1$)

According to Eq. (2.140), if $\xi = 1$, then $p_1 = p_2$. Repeated roots lead to a ***critically damped*** step response, so called because it provides the fastest path to the steady state that the system approaches as $t \to \infty$. With $p_1 = p_2$, the expression for $h(t)$ reduces to a single exponential, which is not a viable solution, because a second-order LCCDE should include two time-dependent functions in its solution. To derive an appropriate expression for $h(t)$, we take the following indirect approach.

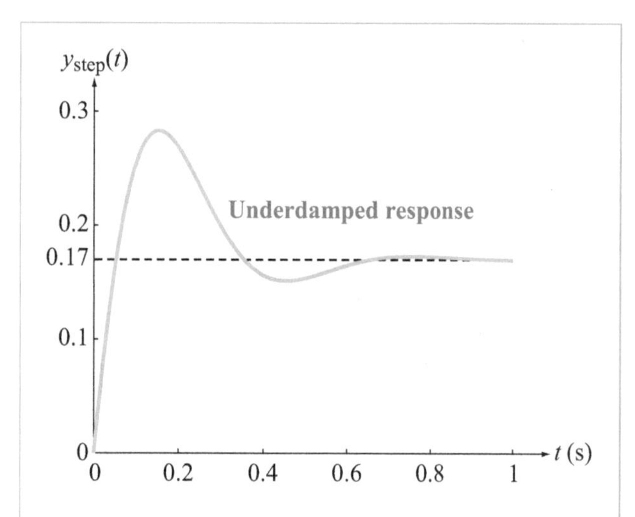

Figure 2-24: Underdamped response of Example 2-15. Note the oscillations.

Step 1: Start with a slightly underdamped system:

Suppose we have a slightly underdamped system with a very small damped natural frequency $\omega_{\text{d}} = \epsilon$ and roots

$$p_1 = -\alpha + j\epsilon \quad \text{and} \quad p_2 = -\alpha - j\epsilon.$$

According to Eq. (2.131), the impulse response $h_{\text{c}}(t)$ of a system with no input derivative is

$$\begin{aligned} h_{\text{c}}(t) &= \left[\frac{1}{p_1 - p_2}\right] [e^{p_1 t} - e^{p_2 t}]\, u(t) \\ &= \left[\frac{e^{j\epsilon t} - e^{-j\epsilon t}}{j2\epsilon}\right] e^{-\alpha t}\, u(t) = \frac{\sin \epsilon t}{\epsilon}\, e^{-\alpha t}\, u(t). \end{aligned} \tag{2.151}$$

Since ϵ is infinitesimally small, $\epsilon \ll \alpha$. Hence, $e^{-\alpha t}$ will decay to approximately zero long before ϵt becomes significant in magnitude, which means that for all practical purposes, the function $\sin \epsilon t$ is relevant only when ϵt is very small, in which case the approximation $\sin(\epsilon t) \approx \epsilon t$ is valid. Accordingly, $h_{\text{c}}(t)$ becomes

$$h_{\text{c}}(t) = te^{-\alpha t}\, u(t), \qquad \text{as } \epsilon \to 0. \tag{2.152}$$

Step 2: Obtain impulse response $h(t)$:

Implementation of Eq. (2.135) to obtain $h(t)$ of the system containing an input derivative dx/dt from that of the same system without the dx/dt term leads to

$$\begin{aligned} h(t) &= b_1 \frac{dh_{\text{c}}}{dt} + b_2 h_{\text{c}}(t) \\ &= (C_1 + C_2 t)e^{-\alpha t}\, u(t) \end{aligned} \tag{2.153}$$

(critically damped impulse response)

with

$$C_1 = b_1 \tag{2.154a}$$

and

$$C_2 = b_2 - \alpha b_1. \tag{2.154b}$$

Example 2-16: Critically Damped Response

Compute and plot the step response of a system described by

$$\frac{d^2y}{dt^2} + 12\,\frac{dy}{dt} + 36y(t) = 4\,\frac{dx}{dt} + 25x(t).$$

Solution: From the LCCDE, $a_1 = 12$, $a_2 = 36$, $b_1 = 4$, and $b_2 = 25$. The damping coefficient is

$$\xi = \frac{a_1}{2\sqrt{a_2}} = \frac{12}{2\sqrt{36}} = 1.$$

Hence, this is a critically damped system. The relevant constants are

$$\alpha = \frac{a_1}{2} = 6 \text{ Np/s}, \; C_1 = 4, \text{ and } C_2 = 1,$$

and the impulse response is

$$h(t) = (4 + t)e^{-6t}\, u(t).$$

The corresponding step response is

$$\begin{aligned} y_{\text{step}}(t) = h(t) * u(t) &= \left[\int_0^t (4 + \tau)e^{-6\tau}\, d\tau\right] u(t) \\ &= \left[\frac{25}{36}(1 - e^{-6t}) - \frac{1}{6}\, te^{-6t}\right] u(t), \end{aligned}$$

and its profile is displayed in Fig. 2-25. The step response starts at zero and approaches a final value of $25/36 = 0.69$ as $t \to \infty$. It exhibits the fastest damping rate possible without oscillation.

The impulse and step responses of the second-order LCCDE, namely Eq. (2.132), are summarized in Table 2-3 for each of the three damping conditions.

Table 2-3: **Impulse and step responses of second-order LCCDE.**

LCCDE $$\frac{d^2y}{dt^2} + a_1 \frac{dy}{dt} + a_2 y = b_1 \frac{dx}{dt} + b_2 x$$

$$\alpha = \frac{a_1}{2}, \qquad \omega_0 = \sqrt{a_2}, \qquad \xi = \frac{\alpha}{\omega_0}, \qquad p_1 = \omega_0[-\xi + \sqrt{\xi^2 - 1}], \qquad p_2 = \omega_0[-\xi - \sqrt{\xi^2 - 1}]$$

Overdamped Case $\xi > 1$

$$h(t) = A_1 e^{p_1 t}\, u(t) + A_2 e^{p_2 t}\, u(t) \qquad y_{\text{step}}(t) = \left[\frac{A_1}{p_1}(e^{p_1 t} - 1) + \frac{A_2}{p_2}(e^{p_2 t} - 1)\right] u(t)$$

$$A_1 = \frac{b_1 p_1 + b_2}{p_1 - p_2}, \qquad A_2 = \frac{-(b_1 p_2 + b_2)}{p_1 - p_2}$$

Underdamped Case $\xi < 1$

$$h(t) = [B_1 \cos \omega_d t + B_2 \sin \omega_d t] e^{-\alpha t}\, u(t)$$

$$y_{\text{step}}(t) = \frac{1}{\alpha^2 + \omega_d^2} \left\{[-(B_1\alpha + B_2\omega_d)\cos\omega_d t + (B_1\omega_d + B_2\alpha)\sin\omega_d t]e^{-\alpha t} + (B_1\alpha + B_2\omega_d)\right\} u(t)$$

$$B_1 = b_1, \qquad B_2 = \frac{b_2 - b_1\alpha}{\omega_d}, \qquad \omega_d = \omega_0\sqrt{1 - \xi^2}$$

Critically Damped Case $\xi = 1$

$$h(t) = (C_1 + C_2 t)e^{-\alpha t}\, u(t) \qquad y_{\text{step}}(t) = \left[\left(\frac{C_1}{\alpha} + \frac{C_2}{\alpha^2}\right)(1 - e^{-\alpha t}) - \frac{C_2}{\alpha}\, t e^{-\alpha t}\right] u(t)$$

$$C_1 = b_1, \qquad C_2 = b_2 - \alpha b_1$$

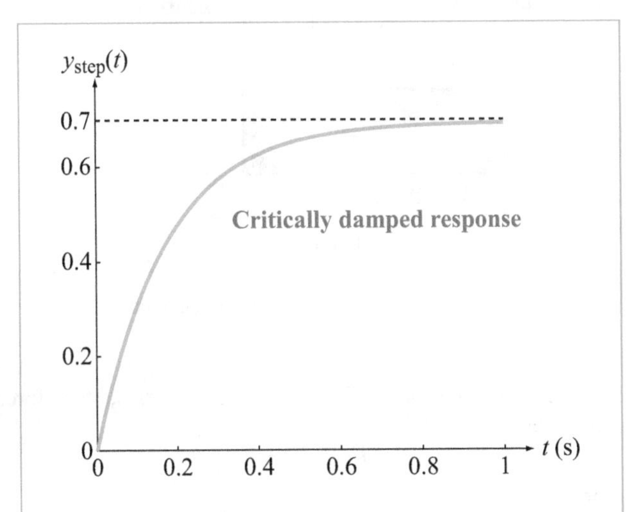

Figure 2-25: Critically damped response of Example 2-16.

Concept Question 2-14: What are the three damping conditions of the impulse response? (See S²)

Concept Question 2-15: How do input derivatives affect impulse responses? (See S²)

Exercise 2-15: Which damping condition is exhibited by $h(t)$ of

$$\frac{d^2y}{dt^2} + 5\frac{dy}{dt} + 4y(t) = 2\frac{dx}{dt}\ ?$$

Answer: Overdamped, because $\xi = 1.25 > 1$. (See S²)

Exercise 2-16: For what constant a_1 is

$$\frac{d^2y}{dt^2} + a_1 \frac{dy}{dt} + 9y(t) = 2 \frac{dx}{dt}$$

critically damped?

Answer: $a_1 = 6$. (See S²)

2-9 Car Suspension System

The results of the preceding section will now be used to analyze a car suspension system, which is selected in part because it offers a nice demonstration of how to model both the car suspension system and several examples of input excitations, including driving over a curb, over a pothole, and on a wavy pavement.

2-9.1 Spring-Mass-Damper System

The basic elements of an automobile suspension system are depicted in Fig. 2-26.

- $x(t)$ = input = vertical displacement of the pavement, defined relative to a reference ground level.
- $y(t)$ = output = vertical displacement of the car chassis from its equilibrium position.
- m = *one-fourth* of the car's mass, because the car has four wheels.
- k = ***spring constant*** or ***stiffness*** of the coil.
- b = ***damping coefficient*** of the shock absorber.

The forces exerted by the spring and shock absorber, which act on the car mass in parallel, depend on the relative displacement $(y - x)$ of the car relative to the pavement. They act to oppose the upward inertial force $\mathbf{F}_c$ on the car, which depends on only the car displacement $y(t)$. When $(y - x)$ is positive (car mass moving away from the pavement), the ***spring force*** $\mathbf{F}_s$ is directed downward. Hence, $\mathbf{F}_s$ is given by

$$\mathbf{F}_s = -k(y - x). \tag{2.155}$$

The ***damper force*** $\mathbf{F}_d$ exerted by the shock absorber is governed by viscous compression. It also is pointed downward, but it opposes the *change* in $(y - x)$. Therefore it opposes the derivative of $(y - x)$ rather than $(y - x)$ itself:

$$\mathbf{F}_d = -b \frac{d}{dt}(y - x). \tag{2.156}$$

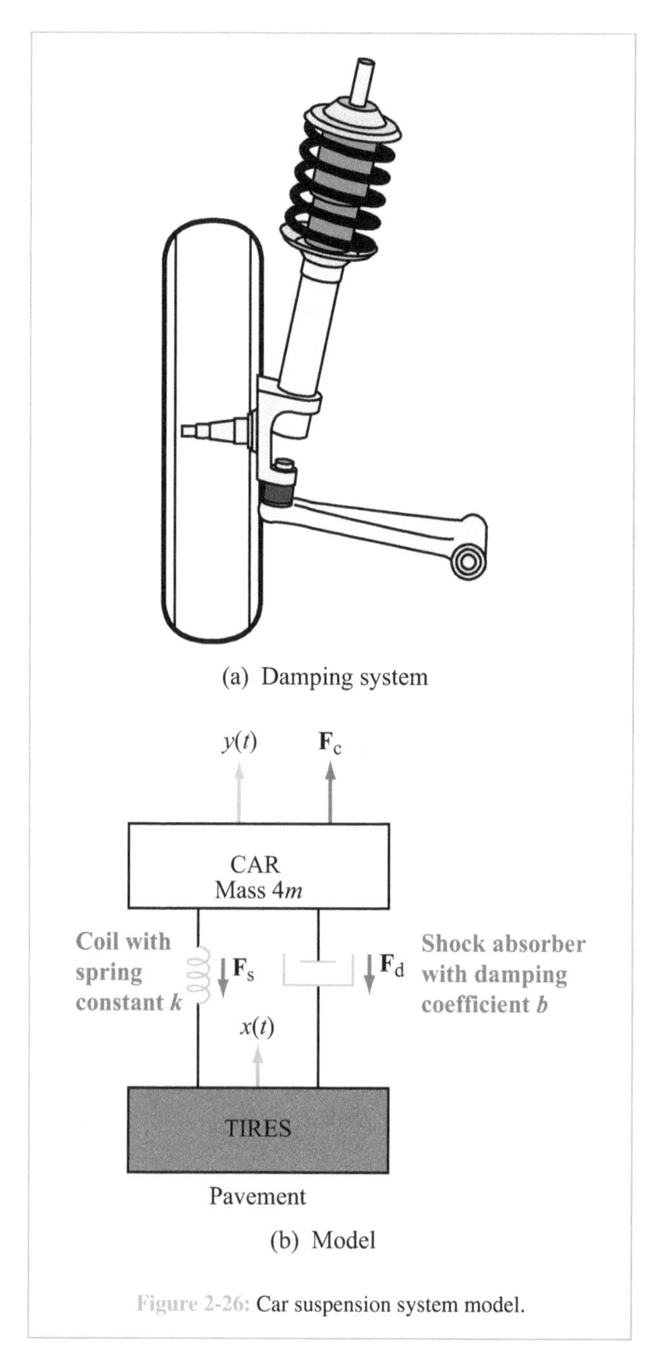

Figure 2-26: Car suspension system model.

Using Newton's law, $\mathbf{F}_c = ma = m(d^2y/dt^2)$, the force equation is

$$\mathbf{F}_c = \mathbf{F}_s + \mathbf{F}_d \tag{2.157}$$

or

$$m \frac{d^2y}{dt^2} = -k(y - x) - b \frac{d}{dt}(y - x),$$

which can be recast as

$$\frac{d^2y}{dt^2}+\frac{b}{m}\frac{dy}{dt}+\frac{k}{m}\,y=\frac{b}{m}\frac{dx}{dt}+\frac{k}{m}\,x. \quad (2.158)$$

The form of Eq. (2.158) is identical with that of the second-order LCCDE given by Eq. (2.132). Hence, all of the results we derived in the preceding section become directly applicable to the automobile suspension system upon setting $a_1 = b_1 = b/m$ and $a_2 = b_2 = k/m$.

Typical values for a small automobile are:

- $m = 250$ kg for a car with a total mass of one metric ton (1000 kg); each wheel supports one-fourth of the car's mass.
- $k = 10^5$ N/m; it takes a force of 1000 N to compress the spring by 1 cm.
- $b = 10^4$ N·s/m; a vertical motion of 1 m/s incurs a resisting force of 10^4 N.

2-9.2 Pavement Models

Driving on a curb

A car driving over a curb can be modeled as a step in $x(t)$ given by

$$x_1(t) = A_1\, u(t), \quad (2.159)$$

where A_1 is the height of the curb (Fig. 2-27(a)).

Driving over a pothole

For a car moving at speed s over a pothole of length d, the pothole represents a depression of duration $T = d/s$. Hence, driving over the pothole can be modeled (Fig. 2-27(b)) as

$$x_2(t) = A_2[-u(t) + u(t-T)], \quad (2.160)$$

where A_2 is the depth of the pothole.

Driving over wavy pavement

Figure 2-27(c) depicts a wavy pavement whose elevation is a sinusoid of amplitude A_3 and period T_0. Input $x_3(t)$ is then

$$x_3(t) = A_3 \cos\frac{2\pi t}{T_0}\,. \quad (2.161)$$

Example 2-17: Car Response to a Curb

A car with a mass of 1,000 kg is driven over a 10 cm high curb. Each wheel is supported by a coil with spring constant $k = 10^5$ N/m. Determine the car's response to driving over the curb for each of the following values of b, the damping constant of the shock absorber: (a) 2×10^4 N·s/m, (b) 10^4 N·s/m, and (c) 5000 N·s/m.

Solution: (a) The mass per wheel is $m = 1000/4 = 250$ kg. Comparison of the constant coefficients in Eq. (2.158) with those in Eq. (2.132) for the LCCDE of Section 2-8 leads to

LCCDE Suspension System

$$a_1 = \frac{b}{m} = \frac{2\times10^4}{250} = 80\ \text{s}^{-1},$$

$$a_2 = \frac{k}{m} = \frac{10^5}{250} = 400\ \text{s}^{-2},$$

$$b_1 = \frac{b}{m} = 80\ \text{s}^{-1},$$

$$b_2 = \frac{k}{m} = 400\ \text{s}^{-2},$$

$$\omega_0 = \sqrt{a_2} = 20 \text{ rad/s},$$

and

$$\alpha = \frac{a_1}{2} = 40 \text{ Np/s}.$$

The damping coefficient is

$$\xi = \frac{\alpha}{\omega_0} = \frac{40}{20} = 2.$$

Since $\xi > 1$, the car suspension is an overdamped system. As was noted earlier in Section 2-9.2, the curb is modeled as a step function with an amplitude $A = 10$ cm $= 0.1$ m. From Table 2-3, the step response of an overdamped system scaled by a factor of 0.1 is

$$\begin{aligned} y_1(t) &= 0.1\, y_{\text{step}}(t) \\ &= 0.1\left[\frac{A_1}{p_1}\,(e^{p_1 t}-1)+\frac{A_2}{p_2}\,(e^{p_2 t}-1)\right] u(t). \end{aligned} \quad (2.162)$$

From Table 2-3, we have

$$p_1 = \omega_0[-\xi + \sqrt{\xi^2-1}] = -5.36 \text{ Np/s},$$

$$p_2 = \omega_0[-\xi - \sqrt{\xi^2-1}] = -74.64 \text{ Np/s},$$

$$A_1 = \frac{b_1 p_1 + b_2}{p_1 - p_2} = \frac{80(-5.36)+400}{-5.36+74.64} = -0.42\ \text{s}^{-1},$$

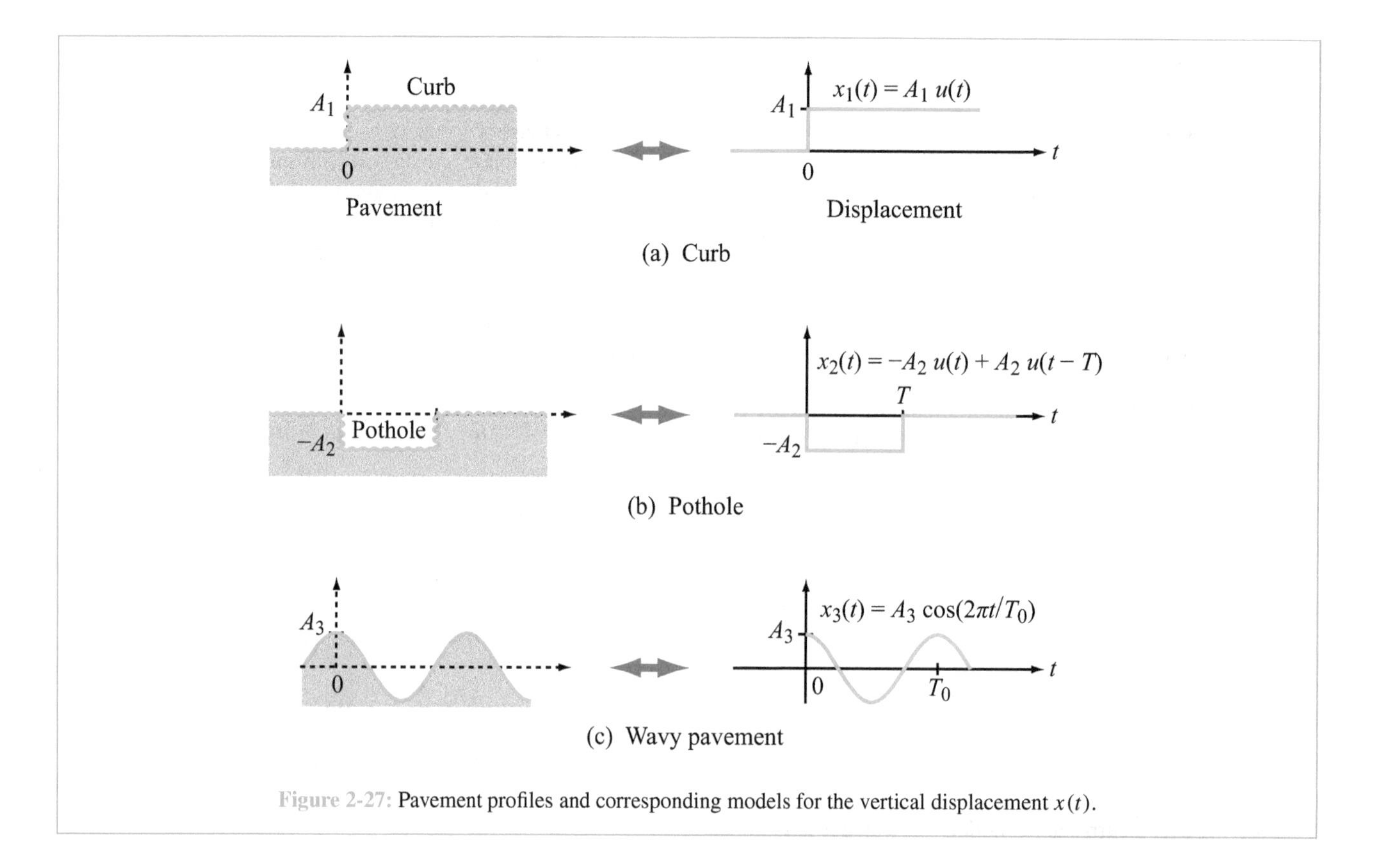

Figure 2-27: Pavement profiles and corresponding models for the vertical displacement $x(t)$.

and

$$A_2 = \frac{-(b_1 p_2 + b_2)}{p_1 - p_2} = \frac{-[80(-74.64) + 400]}{-5.36 + 74.64} = 80.42 \text{ s}^{-1}.$$

Hence, $y_1(t)$ in meters becomes

$$y_1(t) = [0.108(1 - e^{-74.64t}) - 0.008(1 - e^{-5.36t})]\, u(t) \text{ m}, \tag{2.163}$$

A plot of $y_1(t)$ is displayed in Fig. 2-28.

(b) Changing the value of b to 10^4 N·s/m leads to

$$a_1 = b_1 = \frac{10^4}{250} = 40 \text{ s}^{-1},$$

$$a_2 = b_2 = \frac{k}{m} = 400 \text{ s}^{-2} \text{ (unchanged)},$$

$$\omega_0 = \sqrt{a_2} = 20 \text{ rad/s (unchanged)},$$

$$\alpha = \frac{a_1}{2} = 20 \text{ Np/s},$$

and

$$\xi = \frac{\alpha}{\omega_0} = \frac{20}{20} = 1.$$

For this critically damped case, the expressions given in Table 2-3 lead to

$$y_2(t) = 0.1[(1 - e^{-20t}) + 20te^{-20t}]\, u(t) \text{ m}. \tag{2.164}$$

From the plot of $y_2(t)$ in Fig. 2-28, it is easy to see that it approaches its final destination of 0.1 m (height of the curb) much sooner than the overdamped response exhibited by $y_1(t)$.

(c) For an old shock absorber with $b = 5000$ N·s/m, the parameter values are

$$a_1 = b_1 = \frac{5000}{250} = 20 \text{ s}^{-1},$$

$$a_2 = b_2 = \frac{k}{m} = 400 \text{ s}^{-2} \text{ (unchanged)},$$

$$\omega_0 = \sqrt{a_2} = 20 \text{ rad/s (unchanged)},$$

$$\alpha = \frac{a_1}{2} = \frac{20}{2} = 10 \text{ Np/s},$$

and

$$\xi = \frac{\alpha}{\omega_0} = \frac{10}{20} = 0.5.$$

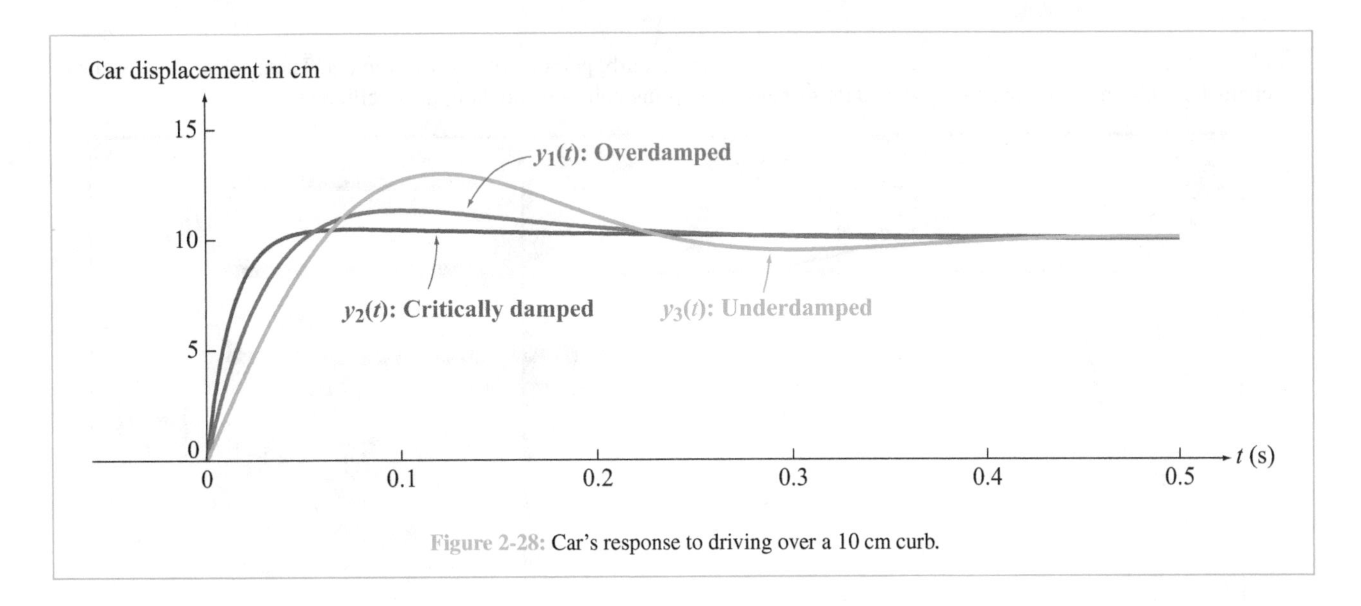

Figure 2-28: Car's response to driving over a 10 cm curb.

Since $\xi < 1$, the system is underdamped, which means that the car response to the curb will include some oscillations. Using Table 2-3, we have

$$\omega_d = \omega_0\sqrt{1-\xi^2} = 20\sqrt{1-0.25} = 17.32 \text{ rad/s}$$

and

$$y_3(t) = 0.1\{[-\cos 17.32t + 1.15 \sin 17.32t]e^{-10t} + 1\}\, u(t) \text{ m}.$$

The oscillatory behavior of $y_3(t)$ is clearly evident in the plot of its profile in Fig. 2-28 (see S^2 for details).

Example 2-18: Car Response to a Pothole

Simulate the response of a car driven at 5 m/s over a 1 m long, 10 cm deep pothole if the damping constant of its shock absorber is (a) 10^4 N·s/m or (b) 2000 N·s/m. All other attributes are the same as those in Example 2-17, namely, $m = 250$ kg and $k = 10^5$ N/m.

Solution: (a) The travel time across the pothole is $T = \frac{1}{5} = 0.2$ s. According to part (b) of the solution of Example 2-17, $\xi = 1$ when $b = 10^4$ N·s/m, representing a critically damped system with the response given by Eq. (2.164) as

$$y_2(t) = 0.1[(1 - e^{-20t}) + 20te^{-20t}]\, u(t) \text{ m}. \tag{2.165}$$

The car's vertical displacement $y_2(t)$ is in response to a 0.1 m vertical step (curb). For the pothole model shown in Fig. 2-27, the response $y_4(t)$ can be synthesized as

$$\begin{aligned} y_4(t) &= -y_2(t) + y_2(t-0.2) \\ &= -0.1[(1-e^{-20t}) + 20te^{-20t}]\, u(t) \\ &\quad + 0.1(1-e^{-20(t-0.2)})\, u(t-0.2) \\ &\quad + 2(t-0.2)e^{-20(t-0.2)}\, u(t-0.2) \text{ m}. \end{aligned} \tag{2.166}$$

Because the height of the curb and the depth of the pothole are both 0.1 m, no scaling was necessary in this case. For a pothole of depth A, the multiplying coefficient (0.1) should be replaced with A.

(b) For $b = 2000$ N·s/m, we have

$$a_1 = b_1 = \frac{2000}{250} = 8 \text{ s}^{-1},$$

$$a_2 = b_2 = \frac{k}{m} = 400 \text{ s}^{-2},$$

$$\omega_0 = \sqrt{a_2} = 20 \text{ rad/s},$$

$$\alpha = \frac{a_1}{2} = \frac{8}{2} = 4 \text{ Np/s},$$

and

$$\xi = \frac{\alpha}{\omega_0} = \frac{4}{20} = 0.2.$$

Module 2.2 Automobile Suspension Response Select curb, pothole, or wavy pavement. Then, select the pavement characteristics, the automobile's mass, and its suspension's spring constant and damping coefficient.

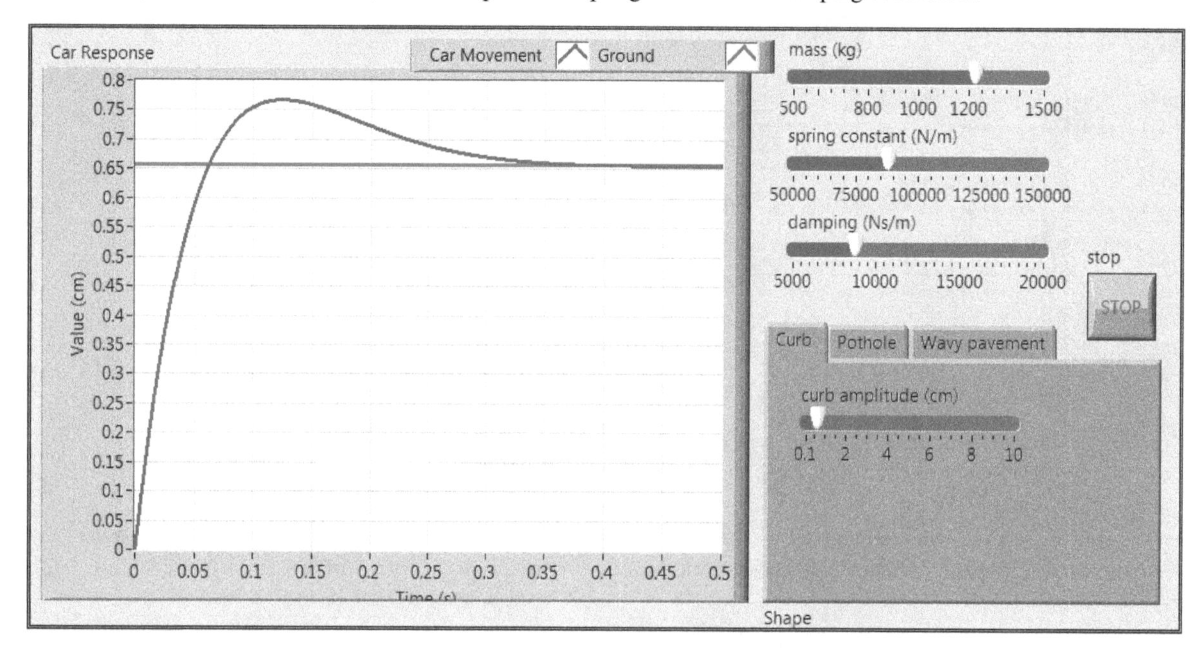

For this underdamped case, the expressions in Table 2-3 lead to

$$\omega_d = \omega_0\sqrt{1-\xi^2} = 20\sqrt{1-0.2^2} = 19.6 \text{ rad/s}$$

and a unit-step response given by

$$y(t) = \{[-\cos 19.6t + 0.58 \sin 19.6t]e^{-4t} + 1\}\, u(t) \text{ m.} \tag{2.167}$$

For the pothole response, we have

$$\begin{aligned} y_5(t) &= -0.1\, y(t) + 0.1\, y(t-0.2) \\ &= 0.1\{[\cos 19.6t - 0.58 \sin 19.6t]e^{-4t} - 1\}\, u(t) \\ &\quad - 0.1\{[\cos(19.6(t-0.2)) \\ &\quad - 0.58 \sin(19.6(t-0.2))]e^{-4(t-0.2)} \\ &\quad - 1\}\, u(t-0.2) \text{ m.} \end{aligned} \tag{2.168}$$

Plots for $y_4(t)$ and $y_5(t)$ are displayed in Fig. 2-29 [see S^2 for details].

Example 2-19: Driving over Wavy Pavement

A 1,000 kg car is driven over a wavy pavement (Fig. 2-27(c)) of amplitude $A_3 = 5$ cm and a period $T_0 = 0.314$ s. The suspension system has a spring constant $k = 10^5$ N/m and a damping constant $b = 10^4$ N·s/m. Simulate the car displacement as a function of time.

Solution: The car suspension parameter values are

$$\frac{b}{m} = \frac{10^4}{250} = 40 \text{ s}^{-1}$$

and

$$\frac{k}{m} = \frac{10^5}{250} = 400 \text{ s}^{-2}.$$

Using these values in Eq. (2.158) gives

$$\frac{d^2y}{dt^2} + 40\frac{dy}{dt} + 400y = 40\frac{dx}{dt} + 400x. \tag{2.169}$$

Following the recipe outlined in Section 2-7.3, wherein we set $x(t) = e^{j\omega t}$ and $y(t) = \mathbf{H}(\omega)\, e^{j\omega t}$, Eq. (2.169) leads to the following expression for the frequency response function $\mathbf{H}(\omega)$:

$$\mathbf{H}(\omega) = \frac{400 + j40\omega}{(j\omega)^2 + j40\omega + 400} = \frac{400 + j40\omega}{(400-\omega^2) + j40\omega}. \tag{2.170}$$

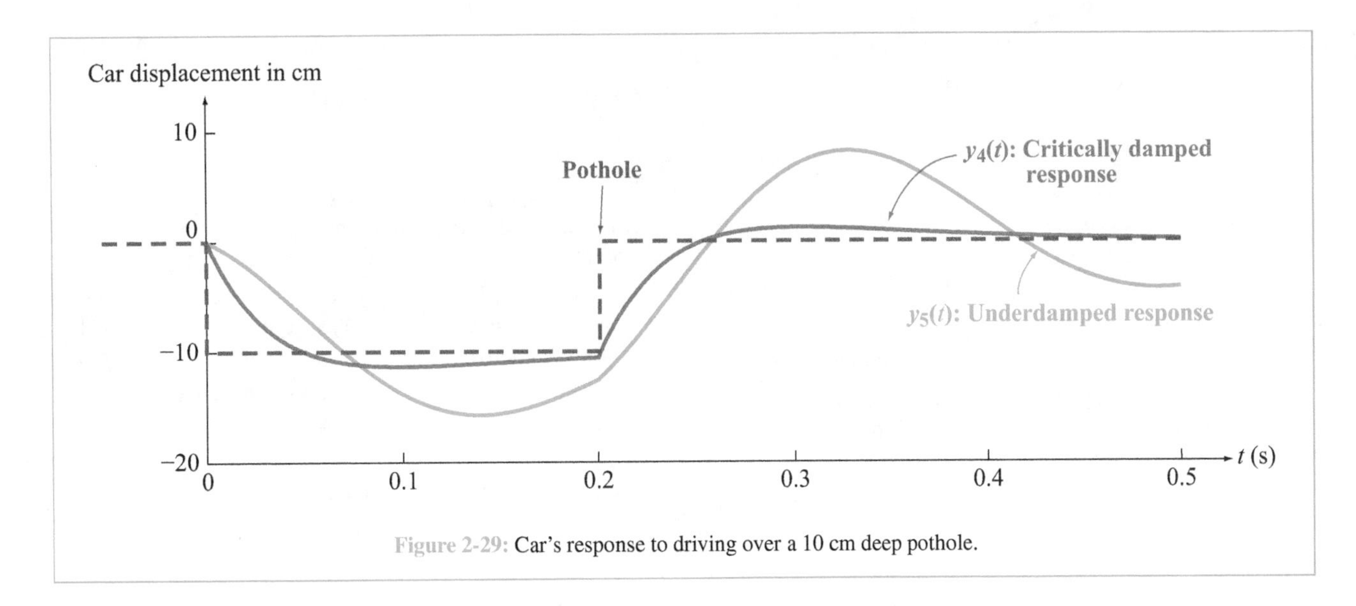

Figure 2-29: Car's response to driving over a 10 cm deep pothole.

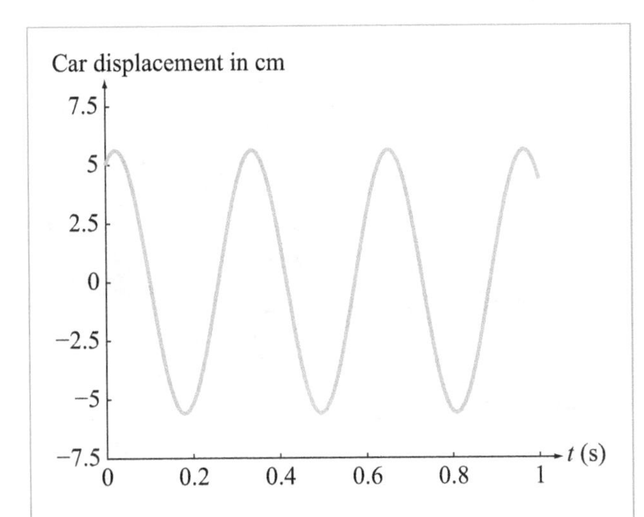

Figure 2-30: Car's response to driving over a wavy pavement with a 5 cm amplitude.

The angular frequency ω of the wavy pavement is

$$\omega_0 = \frac{2\pi}{T_0} = \frac{2\pi}{0.314} = 20 \text{ rad/s}. \tag{2.171}$$

Evaluating $\mathbf{H}(\omega)$ at $\omega_0 = 20$ rad/s gives

$$\mathbf{H}(20) = \frac{400 + j800}{(400 - 400) + j800} = 1 - j0.5 = 1.12e^{-j26.6^\circ}. \tag{2.172}$$

Application of Eq. (2.118) with a scaling amplitude of 0.05 m yields

$$\begin{aligned} y_6(t) &= 0.05|\mathbf{H}(20)| \cos(\omega_0 t + \theta) \\ &= 5.6 \cos(20t - 26.6^\circ) \times 10^{-2} \text{ m}. \end{aligned} \tag{2.173}$$

Note that the amplitude of $y_6(t)$ in Fig. 2-30 is slightly greater than the amplitude of the pavement displacement [5.6 cm compared with 5 cm for $x(t)$]. This is an example of ***resonance***.

Exercise 2-17: Use LabVIEW Module 2.2 to compute the wavy pavement response in Example 2-19 and shown in Fig. 2-30.

Answer:

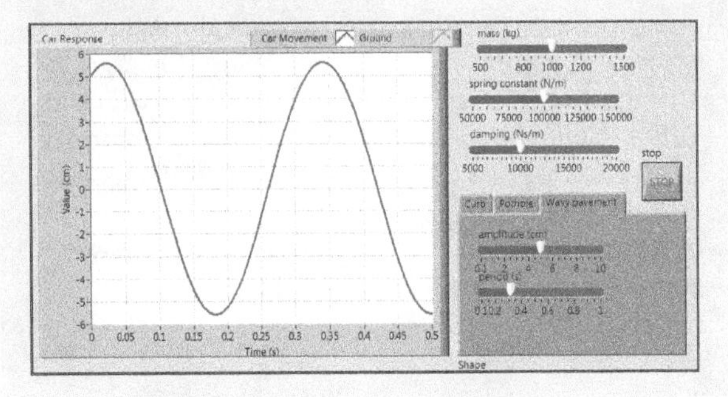

Summary

Concepts

- A system is linear if it has both the scaling and additivity properties.
- A system is time-invariant if delaying the input delays the output by the same amount.
- A system is LTI if it is both linear and time-invariant.
- The impulse response $h(t)$ is the response to an impulse $\delta(t)$. The step response $y_{\text{step}}(t)$ is the response to a step function input $u(t)$.
- If a system is LTI and its impulse response is known, its response to any other input can be computed using convolution, if all initial conditions are zero.
- If a system is LTI, its response to a complex exponential time signal is another complex exponential time signal at the same frequency.
- If a system is LTI, its response to a sinusoid is another sinusoid at the same frequency.
- The frequency response function of a system can be computed from the system's LCCDE.

Mathematical and Physical Models

Computation of impulse response $h(t)$

$$h(t) = \frac{dy_{\text{step}}}{dt}, \text{ where } y_{\text{step}}(t) = \text{step response}$$

Convolution

$$y(t) = \int_{-\infty}^{\infty} x(\tau)\, h(t-\tau)\, d\tau,$$

where $x(t) =$ input and $y(t) =$ output

LTI Causality $\quad h(t) = 0 \text{ for } t < 0$

LTI Stability $\quad \int_{-\infty}^{\infty} |h(t)|\, dt = \text{finite}$

Frequency response $\mathbf{H}(\omega)$

For $\displaystyle\sum_{i=0}^{n} a_{n-i} \frac{d^i y}{dt^i} = \sum_{i=0}^{m} b_{m-i} \frac{d^i x}{dt^i}$

$$\mathbf{H}(\omega) = \frac{\sum_{i=0}^{m} b_{m-i}(j\omega)^i}{\sum_{i=0}^{n} a_{n-i}(j\omega)^i}$$

Important Terms

Provide definitions or explain the meaning of the following terms:

additivity
anticipatory
area property
associative property
attenuation coefficient
BIBO stability
bounded
causal
causality
commutative property
contrapositive
convolution
critically damped
damper force
damping coefficient
distributive property
frequency response function
impulse response
LCCDE
LDE
linear
marginally stable
memoryless
natural undamped frequency
overdamped
rational function
scalability
spring force
step response
superposition property
time constant
time invariant
triangle inequality
underdamped
width property

PROBLEMS

Section 2-1 and 2-2: LTI Systems

2.1 For each of the following systems, specify whether or not the system is: (i) linear and/or (ii) time-invariant.

(a) $y(t) = 3x(t) + 1$

*(b) $y(t) = 3\sin(t)\, x(t)$

(c) $\frac{dy}{dt} + t\, y(t) = x(t)$

(d) $\frac{dy}{dt} + 2y(t) = 3\,\frac{dx}{dt}$

(e) $y(t) = \int_{-\infty}^{t} x(\tau)\, d\tau$

(f) $y(t) = \int_{0}^{t} x(\tau)\, d\tau$

(g) $y(t) = \int_{t-1}^{t+1} x(\tau)\, d\tau$

2.2 For each of the following systems, specify whether or not the system is: (i) linear and/or (ii) time-invariant.

(a) $y(t) = 3x(t-1)$

(b) $y(t) = t\, x(t)$

(c) $\frac{dy}{dt} + y(t-1) = x(t)$

(d) $\frac{dy}{dt} + 2y(t) = \int_{-\infty}^{t} x(\tau)\, d\tau$

(e) $y(t) = x(t)\, u(t)$

(f) $y(t) = \int_{t}^{\infty} x(\tau)\, d\tau$

(g) $y(t) = \int_{t}^{2t} x(\tau)\, d\tau$

2.3 Compute the impulse response of the LTI system whose step response is

2.4 Compute the impulse response of the LTI system whose step response is

2.5 The step response of an LTI system is

Compute the response of the system to the following inputs.

(a)

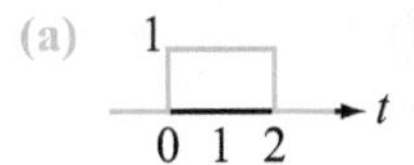

*(b)

(c)

(d)

2.6 Compute the response $y(t)$ of an initially uncharged RC circuit to a pulse $x(t)$ of duration ϵ, height $\frac{1}{\epsilon}$, and area $\epsilon\,\frac{1}{\epsilon} = 1$ for $\epsilon \ll 1$ (Fig. P2.6).

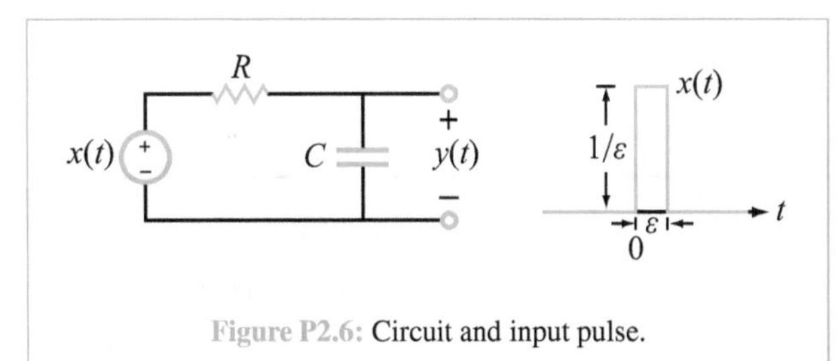

Figure P2.6: Circuit and input pulse.

The power series for e^{ax} truncated to two terms is $e^{ax} \approx 1 + ax$ and is valid for $ax \ll 1$. Set $a = \frac{\epsilon}{RC}$ and substitute the result in your answer. Show that $y(t)$ simplifies to Eq. (2.17).

2.7 Plot the response of the RC circuit shown in Fig. P2.6 to the input shown in Fig. P2.7, given that $RC = 1$ s.

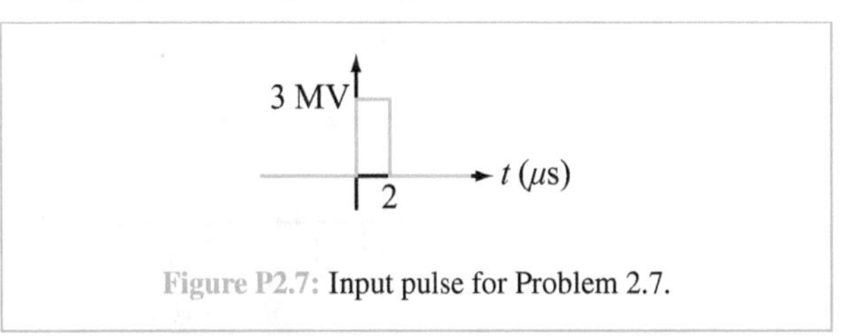

Figure P2.7: Input pulse for Problem 2.7.

2.8 For the RC circuit shown in Fig. 2-5(a), apply the superposition principle to obtain the response $y(t)$ to input:

(a) $x_1(t)$ in Fig. P1.23(a) (in Chapter 1)

(b) $x_2(t)$ in Fig. P1.23(b) (in Chapter 1)

(c) $x_3(t)$ in Fig. P1.23(c) (in Chapter 1)

2.9 For the RC circuit shown in Fig. 2-5(a), obtain the response $y(t)$ to input:

*Answer(s) in Appendix F.

(a) $x_1(t)$ in Fig. P1.27(a) (in Chapter 1)

(b) $x_2(t)$ in Fig. P1.27(b) (in Chapter 1)

(c) $x_3(t)$ in Fig. P1.27(c) (in Chapter 1)

Section 2-3 to 2-5: Convolution

2.10 Functions $x(t)$ and $h(t)$ are both rectangular pulses, as shown in Fig. P2.10. Apply graphical convolution to determine $y(t) = x(t) * h(t)$ given the following data.

(a) $A = 1,\ B = 1,\ T_1 = 2$ s, $T_2 = 4$ s

(b) $A = 2,\ B = 1,\ T_1 = 4$ s, $T_2 = 2$ s

*(c) $A = 1,\ B = 2,\ T_1 = 4$ s, $T_2 = 2$ s.

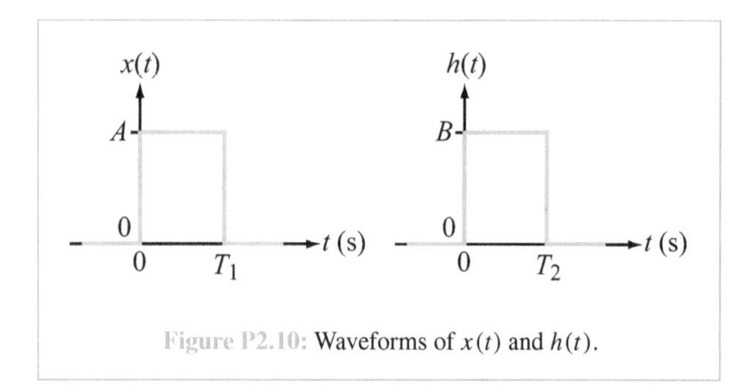

Figure P2.10: Waveforms of $x(t)$ and $h(t)$.

2.11 Apply graphical convolution to the waveforms of $x(t)$ and $h(t)$ shown in Fig. P2.11 to determine $y(t) = x(t) * h(t)$.

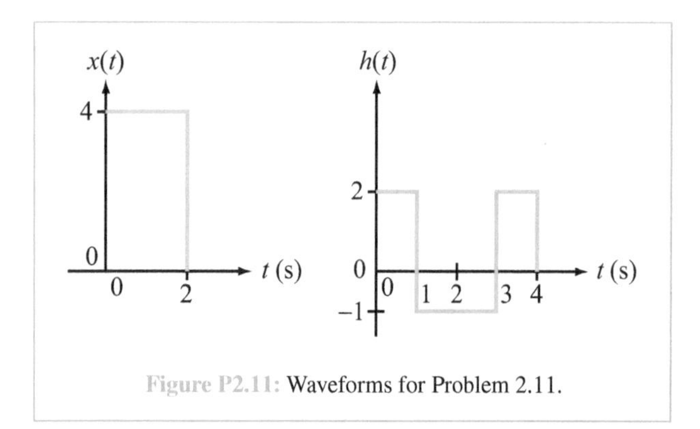

Figure P2.11: Waveforms for Problem 2.11.

2.12 Functions $x(t)$ and $h(t)$ have the waveforms shown in Fig. P2.12. Determine and plot $y(t) = x(t) * h(t)$ using the following methods.

(a) Integrating the convolution analytically.

(b) Integrating the convolution graphically.

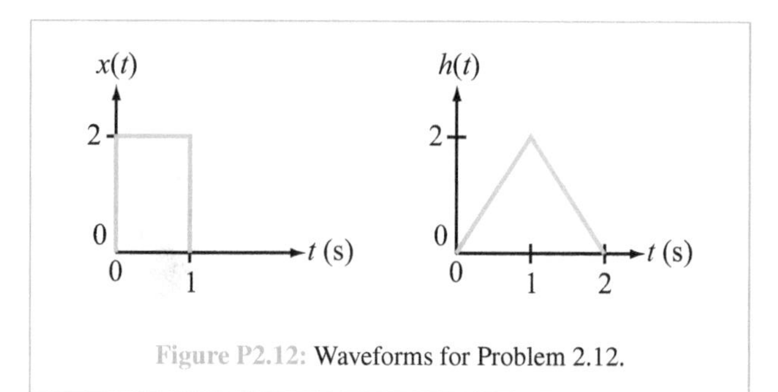

Figure P2.12: Waveforms for Problem 2.12.

2.13 Functions $x(t)$ and $h(t)$ have the waveforms shown in Fig. P2.13. Determine and plot $y(t) = x(t) * h(t)$ using the following methods.

(a) Integrating the convolution analytically.

(b) Integrating the convolution graphically.

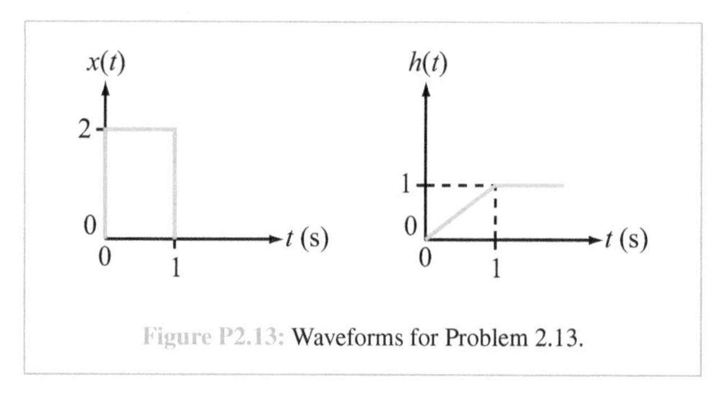

Figure P2.13: Waveforms for Problem 2.13.

2.14 Functions $x(t)$ and $h(t)$ are given by

$$x(t) = \begin{cases} 0, & \text{for } t < 0 \\ \sin \pi t, & \text{for } 0 \le t \le 1 \text{ s} \\ 0, & \text{for } t \ge 1 \text{ s} \end{cases}$$

$$h(t) = u(t).$$

Determine $y(t) = x(t) * h(t)$.

2.15 Compute the following convolutions *without computing any integrals*.

*(a) $u(t) * [\delta(t) - 3\delta(t-1) + 2\delta(t-2)]$

(b) $u(t) * [2u(t) - 2u(t-3)]$

(c) $u(t) * [(t-1)\,u(t-1)]$

2.16 Compute the following convolutions *without computing any integrals*.

(a) $\delta(t-2) * [u(t) - 3u(t-1) + 2u(t-2)]$

(b) $[\delta(t) + 2\delta(t-1) + 3\delta(t-2)] * [4\delta(t) + 5\delta(t-1)]$

(c) $u(t) * [u(t) - u(t-2) - 2\delta(t-2)]$

2.17 Compute the following convolutions.

*(a) $e^{-t}\, u(t) * e^{-2t}\, u(t)$

(b) $e^{-2t}\, u(t) * e^{-3t}\, u(t)$

(c) $e^{-3t}\, u(t) * e^{-3t}\, u(t)$

2.18 Show that the overall impulse response of the interconnected system shown in Fig. P2.18 is $h(t) = 0$!

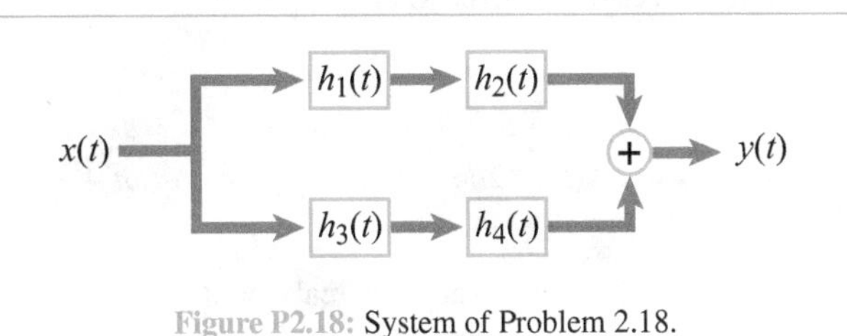

Figure P2.18: System of Problem 2.18.

The impulse responses of the individual systems are

- $h_1(t) = e^{-t}\, u(t) - e^{-2t}\, u(t)$,
- $h_2(t) = e^{-3t}\, u(t)$,
- $h_3(t) = e^{-3t}\, u(t) - e^{-2t}\, u(t)$,
- $h_4(t) = e^{-t}\, u(t)$.

2.19 Prove the following convolution properties.

(a) The convolution of two even functions is an even function.

(b) The convolution of an even function and an odd function is an odd function.

(c) The convolution of two odd functions is an even function.

2.20 Describe in words what this cascade connection of LTI systems does.

$\rightarrow$ $e^{-2t}\, u(t)$ $\rightarrow$ $\delta(t) + 2u(t)$ $\rightarrow$ $\dfrac{d}{dt}$ $\rightarrow$

2.21 Compute the response of an initially uncharged RC circuit with $RC = 1$ to the input voltage shown in Fig. P2.21, using the following methods.

(a) Computing the appropriate convolution

(b) Writing the input in terms of delayed and scaled step and ramp functions

1
0 1 2 t

Figure P2.21: Input signal of Problem 2.21.

Section 2-6: Causality and Stability

2.22 Determine whether or not each of the LTI systems whose impulse responses are specified below are (i) causal and/or (ii) BIBO stable.

(a) $h(t) = e^{-|t|}$

(b) $h(t) = (1 - |t|)[u(t+1) - u(t-1)]$

(c) $h(t) = e^{2t}\, u(-t)$

*(d) $h(t) = e^{2t} u(t)$

(e) $h(t) = \cos(2t) u(t)$

(f) $h(t) = \frac{1}{t+1}\, u(t)$

2.23 Determine whether or not each of the following LTI systems is (i) causal and/or (ii) BIBO stable. If the system is not BIBO stable, provide an example of a bounded input that yields an unbounded output.

(a) $y(t) = \frac{dx}{dt}$

(b) $y(t) = \int_{-\infty}^{t} x(\tau)\, d\tau$

(c) $y(t) = \int_{-\infty}^{t} x(\tau)\, \cos(t-\tau)\, d\tau$

*(d) $y(t) = x(t+1)$

(e) $y(t) = \int_{t-1}^{t+1} x(\tau)\, d\tau$

(f) $y(t) = \int_{t}^{\infty} x(\tau)\, e^{2(t-\tau)}\, d\tau$

2.24 This problem demonstrates the significance of *absolute* integrability of the impulse response for BIBO stability of LTI systems. An LTI system has impulse response

$$h(t) = \sum_{n=1}^{\infty} \frac{(-1)^n}{n}\, \delta(t-n).$$

(a) Show that $h(t)$ is integrable: $\int_{-\infty}^{\infty} h(t)\, dt < \infty$.

(b) Show that $h(t)$ is not *absolutely* integrable:

$$\int_{-\infty}^{\infty} |h(t)|\, dt \to \infty.$$

(c) Provide an example of a bounded input $x(t)$ that yields an unbounded output.

2.25 An LTI system has impulse response

$$h(t) = (1/t^a)\, u(t-1).$$

Show that the system is BIBO stable if $a > 1$.

2.26 Prove the following statements.

(a) Parallel connections of BIBO-stable systems are BIBO stable.

(b) Parallel connections of causal systems are causal.

2.27 Prove the following statements.

(a) Series connection of BIBO-stable systems are BIBO stable.

(b) Series connections of causal systems are causal.

2.28 An LTI system has an impulse response given by

$$h(t) = 2\cos(t)\, u(t).$$

Obtain the response to input $x(t) = 2\cos(t)\, u(t)$ and determine whether or not the system is BIBO-stable.

Section 2-7: LTI Sinusoidal Response

2.29 An LTI system has the frequency response function $\mathbf{H}(\omega) = 1/(j\omega + 3)$. Compute the output if the input is

(a) $x(t) = 3$

(b) $x(t) = 3\sqrt{2}\cos(3t)$

*(c) $x(t) = 5\cos(4t)$

(d) $x(t) = \delta(t)$

(e) $x(t) = u(t)$

(f) $x(t) = 1$

2.30 An LTI system is described by the LCCDE

$$\frac{d^2y}{dt^2} + 2\frac{dy}{dt} + 7y = 5\frac{dx}{dt}.$$

Its input is $x(t) = \cos(\omega t)$.

(a) If $\omega = 2$ rad/s, compute the output $y(t)$.

(b) Find the frequency ω so that $y(t) = A\cos(\omega t)$ for some constant $A > 0$. That is, the input and output sinusoids are in phase.

2.31 An LTI system has the impulse response

$$h(t) = e^{-12t}\, u(t).$$

The input to the system is

$$x(t) = 12 + 26\cos(5t) + 45\cos(9t) + 80\cos(16t).$$

Compute the output $y(t)$.

2.32 An LTI system has the impulse response

$$h(t) = 5e^{-t}u(t) - 16e^{-2t}u(t) + 13e^{-3t}u(t).$$

The input is $x(t) = 7\cos(2t + 25°)$. Compute the output $y(t)$.

2.33 Repeat Problem 2.32 after replacing the input with $x(t) = 6 + 10\cos(t) + 13\cos(2t)$.

2.34 If

$$\cos(\omega t) \rightarrow \boxed{\delta(t) - 2ae^{-at}\, u(t)} \rightarrow A\cos(\omega t + \theta),$$

prove that $A = 1$ for any ω and any real $a > 0$.

2.35 If

$$\cos(\omega t) \rightarrow \boxed{h(t)} \rightarrow \boxed{h(-t)} \rightarrow A\cos(\omega t + \theta),$$

prove that $\theta = 0$ for any ω and any real $h(t)$.

2.36 We observe the following input-output pair for an LTI system:

- $x(t) = 1 + 2\cos(t) + 3\cos(2t)$
- $y(t) = 6\cos(t) + 6\cos(2t)$
- $x(t) \rightarrow \boxed{\text{LTI}} \rightarrow y(t)$

Determine $y(t)$ in response to a new input

$$x(t) = 4 + 4\cos(t) + 2\cos(2t).$$

2.37 We observe the following input-output pair for an LTI system:

- $x(t) = u(t) + 2\cos(2t)$
- $y(t) = u(t) - e^{-2t}\, u(t) + \sqrt{2}\cos(2t - 45°)$
- $x(t) \rightarrow \boxed{\text{LTI}} \rightarrow y(t)$

Determine $y(t)$ in response to a new input

$$x(t) = 5u(t-3) + 3\sqrt{2}\cos(2t - 60°).$$

*2.38 Compute the convolution of the two signals:

- $x(t) = 60\sqrt{2}\cos(3t) + 60\sqrt{2}\cos(4t)$
- $h(t) = e^{-3t}\,u(t) - e^{-4t}\,u(t)$

2.39 Compute the convolution of the two signals:

- $x_1(t) = 60\sqrt{2}\,(e^{j3t} + e^{j4t})$
- $x_2(t) = e^{-3t}\,u(t) - e^{-4t}\,u(t)$

2.40 An inductor is modeled as an ideal inductor in series with a resistor representing the coil resistance. A current sine-wave generator delivers a current $i(t) = \cos(500t) + \cos(900t)$ through the inductor, as shown in Fig. P2.40. The voltage across the inductor is measured to be $\upsilon(t) = 13\cos(500t + \theta_1) + 15\cos(900t + \theta_2)$, where the current is in amps and the voltage in volts and phase angles θ_1 and θ_2 are unknown. Compute the inductance L and coil resistance R.

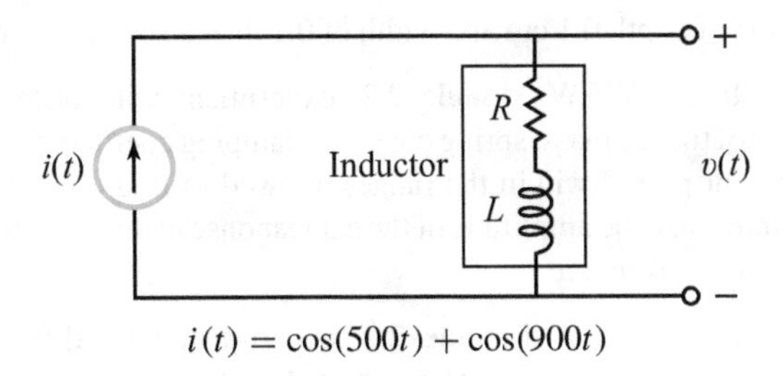

Figure P2.40: Circuit for Problem 2.40.

2.41 A system is modeled as

$$\frac{dy}{dt} + a\,y(t) = b\,\frac{dx}{dt} + c\,x(t),$$

where the constants a, b, and c are all unknown. The response to input $x(t) = 9 + 15\cos(12t)$ is output

$$y(t) = 5 + 13\cos(12t + 0.2487).$$

Determine constants a, b, and c.

2.42 Show that the circuit in Fig. P2.42 is *all pass*. Specifically:

(a) Show that its gain is unity at all frequencies.

(b) Find the frequency ω_0, in terms of R and C, for which if $\upsilon_1(t) = \cos(\omega_0 t)$ then $\upsilon_2(t) = \sin(\omega_0 t)$. This is called a ***90-degree phase shifter***.

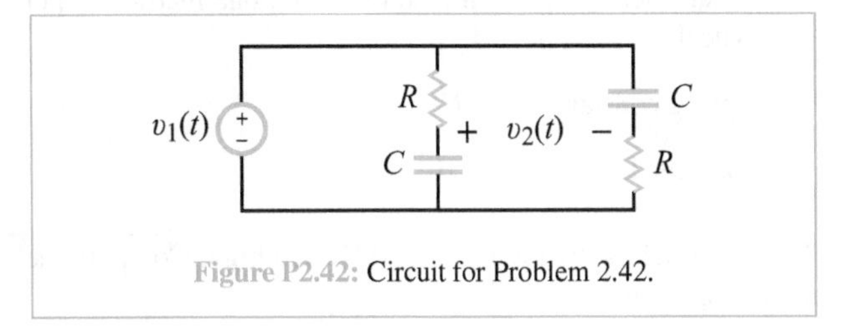

Figure P2.42: Circuit for Problem 2.42.

Section 2-8: Second-Order LCCDE

2.43 An LTI system is described by the LCCDE

$$\frac{d^2y}{dt^2} + B\,\frac{dy}{dt} + 25y(t) = \frac{dx}{dt} + 23x(t).$$

Compute the range of values of constant B so that the system impulse response is

(a) Overdamped (non-oscillatory)

(b) Underdamped (oscillatory)

*(c) Unstable (blows up)

If $B = 26$, compute the impulse response $h(t)$.

2.44 An LTI system is described by the LCCDE

$$\frac{d^2y}{dt^2} + B\,\frac{dy}{dt} + 49y(t) = 21\sqrt{3}\,x(t).$$

Compute the range of values of constant B so that the system impulse response is

(a) Overdamped (non-oscillatory)

(b) Underdamped (oscillatory)

(c) Unstable (blows up)

If $B = 7$, compute the impulse response $h(t)$.

2.45 A series RLC circuit with $L = 10$ mH and $C = 1\ \mu$F is connected to an input voltage source $\upsilon_{in}(t)$. Output voltage $\upsilon_{out}(t)$ is taken across the resistor. For what value of R is the circuit step response critically damped?

2.46 A parallel RLC circuit with $L = 10$ mH and $C = 1\ \mu$F is connected to an input source $i_{in}(t)$. The system output is current $i_{out}(t)$ flowing through the inductor. For what value of R is the circuit step response critically damped?

Section 2-9: Car Suspension System

For each of the following four problems:

- Total truck mass is 4 metric tons (i.e., one metric ton per wheel).
- Spring constant is 10^5 N/m.
- Viscous damper is 5×10^4 Ns/m.

2.47 A truck is driven over a curb 10 cm high. Compute the truck displacement.

2.48 A truck is driven over a pothole 5 cm deep and 1 m wide at 5 m/s. Compute the truck displacement.

2.49 A truck is driven up a ramp 5 m long with a 10% slope at 5 m/s. Compute the truck displacement.

0.5 m
0 5 m distance

2.50 A truck is driven across a wavy pavement with a 1 cm amplitude and 1 cm period. The truck speed is 5 m/s. Compute the truck displacement.

2.51 For a truck with the specified mass and spring constant, what should the viscous damper be for a critically damped response?

LabVIEW Module 2.1

2.52 Use LabVIEW Module 2.1 to compute and display the convolution $e^{-t}\ u(t) * e^{-10t}\ u(t)$. This is the response of an RC circuit with $RC = 1$ to input $e^{-10t}\ u(t)$.

2.53 Use LabVIEW Module 2.1 with one signal given by $e^{-5t}\ u(t)$. Choose a in $e^{-at}\ u(t)$ of the second signal so that $e^{-5t}\ u(t) * e^{-at}\ u(t)$ decays to zero as rapidly as possible. Explain your result.

2.54 In LabVIEW Module 2.1, let the two inputs be $e^{-at}\ u(t)$ and $e^{-bt}\ u(t)$. Choose a and b so that $e^{-at}\ u(t) * e^{-bt}\ u(t)$ decays to zero as rapidly as possible.

2.55 In LabVIEW Module 2.1, compute and display $e^{-5t}\ u(t) * e^{-5t}\ u(t)$. Explain why the analytic plot is blank but the numeric plot shows the correct result.

LabVIEW Module 2.2

2.56 In LabVIEW Module 2.2, for a car mass of 1500 kg, a spring constant of 100000 N/m, and a damping coefficient of 5000 Ns/m, compute and display the response to a curb of amplitude 2 cm.

2.57 In LabVIEW Module 2.2, for a car mass of 1500 kg, a spring constant of 150000 N/m, and a damping coefficient of 5000 Ns/m, compute and display the response to a curb of amplitude 4 cm.

2.58 In LabVIEW Module 2.2, for a car mass of 1000 kg, a spring constant of 100000 N/m, and a curb of height 4 cm, find the value of the damping coefficient that makes the response decay to zero and stay there as quickly as possible.

2.59 In LabVIEW Module 2.2, for a car mass of 1000 kg, a spring constant of 100000 N/m, and a damping coefficient of 5000 Ns/m, compute and display the response to a pothole of depth 10 cm and width of 1 m at a speed of 4 m/s.

2.60 In LabVIEW Module 2.2, for a car mass of 1500 kg, a spring constant of 150000 N/m, and a damping coefficient of 5000 Ns/m, compute and display the response to a (small) pothole of depth 0.1 cm and width of 0.1 m at a speed of 2 m/s.

2.61 In LabVIEW Module 2.2, experiment with different values of the car mass, spring constant, damping coefficient, and pavement period within the ranges allowed to find the values that minimize the amplitude of the car response to the amplitude of a wavy pavement.

2.62 In LabVIEW Module 2.2, experiment with different values of the car mass, spring constant, damping coefficient, and pavement period within the ranges allowed to find the values that make the car response match the wavy pavement as much as possible (similar amplitude and minimal phase difference).

2.63 In LabVIEW Module 2.2, for a car mass of 1500 kg, a spring constant of 150000 N/m, and a damping coefficient of 5000 Ns/m, find the speed that makes the smoothest ride (minimal cusp) over a pothole of depth 10 cm and width of 1 m.

3 Laplace Transform

Contents

Overview, 86
3-1 Definition of the (Unilateral) Laplace Transform, 86
3-2 Poles and Zeros, 89
3-3 Properties of the Laplace Transform, 90
3-4 Circuit Analysis Example, 98
3-5 Partial Fraction Expansion, 99
3-6 Transfer Function **H**(**s**), 106
3-7 Poles and System Stability, 108
3-8 Invertible Systems, 111
3-9 Bilateral Transform for Continuous-Time Sinusoidal Signals, 113
3-10 Interrelating Different Descriptions of LTI Systems, 114
3-11 LTI System Response Partitions, 117
Summary, 123
Problems, 124

Objectives

Learn to:

- Compute the Laplace transform of a signal.
- Apply partial fraction expansion to compute the inverse Laplace transform.
- Perform convolution of two functions using the Laplace transform.
- Relate system stability to the poles of the transfer function.
- Interrelate the six different descriptions of LTI systems.

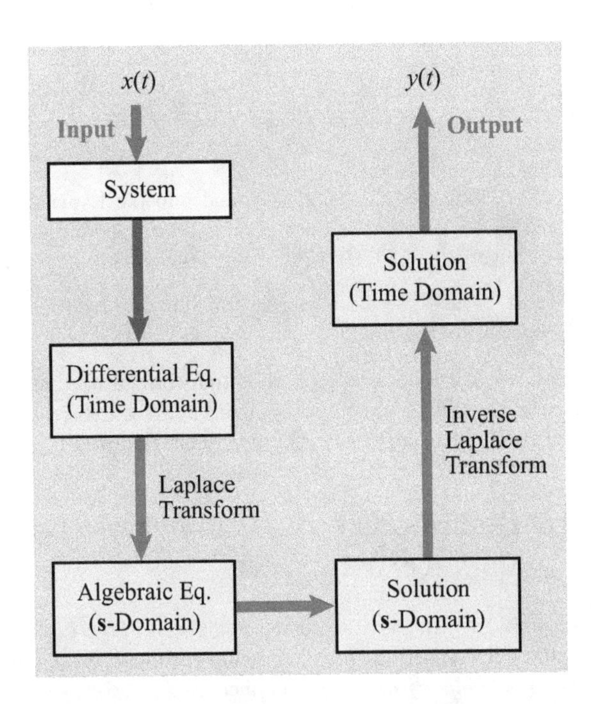

The beauty of the Laplace-transform technique is that it transforms a complicated differential equation into a straightforward algebraic equation. This chapter covers the A-to-Z of how to transform a differential equation from the time domain to the complex frequency **s** domain, solve it, and then inverse transform the solution to the time domain.

Overview

A ***domain transformation*** is a mathematical operation that converts a set of dependent variables from their actual domain, such as the time domain t, into a corresponding set of dependent variables defined in another domain. In this chapter, we explore how and why the Laplace transform is used to convert continuous-time signals and systems from the time domain into corresponding manifestations in a two-dimensional complex domain called the ***complex frequency* s**. In contrast with the answer to the *how* question, which consumes the bulk of the material in this chapter, the answer to the *why* question is very straightforward: *mathematical expediency!* Often, signal/system scenarios are described in terms of one or more LCCDEs. To determine the system's output response to a specified input excitation, the LCCDEs can be solved entirely in the time domain. In general, however, the solution method is mathematically demanding and cumbersome. The Laplace transform technique offers an alternative—and mathematically simpler—approach to arriving at the same solution. It entails a three-step procedure.

Solution Procedure: Laplace Transform

Step 1: The LCCDE is transformed into the Laplace domain—also known as the **s**-domain. The transformation converts the LCCDE into an algebraic equation.

Step 2: The **s**-domain algebraic equation is solved for the variable of interest.

Step 3: The **s**-domain solution is transformed back to the time domain. The solution includes both the transient and steady-state components of the system response.

3-1 Definition of the (Unilateral) Laplace Transform

The symbol $\mathcal{L}[x(t)]$ is a shorthand notation for "the Laplace transform of function $x(t)$." Usually denoted $\mathbf{X}(\mathbf{s})$, the ***unilateral Laplace transform*** is defined as

$$\mathbf{X}(\mathbf{s}) = \mathcal{L}[x(t)] = \int_{0^-}^{\infty} x(t)\, e^{-\mathbf{s}t}\, dt, \tag{3.1}$$

where **s** is a complex variable—with a real part σ and an imaginary part ω—given by

$$\mathbf{s} = \sigma + j\omega. \tag{3.2}$$

In the interest of clarity, complex quantities are represented by bold letters. Note that the lower limit of 0^- in Eq. (3.1) means that $t = 0$ is included in the interval of integration.

Given that the exponent $\mathbf{s}t$ in the integral has to be dimensionless, **s** has the unit of inverse second (which is the same as Hz or rad/s). Moreover, since **s** is a complex quantity, it has the name ***complex frequency***.

In view of the definite limits on the integral in Eq. (3.1), the outcome of the integration will be an expression that depends on a single variable, namely **s**. The transform operation converts a function or signal $x(t)$ defined in the time domain into a function $\mathbf{X}(\mathbf{s})$ defined in the **s**-domain. Functions $x(t)$ and $\mathbf{X}(\mathbf{s})$ constitute a ***Laplace transform pair***.

Because the lower limit on the integral in Eq. (3.1) is 0^-, $\mathbf{X}(\mathbf{s})$ is called a ***unilateral transform*** (or ***one-sided transform***), in contrast with the ***bilateral transform*** or ***two-sided transform***, for which the lower limit is $-\infty$. When we apply the Laplace transform technique to physically realizable systems, we select the start time for the system operation as $t = 0^-$, making the unilateral transform perfectly suitable for handling causal systems with non-zero initial conditions. Should there be a need to examine conceptual scenarios in which the signal is ***everlasting*** (exists over $-\infty < t < \infty$), the bilateral transform should be used instead. The bilateral Laplace transform is discussed in Section 3-9.

► Because we are primarily interested in physically realizable systems excited by causal signals, we shall refer to the ***unilateral Laplace transform*** as simply the ***Laplace transform***. Moreover, unless noted to the contrary, it will be assumed that a signal $x(t)$ always is multiplied by an implicit invisible step function $u(t)$. ◄

3-1.1 Uniqueness Property

The uniqueness property of the Laplace transform states:

► A given $x(t)$ has a unique Laplace transform $\mathbf{X}(\mathbf{s})$, and vice versa. ◄

In symbolic form, the uniqueness property can be expressed as

$$x(t) \longleftrightarrow \mathbf{X}(\mathbf{s}) \tag{3.3a}$$

The **two-way arrow** is a shorthand notation for the combination of the two statements

$$\mathcal{L}[x(t)] = \mathbf{X}(\mathbf{s}) \quad \text{and} \quad \mathcal{L}^{-1}[\mathbf{X}(\mathbf{s})] = x(t). \tag{3.3b}$$

The first statement asserts that $\mathbf{X}(\mathbf{s})$ is the Laplace transform of $x(t)$, and the second statement asserts that $\mathcal{L}^{-1}[\mathbf{X}(\mathbf{s})]$, which is the ***inverse Laplace transform*** of $\mathbf{X}(\mathbf{s})$, is $x(t)$.

3-1.2 Convergence Condition

Depending on the functional form of $x(t)$, the Laplace transform integral given by Eq. (3.1) may or may not converge to a finite value. If it does not, the Laplace transform does not exist. It can be shown that convergence requires that

$$\int_{0-}^{\infty} |x(t)\, e^{-st}|\, dt = \int_{0-}^{\infty} |x(t)||e^{-\sigma t}||e^{-j\omega t}|\, dt$$
$$= \int_{0^-}^{\infty} |x(t)|e^{-\sigma t}\, dt < \infty \tag{3.4}$$

for some real value of σ. We used the fact that $|e^{-j\omega t}| = 1$ for any value of ωt, and since σ is real, $|e^{-\sigma t}| = e^{-\sigma t}$. If σ_c is the smallest value of σ for which the integral converges, then the ***region of convergence*** (*ROC*) is $\sigma > \sigma_c$. Fortunately, this convergence issue is somewhat esoteric to analysts and designers of real systems because the waveforms of the excitation sources usually do satisfy the convergence condition, and hence, their Laplace transforms do exist. *We will not consider ROC further in this book.*

3-1.3 Inverse Laplace Transform

Equation (3.1) allows us to obtain Laplace transform $\mathbf{X(s)}$ corresponding to time function $x(t)$. The inverse process, denoted $\mathcal{L}^{-1}[\mathbf{X(s)}]$, allows us to perform an integration on $\mathbf{X(s)}$ to obtain $x(t)$:

$$x(t) = \mathcal{L}^{-1}[\mathbf{X(s)}] = \frac{1}{2\pi j} \int_{\sigma-j\infty}^{\sigma+j\infty} \mathbf{X(s)}\, e^{\mathbf{s}t}\, d\mathbf{s}, \tag{3.5}$$

where $\sigma > \sigma_c$. The integration, which has to be performed in the two-dimensional complex plane, is rather cumbersome and to be avoided if an alternative approach is available for converting $\mathbf{X(s)}$ into $x(t)$.

Fortunately, there is an alternative approach. Recall from our earlier discussion in the Overview section that the Laplace transform technique entails several steps, with the final step involving a transformation of the solution realized in the **s**-domain to the time domain. Instead of applying Eq. (3.5), we can generate a table of Laplace transform pairs for all of the time functions commonly encountered in real systems and then use it as a look-up table to transform the **s**-domain solution to the time domain. The validity of this approach is supported by the uniqueness property of the Laplace transform, which guarantees a one-to-one correspondence between every $x(t)$ and its corresponding $\mathbf{X(s)}$. The details of the inverse-transform process are covered in Section 3-3. As a result, *we will not apply the complex integral given by Eq. (3.5) in this book.*

Example 3-1: Laplace Transforms of Singularity Functions

Determine the Laplace transforms of the signal waveforms displayed in Fig. 3-1.

Solution: (a) The step function in Fig. 3-1(a) is given by

$$x_1(t) = A\, u(t - T).$$

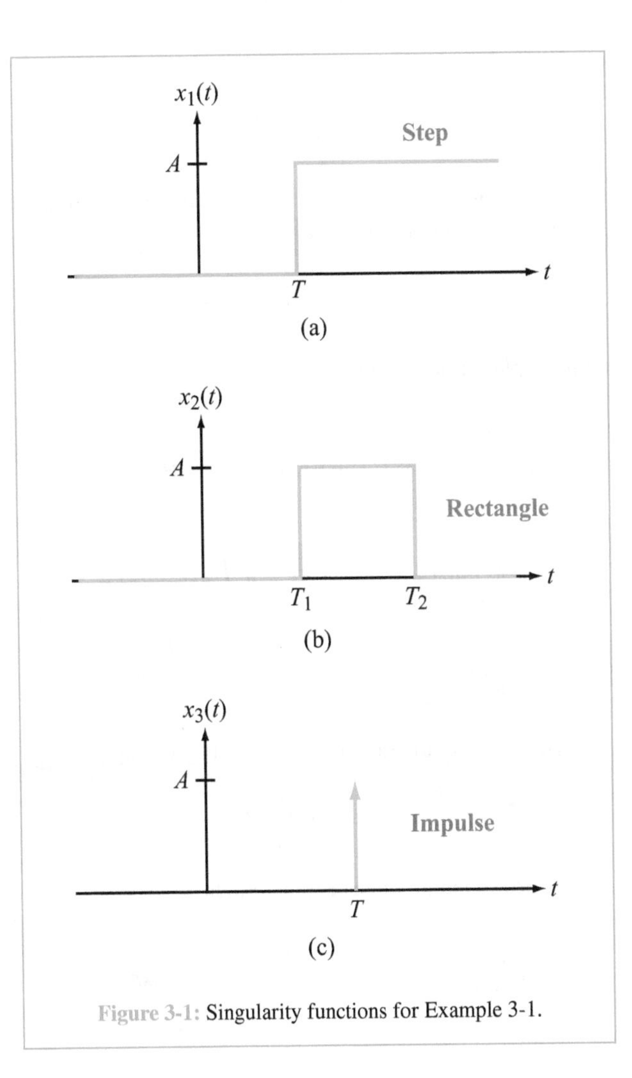

Figure 3-1: Singularity functions for Example 3-1.

Application of Eq. (3.1) gives

$$\begin{aligned}\mathbf{X}_1(\mathbf{s}) &= \int_{0^-}^{\infty} x_1(t)\, e^{-\mathbf{s}t}\, dt \\ &= \int_{0^-}^{\infty} A\, u(t-T)\, e^{-\mathbf{s}t}\, dt \\ &= A \int_{T}^{\infty} e^{-\mathbf{s}t}\, dt = -\frac{A}{\mathbf{s}}\, e^{-\mathbf{s}t}\Big|_{T}^{\infty} = \frac{A}{\mathbf{s}}\, e^{-\mathbf{s}T}.\end{aligned}$$

For the special case where $A = 1$ and $T = 0$ (i.e., the step occurs at $t = 0$), the transform pair becomes

$$u(t) \quad \longleftrightarrow \quad \frac{1}{\mathbf{s}}. \tag{3.6}$$

(b) The rectangle function in Fig. 3-1(b) can be constructed as the sum of two step functions:

$$x_2(t) = A[u(t-T_1) - u(t-T_2)],$$

and its Laplace transform is

$$\begin{aligned}\mathbf{X}_2(\mathbf{s}) &= \int_{0^-}^{\infty} A[u(t-T_1) - u(t-T_2)]e^{-\mathbf{s}t}\, dt \\ &= A \int_{0^-}^{\infty} u(t-T_1)\, e^{-\mathbf{s}t}\, dt - A \int_{0^-}^{\infty} u(t-T_2)\, e^{-\mathbf{s}t}\, dt \\ &= \frac{A}{\mathbf{s}}[e^{-\mathbf{s}T_1} - e^{-\mathbf{s}T_2}].\end{aligned}$$

(c) The impulse function in Fig. 3-1(c) is given by

$$x_3(t) = A\, \delta(t-T),$$

and the corresponding Laplace transform is

$$\mathbf{X}_3(\mathbf{s}) = \int_{0^-}^{\infty} A\, \delta(t-T)\, e^{-\mathbf{s}t}\, dt = Ae^{-\mathbf{s}T},$$

where we used the sampling property of the impulse function defined by Eq. (1.29). For the special case where $A = 1$ and $T = 0$, the Laplace transform pair simplifies to

$$\delta(t) \quad \longleftrightarrow \quad 1. \tag{3.7}$$

Example 3-2: Laplace Transform Pairs

Obtain the Laplace transforms of (a) $x_1(t) = e^{-at}\, u(t)$ and (b) $x_2(t) = [\cos(\omega_0 t)]\, u(t)$.

Solution:

(a) Application of Eq. (3.1) gives

$$\mathbf{X}_1(\mathbf{s}) = \int_{0^-}^{\infty} e^{-at}\, u(t)\, e^{-\mathbf{s}t}\, dt = \frac{e^{-(\mathbf{s}+a)t}}{-(\mathbf{s}+a)}\Bigg|_{0}^{\infty} = \frac{1}{\mathbf{s}+a}.$$

Hence,

$$e^{-at}\, u(t) \quad \longleftrightarrow \quad \frac{1}{\mathbf{s}+a}. \tag{3.8}$$

Note that setting $a = 0$ in Eq. (3.8) yields Eq. (3.6).

(b) We start by expressing $\cos(\omega_0 t)$ in the form

$$\cos(\omega_0 t) = \frac{1}{2}[e^{j\omega_0 t} + e^{-j\omega_0 t}].$$

Next, we take advantage of Eq. (3.8):

$$\begin{aligned}\mathbf{X}_2(\mathbf{s}) &= \mathcal{L}[\cos(\omega_0 t)\, u(t)] \\ &= \frac{1}{2}\, \mathcal{L}[e^{j\omega_0 t}\, u(t)] + \frac{1}{2}\, \mathcal{L}[e^{-j\omega_0 t}\, u(t)] \\ &= \frac{1}{2}\, \frac{1}{\mathbf{s}-j\omega_0} + \frac{1}{2}\, \frac{1}{\mathbf{s}+j\omega_0} = \frac{\mathbf{s}}{\mathbf{s}^2+\omega_0^2}.\end{aligned}$$

Hence,

$$[\cos(\omega_0 t)]\, u(t) \quad \longleftrightarrow \quad \frac{\mathbf{s}}{\mathbf{s}^2+\omega_0^2}. \tag{3.9}$$

Concept Question 3-1: Is the uniqueness property of the Laplace transform uni-directional or bi-directional? Why is that significant? (See S^2)

Concept Question 3-2: Is convergence of the Laplace transform integral an issue when applied to physically realizable systems? If not, why not? (See S^2)

Exercise 3-1: Determine the Laplace transform of (a) $[\sin(\omega_0 t)]\, u(t)$ and (b) $r(t-T)$ (see ramp function in Chapter 1).

Answer: (a) $[\sin(\omega_0 t)]\, u(t) \leftrightarrow \dfrac{\omega_0}{\mathbf{s}^2+\omega_0^2}$,

(b) $r(t-T) \leftrightarrow \dfrac{e^{-\mathbf{s}T}}{\mathbf{s}^2}$. (See S^2)

Exercise 3-2: Determine the Laplace transform of the causal sawtooth waveform shown in Fig. E3-2 (compare with Example 1-4).

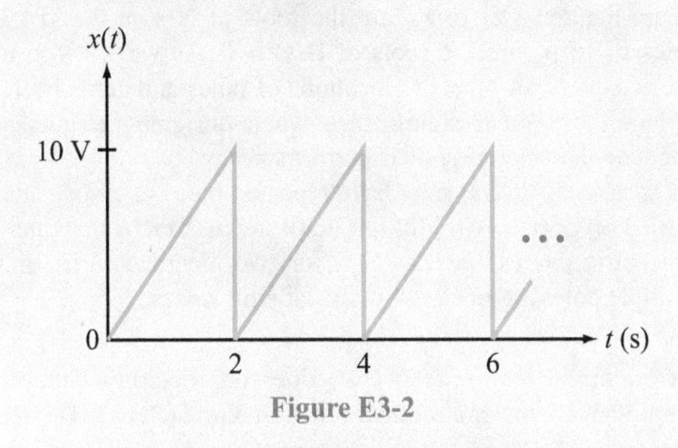

Figure E3-2

Answer:

$$\mathbf{X}(\mathbf{s}) = \mathbf{X}_1(\mathbf{s}) \sum_{n=0}^{\infty} e^{-2n\mathbf{s}} = \frac{\mathbf{X}_1(\mathbf{s})}{1-e^{-2\mathbf{s}}},$$

where

$$\mathbf{X}_1(\mathbf{s}) = \int_0^2 (5t)e^{-\mathbf{s}t}\, dt = \frac{5}{\mathbf{s}^2}\,[1-(2\mathbf{s}+1)e^{-2\mathbf{s}}].$$

Alternatively, we can compute $\mathbf{X}_1(\mathbf{s})$ by applying the answer to part (b) of Exercise 3-1 to the expression developed in Example 1-4 for the first cycle of the sawtooth:

$$x_1(t) = 5r(t) - 5r(t-2) - 10u(t-2)$$

and

$$\begin{aligned}\mathbf{X}_1(\mathbf{s}) &= \mathcal{L}[x_1(t)] \\ &= 5\mathcal{L}[r(t)] - 5\mathcal{L}[r(t-2)] - 10\mathcal{L}[u(t-2)] \\ &= \frac{5}{\mathbf{s}^2} - \frac{5}{\mathbf{s}^2}\, e^{-2\mathbf{s}} - \frac{10}{\mathbf{s}}\, e^{-2\mathbf{s}} \\ &= \frac{5}{\mathbf{s}^2}[1-(2\mathbf{s}+1)e^{-2\mathbf{s}}].\end{aligned}$$

In general, if $x(t)$ is a periodic function of period T, the Laplace transform of $x(t)\, u(t)$ is

$$\mathbf{X}(\mathbf{s}) = \frac{\mathbf{X}_1(\mathbf{s})}{1-e^{-T\mathbf{s}}},$$

where $\mathbf{X}_1(\mathbf{s})$ is the Laplace transform of the cycle starting at $t=0$. (See S^2)

3-2 Poles and Zeros

The Laplace-transform operation defined by Eq. (3.1) converts a one-dimensional function $x(t)$ into a two-dimensional function $\mathbf{X}(\mathbf{s})$, with $\mathbf{s} = \sigma + j\omega$. The $\mathbf{s}$-domain is also known as the ***s-plane***. The transform pairs of the four functions we examined thus far are

$$x_1(t) = u(t) \quad \leftrightarrow \quad \mathbf{X}_1(\mathbf{s}) = \frac{1}{\mathbf{s}}, \tag{3.10a}$$

$$x_2(t) = \delta(t) \quad \leftrightarrow \quad \mathbf{X}_2(\mathbf{s}) = 1, \tag{3.10b}$$

$$x_3(t) = e^{-at}\, u(t) \quad \leftrightarrow \quad \mathbf{X}_3(\mathbf{s}) = \frac{1}{\mathbf{s}+a}, \tag{3.10c}$$

and

$$x_4(t) = [\cos(\omega_0 t)]\, u(t) \quad \leftrightarrow \quad \mathbf{X}_4(\mathbf{s}) = \frac{\mathbf{s}}{\mathbf{s}^2+\omega_0^2}. \tag{3.10d}$$

In the expression for $x_4(t)$, we added a subscript to the angular frequency to distinguish it from the variable ω, representing the imaginary axis in the $\mathbf{s}$-plane.

In general, $\mathbf{X}(\mathbf{s})$ is a ***rational function*** of the form

$$\mathbf{X}(\mathbf{s}) = \frac{\mathbf{N}(\mathbf{s})}{\mathbf{D}(\mathbf{s})},$$

where $\mathbf{N(s)}$ is a numerator polynomial and $\mathbf{D(s)}$ is a denominator polynomial. For $\mathbf{X}_4(\mathbf{s})$, for example, $\mathbf{N(s)} = \mathbf{s}$ and $\mathbf{D(s)} = \mathbf{s}^2 + \omega_0^2$.

▶ The *zeros* of $\mathbf{X(s)}$ are the values of $\mathbf{s}$ that render $\mathbf{N(s)} = 0$, which are also called the *roots* of $\mathbf{N(s)}$. Similarly, the *poles* of $\mathbf{X(s)}$ are the roots of its denominator $\mathbf{D(s)}$. ◀

The Laplace transform of the unit step function, $\mathbf{X}_1(\mathbf{s})$, has no zeros and only one pole, namely, $\mathbf{s} = 0$. In Fig. 3-2(a), the origin in the $\mathbf{s}$-plane is marked with an "✕," indicating that $\mathbf{X}_1(\mathbf{s})$ has a pole at that location. To distinguish a zero from a pole, a zero is marked with a small circle "○."

Laplace transform $\mathbf{X}_2(\mathbf{s}) = 1$ has no poles and no zeros, and $\mathbf{X}_4(\mathbf{s})$, corresponding to $\cos(\omega_0 t)\, u(t)$, has a single zero and two poles (Fig. 3-2(b)). Given that its numerator is simply $\mathbf{N}_4(\mathbf{s}) = \mathbf{s}$, $\mathbf{X}_4(\mathbf{s})$ has its single zero at the origin $\mathbf{s} = 0$, corresponding to coordinates $(\sigma, \omega) = (0, 0)$ in the $\mathbf{s}$-plane. To determine the locations of its two poles, we need to compute the roots of

$$\mathbf{D}_4(\mathbf{s}) = \mathbf{s}^2 + \omega_0^2 = 0,$$

which leads to

$$\mathbf{s}_1 = -j\omega_0 \quad \text{and} \quad \mathbf{s}_2 = j\omega_0.$$

Hence, Eq. (3.10d) can be recast as

$$\mathbf{X}_4(\mathbf{s}) = \frac{\mathbf{s}}{(\mathbf{s} - j\omega_0)(\mathbf{s} + j\omega_0)}.$$

Because the real parts of $\mathbf{s}_1$ and $\mathbf{s}_2$ are zero, the poles of $\mathbf{X}_4(\mathbf{s})$ are located in the $\mathbf{s}$-plane at $(0, j\omega_0)$ and $(0, -j\omega_0)$.

In general, identifying the locations of the poles and zeros of a function $\mathbf{X(s)}$ can be greatly facilitated if $\mathbf{X(s)}$ can be cast in the form

$$\mathbf{X(s)} = \frac{\mathbf{N(s)}}{\mathbf{D(s)}} = \frac{A(\mathbf{s} - \mathbf{z}_1)(\mathbf{s} - \mathbf{z}_2)\dots(\mathbf{s} - \mathbf{z}_m)}{(\mathbf{s} - \mathbf{p}_1)(\mathbf{s} - \mathbf{p}_2)\dots(\mathbf{s} - \mathbf{p}_n)},$$

where the zeros $\mathbf{z}_1$ to $\mathbf{z}_m$ are the roots of $\mathbf{N(s)} = 0$ and the poles $\mathbf{p}_1$ to $\mathbf{p}_n$ are the roots of $\mathbf{D(s)} = 0$. As we will see in later chapters, the specific locations of poles and zeros in the $\mathbf{s}$-plane carry great significance when designing frequency filters or characterizing their performance.

Occasionally, $\mathbf{X(s)}$ may have repeated poles or zeros, such as $\mathbf{z}_1 = \mathbf{z}_2$ or $\mathbf{p}_1 = \mathbf{p}_2$. Multiple zeros are marked by that many concentric circles, such as "◎" for two identical zeros, and multiple poles are marked by overlapping Xs "⨯."

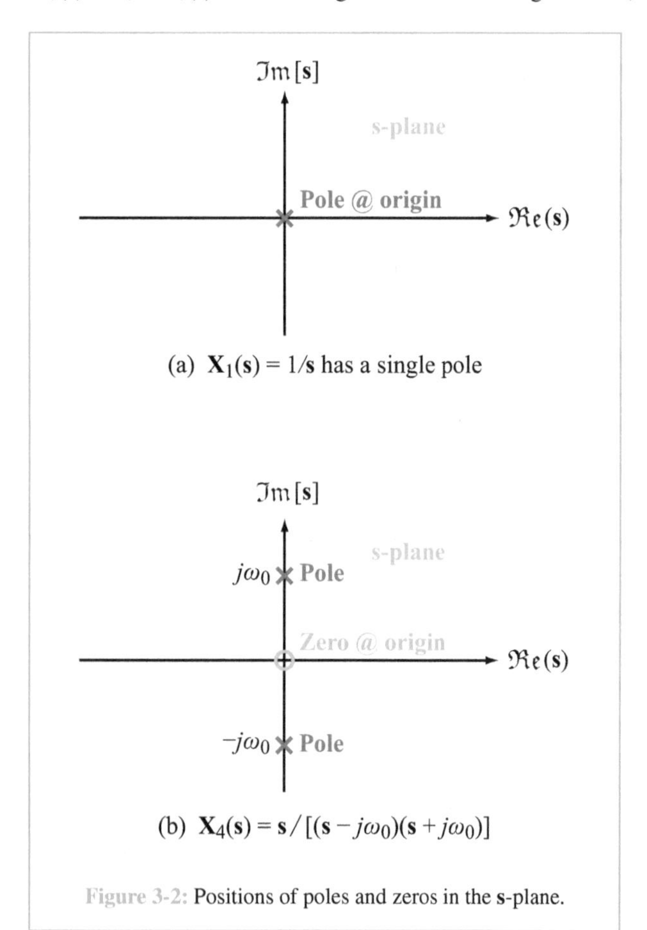

Figure 3-2: Positions of poles and zeros in the $\mathbf{s}$-plane.

Concept Question 3-3: How does one determine the poles and zeros of a rational function $\mathbf{X(s)}$? (See S²)

Exercise 3-3: Determine the poles and zeros of $\mathbf{X(s)} = (\mathbf{s} + a)/[(\mathbf{s} + a)^2 + \omega_0^2]$.

Answer: $\mathbf{z} = (-a + j0)$, $\mathbf{p}_1 = (-a - j\omega_0)$, and $\mathbf{p}_2 = (-a + j\omega_0)$. (See S²)

3-3 Properties of the Laplace Transform

The Laplace transform has a number of useful *universal properties* that apply to any function $x(t)$—greatly facilitating the process of transforming a system from the t-domain to the $\mathbf{s}$-domain. To demonstrate what we mean by a universal property, we consider the ***linearity property*** of the Laplace transform, which states that if

$$x_1(t) \longleftrightarrow \mathbf{X}_1(\mathbf{s})$$

and

$$x_2(t) \longleftrightarrow \mathbf{X}_2(\mathbf{s}),$$

then for any linear combination of $x_1(t)$ and $x_2(t)$,

$$K_1\, x_1(t) + K_2\, x_2(t) \longleftrightarrow K_1\, \mathbf{X}_1(\mathbf{s}) + K_2\, \mathbf{X}_2(\mathbf{s}), \quad (3.11)$$

(linearity property)

where K_1 and K_2 are constant coefficients. Thus, the Laplace transform obeys the linearity property and the principle of superposition holds.

Another example of a universal property of the Laplace transform is the ***time-scaling property***, which states that if $x(t)$ and $\mathbf{X}(\mathbf{s})$ constitute a Laplace transform pair,

$$x(t) \longleftrightarrow \mathbf{X}(\mathbf{s}),$$

then the transform of the time-scaled signal $x(at)$ is

$$x(at) \longleftrightarrow \frac{1}{a}\,\mathbf{X}\left(\frac{\mathbf{s}}{a}\right), \qquad a > 0. \quad (3.12)$$

(time-scaling property)

► A ***universal property*** of the Laplace transform applies to any $x(t)$, regardless of its specific functional form. ◄

This section concludes with Table 3-1, featuring 13 universal properties of the Laplace transform and to which we will be making frequent reference throughout this chapter. Some of these properties are intuitively obvious, while others may require some elaboration. Proof of the time scaling property follows next.

3-3.1 Time Scaling

The time scaling property given by Eq. (3.12) states that shrinking the time axis by a factor a corresponds to stretching the $\mathbf{s}$-domain and the amplitude of $\mathbf{X}(\mathbf{s})$ by the same factor, and vice-versa. Note that $a > 0$.

To prove Eq. (3.12), we start with the standard definition of the Laplace transform given by Eq. (3.1):

$$\mathcal{L}[x(at)] = \int_{0^-}^{\infty} x(at)\, e^{-\mathbf{s}t}\, dt. \quad (3.13)$$

In the integral, if we set $t' = at$ and $dt = \frac{1}{a}\, dt'$, we have

$$\mathcal{L}[x(at)] = \frac{1}{a}\int_{0^-}^{\infty} x(t')\, e^{-(\mathbf{s}/a)t'}\, dt'$$

$$= \frac{1}{a}\int_{0^-}^{\infty} x(t')\, e^{-\mathbf{s}'t'}\, dt' \qquad \text{with } \mathbf{s}' = \frac{\mathbf{s}}{a}. \quad (3.14)$$

The definite integral is identical in form with the Laplace transform definition given by Eq. (3.1), except that the dummy variable is t', instead of t, and the coefficient of the exponent is $\mathbf{s}' = \mathbf{s}/a$, instead of just $\mathbf{s}$. Hence,

$$\mathcal{L}[x(at)] = \frac{1}{a}\,\mathbf{X}(\mathbf{s}') = \frac{1}{a}\,\mathbf{X}\left(\frac{\mathbf{s}}{a}\right), \qquad a > 0. \quad (3.15)$$

3-3.2 Time Shift

If t is shifted by T along the time axis with $T \geq 0$, then

$$x(t-T)\, u(t-T) \longleftrightarrow e^{-T\mathbf{s}}\, \mathbf{X}(\mathbf{s}), \quad T \geq 0. \quad (3.16)$$

(time-shift property)

The validity of this property is demonstrated as follows:

$$\mathcal{L}[x(t-T)\, u(t-T)] = \int_{0^-}^{\infty} x(t-T)\, u(t-T)\, e^{-\mathbf{s}t}\, dt$$

$$= \int_{T}^{\infty} x(t-T)\, e^{-\mathbf{s}t}\, dt$$

$$= \int_{0^-}^{\infty} x(t')\, e^{-\mathbf{s}(t'+T)}\, dt'$$

$$= e^{-T\mathbf{s}} \int_{0^-}^{\infty} x(t')\, e^{-\mathbf{s}t'}\, dt'$$

$$= e^{-T\mathbf{s}}\, \mathbf{X}(\mathbf{s}), \quad (3.17)$$

where we made the substitutions $t - T = t'$ and $dt = dt'$ and then applied the definition for $\mathbf{X}(\mathbf{s})$ given by Eq. (3.1).

To illustrate the utility of the time-shift property, we consider the cosine function of Example 3-2, where it was shown that

$$\cos(\omega_0 t)\, u(t) \longleftrightarrow \frac{\mathbf{s}}{\mathbf{s}^2 + \omega_0^2}. \quad (3.18)$$

Application of Eq. (3.16) yields

$$\cos[\omega_0(t-T)]\, u(t-T) \quad \longleftrightarrow \quad e^{-T\mathbf{s}} \frac{\mathbf{s}}{\mathbf{s}^2+\omega_0^2}. \tag{3.19}$$

Had we analyzed a linear circuit (or system) driven by a sinusoidal source that started at $t=0$ and then wanted to reanalyze it anew, but we wanted to delay both the cosine function and the start time by T, Eq. (3.19) provides an expedient approach to obtaining the transform of the delayed cosine function.

Exercise 3-4: Determine, for $T \geq 0$,

$$\mathcal{L}\{[\sin\omega_0(t-T)]\, u(t-T)\}.$$

Answer: $e^{-T\mathbf{s}} \dfrac{\omega_0}{\mathbf{s}^2+\omega_0^2}$. (See S²)

3-3.3 Frequency Shift

According to the time-shift property, if t is replaced with $(t-T)$ in the time domain, $\mathbf{X}(\mathbf{s})$ gets multiplied by $e^{-T\mathbf{s}}$ in the s-domain. Within a $(-)$ sign, the converse is also true: if $\mathbf{s}$ is replaced with $(\mathbf{s}+\mathbf{a})$ in the s-domain, $x(t)$ gets multiplied by $e^{-\mathbf{a}t}$ in the time domain. Thus,

$$e^{-\mathbf{a}t}\, x(t) \quad \longleftrightarrow \quad \mathbf{X}(\mathbf{s}+\mathbf{a}). \tag{3.20}$$

(frequency shift property)

Proof of Eq. (3.20) is part of Exercise 3-5.

Concept Question 3-4: According to the time scaling property of the Laplace transform, "shrinking the time axis corresponds to stretching the s-domain." What does that mean? (See S²)

Concept Question 3-5: Explain the similarities and differences between the time-shift and frequency-shift properties of the Laplace transform. (See S²)

Exercise 3-5: (a) Prove Eq. (3.20) and (b) apply it to determine $\mathcal{L}[e^{-at}\cos(\omega_0 t)\, u(t)]$.

Answer: (a) (See S²), and

(b) $e^{-at}\cos(\omega_0 t)\, u(t) \longleftrightarrow \dfrac{\mathbf{s}+a}{(\mathbf{s}+a)^2+\omega_0^2}$. (See S²)

3-3.4 Time Differentiation

Differentiating $x(t)$ in the time domain is equivalent to (a) multiplying $\mathbf{X}(\mathbf{s})$ by $\mathbf{s}$ in the s-domain and then (b) subtracting $x(0^-)$ from $\mathbf{s}\,\mathbf{X}(\mathbf{s})$:

$$x' = \frac{dx}{dt} \quad \longleftrightarrow \quad \mathbf{s}\,\mathbf{X}(\mathbf{s}) - x(0^-). \tag{3.21}$$

(time-differentiation property)

To verify Eq. (3.21), we start with the standard definition for the Laplace transform:

$$\mathcal{L}[x'] = \int_{0^-}^{\infty} \frac{dx}{dt}\, e^{-\mathbf{s}t}\, dt. \tag{3.22}$$

Integration by parts with

$$u = e^{-\mathbf{s}t}, \qquad du = -\mathbf{s}e^{-\mathbf{s}t}\, dt,$$

$$dv = \left(\frac{dx}{dt}\right) dt, \qquad \text{and} \qquad v = x$$

gives

$$\begin{aligned}
\mathcal{L}[x'] &= uv\Big|_{0^-}^{\infty} - \int_{0^-}^{\infty} v\, du \\
&= e^{-\mathbf{s}t}\, x(t)\Big|_{0^-}^{\infty} - \int_{0^-}^{\infty} -\mathbf{s}\, x(t)\, e^{-\mathbf{s}t}\, dt \\
&= -x(0^-) + \mathbf{s}\,\mathbf{X}(\mathbf{s}),
\end{aligned} \tag{3.23}$$

which is equivalent to Eq. (3.21).

Higher derivatives can be obtained by repeating the application of Eq. (3.21). For the second derivative of $x(t)$, we have

$$x'' = \frac{d^2x}{dt^2} \longleftrightarrow \mathbf{s}^2\,\mathbf{X}(\mathbf{s}) - \mathbf{s}\, x(0^-) - x'(0^-), \tag{3.24}$$

(second-derivative property)

where $x'(0^-)$ is the derivative of $x(t)$ evaluated at $t = 0^-$.

Example 3-3: Second Derivative Property

Verify the second-derivative property for

$$x(t) = \sin(\omega_0 t)\, u(t)$$

by (a) applying the transform equation to $x''(t)$ and (b) comparing it with the result obtained via Eq. (3.24).

Solution: (a) The first derivative of $x(t)$ is

$$\begin{aligned} x'(t) &= \frac{d}{dt}[\sin(\omega_0 t)\, u(t)] \\ &= \omega_0 \cos(\omega_0 t)\, u(t) + \sin(\omega_0 t)\, \frac{d}{dt}\,[u(t)] \\ &= \omega_0 \cos(\omega_0 t)\, u(t) + \sin(\omega_0 t)\, \delta(t) \\ &= \omega_0 \cos(\omega_0 t)\, u(t) + 0. \end{aligned}$$

The second term is zero because

$$\sin(\omega_0 t)\, \delta(t) = \sin(0)\, \delta(t) = 0.$$

The second derivative of $x(t)$ is

$$\begin{aligned} x''(t) = \frac{d}{dt}[x'(t)] &= \frac{d}{dt}[\omega_0 \cos(\omega_0 t)\, u(t)] \\ &= -\omega_0^2 \sin(\omega_0 t)\, u(t) + \omega_0 \cos(\omega_0 t)\, \delta(t) \\ &= -\omega_0^2 \sin(\omega_0 t)\, u(t) + \omega_0\, \delta(t). \end{aligned}$$

From Exercise 3-1,

$$\mathcal{L}[\sin(\omega_0 t)\, u(t)] = \frac{\omega_0}{\mathbf{s}^2 + \omega_0^2},$$

and from Eq. (3.7),

$$\mathcal{L}[\delta(t)] = 1.$$

Hence, the Laplace transform of $x''(t)$ is

$$\mathcal{L}[x''] = -\omega_0^2 \left(\frac{\omega_0}{\mathbf{s}^2 + \omega_0^2} \right) + \omega_0 = \frac{\omega_0 \mathbf{s}^2}{\mathbf{s}^2 + \omega_0^2}.$$

(b) Application of Eq. (3.24) gives

$$\mathcal{L}[x''] = \mathbf{s}^2\, \mathbf{X}(\mathbf{s}) - \mathbf{s}x(0^-) - x'(0^-).$$

For $x(t) = \sin(\omega_0 t)\, u(t)$, we have

$$\mathbf{X}(\mathbf{s}) = \frac{\omega_0}{\mathbf{s}^2 + \omega_0^2},$$

$$x(0^-) = \sin(0^-)\, u(0^-) = 0,$$

and

$$x'(0^-) = \omega_0 \cos(\omega_0 t)\, u(t)|_{t=0^-} = \omega_0\, u(0^-) = 0,$$

because $u(0^-) = 0$.

Hence,

$$\mathcal{L}[x''] = \mathbf{s}^2 \left(\frac{\omega_0}{\mathbf{s}^2 + \omega_0^2} \right) = \frac{\omega_0 \mathbf{s}^2}{\mathbf{s}^2 + \omega_0^2},$$

which agrees with the result of part (a).

3-3.5 Time Integration

Integration of $x(t)$ in the time domain is equivalent to dividing $\mathbf{X}(\mathbf{s})$ by $\mathbf{s}$ in the $\mathbf{s}$-domain:

$$\int_0^t x(t')\, dt' \quad \longleftrightarrow \quad \frac{1}{\mathbf{s}}\, \mathbf{X}(\mathbf{s}). \tag{3.25}$$

(time-integration property)

Application of the Laplace transform definition gives

$$\mathcal{L}\left[\int_{0^-}^{t} x(t')\, dt' \right] = \int_{0^-}^{\infty} \left[\int_{0^-}^{t} x(t')\, dt' \right] e^{-\mathbf{s}t}\, dt. \tag{3.26}$$

Integration by parts with

$$u = \int_{0^-}^{t} x(t')\, dt', \qquad du = x(t)\, dt,$$

$$dv = e^{-\mathbf{s}t}\, dt, \qquad \text{and} \qquad v = -\frac{e^{-\mathbf{s}t}}{\mathbf{s}}$$

leads to

$$\mathcal{L}\left[\int_{0^-}^{t} x(t')\,dt'\right] = uv\Big|_{0^-}^{\infty} - \int_{0^-}^{\infty} v\,du$$

$$= \left[-\frac{e^{-st}}{s}\int_{0^-}^{t} x(t')\,dt'\right]\Bigg|_{0^-}^{\infty} + \frac{1}{s}\int_{0^-}^{\infty} x(t)\,e^{-st}\,dt = \frac{1}{s}\,\mathbf{X}(s). \tag{3.27}$$

Both limits on the first term on the right-hand side yield zero values.

For example, since

$$\delta(t) \leftrightarrow 1,$$

it follows that

$$u(t) = \int_{0^-}^{t} \delta(t')\,dt' \leftrightarrow \frac{1}{s}$$

and

$$r(t) = \int_{0^-}^{t} u(t')\,dt' \leftrightarrow \frac{1}{s^2}.$$

3-3.6 Initial- and Final-Value Theorems

The relationship between $x(t)$ and $\mathbf{X}(s)$ is such that the initial value $x(0^+)$ and the final value $x(\infty)$ of $x(t)$ can be determined directly from the expression of $\mathbf{X}(s)$—provided certain conditions are satisfied (as discussed later in this subsection).

Consider the derivative property represented by Eq. (3.23) as

$$\mathcal{L}[x'] = \int_{0^-}^{\infty} \frac{dx}{dt}\,e^{-st}\,dt = s\,\mathbf{X}(s) - x(0^-). \tag{3.28}$$

If we take the limit as $s \to \infty$ while recognizing that $x(0^-)$ is independent of s, we get

$$\lim_{s\to\infty}\left[\int_{0^-}^{\infty} \frac{dx}{dt}\,e^{-st}\,dt\right] = \lim_{s\to\infty}[s\,\mathbf{X}(s)] - x(0^-). \tag{3.29}$$

The integral on the left-hand side can be split into two integrals: one over the time segment $(0^-, 0^+)$, for which $e^{-st} = 1$, and another over the segment $(0^+, \infty)$. Thus,

$$\lim_{s\to\infty}\left[\int_{0^-}^{\infty} \frac{dx}{dt}\,e^{-st}\,dt\right] = \lim_{s\to\infty}\left[\int_{0^-}^{0^+} \frac{dx}{dt}\,dt + \int_{0^+}^{\infty} \frac{dx}{dt}\,e^{-st}\,dt\right] = x(0^+) - x(0^-). \tag{3.30}$$

As $s \to \infty$, the exponential function e^{-st} causes the integrand of the last term to vanish. Equating Eqs. (3.29) and (3.30) leads to

$$x(0^+) = \lim_{s\to\infty} s\,\mathbf{X}(s), \tag{3.31}$$

(initial-value theorem)

which is known as the ***initial-value theorem***.

A similar treatment in which s is made to approach 0 (instead of ∞) in Eq. (3.29) leads to the ***final-value theorem***:

$$x(\infty) = \lim_{s\to 0} s\,\mathbf{X}(s). \tag{3.32}$$

(final-value theorem)

We should note that Eq. (3.32) is useful for determining $x(\infty)$, *so long as $x(\infty)$ exists*. Otherwise, application of Eq. (3.32) may lead to an erroneous result. Consider, for example, $x(t) = \cos(\omega_0 t)\,u(t)$, which does not have a unique value as $t \to \infty$. Yet, application of Eq. (3.32) to Eq. (3.9) leads to $x(\infty) = 0$, which is incorrect.

Example 3-4: Initial and Final Values

Determine the initial and final values of a function $x(t)$ whose Laplace transform is given by

$$\mathbf{X}(s) = \frac{25s(s+3)}{(s+1)(s^2+2s+36)}.$$

Solution: Application of Eq. (3.31) gives

$$x(0^+) = \lim_{s\to\infty} s\,\mathbf{X}(s) = \lim_{s\to\infty} \frac{25s^2(s+3)}{(s+1)(s^2+2s+36)}.$$

To avoid the problem of dealing with ∞, it is often more convenient to first apply the substitution $\mathbf{s} = 1/\mathbf{u}$, rearrange the function, and then find the limit as $\mathbf{u} \to 0$. That is,

$$\begin{aligned} x(0^+) &= \lim_{\mathbf{u}\to 0} \frac{25(1/\mathbf{u}^2)(1/\mathbf{u}+3)}{(1/\mathbf{u}+1)(1/\mathbf{u}^2+2/\mathbf{u}+36)} \\ &= \lim_{\mathbf{u}\to 0} \frac{25(1+3\mathbf{u})}{(1+\mathbf{u})(1+2\mathbf{u}+36\mathbf{u}^2)} \\ &= \frac{25(1+0)}{(1+0)(1+0+0)} = 25. \end{aligned}$$

To determine $x(\infty)$, we apply Eq. (3.32):

$$x(\infty) = \lim_{\mathbf{s}\to 0} \mathbf{s}\,\mathbf{X}(\mathbf{s}) = \lim_{\mathbf{s}\to 0} \frac{25\mathbf{s}^2(\mathbf{s}+3)}{(\mathbf{s}+1)(\mathbf{s}^2+2\mathbf{s}+36)} = 0.$$

Exercise 3-6: Determine the initial and final values of $x(t)$ if its Laplace transform is given by

$$\mathbf{X}(\mathbf{s}) = \frac{\mathbf{s}^2+6\mathbf{s}+18}{\mathbf{s}(\mathbf{s}+3)^2}.$$

Answer: $x(0^+) = 1$, $x(\infty) = 2$. (See S²)

3-3.7 Frequency Differentiation

Given the definition of the Laplace transform, namely,

$$\mathbf{X}(\mathbf{s}) = \mathcal{L}[x(t)] = \int_{0^-}^{\infty} x(t)\, e^{-\mathbf{s}t}\, dt, \tag{3.33}$$

if we take the derivative with respect to $\mathbf{s}$ on both sides, we have

$$\begin{aligned} \frac{d\,\mathbf{X}(\mathbf{s})}{d\mathbf{s}} &= \int_{0^-}^{\infty} \frac{d}{d\mathbf{s}}[x(t)\, e^{-\mathbf{s}t}]\, dt \\ &= \int_{0^-}^{\infty} [-t\, x(t)]e^{-\mathbf{s}t}\, dt = \mathcal{L}[-t\, x(t)], \end{aligned} \tag{3.34}$$

where we recognize the integral as the Laplace transform of the function $[-t\, x(t)]$. Rearranging Eq. (3.34) provides the *frequency differentiation relation*:

$$t\, x(t) \quad \longleftrightarrow \quad -\frac{d\,\mathbf{X}(\mathbf{s})}{d\mathbf{s}} = -\mathbf{X}'(\mathbf{s}), \tag{3.35}$$

(frequency differentiation property)

which states that multiplication of $x(t)$ by $-t$ in the time domain is equivalent to differentiating $\mathbf{X}(\mathbf{s})$ in the $\mathbf{s}$-domain.

Example 3-5: Applying the Frequency Differentiation Property

Given that

$$\mathbf{X}(\mathbf{s}) = \mathcal{L}[e^{-at}\, u(t)] = \frac{1}{\mathbf{s}+a},$$

apply Eq. (3.35) to obtain the Laplace transform of $te^{-at}\, u(t)$.

Solution:

$$\mathcal{L}[te^{-at}\, u(t)] = -\frac{d}{d\mathbf{s}}\,\mathbf{X}(\mathbf{s}) = -\frac{d}{d\mathbf{s}}\left[\frac{1}{\mathbf{s}+a}\right] = \frac{1}{(\mathbf{s}+a)^2}.$$

3-3.8 Frequency Integration

Integrating both sides of Eq. (3.33) from $\mathbf{s}$ to ∞ gives

$$\int_{\mathbf{s}}^{\infty} \mathbf{X}(\mathbf{s}')\, d\mathbf{s}' = \int_{\mathbf{s}}^{\infty}\left[\int_{0^-}^{\infty} x(t)\, e^{-\mathbf{s}'t}\, dt\right] d\mathbf{s}'. \tag{3.36}$$

Since t and $\mathbf{s}'$ are independent variables, we can interchange the order of the integration on the right-hand side of Eq. (3.36),

$$\begin{aligned} \int_{\mathbf{s}}^{\infty} \mathbf{X}(\mathbf{s}')\, d\mathbf{s}' &= \int_{0^-}^{\infty}\left[\int_{\mathbf{s}}^{\infty} x(t)\, e^{-\mathbf{s}'t}\, d\mathbf{s}'\right] dt \\ &= \int_{0^-}^{\infty}\left[\frac{x(t)}{-t}\, e^{-\mathbf{s}'t}\Big|_{\mathbf{s}}^{\infty}\right] dt \\ &= \int_{0^-}^{\infty}\left[\frac{x(t)}{t}\right] e^{-\mathbf{s}t}\, dt = \mathcal{L}\left[\frac{x(t)}{t}\right]. \end{aligned} \tag{3.37}$$

This *frequency integration property* can be expressed as

$$\frac{x(t)}{t} \quad \longleftrightarrow \quad \int_{\mathbf{s}}^{\infty} \mathbf{X}(\mathbf{s}')\, d\mathbf{s}'. \tag{3.38}$$

(frequency integration property)

Table 3-1 provides a summary of the principal properties of the Laplace transform. Entry #11 in Table 3-1 is proved in Section 3-6 below. For easy reference, Table 3-2 contains a list of Laplace transform pairs that we are likely to encounter in future sections.

Table 3-1: **Properties of the Laplace transform for causal functions; i.e., $x(t) = 0$ for $t < 0$.**

Property	$x(t)$		$\mathbf{X}(\mathbf{s}) = \mathcal{L}[x(t)]$
1. Multiplication by constant	$K\,x(t)$	$\longleftrightarrow$	$K\,\mathbf{X}(\mathbf{s})$
2. Linearity	$K_1\,x_1(t) + K_2\,x_2(t)$	$\longleftrightarrow$	$K_1\,\mathbf{X}_1(\mathbf{s}) + K_2\,\mathbf{X}_2(\mathbf{s})$
3. Time scaling	$x(at), \quad a > 0$	$\longleftrightarrow$	$\frac{1}{a}\,\mathbf{X}\left(\frac{\mathbf{s}}{a}\right)$
4. Time shift	$x(t-T)\,u(t-T)$	$\longleftrightarrow$	$e^{-T\mathbf{s}}\,\mathbf{X}(\mathbf{s})$
5. Frequency shift	$e^{-\mathbf{a}t}\,x(t)$	$\longleftrightarrow$	$\mathbf{X}(\mathbf{s}+\mathbf{a})$
6. Time 1st derivative	$x' = \frac{dx}{dt}$	$\longleftrightarrow$	$\mathbf{s}\,\mathbf{X}(\mathbf{s}) - x(0^-)$
7. Time 2nd derivative	$x'' = \frac{d^2x}{dt^2}$	$\longleftrightarrow$	$\mathbf{s}^2\mathbf{X}(\mathbf{s}) - \mathbf{s}x(0^-) - x'(0^-)$
8. Time integral	$\int_{0^-}^{t} x(t')\,dt'$	$\longleftrightarrow$	$\frac{1}{\mathbf{s}}\,\mathbf{X}(\mathbf{s})$
9. Frequency derivative	$t\,x(t)$	$\longleftrightarrow$	$-\frac{d}{d\mathbf{s}}\,\mathbf{X}(\mathbf{s}) = -\mathbf{X}'(\mathbf{s})$
10. Frequency integral	$\frac{x(t)}{t}$	$\longleftrightarrow$	$\int_{\mathbf{s}}^{\infty} \mathbf{X}(\mathbf{s}')\,d\mathbf{s}'$
11. Initial value	$x(0^+)$	$=$	$\lim_{\mathbf{s}\to\infty} \mathbf{s}\,\mathbf{X}(\mathbf{s})$
12. Final value	$\lim_{t\to\infty} x(t) = x(\infty)$	$=$	$\lim_{\mathbf{s}\to 0} \mathbf{s}\,\mathbf{X}(\mathbf{s})$
13. Convolution	$x_1(t) * x_2(t)$	$\longleftrightarrow$	$\mathbf{X}_1(\mathbf{s})\,\mathbf{X}_2(\mathbf{s})$

Example 3-6: Laplace Transform

Obtain the Laplace transform of

$$x(t) = t^2 e^{-3t}\cos(4t)\,u(t).$$

Solution: The given function is a product of three functions. We start with the cosine function which we will call $x_1(t)$:

$$x_1(t) = \cos(4t)\,u(t). \tag{3.39}$$

According to entry #11 in Table 3-2, the corresponding Laplace transform is

$$\mathbf{X}_1(\mathbf{s}) = \frac{\mathbf{s}}{\mathbf{s}^2 + 16}. \tag{3.40}$$

Next we define

$$x_2(t) = e^{-3t}\cos(4t)\,u(t) = e^{-3t}\,x_1(t), \tag{3.41}$$

and we apply the frequency shift property (entry #5 in Table 3-1) to obtain

$$\mathbf{X}_2(\mathbf{s}) = \mathbf{X}_1(\mathbf{s}+3) = \frac{\mathbf{s}+3}{(\mathbf{s}+3)^2 + 16}, \tag{3.42}$$

where we replaced $\mathbf{s}$ with $(\mathbf{s}+3)$ everywhere in the expression of Eq. (3.40). Finally, we define

$$x(t) = t^2\,x_2(t) = t^2 e^{-3t}\cos(4t)\,u(t), \tag{3.43}$$

Table 3-2: **Examples of Laplace transform pairs. Note that $x(t) = 0$ for $t < 0^-$ and $T \geq 0$.**

Laplace Transform Pairs

	$x(t)$		$\mathbf{X}(\mathbf{s}) = \mathcal{L}[x(t)]$
1	$\delta(t)$	⟷	1
1a	$\delta(t-T)$	⟷	$e^{-T\mathbf{s}}$
2	$u(t)$	⟷	$\dfrac{1}{\mathbf{s}}$
2a	$u(t-T)$	⟷	$\dfrac{e^{-T\mathbf{s}}}{\mathbf{s}}$
3	$e^{-at}\,u(t)$	⟷	$\dfrac{1}{\mathbf{s}+a}$
3a	$e^{-a(t-T)}\,u(t-T)$	⟷	$\dfrac{e^{-T\mathbf{s}}}{\mathbf{s}+a}$
4	$t\,u(t)$	⟷	$\dfrac{1}{\mathbf{s}^2}$
4a	$(t-T)\,u(t-T)$	⟷	$\dfrac{e^{-T\mathbf{s}}}{\mathbf{s}^2}$
5	$t^2\,u(t)$	⟷	$\dfrac{2}{\mathbf{s}^3}$
6	$te^{-at}\,u(t)$	⟷	$\dfrac{1}{(\mathbf{s}+a)^2}$
7	$t^2e^{-at}\,u(t)$	⟷	$\dfrac{2}{(\mathbf{s}+a)^3}$
8	$t^{n-1}e^{-at}\,u(t)$	⟷	$\dfrac{(n-1)!}{(\mathbf{s}+a)^n}$
9	$\sin(\omega_0 t)\,u(t)$	⟷	$\dfrac{\omega_0}{\mathbf{s}^2+\omega_0^2}$
10	$\sin(\omega_0 t+\theta)\,u(t)$	⟷	$\dfrac{\mathbf{s}\sin\theta+\omega_0\cos\theta}{\mathbf{s}^2+\omega_0^2}$
11	$\cos(\omega_0 t)\,u(t)$	⟷	$\dfrac{\mathbf{s}}{\mathbf{s}^2+\omega_0^2}$
12	$\cos(\omega_0 t+\theta)\,u(t)$	⟷	$\dfrac{\mathbf{s}\cos\theta-\omega_0\sin\theta}{\mathbf{s}^2+\omega_0^2}$
13	$e^{-at}\sin(\omega_0 t)\,u(t)$	⟷	$\dfrac{\omega_0}{(\mathbf{s}+a)^2+\omega_0^2}$
14	$e^{-at}\cos(\omega_0 t)\,u(t)$	⟷	$\dfrac{\mathbf{s}+a}{(\mathbf{s}+a)^2+\omega_0^2}$
15	$2e^{-at}\cos(bt-\theta)\,u(t)$	⟷	$\dfrac{e^{j\theta}}{\mathbf{s}+a+jb}+\dfrac{e^{-j\theta}}{\mathbf{s}+a-jb}$
15a	$e^{-at}\cos(bt-\theta)\,u(t)$	⟷	$\dfrac{(\mathbf{s}+a)\cos\theta+b\sin\theta}{(\mathbf{s}+a)^2+b^2}$
16	$\dfrac{2t^{n-1}}{(n-1)!}\,e^{-at}\cos(bt-\theta)\,u(t)$	⟷	$\dfrac{e^{j\theta}}{(\mathbf{s}+a+jb)^n}+\dfrac{e^{-j\theta}}{(\mathbf{s}+a-jb)^n}$

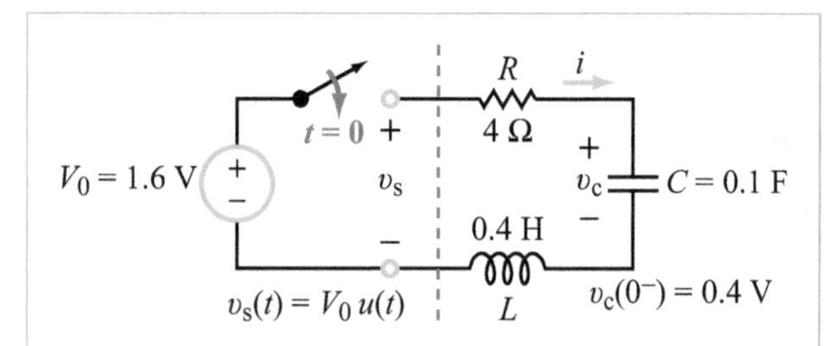

Figure 3-3: RLC circuit. The dc source, in combination with the switch, constitutes an input excitation $\upsilon_s(t) = V_0\, u(t)$.

and we apply the frequency derivative property (entry #9 in Table 3-1), twice:

$$\mathbf{X}(\mathbf{s}) = \mathbf{X}_2''(\mathbf{s}) = \frac{d^2}{d\mathbf{s}^2}\left[\frac{\mathbf{s}+3}{(\mathbf{s}+3)^2+16}\right]$$
$$= \frac{2(\mathbf{s}+3)[(\mathbf{s}+3)^2-48]}{[(\mathbf{s}+3)^2+16]^3}. \quad (3.44)$$

Exercise 3-7: Obtain the Laplace transform of
(a) $x_1(t) = 2(2 - e^{-t})\, u(t)$ and
(b) $x_2(t) = e^{-3t}\cos(2t + 30°)\, u(t)$.

Answer: (a) $\mathbf{X}_1(\mathbf{s}) = \dfrac{2\mathbf{s}+4}{\mathbf{s}(\mathbf{s}+1)}$,
(b) $\mathbf{X}_2(\mathbf{s}) = \dfrac{0.866\mathbf{s}+1.6}{\mathbf{s}^2+6\mathbf{s}+13}$. (See S²)

3-4 Circuit Analysis Example

Having learned how to transform a time-domain signal into its Laplace transform counterpart, we will now demonstrate the utility of the Laplace transform by analyzing a simple electric circuit. Figure 3-3 displays a series RLC circuit connected to a 1.6 V dc voltage source via a *single-pole, single-throw* (*SPST*) switch that closes at $t = 0$. Prior to $t = 0$, the RLC circuit had been connected to another input source, as a result of which the capacitor continued to hold electric charge up until $t = 0^-$ (immediately before closing the switch). The corresponding voltage across the capacitor was $\upsilon_c(0^-) = 0.4$ V. Our objective is to determine $i(t)$ for $t \ge 0$.

Step 1: Apply *Kirchhoff's voltage law* (*KVL*) to obtain an integrodifferential equation for the current $i(t)$ for $t \ge 0$.

The dc source, in combination with the switch, constitutes an input excitation (signal) given by

$$\upsilon_s(t) = V_0\, u(t). \quad (3.45)$$

For the loop, application of KVL for $t \ge 0^-$ gives

$$\upsilon_R + \upsilon_C + \upsilon_L = \upsilon_s,$$

where

$$\upsilon_R = R\, i(t),$$

$$\upsilon_C = \frac{1}{C}\int_{0^-}^{t} i(t')\, dt' + \upsilon_C(0^-),$$

$$\upsilon_L = L\,\frac{di(t)}{dt},$$

and

$$\upsilon_s = V_0\, u(t).$$

Substitution of these expressions in the KVL equation gives

$$R\, i(t) + \left[\frac{1}{C}\int_{0^-}^{t} i(t')\, dt' + \upsilon_C(0^-)\right] + L\,\frac{di(t)}{dt} = V_0\, u(t). \quad (3.46)$$

Step 2: Define $\mathbf{I}(\mathbf{s})$ as the Laplace transform corresponding to the unknown current $i(t)$, obtain $\mathbf{s}$-domain equivalents for each term in Eq. (3.46), and then apply the linearity property of the Laplace transform to transform the entire integrodifferential equation into the $\mathbf{s}$-domain.

The four terms of Eq. (3.46) have the following Laplace transform pairs:

$$R\, i(t) \longleftrightarrow R\, \mathbf{I}(\mathbf{s})$$

$$\frac{1}{C}\int_{0^-}^{t} i(t')\, dt' + \upsilon_C(0^-) \longleftrightarrow \frac{1}{C}\left[\frac{\mathbf{I}(\mathbf{s})}{\mathbf{s}}\right] + \frac{\upsilon_C(0^-)}{\mathbf{s}}$$

(time integral property)

$$L\,\frac{di(t)}{dt} \longleftrightarrow L[\mathbf{s}\,\mathbf{I}(\mathbf{s}) - i(0^-)]$$

(time derivative property)

and

$$V_0\, u(t) \longleftrightarrow \frac{V_0}{\mathbf{s}}$$

(transform of step function).

Hence, the **s**-domain equivalent of Eq. (3.46) is

$$R\,\mathbf{I(s)} + \frac{\mathbf{I(s)}}{C\mathbf{s}} + \frac{\upsilon_C(0^-)}{\mathbf{s}} + L\mathbf{s}\,\mathbf{I(s)} = \frac{V_0}{\mathbf{s}}, \tag{3.47}$$

where we have set $i(0^-) = 0$, because no current could have been flowing through the loop prior to closing the switch. Equation (3.47) is the **s**-domain equivalent of the integrodifferential equation (3.46). Whereas Eq. (3.46) has time-derivatives and integrals, Eq. (3.47) is a simple algebraic equation in **s**.

▶ The beauty of the Laplace transform is that it converts an integrodifferential equation in the time domain into a straightforward algebraic equation in the **s**-domain. ◀

Solving for $\mathbf{I(s)}$ and then replacing R, L, C, V_0, and $\upsilon_C(0^-)$ with their numerical values leads to

$$\mathbf{I(s)} = \frac{V_0 - \upsilon_C(0^-)}{L\left[\mathbf{s}^2 + \frac{R}{L}\mathbf{s} + \frac{1}{LC}\right]} = \frac{1.6 - 0.4}{0.4\left(\mathbf{s}^2 + \frac{4}{0.4}\mathbf{s} + \frac{1}{0.4 \times 0.1}\right)}$$
$$= \frac{3}{\mathbf{s}^2 + 10\mathbf{s} + 25} = \frac{3}{(\mathbf{s}+5)^2}. \tag{3.48}$$

According to entry #6 in Table 3-2, we have

$$\mathcal{L}^{-1}\left[\frac{1}{(\mathbf{s}+a)^2}\right] = te^{-at}\,u(t).$$

Hence,

$$i(t) = 3te^{-5t}\,u(t). \tag{3.49}$$

In this particular example, the expression for $\mathbf{I(s)}$ given by Eq. (3.48) matches one of the entries available in Table 3-2, but what should we do if it does not? We have two options.

(1) We can apply the inverse Laplace transform relation given by Eq. (3.5), which in general involves a rather cumbersome contour integration.

(2) We can apply the ***partial-fraction-expansion*** method to rearrange the expression for $\mathbf{I(s)}$ into a sum of terms, each of which has an appropriate match in Table 3-2. This latter approach is the subject of the next section.

3-5 Partial Fraction Expansion

Let us assume that, after transforming the integrodifferential equation associated with a system of interest to the **s**-domain and then solving it for the output signal whose behavior we wish to examine, we end up with an expression $\mathbf{X(s)}$. Our next step is to inverse transform $\mathbf{X(s)}$ to the time domain, thereby completing our solution. The degree of mathematical difficulty associated with the implementation of the inverse transformation depends on the mathematical form of $\mathbf{X(s)}$.

Consider for example the expression

$$\mathbf{X(s)} = \frac{4}{\mathbf{s}+2} + \frac{6}{(\mathbf{s}+5)^2} + \frac{8}{\mathbf{s}^2 + 4\mathbf{s} + 5}. \tag{3.50}$$

The inverse transform, $x(t)$, is given by

$$x(t) = \mathcal{L}^{-1}[\mathbf{X(s)}]$$
$$= \mathcal{L}^{-1}\left[\frac{4}{\mathbf{s}+2}\right] + \mathcal{L}^{-1}\left[\frac{6}{(\mathbf{s}+5)^2}\right] + \mathcal{L}^{-1}\left[\frac{8}{\mathbf{s}^2+4\mathbf{s}+5}\right]. \tag{3.51}$$

By comparison with the entries in Table 3-2, we note the following:

(a) The first term in Eq. (3.51), $4/(\mathbf{s}+2)$, is functionally the same as entry #3 in Table 3-2 with $a = 2$. Hence,

$$\mathcal{L}^{-1}\left[\frac{4}{\mathbf{s}+2}\right] = 4e^{-2t}\,u(t). \tag{3.52a}$$

(b) The second term, $6/(\mathbf{s}+5)^2$, is functionally the same as entry #6 in Table 3-2 with $a = 5$. Hence,

$$\mathcal{L}^{-1}\left[\frac{6}{(\mathbf{s}+5)^2}\right] = 6te^{-5t}\,u(t). \tag{3.52b}$$

(c) The third term $8/(\mathbf{s}^2+4\mathbf{s}+5)$, is similar (but not identical) in form to entry #13 in Table 3-2. However, it can be rearranged to assume the proper form:

$$\frac{8}{\mathbf{s}^2+4\mathbf{s}+5} = \frac{8}{(\mathbf{s}+2)^2+1}.$$

Consequently,

$$\mathcal{L}^{-1}\left[\frac{8}{(\mathbf{s}+2)^2+1}\right] = 8e^{-2t}\sin t\;u(t). \tag{3.52c}$$

Combining the results represented by Eqs. (3.52a–c) gives

$$x(t) = [4e^{-2t} + 6te^{-5t} + 8e^{-2t} \sin t]\, u(t). \qquad (3.53)$$

3-5.1 Proper Form

The preceding example demonstrated that the implementation of the inverse Laplace transform is a rather painless process, as long as the expression for $\mathbf{X(s)}$ is composed of a series of terms similar to those in Eq. (3.50). More often than not, $\mathbf{X(s)}$ is not in the proper form, so we will need to reconfigure it before we can apply the inverse transform. At the most general level, $\mathbf{X(s)}$ is given by the ratio of a polynomial numerator $\mathbf{N(s)}$ to a polynomial denominator $\mathbf{D(s)}$:

$$\mathbf{X(s)} = \frac{\mathbf{N(s)}}{\mathbf{D(s)}} = \frac{b_m \mathbf{s}^m + b_{m-1}\mathbf{s}^{m-1} + \cdots + b_1\mathbf{s} + b_0}{a_n \mathbf{s}^n + a_{n-1}\mathbf{s}^{n-1} + \cdots + a_1\mathbf{s} + a_0}, \qquad (3.54a)$$

where all of the a_i and b_j coefficients are real and the powers m and n are positive integers.

► The roots of $\mathbf{N(s)}$, namely the values of $\mathbf{s}$ at which $\mathbf{N(s)} = 0$, are called the *zeros* of $\mathbf{X(s)}$, and we designate them $\{\mathbf{z}_i,\ i = 1, 2, \ldots, m\}$. A polynomial of order m has m roots. Similarly, the roots of $\mathbf{D(s)} = 0$ are called the *poles* of $\mathbf{X(s)}$ and are designated $\{\mathbf{p}_i,\ i = 1, 2, \ldots, n\}$. ◄

Function $\mathbf{X(s)}$ can be expressed in terms of its poles and zeros as

$$\mathbf{X(s)} = \mathbf{C}\,\frac{(\mathbf{s}-\mathbf{z}_1)(\mathbf{s}-\mathbf{z}_2)\ldots(\mathbf{s}-\mathbf{z}_m)}{(\mathbf{s}-\mathbf{p}_1)(\mathbf{s}-\mathbf{p}_2)\ldots(\mathbf{s}-\mathbf{p}_n)} = \mathbf{C}\,\frac{\prod_{i=1}^{m}(\mathbf{s}-\mathbf{z}_i)}{\prod_{i=1}^{n}(\mathbf{s}-\mathbf{p}_i)}, \qquad (3.54b)$$

where $\mathbf{C}$ is a constant.

An important attribute of $\mathbf{X(s)}$ is the *degree* of its numerator, m, relative to that of its denominator, n.

(a) If $m < n$, $\mathbf{X(s)}$ is considered a *strictly proper rational function*, in which case it can be expanded into a sum of partial fractions by applying the applicable recipe from among those outlined in Subsections 3-5.2 through 3-5.5 (to follow).

(b) If $m = n$, $\mathbf{X(s)}$ is called a *proper rational function*, and if $m > n$, it is called an *improper rational function*. In both cases, a preparatory step is required prior to the application of partial fraction expansion.

Case 1: $m = n$

Consider the function

$$\mathbf{X}_1(\mathbf{s}) = \frac{2\mathbf{s}^2 + 8\mathbf{s} + 6}{\mathbf{s}^2 + 2\mathbf{s} + 1}.$$

Since $m = n = 2$, this is a proper function, but not a strictly proper function. To convert it to the latter, we use the following steps:

Step 1: Factor out a constant equal to the ratio of the coefficient of the highest term in the numerator to the coefficient of the highest term in the denominator, which in the present case is $2/1 = 2$.

Step 2: Apply the *division relationship*

$$\frac{\mathbf{N(s)}}{\mathbf{D(s)}} = 1 + \frac{\mathbf{N(s)} - \mathbf{D(s)}}{\mathbf{D(s)}}. \qquad (3.55)$$

For $\mathbf{X}_1(\mathbf{s})$, this two-step process leads to

$$\begin{aligned}
\mathbf{X}_1(\mathbf{s}) &= 2\left(\frac{\mathbf{s}^2 + 4\mathbf{s} + 3}{\mathbf{s}^2 + 2\mathbf{s} + 1}\right) && \textbf{(step 1)}\\
&= 2\left[1 + \frac{(\mathbf{s}^2 + 4\mathbf{s} + 3) - (\mathbf{s}^2 + 2\mathbf{s} + 1)}{(\mathbf{s}^2 + 2\mathbf{s} + 1)}\right] && \textbf{(step 2)}\\
&= 2\left[1 + \frac{2\mathbf{s} + 2}{\mathbf{s}^2 + 2\mathbf{s} + 1}\right]\\
&= 2 + \underbrace{\frac{4\mathbf{s} + 4}{\mathbf{s}^2 + 2\mathbf{s} + 1}}_{\text{strictly proper function}}.
\end{aligned}$$

Case 2: $m > n$

Function

$$\mathbf{X}_2(\mathbf{s}) = \frac{6\mathbf{s}^3 + 4\mathbf{s}^2 + 8\mathbf{s} + 6}{\mathbf{s}^2 + 2\mathbf{s} + 1}$$

is an improper rational function. To convert $\mathbf{X}_2(\mathbf{s})$ into a form in which the highest power terms are $\mathbf{s}^3$ in both the numerator and denominator, we factor out $6\mathbf{s}$:

$$\mathbf{X}_2(\mathbf{s}) = 6\mathbf{s}\left[\frac{\mathbf{s}^3 + (2/3)\mathbf{s}^2 + (8/6)\mathbf{s} + 1}{\mathbf{s}^3 + 2\mathbf{s}^2 + \mathbf{s}}\right].$$

Next, we apply the division relationship, which yields

$$\mathbf{X}_2(\mathbf{s}) = 6\mathbf{s}\left[1 + \frac{[\mathbf{s}^3 + (2/3)\mathbf{s}^2 + (8/6)\mathbf{s} + 1] - [\mathbf{s}^3 + 2\mathbf{s}^2 + \mathbf{s}]}{\mathbf{s}^3 + 2\mathbf{s}^2 + \mathbf{s}}\right]$$

$$= 6\mathbf{s}\left[1 + \frac{-(4/3)\mathbf{s}^2 + (2/6)\mathbf{s} + 1}{\mathbf{s}^3 + 2\mathbf{s}^2 + \mathbf{s}}\right]$$

$$= 6\mathbf{s} + \underbrace{\frac{(-8\mathbf{s}^2 + 2\mathbf{s} + 6)}{\mathbf{s}^2 + 2\mathbf{s} + 1}}_{\text{proper function}}.$$

The second term of $\mathbf{X}_2(\mathbf{s})$ is a proper function. We can convert it into a strictly proper function by following the recipe in Case 1. The final outcome is

$$\mathbf{X}_2(\mathbf{s}) = (6\mathbf{s} - 8) + \underbrace{\frac{18\mathbf{s} + 14}{\mathbf{s}^2 + 2\mathbf{s} + 1}}_{\text{strictly proper function}}.$$

3-5.2 Distinct Real Poles

Consider the s-domain function

$$\mathbf{X}(\mathbf{s}) = \frac{\mathbf{s}^2 - 4\mathbf{s} + 3}{\mathbf{s}(\mathbf{s} + 1)(\mathbf{s} + 3)}. \quad (3.56)$$

The poles of $\mathbf{X}(\mathbf{s})$ are $\mathbf{p}_1 = 0$, $\mathbf{p}_2 = -1$, and $\mathbf{p}_3 = -3$. All three poles are ***real*** and ***distinct***. By distinct we mean that no two or more poles are the same. [In $(\mathbf{s} + 4)^2$, for example, the pole $\mathbf{p} = -4$ occurs twice, and therefore, it is not distinct.] The highest power of $\mathbf{s}$ in the numerator of Eq. (3.56) is $m = 2$, and the highest power of $\mathbf{s}$ in the denominator is $n = 3$. Hence, $\mathbf{X}(\mathbf{s})$ is a strictly proper rational function because $m < n$. Given these attributes, $\mathbf{X}(\mathbf{s})$ can be decomposed into partial fractions corresponding to the three factors in the denominator of $\mathbf{X}(\mathbf{s})$:

$$\mathbf{X}(\mathbf{s}) = \frac{A_1}{\mathbf{s}} + \frac{A_2}{(\mathbf{s} + 1)} + \frac{A_3}{(\mathbf{s} + 3)}, \quad (3.57a)$$

where A_1 to A_3 are ***expansion coefficients***, sometimes called ***residues*** (to be determined shortly). Equating the two functional forms of $\mathbf{X}(\mathbf{s})$, we have

$$\frac{A_1}{\mathbf{s}} + \frac{A_2}{(\mathbf{s} + 1)} + \frac{A_3}{(\mathbf{s} + 3)} = \frac{\mathbf{s}^2 - 4\mathbf{s} + 3}{\mathbf{s}(\mathbf{s} + 1)(\mathbf{s} + 3)}. \quad (3.57b)$$

Associated with each expansion coefficient is a ***pole factor***: $\mathbf{s}$, $(\mathbf{s}+1)$, and $(\mathbf{s}+3)$ are the pole factors associated with expansion coefficients A_1, A_2, and A_3, respectively. To determine the value of any expansion coefficient we multiply both sides of Eq. (3.57b) by the pole factor of that expansion coefficient, and then we evaluate them at $\mathbf{s}$ = pole value of that pole factor. The procedure is called the ***residue method***.

To determine A_2, for example, we multiply both sides of Eq. (3.57b) by $(\mathbf{s} + 1)$, we reduce the expressions, and then we evaluate them at $\mathbf{s} = -1$:

$$\left\{(\mathbf{s} + 1)\left[\frac{A_1}{\mathbf{s}} + \frac{A_2}{(\mathbf{s} + 1)} + \frac{A_3}{(\mathbf{s} + 3)}\right]\right\}\Bigg|_{\mathbf{s}=-1}$$

$$= \left[\frac{(\mathbf{s} + 1)(\mathbf{s}^2 - 4\mathbf{s} + 3)}{\mathbf{s}(\mathbf{s} + 1)(\mathbf{s} + 3)}\right]\Bigg|_{\mathbf{s}=-1}. \quad (3.58)$$

After canceling factors of $(\mathbf{s} + 1)$, the expression becomes

$$\left[\frac{A_1(\mathbf{s} + 1)}{\mathbf{s}} + A_2 + \frac{A_3(\mathbf{s} + 1)}{(\mathbf{s} + 3)}\right]\Bigg|_{\mathbf{s}=-1}$$

$$= \left[\frac{(\mathbf{s}^2 - 4\mathbf{s} + 3)}{\mathbf{s}(\mathbf{s} + 3)}\right]\Bigg|_{\mathbf{s}=-1}. \quad (3.59)$$

We note that (a) the presence of $(\mathbf{s} + 1)$ in the numerators of terms 1 and 3 on the left-hand side will force those terms to go to zero when evaluated at $\mathbf{s} = -1$, (b) the middle term has only A_2 in it, and (c) the reduction on the right-hand side of Eq. (3.59) eliminated the pole factor $(\mathbf{s}+1)$ from the expression. Consequently,

$$A_2 = \frac{(-1)^2 + 4 + 3}{(-1)(-1 + 3)} = -4.$$

Similarly,

$$A_1 = \mathbf{s}\,\mathbf{X}(\mathbf{s})|_{\mathbf{s}=0} = \frac{\mathbf{s}^2 - 4\mathbf{s} + 3}{(\mathbf{s} + 1)(\mathbf{s} + 3)}\Bigg|_{\mathbf{s}=0} = 1,$$

and

$$A_3 = (\mathbf{s} + 3)\,\mathbf{X}(\mathbf{s})|_{\mathbf{s}=-3} = \frac{\mathbf{s}^2 - 4\mathbf{s} + 3}{\mathbf{s}(\mathbf{s} + 1)}\Bigg|_{\mathbf{s}=-3} = 4.$$

Having established the values of A_1, A_2, and A_3, we now are ready to apply the inverse Laplace transform to Eq. (3.57a):

$$x(t) = \mathcal{L}^{-1}[\mathbf{X}(\mathbf{s})] = \mathcal{L}\left[\frac{1}{\mathbf{s}} - \frac{4}{\mathbf{s} + 1} + \frac{4}{\mathbf{s} + 3}\right]$$

$$= [1 - 4e^{-t} + 4e^{-3t}]\,u(t). \quad (3.60)$$

Building on this example, we can generalize the process to any strictly proper rational function.

Distinct Real Poles

Given a strictly proper rational function defined by

$$\mathbf{X}(\mathbf{s}) = \frac{\mathbf{N}(\mathbf{s})}{\mathbf{D}(\mathbf{s})} = \frac{\mathbf{N}(\mathbf{s})}{(\mathbf{s}-p_1)(\mathbf{s}-p_2)\dots(\mathbf{s}-p_n)}, \tag{3.61}$$

with distinct real poles p_1 to p_n, such that $p_i \neq p_j$ for all $i \neq j$, and $m < n$ (where m and n are the highest powers of $\mathbf{s}$ in $\mathbf{N}(\mathbf{s})$ and $\mathbf{D}(\mathbf{s})$, respectively), then $\mathbf{X}(\mathbf{s})$ can be expanded into the equivalent form:

$$\begin{aligned}\mathbf{X}(\mathbf{s}) &= \frac{A_1}{\mathbf{s}-p_1} + \frac{A_2}{\mathbf{s}-p_2} + \dots + \frac{A_n}{\mathbf{s}-p_n} \\ &= \sum_{i=1}^{n} \frac{A_i}{\mathbf{s}-p_i}\end{aligned} \tag{3.62}$$

with expansion coefficients A_1 to A_n given by

$$\begin{aligned}A_i &= (\mathbf{s}-p_i)\,\mathbf{X}(\mathbf{s})|_{\mathbf{s}=p_i}, \\ &\quad i = 1, 2, \dots, n.\end{aligned} \tag{3.63}$$

In view of entry #3 in Table 3-2, the inverse Laplace transform of Eq. (3.62) is obtained by replacing a with $-p_i$:

$$\begin{aligned}x(t) &= \mathcal{L}^{-1}[\mathbf{X}(\mathbf{s})] \\ &= [A_1 e^{p_1 t} + A_2 e^{p_2 t} + \dots + A_n e^{p_n t}]\, u(t).\end{aligned} \tag{3.64}$$

Exercise 3-8: Apply the partial fraction expansion method to determine $x(t)$ given that its Laplace transform is

$$\mathbf{X}(\mathbf{s}) = \frac{10\mathbf{s}+16}{\mathbf{s}(\mathbf{s}+2)(\mathbf{s}+4)}.$$

Answer: $x(t) = [2 + e^{-2t} - 3e^{-4t}]\, u(t)$. (See S²)

3-5.3 Repeated Real Poles

We now will consider the case when $\mathbf{X}(\mathbf{s})$ is a strictly proper rational function containing repeated real poles or a combination of distinct and repeated poles. The partial fraction expansion method is outlined by the following steps.

Step 1. We are given a strictly proper rational function $\mathbf{X}(\mathbf{s})$ composed of the product

$$\mathbf{X}(\mathbf{s}) = \mathbf{X}_1(\mathbf{s})\,\mathbf{X}_2(\mathbf{s}) \tag{3.65}$$

with

$$\mathbf{X}_1(\mathbf{s}) = \frac{\mathbf{N}(\mathbf{s})}{(\mathbf{s}-p_1)(\mathbf{s}-p_2)\dots(\mathbf{s}-p_n)} \tag{3.66}$$

and

$$\mathbf{X}_2(\mathbf{s}) = \frac{1}{(\mathbf{s}-p)^m}. \tag{3.67}$$

We note that $\mathbf{X}_1(\mathbf{s})$ is identical in form with Eq. (3.61) and contains only distinct real poles, p_1 to p_n, thereby qualifying it for representation by a series of terms as in Eq. (3.62). The second function, $\mathbf{X}_2(\mathbf{s})$, has an m-repeated pole at $\mathbf{s} = p$, where m is a positive integer. Also, the repeated pole is not a pole of $\mathbf{X}_1(\mathbf{s})$; $p \neq p_i$ for $i = 1, 2, \dots, n$.

Step 2. Partial fraction representation for an m-repeated pole at $\mathbf{s} = p$ consists of m terms:

$$\frac{B_1}{\mathbf{s}-p} + \frac{B_2}{(\mathbf{s}-p)^2} + \dots + \frac{B_m}{(\mathbf{s}-p)^m}. \tag{3.68}$$

Step 3. Partial fraction expansion for the combination of the product $\mathbf{X}_1(\mathbf{s})\,\mathbf{X}_2(\mathbf{s})$ is then given by

$$\begin{aligned}\mathbf{X}(\mathbf{s}) &= \frac{A_1}{\mathbf{s}-p_1} + \frac{A_2}{\mathbf{s}-p_2} + \dots + \frac{A_n}{\mathbf{s}-p_n} \\ &\quad + \frac{B_1}{\mathbf{s}-p} + \frac{B_2}{(\mathbf{s}-p)^2} + \dots + \frac{B_m}{(\mathbf{s}-p)^m} \\ &= \sum_{i=1}^{n} \frac{A_i}{\mathbf{s}-p_i} + \sum_{j=1}^{m} \frac{B_j}{(\mathbf{s}-p)^j}.\end{aligned} \tag{3.69}$$

Step 4. Expansion coefficients A_1 to A_n are determined by applying Eq. (3.63):

$$A_i = (\mathbf{s}-p_i)\,\mathbf{X}(\mathbf{s})|_{\mathbf{s}=p_i}, \qquad i = 1, 2, \dots, n. \tag{3.70}$$

Repeated Real Poles

Expansion coefficients B_1 to B_m are determined through a procedure that involves multiplication by $(\mathbf{s}-p)^m$, differentiation with respect to $\mathbf{s}$, and evaluation at $\mathbf{s}=p$:

$$B_j = \left\{ \frac{1}{(m-j)!} \frac{d^{m-j}}{d\mathbf{s}^{m-j}} [(\mathbf{s}-p)^m \mathbf{X}(\mathbf{s})] \right\}\Bigg|_{\mathbf{s}=p}, \quad j = 1, 2, \ldots, m. \tag{3.71}$$

For the m, $m-1$, and $m-2$ terms, Eq. (3.71) reduces to

$$B_m = (\mathbf{s}-p)^m \mathbf{X}(\mathbf{s})|_{\mathbf{s}=p}, \tag{3.72a}$$

$$B_{m-1} = \left\{ \frac{d}{d\mathbf{s}} [(\mathbf{s}-p)^m \mathbf{X}(\mathbf{s})] \right\}\Bigg|_{\mathbf{s}=p}, \tag{3.72b}$$

$$B_{m-2} = \left\{ \frac{1}{2!} \frac{d^2}{d\mathbf{s}^2} [(\mathbf{s}-p)^m \mathbf{X}(\mathbf{s})] \right\}\Bigg|_{\mathbf{s}=p}. \tag{3.72c}$$

Thus, the evaluation of B_m does not involve any differentiation, that of B_{m-1} involves differentiation with respect to $\mathbf{s}$ only once (and division by 1!), and that of B_{m-2} involves differentiation twice and division by 2!. In practice, it is easiest to start by evaluating B_m first and then evaluating the other expansion coefficients in descending order.

Step 5. Once the values of all of the expansion coefficients of Eq. (3.69) have been determined, transformation to the time domain is accomplished by applying entry #8 of Table 3-2,

$$\mathcal{L}^{-1}\left[\frac{(n-1)!}{(\mathbf{s}+a)^n}\right] = t^{n-1} e^{-at}\, u(t) \tag{3.73}$$

with $a=-p$. The result is

$$x(t) = \mathcal{L}^{-1}[\mathbf{X}(\mathbf{s})] = \left[\sum_{i=1}^{n} A_i e^{p_i t} + \sum_{j=1}^{m} \frac{B_j t^{j-1}}{(j-1)!} e^{pt}\right] u(t). \tag{3.74}$$

Example 3-7: Repeated Poles

Determine the inverse Laplace transform of

$$\mathbf{X}(\mathbf{s}) = \frac{\mathbf{N}(\mathbf{s})}{\mathbf{D}(\mathbf{s})} = \frac{\mathbf{s}^2+3\mathbf{s}+3}{\mathbf{s}^4+11\mathbf{s}^3+45\mathbf{s}^2+81\mathbf{s}+54}.$$

Solution: In theory, any polynomial with real coefficients can be expressed as a product of linear and quadratic factors (of the form $(s+p)$ and (s^2+as+b), respectively). The process involves long division, but it requires knowledge of the roots of the polynomial, which can be determined through the application of numerical techniques. In the present case, numerical evaluation reveals that $\mathbf{s}=-2$ and $\mathbf{s}=-3$ are roots of $\mathbf{D}(\mathbf{s})$. Given that $\mathbf{D}(\mathbf{s})$ is of order four, it should have four roots, including possible duplicates.

Since $\mathbf{s}=-2$ is a root of $\mathbf{D}(\mathbf{s})$, we should be able to factor out $(\mathbf{s}+2)$ from it. Long division gives

$$\begin{aligned}\mathbf{D}(\mathbf{s}) &= \mathbf{s}^4+11\mathbf{s}^3+45\mathbf{s}^2+81\mathbf{s}+54\\ &= (\mathbf{s}+2)(\mathbf{s}^3+9\mathbf{s}^2+27\mathbf{s}+27).\end{aligned}$$

Next, we factor out $(\mathbf{s}+3)$:

$$\mathbf{D}(\mathbf{s}) = (\mathbf{s}+2)(\mathbf{s}+3)(\mathbf{s}^2+6\mathbf{s}+9) = (\mathbf{s}+2)(\mathbf{s}+3)^3.$$

Hence, $\mathbf{X}(\mathbf{s})$ has a distinct real pole at $\mathbf{s}=-2$ and a triple repeated pole at $\mathbf{s}=-3$, and the given expression can be rewritten as

$$\mathbf{X}(\mathbf{s}) = \frac{\mathbf{s}^2+3\mathbf{s}+3}{(\mathbf{s}+2)(\mathbf{s}+3)^3}.$$

Through partial fraction expansion, $\mathbf{X}(\mathbf{s})$ can be decomposed into

$$\mathbf{X}(\mathbf{s}) = \frac{A}{\mathbf{s}+2} + \frac{B_1}{\mathbf{s}+3} + \frac{B_2}{(\mathbf{s}+3)^2} + \frac{B_3}{(\mathbf{s}+3)^3},$$

with

$$A = (\mathbf{s}+2)\,\mathbf{X}(\mathbf{s})|_{\mathbf{s}=-2} = \frac{\mathbf{s}^2+3\mathbf{s}+3}{(\mathbf{s}+3)^3}\Bigg|_{\mathbf{s}=-2} = 1,$$

$$B_3 = (\mathbf{s}+3)^3\,\mathbf{X}(\mathbf{s})|_{\mathbf{s}=-3} = \frac{\mathbf{s}^2+3\mathbf{s}+3}{\mathbf{s}+2}\Bigg|_{\mathbf{s}=-3} = -3,$$

$$B_2 = \frac{d}{d\mathbf{s}}[(\mathbf{s}+3)^3\,\mathbf{X}(\mathbf{s})]\Bigg|_{\mathbf{s}=-3} = 0,$$

and

$$B_1 = \frac{1}{2}\frac{d^2}{d\mathbf{s}^2}[(\mathbf{s}+3)^3\,\mathbf{X}(\mathbf{s})]\Bigg|_{\mathbf{s}=-3} = -1.$$

Hence,

$$\mathbf{X}(\mathbf{s}) = \frac{1}{\mathbf{s}+2} - \frac{1}{\mathbf{s}+3} - \frac{3}{(\mathbf{s}+3)^3},$$

and application of Eq. (3.73) leads to

$$\mathcal{L}^{-1}[\mathbf{X}(\mathbf{s})] = \left[e^{-2t} - e^{-3t} - \frac{3}{2}\, t^2 e^{-3t}\right] u(t).$$

Concept Question 3-6: What purpose does the partial fraction expansion method serve? (See S²)

Concept Question 3-7: When evaluating the expansion coefficients of a function containing repeated poles, is it more practical to start by evaluating the coefficient of the fraction with the lowest-order pole or that with the highest-order pole? Why? (See S²)

Exercise 3-9: Determine the inverse Laplace transform of

$$\mathbf{X}(\mathbf{s}) = \frac{4\mathbf{s}^2 - 15\mathbf{s} - 10}{(\mathbf{s}+2)^3}.$$

Answer: $x(t) = (4 - 31t + 18t^2)e^{-2t}\, u(t)$. (See S²)

3-5.4 Distinct Complex Poles

The Laplace transform of a certain system is given by

$$\mathbf{X}(\mathbf{s}) = \frac{4\mathbf{s}+1}{(\mathbf{s}+1)(\mathbf{s}^2+4\mathbf{s}+13)}. \tag{3.75}$$

In addition to the simple-pole factor, the denominator includes a quadratic-pole factor with roots $\mathbf{p}_1$ and $\mathbf{p}_2$. Solution of $\mathbf{s}^2 + 4\mathbf{s} + 13 = 0$ gives

$$\mathbf{p}_1 = -2 + j3, \qquad \mathbf{p}_2 = -2 - j3. \tag{3.76}$$

The fact that the two roots are complex conjugates of one another is a consequence of an important property of Eq. (3.54a), namely that *if all of the coefficients a_i and b_i are real-valued, then the roots of the numerator and denominator polynomials occur in complex conjugate pairs.*

In view of Eq. (3.76), the quadratic factor is given by

$$\mathbf{s}^2 + 4\mathbf{s} + 13 = (\mathbf{s} - \mathbf{p}_1)(\mathbf{s} - \mathbf{p}_2) = (\mathbf{s}+2-j3)(\mathbf{s}+2+j3), \tag{3.77}$$

and $\mathbf{X}(\mathbf{s})$ can now be expanded into partial fractions:

$$\mathbf{X}(\mathbf{s}) = \frac{A}{\mathbf{s}+1} + \frac{\mathbf{B}_1}{\mathbf{s}+2-j3} + \frac{\mathbf{B}_2}{\mathbf{s}+2+j3}. \tag{3.78}$$

Expansion coefficients $\mathbf{B}_1$ and $\mathbf{B}_2$ are printed in bold letters to signify the fact that they may be complex quantities. Determination of A, $\mathbf{B}_1$, and $\mathbf{B}_2$ follows the same factor-multiplication technique employed in Section 3-5.2:

$$A = (\mathbf{s}+1)\,\mathbf{X}(\mathbf{s})|_{\mathbf{s}=-1} = \left.\frac{4\mathbf{s}+1}{\mathbf{s}^2+4\mathbf{s}+13}\right|_{\mathbf{s}=-1} = -0.3, \tag{3.79a}$$

$$\begin{aligned}\mathbf{B}_1 &= (\mathbf{s}+2-j3)\,\mathbf{X}(\mathbf{s})|_{\mathbf{s}=-2+j3}\\ &= \left.\frac{4\mathbf{s}+1}{(\mathbf{s}+1)(\mathbf{s}+2+j3)}\right|_{\mathbf{s}=-2+j3}\\ &= \frac{4(-2+j3)+1}{(-2+j3+1)(-2+j3+2+j3)}\\ &= \frac{-7+j12}{-18-j6} = 0.73e^{-j78.2^\circ},\end{aligned} \tag{3.79b}$$

and

$$\begin{aligned}\mathbf{B}_2 &= (\mathbf{s}+2+j3)\,\mathbf{X}(\mathbf{s})|_{\mathbf{s}=-2-j3}\\ &= \left.\frac{4\mathbf{s}+1}{(\mathbf{s}+1)(\mathbf{s}+2-j3)}\right|_{\mathbf{s}=-2-j3} = 0.73e^{j78.2^\circ}.\end{aligned} \tag{3.79c}$$

We observe that $\mathbf{B}_2 = \mathbf{B}_1^*$.

> ▶ The expansion coefficients associated with conjugate poles are always conjugate pairs themselves. ◀

The inverse Laplace transform of Eq. (3.78) is

$$\begin{aligned}x(t) &= \mathcal{L}^{-1}[\mathbf{X}(\mathbf{s})]\\ &= \mathcal{L}^{-1}\left(\frac{-0.3}{\mathbf{s}+1}\right) + \mathcal{L}^{-1}\left(\frac{0.73e^{-j78.2^\circ}}{\mathbf{s}+2-j3}\right)\\ &\quad + \mathcal{L}^{-1}\left(\frac{0.73e^{j78.2^\circ}}{\mathbf{s}+2+j3}\right)\\ &= [-0.3e^{-t} + 0.73e^{-j78.2^\circ}e^{-(2-j3)t}\\ &\quad + 0.73e^{j78.2^\circ}e^{-(2+j3)t}]\, u(t).\end{aligned} \tag{3.80}$$

Because complex numbers do not belong in the time domain, our initial reaction to their presence in the solution given by Eq. (3.80) is that perhaps an error was committed somewhere along the way. The truth is that the solution is correct but incomplete. Terms 2 and 3 are conjugate pairs, so by applying

Euler's formula, they can be combined into a single term containing only real quantities:

$$\begin{aligned}[0.73e^{-j78.2^\circ} e^{-(2-j3)t} + 0.73e^{j78.2^\circ} e^{-(2+j3)t}]\, u(t) \\ = 0.73e^{-2t}[e^{j(3t-78.2^\circ)} + e^{-j(3t-78.2^\circ)}]\, u(t) \\ = 2 \times 0.73e^{-2t} \cos(3t - 78.2^\circ)\, u(t) \\ = 1.46e^{-2t} \cos(3t - 78.2^\circ)\, u(t). \qquad (3.81)\end{aligned}$$

This approach is supparized in entry #3 of Table 3-3. Hence, the final time-domain solution is

$$x(t) = [-0.3e^{-t} + 1.46e^{-2t} \cos(3t - 78.2^\circ)]\, u(t). \qquad (3.82)$$

Exercise 3-10: Determine the inverse Laplace transform of

$$\mathbf{X(s)} = \frac{2\mathbf{s} + 14}{\mathbf{s}^2 + 6\mathbf{s} + 25}.$$

Answer: $x(t) = [2\sqrt{2}\, e^{-3t} \cos(4t - 45^\circ)]\, u(t)$. (See s^2)

3-5.5 Repeated Complex Poles

If the Laplace transform $\mathbf{X(s)}$ contains repeated complex poles, we can expand it into partial fractions by using a combination of the tools introduced in Sections 3-5.3 and 3-5.4. The process is illustrated in Example 3-8.

Example 3-8: Five-Pole Function

Determine the inverse Laplace transform of

$$\mathbf{X(s)} = \frac{108(\mathbf{s}^2 + 2)}{(\mathbf{s} + 2)(\mathbf{s}^2 + 10\mathbf{s} + 34)^2}.$$

Solution: The roots of $\mathbf{s}^2 + 10\mathbf{s} + 34 = 0$ are $\mathbf{p}_1 = -5 - j3$ and $\mathbf{p}_2 = -5 + j3$. Hence,

$$\mathbf{X(s)} = \frac{108(\mathbf{s}^2 + 2)}{(\mathbf{s} + 2)(\mathbf{s} + 5 + j3)^2(\mathbf{s} + 5 - j3)^2},$$

and its partial fraction expansion can be expressed as

$$\mathbf{X(s)} = \frac{A}{\mathbf{s} + 2} + \frac{\mathbf{B}_1}{\mathbf{s} + 5 + j3} + \frac{\mathbf{B}_2}{(\mathbf{s} + 5 + j3)^2} + \frac{\mathbf{B}_1^*}{\mathbf{s} + 5 - j3} + \frac{\mathbf{B}_2^*}{(\mathbf{s} + 5 - j3)^2},$$

where $\mathbf{B}_1^*$ and $\mathbf{B}_2^*$ are the complex conjugates of $\mathbf{B}_1$ and $\mathbf{B}_2$, respectively. Coefficients A, $\mathbf{B}_1$, and $\mathbf{B}_2$ are evaluated as

$$A = (\mathbf{s} + 2)\, \mathbf{X(s)}\big|_{\mathbf{s}=-2} = \left.\frac{108(\mathbf{s}^2 + 2)}{(\mathbf{s}^2 + 10\mathbf{s} + 34)^2}\right|_{\mathbf{s}=-2} = 2,$$

$$\begin{aligned}\mathbf{B}_2 &= (\mathbf{s} + 5 + j3)^2\, \mathbf{X(s)}\big|_{\mathbf{s}=-5-j3} \\ &= \left.\frac{108(\mathbf{s}^2 + 2)}{(\mathbf{s} + 2)(\mathbf{s} + 5 - j3)^2}\right|_{\mathbf{s}=-5-j3} \\ &= \frac{108[(-5 - j3)^2 + 2]}{(-5 - j3 + 2)(-5 - j3 + 5 - j3)^2} \\ &= 24 + j6 = 24.74e^{j14^\circ},\end{aligned}$$

and

$$\begin{aligned}\mathbf{B}_1 &= \frac{d}{d\mathbf{s}}[(\mathbf{s} + 5 + j3)^2\, \mathbf{X(s)}]\Big|_{\mathbf{s}=-5-j3} \\ &= \frac{d}{d\mathbf{s}}\left[\frac{108(\mathbf{s}^2 + 2)}{(\mathbf{s} + 2)(\mathbf{s} + 5 - j3)^2}\right]\Bigg|_{\mathbf{s}=-5-j3} \\ &= \left[\frac{108(2\mathbf{s})}{(\mathbf{s} + 2)(\mathbf{s} + 5 - j3)^2} - \frac{108(\mathbf{s}^2 + 2)}{(\mathbf{s} + 2)^2(\mathbf{s} + 5 - j3)^2} \right. \\ &\qquad \left. - \frac{2 \times 108(\mathbf{s}^2 + 2)}{(\mathbf{s} + 2)(\mathbf{s} + 5 - j3)^3}\right]\Bigg|_{\mathbf{s}=-5-j3} \\ &= -(1 + j9) = 9.06e^{-j96.34^\circ}.\end{aligned}$$

The remaining constants are

$$\mathbf{B}_1^* = 9.06e^{j96.34^\circ} \quad \text{and} \quad \mathbf{B}_2^* = 24.74e^{-j14^\circ},$$

and the inverse Laplace transform is

$$\begin{aligned}x(t) &= \mathcal{L}^{-1}[\mathbf{X(s)}] \\ &= \mathcal{L}^{-1}\left[\frac{2}{\mathbf{s} + 2} + \frac{9.06e^{-j96.34^\circ}}{\mathbf{s} + 5 + j3} + \frac{9.06e^{j96.34^\circ}}{\mathbf{s} + 5 - j3} \right. \\ &\qquad \left. + \frac{24.74e^{j14^\circ}}{(\mathbf{s} + 5 + j3)^2} + \frac{24.74e^{-j14^\circ}}{(\mathbf{s} + 5 - j3)^2}\right] \\ &= \big[2e^{-2t} \\ &\quad + 9.06(e^{-j96.34^\circ} e^{-(5+j3)t} + e^{j96.34^\circ} e^{-(5-j3)t}) \\ &\quad + 24.74t(e^{j14^\circ} e^{-(5+j3)t} + e^{-j14^\circ} e^{-(5-j3)t})\big]\, u(t) \\ &= [2e^{-2t} + 18.12e^{-5t} \cos(3t + 96.34^\circ) \\ &\quad + 49.48te^{-5t} \cos(3t - 14^\circ)]\, u(t).\end{aligned}$$

This approach is summarized in entry #4 of Table 3-3.

Table 3-3: **Transform pairs for four types of poles.**

Pole	$\mathbf{X}(\mathbf{s})$	$x(t)$
1. Distinct real	$\dfrac{A}{\mathbf{s}+a}$	$Ae^{-at}\,u(t)$
2. Repeated real	$\dfrac{A}{(\mathbf{s}+a)^n}$	$A\,\dfrac{t^{n-1}}{(n-1)!}\,e^{-at}\,u(t)$
3. Distinct complex	$\left[\dfrac{Ae^{j\theta}}{\mathbf{s}+a+jb}+\dfrac{Ae^{-j\theta}}{\mathbf{s}+a-jb}\right]$	$2Ae^{-at}\cos(bt-\theta)\,u(t)$
4. Repeated complex	$\left[\dfrac{Ae^{j\theta}}{(\mathbf{s}+a+jb)^n}+\dfrac{Ae^{-j\theta}}{(\mathbf{s}+a-jb)^n}\right]$	$\dfrac{2At^{n-1}}{(n-1)!}\,e^{-at}\cos(bt-\theta)\,u(t)$

Example 3-9: Transform Involving e^{-as}

Determine the time-domain equivalent of the Laplace transform

$$\mathbf{X}(\mathbf{s})=\frac{\mathbf{s}e^{-3\mathbf{s}}}{\mathbf{s}^2+4}\,.$$

Solution: We start by separating out the exponential $e^{-3\mathbf{s}}$ from the remaining polynomial fraction. We do so by defining

$$\mathbf{X}(\mathbf{s})=e^{-3\mathbf{s}}\,\mathbf{X}_1(\mathbf{s}),$$

where

$$\mathbf{X}_1(\mathbf{s})=\frac{\mathbf{s}}{\mathbf{s}^2+4}=\frac{\mathbf{s}}{(\mathbf{s}+j2)(\mathbf{s}-j2)}=\frac{\mathbf{B}_1}{\mathbf{s}+j2}+\frac{\mathbf{B}_2}{\mathbf{s}-j2}$$

with

$$\mathbf{B}_1=(\mathbf{s}+j2)\,\mathbf{X}(\mathbf{s})|_{\mathbf{s}=-j2}=\left.\frac{\mathbf{s}}{\mathbf{s}-j2}\right|_{\mathbf{s}=-j2}=\frac{-j2}{-j4}=\frac{1}{2},$$

and

$$\mathbf{B}_2=\mathbf{B}_1^*=\frac{1}{2}\,.$$

Hence,

$$\mathbf{X}(\mathbf{s})=e^{-3\mathbf{s}}\,\mathbf{X}_1(\mathbf{s})=\frac{e^{-3\mathbf{s}}}{2(\mathbf{s}+j2)}+\frac{e^{-3\mathbf{s}}}{2(\mathbf{s}-j2)}\,.$$

By invoking property #3a of Table 3-2, we obtain the inverse Laplace transform

$$\begin{aligned}x(t)=\mathcal{L}^{-1}[\mathbf{X}(\mathbf{s})]&=\mathcal{L}^{-1}\left[\frac{1}{2}\,\frac{e^{-3\mathbf{s}}}{\mathbf{s}+j2}+\frac{1}{2}\,\frac{e^{-3\mathbf{s}}}{\mathbf{s}-j2}\right]\\&=\left[\frac{1}{2}(e^{-j2(t-3)}+e^{j2(t-3)})\right]u(t-3)\\&=[\cos(2t-6)]\,u(t-3).\end{aligned}$$

Alternatively, we could have obtained this result by using properties #4 in Table 3-1 and #11 in Table 3-2.

We conclude this section with Table 3-3, which lists $\mathbf{X}(\mathbf{s})$ and its corresponding inverse transform $x(t)$ for all combinations of real versus complex and distinct versus repeated poles.

▶ Computation of partial fraction expansions can be algebraically arduous. Hence, in Appendix D, we demonstrate how the `residue` command of MATLAB or MathScript can be used to compute the expansion coefficients. ◀

3-6 Transfer Function H(s)

If $y(t)$ is the output response of an LTI system to an input signal $x(t)$, and if $\mathbf{X}(\mathbf{s})$ and $\mathbf{Y}(\mathbf{s})$ are the Laplace transforms of $x(t)$ and $y(t)$, respectively, then the system is said to be characterized by an $\mathbf{s}$-domain ***transfer function*** $\mathbf{H}(\mathbf{s})$ defined as the ratio of $\mathbf{Y}(\mathbf{s})$ to $\mathbf{X}(\mathbf{s})$, *provided that all initial conditions of $y(t)$ are zero at $t=0^-$*. That is,

$$\mathbf{H}(\mathbf{s})=\frac{\mathbf{Y}(\mathbf{s})}{\mathbf{X}(\mathbf{s})}\qquad (\text{with } y(0^-)=0,\ y'(0^-)=0,\ldots). \tag{3.83}$$

As we will demonstrate shortly, $\mathbf{H}(\mathbf{s})$ is the Laplace transform of the system's ***impulse response*** $h(t)$, which was first introduced in Section 2-2.

3-6.1 Relationship to Impulse Response

Consider a system with an impulse response $h(t)$, excited by a causal input signal $x(t)$ and generating an output response $y(t)$. Their corresponding Laplace transforms are $\mathbf{X}(\mathbf{s})$, $\mathbf{H}(\mathbf{s})$, and $\mathbf{Y}(\mathbf{s})$:

$$x(t) \longleftrightarrow \mathbf{X}(\mathbf{s}) \tag{3.84a}$$

$$h(t) \longleftrightarrow \mathbf{H}(\mathbf{s}) \tag{3.84b}$$

$$y(t) \longleftrightarrow \mathbf{Y}(\mathbf{s}) \tag{3.84c}$$

According to Eq. (2.38), the system output $y(t)$ is given by the convolution of $x(t)$ with $h(t)$,

$$y(t) = x(t) * h(t) = \int_{0^-}^{\infty} x(\tau)\, h(t-\tau)\, d\tau, \tag{3.85}$$

where we used 0^- as the lower limit on the integral (as opposed to $-\infty$) because the signals and systems are causal.

Application of the Laplace transform to both sides yields

$$\begin{aligned}\mathcal{L}[x(t) * h(t)] &= \int_{0^-}^{\infty} \left[\int_{0^-}^{\infty} x(\tau)\, h(t-\tau)\, d\tau \right] e^{-\mathbf{s}t}\, dt \\ &= \int_{0^-}^{\infty} x(\tau) \left[\int_{0^-}^{\infty} h(t-\tau)\, e^{-\mathbf{s}t}\, dt \right] d\tau,\end{aligned} \tag{3.86}$$

where we interchanged the order of the two integrals in the second step, which is allowable because t and τ are independent variables.

Because τ in the inner integral represents nothing more than a constant time shift, we can introduce the dummy variable $\mu = t - \tau$ and replace dt with $d\mu$,

$$\begin{aligned}\mathcal{L}[x(t) * h(t)] &= \int_{0^-}^{\infty} x(\tau)\, e^{-\mathbf{s}\tau} \left[\int_{-\tau}^{\infty} h(\mu)\, e^{-\mathbf{s}\mu}\, d\mu \right] d\tau \\ &= \int_{0^-}^{\infty} x(\tau)\, e^{-\mathbf{s}\tau}\, d\tau \int_{0^-}^{\infty} h(\mu)\, e^{-\mathbf{s}\mu}\, d\mu \\ &= \mathbf{X}(\mathbf{s})\, \mathbf{H}(\mathbf{s}).\end{aligned} \tag{3.87}$$

In the middle step, we changed the lower limit on the second integral from $-\tau$ to 0^- because $h(\mu)$, being a causal signal, equals zero when its independent variable $\mu < 0$.

The result given by Eq. (3.87) can be framed as

$$y(t) = x(t) * h(t) \longleftrightarrow \mathbf{Y}(\mathbf{s}) = \mathbf{X}(\mathbf{s})\, \mathbf{H}(\mathbf{s}). \tag{3.88}$$

► Convolution in the time domain corresponds to multiplication in the **s**-domain. ◄

In symbolic form:

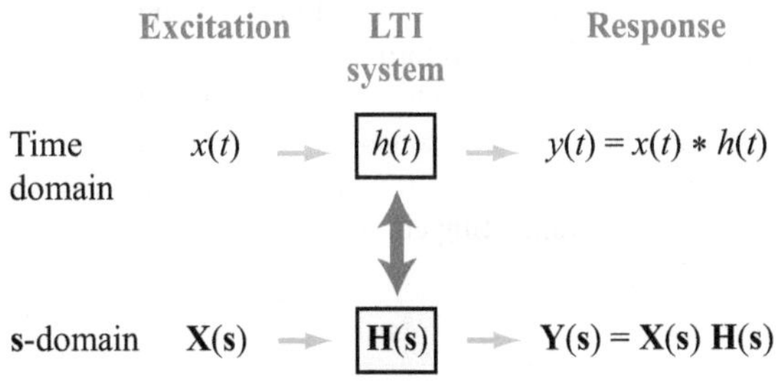

	Excitation	LTI system	Response
Time domain	$x(t)$ →	$h(t)$	→ $y(t) = x(t) * h(t)$
		↕	
s-domain	$\mathbf{X}(\mathbf{s})$ →	$\mathbf{H}(\mathbf{s})$	→ $\mathbf{Y}(\mathbf{s}) = \mathbf{X}(\mathbf{s})\,\mathbf{H}(\mathbf{s})$

3-6.2 Significance of the Transfer Function

For a linear system, $\mathbf{H}(\mathbf{s})$ does not depend on $\mathbf{X}(\mathbf{s})$, so an easy way by which to determine $\mathbf{H}(\mathbf{s})$ is to select an arbitrary excitation $\mathbf{X}(\mathbf{s})$, determine the corresponding response $\mathbf{Y}(\mathbf{s})$, and then form the ratio defined by Eq. (3.83). Of particular note is the excitation $\mathbf{X}(\mathbf{s}) = 1$, because then Eq. (3.83) simplifies to $\mathbf{H}(\mathbf{s}) = \mathbf{Y}(\mathbf{s})$. The inverse Laplace transform of 1 is the ***unit impulse function*** $\delta(t)$. Thus, when a system is excited by $x(t) = \delta(t)$, its **s**-domain output is equal to the system's transfer function $\mathbf{H}(\mathbf{s})$:

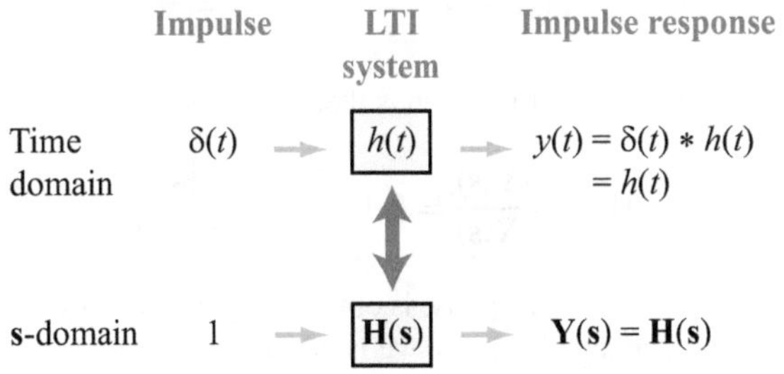

	Impulse	LTI system	Impulse response
Time domain	$\delta(t)$ →	$h(t)$	→ $y(t) = \delta(t) * h(t)$ $= h(t)$
		↕	
s-domain	1 →	$\mathbf{H}(\mathbf{s})$	→ $\mathbf{Y}(\mathbf{s}) = \mathbf{H}(\mathbf{s})$

Once $\mathbf{H}(\mathbf{s})$ of a given system has been established, $y(t)$ can be readily determined for any excitation $x(t)$ using the following steps:

Step 1: Transform $x(t)$ to the s-domain to obtain $\mathbf{X}(\mathbf{s})$.

Step 2: Multiply $\mathbf{X}(\mathbf{s})$ by $\mathbf{H}(\mathbf{s})$ to obtain $\mathbf{Y}(\mathbf{s})$.

Step 3: Express $\mathbf{Y}(\mathbf{s})$ in the form of a sum of partial fractions.

Step 4: Transform the partial-fraction expansion of $\mathbf{Y}(\mathbf{s})$ to the time domain to obtain $y(t)$.

The process is straightforward, as long as $x(t)$ is transformable to the s-domain and $\mathbf{Y}(\mathbf{s})$ is transformable to the time domain (i.e., steps 1 and 4). If $x(t)$ is some irregular or unusual waveform, or if it consists of a series of experimental measurements generated by another system, it may not be possible to transform $x(t)$ to the s-domain analytically. Also, in some cases, the functional form of $\mathbf{Y}(\mathbf{s})$ may be so complicated that it may be very difficult to express it in a form amenable for transformation to the time domain. In such cases, the alternative approach is to determine $y(t)$ by computing the convolution integral, thereby operating entirely in the time domain.

Example 3-10: Transfer Function

The output response of a system excited by a unit step function at $t = 0$ is given by

$$y(t) = [2 + 12e^{-3t} - 6\cos 2t]\, u(t).$$

Determine: (a) the transfer function of the system and (b) its impulse response.

Solution: (a) The Laplace transform of a unit step function is

$$\mathbf{X}(\mathbf{s}) = \frac{1}{\mathbf{s}}.$$

By using entries #2, 3, and 11 in Table 3-2, we obtain the Laplace transform of the output response

$$\mathbf{Y}(\mathbf{s}) = \frac{2}{\mathbf{s}} + \frac{12}{\mathbf{s}+3} - \frac{6\mathbf{s}}{\mathbf{s}^2+4}.$$

The system transfer function is then given by

$$\mathbf{H}(\mathbf{s}) = \frac{\mathbf{Y}(\mathbf{s})}{\mathbf{X}(\mathbf{s})} = 2 + \frac{12\mathbf{s}}{\mathbf{s}+3} - \frac{6\mathbf{s}^2}{\mathbf{s}^2+4}.$$

(b) The impulse response is obtained by transferring $\mathbf{H}(\mathbf{s})$ to the time domain. To do so, we need to have every term in the expression for $\mathbf{H}(\mathbf{s})$ to be in the form of a sum of a constant (which may be zero) and a strictly proper rational function, which can be realized by applying the division relationship discussed in Section 3-5. Such a process leads to

$$\begin{aligned}\mathbf{H}(\mathbf{s}) &= 2 + \left(12 - \frac{36}{\mathbf{s}+3}\right) + \left(-6 + \frac{24}{\mathbf{s}^2+4}\right) \\ &= 8 - \frac{36}{\mathbf{s}+3} + \frac{24}{\mathbf{s}^2+4}.\end{aligned}$$

The corresponding inverse transform is

$$h(t) = 8\delta(t) - 36e^{-3t}\, u(t) + 12\sin(2t)\, u(t).$$

Concept Question 3-8: The transfer function $\mathbf{H}(\mathbf{s})$ is defined under what initial conditions? (See S²)

3-7 Poles and System Stability

In general, a system transfer function $\mathbf{H}(\mathbf{s})$ is given in the form of a ratio of two polynomials,

$$\mathbf{H}(\mathbf{s}) = \frac{\mathbf{N}(\mathbf{s})}{\mathbf{D}(\mathbf{s})}. \tag{3.89}$$

Suppose $\mathbf{N}(\mathbf{s})$ is of degree m (highest power of $\mathbf{s}$) and $\mathbf{D}(\mathbf{s})$ is of degree n. It then follows that $\mathbf{H}(\mathbf{s})$ has m zeros and n poles. The zeros and poles may be real or complex, also distinct or repeated. We will now examine how the poles/zeros and their orders define the stability of the system characterized by $\mathbf{H}(\mathbf{s})$.

3-7.1 Strictly Proper Rational Function ($m < n$)

If $\mathbf{H}(\mathbf{s})$ has n poles, several combinations of distinct and repeated poles may exist.

Case 1: All n poles are distinct

Application of partial fraction expansion (Section 3-5.1) leads to

$$\mathbf{H}(\mathbf{s}) = \sum_{i=1}^{n} \frac{\mathbf{A}_i}{\mathbf{s} - \mathbf{p}_i}, \tag{3.90}$$

where $\mathbf{p}_1$ to $\mathbf{p}_n$ are real or complex poles and $\mathbf{A}_1$ to $\mathbf{A}_n$ are their corresponding residues. According to Section 3-5, the inverse Laplace transform of $\mathbf{H}(\mathbf{s})$ is

$$h(t) = [\mathbf{A}_1 e^{\mathbf{p}_1 t} + \mathbf{A}_2 e^{\mathbf{p}_2 t} + \cdots \mathbf{A}_n e^{\mathbf{p}_n t}]\, u(t). \tag{3.91}$$

In Section 2-6.4, we demonstrated that an LTI system with such an impulse response is BIBO stable if and only if all of the exponential coefficients have negative real parts. That is,

$$\mathfrak{Re}[\mathbf{p}_i] < 0, \qquad i = 1, 2, \ldots, n. \tag{3.92}$$

▶ The condition for BIBO stability defined by Eq. (3.92) requires that $\mathbf{p}_1$ to $\mathbf{p}_n$ reside in the *open left half-plane* (*OLHP*) in the **s**-domain. The imaginary axis is not included in the OLHP; a system with poles along the imaginary axis is not BIBO stable. ◀

Example 3-11: Transfer Function I

Determine if the transfer function

$$\mathbf{H}(\mathbf{s}) = \frac{\mathbf{s}^2 - 4\mathbf{s} + 3}{\mathbf{s}(\mathbf{s}+1)(\mathbf{s}+3)} \tag{3.93}$$

is BIBO stable.

Solution: The expression for $\mathbf{H}(\mathbf{s})$ has poles $\mathbf{p}_1 = 0$, $\mathbf{p}_2 = -1$, and $\mathbf{p}_3 = -3$. Hence, two of the poles are in the OLHP, but the third is at the origin (Fig. 3-4). According to the condition given by Eq. (3.92), the system is not BIBO stable. This conclusion can be further ascertained by considering the time-domain equivalent of Eq. (3.93), which is given by Eq. (3.60) as

$$h(t) = [1 - 4e^{-t} + 4e^{-3t}]\, u(t). \tag{3.94}$$

For $x(t) = u(t)$, the step response $y_{\text{step}}(t)$ is

$$y_{\text{step}}(t) = h(t) * u(t) = \left\{[1 - 4e^{-t} + 4e^{-3t}]\, u(t)\right\} * u(t). \tag{3.95}$$

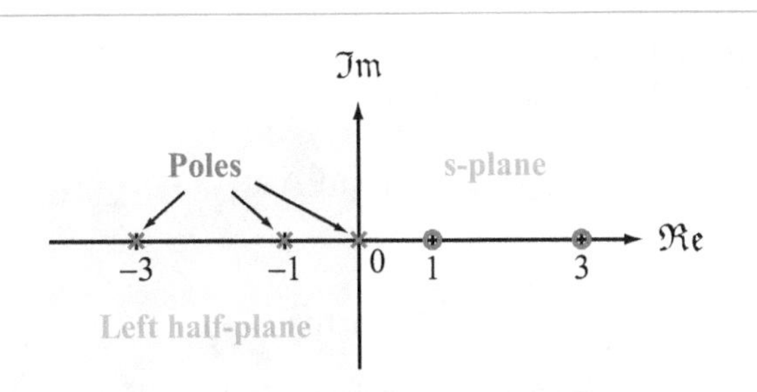

Figure 3-4: The system with poles at $\mathbf{s} = 0, -1$, and -3 is not BIBO stable because not all its poles are in the open left-hand plane (OLHP) of the **s**-domain (the pole at the origin is not in the OLHP).

From Table 2-2, we obtain

$$u(t) * u(t) = t\, u(t)$$

and

$$e^{at}\, u(t) * u(t) = \left[\frac{e^{at} - 1}{a}\right] u(t),$$

which when applied to Eq. (3.95), yields

$$y_{\text{step}}(t) = \left[t + 4(e^{-t} - 1) - \frac{4}{3}(e^{-3t} - 1)\right] u(t). \tag{3.96}$$

As $t \to \infty$, $y_{\text{step}}(t) \approx (t - 8/3)\, u(t) \to \infty$, thereby confirming our earlier conclusion that the system described by Eq. (3.93) is not BIBO stable.

Case 2: **Some poles are repeated**

If $\mathbf{H}(\mathbf{s})$ has a k-repeated pole $\mathbf{p}$, for example, of the form

$$\mathbf{H}(\mathbf{s}) = \frac{\mathbf{N}(\mathbf{s})}{(\mathbf{s} - \mathbf{p})^k}, \tag{3.97}$$

it can be transformed into

$$\mathbf{H}(\mathbf{s}) = \sum_{j=1}^{k} \frac{\mathbf{B}_j}{(\mathbf{s} - \mathbf{p})^j} = \frac{\mathbf{B}_1}{\mathbf{s} - \mathbf{p}} + \frac{\mathbf{B}_2}{(\mathbf{s} - \mathbf{p})^2} + \cdots + \frac{\mathbf{B}_k}{(\mathbf{s} - \mathbf{p})^k}. \tag{3.98}$$

According to Eq. (3.74), its inverse transform is

$$\begin{aligned} h(t) &= \left[\sum_{j=1}^{k} \frac{\mathbf{B}_j t^{j-1}}{(j-1)!}\, e^{\mathbf{p}t}\right] u(t) \\ &= \left[\mathbf{B}_1 + \mathbf{B}_2 t + \frac{\mathbf{B}_3 t^2}{2} + \cdots + \frac{\mathbf{B}_k t^{k-1}}{(k-1)!}\right] e^{\mathbf{p}t}\, u(t). \end{aligned} \tag{3.99}$$

According to Section 2-6.3, a system is BIBO stable if and only if its impulse response is absolutely integrable. That is,

$$\int_{-\infty}^{\infty} |h(t)|\, dt \text{ is finite.}$$

Hence $h(t)$ is absolutely integrable if and only if $\mathfrak{Re}[\mathbf{p}] < 0$. This requirement for repeated poles is identical with the requirement for distinct poles given by Eq. (3.92).

▶ A system whose transfer function $\mathbf{H}(\mathbf{s})$ is a strictly proper rational function is BIBO stable if and only if its distinct and repeated poles, whether real or complex, reside in the OLHP of the **s**-domain, which excludes the imaginary axis. Furthermore, the locations of the zeros of $\mathbf{H}(\mathbf{s})$ have no bearing on the system's stability. ◀

3-7.2 Proper Rational Function ($m = n$)

Consider the function

$$\mathbf{H(s)} = \frac{2\mathbf{s}^2 + 10\mathbf{s} + 6}{\mathbf{s}^2 + 5\mathbf{s} + 6}. \tag{3.100}$$

This is a proper rational function because $m = n = 2$. By applying the division relationship given by Eq. (3.55), $\mathbf{H(s)}$ can be re-expressed as

$$\begin{aligned}\mathbf{H(s)} &= 2 - \frac{6}{\mathbf{s}^2 + 5\mathbf{s} + 6} = 2 - \frac{6}{(\mathbf{s}+2)(\mathbf{s}+3)} \\ &= 2 - \frac{6}{(\mathbf{s}+2)} + \frac{6}{(\mathbf{s}+3)}. \end{aligned} \tag{3.101}$$

The two poles of $\mathbf{H(s)}$ are $\mathbf{p}_1 = -2$ and $\mathbf{p}_2 = -3$, both of which are in the OLHP in the $\mathbf{s}$-domain, but what role will the constant term (2) play with regard to system stability? Let's examine the inverse transform,

$$h(t) = 2\delta(t) - [6e^{-2t} - 6e^{-3t}]\, u(t). \tag{3.102}$$

We already know that the last two terms of Eq. (3.102) satisfy the stability condition, so we will deal with the first term only:

$$\int_{-\infty}^{\infty} 2|\delta(t)|\, dt = 2,$$

which is finite.

> ► A system whose transfer function is a proper rational function obeys the same rules with regard to system stability as a strictly proper rational function. ◄

3-7.3 Improper Rational Function ($m > n$)

The function

$$\mathbf{H(s)} = \frac{6\mathbf{s}^3 + 4\mathbf{s}^2 + 8\mathbf{s} + 6}{\mathbf{s}^2 + 2\mathbf{s} + 1} \tag{3.103}$$

is an improper rational function because $m = 3$ and $n = 2$. By applying the division relationship given by Eq. (3.55) twice, $\mathbf{H(s)}$ can be expressed as

$$\mathbf{H(s)} = 6\mathbf{s} - 8 + \frac{18\mathbf{s} + 14}{(\mathbf{s}+1)^2}. \tag{3.104}$$

Based on our analysis in the preceding two subsections, terms 2 and 3 present no issues with regard to system stability, but the first term is problematic. Let us define the first term as

$$\mathbf{H}_1(\mathbf{s}) = 6\mathbf{s}, \tag{3.105}$$

and let us examine the output response to a bounded input $x(t) = u(t)$. The Laplace transform of $u(t)$ is $1/\mathbf{s}$, so

$$\mathbf{Y(s)} = \mathbf{H}_1(\mathbf{s})\, \mathbf{X(s)} = 6\mathbf{s} \cdot \frac{1}{\mathbf{s}} = 6. \tag{3.106}$$

The corresponding time-domain output is

$$y(t) = 6\,\delta(t), \tag{3.107}$$

which is an unbounded singularity function.

> ► A system whose transfer function is an improper rational function is not BIBO stable, regardless of the locations of its poles and zeros. ◄

Example 3-12: Dangerous Consequences of Unstable Systems

A system has a transfer function $\mathbf{H(s)} = \mathbf{s} + 1$. Because $\mathbf{H(s)}$ is an improper rational function, the system is not BIBO stable. The undesirability of such a system can be demonstrated by examining the consequence of having high-frequency noise accompany an input signal. Suppose the input to the system is

$$x(t) = x_s(t) + x_n(t),$$

where $x_s(t)$ is the intended input signal and $x_n(t)$ is the unintended input noise, respectively. Furthermore, let

$$x_s(t) = 10 \sin(10^3 t)\, u(t)$$

and

$$x_n(t) = 10^{-2} \sin(10^7 t)\, u(t).$$

Note that the amplitude of the noise is 1000 times smaller than that of the signal, but the angular frequency of the signal is 10^3 rad/s, compared with 10^7 rad/s for the noise. Determine the *signal-to-noise ratio* [the ratio of the average power of $x_s(t)$ to that of $x_n(t)$] at the input and output of the system.

Solution:

(a) **At input**

The signal-to-noise ratio is defined in terms of the average power carried by a waveform. Per Eq. (1.38), the average power of a sinusoidal signal of amplitude 10 is

$$P_s(\text{at input}) = \tfrac{1}{2}\,(10)^2 = 50.$$

Similarly, for the noise component, we have

$$P_n(\text{at input}) = \tfrac{1}{2}\,(10^{-2})^2 = 5 \times 10^{-5}.$$

The signal-to-noise ratio is

$$(S/N)_{\text{input}} = \frac{P_s}{P_n} = \frac{50}{5 \times 10^{-5}} = 10^6.$$

Note that since the average power is proportional to the square of amplitude, the amplitude ratio of 10^3 becomes an average power ratio of 10^6. The fact that $S/N = 10^6$ means that the signal at the input to the system is essentially unaffected by the noise.

(b) **At output**

For $\mathbf{H(s)} = \mathbf{s} + 1$, the output is

$$\mathbf{Y(s)} = \mathbf{H(s)}\,\mathbf{X(s)} = \mathbf{s}\,\mathbf{X(s)} + \mathbf{X(s)}.$$

From Table 3-1, the following two Laplace transform properties are of interest:

$$x(t) \quad\longleftrightarrow\quad \mathbf{X(s)}$$
$$\frac{dx}{dt} \quad\longleftrightarrow\quad \mathbf{s}\,\mathbf{X(s)} - \mathbf{s}\,x(0^-).$$

In the present case, $x(t)$ consists of sinusoidal waveforms, so $x(0^-) = 0$. Application of the Laplace transform properties leads to

$$\begin{aligned}
y(t) &= \mathcal{L}^{-1}[\mathbf{Y(s)}] = \mathcal{L}^{-1}[\mathbf{s}\,\mathbf{X(s)} + \mathbf{X(s)}] \\
&= \frac{dx}{dt} + x(t) \\
&= \frac{d}{dt}\,[(x_s + x_n)] + (x_s + x_n) \\
&= \frac{d}{dt}[10\sin(10^3 t) + 10^{-2}\sin(10^7 t)] \\
&\quad + 10\sin(10^3 t) + 10^{-2}\sin(10^7 t) \\
&= \underbrace{[10^4\cos(10^3 t) + 10\sin(10^3 t)]}_{\text{output signal at } 10^3 \text{ rad/s}} \\
&\quad + \underbrace{[10^5\cos(10^7 t) + 10^{-2}\sin(10^7 t)]}_{\text{output noise at } 10^7 \text{ rad/s}}.
\end{aligned}$$

The average powers associated with the signal and noise are

$$P_s(\text{@output}) = \tfrac{1}{2}\,(10^4)^2 + \tfrac{1}{2}\,(10)^2 \approx 5 \times 10^7,$$
$$P_n(\text{@output}) = \tfrac{1}{2}\,(10^5)^2 + \tfrac{1}{2}\,(10^{-2})^2 \approx 5 \times 10^9,$$

and the signal-to-noise ratio is

$$(S/N)_{\text{output}} = \frac{5 \times 10^7}{5 \times 10^9} = 10^{-2}.$$

Whereas the noise component was inconsequential at the input end of the system, propagation through the unstable system led to the undesirable consequence that at the output the noise power became two orders of magnitude greater than that of the signal power.

Concept Question 3-9: Why is it that zeros of the transfer function have no bearing on system stability? (See S²)

Concept Question 3-10: Why is a system with an improper transfer function always BIBO unstable? (See S²)

Exercise 3-11: Is the system with transfer function

$$\mathbf{H(s)} = \frac{\mathbf{s} + 1}{(\mathbf{s} + j3)(\mathbf{s} - j3)}$$

BIBO stable?

Answer: No. It has poles *on* the imaginary axis. A stable system has all its poles in the OLHP. (See S²)

Exercise 3-12: Is the system with transfer function

$$\mathbf{H(s)} = \frac{(\mathbf{s}+1)(\mathbf{s}+2)(\mathbf{s}+3)}{(\mathbf{s}+4)(\mathbf{s}+5)}$$

BIBO stable?

Answer: No. $\mathbf{H(s)}$ is improper. (See S²)

3-8 Invertible Systems

A system with input signal $x(t)$ and a corresponding output response $y(t)$ is said to be ***invertible*** if an operation exists by which $x(t)$ can be determined from $y(t)$. In other words, the system is invertible if an ***inverse system*** exists such that

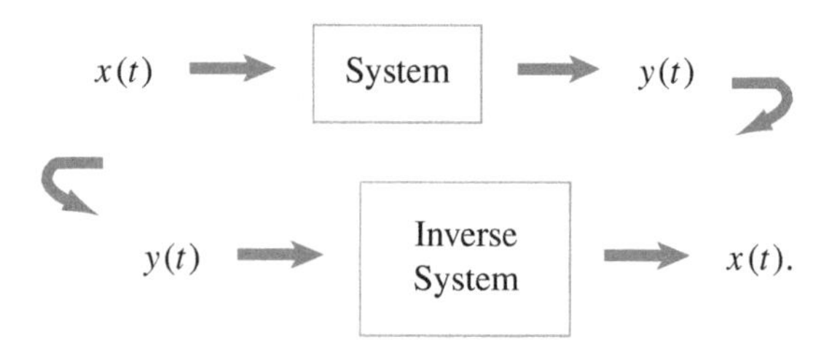

Invertible systems are important in signal processing. One example is when ***equalization*** is applied to the system's frequency response spectrum by boosting the amplitudes of its high-frequency components.

A trivial example of a non-invertible system is

$$y(t) = 2|x(t)|;$$

the sign of $x(t)$ cannot be recovered from $y(t)$.

3-8.1 Inverse System

Consider the following three causal LTI systems and their inverses:

$$y_1(t) = 4x_1(t) \Rightarrow x_1(t) = \frac{1}{4}\, y_1(t), \tag{3.108}$$

$$y_2(t) = [5 + 3\sin 4t]\, x_2(t) \Rightarrow x_2(t) = \frac{1}{5 + 3\sin 4t}\, y_2(t), \tag{3.109}$$

and

$$y_3(t) = 2\,\frac{dx_3}{dt} \Rightarrow x_3(t) = \frac{1}{2}\int_{-\infty}^{t} y_3(\tau)\, d\tau.$$

All three are very simple, easily invertible systems, but most systems are not so simple. Consider, for example, the LCCDE

$$\frac{d^2y}{dt^2} + a_1\,\frac{dy}{dt} + a_2\, y(t) = b_1\,\frac{dx}{dt} + b_2\, x(t). \tag{3.110}$$

Given $x(t)$, we can determine $y(t)$ either by solving the equation in the time domain or by applying the Laplace transform technique. For causal signals with zero initial conditions, transforming Eq. (3.110) to the s-domain entails replacing $x(t)$ with $\mathbf{X(s)}$ and $y(t)$ with $\mathbf{Y(s)}$, and replacing each time derivative with multiplication by $\mathbf{s}$. That is,

$$\mathbf{s}^2\,\mathbf{Y(s)} + a_1\mathbf{s}\,\mathbf{Y(s)} + a_2\,\mathbf{Y(s)} = b_1\mathbf{s}\,\mathbf{X(s)} + b_2\,\mathbf{X(s)},$$

which can be solved to obtain the transfer function

$$\mathbf{H(s)} = \frac{\mathbf{Y(s)}}{\mathbf{X(s)}} = \frac{b_1\mathbf{s} + b_2}{\mathbf{s}^2 + a_1\mathbf{s} + a_2}. \tag{3.111}$$

According to Section 3-8, $\mathbf{H(s)}$ is a proper rational function because the order of its numerator is smaller than that of its denominator, namely, $1 < 2$. Hence, if both poles of $\mathbf{H(s)}$ reside in the OLHP of the s-domain, then $\mathbf{H(s)}$ represents a stable system.

In contrast, the transfer function of the inverse system, $\mathbf{G(s)}$, is given by

$$\mathbf{G(s)} = \frac{1}{\mathbf{H(s)}} = \frac{\mathbf{X(s)}}{\mathbf{Y(s)}} = \frac{\mathbf{s}^2 + a_1\mathbf{s} + a_2}{b_1\mathbf{s} + b_2} \tag{3.112}$$

and is not stable because $\mathbf{G(s)}$ is an improper rational function. This means that in order for both a system $\mathbf{H(s)}$ and its inverse $\mathbf{G(s)}$ to be stable, it is necessary (but not sufficient) that $\mathbf{H(s)}$ be a proper rational function with $m = n$, where m and n are the degrees of its numerator and denominator. To extend this statement to include the ***sufficient*** part, *both the poles and zeros of* $\mathbf{H(s)}$ *should reside in the OLHP of the* s*-domain*. Under those conditions, the poles of $\mathbf{G(s)}$, which are the zeros of $\mathbf{H(s)}$, will also lie in the OLHP, thereby satisfying the stability condition.

► A BIBO stable and causal LTI system has a BIBO stable and causal inverse system if and only if all of its poles and zeros are in the open left half-plane, and they are equal in number (its transfer function is proper). Such a system is called a ***minimum phase*** system. ◄

Example 3-13: Inverse System

Compute the inverse system for the system with

$$h(t) = \delta(t) + 2e^{-3t}\, u(t) - 6e^{-4t}\, u(t). \tag{3.113}$$

Solution: Taking the Laplace transform of $h(t)$ and putting each rational function term over a common denominator gives

$$\begin{aligned}\mathbf{H(s)} &= 1 + \frac{2}{\mathbf{s}+3} - \frac{6}{\mathbf{s}+4} \\ &= \frac{(\mathbf{s}+3)(\mathbf{s}+4) + 2(\mathbf{s}+4) - 6(\mathbf{s}+3)}{(\mathbf{s}+3)(\mathbf{s}+4)} \\ &= \frac{\mathbf{s}^2 + 3\mathbf{s} + 2}{\mathbf{s}^2 + 7\mathbf{s} + 12} = \frac{(\mathbf{s}+1)(\mathbf{s}+2)}{(\mathbf{s}+3)(\mathbf{s}+4)}.\end{aligned} \tag{3.114}$$

We note that $\mathbf{H}(\mathbf{s})$ is proper ($m = n = 2$) and has

- Zeros at $\{-1, -2\}$, both in the open left half-plane
- Poles at $\{-3, -4\}$, both in the open left half-plane

Hence, the inverse system exists, and its transfer function is

$$\mathbf{G}(\mathbf{s}) = \frac{1}{\mathbf{H}(\mathbf{s})} = \frac{\mathbf{s}^2 + 7\mathbf{s} + 12}{\mathbf{s}^2 + 3\mathbf{s} + 2} = 1 + \frac{4\mathbf{s} + 10}{(\mathbf{s}+1)(\mathbf{s}+2)} = 1 + \frac{6}{\mathbf{s}+1} - \frac{2}{\mathbf{s}+2}. \tag{3.115}$$

The impulse response of the inverse system is

$$g(t) = \delta(t) + 6e^{-t}\, u(t) - 2e^{-2t}\, u(t). \tag{3.116}$$

Concept Question 3-11: Why doesn't a strictly proper transfer function have a BIBO stable and causal inverse system? (See S²)

Concept Question 3-12: What role do zeros of transfer functions play in system invertibility? (See S²)

Exercise 3-13: A system has the impulse response $h(t) = \delta(t) - 2e^{-3t}\, u(t)$. Find its inverse system.

Answer: $g(t) = \delta(t) + 2e^{-t}\, u(t)$. (See S²)

3-9 Bilateral Transform for Continuous-Time Sinusoidal Signals

So far, we have limited our examination of the Laplace transform to causal signals that start at $t = 0$, and in order to incorporate initial conditions, we defined the Laplace transform of the system impulse response as

$$\mathbf{H}(\mathbf{s}) = \int_{0^-}^{\infty} h(\tau)\, e^{-\mathbf{s}\tau}\, d\tau. \tag{3.117}$$

This is the ***unilateral Laplace transform*** with a lower integration limit of 0^-.

Let's assume we wish to determine the output response $y(t)$ to a sinusoidal signal

$$x(t) = A\cos(\omega t + \phi), \tag{3.118}$$

and that $x(t)$ has existed for a long time, long enough to be regarded as ***everlasting***; i.e., extending over $(-\infty, \infty)$. In that case, there are no initial conditions, and the solution $y(t)$ we seek is the steady-state response. Because $x(t)$ is everlasting, we need to adopt the ***bilateral Laplace transform*** definition for the Laplace transforms of $x(t)$, $y(t)$, and $h(t)$. For $h(t)$, its bilateral transform is

$$\mathbf{H}_b(\mathbf{s}) = \int_{-\infty}^{\infty} h(\tau)\, e^{-\mathbf{s}\tau}\, d\tau, \tag{3.119}$$

wherein we used $-\infty$ as the lower integration limit and added a subscript "b" to distinguish the bilateral transform $\mathbf{H}_b(\mathbf{s})$ from the unilateral transform $\mathbf{H}(\mathbf{s})$. However, because $h(t)$ is causal, changing the lower integration limit from 0^- to $-\infty$ has no impact on the resulting expression for $\mathbf{H}_b(\mathbf{s})$. *Thus, $\mathbf{H}_b(\mathbf{s}) = \mathbf{H}(\mathbf{s})$ for causal systems.*

In Section 2-7.4, we demonstrated that for an LTI system with impulse response $h(t)$,

$$\cos\omega t \longrightarrow \boxed{h(t)} \longrightarrow y(t) = |\mathbf{H}(\omega)|\cos(\omega t + \theta), \tag{3.120}$$

where

$$\mathbf{H}(\omega) = |\mathbf{H}(\omega)|\angle\theta$$

and

$$\mathbf{H}(\omega) = \int_{-\infty}^{\infty} h(\tau)\, e^{-j\omega\tau}\, d\tau. \tag{3.121}$$

Comparison of Eqs. (3.119) and (3.121) leads to the conclusion that *$\mathbf{H}_b(j\omega)$ reduces to $\mathbf{H}(\omega)$ when the input signal is a sinusoid with angular frequency ω. Hence, if we have an expression for $\mathbf{H}(\mathbf{s})$, $\mathbf{H}(\omega)$ can be obtained by replacing $\mathbf{s}$ with $j\omega$,*

$$\mathbf{H}(\omega) = \mathbf{H}(\mathbf{s})\big|_{\mathbf{s}=j\omega} \qquad \begin{pmatrix}\text{sinusoidal input}\\ \text{causal system}\end{pmatrix}. \tag{3.122}$$

The output $y(t)$ can then be determined by applying Eq. (3.120).

Example 3-14: Sinusoidal Signal

An LTI system with a transfer function

$$\mathbf{H}(\mathbf{s}) = 100/(\mathbf{s}^2 + 15\mathbf{s} + 600)$$

has an input signal

$$x(t) = 10\cos(20t + 30^\circ).$$

Determine the output response $y(t)$.

Solution: At $\omega = 20$ rad/s, we have

$$\begin{aligned}\mathbf{H}(\omega) = \mathbf{H}(\mathbf{s})|_{\mathbf{s}=j20} &= \frac{100}{(j20)^2 + 15(j20) + 600} \\ &= \frac{100}{200 + j300} = 0.28\underline{/-56.31^\circ}\ .\end{aligned}$$

Application of Eq. (3.120) leads to

$$\begin{aligned}y(t) &= A|\mathbf{H}(\omega)|\cos(\omega t + \theta + \phi) \\ &= 10 \times 0.28\cos(20t - 56.31^\circ + 30^\circ) \\ &= 2.8\cos(20t - 26.31^\circ).\end{aligned}$$

3-10 Interrelating Different Descriptions of LTI Systems

An LTI system can be described in six different ways, namely:

- Transfer function $\mathbf{H}(\mathbf{s})$
- Impulse response $h(t)$
- Poles $\{\mathbf{p}_i,\ i = 1, \ldots, n\}$ and zeros $\{\mathbf{z}_i,\ i = 1, \ldots, m\}$
- Specific input-output pair, $x(t)$ and $y(t)$, where

$x(t)$ ⟶ LTI ⟶ $y(t)$

- Differential Equation (LCCDE)
- Frequency response $\mathbf{H}(\omega)$

Given any one of these descriptions, we can easily determine the other five (Fig. 3-5). Specifically, the transfer function $\mathbf{H}(\mathbf{s})$ can be related to the other descriptions as follows.

Impulse response $h(t)$

- Given $h(t)$, its corresponding transfer function can be determined by applying the transformation

$$\mathbf{H}(\mathbf{s}) = \mathcal{L}[h(t)].$$

- Conversely, given $\mathbf{H}(\mathbf{s})$, it follows that

$$h(t) = \mathcal{L}^{-1}[\mathbf{H}(\mathbf{s})].$$

Poles and zeros

- Given $\{\mathbf{p}_i,\ i = 1, \ldots, n\}$ and $\{\mathbf{z}_i,\ i = 1, \ldots, m\}$, $\mathbf{H}(\mathbf{s})$ can be expressed as

$$\mathbf{H}(\mathbf{s}) = \mathbf{C}\,\frac{\prod_{i=1}^{m}(\mathbf{s} - \mathbf{z}_i)}{\prod_{i=1}^{n}(\mathbf{s} - \mathbf{p}_i)}\ . \tag{3.123}$$

where $\mathbf{C}$ is a constant that needs to be specified.

- Conversely, given $\mathbf{H}(\mathbf{s})$, we can recast its expression as the ratio of two polynomials $\mathbf{N}(\mathbf{s})$ and $\mathbf{D}(\mathbf{s})$:

$$\mathbf{H}(\mathbf{s}) = \frac{\mathbf{N}(\mathbf{s})}{\mathbf{D}(\mathbf{s})}\ ,$$

in which case we can determine the poles $\{\mathbf{p}_i,\ i = 1, \ldots, n\}$ by computing the roots of $\mathbf{D}(\mathbf{s}) = 0$ and the zeros $\{\mathbf{z}_i,\ i = 1, \ldots, m\}$ by computing the roots of $\mathbf{N}(\mathbf{s}) = 0$.

Input-output pair

- Given the response $y(t)$ to an input $x(t)$, $\mathbf{H}(\mathbf{s})$ can be determined from the ratio of their Laplace transforms:

$$\mathbf{H}(\mathbf{s}) = \frac{\mathcal{L}[y(t)]}{\mathcal{L}[x(t)]}\ .$$

- Conversely, given $\mathbf{H}(\mathbf{s})$, the output response $y(t)$ to a specific input $x(t)$ is

$$y(t) = \mathcal{L}^{-1}\left[\mathbf{H}(\mathbf{s}) \cdot \mathcal{L}[x(t)]\right].$$

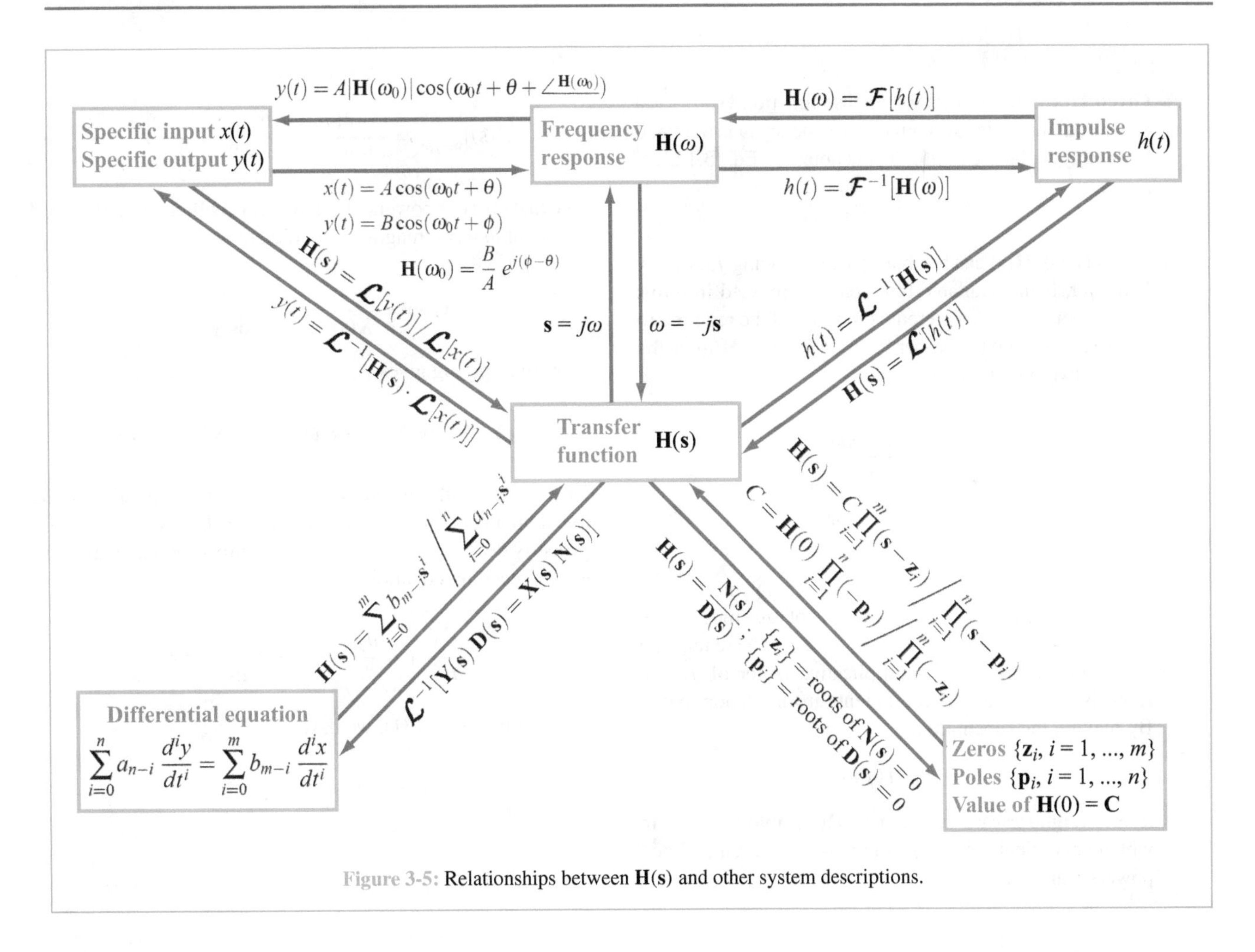

Figure 3-5: Relationships between $\mathbf{H(s)}$ and other system descriptions.

Differential equation (LCCDE)

- Given an LCCDE in the general form given by Eq. (2.6):

$$\sum_{i=0}^{n} a_{n-i} \frac{d^i y}{dt^i} = \sum_{i=0}^{m} b_{m-i} \frac{d^i x}{dt^i} \tag{3.124}$$

taking the Laplace transform of both sides gives

$$\mathcal{L}\left[\sum_{i=0}^{n} a_{n-i} \frac{d^i y}{dt^i}\right] = \mathcal{L}\left[\sum_{i=0}^{m} b_{m-i} \frac{d^i x}{dt^i}\right]. \tag{3.125a}$$

Applying the linearity property of the Laplace transform, we have

$$\sum_{i=0}^{n} a_{n-i}\, \mathcal{L}\left[\frac{d^i y}{dt^i}\right] = \sum_{i=0}^{m} b_{m-i}\, \mathcal{L}\left[\frac{d^i x}{dt^i}\right]. \tag{3.125b}$$

Under zero initial conditions, application of property #6 in Table 3-1 leads to

$$\sum_{i=0}^{n} a_{n-i}\mathbf{s}^i\ \mathbf{Y(s)} = \sum_{i=0}^{m} b_{m-i}\mathbf{s}^i\ \mathbf{X(s)}, \tag{3.126}$$

from which we obtain the transfer function

$$\mathbf{H(s)} = \frac{\mathbf{Y(s)}}{\mathbf{X(s)}} = \frac{\sum_{i=0}^{m} b_{m-i}\mathbf{s}^i}{\sum_{i=0}^{n} a_{n-i}\mathbf{s}^i}. \tag{3.127}$$

- Given $\mathbf{H(s)}$, the LCCDE can be obtained by expressing $\mathbf{H(s)}$ as the ratio of two polynomials and then reversing steps back to Eq. (3.124).

Frequency response $\mathbf{H}(\omega)$

- Given $\mathbf{H}(\mathbf{s})$, the frequency response function $\mathbf{H}(\omega)$ for a sinusoidal input signal at angular frequency ω is obtained by simply replacing $\mathbf{s}$ with $j\omega$ [as noted by Eq. (3.122)]:

$$\mathbf{H}(\omega) = \mathbf{H}(\mathbf{s})\big|_{\mathbf{s}=j\omega}. \tag{3.128}$$

- Given $\mathbf{H}(\omega)$, $\mathbf{H}(\mathbf{s})$ can be obtained by replacing $j\omega$ with $\mathbf{s}$. However, if the available function is expressed in terms of ω rather than $j\omega$, a prerequisite step will be required to convert $\mathbf{H}(\omega)$ into the proper form. In general, $\mathbf{H}(\omega)$ is the ratio of two polynomials:

$$\mathbf{H}(\omega) = \frac{\sum_{i=0}^{m} b_i \omega^i}{\sum_{i=0}^{n} a_i \omega^i}. \tag{3.129}$$

For a real system, even powers of ω must have real coefficients and odd powers of ω must have imaginary coefficients. That is, $b_0, b_2, b_4, \ldots$ are real and $b_1, b_3, b_5, \ldots$ include a multiplicative factor of j. The same is true for the a_i coefficients in the denominator. By making the substitution

$$\omega = -j(j\omega),$$

it is straightforward to convert $\mathbf{H}(\omega)$ into a form in which coefficients of all $(j\omega)$ terms—both even and odd powers—are real.

Example 3-15: Relate $h(t)$ to Other LTI System Descriptions

Given an LTI system with $h(t) = e^{-2t}\, u(t) + e^{-4t}\, u(t)$, determine the following: (a) $\mathbf{H}(\mathbf{s})$, (b) $\mathbf{H}(\omega)$, (c) LCCDE, (d) poles and zeros, and (e) the output response $y(t)$ due to an input $x(t) = \delta(t) + e^{-3t}\, u(t)$. Assume zero initial conditions.

Solution:

(a)

$$\mathbf{H}(\mathbf{s}) = \mathcal{L}[h(t)] = \mathcal{L}[e^{-2t}\, u(t) + e^{-4t}\, u(t)]. \tag{3.130}$$

Application of entry #3 in Table 3-2 leads to

$$\mathbf{H}(\mathbf{s}) = \frac{1}{\mathbf{s}+2} + \frac{1}{\mathbf{s}+4} = \frac{(\mathbf{s}+4)+(\mathbf{s}+2)}{(\mathbf{s}+2)(\mathbf{s}+4)} = \frac{2\mathbf{s}+6}{\mathbf{s}^2+6\mathbf{s}+8}. \tag{3.131}$$

(b)

$$\mathbf{H}(\omega) = \mathbf{H}(\mathbf{s})\big|_{\mathbf{s}=j\omega} = \frac{2(j\omega)+6}{(j\omega)^2 + j6\omega + 8} = \frac{6+j2\omega}{(8-\omega^2)+j6\omega}.$$

Note that all even powers of ω have real coefficients and all odd powers of ω have imaginary coefficients.

(c)

$$\mathbf{H}(\mathbf{s}) = \frac{\mathbf{Y}(\mathbf{s})}{\mathbf{X}(\mathbf{s})} = \frac{2\mathbf{s}+6}{\mathbf{s}^2+6\mathbf{s}+8}.$$

Cross multiplying gives

$$\mathbf{s}^2\, \mathbf{Y}(\mathbf{s}) + 6\mathbf{s}\, \mathbf{Y}(\mathbf{s}) + 8\mathbf{Y}(\mathbf{s}) = 2\mathbf{s}\, \mathbf{X}(\mathbf{s}) + 6\mathbf{X}(\mathbf{s}).$$

For a system with zero initial conditions, differentiation in the time domain corresponds to multiplication by $\mathbf{s}$ in the $\mathbf{s}$-domain (property #6 in Table 3-1). Hence, the time-domain equivalent of the preceding equation is

$$\frac{d^2y}{dt^2} + 6\,\frac{dy}{dt} + 8y = 2\,\frac{dx}{dt} + 6x.$$

(d) From Eq. (3.131), we have

$$2\mathbf{s}+6 = 0 \quad \Rightarrow \quad \text{zero } \{-3\}$$

and

$$\mathbf{s}^2+6\mathbf{s}+8 = 0 \quad \Rightarrow \quad \text{poles } \{-2, -4\}.$$

(e) For $x(t) = \delta(t) + e^{-3t}\, u(t)$, application of entries #1 and 3 in Table 3-2 leads to

$$\mathbf{X}(\mathbf{s}) = 1 + \frac{1}{\mathbf{s}+3} = \frac{\mathbf{s}+4}{\mathbf{s}+3}.$$

Output $\mathbf{Y}(\mathbf{s})$ is then given by

$$\mathbf{Y}(\mathbf{s}) = \mathbf{H}(\mathbf{s})\, \mathbf{X}(\mathbf{s}) = \frac{2(\mathbf{s}+3)}{\mathbf{s}^2+6\mathbf{s}+8} \cdot \frac{\mathbf{s}+4}{\mathbf{s}+3} = \frac{2(\mathbf{s}+3)(\mathbf{s}+4)}{(\mathbf{s}+2)(\mathbf{s}+4)(\mathbf{s}+3)} = \frac{2}{\mathbf{s}+2}.$$

Hence,

$$y(t) = \mathcal{L}^{-1}\left(\frac{2}{\mathbf{s}+2}\right) = 2e^{-2t}\, u(t).$$

Example 3-16: Determine $\mathbf{H}(\omega)$ from LCCDE

Given

$$\frac{d^2y}{dt^2} + 5\,\frac{dy}{dt} + 6y = \frac{d^2x}{dt^2} + 5\,\frac{dx}{dt} + 4x$$

for a system with zero initial conditions, determine the frequency response function $\mathbf{H}(\omega)$.

Solution: Transforming the LCCDE to the **s**-domain gives

$$\mathbf{Y(s)}[\mathbf{s}^2 + 5\mathbf{s} + 6] = \mathbf{X(s)}[\mathbf{s}^2 + 5\mathbf{s} + 4],$$

from which we obtain the transfer function

$$\mathbf{H(s)} = \frac{\mathbf{Y(s)}}{\mathbf{X(s)}} = \frac{\mathbf{s}^2 + 5\mathbf{s} + 4}{\mathbf{s}^2 + 5\mathbf{s} + 6}\,.$$

The frequency response $\mathbf{H}(\omega)$ is then given by

$$\mathbf{H}(\omega) = \mathbf{H(s)}\big|_{\mathbf{s}=j\omega} = \frac{(j\omega)^2 + j5\omega + 4}{(j\omega)^2 + j5\omega + 6} = \frac{(4-\omega^2) + j5\omega}{(6-\omega^2) + j5\omega}\,.$$

Example 3-17: Determine $\mathbf{H(s)}$ from its Poles and Zeros

Transfer functions $\mathbf{H(s)}$ has zero $\{+1\}$ and pole $\{-3\}$. Also, $\mathbf{H}(0) = -1$. Obtain an expression for $\mathbf{H(s)}$.

Solution: $\mathbf{H(s)}$ has one zero at $\mathbf{s} = 1$ and one pole at $\mathbf{s} = -3$. Hence, for some constant $\mathbf{C}$, $\mathbf{H(s)}$ is given by

$$\mathbf{H(s)} = \mathbf{C}\,\frac{\mathbf{s}-1}{\mathbf{s}+3}\,.$$

At $\mathbf{s} = 0$, $\mathbf{H}(0) = -1$. Hence,

$$-1 = \mathbf{C}\,\frac{(-1)}{3}\,,$$

so $\mathbf{C} = 3$, and then

$$\mathbf{H(s)} = 3\,\frac{\mathbf{s}-1}{\mathbf{s}+3}\,.$$

Concept Question 3-13: Does knowledge of just the poles and zeros completely determine the LCCDE? (See S²)

Concept Question 3-14: How do we convert a function of form $\mathbf{H}(\omega)$ into one of form $\mathbf{H(s)}$? (See S²)

Exercise 3-14: An LTI system has impulse response $h(t) = 3e^{-t}\,u(t) - 2e^{-2t}\,u(t)$. Determine the LCCDE description.

Answer: $\dfrac{d^2y}{dt^2} + 3\,\dfrac{dy}{dt} + 2y = \dfrac{dx}{dt} + 4x$. (See S²)

Exercise 3-15: Compute the impulse response of the system described by LCCDE

$$\frac{d^2y}{dt^2} + 5\,\frac{dy}{dt} + 4y = 3x.$$

Answer: $h(t) = e^{-t}\,u(t) - e^{-4t}\,u(t)$. (See S²)

3-11 LTI System Response Partitions

The beauty of the Laplace transform technique is that it can provide a complete solution for the LCCDE describing an LTI system, whether or not the system has zero initial conditions. The solution may be partitioned (divided or organized) into different formats so as to discern from them different information about the *character* of the system's response. In this section, we will examine three different approaches to partitioning the system response. We will explain what each type of partition means, how to extract it from the system response, and what purpose it serves.

3-11.1 Zero-State / Zero-Input Partition

As a working example, let us consider the RC circuit shown in Fig. 3-6. Prior to closing the switch at $t = 0$, the capacitor voltage was

$$\upsilon(0^-) = A_c. \tag{3.132}$$

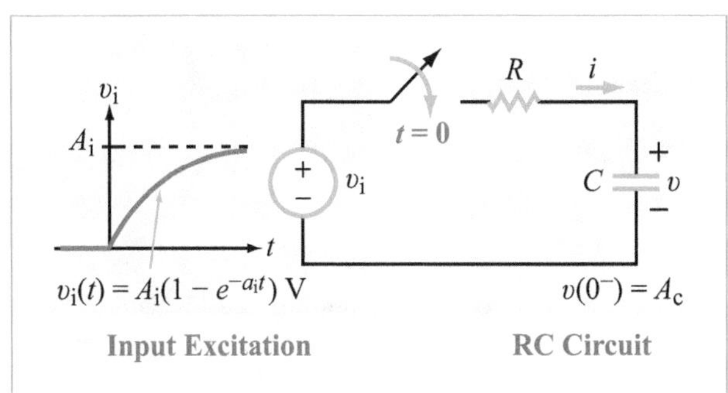

Figure 3-6: The RC circuit and input excitation are treated as separate entities that get connected at $t = 0$.

Application of KVL at $t \geq 0$ gives

$$RI + \upsilon = \upsilon_i(t), \tag{3.133}$$

where υ is the voltage across the capacitor and $\upsilon_i(t)$ is the input voltage excitation introduced at $t = 0$. Since $i = C\,d\upsilon/dt$ for the capacitor, Eq. (3.133) can be converted into a first-order differential equation in one variable, namely,

$$RC\,\frac{d\upsilon}{dt} + \upsilon = \upsilon_{\mathrm{i}}(t). \tag{3.134}$$

So we may easily distinguish between the attributes of the RC circuit and those of the input voltage, we will use the following notation:

- Subscript c refers to attributes that pertain to the circuit alone, independently of the character of the input excitation.
- Subscript i refers to attributes specific to the input excitation.

Accordingly, we denoted in Eq. (3.132) the initial voltage across the capacitor as A_{c}, and we now define the characteristic time constant of the RC circuit as

$$RC = \frac{1}{a_{\mathrm{c}}}. \tag{3.135}$$

Transforming Eq. (3.134) to the **s**-domain entails

$$\upsilon(t) \rightarrow \mathbf{V}(\mathbf{s}),$$

$$\frac{d\upsilon}{dt} \rightarrow \mathbf{s}\,\mathbf{V}(\mathbf{s}) - \upsilon(0^-) = \mathbf{s}\,\mathbf{V}(\mathbf{s}) - A_{\mathrm{c}}$$

(by property #6 in Table 3-1),

$$\upsilon_{\mathrm{i}}(t) \rightarrow \mathbf{V}_{\mathrm{i}}(\mathbf{s}),$$

which leads to

$$\underbrace{\left(\frac{\mathbf{s}}{a_{\mathrm{c}}} + 1\right)\mathbf{V}(\mathbf{s})}_{\text{Zero state}} = \underbrace{\frac{A_{\mathrm{c}}}{a_{\mathrm{c}}}}_{\text{Initial condition}} + \underbrace{\mathbf{V}_{\mathrm{i}}(\mathbf{s})}_{\text{Input excitation}}. \tag{3.136}$$

A system (or RC circuit in the present case) is said to be in *zero state* if it has zero initial conditions, i.e., no stored energy. The term on the left-hand side of Eq. (3.136) represents the zero-state status of the circuit; it is the **s**-domain sum of the voltages across the resistor and the capacitor, excluding the initial voltage $\upsilon(0^-)$. The first term on the right-hand side (with subscript c) represents the initial conditions as an *equivalent excitation source*, and the last term is the real excitation voltage $\mathbf{V}_{\mathrm{i}}(\mathbf{s})$.

Our goal is to solve for $\mathbf{V}(\mathbf{s})$. Dividing all terms in Eq. (3.136) by the factor multiplying $\mathbf{V}(\mathbf{s})$ leads to the expression

$$\underbrace{\mathbf{V}(\mathbf{s})}_{\text{Total response}} = \underbrace{\frac{A_{\mathrm{c}}}{\mathbf{s} + a_{\mathrm{c}}}}_{\text{Zero-input response}} + \underbrace{\frac{a_{\mathrm{c}}\,\mathbf{V}_{\mathrm{i}}(\mathbf{s})}{\mathbf{s} + a_{\mathrm{c}}}}_{\text{Zero-state response}}. \tag{3.137}$$

The total response is a sum of the following:

- **Zero-input response (ZIR)** = response of the system in the absence of an input excitation ($\mathbf{V}_{\mathrm{i}}(\mathbf{s}) = 0$).
- **Zero-state response (ZSR)** = response of the system when it is in zero state (zero initial conditions, $A_{\mathrm{c}} = \upsilon(0^-) = 0$).

In the present case, the RC circuit is excited by an input voltage given by

$$\upsilon_{\mathrm{i}}(t) = A_{\mathrm{i}}[1 - e^{-a_{\mathrm{i}}t}]\,u(t)\ \mathrm{V} \tag{3.138}$$

with amplitude A_{i} and exponential coefficient a_{i}. The corresponding **s**-domain expression is

$$\mathbf{V}_{\mathrm{i}}(\mathbf{s}) = \frac{A_{\mathrm{i}}}{\mathbf{s}} - \frac{A_{\mathrm{i}}}{\mathbf{s} + a_{\mathrm{i}}} = \frac{A_{\mathrm{i}}a_{\mathrm{i}}}{\mathbf{s}(\mathbf{s} + a_{\mathrm{i}})}. \tag{3.139}$$

Inserting Eq. (3.139) into Eq. (3.137) leads to

$$\mathbf{V}(\mathbf{s}) = \frac{A_{\mathrm{c}}}{\mathbf{s} + a_{\mathrm{c}}} + \frac{A_{\mathrm{i}}a_{\mathrm{c}}a_{\mathrm{i}}}{\mathbf{s}(\mathbf{s} + a_{\mathrm{c}})(\mathbf{s} + a_{\mathrm{i}})}. \tag{3.140}$$

In preparation for transforming the expression to the time domain, we use the recipe in Section 3-5.2 to express the second term as a sum of partial fractions. The result is

$$\mathbf{V}(\mathbf{s}) = \frac{A_{\mathrm{c}}}{\mathbf{s} + a_{\mathrm{c}}} + \left[\frac{A_{\mathrm{i}}}{\mathbf{s}} + \frac{A_{\mathrm{i}}a_{\mathrm{i}}}{(a_{\mathrm{c}} - a_{\mathrm{i}})}\left(\frac{1}{\mathbf{s} + a_{\mathrm{c}}}\right) + \frac{A_{\mathrm{i}}a_{\mathrm{c}}}{(a_{\mathrm{i}} - a_{\mathrm{c}})}\left(\frac{1}{\mathbf{s} + a_{\mathrm{i}}}\right)\right]. \tag{3.141}$$

In view of entries #2 and 3 in Table 3-2, the time-domain counterpart of Eq. (3.141) is

$$\upsilon(t) = \underbrace{A_c e^{-a_c t}\, u(t)}_{\text{ZIR}} + \underbrace{\left[A_i + \left(\frac{A_i a_i}{a_c - a_i}\right) e^{-a_c t} + \left(\frac{A_i a_c}{a_i - a_c}\right) e^{-a_i t}\right] u(t).}_{\text{ZSR}} \tag{3.142}$$

For $A_c = \upsilon(0^-) = 2$ V, $a_c = 3\ \text{s}^{-1}$, $A_i = 6$ V, and $a_i = 2\ \text{s}^{-1}$, we have

$$\upsilon(t) = \underbrace{2e^{-3t}\, u(t)}_{\text{ZIR}} + \underbrace{[6 + 12e^{-3t} - 18e^{-2t}]\, u(t).}_{\text{ZSR}} \tag{3.143}$$

The function $e^{-a_c t} = e^{-3t}$ is called a *mode* of the system. It is independent of the input function because $a_c = 1/(RC)$ is characteristic of the RC circuit alone. The zero-input response is composed exclusively of modes, whereas the zero-state response includes both the modes of the system as well as other modes that mimic the character of the excitation function. The second term in Eq. (3.143) includes a term that varies as e^{-3t} and two others with functional forms similar to those in the expression for $\upsilon_i(t)$ given by Eq. (3.138).

3-11.2 Natural / Forced Partition

If we partition the total response into a

- **Natural response** = all terms that vary with t according to the system's modes (in the present case, the system has only one mode, namely e^{-3t}), and a
- **Forced response** = all remaining terms,

then Eq. (3.143) would become

$$\begin{aligned} \upsilon(t) &= [2e^{-3t} + 12e^{-3t}]\, u(t) + [6 - 18e^{-2t}]\, u(t) \\ &= \underbrace{14e^{-3t}\, u(t)}_{\text{Natural response}} + \underbrace{[6 - 18e^{-2t}]\, u(t).}_{\text{Forced response}} \end{aligned} \tag{3.144}$$

Recall that the exponent $a_c = 3\ \text{s}^{-1}$ is $(RC)^{-1}$, which reflects the *natural* identity of the circuit. Hence, all terms of the total response that vary as e^{-3t} are regarded as elements of the natural response. Remaining terms are part of the forced response, which includes modes that mimic the character of the excitation function.

3-11.3 Transient / Steady-State Partition

In many practical situations, the system analyst or designer may be particularly interested in partitioning the response into a

- **Transient response** = all terms that decay to zero as $t \to \infty$, and a
- **Steady-state response** = all terms that remain after the demise of the transient response.

Partitioning the expression for $\upsilon(t)$ along those lines gives

$$\upsilon(t) = \underbrace{[14e^{-3t} - 18e^{-2t}]\, u(t)}_{\text{Transient response}} + \underbrace{6u(t).}_{\text{Steady-state response}} \tag{3.145}$$

The terms included in the first square bracket decay exponentially to zero as $t \to \infty$, leaving behind the steady-state component $6u(t)$. The steady-state component need not be a constant (dc value). In fact, if the input function is a sinusoid, the steady-state response will also be a sinusoid at the same frequency.

Example 3-18: RC Circuit with $\upsilon_i(t) = 6\sin^2(2t)\, u(t)$

Analyze the RC circuit of Fig. 3-6 to determine $\upsilon(t)$ for $t \geq 0$ given that $\upsilon(0^-) = 2$ V, $RC = (1/3)$ s, and

$$\upsilon_i(t) = 6\sin^2(2t)\, u(t)\ \text{V}.$$

Divide the total response along each of the three partitions discussed in this section.

Solution: Using the identity $\sin^2 x = (1-\cos 2x)/2$, the input voltage can be rewritten in the form

$$\upsilon_i(t) = 6\sin^2(2t)\, u(t) = 3[1 - \cos 4t]\, u(t). \tag{3.146}$$

Its s-domain counterpart is

$$\mathbf{V}_i(\mathbf{s}) = \frac{3}{\mathbf{s}} - \frac{3\mathbf{s}}{\mathbf{s}^2 + 16}. \tag{3.147}$$

Using Eq. (3.147) in Eq. (3.137) leads to

$$\begin{aligned}\mathbf{V}(\mathbf{s}) &= \frac{2}{\mathbf{s}+3} + \frac{9}{\mathbf{s}(\mathbf{s}+3)} - \frac{9\mathbf{s}}{(\mathbf{s}+3)(\mathbf{s}^2+16)} \\ &= \frac{2}{\mathbf{s}+3} + \frac{9}{\mathbf{s}(\mathbf{s}+3)} - \frac{9\mathbf{s}}{(\mathbf{s}+3)(\mathbf{s}+j4)(\mathbf{s}-j4)},\end{aligned} \tag{3.148}$$

where we used the numerical values specified in the problem ($A_c = \upsilon(0^-) = 2$ V and $a_c = 3$ s^{-1}) and expanded ($\mathbf{s}^2 + 16$) into the product of two conjugate terms.

The first term of Eq. (3.148) is already in a form amenable to direct transformation to the time domain, but the second and third terms require partial fraction expansion:

$$\frac{9}{\mathbf{s}(\mathbf{s}+3)} = \frac{3}{\mathbf{s}} - \frac{3}{(\mathbf{s}+3)},$$

$$\frac{-9\mathbf{s}}{(\mathbf{s}+3)(\mathbf{s}+j4)(\mathbf{s}-j4)} = \frac{B_1}{\mathbf{s}+3} + \frac{\mathbf{B}_2}{\mathbf{s}+j4} + \frac{\mathbf{B}_2^*}{\mathbf{s}-j4},$$

with

$$B_1 = \left.\frac{-9\mathbf{s}}{(\mathbf{s}+j4)(\mathbf{s}-j4)}\right|_{\mathbf{s}=-3} = \frac{27}{25} = 1.08,$$

$$\mathbf{B}_2 = \left.\frac{-9\mathbf{s}}{(\mathbf{s}+3)(\mathbf{s}-j4)}\right|_{\mathbf{s}=-j4} = -0.9e^{j53.1^\circ},$$

$$\mathbf{B}_2^* = -0.9e^{-j53.1^\circ}.$$

Hence,

$$\mathbf{V}(\mathbf{s}) = \frac{2}{\mathbf{s}+3} + \frac{3}{\mathbf{s}} - \frac{3}{\mathbf{s}+3} + \frac{1.08}{\mathbf{s}+3} - 0.9\left[\frac{e^{j53.1^\circ}}{\mathbf{s}+j4} + \frac{e^{-j53.1^\circ}}{\mathbf{s}-j4}\right]. \tag{3.149}$$

The corresponding time-domain response is

$$\begin{aligned}\upsilon(t) &= 2e^{-3t}\,u(t) + 3(1-e^{-3t})\,u(t) \\ &\quad + 1.08e^{-3t}\,u(t) - 1.8\cos(4t - 53.1^\circ)\,u(t).\end{aligned} \tag{3.150}$$

- ZIR / ZSR partition:

$$\upsilon = \underbrace{2e^{-3t}\,u(t)}_{\text{ZIR}} + \underbrace{3(1-e^{-3t})\,u(t) + 1.08e^{-3t}\,u(t) - 1.8\cos(4t-53.1^\circ)\,u(t)}_{\text{ZSR}}. \tag{3.151}$$

- Natural / forced partition:

$$\begin{aligned}\upsilon(t) &= [2e^{-3t} - 3e^{-3t} + 1.08e^{-3t}]\,u(t) \\ &\quad + [3 - 1.8\cos(4t - 53.1^\circ)]\,u(t) \\ &= \underbrace{0.08e^{-3t}\,u(t)}_{\text{Natural response}} \\ &\quad + \underbrace{[3 - 1.8\cos(4t - 53.1^\circ)]\,u(t)}_{\text{Forced response}}.\end{aligned} \tag{3.152}$$

- Transient / steady-state partition:

$$\begin{aligned}\upsilon(t) = &\underbrace{0.08e^{-3t}\,u(t)}_{\text{Transient response}} \\ &+ \underbrace{[3 - 1.8\cos(4t - 53.1^\circ)]\,u(t)}_{\text{Steady-state response}},\end{aligned}$$

which is the same as the natural/forced partition.

Concept Question 3-15: For stable systems, the transient response includes what other responses? (See S²)

Concept Question 3-16: For stable systems, the zero-state response includes what other responses? (See S²)

Exercise 3-16: Compute the poles and modes of the system with LCCDE

$$\frac{d^2y}{dt^2} + 3\frac{dy}{dt} + 2y = \frac{dx}{dt} + 2x.$$

Answer: Modes: $\{-1, -2\}$. Poles: $\{-1\}$. (See S²)

Exercise 3-17: Compute the zero-input response of

$$\frac{dy}{dt} + 2y = 3\,\frac{dx}{dt} + 4x$$

with $y(0) = 5$.

Answer: $y(t) = 5e^{-2t}\,u(t)$. (See S²)

We conclude this chapter with a long example that brings together many of the concepts presented in the last half of the chapter, and shows how they relate to each other.

Example 3-19: System Responses

An LTI system is described by the LCCDE

$$\frac{d^2y}{dt^2} + 3\,\frac{dy}{dt} + 2y(t) = \frac{dx}{dt} + 2x(t). \tag{3.153}$$

The input is $x(t) = 4\cos(t)\,u(t)$, and the initial conditions are $y(0^-) = 1$ and $\frac{dy}{dt}\,(0^-) = 0$. Compute the following:

(a) The transfer function, poles, zeros, and modes
(b) The zero-state and zero-input responses
(c) The transient and steady-state responses
(d) The forced and natural responses
(e) The choice of initial conditions for which the transient response is zero

Solution: (a) The transfer function does not depend on the specific input $x(t) = 4\cos(t)\,u(t)$, and it assumes zero initial conditions. The Laplace transform of the LCCDE is

$$[\mathbf{s}^2\,\mathbf{Y}(\mathbf{s}) - \mathbf{s}\,y(0^-) - \frac{dy}{dt}\,(0^-)] + 3[\mathbf{s}\,\mathbf{Y}(\mathbf{s}) - y(0^-)] + 2\mathbf{Y}(\mathbf{s})$$
$$= [\mathbf{s}\,\mathbf{X}(\mathbf{s}) - x(0^-)] + 2\mathbf{X}(\mathbf{s}). \tag{3.154}$$

Setting all of the initial conditions to zero gives

$$\mathbf{s}^2\,\mathbf{Y}(s) + 3\mathbf{s}\,\mathbf{Y}(\mathbf{s}) + 2\mathbf{Y}(\mathbf{s}) = \mathbf{s}\,\mathbf{X}(\mathbf{s}) + \mathbf{X}(\mathbf{s}). \tag{3.155}$$

The transfer function is then

$$\mathbf{H}(\mathbf{s}) = \frac{\mathbf{Y}(\mathbf{s})}{\mathbf{X}(\mathbf{s})} = \frac{\mathbf{s}+2}{\mathbf{s}^2+3\mathbf{s}+2} = \frac{\mathbf{s}+2}{(\mathbf{s}+1)(\mathbf{s}+2)} = \frac{1}{\mathbf{s}+1}. \tag{3.156}$$

The zeros are the roots of the numerator polynomial set equal to zero, so there are no zeros. The poles are the roots of the denominator polynomial set equal to zero, so the only pole is -1. The system is BIBO stable.

The modes are the roots of the characteristic polynomial set equal to zero. From Eq. (2.123), the characteristic equation is the polynomial whose coefficients are the coefficients of the left side of the LCCDE. In this example, the characteristic equation is

$$\mathbf{s}^2 + 3\mathbf{s} + 2 = 0, \tag{3.157}$$

so the modes are $\{-1, -2\}$. Note that modes and poles are *not* the same in this example, due to the cancellation of the factors $(\mathbf{s}+2)$ in the numerator and denominator of Eq. (3.156). This is called pole-zero cancellation.

(b) The zero-state response is found by setting all initial conditions to zero and setting the input to $x(t) = 4\cos(t)\,u(t)$. The Laplace transform of $x(t)$ is

$$\mathbf{X}(\mathbf{s}) = \mathcal{L}[4\cos(t)\,u(t)] = \frac{4\mathbf{s}}{\mathbf{s}^2+1}, \tag{3.158}$$

and the Laplace transform of the zero-state response is

$$\mathbf{Y}_{\mathrm{ZSR}}(\mathbf{s}) = \mathbf{H}(\mathbf{s})\,\mathbf{X}(\mathbf{s}) = \left[\frac{1}{\mathbf{s}+1}\right]\left[\frac{4\mathbf{s}}{\mathbf{s}^2+1}\right]. \tag{3.159}$$

Partial fraction expansion of $\mathbf{Y}_{\mathrm{ZSR}}(\mathbf{s})$ gives

$$\mathbf{Y}_{\mathrm{ZSR}}(\mathbf{s}) = \frac{1-j}{\mathbf{s}-j} + \frac{1+j}{\mathbf{s}+j} - \frac{2}{\mathbf{s}+1}, \tag{3.160}$$

and the inverse Laplace transform of $\mathbf{Y}_{\mathrm{ZSR}}(\mathbf{s})$ is

$$y_{\mathrm{zsr}}(t) = (1-j)e^{jt}\,u(t) + (1+j)e^{-jt}\,u(t) - 2e^{-t}\,u(t). \tag{3.161}$$

Noting that $(1 \pm j) = \sqrt{2}\,e^{\pm j45^\circ}$, Eq. (3.161) can be put into trigonometric form using entry #3 in Table 3-3. The result is

$$y_{\mathrm{zsr}}(t) = 2\sqrt{2}\cos(t - 45^\circ)\,u(t) - 2e^{-t}\,u(t). \tag{3.162}$$

The zero-input response is found by setting the input $x(t)$ to zero. The Laplace transform of the zero-input response can then be computed from Eq. (3.153) as

$$\mathbf{Y}_{\text{ZIR}}(\mathbf{s}) = \frac{\mathbf{s}\, y(0^-) + \frac{dy}{dt}(0^-) + 3y(0^-)}{\mathbf{s}^2 + 3\mathbf{s} + 2} = \frac{\mathbf{s}+3}{\mathbf{s}^2+3\mathbf{s}+2}. \tag{3.163}$$

Partial fraction expansion of $\mathbf{Y}_{\text{ZIR}}(\mathbf{s})$ gives

$$\mathbf{Y}_{\text{ZIR}}(\mathbf{s}) = \frac{2}{\mathbf{s}+1} - \frac{1}{\mathbf{s}+2}, \tag{3.164}$$

and its inverse Laplace transform is

$$y_{\text{zir}}(t) = 2e^{-t}\, u(t) - e^{-2t}\, u(t). \tag{3.165}$$

(c) The complete response is then

$$\begin{aligned} y(t) &= y_{\text{zsr}}(t) + y_{\text{zir}}(t) \\ &= 2\sqrt{2}\cos(t - 45^\circ)\, u(t) - 2e^{-t}\, u(t) \\ &\quad + 2e^{-t}\, u(t) - e^{-2t}\, u(t) \\ &= 2\sqrt{2}\cos(t - 45^\circ)\, u(t) - e^{-2t}\, u(t). \end{aligned} \tag{3.166}$$

The given choice of initial conditions has produced a cancellation of two $2e^{-t}\, u(t)$ terms.

The transient response is the part of the complete response that decays to zero:

$$y_{\text{trans}}(t) = -e^{-2t}\, u(t). \tag{3.167}$$

The steady-state response is the part of the complete response that does not decay to zero:

$$y_{\text{ss}}(t) = 2\sqrt{2}\cos(t - 45^\circ)\, u(t). \tag{3.168}$$

Note that except for the step function $u(t)$, we could have computed the steady-state response using phasors, without the Laplace transform. The frequency response function for the LCCDE can be read off as

$$\mathbf{H}(\omega) = \frac{(j\omega)+2}{(j\omega)^2 + 3(j\omega) + 2} = \frac{1}{j\omega+1}. \tag{3.169}$$

Inserting $\omega = 1$ gives

$$\mathbf{H}(1) = \frac{1}{j\omega+1} = \frac{1}{\sqrt{2}}\, e^{-j45^\circ}. \tag{3.170}$$

The response to $4\cos(t)$ in the sinusoidal steady-state is

$$y_{\text{phasors}}(t) = \frac{4}{\sqrt{2}}\cos(t - 45^\circ) = 2\sqrt{2}\cos(t - 45^\circ) \tag{3.171}$$

which agrees with $y_{\text{ss}}(t)$ for $t > 0$.

(d) The forced response is the part of the complete response that resembles the input:

$$y_{\text{forced}}(t) = 2\sqrt{2}\cos(t - 45^\circ)\, u(t). \tag{3.172}$$

The input and the forced response are both sinusoids with frequency $\omega = 1$ rad/s.

The natural response is the part of the complete response that resembles the zero-input response:

$$y_{\text{natural}}(t) = -e^{-2t}\, u(t). \tag{3.173}$$

Both have exponential time functions proportional to $e^{-2t}\, u(t)$. But the $e^{-t}\, u(t)$ term in the zero-input response does not appear in the natural response (actually, it does appear, but with a coefficient of zero).

(e) The former initial conditions associated with parts (a) to (d) no longer apply, so Eqs. (3.163)–(3.167) are no longer valid. To cancel the $-2e^{-t}\, u(t)$ term in the zero-state response, we now require

$$y_{\text{zir}}(t) = 2e^{-t}\, u(t) \rightarrow \mathbf{Y}_{\text{ZIR}}(\mathbf{s}) = \frac{2}{\mathbf{s}+1}. \tag{3.174}$$

So we must choose the initial conditions $y(0^-)$ and $\frac{dy}{dt}(0^-)$ in Eq. (3.163) so that

$$\begin{aligned} \mathbf{Y}_{\text{ZIR}}(\mathbf{s}) &= \frac{\mathbf{s}y(0^-) + \frac{dy}{dt}(0^-) + 3y(0^-)}{\mathbf{s}^2 + 3\mathbf{s} + 2} \\ &= \frac{2}{\mathbf{s}+1} = \frac{2(\mathbf{s}+2)}{\mathbf{s}^2+3\mathbf{s}+2}. \end{aligned} \tag{3.175}$$

Equating coefficients of $\mathbf{s}$ in the numerator of $\mathbf{Y}_{\text{ZIR}}(\mathbf{s})$, leads to

$$y(0^-) = 2 \quad \text{and} \quad \frac{dy}{dt}(0^-) = -2. \tag{3.176}$$

Summary

Concepts

- The unilateral Laplace transform converts a causal time-domain signal $x(t)$ into an $\mathbf{s}$-domain counterpart $\mathbf{X}(\mathbf{s})$. A major advantage of the Laplace transform is that it converts differential equations into algebraic equations, thereby facilitating their solutions considerably.
- The Laplace transform has many useful properties that can be applied to streamline the process of finding the Laplace transform of a time function. Not all time-domain functions have Laplace transforms.
- The $\mathbf{s}$-domain counterpart of a system's impulse response $h(t)$ is the system transfer function $\mathbf{H}(\mathbf{s})$.
- The convolution-integral method allows us to determine the output response of a system by convolving the input excitation with the impulse response of the system. This approach is particularly useful when the excitation is in the form of experimental measurements that may be difficult to characterize in the form of a mathematical function.

Mathematical and Physical Models

Laplace Transform Pair (Unilateral)

$$\mathbf{X}(\mathbf{s}) = \mathcal{L}[x(t)] = \int_{0^-}^{\infty} x(t)\, e^{-\mathbf{s}t}\, dt$$

$$x(t) = \mathcal{L}^{-1}[\mathbf{X}(\mathbf{s})] = \frac{1}{2\pi j} \int_{\sigma-j\infty}^{\sigma+j\infty} \mathbf{X}(\mathbf{s})\, e^{\mathbf{s}t}\, d\mathbf{s}$$

Properties of Laplace Transform

(comprehensive list in Table 3-1)

$$x(at), \quad a > 0 \quad \longleftrightarrow \quad \frac{1}{a}\,\mathbf{X}\!\left(\frac{\mathbf{s}}{a}\right)$$

$$e^{-at}\, x(t) \quad \longleftrightarrow \quad \mathbf{X}(\mathbf{s}+a)$$

$$x' = \frac{dx}{dt} \quad \longleftrightarrow \quad \mathbf{s}\,\mathbf{X}(\mathbf{s}) - x(0^-)$$

$$\int_{0^-}^{t} x(t')\, dt' \quad \longleftrightarrow \quad \frac{1}{\mathbf{s}}\,\mathbf{X}(\mathbf{s})$$

Examples of Laplace Transform Pairs

(longer list in Table 3-2)

$$\delta(t) \quad \longleftrightarrow \quad 1$$

$$u(t) \quad \longleftrightarrow \quad \frac{1}{\mathbf{s}}$$

$$e^{-at}\, u(t) \quad \longleftrightarrow \quad \frac{1}{\mathbf{s}+a}$$

$$t\, u(t) \quad \longleftrightarrow \quad \frac{1}{\mathbf{s}^2}$$

Transfer Function

$$\mathbf{H}(\mathbf{s}) = \frac{\mathbf{Y}(\mathbf{s})}{\mathbf{X}(\mathbf{s})}$$

Convolution

$$x(t) * h(t) \quad \longleftrightarrow \quad \mathbf{X}(\mathbf{s})\,\mathbf{H}(\mathbf{s})$$

Important Terms

Provide definitions or explain the meaning of the following terms:

complex frequency
convergence condition
convolution
Dirac delta function
expansion coefficients
final-value theorem
frequency shift
improper rational function
impulse response
initial-value theorem
Laplace transform
partial fraction expansion
pole
pole factor
proper rational function
residue method
sampling property
singularity function
time invariance
time scaling
time shift
transfer function
uniqueness property
unit impulse function
unit step function
zero (of a polynomial)

PROBLEMS

Sections 3-1 to 3-3: Laplace Transform and Its Properties

3.1 Express each of the waveforms in Fig. P3.1 in terms of step functions and then determine its Laplace transform. (Recall that the ramp function is related to the step function by $r(t-T) = (t-T)\, u(t-T)$.) Assume that all waveforms are zero for $t < 0$.

*(a) Staircase
(b) Square wave
(c) Top hat
(d) Mesa
(e) Negative ramp
(f) Triangular wave

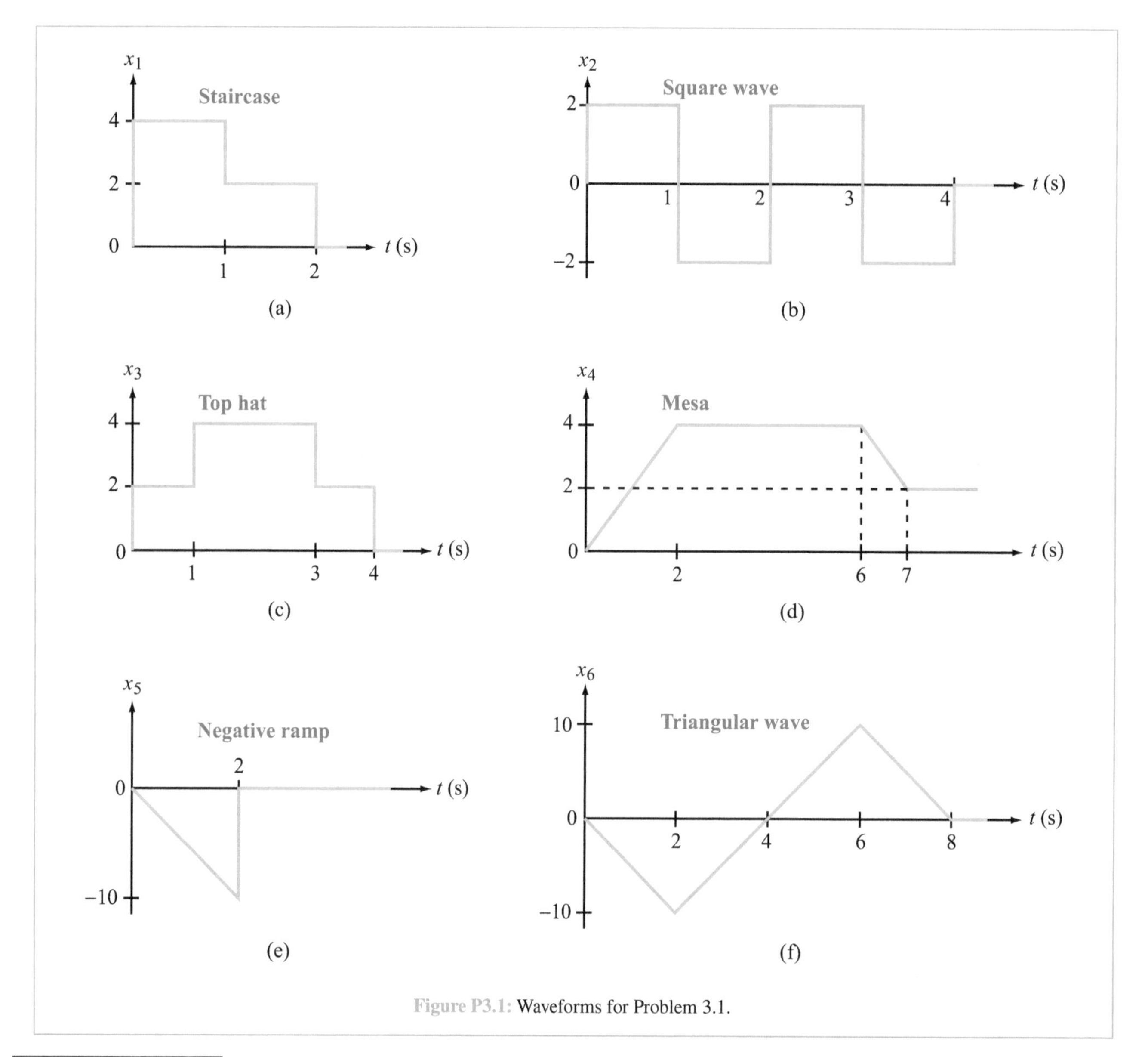

Figure P3.1: Waveforms for Problem 3.1.

*Answer(s) in Appendix F.

3.2 Determine the Laplace transform of each of the *periodic* waveforms shown in Fig. P3.2. (Hint: See Exercise 3.2.)

(a) Sawtooth

(b) Interrupted ramps

(c) Impulses

(d) Periodic exponentials

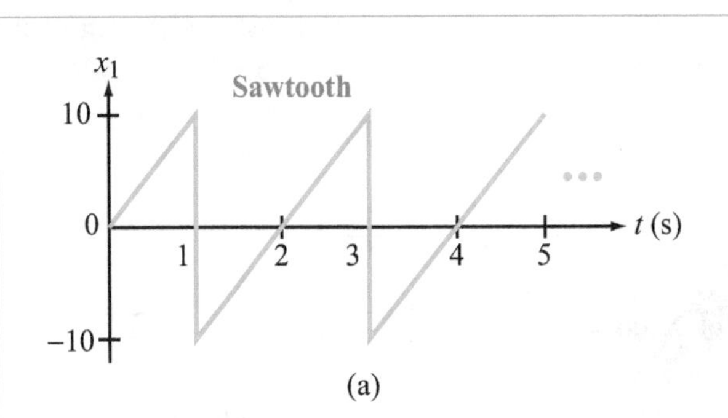

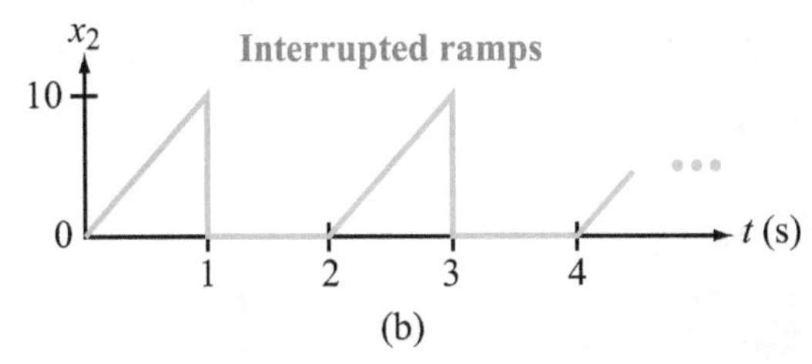

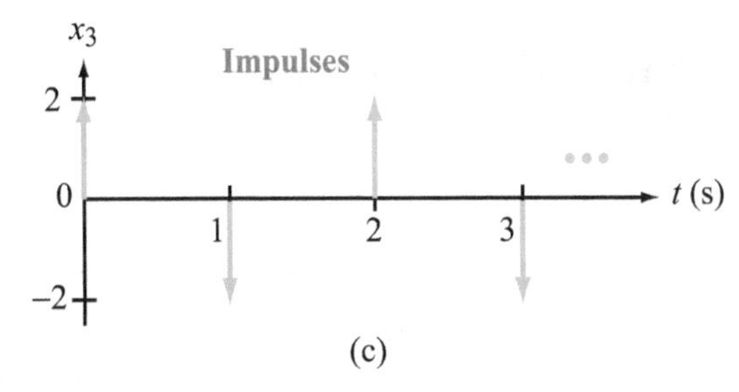

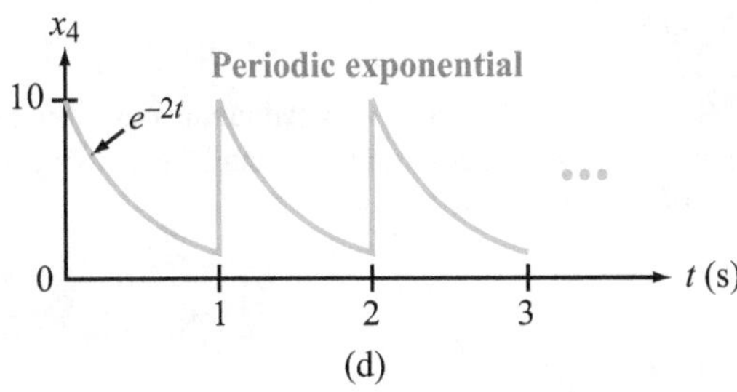

Figure P3.2: Periodic waveforms for Problem 3.2.

3.3 Determine the Laplace transform of each of the following functions by applying the properties given in Tables 3-1 and 3-2.

(a) $x_1(t) = 4te^{-2t}\ u(t)$

(b) $x_2(t) = 10\cos(12t + 60°)\ u(t)$

(c) $x_3(t) = 12e^{-3(t-4)}\ u(t-4)$

3.4 Determine the Laplace transform of each of the following functions by applying the properties given in Tables 3-1 and 3-2.

(a) $x_1(t) = 12te^{-3(t-4)}\ u(t-4)$

(b) $x_2(t) = 27t^2 \sin(6t - 60°)\ u(t)$

*(c) $x_3(t) = 10t^3 e^{-2t}\ u(t)$

3.5 Determine the Laplace transform of each of the following functions by applying the properties given in Tables 3-1 and 3-2.

(a) $x_1(t) = 16e^{-2t} \cos 4t\ u(t)$

(b) $x_2(t) = 20te^{-2t} \sin 4t\ u(t)$

(c) $x_3(t) = 10e^{-3t}\ u(t-4)$

3.6 Determine the Laplace transform of each of the following functions by applying the properties given in Tables 3-1 and 3-2.

(a) $x_1(t) = 30(e^{-3t} + e^{3t})\ u(t)$

(b) $x_2(t) = 5(t-6)\ u(t-3)$

(c) $x_3(t) = 4e^{-2(t-3)}\ u(t-4)$

3.7 Determine the Laplace transform of the following functions:

*(a) $x_1(t) = 25\cos(4\pi t + 30°)\ \delta(t)$

(b) $x_2(t) = 25\cos(4\pi t + 30°)\ \delta(t - 0.2)$

(c) $x_3(t) = 10\ \dfrac{\sin(3t)}{t}\ u(t)$

(d) $x_4(t) = \dfrac{d^2}{dt^2}\ [e^{-4t}\ u(t)]$

3.8 Determine the Laplace transform of the following functions:

(a) $x_1(t) = \dfrac{d}{dt}\ [4te^{-2t} \cos(4\pi t + 30°)\ u(t)]$

(b) $x_2(t) = e^{-3t} \cos(4t + 30°)\ u(t)$

(c) $x_3(t) = t^2[u(t) - u(t-4)]$

(d) $x_4(t) = 10\cos(6\pi t + 30°)\ \delta(t - 0.2)$

3.9 Determine $x(0^+)$ and $x(\infty)$ given that

$$\mathbf{X(s)} = \frac{4\mathbf{s}^2 + 28\mathbf{s} + 40}{\mathbf{s(s+3)(s+4)}}.$$

3.10 Determine $x(0^+)$ and $x(\infty)$ given that

$$\mathbf{X}(\mathbf{s}) = \frac{\mathbf{s}^2+4}{2\mathbf{s}^3+4\mathbf{s}^2+10\mathbf{s}}.$$

***3.11** Determine $x(0^+)$ and $x(\infty)$ given that

$$\mathbf{X}(\mathbf{s}) = \frac{12e^{-2\mathbf{s}}}{\mathbf{s}(\mathbf{s}+2)(\mathbf{s}+3)}.$$

3.12 Determine $x(0^+)$ and $x(\infty)$ given that

$$\mathbf{X}(\mathbf{s}) = \frac{19-e^{-\mathbf{s}}}{\mathbf{s}(\mathbf{s}^2+5\mathbf{s}+6)}.$$

Section 3-4 and 3-5: Partial Fractions and Circuit Examples

3.13 Obtain the inverse Laplace transform of each of the following functions, by first applying the partial-fraction-expansion method.

(a) $\mathbf{X}_1(\mathbf{s}) = \dfrac{6}{(\mathbf{s}+2)(\mathbf{s}+4)}$

(b) $\mathbf{X}_2(\mathbf{s}) = \dfrac{4}{(\mathbf{s}+1)(\mathbf{s}+2)^2}$

(c) $\mathbf{X}_3(\mathbf{s}) = \dfrac{3\mathbf{s}^3+36\mathbf{s}^2+131\mathbf{s}+144}{\mathbf{s}(\mathbf{s}+4)(\mathbf{s}^2+6\mathbf{s}+9)}$

3.14 Obtain the inverse Laplace transform of each of the following functions:

(a) $\mathbf{X}_1(\mathbf{s}) = \dfrac{\mathbf{s}^2+17\mathbf{s}+20}{\mathbf{s}(\mathbf{s}^2+6\mathbf{s}+5)}$

(b) $\mathbf{X}_2(\mathbf{s}) = \dfrac{2\mathbf{s}^2+10\mathbf{s}+16}{(\mathbf{s}+2)(\mathbf{s}^2+6\mathbf{s}+10)}$

(c) $\mathbf{X}_3(\mathbf{s}) = \dfrac{4}{(\mathbf{s}+2)^3}$

3.15 Obtain the inverse Laplace transform of each of the following functions:

(a) $\mathbf{X}_1(\mathbf{s}) = \dfrac{(\mathbf{s}+2)^2}{\mathbf{s}(\mathbf{s}+1)^3}$

(b) $\mathbf{X}_2(\mathbf{s}) = \dfrac{1}{(\mathbf{s}^2+4\mathbf{s}+5)^2}$

*(c) $\mathbf{X}_3(\mathbf{s}) = \dfrac{\sqrt{2}(\mathbf{s}+1)}{\mathbf{s}^2+6\mathbf{s}+13}$

3.16 Obtain the inverse Laplace transform of each of the following functions:

(a) $\mathbf{X}_1(\mathbf{s}) = \dfrac{2\mathbf{s}^2+4\mathbf{s}-16}{(\mathbf{s}+6)(\mathbf{s}+2)^2}$

(b) $\mathbf{X}_2(\mathbf{s}) = \dfrac{2(\mathbf{s}^3+12\mathbf{s}^2+16)}{(\mathbf{s}+1)(\mathbf{s}+4)^3}$

(c) $\mathbf{X}_3(\mathbf{s}) = \dfrac{-2(\mathbf{s}^2+20)}{\mathbf{s}(\mathbf{s}^2+8\mathbf{s}+20)}$

3.17 Obtain the inverse Laplace transform of each of the following functions:

(a) $\mathbf{X}_1(\mathbf{s}) = 2 + \dfrac{4(\mathbf{s}-4)}{\mathbf{s}^2+16}$

(b) $\mathbf{X}_2(\mathbf{s}) = \dfrac{4}{\mathbf{s}} + \dfrac{4\mathbf{s}}{\mathbf{s}^2+9}$

(c) $\mathbf{X}_3(\mathbf{s}) = \dfrac{(\mathbf{s}+5)e^{-2\mathbf{s}}}{(\mathbf{s}+1)(\mathbf{s}+3)}$

3.18 Obtain the inverse Laplace transform of each of the following functions:

(a) $\mathbf{X}_1(\mathbf{s}) = \dfrac{(1-e^{-4\mathbf{s}})(24\mathbf{s}+40)}{(\mathbf{s}+2)(\mathbf{s}+10)}$

*(b) $\mathbf{X}_2(\mathbf{s}) = \dfrac{\mathbf{s}(\mathbf{s}-8)e^{-6\mathbf{s}}}{(\mathbf{s}+2)(\mathbf{s}^2+16)}$

(c) $\mathbf{X}_3(\mathbf{s}) = \dfrac{4\mathbf{s}(2-e^{-4\mathbf{s}})}{\mathbf{s}^2+9}$

3.19 Solve the following two simultaneous differential equations by taking Laplace transforms and then solving a 2×2 linear system of equations:

$$\frac{dx}{dt} = 2x(t) - 3y(t), \qquad \text{with } x(0^-) = 8,$$

$$\frac{dy}{dt} = -2x(t) + y(t), \qquad \text{with } y(0^-) = 3.$$

3.20 Solve the following two simultaneous differential equations by taking Laplace transforms and then solving a 2×2 linear system of equations:

$$\frac{dx}{dt} + 3x(t) + 8y(t) = 0, \qquad \text{with } x(0^-) = 1,$$

$$\frac{dy}{dt} + 3y(t) - 2x(t) = 0, \qquad \text{with } y(0^-) = 2.$$

3.21 Determine $\upsilon(t)$ in the circuit of Fig. P3.21 given that $\upsilon_s(t) = 2u(t)$ V, $R_1 = 1\ \Omega$, $R_2 = 3\ \Omega$, $C = 0.3689$ F, and $L = 0.2259$ H.

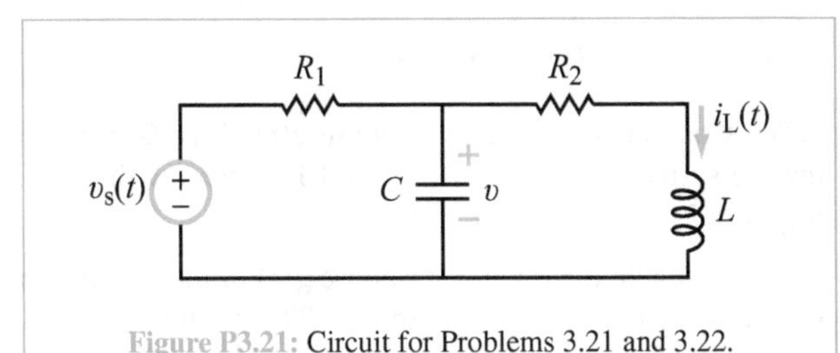

Figure P3.21: Circuit for Problems 3.21 and 3.22.

3.22 Determine $i_L(t)$ in the circuit in Fig. P3.21 given that $\upsilon_s(t) = 2u(t)$, $R_1 = 2\ \Omega$, $R_2 = 6\ \Omega$, $L = 2.215$ H, and $C = 0.0376$ F.

3.23 Determine $\upsilon_{out}(t)$ in the circuit in Fig. P3.23 given that $\upsilon_s(t) = 35u(t)$ V, $\upsilon_{C_1}(0^-) = 20$ V, $R_1 = 1\ \Omega$, $C_1 = 1$ F, $R_2 = 0.5\ \Omega$, and $C_2 = 2$ F.

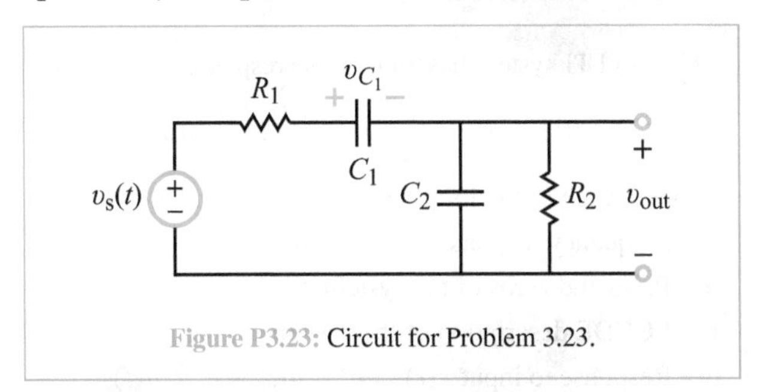

Figure P3.23: Circuit for Problem 3.23.

Section 3-6: Transfer Function

3.24 A system is characterized by a transfer function given by

$$\mathbf{H(s)} = \frac{18\mathbf{s} + 10}{\mathbf{s}^2 + 6\mathbf{s} + 5}.$$

Determine the output response $y(t)$, if the input excitation is given by the following:

*(a) $x_1(t) = u(t)$

(b) $x_2(t) = 2t\ u(t)$

(c) $x_3(t) = 2e^{-4t}\ u(t)$

(d) $x_4(t) = [4\cos(4t)]\ u(t)$

3.25 When excited by a unit step function at $t = 0$, a system generates the output response

$$y(t) = [5 - 10t + 20\sin(2t)]\ u(t).$$

Determine (a) the system transfer function and (b) the impulse response.

3.26 When excited by a unit step function at $t = 0$, a system generates the output response

$$y(t) = 10\,\frac{\sin(5t)}{t}\ u(t).$$

Determine (a) the system transfer function and (b) the impulse response.

3.27 When excited by a unit step function at $t = 0$, a system generates the output response

$$y(t) = 10t^2e^{-3t}\ u(t).$$

Determine (a) the system transfer function and (b) the impulse response.

3.28 When excited by a unit step function at $t = 0$, a system generates the output response

$$y(t) = 9t^2\sin(6t - 60^\circ)\ u(t).$$

Determine (a) the system transfer function and (b) the impulse response.

3.29 A system has $2N$ OLHP poles $\{\,\mathbf{p}_i, \mathbf{p}_i^*, i = 1,\dots,N\,\}$ in complex conjugate pairs and $2N$ zeros $\{\,-\mathbf{p}_i, -\mathbf{p}_i^*, i = 1,\dots,N\,\}$ in complex conjugate pairs in the right half-plane. Show that the gain $|\mathbf{H}(j\omega)|$ of the system is constant for all ω. This is an *all-pass system.* All-pass systems are used to alter a filter's phase response without affecting its gain. *Hint:* For any complex number $\mathbf{z} = |\mathbf{z}|e^{j\theta}$, the ratio

$$\frac{\mathbf{z}}{\mathbf{z}^*} = \frac{|\mathbf{z}|e^{j\theta}}{|\mathbf{z}|e^{-j\theta}} = e^{j2\theta}$$

has magnitude one.

3.30 For the circuit shown in Fig. P3.30, determine (a) $\mathbf{H(s)} = \mathbf{V_o}/\mathbf{V_i}$ and (b) $h(t)$ given that $R_1 = 1\ \Omega$, $R_2 = 2\ \Omega$, $C_1 = 1\ \mu$F, and $C_2 = 2\ \mu$F.

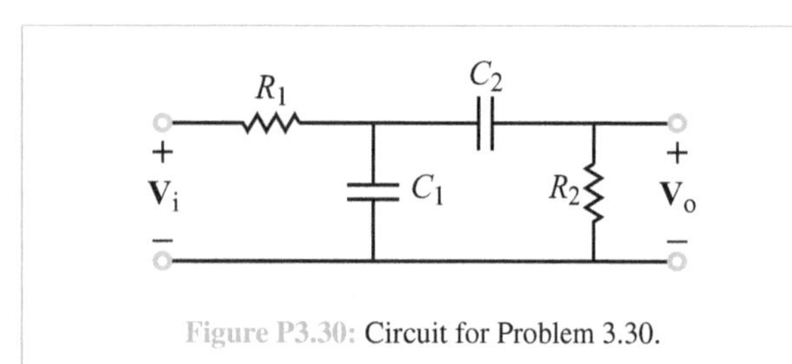

Figure P3.30: Circuit for Problem 3.30.

3.31 For the circuit shown in Fig. P3.31, determine (a) $\mathbf{H(s)} = \mathbf{V_o}/\mathbf{V_i}$ and (b) $h(t)$ given that $R_1 = 1\ \Omega$, $R_2 = 2\ \Omega$, $L_1 = 1$ mH, and $L_2 = 2$ mH.

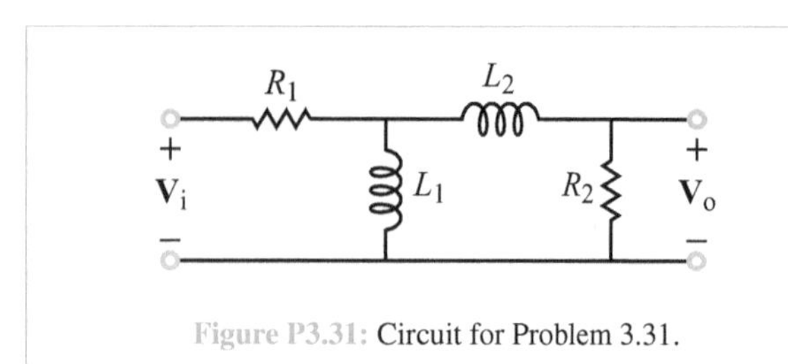

Figure P3.31: Circuit for Problem 3.31.

*3.32 For the circuit shown in Fig. P3.32, determine (a) $\mathbf{H(s)} = \mathbf{V_o}/\mathbf{V_i}$ and (b) $h(t)$ given that $R = 5\ \Omega$, $L = 0.1$ mH, and $C = 1\ \mu$F.

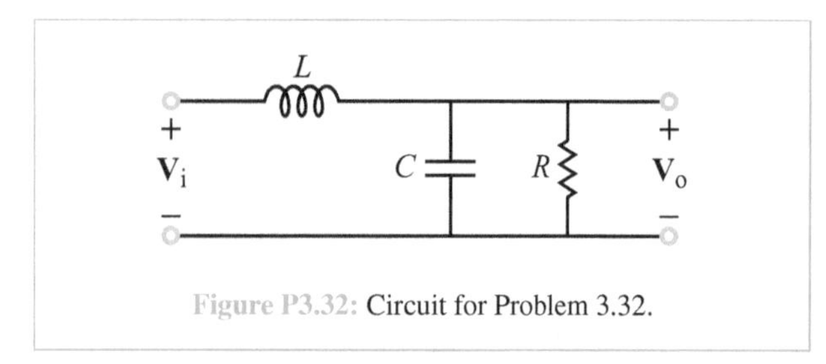

Figure P3.32: Circuit for Problem 3.32.

Section 3-7: LTI System Stability

3.33 An LTI system is described by the LCCDE

$$\frac{d^2y}{dt^2} - 7\frac{dy}{dt} + 12y = \frac{dx}{dt} + 5x.$$

Is the system BIBO stable?

3.34 The response of an LTI system to input

$$x(t) = \delta(t) - 4e^{-3t}\ u(t)$$

is output $y(t) = e^{-2t}\ u(t)$. Is the system BIBO stable?

3.35 An LTI system has impulse response

$$h(t) = (4 + j5)e^{(2+j3)t}\ u(t) + (4 - j5)e^{(2-j3)t}\ u(t).$$

Is the system BIBO stable?

3.36 An LTI system has transfer function

$$\mathbf{H(s)} = \frac{(\mathbf{s}+1)(\mathbf{s}+2)}{\mathbf{s}(\mathbf{s}+3)}.$$

Is it BIBO stable?

Section 3-8: Invertible Systems

*3.37 Compute the impulse response $g(t)$ of the BIBO stable inverse system corresponding to the LTI system with impulse response $h(t) = \delta(t) - 2e^{-3t}\ u(t)$.

3.38 Compute the impulse response $g(t)$ of the BIBO stable inverse system corresponding to the LTI system with impulse response $h(t) = \delta(t) + te^{-t}\ u(t)$.

3.39 Show that the LTI system with impulse response $h(t) = \delta(t) - 4e^{-3t}\ u(t)$ does not have a BIBO stable inverse system.

3.40 Show that the LTI system with impulse response $h(t) = e^{-t}\ u(t)$ does not have a BIBO stable inverse system.

Section 3-10: Interrelating Descriptions

3.41 An LTI system has an impulse response

$$h(t) = \delta(t) + 6e^{-t}\ u(t) - 2e^{-2t}\ u(t).$$

Compute each of the following:

(a) Frequency response function $\mathbf{H}(\omega)$

*(b) Poles and zeros of the system

(c) LCCDE description of the system

(d) Response to input $x(t) = e^{-3t}\ u(t) - e^{-4t}\ u(t)$

3.42 An LTI system has an impulse response

$$h(t) = \delta(t) + 4e^{-3t}\cos(2t)\ u(t).$$

Compute each of the following:

(a) Frequency response function $\mathbf{H}(\omega)$

(b) Poles and zeros of the system

(c) LCCDE description of the system

(d) Response to input $x(t) = 2te^{-5t}\ u(t)$

3.43 An LTI system has the LCCDE description

$$\frac{d^2y}{dt^2} + 7\frac{dy}{dt} + 12y = \frac{dx}{dt} + 2x.$$

Compute each of the following:

(a) Frequency response function $\mathbf{H}(\omega)$

(b) Poles and zeros of the system

(c) Impulse response $h(t)$

*(d) Response to input $x(t) = e^{-2t}\ u(t)$

3.44 An LTI system has the LCCDE description

$$\frac{d^2y}{dt^2} + 4\frac{dy}{dt} + 13y = \frac{dx}{dt} + 2x.$$

Compute each of the following:

(a) Frequency response function $\mathbf{H}(\omega)$

(b) Poles and zeros of the system

(c) Impulse response $h(t)$

(d) Response to input $x(t) = e^{-2t}\ u(t)$

3.45 The response of an LTI system to input

$$x(t) = \delta(t) - 2e^{-3t}\ u(t)$$

is output $y(t) = e^{-2t}\ u(t)$. Compute each of the following:

(a) Frequency response function $\mathbf{H}(\omega)$

(b) Poles and zeros of the system

(c) LCCDE description of the system

(d) Impulse response $h(t)$

3.46 An LTI system has $\mathbf{H}(0) = 1$ and zeros: $\{-1\}$; poles: $\{-3, -5\}$. Compute each of the following:

(a) Frequency response function $\mathbf{H}(\omega)$

(b) LCCDE description of the system

*(c) Impulse response $h(t)$

(d) Response to input $x(t) = e^{-t}\ u(t)$

3.47 An LTI system has $\mathbf{H}(0) = 15$ and zeros: $\{-3 \pm j4\}$; poles: $\{-1 \pm j2\}$. Compute each of the following:

(a) Frequency response function $\mathbf{H}(\omega)$

(b) LCCDE description of the system

(c) Impulse response $h(t)$

(d) Response to input $x(t) = e^{-3t} \sin(4t)\ u(t)$

3.48 The response of an LTI system to input

$$x(t) = e^{-3t} \cos(4t)\ u(t)$$

is output $y(t) = e^{-3t} \sin(4t)\ u(t)$. Compute each of the following:

(a) Frequency response $\mathbf{H}(\omega)$

(b) Poles and zeros of the system

(c) LCCDE description of the system

(d) Impulse response $h(t)$

Section 3-11: Partitions of Responses

3.49 Compute the following system responses for the circuit shown in Fig. P3.49 given that $x(t) = 25\cos(3t)\ u(t)$, $R = 250$ kΩ, $C = 1\ \mu$F, and the capacitor was initially charged to 2 V:

(a) Zero-input response.

(b) Zero-state response.

(c) Transient response.

(d) Steady-state response.

(e) Natural response.

*(f) Forced response.

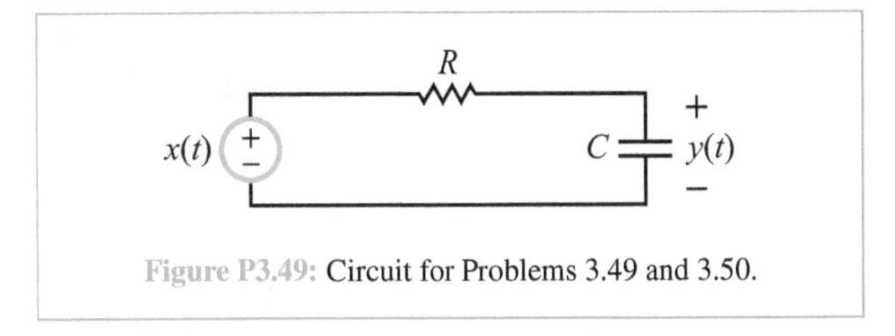

Figure P3.49: Circuit for Problems 3.49 and 3.50.

*3.50 If the capacitor in the circuit of Fig. P3.49 is initially charged to $y(0)$ volts, instead of 2 V, for what value of $y(0)$ is the transient response identically equal to zero (i.e., no transient)?

3.51 For the circuit in Fig. P3.51, compute the steady-state unit-step response for $i(t)$ in terms of R.

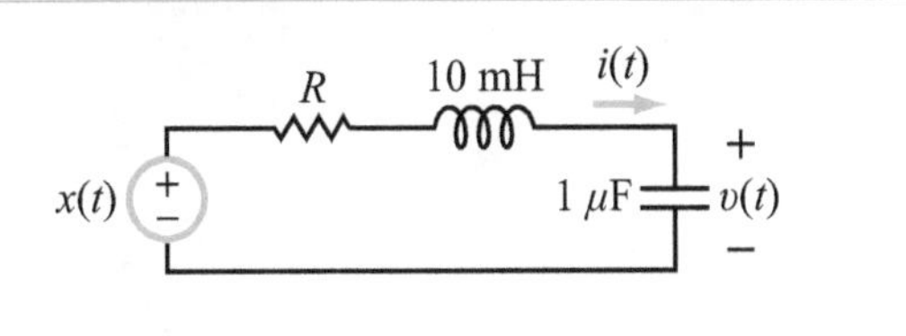

Figure P3.51: Circuit for Problems 3.51 to 3.53.

3.52 In the circuit of Fig. P3.51, let $i(0) = 1$ mA.

(a) Compute the resistance R so that the zero-input response has the form

$$i(t) = Ae^{-20000t}\, u(t) + Be^{-5000t}\, u(t)$$

for some constants A and B.

(b) Using the resistance R from (a), compute the initial capacitor voltage $\upsilon(0)$ so that the zero-input response is $i(t) = Be^{-5000t}\, u(t)$ for some constant B.

3.53 In the circuit of Fig. P3.51, $i(0) = 0$.

(a) Compute the resistance R so that the zero-input response has the form

$$i(t) = Ae^{-2000t}\, u(t) + Be^{-50000t}\, u(t)$$

for some constants A and B.

(b) Using the resistance R from (a), show that no initial capacitor voltage $\upsilon(0)$ will make the zero-input response be $i(t) = Be^{-50000t}\, u(t)$ for some constant B.

3.54 An LTI system is described by the LCCDE

$$\sum_{i=0}^{N} a_{N-i} \frac{d^i y}{dt^i} = K \sum_{j=0}^{M} b_{M-j} \frac{d^j x}{dt^j}$$

with $a_0 = b_0 = 1$ and initial conditions

$$y(0) = y_0;\quad \left.\frac{dy}{dt}\right|_{t=0} = y_1;\ \ldots\ \left.\frac{dy^{N-1}}{dt^{N-1}}\right|_{t=0} = y_{N-1}.$$

(a) Derive an explicit expression for $y(t)$ that exhibits the Laplace transforms of the zero-state response and zero-input response as two separate terms.

(b) Apply your result to

$$\frac{d^2y}{dt^2} + 4\frac{dy}{dt} + 3y(t) = 0$$

with initial conditions $y(0) = 2$ and $\left.\frac{dy}{dt}\right|_{t=0} = -12$.

3.55 The impulse response of a strictly proper LTI system with N distinct poles $\{\mathbf{p}_i\}$ has the form

$$h(t) = \sum_{i=1}^{N} \mathbf{C}_i e^{\mathbf{p}_i t}\, u(t)$$

for some constants $\{\mathbf{C}_i\}$. Its input $x(t)$ has the form

$$x(t) = \sum_{i=1}^{n} \mathbf{B}_i e^{\mathbf{q}_i t}\, u(t)$$

for some constants $\{\mathbf{B}_i\}$ and $\{\mathbf{q}_i\}$. If all of the $\{\mathbf{p}_i, \mathbf{q}_i\}$ are distinct, show that the output $y(t)$ has the form

$$y(t) = \underbrace{\sum_{i=1}^{n} \mathbf{B}_i\ \mathbf{H}(\mathbf{q}_i)\ e^{\mathbf{q}_i t}\ u(t)}_{\text{FORCED: Like } x(t)} + \underbrace{\sum_{i=1}^{N} \mathbf{C}_i\ \mathbf{X}(\mathbf{p}_i)\ e^{\mathbf{p}_i t}\ u(t)}_{\text{NATURAL: Like } h(t)}.$$

This exhibits the forced and natural responses of the system as two separate sums of terms.

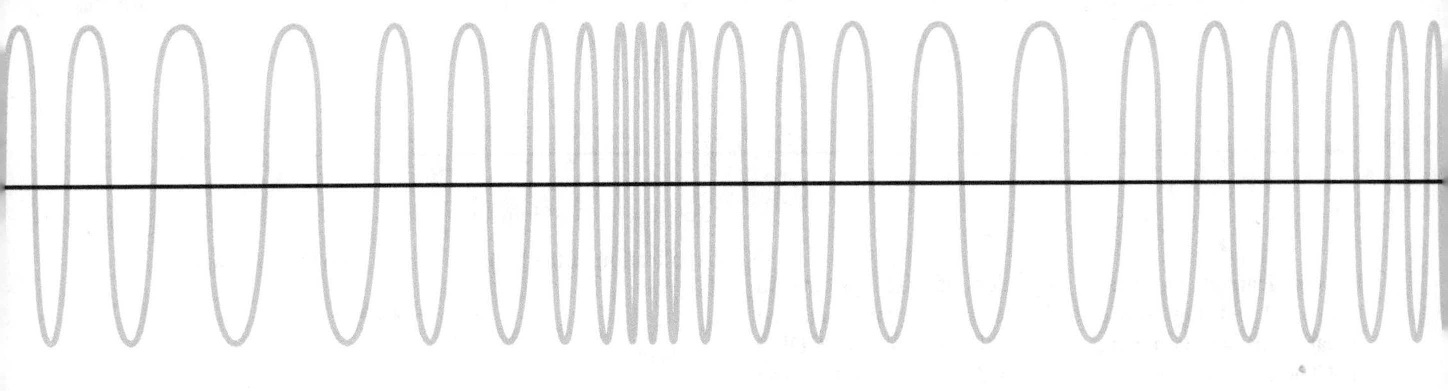

4 Applications of the Laplace Transform

Contents

Overview, 132
4-1 s-Domain Circuit Element Models, 132
4-2 s-Domain Circuit Analysis, 134
4-3 Electromechanical Analogues, 140
4-4 Biomechanical Model of a Person Sitting in a Moving Chair, 146
4-5 Op-Amp Circuits, 149
4-6 Configurations of Multiple Systems, 154
4-7 System Synthesis, 157
4-8 Basic Control Theory, 160
4-9 Temperature Control System, 167
4-10 Amplifier Gain-Bandwidth Product, 171
4-11 Step Response of a Motor System, 174
4-12 Control of a Simple Inverted Pendulum on a Cart, 178
Summary, 183
Problems, 183

Objectives

Learn to:

- Use **s**-domain circuit element models to analyze electric circuits.
- Use electromechanical analogues to simulate and analyze mechanical systems.
- Use op-amp circuits to implement systems.
- Develop system realizations that conform to specified transfer functions.
- Employ feedback control techniques to improve system performance and stability

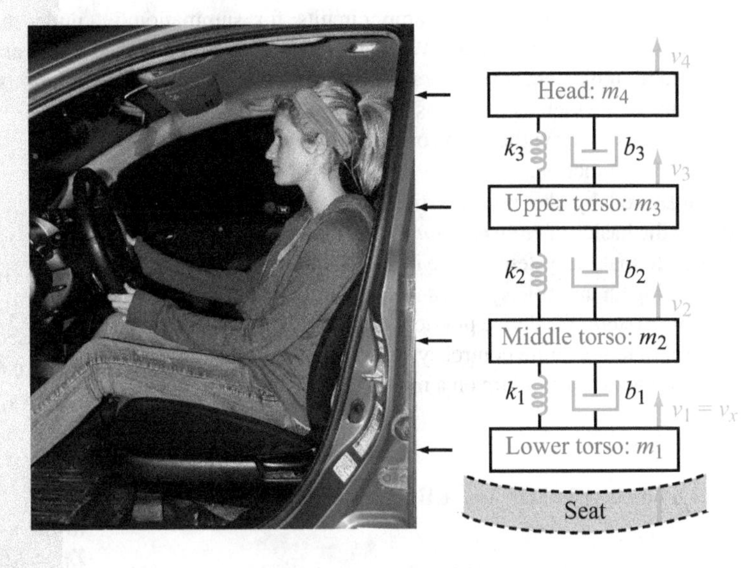

The Laplace-transform tools learned in the previous chapter are now applied to model and solve a wide variety of ***mechanical and thermal systems***, including how to compute the movement of a passenger's head as the car moves over curbs and other types of pavements, and how to design ***feedback loops*** to control ***motors*** and heating systems.

Overview

Having learned the basic properties of LTI systems in Chapter 2 and how to use the Laplace transform to analyze linear, constant-coefficient differential equations (LCCDEs) in Chapter 3, we are now ready to demonstrate how these mathematical techniques can be applied to analyze the behavior of specified systems or to design a system's configuration so it will behave in a desired way.

First, we develop the concept of ***s-domain circuits***. These are like phasors in circuit analysis, except that they can be used with any type of input (not just sinusoidal sources), and they can incorporate non-zero initial conditions. Next, we show that ***spring-mass-damper***-based mechanical systems can be modeled as electrical circuits using ***electromechanical analogues***. By so doing, circuit analysis techniques can be applied to analyze and synthesize mechanical systems. By way of an example, we use electromechanical analogues to analyze a biomechanical model of a sitting person subjected to a sinusoidal chair motion.

Next, we review basic op-amp circuits for summation, differentiation, and integration. We then show how op-amps can be used to implement series, parallel, and feedback connection of systems. We then use op-amps for system ***synthesis***: Given a transfer function, synthesize an op-amp circuit that implements the transfer function.

Motivated by the non-invertibility of many systems, we derive the basics of ***control theory***, in which negative feedback is used to make a system behave in a desired way. This includes stabilizing an unstable system and speeding up the response of a stable system. We derive physical models and apply feedback control to temperature control systems, motor control systems, and an inverted pendulum on a moving cart.

4-1 s-Domain Circuit Element Models

The **s**-domain technique can be used to analyze circuits excited by sources with many types of waveforms—including pulse, step, ramp, and exponential—and provides a complete solution that incorporates both the steady-state and transient components of the overall response.

> ▶ The **s**-domain transformation of circuit elements incorporates initial conditions associated with any energy storage that may have existed in capacitors and inductors at $t = 0^-$. ◀

Resistor in the s-domain

Application of the Laplace transform to Ohm's law,

$$\mathcal{L}[\upsilon] = \mathcal{L}[Ri], \tag{4.1}$$

leads to

$$\mathbf{V} = R\mathbf{I}, \tag{4.2}$$

where by definition,

$$\mathbf{V} = \mathcal{L}[\upsilon] \quad \text{and} \quad \mathbf{I} = \mathcal{L}[i]. \tag{4.3}$$

Hence, for the resistor, the correspondence between the time and **s**-domains is

$$\upsilon = Ri \quad \longleftrightarrow \quad \mathbf{V} = R\mathbf{I}. \tag{4.4}$$

V and **I** are functions of **s**; the **s**-dependence is not expressed explicitly to keep the notation simple.

Inductor in the s-domain

For R, the form of the i–υ relationship remained invariant under the transformation to the **s**-domain. That is not the case for L and C. Application of the Laplace transform to the i–υ relationship of the inductor,

$$\mathcal{L}[\upsilon] = \mathcal{L}\left[L\,\frac{di}{dt}\right], \tag{4.5}$$

gives

$$\mathbf{V} = L[\mathbf{sI} - i(0^-)], \tag{4.6}$$

where $i(0^-)$ is the current that was flowing through the inductor at $t = 0^-$. The time-differentiation property (#6 in Table 3-1) was used in obtaining Eq. (4.6). The correspondence between the two domains is expressed as

$$\upsilon = L\,\frac{di}{dt} \quad \longleftrightarrow \quad \mathbf{V} = \mathbf{s}L\mathbf{I} - L\,i(0^-). \tag{4.7}$$

In the **s**-domain, an inductor is represented by an impedance $\mathbf{Z}_L = \mathbf{s}L$, in series with a dc voltage source given by $L\,i(0^-)$ or (through source transformation) in parallel with a dc current source $i(0^-)/\mathbf{s}$, as shown in Table 4-1. Note that the current **I** flows from $(-)$ to $(+)$ through the dc voltage source (if $i(0^-)$ is positive).

Capacitor in the s-domain

Similarly,

$$i = C\,\frac{d\upsilon}{dt} \quad \longleftrightarrow \quad \mathbf{I} = \mathbf{s}C\mathbf{V} - C\,\upsilon(0^-), \tag{4.8}$$

Table 4-1: Circuit models for R, L, and C in the s-domain.

Time-Domain	s-Domain		
Resistor $v = Ri$	$\mathbf{V} = R\mathbf{I}$		
Inductor $v_L = L \frac{di_L}{dt}$; $i_L = \frac{1}{L}\int_{0^-}^{t} v_L \, dt' + i_L(0^-)$	$\mathbf{V}_L = \mathbf{s}L\mathbf{I}_L - L\, i_L(0^-)$	OR	$\mathbf{I}_L = \frac{\mathbf{V}_L}{\mathbf{s}L} + \frac{i_L(0^-)}{\mathbf{s}}$
Capacitor $i_C = C \frac{dv_C}{dt}$; $v_C = \frac{1}{C}\int_{0^-}^{t} i_C \, dt + v_C(0^-)$	$\mathbf{V}_C = \frac{\mathbf{I}_C}{\mathbf{s}C} + \frac{v_C(0^-)}{\mathbf{s}}$	OR	$\mathbf{I}_C = \mathbf{s}C\mathbf{V}_C - C\, v_C(0^-)$

where $v(0^-)$ is the initial voltage across the capacitor. The s-domain circuit models for the capacitor are available in Table 4-1.

Impedances $\mathbf{Z}_R$, $\mathbf{Z}_L$, and $\mathbf{Z}_C$ are defined in the s-domain in terms of voltage to current ratios, under zero initial conditions $[i(0^-) = v(0^-) = 0]$:

$$\mathbf{Z}_R = R, \qquad \mathbf{Z}_L = \mathbf{s}L, \qquad \mathbf{Z}_C = \frac{1}{\mathbf{s}C}. \tag{4.9}$$

Concept Question 4-1: In the **s**-domain, initial conditions associated with capacitors and inductors are represented by equivalent voltage sources or current sources. Are such voltage sources connected in series with the circuit element or in parallel with it? What about equivalent current sources, how are they connected? (See S²)

Concept Question 4-2: The **s**-domain circuit model for an inductor accounts for the initial current flowing through the inductor, but not the initial voltage across it. Is that an issue? (See S²)

Exercise 4-1: Convert the circuit in Fig. E4-1 into the **s**-domain.

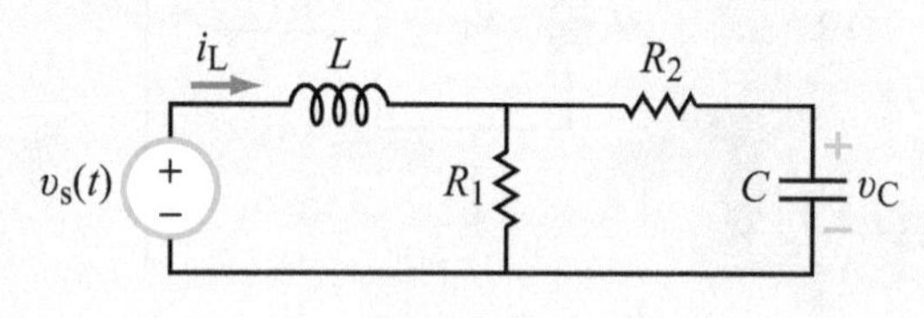

Figure E4-1

Answer:

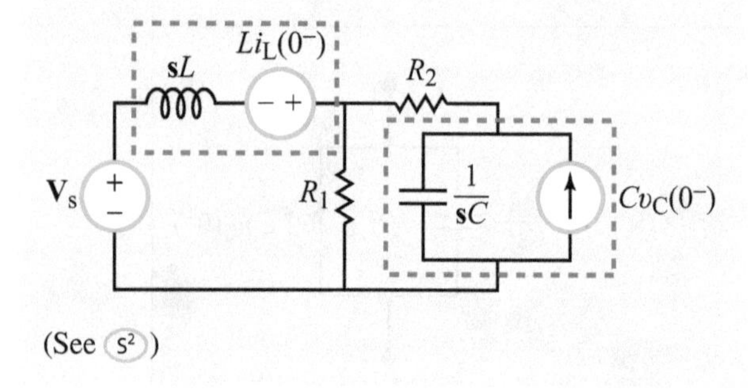

(See S²)

4-2 s-Domain Circuit Analysis

Circuit laws and analysis tools used in the time domain are equally applicable in the **s**-domain. They include Kirchhoff voltage law (KVL) and Kirchhoff current law (KCL); voltage and current division; source transformation; source superposition; and Thévenin and Norton equivalent circuits. Execution of the **s**-domain analysis technique entails the following four steps:

Solution Procedure: s-Domain Technique

Step 1: Transform the circuit from the time domain to the **s**-domain.

Step 2: Apply KVL, KCL, and the other circuit tools to obtain an explicit expression for the voltage or current of interest.

Step 3: If necessary, expand the expression into partial fractions (Section 3-5).

Step 4: Use the list of transform pairs given in Tables 3-2 and 3-3 and the list of properties in Table 3-1 (if needed) to transform the partial fraction to the time domain.

This process is illustrated through Examples 4-1 to 4-4, involving circuits excited by a variety of different waveforms.

Example 4-1: Interrupted Voltage Source

The circuit shown in Fig. 4-1(a) is driven by an input voltage source $\upsilon_{in}(t)$, and its output is taken across the 3-Ω resistor. The input waveform is depicted in Fig. 4-1(b): It starts out as a 15 V dc level that had existed for a long time prior to $t = 0$, it then experiences a momentary drop down to zero volts at $t = 0$, followed by a slow recovery towards its earlier level. The waveform of the output voltage is shown in Fig. 4-1(c). Analyze the circuit to obtain an expression for $\upsilon_{out}(t)$, in order to confirm that the waveform in Fig. 4-1(c) is indeed correct.

Solution: Before transforming the circuit to the **s**-domain, we should always evaluate it at $t = 0^-$, to determine the voltages across all capacitors and currents through all inductors at $t = 0^-$. Until $t = 0$, the circuit was in a *static* state because $\upsilon_{in}(t)$ had been a constant 15 V dc source for a long time. Hence, the voltage across the capacitor is constant, and therefore, the current through it is zero. Similarly, the current through the inductor is constant and the voltage across it is zero. Accordingly, the circuit condition at $t = 0^-$ is as depicted is Fig. 4-1(d), where C is replaced with an open circuit and L replaced with a short circuit. A simple examination of the circuit reveals that

$$\upsilon_C(0^-) = 9\text{ V}, \quad i_L(0^-) = 3\text{ A}, \quad \text{and } \upsilon_{out}(0^-) = 9\text{ V}. \tag{4.10}$$

Next, we need to transform the circuit in Fig. 4-1(a) to the **s**-domain. The voltage waveform is given by

$$\upsilon_{in}(t) = \begin{cases} 15\text{ V} & \text{for } t < 0, \\ 15(1 - e^{-2t})\, u(t)\text{ V} & \text{for } t \geq 0. \end{cases} \tag{4.11}$$

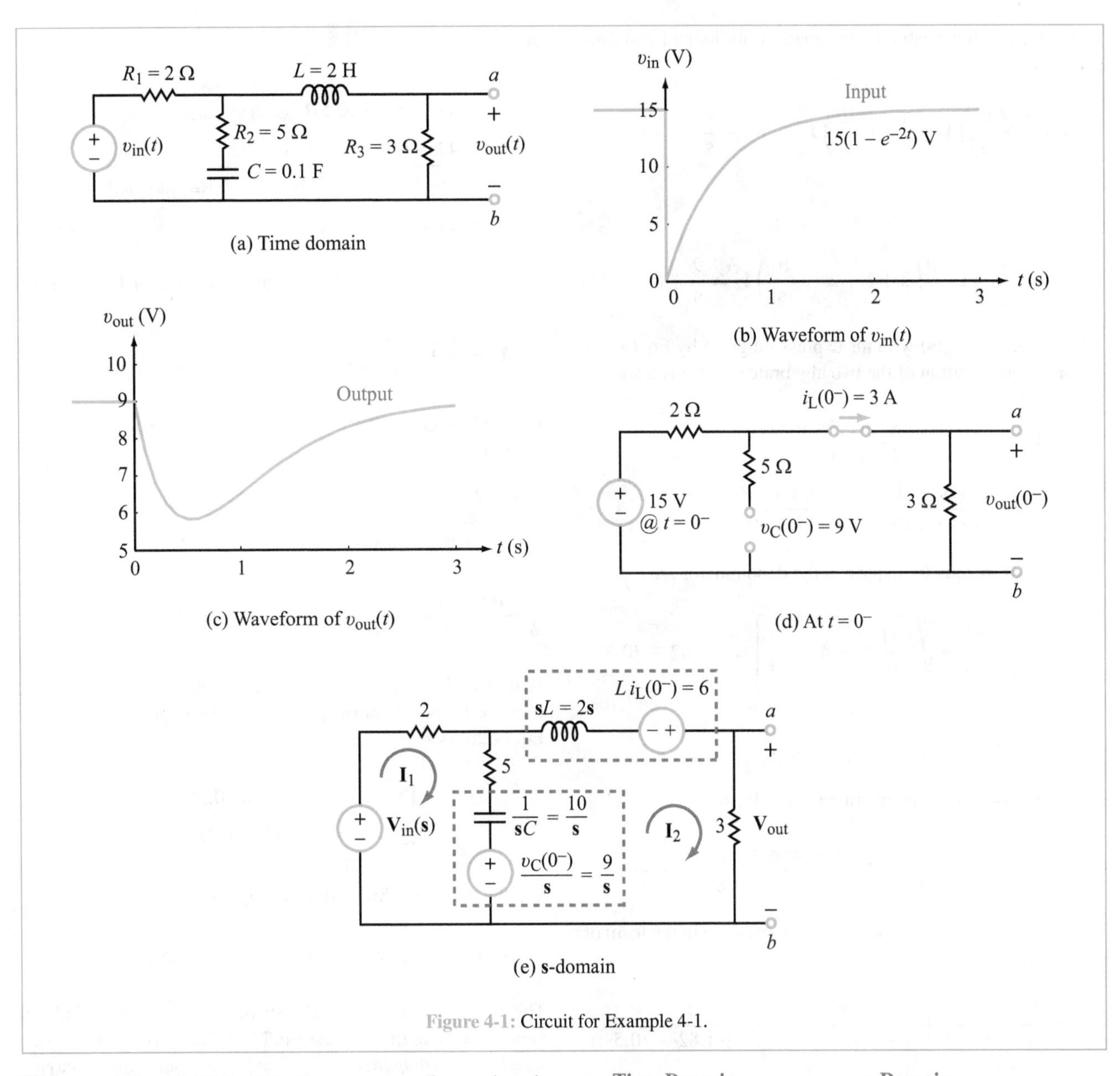

Figure 4-1: Circuit for Example 4-1.

With the help of Table 3-2, the corresponding **s**-domain function for $t \geq 0$ is given by

$$\mathbf{V}_{in}(\mathbf{s}) = \frac{15}{\mathbf{s}} - \frac{15}{\mathbf{s}+2}. \tag{4.12}$$

The **s**-domain circuit is shown in Fig. 4-1(e), in which L and C are represented by their **s**-domain models in accordance with Table 4-1, namely,

Time Domain **s-Domain**

i_L L → $\mathbf{I}_L$ $\mathbf{s}L$ $L\, i_L(0^-)$ $=$ $2\mathbf{s}$ 6 V,

and

i_C C → $\mathbf{I}_C$ $1/\mathbf{s}C$ $v_C(0^-)/\mathbf{s}$ $=$ $10/\mathbf{s}$ $9/\mathbf{s}$.

By inspection, the mesh-current equations for loops 1 and 2 are given by

$$\left(2+5+\frac{10}{\mathbf{s}}\right)\mathbf{I}_1-\left(5+\frac{10}{\mathbf{s}}\right)\mathbf{I}_2=\mathbf{V}_{\text{in}}-\frac{9}{\mathbf{s}} \tag{4.13}$$

and

$$-\left(5+\frac{10}{\mathbf{s}}\right)\mathbf{I}_1+\left(3+5+2\mathbf{s}+\frac{10}{\mathbf{s}}\right)\mathbf{I}_2=\frac{9}{\mathbf{s}}+6. \tag{4.14}$$

After replacing $\mathbf{V}_{\text{in}}(\mathbf{s})$ with the expression given by Eq. (4.12), simultaneous solution of the two algebraic equations leads to

$$\begin{aligned}\mathbf{I}_2&=\frac{42\mathbf{s}^3+162\mathbf{s}^2+306\mathbf{s}+300}{\mathbf{s}(\mathbf{s}+2)(14\mathbf{s}^2+51\mathbf{s}+50)}\\&=\frac{42\mathbf{s}^3+162\mathbf{s}^2+306\mathbf{s}+300}{14\mathbf{s}(\mathbf{s}+2)(\mathbf{s}^2+51\mathbf{s}/14+50/14)}.\end{aligned} \tag{4.15}$$

The roots of the quadratic term in the denominator are

$$\mathbf{s}_1=\frac{1}{2}\left[-\frac{51}{14}-\sqrt{\left(\frac{51}{14}\right)^2-4\times\frac{50}{14}}\right]=-1.82-j0.5, \tag{4.16a}$$

and

$$\mathbf{s}_2=-1.82+j0.5. \tag{4.16b}$$

Hence, Eq. (4.15) can be rewritten in the form

$$\mathbf{I}_2=\frac{42\mathbf{s}^3+162\mathbf{s}^2+306\mathbf{s}+300}{14\mathbf{s}(\mathbf{s}+2)(\mathbf{s}+1.82+j0.5)(\mathbf{s}+1.82-j0.5)}. \tag{4.17}$$

The expression for $\mathbf{I}_2$ is now ready for expansion in the form of partial fractions:

$$\mathbf{I}_2=\frac{A_1}{\mathbf{s}}+\frac{A_2}{\mathbf{s}+2}+\frac{\mathbf{B}}{\mathbf{s}+1.82+j0.5}+\frac{\mathbf{B}^*}{\mathbf{s}+1.82-j0.5}, \tag{4.18}$$

with

$$\begin{aligned}A_1&=\mathbf{s}\mathbf{I}_2|_{\mathbf{s}=0}\\&=\left.\frac{42\mathbf{s}^3+162\mathbf{s}^2+306\mathbf{s}+300}{14(\mathbf{s}+2)(\mathbf{s}^2+51\mathbf{s}/14+50/14)}\right|_{\mathbf{s}=0}=3,\end{aligned} \tag{4.19a}$$

$$\begin{aligned}A_2&=(\mathbf{s}+2)\mathbf{I}_2|_{\mathbf{s}=-2}\\&=\left.\frac{42\mathbf{s}^3+162\mathbf{s}^2+306\mathbf{s}+300}{14\mathbf{s}(\mathbf{s}^2+51\mathbf{s}/14+50/14)}\right|_{\mathbf{s}=-2}=0,\end{aligned} \tag{4.19b}$$

and

$$\begin{aligned}\mathbf{B}&=(\mathbf{s}+1.82+j0.5)\mathbf{I}_2|_{\mathbf{s}=-1.82-j0.5}\\&=\left.\frac{42\mathbf{s}^3+162\mathbf{s}^2+306\mathbf{s}+300}{14\mathbf{s}(\mathbf{s}+2)(\mathbf{s}+1.82-j0.5)}\right|_{\mathbf{s}=-1.82-j0.5}\\&=5.32e^{-j90^\circ}.\end{aligned} \tag{4.19c}$$

Inserting the values of A_1, A_2, and $\mathbf{B}$ into Eq. (4.18) leads to

$$\mathbf{I}_2=\frac{3}{\mathbf{s}}+\frac{5.32e^{-j90^\circ}}{\mathbf{s}+1.82+j0.5}+\frac{5.32e^{j90^\circ}}{\mathbf{s}+1.82-j0.5}. \tag{4.20}$$

For the first term, entry #2 in Table 3-2 leads to inverse Laplace transform

$$\frac{3}{\mathbf{s}} \longleftrightarrow 3\,u(t),$$

and from property #3 of Table 3-3, we have

$$\frac{Ae^{j\theta}}{\mathbf{s}+a+jb}+\frac{Ae^{-j\theta}}{\mathbf{s}+a-jb} \longleftrightarrow 2Ae^{-at}\cos(bt-\theta)\,u(t). \tag{4.21}$$

With $A=5.32$, $\theta=-90^\circ$, $a=1.82$, and $b=0.5$, the inverse Laplace transform corresponding to the expression given by Eq. (4.20) is

$$\begin{aligned}i_2(t)&=[3+10.64e^{-1.82t}\cos(0.5t+90^\circ)]\,u(t)\\&=[3-10.64e^{-1.82t}\sin 0.5t]\,u(t)\text{ A},\end{aligned} \tag{4.22}$$

and the corresponding output voltage is

$$\upsilon_{\text{out}}(t)=3i_2(t)=[9-31.92e^{-1.82t}\sin 0.5t]\,u(t)\text{ V}. \tag{4.23}$$

The waveform of $\upsilon_{\text{out}}(t)$ shown in Fig. 4-1(c) was (indeed) generated using this expression. Note that $\upsilon_{\text{out}}(t)$ consists of a *steady-state component* of 9 V and an exponentially decaying *transient component* that goes to zero as $t\to\infty$.

Example 4-2: ac Source with a dc Bias

Repeat the analysis of the circuit shown in Fig. 4-1(a), but change the waveform of the voltage source into a 15 V dc level that has existed for a long time in combination with a 20 V ac signal that starts at $t=0$, as shown in Fig. 4-2(a).

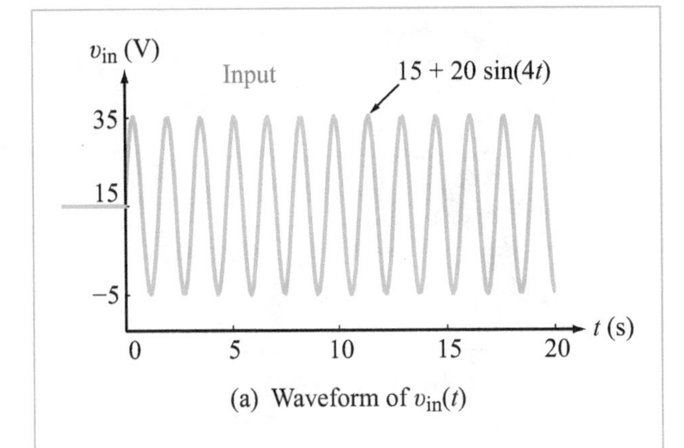

(a) Waveform of $v_{in}(t)$

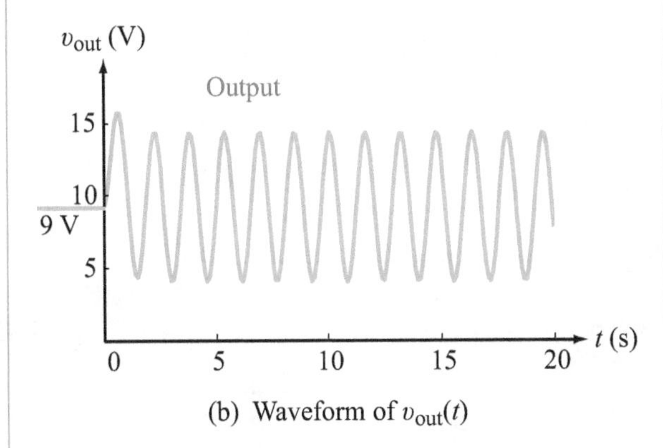

(b) Waveform of $v_{out}(t)$

Figure 4-2: Input and output waveforms for the circuit in Fig. 4-1 (Example 4-2).

Solution*:

$$v_{in}(t) = \begin{cases} 15\text{ V}, & \text{for } t < 0 \\ [15 + 20\sin 4t]\, u(t)\text{ V}, & \text{for } t \ge 0. \end{cases} \tag{4.24}$$

In view of entry #9 in Table 3-2, the **s**-domain counterpart of $v_{in}(t)$ is

$$\mathbf{V}_{in}(\mathbf{s}) = \frac{15}{\mathbf{s}} + \frac{20\omega}{\mathbf{s}^2 + \omega^2} = \frac{15}{\mathbf{s}} + \frac{80}{\mathbf{s}^2 + 16}, \tag{4.25}$$

where we replaced ω with 4 rad/s. For $t \le 0^-$, the voltage waveform is the same as it was before in Example 4-1, namely 15 V, so initial conditions remain as before (Fig. 4-1(d)), as does the circuit configuration in the **s**-domain (Fig. 4-1(e)). The only quantity that has changed is the expression for $\mathbf{V}_{in}(\mathbf{s})$. Rewriting Eq. (4.13), with $\mathbf{V}_{in}(\mathbf{s})$ as given by Eq. (4.25), leads to

$$\begin{aligned}\left(7 + \frac{10}{\mathbf{s}}\right)\mathbf{I}_1 - \left(5 + \frac{10}{\mathbf{s}}\right)\mathbf{I}_2 &= \mathbf{V}_{in} - \frac{9}{\mathbf{s}} \\ &= \frac{15}{\mathbf{s}} + \frac{80}{\mathbf{s}^2 + 16} - \frac{9}{\mathbf{s}} \\ &= \frac{6\mathbf{s}^2 + 80\mathbf{s} + 96}{\mathbf{s}(\mathbf{s}^2 + 16)}.\end{aligned} \tag{4.26}$$

Equation (4.14) remains unchanged as

$$-\left(5 + \frac{10}{\mathbf{s}}\right)\mathbf{I}_1 + \left(3 + 5 + 2\mathbf{s} + \frac{10}{\mathbf{s}}\right)\mathbf{I}_2 = \frac{9}{\mathbf{s}} + 6. \tag{4.27}$$

Simultaneous solution of Eqs. (4.26) and (4.27) leads to

$$\begin{aligned}\mathbf{I}_2 &= \frac{42\mathbf{s}^4 + 153\mathbf{s}^3 + 1222\mathbf{s}^2 + 3248\mathbf{s} + 2400}{14\mathbf{s}(\mathbf{s}^2 + 16)(\mathbf{s}^2 + 51\mathbf{s}/14 + 50/14)} = \\ &\frac{42\mathbf{s}^4 + 153\mathbf{s}^3 + 1222\mathbf{s}^2 + 3248\mathbf{s} + 2400}{14\mathbf{s}(\mathbf{s} + j4)(\mathbf{s} - j4)(\mathbf{s} + 1.82 + j0.5)(\mathbf{s} + 1.82 - j0.5)},\end{aligned} \tag{4.28}$$

where we expanded the two quadratic terms in the denominator into a product of four simple-pole factors.

Partial fraction representation takes the form:

$$\begin{aligned}\mathbf{I}_2 = \frac{A_1}{\mathbf{s}} + \frac{\mathbf{B}_1}{\mathbf{s} + j4} + \frac{\mathbf{B}_1^*}{\mathbf{s} - j4} + \frac{\mathbf{B}_2}{\mathbf{s} + 1.82 + j0.5} \\ + \frac{\mathbf{B}_2^*}{\mathbf{s} + 1.82 - j0.5},\end{aligned} \tag{4.29}$$

with

$$\begin{aligned}A_1 &= \mathbf{s}\mathbf{I}_2|_{\mathbf{s}=0} = \\ &\left.\frac{42\mathbf{s}^4 + 153\mathbf{s}^3 + 1222\mathbf{s}^2 + 3248\mathbf{s} + 2400}{14(\mathbf{s} + j4)(\mathbf{s} - j4)(\mathbf{s} + 1.82 + j0.5)(\mathbf{s} + 1.82 - j0.5)}\right|_{\mathbf{s}=0} \\ &= 3,\end{aligned} \tag{4.30a}$$

$$\mathbf{B}_1 = (\mathbf{s} + j4)\mathbf{I}_2|_{\mathbf{s}=-j4} = 0.834\underline{/157.0^\circ}, \tag{4.30b}$$

and

$$\mathbf{B}_2 = (\mathbf{s} + 1.82 + j0.5)\mathbf{I}_2|_{\mathbf{s}=-1.82-j0.5} = 0.79\underline{/14.0^\circ}. \tag{4.30c}$$

*MATLAB/MathScript solution is available on book website.

Hence, $\mathbf{I}_2$ becomes

$$\mathbf{I}_2 = \frac{3}{\mathbf{s}} + \frac{0.834e^{j157^\circ}}{\mathbf{s}+j4} + \frac{0.834e^{-j157^\circ}}{\mathbf{s}-j4} + \frac{0.79e^{j14^\circ}}{\mathbf{s}+1.82+j0.5} + \frac{0.79e^{-j14^\circ}}{\mathbf{s}+1.82-j0.5}. \quad (4.31)$$

With the help of entry #3 in Table 3-3, conversion to the time domain gives

$$i_2(t) = [3 + 1.67\cos(4t - 157^\circ) + 1.58e^{-1.82t}\cos(0.5t - 14^\circ)]\,u(t) \text{ A}, \quad (4.32)$$

and the corresponding output voltage is

$$\begin{aligned} \upsilon_{\text{out}}(t) &= 3\,i_2(t) \\ &= [9 + 5\cos(4t - 157^\circ) \\ &\quad + 4.73e^{-1.82t}\cos(0.5t - 14^\circ)]\,u(t) \text{ V}. \end{aligned} \quad (4.33)$$

The profile of $\upsilon_{\text{out}}(t)$ is shown in Fig. 4-2(b): In response to the introduction of the ac signal at the input at $t = 0$, the output consists of a dc term equal to 9 V, plus an *oscillatory transient component* (the last term in Eq. (4.33)) that decays down to zero over time, and a *steady-state oscillatory component* that continues indefinitely.

Another approach for analyzing the circuit is by applying the source-superposition method, wherein the circuit is analyzed twice: once with the 15 V dc component alone and a second time with only the ac component that starts at $t = 0$. The first solution would lead to the first term in Eq. (4.33). For the ac source, we can apply phasor-analysis, but the solution would have yielded the steady-state component only.

► The advantage of the Laplace transform technique is that it can provide a complete solution that automatically includes both the transient and steady-state components and would do so for any type of excitation. ◄

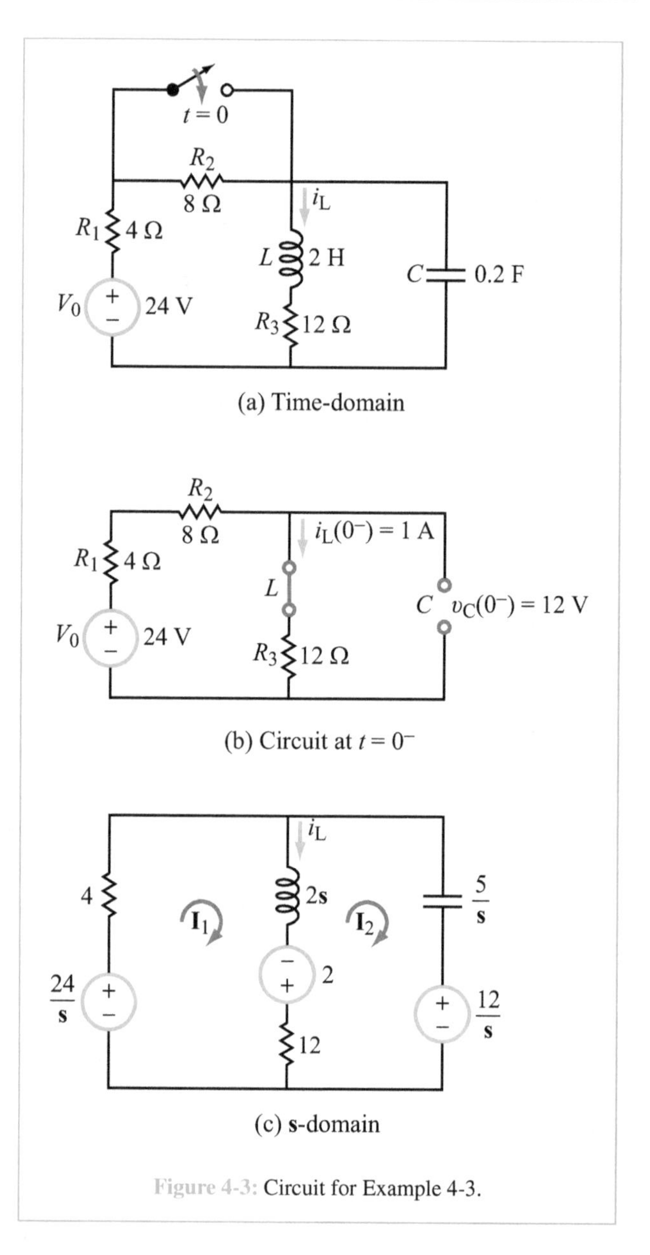

Figure 4-3: Circuit for Example 4-3.

Example 4-3: Circuit with a Switch

Determine $i_L(t)$ in the circuit of Fig. 4-3 for $t \geq 0$.

Solution: We start by examining the state of the circuit at $t = 0^-$ (before closing the switch). Upon replacing L with a short circuit and C with an open circuit, as portrayed by the configuration in Fig. 4-3(b), we establish that

$$i_L(0^-) = 1 \text{ A}, \qquad \upsilon_C(0^-) = 12 \text{ V}. \quad (4.34)$$

For $t \geq 0$, the $\mathbf{s}$-domain equivalent of the original circuit is shown in Fig. 4-3(c), where we have replaced R_2 with a short circuit, converted the dc source into its $\mathbf{s}$-domain equivalent, and in accordance with the circuit models given in Table 4-1, converted L and C into impedances, each with its own appropriate voltage source. By inspection, the two mesh current equations are given by

$$(4 + 12 + 2\mathbf{s})\mathbf{I}_1 - (12 + 2\mathbf{s})\mathbf{I}_2 = \frac{24}{\mathbf{s}} + 2, \tag{4.35}$$

$$-(12 + 2\mathbf{s})\mathbf{I}_1 + \left(12 + 2\mathbf{s} + \frac{5}{\mathbf{s}}\right)\mathbf{I}_2 = -2 - \frac{12}{\mathbf{s}}. \tag{4.36}$$

Simultaneous solution of the two equations leads to

$$\mathbf{I}_1 = \frac{12\mathbf{s}^2 + 77\mathbf{s} + 60}{\mathbf{s}(4\mathbf{s}^2 + 29\mathbf{s} + 40)} \tag{4.37a}$$

and

$$\mathbf{I}_2 = \frac{8(\mathbf{s} + 6)}{4\mathbf{s}^2 + 29\mathbf{s} + 40}. \tag{4.37b}$$

The associated inductor current $\mathbf{I}_L$ is

$$\mathbf{I}_L = \mathbf{I}_1 - \mathbf{I}_2 = \frac{4\mathbf{s}^2 + 29\mathbf{s} + 60}{\mathbf{s}(4\mathbf{s}^2 + 29\mathbf{s} + 40)} = \frac{4\mathbf{s}^2 + 29\mathbf{s} + 60}{4\mathbf{s}(\mathbf{s} + 1.85)(\mathbf{s} + 5.4)}, \tag{4.38}$$

which can be represented by the partial fraction expansion

$$\mathbf{I}_L = \frac{A_1}{\mathbf{s}} + \frac{A_2}{\mathbf{s} + 1.85} + \frac{A_3}{\mathbf{s} + 5.4}. \tag{4.39}$$

The values of A_1 to A_3 are obtained from

$$A_1 = \mathbf{s}\mathbf{I}_L|_{\mathbf{s}=0} = \frac{60}{40} = 1.5, \tag{4.40a}$$

$$A_2 = (\mathbf{s} + 1.85)\mathbf{I}_L|_{\mathbf{s}=-1.85} = \left.\frac{4\mathbf{s}^2 + 29\mathbf{s} + 60}{4\mathbf{s}(\mathbf{s} + 5.4)}\right|_{\mathbf{s}=-1.85} = -0.76, \tag{4.40b}$$

and

$$A_3 = (\mathbf{s} + 5.4)\mathbf{I}_L|_{\mathbf{s}=-5.4} = 0.26. \tag{4.40c}$$

Hence,

$$\mathbf{I}_L = \frac{1.5}{\mathbf{s}} - \frac{0.76}{\mathbf{s} + 1.85} + \frac{0.26}{\mathbf{s} + 5.4}, \tag{4.41}$$

and the corresponding time-domain current is

$$i_L(t) = [1.5 - 0.76e^{-1.85t} + 0.26e^{-5.4t}]\, u(t) \text{ A}. \tag{4.42}$$

Example 4-4: Lowpass Filter Response to a Rectangular Pulse

Given the RC circuit shown in Fig. 4-4(a), determine the output response to a 1 s long rectangular pulse. The pulse amplitude is 1 V. This is a repeat of Example 2-4, which was analyzed in Chapter 2 by applying the convolution method in the time domain.

Solution: With $R = 0.5$ MΩ and $C = 1$ μF, the product is $RC = 0.5$ s. Voltage division in the s-domain (Fig. 4-4(b)) leads to

$$\mathbf{H}(\mathbf{s}) = \frac{\mathbf{V}_{\text{out}}(\mathbf{s})}{\mathbf{V}_{\text{in}}(\mathbf{s})} = \frac{1/\mathbf{s}C}{R + 1/\mathbf{s}C} = \frac{1/RC}{\mathbf{s} + 1/RC} = \frac{2}{\mathbf{s} + 2}. \tag{4.43}$$

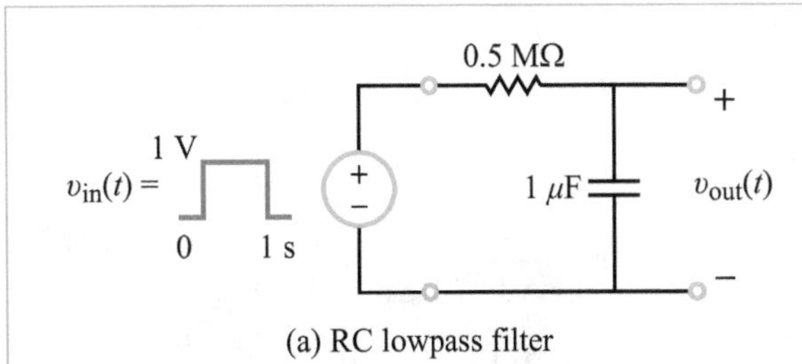

(a) RC lowpass filter

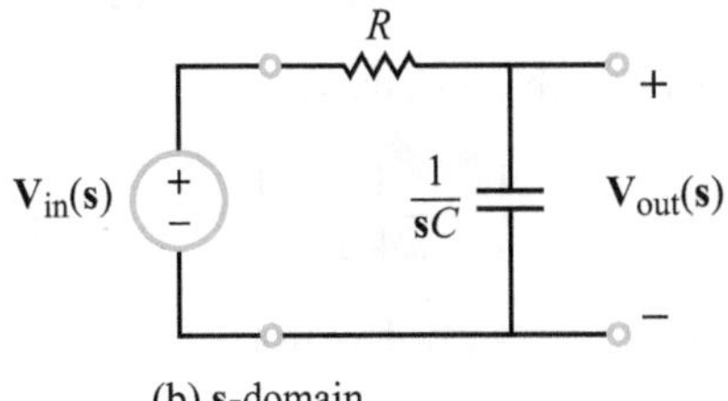

(b) s-domain

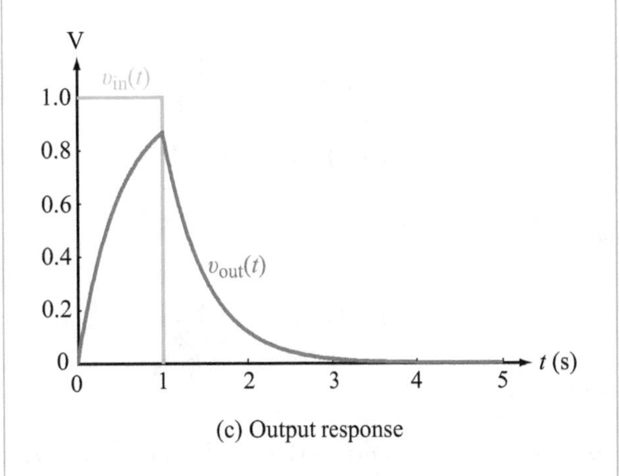

(c) Output response

Figure 4-4: Example 4-4.

The rectangular pulse is given by

$$\upsilon_{\text{in}}(t) = [u(t) - u(t-1)]\text{ V}, \tag{4.44}$$

and with the help of Table 3-2, its s-domain counterpart is

$$\mathbf{V}_{\text{in}}(\mathbf{s}) = \left[\frac{1}{\mathbf{s}} - \frac{1}{\mathbf{s}}\,e^{-\mathbf{s}}\right]\text{ V}. \tag{4.45}$$

Hence,

$$\mathbf{V}_{\text{out}}(\mathbf{s}) = \mathbf{H}(\mathbf{s})\,\mathbf{V}_{\text{in}}(\mathbf{s}) = 2(1-e^{-\mathbf{s}})\left[\frac{1}{\mathbf{s}(\mathbf{s}+2)}\right]. \tag{4.46}$$

In preparation for transformation to the time domain, we expand the function inside the square bracket into partial fractions:

$$\frac{1}{\mathbf{s}(\mathbf{s}+2)} = \frac{A_1}{\mathbf{s}} + \frac{A_2}{\mathbf{s}+2}, \tag{4.47}$$

with

$$A_1 = \mathbf{s}\left[\frac{1}{\mathbf{s}(\mathbf{s}+2)}\right]\Bigg|_{\mathbf{s}=0} = \frac{1}{2},$$

$$A_2 = (\mathbf{s}+2)\left[\frac{1}{\mathbf{s}(\mathbf{s}+2)}\right]\Bigg|_{\mathbf{s}=-2} = -\frac{1}{2}.$$

Incorporating these results gives

$$\mathbf{V}_{\text{out}}(\mathbf{s}) = \frac{1}{\mathbf{s}} - \frac{1}{\mathbf{s}+2} - \frac{1}{\mathbf{s}}\,e^{-\mathbf{s}} + \frac{1}{\mathbf{s}+2}\,e^{-\mathbf{s}}. \tag{4.48}$$

From Table 3-2, we deduce that

$$u(t) \;\leftrightarrow\; \frac{1}{\mathbf{s}},$$

$$e^{-2t}\,u(t) \;\leftrightarrow\; \frac{1}{\mathbf{s}+2},$$

$$u(t-1) \;\leftrightarrow\; \frac{1}{\mathbf{s}}\,e^{-\mathbf{s}},$$

and

$$e^{-2(t-1)}\,u(t-1) \;\leftrightarrow\; \frac{1}{\mathbf{s}+2}\,e^{-\mathbf{s}}.$$

Hence,

$$\upsilon_{\text{out}}(t) = \Big[[1-e^{-2t}]\,u(t) - [1-e^{-2(t-1)}]\,u(t-1)\Big]\text{ V}. \tag{4.49}$$

Figure 4-4(c) displays the temporal response of $\upsilon_{\text{out}}(t)$.

Exercise 4-2: Compute the s-domain impedance of a series RLC circuit with zero initial conditions. Simplify the expression to a ratio of polynomials.

Answer:

$$\mathbf{Z}(\mathbf{s}) = R + \mathbf{s}L + \frac{1}{\mathbf{s}C} = \frac{\mathbf{s}^2 + (R/L)\mathbf{s} + 1/(LC)}{\mathbf{s}/L}.$$

(See S²)

4-3 Electromechanical Analogues

In this section, we show that mechanical systems consisting of springs, masses, and dampers can be mapped to an equivalent electrical system consisting of inductors, capacitors, and resistors. The electrical system can then be solved using s-domain circuit analysis. We then apply this to a biomechanical model of a human sitting on a chair subjected to an applied motion (e.g., a vehicle moving on an uneven pavement).

4-3.1 A Revealing Example

Let us revisit the car spring-mass-damper system analyzed earlier in Section 2-9 and diagrammed again in Fig. 4-5(a). We will shortly demonstrate that this shock-absorber system is mathematically equivalent to the RLC circuit shown in Fig. 4-5(b). Equivalency stems from the fact that the mechanical and electrical systems are characterized by LCCDEs of identical form.

Figure 4-6 depicts a car moving horizontally at a constant speed v_z over a pavement with height profile $x(z)$, where z is the direction of travel. The height profile appears to the car's tires as a time-dependent vertical displacement $x(t)$ with associated vertical velocity

$$v_x = \frac{dx}{dt}. \tag{4.50}$$

We shall call $v_x(t)$ the *input vertical velocity* experienced by the tire as it moves over the pavement. The car, including the shock absorber, is the *system*, and the *vertical velocity of the car's body*, $v_y(t)$, is the output response. To keep the analysis simple, we are assuming that all four tires experience the same pavement profile, which obviously is not always true for the front and back tires.

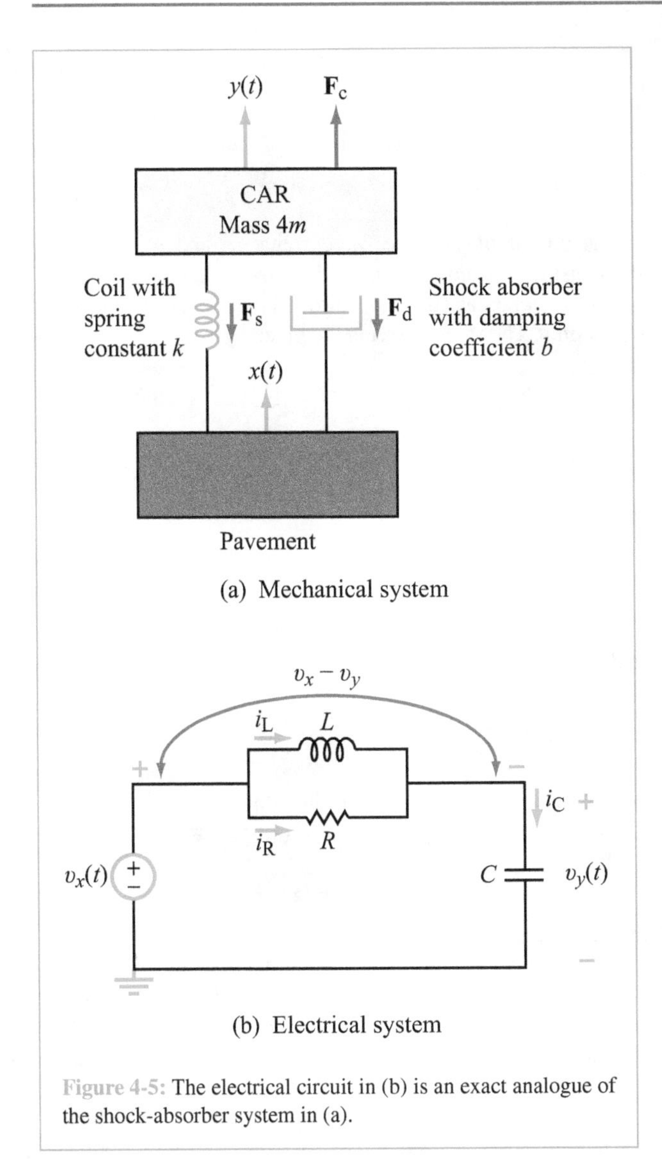

Figure 4-5: The electrical circuit in (b) is an exact analogue of the shock-absorber system in (a).

For the purpose of our electromechanical analogy, we offer the following specifications:

Mechanical system

- $v_x(t)$ is the vertical velocity of the tire (input signal).
- $v_y(t)$ is the vertical velocity of the car (output response).

Electrical system

- $\upsilon_x(t)$ is the voltage of voltage source (input signal).
- $\upsilon_y(t)$ is the voltage across the capacitor (output response).

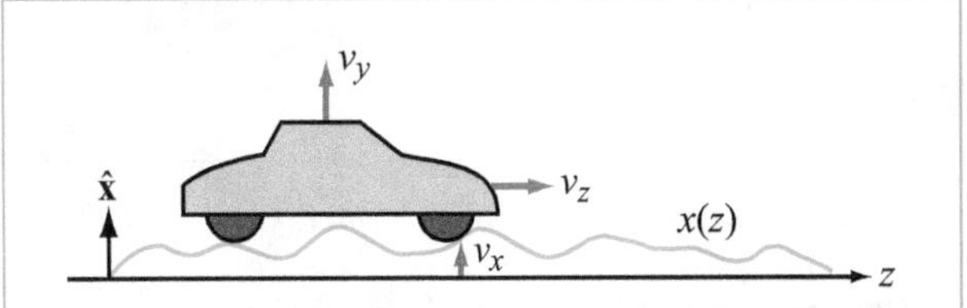

Figure 4-6: Car moving with horizontal speed v_z along the z direction over a pavement with height profile $x(z)$. The motion of the car over the pavement imparts a vertical velocity $v_x(t)$ on the car wheel. In response, the car's body experiences a vertical velocity $v_y(t)$.

Consistent with our earlier treatment in Section 2-9.1, we ascribe a positive polarity to a force when it is pointing upwards.

Comparison

The two derivations leading to the LCCDEs given by Eqs. (4.54) and (4.58) are *mathematically identical*. Accordingly, we draw the parallels outlined in Table 4-2. By applying these parallels, we can construct electrical circuits to represent mechanical systems and then apply circuit analysis techniques to determine the output response of the mechanical systems to specified input excitations.

4-3.2 Analogue Models

Building on the simple example of the preceding subsection, we will now generalize the analogy to more complex systems. We start with the following definitions:

- An *SMD system* is a mechanical system consisting entirely of ***springs***, ***masses***, and ***dampers***.
- An *RLC system* is an electrical system consisting entirely of ***resistors***, ***inductors***, and ***capacitors*** (no dependent sources or op amps).

► SMD systems can be modeled in terms of equivalent RLC systems. "Equivalent" means their mathematical behavior is the same, even though the physical quantities are different. So in analyzing an RLC system, we are also implicitly analyzing an SMD system. ◄

Mechanical System

The forces exerted on the car's body (Fig. 4-5(a)) by the spring and damper are $\mathbf{F}_\mathrm{s}$ and $\mathbf{F}_\mathrm{d}$, respectively; both depend on the *relative* velocity $(v_y - v_x)$. The inertial force $\mathbf{F}_\mathrm{c}$ depends only on the velocity of the car's body, v_y. With $\hat{\mathbf{x}}$ denoting a unit vector pointing upwards, the three forces are given by

$$\mathbf{F}_\mathrm{s} = -k(y - x)\hat{\mathbf{x}} = \left\{ k \int_0^t [v_x(\tau) - v_y(\tau)]\, d\tau \right\} \hat{\mathbf{x}}, \tag{4.51a}$$

$$\mathbf{F}_\mathrm{d} = -b\, \frac{d}{dt}\, (y - x)\hat{\mathbf{x}} = b(v_x - v_y)\hat{\mathbf{x}}, \tag{4.51b}$$

and

$$\mathbf{F}_\mathrm{c} = m\, \frac{d^2 y}{dt^2}\, \hat{\mathbf{x}} = m\, \frac{dv_y}{dt}\, \hat{\mathbf{x}}, \tag{4.51c}$$

where k is the spring constant, b is the damping coefficient of the shock absorber, and m is one-fourth of the car's mass. The force balance equation is

$$\mathbf{F}_\mathrm{c} = \mathbf{F}_\mathrm{s} + \mathbf{F}_\mathrm{d} \tag{4.52}$$

or, equivalently,

$$-k \int_0^t [v_x(\tau) - v_y(\tau)]\, d\tau - b(v_x - v_y) + m\, \frac{dv_y}{dt} = 0. \tag{4.53}$$

Upon differentiating all terms with respect to t and then rearranging them so that those involving the input $v_x(t)$ appear on the right-hand side and those involving the output response $v_y(t)$ appear on the left, we have

$$m\, \frac{d^2 v_y}{dt^2} + b\, \frac{dv_y}{dt} + kv_y = b\, \frac{dv_x}{dt} + kv_x. \tag{4.54}$$

Electrical System

In the circuit of Fig. 4-5(b), currents i_R and i_L through the resistor and inductor depend on the voltage difference $(\upsilon_x - \upsilon_y)$, whereas current i_C through the capacitor depends on υ_y only. The three currents are given by

$$i_\mathrm{L} = \frac{1}{L} \int_0^t [\upsilon_x(\tau) - \upsilon_y(\tau)]\, d\tau, \tag{4.55a}$$

$$i_\mathrm{R} = \frac{1}{R}\, (\upsilon_x - \upsilon_y), \tag{4.55b}$$

and

$$i_\mathrm{C} = C\, \frac{d\upsilon_y}{dt}\, , \tag{4.55c}$$

Note the symmetry between the expressions given by Eq. (4.51) and those given by Eq. (4.55). From the KCL, we have

$$i_\mathrm{C} = i_\mathrm{L} + i_\mathrm{R} \tag{4.56}$$

or, equivalently,

$$-\frac{1}{L} \int_0^t [\upsilon_x(\tau) - \upsilon_y(\tau)]\, d\tau - \frac{1}{R}\, (\upsilon_x - \upsilon_y) + C\, \frac{d\upsilon_y}{dt} = 0. \tag{4.57}$$

Upon differentiating all terms with respect to t and then rearranging them so that those involving the input source $\upsilon_x(t)$ appears on the right-hand side and those involving the output response $\upsilon_y(t)$ appear on the left, we have

$$C\, \frac{d^2 \upsilon_y}{dt^2} + \frac{1}{R}\, \frac{d\upsilon_y}{dt} + \frac{1}{L}\, \upsilon_y = \frac{1}{R}\, \frac{d\upsilon_x}{dt} + \frac{1}{L}\, \upsilon_x. \tag{4.58}$$

Table 4-2: Mechanical-electrical analogue.

Mechanical		Electrical
• Force **F** **F** is positive when pointing upwards	⟷	Current i i is positive when entering positive voltage terminal of device
• Vertical velocity v v is positive when car or tire is moving upwards	⟷	Voltage υ υ's positive terminal is where i enters device
• Mass m (1/4 of car's mass) $F_c = m\,\dfrac{dv_y}{dt}$	⟷	Capacitance C $i_C = C\,\dfrac{d\upsilon_y}{dt}$
• Spring constant k $F_s = k\int_0^t (v_x - v_y)\,d\tau$	⟷	$1/L$: Inverse of inductance $i_L = \dfrac{1}{L}\int_0^t (\upsilon_x - \upsilon_y)\,d\tau$
• Damping coefficient b $F_d = b(v_x - v_y)$	⟷	$1/R$: Inverse of resistance (conductance) $i_R = \dfrac{1}{R}(\upsilon_x - \upsilon_y)$
• $F_c = F_s + F_d$	⟷	$i_C = i_L + i_R$

Example 4-5: Mechanical System

The mechanical system in Fig. 4-7(a) consists of a spring-damper-mass sequence connected to a platform that sits over a hydraulic lift. The lift was used to raise the platform by 4 m at a constant speed of 0.5 m/s. Determine the corresponding vertical speed and displacement of the mass m, given that $m = 150$ kg, $k = 1200$ N/m, and $b = 200$ N·s/m.

Solution: We start by constructing an equivalent electrical circuit; the series spring-damper-mass is represented by a resistor, inductor, and capacitor in series, with the combination driven by a voltage source $\upsilon_x(t)$. The output is voltage $\upsilon_y(t)$ across the capacitor (Fig. 4-7(b)).

We then transform the time-domain circuit into the **s**-domain as shown in Fig. 4-7(c). Voltage division gives

$$\mathbf{V}_y = \frac{\mathbf{V}_x \mathbf{Z}_C}{\mathbf{Z}_R + \mathbf{Z}_L + \mathbf{Z}_C} = \frac{1/(\mathbf{s}m)}{\frac{1}{b} + \frac{\mathbf{s}}{k} + \frac{1}{\mathbf{s}m}}\,\mathbf{V}_x = \frac{k/m}{\mathbf{s}^2 + \frac{k}{b}\,\mathbf{s} + \frac{k}{m}}\,\mathbf{V}_x = \frac{8}{\mathbf{s}^2 + 6\mathbf{s} + 8}\,\mathbf{V}_x. \tag{4.59}$$

The input excitation $\mathbf{V}_x$ is the Laplace transform of the input velocity $v_x(t)$. Given that the lift moved at a constant speed of 0.5 m/s over a distance of 4 m, which corresponds to a travel time of $4/0.5 = 8$ s, $v_x(t)$ is a rectangle waveform given by

$$v_x(t) = 0.5[u(t) - u(t-8)] \text{ m/s}. \tag{4.60}$$

Using entries #2 and #2a in Table 3-2, the Laplace transform of $v_x(t)$ is

$$\mathbf{V}_x = \frac{0.5}{\mathbf{s}} - \frac{0.5}{\mathbf{s}}\,e^{-8\mathbf{s}}. \tag{4.61}$$

Hence,

$$\mathbf{V}_y = \frac{4}{\mathbf{s}(\mathbf{s}^2 + 6\mathbf{s} + 8)} - \frac{4e^{-8\mathbf{s}}}{\mathbf{s}(\mathbf{s}^2 + 6\mathbf{s} + 8)} = \frac{4}{\mathbf{s}(\mathbf{s}+2)(\mathbf{s}+4)} - \frac{4e^{-8\mathbf{s}}}{\mathbf{s}(\mathbf{s}+2)(\mathbf{s}+4)}. \tag{4.62}$$

SMD-RLC Analysis Procedure

Step 1: Replace each mass with a capacitor with one terminal connected to a node and the other to ground.

Step 2: Replace each spring with an inductor with $L = 1/k$, where k is the spring's stiffness coefficient.

- If the spring connects two masses, its equivalent inductor connects to their equivalent capacitors at their non-ground terminals.
- If the spring connects a mass to a stationary surface, its equivalent inductor should be connected between the capacitor's non-ground terminal and ground.
- If one end of the spring connects to a moving surface, the corresponding terminal of its equivalent inductor should be connected to a voltage source.

Step 3: Replace each damper with a resistor with $R = 1/b$, where b is the damper's damping coefficient. Connection rules are the same as for springs.

Step 4: Analyze the RLC circuit using the **s**-domain technique described in Section 4-2.

The solution of the RLC circuit provides expressions for the voltages across capacitors, corresponding to the velocities of their counterpart masses in the mechanical system. Displacement of a mass or its acceleration can be obtained by integrating or differentiating its velocity $v(t)$, respectively.

We will label the first term $\mathbf{V}_{y_1}$ and expand it into partial fractions:

$$\mathbf{V}_{y_1} = \frac{4}{\mathbf{s}(\mathbf{s}+2)(\mathbf{s}+4)} = \frac{A_1}{\mathbf{s}} + \frac{A_2}{\mathbf{s}+2} + \frac{A_3}{\mathbf{s}+4}. \tag{4.63}$$

Application of the procedure outlined in Section 3-5.1 leads to

$$A_1 = \mathbf{s}\,\mathbf{V}_{y_1}\big|_{\mathbf{s}=0} = \frac{4}{2 \times 4} = 0.5,$$

$$A_2 = (\mathbf{s}+2)\,\mathbf{V}_{y_1}\big|_{\mathbf{s}=-2} = \frac{4}{(-2) \times 2} = -1,$$

and

$$A_3 = (\mathbf{s}+4)\,\mathbf{V}_{y_1}\big|_{\mathbf{s}=-4} = \frac{4}{(-4) \times (-2)} = 0.5.$$

Hence,

$$\mathbf{V}_{y_1} = \frac{0.5}{\mathbf{s}} - \frac{1}{\mathbf{s}+2} + \frac{0.5}{\mathbf{s}+4}, \tag{4.64}$$

and its inverse Laplace transform is

$$v_{y_1}(t) = [0.5 - e^{-2t} + 0.5e^{-4t}]\, u(t). \tag{4.65}$$

The second term in Eq. (4.62) is identical with the first term except for the multiplication factor $(-e^{-8\mathbf{s}})$. From property #3a in Table 3-2 we surmise that its time-domain equivalent is the same as that of the first term, but delayed by 8 s and multiplied by (-1). Accordingly,

$$v_{y_2}(t) = -[0.5 - e^{-2(t-8)} + 0.5e^{-4(t-8)}]\, u(t-8), \tag{4.66}$$

and

$$v_y(t) = v_{y_1}(t) + v_{y_2}(t). \tag{4.67}$$

Figure 4-7 displays plots of $v_y(t)$, the vertical velocity of mass m, and its vertical displacement

$$y(t) = \int_0^t v_y(t)\, d\tau. \tag{4.68}$$

Concept Question 4-3: What are the electrical analogues of springs, masses, and dampers? (See S²)

Concept Question 4-4: After computing the node voltages of the equivalent RLC system, what do you need to do to complete the analysis of the mechanical system? (See S²)

Exercise 4-3: A mass is connected by a spring to a moving surface. What is its electrical analogue? **Answer:**

A series LC circuit driven by a voltage source, the same as the circuit in Fig. 4-7(b), but without the resistor. (See S²)

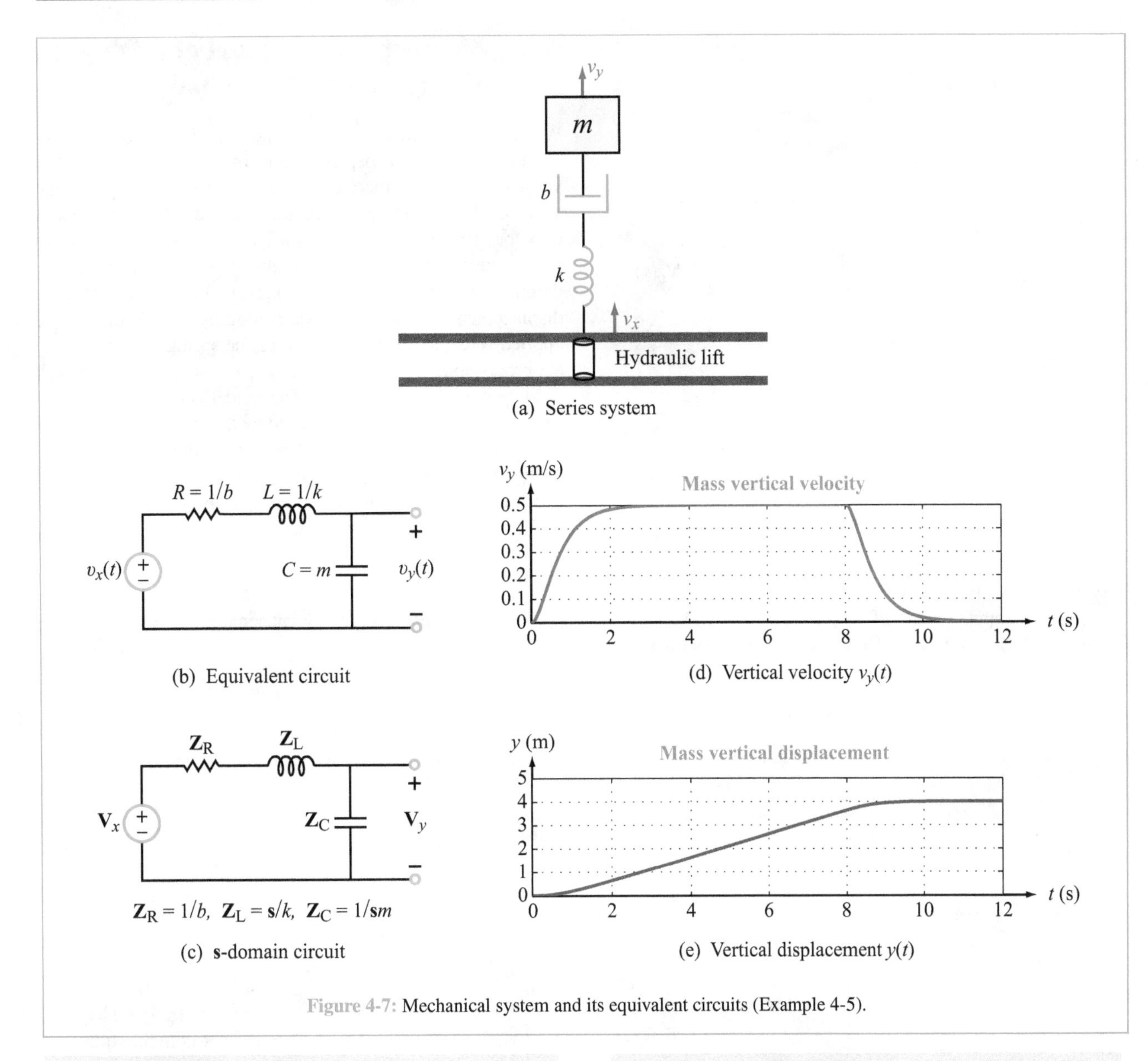

Figure 4-7: Mechanical system and its equivalent circuits (Example 4-5).

Exercise 4-4: What do you expect the impulse response of the system in Exercise 4-3 to be like?

Answer: Oscillatory, with no damping (no resistor). (See S^2)

Exercise 4-5: In the SMD system shown in Fig. E4-5, $v_x(t)$ is the input velocity of the platform and $v_y(t)$ is the output velocity of mass m. Draw the equivalent **s**-domain circuit.

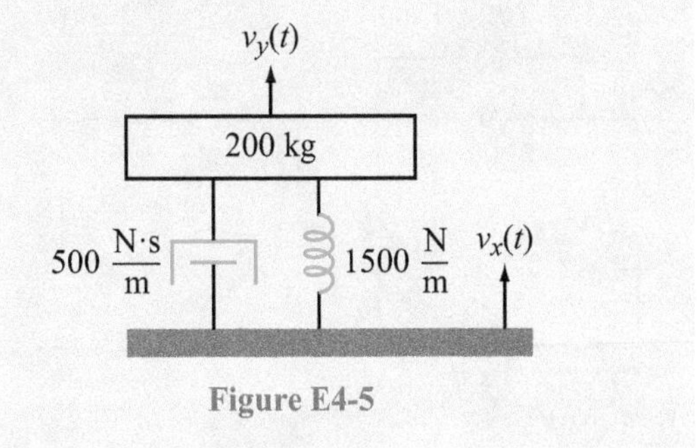

Figure E4-5

Answer:

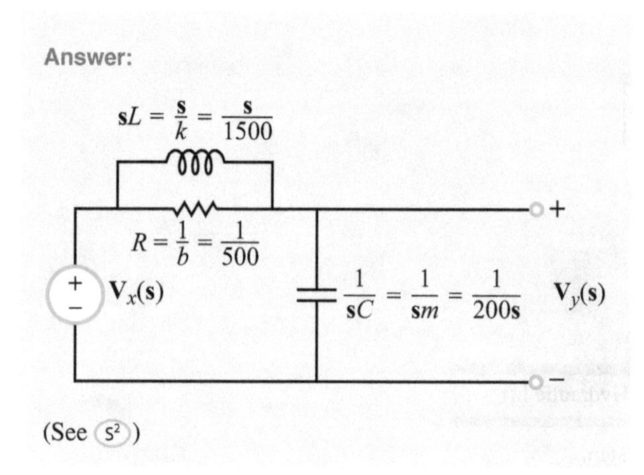

(See S²)

Exercise 4-6: In the SMD system shown in Fig. E4-6, $v_x(t)$ is the input velocity of the platform and $v_y(t)$ is the output velocity of mass m. Draw the equivalent **s**-domain circuit.

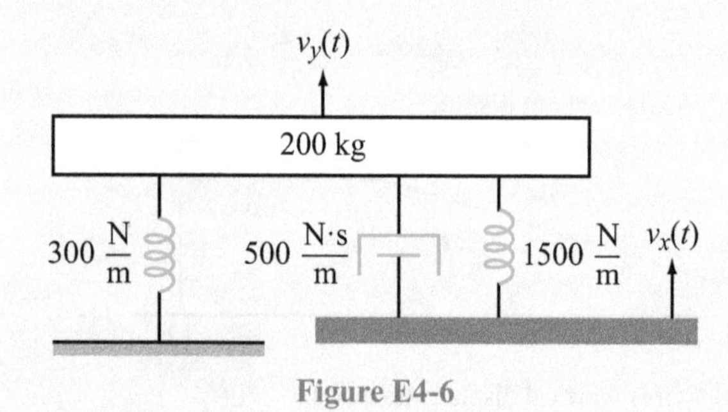

Figure E4-6

Answer:

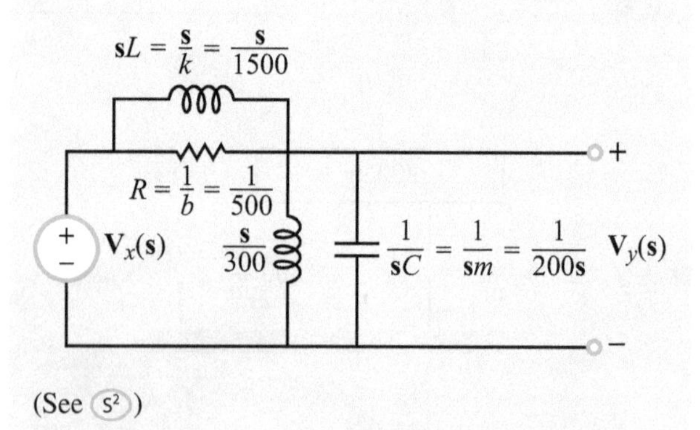

(See S²)

4-4 Biomechanical Model of a Person Sitting in a Moving Chair

We now apply an electromechanical analogue to a biomechanical model of a person sitting in a car chair subject to vertical motions as the car moves over an uneven pavement. The model shown in Fig. 4-8 divides the body above the car seat into four sections with masses m_1 to m_4.† Adjacent sections are connected by a parallel combination of a spring and a damper. The springs represent bones, including the spine; the displacement allowed by a bone is roughly proportional to the applied force, and the stiffness k of the spring corresponds to the proportionality factor. Soft tissue, which basically is a thick fluid, is modeled as a damper. The model is used to assess the risk of lower back disorders caused by driving trucks or similar heavy equipment. The input to the system is the vertical velocity of the chair, and the output is the vertical velocity of the head (if it exceeds certain limits, spinal injury may occur).

We use the following notation:

Biomechanical system

- $x_i(t)$ is the vertical displacement of mass i, with $i = 1, 2, 3, 4$.
- $v_i(t) = dx_i/dt$ is the vertical velocity of mass i.
- $v_1(t) = v_x(t)$, where $v_x(t)$ is the vertical velocity of the car seat (lower torso section is connected directly to car seat).
- k_i is the stiffness of the spring connecting mass i to mass $i + 1$, with $i = 1, 2, 3$.
- b_i is the damping factor of the damper connecting mass i to mass $i + 1$, with $i = 1, 2, 3$.

Electrical analogue

- Voltages $\upsilon_i(t)$ correspond to $v_i(t)$ with $\upsilon_1(t) = \upsilon_x(t)$ where $\upsilon_x(t)$ is a voltage source; its **s**-domain equivalent is $\mathbf{V}_x$,
- Spring with k_i → Inductor with $L_i = 1/k_i$,
- Damper with b_i → Resistor with $R_i = 1/b_i$,
- Mass m_i → Capacitor with $C_i = m_i$.

†T. R. Waters, F. Li, R. L. Huston and N. K. Kittusamy, "Biomechanical Modeling of Spinal Loading Due to Jarring and Jolting for Heavy Equipment Operators," *Proceedings of the XVth Triennial Congress of the International Ergonomics Association and 7th Joint Conference of Ergonomics Society of Korea/Japan Ergonomics Society*, Seoul, Korea, Aug. 24–29, 2003.

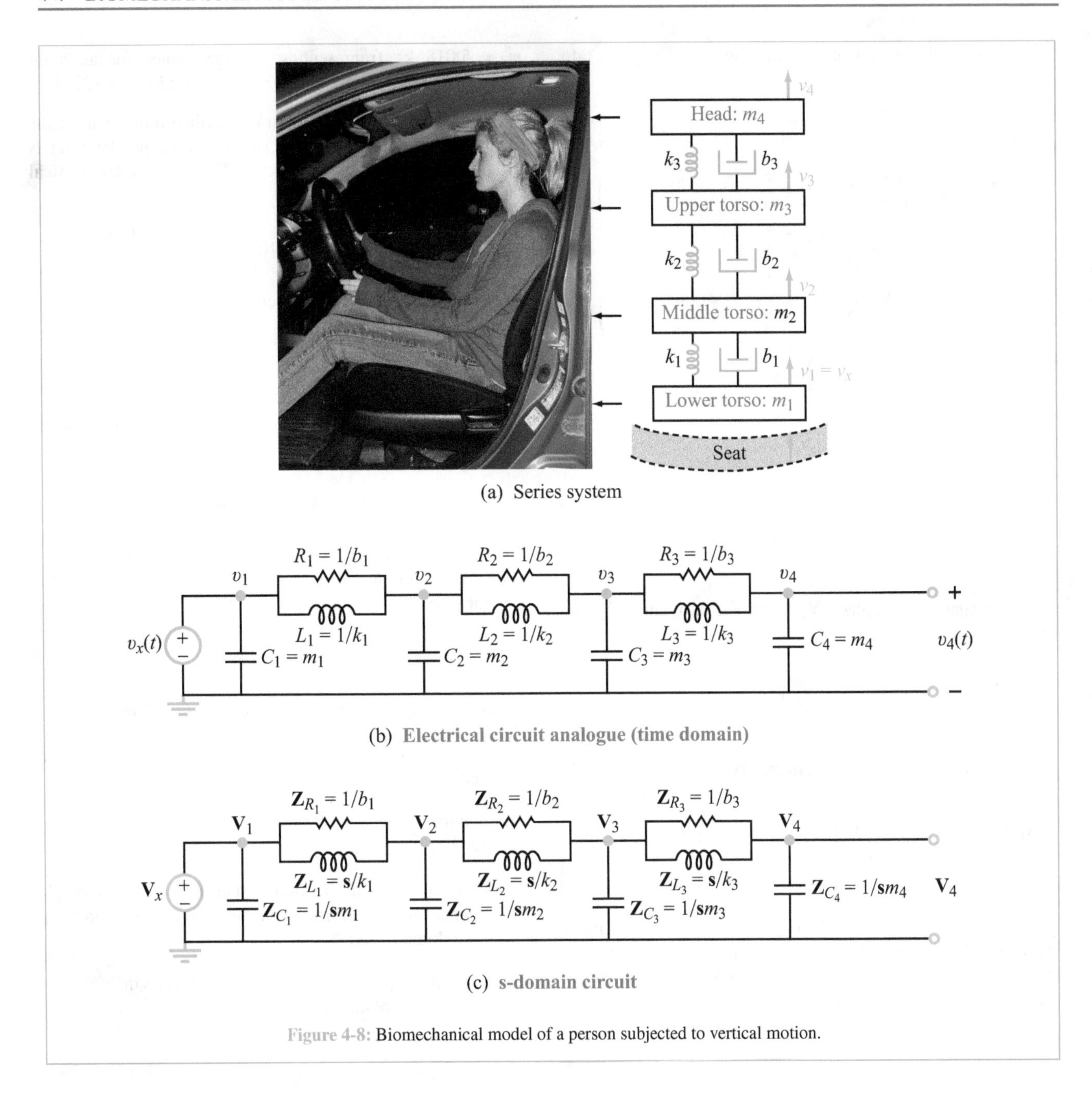

Figure 4-8: Biomechanical model of a person subjected to vertical motion.

The circuit in Fig. 4-8(b) is an analogue of the biomechanical model in part (a). Each mass is represented by a capacitor with one of its terminals connected to ground; adjoining masses are connected by a parallel combination of a resistor and an inductor; and the excitation is provided by voltage source $\upsilon_x(t)$, representing the vertical velocity of the car seat $v_x(t)$.

The **s**-domain equivalent circuit is shown in Fig. 4-8(c). Its node equations are

$$\mathbf{V}_1 = \mathbf{V}_x, \tag{4.69a}$$

$$-\mathbf{V}_1\left(b_1+\frac{k_1}{\mathbf{s}}\right)+\mathbf{V}_2\left(b_1+\frac{k_1}{\mathbf{s}}+b_2+\frac{k_2}{\mathbf{s}}+\mathbf{s}m_2\right) -\mathbf{V}_3\left(b_2+\frac{k_2}{\mathbf{s}}\right)=0, \tag{4.69b}$$

$$-\mathbf{V}_2\left(b_2+\frac{k_2}{\mathbf{s}}\right)+\mathbf{V}_3\left(b_2+\frac{k_2}{\mathbf{s}}+b_3+\frac{k_3}{\mathbf{s}}+\mathbf{s}m_3\right) -\mathbf{V}_4\left(b_3+\frac{k_3}{\mathbf{s}}\right)=0, \tag{4.69c}$$

and

$$-\mathbf{V}_3\left(b_3+\frac{k_3}{\mathbf{s}}\right)+\mathbf{V}_4\left(b_3+\frac{k_3}{\mathbf{s}}+\mathbf{s}m_4\right)=0. \tag{4.69d}$$

To save space, we replace $\mathbf{V}_1$ with $\mathbf{V}_x$, and introduce the abbreviations

$$\mathbf{a}_i = b_i + \frac{k_i}{\mathbf{s}}, \qquad \text{for } i = 1, 2, 3. \tag{4.70}$$

We then cast the node equation in matrix form as

$$\begin{bmatrix} \mathbf{a}_1+\mathbf{a}_2+\mathbf{s}m_2 & -\mathbf{a}_2 & 0 \\ -\mathbf{a}_2 & \mathbf{a}_2+\mathbf{a}_3+\mathbf{s}m_3 & -\mathbf{a}_3 \\ 0 & -\mathbf{a}_3 & \mathbf{a}_3+\mathbf{s}m_4 \end{bmatrix} \begin{bmatrix} \mathbf{V}_2 \\ \mathbf{V}_3 \\ \mathbf{V}_4 \end{bmatrix} = \begin{bmatrix} \mathbf{a}_1\mathbf{V}_x \\ 0 \\ 0 \end{bmatrix}. \tag{4.71}$$

In Eq. (4.71), $\mathbf{V}_x$ represents the Laplace transform of the input vertical velocity imparted by the car seat onto mass m_1 (lower torso), and $\mathbf{V}_4$, which is the Laplace transform of the output response, represents the vertical velocity of the head.

Example 4-6: Car Driving over a Curb

For the biomechanical model represented by Eq. (4.71), compute and plot the head's vertical velocity as a function of time in response to the car going over a curb 1 cm in height. Assume $m_1 = 8.164$ kg, $m_2 = 11.953$ kg, $m_3 = 11.654$ kg, $m_4 = 5.018$ kg (representing average values for an adult human), $b_i = 90$ N·s/m, and $k_i = 3500$ N/m for $i = 1, 2, 3$.‡

Solution: We will assume that the vertical movement of the chair seat mimics that of the car tires (i.e., no damping is provided by the shock absorbers). The seat's 1 cm vertical displacement is then

$$x(t) = 0.01\, u(t) \qquad \text{(m)},$$

and the associated vertical velocity is

$$v_x(t) = \frac{dx}{dt} = 0.01\, \delta(t).$$

This input signal corresponds to a voltage source $\upsilon_x(t)$ with the same expression and a Laplace transform

$$\mathbf{V}_x = \mathcal{L}[v_x(t)] = 0.01.$$

When the input to a system is an impulse function, its output is equal to the system's impulse response $h(t)$. Hence, in the present case, the desired output, $v_4(t)$, is

$$v_4(t) = 0.01\, h(t) \tag{4.72}$$

with $h(t)$ being the solution based on Eq. (4.71) with $\mathbf{V}_x = 1$. Upon:

(a) setting $\mathbf{V}_x = 1$ in Eq. (4.71),

(b) solving the system of equations to obtain an expression for $\mathbf{V}_4$,

(c) applying partial fraction expansion to convert the expression into a form amenable to transformation to the time domain, and

(d) using the applicable Laplace transform pairs in Tables 3-2 and 3-3, we obtain the result

$$h(t) = 3.22e^{-15.2t}\cos(30.83t - 37.5^\circ)\, u(t) + 8.65e^{-7.95t}\cos(23.56t + 127.3^\circ)\, u(t) + 11.5e^{-1.08t}\cos(9.10t - 76.5^\circ)\, u(t). \tag{4.73}$$

‡V. M. Zatsiorsky, *Kinetics of Human Motion*, Human Kinetics, Champaign IL, 2002, p. 304. M. Fritz, "An Improved Biomechanical Model for Simulating the Strain of the Hand-Arm System under Vibration Stress," *J. Biomechanics* 24(12), 1165–1171, 1991.

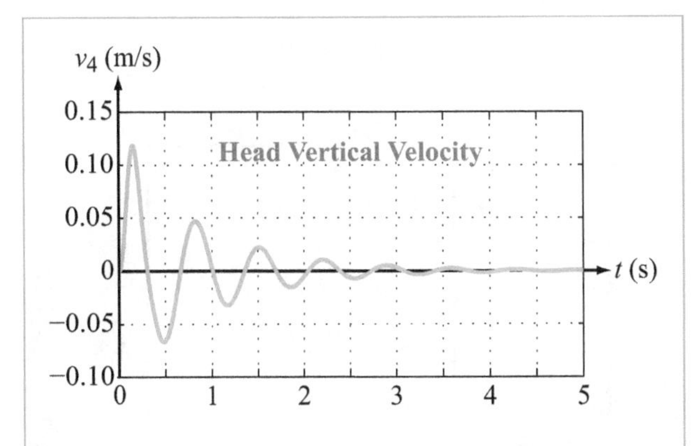

Figure 4-9: Vertical velocity of the head in response to a car driving over a curb (Example 4-6).

A plot of the head's vertical velocity, $v_4(t) = 0.01\, h(t)$, is displayed in Fig. 4-9. It is approximately a decaying sinusoid with a first peak at 0.12 m/s (see S² for more details).

Example 4-7: Wavy Pavement Response

Use the biomechanical model of Example 4-6 to determine the time variation of the head's vertical velocity when the seated person is riding in a vehicle over a wavy surface that imparts a vertical velocity on the seat given by $v_x(t) = \cos 10t$ (m/s).

Solution: According to Eq. (2.118), a sinusoidal input signal $A\cos(\omega t + \phi)$ generates an output response

$$y(t) = A|\mathbf{H}(j\omega)| \cos(\omega t + \theta + \phi).$$

In the present case, $A = 1$ m/s, $\phi = 0$, $\omega = 10$ rad/s, $y(t)$ is the output $v_4(t)$, and $|\mathbf{H}(j\omega)|$ and θ are the magnitude and phase angle of $\mathbf{H}(\mathbf{s})$, when evaluated at $\mathbf{s} = j\omega$. After replacing $\mathbf{s}$ with $j\omega$ in Eqs. (4.70) and (4.71), solving the system of equations leads to

$$\mathbf{H}(j\omega) = \mathbf{V}_4(j\omega) = 4.12\angle{-116.1^\circ}.$$

Hence,

$$v_4(t) = 4.12\cos(10t - 116.1^\circ) \qquad \text{m/s}.$$

Note that the head's vertical velocity has an amplitude 4.12 times that of the chair's vertical velocity! This means that the body is a poor damping system at this angular frequency.

Exercise 4-7: What is the amplitude of the head *displacement* for the person in Example 4-7, if the seat *displacement* is $x_1(t) = 0.02\cos(10t)$ (m)?

Answer:

$$\begin{aligned} v_1(t) &= dx_1/dt \\ &= -0.2\sin(10t) = 0.2\cos(10t + 90^\circ) \text{ (m/s)}; \\ v_4(t) &= 0.2 \times 4.12\cos(10t + 90^\circ - 116.1^\circ) \\ &= 0.824\cos(10t - 26.1^\circ) \text{ (m/s)}; \\ x_4(t) &= \int_{-\infty}^{t} v_4(\tau)\, d\tau = 0.0824\sin(10t - 26.1^\circ) \text{ (m)}; \end{aligned}$$

amplitude = 8.24 cm. (See S²)

4-5 Op-Amp Circuits

A system design process usually starts by defining the desired performance specifications of the system, such as its gain, frequency response, sensitivity to certain parameters, and immunity to noise interference, among others. As we will see in this and later chapters, these specifications can be used to establish a functional form for the transfer function of the desired system, $\mathbf{H}(\mathbf{s})$. The next step is to construct a physical system with an input-output response that matches the desired $\mathbf{H}(\mathbf{s})$. The physical system may be electrical, mechanical, chemical, or some combination of different technologies. When electrical systems are constructed to process signals or control other systems, it is common practice to use *operational-amplifier* circuits whenever possible, because they can be used in a modular mode to design a wide variety of system transfer functions.

Operational amplifiers (*op amps* for short) usually are covered in introductory circuits courses. An op amp is represented by the triangular symbol shown in Fig. 4-10 where five voltage node terminals are connected.

Op-Amp Terminals

υ_n	inverting (or negative) input voltage
υ_p	noninverting (or positive) input voltage
$-V_{dc}$	negative dc power supply voltage
$+V_{dc}$	positive dc power supply voltage
υ_o	output voltage

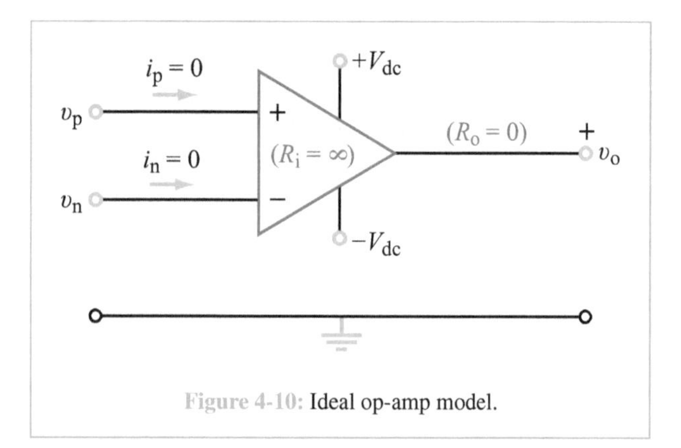

Figure 4-10: Ideal op-amp model.

By way of a review, the *ideal op-amp model* is characterized by the following properties:

(a) **Infinite input resistance:** $R_i \approx \infty$

(b) **Zero output resistance:** $R_o \approx 0$ (so long as any load that gets connected between υ_o and ground is at least 1 kΩ or greater)

(c) **Zero input-current constraint:** $i_p = i_n = 0$

(d) **Zero input-voltage constraint:** $\upsilon_p - \upsilon_n = 0$ (or, equivalently, $\upsilon_p = \upsilon_n$).

(e) **Output voltage saturation constraint:** υ_o is bounded between $-V_{dc}$ and $+V_{dc}$.

Application of the ideal op-amp constraints greatly facilitates the analysis and design of op-amp circuits.

► In forthcoming discussions involving op amps, we shall assume that the positive and negative power supply voltages have equal magnitudes (which is not universally the case), so a single voltage rail is sufficient in op-amp diagrams. Moreover, in op-amp circuits where no voltage rails are indicated at all, the assumption is that the signal magnitudes at the op-amp output terminals are well below the saturation level $|V_{dc}|$, so there is no need to show a voltage rail explicitly. ◄

4-5.1 Basic Inverting Amplifier

The **s**-domain op-amp circuit shown in Fig. 4-11 includes an input impedance $\mathbf{Z}_i$ and a feedback impedance $\mathbf{Z}_f$. At node $\mathbf{V}_n$, using the KCL gives

$$\mathbf{I}_1 + \mathbf{I}_2 + \mathbf{I}_n = 0$$

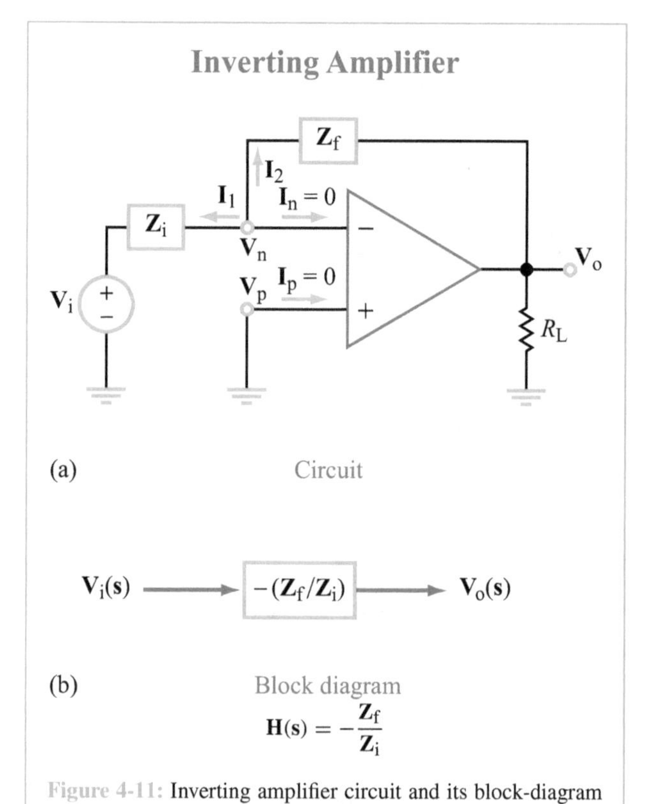

Figure 4-11: Inverting amplifier circuit and its block-diagram equivalent in the **s**-domain.

or, equivalently,

$$\frac{\mathbf{V}_n - \mathbf{V}_i}{\mathbf{Z}_i} + \frac{\mathbf{V}_n - \mathbf{V}_o}{\mathbf{Z}_f} + \mathbf{I}_n = 0. \tag{4.74}$$

In view of the current constraint ($\mathbf{I}_n = 0$), Eq. (4.74) leads to

$$\mathbf{H}(\mathbf{s}) = \frac{\mathbf{V}_o(\mathbf{s})}{\mathbf{V}_i(\mathbf{s})} = -\frac{\mathbf{Z}_f}{\mathbf{Z}_i}. \tag{4.75a}$$

Because of the minus sign in Eq. (4.75a), this circuit is called an *inverting amplifier*, and its gain is equal to the ratio of the feedback impedance to that of the input impedance.

For the special case where both $\mathbf{Z}_i$ and $\mathbf{Z}_f$ are simply resistors R_i and R_f, the circuit becomes a simple inverting amplifier with

$$\mathbf{H}(\mathbf{s}) = -\frac{R_f}{R_i} \quad \textbf{(resistive impedances)}. \tag{4.75b}$$

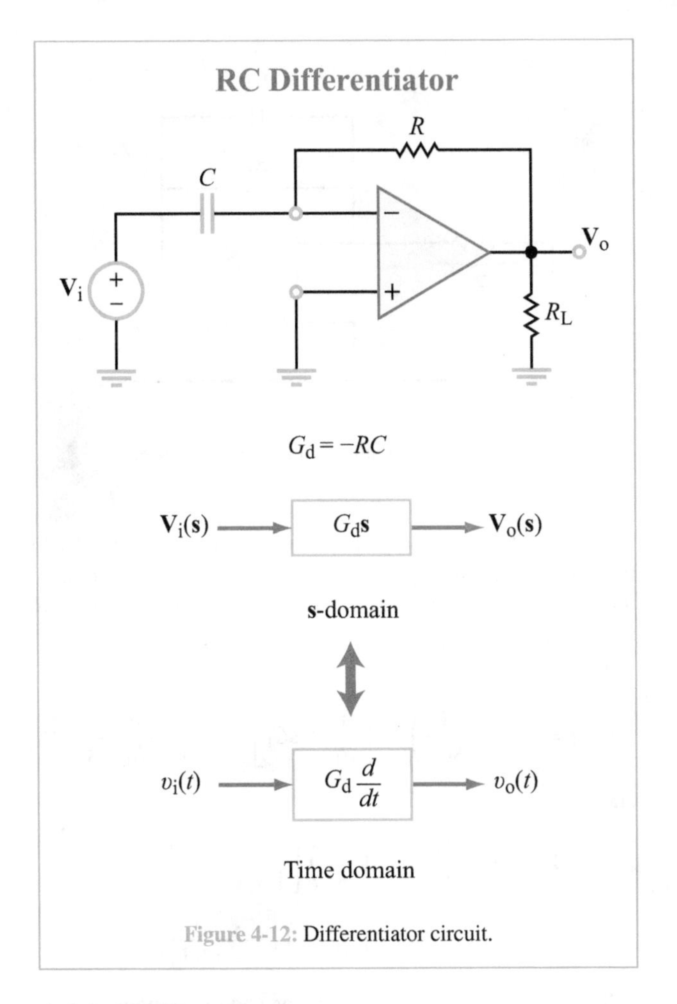

Figure 4-12: Differentiator circuit.

4-5.2 Differentiator

If the input impedance is a capacitor C and the feedback impedance is a resistor R, the inverting amplifier circuit becomes as shown in Fig. 4-12, and its transfer function simplifies to

$$\mathbf{H}_\mathrm{d}(\mathbf{s}) = -\frac{R}{1/\mathbf{s}C} = -RC\mathbf{s}.$$

This can be rearranged into the compact form

$$\mathbf{H}_\mathrm{d}(\mathbf{s}) = G_\mathrm{d}\mathbf{s} \qquad \textbf{(differentiator)}, \tag{4.76}$$

where G_d is a differentiator gain-constant given by

$$G_\mathrm{d} = -RC. \tag{4.77}$$

This circuit is called a ***differentiator*** because multiplication by $\mathbf{s}$ in the $\mathbf{s}$-domain is equivalent to differentiation in the time domain (under zero initial conditions).

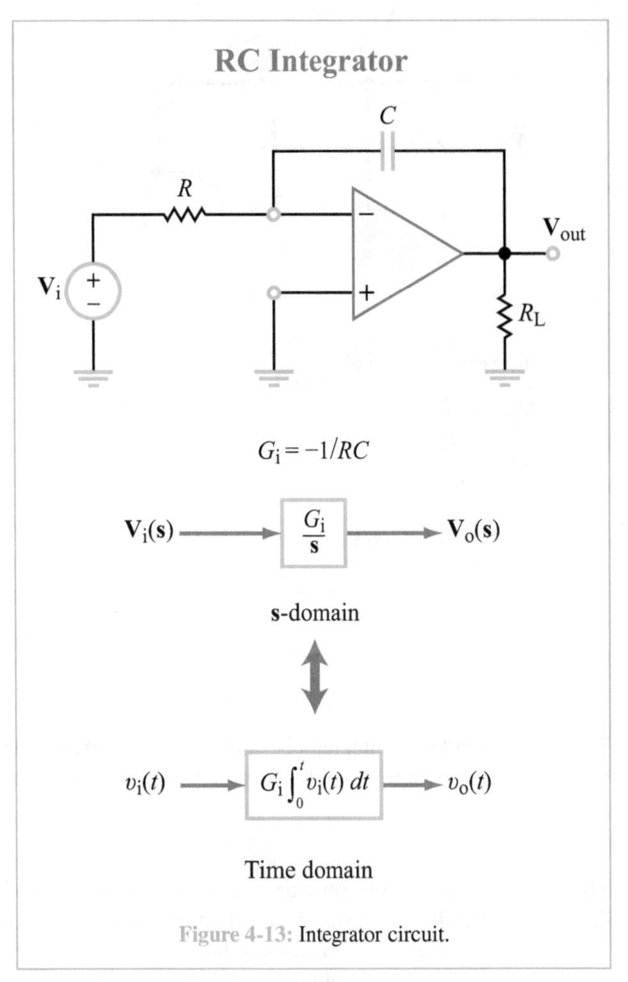

Figure 4-13: Integrator circuit.

4-5.3 Integrator

Reversing the locations of R and C in the differentiator circuit leads to the ***integrator circuit*** shown in Fig. 4-13. Its transfer function is

$$\mathbf{H}(\mathbf{s}) = \frac{-1/\mathbf{s}C}{R} = \left(-\frac{1}{RC}\right)\frac{1}{\mathbf{s}}$$

or, equivalently,

$$\mathbf{H}_\mathrm{i}(\mathbf{s}) = G_\mathrm{i}\left(\frac{1}{\mathbf{s}}\right) \qquad \textbf{(integrator)} \tag{4.78}$$

with an integrator gain constant

$$G_\mathrm{i} = -\frac{1}{RC}. \tag{4.79}$$

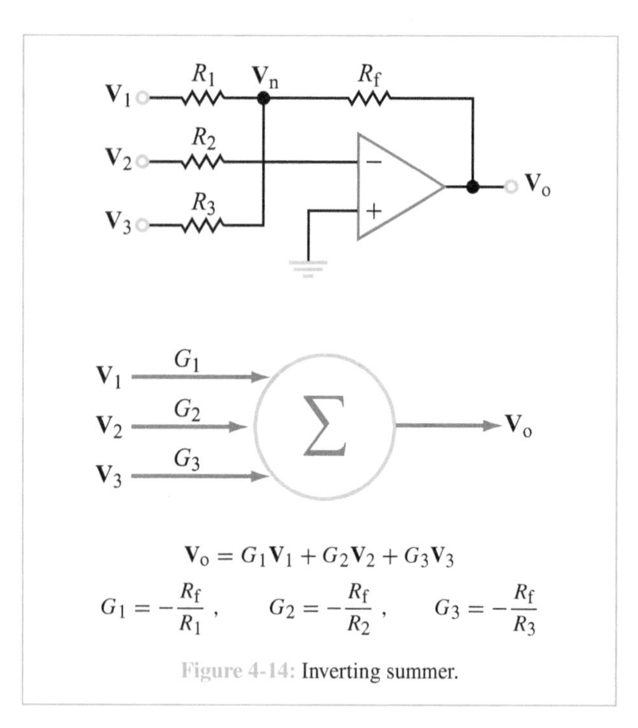

Figure 4-14: Inverting summer.

This circuit is called an integrator because integration in the time domain is equivalent to multiplication by $1/\mathbf{s}$ in the $\mathbf{s}$-domain.

4-5.4 Summer

The two preceding circuits provide tools for multiplication and division by $\mathbf{s}$ in the $\mathbf{s}$-domain. We need a third circuit in order to construct transfer functions, namely, the ***summing amplifier*** shown in Fig. 4-14. Application of KCL analysis at node $\mathbf{V}_n$ leads to

$$\begin{aligned}\mathbf{V}_o(\mathbf{s}) &= -\frac{R_f}{R_1}\,\mathbf{V}_1 - \frac{R_f}{R_2}\,\mathbf{V}_2 - \frac{R_f}{R_3}\,\mathbf{V}_3 \\ &= G_1\mathbf{V}_1 + G_2\mathbf{V}_2 + G_3\mathbf{V}_3 \end{aligned} \tag{4.80}$$

with

$$G_1 = -\frac{R_f}{R_1}, \quad G_2 = -\frac{R_f}{R_2}, \quad \text{and} \quad G_3 = -\frac{R_f}{R_3}. \tag{4.81}$$

The summing amplifier circuit is extendable to any number of input voltages.

Example 4-8: One-Pole Transfer Function

Derive the transfer function $\mathbf{H}(\mathbf{s}) = \mathbf{V}_o(\mathbf{s})/\mathbf{V}_i(\mathbf{s})$ for the op-amp circuit shown in Fig. 4-15, given that $R_i = 10$ kΩ, $R_f = 20$ kΩ, and $C_f = 25$ μF.

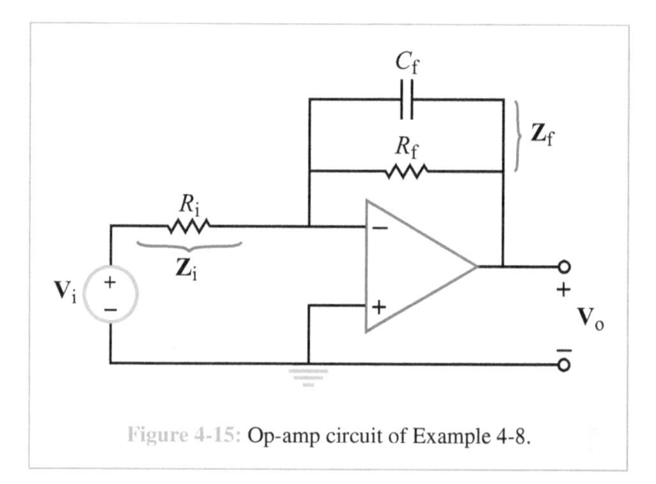

Figure 4-15: Op-amp circuit of Example 4-8.

Solution: The basic structure of the circuit is identical with the generic inverting op-amp circuit of Fig. 4-11, except that in the present case $\mathbf{Z}_f$ represents the parallel combination of R_f and C_f and $\mathbf{Z}_i = R_i$.

Hence,

$$\mathbf{Z}_f = R_f \parallel \left(\frac{1}{\mathbf{s}C_f}\right) = \left[\frac{1}{R_f} + \mathbf{s}C_f\right]^{-1} = \frac{R_f}{1 + R_f C_f \mathbf{s}},$$

and

$$\begin{aligned}\mathbf{H}(\mathbf{s}) = -\frac{\mathbf{Z}_f}{\mathbf{Z}_i} &= -\left(\frac{R_f}{R_i}\right)\left[\frac{1}{1 + R_f C_f \mathbf{s}}\right] \\ &= -\left(\frac{1}{R_i C_f}\right)\left[\frac{1}{\mathbf{s} + 1/(R_f C_f)}\right].\end{aligned}$$

The transfer function has no zeros and only one pole given by

$$p = -\frac{1}{R_f C_f}. \tag{4.82a}$$

In terms of p, $\mathbf{H}(\mathbf{s})$ can be written as

$$\mathbf{H}(\mathbf{s}) = G\left(\frac{1}{\mathbf{s} - p}\right) \tag{4.82b}$$

with

$$G = -\frac{1}{R_i C_f}. \tag{4.82c}$$

For the specified element values, we have

$$p = -\frac{1}{2 \times 10^4 \times 25 \times 10^{-6}} = -2,$$

$$G = -\frac{1}{10^4 \times 25 \times 10^{-6}} = -4,$$

Table 4-3: Op-amp circuits and their block-diagram representations.

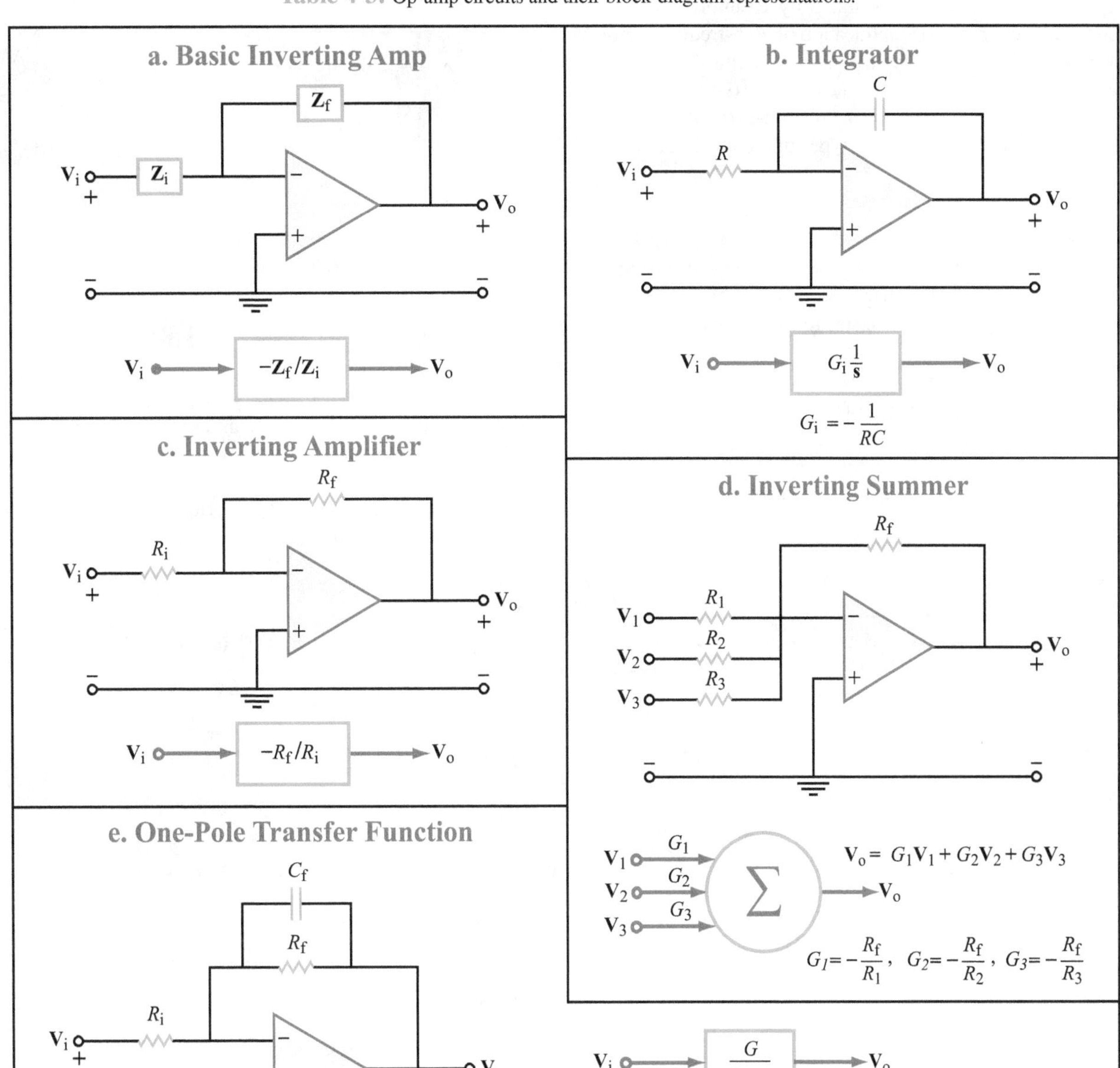

and

$$\mathbf{H}(\mathbf{s}) = -4\left(\frac{1}{\mathbf{s}+2}\right).$$

System performance often is specified in terms of the locations of the poles and zeros of its transfer function in the **s**-domain. The op-amp circuits of Table 4-3 are examples of basic modules that can be combined together to form the desired transfer function. In practice, *it is advisable to avoid circuits* *whose transfer function corresponds to differentiation in the time domain*, because (as illustrated by Example 3-12) such circuits are vulnerable to amplification of high-frequency noise that may accompany the input signal. A circuit that performs time differentiation has a term of the form $\mathbf{s}$ or $(\mathbf{s}+a)$ in its transfer function. Hence, in forthcoming sections, we will synthesize transfer functions using only the types of circuits in Table 4-3.

Concept Question 4-5: Why is it preferable to use integrators rather than differentiators when using op-amp circuits to model desired transfer functions? (See S²)

Exercise 4-8: Obtain the transfer function of the op-amp circuit shown in Fig. E4-8. (The dc supply voltage $V_{\mathrm{dc}} = \pm 10$ V means that an op amp has a positive dc voltage of 10 V and a separate negative dc voltage of -10 V.)

v_i 5 kΩ, 4 μF, v_{out_1} 1 MΩ, 5 μF, v_{out_2}, $V_{\mathrm{dc}} = \pm 10$ V, $V_{\mathrm{dc}} = \pm 10$ V

Figure E4-8

Answer: $\mathbf{H}(\mathbf{s}) = 10/\mathbf{s}^2$. (See S²)

4-6 Configurations of Multiple Systems

More often than not, real systems are composed of several subsystems—each characterized by its own transfer function. The linear feature of LTI systems allows us to combine multiple systems in various ways, including the three common configurations shown in Fig. 4-16.

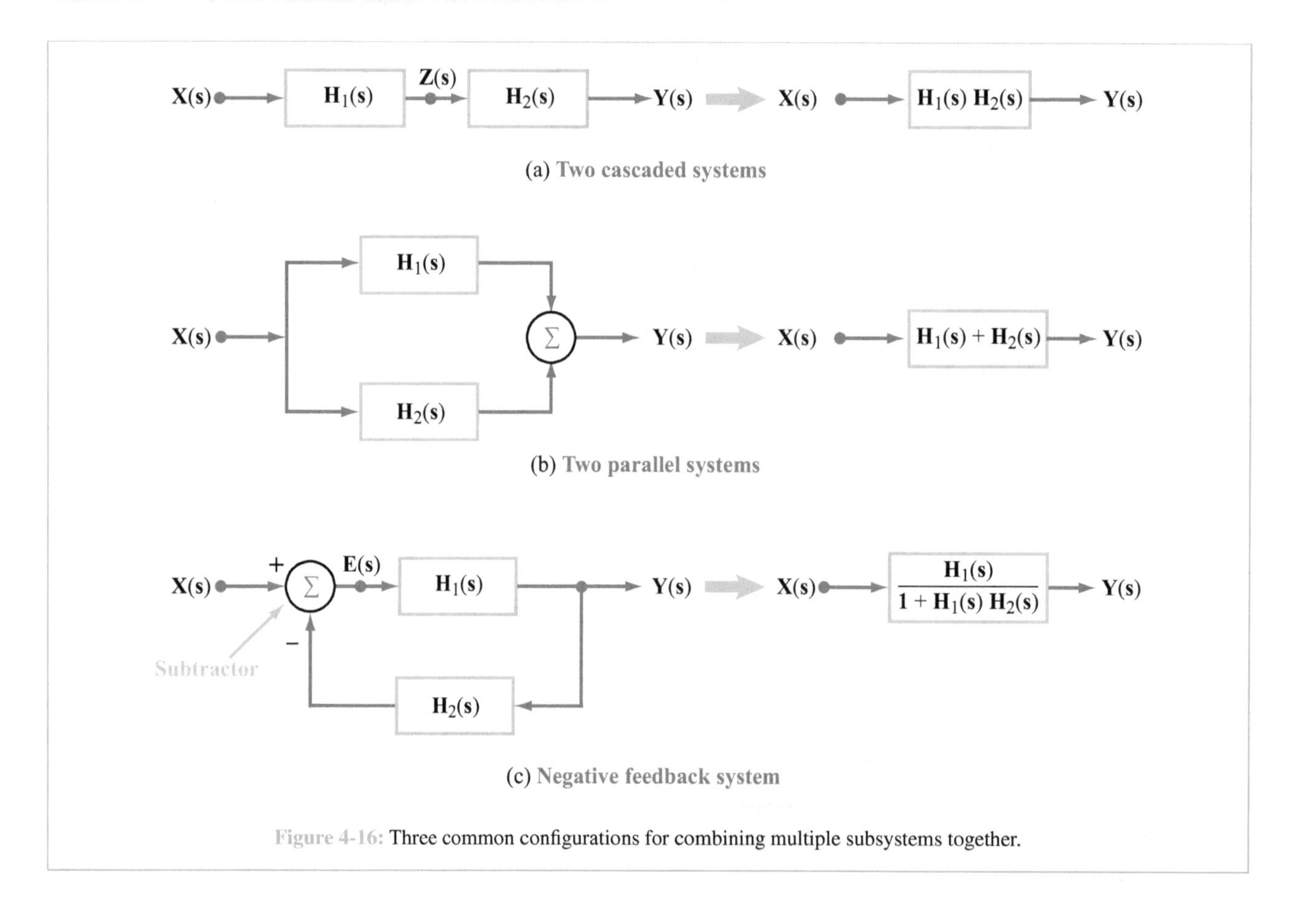

Figure 4-16: Three common configurations for combining multiple subsystems together.

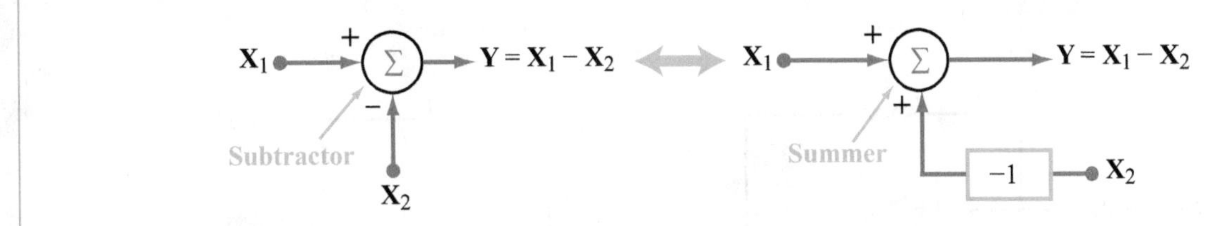

Figure 4-17: A subtractor is equivalent to a summer with one of its inputs preceded by an inverter.

(a) **Cascade configuration**

The combined transfer function $\mathbf{H}(\mathbf{s})$ of two cascaded systems (connected in series) is

$$\mathbf{H}(\mathbf{s}) = \frac{\mathbf{Y}(\mathbf{s})}{\mathbf{X}(\mathbf{s})} = \frac{\mathbf{Y}(\mathbf{s})}{\mathbf{Z}(\mathbf{s})} \cdot \frac{\mathbf{Z}(\mathbf{s})}{\mathbf{X}(\mathbf{s})} = \mathbf{H}_1(\mathbf{s})\,\mathbf{H}_2(\mathbf{s}), \tag{4.83}$$

(series configuration)

where $\mathbf{Z}(\mathbf{s})$ defines an intermediary signal in between the two systems (Fig. 4-16(a)).

(b) **Parallel configuration**

In Fig. 4-16(b), input $\mathbf{X}(\mathbf{s})$ serves as the input to two separate systems, $\mathbf{H}_1(\mathbf{s})$ and $\mathbf{H}_2(\mathbf{s})$. The sum of their outputs is $\mathbf{Y}(\mathbf{s})$. Hence,

$$\mathbf{H}(\mathbf{s}) = \frac{\mathbf{Y}(\mathbf{s})}{\mathbf{X}(\mathbf{s})} = \frac{\mathbf{H}_1(\mathbf{s})\,\mathbf{X}(\mathbf{s}) + \mathbf{H}_2(\mathbf{s})\,\mathbf{X}(\mathbf{s})}{\mathbf{X}(\mathbf{s})} = \mathbf{H}_1(\mathbf{s}) + \mathbf{H}_2(\mathbf{s}). \tag{4.84}$$

(parallel configuration)

(c) **Feedback configuration**

If a sample of the output $\mathbf{Y}(\mathbf{s})$ is fed back (returned) through a system with transfer function $\mathbf{H}_2(\mathbf{s})$ to a summer that subtracts it from input $\mathbf{X}(\mathbf{s})$, the output of the summer is the intermediary (error) signal

$$\mathbf{E}(\mathbf{s}) = \mathbf{X}(\mathbf{s}) - \mathbf{Y}(\mathbf{s})\,\mathbf{H}_2(\mathbf{s}). \tag{4.85}$$

Moreover, $\mathbf{E}(\mathbf{s})$ is related to $\mathbf{Y}(\mathbf{s})$ by

$$\mathbf{Y}(\mathbf{s}) = \mathbf{E}(\mathbf{s})\,\mathbf{H}_1(\mathbf{s}). \tag{4.86}$$

Solving the two equations for $\mathbf{Y}(\mathbf{s})$ and then dividing by $\mathbf{X}(\mathbf{s})$ leads to

$$\mathbf{H}(\mathbf{s}) = \frac{\mathbf{Y}(\mathbf{s})}{\mathbf{X}(\mathbf{s})} = \frac{\mathbf{H}_1(\mathbf{s})}{1 + \mathbf{H}_1(\mathbf{s})\,\mathbf{H}_2(\mathbf{s})}. \tag{4.87}$$

(negative feedback)

The configuration depicted by the diagram in Fig. 4-16(c) is called ***negative feedback*** because it involves *subtracting* a copy of the output from the input. Subtraction can be realized by using a difference amplifier circuit (which we have not covered) or by using the combination of a summer and inverter as shown in Fig. 4-17.

Example 4-9: Circuit Transfer Function

Determine the overall transfer function $\mathbf{H}(\mathbf{s}) = \mathbf{Y}(\mathbf{s})/\mathbf{X}(\mathbf{s})$ of the circuit in Fig. 4-18. Element values are $R = 100$ kΩ and $C = 5$ μF.

Solution: Through comparison with the circuit configurations we examined earlier, we recognize that the circuit involving Op amp #1 in Fig. 4-18 is an inverting summer with two input channels, both with absolute gain constants of 1 (because the input and feedback resistors are the same). Hence, in the block diagram of Fig. 4-18(b), the inverting summer is shown with gains of -1. The circuit that uses Op amp #2 is an integrator with

$$\mathbf{H}_1(\mathbf{s}) = \left(-\frac{1}{RC}\right)\frac{1}{\mathbf{s}} = \left(-\frac{1}{10^5 \times 5 \times 10^{-6}}\,\frac{1}{\mathbf{s}}\right) = -\frac{2}{\mathbf{s}}.$$

The last op-amp circuit is an inverter with

$$\mathbf{H}_2(\mathbf{s}) = -\frac{0.2R}{R} = -0.2.$$

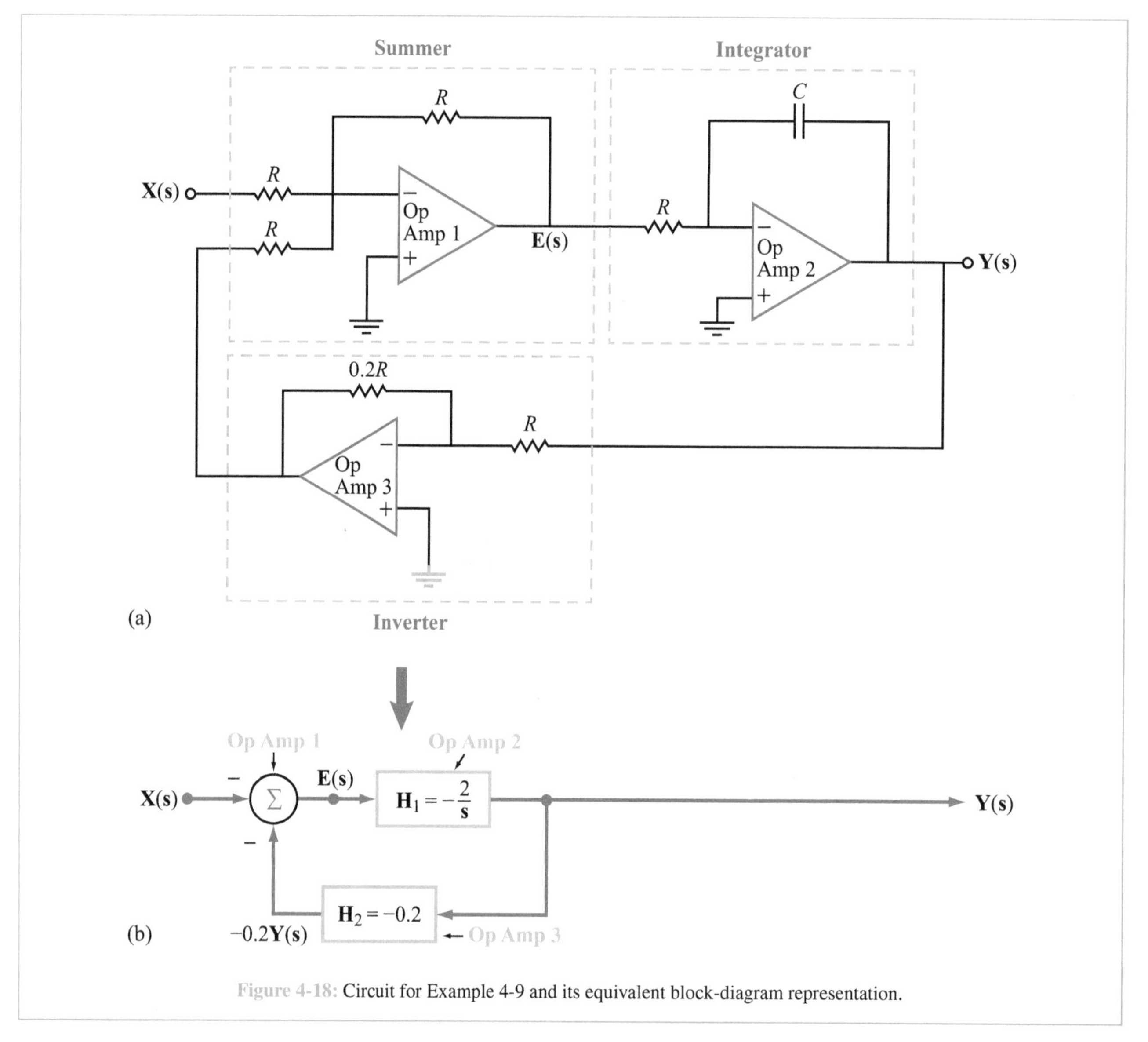

Figure 4-18: Circuit for Example 4-9 and its equivalent block-diagram representation.

The output of the inverter is $[-0.2\mathbf{Y}(\mathbf{s})]$, and the output of the inverting summer is

$$\mathbf{E}(\mathbf{s}) = -\mathbf{X}(\mathbf{s}) - [-0.2\mathbf{Y}(\mathbf{s})] = -\mathbf{X}(\mathbf{s}) + 0.2\mathbf{Y}(\mathbf{s}).$$

Also,

$$\mathbf{Y}(\mathbf{s}) = \mathbf{H}_1(\mathbf{s})\,\mathbf{E}(\mathbf{s}) = -\frac{2}{\mathbf{s}}\,[-\mathbf{X}(\mathbf{s}) + 0.2\mathbf{Y}(\mathbf{s})].$$

Solving for $\mathbf{Y}(\mathbf{s})$ and then dividing by $\mathbf{X}(\mathbf{s})$ leads to

$$\mathbf{H}(\mathbf{s}) = \frac{\mathbf{Y}(\mathbf{s})}{\mathbf{X}(\mathbf{s})} = \frac{2}{\mathbf{s}+0.4}\,.$$

Exercise 4-9: How many op amps are needed (as a minimum) to implement a system with transfer function $\mathbf{H}(\mathbf{s}) = b/(\mathbf{s}+a)$ where $a, b > 0$?

Answer: Two are needed: a one-pole configuration with $a = 1/(R_f C_f)$ and $b = 1/(R_i C_f)$ and an inverter with $G = -1$. (See S^2)

4-7 System Synthesis

Consider a scenario where we are asked to design a *notch filter* to remove signals at a specific frequency because such signals cause undesirable interference. In another scenario, we may be asked to design a feedback control system to maintain the speed of a motor constant or to control a room's heating system so as to keep the temperature at the value selected by the user. In these and many other system design scenarios, we translate the specifications into poles and zeros, and then we use the locations of those poles and zeros to synthesize a transfer function $\mathbf{H(s)}$. (This latter step was covered in Section 3-10.) The prescribed transfer function $\mathbf{H(s)}$ can then guide the construction of an analog system using op-amp circuits.

▶ The formal procedure for creating a modular block diagram representing $\mathbf{H(s)}$ wherein the individual modules are summers, integrators, and inverters is called *system realization*. ◀

4-7.1 Direct Form I (DFI) Topology

Several topologies are available to *realize* a transfer function $\mathbf{H(s)}$ [i.e., to convert it into a real system], of which the most straightforward is called the *direct form I* (*DFI*) realization. Even though the system realized by implementing the DFI is not as *efficient* (in terms of circuit complexity) as that realized by the DFII topology covered in the next subsection, we will examine it anyway, because it serves as a useful reference for comparison with other realization topologies.

Recall from Eq. (3.127) that a transfer function can be expressed as the ratio of two polynomials. Hence, the most general form of a transfer function $\mathbf{H(s)}$ is

$$\mathbf{H(s)} = \frac{\sum_{i=0}^{m} b_{m-i}\, \mathbf{s}^i}{\sum_{i=0}^{n} a_{n-i}\, \mathbf{s}^i},$$

where some of the constant coefficients may be zero. To illustrate the DFI procedure, we will implement it for a function with $m = n = 3$ and $b_3 = 0$ as in

$$\mathbf{H(s)} = \frac{b_0\mathbf{s}^3 + b_1\mathbf{s}^2 + b_2\mathbf{s}}{\mathbf{s}^3 + a_1\mathbf{s}^2 + a_2\mathbf{s} + a_3}, \tag{4.88}$$

where we have already divided all terms by a_0 so as to make the coefficient of $\mathbf{s}^3$ in the denominator unity. The first step in the procedure entails rewriting the expression in terms of inverse powers of $\mathbf{s}$:

$$\begin{aligned}\mathbf{H(s)} &= \frac{b_0\mathbf{s}^3 + b_1\mathbf{s}^2 + b_2\mathbf{s}}{\mathbf{s}^3 + a_1\mathbf{s}^2 + a_2\mathbf{s} + a_3} \cdot \frac{1/\mathbf{s}^3}{1/\mathbf{s}^3} \\ &= \left(b_0 + \frac{b_1}{\mathbf{s}} + \frac{b_2}{\mathbf{s}^2}\right)\left(\frac{1}{1 + \frac{a_1}{\mathbf{s}} + \frac{a_2}{\mathbf{s}^2} + \frac{a_3}{\mathbf{s}^3}}\right) \\ &= \mathbf{H}_1(\mathbf{s})\,\mathbf{H}_2(\mathbf{s}),\end{aligned} \tag{4.89}$$

with

$$\mathbf{H}_1(\mathbf{s}) = b_0 + \frac{b_1}{\mathbf{s}} + \frac{b_2}{\mathbf{s}^2} \tag{4.90a}$$

and

$$\mathbf{H}_2(\mathbf{s}) = \left(1 + \frac{a_1}{\mathbf{s}} + \frac{a_2}{\mathbf{s}^2} + \frac{a_3}{\mathbf{s}^3}\right)^{-1}. \tag{4.90b}$$

Because $\mathbf{H(s)}$ represents an LTI system, the commutative property allows us to realize $\mathbf{H(s)}$ in either of two sequences: $\mathbf{H}_1(\mathbf{s})$ followed by $\mathbf{H}_2(\mathbf{s})$ or in reverse order. The two sequences are outlined in Fig. 4-19; DFI starts by realizing $\mathbf{H}_1(\mathbf{s})$ first, whereas DFII starts by realizing $\mathbf{H}_2(\mathbf{s})$ first. The difference may seem inconsequential, but as we will see when we compare the two topologies in Section 4-7.2, the number of circuit modules required by DFII is about half of that required by DFI.

In Fig. 4-19(a), $\mathbf{Z}_1(\mathbf{s})$ is an intermediate signal representing the output of $\mathbf{H}_1(\mathbf{s})$. Construction of the expression defining $\mathbf{H}_1(\mathbf{s})$ requires three types of algebraic operations, namely, addition, multiplication by constant coefficients, and division by $\mathbf{s}$. Figure 4-20 shows a configuration that utilizes these

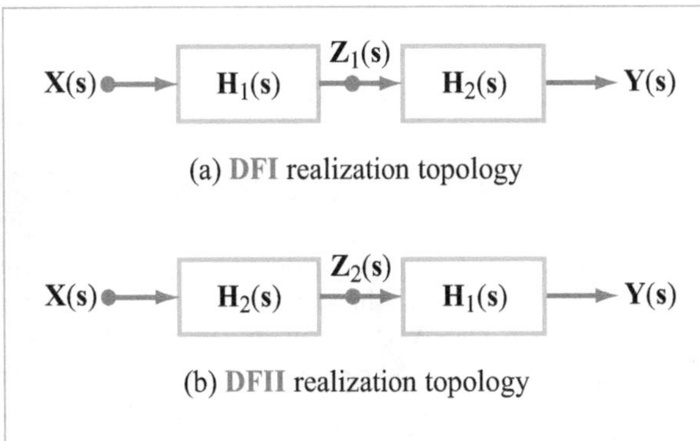

Figure 4-19: In the DFI process, $\mathbf{H}_1(\mathbf{s})$ is realized ahead of $\mathbf{H}_2(\mathbf{s})$, whereas the reverse is the case for the DFII process.

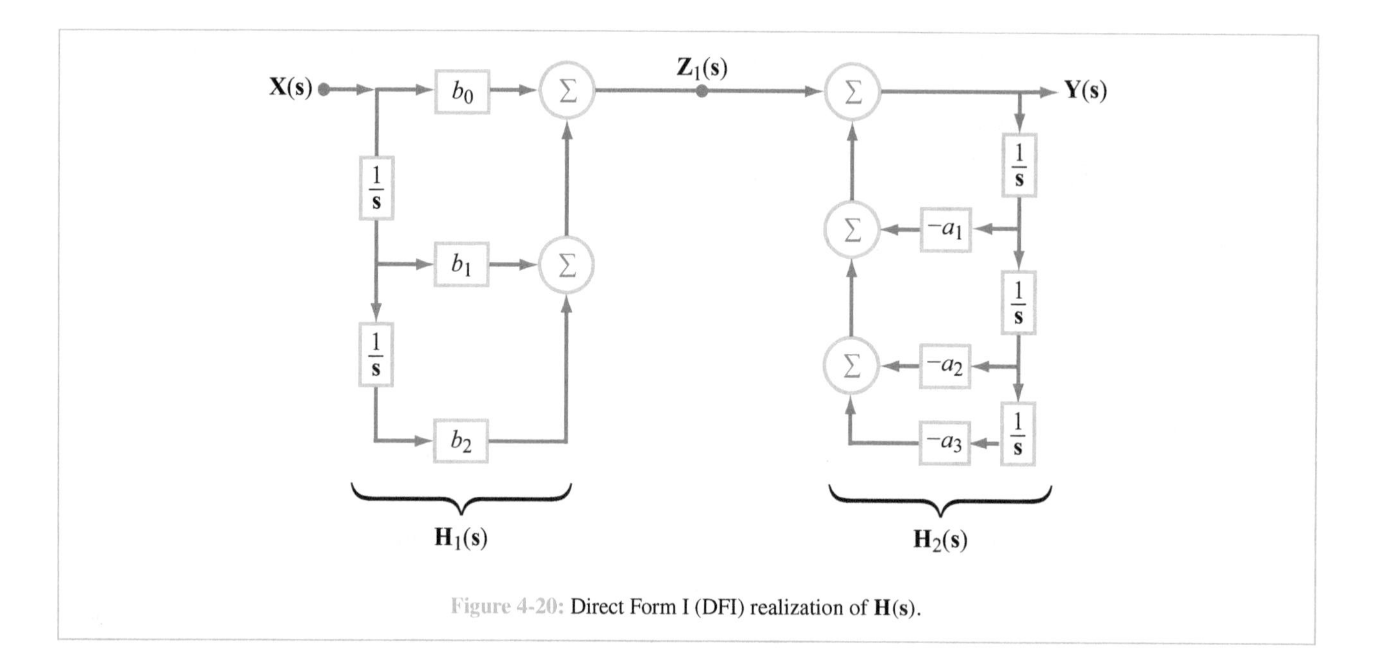

Figure 4-20: Direct Form I (DFI) realization of $\mathbf{H}(\mathbf{s})$.

algebraic steps to generate $\mathbf{Z}_1(\mathbf{s})$ from $\mathbf{X}(\mathbf{s})$, thereby realizing transfer function $\mathbf{H}_1(\mathbf{s})$. Note that generating term $b_2/\mathbf{s}^2$ involves multiplication by $(1/\mathbf{s})$ twice and is followed with multiplication by b_2.

The form of $\mathbf{H}_1(\mathbf{s})$ given by Eq. (4.90a) made its realization straightforward; all that was needed was addition, multiplication, and division by $\mathbf{s}$ (which is equivalent to integration in the time domain). That is not the case for the expression of $\mathbf{H}_2(\mathbf{s})$ given by Eq. (4.90b), so we need to manipulate the realization process to create a form similar to that of $\mathbf{H}_1(\mathbf{s})$. To that end, we use $\mathbf{Y}(\mathbf{s}) = \mathbf{H}_2(\mathbf{s})\,\mathbf{Z}_1(\mathbf{s})$ to solve for $\mathbf{Z}_1(\mathbf{s})$:

$$\mathbf{Z}_1(\mathbf{s}) = \frac{\mathbf{Y}(\mathbf{s})}{\mathbf{H}_2(\mathbf{s})} = \left(1 + \frac{a_1}{\mathbf{s}} + \frac{a_2}{\mathbf{s}^2} + \frac{a_3}{\mathbf{s}^3}\right) \mathbf{Y}(\mathbf{s}). \tag{4.91}$$

Next, we rearrange the expression into

$$\mathbf{Y}(\mathbf{s}) = \mathbf{Z}_1(\mathbf{s}) - \left(\frac{a_1}{\mathbf{s}} + \frac{a_2}{\mathbf{s}^2} + \frac{a_3}{\mathbf{s}^3}\right) \mathbf{Y}(\mathbf{s}). \tag{4.92}$$

Even though $\mathbf{Y}(\mathbf{s})$ appears on both sides of Eq. (4.92), its form allows us to realize $\mathbf{H}_2(\mathbf{s})$, as shown by the right-hand segment of Fig. 4-20. The fact that $\mathbf{Y}(\mathbf{s})$ is needed on the right-hand side of Eq. (4.92) to calculate $\mathbf{Y}(\mathbf{s})$ on the left-hand side means that it is a feedback process similar to that used in most op-amp circuits. We should note that the operations represented in the flow diagram of Fig. 4-20 include five summers, six amplifiers, and five integrators.

4-7.2 Direct Form II (DFII)

Reversing the implementation order of $\mathbf{H}_1(\mathbf{s})$ and $\mathbf{H}_2(\mathbf{s})$ and defining an intermediate signal $\mathbf{Z}_2(\mathbf{s})$ in between them gives

$$\mathbf{Z}_2(\mathbf{s}) = \mathbf{H}_2(\mathbf{s})\,\mathbf{X}(\mathbf{s}), \tag{4.93a}$$

$$\mathbf{Y}(\mathbf{s}) = \mathbf{H}_1(\mathbf{s})\,\mathbf{Z}_2(\mathbf{s}), \tag{4.93b}$$

which combine into

$$\mathbf{Y}(\mathbf{s}) = \mathbf{H}_1(\mathbf{s})\,\mathbf{H}_2(\mathbf{s})\,\mathbf{X}(\mathbf{s}).$$

Following the recipes introduced in Section 4-7.1 in connection with the DFI process, $\mathbf{H}_1(\mathbf{s})$ and $\mathbf{H}_2(\mathbf{s})$ can be implemented as shown in Fig. 4-21(a). The configuration uses separate integrators for $\mathbf{H}_1(\mathbf{s})$ and $\mathbf{H}_2(\mathbf{s})$. The number of required operations is identical with those needed to implement the DFI process shown in Fig. 4-20. However, since the integrators for both $\mathbf{H}_1(\mathbf{s})$ and $\mathbf{H}_2(\mathbf{s})$ are fed by $\mathbf{Z}_2(\mathbf{s})$, they can serve to supply the necessary signals to both $\mathbf{H}_1(\mathbf{s})$ and $\mathbf{H}_2(\mathbf{s})$, as shown in Fig. 4-21(b). Consequently, only three integrators are needed instead of five.

Example 4-10: One-Pole Transfer Function

Develop a realization and implement it using op-amp circuits for the transfer function

$$\mathbf{H}(\mathbf{s}) = \frac{20}{\mathbf{s}+5}. \tag{4.94}$$

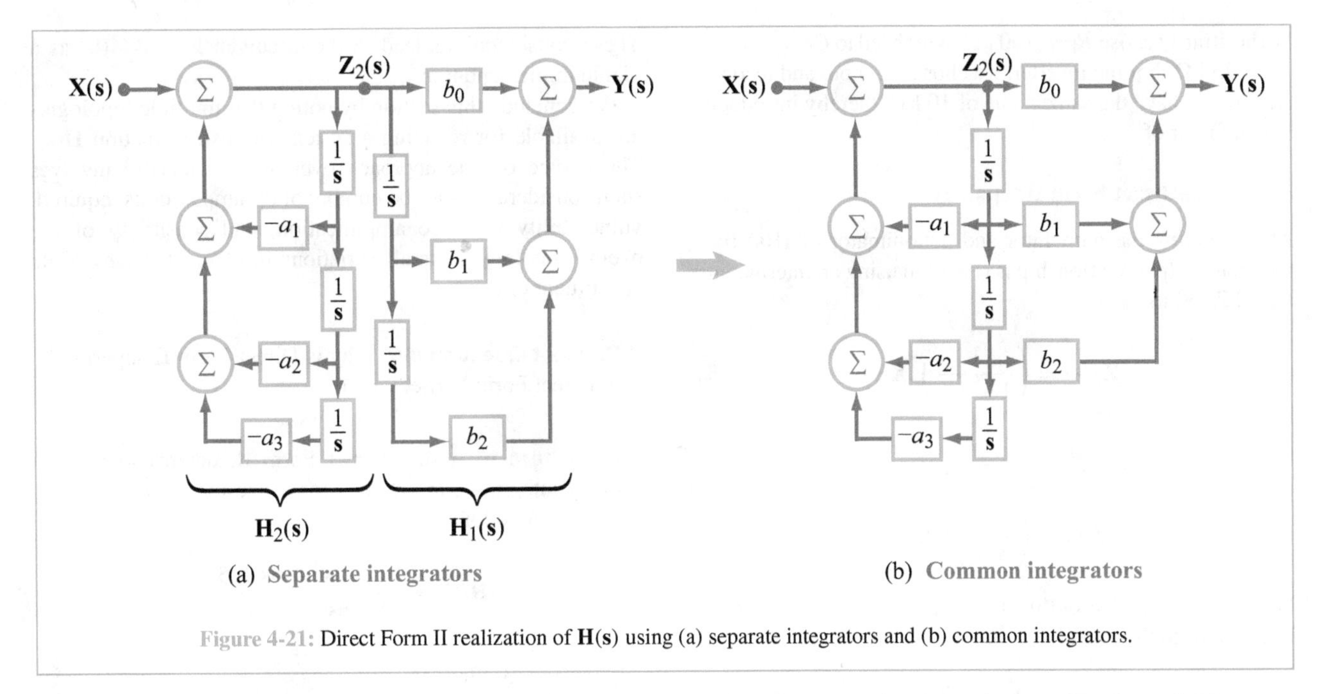

Figure 4-21: Direct Form II realization of $\mathbf{H}(\mathbf{s})$ using (a) separate integrators and (b) common integrators.

Solution:

Method 1: One-pole circuit

The form of $\mathbf{H}(\mathbf{s})$ matches Eq. (4.82b) of the one-pole op-amp circuit shown in Table 4-3(e). The gain of the op-amp circuit is negative, however, so we need to add an inverter either ahead or after the one-pole circuit. Our solution is displayed in Fig. 4-22.

The circuit contains three elements (R_i, R_f, and C_f), yet we need to satisfy only the two conditions given by Eqs. (4.82a and c):

$$p = -\frac{1}{R_\mathrm{f} C_\mathrm{f}} = -5,$$

and

$$\frac{1}{R_\mathrm{i} C_\mathrm{f}} = 20.$$

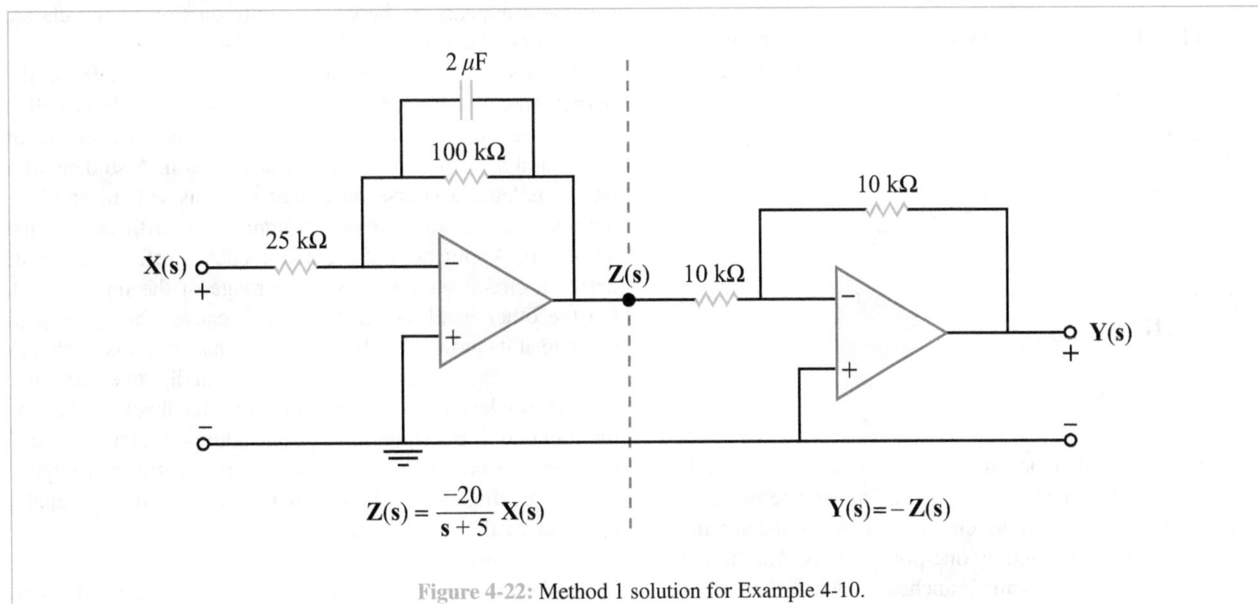

Figure 4-22: Method 1 solution for Example 4-10.

We arbitrarily chose $R_f = 100$ kΩ, which led to $C_f = 2\ \mu$F and $R_i = 25$ kΩ. For the inverter, we chose the input and feedback resistors to have the same value of 10 kΩ, thereby introducing a gain factor of −1.

Method 2: Direct Form II topology

After dividing the numerator and denominator of $\mathbf{H(s)}$ by $\mathbf{s}$, the input-output relationship is rewritten using an intermediary signal $\mathbf{Z}_2(\mathbf{s})$ as

$$\mathbf{Z}_2(\mathbf{s}) = \left(\frac{1}{1+5/\mathbf{s}}\right) \mathbf{X(s)} \tag{4.95a}$$

and

$$\mathbf{Y(s)} = \left(\frac{20}{\mathbf{s}}\right) \mathbf{Z}_2(\mathbf{s}). \tag{4.95b}$$

Following the recipe outlined in Section 4-7.2, we rearrange Eq. (4.95a) to the form

$$\mathbf{Z}_2(\mathbf{s}) = \mathbf{X(s)} - \frac{5}{\mathbf{s}}\,\mathbf{Z}_2(\mathbf{s}). \tag{4.96}$$

Implementation of Eqs. (4.95b) and (4.96) leads to the flow diagram shown in Fig. 4-23(a). Parts (b) and (c) of the figure depict the op-amp symbolic diagram and the detailed circuit implementation, respectively.

4-7.3 Parallel Realization Process

The DFI and DFII processes are two among several different approaches used to realize system transfer functions. The ***parallel realization*** process relies on expressing the transfer function as a sum of partial fractions. Consider, for example, the transfer function

$$\mathbf{H(s)} = \frac{\mathbf{s}^2 + 17\mathbf{s} + 20}{\mathbf{s}(\mathbf{s}^2 + 6\mathbf{s} + 5)} . \tag{4.97}$$

Application of partial fraction expansion (Chapter 3) allows us to express $\mathbf{H(s)}$ as

$$\mathbf{H(s)} = \frac{4}{\mathbf{s}} - \frac{1}{\mathbf{s}+1} - \frac{2}{\mathbf{s}+5} . \tag{4.98}$$

The three terms can be added together using the parallel configuration shown in Fig. 4-24(a). The first term can be implemented by an integrator circuit, and the second and third terms can be implemented by one-pole circuits. Additionally, inverters are needed in some branches to switch the polarity. These considerations lead to the circuit in Fig. 4-24(b) as a viable implementation.

We conclude this section by noting that multiple topologies are available for realizing a system's transfer function $\mathbf{H(s)}$. The choice of one approach over another usually involves such considerations as the number of op-amp circuits required, vulnerability to noise amplification, and sensitivity of the overall topology to small variations in element values of its constituent circuits.

Concept Question 4-6: Why is Direct Form II superior to Direct Form I? (See S²)

Exercise 4-10: Using Direct Form II, determine how many integrators are needed to realize the system with transfer function

$$\mathbf{H(s)} = \frac{2\mathbf{s}^3 + 3\mathbf{s}^2 + 4\mathbf{s} + 5}{\mathbf{s}^3 + 8\mathbf{s}^2 + 7\mathbf{s} + 6} .$$

Answer: 3. (See S²)

4-8 Basic Control Theory

"Without control systems there could be no manufacturing, no vehicles, no computers, no regulated environment—in short, no technology. Control systems are what make machines, in the broadest sense of the term, function as intended." This statement appears in the opening introduction to the classic book on *Feedback Control Theory* by Doyle et al.

The basic idea of feedback is to take a sample of the output signal and feed it back into the input. It is called ***positive feedback*** if it increases the strength of the input signal and ***negative feedback*** if it decreases it. A student who has completed a course on circuit analysis will most likely have learned about feedback in connection with operational amplifiers. Application of negative feedback offers a trade-off between circuit gain and dynamic range of the input signal. On the other hand, positive feedback causes the op amp to saturate at its power supply voltage. Such a circuit is known as a ***Schmitt trigger*** and is used extensively in digital electronics. Another (undesirable) example of positive feedback is when the microphone (input signal) in a public address system is placed too close to one or more of the system's speakers (output), thereby picking up part of the output signal, amplifying it again, and causing the sound to squeal!

Our discussion will be limited to the use of negative feedback to regulate the behavior of a system or improve its stability.

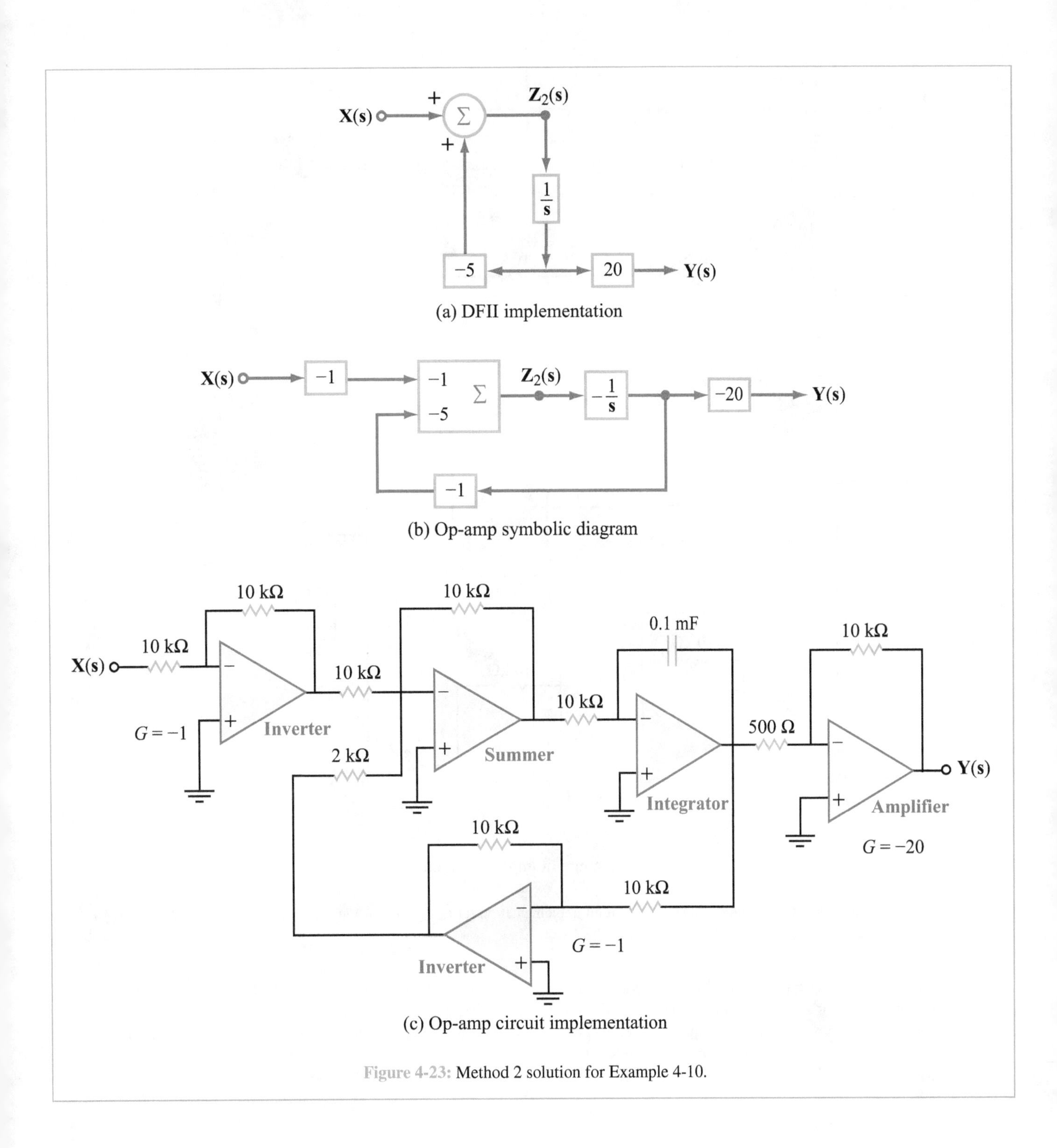

Figure 4-23: Method 2 solution for Example 4-10.

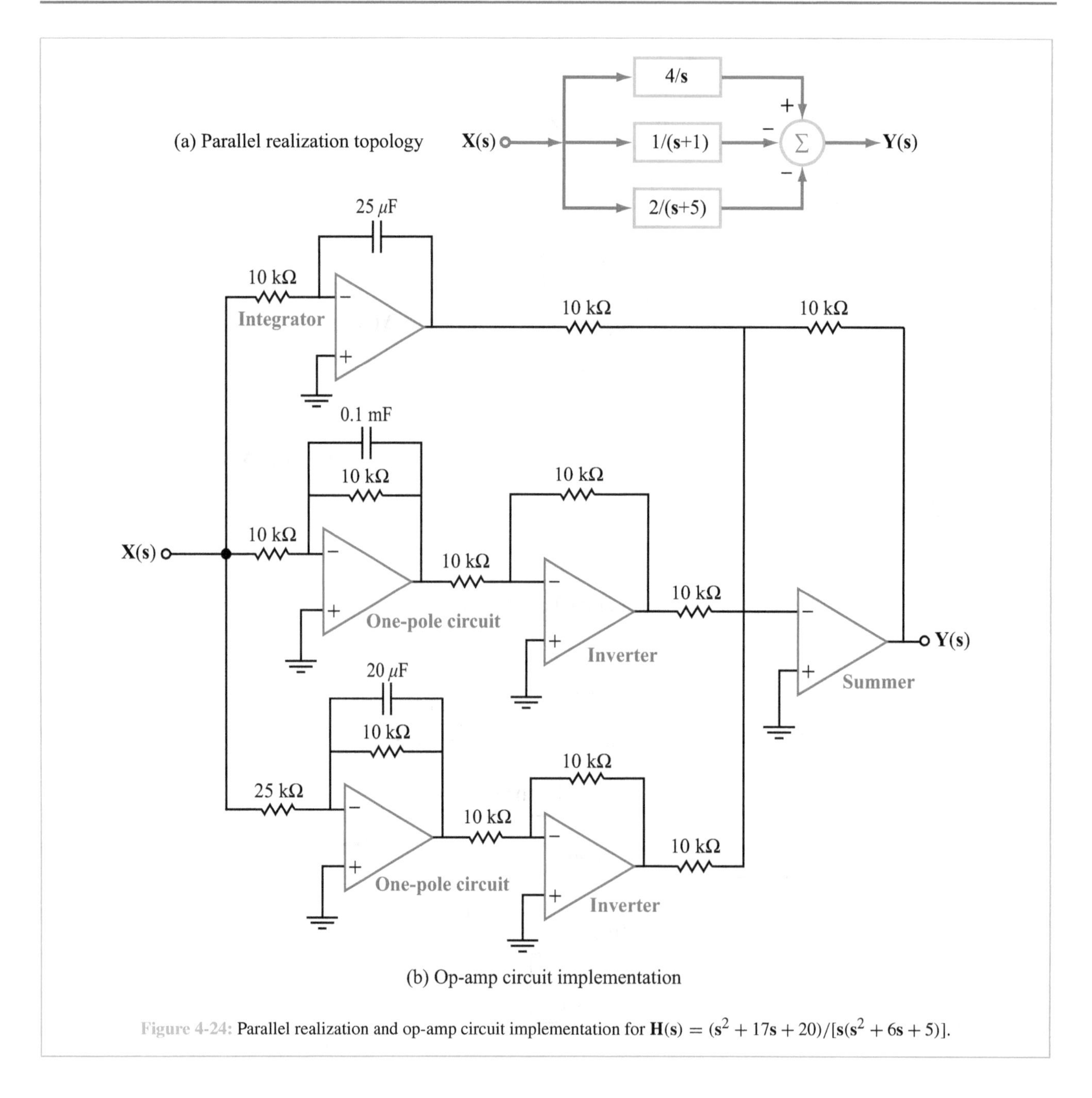

Figure 4-24: Parallel realization and op-amp circuit implementation for $\mathbf{H}(\mathbf{s}) = (\mathbf{s}^2 + 17\mathbf{s} + 20)/[\mathbf{s}(\mathbf{s}^2 + 6\mathbf{s} + 5)]$.

Common examples include cruise control in cars, controlling the operation of a heater or air conditioning system to maintain the temperature in a room or house at a specified level, and rudder control in aircraft, as well as a seemingly endless list of applications to mechanical, electrical, chemical, and biological systems.

The block diagram in Fig. 4-25 illustrates the basic operation of a feedback system configured to maintain a room's temperature at T_{ref}, which is the temperature selected by a human. The room temperature measured by a thermistor, T_{m}, is compared with and subtracted from T_{ref} by a summer. If the difference $(T_{\text{ref}} - T_{\text{m}})$ exceeds a minimum threshold (such as

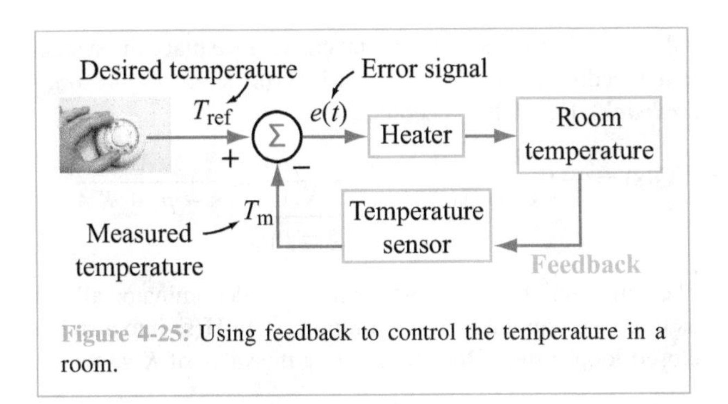

Figure 4-25: Using feedback to control the temperature in a room.

0.5 °C), the corresponding error signal $e(t)$ will instruct the heater to turn on. When $(T_{ref} - T_m)$ drops below 0.5 °C, the error signal will instruct the heater to stop generating hot air.

4-8.1 Closed-Loop Transfer Function

In the absence of feedback, a system is said to be operating in an ***open-loop*** mode; in Fig. 4-26(a), **H**(**s**) is called the ***open-loop transfer function***. In a ***closed-loop*** mode, a copy of output signal **Y**(**s**) is used to modify input signal **X**(**s**), as diagrammed in Fig. 4-26(b). The feedback path includes a system characterized by a ***feedback transfer function*** **G**(**s**) with input **Y**(**s**) and output **G**(**s**) **Y**(**s**). The output of the summer is an ***error signal*** **E**(**s**) given by

$$\mathbf{E(s)} = \mathbf{X(s)} - \mathbf{G(s)}\,\mathbf{Y(s)}\,, \tag{4.99}$$

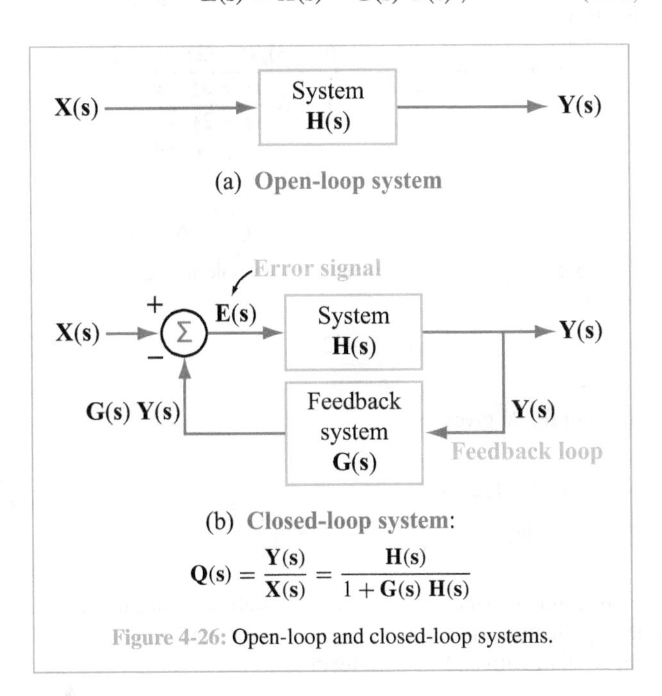

Figure 4-26: Open-loop and closed-loop systems.

and the overall system output is

$$\mathbf{Y(s)} = \mathbf{H(s)}\,\mathbf{E(s)} = \mathbf{H(s)}[\mathbf{X(s)} - \mathbf{G(s)}\,\mathbf{Y(s)}],$$

which leads to

$$\mathbf{Y(s)} = \frac{\mathbf{H(s)}\,\mathbf{X(s)}}{1 + \mathbf{G(s)}\,\mathbf{H(s)}}\,. \tag{4.100}$$

The output-to-input ratio of the closed-loop system is called the ***closed-loop transfer function*** **Q**(**s**). From Eq. (4.100), we obtain

$$\mathbf{Q(s)} = \frac{\mathbf{Y(s)}}{\mathbf{X(s)}} = \frac{\mathbf{H(s)}}{1 + \mathbf{G(s)}\,\mathbf{H(s)}}\,. \tag{4.101}$$

Negative feedback requires that **G**(**s**) be a positive quantity, but it imposes no restrictions on its magnitude or functional form. In the absence of feedback [**G**(**s**) = 0], Eq. (4.101) reduces to **Q**(**s**) = **H**(**s**).

We also should note that, whereas the introduction of feedback can alter the poles of the system, it has no effect on its zeros unless **G**(**s**) has poles. In its most general form, **H**(**s**) is the ratio of two polynomials:

$$\mathbf{H(s)} = \frac{\mathbf{N(s)}}{\mathbf{D(s)}}\,. \tag{4.102}$$

Its zeros and poles are the roots of **N**(**s**) = 0 and **D**(**s**) = 0, respectively. Inserting Eq. (4.102) into Eq. (4.101) leads to

$$\mathbf{Q(s)} = \frac{\mathbf{N(s)}/\mathbf{D(s)}}{1 + \mathbf{G(s)}\,\mathbf{N(s)}/\mathbf{D(s)}} = \frac{\mathbf{N(s)}}{\mathbf{D(s)} + \mathbf{G(s)}\,\mathbf{N(s)}}\,. \tag{4.103}$$

► We observe that the zeros of **Q**(**s**) are determined by **N**(**s**), which is the same as for **H**(**s**), but the presence of **G**(**s**) in the denominator of Eq. (4.103) allows us to alter the locations of the poles of **H**(**s**). As we shall demonstrate in future sections, this ability to move the locations of the poles of a system's transfer function is a powerful tool in many practical applications. ◄

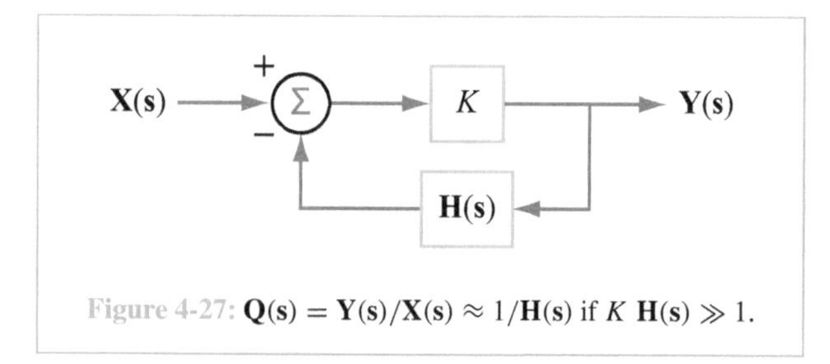

Figure 4-27: $\mathbf{Q}(\mathbf{s}) = \mathbf{Y}(\mathbf{s})/\mathbf{X}(\mathbf{s}) \approx 1/\mathbf{H}(\mathbf{s})$ if $K\,\mathbf{H}(\mathbf{s}) \gg 1$.

4-8.2 Constructing Inverse System

Suppose we have a physical system with transfer function $\mathbf{H}(\mathbf{s})$ and we want to construct its inverse. Mathematically, we simply invert the expression for $\mathbf{H}(\mathbf{s})$, but how do we physically construct a system with transfer function $\mathbf{H}'(\mathbf{s}) = 1/\mathbf{H}(\mathbf{s})$ from the physical system $\mathbf{H}(\mathbf{s})$? One way to realize the inverse system is to construct the feedback loop shown in Fig. 4-27 in which the original system $\mathbf{H}(\mathbf{s})$ is placed in the negative feedback branch of the loop and a constant multiplier K is placed in the forward branch. Replacing $\mathbf{H}(\mathbf{s})$ with K and $\mathbf{G}(\mathbf{s})$ with $\mathbf{H}(\mathbf{s})$ in Eq. (4.101) leads to

$$\mathbf{Q}(\mathbf{s}) = \frac{K}{1 + K\,\mathbf{H}(\mathbf{s})}. \tag{4.104}$$

If we choose the constant multiplier to have a large gain such that $K\,\mathbf{H}(\mathbf{s}) \gg 1$, Eq. (4.104) reduces to

$$\mathbf{Q}(\mathbf{s}) \approx \frac{1}{\mathbf{H}(\mathbf{s})} = \mathbf{H}'(\mathbf{s}). \tag{4.105}$$

Thus, with the proper choice of K, the overall feedback system behaves like the inverse of the open-loop system $\mathbf{H}(\mathbf{s})$.

4-8.3 System Stabilization

As we discussed earlier in Section 3-8, a system is unstable unless all the poles of its transfer function $\mathbf{H}(\mathbf{s})$ are in the open left-hand plane (OLHP) of the $\mathbf{s}$-domain. Feedback can be used to stabilize an otherwise unstable system. We will demonstrate the stabilization process through two examples.

(a) Proportional feedback in first-order system

Consider the first-order system characterized by

$$\mathbf{H}(\mathbf{s}) = \frac{A}{\mathbf{s} - p_1} \quad \text{with } p_1 > 0. \tag{4.106}$$

The open-loop transfer function $\mathbf{H}(\mathbf{s})$ has a single pole at $\mathbf{s} = p_1$, and since $p_1 > 0$, the pole resides in the RHP of the $\mathbf{s}$-domain. Hence, the system is unstable. If we place the system in the feedback loop of Fig. 4-26 with $\mathbf{G}(\mathbf{s}) = K$ (*proportional feedback*), Eq. (4.101) becomes

$$\mathbf{Q}(\mathbf{s}) = \frac{\mathbf{H}(\mathbf{s})}{1 + \mathbf{G}(\mathbf{s})\,\mathbf{H}(\mathbf{s})} = \frac{A/(\mathbf{s} - p_1)}{1 + \dfrac{KA}{\mathbf{s} - p_1}} = \frac{A}{\mathbf{s} - p_1 + KA}. \tag{4.107}$$

The introduction of the last term in the denominator allows us to convert an unstable open-loop system $\mathbf{H}(\mathbf{s})$ into a stable closed-loop system $\mathbf{Q}(\mathbf{s})$ by selecting the value of K so that

$$KA > p_1.$$

An important feature of this condition is that we do not need to know the exact value of the pole p_1 to stabilize the system. In practice, the value of the pole p_1 is often not known exactly, so an unstable pole p_1 cannot be canceled by a zero at p_1. Using feedback, this is not a problem: We need merely to choose K conservatively to ensure that $KA > p_1$, even if we do not know p_1 or A exactly.

Example 4-11: First-Order Stabilization

Choose the value of the feedback-loop constant K so that the pole of $\mathbf{H}(\mathbf{s}) = (\mathbf{s}+3)/(\mathbf{s}-2)$ moves from 2 to its diametrically opposite location in the OLHP, namely, -2.

Solution: With *proportional feedback*, $\mathbf{G}(\mathbf{s}) = K$, which leads to

$$\mathbf{Q}(\mathbf{s}) = \frac{\mathbf{H}(\mathbf{s})}{1 + K\,\mathbf{H}(\mathbf{s})} = \frac{(\mathbf{s}+3)/(\mathbf{s}-2)}{1 + K\dfrac{(\mathbf{s}+3)}{(\mathbf{s}-2)}} \cdot \frac{\mathbf{s}-2}{\mathbf{s}-2} = \frac{\mathbf{s}+3}{(K+1)\left[\mathbf{s} + \dfrac{3K-2}{K+1}\right]}.$$

To move the pole of $\mathbf{H}(\mathbf{s})$ from $+2$ to a pole at -2 (Fig. 4-28) for $\mathbf{Q}(\mathbf{s})$, it is necessary that

$$\frac{3K-2}{K+1} = 2.$$

Solving for K gives

$$K = 4,$$

which in turn leads to

$$\mathbf{Q}(\mathbf{s}) = 0.2\,\frac{\mathbf{s}+3}{\mathbf{s}+2}.$$

To appreciate the significance of the stabilization gained from the application of feedback, let us examine the step response of the system with and without feedback.

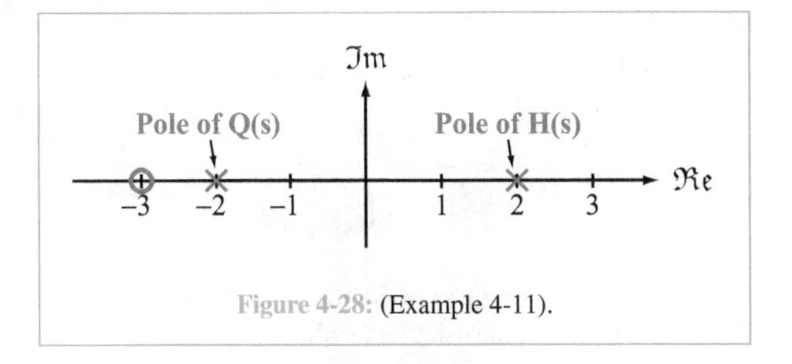

Figure 4-28: (Example 4-11).

1. Without feedback

For an input step function $x(t) = u(t)$ and corresponding Laplace transform $\mathbf{X(s)} = 1/\mathbf{s}$, the step response is $y_1(t)$ and the Laplace transform of $y_1(t)$ is

$$\mathbf{Y}_1(\mathbf{s}) = \mathbf{H}(\mathbf{s})\ \mathbf{X}(\mathbf{s}) = \frac{1}{\mathbf{s}} \frac{(\mathbf{s}+3)}{(\mathbf{s}-2)} .$$

Through application of partial fraction expansion, $\mathbf{Y}_1(\mathbf{s})$ becomes

$$\mathbf{Y}_1(\mathbf{s}) = -\frac{1.5}{\mathbf{s}} + \frac{2.5}{\mathbf{s}-2} .$$

Conversion to the time domain, facilitated by entries #2 and 3 in Table 3-2, gives

$$y_1(t) = -1.5u(t) + 2.5e^{2t}\ u(t).$$

As $t \to \infty$, so does $y_1(t)$. *This result confirms that* $\mathbf{H(s)}$ *represents a BIBO unstable system.*

2. With feedback

For a feedback system with closed-loop transfer function $\mathbf{Q(s)}$, repetition of the earlier steps leads to

$$\mathbf{Y}_2(\mathbf{s}) = \mathbf{Q}(\mathbf{s})\ \mathbf{X}(\mathbf{s}) = \frac{1}{\mathbf{s}} \cdot 0.2\ \frac{\mathbf{s}+3}{\mathbf{s}+2} = \frac{0.3}{\mathbf{s}} - \frac{0.1}{\mathbf{s}+2} ,$$

and

$$y_2(t) = 0.3u(t) - 0.1e^{-2t}\ u(t).$$

Because of the feedback loop, the exponent of the second term is negative, so as $t \to \infty$, $y_2(t) \to 0.3$. *Hence, the system is BIBO stable.*

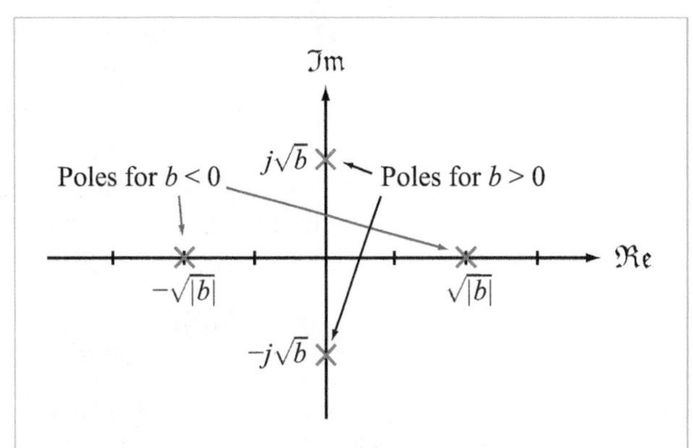

Figure 4-29: Poles of $\mathbf{H(s)} = (\mathbf{s}+a)/(\mathbf{s}^2+b)$ for $b > 0$ and $b < 0$ (zero at $-a$ not shown).

(b). **Proportional feedback in second-order system**

1. Open loop (without feedback)

Suppose we are given a second-order system with

$$\mathbf{H}(\mathbf{s}) = \frac{\mathbf{s}+a}{\mathbf{s}^2+b} \qquad \textbf{(open-loop)}. \qquad (4.108)$$

The poles of $\mathbf{H(s)}$ are the roots of its denominator. Setting

$$\mathbf{s}^2 + b = 0 \qquad (4.109)$$

gives

$$\begin{aligned} \mathbf{s} &= \pm\sqrt{|b|} \qquad \text{for } b < 0, \\ \mathbf{s} &= \pm j\sqrt{b} \qquad \text{for } b > 0. \end{aligned} \qquad (4.110)$$

If $b > 0$, the poles reside on the imaginary axis, as shown in Fig. 4-29, which makes the system BIBO unstable. If $b < 0$, the poles reside on the real axis, with one of them in the RHP. Hence, the system is again unstable. In summary, $\mathbf{H(s)}$ *is unstable for all values of* b.

2. Closed loop (with feedback)

Our task is to use feedback control to stabilize the system given by Eq. (4.108). We will consider four cases of the real-valued constants a and b, each of which may be either positive or negative. Employing a feedback system with a constant K gives

$$\begin{aligned} \mathbf{Q}(\mathbf{s}) &= \frac{\mathbf{H}(\mathbf{s})}{1+K\ \mathbf{H}(\mathbf{s})} \\ &= \frac{(\mathbf{s}+a)/(\mathbf{s}^2+b)}{1+K\ \dfrac{\mathbf{s}+a}{\mathbf{s}^2+b}} \\ &= \frac{\mathbf{s}+a}{\mathbf{s}^2+K\mathbf{s}+(Ka+b)} \qquad \textbf{(closed-loop)}. \qquad (4.111) \end{aligned}$$

The poles of $\mathbf{Q(s)}$ are the roots of its denominator:

$$s^2 + Ks + (Ka + b) = 0. \tag{4.112}$$

The two roots of this quadratic equation will reside in the open left half-plane if and only if the linear and constant terms are positive. That is, if and only if

$$K > 0 \quad \text{and} \quad Ka + b > 0. \tag{4.113}$$

We therefore require $K > 0$ regardless of the values of a and b. Additional constraints apply to satisfy the second condition, as follows:

(1) $a > 0$ and $b > 0$

If a and b are greater than zero, the second condition of Eq. (4.113) is automatically satisfied, and the closed-loop system is BIBO stable.

(2) $a < 0$ and $b > 0$

If only b is greater than zero, the closed-loop system is BIBO stable if (in addition to $K > 0$) $b > -aK$. Equivalently, the conditions can be combined into

$$0 < K < \frac{b}{|a|}.$$

(3) $a > 0$ and $b < 0$

If only a is greater than zero, the closed-loop system is BIBO stable if

$$K > \frac{|b|}{a}.$$

(4) $a < 0$ and $b < 0$

If both a and b have negative values, the closed-loop system is always unstable, since $Ka + b < 0$.

Example 4-12: Second-Order Stabilization

Given a system with $\mathbf{H(s)} = 3/(s^2 + 4)$, determine if/how it can be stabilized by a feedback loop with (a) *proportional feedback* $[\mathbf{G(s)} = K]$, and (b) *proportional-plus-derivative (PD)* feedback $[\mathbf{G(s)} = K_1 + K_2 s]$.

Solution:
(a) For $\mathbf{G(s)} = K$, we obtain

$$\mathbf{Q(s)} = \frac{\mathbf{H(s)}}{1 + K\,\mathbf{H(s)}} = \frac{3/(s^2+4)}{1 + K\left(\dfrac{3}{s^2+4}\right)} = \frac{3}{s^2 + (3K + 4)}.$$

The denominator of $\mathbf{Q(s)}$ is identical in form with that of Eq. (4.108). Hence, the use of a feedback loop with a constant-gain function K is insufficient to realize stabilization.

(b) Using PD feedback with $\mathbf{G(s)} = K_1 + K_2 s$ leads to

$$\mathbf{Q(s)} = \frac{\mathbf{H(s)}}{1 + (K_1 + K_2 s)\,\mathbf{H(s)}} = \frac{3/(s^2+4)}{1 + (K_1 + K_2 s)\left(\dfrac{3}{s^2+4}\right)} = \frac{3}{s^2 + 3K_2 s + (3K_1 + 4)}.$$

The poles of $\mathbf{Q(s)}$ will have negative real parts (i.e., they will reside in the OLHP) only if all constant coefficients in the denominator of $\mathbf{Q(s)}$ are real and positive. Equivalently, K_1 and K_2 must satisfy the conditions

$$K_1 > -\frac{4}{3}, \qquad K_2 > 0,$$

in order for the closed-loop system to be stable.

Concept Question 4-7: In control applications, why do we usually use negative, not positive, feedback? (See S²)

Concept Question 4-8: Why is using a zero to cancel an unstable pole a bad idea? (See S²)

Exercise 4-11: What is the minimum value of the feedback factor K needed to stabilize a system with transfer function $\mathbf{H(s)} = 1/[(s + 3)(s - 2)]$?

Answer: $K > 6$. (See S²)

Exercise 4-12: What values of K can stabilize a system with transfer function

$$\mathbf{H(s)} = \frac{1}{(s-3)(s+2)}?$$

Answer: None. (See S²)

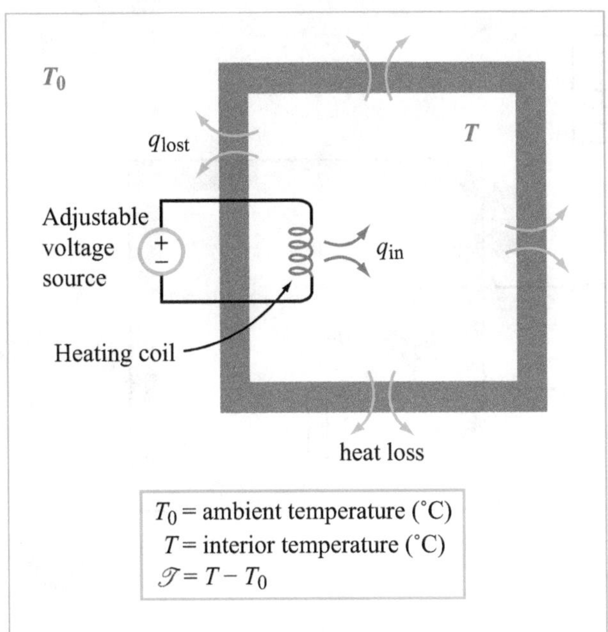

Figure 4-30: Heating coil used to raise the temperature of the interior space of a container relative to the temperature of its outside surroundings, T_0.

4-9 Temperature Control System

The box shown in Fig. 4-30 represents a container whose initial interior temperature is the same as the ambient temperature T_0 of the exterior space surrounding the container. The box might contain a bacterial culture whose growth rate depends on its temperature, a chemical compound whose synthesis requires careful temperature control, or any space whose temperature is to be monitored and controlled. The goal is to design an actuator system that uses an adjustable heat source affixed to an interior wall of the container to raise the container's interior temperature from T_0 to a selectable final value T_∞ and to do so fairly quickly. We will examine two system configurations for realizing the stated goal: one without feedback (open-loop mode) and another with feedback (closed-loop mode). Before we do so, however, we will derive the LCCDE relating the heat flow supplied by the heater to the temperature of the container's interior, $T(t)$.

4-9.1 Heat Transfer Model

The heat transfer quantities of interest are the following:

- T_0 = ambient temperature of the box's exterior space, and also the initial temperature of its interior space.
- T_∞ = the final temperature to which the interior space will be raised (selected by the operator).
- $\Delta T = T_\infty - T_0$ = desired temperature rise.
- $T(t)$ = temperature of the container's interior space as a function of time with $T(0) = T_0$ and $T(\infty) = T_\infty$.
- $\mathcal{T}(t) = T(t) - T_0$ = temperature of container's interior *relative* to the ambient temperature. That is, $\mathcal{T}(t)$ is a ***temperature deviation*** with initial and final values $\mathcal{T}(0) = 0$ and $\mathcal{T}(\infty) = T_\infty - T_0$.
- Q = ***enthalpy*** (heat content in joules) of the interior space of the container (including its contents), defined *relative* to that of the exterior space. That is, $Q = 0$ represents the equilibrium condition when $T = T_0$ or, equivalently, $\mathcal{T} = 0$.
- C = ***heat capacity*** of the container's interior space, measured in joules/ °C.
- R = thermal resistance of the interface (walls) between the container's interior and exterior measured in °C per watt (°C/W).
- $q = \frac{dQ}{dt}$ = the rate of heat flow in joules/s or, equivalently, watts (W).

The amount of heat required to raise the temperature of the container's interior by $\mathcal{T}$ is

$$Q = C\mathcal{T}. \tag{4.114}$$

The rate of heat flow q is analogous to electric current i in electric circuits. For the scenario depicted in Fig. 4-31:

- q_{in} = rate of heat flow supplied by the source.
- q_{lost} = rate of heat flow loss to the outside through the walls.
- q_{abs} = rate of heat flow absorbed by the air and contents of the container's interior, raising their temperature.

For an interface with thermal resistance R,

$$q_{\text{lost}} = \frac{T - T_0}{R} = \frac{\mathcal{T}}{R} \quad \longleftrightarrow \quad i = \frac{\upsilon}{R}. \tag{4.115}$$

In an electrothermal analogue, temperature deviation $\mathcal{T}$ is equivalent to voltage υ (referenced to ground) in an electric circuit. Hence, Eq. (4.115) is the analogue of Ohm's law.

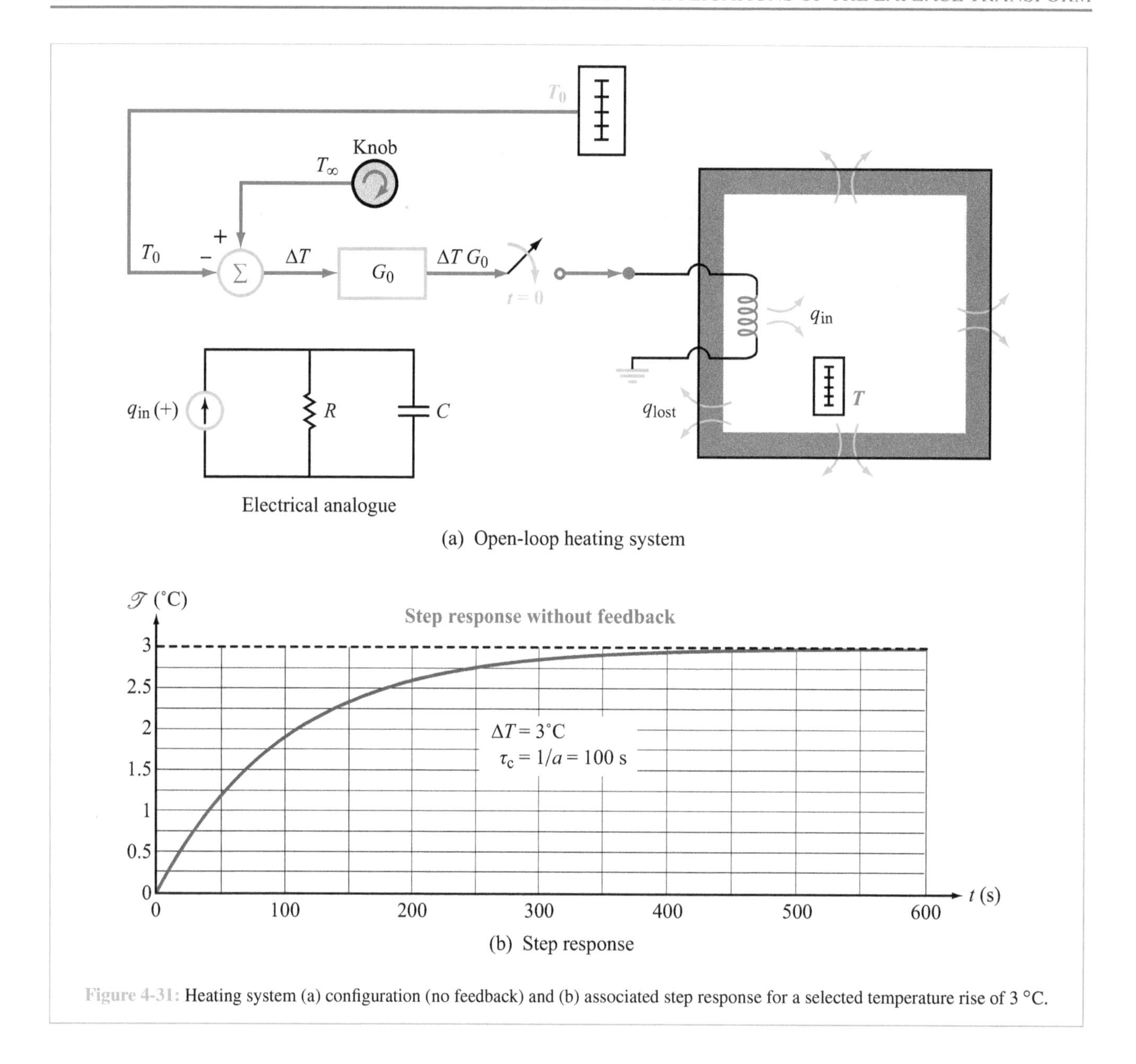

Figure 4-31: Heating system (a) configuration (no feedback) and (b) associated step response for a selected temperature rise of 3 °C.

Differentiating Eq. (4.114) shows that the rate of heat flow absorbed by the container's interior is equal to the increase of its heat content:

$$q_{\text{abs}} = \frac{dQ}{dT} = C\,\frac{d\mathcal{T}}{dt} \quad \longleftrightarrow \quad i = C\,\frac{dv}{dt}, \tag{4.116}$$

which is analogous to the i-v relationship for a capacitor. Conservation of energy dictates that, analogous to KCL,

$$q_{\text{abs}} + q_{\text{lost}} = q_{\text{in}}$$

or, equivalently,

$$C\,\frac{d\mathcal{T}}{dt} + \frac{\mathcal{T}}{R} = q_{\text{in}}(t). \tag{4.117}$$

The electrical analogue of the ***thermal circuit*** consists of a parallel RC circuit connected to a current source $q_{\text{in}}(t)$, as shown in Fig. 4-31(a).

4-9.2 Step Response without Feedback

In Fig. 4-31(a), the exterior ambient temperature T_0 is measured by a temperature sensor, and the desired final temperature T_∞ is selected by the user. The difference between the two temperatures, $\Delta T = (T_\infty - T_0)$, is multiplied by a constant scaling factor G_0 whose output provides the electric current necessary to heat the coil. As we will see shortly, the value of G_0 is selected such that after closing the switch the temperature $T(t)$ will rise from T_0 to an asymptotic level T_∞. Closing the switch is equivalent to establishing a step-function input:

$$q_{\text{in}}(t) = \Delta T \; G_0 \; u(t). \tag{4.118}$$

If we divide all terms in Eq. (4.117) by C and define

$$a = \frac{1}{RC},$$

the differential equation assumes the form

$$\frac{d\mathcal{T}}{dt} + a\mathcal{T} = \frac{1}{C}\, q_{\text{in}}(t). \tag{4.119}$$

For $q_{\text{in}}(t)$ as given by Eq. (4.118), the solution of this first-order differential equation with initial condition T_0 is

$$\mathcal{T}(t) = \frac{\Delta T \; G_0}{aC}\, [1 - e^{-at}]\, u(t).$$

By selecting the scaling factor to be

$$G_0 = aC,$$

and making the substitutions $\mathcal{T}(t) = T(t) - T_0$ and $\Delta T = T_\infty - T_0$, the expression for $\mathcal{T}(t)$ reduces to

$$T(t) - T_0 = (T_\infty - T_0)[1 - e^{-at}]\, u(t). \tag{4.120}$$

(without feedback)

As $t \to \infty$, $T(t) \to T_\infty$, which satisfies the requirement that the temperature inside the container should rise from T_0 to T_∞. Figure 4-31(b) displays a plot of Eq. (4.120) for $\Delta T = 3$ °C and $a = 10^{-2}$ s^{-1}. The constant a, which is the reciprocal of the product of C (the heat capacity of the air and interior of the container) and R (the thermal resistance of the container's walls), is the inverse of the time constant of the circuit, τ_c. Hence, $\tau_c = 100$ s, which means that it takes about $4.6\tau_c = 460$ s $= 7.67$ minutes for the temperature to rise to 99% of the ultimate rise of 3 °C. Since the value of a is constrained by physical attributes of the system, it is not possible to speed up the heating response time of the system as currently configured. Next, we will show how to shorten the effective time constant of the system through the application of feedback.

4-9.3 Step Response with Feedback

Through the use of a temperature sensor placed inside the container, we can measure the inside temperature $T(t)$, as shown in Fig. 4-32. Upon subtracting T_0 from $T(t)$, we generate the temperature deviation $\mathcal{T}(t)$, to serve as input to a proportional feedback transfer function K. In preparation for incorporating the feedback loop, we convert symbols to our standard signals and systems notation and transform them to the **s**-domain:

$$\mathcal{T}(t) = y(t) \;\longrightarrow\; \mathbf{Y(s)},$$

$$\frac{1}{C}\, q_{\text{in}}(t) = x(t) \;\longrightarrow\; \mathbf{X(s)}.$$

The **s**-domain equivalent of Eq. (4.119) is then

$$\mathbf{s}\,\mathbf{Y(s)} + a\,\mathbf{Y(s)} = \mathbf{X(s)}$$

from which we obtain the transfer function

$$\mathbf{H(s)} = \frac{\mathbf{Y(s)}}{\mathbf{X(s)}} = \frac{1}{\mathbf{s} + a} \quad \textbf{(without feedback)}. \tag{4.121}$$

This is the transfer function of the heating system without feedback. Upon incorporating the heating system into a negative feedback loop, as shown in Fig. 4-32(a), we obtain the closed-loop transfer function

$$\mathbf{Q(s)} = \frac{\mathbf{H(s)}}{1 + K\,\mathbf{H(s)}} = \frac{1/(\mathbf{s}+a)}{1 + K/(\mathbf{s}+a)} = \frac{1}{\mathbf{s} + (a+K)} = \frac{1}{\mathbf{s}+b}, \tag{4.122a}$$

(with proportional feedback)

where

$$b = a + K. \tag{4.122b}$$

For the step-function input given by Eq. (4.118), the system input is

$$x(t) = \frac{1}{C}\, q_{\text{in}}(t) = \left(\frac{\Delta T\, G_0}{C}\right) u(t).$$

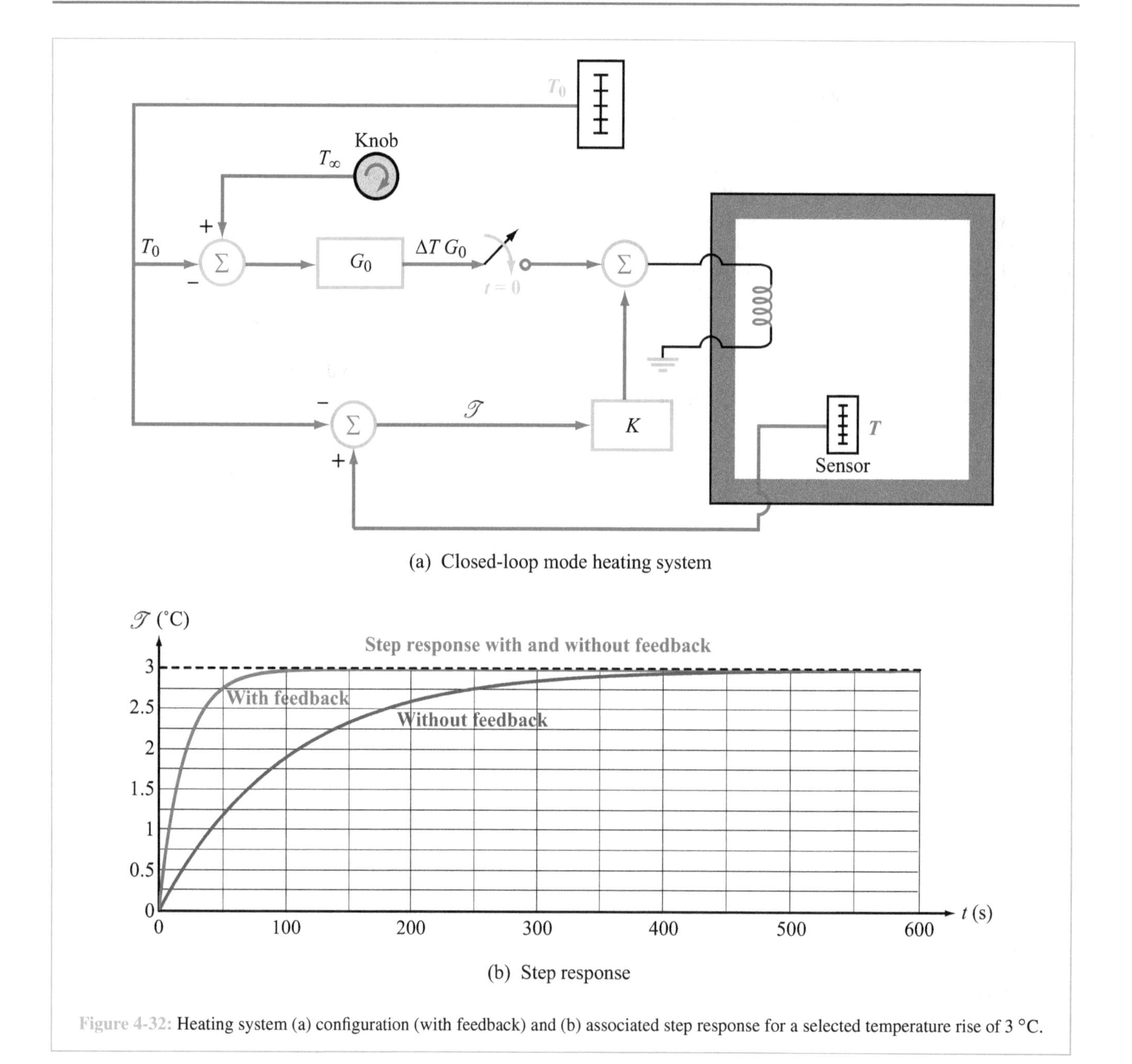

Figure 4-32: Heating system (a) configuration (with feedback) and (b) associated step response for a selected temperature rise of 3 °C.

The Laplace transform of $x(t)$ is

$$\mathbf{X(s)} = \left(\frac{\Delta T\ G_0}{C}\right)\frac{1}{\mathbf{s}},$$

and the output step response is

$$\mathbf{Y(s)} = \mathbf{Q(s)}\ \mathbf{X(s)} = \left(\frac{\Delta T\ G_0}{C}\right)\frac{1}{\mathbf{s}(\mathbf{s}+b)}.$$

Partial fraction expansion leads to

$$\mathbf{Y(s)} = \left(\frac{\Delta T\ G_0}{bC}\right)\left[\frac{1}{\mathbf{s}} - \frac{1}{\mathbf{s}+b}\right], \tag{4.123}$$

which has the inverse Laplace transform

$$\mathcal{T}(t) = \frac{\Delta T\ G_0}{bC}\ [1 - e^{-bt}]\ u(t). \tag{4.124}$$

By selecting the scaling factor to be

$$G_0 = bC,$$

and making the substitution $\mathcal{T}(t) = T(t) - T_0$ and $\Delta T = T_\infty - T_0$, the expression for $\mathcal{T}(t)$ reduces to

$$T(t) - T_0 = (T_\infty - T_0)[1 - e^{-bt}]\, u(t). \tag{4.125}$$

(with feedback)

We observe that Eq. (4.125) is identical in form with Eq. (4.120), except for an important difference, namely the coefficient of the exponential term. The introduction of the feedback loop changes the coefficient from a to $b = a + K$, as evident in Eq. (4.122a).

Figure 4-32(b) displays $\mathcal{T}(t)$ for both the open-loop (no-feedback) and closed-loop (with-feedback) modes, with $a = 10^{-2}$ s^{-1} and $K = 4 \times 10^{-2}$ s^{-1}. Whereas without feedback it takes 7.67 minutes for $T(t)$ to rise to within 1% of 3 °C, it takes only 20% of that time (1.53 min) to reach that level when feedback is used with $K = 4a$.

Can we make K larger and speed the process even more? The answer is yes, as long as the coil can tolerate the initial current (at $t = 0$) flowing through it, which is proportional to the amplitude of the input step function. That amplitude is $\Delta T a$ for the open-loop mode, as compared with $\Delta T b = \Delta T\ (a+K)$ for the closed-loop mode.

Concept Question 4-9: What is the thermal analogue of capacitance? What is the electrical analogue of an oven warming up? (See S²)

Exercise 4-13: What is the time constant of an oven whose heat capacity is 20 J/ °C and thermal resistance is 5 °C/W?

Answer: 100 s. (See S²)

Exercise 4-14: What is the closed-loop time constant when feedback with $K = 0.04$ s^{-1} is used on the oven of Exercise 4-13?

Answer: 20 s. (See S²)

Exercise 4-15: Use LabVIEW Module 4.1 to compute the oven temperature responses shown in Fig. 4-32, using values given in the text.

Answer:

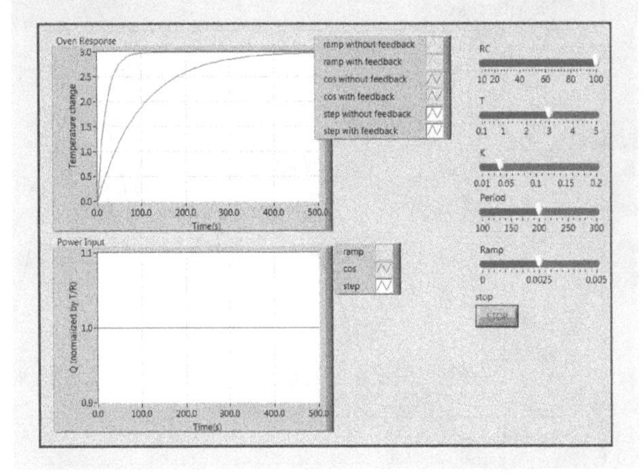

4-10 Amplifier Gain-Bandwidth Product

4-10.1 Open-Loop Mode

According to the manufacturer's specification sheet, the popular μA741 op amp has a dc gain of 106 dB (equivalently, $G_0 = 2 \times 10^5$) and a half-power bandwidth of 8 Hz (equivalently, an angular frequency $\omega_c = 2\pi \times 8 \approx 50$ rad/s). This means that, when operated in open-loop mode, the op amp's gain $|\mathbf{G}(\omega)|$ has a spectrum that starts at $G_0 = 2 \times 10^5$ at $\omega = 0$ and then decreases asymptotically with increasing ω in a manner similar to that shown in Fig. 4-33. The gain function $|\mathbf{G}(\omega)|$ represents the ratio of the output voltage $\mathbf{V}_o$ to the input voltage $\mathbf{V}_i$, so the power ratio is equal to $|\mathbf{G}(\omega)|^2$. The specified value of 50 rad/s for the half-power angular frequency ω_c means that

$$\frac{|\mathbf{G}(50)|^2}{G_0^2} = \frac{1}{2} \tag{4.126}$$

or

$$|\mathbf{G}(50)| = \frac{G_0}{\sqrt{2}} = 0.707 G_0, \tag{4.127}$$

as noted in Fig. 4-33. The frequency response of $\mathbf{G}(\omega)$ is modeled by the function

$$\mathbf{G}(\omega) = \frac{\mathbf{V}_o}{\mathbf{V}_i} = \frac{G_0}{1 + j\omega/\omega_c} \tag{4.128}$$

Module 4.1 Oven Temperature Response The input can be a step, ramp, or sinusoid. The RC time constant of the oven, the desired temperature change T, the period of the sinusoidal input, the slope of the ramp input, and the gain K of the feedback are all selectable parameters.

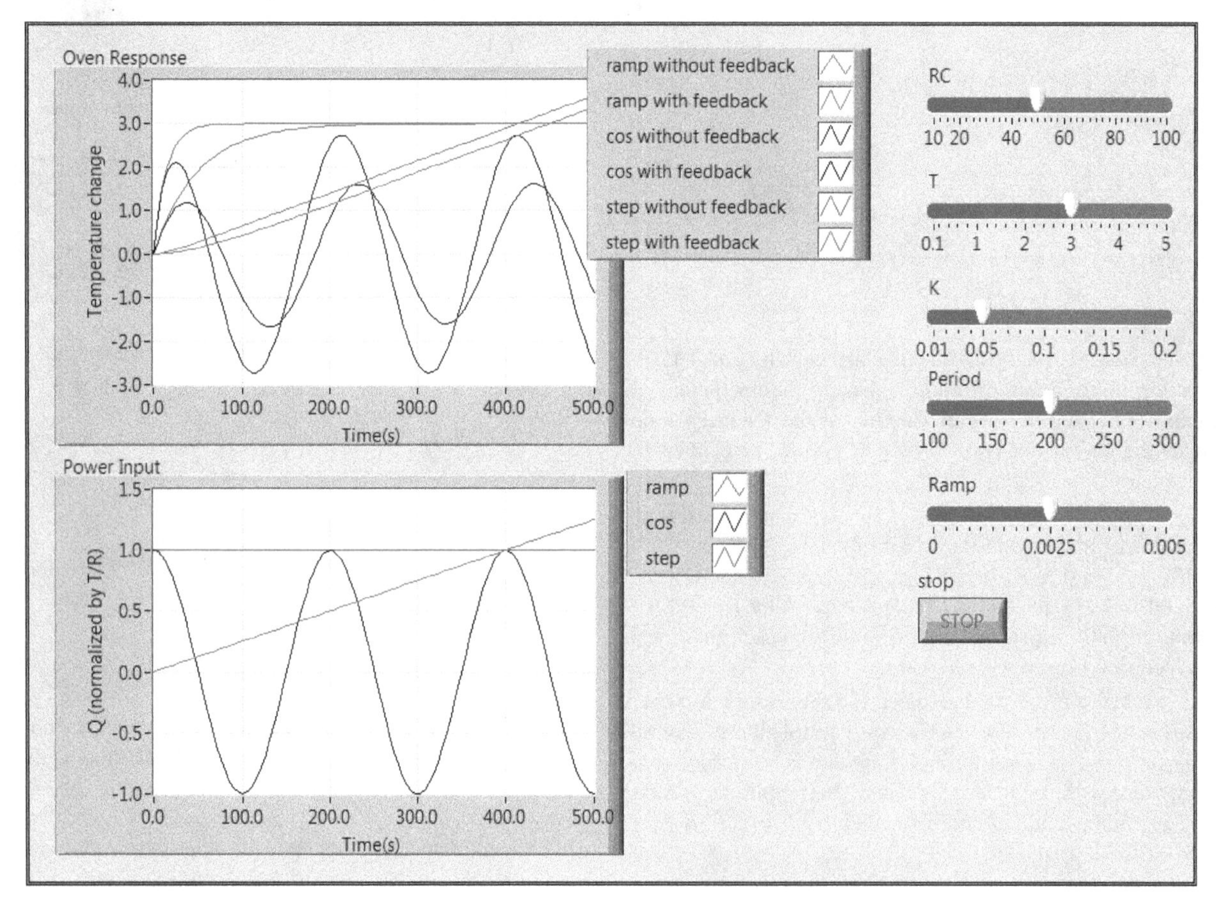

with $G_0 = 2 \times 10^5$ and $\omega_c = 50$ rad/s. At $\omega = \omega_c$, Eq. (4.128) satisfies the half-power condition given by Eq. (4.126), and at $\omega = 0$, it satisfies $\mathbf{G}(0) = G_0$.

The product of G_0 and ω_c is called the amplifier's *gain-bandwidth product*:

$$\text{GBP} = G_0\omega_c = 2 \times 10^5 \times 50 = 10^7 \qquad \textbf{(open-loop mode)}.$$

4-10.2 Closed-Loop Mode

When part of the output signal $\mathbf{V}_o$ is fed back to the negative terminal of the input side of the op amp, the circuit operation is equivalent to the feedback configuration shown in Fig. 4-33(b) with

$$K = \frac{R_i}{R_f}. \tag{4.129}$$

The frequency response function of the closed-loop system is

$$\begin{aligned}\mathbf{Q}(\omega) = \frac{\mathbf{G}(\omega)}{1 + K\,\mathbf{G}(\omega)} &= \frac{G_0/(1 + j\omega/\omega_c)}{1 + \dfrac{KG_0}{1 + j\omega/\omega_c}} \\ &= \frac{G_0\omega_c}{j\omega + (\omega_c + G_0\omega_c K)} \\ &= \frac{10^7}{j\omega + (50 + 10^7 K)}. \end{aligned} \tag{4.130}$$

If $K = 0$, which corresponds to no feedback, the closed-loop expression given by Eq. (4.130) reduces to the open-loop mode described by Eq. (4.128).

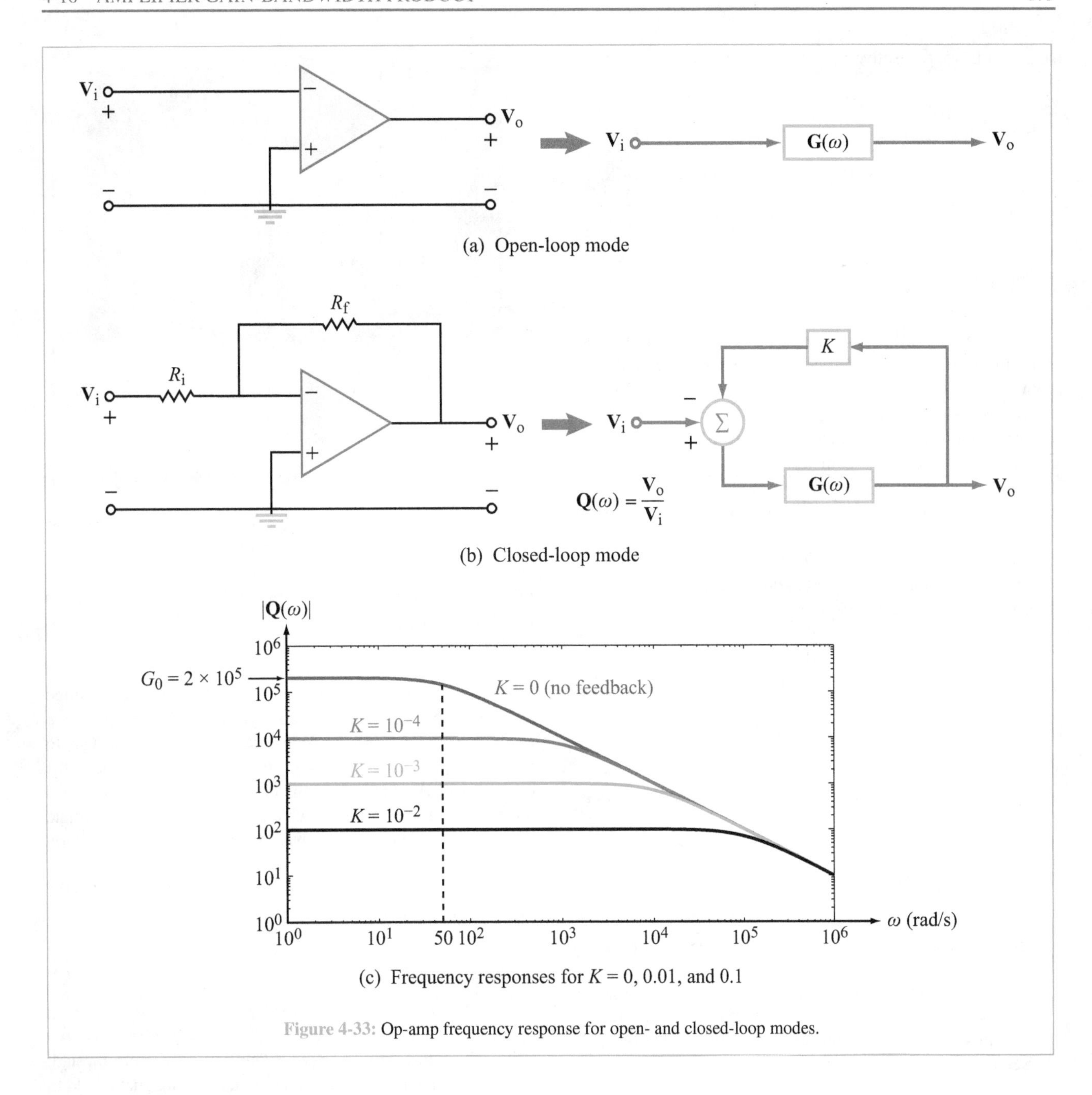

Figure 4-33: Op-amp frequency response for open- and closed-loop modes.

To observe the effect of feedback, we present in Fig. 4-33(c) plots of $|\mathbf{Q}(\omega)|$ for four values of K, namely $K = 0, 0.01, 0.1$, and 1. We observe that as the value of K is increased from 0 (no feedback) to 1, the dc gain decreases, but the bandwidth increases. By rewriting Eq. (4.130) in a form similar to that of Eq. (4.128), namely,

$$\mathbf{Q}(\omega) = \frac{10^7/(50 + 10^7 K)}{1 + j\omega/(50 + 10^7 K)}, \tag{4.131}$$

we deduce from the numerator and the second term in the denominator that the closed-loop dc gain and half-power bandwidth are, respectively,

$$G'_0 = \frac{10^7}{50 + 10^7 K} \approx \frac{1}{K} \qquad \text{(closed-loop dc gain)},$$

$$\omega'_c = (50 + 10^7 K) \approx 10^7 K \quad \text{(closed-loop bandwidth)}.$$

Hence, the gain-bandwidth product remains

$$\text{GBP} = G'_0 \omega'_c = 10^7 \qquad \text{(closed-loop mode)}.$$

The implication of this result (the gain-bandwidth product is a constant) is that feedback provides a mechanism for trading off gain for bandwidth, and vice versa. The degree of trade-off is controlled by the value of K (equivalently, the ratio R_i/R_f).

Another advantage accrued by the use of feedback is the reduced sensitivity to dc gain variations. If changes in the op amp's environmental temperature were to cause its dc gain G_0 to increase from 2×10^5 to 4×10^5, for example, in open-loop mode the $\times 2$ increase will be reflected in the level of the output voltage $\mathbf{V}_o$, which may distort the signal passing through the op-amp circuit or cause problems in succeeding circuits. In contrast, when feedback is used to increase the circuit's bandwidth, which means that $K \gg 50 \times 10^{-7}$, the expression given by Eq. (4.130) simplifies to

$$\mathbf{Q}(\omega) \approx \frac{10^7}{j\omega + 10^7 K}.$$

At the usable part of the spectrum corresponding to $\omega \ll 10^7 K$, $\mathbf{Q}(\omega)$ becomes

$$\mathbf{Q}(\omega) \approx \frac{1}{K} \qquad (\omega \ll 10^7 K),$$

which is independent of the dc gain G_0, thereby providing immunity to dc gain variations with temperature.

Concept Question 4-10: How does feedback provide immunity to op-amp dc gain variations with temperature? (See S²)

Exercise 4-16: In open-loop mode, an op-amp circuit has a gain of 100 dB and half-power bandwidth of 32 Hz. What will the gain and bandwidth be in closed-loop mode with $K = 0.01$?

Answer: 40 dB and 32 kHz. (See S²)

Figure 4-34: Motor with a shaft rotating at angular velocity ω.

4-11 Step Response of a Motor System

Motors are used in a seemingly endless list of applications, from power tools and household appliances, to motorbikes, automobiles, and jet engines. Their basic principle of operation relies on converting electrical energy into mechanical energy to rotate a shaft that, in turn, rotates an object of interest, such as a wheel or a drill bit (Fig. 4-34).

In this section, we examine two different types of motor applications, which from a system perspective can be classified as open-loop and closed-loop configurations. In the open-loop configuration, an applied voltage causes the motor's shaft to rotate at a proportional angular velocity ω. An example is the electric drill. An example of the closed-loop configuration is a telescope pointing system in which the objective is not to rotate the telescope continuously, but rather to rotate its direction *quickly* and *smoothly* from whatever initial direction it was pointing along to a new specified direction associated with the location of a particular star of interest.

4-11.1 Motor Model

Figure 4-35 is a simplified equivalent circuit of a motor driven by an external voltage source $\upsilon(t)$. The motor consists of two coils called its ***stator*** and ***rotor***. The flow of current through its stator, which remains stationary, induces a magnetic field in the space occupied by the rotor, which in turn induces a ***torque*** $\tau(t)$, causing the rotor to rotate. A load attached to the rotor shaft rotates at the ***angular velocity of the shaft***, ω.

In the equivalent circuit of Fig. 4-35, resistance R accounts for the resistances of the voltage source and the motor coils, and $\upsilon_{\text{emf}}(t)$ is an induced electromotive force (voltage) called "back emf" because it acts in opposition to the applied voltage

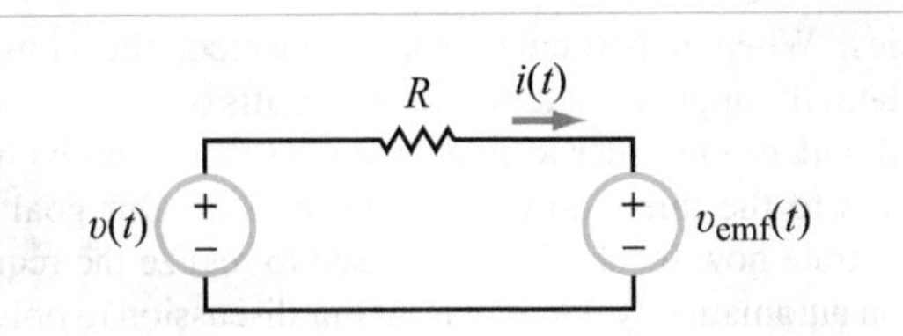

Figure 4-35: Equivalent circuit of a motor driven by a voltage source $\upsilon(t)$.

$\upsilon(t)$. The inductance associated with the coils has been ignored. The loop equation is

$$\upsilon(t) - \upsilon_{\rm emf} - R\, i(t) = 0. \tag{4.132}$$

By Faraday's law, the induced emf is directly proportional to the angular velocity ω, which by definition is the time derivative of the ***shaft rotation angle*** θ. That is,

$$\upsilon_{\rm emf}(t) = c_1 \omega = c_1 \frac{d\theta}{dt}, \tag{4.133}$$

where c_1 is a motor-specific constant.

By Ampère's law, the induced magnetic field is proportional to the current $i(t)$. Moreover, the torque is proportional to the magnetic-field strength. Hence, current and torque are linearly related as

$$\tau(t) = c_2\, i(t), \tag{4.134}$$

where c_2 is another motor-specific constant. Also, the rotational version of the force equation ($f = ma$) is

$$\tau(t) = J \frac{d^2\theta}{dt^2}, \tag{4.135}$$

where J is the ***moment of inertia*** of the load attached to the rotor shaft. Combining Eqs. (4.134) and (4.135) leads to

$$i(t) = \frac{J}{c_2} \frac{d^2\theta}{dt^2}. \tag{4.136}$$

Inserting Eqs. (4.133) and (4.136) into Eq. (4.132) leads to

$$\frac{d^2\theta}{dt^2} + a \frac{d\theta}{dt} = b\, \upsilon(t), \tag{4.137}$$

where $a = (c_1 c_2)/(RJ)$ and $b = c_2/(RJ)$. From here on forward, we treat a and b as positive-valued system constants related to the physical characteristics of the motor. Our task is to examine the second-order LCCDE given by Eq. (4.137) with

$$\upsilon(t) = \text{ input signal}$$

and

$$\theta(t) = \text{ output response (measured in radians relative to a reference plane (Fig. 4-34)).}$$

4-11.2 Open-Loop Configuration

The **s**-domain counterpart of Eq. (4.137) is

$$\mathbf{s}^2\, \boldsymbol{\theta}(\mathbf{s}) + a\mathbf{s}\, \boldsymbol{\theta}(\mathbf{s}) = b\mathbf{V}(\mathbf{s}), \tag{4.138}$$

and the ***motor transfer function*** is

$$\mathbf{H}(\mathbf{s}) = \frac{\boldsymbol{\theta}(\mathbf{s})}{\mathbf{V}(\mathbf{s})} = \frac{b}{\mathbf{s}^2 + a\mathbf{s}} = \frac{b}{\mathbf{s}(\mathbf{s}+a)}. \tag{4.139}$$

The transfer function has a pole at $\mathbf{s} = 0$ and another at $\mathbf{s} = -a$. Consequently, a motor is an unstable system because according to our discussion in Section 3-7, a system is BIBO unstable if its transfer function has a pole at the origin in the **s**-domain. Physically, this conclusion makes sense, because applying a constant voltage to the motor will cause it to rotate its shaft at a constant rate, increasing its angle $\theta(t)$ indefinitely. This assertion can be verified by calculating $y(t)$ in response to a step input $\upsilon(t) = u(t)$ or, equivalently, $\mathbf{V}(\mathbf{s}) = 1/\mathbf{s}$. From Eq. (4.139), the step response is

$$\boldsymbol{\theta}(\mathbf{s}) = \mathbf{H}(\mathbf{s})\, \mathbf{V}(\mathbf{s}) = \frac{b}{\mathbf{s}(\mathbf{s}+a)} \cdot \frac{1}{\mathbf{s}} = \frac{b}{\mathbf{s}^2(\mathbf{s}+a)}. \tag{4.140}$$

Partial fraction expansion gives

$$\boldsymbol{\theta}(\mathbf{s}) = \frac{b/a}{\mathbf{s}^2} - \frac{b/a^2}{\mathbf{s}} + \frac{b/a^2}{\mathbf{s}+a}, \tag{4.141}$$

and using Table 3-2, the time-domain counterpart of $\boldsymbol{\theta}(\mathbf{s})$ is

$$\theta(t) = \frac{b}{a^2}\, [at - 1 + e^{-at}]\, u(t) \qquad \text{(radians)}. \tag{4.142}$$

The rotational angular velocity of the shaft is

$$\omega_{\rm rad/s} = \frac{d\theta}{dt} = \frac{b}{a}\, (1 - e^{-at})\, u(t) \qquad \text{(rad/s)}. \tag{4.143}$$

Converting ω in (rad/s) to revolutions per minute (with 1 revolution = 2π radians) gives

$$\omega_{\rm rev/m} = \frac{60}{2\pi}\, \omega_{\rm rad/s} = \frac{30b}{\pi a}\, (1 - e^{-at})\, u(t). \tag{4.144}$$

If the motor is designed such that $a \approx 3$ or greater, the second term becomes negligible compared to the first within 1 second or less, thereby simplifying the result to

$$\omega_{\rm rev/m} \approx \frac{30b}{\pi a} \qquad (t > 3/a \text{ seconds}). \tag{4.145}$$

Through proper choice of the motor physical parameters, the ratio b/a can be selected to yield the desired value for ω.

In summary, when a motor is operated in an open-loop configuration, the application of a step voltage causes the shaft to reach a constant angular velocity ω within a fraction of a second (if $a > 3$).

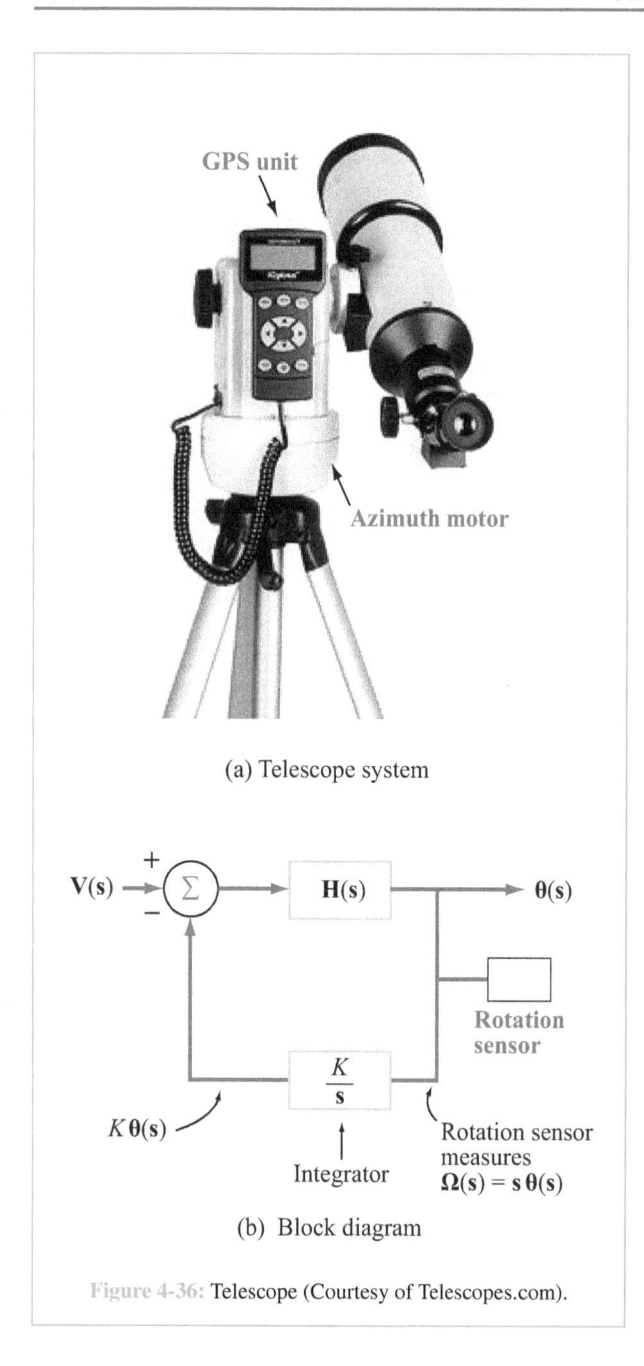

(a) Telescope system

(b) Block diagram

Figure 4-36: Telescope (Courtesy of Telescopes.com).

4-11.3 Closed-Loop Configuration

The telescope automatic pointing system shown in Fig. 4-36(a) uses two motors: one to rotate the telescope in elevation and another to rotate it in azimuth. Its global positioning system (GPS) determines the telescope's coordinates, and its computer contains celestial coordinates for a large number of stars and galaxies. When a particular star is selected, the computer calculates the angular rotations that the shafts of the two motors should undergo in order to move the telescope from its initial direction to the direction of the selected star. Our goal is to demonstrate how feedback can be used to realize the required rotation automatically. We will limit our discussion to only one of the motors.

The feedback loop is shown in Fig. 4-36(b). A magnetic sensor positioned close to the motor's shaft measures the angular velocity ω. Since $\omega = d\theta/dt$ and differentiation in the time domain is equivalent to multiplication by **s** in the **s**-domain, the **s**-domain quantity measured by the sensor is

$$\mathbf{\Omega}(\mathbf{s}) = \mathbf{s}\,\boldsymbol{\theta}(\mathbf{s}). \tag{4.146}$$

To perform the desired feedback, we need to feed $K\,\boldsymbol{\theta}(\mathbf{s})$ into the negative terminal of the summer. Hence, an integrator circuit with transfer function $K/\mathbf{s}$ is used in the feedback arm of the loop. The net result of the process is equivalent to using a feedback loop with $\mathbf{G}(\mathbf{s}) = K$. Consequently, the closed-loop transfer function is

$$\mathbf{Q}(\mathbf{s}) = \frac{\mathbf{H}(\mathbf{s})}{1 + K\,\mathbf{H}(\mathbf{s})}. \tag{4.147}$$

Use of Eq. (4.139) for $\mathbf{H}(\mathbf{s})$ leads to

$$\mathbf{Q}(\mathbf{s}) = \frac{b/(\mathbf{s}^2 + a\mathbf{s})}{1 + Kb/(\mathbf{s}^2 + a\mathbf{s})} = \frac{b}{\mathbf{s}^2 + a\mathbf{s} + bK}. \tag{4.148}$$

The expression for $\mathbf{Q}(\mathbf{s})$ can be rewritten in the form

$$\mathbf{Q}(\mathbf{s}) = \frac{b}{(\mathbf{s} - p_1)(\mathbf{s} - p_2)}, \tag{4.149}$$

where poles p_1 and p_2 are given by

$$p_1 = -\frac{a}{2} + \sqrt{\left(\frac{a}{2}\right)^2 - bK}, \tag{4.150a}$$

$$p_2 = -\frac{a}{2} - \sqrt{\left(\frac{a}{2}\right)^2 - bK}. \tag{4.150b}$$

By choosing $K > 0$, we ensure that the poles p_1 and p_2 are in the open left half-plane, thereby guaranteeing BIBO stability.

The step response is obtained by calculating $\boldsymbol{\theta}(\mathbf{s})$ for $\upsilon(t) = u(t)$ or, equivalently, $\mathbf{V}(\mathbf{s}) = 1/\mathbf{s}$:

$$\boldsymbol{\theta}(\mathbf{s}) = \mathbf{Q}(\mathbf{s})\,\mathbf{V}(\mathbf{s}) = \frac{b}{\mathbf{s}(\mathbf{s} - p_1)(\mathbf{s} - p_2)}. \tag{4.151}$$

Partial fraction expansion leads to the following equivalent expression for $\boldsymbol{\theta}(\mathbf{s})$:

$$\boldsymbol{\theta}(\mathbf{s}) = \left(\frac{b}{p_1 p_2}\right)\frac{1}{\mathbf{s}} + \frac{b}{p_1(p_1-p_2)} \cdot \frac{1}{(\mathbf{s}-p_1)} + \frac{b}{p_2(p_2-p_1)} \cdot \frac{1}{(\mathbf{s}-p_2)}. \tag{4.152}$$

Its time-domain equivalent is

$$\theta(t) = \left[\frac{b}{p_1 p_2} + \frac{b}{p_1(p_1-p_2)}\, e^{p_1 t} + \frac{b}{p_2(p_2-p_1)}\, e^{p_2 t}\right] u(t). \tag{4.153}$$

If p_1 and p_2 are both in the open left half-plane, the second and third term will decay to zero as a function of time, leaving behind

$$\lim_{t\to\infty} \theta(t) = \frac{b}{p_1 p_2}. \tag{4.154}$$

Use of the expressions for p_1 and p_2 given by Eq. (4.150) in Eq. (4.154) leads to

$$\lim_{t\to\infty} \theta(t) = \frac{1}{K}. \tag{4.155}$$

If the magnitude of K is on the order of inverse seconds, the final value of $\theta(t)$ is approached very quickly. This final value is equal to the total angular rotation required to move the telescope from its original direction to the designated new direction.

Note that Eq. (4.154) can also be obtained directly from Eq. (4.148) using the final value theorem of the Laplace transform. Application of property #12 in Table 3-1 gives

$$\lim_{t\to\infty} \theta(t) = \lim_{\mathbf{s}\to 0} \mathbf{s}\,\boldsymbol{\theta}(\mathbf{s}) = \mathbf{s}\left.\frac{b}{\mathbf{s}(\mathbf{s}^2+a\mathbf{s}+bK)}\right|_{\mathbf{s}=0} = \frac{1}{K} \tag{4.156}$$

provided the system is BIBO stable. The open-loop system given by Eq. (4.139) is BIBO-unstable because one of the poles is at the origin. But the closed-loop system is stable if $K > 0$. Moreover, the closed-loop steady-state unit step response is $\theta(\infty) = 1/K$, independent of a and b.

Example 4-13: Stabilization of Telescope Pointing System

The system parameters of a telescope pointing system given by Eq. (4.139) are $a = 3$ and $b = 6$. Using a *proportional-plus-derivative (PD) feedback* transfer function $\mathbf{G}(\mathbf{s}) = K_1 + K_2\mathbf{s}$, choose K_1 and K_2 so that the closed-loop system is critically damped with both poles at -6. Also determine the size of the step input required for $\theta(\infty) = 28.65° = 0.5$ rad.

Solution: For $\mathbf{G}(\mathbf{s}) = K_1 + K_2\mathbf{s}$, the closed-loop transfer function is

$$\begin{aligned}\mathbf{Q}(\mathbf{s}) &= \frac{\mathbf{H}(\mathbf{s})}{1+(K_1+K_2\mathbf{s})\,\mathbf{H}(\mathbf{s})} \\ &= \frac{6/(\mathbf{s}^2+3\mathbf{s})}{1+(K_1+K_2\mathbf{s})\left(\dfrac{6}{\mathbf{s}^2+3\mathbf{s}}\right)} \cdot \left[\frac{\mathbf{s}^2+3\mathbf{s}}{\mathbf{s}^2+3\mathbf{s}}\right] \\ &= \frac{6}{\mathbf{s}^2+3\mathbf{s}+6(K_1+K_2\mathbf{s})} \\ &= \frac{6}{\mathbf{s}^2+(3+6K_2)\mathbf{s}+(6K_1)}.\end{aligned} \tag{4.157}$$

The polynomial in the closed-loop system is $\mathbf{s}^2 + (3+6K_2)\mathbf{s} + 6K_1$. The problem statement specifies that the poles of the closed-loop system should be -6 and -6, which means the polynomial should be $(\mathbf{s}+6)^2$. Hence, we must choose K_1 and K_2 so that

$$(\mathbf{s}+6)^2 = \mathbf{s}^2+12\mathbf{s}+36 = \mathbf{s}^2+(3+6K_2)\mathbf{s}+(6K_1). \tag{4.158}$$

Equating coefficients leads to $K_1 = 6$ and $K_2 = 1.5$. An analysis similar to Eq. (4.156) shows that the steady-state unit-step response is $1/K_1 = 1/6$, independent of a and b. Since $x(t) = u(t)$ leads to $\theta(\infty) = 1/6$, the scaling property of LTI systems shows that $x(t) = 3u(t)$ would lead to $\theta(\infty) = 3/6 = 0.5$ rad, as required.

Concept Question 4-11: How do we compute the steady-state step response without performing partial fraction expansion? (See S²)

Exercise 4-17: Compute the steady-state step response $\lim_{t\to\infty} y_{\text{step}}(t)$ for the BIBO stable system with transfer function

$$\mathbf{H}(\mathbf{s}) = \frac{2\mathbf{s}^2+3\mathbf{s}+4}{5\mathbf{s}^3+6\mathbf{s}^2+7\mathbf{s}+8}.$$

Answer: $\lim_{t\to\infty} y_{\text{step}}(t) = 0.5$. (See S²)

Exercise 4-18: In Example 4-13, suppose $a = 101$, $b = 100$ and $K_1 = 1$. Compute K_2 so that the closed-loop system is critically damped using PD feedback.

Answer: $K_2 = -0.81$; $h(t) = 100te^{-10t}\, u(t)$. (See S²)

4-12 Control of a Simple Inverted Pendulum on a Cart

The inverted pendulum is a good illustration of how the feedback control theory developed in this chapter is applied to a (simplified) real-world problem. Topics involved in the control of an inverted pendulum include: (1) linearization of a nonlinear system into an LTI system, so that LTI system theory can be applied to solving the problem; (2) stability analysis of the open-loop system, which turns out to be unstable; (3) realization of the inadequacy of proportional and PD feedback for this particular problem, and (4) introduction of ***PI (proportional-plus-integral)*** control to secure the necessary stability.

The inverted pendulum on a cart is a very crude model of a Segway® people mover. Hence, the analysis to follow illustrates some of the control and stability issues associated with the operation of the Segway system.

(a) Inverted pendulum on wheels

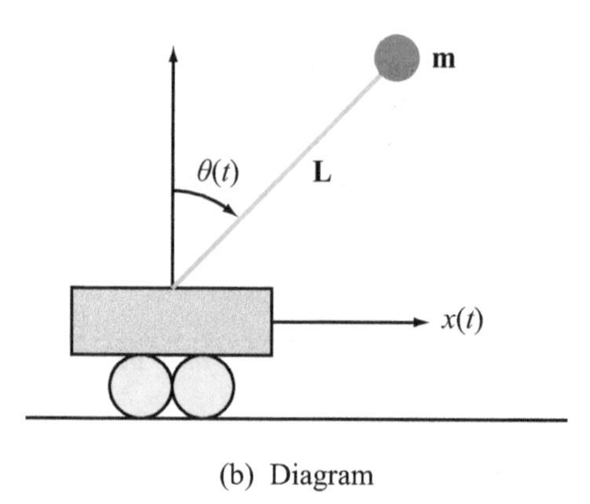

(b) Diagram

Figure 4-37: Inverted pendulum on a cart and its (simplified) equivalent diagram (Courtesy of Chalmers University of Technology).

4-12.1 Basic Physics of Inverted Pendulum

The simplest form of the inverted pendulum is depicted in Fig. 4-37, where we define the following variables:

- Input $x(t)$ is the back-and-forth position of the cart.
- Output $\theta(t)$ is the angle made with the vertical.
- Pendulum has length L and mass m, all at its end. The pendulum thus has moment of inertia mL^2.
- The pendulum has a frictionless hinge at its base.

We note in passing that a pendulum having its mass distributed evenly over its length (instead of all of it at its end) has moment of inertia $\frac{1}{3}mL^2$, not mL^2.

There are three torques around the pendulum hinge acting on the mass m at the end of the pendulum.

(a) The torque $mLg \sin(\theta)$ due to gravity on the mass.

(b) The torque $-mL\cos(\theta)\,(d^2x/dt^2)$ due to the motion of the cart. Note that forward motion of the cart makes the pendulum tend to swing backwards.

(c) The torque $mL^2(d^2\theta/dt^2)$ due to angular acceleration of the pendulum.

Equating these torques around the pendulum hinge gives

$$mL^2\,\frac{d^2\theta}{dt^2} = mLg\sin(\theta) - mL\cos(\theta)\,\frac{d^2x}{dt^2}, \qquad (4.159)$$

where $g = 9.8\text{ m}\cdot\text{s}^{-2}$ is the acceleration of gravity, and the angle θ has units of radians.

4-12.2 Linearization

Equation (4.159) is clearly a non-linear system. We ***linearize*** the system about $\theta = 0$, so that we can apply the techniques for LTI systems. Position $\theta = 0$ means that the pendulum is vertical, so the assumption that θ is small is equivalent to the assumption that the pendulum is close to vertical.

The power series expansions for sine and cosine are given by

$$\sin(\theta) = \theta - \frac{\theta^3}{3!} + \cdots \approx \theta \qquad \text{if } \theta \ll 1 \tag{4.160a}$$

and

$$\cos(\theta) = 1 - \frac{\theta^2}{2!} + \cdots \approx 1 \qquad \text{if } \theta \ll 1. \tag{4.160b}$$

Note that θ is in radians. To illustrate that these are indeed good approximations, consider the case $\theta = 0.2$ rad, which corresponds to a deviation of 11° from the vertical. The values of $\sin(0.2)$ and $\cos(0.2)$ are

$$\sin(0.2) = 0.1987 \approx 0.2,$$

$$\cos(0.2) = 0.9801 \approx 1.$$

Substituting the truncated power series given by Eq. (4.160) into the non-linear system given by Eq. (4.159) gives

$$mL^2 \frac{d^2\theta}{dt^2} = mLg\,\theta(t) - mL(1)\frac{d^2x}{dt^2}. \tag{4.161}$$

Finally, dividing by mL^2 leads to the linearized equation

$$\frac{d^2\theta}{dt^2} = \frac{g}{L}\,\theta(t) - \frac{1}{L}\frac{d^2x}{dt^2}. \tag{4.162}$$

This system model is LTI; in fact, it is an LCCDE. To compute its transfer function, we transform the differential equation to the **s**-domain:

$$\mathbf{s}^2\,\boldsymbol{\theta}(\mathbf{s}) = \frac{g}{L}\,\boldsymbol{\theta}(\mathbf{s}) - \frac{1}{L}\,\mathbf{s}^2\,\mathbf{X}(\mathbf{s}). \tag{4.163}$$

Solving for $\boldsymbol{\theta}(\mathbf{s})$ and then dividing by $\mathbf{X}(\mathbf{s})$ gives

$$\mathbf{H}(\mathbf{s}) = \frac{\boldsymbol{\theta}(\mathbf{s})}{\mathbf{X}(\mathbf{s})} = -\frac{\mathbf{s}^2/L}{\mathbf{s}^2 - g/L}. \tag{4.164}$$

The system has poles at $\pm\sqrt{g/L}$. Since one of these is in the right half-plane, the system is unstable. Physically, this makes sense: Any departure of the pendulum from the vertical will make it swing farther away from the vertical, until it crashes into a horizontal position on the cart. This initial perturbation could come from a tiny vibration or a tiny breeze that moves the cart slightly, as the following example demonstrates.

Example 4-14: Open-Loop Inverted Pendulum

A cart carrying an inverted pendulum of length 61.25 cm is suddenly moved 1 mm. Compute its response.

Solution: Inserting $L = 61.25$ cm and $g = 9.8\ \text{m}\cdot\text{s}^{-2}$ into Eq. (4.164) gives

$$\mathbf{H}(\mathbf{s}) = \frac{-\mathbf{s}^2/0.6125}{\mathbf{s}^2 - 9.8/0.6125} = \frac{-1.633\mathbf{s}^2}{\mathbf{s}^2 - 16}. \tag{4.165}$$

A sudden movement of length 1 mm means that the input $x(t) = 0.001u(t)$. Its Laplace transform is $\mathbf{X}(\mathbf{s}) = 0.001/\mathbf{s}$, and the corresponding response is

$$\boldsymbol{\theta}(\mathbf{s}) = \mathbf{H}(\mathbf{s})\,\mathbf{X}(\mathbf{s}) = \frac{-1.633\mathbf{s}^2}{\mathbf{s}^2 - 16}\,\frac{0.001}{\mathbf{s}} = \frac{-0.001633\mathbf{s}}{\mathbf{s}^2 - 16}. \tag{4.166}$$

Partial fraction expansion gives

$$\boldsymbol{\theta}(\mathbf{s}) = \frac{-0.001633\mathbf{s}}{(\mathbf{s}-4)(\mathbf{s}+4)} = \frac{-0.000817}{\mathbf{s}-4} + \frac{-0.000817}{\mathbf{s}+4}. \tag{4.167}$$

The inverse Laplace transform of $\boldsymbol{\theta}(\mathbf{s})$ is

$$\theta(t) = -0.000817(e^{4t} + e^{-4t})\,u(t). \tag{4.168}$$

The pendulum topples over in less than two seconds, since $-0.000817e^8 = -2.435$ rad $= -139.5° < -90°$.

Note that forward motion of the cart makes the pendulum fall backwards. That is why the angle $\theta(t) < 0$.

4-12.3 Control Using Proportional Feedback

We try to stabilize the system using feedback: The cart position $x(t)$ is moved back and forth in an attempt to keep the pendulum from tipping over, like balancing a broomstick on a finger by moving the finger in response to the perceived angle the broomstick makes with the vertical. "Proportional" feedback means feeding back $K\,\theta(t)$ to the input, as we have seen before in Section 4-11.

Physically, the output $\theta(t)$ can be measured by placing one terminal of a rheostat on the hinge and the other terminal on the pendulum. The resistance of the tapped rheostat is directly proportional to $\theta(t) + \frac{\pi}{2}$. If $\theta(t) = -\frac{\pi}{2}$, then the resistance is zero, and if $\theta(t) = \frac{\pi}{2}$ the resistance is maximum. The resistance can then be used to control a motor attached to the cart wheels, to control the cart position $x(t)$.

Recall that the equation for proportional feedback control is

$$\mathbf{Q}(\mathbf{s}) = \frac{\mathbf{H}(\mathbf{s})}{1 + K\,\mathbf{H}(\mathbf{s})}. \tag{4.169}$$

Substituting $\mathbf{H}(\mathbf{s})$ into this feedback equation gives

$$\begin{aligned}\mathbf{Q}(\mathbf{s}) &= \frac{\mathbf{H}(\mathbf{s})}{1 + K\,\mathbf{H}(\mathbf{s})} \\ &= -\frac{\mathbf{s}^2/L}{\mathbf{s}^2 - g/L} \Big/ \left(1 - K\frac{\mathbf{s}^2/L}{\mathbf{s}^2 - g/L}\right) \\ &= -\frac{\mathbf{s}^2/L}{\mathbf{s}^2 - g/L - K\mathbf{s}^2/L} = \frac{-\mathbf{s}^2}{(L-K)\mathbf{s}^2 - g}. \end{aligned} \tag{4.170}$$

Note that in proportional feedback, $K\theta$ is added to input $x(t)$, and since $x(t)$ is measured in units of length, K has units of length/radian. If $K < L$ the system will continue to be unstable. But if $K > L$, the poles are pure imaginary, namely $\pm j\sqrt{g/(K-L)}$, in which case the natural response of the system will be oscillatory: The best we can do is to keep the pendulum swinging back and forth, while keeping the amplitude of its swings from getting too large!

Example 4-15: Inverted Pendulum Control Using Proportional Feedback

This is a repeat of Example 4-14, but with proportional feedback. A cart carrying an inverted pendulum of length 61.25 cm is suddenly moved 1 mm. Compute its response if proportional feedback is used with $K = 122.5$ cm/rad.

Solution: Inserting $L = 61.25$ cm, $K = 122.5$ cm/radian, and $g = 9.8\ \text{m}\cdot\text{s}^{-2}$ into the closed-loop transfer function $\mathbf{Q}(\mathbf{s})$ gives

$$\begin{aligned}\mathbf{Q}(\mathbf{s}) &= \frac{-\mathbf{s}^2}{(L-K)\mathbf{s}^2 - g} \\ &= \frac{-\mathbf{s}^2}{(0.6125 - 1.225)\mathbf{s}^2 - 9.8} = \frac{1.633\mathbf{s}^2}{\mathbf{s}^2 + 16}. \end{aligned} \tag{4.171}$$

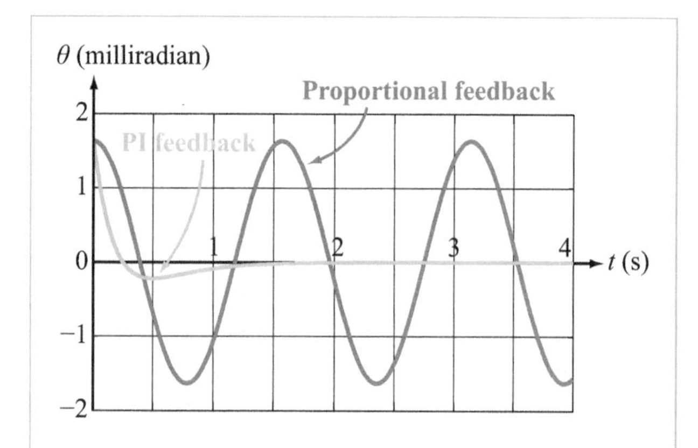

Figure 4-38: Inverted pendulum response to input $0.001u(t)$ using proportional and proportional-plus-integral (PI) feedback.

A sudden movement of length 1 mm means that the input $x(t) = 0.001u(t)$. Its Laplace transform is $\mathbf{X}(\mathbf{s}) = 0.001/\mathbf{s}$, and

$$\boldsymbol{\theta}(\mathbf{s}) = \mathbf{Q}(\mathbf{s})\,\mathbf{X}(\mathbf{s}) = \frac{1.633\mathbf{s}^2}{\mathbf{s}^2 + 16} \times \frac{0.001}{\mathbf{s}} = \frac{0.001633\mathbf{s}}{\mathbf{s}^2 + 16}. \tag{4.172}$$

The inverse Laplace transform of $\boldsymbol{\theta}(\mathbf{s})$ is

$$\theta(t) = 0.001633\cos(4t)\,u(t). \tag{4.173}$$

The pendulum swings back and forth with period $2\pi/4 = 1.571$ s and amplitude 0.001633 radians $= 0.094°$, which is small enough to support the assumption that the linear model is applicable.

The response $\theta(t)$ is plotted in Fig. 4-38.

4-12.4 Control Using PD Feedback

Having seen the results of using proportional feedback, we now introduce ***proportional-plus-derivative (PD)*** feedback with

$$\mathbf{G}(\mathbf{s}) = K_1 + K_2\mathbf{s}, \tag{4.174}$$

so that the feedback depends on both the angle $\theta(t)$ and its derivative $d\theta/dt$ (angular velocity). The angular velocity $d\theta/dt$ can be measured directly using a magnetic sensor, as was done previously with the motor. The closed-loop transfer function $\mathbf{Q}(\mathbf{s})$ is then

$$\begin{aligned}\mathbf{Q}(\mathbf{s}) &= \frac{\mathbf{H}(\mathbf{s})}{1 + \mathbf{G}(\mathbf{s})\,\mathbf{H}(\mathbf{s})} \\ &= -\frac{\mathbf{s}^2/L}{\mathbf{s}^2 - g/L} \Big/ \left(1 - (K_1 + K_2\mathbf{s})\,\frac{\mathbf{s}^2/L}{\mathbf{s}^2 - g/L}\right) \\ &= \frac{-\mathbf{s}^2}{(L-K_1)\mathbf{s}^2 - K_2\mathbf{s}^3 - g}. \end{aligned} \tag{4.175}$$

The denominator of $\mathbf{Q}(\mathbf{s})$ is a cubic polynomial, so there will be three poles. With two arbitrary constants, K_1 and K_2, to determine three poles, one might think that the closed-loop system can always be stabilized. But this is not true.

▶ A cubic polynomial in which the coefficient of the $\mathbf{s}$-term is zero and the coefficients of other terms are real and non-zero must have at least one root with positive real part. ◀

Hence the closed-loop system can never be stabilized using PD control!

4-12.5 Control Using PI Feedback

Since neither proportional nor PD control can stabilize the inverted pendulum, let us now try something new: ***proportional-plus-integral (PI)*** feedback, wherein

$$\mathbf{G}(\mathbf{s}) = K_1 + \frac{K_2}{\mathbf{s}} . \tag{4.176}$$

Now the feedback depends on both the angle $\theta(t)$ and its integral $\int_0^t \theta(t')\,dt'$. The integral can be computed using an op-amp integrator circuit. The closed-loop transfer function $\mathbf{Q}(\mathbf{s})$ is then

$$\begin{aligned}\mathbf{Q}(\mathbf{s}) &= \frac{\mathbf{H}(\mathbf{s})}{1+\mathbf{G}(\mathbf{s})\,\mathbf{H}(\mathbf{s})} \\ &= -\frac{\mathbf{s}^2/L}{\mathbf{s}^2 - g/L} \Bigg/ \left(1 - (K_1 + K_2/\mathbf{s})\,\frac{\mathbf{s}^2/L}{\mathbf{s}^2 - g/L}\right) \\ &= -\frac{\mathbf{s}^2}{(L-K_1)\mathbf{s}^2 - K_2\mathbf{s} - g} .\end{aligned} \tag{4.177}$$

The denominator of $\mathbf{Q}(\mathbf{s})$ is now a quadratic polynomial, so it has only two poles. With two arbitrary constants, K_1 and K_2, to determine two poles, we can not only stabilize the closed-loop system, but we can also place its poles anywhere we want!

Example 4-16: Inverted Pendulum Control Using PI Feedback

A cart carrying an inverted pendulum of length 61.25 cm is suddenly moved 1 mm. Compute its response if PI feedback is used with $K_1 = 122.5$ cm/rad and $K_2 = 490$ cm s^{-1}/rad.

Solution: Inserting $L = 61.25$ cm, $K_1 = 122.5$ cm/rad, $K_2 = 490$ cm s^{-1}/rad, and $g = 9.8$ m s^{-2} into the closed-loop transfer function $\mathbf{Q}(\mathbf{s})$ gives

$$\begin{aligned}\mathbf{Q}(\mathbf{s}) &= \frac{-\mathbf{s}^2}{(L-K_1)\mathbf{s}^2 - K_2\mathbf{s} - g} \\ &= \frac{-\mathbf{s}^2}{(0.6125 - 1.225)\mathbf{s}^2 - 4.9\mathbf{s} - 9.8} = \frac{1.633\mathbf{s}^2}{\mathbf{s}^2 + 8\mathbf{s} + 16} .\end{aligned} \tag{4.178}$$

A sudden movement of length 1 mm means that the input $x(t) = 0.001u(t)$. Its Laplace transform is $\mathbf{X}(\mathbf{s}) = 0.001/\mathbf{s}$, and

$$\begin{aligned}\boldsymbol{\theta}(\mathbf{s}) &= \mathbf{Q}(\mathbf{s})\,\mathbf{X}(\mathbf{s}) \\ &= \frac{1.633\mathbf{s}^2}{\mathbf{s}^2 + 8\mathbf{s} + 16} \times \frac{0.001}{\mathbf{s}} = \frac{0.001633\mathbf{s}}{(\mathbf{s}+4)^2} .\end{aligned} \tag{4.179}$$

To enable obtaining the inverse Laplace transform, we rewrite $\boldsymbol{\theta}(\mathbf{s})$ by expressing $\mathbf{s}$ as $\mathbf{s} = (\mathbf{s}+4) - 4$:

$$\boldsymbol{\theta}(\mathbf{s}) = \frac{0.001633\mathbf{s}}{(\mathbf{s}+4)^2} = \frac{0.001633(\mathbf{s}+4)}{(\mathbf{s}+4)^2} - \frac{0.006532}{(\mathbf{s}+4)^2} .$$

The inverse Laplace transform of $\boldsymbol{\theta}(\mathbf{s})$ is

$$\theta(t) = 0.001633(1 - 4t)e^{-4t}\,u(t). \tag{4.180}$$

The response (shown in Fig. 4-38) not only decays to zero, but it is critically damped for this choice of K_2.

Concept Question 4-12: What assumption about angle θ was made in order to linearize the inverted pendulum system? (See S²)

Concept Question 4-13: What is the difference between PD and PI feedback? (See S²)

Exercise 4-19: Using proportional feedback with $K = L + 0.2$, compute the response to input

$$x(t) = 0.01\,u(t).$$

Answer: $y(t) = 0.05\cos(7t)\,u(t)$. (See S²)

Exercise 4-20: Using PI feedback, show that the closed-loop system is stable if $K_1 > L$ and $K_2 > 0$.

Answer: (See S²).

Module 4.2 Inverted Pendulum Response The input can be a step, ramp, or sinusoid. The period of the sinusoidal input and the gain of the feedback (K for proportional control, and K_1 and K_2 for PI control), are selectable parameters. Waveforms of the input and angular response are displayed.

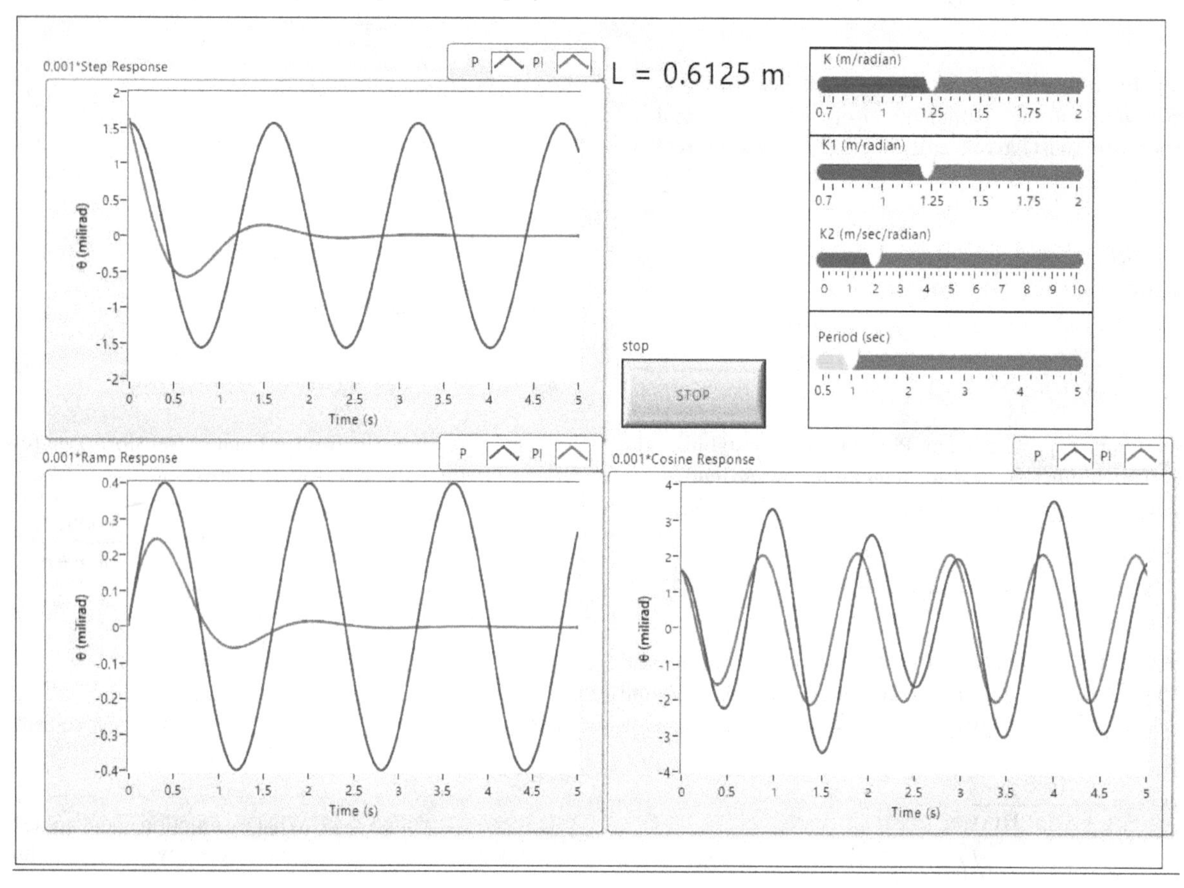

Exercise 4-21: Using PI feedback with

$$K_1 = L + 0.2,$$

select the value of K_2 so that the closed-loop system is critically damped.

Answer: $K_2 = 2.8$. (See S²)

Exercise 4-22: Use LabVIEW Module 4.2 to compute the inverted pendulum responses shown in Fig. 4-38.

Answer:

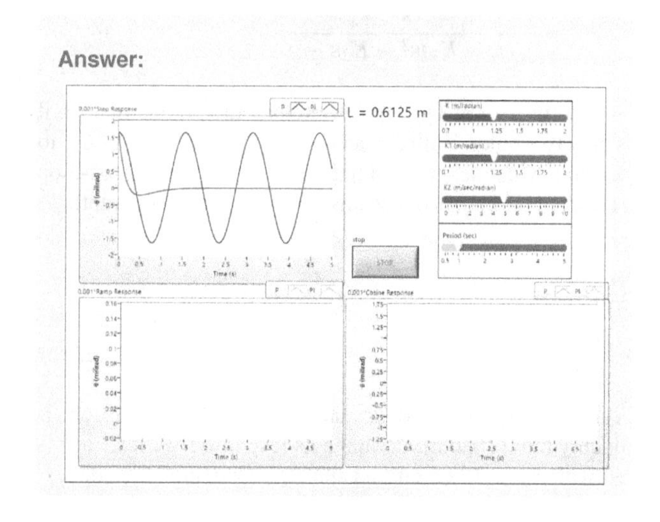

Summary

Concepts

- The Laplace transform can be used to analyze any causal LTI system, even if the system has non-zero initial conditions.
- Many mechanical systems can be modeled in terms of electromechanical analogues.
- Op-amp circuits can be used to develop system realizations that conform to specified transfer functions.
- Feedback control can be employed to improve system performance and stability.

Mathematical and Physical Models

s-Domain Impedance (no initial conditions)

$$\mathbf{Z}_{\mathrm{R}} = R, \qquad \mathbf{Z}_{\mathrm{L}} = \mathbf{s}L, \qquad \mathbf{Z}_{\mathrm{C}} = \frac{1}{\mathbf{s}C}$$

Configurations of Multiple Systems

Series $\quad \mathbf{H}(\mathbf{s}) = \mathbf{H}_1(\mathbf{s})\,\mathbf{H}_2(\mathbf{s})$

Parallel $\quad \mathbf{H}(\mathbf{s}) = \mathbf{H}_1(\mathbf{s}) + \mathbf{H}_2(\mathbf{s})$

Negative Feedback $\quad \mathbf{H}(\mathbf{s}) = \dfrac{\mathbf{H}_1(\mathbf{s})}{1 + \mathbf{H}_1(\mathbf{s})\,\mathbf{H}_2(\mathbf{s})}$

Motor s-domain Model

$$\mathbf{s}^2\,\boldsymbol{\theta}(\mathbf{s}) + a\mathbf{s}\,\boldsymbol{\theta}(\mathbf{s}) = b\,\mathbf{V}(\mathbf{s})$$

Important Terms

Provide definitions or explain the meaning of the following terms:

closed loop
damper
DFI
DFII
electromechanical analogue
feedback control
gain-bandwidth product
inverse system
inverted pendulum
inverter
negative feedback
open loop
parallel configuration
proportional feedback
proportional-plus-derivative (PD) feedback
proportional-plus-integral (PI) feedback
Schmitt trigger
series configuration
SMD
subtractor
summer
system realization
transfer function

PROBLEMS

Sections 4-1 and 4-2: s-domain Analysis

*4.1 Determine $v(t)$ in the circuit of Fig. P4.1 given that $v_{\mathrm{s}}(t) = 2u(t)$ V, $R_1 = 1\ \Omega$, $R_2 = 3\ \Omega$, $C = 0.3689$ F, and $L = 0.2259$ H.

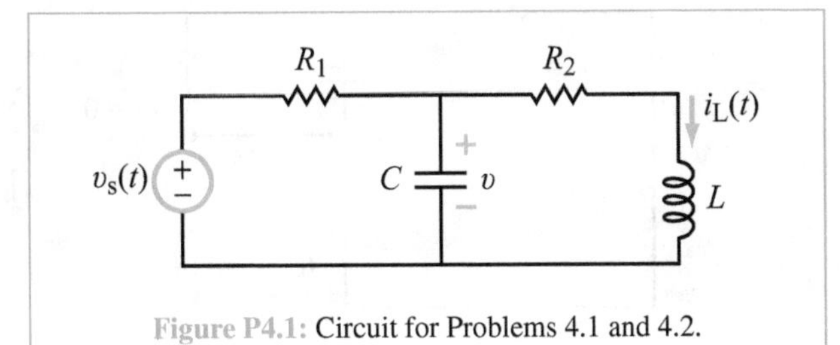

Figure P4.1: Circuit for Problems 4.1 and 4.2.

*Answer(s) in Appendix F.

4.2 Determine $i_{\mathrm{L}}(t)$ in the circuit in Fig. P4.1 given that $v_{\mathrm{s}}(t) = 2u(t)$, $R_1 = 2\ \Omega$, $R_2 = 6\ \Omega$, $L = 2.215$ H, and $C = 0.0376$ F.

4.3 Determine $v_{\mathrm{out}}(t)$ in the circuit in Fig. P4.3 given that $v_{\mathrm{s}}(t) = 35u(t)$ V, $v_{C_1}(0^-) = 20$ V, $R_1 = 1\ \Omega$, $C_1 = 1$ F, $R_2 = 0.5\ \Omega$, and $C_2 = 2$ F.

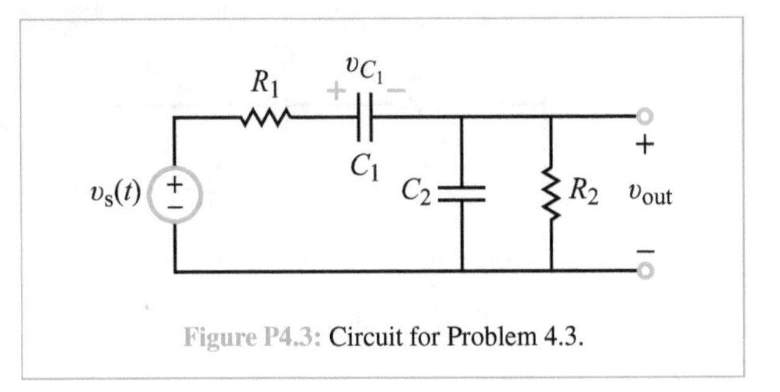

Figure P4.3: Circuit for Problem 4.3.

4.4 Determine $i_L(t)$ in the circuit of Fig. P4.4 for $t \geq 0$ given that the switch was opened at $t = 0$, after it had been closed for a long time, $v_s = 12$ mV, $R_0 = 5\ \Omega$, $R_1 = 10\ \Omega$, $R_2 = 20\ \Omega$, $L = 0.2$ H, and $C = 6$ mF.

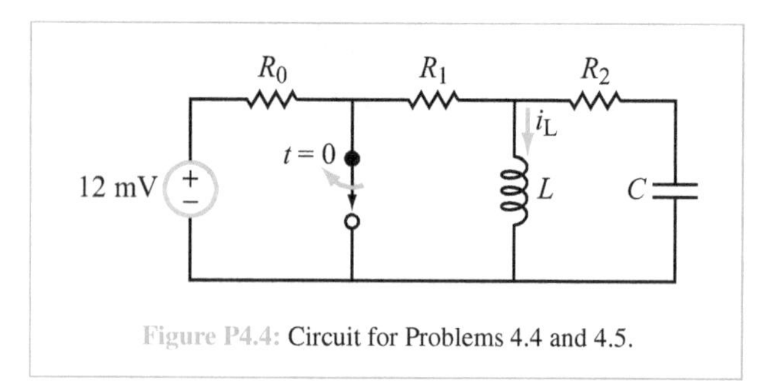

Figure P4.4: Circuit for Problems 4.4 and 4.5.

4.5 Repeat Problem 4.4, but assume that the switch had been open for a long time and then closed at $t = 0$. Retain the dc source at 12 mV and the resistors at $R_0 = 5\ \Omega$, $R_1 = 10\ \Omega$, and $R_2 = 20\ \Omega$, but change L to 2 H and C to 0.4 F.

4.6 Determine $i_L(t)$ in the circuit of Fig. P4.6 given that $R_1 = 2\ \Omega$, $R_2 = 1/6\ \Omega$, $L = 1$ H, and $C = 1/13$ F. Assume no energy was stored in the circuit segment to the right of the switch prior to $t = 0$.

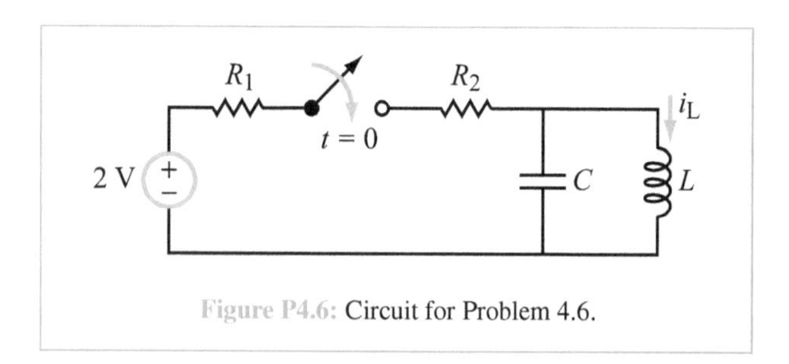

Figure P4.6: Circuit for Problem 4.6.

*__4.7__ Determine $v_{C_2}(t)$ in the circuit of Fig. P4.7 given that $R = 200\ \Omega$, $C_1 = 1$ mF, and $C_2 = 5$ mF.

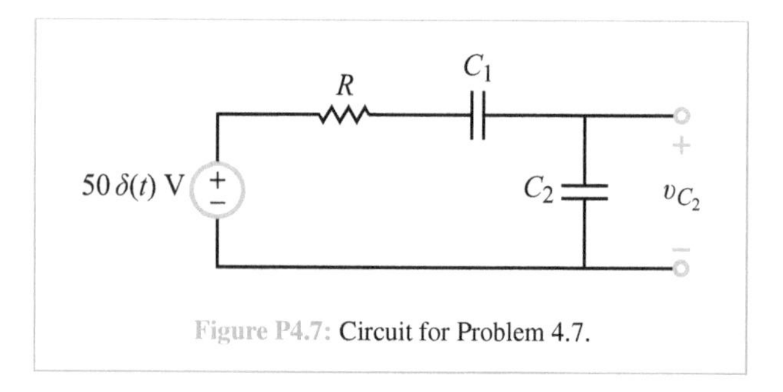

Figure P4.7: Circuit for Problem 4.7.

4.8 Determine $i_L(t)$ in the circuit of Fig. P4.8 given that before closing the switch, $v_C(0^-) = 24$ V. Also, the element values are $R = 1\ \Omega$, $L = 0.8$ H, and $C = 0.25$ F.

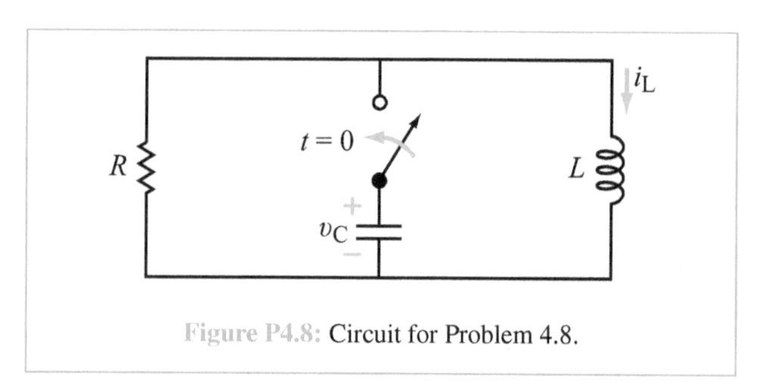

Figure P4.8: Circuit for Problem 4.8.

4.9 Determine $v_{out}(t)$ in the circuit of Fig. P4.9 given that $v_s(t) = 11u(t)$ V, $R_1 = 2\ \Omega$, $R_2 = 4\ \Omega$, $R_3 = 6\ \Omega$, $L = 1$ H, and $C = 0.5$ F.

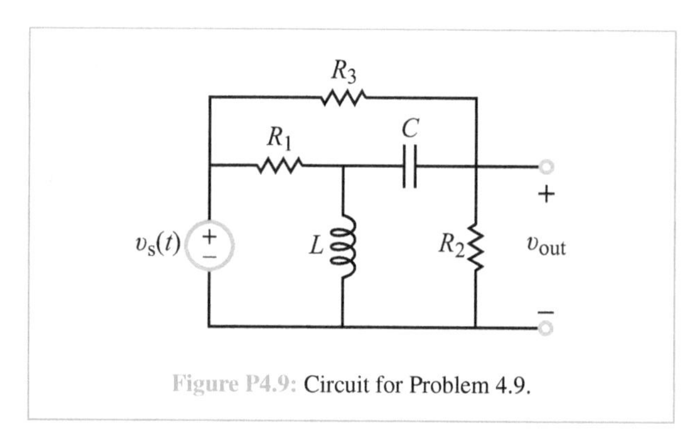

Figure P4.9: Circuit for Problem 4.9.

4.10 Determine $i_L(t)$ in the circuit of Fig. P4.10 for $t \geq 0$ given that $R = 3.5\ \Omega$, $L = 0.5$ H, and $C = 0.2$ F.

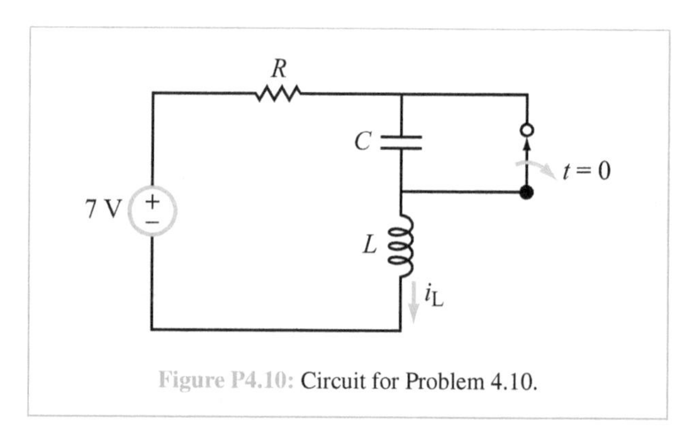

Figure P4.10: Circuit for Problem 4.10.

4.11 Apply mesh-current analysis in the **s**-domain to determine $i_L(t)$ in the circuit of Fig. P4.11 given that $\upsilon_s(t) = 44u(t)$ V, $R_1 = 2\,\Omega$, $R_2 = 4\,\Omega$, $R_3 = 6\,\Omega$, $C = 0.1$ F, and $L = 4$ H.

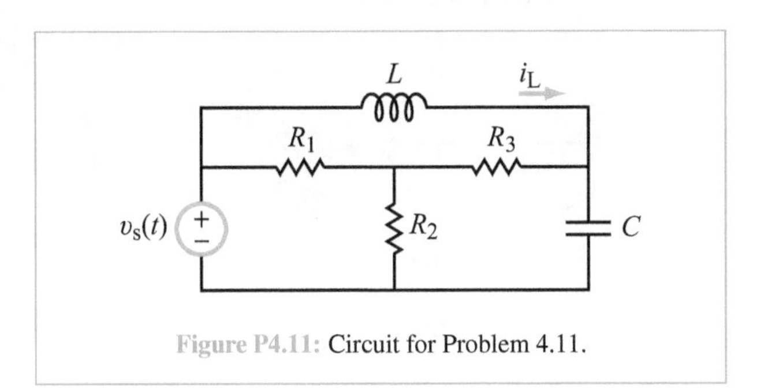

Figure P4.11: Circuit for Problem 4.11.

4.12 The voltage source in the circuit of Fig. P4.12 is given by $\upsilon_s(t) = [10 - 5u(t)]$ V. Determine $i_L(t)$ for $t \geq 0$ given that $R_1 = 1\,\Omega$, $R_2 = 3\,\Omega$, $L = 2$ H, and $C = 0.5$ F.

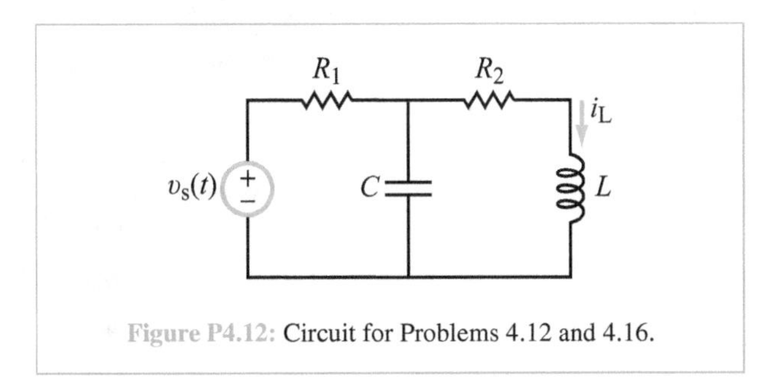

Figure P4.12: Circuit for Problems 4.12 and 4.16.

*__4.13__ The current source in the circuit of Fig. P4.13 is given by $i_s(t) = [10u(t) + 20\delta(t)]$ mA. Determine $\upsilon_C(t)$ for $t \geq 0$ given that $R_1 = R_2 = 1$ kΩ and $C = 0.5$ mF.

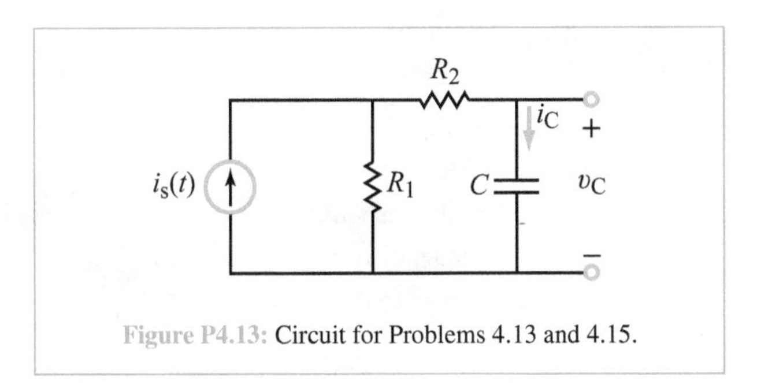

Figure P4.13: Circuit for Problems 4.13 and 4.15.

4.14 The circuit in Fig. P4.14 is excited by a 10 V, 1 s long rectangular pulse. Determine $i(t)$ given that $R_1 = 1\,\Omega$, $R_2 = 2\,\Omega$, and $L = 1/3$ H.

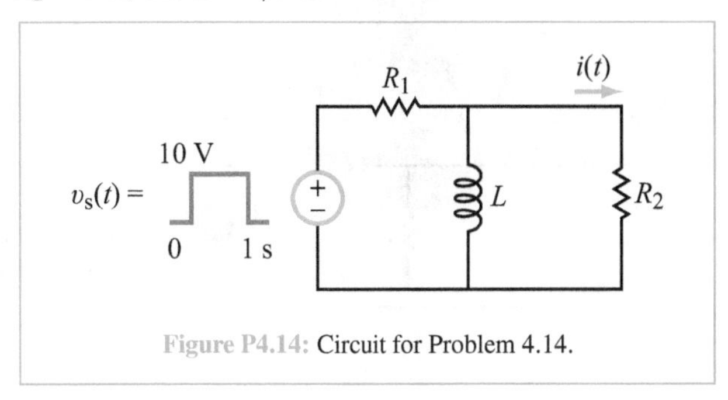

Figure P4.14: Circuit for Problem 4.14.

4.15 Repeat Problem 4.13, after replacing the current source with a 10 mA, 2 s long rectangular pulse.

4.16 Analyze the circuit shown in Fig. P4.12 to determine $i_L(t)$, in response to a voltage excitation $\upsilon_s(t)$ in the form of a 10 V rectangular pulse that starts at $t = 0$ and ends at $t = 5$ s. The element values are $R_1 = 1\,\Omega$, $R_2 = 3\,\Omega$, $L = 2$ H, and $C = 0.5$ F.

4.17 The current source in the circuit of Fig. P4.17 is given by $i_s(t) = 6e^{-2t}\,u(t)$ A. Determine $i_L(t)$ for $t \geq 0$ given that $R_1 = 10\,\Omega$, $R_2 = 5\,\Omega$, $L = 0.6196$ H, and $LC = 1/15$ s.

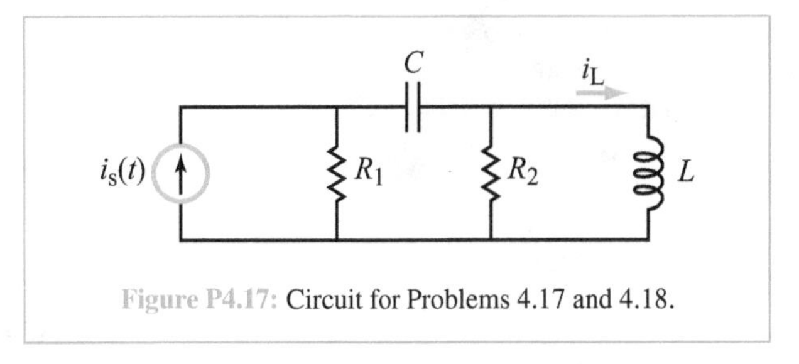

Figure P4.17: Circuit for Problems 4.17 and 4.18.

4.18 Given the current-source waveform displayed in Fig. P4.18, determine $i_L(t)$ in the circuit of Fig. P4.17 given that $R_1 = 10\,\Omega$, $R_2 = 5\,\Omega$, $L = 0.6196$ H, and $LC = 1/15$ s.

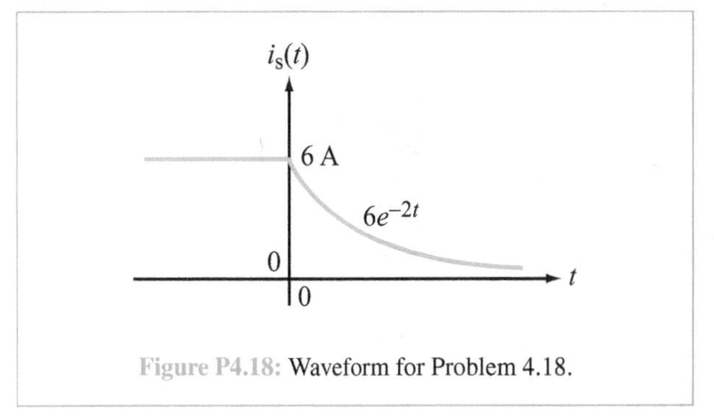

Figure P4.18: Waveform for Problem 4.18.

*4.19 The current source shown in the circuit of Fig. P4.19 is given by the displayed waveform. Determine $\upsilon_{out}(t)$ for $t \geq 0$ given that $R_1 = 1\ \Omega$, $R_2 = 0.5\ \Omega$, and $L = 0.5$ H.

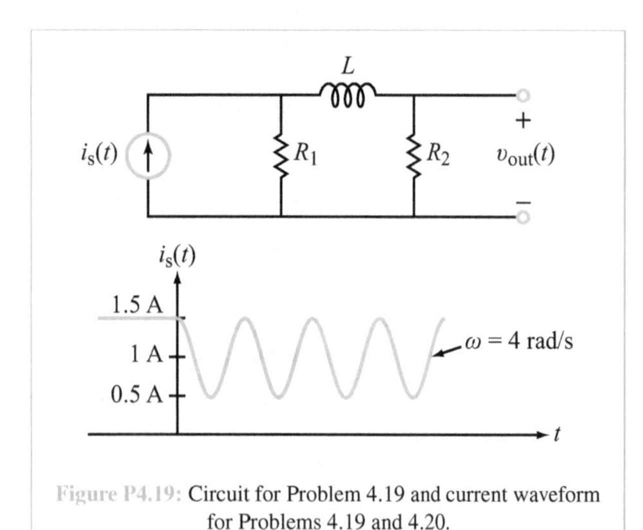

Figure P4.19: Circuit for Problem 4.19 and current waveform for Problems 4.19 and 4.20.

4.20 If the circuit shown in Fig. P4.20 is excited by the current waveform $i_s(t)$ shown in Fig. P4.19, determine $i(t)$ for $t \geq 0$ given that $R_1 = 10\ \Omega$, $R_2 = 5\ \Omega$, and $C = 0.02$ F.

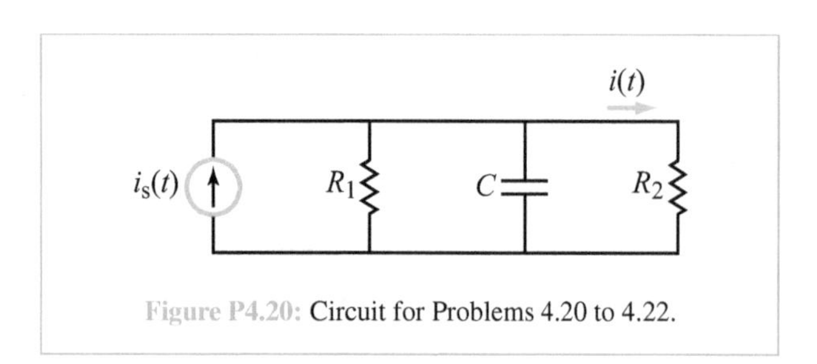

Figure P4.20: Circuit for Problems 4.20 to 4.22.

4.21 If the circuit shown in Fig. P4.20 is excited by current waveform $i_s(t) = 36te^{-6t}\ u(t)$ mA, determine $i(t)$ for $t \geq 0$ given that $R_1 = 2\ \Omega$, $R_2 = 4\ \Omega$, and $C = (1/8)$ F.

4.22 If the circuit shown in Fig. P4.20 is excited by a current waveform given by $i_s(t) = 9te^{-3t}\ u(t)$ mA, determine $i(t)$ for $t \geq 0$ given that $R_1 = 1\ \Omega$, $R_2 = 3\ \Omega$, and $C = 1/3$ F.

Sections 4-3 and 4-4: Electromechanical Analogues

4.23 In the SMD system shown in Fig. P4.23, $v_x(t)$ and $v_y(t)$ are the input velocity of the platform surface and output velocity of the 100 kg mass, respectively.

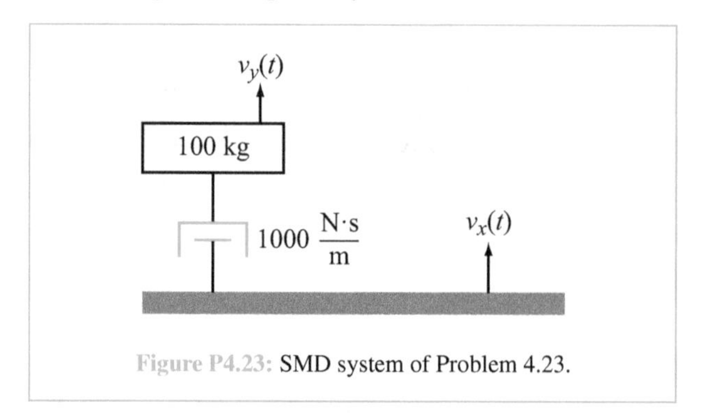

Figure P4.23: SMD system of Problem 4.23.

(a) Draw the equivalent s-domain circuit.

*(b) Determine the system transfer function.

(c) Compute the impulse response.

(d) Determine the frequency response.

(e) Compute the response to the rectangular pulse $v_x(t) = [5u(t) - 5u(t - 0.2)]$ (m/s).

4.24 In the SMD system shown in Fig. P4.24, $v_x(t)$ is the input velocity of the platform and $v_y(t)$ is the output velocity of the 200 kg mass.

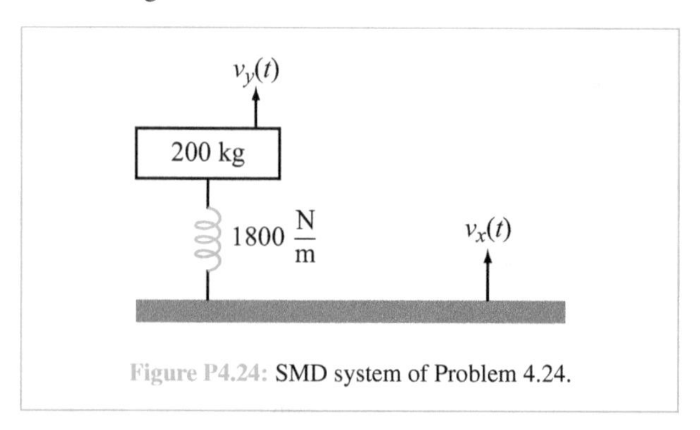

Figure P4.24: SMD system of Problem 4.24.

(a) Draw the equivalent s-domain circuit.

(b) Determine the system transfer function.

*(c) Compute the impulse response.

(d) Determine the frequency response.

(e) Compute the response to the sinusoid

$$v_x(t) = \cos(3t)\ u(t)\ \text{(m/s)}.$$

4.25 In the SMD system shown in Fig. P4.25, $v_x(t)$ is the input velocity of the platform and $v_y(t)$ is the output velocity of the 200 kg mass.

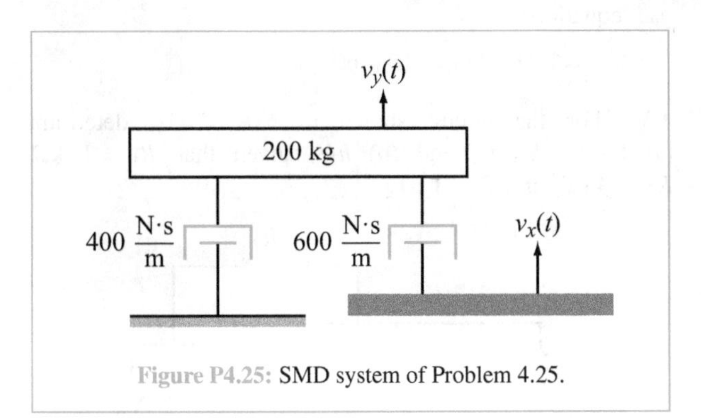

Figure P4.25: SMD system of Problem 4.25.

(a) Draw the equivalent **s**-domain circuit.

(b) Determine the system transfer function.

(c) Compute the impulse response.

(d) Determine the frequency response.

(e) Compute the response to the decaying exponential $v_x(t) = e^{-5t}\, u(t)$ (m/s).

4.26 In the SMD system shown in Fig. P4.26, $v_x(t)$ is the input velocity of the platform and $v_y(t)$ is the output velocity of the 200 kg mass.

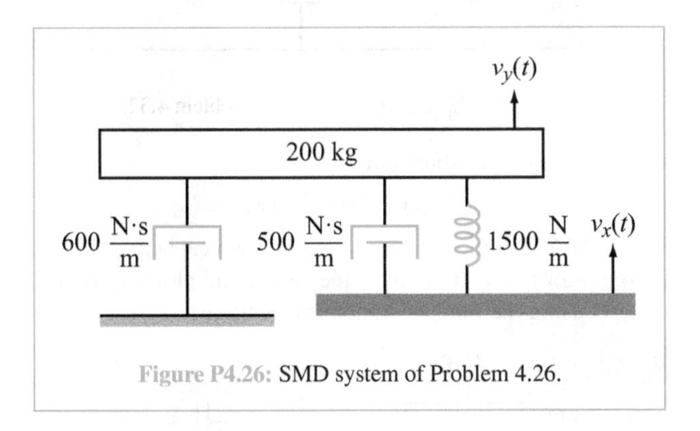

Figure P4.26: SMD system of Problem 4.26.

(a) Draw the equivalent **s**-domain circuit.

(b) Determine the system transfer function.

(c) Compute the impulse response.

(d) Determine the frequency response.

(e) Compute the response to the step $v_x(t) = u(t)$ (m/s).

4.27 In the SMD system shown in Fig. P4.27, $v_x(t)$ is the input velocity of the platform and $v_y(t)$ is the output velocity of the 200 kg mass.

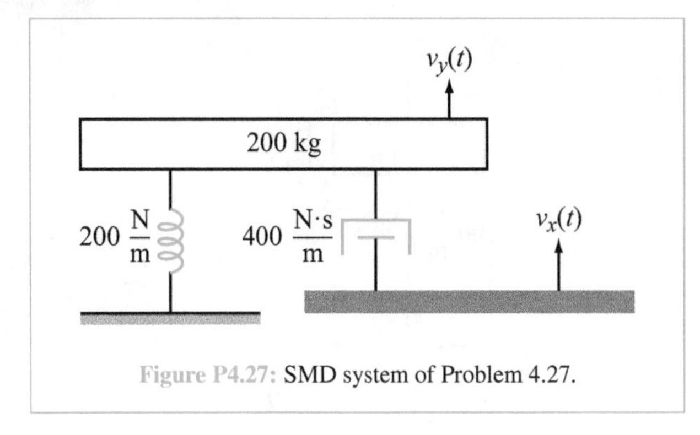

Figure P4.27: SMD system of Problem 4.27.

(a) Draw the equivalent **s**-domain circuit.

(b) Determine the system transfer function.

(c) Compute the impulse response.

(d) Determine the frequency response.

(e) Let $v_x(t) = \cos(\omega t)$. At what frequency ω does $v_y(t) = v_x(t)$, i.e., output = input?

4.28 In the SMD system shown in Fig. P4.28, $v_x(t)$ is the input velocity of the platform and $v_y(t)$ is the output velocity of the 300 kg mass.

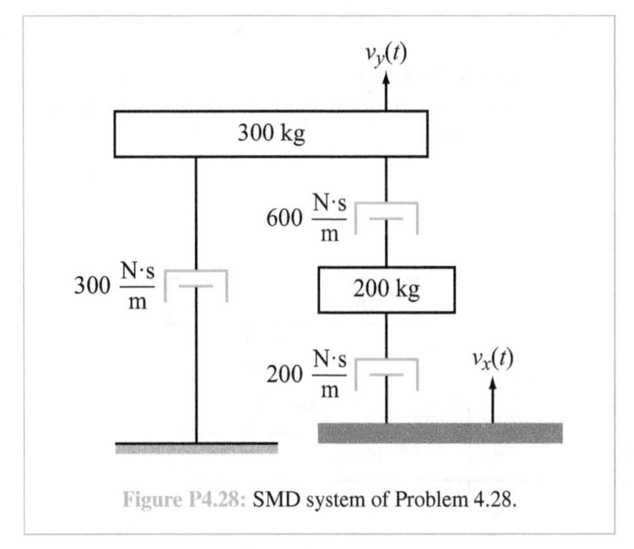

Figure P4.28: SMD system of Problem 4.28.

(a) Draw the equivalent **s**-domain circuit.

*(b) Determine the system transfer function.

(c) Compute the impulse response.

(d) Determine the frequency response.

4.29 In the SMD system shown in Fig. P4.29, $v_x(t)$ is the input velocity of the platform and $v_y(t)$ is the output velocity of the 100 kg mass.

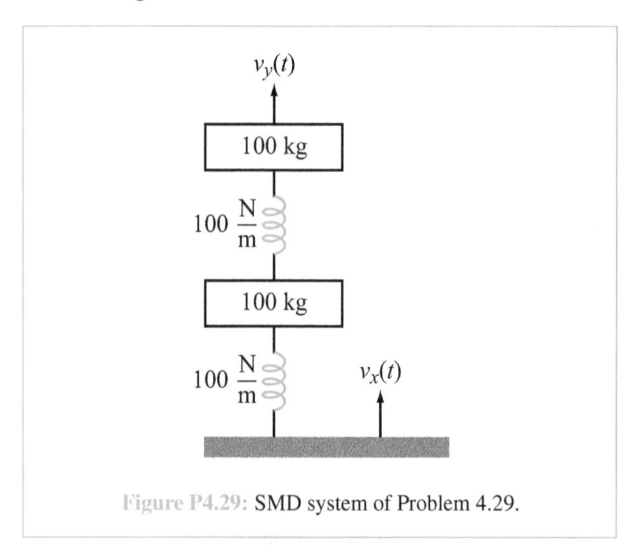

Figure P4.29: SMD system of Problem 4.29.

(a) Draw the equivalent **s**-domain circuit.

(b) Determine the system transfer function.

(c) Compute the impulse response.

(d) Determine the frequency response.

(e) Let $v_x(t) = \cos(\omega t)$. At what frequency ω does $v_y(t)$ blow up?

4.30 In the SMD system shown in Fig. P4.30, $v_x(t)$ is the input velocity of the platform and $v_y(t)$ is the output velocity of the 100 kg mass.

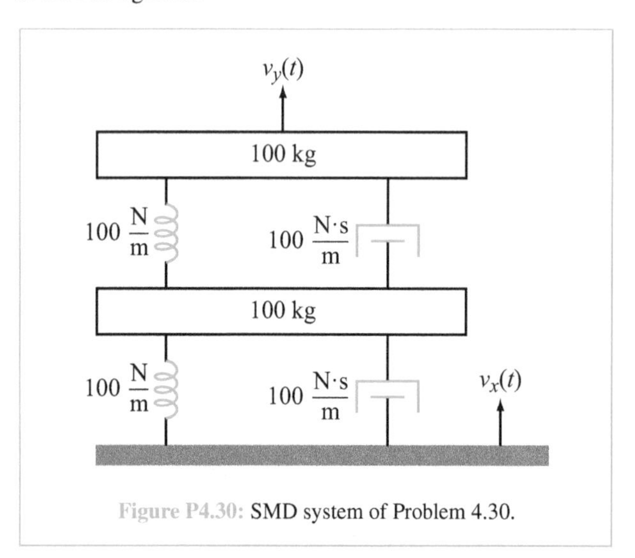

Figure P4.30: SMD system of Problem 4.30.

(a) Draw the equivalent **s**-domain circuit.

(b) Determine the system transfer function.

(c) Determine the frequency response. *Hint:* Use two node equations.

Section 4-5: Op-Amp Circuits

4.31 For the circuit shown in Fig. P4.31, determine (a) $\mathbf{H(s)} = \mathbf{V}_o/\mathbf{V}_s$ and (b) $h(t)$ given that $R_1 = 1$ kΩ, $R_2 = 4$ kΩ, and $C = 1\ \mu$F.

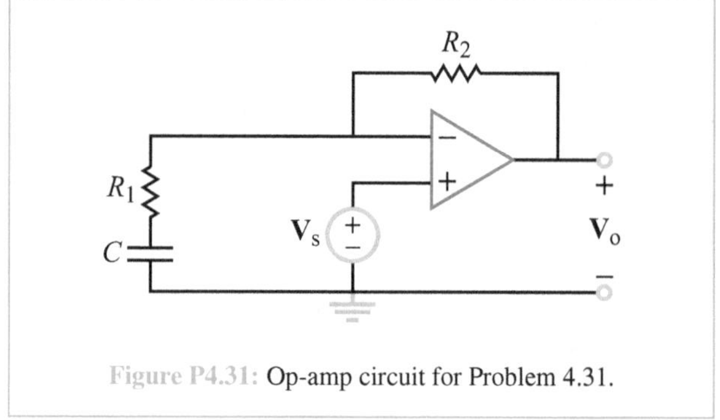

Figure P4.31: Op-amp circuit for Problem 4.31.

*4.32 For the circuit shown in Fig. P4.32, determine (a) $\mathbf{H(s)} = \mathbf{V}_o/\mathbf{V}_s$ and (b) $h(t)$ given that $R_1 = R_2 = 100\ \Omega$ and $C_1 = C_2 = 1\ \mu$F.

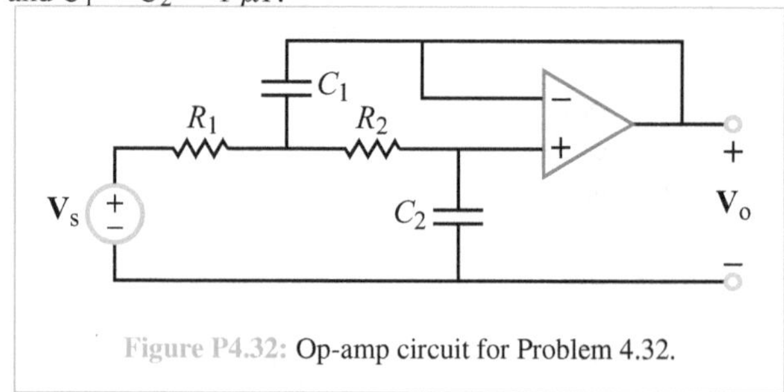

Figure P4.32: Op-amp circuit for Problem 4.32.

4.33 In the circuit shown in Fig. P4.32,

$$\upsilon_i(t) = 10u(t) \text{ mV},$$

$V_{CC} = 10$ V for both op amps, and the two capacitors had no charge prior to $t = 0$. Analyze the circuit and plot $\upsilon_{out_1}(t)$ and $\upsilon_{out_2}(t)$ using the Laplace transform technique.

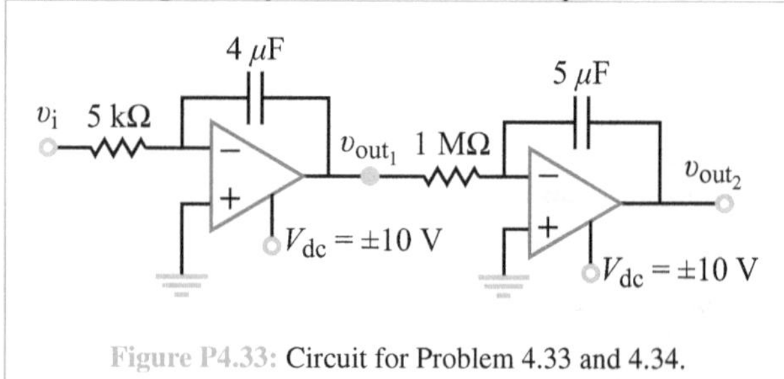

Figure P4.33: Circuit for Problem 4.33 and 4.34.

4.34 Repeat Problem 4.33 retaining all element values and conditions but changing the input voltage to $\upsilon_i(t) = 0.4te^{-2t}\,u(t)$.

Sections 4-6 and 4-7: System Configurations

*4.35 Draw the Direct Form II realization of the LTI system

$$\frac{d^2y}{dt^2} + 2\,\frac{dy}{dt} + 3y(t) = 4\,\frac{d^2x}{dt^2} + 5\,\frac{dx}{dt} + 6x(t).$$

4.36 Draw the Direct Form II realization of the LTI system with transfer function

$$\mathbf{H}(\mathbf{s}) = \frac{3\mathbf{s}^2 + 4\mathbf{s} + 5}{\mathbf{s}^2 + 6\mathbf{s} + 7}.$$

4.37 Draw the Direct Form II realization of the LTI system with poles $\{-1, -4\}$, zeros $\{-2, -3\}$, and $\mathbf{H}(0) = 1.5$, where $\mathbf{H}(\mathbf{s})$ is its transfer function.

4.38 Draw the Direct Form II realization of the LTI system that has the impulse response

$$h(t) = \delta(t) + e^{-t}\,u(t) + e^{-2t}\,u(t).$$

4.39 Draw the Direct Form II realization of the LTI system described by

$$\frac{d^2y}{dt^2} + 2\,\frac{dy}{dt} + 3y(t) = x(t).$$

4.40 Draw the Direct Form II realization of an LTI system whose frequency response is

$$\hat{\mathbf{H}}(\omega) = \frac{(4 - 6\omega^2) + j5\omega}{(2 - \omega^2) + j3\omega}.$$

Section 4-8: Control Systems

4.41 An unstable LTI system has the impulse response $h(t) = e^t\,u(t) - e^{-5t}\,u(t)$. For proportional feedback with $\mathbf{G}(\mathbf{s}) = K$, compute K to realize the following:

(a) The closed-loop system has poles $\{-1, -3\}$.

(b) The closed-loop system is critically damped.

4.42 An unstable LTI system has the impulse response $h(t) = \cos(4t)\,u(t)$. For proportional feedback with $\mathbf{G}(\mathbf{s}) = K$, compute K to realize the following:

*(a) The closed-loop system has poles $\{-2, -8\}$.

(b) The closed-loop system is critically damped.

4.43 An unstable LTI system has the impulse response $h(t) = \sin(4t)\,u(t)$.

(a) Show that proportional feedback ($\mathbf{G}(\mathbf{s}) = K$) cannot BIBO-stabilize the system.

(b) Show that derivative feedback ($\mathbf{G}(\mathbf{s}) = K\mathbf{s}$) can stabilize the system.

(c) Using derivative control, choose K so that the closed-loop system is critically damped.

4.44 An unstable LTI system has the impulse response $h(t) = e^{3t}\,u(t) - e^{2t}\,u(t)$.

(a) Show that proportional feedback (using $\mathbf{G}(\mathbf{s}) = K$) cannot BIBO-stabilize the system.

(b) Show that derivative feedback ($\mathbf{G}(\mathbf{s}) = K\mathbf{s}$) can stabilize the system, but the closed-loop system has a pole $p \geq -\sqrt{6}$, so its step response is slow.

(c) Show that proportional-plus-derivative feedback control, ($\mathbf{G}(\mathbf{s}) = K_1 + K_2\mathbf{s}$), can be used to put both closed-loop poles at -10. Compute K_1 and K_2 to accomplish this.

4.45 The impulse response of a first-order actuator is $h(t) = ae^{-at}\,u(t)$. The 3 dB frequency is the frequency at which the gain is down 3 dB. Show that with proportional feedback the following occurs:

(a) The product of the time constant and the 3 dB frequency is independent of K.

(b) The product of the dc gain and the 3 dB frequency is independent of K.

4.46 Show that derivative control $\mathbf{G}(\mathbf{s}) = K\mathbf{s}$ does not alter the steady-state step response.

4.47 A BIBO-stable LTI system with transfer function $\mathbf{H}(\mathbf{s})$ has a step response $y_{\text{step}}(t)$. The steady-state step response is $y_{\text{ss}} = \lim_{t\to\infty} y_{\text{step}}(t)$. Show that $y_{\text{ss}} = \mathbf{H}(0)$.

4.48 An unstable LTI system is described by the LCCDE

$$\frac{d^2y}{dt^2} - 2\,\frac{dy}{dt} + 5y(t) = x(t).$$

Using proportional plus derivative feedback with $\mathbf{G(s)} = K_1 + K_2\mathbf{s}$, compute K_1 and K_2 to realize the following:

(a) The closed-loop system is BIBO stable and critically damped.

(b) The closed-loop steady state step response

$$\lim_{t\to\infty} y_{\text{step}}(t) = 0.01.$$

4.49 The feedback system configuration shown in Fig. P4.49 is called the compensator feedback configuration. It is used often in control systems. Note that $\mathbf{G(s)}$ now appears just before the system, rather than in the feedback loop.

(a) Show that the closed-loop transfer function is

$$\mathbf{Q(s)} = \frac{\mathbf{G(s)\ H(s)}}{1 + \mathbf{G(s)\ H(s)}}.$$

(b) What happens to $\mathbf{Q(s)}$ when $\mathbf{G(s)} = K \to \infty$?

(c) Let $\mathbf{H(s)} = \mathbf{N(s)}/\mathbf{D(s)}$. Show that the closed-loop poles are still the roots of

$$\mathbf{D(s)} + \mathbf{G(s)\ N(s)} = 0,$$

in agreement with Eq. (4.104), if $\mathbf{G(s)}$ is a polynomial.

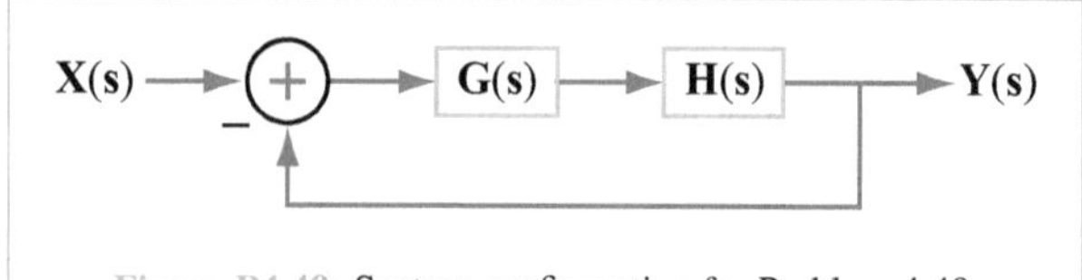

Figure P4.49: System configuration for Problem 4.49.

Section 4-9: Temperature Control

4.50 An oven has the following specifications: heat capacity = 2000 J/°C, thermal resistance = 0.1 °C/watt, ambient temperature = 20 °C, and desired temperature = 350 °C.

(a) How long will it take the oven to reach 340 °C, without feedback?

(b) If feedback control is used, what value of K is required to reduce that time to 1 minute?

4.51 An oven has the following specifications: heat capacity = 2000 J/ °C, thermal resistance = 0.15 °C/watt, ambient temperature = 20 °C, and desired temperature = 300 °C.

(a) How long will it take the oven to reach 290 °C, without feedback?

(b) If feedback control is used, what value of K is required to reduce that time to 1 minute?

*4.52 An oven is to be heated from 20 °C to 340 °C in 60 s. If its time constant is 300 s and no feedback is used, to what T_∞ should its input be set?

4.53 An oven is to be heated from 20 °C to 340 °C in 1 s. If its time constant is 300 s and no feedback is used, to what T_∞ should its input be set?

4.54 The desired temperature for an oven is 400 °C. However, to minimize the amount of time the oven is on, it is made to switch on and off, as shown in Fig. P4.54, so as to maintain the temperature between 380 °C and 420 °C. When the oven is on, its heat source would heat it up to $T_\infty = 550$ °C, if it were to stay on forever. The oven's time constant is 190 s and its ambient temperature is 20 °C. Compute the time durations T_{ON} and T_{OFF} so that the oven temperature stays between 380 °C and 420 °C.

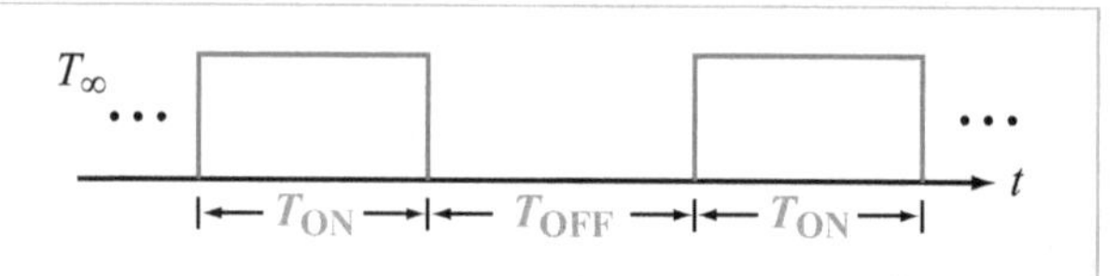

Figure P4.54: Oven temperature control (Problems 4.54 and 4.55).

4.55 The desired temperature for an oven is 300 °C. However, to minimize the amount of time the oven is on, it is made to switch on and off, as shown in Fig. P4.54, with $T_{\text{ON}} = T_{\text{OFF}}$, so as to maintain the temperature between 290 °C and 310 °C. The oven's time constant is 280 s and its ambient temperature is 20 °C. When the oven is on, its heat source would heat it up to T_∞, if it were to stay on forever. Compute the oven setting T_∞ so that the oven temperature stays between 290 °C and 310 °C.

Section 4-11: Motor System Control

4.56 A motor has transfer function $\mathbf{H(s)} = b/[\mathbf{s}(\mathbf{s} + a)]$ for known constants $a, b > 0$.

(a) Show that the step response blows up.

(b) Show that by using proportional feedback, the motor can be BIBO-stabilized.

(c) Show that if and only if $K = 1$, then the steady-state response error is zero.

*4.57 A motor has transfer function $\mathbf{H(s)} = 3/[\mathbf{s}(\mathbf{s}+3)]$. Show that PD feedback control $\mathbf{G}(\mathbf{s}) = K_1 + K_2\mathbf{s}$ can be used to place both closed-loop poles at -30. Compute K_1 and K_2 to accomplish this.

4.58 An unstable system has transfer function

$$\mathbf{H(s)} = \frac{1}{(\mathbf{s}-1)(\mathbf{s}-2)(\mathbf{s}+6)} = \frac{1}{\mathbf{s}^3 + 3\mathbf{s}^2 - 16\mathbf{s} + 12}.$$

Show that PD feedback control $\mathbf{G(s)} = K_1 + K_2\mathbf{s}$ can be used to place all three closed-loop poles at -1. Compute K_1 and K_2 to accomplish this.

4.59 Show that the system described by the LCCDE

$$\frac{d^3y}{dt^3} + a\,\frac{d^2y}{dt^2} + b\,\frac{dy}{dt} + c\;y(t) = x(t)$$

cannot be stabilized by PD control unless $a > 0$.

Section 4-12: Inverted Pendulum

4.60 A cart carrying an inverted pendulum of length 20 cm suddenly moves one Angstrom (10^{-10} m). How long does it take it to topple over?

4.61 A cart carrying an inverted pendulum of length 20 cm suddenly moves 1 cm. Compute its response if PI control $\mathbf{G(s)} = K_1 + K_2/\mathbf{s}$ is used with $K_1 = 40$ cm/rad and $K_2 = 1000$ cm $\cdot$ s^{-1}/rad.

4.62 A cart carries an inverted pendulum of length 20 cm. PI control $\mathbf{G(s)} = K_1 + K_2/\mathbf{s}$ with $K_1 = 40$ cm/rad is used. Compute K_2 in cm$\cdot$s^{-1}/rad so that its impulse response is critically damped.

4.63 A cart carries an inverted pendulum of length 20 cm. PI control $\mathbf{G(s)} = K_1 + K_2/\mathbf{s}$ is used. Compute K_1 in cm/rad and K_2 in cm$\cdot$s^{-1}/rad so that its impulse response has a double pole at -1.

LabVIEW Module 4.1

4.64

(a) Use LabVIEW Module 4.1 with a step input, a desired temperature change (T) of 5.0, a thermal time constant $RC = 100$, and no feedback. Plot the oven temperature response.

(b) Repeat part (a) using feedback with $K = 0.1$. What is the most significant difference between your answers to parts (a) and (b)?

4.65

(a) Use LabVIEW Module 4.1 with a sinusoidal input of period 300, desired temperature change (T) of 5.0, thermal time constant $RC = 100$, and no feedback. Plot the oven temperature response.

(b) Repeat part (a) using feedback with $K = 0.2$. What are the differences in gain and phase between your answers to parts (a) and (b)?

(c) Repeat part (a) but change the period to 100.

(d) Repeat part (c) using feedback with $K = 0.2$. What are the differences in gain and phase between your answers to parts (c) and (d)?

LabVIEW Module 4.2

4.66 For a step input, display the response using PI control with $K_1 = 1$ and $K_2 = 0.7$. Describe the response. What would the response be if $K_2 = 0$?

4.67 For a step input, using PI control, choose from among the allowed K_1 and K_2 values that make the response return to zero most quickly. Display the response.

4.68 For a step input, using PI control and $K_1 = 2$, choose the smallest value of K_2 that eliminates the overshoot. Display the response.

4.69 For a sinusoidal input with period 0.5, display the response using P control with $K = 0.7$. Explain the form of the response as forced plus natural responses.

4.70 For a sinusoidal input with period 0.5, display the response using PI control with $K_1 = 0.7$ and $K_2 = 10$. Explain the form of the response as forced plus natural responses. *Hint:* Also display the step response in the same window.

4.71 For a sinusoidal input with period 0.5, using PI control, choose the allowed K_1 and K_2 values that minimize the transient response. Display the response.

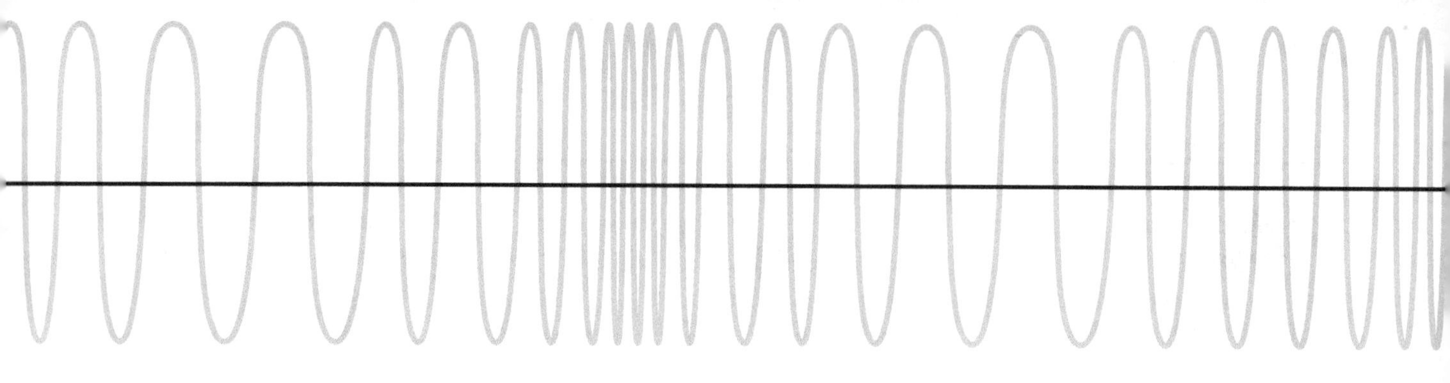

5 Fourier Analysis Techniques

Contents

Overview, 193
5-1 Phasor-Domain Technique, 193
5-2 Fourier Series Analysis Technique, 195
5-3 Fourier Series Representations, 197
5-4 Computation of Fourier Series Coefficients, 198
5-5 Circuit Analysis with Fourier Series, 213
5-6 Parseval's Theorem for Periodic Waveforms, 216
5-7 Fourier Transform, 218
5-8 Fourier Transform Properties, 223
5-9 Parseval's Theorem for Fourier Transforms, 230
5-10 Additional Attributes of the Fourier Transform, 232
5-11 Phasor vs. Laplace vs. Fourier, 235
5-12 Circuit Analysis with Fourier Transform, 236
5-13 The Importance of Phase Information, 238
Summary, 243
Problems, 244

Objectives

Learn to:

- Apply the phasor-domain technique to analyze systems driven by sinusoidal excitations.
- Express periodic signals in terms of Fourier series.
- Use Fourier series to analyze systems driven by continuous periodic signals.
- Apply Parseval's theorem to compute the power or energy contained in a signal.
- Compute the Fourier transform of nonperiodic signals and use it to analyze the system response to nonperiodic excitations.

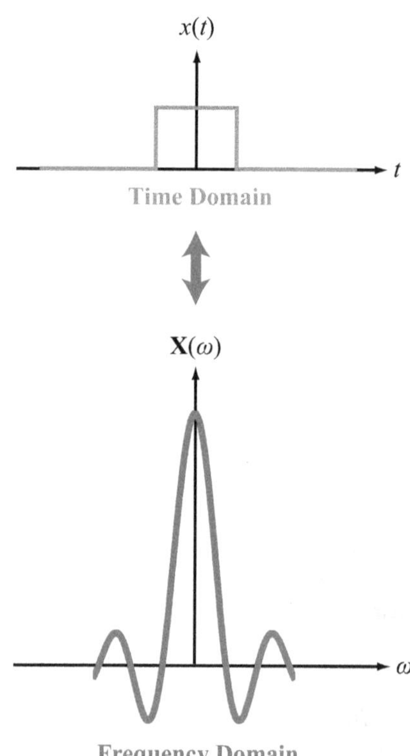

Time-domain signals have ***frequency domain spectra***. Because many analysis and design projects are easier to work with in the frequency domain, the ability to easily ***transform signals and systems*** back and forth between the two domains will prove invaluable in succeeding chapters.

Overview

Akin to the Laplace transform technique of Chapters 3 and 4, the ***phasor-domain technique*** is particularly suitable for analyzing systems when excited by sinusoidal waveforms. Through a phasor-domain transformation, the time-domain differential equations describing the system get transformed into algebraic equations. The solution is then transformed back to the time domain, yielding the same solution that would have been obtained had the original time-domain differential equations been solved entirely in the time domain. Even though the procedure involves multiple steps, it avoids the complexity of solving differential equations containing sinusoidal forcing functions.

As a periodic function with period T_0, a sinusoidal signal shares a distinctive property with all other members of the family of ***periodic functions***, namely the ***periodicity property*** given by

$$x(t) = x(t + nT_0), \tag{5.1}$$

where n is any integer. Given this natural connection between sinusoids and other periodic functions, can we somehow extend the phasor-domain solution technique to non-sinusoidal periodic excitations? The answer is yes, and the process for realizing it is facilitated by two enabling mechanisms: the ***Fourier theorem*** and the superposition principle. The Fourier theorem makes it possible to mathematically characterize any periodic excitation in the form of a sum of multiple sinusoidal ***harmonics***, and the superposition principle allows us to apply phasor analysis to calculate the circuit response due to each harmonic and then to add all of the responses together, thereby realizing the response to the original periodic excitation. The first half of this chapter aims to demonstrate the mechanics of the solution process, as well as to explain the physics associated with the system response to the different harmonics.

The second half of the chapter is devoted to the ***Fourier transform***, which is particularly useful for analyzing systems excited by nonperiodic waveforms, such as single pulses or step functions. As we will see in Section 5-11, the Fourier transform is related to the Laplace transform of Chapter 3 and has the same form under certain circumstances, but the two techniques are generally distinct (as are their conditions of applicability).

5-1 Phasor-Domain Technique

Any ***sinusoidally time-varying function*** $x(t)$ of angular frequency ω, representing a system excitation or response, can be expressed in the form

$$x(t) = \mathfrak{Re}[\underbrace{\mathbf{X}}_{\text{phasor}}\, e^{j\omega t}], \tag{5.2}$$

where $\mathbf{X}$ is a *time-independent* constant called the ***phasor counterpart*** of $x(t)$. Thus, $x(t)$ is defined in the time domain, while its counterpart $\mathbf{X}$ is defined in the phasor domain. *To distinguish phasor quantities from their time-domain counterparts, phasors are always represented by* **bold** *letters in this book.*

In general, the phasor-domain quantity $\mathbf{X}$ is complex, consisting of a magnitude $|\mathbf{X}|$ and a phase angle ϕ:

$$\mathbf{X} = |\mathbf{X}|e^{j\phi}. \tag{5.3}$$

Using this expression in Eq. (5.2) gives

$$\begin{aligned} x(t) &= \mathfrak{Re}[|\mathbf{X}|e^{j\phi}e^{j\omega t}] \\ &= \mathfrak{Re}[|\mathbf{X}|e^{j(\omega t+\phi)}] \\ &= |\mathbf{X}|\cos(\omega t + \phi). \end{aligned} \tag{5.4}$$

Application of the $\mathfrak{Re}$ operator allows us to transform sinusoids from the phasor domain to the time domain. The reverse operation, namely to specify the phasor-domain equivalent of a sinusoid, can be ascertained by comparing the two sides of Eq. (5.4). Thus, for a voltage $\upsilon(t)$ with phasor counterpart $\mathbf{V}$, the correspondence between the two domains is as follows:

Time Domain		**Phasor Domain**	
$\upsilon(t) = V_0 \cos \omega t$	$\longleftrightarrow$	$\mathbf{V} = V_0,$	(5.5a)
$\upsilon(t) = V_0 \cos(\omega t + \phi)$	$\longleftrightarrow$	$\mathbf{V} = V_0 e^{j\phi}.$	(5.5b)

If $\phi = -\pi/2$,

$$\upsilon(t) = V_0 \cos(\omega t - \pi/2) \longleftrightarrow \mathbf{V} = V_0 e^{-j\pi/2}. \tag{5.6}$$

Since $\cos(\omega t - \pi/2) = \sin \omega t$ and $e^{-j\pi/2} = -j$, it follows that

$$\upsilon(t) = V_0 \sin \omega t \longleftrightarrow \mathbf{V} = -jV_0. \tag{5.7}$$

Differentiation

Given a sinusoidal signal $i(t)$ with a corresponding phasor $\mathbf{I}$,

$$i(t) = \mathfrak{Re}[\mathbf{I}e^{j\omega t}], \tag{5.8}$$

the derivative di/dt is given by

$$\begin{aligned} \frac{di}{dt} &= \frac{d}{dt}[\mathfrak{Re}(\mathbf{I}e^{j\omega t})] \\ &= \mathfrak{Re}\left[\frac{d}{dt}(\mathbf{I}e^{j\omega t})\right] \\ &= \mathfrak{Re}[\underbrace{j\omega\mathbf{I}}_{\text{phasor}}\, e^{j\omega t}], \end{aligned} \tag{5.9}$$

where in the second step we interchanged the order of the two operators, $\mathfrak{Re}$ and d/dt, which is justified by the fact that the two operators are independent of one another, meaning that *taking the real part* of a quantity has no influence on taking its time derivative and vice versa. We surmise from Eq. (5.9) that

$$\frac{di}{dt} \longleftrightarrow j\omega\mathbf{I}. \tag{5.10}$$

▶ Differentiation of a sinusoidal time function $i(t)$ in the time domain is equivalent to multiplication of its phasor counterpart **I** by $j\omega$ in the phasor domain. ◀

Integration

Similarly,

$$\begin{aligned}\int i\,dt' &= \int \mathfrak{Re}[\mathbf{I}e^{j\omega t'}]\,dt' \\ &= \mathfrak{Re}\left[\left[\int \mathbf{I}e^{j\omega t'}\,dt'\right]\right] \\ &= \mathfrak{Re}\left[\underbrace{\frac{\mathbf{I}}{j\omega}}_{\text{phasor}}\,e^{j\omega t}\right],\end{aligned} \tag{5.11}$$

or

$$\int i\,dt' \longleftrightarrow \frac{\mathbf{I}}{j\omega}. \tag{5.12}$$

▶ Integration of $i(t)$ in the time domain is equivalent to dividing its phasor **I** by $j\omega$ in the phasor domain. ◀

Table 5-1 provides a summary of some time functions and their phasor-domain counterparts.

Example 5-1: Sinusoidal Excitation

Given a system characterized by the differential equation

$$\frac{d^2y}{dt^2} + a\,\frac{dy}{dt} + by = x(t),$$

with $a = 300,\ b = 5 \times 10^4$, and

$$x(t) = 10\sin(100t + 60°),$$

use the phasor-domain technique to determine $y(t)$.

Table 5-1: Time-domain sinusoidal functions $x(t)$ and their cosine-reference phasor-domain counterparts **X**, where $x(t) = \mathfrak{Re}\ [\mathbf{X}e^{j\omega t}]$.

$x(t)$		$\mathbf{X}$
$A\cos\omega t$	$\longleftrightarrow$	A
$A\cos(\omega t + \phi)$	$\longleftrightarrow$	$Ae^{j\phi}$
$-A\cos(\omega t + \phi)$	$\longleftrightarrow$	$Ae^{j(\phi \pm \pi)}$
$A\sin\omega t$	$\longleftrightarrow$	$Ae^{-j\pi/2} = -jA$
$A\sin(\omega t + \phi)$	$\longleftrightarrow$	$Ae^{j(\phi - \pi/2)}$
$-A\sin(\omega t + \phi)$	$\longleftrightarrow$	$Ae^{j(\phi + \pi/2)}$
$\frac{d}{dt}[A\cos(\omega t + \phi)]$	$\longleftrightarrow$	$j\omega Ae^{j\phi}$
$\int A\cos(\omega t' + \phi)\,dt'$	$\longleftrightarrow$	$\frac{1}{j\omega}\,Ae^{j\phi}$

Solution:

Step 1: **Convert $x(t)$ to cosine format**

$$\begin{aligned}x(t) &= 10\sin(100t + 60°) \\ &= 10\cos(100t + 60° - 90°) \\ &= 10\cos(100t - 30°).\end{aligned}$$

Step 2: **Define phasor-domain counterparts for all time-dependent variables**

$$\begin{aligned}x(t) = 10\cos(100t - 30°) &\longrightarrow \mathbf{X} = 10e^{-j30°} \\ y(t) &\longrightarrow \mathbf{Y} \\ \frac{dy}{dt} &\longrightarrow j\omega\mathbf{Y} \\ \frac{d^2y}{dt^2} &\longrightarrow (j\omega)^2\mathbf{Y}\end{aligned}$$

Step 3: **Transform differential equation to the phasor domain**

$$(j\omega)^2\mathbf{Y} + j\omega a\mathbf{Y} + b\mathbf{Y} = 10e^{-j30°}.$$

Step 4: **Solve for quantity of interest**

$$\begin{aligned}\mathbf{Y} &= \frac{10e^{-j30^\circ}}{b-\omega^2+j\omega a} \\ &= \frac{10e^{-j30^\circ}}{5\times10^4-10^4+j10^2\times300} \\ &= \frac{10e^{-j30^\circ}}{10^4(4+j3)} \\ &= \frac{10^{-3}e^{-j30^\circ}}{5e^{j36.87^\circ}} = 0.2\times10^{-3}e^{-j66.87^\circ}.\end{aligned}$$

Step 5: **Transform solution to the time domain**

$$\begin{aligned}y(t) &= \mathfrak{Re}[\mathbf{Y}e^{j\omega t}] \\ &= \mathfrak{Re}[0.2\times10^{-3}e^{-j66.87^\circ}e^{j100t}] \\ &= 0.2\times10^{-3}\cos(100t-66.87^\circ).\end{aligned}$$

5-2 Fourier Series Analysis Technique

By way of introducing the Fourier series analysis technique, let us consider the RL circuit shown in Fig. 5-1(a), which is excited by the square-wave voltage waveform shown in Fig. 5-1(b). The waveform amplitude is 3 V and its period $T_0 = 2$ s. Our goal is to determine the output voltage response, $\upsilon_{\text{out}}(t)$. The solution procedure consists of three basic steps.

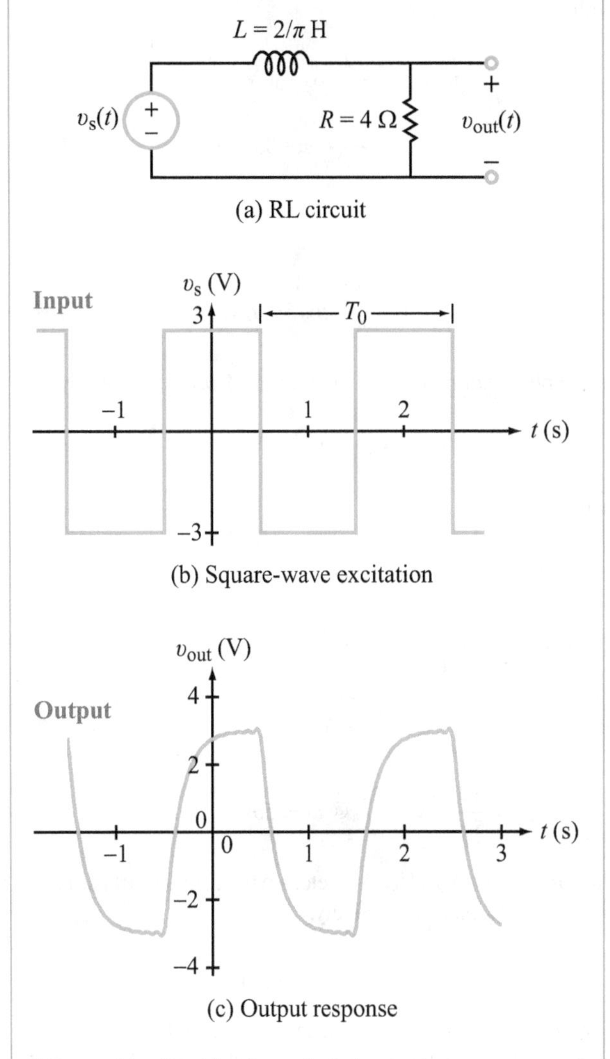

Figure 5-1: RL circuit excited by a square wave, and corresponding output response.

Step 1: **Express the periodic excitation in terms of Fourier harmonics**

According to the Fourier theorem (which we will introduce and examine in detail in Section 5-3), the waveform shown in Fig. 5-1(b) can be represented by the series

$$\upsilon_s(t) = \frac{12}{\pi}\left(\cos\omega_0 t - \frac{1}{3}\cos3\omega_0 t + \frac{1}{5}\cos5\omega_0 t - \cdots\right), \quad (5.13)$$

where $\omega_0 = 2\pi/T_0 = 2\pi/2 = \pi$ (rad/s) is the ***fundamental angular frequency*** of the waveform. Since our present objective is to outline the solution procedure, we will accept that the infinite-series representation given by Eq. (5.13) is indeed equivalent to the square wave of Fig. 5-1(b). The series consists of cosine functions of the form $\cos m\omega_0 t$, with m assuming only odd values (1, 3, 5, etc.). Thus, the series contains only odd ***harmonics*** of ω_0. The coefficient of the mth harmonic is equal to $1/m$ (relative to the coefficient of the fundamental), and its polarity is negative if $m = 3, 7, \ldots$, and positive if $m = 5, 9, \ldots$. In view of these properties, we can replace m with $(2n-1)$ and cast $\upsilon_s(t)$ in the form

$$\upsilon_s(t) = \frac{12}{\pi}\sum_{n=1}^{\infty}(-1)^{n+1}\frac{1}{2n-1}\cos(2n-1)\pi t \text{ V}. \quad (5.14)$$

In terms of its first few components, $\upsilon_s(t)$ is given by

$$\upsilon_s(t) = \upsilon_{s_1}(t) + \upsilon_{s_2}(t) + \upsilon_{s_3}(t) + \cdots \quad (5.15)$$

with

$$\upsilon_{s_1}(t) = \frac{12}{\pi}\cos\omega_0 t \text{ V}, \quad (5.16a)$$

$$\upsilon_{s_2}(t) = -\frac{12}{3\pi}\cos 3\omega_0 t \text{ V}, \quad (5.16b)$$

and

$$\upsilon_{s_3}(t) = \frac{12}{5\pi}\cos 5\omega_0 t \text{ V}, \quad \text{etc.} \quad (5.16c)$$

In the phasor domain, the counterpart of $\upsilon_s(t)$ is given by:

$$\mathbf{V}_s = \mathbf{V}_{s_1} + \mathbf{V}_{s_2} + \mathbf{V}_{s_3} + \cdots \quad (5.17)$$

with

$$\mathbf{V}_{s_1} = \frac{12}{\pi} \text{ V}, \quad @\ \omega = \omega_0, \quad (5.18a)$$

$$\mathbf{V}_{s_2} = -\frac{12}{3\pi} \text{ V}, \quad @\ \omega = 3\omega_0, \quad (5.18b)$$

and

$$\mathbf{V}_{s_3} = \frac{12}{5\pi} \text{ V}, \quad @\ \omega = 5\omega_0, \quad \text{etc.} \quad (5.18c)$$

Phasor voltages $\mathbf{V}_{s_1}, \mathbf{V}_{s_2}, \mathbf{V}_{s_3}$, etc. are the counterparts of $\upsilon_{s_1}(t)$, $\upsilon_{s_2}(t)$, $\upsilon_{s_3}(t)$, etc., respectively.

Step 2: **Determine output responses to input harmonics**

For the circuit in Fig. 5-1(a), input voltage $\mathbf{V}_{s_1}$ acting alone would generate a corresponding output voltage $\mathbf{V}_{\text{out}_1}$. In the phasor domain, the impedance of the inductor is $\mathbf{Z}_L = j\omega L$. Keeping in mind that $\mathbf{V}_{s_1}$ corresponds to $\upsilon_{s_1}(t)$ at $\omega = \omega_0 = \pi$, voltage division gives

$$\begin{aligned}\mathbf{V}_{\text{out}_1} &= \left(\frac{R}{R + j\omega_0 L}\right)\mathbf{V}_{s_1} \\ &= \frac{4}{4 + j\pi \times \frac{2}{\pi}} \cdot \frac{12}{\pi} = 3.42\underline{/-26.56^\circ}, \end{aligned} \quad (5.19)$$

with a corresponding time-domain voltage

$$\begin{aligned}\upsilon_{\text{out}_1}(t) &= \mathfrak{Re}[\mathbf{V}_{\text{out}_1} e^{j\omega_0 t}] \\ &= 3.42\cos(\omega_0 t - 26.56^\circ) \quad \text{(V)}. \end{aligned} \quad (5.20)$$

Similarly, at $\omega = 3\omega_0 = 3\pi$,

$$\begin{aligned}\mathbf{V}_{\text{out}_2} &= \frac{R}{R + j3\omega_0 L}\mathbf{V}_{s_2} \\ &= \frac{4}{4 + j3\pi \times \frac{2}{\pi}} \cdot \left(-\frac{12}{3\pi}\right) = -0.71\underline{/-56.31^\circ} \quad \text{(V)}, \end{aligned} \quad (5.21)$$

and

$$\begin{aligned}\upsilon_{\text{out}_2}(t) &= \mathfrak{Re}[\mathbf{V}_{\text{out}_2} e^{j3\omega_0 t}] \\ &= -0.71\cos(3\omega_0 t - 56.31^\circ) \text{ V}. \end{aligned} \quad (5.22)$$

In view of the harmonic pattern expressed in the form of Eq. (5.14), for the harmonic at angular frequency $\omega = (2n-1)\omega_0$,

$$\begin{aligned}\mathbf{V}_{\text{out}_n} &= \frac{4}{4 + j(2n-1)\pi \times \frac{2}{\pi}} \cdot (-1)^{n+1}\frac{12}{\pi(2n-1)} \\ &= (-1)^{n+1}\frac{24}{\pi(2n-1)\sqrt{4+(2n-1)^2}} \\ &\quad \cdot \underline{/-\tan^{-1}[(2n-1)/2]} \text{ V}. \end{aligned} \quad (5.23)$$

The corresponding time domain voltage is

$$\begin{aligned}\upsilon_{\text{out}_n}(t) &= \mathfrak{Re}[\mathbf{V}_{\text{out}_n} e^{j(2n-1)\omega_0 t}] \\ &= (-1)^{n+1}\frac{24}{\pi(2n-1)\sqrt{4+(2n-1)^2}} \\ &\quad \cdot \cos\left[(2n-1)\omega_0 t - \tan^{-1}\left(\frac{2n-1}{2}\right)\right] \text{ V}, \end{aligned} \quad (5.24)$$

with $\omega_0 = \pi$ rad/s.

Step 3: **Apply the superposition principle to determine $\upsilon_{\text{out}}(t)$**

According to the superposition principle, if υ_{out_1} is the output generated by a linear circuit when excited by an input voltage υ_{s_1} acting alone, and if similarly υ_{out_2} is the output due to υ_{s_2} acting alone, then the output due to the combination of υ_{s_1} and υ_{s_2} acting simultaneously is simply the sum of υ_{out_1} and υ_{out_2}. Moreover, the principle is extendable to any number of sources. In the present case, the square-wave excitation is equivalent to a series of sinusoidal sources $\upsilon_{s_1}, \upsilon_{s_2}, \ldots,$

generating corresponding output voltages $\upsilon_{\text{out}_1}, \upsilon_{\text{out}_2}, \ldots$. Consequently,

$$\begin{aligned}\upsilon_{\text{out}}(t) &= \sum_{n=1}^{\infty} \upsilon_{\text{out}_n}(t) \\ &= \sum_{n=1}^{\infty} (-1)^{n+1} \frac{24}{\pi(2n-1)\sqrt{4+(2n-1)^2}} \\ &\quad \cdot \cos\left[(2n-1)\omega_0 t - \tan^{-1}\left(\frac{2n-1}{2}\right)\right] \\ &= 3.42\cos(\omega_0 t - 26.56^\circ) \\ &\quad - 0.71\cos(3\omega_0 t - 56.31^\circ) \\ &\quad + 0.28\cos(5\omega_0 t - 68.2^\circ) + \cdots \text{ V}, \end{aligned} \tag{5.25}$$

with $\omega_0 = \pi$ rad/s.

We note that the fundamental component of $\upsilon_{\text{out}}(t)$ has the dominant amplitude and that the higher the harmonic, the smaller its amplitude. This allows us to approximate $\upsilon_{\text{out}}(t)$ by retaining only a few terms, such as up to $n = 10$, depending on the level of desired accuracy. The plot of $\upsilon_{\text{out}}(t)$ displayed in Fig. 5-1(c), which is based only on the first 10 terms, is sufficiently accurate for most practical applications.

The foregoing three-step procedure, which is equally applicable to any linear system excited by any realistic periodic function, relied on the use of the Fourier theorem to express the square-wave pattern in terms of sinusoids. In the next section, we will examine the attributes of the Fourier theorem and how we may apply it to any periodic function.

Concept Question 5-1: The Fourier-series technique is applied to analyze circuits excited by what type of functions? (See S²)

Concept Question 5-2: How is the angular frequency of the nth harmonic related to that of the fundamental ω_0? How is ω_0 related to the period T of the periodic function? (See S²)

Concept Question 5-3: What steps constitute the Fourier-series solution procedure? (See S²)

5-3 Fourier Series Representations

In 1822, the French mathematician Jean Baptiste Joseph Fourier developed an elegant formulation for representing periodic functions in terms of a series of sinusoidal harmonics. The representation is known today as the ***Fourier series***, and the formulation is called the ***Fourier theorem***. To guarantee that a periodic function $x(t)$ has a realizable Fourier series, it should satisfy a set of requirements known as the ***Dirichlet conditions***, which we shall discuss in Section 5-10.3. Fortunately, any periodic function generated by a real system will automatically meet these conditions, and therefore we are assured that its Fourier series does indeed exist.

The ***Fourier theorem*** states that a periodic function $x(t)$ of ***period*** T_0 can be expanded in any of the following three forms:

$$x(t) = a_0 + \sum_{n=1}^{\infty} [a_n \cos(n\omega_0 t) + b_n \sin(n\omega_0 t)] \tag{5.26a}$$

(sine/cosine representation)

$$= c_0 + \sum_{n=1}^{\infty} c_n \cos(n\omega_0 t + \phi_n) \tag{5.26b}$$

(amplitude/phase representation)

$$= \sum_{n=-\infty}^{\infty} \mathbf{x}_n e^{jn\omega_0 t}, \tag{5.26c}$$

(exponential representation)

where $\omega_0 = 2\pi/T_0$ is the ***fundamental angular frequency*** in rad/s. The ***fundamental frequency***, measured in Hz, is $f_0 = 1/T_0$. The first two representations can be used only if $x(t)$ is real-valued, whereas the exponential representation can be used for complex-valued $\mathbf{x}(t)$ as well.

- The constant terms a_0 and c_0 are equal and both are called dc (direct current) or ***average*** terms, since they have a frequency of zero Hz and are the average value of $x(t)$.
- The $n = 1$ terms $[a_1 \cos(\omega_0 t) + b_1 \sin(\omega_0 t)$ and $c_1 \cos(\omega_0 t + \phi_1)]$ are called ***fundamentals***. The period of the fundamental term is T_0, the same as the period of $x(t)$.
- The terms

  $$a_n \cos(n\omega_0 t) + b_n \sin(n\omega_0 t)$$

  and

  $$c_n \cos(n\omega_0 t + \phi_n)$$

  are called the ***nth harmonics***. The frequency of the nth harmonic is n/T_0 Hz.
- In music theory, harmonics are called ***overtones***, since they add richness to the single-frequency sound of a simple tone.
- The $\{a_n, b_n, c_n, \mathbf{x}_n\}$ are called ***Fourier coefficients***.

Here are three ways to think about Fourier series:

1. A Fourier series is a mathematical version of a *prism*; it breaks up a signal into different frequencies, just as a prism (or diffraction grating) breaks up light into different colors (which are light at different frequencies).

2. A Fourier series is a mathematical depiction of *adding overtones* to a basic note to give a richer and fuller sound. It can also be used as a formula for *synthesis* of sounds and tones.

3. A Fourier series is a representation of $x(t)$ in terms of *orthogonal functions*.

5-4 Computation of Fourier Series Coefficients

5-4.1 Sine / Cosine Representation

According to Eq. (5.26a), the sine/cosine representation of a periodic function $x(t)$ is given by

$$x(t) = a_0 + \sum_{n=1}^{\infty} [a_n \cos(n\omega_0 t) + b_n \sin(n\omega_0 t)].$$

(sine/cosine representation) (5.27)

Its *Fourier coefficients* are determined by evaluating integral expressions involving $x(t)$, namely,

$$a_0 = \frac{1}{T_0} \int_0^{T_0} x(t)\, dt, \quad (5.28a)$$

$$a_n = \frac{2}{T_0} \int_0^{T_0} x(t) \cos(n\omega_0 t)\, dt, \quad (5.28b)$$

and $$b_n = \frac{2}{T_0} \int_0^{T_0} x(t) \sin(n\omega_0 t)\, dt. \quad (5.28c)$$

Table 5-2: Trigonometric integral properties for any integers m and n with $n \neq 0$ and $m \neq 0$. The integration period $T_0 = 2\pi/\omega_0$, and angles ϕ, ϕ_1, and ϕ_2 are any time-independent constants.

Property	Integral
1	$\int_0^{T_0} \sin(n\omega_0 t + \phi)\, dt = 0$
2	$\int_0^{T_0} \cos(n\omega_0 t + \phi)\, dt = 0$
3	$\int_0^{T_0} \sin(n\omega_0 t + \phi_1) \sin(m\omega_0 t + \phi_2)\, dt = 0$ $n \neq m$
4	$\int_0^{T_0} \cos(n\omega_0 t + \phi_1) \cos(m\omega_0 t + \phi_2)\, dt = 0$ $n \neq m$
5	$\int_0^{T_0} \sin(n\omega_0 t + \phi_1) \cos(m\omega_0 t + \phi_2)\, dt = 0$
6	$\int_0^{T_0} \sin^2(n\omega_0 t + \phi)\, dt = T_0/2$
7	$\int_0^{T_0} \cos^2(n\omega_0 t + \phi)\, dt = T_0/2$

Even though the indicated limits of integration are from 0 to T_0, the expressions are equally valid if the lower limit is changed to t_0 and the upper limit to $(t_0 + T_0)$, for any value of t_0. In some cases, the evaluation is easier to perform by integrating from $-T_0/2$ to $T_0/2$.

To verify the validity of the expressions given by Eq. (5.28), we will make use of the trigonometric integral properties listed in Table 5-2.

dc Fourier Component a_0

The average value of a periodic function is obtained by integrating it over a complete period T_0 and then dividing the integral by T_0. Applying the definition to Eq. (5.27) gives

$$\begin{aligned}\frac{1}{T_0}\int_0^{T_0} x(t)\,dt &= \frac{1}{T_0}\int_0^{T_0} a_0\,dt \\ &+ \frac{1}{T_0}\int_0^{T_0}\left[\sum_{n=1}^{\infty}[a_n\cos(n\omega_0 t)+b_n\sin(n\omega_0 t)]\right]dt \\ &= a_0 + \frac{1}{T_0}\int_0^{T_0} a_1\cos(\omega_0 t)\,dt \\ &+ \frac{1}{T_0}\int_0^{T_0} a_2\cos(2\omega_0 t)\,dt + \cdots \\ &+ \frac{1}{T_0}\int_0^{T_0} b_1\sin(\omega_0 t)\,dt \\ &+ \frac{1}{T_0}\int_0^{T_0} b_2\sin(2\omega_0 t)\,dt + \cdots . \end{aligned} \tag{5.29}$$

According to property #1 in Table 5-2, the average value of a sine function is zero, and the same is true for a cosine function (property #2). Hence, all of the terms in Eq. (5.29) containing $\cos n\omega_0 t$ or $\sin n\omega_0 t$ will vanish, leaving behind

$$\frac{1}{T_0}\int_0^{T_0} x(t)\,dt = a_0, \tag{5.30}$$

which is identical with the definition given by Eq. (5.28a).

a_n Fourier Coefficients

Multiplication of both sides of Eq. (5.27) by $\cos m\omega_0 t$ (with m being any integer value equal to or greater than 1) followed with integration over $[0, T_0]$ yields

$$\begin{aligned}\int_0^{T_0} x(t)\,\cos(m\omega_0 t)\,dt &= \int_0^{T_0} a_0\cos(m\omega_0 t)\,dt \\ &+ \int_0^{T_0}\sum_{n=1}^{\infty} a_n\cos(n\omega_0 t)\cos(m\omega_0 t)\,dt \\ &+ \int_0^{T_0}\sum_{n=1}^{\infty} b_n\sin(n\omega_0 t)\cos(m\omega_0 t)\,dt. \\ &= \sum_{n=1}^{\infty} a_n\int_0^{T_0}\cos(n\omega_0 t)\cos(m\omega_0 t)\,dt \\ &+ \sum_{n=1}^{\infty} b_n\int_0^{T_0}\sin(n\omega_0 t)\cos(m\omega_0 t)\,dt. \end{aligned} \tag{5.31}$$

On the right-hand side of Eq. (5.31):

(1) the term containing a_0 is equal to zero (property #2 in Table 5-2),

(2) all terms containing b_n are equal to zero (property #5), and

(3) all terms containing a_n are equal to zero (property #4), except when $m = n$, in which case property #7 applies.

Hence, after eliminating all of the zero-valued terms and then setting $m = n$ in the two remaining terms, we have

$$\int_0^{T_0} x(t)\,\cos(n\omega_0 t)\,dt = a_n\,\frac{T_0}{2}, \tag{5.32}$$

which proves Eq. (5.28b).

b_n Fourier Coefficients

Similarly, if we were to repeat the preceding process, after multiplication of Eq. (5.27) by $\sin m\omega_0 t$ (instead of $\cos m\omega_0 t$), we would conclude with a result affirming the validity of Eq. (5.28c).

To develop an appreciation for how the components of the Fourier series add up to represent the periodic waveform, let us consider the square-wave voltage waveform shown in Fig. 5-2(a). Over the period extending from $-T_0/2$ to $T_0/2$, $x(t)$ is given by

$$x(t) = \begin{cases} -A & \text{for } -T_0/2 < t < -T_0/4, \\ A & \text{for } -T_0/4 < t < T_0/4, \\ -A & \text{for } T_0/4 < t < T_0/2. \end{cases}$$

If we apply Eq. (5.28) with integration limits $[-T_0/2, T_0/2]$ to evaluate the Fourier coefficients and then use them in Eq. (5.27), we end up with the series

$$\begin{aligned} x(t) &= \sum_{n=1}^{\infty}\frac{4A}{n\pi}\sin\left(\frac{n\pi}{2}\right)\cos\left(\frac{2n\pi t}{T_0}\right) \\ &= \frac{4A}{\pi}\cos\left(\frac{2\pi t}{T_0}\right) - \frac{4A}{3\pi}\cos\left(\frac{6\pi t}{T_0}\right) \\ &+ \frac{4A}{5\pi}\cos\left(\frac{10\pi t}{T_0}\right) - \cdots . \end{aligned}$$

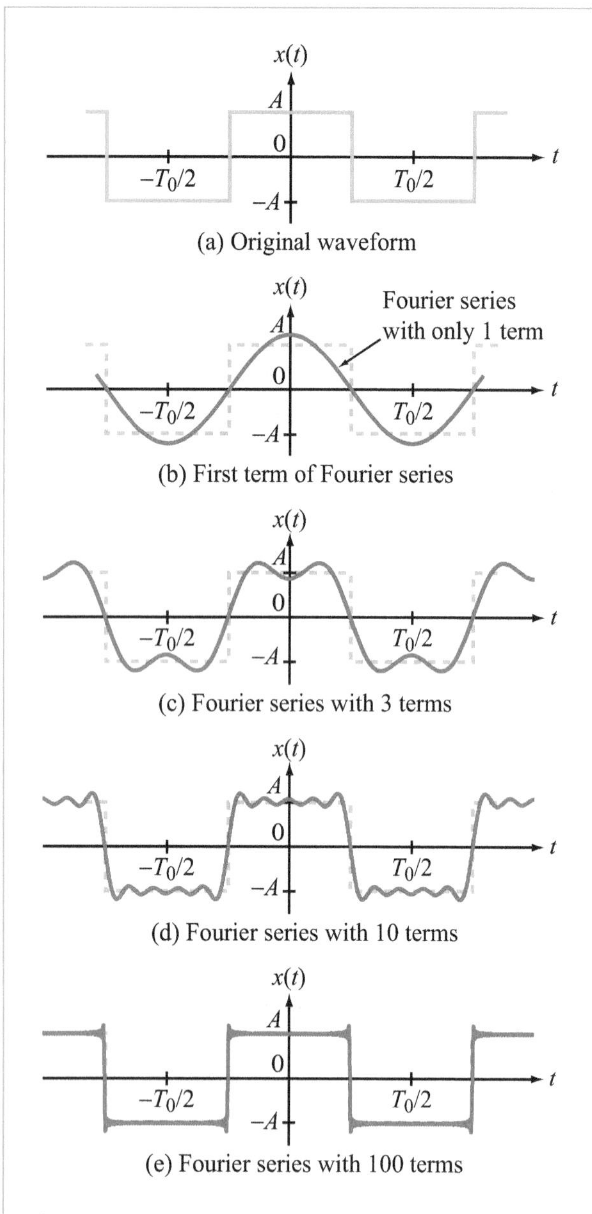

Figure 5-2: Comparison of the square-wave waveform with its Fourier series representation using only the first term (b), the sum of the first three (c), ten (d), and 100 terms (e).

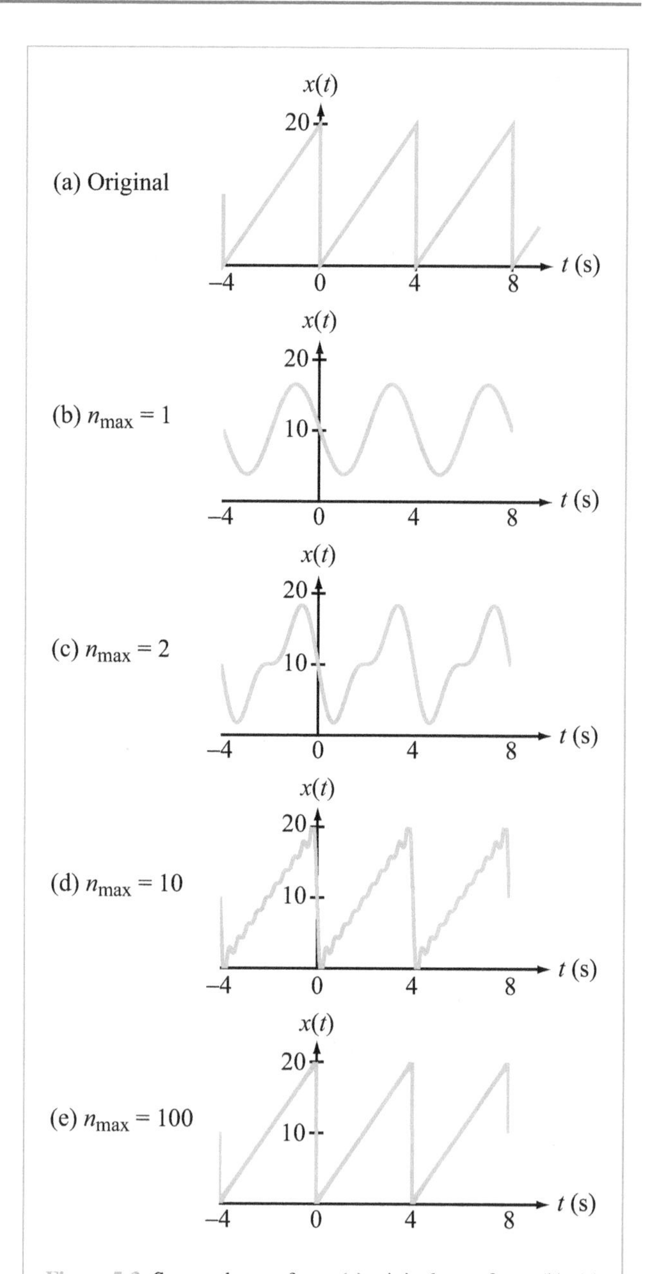

Figure 5-3: Sawtooth waveform: (a) original waveform, (b)–(e) representation by a truncated Fourier series with $n_{\max} = 1, 2, 10$, and 100, respectively.

Alone, the first term of the series provides a crude approximation of the square wave (Fig. 5-2(b)), but as we add more and more terms, the sum starts to better resemble the general shape of the square wave, as demonstrated by the waveforms in Figs. 5-2(c) to (e).

Example 5-2: Sawtooth Waveform

Express the sawtooth waveform shown in Fig. 5-3(a) in terms of a Fourier series, and then evaluate how well the original waveform is represented by a truncated series in which the summation stops when n reaches a specified truncation number $n_{\max}$. Generate plots for $n_{\max} = 1$, 2, 10, and 100.

Solution: The sawtooth waveform is characterized by a period $T_0 = 4$ s and $\omega_0 = 2\pi/T_0 = \pi/2$ (rad/s). Over the waveform's first cycle ($t = 0$ to $t = 4$ s), its amplitude variation is given by

$$x(t) = 5t, \qquad \text{for } 0 \le t \le 4 \text{ s}.$$

Application of Eq. (5.28) yields

$$a_0 = \frac{1}{T_0} \int_0^{T_0} x(t)\, dt = \frac{1}{4} \int_0^4 5t\, dt = 10,$$

$$a_n = \frac{2}{T_0} \int_0^{T_0} x(t)\, \cos(n\omega_0 t)\, dt$$

$$= \frac{2}{4} \int_0^4 5t \cos\left(\frac{n\pi}{2}\, t\right)\, dt = 0,$$

and

$$b_n = \frac{2}{T_0} \int_0^{T_0} x(t)\, \sin(n\omega_0 t)\, dt$$

$$= \frac{2}{4} \int_0^4 5t \sin\left(\frac{n\pi}{2}\, t\right)\, dt = -\frac{20}{n\pi}.$$

Upon inserting these results into Eq. (5.27), we obtain the following *complete* Fourier series representation for the sawtooth waveform:

$$x(t) = 10 - \frac{20}{\pi} \sum_{n=1}^{\infty} \frac{1}{n} \sin\left(\frac{n\pi}{2}\, t\right).$$

The $n_{\max}$-*truncated series* is identical in form with the complete series, except that the summation is terminated after the index n reaches $n_{\max}$. Figures 5-3(b) through (e) display the waveforms calculated using the truncated series with $n_{\max} = 1$, 2, 10, and 100. As expected, the addition of more terms improves the accuracy of the Fourier-series representation, but even with only 10 terms (in addition to the dc component), the truncated series appears to provide a reasonable approximation of the original waveform.

Concept Question 5-4: Is the Fourier-series representation given by Eq. (5.27) applicable to a periodic function that starts at $t = 0$ (and is zero for $t < 0$)? (See S²)

Concept Question 5-5: What is a truncated series? (See S²)

Exercise 5-1: Obtain the Fourier-series representation for the waveform shown in Fig. E5-1.

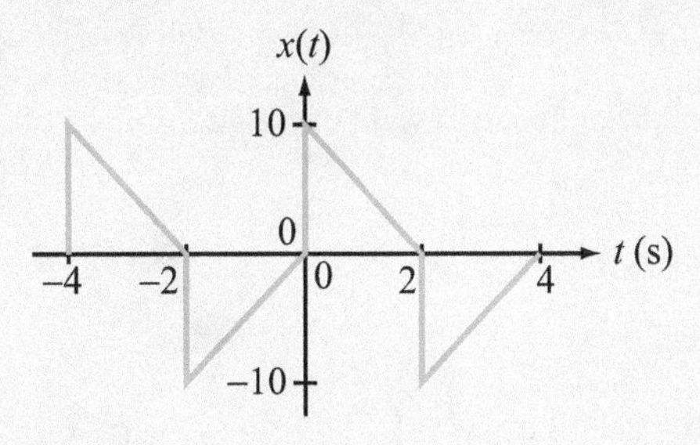

Figure E5-1

Answer:

$$x(t) = \sum_{n=1}^{\infty} \left[\frac{20}{n^2\pi^2} (1 - \cos n\pi) \cos \frac{n\pi t}{2} + \frac{10}{n\pi} (1 - \cos n\pi) \sin \frac{n\pi t}{2} \right].$$

(See S²)

5-4.2 Amplitude and Phase Representation

In the sine/cosine Fourier-series representation given by Eq. (5.27), at each value of the integer index n, the summation contains the sum of a sine term and a cosine term, both at angular frequency $n\omega_0$. The sum can be converted into a single sinusoid as follows. For $n \neq 0$,

$$a_n \cos n\omega_0 t + b_n \sin n\omega_0 t = c_n \cos(n\omega_0 t + \phi_n), \tag{5.33}$$

where c_n is called the ***amplitude of the nth harmonic*** and ϕ_n is its associated ***phase***. The relationships between (c_n, ϕ_n) and (a_n, b_n) are obtained by expanding the right-hand side of Eq. (5.33) in accordance with the trigonometric identity

$$\cos(x + y) = \cos x \cos y - \sin x \sin y. \tag{5.34}$$

Thus,

$$a_n \cos(n\omega_0 t) + b_n \sin(n\omega_0 t) = c_n \cos\phi_n \cos(n\omega_0 t) - c_n \sin\phi_n \sin(n\omega_0 t). \tag{5.35}$$

Upon equating the coefficients of $\cos(n\omega_0 t)$ and $\sin(n\omega_0 t)$ on one side of the equation to their respective counterparts on the other side, we have for $n \neq 0$

$$a_n = c_n \cos\phi_n \quad \text{and} \quad b_n = -c_n \sin\phi_n, \tag{5.36}$$

which can be combined to yield the relationships

$$c_n = \sqrt{a_n^2 + b_n^2}$$

and

$$\phi_n = \begin{cases} -\tan^{-1}\left(\dfrac{b_n}{a_n}\right), & a_n > 0 \\ \pi - \tan^{-1}\left(\dfrac{b_n}{a_n}\right), & a_n < 0 \end{cases} \tag{5.37}$$

In complex vector form, we have

$$c_n\underline{/\phi_n} = a_n - jb_n. \tag{5.38}$$

In view of Eq. (5.33), the sine/cosine Fourier-series representation of $x(t)$ can be rewritten in the alternative ***amplitude/phase*** format

$$x(t) = c_0 + \sum_{n=1}^{\infty} c_n \cos(n\omega_0 t + \phi_n), \tag{5.39}$$

(amplitude/phase representation)

where we renamed a_0 as c_0 for the sake of notational consistency. Associated with each discrete frequency harmonic $n\omega_0$ is an amplitude c_n and phase ϕ_n.

A ***line spectrum*** of a periodic signal $x(t)$ is a visual depiction of its Fourier coefficients, c_n and ϕ_n. Its ***amplitude spectrum*** consists of vertical lines located at discrete values along the ω-axis, with a line of height c_0 located at dc ($\omega = 0$), another of height c_1 at $\omega = \omega_0$, a third of height c_2 at $\omega = 2\omega_0$, and so on. Similarly, the ***phase spectrum*** of $x(t)$ consists of lines of lengths proportional to the values of ϕ_n with each located at its corresponding harmonic $n\omega_0$. Line spectra show at a glance which frequencies in the spectrum of $x(t)$ are most significant and which are not.

The line spectra associated with c_n and ϕ_n are called ***one-sided line spectra*** because they are defined for only non-negative values of $n\omega_0$, which follows from the definition of the amplitude/phase representation given by Eq. (5.39), wherein the summation is over only positive values of n. This is to distinguish it from ***two-sided line spectra*** associated with the exponential representation introduced later in Section 5-4.3.

► Note that $c_0 < 0$ if $a_0 < 0$. Some books define $c_0 = |a_0|$ and $\phi_0 = \pi$ if $a_0 < 0$. In this book we define $c_0 = a_0$ for simplicity. ◄

Example 5-3: Line Spectra

Generate and plot the amplitude and phase spectra of the periodic waveform displayed in Fig. 5-4(a).

Solution: The periodic waveform has a period $T_0 = 2$ s. Hence, $\omega_0 = 2\pi/T_0 = 2\pi/2 = \pi$ rad/s, and the functional expression for $x(t)$ over its first cycle along the positive t-axis is

$$x(t) = \begin{cases} 1 - t, & \text{for } 0 < t \leq 1 \text{ s} \\ 0, & \text{for } 1 \leq t \leq 2 \text{ s}. \end{cases}$$

The dc component of $x(t)$ is given by

$$c_0 = a_0 = \frac{1}{T_0}\int_0^{T_0} x(t)\, dt = \frac{1}{2}\int_0^1 (1-t)\, dt = 0.25,$$

which is equal to the area under a single triangle divided by the period $T_0 = 2$ s.

For the other Fourier coefficients, evaluation of the expressions given by Eqs. (5.28b and c) leads to

$$\begin{aligned} a_n &= \frac{2}{T_0}\int_0^{T_0} x(t)\cos(n\omega_0 t)\, dt \\ &= \frac{2}{2}\int_0^1 (1-t)\cos(n\pi t)\, dt \\ &= \frac{1}{n\pi}\sin(n\pi t)\Big|_0^1 \\ &\quad - \left[\frac{1}{n^2\pi^2}\cos(n\pi t) + \frac{t}{n\pi}\sin(n\pi t)\right]\Bigg|_0^1 \\ &= \frac{1}{n^2\pi^2}[1 - \cos n\pi], \end{aligned}$$

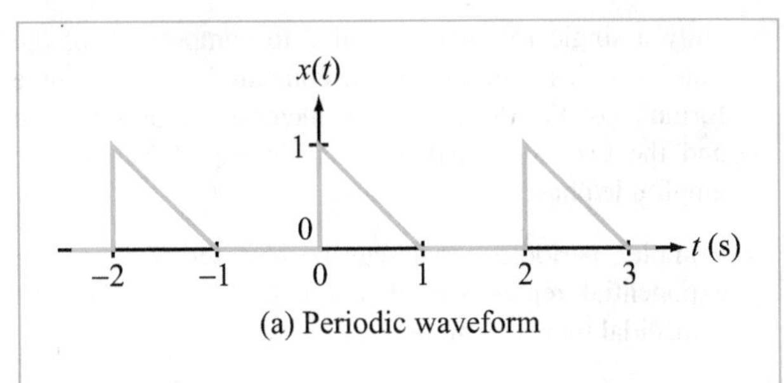

(a) Periodic waveform

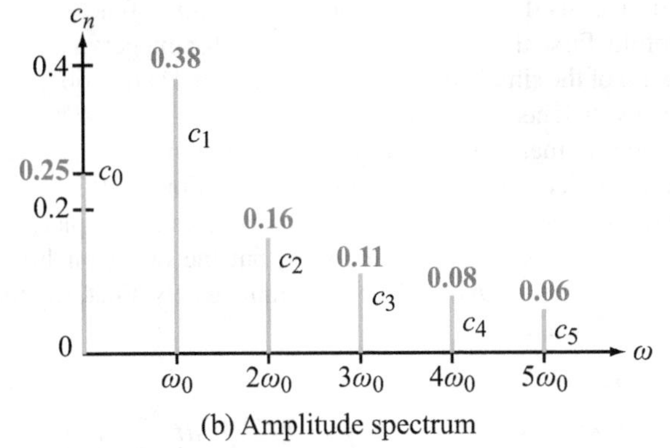

(b) Amplitude spectrum

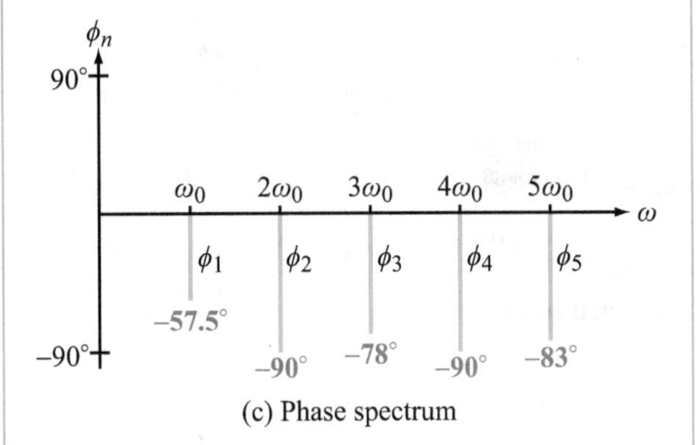

(c) Phase spectrum

Figure 5-4: Periodic waveform of Example 5-3 with its associated line spectra.

and

$$b_n = \frac{2}{T_0} \int_0^{T_0} x(t) \sin(n\omega_0 t)\, dt$$

$$= \frac{2}{2} \int_0^1 (1 - t) \sin(n\pi t)\, dt$$

$$= -\frac{1}{n\pi} \cos(n\pi t)\Big|_0^1 - \left[\frac{1}{n^2\pi^2} \sin(n\pi t) - \frac{t}{n\pi} \cos(n\pi t)\right]\Bigg|_0^1$$

$$= \frac{1}{n\pi} .$$

By Eq. (5.37), the harmonic amplitudes and phases are given by

$$c_n = \sqrt{a_n^2 + b_n^2}$$

$$= \left[\left(\frac{1}{n^2\pi^2}\,[1 - \cos n\pi]\right)^2 + \left(\frac{1}{n\pi}\right)^2\right]^{1/2}$$

$$= \begin{cases} \left(\dfrac{4}{n^4\pi^4} + \dfrac{1}{n^2\pi^2}\right)^{1/2} & \text{for } n = \text{odd}, \\ \dfrac{1}{n\pi} & \text{for } n = \text{even}, \end{cases}$$

and

$$\phi_n = -\tan^{-1} \frac{b_n}{a_n}$$

$$= -\tan^{-1}\left(\frac{n\pi}{[1 - \cos n\pi]}\right)$$

$$= \begin{cases} -\tan^{-1}\left(\dfrac{n\pi}{2}\right) & \text{for } n = \text{odd} \\ -90^\circ & \text{for } n = \text{even}. \end{cases}$$

The values of c_n and ϕ_n for the first three terms are

$$c_1 = 0.38, \qquad \phi_1 = -57.5^\circ,$$
$$c_2 = 0.16, \qquad \phi_2 = -90^\circ,$$
$$c_3 = 0.11, \quad \text{and} \quad \phi_3 = -78^\circ.$$

Spectral plots of c_n and ϕ_n are shown in Figs. 5-4(b) and (c), respectively.

Exercise 5-2: Obtain the line spectra associated with the periodic function of Exercise 5-1.

Answer:

$$c_n = [1 - \cos(n\pi)] \frac{20}{n^2\pi^2} \sqrt{1 + \frac{n^2\pi^2}{4}}$$

and

$$\phi_n = -\tan^{-1}\left(\frac{n\pi}{2}\right).$$

(See S²)

5-4.3 Exponential Representation

Another Fourier series representation of periodic signals $x(t)$ uses complex exponential functions $e^{jn\omega_0 t}$. As is often the case in this book, using complex exponential functions instead of sinusoids reduces the amount of algebra.

The exponential representation of a periodic signal $x(t)$ is

$$x(t) = \sum_{n=-\infty}^{\infty} \mathbf{x}_n e^{jn\omega_0 t}. \quad (5.40)$$

(exponential representation)

where the Fourier series coefficients $\mathbf{x}_n$ are now complex numbers computed using

$$\mathbf{x}_n = \frac{1}{T_0} \int_0^{T_0} x(t)\, e^{-jn\omega_0 t}\, dt. \quad (5.41)$$

As with the other two representations, any interval of length T_0 may be used as the interval of integration, since the integrand is periodic with period T_0. If $x(t)$ is real-valued, the following ***conjugate symmetry*** relation is applicable:

$$\mathbf{x}_{-n} = \mathbf{x}_n^* \quad (x(t) \text{ real}). \quad (5.42)$$

The exponential representation is considerably simpler than the previous representations:

- Only a single formula is needed to compute all of the Fourier series coefficients $\mathbf{x}_n$, compared with the three formulas of Eq. (5.28) for the sine/cosine representation and the two additional formulas in Eq. (5.37) for the amplitude/phase representation.

- Complex periodic signals can be represented using the exponential representation, but not with either of the sinusoidal representations.

The reasons for using complex exponential functions are twofold. First, the ***orthogonality*** of $e^{jn\omega_0 t}$ for integers n enables the use of the simple formula given by Eq. (5.41) for computing the coefficients $\mathbf{x}_n$. Second, the ***completeness*** of $e^{jn\omega_0 t}$ for integers n means that any periodic function $x(t)$ satisfying the Dirichlet conditions discussed in Section 5-10.3 can be represented using Eq. (5.40). The mathematics of completeness are beyond the scope of this book, but the orthogonality of $e^{jn\omega_0 t}$ for integers n can be derived immediately. First, suppose $m \neq n$. Then

$$\int_0^{T_0} e^{jm\omega_0 t} e^{-jn\omega_0 t}\, dt = \int_0^{T_0} e^{j(m-n)\omega_0 t}\, dt$$

$$= \left. \frac{e^{j(m-n)\omega_0 t}}{j(m-n)\omega_0} \right|_0^{T_0} = 0, \quad (5.43)$$

because $e^{j(m-n)\omega_0 0} = 1$ and

$$e^{j(m-n)\omega_0 T_0} = e^{j2\pi(m-n)} = 1.$$

If $m = n$, then we have

$$\int_0^{T_0} e^{jn\omega_0 t} e^{-jn\omega_0 t}\, dt = \int_0^{T_0} 1\, dt = T_0. \quad (5.44)$$

Combining these two results, we have

$$\int_0^{T_0} e^{jm\omega_0 t} e^{-jn\omega_0 t}\, dt = \begin{cases} T_0 & \text{if } m = n, \\ 0 & \text{if } m \neq n. \end{cases} \quad (5.45)$$

The formula given by Eq. (5.41) for the coefficients $\mathbf{x}_n$ then follows from multiplying the complex exponential representation in Eq. (5.40) by $e^{-jm\omega_0 t}$ and integrating from 0 to T_0. The process leads to

$$\int_0^{T_0} x(t)\, e^{-jm\omega_0 t}\, dt = \int_0^{T_0} \sum_{n=-\infty}^{\infty} \mathbf{x}_n e^{jn\omega_0 t} e^{-jm\omega_0 t}\, dt. \quad (5.46)$$

Interchanging the order of integration and summation gives

$$\int_0^{T_0} x(t)\, e^{-jm\omega_0 t}\, dt = \sum_{n=-\infty}^{\infty} \mathbf{x}_n \int_0^{T_0} e^{j(n-m)\omega_0 t}\, dt. \qquad (5.47)$$

In view of Eq. (5.45), Eq. (5.47) simplifies to

$$\int_0^{T_0} x(t)\, e^{-jm\omega_0 t}\, dt = T_0 \mathbf{x}_m, \qquad (5.48)$$

because the only nonzero term in the summation is the one with $n = m$. Dividing both sides by T_0 and replacing m with n leads to Eq. (5.41).

Example 5-4: Complex Exponential Representation of Square Wave

Define the periodic signal $x(t)$ for all integers k as

$$x(t) = \begin{cases} \pi/4 & \text{for } k\pi < t < (k+1)\pi, \\ -\pi/4 & \text{for } (k-1)\pi < t < k\pi. \end{cases} \qquad (5.49)$$

Solution: Signal $x(t)$ is a square wave (Fig. 5-6(e)) that jumps between $\frac{\pi}{4}$ and $-\frac{\pi}{4}$ at times that are integer multiples of π. Its amplitude is $A = \frac{\pi}{4}$ and its period $T_0 = 2\pi$, so

$$\omega_0 = \frac{2\pi}{T_0} = \frac{2\pi}{2\pi} = 1.$$

The coefficients $\mathbf{x}_n$ of its exponential representation can be computed using Eq. (5.41) as

$$\mathbf{x}_n = \frac{1}{2\pi}\int_0^{\pi} \frac{\pi}{4}\, e^{-jnt}\, dt + \frac{1}{2\pi}\int_{\pi}^{2\pi} \frac{-\pi}{4}\, e^{-jnt}\, dt$$
$$= \begin{cases} -j/(2n) & \text{for } n \text{ odd}, \\ 0 & \text{for } n \text{ even}. \end{cases} \qquad (5.50)$$

This is because $e^{-jn\pi} = 1$ for n even and -1 for n odd. The exponential representation can be written out as

$$x(t) = \frac{-j}{2}\, e^{jt} + \frac{-j}{6}\, e^{j3t} + \frac{-j}{10}\, e^{j5t} + \cdots$$
$$+ \frac{j}{2}\, e^{-jt} + \frac{j}{6}\, e^{-j3t} + \frac{j}{10}\, e^{-j5t} + \cdots \qquad (5.51)$$

The top row includes terms for $n = 1, 3, 5 \ldots$ and the bottom row includes terms for $n = -1, -3, -5 \ldots$. The dc ($n = 0$) term is zero, as expected, since $x(t)$ clearly has an average value of zero.

Recalling that

$$\sin(x) = \frac{e^{jx} - e^{-jx}}{2j} = -j\frac{e^{jx} - e^{-jx}}{2},$$

the exponential representation can be rewritten in sine/cosine form as

$$x(t) = \sin(t) + \frac{1}{3}\sin(3t) + \frac{1}{5}\sin(5t) + \cdots, \qquad (5.52)$$

which in this case is simpler than the exponential representation, as it takes advantage of $x(t)$ being an odd function.

The exponential representation can also be derived from the sine/cosine representation. According to Eq. (5.27), a periodic function of period T_0 and corresponding fundamental frequency $\omega_0 = 2\pi/T_0$ can be represented by the series

$$x(t) = a_0 + \sum_{n=1}^{\infty} [a_n \cos(n\omega_0 t) + b_n \sin(n\omega_0 t)]. \qquad (5.53)$$

Sine and cosine functions can be converted into complex exponentials via Euler's identity:

$$\cos(n\omega_0 t) = \frac{1}{2}(e^{jn\omega_0 t} + e^{-jn\omega_0 t}), \qquad (5.54a)$$

and

$$\sin(n\omega_0 t) = \frac{1}{j2}(e^{jn\omega_0 t} - e^{-jn\omega_0 t}). \qquad (5.54b)$$

Upon inserting Eqs. (5.54a and b) into Eq. (5.53), we have

$$x(t) =$$
$$a_0 + \sum_{n=1}^{\infty}\left[\frac{a_n}{2}(e^{jn\omega_0 t} + e^{-jn\omega_0 t}) + \frac{b_n}{j2}(e^{jn\omega_0 t} - e^{-jn\omega_0 t})\right]$$
$$= a_0 + \sum_{n=1}^{\infty}\left[\left(\frac{a_n - jb_n}{2}\right) e^{jn\omega_0 t} + \left(\frac{a_n + jb_n}{2}\right) e^{-jn\omega_0 t}\right]$$
$$= a_0 + \sum_{n=1}^{\infty} [\mathbf{x}_n e^{jn\omega_0 t} + \mathbf{x}_{-n} e^{-jn\omega_0 t}], \qquad (5.55)$$

Table 5-3: Fourier series representations for a real-valued periodic function $x(t)$.

Cosine/Sine	Amplitude/Phase	Exponential
$x(t) = a_0 + \sum_{n=1}^{\infty} [a_n \cos(n\omega_0 t) + b_n \sin(n\omega_0 t)]$	$x(t) = c_0 + \sum_{n=1}^{\infty} c_n \cos(n\omega_0 t + \phi_n)$	$x(t) = \sum_{n=-\infty}^{\infty} \mathbf{x}_n e^{jn\omega_0 t}$
$a_0 = \frac{1}{T_0} \int_0^{T_0} x(t)\,dt$	$c_n e^{j\phi_n} = a_n - jb_n$	$\mathbf{x}_n = \lvert\mathbf{x}_n\rvert e^{j\phi_n};\ \mathbf{x}_{-n} = \mathbf{x}_n^*;\ \phi_{-n} = -\phi_n$
$a_n = \frac{2}{T_0} \int_0^{T_0} x(t) \cos n\omega_0 t\,dt$	$c_n = \sqrt{a_n^2 + b_n^2}$	$\lvert\mathbf{x}_n\rvert = c_n/2;\ \mathbf{x}_0 = c_0$
$b_n = \frac{2}{T_0} \int_0^{T_0} x(t) \sin n\omega_0 t\,dt$	$\phi_n = \begin{cases} -\tan^{-1}(b_n/a_n), & a_n > 0 \\ \pi - \tan^{-1}(b_n/a_n), & a_n < 0 \end{cases}$	$\mathbf{x}_n = \frac{1}{T_0} \int_0^{T_0} x(t)\, e^{-jn\omega_0 t}\,dt$
$a_0 = c_0 = \mathbf{x}_0;\ a_n = c_n \cos\phi_n;\ b_n = -c_n \sin\phi_n;\ \mathbf{x}_n = \frac{1}{2}(a_n - jb_n)$		

where we introduced the complex coefficients

$$\mathbf{x}_n = \frac{a_n - jb_n}{2} \quad \text{and} \quad \mathbf{x}_{-n} = \frac{a_n + jb_n}{2} = \mathbf{x}_n^*. \tag{5.56}$$

As the index n is incremented from 1 to ∞, the second term in Eq. (5.55) generates the series

$$\mathbf{x}_{-1} e^{-j\omega_0 t} + \mathbf{x}_{-2} e^{-j2\omega_0 t} + \cdots,$$

which can also be generated by $\mathbf{x}_n e^{jn\omega_0 t}$ with n decremented from -1 to $-\infty$. This equivalence allows us to express $x(t)$ in the compact exponential form:

$$x(t) = \sum_{n=-\infty}^{\infty} \mathbf{x}_n e^{jn\omega_0 t}, \tag{5.57}$$

where

$$\mathbf{x}_0 = a_0 = c_0 = |\mathbf{x}_0| e^{j\phi_0}, \tag{5.58}$$

and the range of n has been expanded to $(-\infty, \infty)$. For all coefficients $\mathbf{x}_n$, including $\mathbf{x}_0$, it is easy to show that

$$\mathbf{x}_n = \frac{1}{T_0} \int_{-T_0/2}^{T_0/2} x(t)\, e^{-jn\omega_0 t}\, dt. \tag{5.59}$$

► Even though the integration limits indicated in Eq. (5.59) are from $-T_0/2$ to $T_0/2$, they can be chosen arbitrarily so long as the upper limit exceeds the lower limit by exactly T_0. ◄

For easy reference, Table 5-3 provides a summary of the relationships associated with all three Fourier-series representations introduced in this chapter, namely the sine/cosine, amplitude/phase, and complex exponential.

The exponential representation given by Eq. (5.57) is characterized by ***two-sided line spectra*** because in addition to the dc value at $\omega = 0$, the magnitudes $|\mathbf{x}_n|$ and phases ϕ_n are defined at both positive and negative values of ω_0 and its harmonics.

It is easy to understand what a positive value of the angular frequency ω means, but what does a negative angular frequency mean? It does not have a physical meaning; defining ω along the negative axis is purely a mathematical convenience.

The single-sided line spectra of the amplitude/phase representation given by Eq. (5.39) and the two-sided exponential representation given by Eq. (5.57) are interrelated. For a real-valued periodic function $x(t)$ with fundamental angular frequency ω_0:

(a) The ***amplitude line spectrum*** is a plot of amplitudes c_n at $\omega = 0$ (dc value) and at positive, discrete values of ω, namely $n\omega_0$ with n a positive integer. The outcome is a single-sided spectrum similar to that displayed in Fig. 5-5(a).

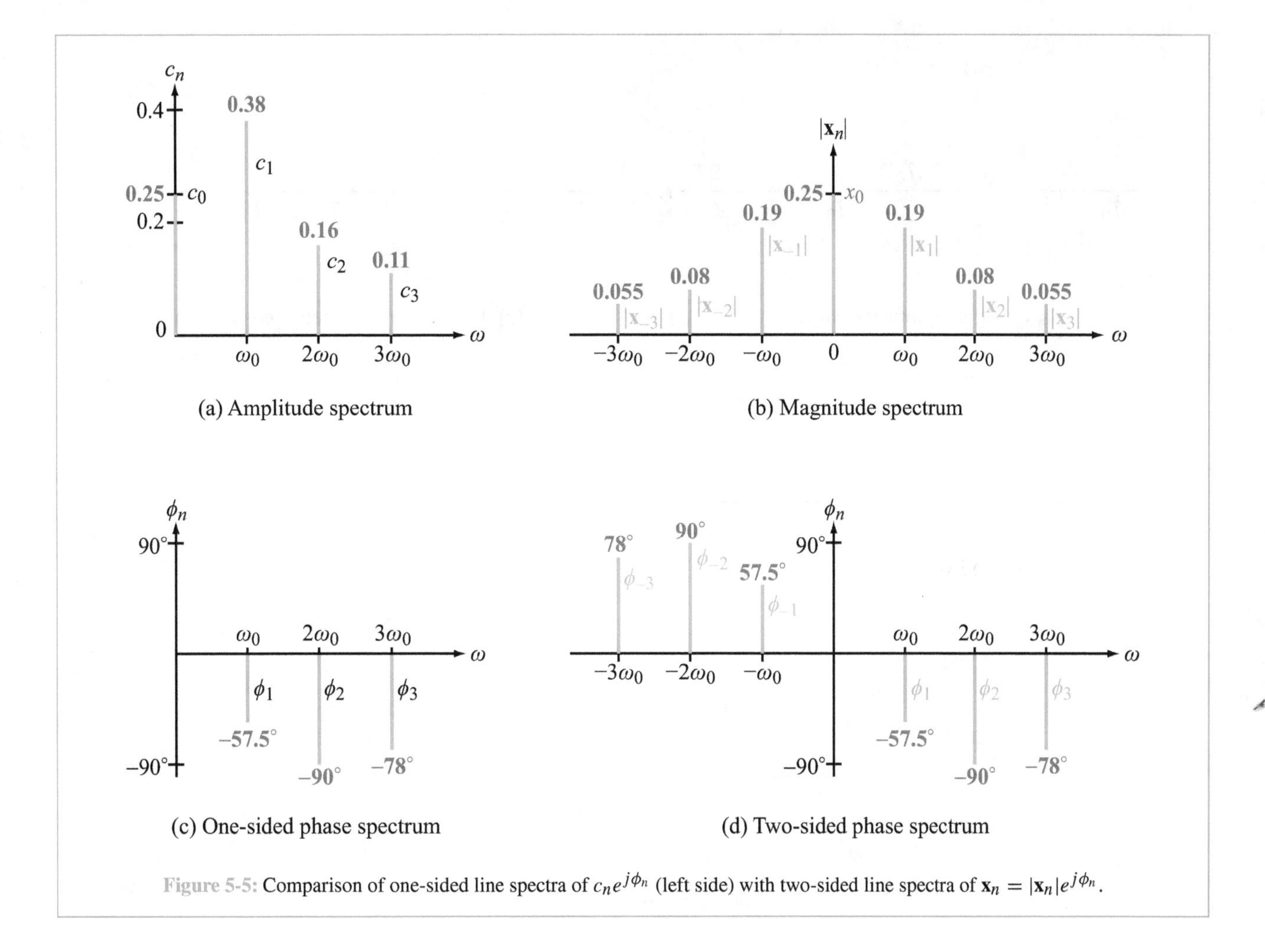

Figure 5-5: Comparison of one-sided line spectra of $c_n e^{j\phi_n}$ (left side) with two-sided line spectra of $\mathbf{x}_n = |\mathbf{x}_n| e^{j\phi_n}$.

(b) The ***magnitude line spectra*** is a plot of magnitudes $|\mathbf{x}_n|$ at $\omega = 0$ (dc value) and at both positive and negative values of ω_0 and its harmonics. The outcome is a two-sided spectrum as shown in Fig. 5-5(b).

(c) The two-sided magnitude spectrum has even symmetry about $n = 0$ (i.e., $|\mathbf{x}_n| = |\mathbf{x}_{-n}|$), and the magnitudes of $\mathbf{x}_n$ are half the corresponding amplitudes c_n. That is, $|\mathbf{x}_n| = c_n/2$, except for $n = 0$ in which case $|\mathbf{x}_0| = |c_0|$.

(d) The ***phase line spectrum*** ϕ_n is single-sided, whereas the phase line spectrum of $\mathbf{x}_n$ is two-sided. The right half of the two-sided phase spectrum (Fig. 5-5(d)) is identical with the one-sided phase spectrum of the amplitude/phase representation (Fig. 5-5(c)).

(e) The two-sided phase spectrum has odd symmetry about $n = 0$ (because $\phi_{-n} = -\phi_n$), as illustrated by Fig. 5-5(d). However, if $a_0 < 0$, then $\phi_0 = \pi$, in which case ϕ_n would no longer be an odd function of n.

Exercise 5-3: A periodic signal $x(t)$ has the exponential Fourier series

$$x(t) = (-2 + j0) + (3 + j4)e^{j2t} + (1 + j)e^{j4t} + (3 - j4)e^{-j2t} + (1 - j)e^{-j4t}.$$

Compute its cosine/sine and amplitude/phase Fourier series representations.

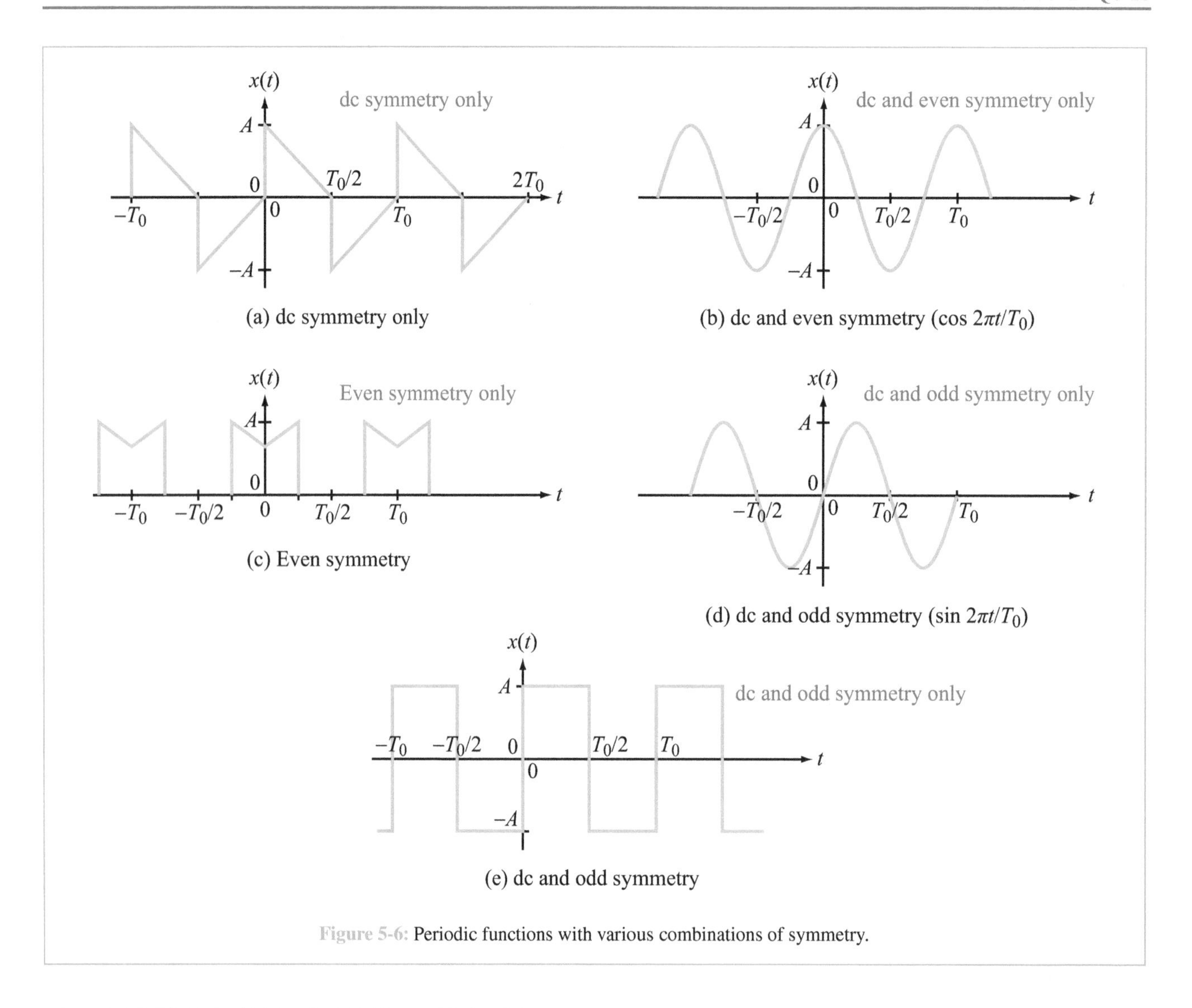

Figure 5-6: Periodic functions with various combinations of symmetry.

Answer: Amplitude/phase representation:

$$x(t) = -2 + 10\cos(2t + 53^\circ) + 2\sqrt{2}\cos(4t + 45^\circ).$$

Cosine/sine representation:

$$x(t) = -2 + 6\cos(2t) - 8\sin(2t) + 2\cos(4t) - 2\sin(4t).$$

(See S²)

5-4.4 Symmetry Considerations

Some functions may exhibit dc symmetry, wherein the area under their waveforms above the t-axis is equal to that below the t-axis. Consequently, $a_0 = c_0 = 0$. Examples include the functions in parts (a), (b), (d), and (e) of Fig. 5-6. Other forms of symmetry include even and odd symmetry, as displayed by some of the periodic functions in Fig. 5-6.

5-4.5 Fourier Coefficients of Even Functions

Recall from Chapter 1 that an even function has the property that $x(t) = x(-t)$. Even symmetry allows us to simplify Eqs. (5.28) to the following expressions:

Even Symmetry: $x(t) = x(-t)$

$$a_0 = \frac{2}{T_0} \int_0^{T_0/2} x(t)\, dt$$

$$a_n = \frac{4}{T_0} \int_0^{T_0/2} x(t) \cos(n\omega_0 t)\, dt \qquad (5.60)$$

$$b_n = 0$$

$$c_n = |a_n|, \qquad \phi_n = \begin{cases} 0 & \text{if } a_n > 0 \\ 180^\circ & \text{if } a_n < 0 \end{cases}$$

The expressions for a_0 and a_n are the same as given earlier by Eq. (5.28a and b), except that the integration limits are now over half of a period and the integral has been multiplied by a factor of 2. The simplification is justified by the even symmetry of $x(t)$. As was stated in connection with Eq. (5.28), the only restriction associated with the integration limits is that the upper limit has to be greater than the lower limit by exactly T_0. Hence, by choosing the limits to be $[-T_0/2, T_0/2]$ and then recognizing that the integral of $x(t)$ over $[-T_0/2, 0]$ is equal to the integral over $[0, T_0/2]$, we justify the changes reflected in the expression for a_0. A similar argument applies to the expression for a_n based on the fact that multiplication of an even function $x(t)$ by $\cos n\omega_0 t$, which itself is an even function, yields an even function.

The rationale for setting $b_n = 0$ for all n relies on the fact that multiplication of an even function $x(t)$ by $\sin n\omega_0 t$, which is an odd function, yields an odd function, and integration of an odd function over $[-T_0/2, T_0/2]$ is always equal to zero. This is because the integral of an odd function over $[-T_0/2, 0]$ is equal in magnitude, but opposite in sign, to the integral over $[0, T_0/2]$.

5-4.6 Fourier Coefficients of Odd Functions

An odd function has the property that $x(-t) = -x(t)$. Odd symmetry allows us to simplify the expressions for the Fourier coefficients.

Odd Symmetry: $x(t) = -x(-t)$

$$a_0 = 0 \qquad a_n = 0$$

$$b_n = \frac{4}{T_0} \int_0^{T_0/2} x(t) \sin(n\omega_0 t)\, dt \qquad (5.61)$$

$$c_n = |b_n| \qquad \phi_n = \begin{cases} -90^\circ & \text{if } b_n > 0 \\ 90^\circ & \text{if } b_n < 0 \end{cases}$$

Selected waveforms are displayed in Table 5-4 together with their corresponding Fourier series expressions.

Example 5-5: M-Periodic Waveform

Evaluate the Fourier coefficients of the M-periodic waveform shown in Fig. 5-7(a),

Solution: The M waveform is even-symmetrical, its period is $T_0 = 4$ s, $\omega_0 = 2\pi/T_0 = \pi/2$ rad/s, and its functional form over the positive half period is

$$x(t) = \begin{cases} \frac{1}{2}(1+t) & 0 \le t \le 1 \text{ s}, \\ 0 & 1 < t \le 2 \text{ s}. \end{cases}$$

Application of Eq. (5.60) yields

$$a_0 = \frac{2}{T_0} \int_0^{T_0/2} x(t)\, dt$$

$$= \frac{2}{4} \int_0^1 \frac{1}{2}(1+t)\, dt, = 0.375,$$

$$a_n = \frac{4}{T_0} \int_0^{T_0/2} x(t) \cos(n\omega_0 t)\, dt$$

$$= \frac{4}{4} \int_0^1 \frac{1}{2}(1+t) \cos(n\omega_0 t)\, dt$$

$$= \frac{2}{n\pi} \sin\frac{n\pi}{2} + \frac{2}{n^2\pi^2}\left(\cos\frac{n\pi}{2} - 1\right),$$

and

$$b_n = 0.$$

Table 5-4: Fourier series expressions for a select set of periodic waveforms.

Waveform	Fourier Series
1. Square Wave	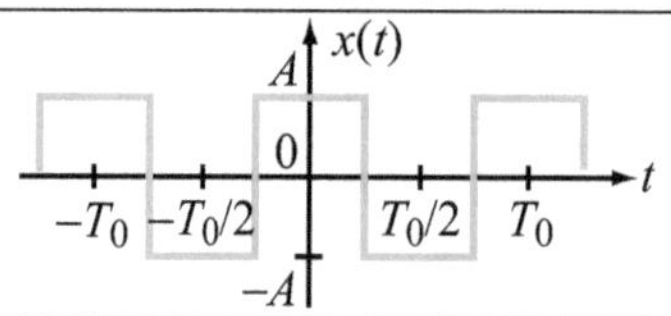 $x(t) = \sum_{n=1}^{\infty} \frac{4A}{n\pi} \sin\left(\frac{n\pi}{2}\right) \cos\left(\frac{2n\pi t}{T_0}\right)$
2. Time-Shifted Square Wave	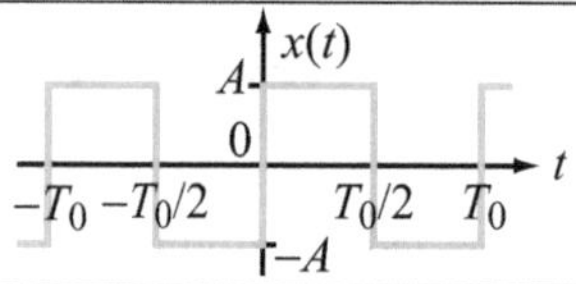 $x(t) = \sum_{\substack{n=1 \\ n=\text{odd}}}^{\infty} \frac{4A}{n\pi} \sin\left(\frac{2n\pi t}{T_0}\right)$
3. Pulse Train	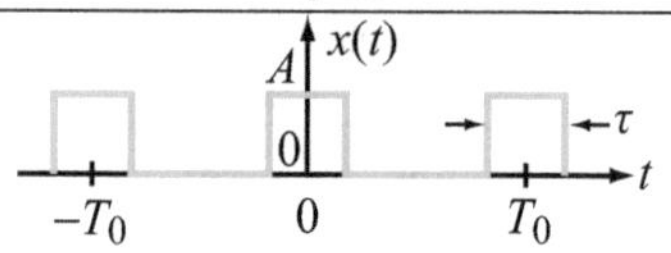 $x(t) = \frac{A\tau}{T_0} + \sum_{n=1}^{\infty} \frac{2A}{n\pi} \sin\left(\frac{n\pi\tau}{T_0}\right) \cos\left(\frac{2n\pi t}{T_0}\right)$
4. Triangular Wave	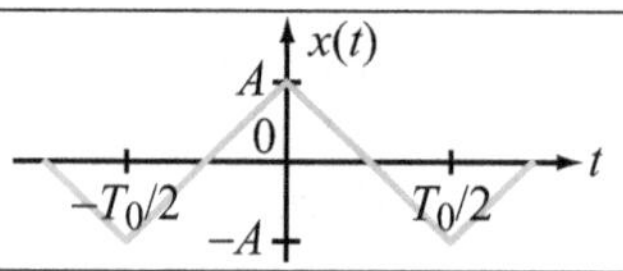 $x(t) = \sum_{\substack{n=1 \\ n=\text{odd}}}^{\infty} \frac{8A}{n^2\pi^2} \cos\left(\frac{2n\pi t}{T_0}\right)$
5. Shifted Triangular Wave	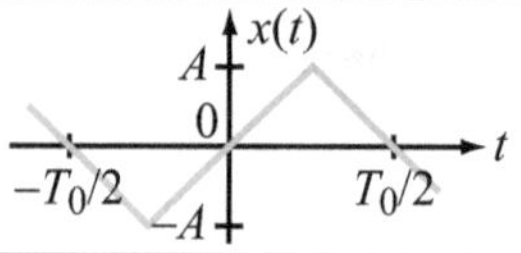 $x(t) = \sum_{\substack{n=1 \\ n=\text{odd}}}^{\infty} \frac{8A}{n^2\pi^2} \sin\left(\frac{n\pi}{2}\right) \sin\left(\frac{2n\pi t}{T_0}\right)$
6. Sawtooth	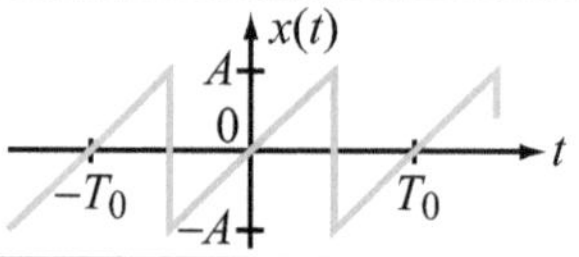 $x(t) = \sum_{n=1}^{\infty} (-1)^{n+1} \frac{2A}{n\pi} \sin\left(\frac{2n\pi t}{T_0}\right)$
7. Backward Sawtooth	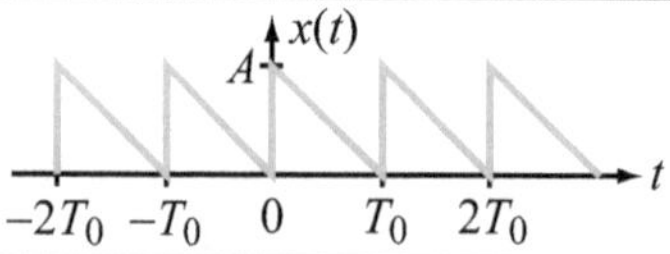 $x(t) = \frac{A}{2} + \sum_{n=1}^{\infty} \frac{A}{n\pi} \sin\left(\frac{2n\pi t}{T_0}\right)$
8. Full-Wave Rectified Sinusoid	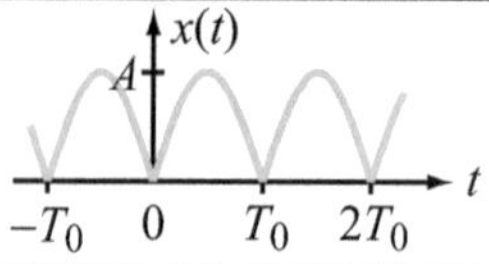 $x(t) = \frac{2A}{\pi} + \sum_{n=1}^{\infty} \frac{4A}{\pi(1-4n^2)} \cos\left(\frac{2n\pi t}{T_0}\right)$
9. Half-Wave Rectified Sinusoid	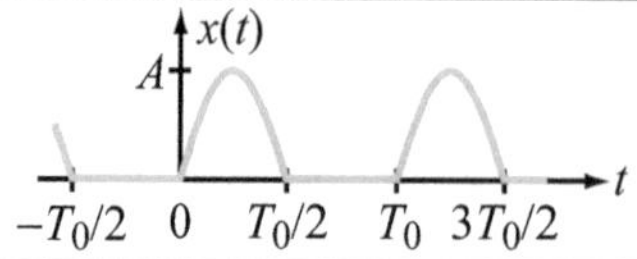 $x(t) = \frac{A}{\pi} + \frac{A}{2} \sin\left(\frac{2\pi t}{T_0}\right) + \sum_{\substack{n=2 \\ n=\text{even}}}^{\infty} \frac{2A}{\pi(1-n^2)} \cos\left(\frac{2n\pi t}{T_0}\right)$

Since $b_n = 0$, we have for $n \neq 0$

$$c_n = |a_n|, \qquad \phi_n = \begin{cases} 0 & \text{if } a_n > 0, \\ 180^\circ & \text{if } a_n < 0. \end{cases}$$

Figures 5-7(b) and (c) display the amplitude and phase line spectra of the M-periodic waveform, and parts (d) through (f) display, respectively, the waveforms based on the first five

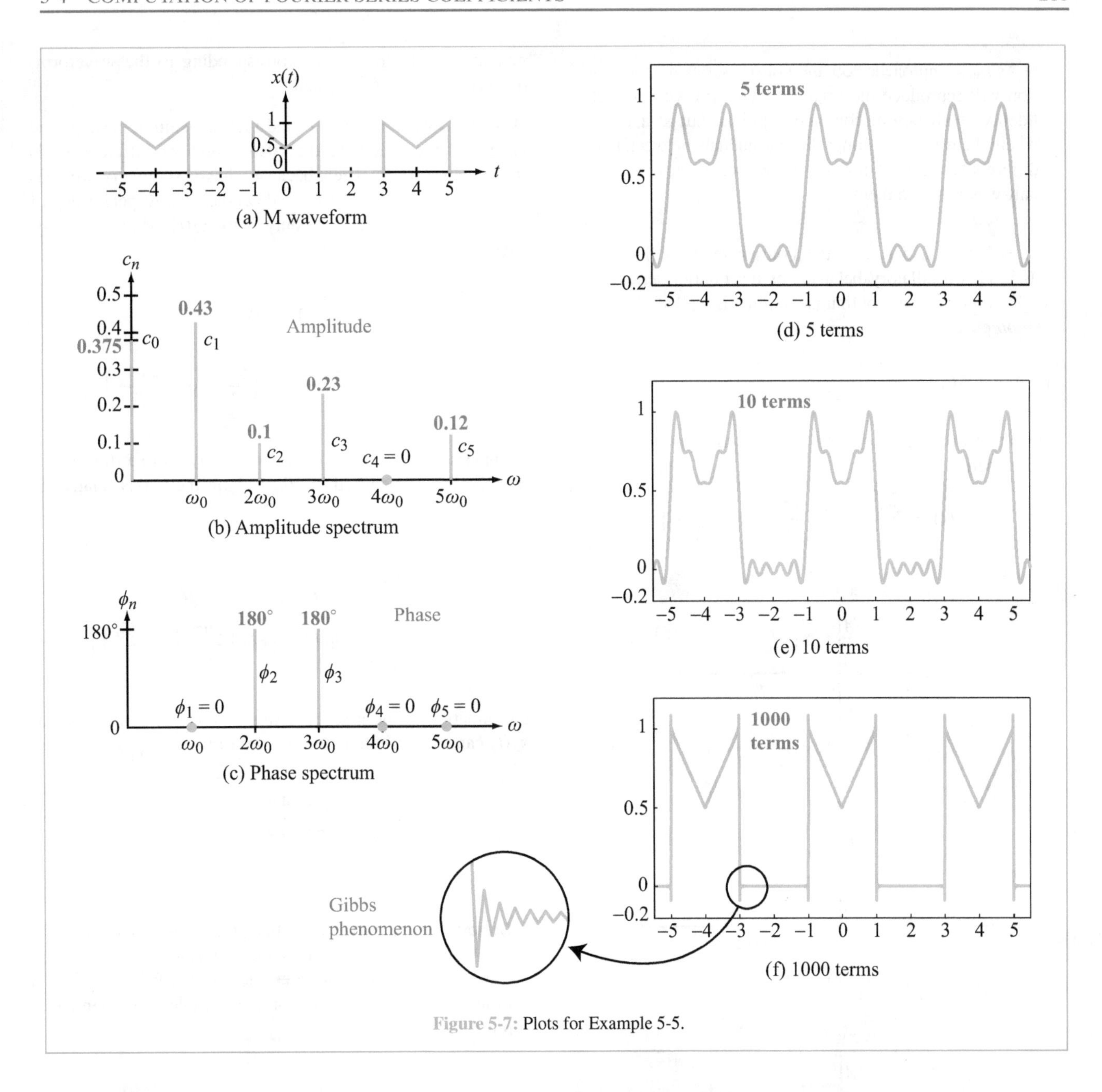

Figure 5-7: Plots for Example 5-5.

terms, the first ten terms, and the first 1000 terms of the Fourier series.

As expected, the addition of more terms in the Fourier series improves the overall fidelity of the reproduced waveform. However, no matter how many terms are included in the series representation, *the reproduction cannot duplicate the original M-waveform at points of discontinuity,* such as when the waveform jumps from zero to one. *Discontinuities generate oscillations.* Increasing the number of terms (adding more harmonics) reduces the period of the oscillation, and ultimately the oscillations fuse into a solid line, except at the discontinuities [see expanded view of the discontinuity at $t = -3$ s in Fig. 5-7(f)].

► As $n_{\max}$ approaches ∞, the Fourier series representation will reproduce the original waveform with perfect fidelity at all non-discontinuous points, but at a point where the waveform jumps discontinuously between two different levels, the Fourier series will converge to a level half-way between them. ◄

At $t = 1$ s, 3 s, 5 s, ..., the Fourier series will converge to 0.5. This oscillatory behavior of the Fourier series in the neighborhood of discontinuous points is called the *Gibbs phenomenon*.

Example 5-6: Waveform Synthesis

Given that waveform $x_1(t)$ in Fig. 5-8(a) is represented by the Fourier series

$$x_1(t) = \sum_{n=1}^{\infty} \frac{4A}{n\pi} \sin\left(\frac{n\pi}{2}\right) \cos\left(\frac{2n\pi t}{T_0}\right),$$

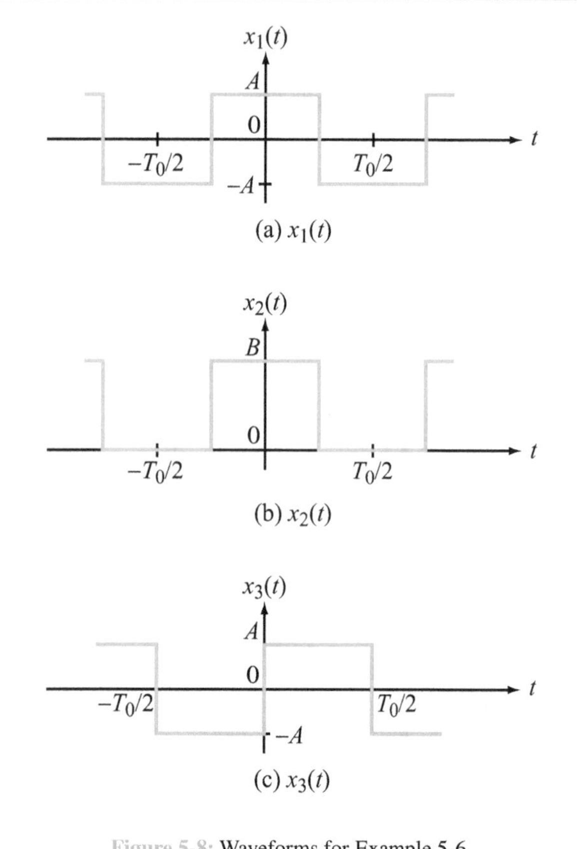

Figure 5-8: Waveforms for Example 5-6.

generate the Fourier series corresponding to the waveforms displayed in Figs. 5-8(b) and (c).

Solution: Waveforms $x_1(t)$ and $x_2(t)$ are similar in shape and have the same period, but they also exhibit two differences: (1) the dc value of $x_1(t)$ is zero because it has dc symmetry, whereas the dc value of $x_2(t)$ is $B/2$, and (2) the peak-to-peak value of $x_1(t)$ is $2A$, compared with only B for $x_2(t)$. Mathematically, $x_2(t)$ is related to $x_1(t)$ by

$$\begin{aligned} x_2(t) &= \frac{B}{2} + \left(\frac{B}{2A}\right) x_1(t) \\ &= \frac{B}{2} + \sum_{n=1}^{\infty} \frac{2B}{n\pi} \sin\left(\frac{n\pi}{2}\right) \cos\left(\frac{2n\pi t}{T_0}\right). \end{aligned}$$

Comparison of waveform $x_1(t)$ with waveform $x_3(t)$ reveals that the latter is shifted by $T_0/4$ along the t-axis relative to $x_1(t)$. That is,

$$\begin{aligned} x_3(t) &= x_1\left(t - \frac{T_0}{4}\right) \\ &= \sum_{n=1}^{\infty} \frac{4A}{n\pi} \sin\left(\frac{n\pi}{2}\right) \cos\left[\frac{2n\pi}{T_0}\left(t - \frac{T_0}{4}\right)\right]. \end{aligned}$$

Examination of the first few terms of $x_3(t)$ demonstrates that $x_3(t)$ can be rewritten in the simpler form as

$$x_3(t) = \sum_{\substack{n=1 \\ n \text{ is odd}}}^{\infty} \frac{4A}{n\pi} \sin\left(\frac{2n\pi t}{T_0}\right).$$

Concept Question 5-6: For the cosine/sine and amplitude/phase Fourier series representations, the summation extends from $n = 1$ to $n = \infty$. What are the limits on the summation for the complex exponential representation? (See S²)

Concept Question 5-7: What purpose is served by the symmetry properties of a periodic function? (See S²)

Concept Question 5-8: What distinguishes the phase angles ϕ_n of an even-symmetrical function from those of an odd-symmetrical function? (See S²)

Concept Question 5-9: What is the Gibbs phenomenon? (See S²)

Exercise 5-4: (a) Does the waveform $x(t)$ shown in Fig. E5-4 exhibit either even or odd symmetry? (b) What is the value of a_0? (c) Does the function $y(t) = x(t) - a_0$ exhibit either even or odd symmetry?

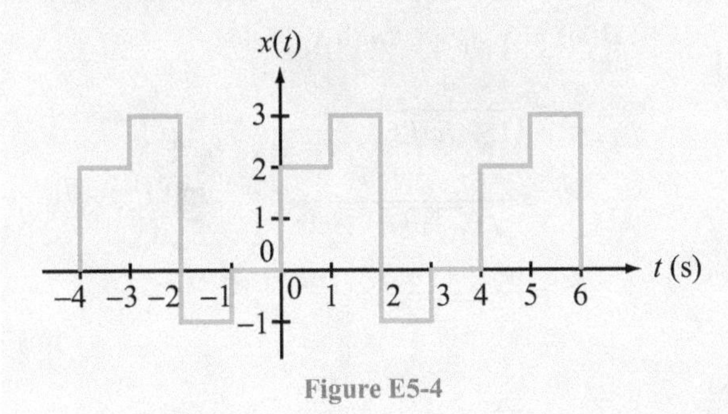

Figure E5-4

Answer: (a) Neither even nor odd symmetry, (b) $a_0 = 1$, (c) odd symmetry. (See S²)

5-5 Circuit Analysis with Fourier Series

Given the tools we developed in the preceding section for how to express a periodic function in terms of a Fourier series, we will now examine how to analyze linear circuits when excited by periodic voltage or current sources. The method of solution relies on the application of the phasor-domain technique that we introduced in Section 5-1. A periodic function can be expressed as the sum of cosine and sine functions with coefficients a_n and b_n and zero phase angles, or expressed as the sum of only cosine functions with amplitudes c_n and phase angles ϕ_n. The latter form is amenable to direct application of the phasor-domain technique, whereas the former will require converting all $\sin(n\omega_0 t)$ terms into $\cos(n\omega_0 t - 90°)$ before implementation of the phasor-domain technique.

Even though the basic solution procedure was outlined earlier in Section 5-2, it is worth repeating it in a form that incorporates the concepts and terminology introduced in Section 5-4. To that end, we shall use $\upsilon_s(t)$ [or $i_s(t)$ if it is a current source] to denote the input excitation and $\upsilon_{out}(t)$ [or $i_{out}(t)$] to denote the output response for which we seek a solution.

Solution Procedure: Fourier Series Analysis

Step 1: Express $\upsilon_s(t)$ in terms of an amplitude/phase Fourier series:

$$\upsilon_s(t) = a_0 + \sum_{n=1}^{\infty} c_n \cos(n\omega_0 t + \phi_n) \tag{5.62}$$

with $c_n \angle \phi_n = a_n - jb_n$.

Step 2: Determine the *generic transfer function* of the circuit at frequency ω:

$$\mathbf{H}(\omega) = \mathbf{V}_{out} \quad \text{when } \upsilon_s = 1\cos\omega t. \tag{5.63}$$

Step 3: Write down the time-domain solution:

$$\upsilon_{out}(t) = a_0\,\mathbf{H}(\omega = 0) + \sum_{n=1}^{\infty} c_n\, \mathfrak{Re}\{\mathbf{H}(\omega = n\omega_0)\, e^{j(n\omega_0 t + \phi_n)}\}. \tag{5.64}$$

For each value of n, coefficient $c_n e^{j\phi_n}$ is associated with frequency harmonic $n\omega_0$. Hence, in Step 3, each harmonic amplitude is multiplied by its corresponding $e^{jn\omega_0 t}$ before application of the $\mathfrak{Re}\{\,\}$ operator.

Example 5-7: RC Circuit

Determine $\upsilon_{out}(t)$ when the circuit in Fig. 5-9(a) is excited by the voltage waveform shown in Fig. 5-9(b). The element values are $R = 20$ kΩ and $C = 0.1$ mF.

Solution:

Step 1: The period of $\upsilon_s(t)$ is 4 s. Hence, $\omega_0 = 2\pi/4 = \pi/2$ rad/s, and by Eq. (5.28), we have

$$a_0 = \frac{1}{T}\int_0^T \upsilon_s(t)\,dt = \frac{1}{4}\int_0^1 10\,dt = 2.5\text{ V},$$

$$a_n = \frac{2}{4}\int_0^1 10\cos\frac{n\pi}{2}t\,dt = \frac{10}{n\pi}\sin\frac{n\pi}{2}\text{ V},$$

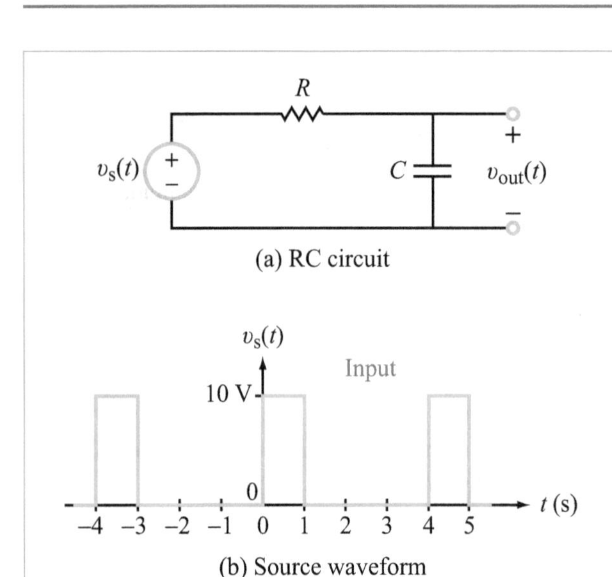

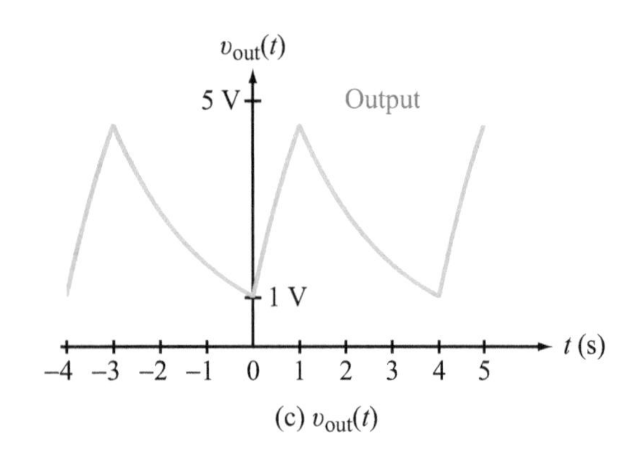

$v_{out}(t)$ 5 V Output 1 V t (s) −4 −3 −2 −1 0 1 2 3 4 5

(c) $v_{out}(t)$

Figure 5-9: Circuit response to periodic pulses.

$$b_n = \frac{2}{4} \int_0^1 10 \sin \frac{n\pi}{2} t \, dt = \frac{10}{n\pi} \left(1 - \cos \frac{n\pi}{2}\right) \text{ V},$$

and

$$c_n \angle \phi_n = a_n - jb_n = \frac{10}{n\pi} \left[\sin \frac{n\pi}{2} - j\left(1 - \cos \frac{n\pi}{2}\right)\right].$$

The values of $c_n \angle \phi_n$ for the first four terms are

$$c_1 \angle \phi_1 = \frac{10\sqrt{2}}{\pi} \angle -45^\circ,$$

$$c_2 \angle \phi_2 = \frac{10}{\pi} \angle -90^\circ,$$

$$c_3 \angle \phi_3 = \frac{10\sqrt{2}}{3\pi} \angle -135^\circ,$$

and

$$c_4 \angle \phi_4 = 0.$$

Step 2: In the phasor domain, the impedance of a capacitor is $\mathbf{Z}_C = 1/(j\omega C)$. By voltage division, the generic phasor-domain transfer function of the circuit is

$$\begin{aligned}
\mathbf{H}(\omega) &= \mathbf{V}_{out} \qquad (\text{with } \mathbf{V}_s = 1) \\
&= \frac{1}{1 + j\omega RC} \\
&= \frac{1}{\sqrt{1 + \omega^2 R^2 C^2}} \, e^{-j \tan^{-1}(\omega RC)} \\
&= \frac{1}{\sqrt{1 + 4\omega^2}} \, e^{-j \tan^{-1}(2\omega)},
\end{aligned}$$

where we used $RC = 2 \times 10^4 \times 10^{-4} = 2$ s.

Step 3: The time-domain output voltage is

$$v_{out}(t) = 2.5 + \sum_{n=1}^{\infty} \mathfrak{Re} \left\{ c_n \frac{1}{\sqrt{1 + 4n^2\omega_0^2}} \, e^{j[n\omega_0 t + \phi_n - \tan^{-1}(2n\omega_0)]} \right\}.$$

Using the values of $c_n \angle \phi_n$ determined earlier for the first four terms and replacing ω_0 with its numerical value of $\pi/2$ rad/s, the expression becomes

$$\begin{aligned}
v_{out}(t) = 2.5 &+ \frac{10\sqrt{2}}{\pi\sqrt{1+\pi^2}} \cos\left[\frac{\pi t}{2} - 45^\circ - \tan^{-1}(\pi)\right] \\
&+ \frac{10}{\pi\sqrt{1+4\pi^2}} \cos[\pi t - 90^\circ - \tan^{-1}(2\pi)] \\
&+ \frac{10\sqrt{2}}{3\pi\sqrt{1+9\pi^2}} \cos\left[\frac{3\pi t}{2} - 135^\circ - \tan^{-1}(3\pi)\right] \cdots \\
= 2.5 &+ 1.37 \cos\left(\frac{\pi t}{2} - 117^\circ\right) + 0.5 \cos(\pi t - 171^\circ) \\
&+ 0.16 \cos\left(\frac{3\pi t}{2} + 141^\circ\right) \cdots \text{ V}.
\end{aligned}$$

The voltage response $v_{out}(t)$ is displayed in Fig. 5-9(c), which was computed using the series solution given by the preceding expression with $n_{max} = 1000$.

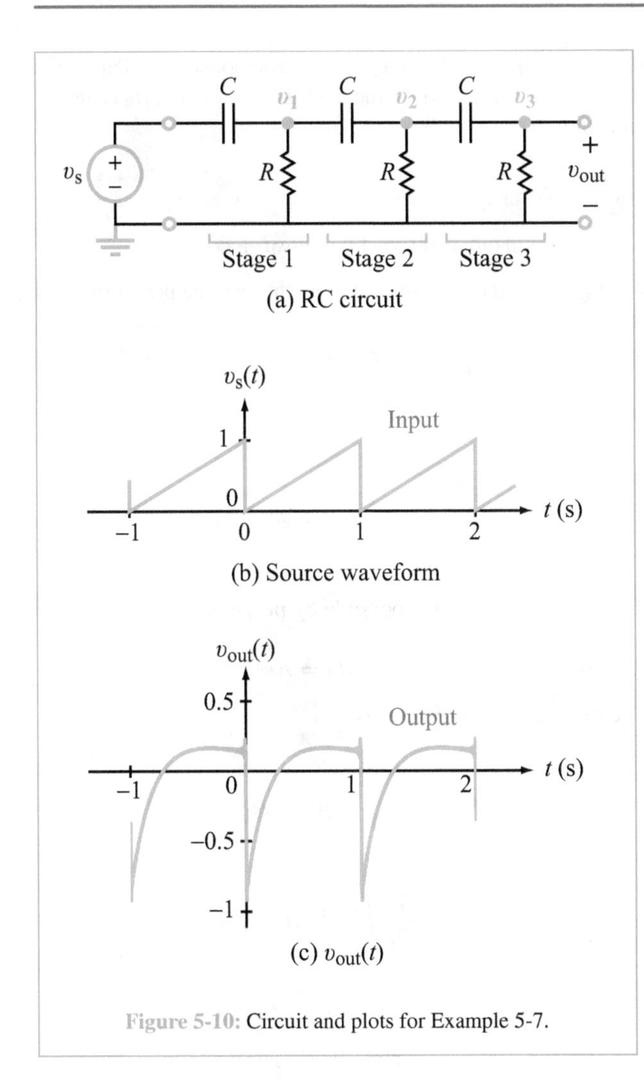

Figure 5-10: Circuit and plots for Example 5-7.

Example 5-8: Three-Stage RC Circuit

Application of KCL to the three-stage RC circuit of Fig. 5-10(a) and defining $x = \omega RC$ leads to

$$\mathbf{H}(\omega) = \frac{\mathbf{V}_{\text{out}}}{\mathbf{V}_{\text{s}}} = \frac{x^3}{(x^3 - 5x) + j(1 - 6x^2)}.$$

Determine the output response to the periodic waveform shown in Fig. 5-10(b), given that $RC = 1$ s.

Solution:

Step 1: With $T = 1$ s, $\omega_0 = 2\pi/T = 2\pi$ rad/s, and $\upsilon_s(t) = t$ over [0, 1], we have

$$a_0 = \frac{1}{T}\int_0^T \upsilon_s(t)\, dt = \int_0^1 t\, dt = 0.5,$$

$$a_n = \frac{2}{1}\int_0^1 t\cos 2n\pi t\, dt$$

$$= 2\left[\frac{1}{(2n\pi)^2}\cos 2n\pi t + \frac{t}{2n\pi}\sin 2n\pi t\right]\Bigg|_0^1 = 0,$$

$$b_n = \frac{2}{1}\int_0^1 t\sin 2n\pi t\, dt$$

$$= 2\left[\frac{1}{(2n\pi)^2}\sin 2n\pi t - \frac{t}{2n\pi}\cos 2n\pi t\right]\Bigg|_0^1$$

$$= -\frac{1}{n\pi},$$

and

$$c_n\underline{/\phi_n} = 0 - jb_n = 0 + j\frac{1}{n\pi} = \frac{1}{n\pi}\underline{/90^\circ} \text{ V}.$$

Step 2: With $RC = 1$ and $x = \omega RC = \omega$, $\mathbf{H}(\omega)$ becomes

$$\mathbf{H}(\omega) = \frac{\omega^3}{(\omega^3 - 5\omega) + j(1 - 6\omega^2)}.$$

Step 3: With $\omega_0 = 2\pi$ rad/s, $\mathbf{H}(\omega = 0) = 0$, and $c_n = [1/(n\pi)]e^{j90^\circ}$, the time-domain voltage is obtained by multiplying each term in the summation by its corresponding $e^{jn\omega_0 t} = e^{j2n\pi t}$, and then taking the real part of the entire expression:

$$\upsilon_{\text{out}}(t) = \sum_{n=1}^{\infty} \mathfrak{Re}\left\{\frac{8n^2\pi^2}{[(2n\pi)((2n\pi)^2 - 5) + j(1 - 24n^2\pi^2)]} \cdot e^{j(2n\pi t + 90^\circ)}\right\}.$$

Evaluating the first few terms of $\upsilon_{\text{out}}(t)$ leads to

$$\upsilon_{\text{out}}(t) = 0.25\cos(2\pi t + 137^\circ) + 0.15\cos(4\pi t + 116^\circ) + 0.10\cos(6\pi t + 108^\circ) + \cdots.$$

A plot of $\upsilon_{\text{out}}(t)$ with 100 terms is displayed in Fig. 5-10(c).

Concept Question 5-10: What is the connection between the Fourier-series solution method and the phasor-domain solution technique? (See S^2)

Concept Question 5-11: Application of the Fourier-series method in circuit analysis relies on which fundamental property of the circuit? (See S^2)

Exercise 5-5: The RL circuit shown in Fig. E5-5(a) is excited by the square-wave voltage waveform of Fig. E5-5(b). Determine $v_{out}(t)$.

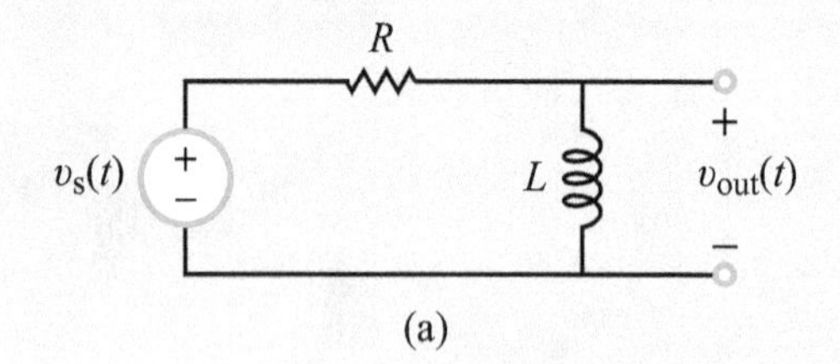

(a)

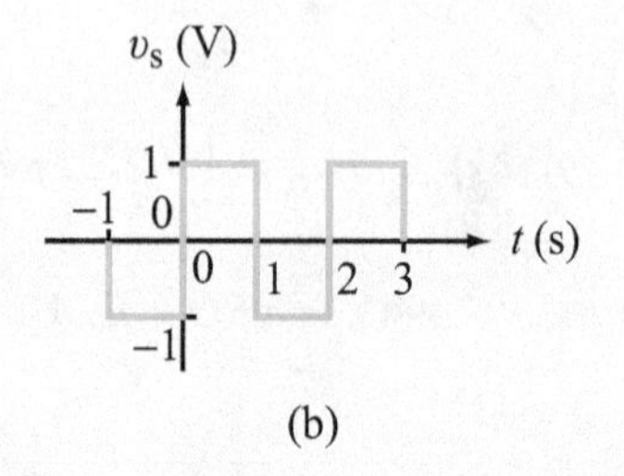

(b)

Figure E5-5

Answer:

$$v_{out}(t) = \sum_{\substack{n=1 \\ n \text{ is odd}}}^{\infty} \frac{4L}{\sqrt{R^2 + n^2\pi^2 L^2}} \cos(n\pi t + \theta_n);$$

$$\theta_n = -\tan^{-1}\left(\frac{n\pi L}{R}\right).$$

(See S^2)

5-6 Parseval's Theorem for Periodic Waveforms

As was noted in Section 1-5, power and energy are standard properties ascribed to signals, whether electrical or not. *Parseval's theorem* states that we can compute the average power of a signal in either the time or frequency domains and obtain the same result. As a prelude to demonstrating Parseval's theorem, we will review the requisite integral properties of sinusoids.

5-6.1 Average Power of Sinusoid

(a) **Real Sinusoid $x_1(t) = A\cos(n\omega_0 t + \phi)$**

For $\omega_0 \neq 0$ and n a positive integer, the average power of $x_1(t)$ is

$$P_{x_1} = \frac{1}{T_0}\int_0^{T_0} |x_1(t)|^2 \, dt$$

$$= \frac{1}{T_0}\int_0^{T_0} A^2 \cos^2(n\omega_0 t + \phi)\, dt = \frac{A^2}{2}. \tag{5.65}$$

The final step was made possible by property #7 in Table 5-2.

(b) **Complex Exponential $x_2(t) = Ae^{j(\omega_0 t + \phi)}$**

The average power of $x_2(t)$ is

$$P_{x_2} = \frac{1}{T_0}\int_0^{T_0} |x_2(t)|^2 \, dt$$

$$= \frac{1}{T_0}\int_0^{T_0} |A^2|\, dt = |A|^2. \tag{5.66}$$

5-6.2 Average Power of Sum of Sinusoids

Let $x(t)$ and $y(t)$ be two periodic signals given by

$$x(t) = c_n \cos(n\omega_0 t + \phi_n) \tag{5.67a}$$

and

$$y(t) = c_m \cos(m\omega_0 t + \phi_m), \tag{5.67b}$$

where m and n are dissimilar integers. From the arguments of their cosine functions, we deduce that $x(t)$ has a period $T_1 = T_0/n$ and $y(t)$ has a period $T_2 = T_0/m$, where T_0 is the period associated with the fundamental angular frequency ω_0 (i.e., $T_0 = 2\pi/\omega_0$). Hence, the average power of $x(t)$ is

$$P_x = \frac{1}{T_1}\int_0^{T_1} |x(t)|^2 \, dt. \tag{5.68}$$

Since the average power of a periodic function over its period T_1 is the same as that performed over multiple periods, we can replace T_1 with the fundamental period $T_0 = nT_1$. Hence, Eq. (5.68) becomes

$$P_x = \frac{1}{T_0}\int_0^{T_0} |x(t)|^2 \, dt$$

$$= \frac{1}{T_0}\int_0^{T_0} c_n^2 \cos^2(n\omega_0 t + \phi_n) \, dt = \frac{c_n^2}{2}. \quad (5.69a)$$

Similarly, for signal $y(t)$ alone,

$$P_y = \frac{1}{T_0}\int_0^{T_0} c_m^2 \cos^2(m\omega_0 t + \phi_m) \, dt = \frac{c_m^2}{2}. \quad (5.69b)$$

For the sum of the two periodic signals, $x(t)+y(t)$, the average can be performed over the period T_0 since it is a multiple of both T_1 and T_2. Thus,

$$P_{x+y} = \frac{1}{T_0}\int_0^{T_0} |x(t) + y(t)|^2 \, dt$$

$$= \frac{1}{T_0}\int_0^{T_0} [x(t) + y(t)][x(t) + y(t)]^* \, dt$$

$$= \frac{1}{T_0}\int_0^{T_0} |x(t)|^2 \, dt + \frac{1}{T_0}\int_0^{T_0} |y(t)|^2 \, dt$$

$$+ \frac{1}{T_0}\int_0^{T_0} [x(t)\, y^*(t)] + [x^*(t)\, y(t)] \, dt.$$

The first two terms are both known from Eq. (5.69). By property #4 in Table 5-2, the third term becomes

$$\frac{1}{T_0}\int_0^{T_0} 2c_n c_m \cos(n\omega_0 t + \phi_n)\cos(m\omega_0 t + \phi_m) \, dt = 0.$$

Hence,

$$P_{x+y} = \frac{c_n^2}{2} + \frac{c_m^2}{2}. \quad (5.70)$$

► The average power of the sum of two periodic signals is equal to the sum of their individual powers, if the sum also is periodic. ◄

5-6.3 Application to Fourier Series

Extending the preceding results to the three Fourier-series representations (Table 5-3) of a periodic signal $x(t)$ leads to the following three formulas for P_x:

$$P_x = \frac{1}{T_0}\int_0^{T_0} |x(t)|^2 \, dt = a_0^2 + \sum_{n=1}^{\infty} (a_n^2 + b_n^2)/2, \quad (5.71a)$$

$$P_x = \frac{1}{T_0}\int_0^{T_0} |x(t)|^2 \, dt = c_0^2 + \sum_{n=1}^{\infty} c_n^2/2, \quad (5.71b)$$

$$P_x = \frac{1}{T_0}\int_0^{T_0} |x(t)|^2 \, dt = \sum_{n=-\infty}^{\infty} |\mathbf{x}_n|^2. \quad (5.71c)$$

The three expressions convey equivalent statements of ***Parseval's theorem***, which asserts that the total average power of a periodic signal is equal to its dc power ($a_0^2 = c_0^2 = |\mathbf{x}_0|^2$) plus the average ac power associated with its fundamental frequency ω_0 and its harmonic multiples. The ***ac power fraction*** is the ratio of the average ac power to the total average power (dc + ac).

Example 5-9: Sawtooth Waveform

Verify Parseval's theorem for the periodic sawtooth waveform shown in Fig. 5-3(a), whose first cycle is given by $x(t) = 5t$ for $0 \le t \le 4$ s.

Solution: (a) Direct integration gives

$$P_x = \frac{1}{4}\int_0^4 (5t)^2 \, dt = \frac{400}{3}.$$

(b) From Example 5-2, the Fourier-series representation of the sawtooth waveform is

$$x(t) = 10 - \frac{20}{\pi}\sum_{n=1}^{\infty} \frac{1}{n} \sin\left(\frac{n\pi}{2}\, t\right).$$

With $a_0 = 10$, $a_n = 0$ for all n, and $b_n = -\frac{20}{n\pi}$, application of Eq. (5.71a) leads to

$$\begin{aligned} P_x &= a_0^2 + \sum_{n=1}^{\infty} (a_n^2 + b_n^2)/2 \\ &= 100 + \sum_{n=1}^{\infty} \frac{400}{2n^2\pi^2} \\ &= 100 + \frac{200}{\pi^2} \sum_{n=1}^{\infty} \frac{1}{n^2} = \frac{400}{3}, \end{aligned}$$

where we used the infinite series

$$\sum_{n=1}^{\infty} \frac{1}{n^2} = \frac{\pi^2}{6}.$$

The direct-computation result for P_x is identical with the value provided by Parseval's theorem.

5-6.4 Inner-Product Version of Parseval's Theorem

Suppose periodic signals $x(t)$ and $y(t)$ have exponential Fourier-series representations given by

$$x(t) = \sum_{n=-\infty}^{\infty} \mathbf{x}_n e^{jn\omega_0 t} \tag{5.72a}$$

and

$$y(t) = \sum_{m=-\infty}^{\infty} \mathbf{y}_m e^{jm\omega_0 t}, \tag{5.72b}$$

where different indices n and m were used for the sake of distinction. The average value of the inner product $x(t)\ y^*(t)$ is

$$\begin{aligned} P_{xy^*} &= \frac{1}{T_0} \int_0^{T_0} x(t)\ y^*(t)\ dt \\ &= \frac{1}{T_0} \int_0^{T_0} \sum_{n=-\infty}^{\infty} \mathbf{x}_n e^{jn\omega_0 t} \sum_{m=-\infty}^{\infty} \mathbf{y}_m^* e^{-jm\omega_0 t}\ dt \\ &= \sum_{n=-\infty}^{\infty} \sum_{m=-\infty}^{\infty} \mathbf{x}_n \mathbf{y}_m^* \frac{1}{T_0} \int_0^{T_0} e^{j(n-m)\omega_0 t}\ dt. \end{aligned} \tag{5.73}$$

Application of Euler's identity to the complex exponential inside the integral leads to

$$\begin{aligned} \frac{1}{T_0} \int_0^{T_0} e^{j(n-m)\omega_0 t}\ dt &= \frac{1}{T_0} \int_0^{T_0} \cos[(n-m)\omega_0 t]\ dt \\ &\quad + \frac{j}{T_0} \int_0^{T_0} \sin[(n-m)\omega_0 t]\ dt \\ &= \begin{cases} 0 & \text{if } m \neq n, \\ 1 & \text{if } m = n. \end{cases} \end{aligned} \tag{5.74}$$

Hence, in the double summation, only the terms $n = m$ are nonzero, in which case Eq. (5.73) reduces to the ***inner-product version of Parseval's theorem***:

$$P_{xy^*} = \frac{1}{T_0} \int_0^{T_0} x(t)\ y^*(t)\ dt = \sum_{n=-\infty}^{\infty} \mathbf{x}_n \mathbf{y}_n^*. \tag{5.75}$$

For the particular case where $\mathbf{x}_n = \mathbf{y}_n$, Eq. (5.75) reduces to the average power version of Parseval's theorem given by Eq. (5.71c).

5-7 Fourier Transform

The Fourier series is a perfectly suitable construct for representing periodic functions, but what about nonperiodic functions? The pulse-train waveform shown in Fig. 5-11(a) consists of a sequence of rectangular pulses, each of width $\tau = 2$ s. The period $T_0 = 4$ s. In part (b) of the same figure, the individual pulses have the same shape as before, but T_0 has been increased to 7 s. So long as T_0 is finite, both waveforms are amenable to representation by Fourier series, but what would happen if we let $T_0 \to \infty$, ending up with the single pulse shown in Fig. 5-11(c)? Can we then represent the no-longer periodic pulse by a Fourier series? We will shortly discover that as $T_0 \to \infty$, the summation in the Fourier series evolves into a continuous integral, which we call the ***Fourier transform***, or ***frequency transform***.

▶ When representing signals, we apply the Fourier-series representation if the signal is periodic, and we use the Fourier transform representation if it is nonperiodic. ◀

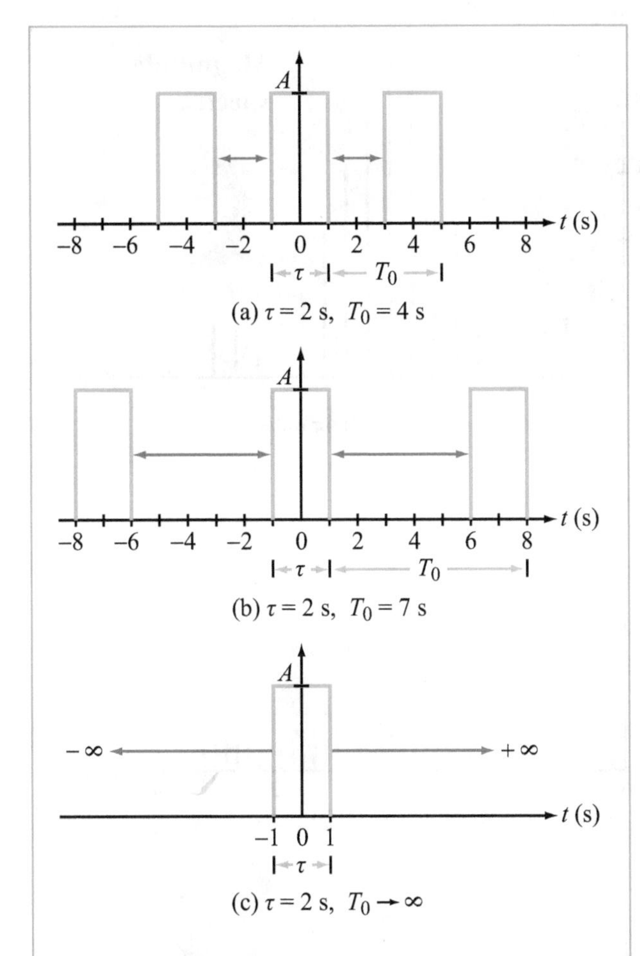

Figure 5-11: The single pulse in (c) is equivalent to a periodic pulse train with $T_0 = \infty$.

Does that mean that we can use both the Laplace-transform (Chapter 3) and the Fourier-transform techniques to analyze circuits containing nonperiodic sources, and if so, which of the two transforms should we use, and why? We will address these questions later (Section 5-11), after formally introducing the Fourier transform and discussing some of its salient features.

5-7.1 Line Spectrum of Pulse Train

As a prelude to introducing the Fourier transform, we shall examine the character of the magnitude line spectrum of a train of pulses as we increase the separation between adjacent pulses. The three waveforms of interest are displayed on the left-hand side of Fig. 5-12.

Over a period T extending from $-T_0/2$ to $T_0/2$,

$$x(t) = \begin{cases} A & \text{for } -\tau/2 \le t \le \tau/2, \\ 0 & \text{otherwise.} \end{cases}$$

With the integration domain chosen to be from $-T_0/2$ to $T_0/2$, Eq. (5.59) gives

$$\begin{aligned} \mathbf{x}_n &= \frac{1}{T_0} \int_{-T_0/2}^{T_0/2} x(t)\, e^{-jn\omega_0 t}\, dt \\ &= \frac{1}{T_0} \int_{-\tau/2}^{\tau/2} A e^{-jn\omega_0 t}\, dt \\ &= \frac{A}{-jn\omega_0 T_0}\, e^{-jn\omega_0 t} \Big|_{-\tau/2}^{\tau/2} \\ &= \frac{2A}{n\omega_0 T_0} \left[\frac{e^{jn\omega_0\tau/2} - e^{-jn\omega_0\tau/2}}{2j} \right]. \end{aligned} \tag{5.76}$$

The quantity inside the square bracket matches the form of one of Euler's formulas, namely,

$$\sin\theta = \frac{e^{j\theta} - e^{-j\theta}}{2j}. \tag{5.77}$$

Hence, Eq. (5.76) can be rewritten in the form

$$\begin{aligned} \mathbf{x}_n &= \frac{2A}{n\omega_0 T_0} \sin(n\omega_0\tau/2) \\ &= \frac{A\tau}{T_0} \frac{\sin(n\omega_0\tau/2)}{(n\omega_0\tau/2)} = \frac{A\tau}{T_0} \operatorname{sinc}(n\omega_0\tau/2), \end{aligned} \tag{5.78}$$

where in the last step we introduced the ***sinc function***, defined as*

$$\operatorname{sinc}(\theta) = \frac{\sin\theta}{\theta}. \tag{5.79}$$

Among the important properties of the sinc function are the following:

(a) When its argument is zero, the sinc function is equal to 1;

$$\operatorname{sinc}(0) = \frac{\sin(\theta)}{\theta}\Big|_{\theta=0} = 1. \tag{5.80}$$

Verification of this property can be established by applying l'Hopital's rule to Eq. (5.79) and then setting $\theta = 0$.

*An alternative definition for the sinc function is

$$\operatorname{sinc}(\theta) = \sin(\pi\theta)/(\pi\theta),$$

and it is used in signal processing and MATLAB. In this book, we use the definition given by Eq. (5.79).

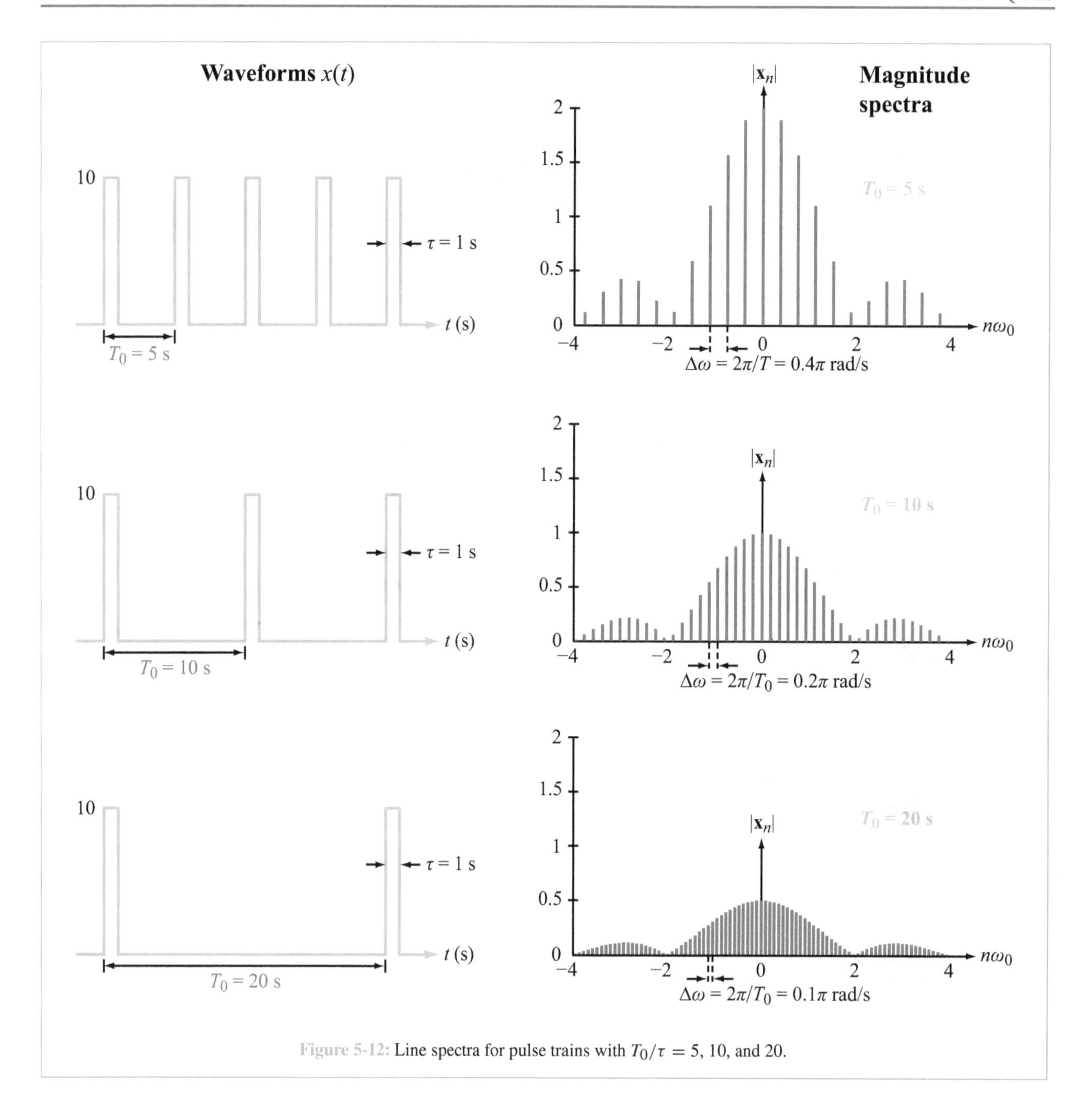

Figure 5-12: Line spectra for pulse trains with $T_0/\tau = 5$, 10, and 20.

(b) Since $\sin(m\pi) = 0$ for any integer value of m, the same is true for the sinc function. That is,

$$\operatorname{sinc}(m\pi) = 0, \qquad m \neq 0. \tag{5.81}$$

(c) Because both $\sin\theta$ and θ are odd functions, their ratio is an even function. Hence, the sinc function possesses even symmetry relative to the vertical axis. Consequently,

$$\mathbf{x}_n = \mathbf{x}_{-n}. \tag{5.82}$$

Evaluation of Eq. (5.78), with $A = 10$, leads to the line spectra displayed on the right-hand side of Fig. 5-12. The general shape of the envelope is dictated by the sinc function, exhibiting a symmetrical pattern with a peak at $n = 0$, a major lobe extending between $n = -T_0/\tau$ and $n = T_0/\tau$, and progressively smaller-amplitude lobes on both sides. The density of spectral lines depends on the ratio of T_0/τ, so in the limit as $T_0 \to \infty$, the line spectrum becomes a continuum.

5-7.2 Nonperiodic Waveforms

In connection with the line spectra displayed in Fig. 5-12, we noted that as the period $T_0 \to \infty$, the periodic function becomes nonperiodic and the associated line spectrum evolves from one containing discrete lines into a continuum. We will now explore this evolution in mathematical terms, culminating in a definition for the Fourier transform of a nonperiodic function. To that end, we begin with the pair of expressions given by Eqs. (5.57) and (5.59), namely,

$$x(t) = \sum_{n=-\infty}^{\infty} \mathbf{x}_n e^{jn\omega_0 t} \tag{5.83a}$$

and

$$\mathbf{x}_n = \frac{1}{T_0} \int_{-T_0/2}^{T_0/2} x(t')\, e^{-jn\omega_0 t'}\, dt'. \tag{5.83b}$$

These two quantities form a complementary pair, with $x(t)$ defined in the continuous time domain and $\mathbf{x}_n$ defined in the discrete frequency domain $n\omega_0$, with $\omega_0 = 2\pi/T_0$. For a given value of T_0, the nth frequency harmonic is at $n\omega_0$ and the next harmonic after that is at $(n+1)\omega_0$. Hence, the ***spacing between adjacent harmonics*** is

$$\Delta\omega = (n+1)\omega_0 - n\omega_0 = \omega_0 = \frac{2\pi}{T_0}\,. \tag{5.84}$$

If we insert Eq. (5.83b) into Eq. (5.83a) and replace $1/T_0$ with $\Delta\omega/2\pi$, we get

$$x(t) = \sum_{n=-\infty}^{\infty} \left[\frac{1}{2\pi} \int_{-T_0/2}^{T_0/2} x(t')\, e^{-jn\omega_0 t'}\, dt' \right] e^{jn\omega_0 t}\, \Delta\omega. \tag{5.85}$$

As $T_0 \to \infty$, $\Delta\omega \to d\omega$, $n\omega_0 \to \omega$, and the sum becomes a continuous integral:

$$x(t) = \frac{1}{2\pi} \int_{-\infty}^{\infty} \left[\int_{-\infty}^{\infty} x(t')\, e^{-j\omega t'}\, dt' \right] e^{j\omega t}\, d\omega. \tag{5.86}$$

Given this new arrangement, we are now ready to offer formal definitions for the Fourier transform $\mathbf{X}(\omega)$ and its inverse transform $x(t)$:

$$\mathbf{X}(\omega) = \mathcal{F}[x(t)] = \int_{-\infty}^{\infty} x(t)\, e^{-j\omega t}\, dt \tag{5.87a}$$

and

$$x(t) = \mathcal{F}^{-1}[\mathbf{X}(\omega)] = \frac{1}{2\pi} \int_{-\infty}^{\infty} \mathbf{X}(\omega)\, e^{j\omega t}\, d\omega, \tag{5.87b}$$

where $\mathcal{F}[x(t)]$ is a short-hand notation for "***the Fourier transform of*** $x(t)$," and similarly $\mathcal{F}^{-1}[\mathbf{X}(\omega)]$ represents the inverse operation. Occasionally, we may also use the symbolic form

$$x(t) \longleftrightarrow \mathbf{X}(\omega).$$

▶ The reader should be aware that Fourier transforms are defined differently in fields other than electrical engineering. Mathematicians define the Fourier transform using the integrand $x(t)\, e^{j\omega t}$, instead of $x(t)\, e^{-j\omega t}$. You will encounter this definition when you take a course in probability, since characteristic functions, which are Fourier transforms of probability density functions, are defined using $x(t)\, e^{j\omega t}$. Sometimes, mathematicians use the integrand $(1/\sqrt{2\pi})\, x(t)\, e^{j\omega t}$, so that the factor of $\frac{1}{2\pi}$ in the inverse Fourier transform is split evenly between the Fourier transform and its inverse. The computer program Mathematica uses this definition to compute Fourier transforms symbolically. Seismologists and geophysicists use different definitions for time and for frequency. For them, the 2-D Fourier transform of the 2-D signal $z(x, t)$, where z is displacement, x is depth, and t is time, is defined using the integrand $z(x, t)\, e^{j(\omega t - kx)}$, where k is wavenumber (spatial frequency). This book will use the integrand $x(t)\, e^{-j\omega t}$ exclusively. ◀

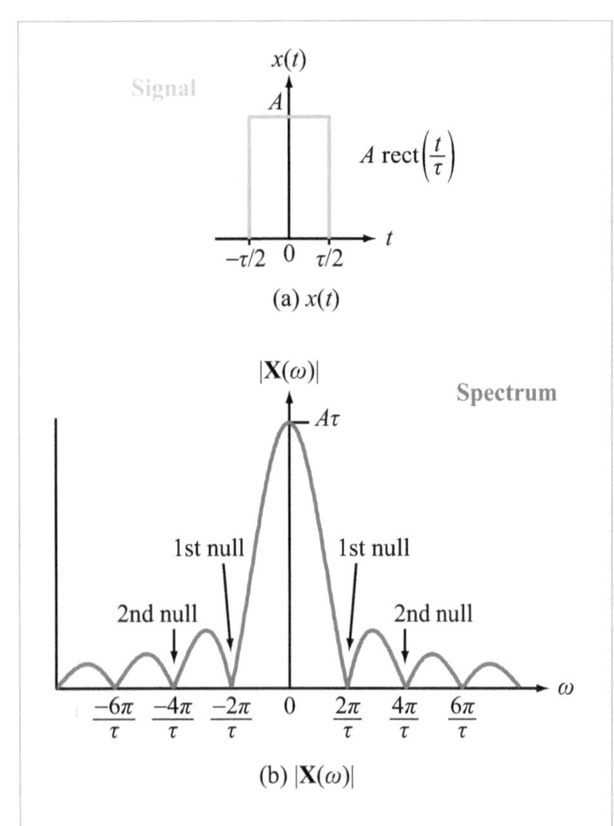

Figure 5-13: (a) Rectangular pulse of amplitude A and width τ; (b) frequency spectrum of $|\mathbf{X}(\omega)|$ for $A = 5$ and $\tau = 1$ s.

Example 5-10: Rectangular Pulse

Determine the Fourier transform of the solitary rectangular pulse shown in Fig. 5-13(a), and then plot its *magnitude spectrum*, $|\mathbf{X}(\omega)|$, for $A = 5$ and $\tau = 1$ s.

Solution: Application of Eq. (5.87a) with

$$x(t) = A \operatorname{rect}(t/\tau)$$

over the integration interval $[-\tau/2, \tau/2]$ leads to

$$\begin{aligned} \mathbf{X}(\omega) &= \int_{-\tau/2}^{\tau/2} Ae^{-j\omega t}\, dt \\ &= \frac{A}{-j\omega} e^{-j\omega t}\Big|_{-\tau/2}^{\tau/2} \\ &= A\tau \frac{\sin(\omega\tau/2)}{(\omega\tau/2)} = A\tau \operatorname{sinc}\left(\frac{\omega\tau}{2}\right). \end{aligned} \quad (5.88)$$

The sinc function was defined earlier in Eq. (5.79).

The *frequency spectrum* of $|\mathbf{X}(\omega)|$ is displayed in Fig. 5-13(b) for the specified values of $A = 5$ and $\tau = 1$ s. The *nulls* in the spectrum occur when the argument of the sinc function is a multiple of $\pm\pi$ (rad/s), which in this specific case correspond to ω equal to multiples of 2π (rad/s).

Concept Question 5-12: What is a sinc function and what are its primary properties? Why is $\operatorname{sinc}(0) = 1$? (See S²)

Concept Question 5-13: What is the functional form for the Fourier transform $\mathbf{X}(\omega)$ of a rectangular pulse of amplitude 1 and duration τ? (See S²)

Exercise 5-6: For a single rectangular pulse of width τ, what is the spacing $\Delta\omega$ between first nulls? If τ is very wide, will its frequency spectrum be narrow and peaked or wide and gentle?

Answer: $\Delta\omega = 4\pi/\tau$. Wide τ leads to narrow spectrum. (See S²)

5-7.3 Convergence of the Fourier Integral

Not every function $x(t)$ has a Fourier transform. The Fourier transform $\mathbf{X}(\omega)$ exists if the Fourier integral given by Eq. (5.87a) converges to a finite number, or to an equivalent expression, but, as we shall discuss shortly, it may also exist even if the Fourier integral does not converge. Convergence depends on the character of $x(t)$ over the integration range $(-\infty, \infty)$. By character, we mean (1) whether or not $x(t)$ exhibits bounded discontinuities and (2) how $x(t)$ behaves as $|t|$ approaches ∞. As a general rule, the Fourier integral does converge if $x(t)$ has no discontinuities and is *absolutely integrable*. That is,

$$\int_{-\infty}^{\infty} |x(t)|\, dt < \infty. \quad (5.89)$$

A function $x(t)$ can still have a Fourier transform even if it has discontinuities, so long as those discontinuities are bounded. The step function $A\, u(t)$ exhibits a bounded discontinuity at $t = 0$ if A is finite.

The stated conditions for the existence of the Fourier transform are sufficient but not necessary conditions. In other words, some functions may still have transforms even though their Fourier integrals do not converge. Among such functions are the constant $x(t) = A$ and the unit step function

$x(t) = A\, u(t)$, both of which represent important excitation waveforms in linear systems. To find the Fourier transform of a function whose transform exists but its Fourier integral does not converge, we need to employ an indirect approach. The approach entails the following ingredients:

(a) If $x(t)$ is a function whose Fourier integral does not converge, we select a second function $x_\epsilon(t)$ whose functional form includes a parameter ϵ, which if allowed to approach a certain limit makes $x_\epsilon(t)$ identical with $x(t)$.

(b) The choice of function $x_\epsilon(t)$ should be such that its Fourier integral does converge, and therefore, $x_\epsilon(t)$ has a definable Fourier transform $\mathbf{X}_\epsilon(\omega)$.

(c) By taking parameter ϵ in the expression for $\mathbf{X}_\epsilon(\omega)$ to its limit, $\mathbf{X}_\epsilon(\omega)$ reduces to the transform $\mathbf{X}(\omega)$ corresponding to the original function $x(t)$.

The procedure is illustrated through an example in Section 5-8.7.

5-8 Fourier Transform Properties

In this section, we shall develop fluency in how to move back and forth between the time domain and the ω-domain. We will learn how to circumvent the convergence issues we noted in Section 5-7.3, and in the process, we will identify a number of useful properties of the Fourier transform.

5-8.1 Linearity Property

If

$$x_1(t) \longleftrightarrow \mathbf{X}_1(\omega)$$

and

$$x_2(t) \longleftrightarrow \mathbf{X}_2(\omega),$$

then

$$K_1\, x_1(t) + K_2\, x_2(t) \longleftrightarrow K_1\, \mathbf{X}_1(\omega) + K_2\, \mathbf{X}_2(\omega), \tag{5.90}$$

(linearity property)

where K_1 and K_2 are constants. Proof of Eq. (5.90) is easily ascertained through the application of Eq. (5.87a).

5-8.2 Scaling Property

If

$$x(t) \longleftrightarrow \mathbf{X}(\omega),$$

and t is scaled by a real constant a, then

$$x(at) \longleftrightarrow \frac{1}{|a|}\, \mathbf{X}\left(\frac{\omega}{a}\right), \qquad \text{for any } a. \tag{5.91}$$

(scaling property)

To prove Eq. (5.91), we replace $x(t)$ with $x(at)$ in Eq. (5.87a), which gives

$$\begin{aligned} \mathcal{F}[x(at)] &= \int_{-\infty}^{\infty} x(at)\, e^{-j\omega t}\, dt \\ &= \int_{-\infty}^{\infty} x(\tau)\, e^{-j(\omega/a)\tau} \cdot \frac{1}{a}\, d\tau \\ &= \frac{1}{a}\, \mathbf{X}\left(\frac{\omega}{a}\right), \qquad \text{for } a > 0, \end{aligned}$$

where we made the substitution $\tau = at$. This result is valid only if a is positive. Repetition of the process for $a < 0$ leads to

$$\mathcal{F}[x(at)] = -\frac{1}{a}\, \mathbf{X}\left(\frac{\omega}{a}\right), \qquad \text{for } a < 0.$$

The two results are combined in the form of Eq. (5.91).

The scaling property is illustrated graphically by the example shown in Fig. 5-14. Part (a) of the figure displays a rectangular pulse of unit amplitude and duration τ. Its corresponding Fourier transform is given by Eq. (5.88) and labeled $\mathbf{X}_1(\omega)$. A similar rectangular pulse (also of unit amplitude) is shown in part (b) of Fig. 5-14, except that its duration is three times longer ($\tau_2 = 3\tau$). Comparison of the waveforms and corresponding Fourier transforms of the two pulses leads to the observations listed in Table 5-5. The second pulse is a stretched-out version of the first one with a scaling factor $a = 3$. Stretching the time dimension by a factor of 3 leads to compression of the angular frequency domain by the same factor. Moreover, preservation of energy (as discussed in Section 5-9) leads to a three-fold increase in the amplitude of the Fourier transform. These conclusions affirm the mathematical statement embodied by Eq. (5.91), which states the following:

▶ Time expansion of a signal $x(t)$ to $x(at)$ by an expansion factor $a < 1$, leads to compression of its spectrum $\mathbf{X}(\omega)$ to $\mathbf{X}(\omega/a)$; conversely, time compression by a compression factor $a > 1$ leads to expansion of the corresponding spectrum. ◀

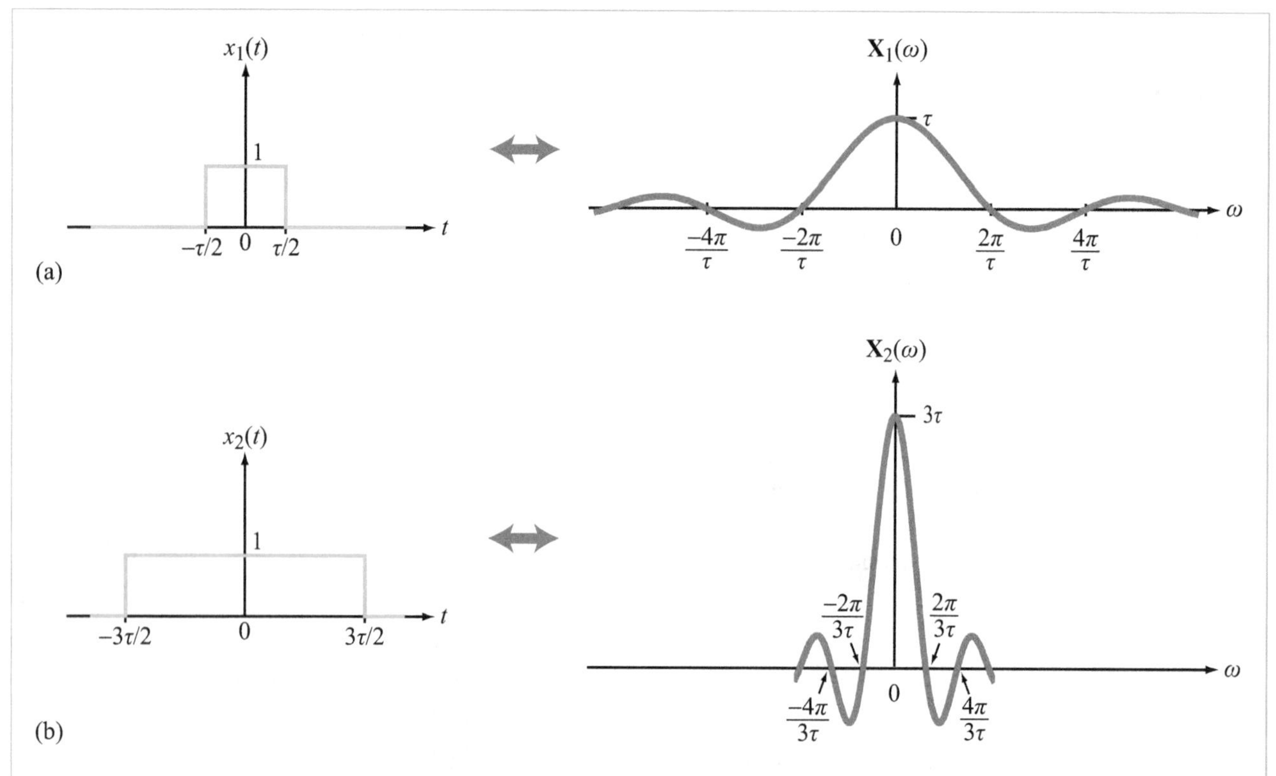

Figure 5-14: Stretching $x_1(t)$ to get $x_2(t)$ entails stretching t by a factor of 3. The corresponding spectrum $\mathbf{X}_2(\omega)$ is a 3-times compressed version of $\mathbf{X}_1(\omega)$.

5-8.3 Fourier Transform of $\delta(t - t_0)$

By Eq. (5.87a), the Fourier transform of $\delta(t - t_0)$ is given by

$$\mathbf{X}(\omega) = \mathcal{F}[\delta(t - t_0)] = \int_{-\infty}^{\infty} \delta(t - t_0) e^{-j\omega t}\, dt$$

$$= e^{-j\omega t}\Big|_{t=t_0} = e^{-j\omega t_0}. \qquad (5.92)$$

Hence,

$$\delta(t - t_0) \longleftrightarrow e^{-j\omega t_0}, \qquad (5.93)$$

and

$$\delta(t) \longleftrightarrow 1. \qquad (5.94)$$

Table 5-5: Comparison of two rectangular pulses.

	Pulse $x_1(t)$	Pulse $x_2(t)$
Amplitude	1	1
Pulse length	$\tau_1 = \tau$	$\tau_2 = 3\tau$
Fourier transform	$\mathbf{X}_1(\omega) = \tau \operatorname{sinc}\left(\frac{\omega\tau}{2}\right)$	$\mathbf{X}_2(\omega) = 3\tau \operatorname{sinc}\left(\frac{3\omega\tau}{2}\right)$
• **Peak value**	τ	3τ
• **null-null width**	$\frac{4\pi}{\tau}$	$\frac{4\pi}{3\tau}$

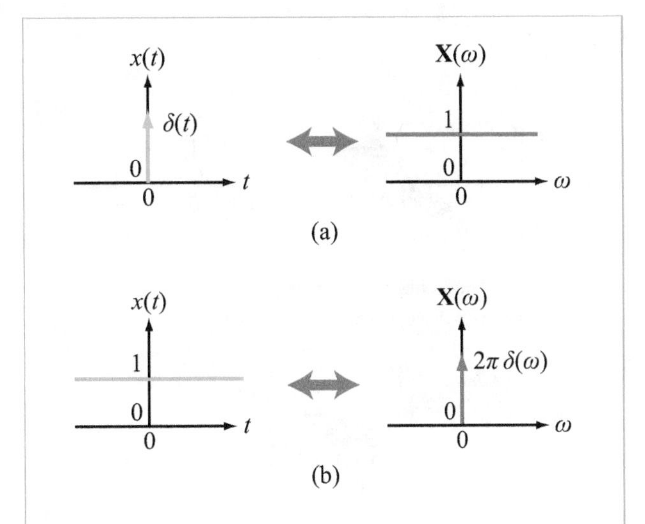

Figure 5-15: The Fourier transform of $\delta(t)$ is 1 and the Fourier transform of 1 is $2\pi\ \delta(\omega)$.

Thus, a unit impulse function $\delta(t)$ generates a constant of unit amplitude that extends over $(-\infty, \infty)$ in the ω-domain, as shown in Fig. 5-15(a),

5-8.4 Shift Properties

By Eq. (5.87b), the inverse Fourier transform of $\mathbf{X}(\omega) = \delta(\omega - \omega_0)$ is

$$x(t) = \mathcal{F}^{-1}[\delta(\omega - \omega_0)]$$
$$= \frac{1}{2\pi} \int_{-\infty}^{\infty} \delta(\omega - \omega_0)\, e^{j\omega t}\, d\omega = \frac{e^{j\omega_0 t}}{2\pi}.$$

Hence,

$$e^{j\omega_0 t} \quad \longleftrightarrow \quad 2\pi\ \delta(\omega - \omega_0) \tag{5.95a}$$

and

$$1 \quad \longleftrightarrow \quad 2\pi\ \delta(\omega). \tag{5.95b}$$

Comparison of the plots in Fig. 5-15(a) and (b) demonstrates the ***duality*** between the time domain and the ω-domain: An impulse $\delta(t)$ in the time domain generates a uniform spectrum in the frequency domain; conversely, a uniform (constant) waveform

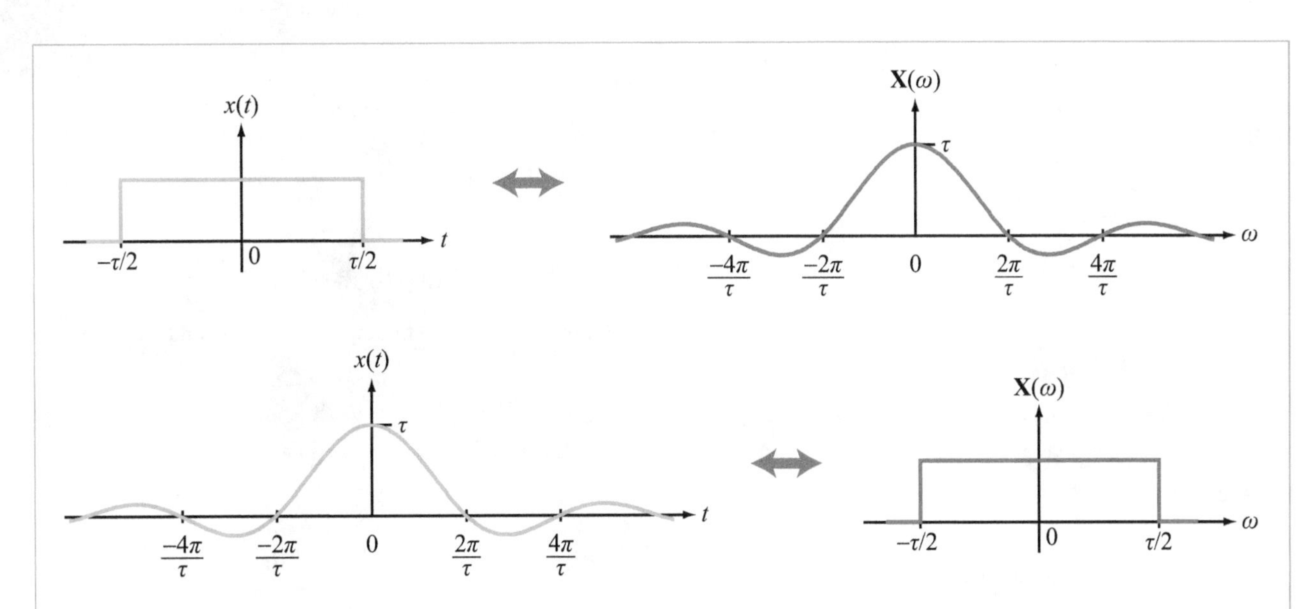

Figure 5-16: Time-frequency duality: a rectangular pulse generates a sinc spectrum and, conversely, a sinc-pulse generates a rectangular spectrum.

in the time domain generates an impulse $\delta(\omega)$ in the frequency domain. By the same token, a rectangular pulse in the time domain generates a sinc pattern in the frequency domain, and a sinc pulse in the time domain generates a rectangular spectrum in the frequency domain (Fig. 5-16).

It is straightforward to show that the result given by Eq. (5.95) can be generalized to

$$e^{j\omega_0 t}\, x(t) \quad \longleftrightarrow \quad \mathbf{X}(\omega - \omega_0), \tag{5.96}$$

(frequency-shift property)

which is known as the ***frequency-shift property*** of the Fourier transform. It states that multiplication of a function $x(t)$ by $e^{j\omega_0 t}$ in the time domain corresponds to shifting the Fourier transform of $x(t)$, $\mathbf{X}(\omega)$, by ω_0 along the ω-axis.

The dual of the frequency-shift property is the ***time-shift property*** given by

$$x(t - t_0) \quad \longleftrightarrow \quad e^{-j\omega t_0}\, \mathbf{X}(\omega). \tag{5.97}$$

(time-shift property)

5-8.5 Fourier Transform of $\cos \omega_0 t$

By Euler's identity,

$$\cos \omega_0 t = \frac{e^{j\omega_0 t} + e^{-j\omega_0 t}}{2}.$$

In view of Eq. (5.95a),

$$\begin{aligned}\mathbf{X}(\omega) &= \mathcal{F}\left[\frac{e^{j\omega_0 t}}{2} + \frac{e^{-j\omega_0 t}}{2}\right] \\ &= \pi\ \delta(\omega - \omega_0) + \pi\ \delta(\omega + \omega_0).\end{aligned}$$

Hence,

$$\cos \omega_0 t \quad \longleftrightarrow \quad \pi[\delta(\omega - \omega_0) + \delta(\omega + \omega_0)], \tag{5.98}$$

and similarly,

$$\sin \omega_0 t \quad \longleftrightarrow \quad j\pi[\delta(\omega + \omega_0) - \delta(\omega - \omega_0)]. \tag{5.99}$$

As shown in Fig. 5-17, the Fourier transform of $\cos \omega_0 t$ consists of impulse functions at $\pm\omega_0$.

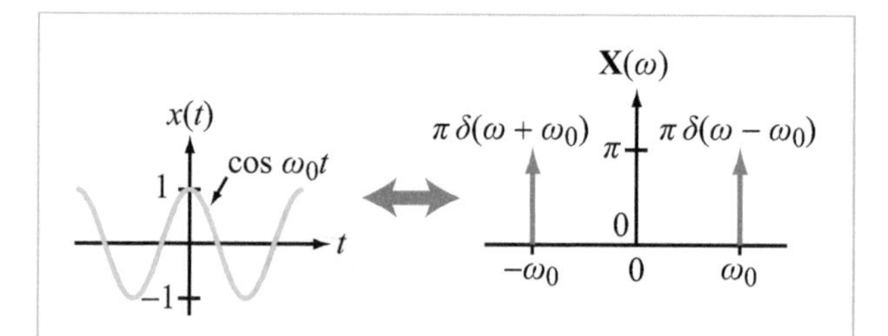

Figure 5-17: The Fourier transform of $\cos \omega_0 t$ is equal to two impulse functions—one at ω_0 and another at $-\omega_0$.

5-8.6 Fourier Transform of $Ae^{-at}\, u(t)$, with $a > 0$

The Fourier transform of an exponentially decaying function that starts at $t = 0$ is

$$\begin{aligned}\mathbf{X}(\omega) = \mathcal{F}[Ae^{-at}\, u(t)] &= \int_0^\infty Ae^{-at} e^{-j\omega t}\, dt \\ &= A\, \frac{e^{-(a+j\omega)t}}{-(a + j\omega)}\bigg|_0^\infty \\ &= \frac{A}{a + j\omega}.\end{aligned}$$

Hence,

$$Ae^{-at}\, u(t) \quad \longleftrightarrow \quad \frac{A}{a + j\omega} \quad \text{for } a > 0. \tag{5.100}$$

5-8.7 Fourier Transform of $u(t)$

The direct approach to finding $\mathbf{X}(\omega)$ for the unit step function leads to

$$\begin{aligned}\mathbf{X}(\omega) = \mathcal{F}[u(t)] &= \int_{-\infty}^\infty u(t)\, e^{-j\omega t}\, dt \\ &= \int_0^\infty e^{-j\omega t}\, dt \\ &= \frac{e^{-j\omega t}}{-j\omega}\bigg|_0^\infty = \frac{j}{\omega}(e^{-j\infty} - 1),\end{aligned}$$

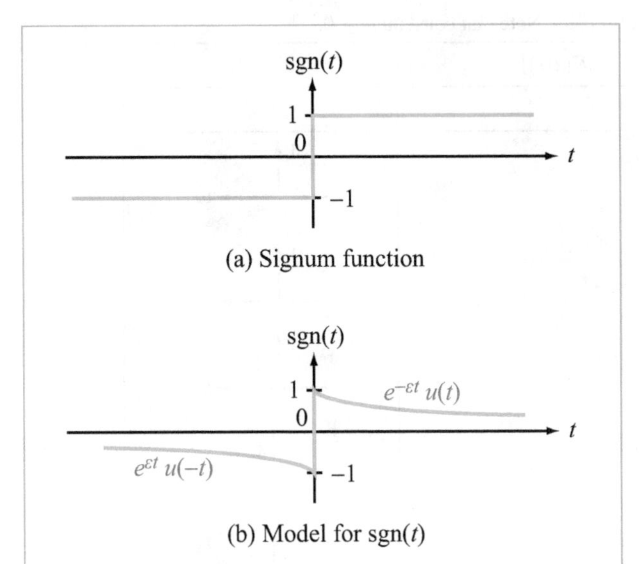

Figure 5-18: The model shown in (b) approaches the exact definition of $\text{sgn}(t)$ as $\epsilon \to 0$.

which is problematic, because $e^{-j\infty}$ does not converge. To avoid the convergence problem, we can pursue an alternative approach that involves the ***signum function***, defined by

$$\text{sgn}(t) = u(t) - u(-t). \tag{5.101}$$

Shown graphically in Fig. 5-18(a), the signum function resembles a step-function waveform (with an amplitude of two units) that has been slid downward by one unit. Looking at the waveform, it is easy to see that one can generate a step function from the signum function as

$$u(t) = \frac{1}{2} + \frac{1}{2}\,\text{sgn}(t). \tag{5.102}$$

The corresponding Fourier transform is given by

$$\begin{aligned}\mathcal{F}[u(t)] &= \mathcal{F}\left[\frac{1}{2}\right] + \frac{1}{2}\,\mathcal{F}[\text{sgn}(t)] \\ &= \pi\,\delta(\omega) + \frac{1}{2}\,\mathcal{F}[\text{sgn}(t)],\end{aligned} \tag{5.103}$$

where in the first term we used the relationship given by Eq. (5.95b). Next, we will obtain $\mathcal{F}[\text{sgn}(t)]$ by modeling the signum function as

$$\text{sgn}(t) = \lim_{\epsilon\to 0}[e^{-\epsilon t}\,u(t) - e^{\epsilon t}\,u(-t)], \tag{5.104}$$

with $\epsilon > 0$. The shape of the modeled waveform is shown in Fig. 5-18(b) for a small value of ϵ.

Now we are ready to apply the formal definition of the Fourier transform given by Eq. (5.87a):

$$\begin{aligned}\mathcal{F}[\text{sgn}(t)] &= \int_{-\infty}^{\infty} \lim_{\epsilon\to 0}[e^{-\epsilon t}\,u(t) - e^{\epsilon t}\,u(-t)]e^{-j\omega t}\,dt \\ &= \lim_{\epsilon\to 0}\left[\int_{0}^{\infty} e^{-(\epsilon+j\omega)t}\,dt - \int_{-\infty}^{0} e^{(\epsilon-j\omega)t}\,dt\right] \\ &= \lim_{\epsilon\to 0}\left[\left.\frac{e^{-(\epsilon+j\omega)t}}{-(\epsilon+j\omega)}\right|_{0}^{\infty} - \left.\frac{e^{(\epsilon-j\omega)t}}{\epsilon-j\omega}\right|_{-\infty}^{0}\right] \\ &= \lim_{\epsilon\to 0}\left[\frac{1}{\epsilon+j\omega} - \frac{1}{\epsilon-j\omega}\right] = \frac{2}{j\omega}.\end{aligned} \tag{5.105}$$

Use of Eq. (5.105) in Eq. (5.103) gives

$$\mathcal{F}[u(t)] = \pi\,\delta(\omega) + \frac{1}{j\omega}.$$

Equivalently, the preceding result can be expressed in the form

$$u(t) \quad \longleftrightarrow \quad \pi\,\delta(\omega) + \frac{1}{j\omega}. \tag{5.106}$$

Table 5-6 provides a list of commonly used time functions, together with their corresponding Fourier transforms, and Table 5-7 offers a summary of the major properties of the Fourier transform, many of which resemble those we encountered earlier in Chapter 3 in connection with the Laplace transform.

Example 5-11: Fourier Transform Properties

Establish the validity of the time derivative and modulation properties of the Fourier transform (#6 and #10 in Table 5-7).

Table 5-6: Examples of Fourier transform pairs. Note that constant $a \geq 0$.

	$x(t)$	$\mathbf{X}(\omega) = \mathcal{F}[x(t)]$	$\lvert\mathbf{X}(\omega)\rvert$
		BASIC FUNCTIONS	
1.	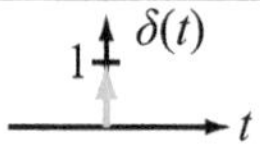	$\delta(t) \leftrightarrow 1$	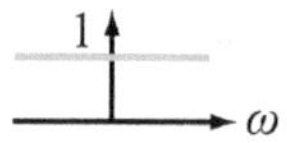
1a.	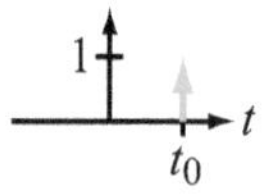	$\delta(t - t_0) \leftrightarrow e^{-j\omega t_0}$	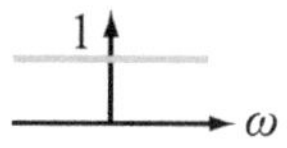
2.	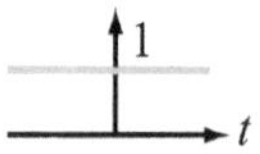	$1 \leftrightarrow 2\pi\ \delta(\omega)$	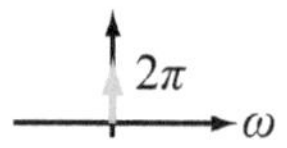
3.	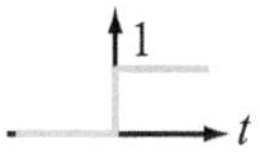	$u(t) \leftrightarrow \pi\ \delta(\omega) + 1/j\omega$	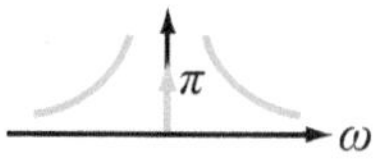
4.	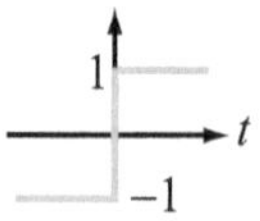	$\text{sgn}(t) \leftrightarrow 2/j\omega$	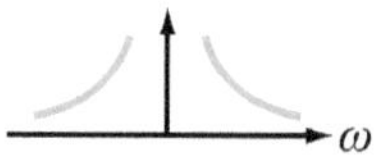
5.	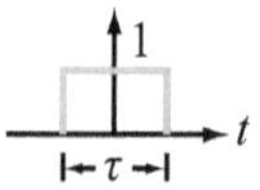	$\text{rect}(t/\tau) \leftrightarrow \tau\ \text{sinc}(\omega\tau/2)$	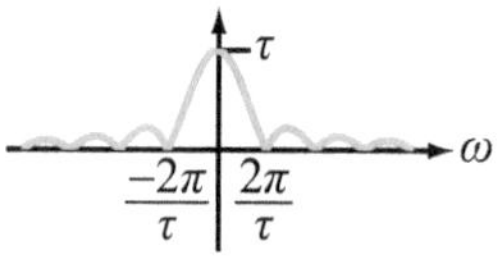
6.	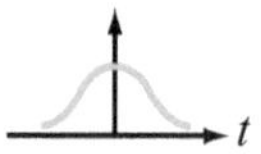	$\dfrac{e^{-t^2/(2\sigma^2)}}{\sqrt{2\pi\sigma^2}} \leftrightarrow e^{-\omega^2\sigma^2/2}$	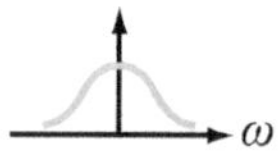
7a.	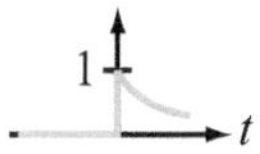	$e^{-at}\ u(t) \leftrightarrow 1/(a + j\omega)$	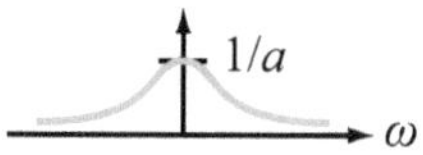
7b.	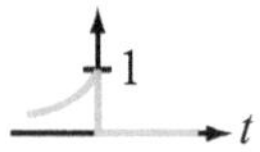	$e^{at}\ u(-t) \leftrightarrow 1/(a - j\omega)$	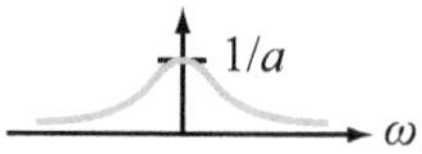
8.	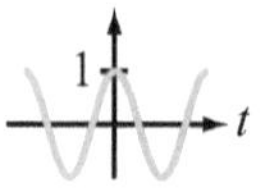	$\cos\omega_0 t \leftrightarrow \pi[\delta(\omega - \omega_0) + \delta(\omega + \omega_0)]$	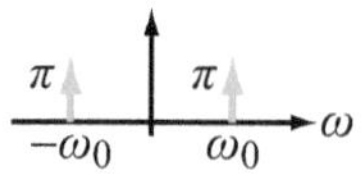
9.	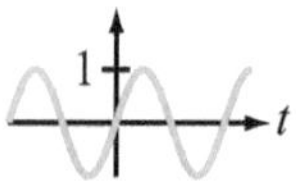	$\sin\omega_0 t \leftrightarrow j\pi[\delta(\omega + \omega_0) - \delta(\omega - \omega_0)]$	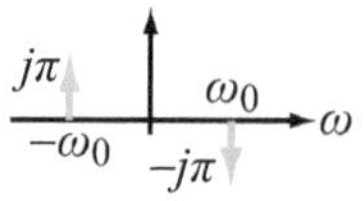
		ADDITIONAL FUNCTIONS	
10.		$e^{j\omega_0 t} \leftrightarrow 2\pi\ \delta(\omega - \omega_0)$	
11.		$te^{-at}\ u(t) \leftrightarrow 1/(a + j\omega)^2$	
12a.		$[e^{-at}\sin\omega_0 t]\ u(t) \leftrightarrow \omega_0/[(a + j\omega)^2 + \omega_0^2]$	
12b.	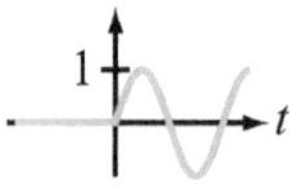	$[\sin\omega_0 t]\ u(t) \leftrightarrow (\pi/2j)[\delta(\omega - \omega_0) - \delta(\omega + \omega_0)] + [\omega_0^2/(\omega_0^2 - \omega^2)]$	
13a.		$[e^{-at}\cos\omega_0 t]\ u(t) \leftrightarrow (a + j\omega)/[(a + j\omega)^2 + \omega_0^2]$	
13b.	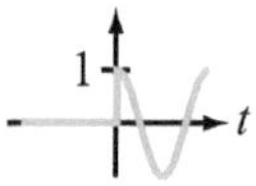	$[\cos\omega_0 t]\ u(t) \leftrightarrow (\pi/2)[\delta(\omega - \omega_0) + \delta(\omega + \omega_0)] + [j\omega/(\omega_0^2 - \omega^2)]$	

Table 5-7: Major properties of the Fourier transform.

Property	$x(t)$		$\mathbf{X}(\omega) = \mathcal{F}[x(t)] = \int_{-\infty}^{\infty} x(t)\, e^{-j\omega t}\, dt$
1. Multiplication by a constant	$K\, x(t)$	$\longleftrightarrow$	$K\, \mathbf{X}(\omega)$
2. Linearity	$K_1\, x_1(t) + K_2\, x_2(t)$	$\longleftrightarrow$	$K_1\, \mathbf{X}_1(\omega) + K_2\, \mathbf{X}_2(\omega)$
3. Time scaling	$x(at)$	$\longleftrightarrow$	$\frac{1}{\lvert a\rvert}\, \mathbf{X}\left(\frac{\omega}{a}\right)$
4. Time shift	$x(t - t_0)$	$\longleftrightarrow$	$e^{-j\omega t_0}\, \mathbf{X}(\omega)$
5. Frequency shift	$e^{j\omega_0 t}\, x(t)$	$\longleftrightarrow$	$\mathbf{X}(\omega - \omega_0)$
6. Time 1st derivative	$x' = \frac{dx}{dt}$	$\longleftrightarrow$	$j\omega\, \mathbf{X}(\omega)$
7. Time nth derivative	$\frac{d^n x}{dt^n}$	$\longleftrightarrow$	$(j\omega)^n\, \mathbf{X}(\omega)$
8. Time integral	$\int_{-\infty}^{t} x(\tau)\, d\tau$	$\longleftrightarrow$	$\frac{\mathbf{X}(\omega)}{j\omega} + \pi\, \mathbf{X}(0)\, \delta(\omega)$, where $\mathbf{X}(0) = \int_{-\infty}^{\infty} x(t)\, dt$
9. Frequency derivative	$t^n\, x(t)$	$\longleftrightarrow$	$(j)^n \frac{d^n \mathbf{X}(\omega)}{d\omega^n}$
10. Modulation	$x(t)\, \cos\omega_0 t$	$\longleftrightarrow$	$\frac{1}{2}[\mathbf{X}(\omega - \omega_0) + \mathbf{X}(\omega + \omega_0)]$
11. Convolution in t	$x_1(t) * x_2(t)$	$\longleftrightarrow$	$\mathbf{X}_1(\omega)\, \mathbf{X}_2(\omega)$
12. Convolution in ω	$x_1(t)\, x_2(t)$	$\longleftrightarrow$	$\frac{1}{2\pi}\, \mathbf{X}_1(\omega) * \mathbf{X}_2(\omega)$
13. Conjugate symmetry			$\mathbf{X}(-\omega) = \mathbf{X}^*(\omega)$

Solution:

Time Derivative Property

From Eq. (5.87b),

$$x(t) = \frac{1}{2\pi} \int_{-\infty}^{\infty} \mathbf{X}(\omega)\, e^{j\omega t}\, d\omega. \tag{5.107}$$

Differentiating both sides with respect to t gives

$$x'(t) = \frac{dx}{dt} = \frac{1}{2\pi} \int_{-\infty}^{\infty} j\omega\, \mathbf{X}(\omega)\, e^{j\omega t}\, d\omega.$$

▶ Differentiating $x(t)$ in the time domain is equivalent to multiplying $\mathbf{X}(\omega)$ by $j\omega$ in the frequency domain. ◀

Thus,

$$x'(t) \longleftrightarrow j\omega\, \mathbf{X}(\omega). \tag{5.108}$$

(derivative property)

Time Modulation Property

We start by multiplying both sides of Eq. (5.107) by $\cos\omega_0 t$ and, for convenience, we change the dummy variable ω to ω':

$$x(t)\, \cos\omega_0 t = \frac{1}{2\pi} \int_{-\infty}^{\infty} \cos\omega_0 t\, \mathbf{X}(\omega')\, e^{j\omega' t}\, d\omega'.$$

Applying Euler's identity to $\cos\omega_0 t$ on the right-hand side leads to

$$
\begin{aligned}
x(t)\ \cos\omega_0 t \\
&= \frac{1}{2\pi}\int_{-\infty}^{\infty}\left(\frac{e^{j\omega_0 t}+e^{-j\omega_0 t}}{2}\right)\mathbf{X}(\omega')\ e^{j\omega' t}\ d\omega' \\
&= \frac{1}{4\pi}\left[\int_{-\infty}^{\infty}\mathbf{X}(\omega')\ e^{j(\omega'+\omega_0)t}\ d\omega'\right. \\
&\quad \left. + \int_{-\infty}^{\infty}\mathbf{X}(\omega')\ e^{j(\omega'-\omega_0)t}\ d\omega'\right].
\end{aligned}
$$

Upon making the substitution ($\omega = \omega'+\omega_0$) in the first integral, and independently making the substitution ($\omega = \omega'-\omega_0$) in the second integral, we have

$$
\begin{aligned}
x(t)\ \cos\omega_0 t = \frac{1}{2}\left[\frac{1}{2\pi}\int_{-\infty}^{\infty}\mathbf{X}(\omega-\omega_0)\ e^{j\omega t}\ d\omega\right. \\
\left. + \frac{1}{2\pi}\int_{-\infty}^{\infty}\mathbf{X}(\omega+\omega_0)\ e^{j\omega t}\ d\omega\right],
\end{aligned}
$$

which can be cast in the abbreviated form

$$
x(t)\cos(\omega_0 t) \quad \longleftrightarrow \quad \frac{1}{2}[\mathbf{X}(\omega-\omega_0)+\mathbf{X}(\omega+\omega_0)].
$$

(modulation property) (5.109)

Of course, this result can also be obtained by applying Eq. (5.96) twice, once with ω_0 and another with $-\omega_0$. The modulation property is the foundational cornerstone of ***frequency division multiplexing*** (*FDM*), which allows the simultaneous transmission of several signals over the same channel by allocating part of the available spectrum to each of them. The details are covered in Section 6-12 on amplitude modulation.

Concept Question 5-14: What is the Fourier transform of a dc voltage? (See S²)

Concept Question 5-15: "An impulse in the time domain is equivalent to an infinite number of sinusoids, all with equal amplitude." Is this a true statement? Can one construct an ideal impulse function? (See S²)

Exercise 5-7: Use the entries in Table 5-6 to determine the Fourier transform of $u(-t)$.

Answer: $\mathbf{X}(\omega) = \pi\ \delta(\omega) - 1/j\omega$. (See S²)

Exercise 5-8: Verify the Fourier transform expression for entry #10 in Table 5-6.

Answer: (See S²)

5-9 Parseval's Theorem for Fourier Transforms

5-9.1 Signal Energy

Recall that Parseval's theorem for the Fourier series stated that the *average power* of a signal $x(t)$ can be computed in either the time or frequency domains. Parseval's theorem for the Fourier transform states that the *energy* E of a signal can be computed in either the time or frequency domains. Specifically, we have

$$
E = \int_{-\infty}^{\infty} |x(t)|^2\ dt = \frac{1}{2\pi}\int_{-\infty}^{\infty}|\mathbf{X}(\omega)|^2\ d\omega. \tag{5.110}
$$

To demonstrate the validity of Eq. (5.110), we will perform a few steps of mathematical manipulations, starting with

$$
\begin{aligned}
E &= \int_{-\infty}^{\infty} x(t)\ x^*(t)\ dt \\
&= \int_{-\infty}^{\infty} x(t)\left[\frac{1}{2\pi}\int_{-\infty}^{\infty}\mathbf{X}^*(\omega)\ e^{-j\omega t}\ d\omega\right] dt, \qquad (5.111)
\end{aligned}
$$

where $x^*(t)$ was replaced with the inverse Fourier transform relationship given by Eq. (5.87b). By reversing the order of $x(t)$ and $\mathbf{X}(\omega)$, and reversing the order of integration, we have

$$
\begin{aligned}
E &= \frac{1}{2\pi}\int_{-\infty}^{\infty}\mathbf{X}^*(\omega)\left[\int_{-\infty}^{\infty} x(t)\ e^{-j\omega t}\ dt\right] d\omega \\
&= \frac{1}{2\pi}\int_{-\infty}^{\infty}\mathbf{X}^*(\omega)\ \mathbf{X}(\omega)\ d\omega, \qquad (5.112)
\end{aligned}
$$

where we used the definition of the Fourier transform given by Eq. (5.87a). The combination of Eqs. (5.111) and (5.112) can be written as

$$E = \int_{-\infty}^{\infty} |x(t)|^2 \, dt = \frac{1}{2\pi} \int_{-\infty}^{\infty} |\mathbf{X}(\omega)|^2 \, d\omega. \tag{5.113}$$

(Parseval's theorem)

The inner-product version of Parseval's theorem for the Fourier transform is

$$\int_{-\infty}^{\infty} x(t) \, y^*(t) \, dt = \frac{1}{2\pi} \int_{-\infty}^{\infty} \mathbf{X}(\omega) \, \mathbf{Y}^*(\omega) \, d\omega. \tag{5.114}$$

Its derivation follows the same steps leading to Eq. (5.113).

Example 5-12: Energy of Decaying Exponential

Compute the energy of the signal given by $x(t) = e^{-at} \, u(t)$ in both the time and frequency domains, and show that they agree. The exponent a is a positive real number.

Solution: From Table 5-4, the Fourier transform of $x(t) = e^{-at} \, u(t)$ is $\mathbf{X}(\omega) = 1/(a + j\omega)$.

Energy in time domain:

$$\int_{0}^{\infty} |e^{-at}|^2 \, dt = \int_{0}^{\infty} e^{-2at} \, dt = \frac{1}{2a} \, .$$

Energy in frequency domain:

$$\frac{1}{2\pi} \int_{-\infty}^{\infty} \left| \frac{1}{a + j\omega} \right|^2 d\omega = \frac{1}{2\pi} \int_{-\infty}^{\infty} \frac{1}{a^2 + \omega^2} \, d\omega = \frac{1}{2a} \, .$$

Hence, the energy is $1/(2a)$ in both domains.

5-9.2 Energy Spectral Density

Average power of periodic signal

If $x(t)$ is a periodic signal of period T_0, it can be expressed by the Fourier series representation given by Eq. (5.40) as

$$x(t) = \sum_{n=-\infty}^{\infty} \mathbf{x}_n e^{jn\omega_0 t}, \tag{5.115}$$

where $\omega_0 = 2\pi/T_0$ and $\mathbf{x}_n$ are the Fourier series coefficients given by Eq. (5.41). The *one-sided average power* of $x(t)$ at frequency harmonic $\omega_n = n\omega_0$ is

$$P_{n_1} = |\mathbf{x}_n|^2. \tag{5.116}$$

(one-sided average power)

If $x(t)$ is real-valued, it can also be expressed in terms of the trigonometric Fourier series given by Eq. (5.39), namely

$$x(t) = c_0 + \sum_{n=1}^{\infty} c_n \cos(n\omega_0 t + \phi_n). \tag{5.117}$$

The average dc power is

$$P_0 = c_0^2, \tag{5.118a}$$

and the *two-sided average power* of $x(t)$ at frequency harmonic $\omega_n = n\omega_0$ (for $n \neq 0$) is

$$P_{n_2} = \frac{c_n^2}{2} = 2|\mathbf{x}_n|^2 \qquad \text{for } n \neq 0. \tag{5.118b}$$

(two-sided average power for $x(t)$ = real-valued)

The factor of 2 accounts for the fact that a sinusoid $c_n \cos(n\omega_0 t + \phi_n)$ has complex exponential components at $\pm n\omega_0$, so the two-sided average power is the sum of the two one-sided powers, which are identical because $x(t)$ is real-valued. From Table 5-3, $|x_n| = c_n/2$, so $2|x_n|^2 = c_n^2/2$.

Energy spectral density of nonperiodic signal

As noted in Section 5-7.2, if $x(t)$ is nonperiodic, we should use the Fourier transform (instead of the Fourier series) to represent it. From Eq. (5.87a), the Fourier transform of $x(t)$ is given by

$$\mathbf{X}(\omega) = \int_{-\infty}^{\infty} x(t)\, e^{-j\omega t}\, dt. \tag{5.119}$$

It may be tempting to apply a definition analogous to that given by Eq. (5.116) to describe the power or energy of a nonperiodic signal at a given frequency ω_0. Such a definition would assume the form $|\mathbf{X}(\omega_0)|^2$, but it would not be correct. The energy at a specific frequency is zero, because the total energy of $x(t)$ is an integral, not a sum, of components at various frequencies.

Instead, we define the ***one-sided energy spectral density*** of $x(t)$ at frequency ω_0 as

$$\mathcal{E}_1(\omega_0) = \frac{1}{2\pi}\, |\mathbf{X}(\omega_0)|^2. \tag{5.120}$$

(one-sided energy spectral density)

Accordingly, the energy E carried by signal $x(t)$ in the frequency range between ω_0 and $\omega_0 + \delta\omega$, in the limit as $\delta\omega \to 0$, is, by Parseval's theorem, given by

$$E = \int_{\omega_0}^{\omega_0+\delta\omega} \mathcal{E}_1(\omega_0)\, d\omega = \frac{1}{2\pi} \int_{\omega_0}^{\omega_0+\delta\omega} |\mathbf{X}(\omega_0)|^2\, d\omega = \frac{1}{2\pi}\, |\mathbf{X}(\omega_0)|^2\, \delta\omega. \tag{5.121}$$

The energy E is equal to the product of the energy spectral density and the small frequency interval $\delta\omega$. It is analogous to the definition of the mass m contained in a small volume δV as $m = \rho\,\delta V$, where ρ is the mass density.

If $x(t)$ is real-valued, then $|\mathbf{X}(\omega_0)| = |\mathbf{X}(-\omega_0)|$, in which case we can define the ***two-sided energy spectral density*** as

$$\mathcal{E}_2(\omega_0) = 2\mathcal{E}_1(\omega_0) = \frac{1}{\pi}\, |\mathbf{X}(\omega_0)|^2. \tag{5.122}$$

(two-sided energy spectral density for $x(t)$ = real-valued)

Example 5-13: Energy Spectral Density

Compute the two-sided energy spectral density of the signal

$$x(t) = e^{-3t}\, u(t)$$

at $\omega = 4$ rad/s. Also compute the energy of $x(t)$ contained in the interval $4 < |\omega| < 4.01$ rad/s.

Solution: Application of the Fourier transform relation given by Eq. (5.119) to $x(t)$ leads to

$$\mathbf{X}(\omega) = \frac{1}{3 + j\omega}.$$

Since $x(t)$ is real-valued, the two-sided energy at $\omega = 4$ rad/s is

$$\mathcal{E}_2 = \frac{1}{\pi}\, |\mathbf{X}(4)|^2 = \frac{1}{\pi} \left| \frac{1}{3 + j4} \right|^2 = 1.27 \times 10^{-2}.$$

The total energy contained in the designated interval is

$$E = \mathcal{E}_2\, \delta\omega = 1.27 \times 10^{-2} \times 0.01 = 1.27 \times 10^{-4}.$$

Alternatively, we could have applied Parseval's theorem to compute E:

$$E = \frac{1}{2\pi} \int_{-4.01}^{-4} \left| \frac{1}{3 + j\omega} \right|^2 d\omega + \frac{1}{2\pi} \int_{4}^{4.01} \left| \frac{1}{3 + j\omega} \right|^2 d\omega$$
$$= 1.27 \times 10^{-4}.$$

5-10 Additional Attributes of the Fourier Transform

5-10.1 Fourier Transforms of Periodic Signals

In filtering applications, we sometimes need to compute the Fourier transform of a periodic signal expressed in terms of its Fourier series. Consider a periodic signal $x(t)$ characterized by an exponential Fourier-series representation as

$$x(t) = \sum_{n=-\infty}^{\infty} \mathbf{x}_n e^{jn\omega_0 t}. \tag{5.123}$$

Taking the Fourier transform of both sides gives

$$\mathcal{F}\{x(t)\} = \sum_{n=-\infty}^{\infty} \mathbf{x}_n\, \mathcal{F}\left\{e^{jn\omega_0 t}\right\}. \tag{5.124}$$

According to entry #10 in Table 5-6, we have

$$\mathcal{F}\left\{e^{j\omega_0 t}\right\} = 2\pi\ \delta(\omega - \omega_0).$$

By extension, replacing ω_0 with $n\omega_0$ gives

$$\mathcal{F}\left\{e^{jn\omega_0 t}\right\} = 2\pi\ \delta(\omega - n\omega_0). \tag{5.125}$$

Using Eq. (5.125) in Eq. (5.124) leads to

$$\mathcal{F}\{x(t)\} = \sum_{n=-\infty}^{\infty} \mathbf{x}_n 2\pi\ \delta(\omega - n\omega_0). \tag{5.126}$$

▶ This result states that the Fourier transform of a periodic signal with Fourier series coefficients $\mathbf{x}_n$ and associated frequencies $\omega = n\omega_0$ consists of impulses of areas $2\pi\mathbf{x}_n$ at those frequencies. ◀

Example 5-14: Cosine Waveform

Compute the Fourier transform of

$$x(t) = 8\cos(3t + 2).$$

Solution:

(a) Method 1: Application of Eq. (5.126)

The fundamental angular frequency of $x(t)$ is $\omega_0 = 3$ rad/s, and its period is $T_0 = 2\pi/\omega_0 = 2\pi/3$ s. Given that $x(t)$ is a simple cosine function, we will initially represent it by the amplitude/phase format

$$x(t) = c_0 + \sum_{n=1}^{\infty} c_n \cos(n\omega_0 t + \phi_n).$$

Comparison of $x(t) = 8\cos(3t+2)$ with the infinite summation leads to the obvious conclusion that $c_0 = 0$ and all terms of the summation also are zero, except for $n = 1$. Moreover, $c_1 = 8$, $\omega_0 = 3$ rad/s, and $\phi_1 = 2$ rad. Using Table 5-4, we can convert the representation to the exponential format by noting that

$$\mathbf{x}_1 = \frac{c_1}{2}\ e^{j\phi_1} = \frac{8}{2}\ e^{j2} = 4e^{j2},$$

$$\mathbf{x}_{-1} = \mathbf{x}_1^* = 4e^{-j2}.$$

Hence, the exponential Fourier series of $x(t)$ is

$$x(t) = \mathbf{x}_1 e^{j3t} + \mathbf{x}_{-1} e^{-j3t} = 4e^{j2}e^{j3t} + 4e^{-j2}e^{-j3t},$$

and by application of Eq. (5.126), the Fourier transform of $x(t)$ is

$$\mathcal{F}[8\cos(3t + 2)] = 8\pi e^{j2}\ \delta(\omega - 3) + 8\pi e^{-j2}\ \delta(\omega + 3),$$

which states that in the Fourier frequency domain a cosine waveform at angular frequency ω_0 is represented by two impulses: one at ω_0 and another at $-\omega_0$, in agreement with Eqs. (5.98) and (5.99).

(b) Method 2: Application of Eq. (5.95a)

Using the relation $\cos x = (e^{jx}+e^{-jx})/2$ and Eq. (5.95a) leads to

$$\begin{aligned}\mathcal{F}[8\cos(3t + 2)] &= \mathcal{F}[4e^{j2}e^{j3t} + 4e^{-j2}e^{-j3t}] \\ &= 4e^{j2}\ \mathcal{F}[e^{j3t}] + 4e^{-j2}\ \mathcal{F}[e^{-j3t}] \\ &= 8\pi e^{j2}\ \delta(\omega - 3) + 8\pi e^{-j2}\ \delta(\omega + 3).\end{aligned}$$

5-10.2 Conjugate Symmetry

According to the definition given by Eq. (5.87a), the Fourier transform of a signal $x(t)$ is given by

$$\mathbf{X}(\omega) = \int_{-\infty}^{\infty} x(t)\ e^{-j\omega t}\ dt. \tag{5.127}$$

If $x(t)$ is real valued, conjugating both sides gives

$$\mathbf{X}^*(\omega) = \int_{-\infty}^{\infty} x(t)\ e^{j\omega t}\ dt. \tag{5.128}$$

Alternatively, if we replace ω with $-\omega$ in Eq. (5.127), we have

$$\mathbf{X}(-\omega) = \int_{-\infty}^{\infty} x(t)\ e^{j\omega t}\ dt, \tag{5.129}$$

from which we conclude that

$$\mathbf{X}(-\omega) = \mathbf{X}^*(\omega). \tag{5.130}$$

(reversal property)

Consequently, for a *real-valued* $x(t)$, its Fourier transform $\mathbf{X}(\omega)$ exhibits the following symmetry properties:

(a) $|\mathbf{X}(\omega)|$ is an even function ($|\mathbf{X}(\omega)| = |\mathbf{X}(-\omega)|$).

(b) $\arg[\mathbf{X}(\omega)]$ is an odd function ($\arg[\mathbf{X}(\omega)] = -\arg[\mathbf{X}(-\omega)]$).

(c) If $x(t)$ is real and an even function of time, $\mathbf{X}(\omega)$ will be purely real and an even function.

(d) If $x(t)$ is real and an odd function of time, $\mathbf{X}(\omega)$ will be purely imaginary and an odd function.

▶ Conjugate symmetry has important implications about the properties of $x(t)$ and its Fourier transform $\mathbf{X}(\omega)$:

$x(t)$	$\mathbf{X}(\omega)$
Real and even	Real and even
Real and odd	Imaginary and odd
Imaginary and even	Imaginary and even
Imaginary and odd	Real and odd

◀

5-10.3 Dirichlet Conditions

At a meeting of the Paris Academy in 1807, Jean-Baptiste Joseph Fourier first made the claim that any periodic function can be expressed in terms of sinusoids. In response, Joseph Lagrange stood up and said Fourier was wrong, which led to a protracted argument between the two French mathematicians. Eventually, the issue was settled by agreeing that a periodic function $x(t)$ of period T_0 can be expanded in terms of sinusoids if the following ***Dirichlet conditions***, named after Peter Gustav Lejeune Dirichlet, are fulfilled (these are *sufficient, not necessary*):

- If over the period of length T_0, $x(t)$ has a *finite number* of discontinuities, maxima, and minima,

- and $x(t)$ is ***absolutely integrable***, that is,

$$\int_0^{T_0} |x(t)|\, dt < \infty, \tag{5.131}$$

- then $x(t)$ can be expanded in a Fourier series, and the summation representing $x(t)$ is ***pointwise convergent***:

$$\lim_{N\to\infty} \left| x(t) - \sum_{n=-N}^{N} \mathbf{x}_n e^{jn\omega_0 t} \right| = 0 \qquad \text{for all } t, \tag{5.132}$$

  which means that the summation converges to the true value of $x(t)$ at every point t where it is not discontinuous.

- Furthermore, at any discontinuity t_i of $x(t)$, the summation converges to $\frac{1}{2}[x(t_i^+) + x(t_i^-)]$.

These sufficient (but not necessary) conditions for the existence of a Fourier series of a periodic function are analogous to those associated with the existence of the Fourier transform of a nonperiodic function (Section 5-7.3). Consider the periodic function

$$x(t) = \sum_{n=-\infty}^{\infty} \tilde{x}(t - nT_0),$$

where T_0 is the period of $x(t)$ and $\tilde{x}(t)$ is a single period of $x(t)$. If $\tilde{x}(t)$ fails any of the conditions stated earlier, its periodic parent, $x(t)$, cannot be represented by a Fourier series. One example is

$$\tilde{x}(t) = \frac{1}{\frac{T_0}{2} - t} \qquad \text{for } 0 < t < T_0.$$

Because $\tilde{x}(t)$ is not absolutely integrable and has an unbounded discontinuity at $t = T_0/2$, the Fourier series of $x(t)$ does not exist.

Another example is

$$\tilde{x}(t) = \sin\left(\frac{1}{t}\right) \qquad \text{for } |t| < T_0/2.$$

In this case, because $\tilde{x}(t)$ has an infinite number of maxima and minima, $x(t)$ does not have a Fourier-series representation.

5-10.4 Gibbs Phenomenon

The Gibbs phenomenon refers to the ***overshoot*** behavior of the Fourier series in the neighborhood of discontinuities. We discussed it briefly in connection with Fig. 5-7, but it is deserving of further elaboration. To that end, we reintroduce in Fig. 5-19 the square-wave waveform of Fig. 5-2(a), along with a plot of its Fourier-series representation computed for $n_{\max} = 100$ terms. A close examination of the Gibbs phenomenon at the points in time at which the slope of $x(t)$ is discontinuous reveals that no matter how large $n_{\max}$ gets, the height of the *overshoot* stays constant at 8.9% of the magnitude of the jump at the discontinuity. However, the width (in t) over which the overshoot is significant goes to zero as $n_{\max} \to \infty$. This behavior satisfies the ***mean-square convergence*** condition which pertains to the energy of the function rather than its magnitude. The convergence condition requires that

$$\lim_{N\to\infty} \int_0^{T_0} \left| x(t) - \sum_{n=-N}^{N} \mathbf{x}_n e^{jn\omega_0 t} \right|^2 dt = 0. \tag{5.133}$$

The pointwise convergence condition applies only at points where $x(t)$ is not discontinuous, whereas the mean-square convergence condition integrates over the entire period, including the discontinuities. Even though the overshoot itself never goes to zero, the *energy* it represents does go to zero as $n_{\max} \to \infty$ because the time duration over which the overshoot exists goes to zero.

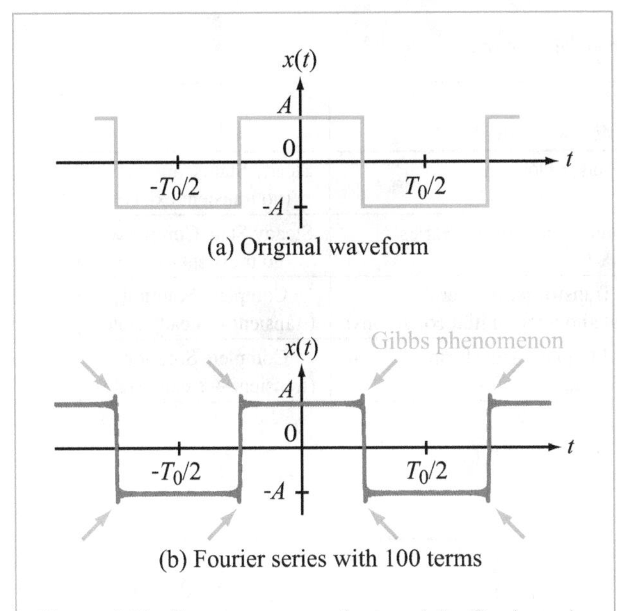

Figure 5-19: Square-wave waveform and its Fourier-series representation computed for $n_{\max} = 100$ terms.

5-11 Phasor vs. Laplace vs. Fourier

Consider an LTI system characterized by an LCCDE with input excitation $x(t)$ and output response $y(t)$. Beyond the time-domain differential equation solution method, which in practice can accommodate only first- and second-order systems, we have available to us three techniques by which to determine $y(t)$.

(a) The phasor-domain technique (Section 5-1).

(b) The Laplace transform technique (Chapters 3 and 4).

(c) The Fourier series and transform techniques (Chapter 5).

The applicability conditions for the three techniques, summarized in Table 5-8, are governed by the duration and shape of the waveform of the input excitation. Based on its duration, an input signal $x(t)$ is said to be:

(1) *everlasting*: if it exists over all time $(-\infty, \infty)$,

(2) *noncausal*: if it starts before $t = 0$,

(3) *causal*: if it starts at or after $t = 0$.

Through a change of variables, it is always possible to time-shift a noncausal signal that starts at a specific time $t = -T$ (where $T > 0$), so as to convert it into a causal signal. Hence, in essence, we have only two time-duration categories, namely, everlasting and causal.

In real life, there is no such thing as an everlasting signal. When we deal with real signals and systems, there is always a starting point in time for both the input and output signals. In general, an output signal consists of two components, a ***transient component*** associated with the initial onset of the input signal, and a ***steady state component*** that alone remains after the decay of the transient component to zero. If the input signal is sinusoidal and we are interested in only the steady state component of the output response, it is often convenient to regard the input signal as everlasting, even though, strictly speaking, it cannot be so. We regard it as such because we can then apply the phasor-domain technique, which is easier to implement than the other two techniques.

According to the summary provided in Table 5-8:

- If $x(t)$ is an ***everlasting sinusoid***, the phasor-domain technique is the solution method of choice.

- If $x(t)$ is an ***everlasting periodic signal***, such as a square wave or any repetitive waveform that can be represented by a Fourier series, then by virtue of the superposition principle, the phasor-domain technique can be used to compute the output responses corresponding to the individual Fourier components of the input signal, and then all of the output components can be added up to generate the total output.

- If $x(t)$ is a ***causal signal***, the Laplace transform technique is the preferred solution method. An important feature of the technique is that it can accommodate non-zero initial conditions of the system, if they exist.

- If $x(t)$ is everlasting and its waveform is nonperiodic, we can obtain $y(t)$ by applying either the bilateral Laplace transform (Section 3-9) or the Fourier transform. For input signals $x(t)$ whose Laplace transforms do not exist but their Fourier transforms do, the Fourier transform approach becomes the only viable option, and the converse is true for signals whose Fourier transforms do not exist but their Laplace transforms do.

The Laplace transform operates in the $\mathbf{s}$-domain, wherein an LTI system is described in terms of its ***transfer function*** $\mathbf{H}(\mathbf{s})$, where $\mathbf{s} = \sigma + j\omega$ is represented by a complex plane with real and imaginary axes along σ and $j\omega$. The Fourier transform operates along the $j\omega$-axis of the $\mathbf{s}$-plane, corresponding to $\sigma = 0$, and an LTI system is described by its ***frequency response*** $\mathbf{H}(\omega)$. In Section 3-9, we demonstrated that for causal systems excited by everlasting sinusoidal signals,

$$\mathbf{H}(\omega) = \mathbf{H}(\mathbf{s})|_{\mathbf{s}=j\omega}. \tag{5.134}$$

Table 5-8: Methods of solution.

Input $x(t)$			
Duration	Waveform	Solution Method	Output $y(t)$
Everlasting	Sinusoid	Phasor Domain	Steady State Component (no transient exists)
Everlasting	Periodic	Phasor Domain and Fourier Series	Steady State Component (no transient exists)
Causal, $x(t) = 0$, for $t < 0$	Any	Laplace Transform (unilateral) (can accommodate non-zero initial conditions)	Complete Solution (transient + steady state)
Everlasting	Any	Bilateral Laplace Transform or Fourier Transform	Complete Solution (transient + steady state)

The Fourier Transform

(a) For a signal $x(t)$, its Fourier transform is its frequency spectrum $\mathbf{X}(\omega)$.

(b) For a system, the Fourier transform of its impulse response $h(t)$ is the system's frequency response $\mathbf{H}(\omega)$.

5-12 Circuit Analysis with Fourier Transform

As was mentioned earlier, the Fourier transform technique can be used to analyze circuits excited by either one-sided or two-sided nonperiodic waveforms, so long as the circuit has no initial conditions. The procedure, which is analogous to the Laplace transform technique, with $\mathbf{s}$ replaced by $j\omega$, is demonstrated through Example 5-15.

Example 5-15: RC Circuit

The RC circuit shown in Fig. 5-20(a) is excited by a voltage source $\upsilon_s(t)$. Apply Fourier analysis to determine $i_C(t)$ if: (a) $\upsilon_s = 10u(t)$, (b) $\upsilon_s(t) = 10e^{-2t}\, u(t)$, and (c) $\upsilon_s(t) = 10+5\cos 4t$, all measured in volts. The element values are $R_1 = 2$ kΩ, $R_2 = 4$ kΩ, and $C = 0.25$ mF. The intent of this example is to demonstrate the solution procedure when using the Fourier transform technique. For all three excitations, the same results can be obtained using the Laplace transform technique, and for excitation (c) the phasor-domain technique is also applicable and easy to implement.

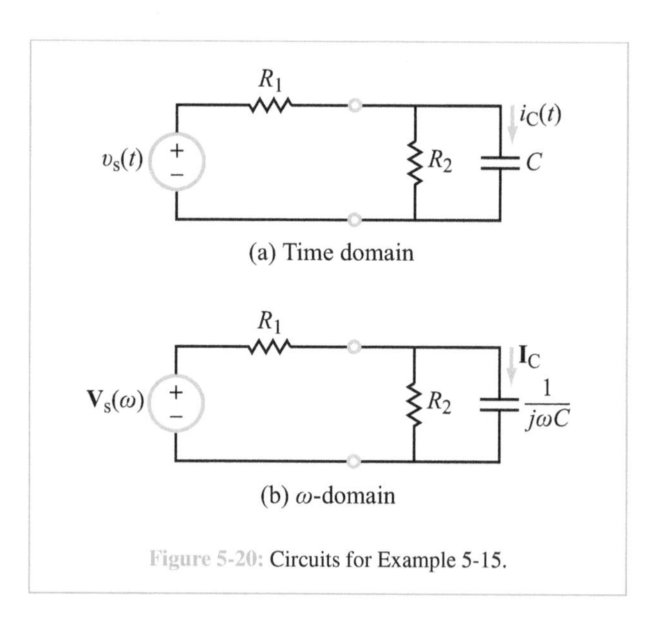

Figure 5-20: Circuits for Example 5-15.

Solution:

Step 1: **Transfer Circuit to ω-Domain**

In the frequency domain circuit shown in Fig. 5-20(b), $\mathbf{V}_s(\omega)$ is the Fourier transform of $\upsilon_s(t)$.

Step 2: **Determine $\mathbf{H}(\omega) = \mathbf{I}_C(\omega)/\mathbf{V}_s(\omega)$**

Application of source transformation to the circuit in Fig. 5-20(b), followed with current division, leads to

$$\mathbf{H}(\omega) = \frac{\mathbf{I}_C(\omega)}{\mathbf{V}_s(\omega)} = \frac{j\omega/R_1}{\dfrac{R_1+R_2}{R_1R_2C} + j\omega} = \frac{j0.5\omega \times 10^{-3}}{3+j\omega}. \tag{5.135}$$

Step 3: **Solve for $\mathbf{I}_C(\omega)$ and $i_C(t)$**

(a) Input $\boldsymbol{v_s(t) = 10u(t)}$:

The corresponding Fourier transform, per entry #3 in Table 5-6, is

$$\mathbf{V}_s(\omega) = 10\pi\ \delta(\omega) + \frac{10}{j\omega},$$

and the corresponding current is

$$\mathbf{I}_C(\omega) = \mathbf{H}(\omega)\ \mathbf{V}_s(\omega)$$
$$= \frac{j5\pi\omega\ \delta(\omega) \times 10^{-3}}{3 + j\omega} + \frac{5 \times 10^{-3}}{3 + j\omega}.$$

The inverse Fourier transform of $\mathbf{I}_C(\omega)$ is given by

$$i_C(t) = \frac{1}{2\pi} \int_{-\infty}^{\infty} \frac{j5\pi\omega\ \delta(\omega) \times 10^{-3}}{3 + j\omega}\ e^{j\omega t}\ d\omega + \mathcal{F}^{-1}\left[\frac{5 \times 10^{-3}}{3 + j\omega}\right],$$

where we applied the formal definition of the inverse Fourier transform to the first term—because it includes an impulse—and the functional form to the second term, because we intend to use look-up entry #7 in Table 5-6. Accordingly,

$$i_C(t) = 0 + 5e^{-3t}\ u(t) \text{ mA}. \tag{5.136}$$

(b) Input $\boldsymbol{v_s(t) = 10e^{-2t}\ u(t)}$:

By entry #7 in Table 5-6, we have

$$\mathbf{V}_s(\omega) = \frac{10}{2 + j\omega}.$$

and the corresponding current $\mathbf{I}_C(\omega)$ is given by

$$\mathbf{I}_C(\omega) = \mathbf{H}(\omega)\ \mathbf{V}_s(\omega)$$
$$= \frac{j5\omega \times 10^{-3}}{(2 + j\omega)(3 + j\omega)}.$$

Application of partial fraction expansion (Section 3-5) gives

$$\mathbf{I}_C(\omega) = \frac{A_1}{2 + j\omega} + \frac{A_2}{3 + j\omega},$$

with

$$A_1 = (2 + j\omega)\ \mathbf{I}_C(\omega)\big|_{j\omega=-2}$$
$$= \left.\frac{j5\omega \times 10^{-3}}{3 + j\omega}\right|_{j\omega=-2} = -10 \times 10^{-3}$$

and

$$A_2 = (3 + j\omega)\ \mathbf{I}_C(\omega)\big|_{j\omega=-3}$$
$$= \left.\frac{j5\omega \times 10^{-3}}{2 + j\omega}\right|_{j\omega=-3} = 15 \times 10^{-3}.$$

Hence, we obtain

$$\mathbf{I}_C(\omega) = \left(\frac{-10}{2 + j\omega} + \frac{15}{3 + j\omega}\right) \times 10^{-3}$$

and

$$i_C(t) = (15e^{-3t} - 10e^{-2t})\ u(t) \text{ mA}. \tag{5.137}$$

(c) Input $\boldsymbol{v_s(t) = 10 + 5\cos 4t}$:

By entries #2 and #8 in Table 5-6,

$$\mathbf{V}_s(\omega) = 20\pi\ \delta(\omega) + 5\pi[\delta(\omega - 4) + \delta(\omega + 4)],$$

and the capacitor current is

$$\mathbf{I}_C(\omega) = \mathbf{H}(\omega)\ \mathbf{V}_s(\omega)$$
$$= \frac{j10\pi\omega\ \delta(\omega) \times 10^{-3}}{3 + j\omega} + j2.5\pi \times 10^{-3}\left[\frac{\omega\ \delta(\omega - 4)}{3 + j\omega} + \frac{\omega\ \delta(\omega + 4)}{3 + j\omega}\right].$$

The corresponding time-domain current is obtained by applying Eq. (5.87b):

$$i_C(t) = \frac{1}{2\pi} \int_{-\infty}^{\infty} \frac{j10\pi\omega\ \delta(\omega) \times 10^{-3} e^{j\omega t}\ d\omega}{3 + j\omega}$$
$$+ \frac{1}{2\pi} \int_{-\infty}^{\infty} \frac{j2.5\pi\omega \times 10^{-3}}{3 + j\omega}\ \delta(\omega - 4)\ e^{j\omega t}\ d\omega$$
$$+ \frac{1}{2\pi} \int_{-\infty}^{\infty} \frac{j2.5\pi\omega \times 10^{-3}}{3 + j\omega}\ \delta(\omega + 4)\ e^{j\omega t}\ d\omega$$
$$= 0 + \frac{j5 \times 10^{-3} e^{j4t}}{3 + j4} - \frac{j5 \times 10^{-3} e^{-j4t}}{3 - j4}$$
$$= 5 \times 10^{-3}\left(\frac{e^{j4t} e^{j36.9^\circ}}{5} + \frac{e^{-j4t} e^{-j36.9^\circ}}{5}\right)$$
$$= 2\cos(4t + 36.9^\circ) \text{ mA}. \tag{5.138}$$

Exercise 5-9: Determine the voltage across the capacitor, $\upsilon_C(t)$, in Fig. 5-20(a) of Example 5-15, for each of the three voltage waveforms given in its example statement.

Answer: (a) $\upsilon_C(t) = \frac{10}{3} + \frac{20}{3}(1 - e^{-3t})\, u(t)$ V,
(b) $\upsilon_C(t) = 20(e^{-2t} - e^{-3t})\, u(t)$ V,
(c) $\upsilon_C(t) = \left[\frac{20}{3} + 2\cos(4t - 36.9°)\right]$ V. (See S²)

5-13 The Importance of Phase Information

A sinusoidal waveform given by

$$x(t) = A\cos(\omega_0 t + \phi) \tag{5.139}$$

is characterized by three parameters, namely its amplitude A, its angular frequency ω_0 (with $\omega_0 = 2\pi f_0$), and its phase angle ϕ. The role of A is straightforward; it determines the peak-to-peak swing of the waveform, and similarly, the role of f_0 also is easy to understand, as it defines the number of oscillations that the waveform goes through in 1 second. What about ϕ? At first glance, we might assign to ϕ a rather trivial role, because its only impact on the waveform is to specify in what direction and by how much the waveform is shifted in time relative to the waveform of $A\cos\omega_0 t$. Whereas such an assignment may be quite reasonable in the case of the simple sinusoidal waveform, we will demonstrate in this section that in the general case of a more elaborate waveform, the phase part of the waveform carries information that is equally important as that contained in the waveform's amplitude.

Let us consider the rectangular pulse shown in Fig. 5-21(a). According to entry #5 in Table 5-6, the pulse and its Fourier transform are given by

$$x(t) = \text{rect}\left(\frac{t}{\tau}\right) \quad \longleftrightarrow \quad \mathbf{X}(\omega) = \tau\,\text{sinc}\left(\frac{\omega\tau}{2}\right), \tag{5.140}$$

where τ is the pulse length and the sinc function is defined by Eq. (5.79). By defining $\mathbf{X}(\omega)$ as

$$\mathbf{X}(\omega) = |\mathbf{X}(\omega)|e^{j\,\phi(\omega)}, \tag{5.141}$$

we determine that the ***phase spectrum*** $\phi(\omega)$ can be ascertained from

$$e^{j\,\phi(\omega)} = \frac{\mathbf{X}(\omega)}{|\mathbf{X}(\omega)|} = \frac{\text{sinc}(\omega\tau/2)}{|\,\text{sinc}(\omega\tau/2)|}. \tag{5.142}$$

The quantity on the right-hand side of Eq. (5.142) is always equal to $+1$ or -1. Hence, $\phi(\omega) = 0°$ when $\text{sinc}(\omega\tau/2)$ is positive and 180° when $\text{sinc}(\omega\tau/2)$ is negative. The magnitude and phase spectra of the rectangular pulse are displayed in Figs. 5-21(b) and (c), respectively.

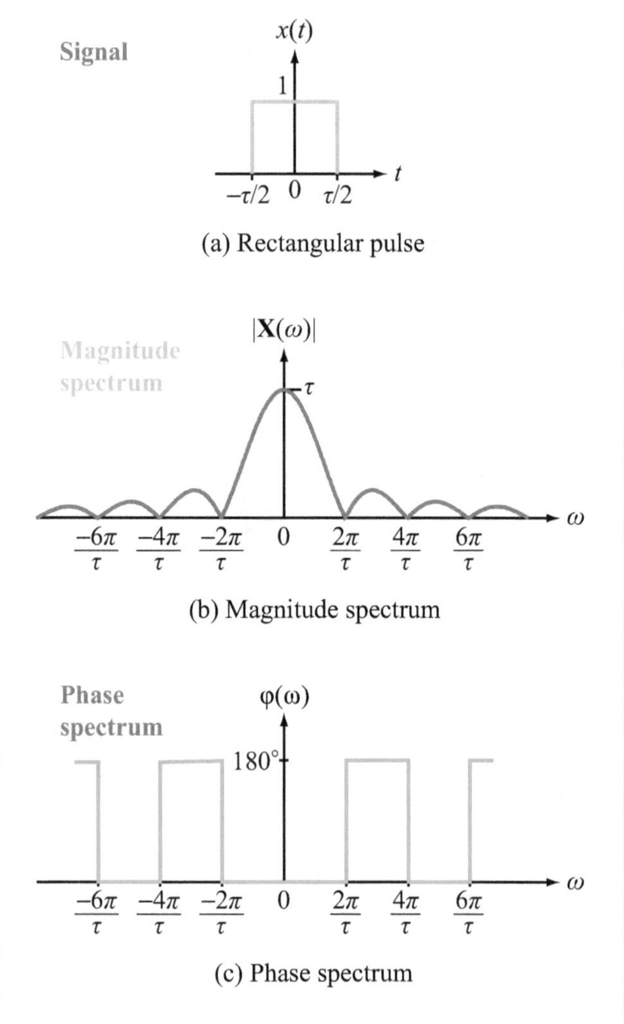

Figure 5-21: (a) Rectangular pulse, and corresponding (b) magnitude spectrum and (c) phase spectrum.

5-13.1 2-D Spatial Transform

When the independent variable is t, measured in seconds, the corresponding independent variable in the Fourier-transform domain is ω, measured in rad/s. The (t, ω) correspondence has analogies in other domains, such as the spatial domain. In fact, in the case of planar images, we deal with two spatial dimensions—rather than just one—which we shall label ξ and η. Accordingly, the image intensity may vary with both ξ and η and will be denoted $x(\xi, \eta)$. Moreover, since $x(\xi, \eta)$ is a function of two variables, so is its Fourier transform, which we will call the ***two-dimensional Fourier transform*** $\mathbf{X}(\omega_1, \omega_2)$, where ω_1 and ω_2 are called ***spatial frequencies***. If ξ and η are measured in meters, ω_1 and ω_2 will have units of rad/m. With digital images, ξ and η are measured in pixels, in which case ω_1 and ω_2 will have units of rad/pixel. Upon extending the Fourier-transform definition given by Eq. (5.87) to the 2-D case—as well as replacing the time dimension with spatial dimensions—we have

$$\mathbf{X}(\omega_1, \omega_2) = \mathcal{F}[x(\xi, \eta)] = \int_{-\infty}^{\infty}\int_{-\infty}^{\infty} x(\xi, \eta)\, e^{-j\omega_1\xi} e^{-j\omega_2\eta}\, d\xi\, d\eta$$

(2-D Fourier transform) (5.143a)

and

$$x(\xi, \eta) = \frac{1}{(2\pi)^2}\int_{-\infty}^{\infty}\int_{-\infty}^{\infty} \mathbf{X}(\omega_1, \omega_2)\, e^{j\omega_1\xi} e^{j\omega_2\eta}\, d\omega_1\, d\omega_2.$$

(2-D inverse Fourier transform) (5.143b)

By way of an example, let us consider the white square shown in Fig. 5-22(a). If we assign an amplitude of 1 to the white part of the image and 0 to the black part, the variation across the image along the ξ-direction is analogous to that representing the time-domain pulse of Fig. 5-21(a), and the same is true along η. Hence, the white square represents the product of two pulses, one along ξ and another along η, and is given by

$$x(\xi, \eta) = \text{rect}\left(\frac{\xi}{\ell}\right) \text{rect}\left(\frac{\eta}{\ell}\right), \quad (5.144)$$

where ℓ is the length of the square sides. Application of Eq. (5.143a) leads to

$$\mathbf{X}(\omega_1, \omega_2) = \ell^2 \,\text{sinc}\left(\frac{\omega_1\ell}{2}\right) \text{sinc}\left(\frac{\omega_2\ell}{2}\right). \quad (5.145)$$

The magnitude and phase spectra associated with the expression given by Eq. (5.145) are displayed in grayscale format in Fig. 5-22(b) and (c), respectively. For the magnitude spectrum, white represents the peak value of $|\mathbf{X}(\omega_1, \omega_2)|$ and black represents $|\mathbf{X}(\omega_1, \omega_2)| = 0$. The phase spectrum $\phi(\omega_1, \omega_2)$ varies between $-180°$ and $180°$, so the grayscale was defined such that white corresponds to $+180°$ and black to $-180°$. The tonal variations along ω_1 and ω_2 are equivalent to the patterns depicted in Figs. 5-21(b) and (c) for the rectangular pulse.

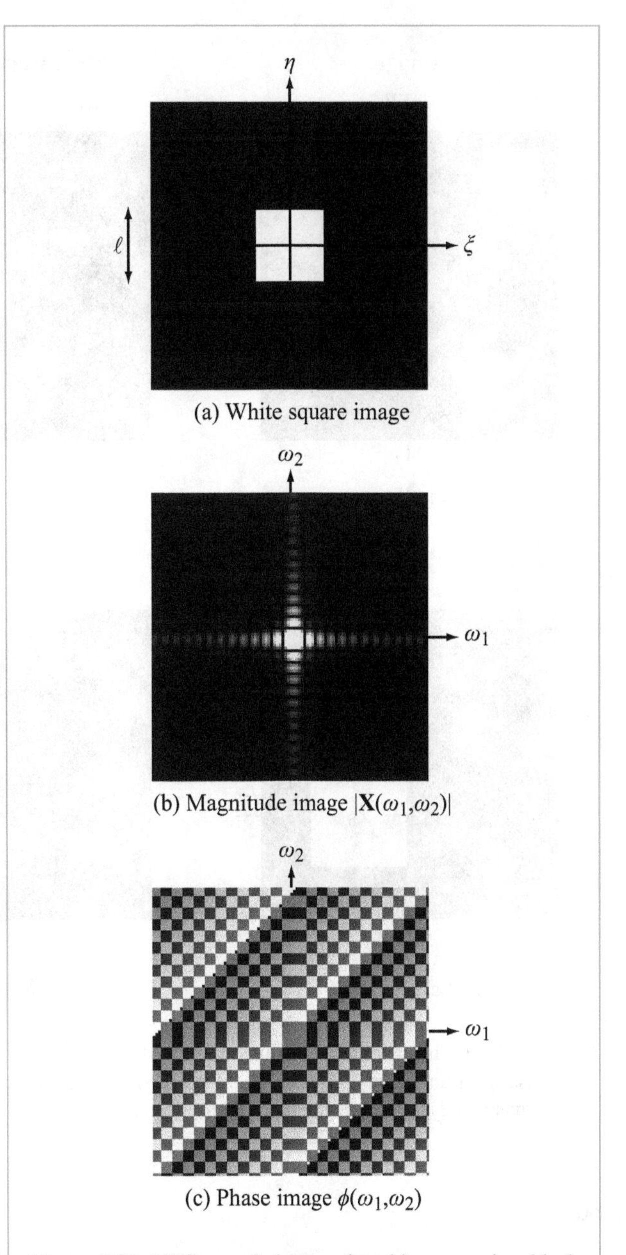

Figure 5-22: (a) Grayscale image of a white square in a black background, (b) magnitude spectrum, and (c) phase spectrum.

5-13.2 Magnitude and Phase Spectra

Next, we consider two white squares, of different sizes and at different locations, as shown in Fig. 5-23(a) and (d). The small square is of side ℓ_s and its center is at $(-L_s, +L_s)$, relative to the center of the image. In contrast, the big square is of side ℓ_b and located in the fourth quadrant with its center at $(+L_b, -L_b)$. Their corresponding functional expressions are

$$x_s(\xi, \eta) = \text{rect}\left(\frac{\xi + L_s}{\ell_s}\right) \text{rect}\left(\frac{\eta - L_s}{\ell_s}\right) \quad (5.146a)$$

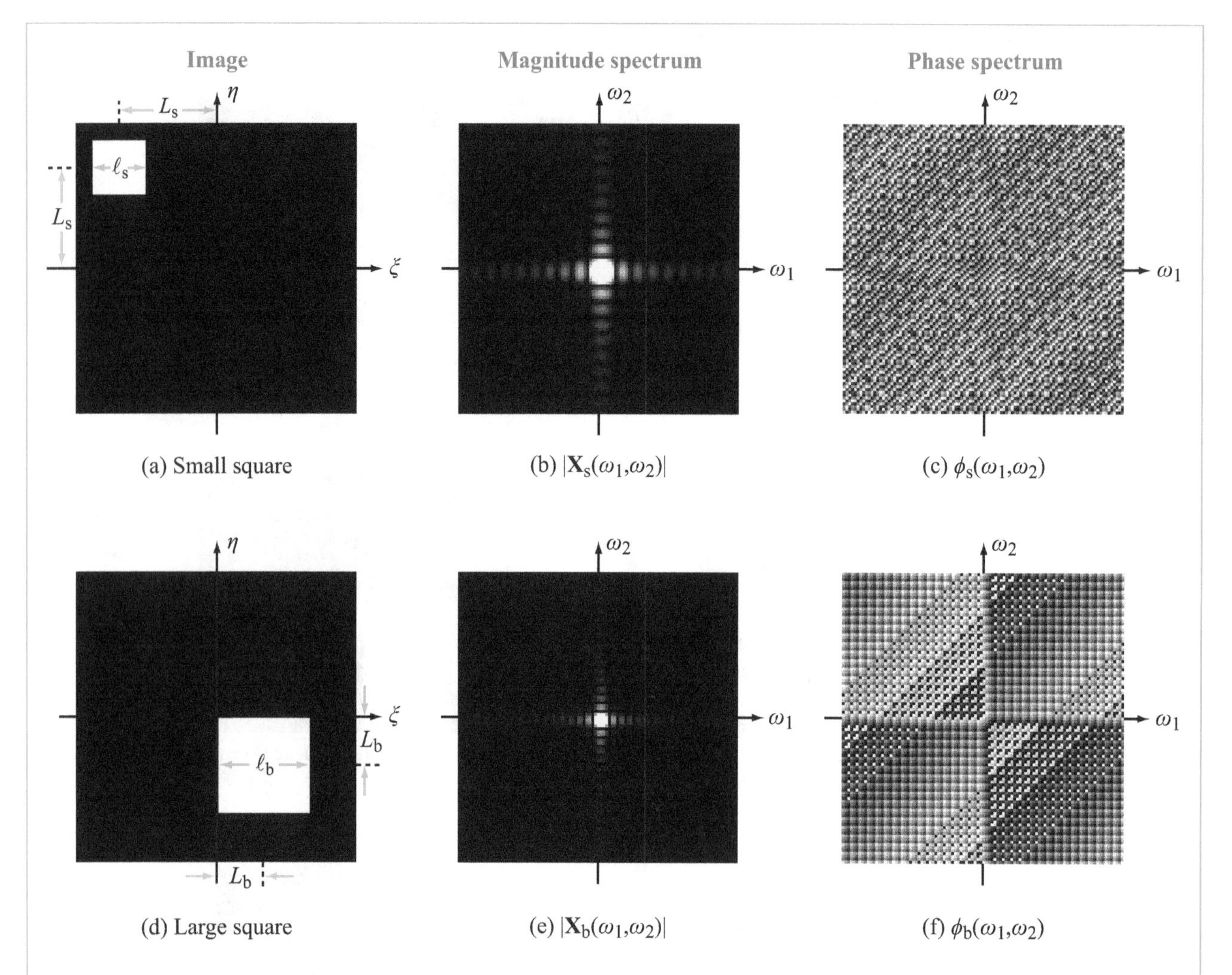

Figure 5-23: (a) Image of a small white square of dimension ℓ_s and center at $(\xi, \eta) = (-L_s, L_s)$, (b) magnitude spectrum of small square, (c) phase spectrum of small square, (d) image of a big white square of dimension ℓ_b and center at $(L_b, -L_b)$, (e) magnitude spectrum of large square, and (f) phase spectrum of large square.

and

$$x_b(\xi, \eta) = \text{rect}\left(\frac{\xi - L_b}{\ell_b}\right) \text{rect}\left(\frac{\eta + L_b}{\ell_b}\right). \quad (5.146b)$$

In view of property #4 in Table 5-7, the corresponding 2-D transforms are given by

$$\mathbf{X}_s(\omega_1, \omega_2) = \ell_s^2 e^{j\omega_1 L_s} \text{sinc}\left(\frac{\omega_1 \ell_s}{2}\right) e^{-j\omega_2 L_s} \text{sinc}\left(\frac{\omega_2 \ell_s}{2}\right) \quad (5.147a)$$

and

$$\mathbf{X}_b(\omega_1, \omega_2) = \ell_b^2 e^{-j\omega_1 L_b} \text{sinc}\left(\frac{\omega_1 \ell_b}{2}\right) e^{j\omega_2 L_b} \text{sinc}\left(\frac{\omega_2 \ell_b}{2}\right). \quad (5.147b)$$

Associated with $\mathbf{X}_s(\omega_1, \omega_2)$ are magnitude and phase spectra defined by

$$|\mathbf{X}_s(\omega_1, \omega_2)|$$

and

$$e^{j\phi_s(\omega_1, \omega_2)} = \frac{\mathbf{X}_s(\omega_1, \omega_2)}{|\mathbf{X}_s(\omega_1, \omega_2)|}, \quad (5.148)$$

and similar definitions apply to the magnitude and phase of $\mathbf{X}_b(\omega_1, \omega_2)$. The four 2-D spectra are displayed in Fig. 5-23.

5-13.3 Image Reconstruction

Figure 5-23 contains 2-D spectra $|\mathbf{X}_s|$, ϕ_s, $|\mathbf{X}_b|$, and ϕ_b. If we were to apply the inverse Fourier transform to $|\mathbf{X}_s|e^{j\phi_s}$, we would reconstruct the original image of the small square, and similarly, application of the inverse transform to $|\mathbf{X}_b|e^{j\phi_b}$ would generate the image of the big square. Neither result would be a surprise, but what if we were to "mix" magnitude and phase spectra? That is, what would we get if in the reconstruction process we were to apply the inverse Fourier transform to $|\mathbf{X}_b|e^{j\phi_s}$, which contains the magnitude spectrum of the big square and the phase spectrum of the small square. Would we still obtain a square, what size would it be, and where will it be located? The result of such an experiment is displayed in Fig. 5-24(a). We observe that the dominant feature in the reconstructed image is still a square, but neither its size nor its location match those of the square in Fig. 5-23(d). In fact, the location corresponds to that of the small square in Fig. 5-23(a). Similarly, in Fig. 5-24(b) we display the image reconstructed by applying the inverse Fourier transform to $|\mathbf{X}_s|e^{j\phi_b}$. In both cases, the location of the square in the reconstructed image is governed primarily by the phase spectrum.

Instead of squares, let us explore what happens when we use more complex images. Figures 5-25(a) and (b) are images of Albert Einstein and the Mona Lisa. The other two images are reconstructions based on mixed amplitude/phase spectra. The image in Fig. 5-25(c) was constructed using the magnitude spectrum of the Einstein image and the phase spectrum of the Mona Lisa image. Even though it contains the magnitude spectrum of the Einstein image, it fails to reproduce an image that resembles the original but successfully reproduces an image with the likeness of the original Mona Lisa image. Thus, in this case, the phase spectrum has proven to be not only important but even more important than the magnitude information. Further confirmation of the importance of phase information is evidenced by the image in Fig. 5-25(d), which reproduces a likeness of the Einstein image even though only the phase spectrum of the Einstein image was used in the reconstruction process.

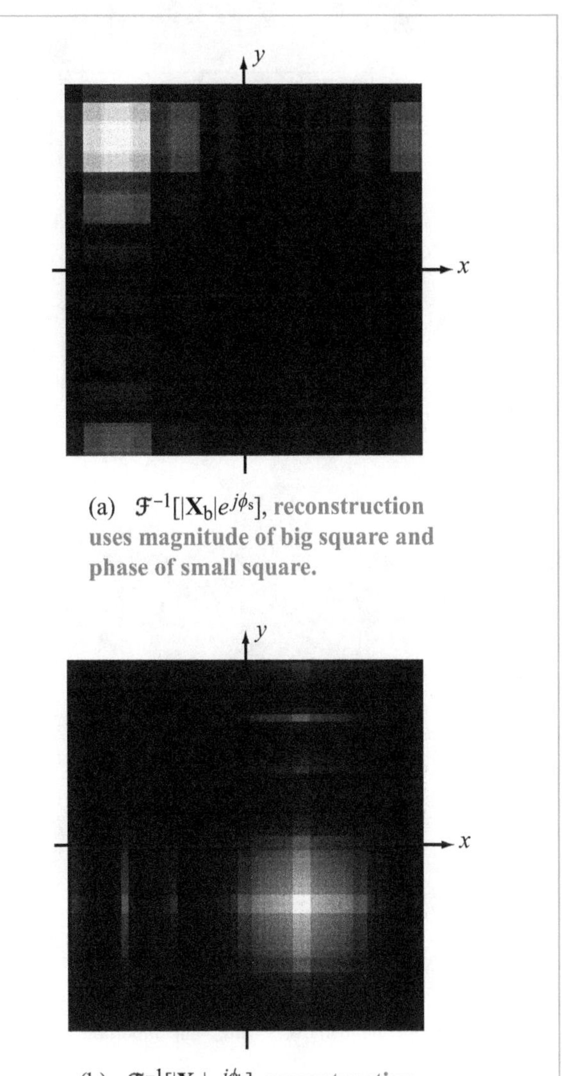

(a) $\mathfrak{F}^{-1}[|\mathbf{X}_b|e^{j\phi_s}]$, **reconstruction uses magnitude of big square and phase of small square.**

(b) $\mathfrak{F}^{-1}[|\mathbf{X}_s|e^{j\phi_b}]$, **reconstruction uses magnitude of small square and phase of big square.**

Figure 5-24: Reconstructed "mixed" images: (a) Although not sharp, the location and size of the small square are governed primarily by the phase information associated with the small square; (b) the phase information of the big square defines its location and approximate shape in the reconstructed image.

5-13.4 Image Reconstruction Recipe

To perform a 2-D Fourier transform or an inverse transform on a black-and-white image, you can use MATLAB®, Mathematica, MathScript, or similar software. The process entails the following steps:

1. Your starting point has to be a digital image, so if your image is a hard copy print, you will need to scan it to convert it into digital format. Consisting of $M \times N$ pixels, the digital image is equivalent to an $M \times N$ matrix. Associated with an individual

(a) Original Einstein image

(b) Original Mona Lisa image

(c) Reconstructed image based on *Einstein magnitude* and *Mona Lisa phase*

(d) Reconstructed image based on *Mona Lisa magnitude* and *Einstein phase*

Figure 5-25: In the reconstructed images, the phase spectrum exercised the dominant role, even more so than the magnitude spectrum. [From A. V. Oppenheim and J. S. Lim, "The importance of phase in signals," *Proceedings of the IEEE,* v. 69, no. 5, May 1981, pp. 529–541.]

pixel (m, n) is a grayscale intensity $I(m, n)$. Indices m, n define the location of a pixel along the vertical and horizontal directions, respectively, which is the converse of the traditional order used in defining 2-D continuous functions.

2. In MATLAB® software, the command `fft(x)` generates the one-dimensional Fast Fourier transform of vector x, and similarly, `fft2(I)` generates the two-dimensional fast Fourier transform of matrix `I`. The outcome of the 2-D FFT is an $M \times N$ matrix whose elements we will designate as

$$\mathbf{X}(m, n) = A(m, n) + jB(m, n),$$

where $A(m, n)$ and $B(m, n)$ are the real and imaginary parts of $\mathbf{X}(m, n)$, respectively. In the frequency domain, coordinates (m, n) represent frequencies ω_1 and ω_2, respectively.

3. The magnitude and phase matrices of the 2-D FFT can be generated from

$$|\mathbf{X}(m, n)| = [A^2(m, n) + B^2(m, n)]^{1/2},$$

and

$$\phi(m, n) = \tan^{-1}\left[\frac{B(m, n)}{A(m, n)}\right].$$

4. To display the matrices associated with $|\mathbf{X}(m, n)|$ and $\phi(m, n)$ as grayscale images with zero frequency located at the center of the image (as opposed to having the dc component located at the upper left corner of the matrix), it is necessary to

apply the command fftshift to each of the images before displaying it.

5. Reconstruction back to the spatial domain $x(m, n)$ entails using the command ifft2 on $\mathbf{X}(m, n)$.

6. If two images are involved, $I_1(m, n)$ and $I_2(m, n)$, with corresponding 2-D FFTs $\mathbf{X}_1(m, n)$ and $\mathbf{X}_2(m, n)$, respectively, reconstruction of a mixed transform composed of the magnitude of one of the transforms and the phase of the other one will require a prerequisite step prior to applying ifft2. The artificial FFT composed of the magnitude of $\mathbf{X}_1(m, n)$ and the phase of $\mathbf{X}_2(m, n)$, for example, is given by

$$\mathbf{X}_3(m, n) = |\mathbf{X}_1(m, n)| \cos \phi_2(m, n) + j|\mathbf{X}_2(m, n)| \sin \phi_2(m, n).$$

Summary

Concepts

- A periodic waveform of period T_0 can be represented by a Fourier series consisting of a dc term and sinusoidal terms that are harmonic multiples of $\omega_0 = 2\pi / T_0$.
- The Fourier series can be represented in terms of a cosine/sine form, amplitude/phase form, and a complex exponential form.
- Circuits excited by a periodic waveform can be analyzed by applying the superposition theorem to the individual terms of the harmonic series.
- Nonperiodic waveforms can be represented by a Fourier transform.
- Upon transforming the circuit to the frequency domain, the circuit can be analyzed for the desired voltage or current of interest and then the result can be inverse transformed to the time domain.
- The Fourier transform technique can be extended to two-dimensional spatial images.
- The phase part of a signal contains vital information, particularly with regard to timing or spatial location.

Mathematical and Physical Models

Fourier Series Table 5-3

Fourier Transform

$$\mathbf{X}(\omega) = \mathcal{F}[x(t)] = \int_{-\infty}^{\infty} x(t)\, e^{-j\omega t}\, dt$$

$$x(t) = \mathcal{F}^{-1}[\mathbf{X}(\omega)] = \frac{1}{2\pi} \int_{-\infty}^{\infty} \mathbf{X}(\omega)\, e^{j\omega t}\, d\omega$$

sinc Function $\quad \operatorname{sinc}(x) = \dfrac{\sin x}{x}$

Properties of Fourier Transform Table 5-7

2-D Fourier Transform

$$\mathbf{X}(\omega_1, \omega_1) = \mathcal{F}[x(\xi, \eta)] = \int_{-\infty}^{\infty}\int_{-\infty}^{\infty} x(\xi, \eta)\, e^{-j\omega_1 \xi} e^{-j\omega_2 \eta}\, dx\, dy$$

$$x(\xi, \eta) = \frac{1}{(2\pi)^2} \int_{-\infty}^{\infty}\int_{-\infty}^{\infty} \mathbf{X}(\omega_1, \omega_2)\, e^{j\omega_1 \xi} e^{j\omega_2 \eta}\, d\omega_1\, d\omega_2$$

Important Terms

Provide definitions or explain the meaning of the following terms:

2-D Fourier transform	Fourier coefficient	harmonic	periodic waveform
amplitude spectrum	Fourier series	line spectra	phase spectrum
dc component	Fourier transform	mixed signal circuit	signum function
even symmetry	frequency spectrum	nulls	sinc function
fft	fundamental angular frequency	odd symmetry	spatial frequency
fftshift	Gibbs phenomenon	periodicity property	truncated series

PROBLEMS

Section 5-1: Phasor-Domain Technique

5.1 A system is characterized by the differential equation

$$c_1 \frac{dy}{dt} + c_2 y = 10\cos(400t - 30°).$$

*(a) Determine $y(t)$, given that $c_1 = 10^{-2}$ and $c_2 = 3$.

(b) Determine $y(t)$, given that $c_1 = 10^{-2}$ and $c_2 = 0.3$.

5.2 A system is characterized by the differential equation

$$c_1 \frac{d^2y}{dt^2} + c_2 \frac{dy}{dt} + c_3 y = A\cos(\omega t + \phi).$$

Determine $y(t)$ for the following:

(a) $c_1 = 10^{-6}$, $c_2 = 3 \times 10^{-3}$, $c_3 = 3$, $A = 12$, $\omega = 10^3$ rad/s, and $\phi = 60°$.

(b) $c_1 = 5 \times 10^{-4}$, $c_2 = 10^{-2}$, $c_3 = 1$, $A = 16$, $\omega = 200$ rad/s, and $\phi = -30°$.

(c) $c_1 = 5 \times 10^{-6}$, $c_2 = 1$, $c_3 = 10^6$, $A = 4$, $\omega = 10^6$ rad/s, and $\phi = -60°$.

5.3 Repeat part (a) of Problem 5.2 after replacing the cosine with a sine.

5.4 A system is characterized by

$$c_1 \frac{d^2y}{dt^2} + c_2 \frac{dy}{dt} + c_3 y = A_1\cos(\omega t + \phi_1) + A_2\sin(\omega t + \phi_2).$$

Determine $y(t)$, given that $c_1 = 10^{-6}$, $c_2 = 3 \times 10^{-3}$, $c_3 = 3$, $A_1 = 10$, $A_2 = 20$, $\omega = 10^3$ rad/s, $\phi_1 = 30°$, and $\phi_2 = 30°$. (*Hint:* Apply the superposition property of LTI systems.)

***5.5** A system is characterized by

$$4 \times 10^{-3} \frac{dy}{dt} + 3y = 5\cos(1000t) - 10\cos(2000t).$$

Determine $y(t)$. (*Hint:* Apply the superposition property of LTI systems.)

*Answer(s) in Appendix F.

Sections 5-3 and 5-4: Fourier Series

Follow these instructions for each of the waveforms in Problems 5.6 through 5.15.

(a) Determine if the waveform has dc, even, or odd symmetry.

(b) Obtain its cosine/sine Fourier series representation.

(c) Convert the representation to amplitude/phase format and plot the line spectra for the first five non-zero terms.

(d) Convert the representation to complex exponential format and plot the line spectra for the first five non-zero terms.

(e) Use MATLAB or MathScript to plot the waveform using a truncated Fourier series representation with $n_{max} = 100$.

5.6 Waveform in Fig. P5.6 with $A = 10$.

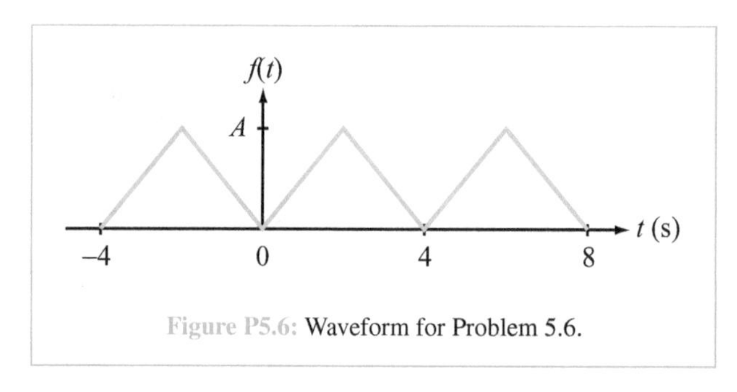

Figure P5.6: Waveform for Problem 5.6.

5.7 Waveform in Fig. P5.7 with $A = 4$.

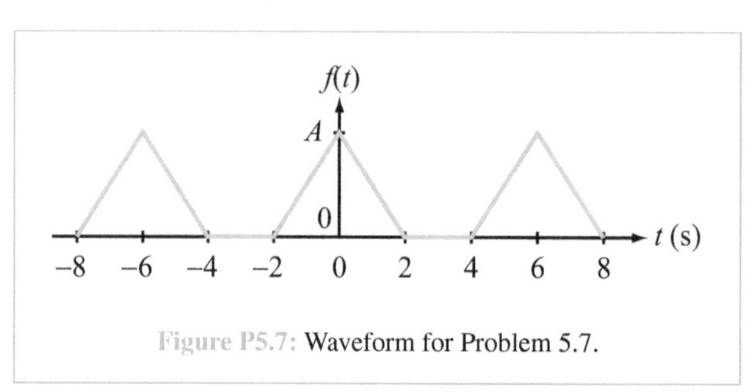

Figure P5.7: Waveform for Problem 5.7.

5.8 Waveform in Fig. P5.8 with $A = 6$.

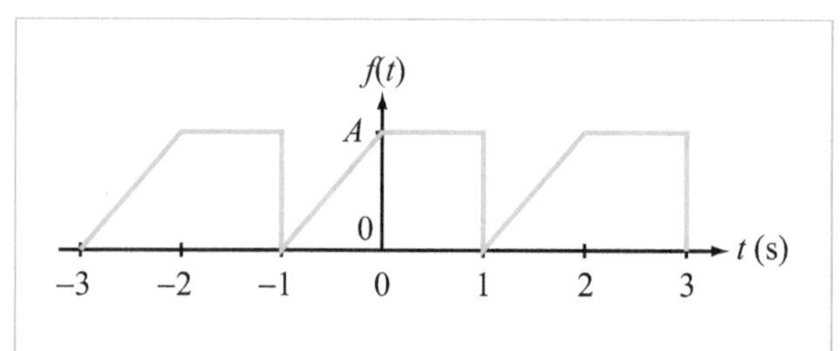

Figure P5.8: Waveform for Problem 5.8.

*5.9 Waveform in Fig. P5.9 with $A = 10$.

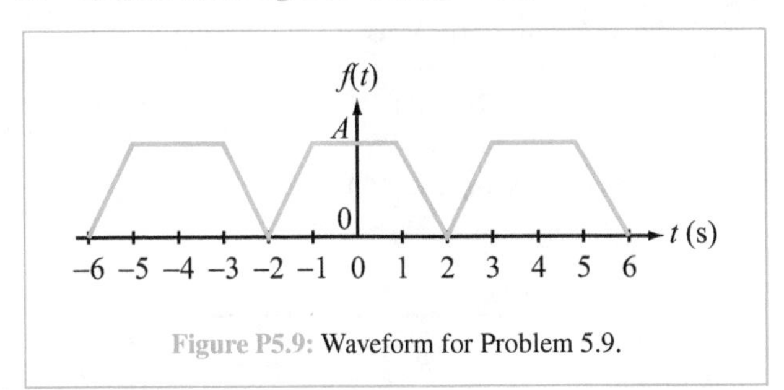

Figure P5.9: Waveform for Problem 5.9.

5.10 Waveform in Fig. P5.10 with $A = 20$.

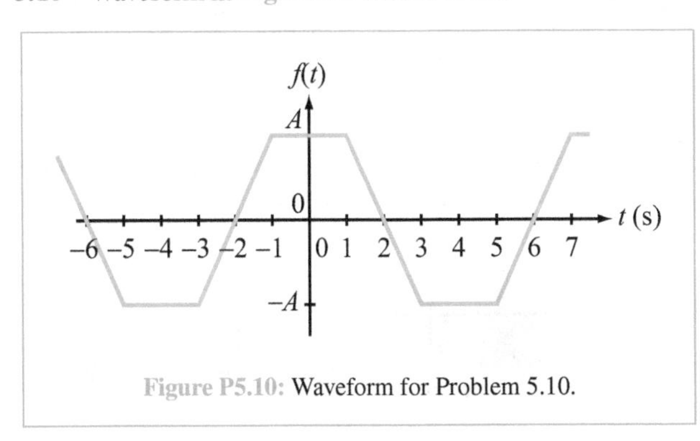

Figure P5.10: Waveform for Problem 5.10.

5.11 Waveform in Fig. P5.11 with $A = 100$.

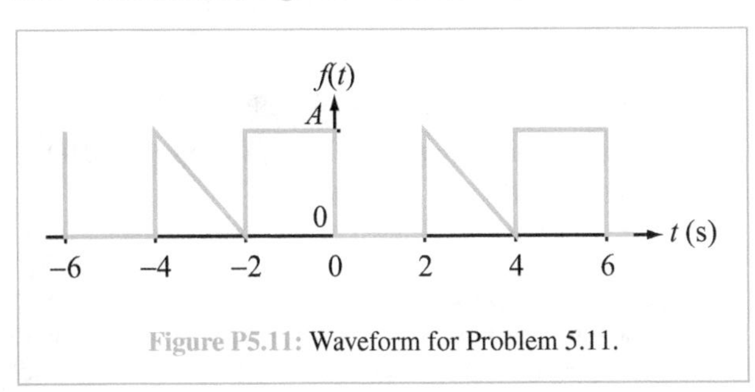

Figure P5.11: Waveform for Problem 5.11.

5.12 Waveform in Fig. P5.12 with $A = 4$.

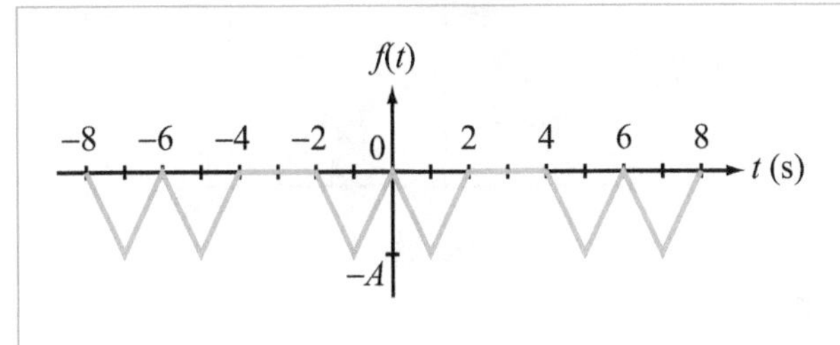

Figure P5.12: Waveform for Problem 5.12.

5.13 Waveform in Fig. P5.13 with $A = 10$.

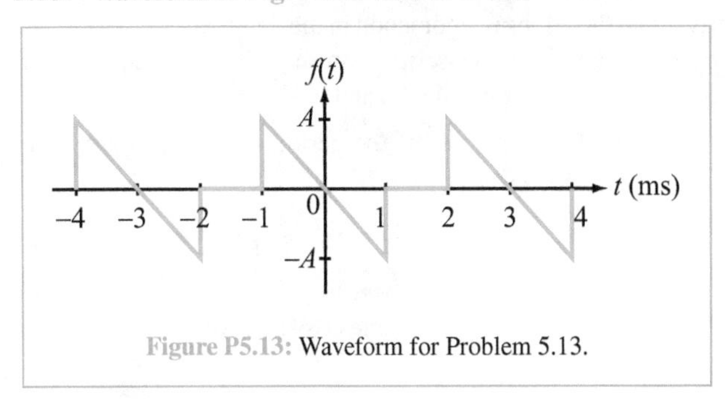

Figure P5.13: Waveform for Problem 5.13.

5.14 Waveform in Fig. P5.14 with $A = 10$.

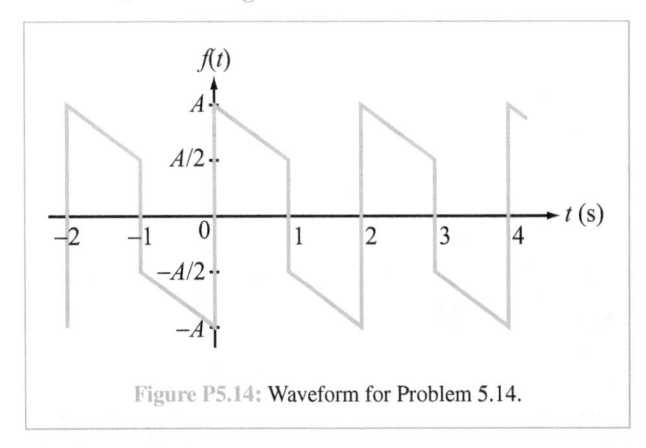

Figure P5.14: Waveform for Problem 5.14.

5.15 Waveform in Fig. P5.15 with $A = 20$.

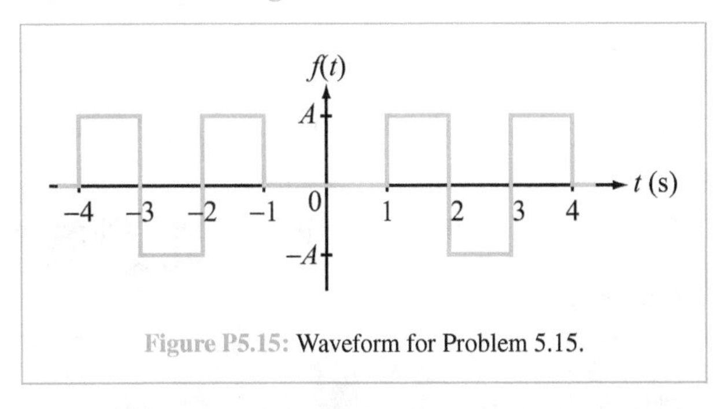

Figure P5.15: Waveform for Problem 5.15.

5.16 Obtain the cosine/sine Fourier-series representation for $f(t) = \cos^2(4\pi t)$, and use MATLAB/MathScript software to plot it with $n_{\max} = 100$.

5.17 Repeat Problem 5.16 for $f(t) = \sin^2(4\pi t)$.

*5.18 Repeat Problem 5.16 for $f(t) = |\sin(4\pi t)|$.

5.19 Which of the six waveforms shown in Figs. P5.6 through P5.11 will exhibit the Gibbs oscillation phenomenon when represented by a Fourier series? Why?

5.20 Consider the sawtooth waveform shown in Fig. 5-3(a). Evaluate the Gibbs phenomenon in the neighborhood of $t = 4$ s by plotting the Fourier-series representation with $n_{\max} = 100$ over the range between 4.01 s and 4.3 s.

5.21 The Fourier series of the periodic waveform shown in Fig. P5.21(a) is given by

$$f_1(t) = 10 - \frac{20}{\pi} \sum_{n=1}^{\infty} \frac{1}{n} \sin\left(\frac{n\pi t}{2}\right).$$

Determine the Fourier series of waveform $f_2(t)$ in Fig. P5.21(b).

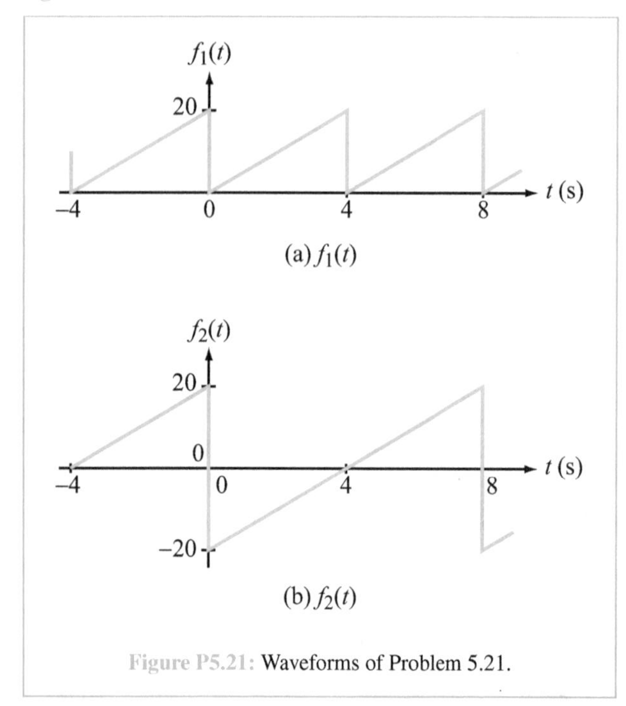

Figure P5.21: Waveforms of Problem 5.21.

5.22 Let $x(t)$ be a real-valued and periodic signal with period T_0 s and maximum frequency B Hz. The signal is known to have a nonzero component at exactly B Hz.

(a) Show that $x(t)$ can be expressed using an expression with only $2BT_0 + 1$ constants.

(b) $x(t)$ has period 0.1 s and maximum frequency 20 Hz. It has the following known values: $x(0.00) = 5$, $x(0.02) = 1$, $x(0.04) = 3$, $x(0.06) = 2$, and $x(0.08) = 4$. Determine an explicit formula for $x(t)$.

Section 5-5: Circuit Applications

5.23 The voltage source $\upsilon_s(t)$ in the circuit of Fig. P5.23 generates a square wave (waveform #1 in Table 5-4) with $A = 10$ V and $T = 1$ ms.

(a) Derive the Fourier series representation of $\upsilon_{\text{out}}(t)$.

(b) Calculate the first five terms of $\upsilon_{\text{out}}(t)$ using

$$R_1 = R_2 = 2 \text{ k}\Omega, \qquad C = 1\ \mu\text{F}.$$

(c) Plot $\upsilon_{\text{out}}(t)$ using $n_{\max} = 100$.

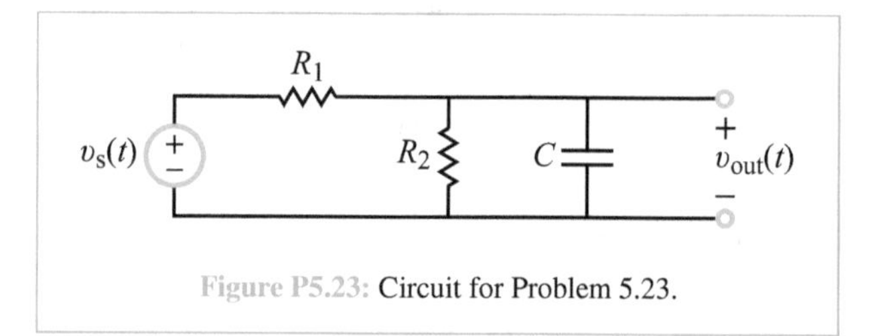

Figure P5.23: Circuit for Problem 5.23.

5.24 The current source $i_s(t)$ in the circuit of Fig. P5.24 generates a sawtooth wave (waveform in Fig. 5-3(a)) with a peak amplitude of 20 mA and a period $T = 5$ ms.

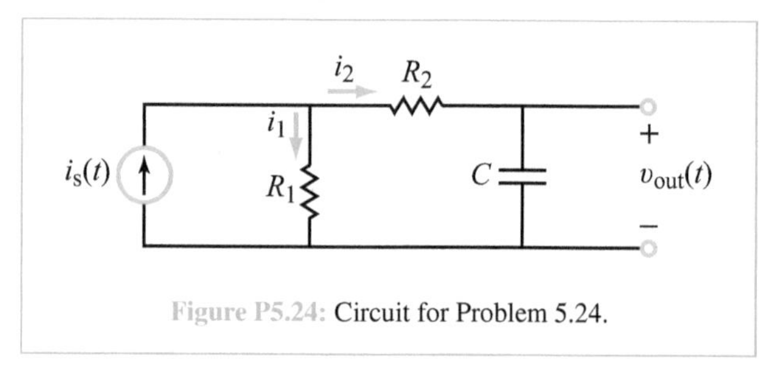

Figure P5.24: Circuit for Problem 5.24.

(a) Derive the Fourier series representation of $\upsilon_{\text{out}}(t)$.

(b) Calculate the first five terms of $\upsilon_{\text{out}}(t)$ using $R_1 = 500\ \Omega$, $R_2 = 2$ kΩ, and $C = 0.33\ \mu$F.

(c) Plot $\upsilon_{\text{out}}(t)$ and $i_s(t)$ using $n_{\max} = 100$.

***5.25** The current source $i_s(t)$ in the circuit of Fig. P5.25 generates a train of pulses (waveform #3 in Table 5-4) with $A = 6$ mA, $\tau = 1\ \mu$s, and $T = 10\ \mu$s.

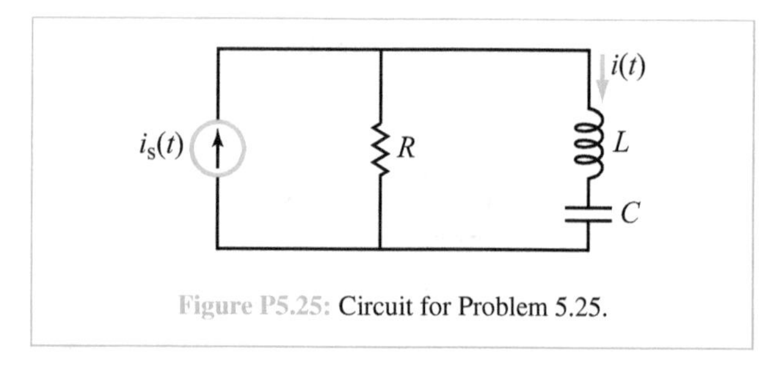

Figure P5.25: Circuit for Problem 5.25.

(a) Derive the Fourier series representation of $i(t)$.

(b) Calculate the first five terms of $i(t)$ using $R = 1$ kΩ, $L = 1$ mH, and $C = 1$ μF.

(c) Plot $i(t)$ and $i_s(t)$ using $n_{max} = 100$.

5.26 Voltage source $v_s(t)$ in the circuit of Fig. P5.26(a) has the waveform displayed in Fig. P5.26(b).

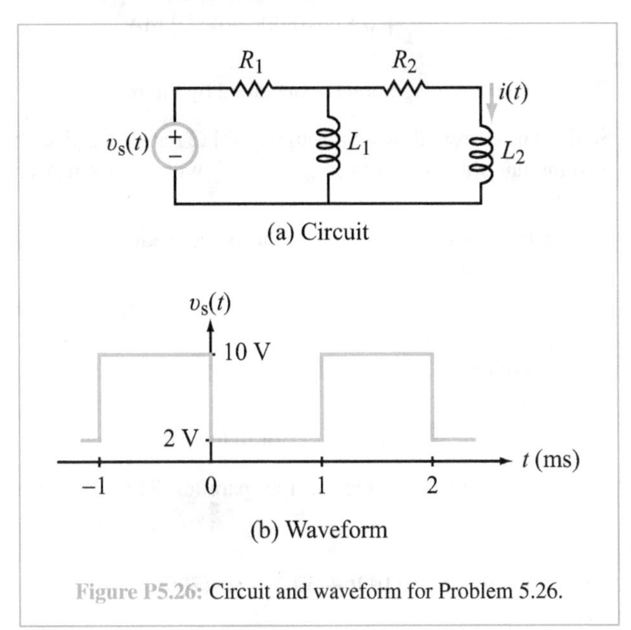

Figure P5.26: Circuit and waveform for Problem 5.26.

(a) Derive the Fourier series representation of $i(t)$.

(b) Calculate the first five terms of $i(t)$ using

$$R_1 = R_2 = 10\ \Omega \quad \text{and} \quad L_1 = L_2 = 10 \text{ mH}.$$

(c) Plot $i(t)$ and $v_s(t)$ using $n_{max} = 100$.

5.27 Determine the output voltage $v_{out}(t)$ in the circuit of Fig. P5.27, given that the input voltage $v_{in}(t)$ is a full-wave rectified sinusoid (waveform #8 in Table 5-4) with $A = 120$ V and $T = 1$ μs.

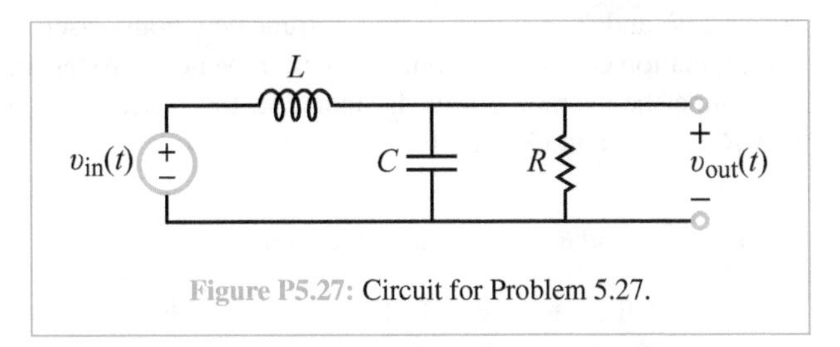

Figure P5.27: Circuit for Problem 5.27.

(a) Derive the Fourier series representation of $v_{out}(t)$.

(b) Calculate the first five terms of $v_{out}(t)$ using $R = 1$ kΩ, $L = 1$ mH, and $C = 1$ nF.

(c) Plot $v_{out}(t)$ and $v_{in}(t)$ using $n_{max} = 100$.

5.28

(a) Repeat Example 5-6, after replacing the capacitor with an inductor $L = 0.1$ H and reducing the value of R to 1 Ω.

(b) Calculate the first five terms of $v_{out}(t)$.

(c) Plot $v_{out}(t)$ and $v_s(t)$ using $n_{max} = 100$.

5.29 Determine $v_{out}(t)$ in the circuit of Fig. P5.29, given that the input excitation is characterized by a triangular waveform (#4 in Table 5-4) with $A = 24$ V and $T = 20$ ms.

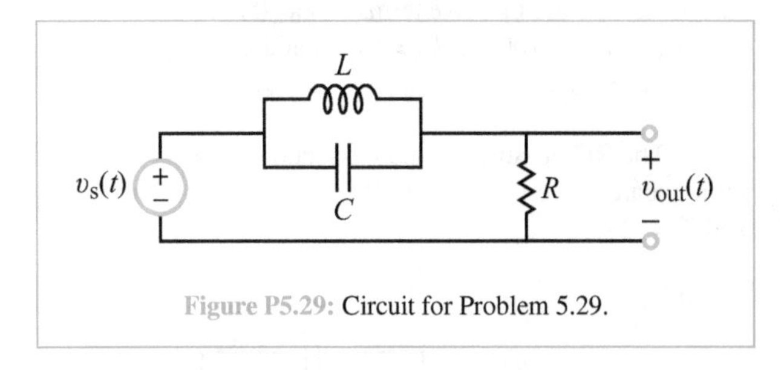

Figure P5.29: Circuit for Problem 5.29.

(a) Derive Fourier series representation of $v_{out}(t)$.

(b) Calculate first five terms of $v_{out}(t)$ using $R = 470$ Ω, $L = 10$ mH, and $C = 10$ μF.

(c) Plot $v_{out}(t)$ and $v_s(t)$ using $n_{max} = 100$.

5.30 A backward-sawtooth waveform (#7 in Table 5-4) with $A = 100$ V and $T = 1$ ms is used to excite the circuit in Fig. P5.30.

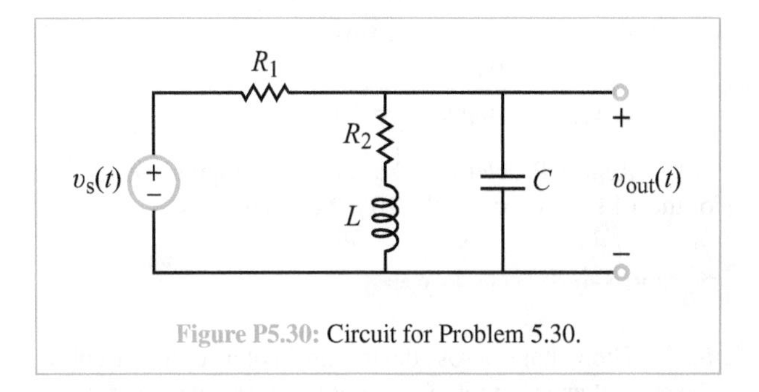

Figure P5.30: Circuit for Problem 5.30.

(a) Derive Fourier series representation of $v_{out}(t)$.

(b) Calculate the first five terms of $v_{out}(t)$ using $R_1 = 1$ kΩ, $R_2 = 100$ Ω, $L = 1$ mH, and $C = 1$ μF.

(c) Plot $v_{out}(t)$ and $v_s(t)$ using $n_{max} = 100$.

*5.31 The circuit in Fig. P5.31 is excited by the source waveform shown in Fig. P5.26(b).

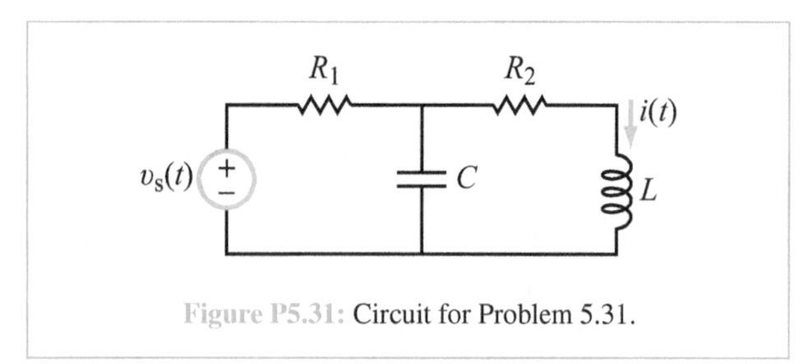

Figure P5.31: Circuit for Problem 5.31.

(a) Derive Fourier series representation of $i(t)$.

(b) Calculate the first five terms of $\upsilon_{out}(t)$ using $R_1 = R_2 = 100\ \Omega$, $L = 1$ mH, and $C = 1\ \mu$F.

(c) Plot $i(t)$ and $\upsilon_s(t)$ using $n_{max} = 100$.

5.32 The RC op-amp integrator circuit of Fig. P5.32 excited by a square wave (waveform #1 in Table 5-4) with $A = 4$ V and $T = 2$ s.

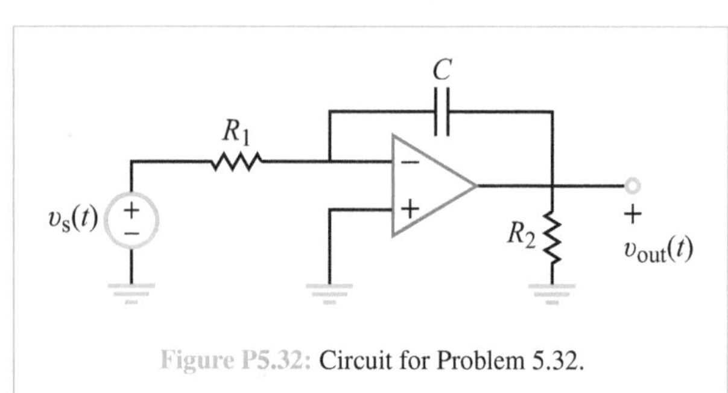

Figure P5.32: Circuit for Problem 5.32.

(a) Derive Fourier series representation of $\upsilon_{out}(t)$.

(b) Calculate the first five terms of $\upsilon_{out}(t)$ using $R_1 = 1\ \text{k}\Omega$ and $C = 10\ \mu$F.

(c) Plot $\upsilon_{out}(t)$ using $n_{max} = 100$.

5.33 Repeat Problem 5.32 after interchanging the locations of the 1 kΩ resistor and the 10 μF capacitor.

Section 5-6: Average Power

5.34 The voltage across the terminals of a certain circuit and the current entering into its (+) voltage terminal are given by

$$\upsilon(t) = [4 + 12\cos(377t + 60^\circ) - 6\cos(754t - 30^\circ)]\ \text{V},$$

$$i(t) = [5 + 10\cos(377t + 45^\circ) + 2\cos(754t + 15^\circ)]\ \text{mA}.$$

Determine the average power consumed by the circuit, and the ac power fraction.

5.35 The current flowing through a 2 kΩ resistor is given by

$$i(t) = [5 + 2\cos(400t + 30^\circ) + 0.5\cos(800t - 45^\circ)]\ \text{mA}.$$

Determine the average power consumed by the resistor.

5.36 The current flowing through a 10 kΩ resistor is given by a triangular waveform (#4 in Table 5-4) with $A = 4$ mA and $T = 0.2$ s.

(a) Determine the exact value of the average power consumed by the resistor.

(b) Using a truncated Fourier-series representation of the waveform with only the first four terms, obtain an approximate value for the average power consumed by the resistor.

(c) What is the percentage of error in the value given in (b)?

*5.37 The current source in the parallel RLC circuit of Fig. P5.37 is given by

$$i_s(t) = [10 + 5\cos(100t + 30^\circ) - \cos(200t - 30^\circ)]\ \text{mA}.$$

Determine the average power dissipated in the resistor given that $R = 1\ \text{k}\Omega$, $L = 1$ H, and $C = 1\ \mu$F.

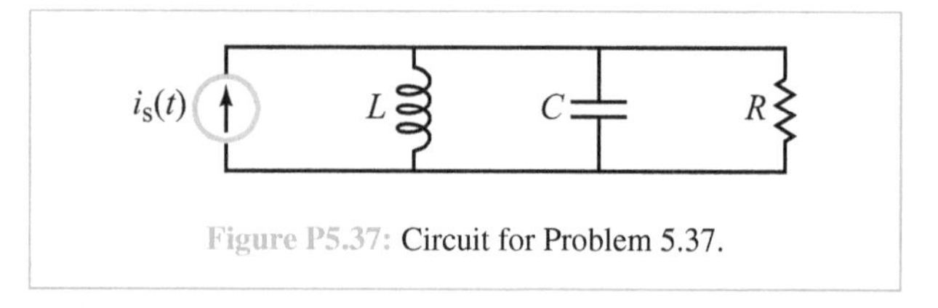

Figure P5.37: Circuit for Problem 5.37.

5.38 A series RC circuit is connected to a voltage source whose waveform is given by waveform #5 in Table 5-4, with $A = 12$ V and $T = 1$ ms. Using a truncated Fourier-series representation composed of only the first three non-zero terms, determine the average power dissipated in the resistor, given that $R = 2\ \text{k}\Omega$ and $C = 1\ \mu$F.

Sections 5-7 and 5-8: Fourier Transform

For each of the waveforms in Problems 5.39 through 5.48, determine the Fourier transform.

5.39 Waveform in Fig. P5.39 with $A = 5$ and $T = 3$ s.

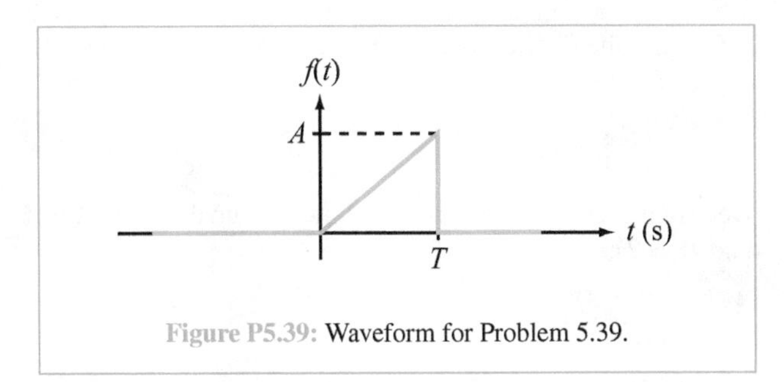

Figure P5.39: Waveform for Problem 5.39.

5.40 Waveform in Fig. P5.40 with $A = 10$ and $T = 6$ s.

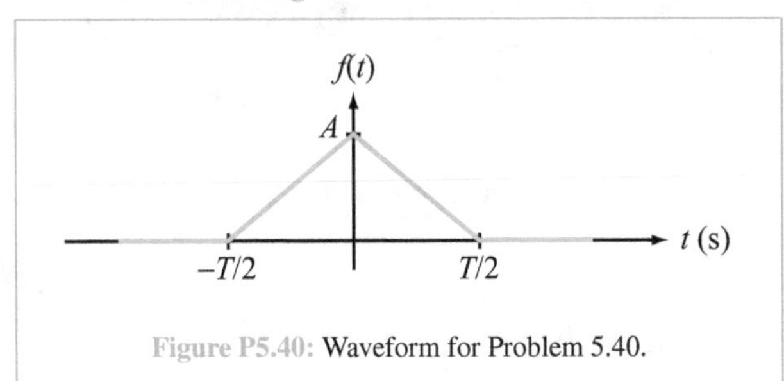

Figure P5.40: Waveform for Problem 5.40.

5.41 Waveform in Fig. P5.41 with $A = 12$ and $T = 3$ s.

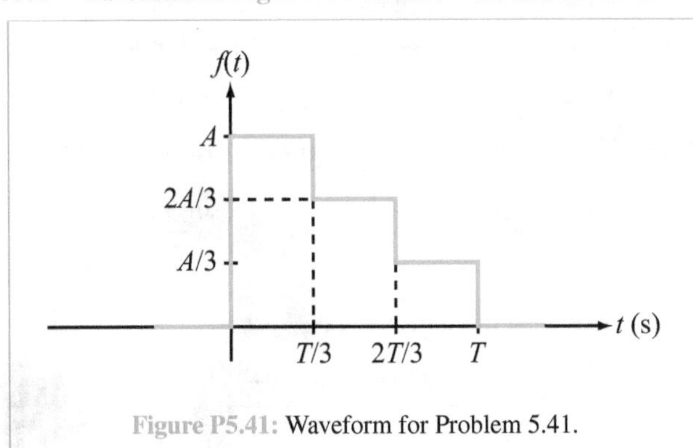

Figure P5.41: Waveform for Problem 5.41.

5.42 Waveform in Fig. P5.42 with $A = 2$ and $T = 12$ s.

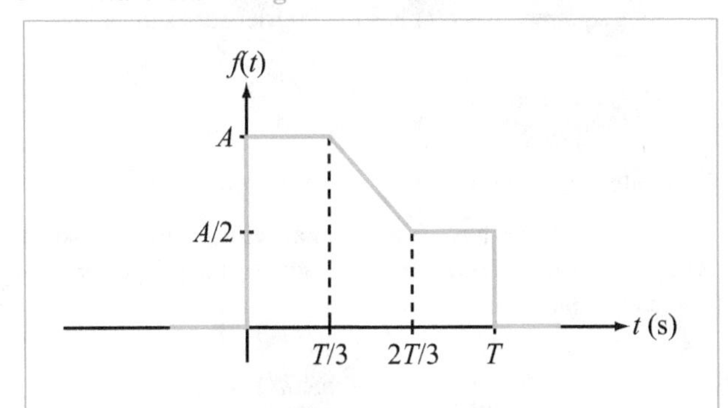

Figure P5.42: Waveform for Problem 5.42.

5.43 Waveform in Fig. P5.43 with $A = 1$ and $T = 3$ s.

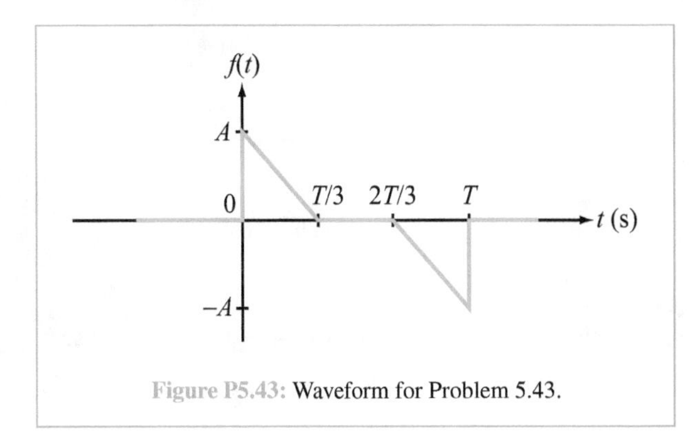

Figure P5.43: Waveform for Problem 5.43.

5.44 Waveform in Fig. P5.44 with $A = 1$ and $T = 2$ s.

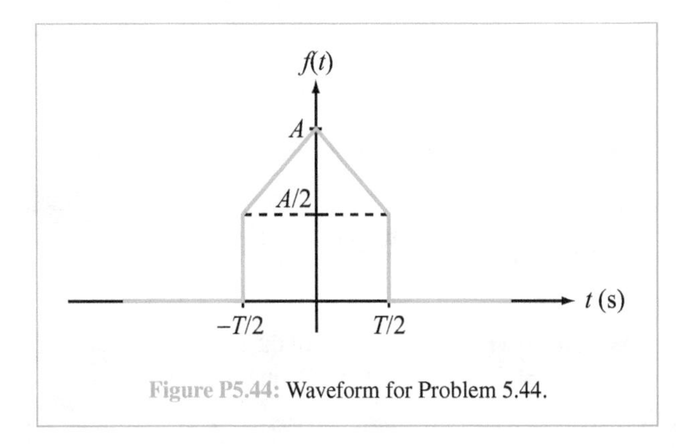

Figure P5.44: Waveform for Problem 5.44.

5.45 Waveform in Fig. P5.45 with $A = 3$ and $T = 1$ s.

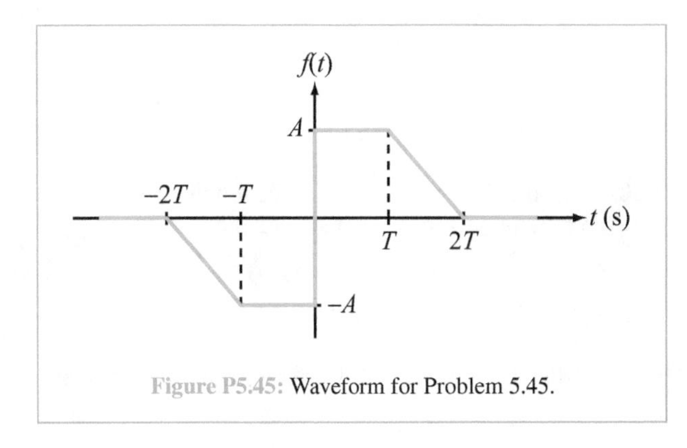

Figure P5.45: Waveform for Problem 5.45.

5.46 Waveform in Fig. P5.46 with $A = 5$, $T = 1$ s, and $\alpha = 10\ \text{s}^{-1}$.

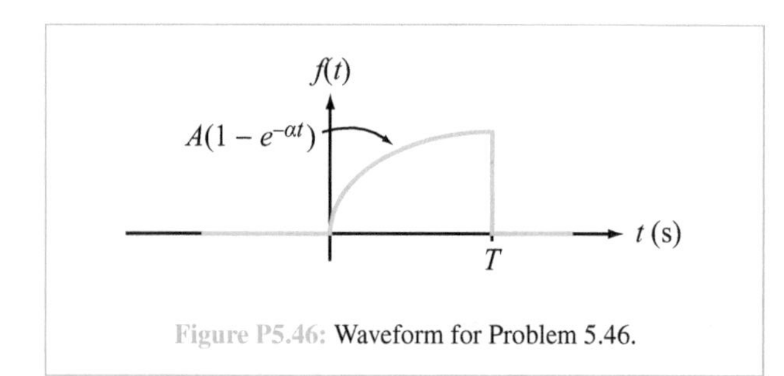

Figure P5.46: Waveform for Problem 5.46.

*5.47 Waveform in Fig. P5.47 with $A = 10$ and $T = 2$ s.

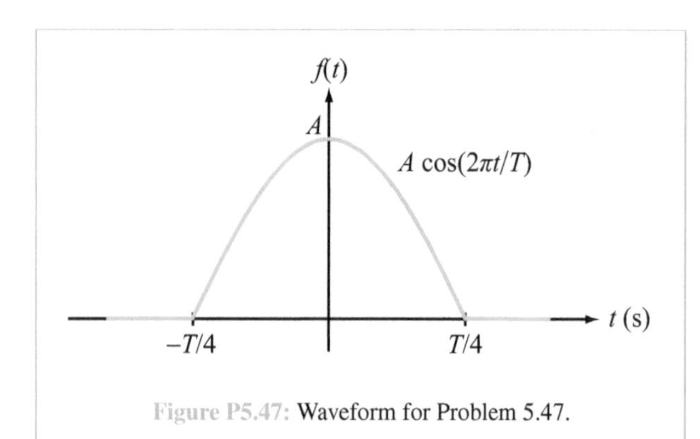

Figure P5.47: Waveform for Problem 5.47.

5.48 Find the Fourier transform of the following signals with $A = 2$, $\omega_0 = 5$ rad/s, $\alpha = 0.5\ \text{s}^{-1}$, and $\phi_0 = \pi/5$.

(a) $f(t) = A\cos(\omega_0 t - \phi_0), \quad -\infty < t < \infty$

(b) $g(t) = e^{-\alpha t}\cos(\omega_0 t)\, u(t)$

5.49 Find the Fourier transform of the following signals with $A = 3$, $B = 2$, $\omega_1 = 4$ rad/s, and $\omega_2 = 2$ rad/s.

(a) $f(t) = [A + B\sin(\omega_1 t)]\sin(\omega_2 t)$

(b) $g(t) = A|t|, \quad |t| < (2\pi/\omega_1)$

5.50 Find the Fourier transform of the following signals with $\alpha = 0.5\ \text{s}^{-1}$, $\omega_1 = 4$ rad/s, and $\omega_2 = 2$ rad/s.

*(a) $f(t) = e^{-\alpha t}\sin(\omega_1 t)\cos(\omega_2 t)\, u(t)$

(b) $g(t) = te^{-\alpha t}, \quad 0 \le t \le 10\alpha$

5.51 Using the definition of Fourier transform, prove that

$$\mathcal{F}[t\ f(t)] = j\,\frac{d}{d\omega}\,\mathcal{F}(\omega).$$

5.52 Let the Fourier transform of $f(t)$ be

$$\mathbf{F}(\omega) = \frac{A}{(B + j\omega)}\ .$$

Determine the transforms of the following signals (using $A = 5$ and $B = 2$):

(a) $f(3t - 2)$

(b) $t\ f(t)$

(c) $d\,f(t)/dt$

5.53 Let the Fourier transform of $f(t)$ be

$$\mathbf{F}(\omega) = \frac{1}{(A + j\omega)}\, e^{-j\omega} + B.$$

Determine the Fourier transforms of the following signals (set $A = 2$ and $B = 1$).

(a) $f\left(\frac{5}{8}\, t\right)$

(b) $f(t)\cos(At)$

(c) $d^3 f/dt^3$

5.54 Prove the following two Fourier transform pairs.

(a) $\cos(\omega T)\ \mathbf{F}(\omega) \quad \longleftrightarrow \quad \frac{1}{2}[f(t - T) + f(t + T)]$

(b) $\sin(\omega T)\ \mathbf{F}(\omega) \quad \longleftrightarrow \quad \frac{1}{2j}[f(t + T) - f(t - T)]$

5.55 Using only Fourier transform properties, show that

$$\frac{\sin(10\pi t)}{\pi t}\,[1 + 2\cos(20\pi t)] = \frac{\sin(30\pi t)}{\pi t}\ .$$

5.56 Show that the spectrum of

$$\frac{\sin(20\pi t)}{\pi t}\ \frac{\sin(10\pi t)}{\pi t}$$

is zero for $|\omega| > 30\pi$.

*5.57 A square wave $x(t)$ has the Fourier series given by

$$x(t) = \sin(t) + \frac{1}{3}\sin(3t) + \frac{1}{5}\sin(5t) + \cdots\ .$$

Compute $y(t) = x(t) * 3e^{-|t|} * [\sin(4t)/(\pi t)]$.

5.58 Let $x(t)$ be any causal signal with Fourier transform $\mathbf{X}(\omega) = R(\omega) + jI(\omega)$. Show that given either $R(\omega)$ or $I(\omega)$, we can compute the other using the formulas:

$$I(\omega) = -\frac{1}{\pi}\int_{-\infty}^{\infty} \frac{R(\omega')}{\omega - \omega'}\, d\omega'$$

and

$$R(\omega) = \frac{1}{\pi} \int_{-\infty}^{\infty} \frac{I(\omega')}{\omega - \omega'}\, d\omega'.$$

Hints: Conjugate symmetry, entry #4 in Table 5-6, and Fig. 1-11. These are called the Kramers-Kronig relations for a dielectric material with complex frequency-dependent electrical permittivity $\mathbf{X}(\omega)$.

5.59 A signal $x(t)$ is ***narrowband*** if its spectrum $\mathbf{X}(\omega) = 0$ unless $\omega_0 - \delta < |\omega| < \omega_0 + \delta$ for some center frequency ω_0 and bandwidth $2\delta \ll \omega_0$. Narrowband signals are used in radar, sonar, and ultrasound.

This problem examines the effect of a system with frequency response $\mathbf{H}(\omega)$ on the narrowband signal $x(t)$. Throughout this problem, we assume $\omega_0 - \delta < |\omega| < \omega_0 + \delta$. We linearize $\mathbf{H}(\omega)$ by writing $\mathbf{H}(\omega)$ as

$$\mathbf{H}(\omega) = |\mathbf{H}(\omega)| e^{j\theta(\omega)} \approx |\mathbf{H}(\omega_0)| e^{-j(\theta_0 + \theta_1 \omega)}.$$

According to this model $\mathbf{H}(\omega)$ has:

(1) Constant gain, $|\mathbf{H}(\omega)| \approx |\mathbf{H}(\omega_0)|$.

(2) Linear phase, $\theta(\omega) = \theta_0 + \theta_1 \omega$, where

$$\theta_1 = -\left.\frac{d\theta}{d\omega}\right|_{\omega=\omega_0}$$

is defined to be the ***group delay*** of $\mathbf{H}(\omega)$.

(a) Show that $x(t)$ can be written as

$$x(t) = s(t)\cos(\omega_0 t),$$

where the spectrum of $s(t)$ has a maximum frequency $\delta \ll \omega_0$. Thus, $s(t)$ varies slowly compared to $\cos(\omega_0 t)$.

(b) Plot the waveform of $x(t)$ for $s(t) = \cos(2\pi t)$ and $\omega_0 = 20\pi$ rad/s.

(c) Write $\mathbf{H}(\omega) \approx \mathbf{H}_1(\omega)\, \mathbf{H}_2(\omega)$, where

$$\mathbf{H}_1(\omega) = |\mathbf{H}(\omega_0)| e^{-j\theta_0}$$

and $\mathbf{H}_2(\omega) = e^{-j\theta_1 \omega}$. Compute the response $y_1(t)$ of $\mathbf{H}_1(\omega)$ to the input $x_1(t) = s(t)\, e^{j\omega_0 t}$.

(d) Compute the response $y_2(t)$ of $\mathbf{H}_2(\omega)$ to the input $y_1(t)$, the output from (c).

(e) Using (c) and (d), compute the response $y(t)$ of $\mathbf{H}(\omega)$ to the original input $x(t)$. Describe what the system does to $s(t)$ and to $\cos(\omega_0 t)$ if input $x(t) = s(t)\cos(\omega_0 t)$.

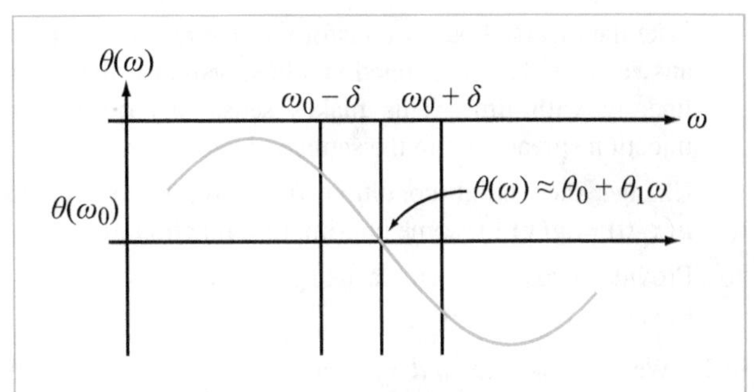

Figure P5.59: Phase response $\theta(\omega)$ is approximately linear for $\omega_0 - \delta < \omega < \omega_0 + \delta$.

5.60 The modulated Gaussian pulse

$$x(t) = \frac{1}{\sqrt{0.2\pi}}\, e^{-(t-1)^2/0.2} \cos(100t)$$

is input into the RC circuit in Fig. 2-5(a), for which $RC = 0.01$. Compute the output $y(t)$ using:

(a) the result of Problem 5.59, and

(b) numerical computation of $h(t) * x(t)$.

5.61 We use Laplace *and* Fourier transforms to solve the partial differential equation

$$\frac{\partial^2 u}{\partial x^2} = \frac{1}{a}\frac{\partial u}{\partial t}.$$

This is the ***Fokker-Planck equation***, also known as the ***heat*** or ***diffusion equation***. The goal is to compute the density $u(x,t)$ at position x and time t for a thin slice of material injected into a semiconductor base at $x = 0$ at $t = 0$. The material diffuses out into the semiconductor base according to the diffusion equation with initial condition (in t) $u(x,0) = \delta(x)$ (initial slice) and diffusion coefficient a.

(a) Take the Laplace transform in t of

$$\frac{\partial^2 u}{\partial x^2} = \frac{1}{a}\frac{\partial u}{\partial t},$$

with $\mathbf{U}(\mathbf{s}, x) = \mathcal{L}\{u(x,t)\}$.

(b) Take the Fourier transform in x of the result. Use wavenumber k instead of ω.

(c) Solve the result for

$$\mathcal{L}\{\mathcal{F}\{u(x,t)\}\} = \mathbf{U}(k, s).$$

(d) Take the inverse Laplace transform of $\mathbf{U}(k, s)$.

(e) Take the inverse Fourier transform of the result. The final answer is a Gaussian function whose variance increases linearly with time. This makes sense physically: the injection spreads out in the semiconductor base.

(f) Change the initial condition from $u(x,0) = \delta(x)$ to $u(x,0) = g(x)$ for some non-impulse function $g(x)$.

(g) Provide an explicit formula, using an integral, for the new $u(x,t)$ in terms of $g(x)$.

5.62 We use Laplace *and* Fourier transforms to solve the partial differential equation

$$\frac{\partial^2 u}{\partial x^2} - \frac{\partial^2 u}{\partial t^2} = 0,$$

with initial conditions (in t) $u(x,0) = \delta(x)$ and $\frac{\partial u}{\partial t}(x,0) = 0$. This is the ***wave equation***. The impulse at position $x = 0$ propagates in both directions. The goal is to compute the field $u(x,t)$, at position x and time t.

(a) Take the Laplace transform in t of

$$\frac{\partial^2 u}{\partial x^2} - \frac{\partial^2 u}{\partial t^2} = 0,$$

with $\mathbf{U}(\mathbf{s}, x) = \mathcal{L}\{u(x,t)\}$.

(b) Take the Fourier transform in x of the result. Use wavenumber k instead of ω.

(c) Solve the result for

$$\mathcal{L}\{\mathcal{F}\{u(x,t)\}\} = \mathbf{U}(k,s).$$

(d) Take the inverse Laplace transform of $\mathbf{U}(k,s)$.

(e) Take the inverse Fourier transform of the result. The final answer is an impulse that propagates in both the x and $-x$ directions. This makes sense physically: the impulse at $x = t = 0$ spreads out as t increases.

(f) Change the initial condition from $u(x,0) = \delta(x)$ to $u(x,0) = g(x)$ for some non-impulse function $g(x)$.

(g) Provide an explicit formula, using an integral, for the new $u(x,t)$ in terms of $g(x)$.

5.63 Let $p(t)$ be a pulse whose spectrum $\mathbf{P}(\omega) \neq 0$ for $|\omega| < \pi B$. Let $\{x[n]\}$ be a sequence of numbers to be transmitted over a communications channel. The sequence of weighted and delayed pulses $x(t) = \sum_{n=-\infty}^{\infty} x[n]\ p(t - nT_s)$ is used to transmit a value of $x[n]$ every T_s seconds over the channel. If the frequency response of the channel $\mathbf{H}(\omega) \neq 0$ for $|\omega| < \pi B$, show that $\{x[n]\}$ can be recovered from the channel output if $BT_s < 1$. *Hint:* Use Fourier series and Fourier transform.

Section 5-9: Parseval's Theorem for Fourier Integral

5.64 If $x(t) = \sin(2t)/(\pi t)$, compute the energy of d^2x/dt^2.

5.65 Compute the energy of $e^{-t}\ u(t) * \sin(t)/(\pi t)$.

5.66 Show that

$$\int_{-\infty}^{\infty} \frac{\sin^2(at)}{(\pi t)^2}\, dt = \frac{a}{\pi}$$

if $a > 0$.

Section 5-12: Circuit Analysis with Fourier Transform

5.67 The circuit in Fig. P5.24 is excited by the source waveform shown in Fig. P5.39.

(a) Derive the expression for $\upsilon_{\text{out}}(t)$ using Fourier analysis.

(b) Plot $\upsilon_{\text{out}}(t)$ using $A = 5$ V, $T = 3$ ms, $R_1 = 500\ \Omega$, $R_2 = 2$ kΩ, and $C = 0.33\ \mu$F.

(c) Repeat part (b) with $C = 0.33$ mF and comment on the results.

5.68 The circuit in Fig. P5.24 is excited by the source waveform shown in Fig. P5.40.

(a) Derive the expression for $\upsilon_{\text{out}}(t)$ using Fourier analysis.

(b) Plot $\upsilon_{\text{out}}(t)$ using $A = 5$ mA, $T = 3$ s, $R_1 = 500\ \Omega$, $R_2 = 2$ kΩ, and $C = 0.33$ mF.

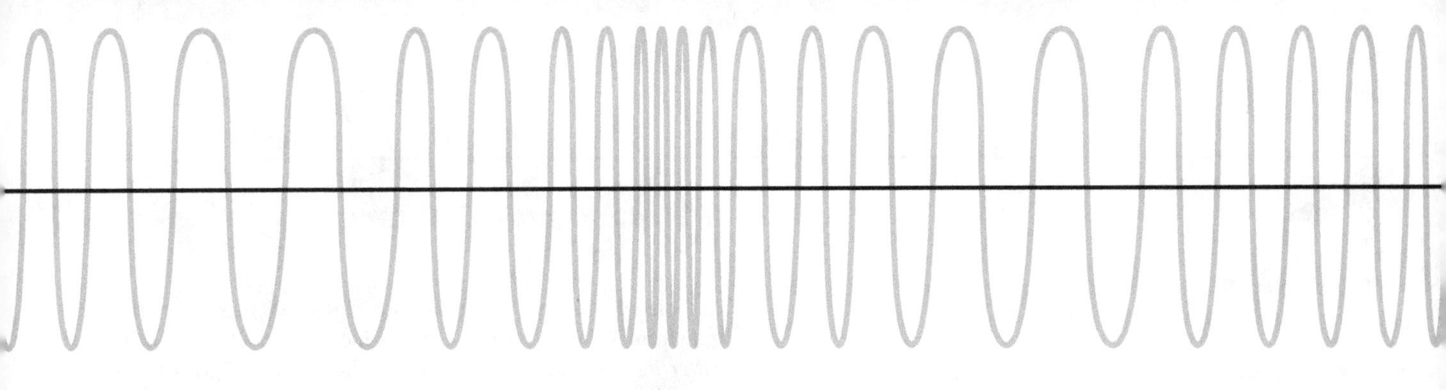

6 Applications of the Fourier Transform

Contents

Overview, 254
6-1 Filtering a 2-D Image, 254
6-2 Types of Filters, 256
6-3 Passive Filters, 263
6-4 Active Filters, 272
6-5 Ideal Brick-Wall Filters, 275
6-6 Filter Design by Poles and Zeros, 278
6-7 Frequency Rejection Filters, 281
6-8 Spectra of Musical Notes, 287
6-9 Butterworth Filters, 289
6-10 Denoising a Trumpet Signal, 298
6-11 Resonator Filter, 300
6-12 Modulation, 303
6-13 Sampling Theorem, 319
Summary, 334
Problems, 336

Objectives

Learn to:

- Design lowpass, bandpass, highpass, bandreject, notch, comb, Butterworth, and resonator filters to remove noise or unwanted interference from signals.
- Compute the spectra of modulated signals.
- Compute the sampling rates necessary to avoid aliasing.

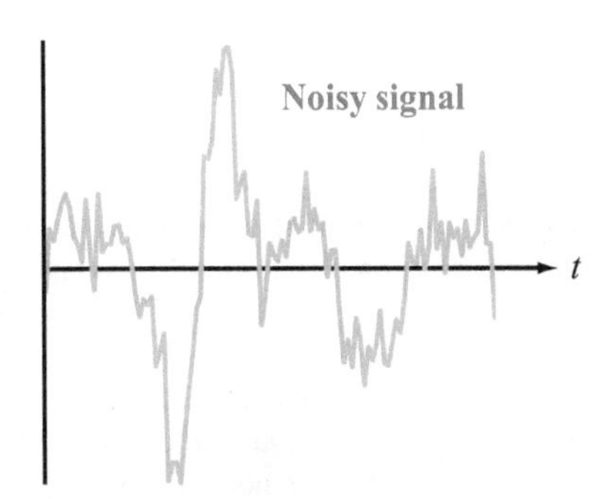

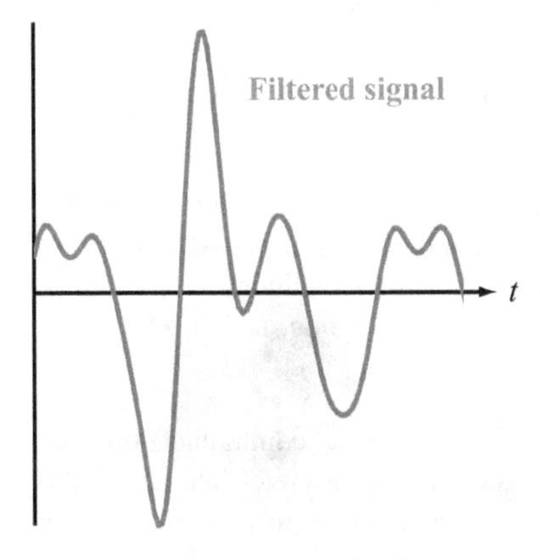

Noise filtering, ***modulation***, frequency division multiplexing, ***signal sampling***, and many related topics are among those treated in this chapter. These are examples of applications that rely on the properties of the ***Fourier transform*** introduced in Chapter 5.

Overview

This chapter offers several examples of how Fourier series and transforms are used to solve engineering problems.

- A *musical instrument* playing a single note is a good example of the physical meaning (fundamental and overtones) of a Fourier series. We show how a Fourier series can be used to represent the sound of a trumpet. We use an actual trumpet signal in this chapter and a discrete-time version of it again in Chapter 8.

- ***Notch filters*** are systems used to remove *sinusoidal interference* (an unwanted sinusoidal signal) from a desired signal. We will design a system that eliminates an unwanted tone from the trumpet signal. We discuss the trade-off between frequency selectivity and impulse response duration, and implementation of filters using op amps.

- ***Comb filters*** are systems that remove *periodic interference* (an unwanted periodic signal) from a desired signal. We will design a system that eliminates one of two notes (one of which is undesired) being played simultaneously by two trumpets.

- ***Lowpass filtering*** is commonly used to reduce noise attached to signals. We design a Butterworth filter to remove some of the noise from a noisy trumpet signal.

- The trumpet signal has a ***line spectrum***, not a continuous spectrum. We use this information to design a ***resonator filter*** that does a much better job of removing noise from the noisy trumpet signal.

- ***Radio*** is familiar to all readers. We use a single property (modulation) of the Fourier transform to derive the analog communication techniques of ***frequency domain multiplexing***, ***amplitude modulation***, ***SSB***, and ***envelope detection***.

- In today's world, most communication and signal processing systems use discrete-time signal processing. We apply the basic properties of Fourier series and transforms to derive the ***sampling theorem***, which demonstrates that a continuous-time ***bandlimited*** signal can be recovered from its samples if the sampling rate is fast enough. We quickly discuss ***aliasing***, ***non-ideal*** sampling and reconstruction and the desirability of ***oversampling***. Chapters 7 to 10 examine discrete-time signal processing in greater detail.

6-1 Filtering a 2-D Image

Associated with any time-varying signal $x(t)$ is a Fourier transform $\mathbf{X}(\omega)$. Obvious examples include speech and music. ***Filtering*** a signal entails passing it through an LTI system—called a ***frequency filter***, or ***filter*** for short—to modify its spectrum so as to realize a desired outcome, such as removing noise accompanying the signal or smoothing out its fast (high frequency) variations. The spectrum may be discrete (Fourier series) or continuous (Fourier transform), depending on whether $x(t)$ is periodic or not. The correspondence between $x(t)$ and $\mathbf{X}(\omega)$ was the subject of Chapter 5.

As was noted earlier in Section 5-13, a two-dimensional image with intensity variation $x(\xi, \eta)$ is also a signal and it has a corresponding spectrum $\mathbf{X}(\omega_1, \omega_2)$, where ω_1 and ω_2 are ***spatial frequencies*** along the ξ and η directions. The purpose of the present section is to develop some degree of intuitive understanding for what actually happens to a signal when we *filter* it.

To that end, we plan to compare a signal before filtering it and after lowpass filtering, and also after highpass-filtering it. Visually, it is much easier to demonstrate the consequences of lowpass and highpass filtering using a spatial signal (i.e., an image) rather than an audio signal or other types of signals. Our test image is shown in Fig. 6-1(a), and its 2-D spectrum $\mathbf{X}(\omega_1, \omega_2)$ is displayed in Fig. 6-1(b). The center of the spectrum corresponds to $\omega_1 = \omega_2 = 0$, and represents the average intensity (dc value) across the entire image. The immediate region surrounding the center of $\mathbf{X}(\omega_1, \omega_2)$ represents the low-frequency variations, while the outer region represents high-frequency variations. Low frequencies are associated with basic structure in an image, while high frequencies are associated with edges and sudden changes.

6-1.1 Lowpass Filtering

A lowpass filter retains low-frequency components and rejects high-frequency components. By applying a window to the spectrum of the original image, as depicted in Fig. 6-1(d), we replace all spectral components outside the window with zero. Upon performing an inverse Fourier transform on the spectrum in Fig. 6-1(d), we obtain the image shown in Fig. 6-1(c). The filtered image no longer contains the frequency components responsible for fast variations, such as the edges of the letters. Consequently, the image looks blurry. It continues to display the basic structures of the letters, but not their edges.

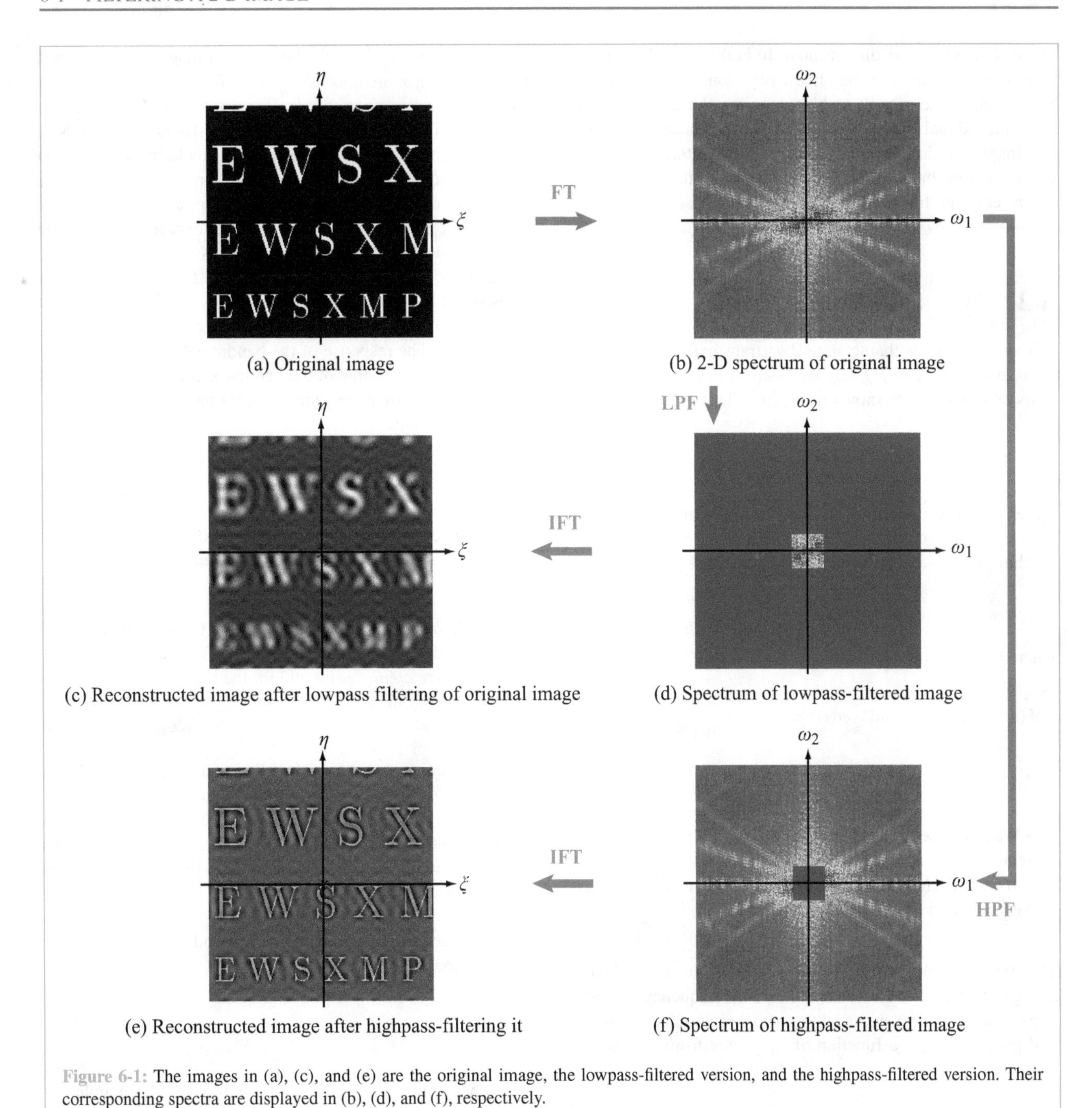

Figure 6-1: The images in (a), (c), and (e) are the original image, the lowpass-filtered version, and the highpass-filtered version. Their corresponding spectra are displayed in (b), (d), and (f), respectively.

6-1.2 Highpass Filtering

If instead of replacing the outer region of the spectrum with zero, we were to replace the *central* region in Fig. 6-1(b) with zero, we would end up implementing a highpass filter to the image. The spectrum is shown in Fig. 6-1(f), and the reconstructed image is shown in Fig. 6-1(e).

In later sections of this chapter, we demonstrate how filters are used to remove noise from a musical signal. The process is similar to what we did with the spatial image, except that the musical signal is one-dimensional. In both cases, the filtering action is performed in the frequency domain after Fourier transforming the signals, which requires the availability of the entire signal before we implement the transformation to the frequency domain and the subsequent filtering and final reconstruction back to the time or spatial domain. Thus, it is *not* a real-time filtering process. *A real-time filtering process requires modifying a time-varying signal in the time domain, instead of in the frequency domain.*

6-2 Types of Filters

A frequency filter is characterized by a transfer function $\mathbf{H}(\omega)$—also called its ***frequency response***—that relates its input $\mathbf{X}(\omega)$ to its output $\mathbf{Y}(\omega)$, as shown in Fig. 6-2. That is,

$$\mathbf{H}(\omega) = \frac{\mathbf{Y}(\omega)}{\mathbf{X}(\omega)}. \tag{6.1}$$

As a complex quantity, the transfer function $\mathbf{H}(\omega)$ has a ***magnitude*** (also called ***gain***)—to which we will assign the symbol $M(\omega)$—and an associated ***phase angle*** $\phi(\omega)$,

$$\mathbf{H}(\omega) = M(\omega)\, e^{j\phi(\omega)}, \tag{6.2}$$

where by definition, we have

$$M(\omega) = |\mathbf{H}(\omega)| \quad \text{and} \quad \phi(\omega) = \tan^{-1}\left\{\frac{\mathfrak{Im}[\mathbf{H}(\omega)]}{\mathfrak{Re}[\mathbf{H}(\omega)]}\right\}. \tag{6.3}$$

6-2.1 Terminology

► We often use the term "frequency" for both the angular frequency ω and the circular frequency $f = \omega/2\pi$. Converting $\mathbf{X}(\omega)$ to $\mathbf{X}(f)$ entails replacing ω with $2\pi f$ everywhere in the expression of $\mathbf{X}(\omega)$. ◄

The four generic types of filters are: ***lowpass***, ***highpass***, ***bandpass***, and ***bandreject***. To visualize the frequency response of a transfer function, we usually generate plots of its magnitude and phase angle as a function of frequency from $\omega = 0$ (dc) to $\omega \to \infty$. Figure 6-3 displays typical magnitude responses for the four aforementioned types of filters. Each of the four filters is characterized by at least one ***passband*** and one ***stopband***. The lowpass filter allows low-frequency signals to pass through essentially unimpeded, but blocks the transmission of high-frequency signals. The qualifiers *low* and *high* are relative to the ***corner frequency*** ω_c (Fig. 6-3(a)), which we shall define shortly. The high-pass filter exhibits the opposite behavior, blocking low-frequency signals while allowing high frequencies to go through. The bandpass filter (Fig. 6-3(c)) is transparent to signals whose frequencies are within a certain range centered at ω_0, but cuts off both very high and very low frequencies. The response of the bandreject filter provides the opposite function to that of the bandpass filter; it is transparent to low and high frequency signals and opaque to intermediate-frequency signals.

Figure 6-2: Filter with transfer function $\mathbf{H}(\omega)$.

Gain factor M_0

All four spectral plots shown in Fig. 6-3 exhibit smooth patterns as a function of ω, and each has a peak value M_0 in its passband.

► If M_0 occurs at dc, as in the case of the lowpass filter, it is called the ***dc gain***; if it occurs at $\omega \to \infty$, it is called the ***high-frequency gain***; and for the bandpass filter, it is called simply the ***gain factor***. ◄

In some cases, the transfer function of a lowpass or highpass filter may exhibit a resonance behavior that manifests itself in the form of a peak in the neighborhood of the resonant frequency of the circuit ω_0, as illustrated in Fig. 6-4. Obviously, the peak value at $\omega = \omega_0$ exceeds M_0, but we will continue to refer to M_0 as the dc gain of $M(\omega)$ because *M_0 is defined as the reference level in the passband of the transfer function*, whereas the behavior of $M(\omega)$ in the neighborhood of ω_0 is specific to that neighborhood.

Corner frequency ω_c

► The corner frequency ω_c is defined as the angular frequency at which $M(\omega)$ is equal to $1/\sqrt{2}$ of the reference peak value:

$$M(\omega_c) = \frac{M_0}{\sqrt{2}} = 0.707 M_0. \tag{6.4}$$

◄

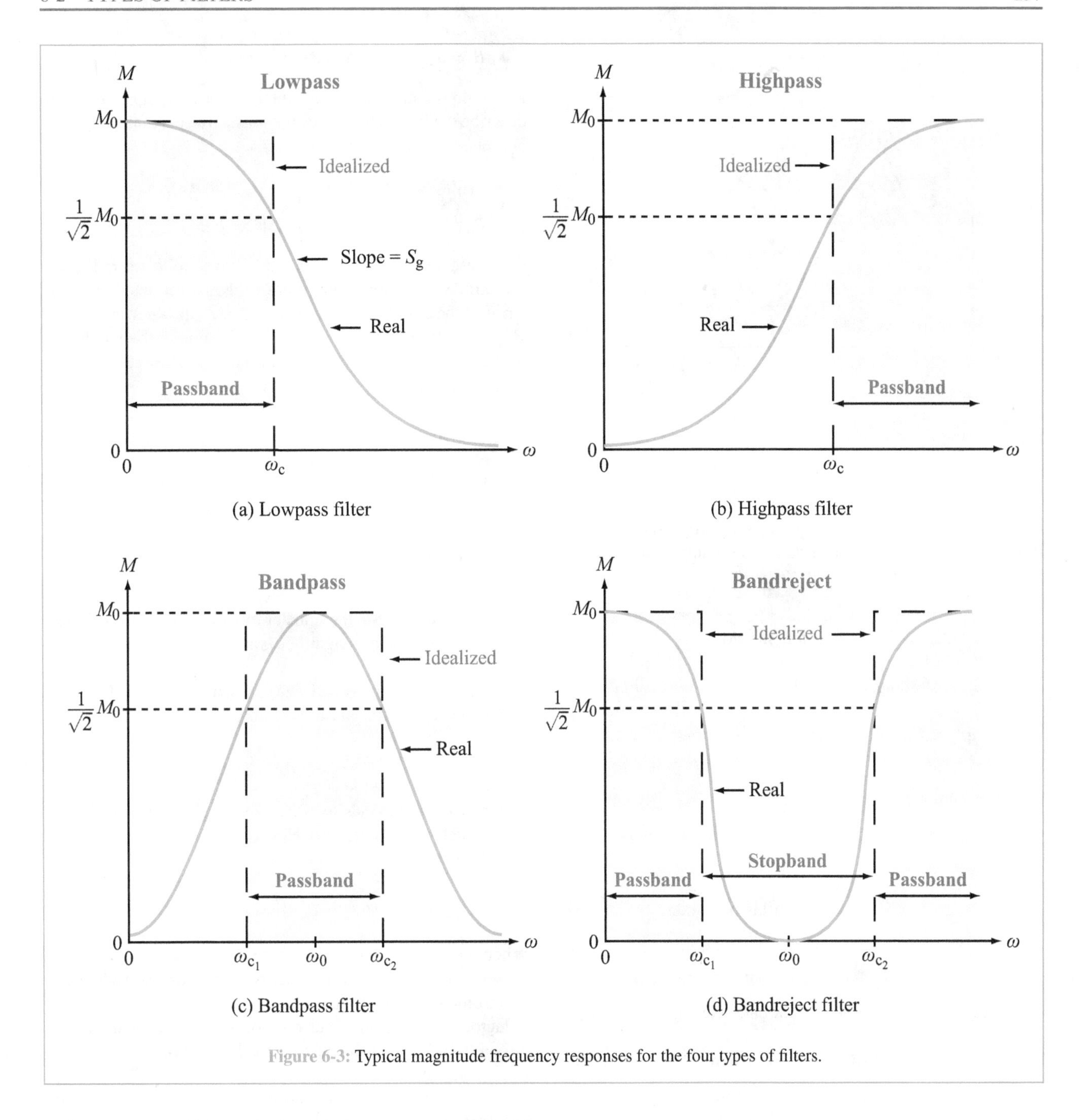

Figure 6-3: Typical magnitude frequency responses for the four types of filters.

Since $M(\omega)$ often is a voltage transfer function, $M^2(\omega)$ is the transfer function for power. The condition described by Eq. (6.4) is equivalent to

$$M^2(\omega_c) = \frac{M_0^2}{2} \qquad \text{or} \qquad P(\omega_c) = \frac{P_0}{2}. \tag{6.5}$$

Hence, ω_c is also called the ***half-power frequency***. The frequency responses of the lowpass and highpass filters shown in Fig. 6-3(a) and (b) have only one half-power frequency each, but the bandpass and bandreject responses have two half-power frequencies each, ω_{c_1} and ω_{c_2}. Even though the

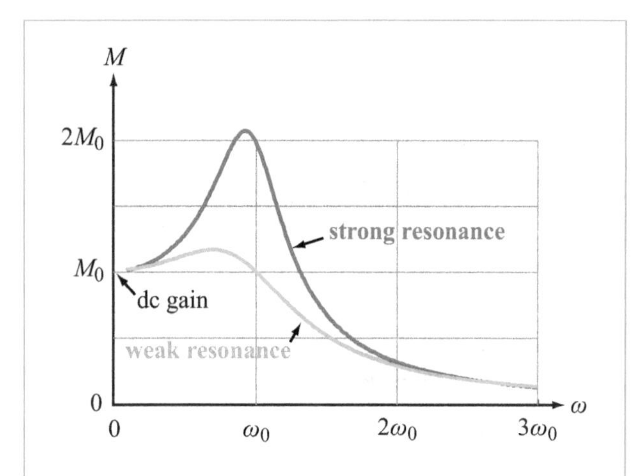

Figure 6-4: Resonant peak in the frequency response of a lowpass filter circuit.

actual frequency response of a filter is a gently varying curve, it is usually approximated to that of an equivalent ***idealized response***, as illustrated in Fig. 6-3.

▶ The idealized version for the lowpass filter has a rectangle-like envelope with a sudden transition at $\omega = \omega_c$. Accordingly, it also is called a ***brick-wall filter*** and ω_c is referred to as its ***cutoff frequency***. These terms also apply to the other three types of filters. ◀

Passband and stopband

The filter *passband is the range of ω over which the filter passes the input signal:*

- $0 \leq \omega < \omega_c$, for lowpass filter,
- $\omega > \omega_c$, for highpass filter,
- $\omega_{c_1} < \omega < \omega_{c_2}$, for bandpass filter,
- $\omega < \omega_{c_1}$ and $\omega > \omega_{c_2}$, for bandreject filter. (6.6)

The ***stopband*** of the bandreject filter extends from ω_{c_1} to ω_{c_2} (Fig. 6-3(d)).

Bandwidth

For lowpass and bandpass filters, the ***bandwidth*** B denotes the extent of the filter's passband.

- $B = \omega_c$ for lowpass filter (Fig. 6-3(a)).
- $B = \omega_{c_2} - \omega_{c_1}$ for bandpass filter (Fig. 6-3(c)).

For highpass and bandreject filters, it is more appropriate to describe the filter's frequency response in terms of the ***rejection bandwidth*** B_{rej}.

- $B_{rej} = \omega_c$ for highpass filter (Fig. 6-3(b)).
- $B_{rej} = \omega_{c_2} - \omega_{c_1}$ for bandreject filter (Fig. 6-3(d)).

Per these definitions, B and B_{rej} have units of rad/s, but sometimes it is more convenient to define the bandwidths in terms of the circular frequency $f = \omega/2\pi$, in which case B and B_{rej} are measured in Hz instead. For example, the bandwidth of a lowpass filter with $\omega_c = 100$ rad/s is $B = 100$ rad/s, or equivalently $100/2\pi = 15.92$ Hz.

Resonant frequency ω_0

▶ ***Resonance*** is a condition that occurs when the input impedance or input admittance of a circuit containing reactive elements is purely real, and the angular frequency at which it occurs is called the ***resonant frequency*** ω_0. ◀

Often, but not always, the transfer function $\mathbf{H}(\omega)$ also is purely real at $\omega = \omega_0$, and its magnitude is at its maximum or minimum value.

Let us consider the two circuits shown in Fig. 6-5. The input impedance of the RL circuit is simply

$$\mathbf{Z}_{in_1} = R + j\omega L. \tag{6.7}$$

Resonance corresponds to when the imaginary part of $\mathbf{Z}_{in_1}$ is zero, which occurs at $\omega = 0$. Hence, the resonant frequency of the RL circuit is

$$\omega_0 = 0 \qquad \text{(RL circuit).} \tag{6.8}$$

When $\omega_0 = 0$ (dc) or ∞, the resonance is regarded as a ***trivial resonance*** because it occurs at the extreme ends of the spectrum. This usually happens when the circuit has either an inductor or a capacitor (but not both). A circuit that exhibits only a trivial resonance, such as the RL circuit in Fig. 6-5(a), is not considered a resonator.

If the circuit contains at least one capacitor and at least one inductor, resonance can occur at intermediate values of ω. A case in point is the series RLC circuit shown in Fig. 6-5(b). Its input impedance is

$$\mathbf{Z}_{in_2} = R + j\left(\omega L - \frac{1}{\omega C}\right). \tag{6.9}$$

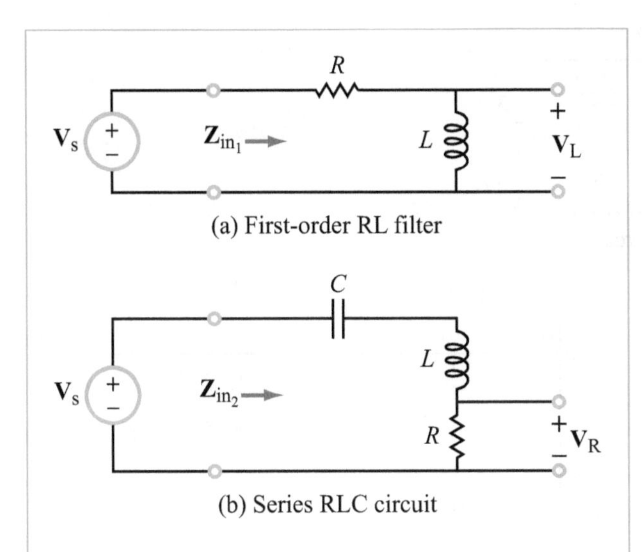

Figure 6-5: Resonance occurs when the imaginary part of the input impedance is zero. For the RL circuit, $\mathfrak{Im}\,[\mathbf{Z}_{\text{in}_1}] = 0$ when $\omega = 0$ (dc), but for the RLC circuit, $\mathfrak{Im}\,[\mathbf{Z}_{\text{in}_2}] = 0$ requires that $\mathbf{Z}_\text{L} = -\mathbf{Z}_C$, or $\omega^2 = 1/LC$.

At resonance ($\omega = \omega_0$), the imaginary part of $\mathbf{Z}_{\text{in}_2}$ is equal to zero. Thus,

$$\omega_0 L - \frac{1}{\omega_0 C} = 0,$$

which leads to

$$\omega_0 = \frac{1}{\sqrt{LC}} \qquad \textbf{(RLC circuit)}. \tag{6.10}$$

So long as neither L nor C is zero or ∞, the transfer function $\mathbf{H}(\omega) = \mathbf{V}_\text{R}/\mathbf{V}_\text{s}$ will exhibit a two-sided spectrum with a peak at ω_0, similar in shape to that of the bandpass filter response shown in Fig. 6-3(c).

Roll-off rate S_g

Outside the passband, the rectangle-shaped idealized responses shown in Fig. 6-3 have infinite slopes, but of course, the actual responses have finite slopes. The steeper the slope, the more discriminating the filter is, and the closer it approaches the idealized response. Hence, the slope S_g outside the passband (called the ***gain roll-off rate***) is an important attribute of the filter response.

6-2.2 RC Circuit Example

To illustrate the transfer-function concept with a concrete example, let us consider the series RC circuit shown in Fig. 6-6(a). Voltage source $\mathbf{V}_\text{s}$ is designated as the input phasor, and on the output side we have designated two voltage phasors, namely, $\mathbf{V}_\text{R}$ and $\mathbf{V}_\text{C}$. We will examine the frequency responses of the transfer functions corresponding to each of those two output voltages.

Lowpass filter

Application of voltage division gives

$$\mathbf{V}_\text{C} = \frac{\mathbf{V}_\text{s}\mathbf{Z}_\text{C}}{R + \mathbf{Z}_\text{C}} = \frac{\mathbf{V}_\text{s}/j\omega C}{R + \frac{1}{j\omega C}}. \tag{6.11}$$

The transfer function corresponding to $\mathbf{V}_\text{C}$ is

$$\mathbf{H}_\text{C}(\omega) = \frac{\mathbf{V}_\text{C}}{\mathbf{V}_\text{s}} = \frac{1}{1 + j\omega RC}, \tag{6.12}$$

where we have multiplied the numerator and denominator of Eq. (6.11) by $j\omega C$ to simplify the form of the expression. In terms of its magnitude $M_\text{C}(\omega)$ and phase angle $\phi_\text{C}(\omega)$, the transfer function is given by

$$\mathbf{H}_\text{C}(\omega) = M_\text{C}(\omega)\, e^{j\phi_\text{C}(\omega)}, \tag{6.13}$$

with

$$M_\text{C}(\omega) = |\mathbf{H}_\text{C}(\omega)| = \frac{1}{\sqrt{1 + \omega^2 R^2 C^2}} \tag{6.14a}$$

and

$$\phi_\text{C}(\omega) = -\tan^{-1}(\omega RC). \tag{6.14b}$$

Frequency response plots for $M_\text{C}(\omega)$ and $\phi_\text{C}(\omega)$ are displayed in Fig. 6-6(b). It is clear from the plot of its magnitude that the expression given by Eq. (6.12) represents the transfer function of a lowpass filter with a dc gain factor $M_0 = 1$. At dc, the capacitor acts like an open circuit, allowing no current to flow through the loop, with the obvious consequence that $\mathbf{V}_\text{C} = \mathbf{V}_\text{s}$. At very high values of ω, the capacitor acts like a short circuit, in which case the voltage across it is approximately zero.

Application of Eq. (6.5) allows us to determine the corner frequency ω_c:

$$M_\text{C}^2(\omega_\text{c}) = \frac{1}{1 + \omega_\text{c}^2 R^2 C^2} = \frac{1}{2}, \tag{6.15}$$

which leads to

$$\omega_\text{c} = \frac{1}{RC}. \tag{6.16}$$

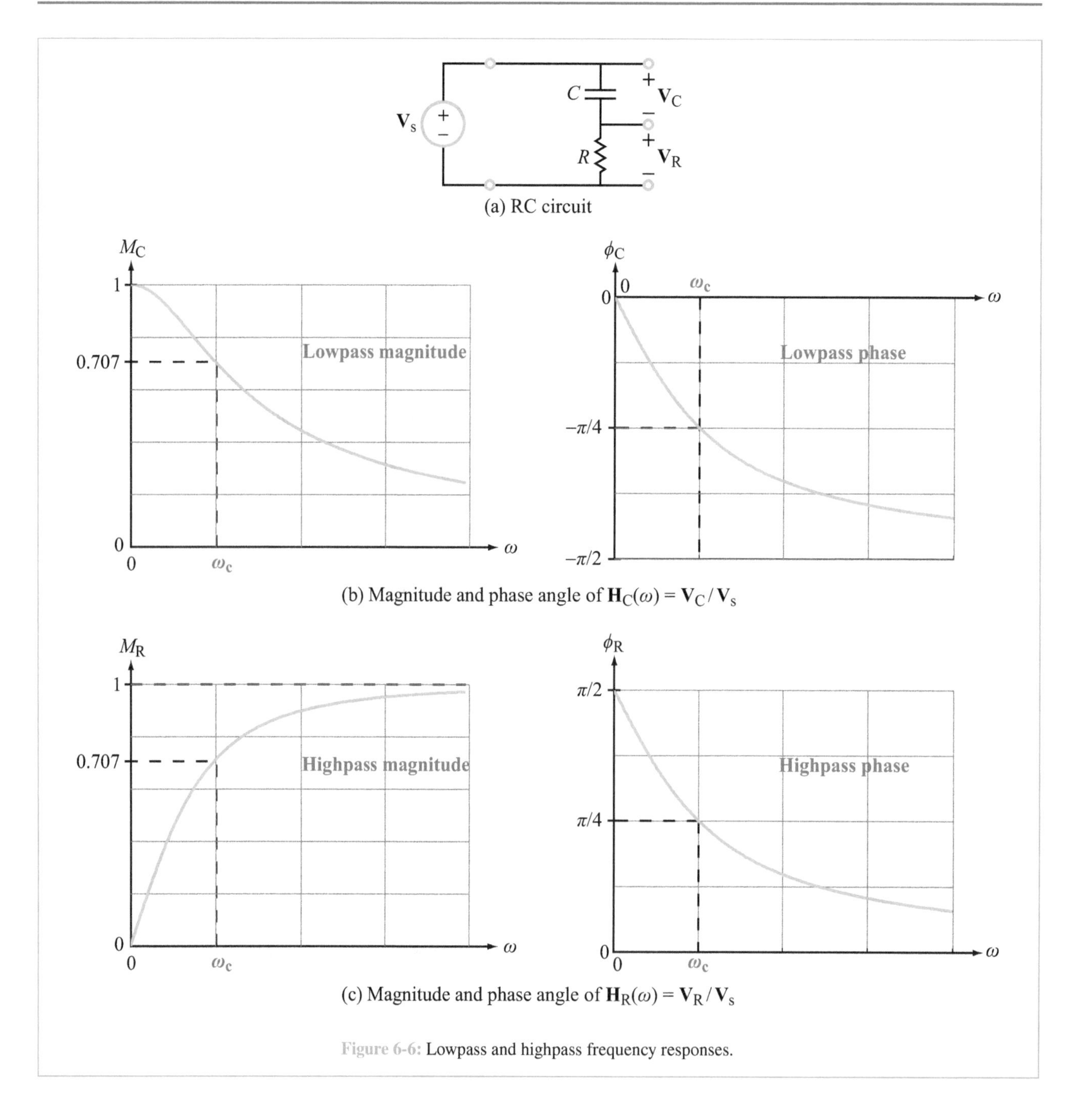

Figure 6-6: Lowpass and highpass frequency responses.

Highpass filter

The output across R in Fig. 6-6(a) leads to

$$\mathbf{H}_R(\omega) = \frac{\mathbf{V}_R}{\mathbf{V}_s} = \frac{j\omega RC}{1 + j\omega RC} . \tag{6.17}$$

The magnitude and phase angle of $\mathbf{H}_R(\omega)$ are given by

$$M_R(\omega) = |\mathbf{H}_R(\omega)| = \frac{\omega RC}{\sqrt{1 + \omega^2 R^2 C^2}} , \tag{6.18a}$$

Table 6-1: Correspondence between M and M [dB].

M	M [dB]
10^N	$20N$ dB
10^3	60 dB
100	40 dB
10	20 dB
4	≈ 12 dB
2	≈ 6 dB
1	0 dB
0.5	≈ -6 dB
0.25	≈ -12 dB
0.1	-20 dB
10^{-N}	$-20N$ dB

and

$$\phi_{\mathrm{R}}(\omega) = \frac{\pi}{2} - \tan^{-1}(\omega RC). \tag{6.18b}$$

Their spectral plots are displayed in Fig. 6-6(c).

6-2.3 Bode Plots

In the late 1930s, inventor Hendrik Bode (pronounced Boh-dee) developed a graphical technique that has since become a standard tool for the analysis and design of resonant circuits, including filters, oscillators, and amplifiers. Bode's technique, which generates what we today call ***Bode plots*** or ***Bode diagrams***, relies on *using a logarithmic scale for ω and on expressing the magnitude of the transfer function in decibels (dB):*

$$M \text{ [dB]} = 20 \log M = 20 \log |\mathbf{H}|. \tag{6.19}$$

The logarithm is in base 10. Note from Table 6-1 that 10^N becomes $20N$ in dB. A useful property of the log operator is that *the log of the product of two numbers is equal to the sum of their logs.* That is:

$$\text{If } G = XY \longrightarrow G \text{ [dB]} = X \text{ [dB]} + Y \text{ [dB]}. \tag{6.20}$$

This result follows from

$$\begin{aligned} G \text{ [dB]} &= 20 \log(XY) = 20 \log X + 20 \log Y \\ &= X \text{ [dB]} + Y \text{ [dB]}. \end{aligned}$$

By the same token,

$$\text{If } G = \frac{X}{Y} \longrightarrow G \text{ [dB]} = X \text{ [dB]} - Y \text{ [dB]}. \tag{6.21}$$

Conversion of products and ratios into sums and differences will prove to be quite useful when constructing the frequency response of a transfer function composed of the product of multiple terms.

Exercise 6-1: Convert the following magnitude ratios to dB: (a) 20, (b) 0.03, (c) 6×10^6.

Answer: (a) 26.02 dB, (b) −30.46 dB, (c) 135.56 dB. (See S²)

Exercise 6-2: Convert the following dB values to magnitude ratios: (a) 36 dB, (b) −24 dB, (c) −0.5 dB.

Answer: (a) 63.1, (b) 0.063, (c) 0.94. (See S²)

Example 6-1: RL Highpass Filter

For the series RL circuit shown in Fig. 6-7(a):

(a) Obtain an expression for the transfer function $\mathbf{H}(\omega) = \mathbf{V}_{\text{out}}/\mathbf{V}_{\text{s}}$ in terms of ω/ω_{c}, where $\omega_{\text{c}} = R/L$.

(b) Determine the magnitude M [dB], and plot it as a function of ω on a log scale, with ω expressed in units of ω_{c}.

(c) Determine and plot the phase angle of $\mathbf{H}(\omega)$.

Solution:

(a) Voltage division gives

$$\mathbf{V}_{\text{out}} = \frac{j\omega L \mathbf{V}_{\text{s}}}{R + j\omega L},$$

which leads to

$$\mathbf{H}(\omega) = \frac{\mathbf{V}_{\text{out}}}{\mathbf{V}_{\text{s}}} = \frac{j\omega L}{R + j\omega L} = \frac{j(\omega/\omega_{\text{c}})}{1 + j(\omega/\omega_{\text{c}})}, \tag{6.22}$$

with $\omega_{\text{c}} = R/L$.

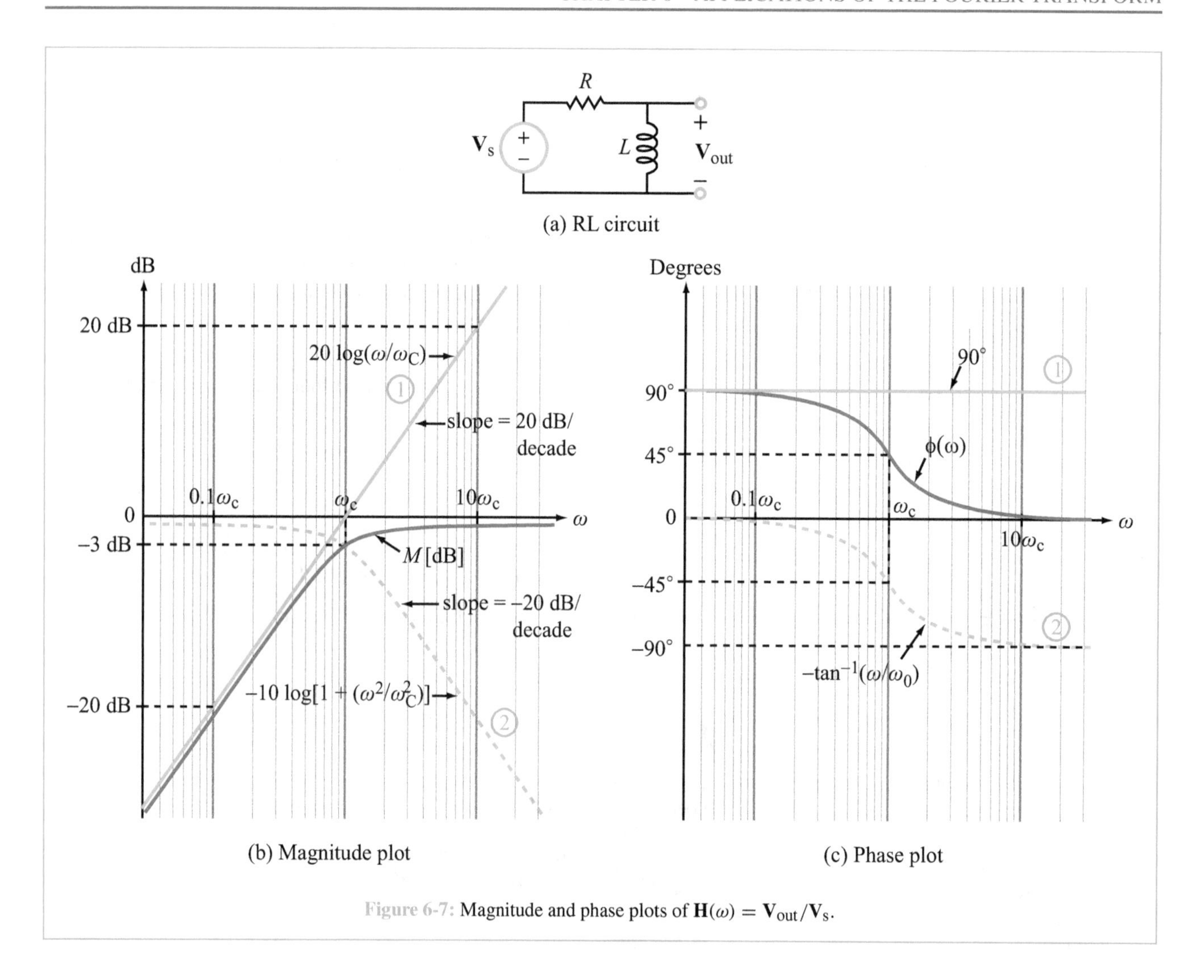

Figure 6-7: Magnitude and phase plots of $\mathbf{H}(\omega) = \mathbf{V}_{\text{out}}/\mathbf{V}_{\text{s}}$.

(b) The magnitude of $\mathbf{H}(\omega)$ is given by

$$M = |\mathbf{H}(\omega)| = \frac{(\omega/\omega_c)}{|1 + j(\omega/\omega_c)|} = \frac{(\omega/\omega_c)}{\sqrt{1 + (\omega/\omega_c)^2}}, \tag{6.23}$$

and

$$\begin{aligned} M\ [\text{dB}] &= 20 \log M \\ &= 20 \log(\omega/\omega_c) - 20 \log[1 + (\omega/\omega_c)^2]^{1/2} \\ &= \underbrace{20 \log(\omega/\omega_c)}_{①} - \underbrace{10 \log[1 + (\omega/\omega_c)^2]}_{②}. \end{aligned} \tag{6.24}$$

In the Bode-diagram terminology, the components of M [dB] are called *factors*, so in the present case M [dB] consists of two factors with the second one having a negative coefficient. A magnitude plot is displayed on semilog graph paper with the vertical axis in dB and the horizontal axis in (rad/s) on a logarithmic scale. Figure 6-7(b) contains individual plots for each of the two factors comprising M [dB], as well as a plot for their sum.

On semilog graph paper, the plot of $\log(\omega/\omega_c)$ is a straight line that crosses the ω-axis at $(\omega/\omega_c) = 1$. This is because $\log 1 = 0$. At $(\omega/\omega_c) = 10$, $20 \log 10 = 20$ dB. Hence,

$$20 \log\left(\frac{\omega}{\omega_c}\right) \quad \rightarrow \quad \text{straight line with slope} = 20 \text{ dB/decade and } \omega\text{-axis crossing at } \omega/\omega_c = 1.$$

The second factor has a nonlinear plot with the following properties:

Low-frequency asymptote

$$\text{As } (\omega/\omega_c) \Rightarrow 0, \qquad -10\log\left[1+\left(\frac{\omega}{\omega_c}\right)^2\right] \Rightarrow 0.$$

High-frequency asymptote

$$\text{As } (\omega/\omega_c) \Rightarrow \infty, \qquad -10\log\left[1+\left(\frac{\omega}{\omega_c}\right)^2\right] \Rightarrow -20\log\left(\frac{\omega}{\omega_c}\right).$$

The plot of M [dB] is obtained by graphically adding together the two plots of its individual factors (Fig. 6-7(b)). At low frequencies ($\omega/\omega_c \ll 1$), M [dB] is dominated by its first factor; at $\omega/\omega_c = 1$, M [dB] $= -10\log 2 \approx -3$ dB; and at high frequencies ($\omega/\omega_c \gg 1$), M [dB] $\to 0$ because its two factors cancel each other out. The overall profile is typical of the spectral response of a highpass filter with a cutoff frequency ω_c.

(c) From Eq. (6.22), the phase angle of $\mathbf{H}(\omega)$ is

$$\phi(\omega) = \underbrace{90^\circ}_{①} \underbrace{- \tan^{-1}\left(\frac{\omega}{\omega_c}\right)}_{②}. \tag{6.25}$$

The 90° component is contributed by j in the numerator and the second term is the phase angle of the denominator. The phase plot is displayed in Fig. 6-7(c).

6-2.4 Standard Form

When generating Bode magnitude and phase plots, the process can be facilitated by casting the expression for $\mathbf{H}(\omega)$ in a *standard form* comprised of factors (either in the denominator or numerator) of the form:

- K, where K is a constant,
- $(j\omega/\omega_{c_1})$, where ω_{c_1} is a constant,
- $[1+(j\omega/\omega_{c_2})]$, where ω_{c_2} is a constant and the real part is 1,
- $[1+(j\omega/\omega_{c_3})^2 + 2\xi(j\omega/\omega_{c_3})]$, where ω_{c_3} and ξ are constants and the non-ω part is 1.

The standard form may include second or higher orders of any of the last three factors, such as $(j\omega/\omega_{c_1})^2$ or $[1+(j\omega/\omega_{c_2})]^3$.

In Eq. (6.22), $\mathbf{H}(\omega)$ was cast in standard form by (a) dividing the numerator and denominator by R so as to make the real part of the denominator equal to 1 and (2) defining the resultant coefficient of ω, namely L/R, as $1/\omega_c$. To illustrate the conversion process for a more complicated expression, consider the following sequence of steps:

$$\begin{aligned}
\mathbf{H}(\omega) &= \frac{(j10\omega+30)^2}{(5+j\omega)(300-3\omega^2+j60\omega)} \\
&= \frac{30^2\left(1+j\dfrac{\omega}{3}\right)^2}{5\left(1+j\dfrac{\omega}{5}\right)300\left(1-\dfrac{\omega^2}{100}+j\dfrac{2\omega}{10}\right)} \\
&= \frac{0.6\left[1+\left(j\dfrac{\omega}{3}\right)\right]^2}{\left(1+j\dfrac{\omega}{5}\right)\left[1+\left(j\dfrac{\omega}{10}\right)^2+2\left(j\dfrac{\omega}{10}\right)\right]}.
\end{aligned}$$

Every factor in the final expression matches one of the standard-form factors listed earlier. In the quadratic term, $\xi = 1$.

6-3 Passive Filters

Analog filters are of two types: passive and active.

> ▶ ***Passive filters*** are resonant circuits that contain only passive elements (resistors, capacitors, and inductors). ◀

In contrast, ***active filters*** contain op amps, transistors, and/or other active devices in addition to the passive elements.

The objective of this section is to examine the basic properties of passive filters by analyzing their transfer functions. To that end, we will use the series RLC circuit shown in Fig. 6-8 in which we have designated four voltage outputs (namely, $\mathbf{V}_R$, $\mathbf{V}_L$, and $\mathbf{V}_C$ across the individual elements and $\mathbf{V}_{LC}$ across the combination of L and C). We will examine the frequency responses of the transfer functions corresponding to all four output voltages.

6-3.1 Bandpass Filter

The current $\mathbf{I}$ flowing through the loop in Fig. 6-9(a) is given by

$$\mathbf{I} = \frac{\mathbf{V}_s}{R+j\left(\omega L - \dfrac{1}{\omega C}\right)} = \frac{j\omega C\mathbf{V}_s}{(1-\omega^2 LC)+j\omega RC}, \tag{6.26}$$

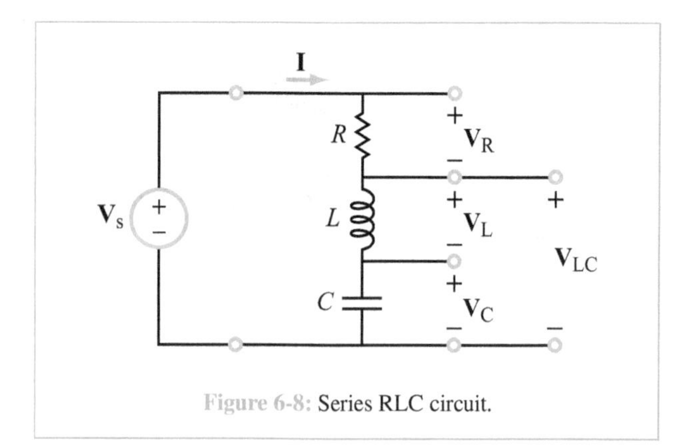

Figure 6-8: Series RLC circuit.

where we multiplied the numerator and denominator by $j\omega C$ to simplify the form of the expression. The transfer function corresponding to $\mathbf{V}_R$ is

$$\mathbf{H}_{BP}(\omega) = \frac{\mathbf{V}_R}{\mathbf{V}_s} = \frac{R\mathbf{I}}{\mathbf{V}_s} = \frac{j\omega RC}{(1-\omega^2 LC) + j\omega RC}, \quad (6.27)$$

where we added the subscript "BP" in anticipation of the fact that $\mathbf{H}_{BP}(\omega)$ is the transfer function of a bandpass filter. Its magnitude and phase angle are given by

$$M_{BP}(\omega) = |\mathbf{H}_{BP}(\omega)| = \frac{\omega RC}{\sqrt{(1-\omega^2 LC)^2 + \omega^2 R^2 C^2}}, \quad (6.28)$$

and

$$\phi_R(\omega) = 90^\circ - \tan^{-1}\left[\frac{\omega RC}{1-\omega^2 LC}\right]. \quad (6.29)$$

According to the plot displayed in Fig. 6-9(b), M_{BP} goes to zero at both extremes of the frequency spectrum and exhibits a maximum across an intermediate range centered at ω_0. Hence, the circuit functions like a bandpass (BP) filter, allowing the transmission (through it) of signals whose angular frequencies are close to ω_0 and discriminating against those with frequencies that are far away from ω_0.

The general profile of $M_{BP}(\omega)$ can be discerned by examining the circuit of Fig. 6-9(a) at specific values of ω. At $\omega = 0$, the capacitor behaves like an open circuit, allowing no current to flow and no voltage to develop across R. As $\omega \to \infty$, it is the inductor that acts like an open circuit, again allowing no current to flow. In the intermediate frequency range when the value of ω is such that $\omega L = 1/(\omega C)$, the impedances of L and C cancel each other out, reducing the total impedance of the RLC circuit to R and the current to $\mathbf{I} = \mathbf{V}_s/R$. Consequently, $\mathbf{V}_R = \mathbf{V}_s$ and $\mathbf{H}_{BP} = 1$. To note the significance of this specific condition, we call it the ***resonance condition*** and we refer to the frequency at

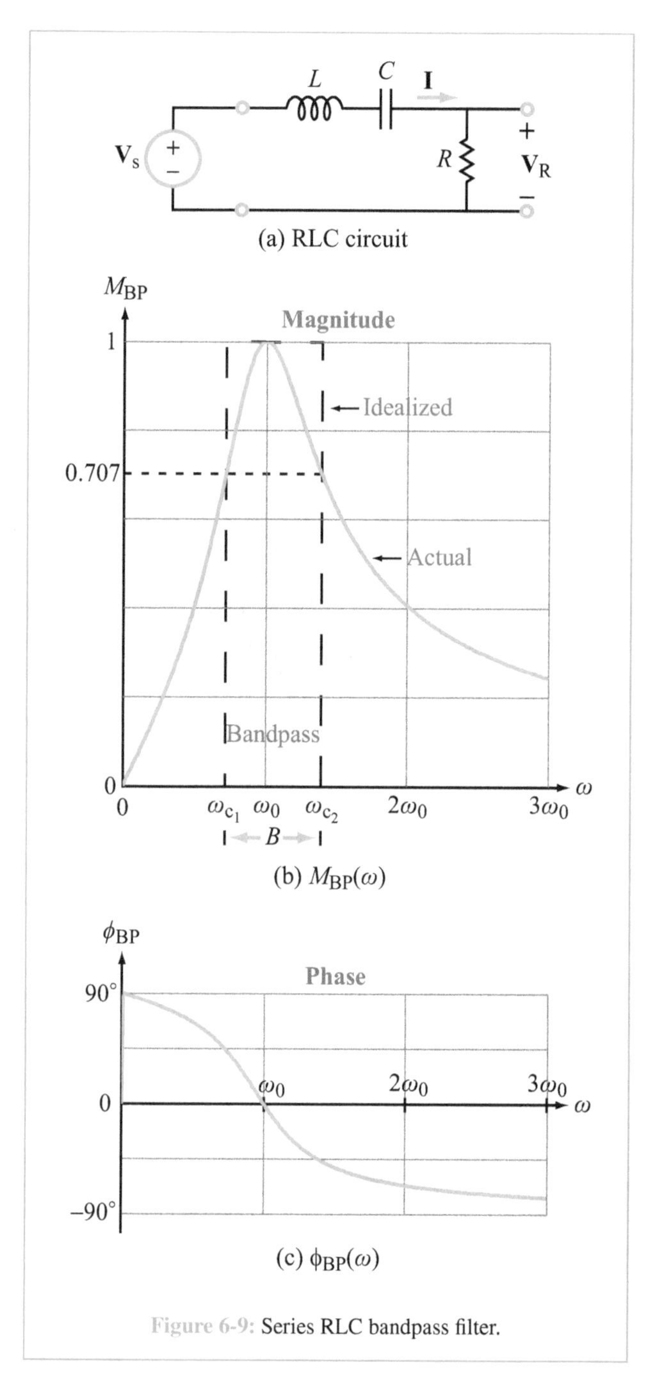

Figure 6-9: Series RLC bandpass filter.

which it occurs as the ***resonant frequency*** ω_0:

$$\omega_0 = \frac{1}{\sqrt{LC}}. \quad (6.30)$$

(resonant frequency)

The phase plot in Fig. 6-9(c) conveys the fact that ϕ_{BP} is dominated by the phase of C at low frequencies and by the phase of L at high frequencies, and $\phi_{BP} = 0$ at $\omega = \omega_0$.

Filter bandwidth

The ***bandwidth*** of the bandpass filter is defined as the frequency range extending between ω_{c_1} and ω_{c_2}, where ω_{c_1} and ω_{c_2} are the values of ω at which $M^2_{BP}(\omega) = 0.5$, or $M_{BP}(\omega) = 1/\sqrt{2} = 0.707$. The quantity M^2_{BP} is proportional to the power delivered to the resistor in the RLC circuit. At resonance, the power is at its maximum, and at ω_{c_1} and ω_{c_2} the power delivered to R is equal to 1/2 of the maximum possible. That is why ω_{c_1} and ω_{c_2} are also referred to as the ***half-power frequencies*** (or the ***3 dB frequencies*** on a dB scale). Thus,

$$M^2_{BP}(\omega) = \frac{1}{2} \qquad @ \ \omega_{c_1} \text{ and } \omega_{c_2}. \tag{6.31}$$

Upon inserting the expression for $M_{BP}(\omega)$ given by Eq. (6.28) and carrying out several steps of algebra, we obtain the solutions

$$\omega_{c_1} = -\frac{R}{2L} + \sqrt{\left(\frac{R}{2L}\right)^2 + \frac{1}{LC}}, \tag{6.32a}$$

and

$$\omega_{c_2} = \frac{R}{2L} + \sqrt{\left(\frac{R}{2L}\right)^2 + \frac{1}{LC}}. \tag{6.32b}$$

The bandwidth is then given by

$$B = \omega_{c_2} - \omega_{c_1} = \frac{R}{L}. \tag{6.33}$$

Quality factor

▶ The ***quality factor*** of a circuit, Q, is an attribute commonly used to characterize the ***degree of selectivity*** of the circuit. ◀

Figure 6-10 displays frequency responses for three circuits, all with the same ω_0. The high-Q circuit exhibits a sharp response with a narrow bandwidth (relative to ω_0), the medium-Q circuit has a broader pattern, and the low-Q circuit has a pattern with limited selectivity.

The formal definition of Q applies to any resonant system and is based on energy considerations. For the RLC bandpass filter,

$$Q = \frac{\omega_0}{B}. \tag{6.34}$$

(RLC circuit)

Thus, Q is the inverse of the bandwidth B, normalized by the center frequency ω_0.

Table 6-2 provides a summary of the salient features of the series RLC bandpass filter. For comparison, the table also includes the corresponding list for the parallel RLC circuit.

Example 6-2: Two-Stage Bandpass Filter

Determine $\mathbf{H}(\omega) = \mathbf{V}_o/\mathbf{V}_s$ for the two-stage BP-filter circuit shown in Fig. 6-11. If $Q_1 = \omega_0 L/R$ is the quality factor of a single stage alone, what is Q_2 for the two stages in combination, given that $R = 2\ \Omega$, $L = 10$ mH, and $C = 1\ \mu$F.

Solution: For each stage alone, we obtain

$$\omega_0 = \frac{1}{\sqrt{LC}} = \frac{1}{\sqrt{10^{-2} \times 10^{-6}}} = 10^4 \text{ rad/s}$$

and

$$Q_1 = \frac{\omega_0 L}{R} = \frac{10^4 \times 10^{-2}}{2} = 50.$$

The loop equations for mesh currents $\mathbf{I}_1$ and $\mathbf{I}_2$ are

$$-\mathbf{V}_s + \mathbf{I}_1\left(j\omega L + \frac{1}{j\omega C} + R\right) - R\mathbf{I}_2 = 0$$

and

$$-R\mathbf{I}_1 + \mathbf{I}_2\left(2R + j\omega L + \frac{1}{j\omega C}\right) = 0.$$

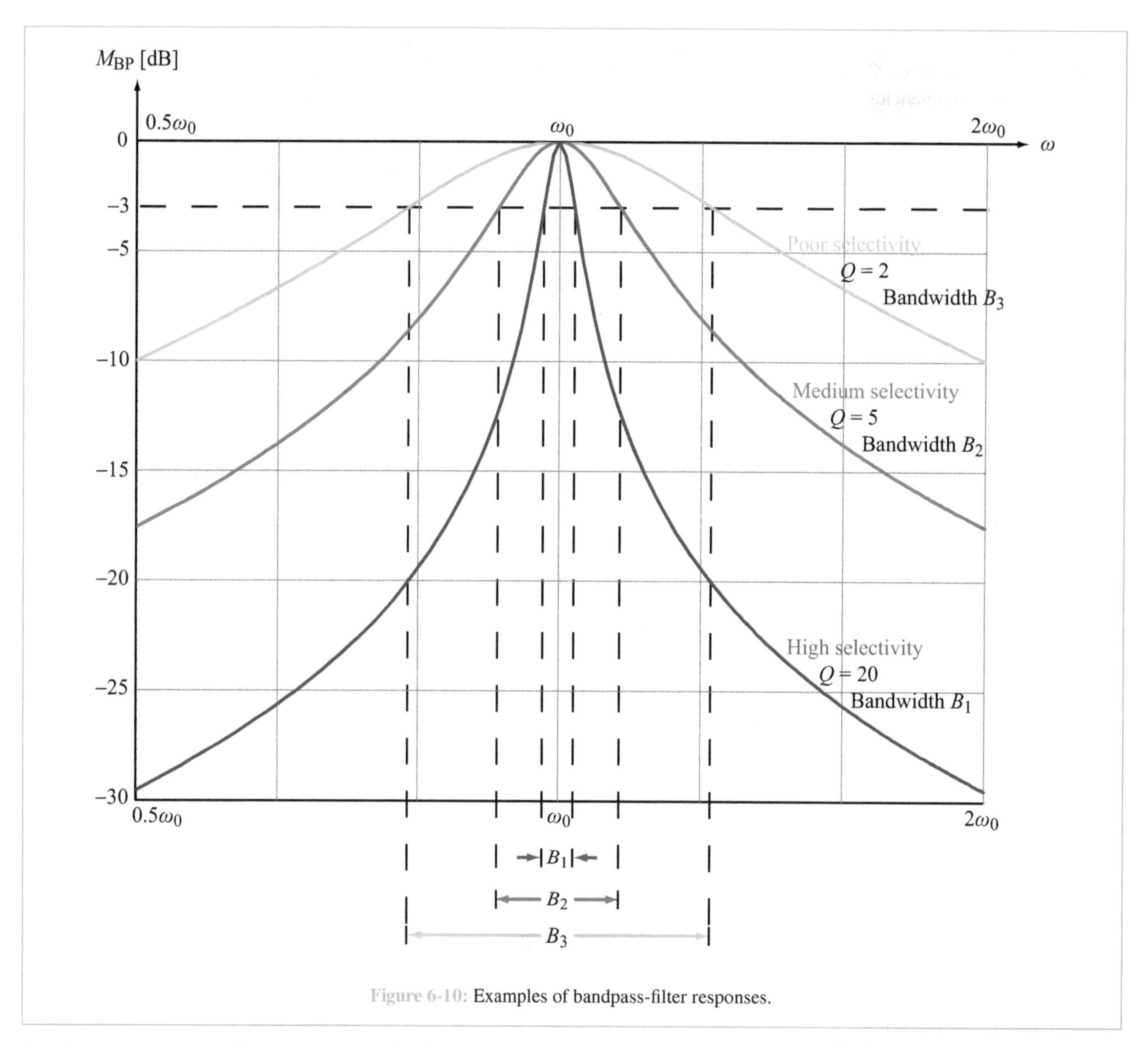

Figure 6-10: Examples of bandpass-filter responses.

Simultaneous solution of the two equations leads to

$$\mathbf{H}(\omega) = \frac{\mathbf{V}_o}{\mathbf{V}_s} =$$

$$\frac{\omega^2 R^2 C^2}{\omega^2 R^2 C^2 - (1 - \omega^2 LC)^2 - j3\omega RC(1 - \omega^2 LC)} =$$

$$\frac{\omega^2 R^2 C^2 [\omega^2 R^2 C^2 - (1 - \omega^2 LC)^2 + j3\omega RC(1 - \omega^2 LC)]}{[\omega^2 R^2 C^2 - (1 - \omega^2 LC)^2]^2 + 9\omega^2 R^2 C^2 (1 - \omega^2 LC)^2}.$$

Resonance occurs when the imaginary part of $\mathbf{H}(\omega)$ is zero, which is satisfied either when $\omega = 0$ (which is a trivial resonance) or when $\omega = 1/\sqrt{LC}$. Hence, the two-stage circuit has the same resonance frequency as a single-stage circuit.

Using the specified values of R, L, and C, we can calculate the magnitude $M(\omega) = |\mathbf{H}(\omega)|$ and plot it as a function of ω. The result is displayed in Fig. 6-11(b). From the spectral plot, we have

$$\omega_{c_1} = 9963 \text{ rad/s}, \qquad \omega_{c_2} = 10037 \text{ rad/s},$$

$$B_2 = \omega_{c_2} - \omega_{c_1} = 10037 - 9963 = 74 \text{ rad/s},$$

and $$Q_2 = \frac{\omega_0}{B_2} = \frac{10^4}{74} = 135,$$

Table 6-2: Attributes of series and parallel RLC bandpass circuits.

	Series	Parallel
RLC circuit	$\mathbf{V}_s$, L, C, R, $\mathbf{V}_R$	$\mathbf{I}_s$, L, C, R, $\mathbf{V}_R$
Transfer function	$\mathbf{H} = \dfrac{\mathbf{V}_R}{\mathbf{V}_s}$	$\mathbf{H} = \dfrac{\mathbf{V}_R}{\mathbf{I}_s}$
Resonant frequency, ω_0	$\dfrac{1}{\sqrt{LC}}$	$\dfrac{1}{\sqrt{LC}}$
Bandwidth, B	$\dfrac{R}{L}$	$\dfrac{1}{RC}$
Quality factor, Q	$\dfrac{\omega_0}{B} = \dfrac{\omega_0 L}{R}$	$\dfrac{\omega_0}{B} = \dfrac{R}{\omega_0 L}$
Lower half-power frequency, ω_{c_1}	$\left[-\dfrac{1}{2Q} + \sqrt{1 + \dfrac{1}{4Q^2}}\right]\omega_0$	$\left[-\dfrac{1}{2Q} + \sqrt{1 + \dfrac{1}{4Q^2}}\right]\omega_0$
Upper half-power frequency, ω_{c_2}	$\left[\dfrac{1}{2Q} + \sqrt{1 + \dfrac{1}{4Q^2}}\right]\omega_0$	$\left[\dfrac{1}{2Q} + \sqrt{1 + \dfrac{1}{4Q^2}}\right]\omega_0$

Notes: (1) The expression for Q of the series RLC circuit is the inverse of that for Q of the parallel circuit. (2) For $Q \geq 10$, $\omega_{c_1} \approx \omega_0 - \frac{B}{2}$, $\omega_{c_2} \approx \omega_0 + \frac{B}{2}$.

where B_2 is the bandwidth of the two-stage BP-filter response. The two-stage combination increases the quality factor from 50 to 135, and reduces the bandwidth from 200 rad/s to 74 rad/s.

6-3.2 Highpass Filter

Transfer function $\mathbf{H}_{\mathrm{HP}}(\omega)$, corresponding to $\mathbf{V}_{\mathrm{L}}$ in the circuit of Fig. 6-12(a), is given by

$$\mathbf{H}_{\mathrm{HP}}(\omega) = \frac{\mathbf{V}_{\mathrm{L}}}{\mathbf{V}_s} = \frac{j\omega L\mathbf{I}}{\mathbf{V}_s} = \frac{-\omega^2 LC}{(1-\omega^2 LC) + j\omega RC} \tag{6.35}$$

with a magnitude and phase angle of

$$M_{\mathrm{HP}}(\omega) = \frac{\omega^2 LC}{[(1-\omega^2 LC)^2 + \omega^2 R^2 C^2]^{1/2}} = \frac{(\omega/\omega_0)^2}{\{[1-(\omega/\omega_0)^2]^2 + [\omega/(Q\omega_0)]^2\}^{1/2}} \tag{6.36a}$$

and

$$\phi_{\mathrm{HP}}(\omega) = 180^\circ - \tan^{-1}\left[\frac{\omega RC}{1-\omega^2 LC}\right] = 180^\circ - \tan^{-1}\left\{\frac{(\omega/\omega_0)}{Q[1-(\omega/\omega_0)^2]}\right\}, \tag{6.36b}$$

where ω_0 and Q are defined by Eqs. (6.30) and (6.34), respectively. Figure 6-12(b) displays logarithmic plots of M_{HP} [dB] for two values of Q. Because $M_{\mathrm{HP}}(\omega)$ has a quadratic pole, its slope in the stopband is 40 dB/decade.

6-3.3 Lowpass Filter

The voltage across the capacitor in Fig. 6-13(a) generates a lowpass-filter transfer function given by

$$\mathbf{H}_{\mathrm{LP}}(\omega) = \frac{\mathbf{V}_{\mathrm{C}}}{\mathbf{V}_s} = \frac{[1/(j\omega C)]\mathbf{I}}{\mathbf{V}_s} = \frac{1}{(1-\omega^2 LC) + j\omega RC} \tag{6.37}$$

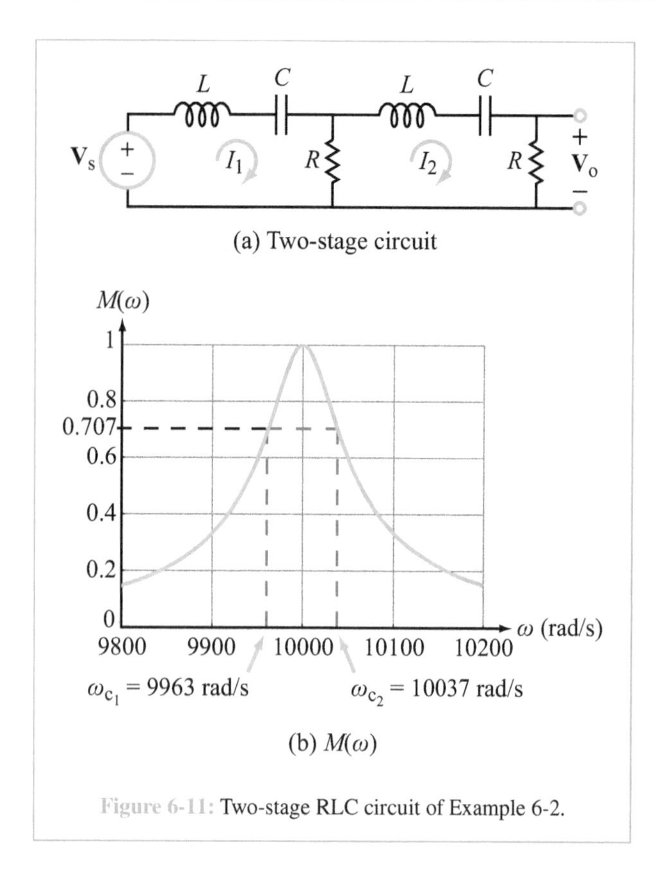

Figure 6-11: Two-stage RLC circuit of Example 6-2.

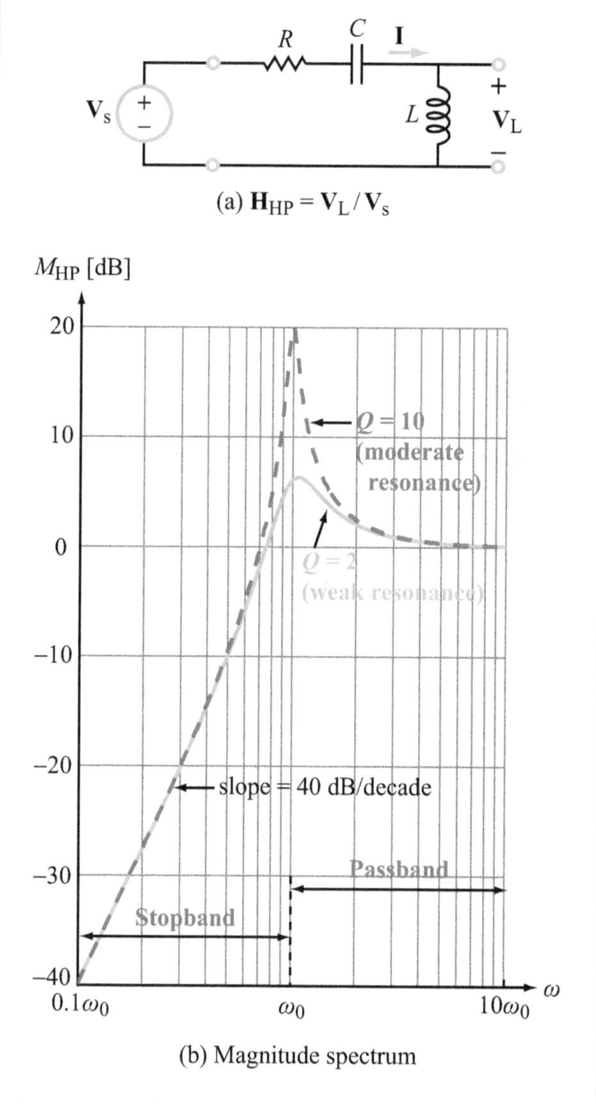

Figure 6-12: Plots of M_{HP} [dB] for $Q = 2$ (weak resonance) and $Q = 10$ (moderate resonance).

with magnitude and phase angle given by

$$M_{\text{LP}}(\omega) = \frac{1}{[(1-\omega^2 LC)^2 + \omega^2 R^2 C^2]^{1/2}}$$
$$= \frac{1}{\{[1-(\omega/\omega_0)^2]^2 + [\omega/(Q\omega_0)]^2\}^{1/2}} \quad (6.38a)$$

and

$$\phi_{\text{LP}}(\omega) = -\tan^{-1}\left(\frac{\omega RC}{1-\omega^2 LC}\right)$$
$$= -\tan^{-1}\left\{\frac{(\omega/\omega_0)}{Q[1-(\omega/\omega_0)^2]}\right\}. \quad (6.38b)$$

The spectral plots of M_{LP} [dB] shown in Fig. 6-13(b) are mirror images of the highpass-filter plots displayed in Fig. 6-12(b).

6-3.4 Bandreject Filter

The output voltage across the combination of L and C in Fig. 6-14(a), which generates a bandreject-filter transfer function, is equal to $\mathbf{V}_s - \mathbf{V}_R$:

$$\mathbf{H}_{\text{BR}}(\omega) = \frac{\mathbf{V}_L + \mathbf{V}_C}{\mathbf{V}_s} = \frac{\mathbf{V}_s - \mathbf{V}_R}{\mathbf{V}_s} = 1 - \mathbf{H}_{\text{BP}}(\omega), \quad (6.39)$$

where $\mathbf{H}_{\text{BP}}(\omega)$ is the bandpass-filter transfer function given by Eq. (6.27). The spectral response of $\mathbf{H}_{\text{BP}}$ passes all frequencies, except for an intermediate band centered at ω_0, as shown in Fig. 6-14(b). The width of the stopband is determined by the values of ω_0 and Q.

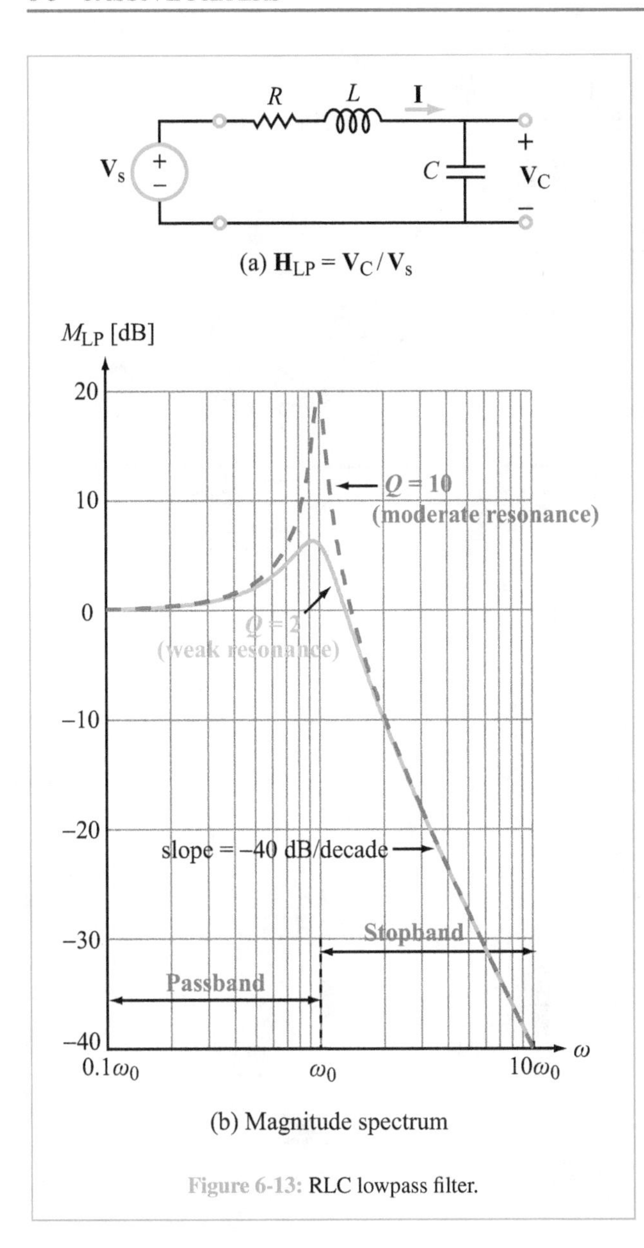

Figure 6-13: RLC lowpass filter.

6-3.5 Filter Order

▶ The order of a filter is equal to the absolute value of the highest power of ω in its transfer function when ω is in the filter's stopband(s). The associated roll-off rate is $S_g = (20 \times \text{filter order})$ in dB/decade. ◀

Let us examine this definition for two circuit configurations.

First-order lowpass RC filter

The transfer function of the RC circuit shown in Fig. 6-15(a) is given by

$$\mathbf{H}_1(\omega) = \frac{\mathbf{V}_C}{\mathbf{V}_s} = \frac{1/(j\omega C)}{R + 1/(j\omega C)} = \frac{1}{1 + j\omega RC} = \frac{1}{1 + j\omega/\omega_{c_1}} \quad \text{(first order)}, \tag{6.40}$$

where we multiplied both the numerator and denominator by $j\omega C$ so as to rearrange the expression into a form in which ω is normalized to the ***corner frequency*** given by

$$\omega_{c_1} = \frac{1}{RC} \quad \textbf{(RC filter)}. \tag{6.41}$$

It is evident from the expression given by Eq. (6.40) that the highest order of ω is 1, and therefore the RC circuit is a first-order filter. Strict application of the definition for the order of a filter requires that we evaluate the power of ω when ω is in the stopband of the filter. In the present case, the stopband covers the range $\omega \geq \omega_{c_1}$. When ω is well into the stopband ($\omega/\omega_{c_1} \gg 1$), Eq. (6.40) simplifies to

$$\mathbf{H}_1(\omega) \approx \frac{-j\omega_{c_1}}{\omega} \quad (\text{for } \omega/\omega_{c_1} \gg 1), \tag{6.42}$$

which confirms the earlier conclusion that the RC circuit is first-order. The corresponding roll-off rate is −20 dB/decade.

Second-order lowpass filter

For the RLC circuit shown in Fig. 6-15(c), we determined in Section 6-2.3 that its transfer function is given by

$$\mathbf{H}_2(\omega) = \frac{\mathbf{V}_C}{\mathbf{V}_s} = \frac{1}{(1 - \omega^2 LC) + j\omega RC}. \tag{6.43}$$

The magnitude spectrum of the RLC lowpass filter was presented earlier in Fig. 6-13(b), where it was observed that the response may exhibit a resonance phenomenon in the neighborhood of $\omega_0 = 1/\sqrt{LC}$, and that it decays with $S_g = -40$ dB/decade in the stopband ($\omega \geq \omega_0$). This is consistent with the fact that the RLC circuit generates a second-order lowpass filter when the output voltage is taken

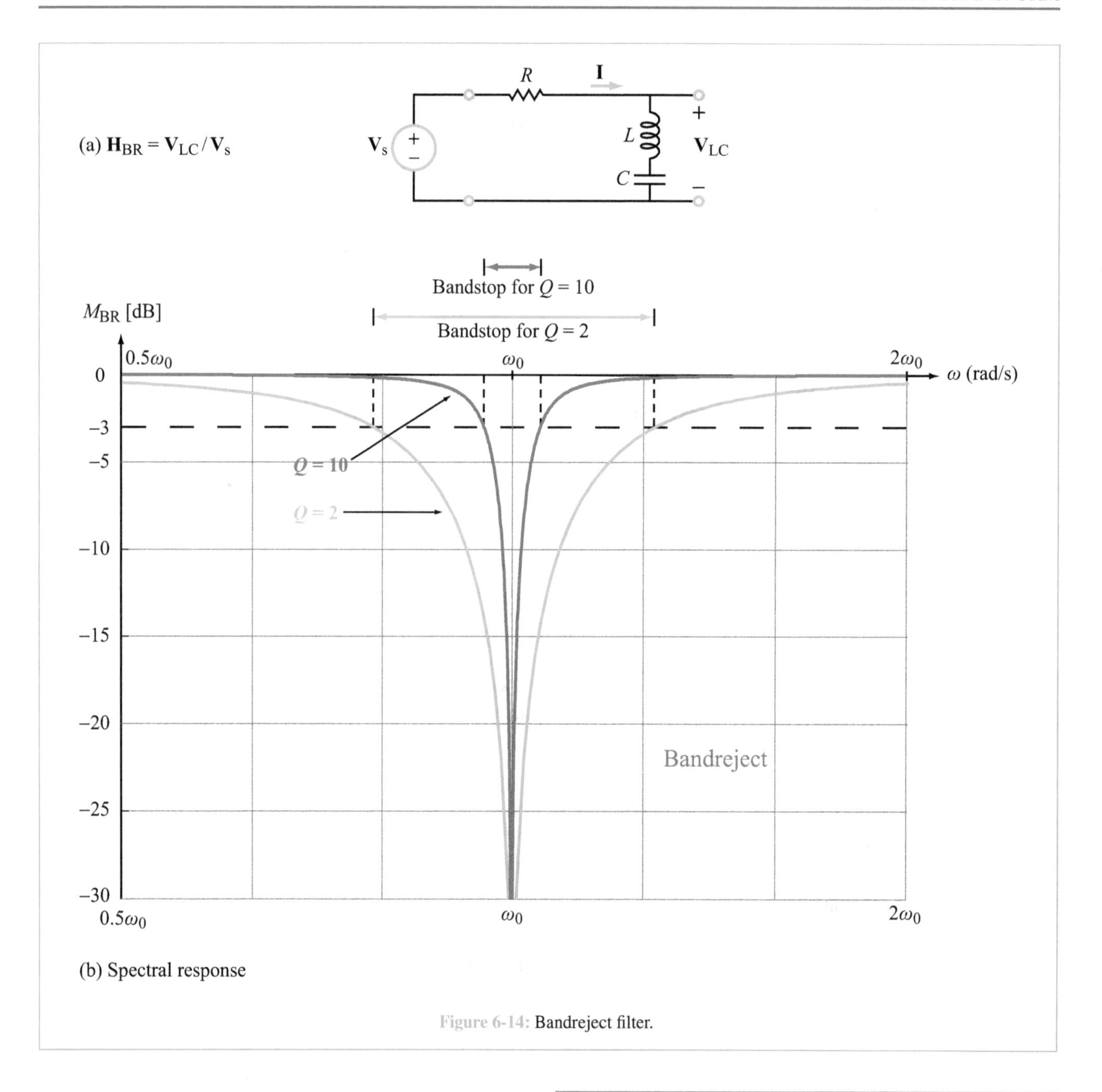

Figure 6-14: Bandreject filter.

across the capacitor. In terms of our definition for the order of a filter, in the stopband ($\omega^2 \gg 1/LC$), Eq. (6.43) reduces to

$$\mathbf{H}_2(\omega) \approx \frac{-1}{\omega^2 LC} \qquad (\text{for } \omega \gg \omega_0), \tag{6.44}$$

which assumes the form of a second-order pole.

Example 6-3: RLC Bandpass Filter

Is the RLC bandpass filter first-order or second-order?

Solution: To answer the question, we start by examining the expression for the transfer function when ω is in the stopbands of the filter. The expression for $\mathbf{H}_{BP}(\omega)$ is given by Eq. (6.27)

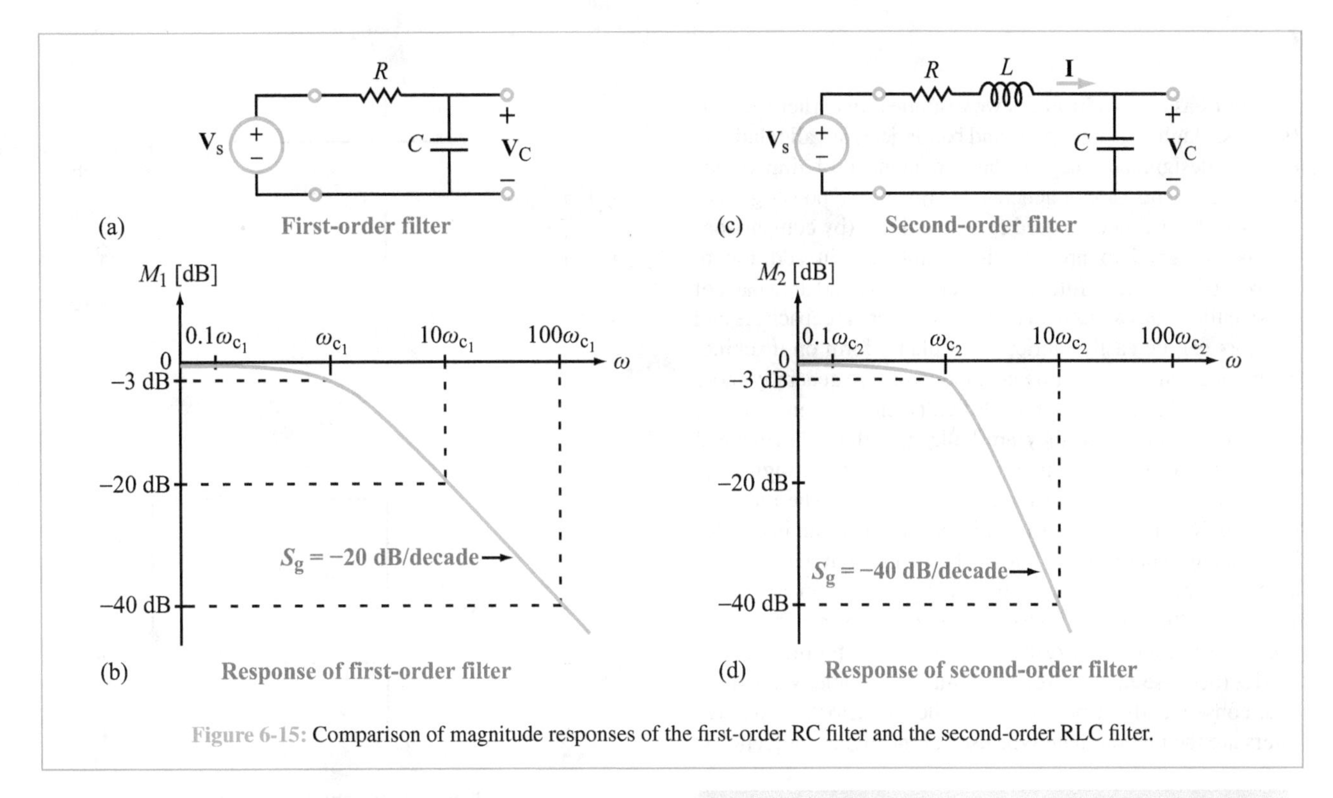

Figure 6-15: Comparison of magnitude responses of the first-order RC filter and the second-order RLC filter.

as

$$\mathbf{H}_{\mathrm{BP}}(\omega) = \frac{j\omega RC}{1 - \omega^2 LC + j\omega RC}. \quad (6.45)$$

For $\omega \ll \omega_0$ and $\omega \gg \omega_0$ (where $\omega_0 = 1/\sqrt{LC}$), the expression simplifies to

$$\mathbf{H}_{\mathrm{BP}}(\omega) \approx \begin{cases} j\omega RC & \text{for } \omega \ll \omega_0 \text{ and } \omega \ll RC, \\ \dfrac{-jR}{\omega L} & \text{for } \omega \gg \omega_0. \end{cases} \quad (6.46)$$

At the low-frequency end, $\mathbf{H}_{\mathrm{BP}}(\omega)$ is proportional to ω, and at the high-frequency end, it is proportional to $1/\omega$. Hence, the RLC bandpass filter is first-order, not second-order.

6-3.6 Real versus Idealized Filters

The higher the order of a real filter is, the steeper is its roll-off rate S_g, and the closer it approaches the idealized ***brick-wall spectrum***. Can idealized filters be realized, and if not, how and why do we use them? Because their spectra are discontinuous at junctions between their passbands and stopbands, idealized filters are not realizable. Nevertheless, they serve as convenient models with which to analyze and/or synthesize applications of interest, as an initial step prior to developing and implementing approximate equivalents in the form of real filters.

Exercise 6-3: Determine the order of

$$\mathbf{H}(\omega) = \mathbf{V}_{\mathrm{out}}/\mathbf{V}_{\mathrm{s}}$$

for the circuit in Fig. E6-3.

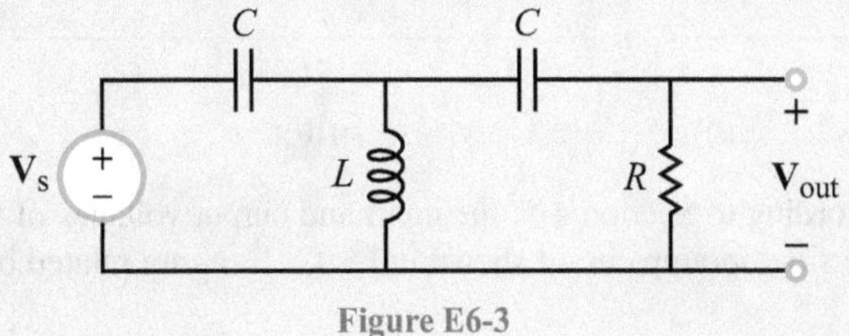

Figure E6-3

Answer:

$$\mathbf{H}(\omega) = \frac{j\omega^3 RLC^2}{\omega^2 LC - (1 - \omega^2 LC)(1 + j\omega RC)},$$

which describes a highpass filter. In the stopband (very small values of ω), $\mathbf{H}(\omega)$ varies as ω^3. Hence, it is third order. (See S²)

6-4 Active Filters

The four basic types of filters we examined in earlier sections (lowpass, highpass, bandpass, and bandreject) are all relatively easy to design, but they do have a number of drawbacks. Passive elements cannot generate energy, so the power gain of a passive filter cannot exceed 1. Active filters (by comparison) can be designed to provide significant gain in addition to realizing the specified filter performance. A second drawback of passive filters has to do with inductors. Whereas capacitors and resistors can be easily fabricated in planar form on machine-assembled printed circuit boards, inductors are generally more expensive to fabricate and more difficult to integrate into the rest of the circuit, because they are bulky and three-dimensional in shape. In contrast, op-amp circuits can be designed to function as filters without the use of inductors. The intended operating-frequency range is an important determining factor in choosing what type of filter is best to design and use. Op amps generally do not perform reliably at frequencies above 1 MHz, so their use as filters is limited to lower frequencies. Fortunately, inductor size becomes less of a problem above 1 MHz (because $\mathbf{Z}_{\mathrm{L}} = j\omega L$, necessitating a smaller value for L and, consequently, a physically smaller inductor), so passive filters are the predominant type used at the higher frequencies.

► One of the major assets of op amp circuits is that they can be easily cascaded together, both in series and in parallel, to realize the intended function. Moreover, by inserting buffer circuits between successive stages, impedance mismatch and loading problems can be minimized or avoided altogether. ◄

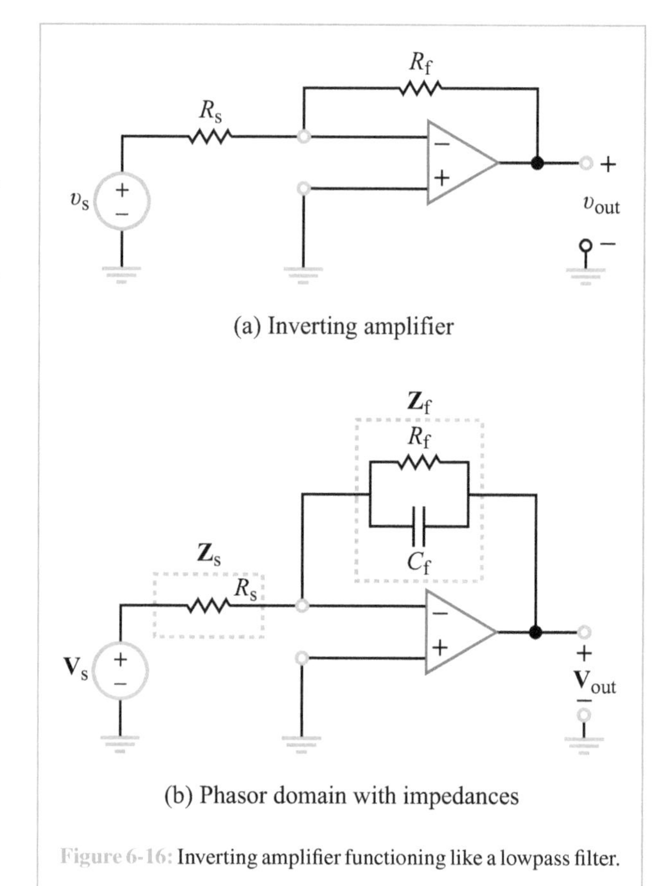

Figure 6-16: Inverting amplifier functioning like a lowpass filter.

6-4.1 Single-Pole Lowpass Filter

According to Section 4-5, the input and output voltages of the inverting op-amp circuit shown in Fig. 6-16(a) are related by

$$\upsilon_{\mathrm{out}} = -\frac{R_{\mathrm{f}}}{R_{\mathrm{s}}}\,\upsilon_{\mathrm{s}}. \tag{6.47}$$

Let us now transform the circuit into the frequency domain and generalize it by replacing resistors R_{s} and R_{f} with impedances $\mathbf{Z}_{\mathrm{s}}$ and $\mathbf{Z}_{\mathrm{f}}$, respectively, as shown in Fig. 6-16(b). Further, let us retain $\mathbf{Z}_{\mathrm{s}}$ as R_{s}, but specify $\mathbf{Z}_{\mathrm{f}}$ as the parallel combination of a resistor R_{f} and a capacitor C_{f}. By analogy with Eq. (6.47), the equivalent relationship for the circuit in Fig. 6-16(b) is

$$\mathbf{V}_{\mathrm{out}} = -\frac{\mathbf{Z}_{\mathrm{f}}}{\mathbf{Z}_{\mathrm{s}}}\,\mathbf{V}_{\mathrm{s}}, \tag{6.48}$$

with

$$\mathbf{Z}_{\mathrm{s}} = R_{\mathrm{s}}, \tag{6.49a}$$

$$\mathbf{Z}_{\mathrm{f}} = R_{\mathrm{f}} \parallel \left(\frac{1}{j\omega C_{\mathrm{f}}}\right) = \frac{R_{\mathrm{f}}}{1 + j\omega R_{\mathrm{f}} C_{\mathrm{f}}}\,. \tag{6.49b}$$

The transfer function of the circuit, which we will soon recognize as that of a lowpass filter, is given by

$$\mathbf{H}_{\mathrm{LP}}(\omega) = \frac{\mathbf{V}_{\mathrm{out}}}{\mathbf{V}_{\mathrm{s}}} = -\frac{\mathbf{Z}_{\mathrm{f}}}{\mathbf{Z}_{\mathrm{s}}} = -\frac{R_{\mathrm{f}}}{R_{\mathrm{s}}}\left(\frac{1}{1 + j\omega R_{\mathrm{f}} C_{\mathrm{f}}}\right) = G_{\mathrm{LP}}\left(\frac{1}{1 + j\omega/\omega_{\mathrm{LP}}}\right), \tag{6.50}$$

where

$$G_{\mathrm{LP}} = -\frac{R_{\mathrm{f}}}{R_{\mathrm{s}}} \quad \text{and} \quad \omega_{\mathrm{LP}} = \frac{1}{R_{\mathrm{f}} C_{\mathrm{f}}}\,. \tag{6.51}$$

The expression for G_{LP} is the same as that of the original inverting amplifier, and ω_{LP} is the cutoff frequency of the lowpass filter. Except for the gain factor, the expression given

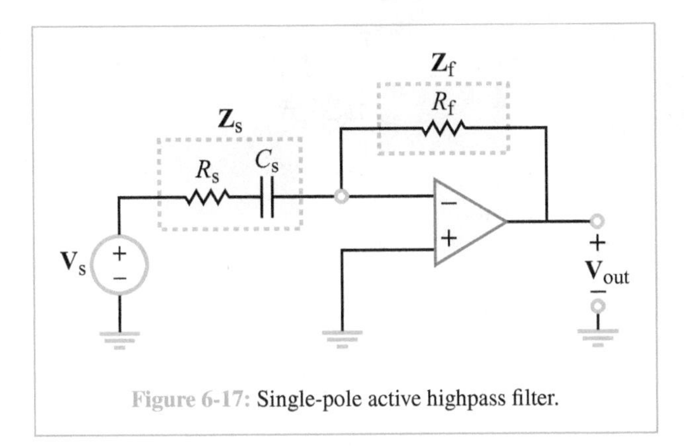

Figure 6-17: Single-pole active highpass filter.

by Eq. (6.50) is identical in form with Eq. (6.40), the transfer function of the RC lowpass filter. A decided advantage of the active lowpass filter over its passive counterpart is that ω_{LP} is independent of both the input resistance R_s and any non-zero load resistance R_L that may be connected across the op amp's output terminals.

6-4.2 Single-Pole Highpass Filter

If in the inverting amplifier circuit, we were to specify the input and feedback impedances as

$$\mathbf{Z}_s = R_s - \frac{j}{\omega C_s} \quad \text{and} \quad \mathbf{Z}_f = R_f, \tag{6.52}$$

as shown in Fig. 6-17, we would obtain the highpass filter transfer function given by

$$\mathbf{H}_{HP}(\omega) = \frac{\mathbf{V}_{out}}{\mathbf{V}_s} = -\frac{\mathbf{Z}_f}{\mathbf{Z}_s} = -\frac{R_f}{R_s - j/(\omega C_s)} = G_{HP}\left[\frac{j\omega/\omega_{HP}}{1 + j\omega/\omega_{HP}}\right], \tag{6.53}$$

where

$$G_{HP} = -\frac{R_f}{R_s} \quad \text{and} \quad \omega_{HP} = \frac{1}{R_s C_s}. \tag{6.54}$$

The expression given by Eq. (6.53) represents a first-order highpass filter with a cutoff frequency ω_{HP} and a high frequency gain factor G_{HP}.

Concept Question 6-1: What are the major advantages of active filters over their passive counterparts? (See S²)

Concept Question 6-2: Are active filters used mostly at frequencies below 1 MHz or above 1 MHz? (See S²)

Exercise 6-4: Choose values for R_s and R_f in the circuit of Fig. 6-16(b) so that the gain magnitude is 10 and the corner frequency is 10^3 rad/s, given that $C_f = 1\ \mu$F.

Answer: $R_s = 100\ \Omega$, $R_f = 1\ \text{k}\Omega$. (See S²)

6-4.3 Cascaded Active Filters

The active lowpass and highpass filters we examined thus far—as well as other op-amp configurations that provide these functions—can be regarded as basic building blocks that can be easily cascaded together to create second- or higher-order lowpass and highpass filters, or to design bandpass and bandreject filters (Fig. 6-18).

▶ The cascading approach allows the designer to work with each stage separately and then combine all of the stages together to achieve the desired specifications. ◀

Moreover, inverting or noninverting amplifier stages can be added to the filter cascade to adjust the gain or polarity of the output signal, and buffer circuits can be inserted in-between stages to provide impedance isolation, if necessary. Throughout the multistage process, it is prudent to compare the positive and negative peak values of the voltage at the output of every stage with the op amp's power supply voltages V_{CC} and $-V_{CC}$ to make sure that the op amp will not go into saturation mode.

Example 6-4: Third-Order Lowpass Filter

For the three-stage active filter shown in Fig. 6-19, generate dB plots for M_1, M_2, and M_3, where $M_1 = |\mathbf{V}_1/\mathbf{V}_s|$, $M_2 = |\mathbf{V}_2/\mathbf{V}_s|$, and $M_3 = |\mathbf{V}_3/\mathbf{V}_s|$.

Solution: Since all three stages have the same values for R_f and C_f, they have the same cutoff frequency

$$\omega_{LP} = \frac{1}{R_f C_f} = \frac{1}{10^4 \times 10^{-9}} = 10^5 \text{ rad/s}.$$

The input resistance of the first stage is 10 Ω, but the input resistances of the second and third stages are 10 kΩ. Hence,

$$G_1 = -\frac{10\text{k}}{10} = -10^3$$

and

$$G_2 = G_3 = -\frac{10\text{k}}{10\text{k}} = -1.$$

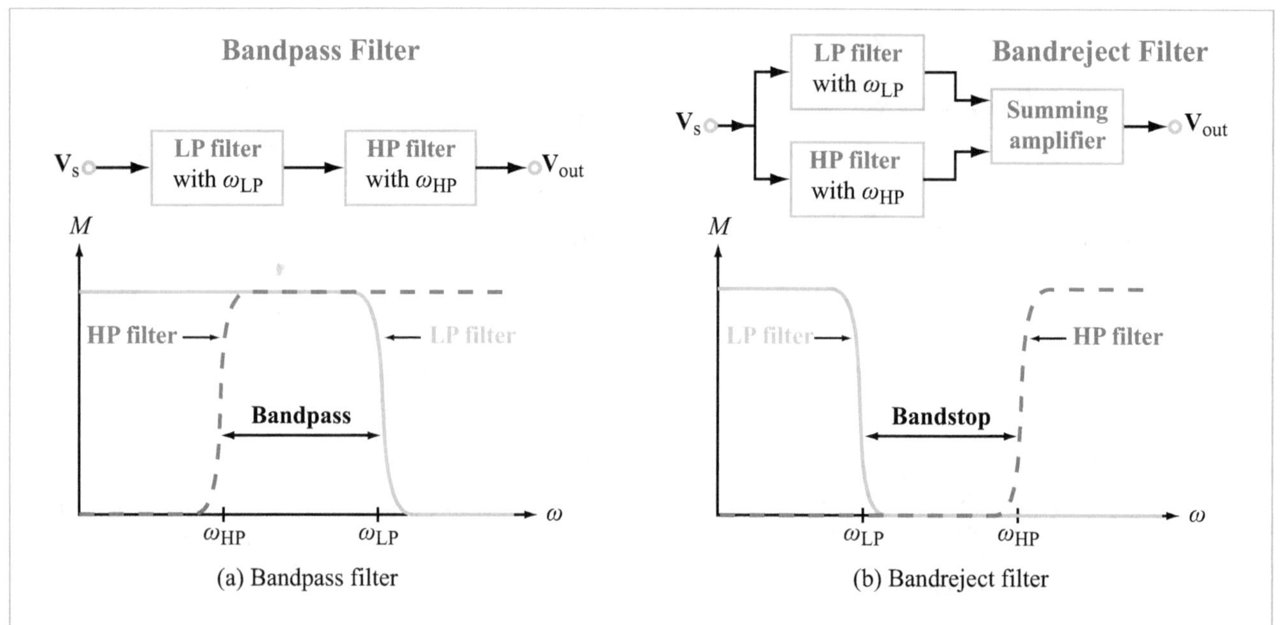

(a) Bandpass filter (b) Bandreject filter

Figure 6-18: In-series cascade of a lowpass and a highpass filter generates a bandpass filter; in-parallel cascading generates a bandreject filter.

Transfer function M_1 is therefore given by

$$M_1 = \left|\frac{\mathbf{V}_1}{\mathbf{V}_s}\right| = \left|\frac{G_1}{1 + j\omega/\omega_{LP}}\right| = \frac{10^3}{\sqrt{1 + (\omega/10^5)^2}},$$

and

$$M_1 \text{ [dB]} = 20 \log \left[\frac{10^3}{\sqrt{1 + (\omega/10^5)^2}}\right] = 60 \text{ dB} - 10 \log[1 + (\omega/10^5)^2].$$

The transfer function corresponding to $\mathbf{V}_2$ is

$$M_2 = \left|\frac{\mathbf{V}_2}{\mathbf{V}_1} \cdot \frac{\mathbf{V}_1}{\mathbf{V}_s}\right| = \left|\frac{G_1}{1 + j\omega/\omega_{LP}}\right| \left|\frac{G_2}{1 + j\omega/\omega_{LP}}\right| = \frac{10^3}{1 + (\omega/10^5)^2},$$

and

$$M_2 \text{ [dB]} = 20 \log \left[\frac{10^3}{1 + (\omega/10^5)^2}\right] = 60 \text{ dB} - 20 \log[1 + (\omega/10^5)^2].$$

Similarly,

$$M_3 \text{ [dB]} = 60 \text{ dB} - 30 \log[1 + (\omega/10^5)^2].$$

The three-stage process is shown in Fig. 6-19(b) in block diagram form, and spectral plots of M_1 [dB], M_2 [dB], and M_3 [dB] are displayed in Fig. 6-19(c). We note that the gain roll-off rate S_g is -20 dB for M_1 [dB], -40 dB for M_2 [dB], and -60 dB for M_3 [dB]. We also note that the -3 dB corner frequencies are not the same for the three stages.

Concept Question 6-3: Why is it more practical to cascade multiple stages of active filters than to cascade multiple stages of passive filters? (See S^2)

Concept Question 6-4: What determines the gain factors of highpass and lowpass op-amp filters? (See S^2)

Exercise 6-5: What are the values of the corner frequencies associated with M_1, M_2, and M_3 of Example 6-4?

Answer: $\omega_{c_1} = 10^5$ rad/s, $\omega_{c_2} = 0.64\omega_{c_1} = 6.4 \times 10^4$ rad/s, $\omega_{c_3} = 0.51\omega_{c_1} = 5.1 \times 10^4$ rad/s. (See S^2)

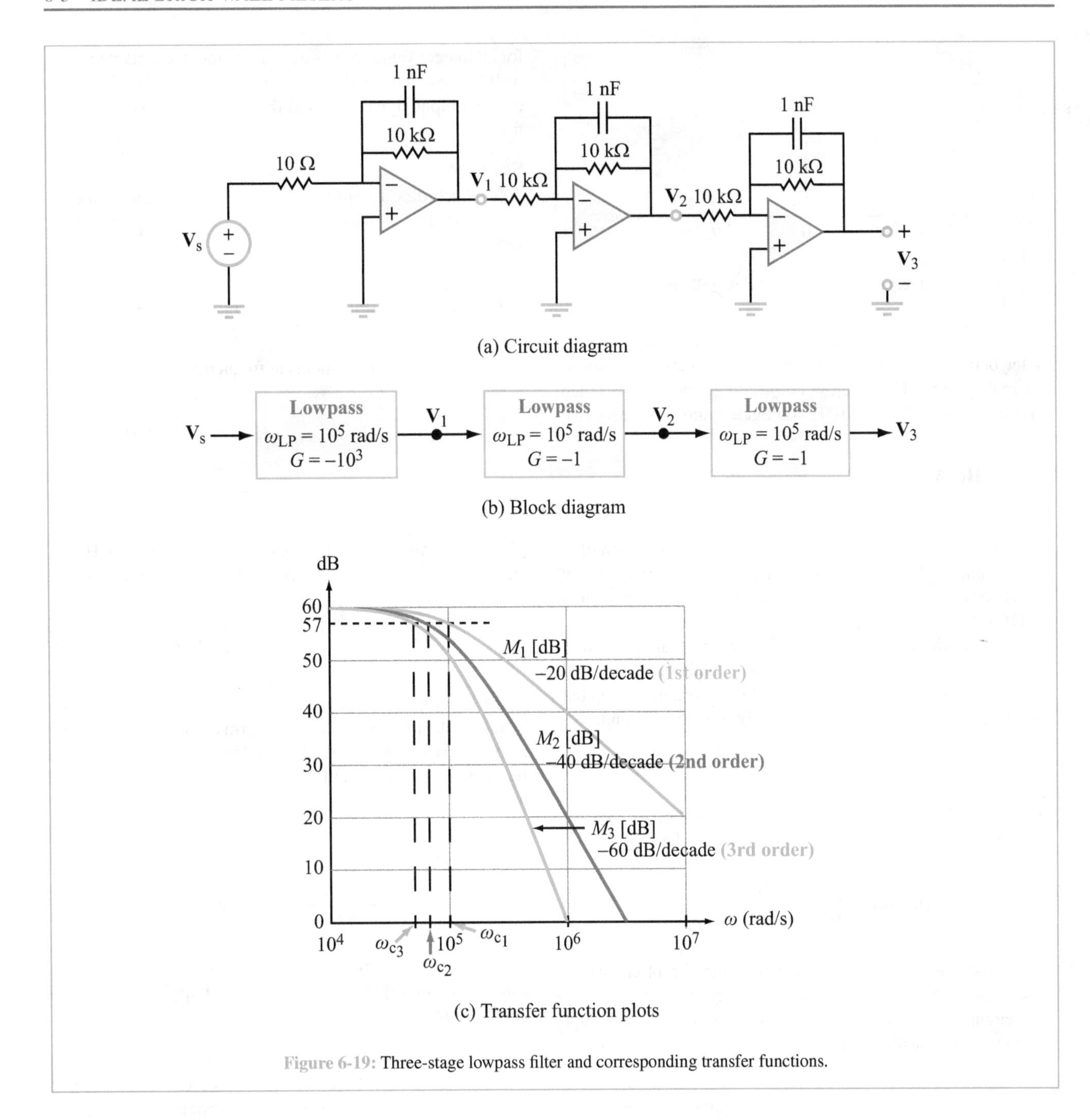

Figure 6-19: Three-stage lowpass filter and corresponding transfer functions.

6-5 Ideal Brick-Wall Filters

In past sections, the frequency responses of filters were plotted as a function of only positive values of ω. Even though "negative frequency" has no physical meaning, we find it mathematically convenient to define ω over $-\infty$ to ∞, just as we do with time t. The Fourier transform of a time-domain signal is a frequency-domain function that extends over both positive and negative frequencies. The same is true for systems; a system with an impulse response $h(t)$ has frequency response $\mathbf{H}(\omega)$, with the

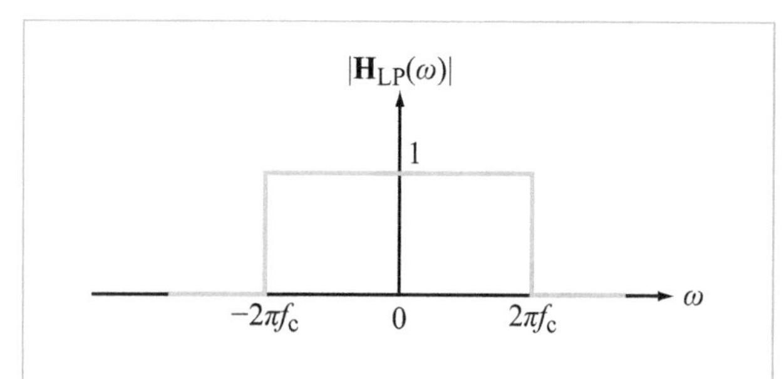

Figure 6-20: Frequency response of brick-wall lowpass filter.

latter being the Fourier transform of the former. Fortunately, for real systems, $\mathbf{H}(\omega)$ enjoys ***conjugate symmetry***. As noted in connection with Eq. (2.109), conjugate symmetry means that

$$\mathbf{H}(-\omega) = \mathbf{H}^*(\omega) \qquad \textbf{(conjugate symmetry)} \tag{6.55}$$

or, equivalently, its magnitude $|\mathbf{H}(\omega)|$ is an even function of ω and its phase $\underline{/\mathbf{H}(\omega)}$ is an odd function. Hence, if we know $\mathbf{H}(\omega)$ for $\omega \geq 0$, we can construct it for $\omega \leq 0$ as well. Plots of $\mathbf{H}(\omega)$ as a function of ω for $\omega \geq 0$ are ***one-sided frequency responses***, whereas plots over $-\infty < \omega < \infty$ are ***two-sided frequency responses***. As we work with both one-sided and two-sided spectra in future sections, it is important to keep the implications of the conjugate-symmetry property in mind.

6-5.1 Brick-Wall Lowpass Filter

A brick-wall lowpass filter with ***cutoff frequency*** f_c exhibits the frequency response displayed in Fig. 6-20. That is,

$$\mathbf{H}_{\mathrm{LP}}(\omega) = \begin{cases} 1 & \text{for } |\omega| < 2\pi f_c, \\ 0 & \text{for } |\omega| > 2\pi f_c. \end{cases} \tag{6.56}$$

The name ***brick wall*** is associated with the abrupt change at $\omega = \pm 2\pi f_c$, between its passband and reject bands. As will be demonstrated in Example 6-5, a filter with such an abrupt response is physically unrealizable.

Example 6-5: Lowpass Filtering of a Square Wave

A square wave given by

$$x(t) = \begin{cases} \pi/4 & \text{for } 2n\pi < t < (2n+1)\pi, \\ -\pi/4 & \text{for } (2n-1)\pi < t < 2n\pi, \end{cases} \tag{6.57}$$

for all integer values of n is used as the input signal into a brick-wall lowpass filter with a cutoff frequency of 1 Hz. Determine (a) the output signal $y(t)$ and (b) the impulse response of the filter.

Solution:

(a) According to entry #2 in Table 5-4, a square wave of amplitude $A = \pi/4$ and period $T = 2\pi$ is given by the Fourier series

$$x(t) = \sin(t) + \frac{1}{3}\sin(3t) + \frac{1}{5}\sin(5t) + \frac{1}{7}\sin(7t) + \cdots, \tag{6.58}$$

which consists of components at frequencies

$$f_1 = \frac{1}{2\pi} = 0.16 \text{ Hz}, \qquad f_3 = \frac{3}{2\pi} = 0.48 \text{ Hz},$$
$$f_5 = \frac{5}{2\pi} = 0.80 \text{ Hz}, \qquad f_7 = \frac{7}{2\pi} = 1.11 \text{ Hz, etc.}$$

The brick-wall filter with a cutoff frequency $f_c = 1$ Hz will allow only the first three components to pass through. Hence,

$$y(t) = \sin(t) + \frac{1}{3}\sin(3t) + \frac{1}{5}\sin(5t). \tag{6.59}$$

(b) The impulse response $h(t)$ of any filter can be computed by taking the inverse Fourier transform of the filter's frequency response $\mathbf{H}(\omega)$. For a brick-wall lowpass filter with cutoff frequency f_c, application of Eq. (5.87b) to Eq. (6.56) leads to

$$\begin{aligned} h_{\mathrm{LP}}(t) &= \mathcal{F}^{-1}\{\mathbf{H}_{\mathrm{LP}}(\omega)\} \\ &= \frac{\sin(2\pi f_c t)}{\pi t} = 2f_c \operatorname{sinc}(2\pi f_c t), \end{aligned} \tag{6.60}$$

where the sinc function was defined by Eq. (5.79) as $\operatorname{sinc}(x) = [\sin(x)]/x$. A plot of the impulse response is displayed in Fig. 6-21.

► Because $h_{\mathrm{LP}}(t)$ is everlasting (exists over $-\infty$ to $+\infty$), it is noncausal, which means that the brick-wall lowpass filter cannot exist physically. ◄

However, *it is possible* to design realizable filters, such as the Butterworth filter of Section 6-8, that can closely approximate the brick-wall filter.

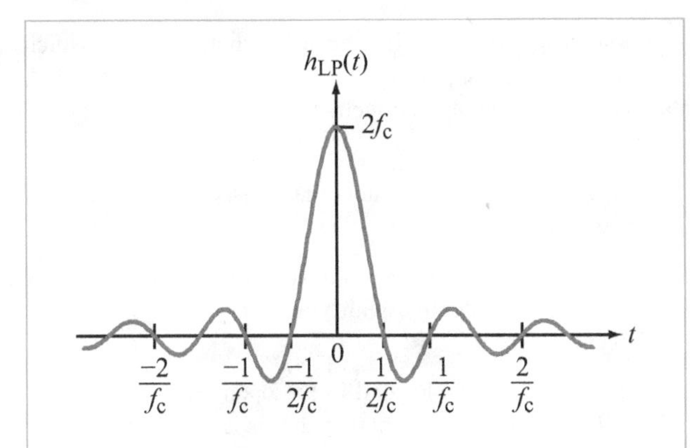

Figure 6-21: Impulse response of brick-wall lowpass filter with cutoff frequency f_c.

6-5.2 Brick-Wall Bandpass Filter

A brick-wall bandpass filter has cutoff frequencies f_{c_1} and f_{c_2} and exhibits the frequency response shown in Fig. 6-22. Its impulse response $h_{BP}(t)$ can be obtained by applying the ***modulation property*** of the Fourier transform (Table 5-7) which, when stated in terms of $h_{LP}(t)$ and its Fourier transform $\mathbf{H}_{LP}(\omega)$, takes the form

$$h_{LP}(t)\cos(\omega_0 t) \quad \longleftrightarrow \quad \frac{1}{2}[\mathbf{H}_{LP}(\omega-\omega_0)+\mathbf{H}_{LP}(\omega+\omega_0)]. \tag{6.61}$$

The bandpass filter spectrum is equivalent to the sum of two lowpass spectra with one shifted to the right by $2\pi f_0 = \pi(f_{c_2}+f_{c_1})$ and another shifted to the left by the same amount. Hence, if we define

$$f_0 = \frac{1}{2}(f_{c_2}+f_{c_1}) \tag{6.62a}$$

and

$$f_c = \frac{1}{2}(f_{c_2}-f_{c_1}), \tag{6.62b}$$

application of the modulation property—with $h_{LP}(t)$ as given by Eq. (6.60)—leads to

$$\begin{aligned} h_{BP}(t) &= 2\frac{\sin(2\pi f_c t)}{\pi t}\cos(2\pi f_0 t) \\ &= 2\frac{\sin[\pi(f_{c_2}-f_{c_1})t]}{\pi t}\cos[\pi(f_{c_2}+f_{c_1})t] \\ &= 2(f_{c_2}-f_{c_1})\operatorname{sinc}[\pi(f_{c_2}-f_{c_1})t]\cos[\pi(f_{c_2}+f_{c_1})t]. \end{aligned} \tag{6.63}$$

If $f_0 = 0$, the bandpass-filter frequency response collapses into the sum of two identical lowpass-filter responses. Accordingly, if we set $f_0 = 0$ in Eq. (6.63), the expression for $h_{BP}(t)$ reduces to $2h_{LP}(t)$.

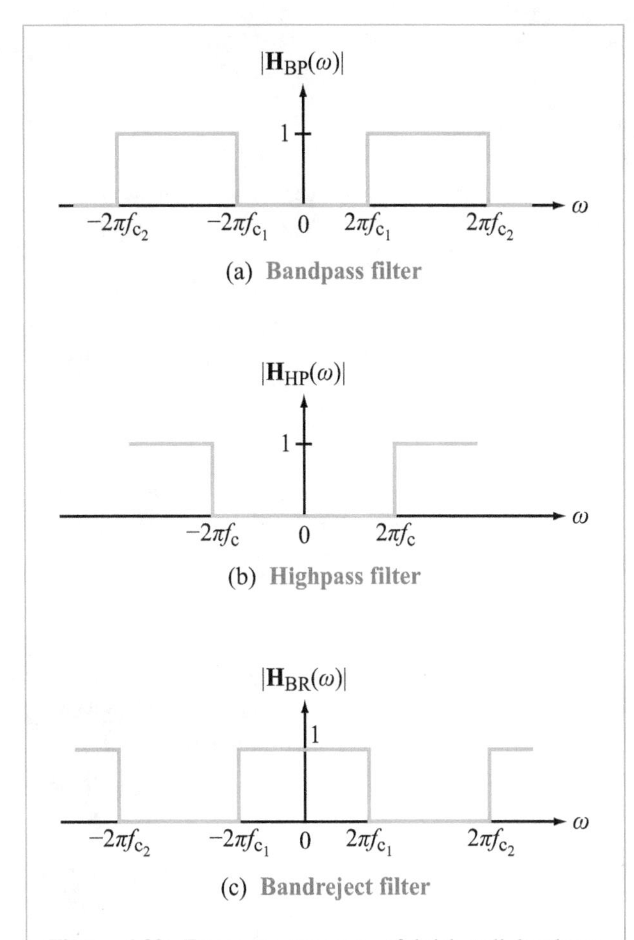

Figure 6-22: Frequency responses of brick-wall bandpass, highpass, and bandreject filters.

6-5.3 Brick-Wall Highpass Filter

The frequency response of the brick-wall highpass filter shown in Fig. 6-22(b) is related to that of the lowpass filter by

$$\mathbf{H}_{HP}(\omega) = 1 - \mathbf{H}_{LP}(\omega).$$

The inverse Fourier transform of 1 is $\delta(t)$. Hence,

$$h_{HP}(t) = \delta(t) - h_{LP}(t) = \delta(t) - 2f_c\operatorname{sinc}(2\pi f_c t).$$

6-5.4 Brick-Wall Bandreject Filter

Similarly, the frequency response of a bandreject filter is

$$\mathbf{H}_{BR}(\omega) = 1 - \mathbf{H}_{BP}(\omega)$$

and

$$h_{\mathrm{BR}}(t) = \delta(t) - h_{\mathrm{BP}}(t),$$

where $h_{\mathrm{BP}}(t)$ is given by Eq. (6.63).

Concept Question 6-5: How can we tell right away that the impulse response of a brick-wall lowpass filter must be noncausal? (See S²)

Concept Question 6-6: What property of the Fourier transform can be used to obtain the impulse response of a brick-wall bandpass filter from that of a brick-wall lowpass filter? (See S²)

Exercise 6-6: Determine the output from a brick-wall lowpass filter with a cutoff frequency of 0.2 Hz, given that the input is the square wave given by Eq. (6.57).

Answer: $y(t) = \sin(t)$. (See S²)

Exercise 6-7: Determine the output from a brick-wall bandpass filter with $f_{c_1} = 0.2$ Hz and $f_{c_2} = 1$ Hz, given that the input is the square wave given by Eq. (6.57).

Answer: $y(t) = \frac{1}{3}\sin 3(t) + \frac{1}{5}\sin 5(t)$. (See S²)

6-6 Filter Design by Poles and Zeros

▶ The frequency response of a filter is governed—within a multiplicative scaling constant—by the locations of the poles and zeros of its transfer function $\mathbf{H}(\mathbf{s})$. ◀

The $\mathbf{s}$-domain transfer function of an LTI filter with n poles $\{\mathbf{p}_i,\ i = 1, 2, \ldots, n\}$ and m zeros $\{\mathbf{z}_i,\ i = 1, 2, \ldots, m\}$ is given by Eq. (3.54b) as

$$\mathbf{H}(\mathbf{s}) = \mathbf{C}\,\frac{\prod_{i=1}^{m}(\mathbf{s} - \mathbf{z}_i)}{\prod_{i=1}^{n}(\mathbf{s} - \mathbf{p}_i)}, \tag{6.64}$$

where $\mathbf{C}$ is a constant. For sinusoidal signals, we replace $\mathbf{s}$ with $j\omega$ to obtain the frequency response:

$$\mathbf{H}(\omega) = \mathbf{H}(\mathbf{s})\big|_{\mathbf{s}=j\omega} = \mathbf{C}\,\frac{\prod_{i=1}^{m}(j\omega - \mathbf{z}_i)}{\prod_{i=1}^{n}(j\omega - \mathbf{p}_i)}. \tag{6.65}$$

As a complex function, $\mathbf{H}(\omega)$ is characterized by two frequency responses, a *magnitude response* (also called *gain*) $M(\omega)$ and a *phase response* $\phi(\omega)$. For the present, we will focus our attention on $M(\omega)$, primarily because it alone controls which frequencies of the input signal's spectrum are emphasized, attenuated, or eliminated altogether.

▶ We will consider only filters that satisfy the following two conditions:

(a) Poles always occur in conjugate pairs, and the same applies to zeros, thereby qualifying the system as a *real LTI system*.

(b) Poles are always located in the open left half-plane (OLHP) in the complex plane; this is a necessary and sufficient condition for *BIBO stability*. ◀

This and succeeding sections aim to address the following corollary questions:

(1) **Filter analysis:** How does the location of a pole or zero in the complex plane affect a filter's magnitude response $M(\omega)$?

(2) **Filter design:** Conversely, given a desired (ideal) frequency response $M_{\mathrm{ideal}}(\omega)$, how should the locations of the poles and zeros be selected so as to realize a filter with a frequency response $M(\omega)$ that closely approximates $M_{\mathrm{ideal}}(\omega)$?

The answer to the first question should provide the insight needed to develop design tools and approaches to address the second question (at least partially).

6-6.1 Single-Zero Transfer Function

By initially analyzing how the spectrum of $M(\omega)$ is influenced by the locations of a single pair of conjugate zeros and then repeating the process for a single pair of conjugate poles, we can extend the results to transfer functions $\mathbf{H}(\omega)$ with any number of poles and zeros.

Consider an LTI system with conjugate zeros given by

$$\{\mathbf{z}, \mathbf{z}^*\} = -\alpha \pm j\omega_0. \tag{6.66}$$

Both α and ω_0 are positive real numbers. In the complex plane (Fig. 6-23(a)), the conjugate zeros are symmetrically located in the OLHP at a distance α from the $j\omega$-axis. Application of Eq. (6.65) with $\mathbf{C} = 1$ gives

$$\begin{aligned} M(\omega) &= |\mathbf{H}(\omega)| \\ &= |j\omega - (-\alpha + j\omega_0)||j\omega - (-\alpha - j\omega_0)| = l_1 l_2, \end{aligned} \tag{6.67}$$

where l_1 and l_2 are the lengths of the two vectors from any point $(0, j\omega)$ to the two zeros at $(-\alpha, j\omega_0)$ and $(-\alpha, -j\omega_0)$. Since $\mathbf{C}$ is just a scaling factor, setting it equal to 1 has no

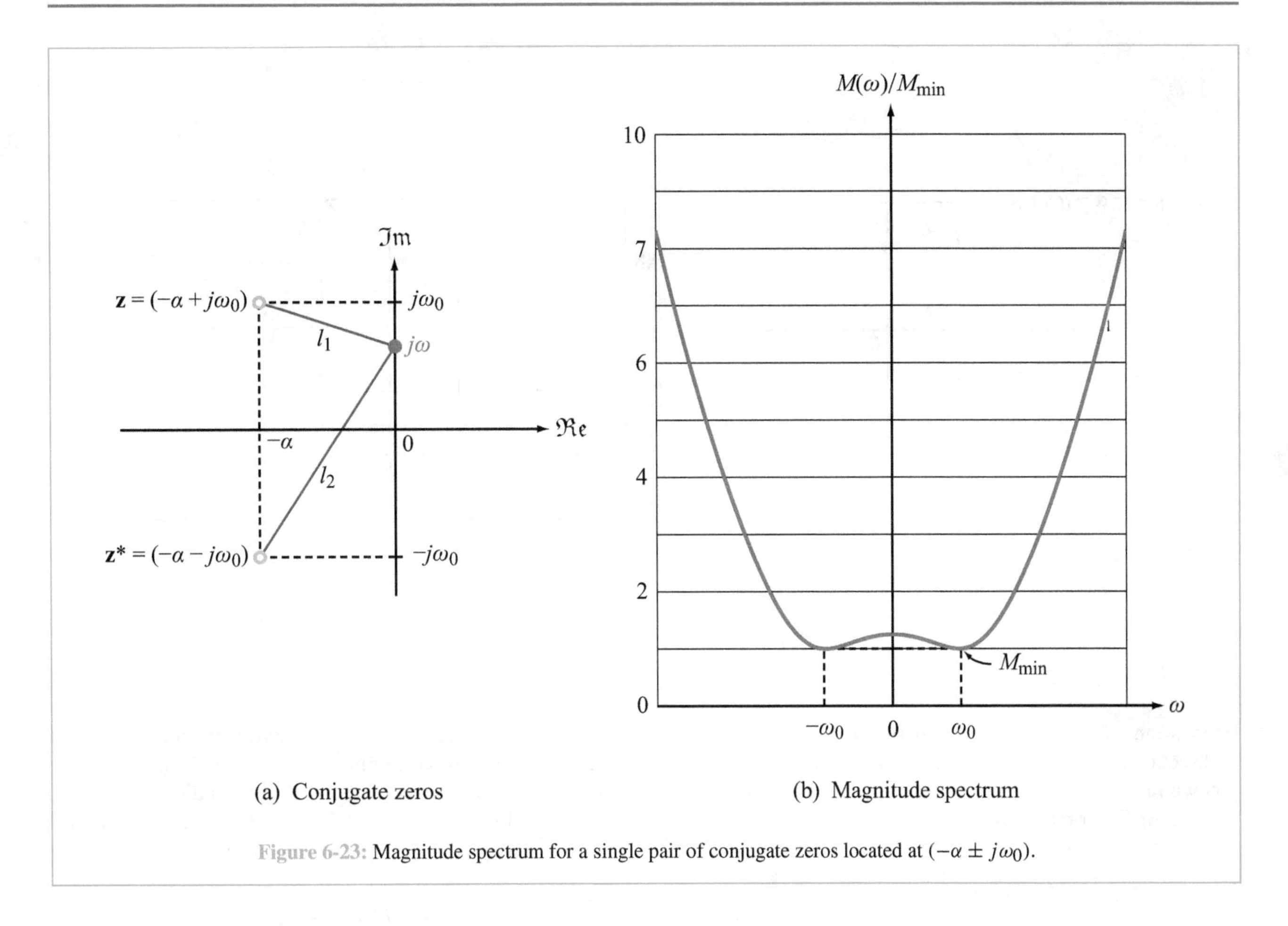

Figure 6-23: Magnitude spectrum for a single pair of conjugate zeros located at $(-\alpha \pm j\omega_0)$.

impact on the shape of the spectrum shown in Fig. 6-23(b). The spectrum exhibits minima at approximately $\omega = \pm\omega_0$, corresponding to the minimum value that l_1 can assume, namely α. As point $(0, j\omega)$ is moved above or below point $(0, j\omega_0)$, while remaining close to $(0, j\omega_0)$ on the $j\omega$-axis, l_1 changes appreciably, but l_2 does not. Consequently, the location of the dip at $\omega = \omega_0$ is governed primarily by the location of the zero in the upper half-plane, and the magnitude of the frequency response at that point is

$$M_{min} \approx M(\pm\omega_0) = \alpha(\alpha^2 + 4\omega_0^2)^{1/2}. \quad (6.68)$$

Since the zeros are conjugate pairs, the magnitude spectrum is symmetrical with respect to the vertical axis, with an identical minimum at $\omega = -\omega_0$.

Of particular interest in filter design is when α is zero or very small in absolute magnitude relative to ω_0, because then $M_{min} \approx 0$. This corresponds to locating the conjugate zeros on or very close to the $j\omega$-axis.

► Since the location of a zero does not impact system stability, α may be positive, negative, or zero. ◄

6-6.2 Single-Pole Transfer Function

Now we consider a transfer function composed of a single pair of conjugate poles given by

$$\{\mathbf{p}, \mathbf{p}^*\} = -\alpha \pm j\omega_0. \quad (6.69)$$

The corresponding magnitude response from Eq. (6.65) with $\mathbf{C} = 1$ is

$$M(\omega) = |\mathbf{H}(\omega)| = \frac{1}{|j\omega - (-\alpha + j\omega_0)|} \cdot \frac{1}{|j\omega - (-\alpha - j\omega_0)|} = \frac{1}{l_1 l_2}. \quad (6.70)$$

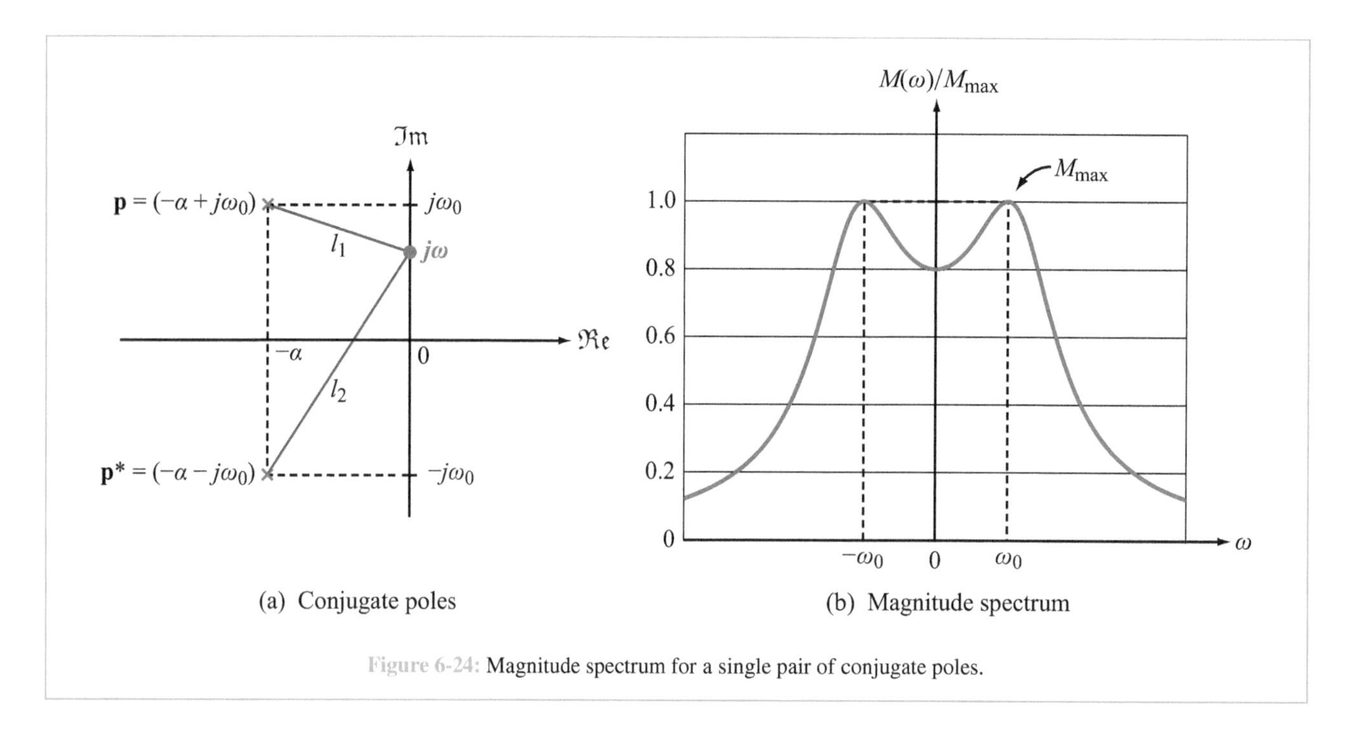

(a) Conjugate poles

(b) Magnitude spectrum

Figure 6-24: Magnitude spectrum for a single pair of conjugate poles.

Repetition of the analysis we performed in the preceding subsection for the pair of conjugate zeros leads to the spectrum shown in Fig. 6-24, where we observe peaks at approximately $\omega = \pm\omega_0$. Their magnitudes are

$$M_{\max} \approx M(\pm\omega_0) = \frac{1}{\alpha(\alpha^2 + 4\omega_0^2)^{1/2}} . \qquad (6.71)$$

A stable system cannot have poles on or to the right of the $j\omega$-axis. Hence, to generate a reasonably selective spectrum, α should be positive and very small relative to ω_0, placing the poles near the $j\omega$-axis in the OLHP.

▶ If a system has zeros $\{a_i \pm jb_i\}$ and poles $\{c_i \pm jd_i\}$, the magnitude of its frequency response, $M(\omega)$, will have peaks at $\omega = \pm d_i$ if $|c_i|$ is small, and dips at $\omega = \pm b_i$ if $|a_i|$ is small. ◀

Example 6-6: Draped-Like Spectrum

Given a filter with zeros $\mathbf{z}_i = \{\pm j2,\ 0.1 \pm j4\}$, poles $\mathbf{p}_i = \{-0.5 \pm j1, -0.5 \pm j3, -0.5 \pm j5\}$, and a scaling factor $\mathbf{C} = 1$, generate and plot its magnitude spectrum $M(\omega)$.

Solution: The locations of the two zero-pairs indicates that the spectrum will exhibit minima at $\omega = \pm 2$ rad/s and ± 4 rad/s. Similarly, the locations of the pole-pairs indicate maxima at $\omega = \pm 1$ rad/s, ± 3 rad/s and ± 5 rad/s. The complete expression for $M(\omega)$ is

$$\begin{aligned} M(\omega) &= |(j\omega - j2)(j\omega + j2)| \\ &\quad \times |(j\omega - 0.1 - j4)(j\omega - 0.1 + j4)| \\ &\quad \times \left| \frac{1}{(j\omega + 0.5 - j1)(j\omega + 0.5 + j1)} \right| \\ &\quad \times \left| \frac{1}{(j\omega + 0.5 - j3)(j\omega + 0.5 + j3)} \right| \\ &\quad \times \left| \frac{1}{(j\omega + 0.5 - j5)(j\omega + 0.5 + j5)} \right| . \end{aligned}$$

The calculated spectrum displayed in Fig. 6-25 resembles a wire draped over electrical poles at $\omega = \pm 1$ rad/s, $\omega = \pm 3$ rad/s, and $\omega = \pm 5$ rad/s and pinned to the ground at $\omega = \pm 2$ rad/s and $\omega = \pm 4$ rad/s. Hence, we use the nomenclature *poles* and *zeros*.

Concept Question 6-7: What effects do the locations of the poles and zeros of a system's transfer function $\mathbf{H}(\mathbf{s})$ have on the system's frequency response? (See S²)

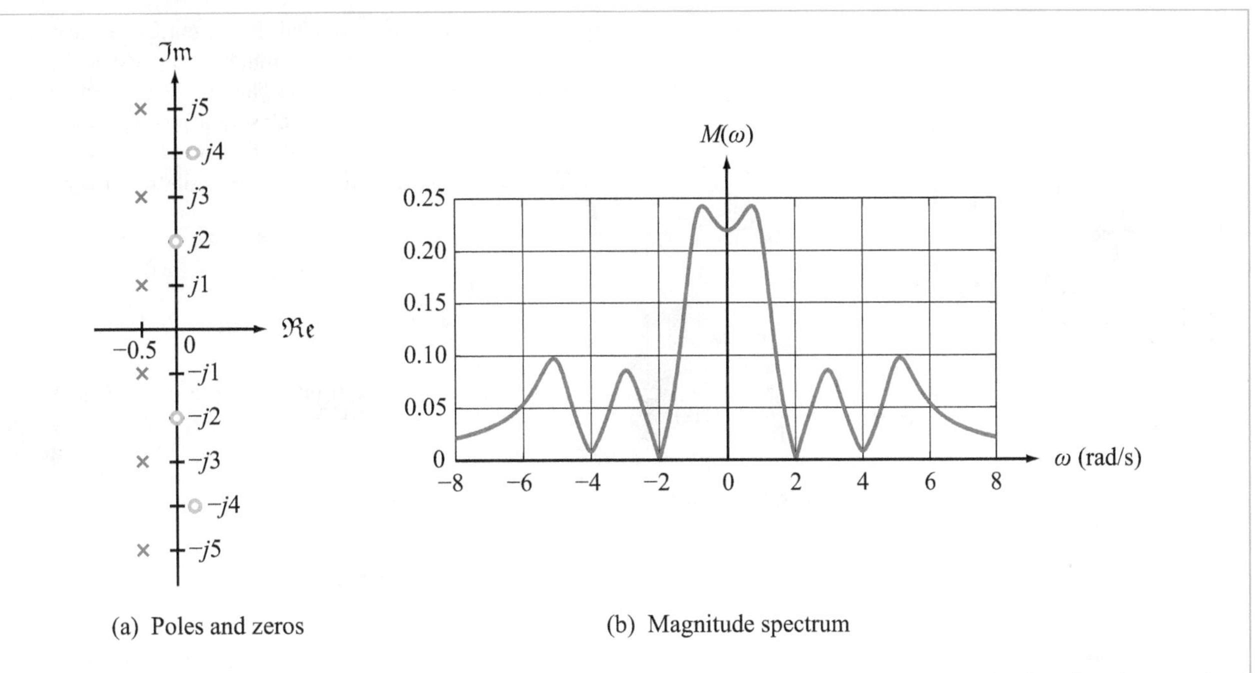

Figure 6-25: Magnitude spectrum of a frequency response composed of 2 pairs of conjugate zeros and 3 pairs of conjugate poles (Example 6-6).

Concept Question 6-8: Why is phase less important than magnitude when designing filters? (See S²)

Exercise 6-8: An LTI system has zeros at $\pm j3$. What sinusoidal signals will it eliminate?

Answer: $A\cos(3t+\theta)$ for any constant A and θ. (See S²)

Exercise 6-9: An LTI system has poles at $-0.1 \pm j4$. What sinusoidal signals will it emphasize?

Answer: $A\cos(4t+\theta)$ for any constant A and θ. (See S²)

6-7 Frequency Rejection Filters

Now that we understand the connection between the locations of poles and zeros in the complex plane and the corresponding general shape of the magnitude spectrum, we will briefly explore how to design filters for practical applications. The emphasis of the present section is on designing filters that *reject* one or more bands of frequencies while passing all others. In contrast, future sections will be oriented along the direction of *passing* signals in specific bands, while rejecting all others.

6-7.1 Notch Filters: Removing Sinusoidal Interference

Power lines outdoors and electrical wires in buildings often act like antennas, radiating 60 Hz sinusoids into their surroundings. Some of the radiated energy may get picked up by wires and conductors in electronic circuits, causing interference with the signals carried by those circuits. While proper circuit grounding can reduce the interference problem, it may still be necessary to block the 60 Hz interference from entering the circuit or to remove it if it has already superimposed itself onto the signal. If the signal spectrum is concentrated at frequencies well below or well above 60 Hz, we can use a lowpass or highpass filter, respectively, to remove the 60 Hz interference, but most realistic signal spectra have frequency components on both sides of 60 Hz. In that case we seek a bandreject filter with a *very narrow* bandstop centered at 60 Hz, as shown in Fig. 6-26. Because its magnitude spectrum exhibits a narrow notch at 60 Hz, it is called a 60 Hz ***notch filter***. The ideal transfer function of a notch filter

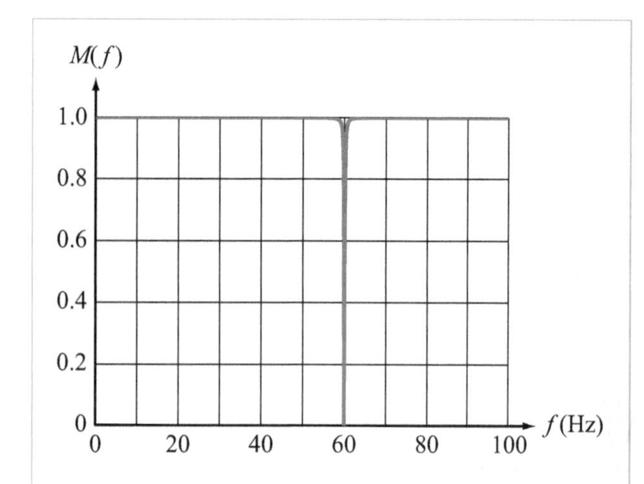

Figure 6-26: Magnitude spectrum of notch filter for removing 60 Hz interference.

is

$$\mathbf{H}_{\text{ideal}}(\omega) = \begin{cases} 1 & \text{for } \omega \neq \pm\omega_0, \\ 0 & \text{for } \omega = \pm\omega_0, \end{cases} \quad (6.72)$$

where $\omega_0 = 2\pi f_0$ and f_0 is the notch frequency.

(a) First attempt: **Zeros only**

To eliminate a sinusoidal input signal at an angular frequency ω_0, the filter's transfer function should have conjugate zeros at $\pm j\omega_0$. In our first design attempt, we will construct a simplistic filter with zeros at $\pm j\omega_0$ (but no other poles or zeros) and we shall denote its transfer function $\mathbf{H}_{\text{zero}}(\omega)$. With $\mathbf{C} = 1$, Eq. (6.65) becomes

$$\mathbf{H}_{\text{zero}}(\omega) = (j\omega - j\omega_0)(j\omega + j\omega_0) = \omega_0^2 - \omega^2. \quad (6.73)$$

From its magnitude spectrum displayed in Fig. 6-27(a), we observe that the filter indeed rejects signals at $\omega = \pm\omega_0$, but the response is anything but constant at other frequencies. In fact $M_{\text{zero}}(\omega) \to \infty$ as $|\omega| \to \infty$. Clearly, this is not a viable notch filter.

(b) Second attempt: **Notch filter with parallel poles and zeros**

In addition to exhibiting nulls at $\omega = \pm\omega_0$, we would like the magnitude $M(\omega)$ to be a constant (flat) at all other frequencies. This can be achieved by placing poles in parallel with and very close to the zeros. To the zero at location $(0, j\omega_0)$ in the complex plane, we add a parallel pole at location $(-\alpha, j\omega_0)$, as shown in Fig. 6-27(b), with α chosen to be positive (so the pole is in the OLHP) and small in magnitude relative to ω_0. In Fig. 6-27(b), $\alpha = 0.1\omega_0$. A similar pole placement is made next to the zero at $(0, -j\omega_0)$. The significance of the ratio α/ω_0, which determines how closely a pole is located to its neighboring zero, will be discussed later in this subsection.

The transfer function of this parallel pole/zero filter is

$$\mathbf{H}_{\text{notch}}(\omega) = \frac{(j\omega - j\omega_0)(j\omega + j\omega_0)}{(j\omega + \alpha - j\omega_0)(j\omega + \alpha + j\omega_0)}. \quad (6.74)$$

The magnitude spectrum $M_{\text{notch}}(\omega)$ is displayed in Fig. 6-27(b) for $\alpha = 0.1\omega_0$. Not surprisingly, $M_{\text{notch}}(\omega)$ does exhibit nulls at $\omega = \pm\omega_0$, but away from the immediate vicinities of $\omega = \pm\omega_0$, the spectrum is essentially flat. This is due to the fact that except in the immediate vicinities of $\omega = \pm\omega_0$, the ratio of each zero/pole magnitude combination is approximately 1, so long as $\alpha \ll \omega_0$.

► In essence, each pole *cancels* the contribution of its parallel zero, except at or near $\omega = \pm\omega_0$, thereby generating the desired notch-filter spectrum. ◄

(c) The role of α

The selectivity of the notch-filter spectrum is dictated by the parameter

$$\beta = \tan^{-1}\left(\frac{\alpha}{\omega_0}\right), \quad (6.75)$$

where β is the angle shown in Fig. 6-28(a). To appreciate the role of β, we show in Fig. 6-28(b) magnitude spectra for several values of β from which it is evident that to generate a highly selective notch filter (very narrow notch) β should be made as small as possible. Since ω_0 is the angular frequency of the interfering signal to be removed by the notch filter, it is not a selectable parameter, but α is. Hence, *from the standpoint of spectral selectivity, α should be selected so that $\alpha/\omega_0 \ll 1$.*

However, α plays another, equally important role that also should be taken into consideration. In fact, this second role, which has to do with the transient component of the impulse response of the filter, favors a value for α large enough to insure that the transient component of the output response decays quickly after the introduction of the input signal (see Section 3-11).

To evaluate the general nature of the transient response to an input signal, we obtain the transfer function of the notch filter

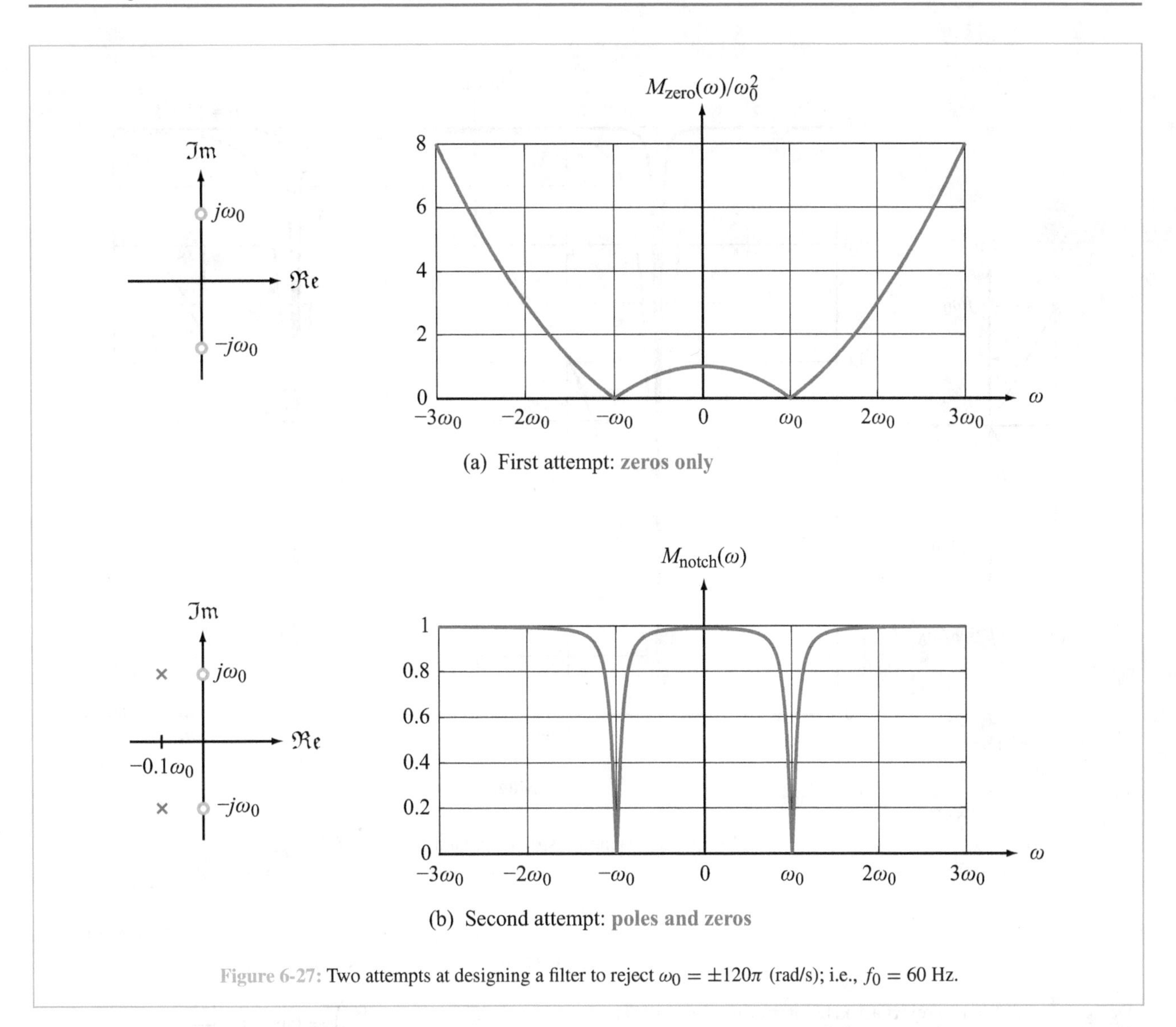

Figure 6-27: Two attempts at designing a filter to reject $\omega_0 = \pm 120\pi$ (rad/s); i.e., $f_0 = 60$ Hz.

by replacing $j\omega$ with $\mathbf{s}$ in Eq. (6.74),

$$\begin{aligned}
\mathbf{H}_{\text{notch}}(\mathbf{s}) &= \mathbf{H}_{\text{notch}}(\omega)|_{j\omega=\mathbf{s}} \\
&= \frac{(\mathbf{s} - j\omega_0)(\mathbf{s} + j\omega_0)}{(\mathbf{s} + \alpha - j\omega_0)(\mathbf{s} + \alpha + j\omega_0)} \\
&= \frac{\mathbf{s}^2 + \omega_0^2}{(\mathbf{s} + \alpha)^2 + \omega_0^2} \\
&= \frac{\mathbf{s}^2 + 2\alpha\mathbf{s} + \alpha^2 + \omega_0^2}{(\mathbf{s} + \alpha)^2 + \omega_0^2} - \frac{(2\alpha\mathbf{s} + \alpha^2)}{(\mathbf{s} + \alpha)^2 + \omega_0^2} \\
&= 1 - \frac{(2\alpha\mathbf{s} + \alpha^2)}{(\mathbf{s} + \alpha)^2 + \omega_0^2}, \qquad (6.76)
\end{aligned}$$

where in the last step, we applied the division relationship given by Eq. (3.55) to convert $\mathbf{H}(\mathbf{s})$ from a ***proper rational function*** (numerator and denominator of the same degree) into a form $\mathbf{H}_{\text{notch}} = 1 + \mathbf{G}(\mathbf{s})$, where $\mathbf{G}(\mathbf{s})$ is a ***strictly*** proper rational function (degree of numerator smaller than that of denominator). Application of partial fraction expansion followed by transformation to the time domain leads to the impulse response:

$$h_{\text{notch}}(t) = \delta(t) - Ae^{-\alpha t}\cos(\omega_0 t + \theta)\,u(t) \qquad (6.77)$$

with

$$A = \left[4\alpha^2 + \frac{\alpha^4}{\omega_0^2}\right]^{1/2}, \quad \text{and} \quad \theta = \tan^{-1}\left(\frac{\alpha}{2\omega_0}\right). \qquad (6.78)$$

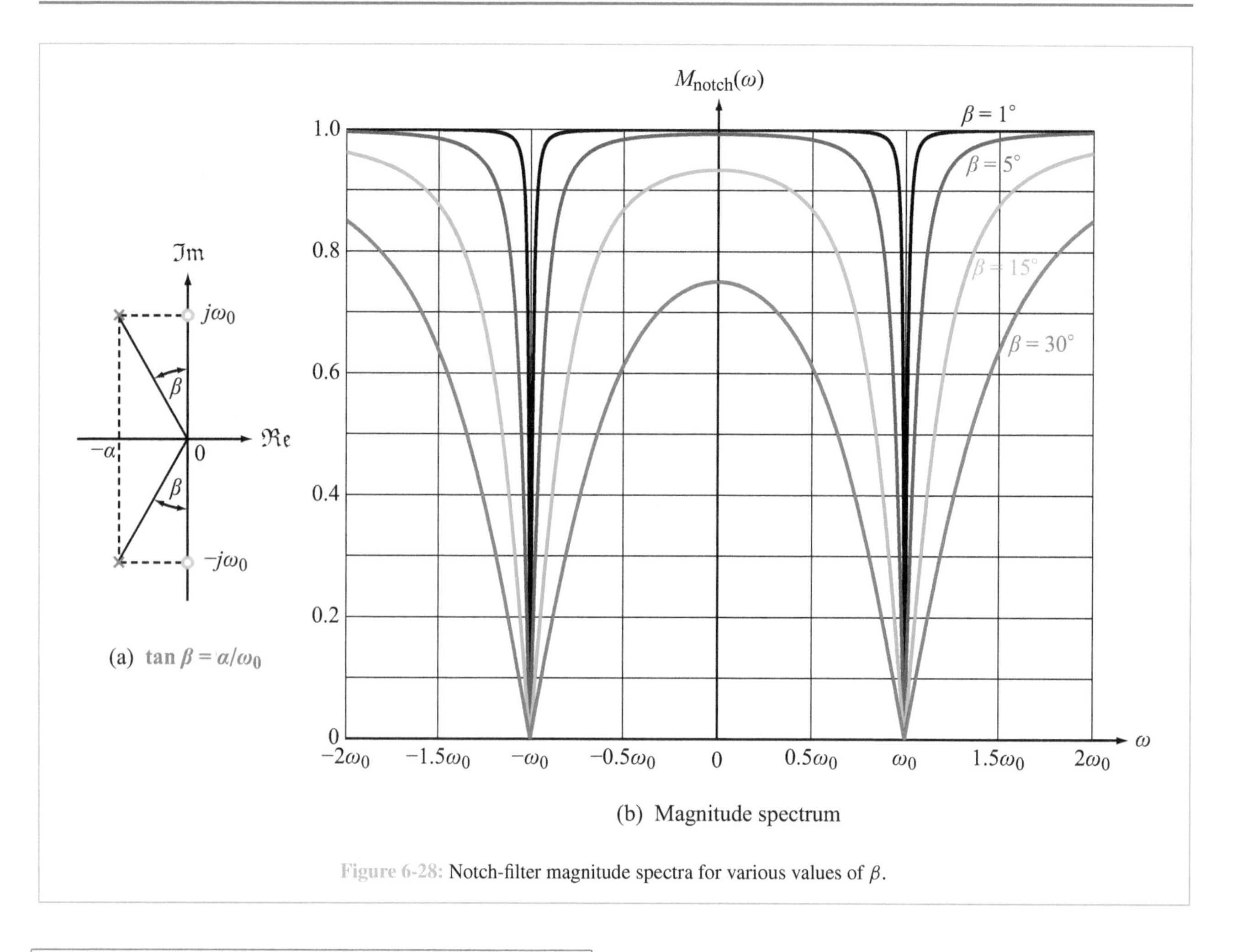

Figure 6-28: Notch-filter magnitude spectra for various values of β.

Example 6-7: Rejecting an Interfering Tone

Design a notch filter to reject a 1 kHz interfering sinusoid. The filter's impulse response should decay to less than 0.01 within 0.1 s.

Solution: At 1 kHz,

$$\omega_0 = 2\pi f_0 = 2\pi \times 10^3 = 6283 \text{ rad/s}.$$

To satisfy the impulse-response decay requirement, the value of α should be such that the coefficient of the cosine term in Eq. (6.77) is no larger than 0.01 at $t = 0.1$ s. That is,

$$\left(4\alpha^2 + \frac{\alpha^4}{\omega_0^2}\right)^{1/2} e^{-0.1\alpha} \leq 0.01.$$

Solution of this inequality (using MATLAB or MathScript) leads to $\alpha \approx 100 \text{ s}^{-1}$ and, in turn, to

$$\beta = \tan^{-1}\left(\frac{\alpha}{\omega_0}\right) = \tan^{-1}\left(\frac{100}{6283}\right) = 0.9°.$$

With both ω_0 and α specified, we now can calculate and plot the impulse response given by Eq. (6.77) (excluding the impulse function), as shown in Fig. 6-29.

Note that although the maximum value of $h_{notch}(t)$, excluding the impulse, is 200, $|h_{notch}(0.1)| = 0.009$, which is smaller than the specified value of 0.01. The magnitude spectrum of this notch filter is essentially the $\beta = 1°$-plot shown in Fig. 6-28(b).

Concept Question 6-9: Why do we need poles in a notch filter? Why not just use zeros only? (See S²)

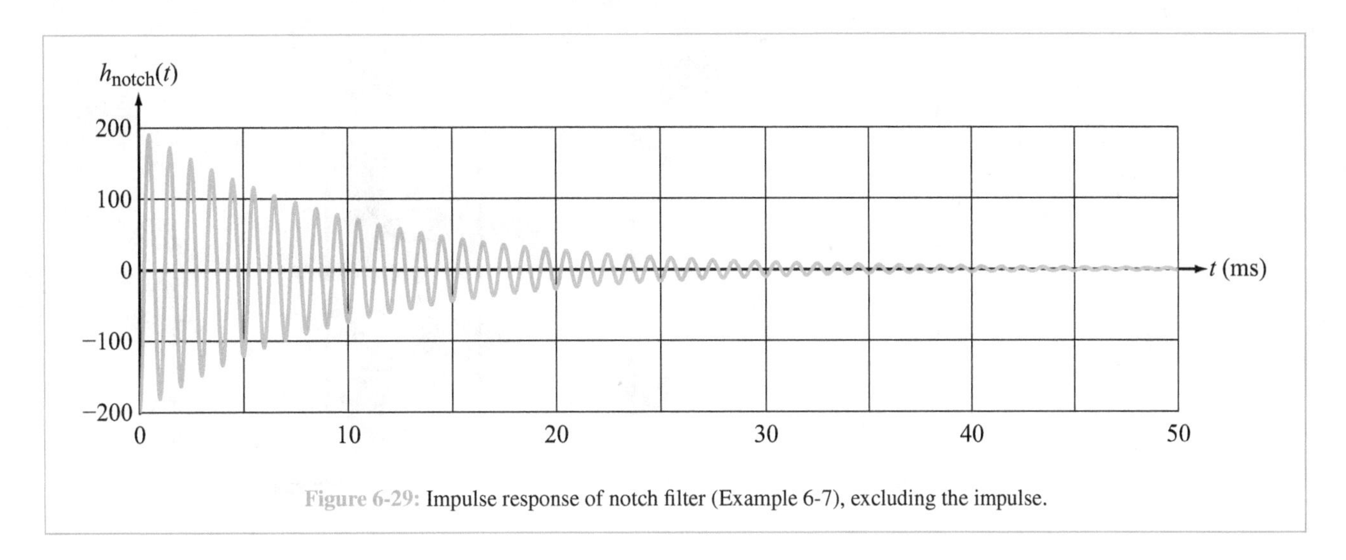

Figure 6-29: Impulse response of notch filter (Example 6-7), excluding the impulse.

Concept Question 6-10: Why is it not advisable in practice to select the real part of the notch filter's poles to be extremely small? (See S²)

Exercise 6-10: Design (specify the transfer function of) a notch filter to reject a 50 Hz sinusoid. The filter's impulse response must decay to 0.005 within 6 seconds.

Answer:

$$\mathbf{H(s)} = 1 - \frac{2\mathbf{s}+1}{\mathbf{s}^2+2\mathbf{s}+98697}.$$

(See S²)

6-7.2 Comb Filters: Removing Periodic Interference

Motors, generators, and air conditioners are examples of systems that generate electromagnetic interference. Often, the interference is periodic, occurring at a fundamental frequency f_0 and its many harmonics. In many cases of practical concern, the amplitude of a harmonic is inversely proportional to its harmonic number, so to remove most of the interference, we need to implement a series connection of notch filters with notches at the fundamental frequency ω_0 and a few of its harmonics. Such a series connection of notch filters is called a ***comb filter*** because its magnitude spectrum resembles the teeth of a comb. An example is shown in Fig. 6-30.

▶ Each notch in the comb filter's spectrum corresponds to a conjugate pair of zeros in parallel with a conjugate pair of poles. ◀

For notches at $k\omega_0$, with $k = 1, 2, \ldots, n$, $\mathbf{H}(\omega)$ of Eq. (6.65) assumes the form

$$\mathbf{H}_{\text{comb}}(\omega) = \prod_{\substack{k=-n \\ k\neq 0}}^{n} \left[\frac{j\omega - jk\omega_0}{j\omega + \alpha - jk\omega_0}\right] \tag{6.79}$$

for $\mathbf{C} = 1$.

Example 6-8: Comb Filter Design

Design a comb filter to reject a 1 kHz interfering signal and its second harmonic. Generate plots of (a) the filter's frequency response and (b) impulse response (excluding the impulse). Use $\alpha = 100\ \text{s}^{-1}$.

Solution:

(a) To filter out $\omega_0 = 2\pi f_0 = 6283$ rad/s and its harmonic $2\omega_0 = 12566$ rad/s, we place:

- zeros at $\{\pm j6283,\ \pm 12566\}$, and parallel
- poles at $\{-100 \pm j6283,\ -100 \pm j12566\}$.

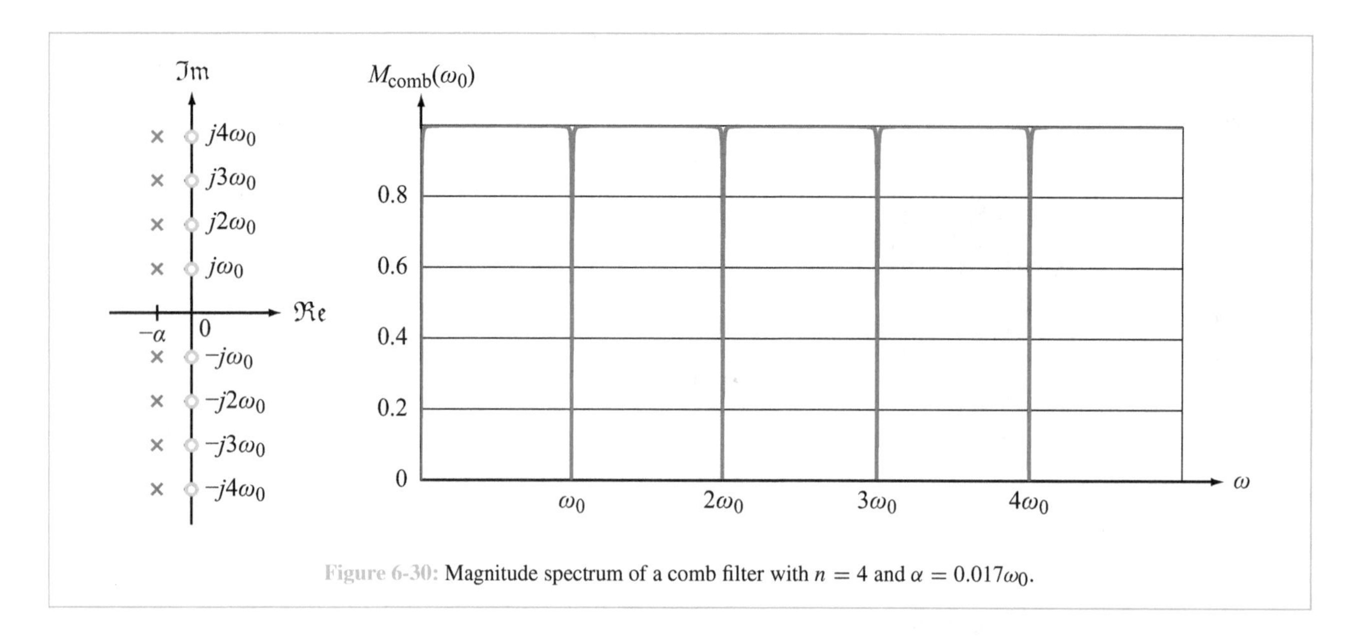

Figure 6-30: Magnitude spectrum of a comb filter with $n = 4$ and $\alpha = 0.017\omega_0$.

Implementation of Eq. (6.79) with $n = 2$ leads to

$$\mathbf{H}_{\text{comb}}(\omega) = \left(\frac{\omega_0^2 - \omega^2}{\omega_0^2 + \alpha^2 - \omega^2 + j2\alpha\omega}\right) \cdot \left(\frac{4\omega_0^2 - \omega^2}{4\omega_0^2 + \alpha^2 - \omega^2 + j2\alpha\omega}\right) \quad (6.80)$$

with $\omega_0 = 6283$ rad/s and $\alpha = 100\text{ s}^{-1}$. A plot of $M_{\text{comb}}(\omega) = |\mathbf{H}_{\text{comb}}(\omega)|$ is displayed in Fig. 6-31(a).

(b) Generating the impulse response of the comb filter requires a straightforward (but rather lengthy) process. Hence, we will only outline its steps.

Step 1: The expression for $\mathbf{H}_{\text{comb}}(\omega)$ should be converted into $\mathbf{H}_{\text{comb}}(\mathbf{s})$ by replacing $j\omega$ with $\mathbf{s}$ everywhere in Eq. (6.80).

Step 2: Because $\mathbf{H}_{\text{comb}}(\mathbf{s})$ is a proper, but not a strictly proper rational function, a step of long division is necessary to convert it into the form $\mathbf{H}_{\text{comb}}(\mathbf{s}) = 1 + \mathbf{G}(\mathbf{s})$.

Step 3: Partial fraction expansion should be applied to prepare $\mathbf{G}(\mathbf{s})$ for transformation to the time domain.

Step 4: Transformation of $\mathbf{H}_{\text{comb}}(\mathbf{s})$ to the time domain yields $h(t)$, which is displayed in Fig. 6-31(b).

Concept Question 6-11: How are comb and notch filters related? (See S²)

Concept Question 6-12: Why do comb filters have that name? (See S²)

Exercise 6-11: Design a comb filter to eliminate periodic interference with period = 1 ms. Assume that harmonics above 2 kHz are negligible. Use $\alpha = 100\text{ s}^{-1}$.

Answer:

$$\mathbf{H}(\mathbf{s}) = \frac{\mathbf{s}^2 + (2000\pi)^2}{\mathbf{s}^2 + 200\mathbf{s} + 10^4 + (2000\pi)^2} \times \frac{\mathbf{s}^2 + (4000\pi)^2}{\mathbf{s}^2 + 200\mathbf{s} + 10^4 + (4000\pi)^2}.$$

(See S²)

6-7.3 Rejecting Bands of Frequencies

Next, we consider how to configure a filter to reject input signals if their angular frequencies are within the range $\omega_L < \omega < \omega_H$, while passing all others. The ideal frequency response of such a filter is displayed in Fig. 6-32.

A reasonable approximation to the ideal response can be realized by extending the results of the single-frequency notch filter to a continuous multi-frequency scenario. Over the range from ω_L to ω_H, we place a string of closely spaced zeros along the $j\omega$-axis, together with a parallel string of poles. Of course, a similar arrangement is needed in the lower-half plane. The process is illustrated in Fig. 6-33 for a stopband extending from 3 rad/s to 5 rad/s. Five equally spaced zeros and parallel poles are placed across the designated stopband, with the poles displaced

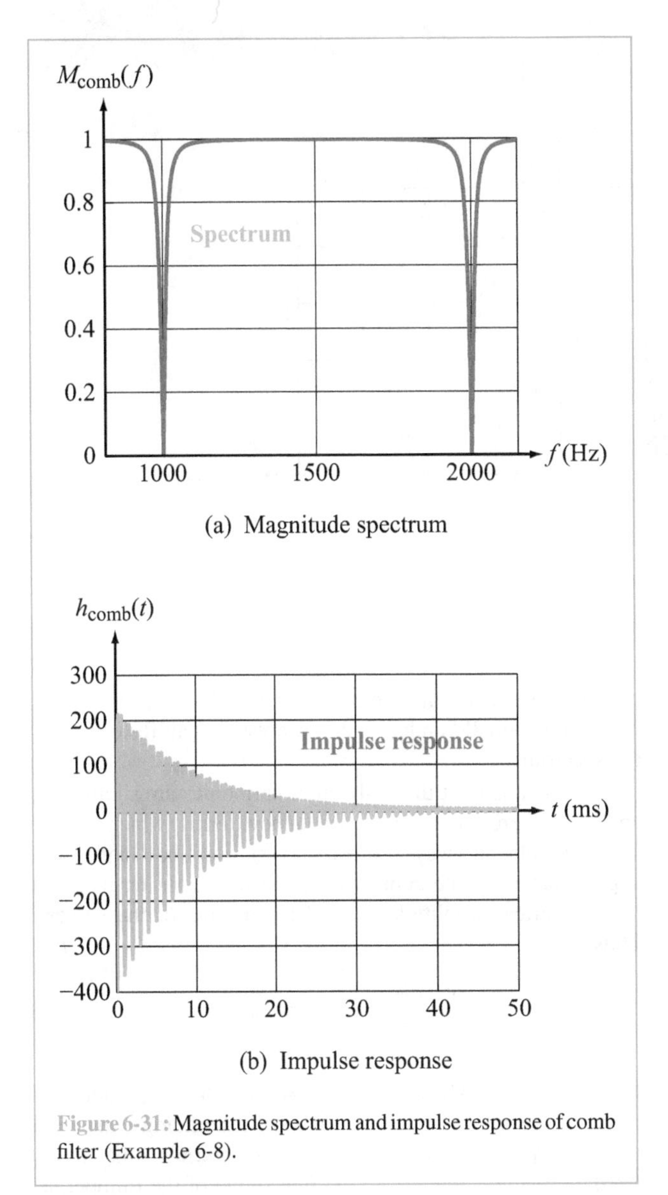

Figure 6-31: Magnitude spectrum and impulse response of comb filter (Example 6-8).

from the $j\omega$-axis by a distance $\alpha = 0.5$. The general shape of the magnitude spectrum does resemble the desired bandreject response, but multiple ripples are observed in the stopbands.

Once ω_L and ω_H have been specified, the filter designer has only two parameters to work with, namely α and n, where n is the number of zero/pole conjugate pairs to be placed over the specified stopband.

► Increasing n leads to better selectivity (steeper roll-off at the stopband/passband boundaries) and lower ripple amplitude in the stopband. In practice, this translates into more cost and complexity. ◄

For a fixed value of n, the value of α is selected in a trade-off between two competing attributes, selectivity and ripple amplitude. Selectivity is inversely proportional to α, and so is ripple amplitude.

6-8 Spectra of Musical Notes

When a musical instrument such as a trumpet or clarinet plays a midrange musical note B, it produces a periodic acoustic signal with a fundamental frequency $f_0 = 494$ Hz, corresponding to a period $T = 1/f_0 \approx 2.0$ ms.

The time waveform of an actual trumpet playing midrange note B is plotted in Fig. 6-34(a). The vertical axis is the intensity of the sound, and the horizontal axis is time in milliseconds. The signal is clearly periodic with a period of about 2 ms. We will refer to this sound as the trumpet signal $x(t)$.

The computed magnitude spectrum of the trumpet signal, $M(\omega)$, is plotted in Fig. 6-34(b). We will learn how to compute this spectrum in Chapter 8. The computed magnitude spectrum is a line spectrum, as the spectrum of a periodic signal should be. However, the trumpet signal is not perfectly periodic, since a human trumpet player cannot repeat the note perfectly. Consequently, the spectrum in Fig. 6-34(b) is not a perfect

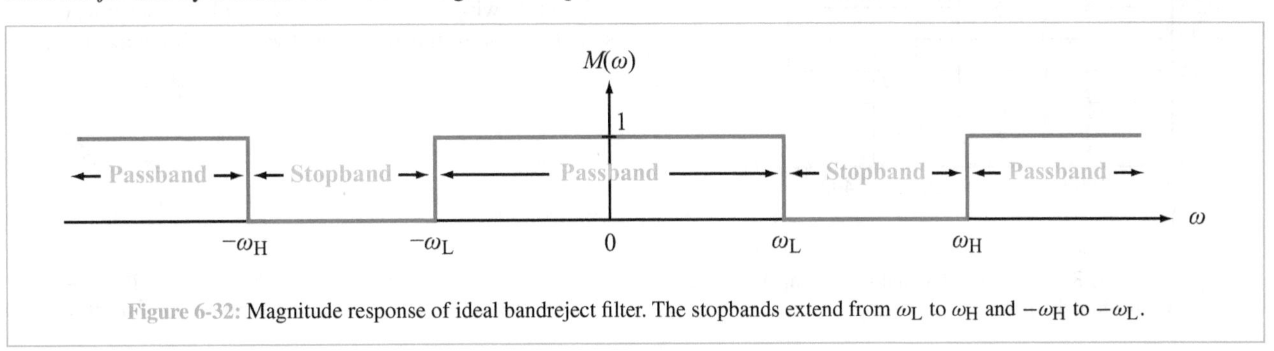

Figure 6-32: Magnitude response of ideal bandreject filter. The stopbands extend from ω_L to ω_H and $-\omega_H$ to $-\omega_L$.

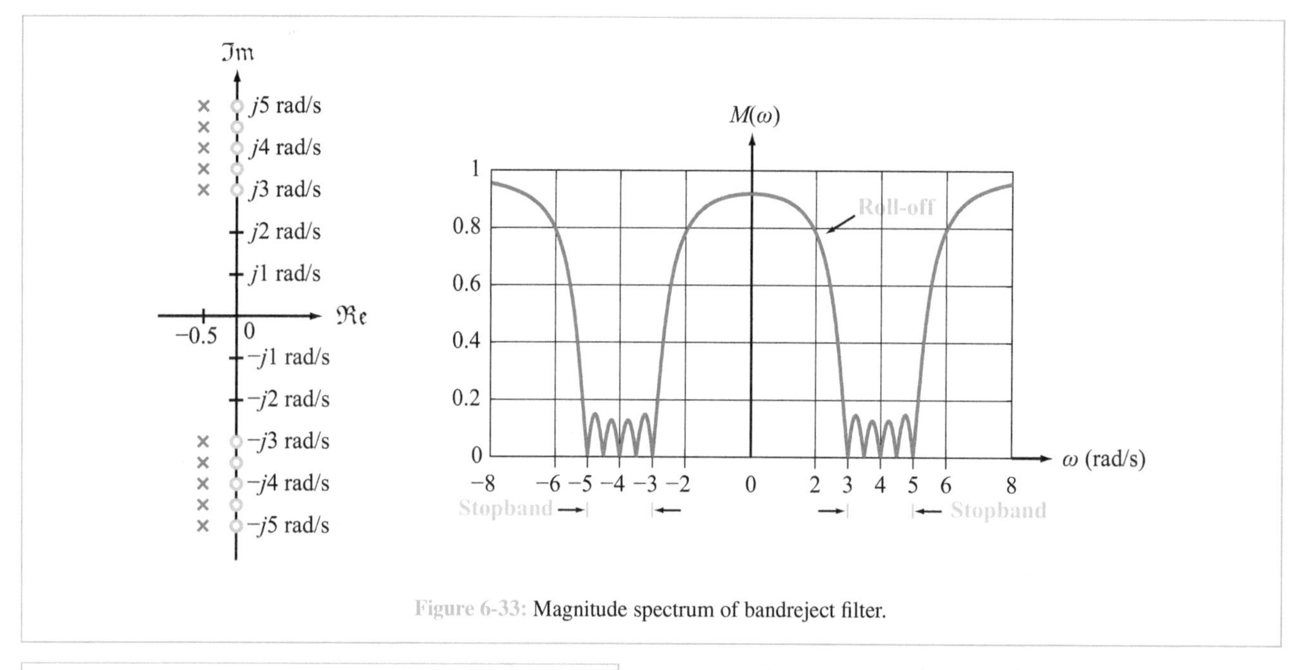

Figure 6-33: Magnitude spectrum of bandreject filter.

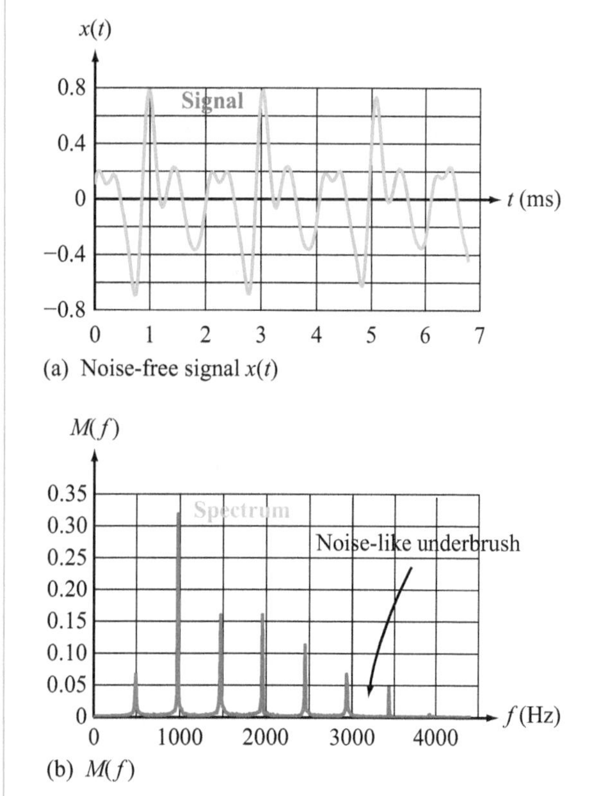

Figure 6-34: (a) Recorded sound signal of a trumpet playing note B; (b) magnitude spectrum.

one-side line spectrum: the peaks have a "base" several hertz wide, and a small "underbrush" of noise is manifested across the spectrum.

It is also apparent that only eight harmonics are significant, and that there is no zero frequency term (which would be inaudible anyway). Hence, $x(t)$ can be reasonably approximated by the sum of only the first eight terms in its Fourier series (amplitudes c_n are the heights of the spikes). Thus,

$$x(t) = \sum_{n=1}^{8} c_n \cos(2\pi n f_0 t + \phi_n), \tag{6.81}$$

where $f_0 = 494$ Hz and c_n and ϕ_n are the amplitude and phase of the nth harmonic. In musical terminology, the term $c_1 \cos(2\pi f_0 t + \phi_1)$ is the ***fundamental***, and the other terms are the ***overtones*** that create the rich sound of the trumpet, as compared with a simple tone.

The following example uses a comb filter to separate out the sounds of two trumpets playing two notes simultaneously.

Example 6-9: Separating Two Simultaneously Played Trumpet Notes

Given signals $x_1(t)$ and $x_2(t)$ of two trumpets playing notes G and A simultaneously, design a comb filter that eliminates

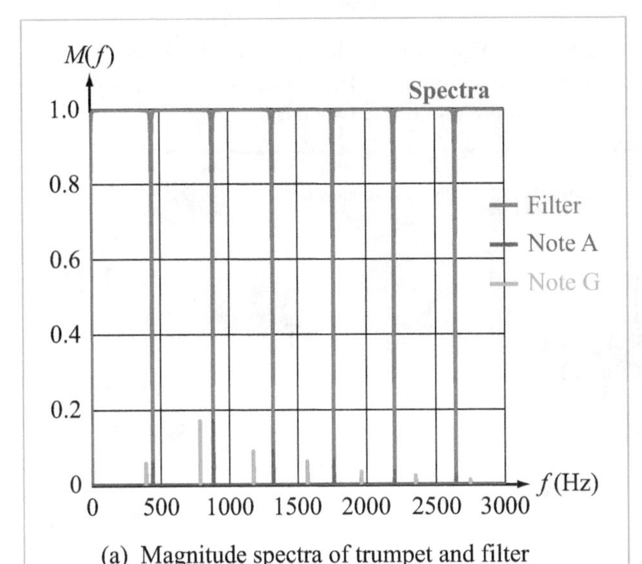

(a) Magnitude spectra of trumpet and filter

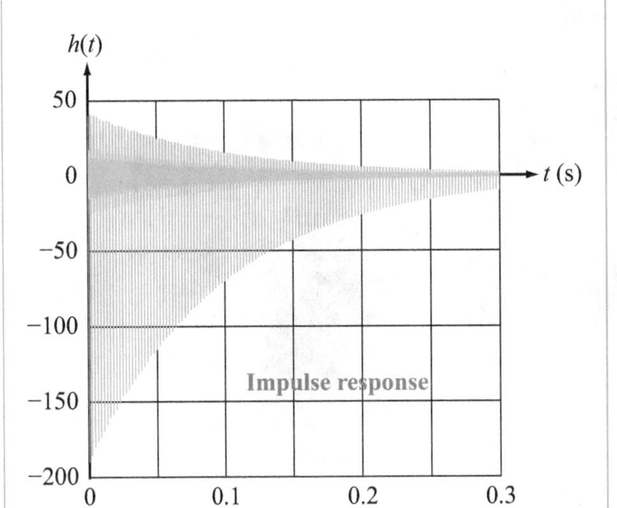

(b) Impulse response

Figure 6-35: (a) Magnitude spectra of trumpet signals and comb filter; (b) impulse response of comb filter (Example 6-9).

the trumpet playing note A, while keeping the trumpet playing note G. Compute and plot the comb filter's impulse and frequency responses. Choose $\alpha = 10\ \text{s}^{-1}$ and $n = 9$. Notes G and A have fundamental frequencies of 392 Hz and 440 Hz, respectively.

Solution: The combined spectrum of $x_1(t)$ and $x_2(t)$ consists of harmonics of 392 Hz and 440 Hz. The frequency response of the desired comb filter is given by Eq. (6.79), with $\omega_0 = 2\pi \times 440 = 2765$ rad/s, $\alpha = 10\ \text{s}^{-1}$, and $n = 9$. MATLAB or MathScript implementation leads to the spectrum displayed in Fig. 6-35(a), and repetition of the procedure outlined in Example 6-8 leads to the impulse response shown in Fig. 6-35(b). The comb filter clearly eliminates the harmonics at 440 Hz, while preserving the harmonics at 392 Hz (note G) (see S² for more details).

6-9 Butterworth Filters

We now switch our emphasis from designing filters that reject bands of frequencies to those that pass bands of frequencies. We begin by designing realizable, near-ideal, lowpass filters. We will then extend our treatment to encompass near-ideal highpass and bandpass filters.

The ideal lowpass filter has the frequency response shown in Fig. 6-36. However, as we had noted earlier in Section 6-4.1, its impulse response is noncausal, and therefore the filter is unrealizable.

► The *Butterworth lowpass filter* is a BIBO-stable and causal LTI system with a frequency response that approximates that of the ideal lowpass filter (Fig. 6-36). ◄

Its design implementation relies on arranging the poles of its transfer function along the perimeter of a semicircle in the complex plane. To gain an appreciation for the advantages offered by such an arrangement, we will precede our presentation of the Butterworth filter with analyses of two elementary arrangements based on the lessons learned from earlier sections.

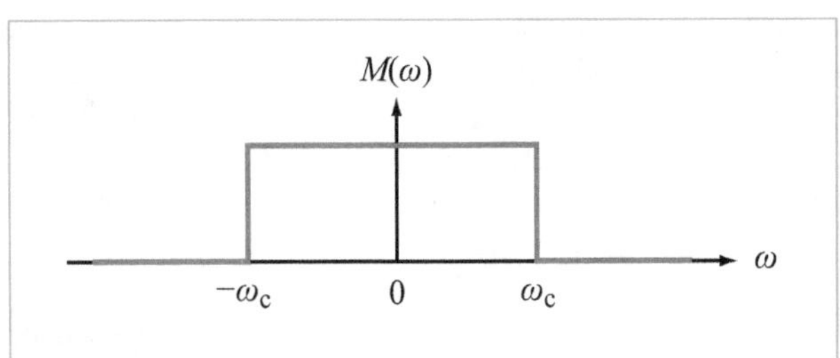

Figure 6-36: Magnitude response of ideal lowpass filter with cutoff frequency ω_c.

Module 6.1 Notch Filter to Remove Sinusoid from Trumpet Signal This module adds a sinusoid to the waveform of a trumpet playing note B. The amplitude and frequency of the sinusoidal interference are selectable, as are the parameters of the notch filter. The module also allows the user to listen to the original, noisy, and filtered signals.

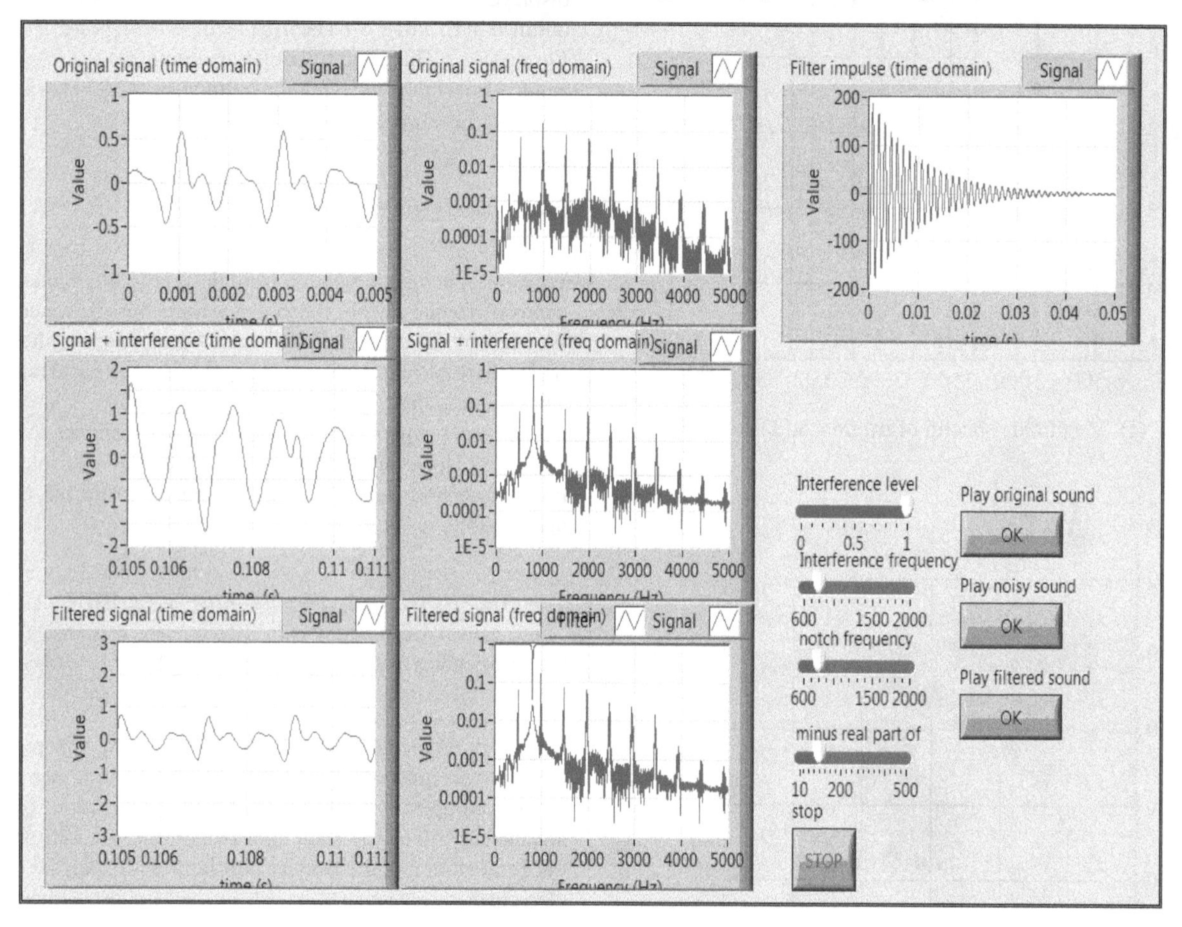

6-9.1 Elementary Arrangement 1: **Vertical String of Poles**

A filter can be viewed either in terms of its passbands or of its stopbands. A lowpass filter with a passband extending from $-\omega_c$ to ω_c is also a rejection filter with stopbands defined by $\omega > \omega_c$ and $\omega < -\omega_c$. If we were to approach the ideal lowpass filter as a rejection filter, we would need to place an infinite number of zeros along the $j\omega$-axis, extending from $j\omega_c$ to $j\infty$, and from $-j\omega_c$ to $-j\infty$. Obviously, that is not a practical solution.

Alternatively, we can approach the filter design by accentuating the magnitude spectrum in the passband. The route to boosting the magnitude of $\mathbf{H}(\omega)$ at a given frequency is to place a pole at that frequency. Hence, it would seem intuitively obvious to place a vertical string of poles in the OLHP between $-\omega_c$ and ω_c, as illustrated by the example shown in Fig. 6-37(a). The vertical string's five equally spaced poles are located at $(-\alpha + j\omega_c)$, $(-\alpha + j\omega_c/2)$, $(-\alpha + j0)$, $(-\alpha - j\omega_c/2)$, and $(-\alpha - j\omega_c)$. From Eq. (6.65), the corresponding frequency response is

$$\mathbf{H}(\omega) = \mathbf{C} \prod_{k=-2}^{2} \left(\frac{1}{j\omega + \alpha - j\,\dfrac{k\omega_c}{2}} \right). \tag{6.82}$$

Even though the associated magnitude spectrum (Fig. 6-37(b)) favors frequencies in the $-\omega_c$ to ω_c range, the pattern includes undesirable resonances at frequencies corresponding to the locations of the poles in the complex plane.

Module 6.2 Comb Filter to Separate Two Trumpet Signals This module adds the waveforms of trumpets playing notes G and A, and uses a comb filter to separate them. The module also allows the user to listen to the original (two trumpets) and filtered (one trumpet) signals.

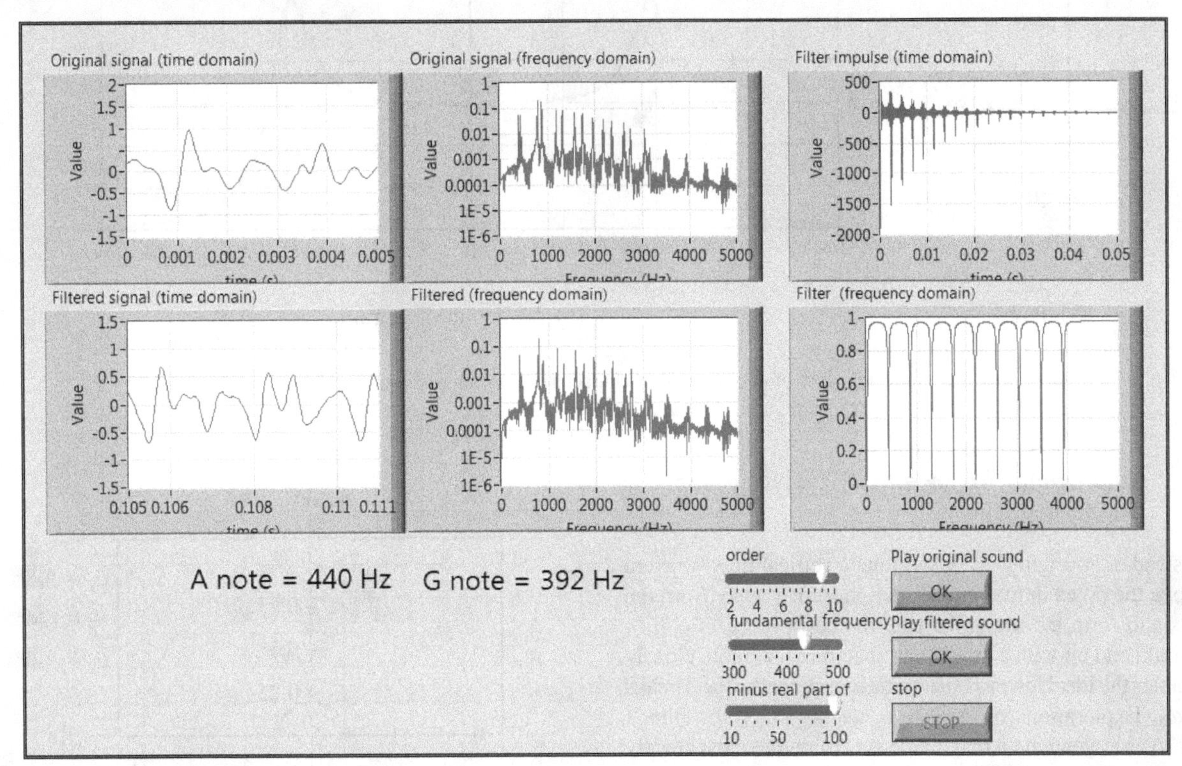

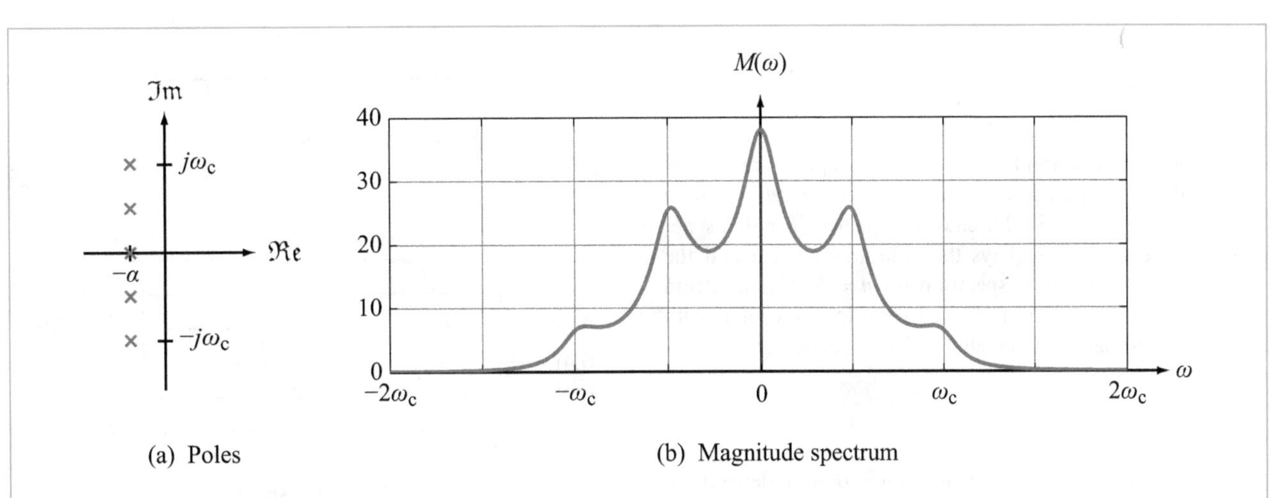

(a) Poles

(b) Magnitude spectrum

Figure 6-37: Magnitude response of a lowpass filter characterized by a vertical string of closely spaced poles displaced from the vertical axis by $\alpha = 0.1\omega_c$. The scaling constant was set at $\mathbf{C} = \omega_c^5$.

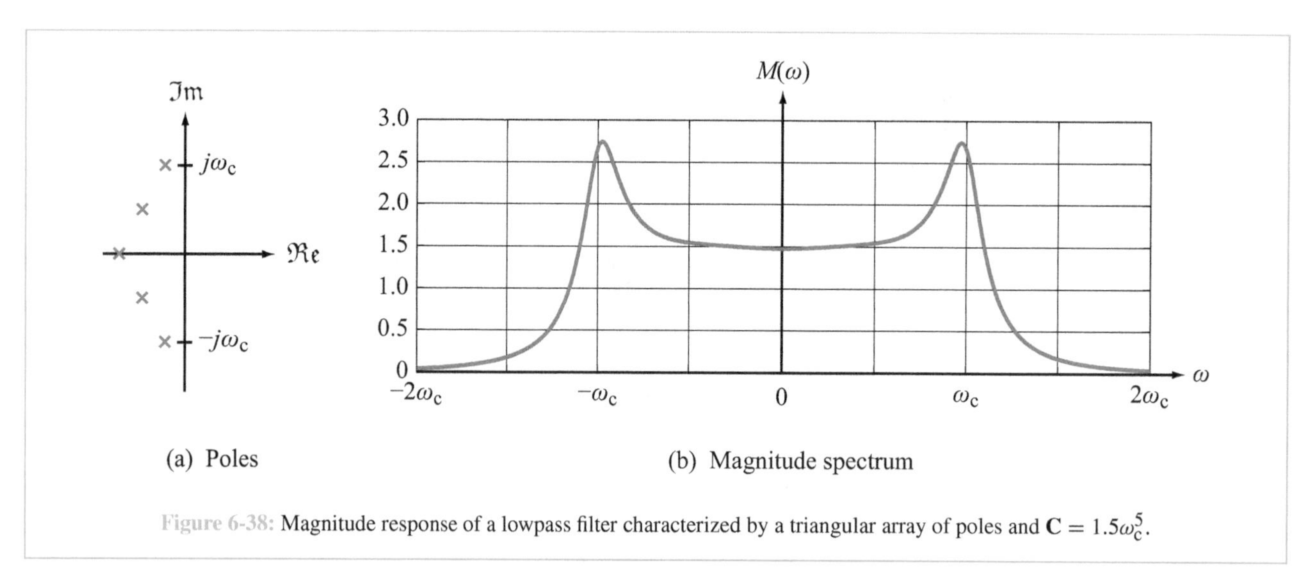

Figure 6-38: Magnitude response of a lowpass filter characterized by a triangular array of poles and $\mathbf{C} = 1.5\omega_c^5$.

6-9.2 Elementary Arrangement 2: **Triangular Pole Pattern**

If, instead of arranging the poles along a vertical line, we were to move the poles in the middle section further away from the $j\omega$-axis to form the triangular pattern shown in Fig. 6-38(a), we would end up with a magnitude spectrum that more closely resembles the ideal lowpass spectrum, except for the local peaks at $\pm\omega_c$. Explorations on how to *optimally* place the filter's poles so as to generate a magnitude spectrum that most closely approximates that of the ideal (*brick-wall*) lowpass filter can lead to multiple specialized designs, including the Butterworth and Chebyshev families of filters. We describe the Butterworth filter next.

6-9.3 Arrangement 3: **Butterworth Pole Placement**

In an n-pole Butterworth lowpass filter, the n poles are equally spaced along the perimeter of a semicircle of radius ω_c in the OLHP, where ω_c is the cutoff frequency in radians per second. Figure 6-39 displays the pole arrangement and the corresponding magnitude spectrum for $n = 5$. The spectrum exhibits the general pattern desired for a lowpass filter, with none of the resonances associated with the spectra in Figs. 6-37 and 6-38.

Pole locations

Pole locations of the Butterworth filter design depend on whether n, the number of poles, is odd or even. An easy recipe follows.

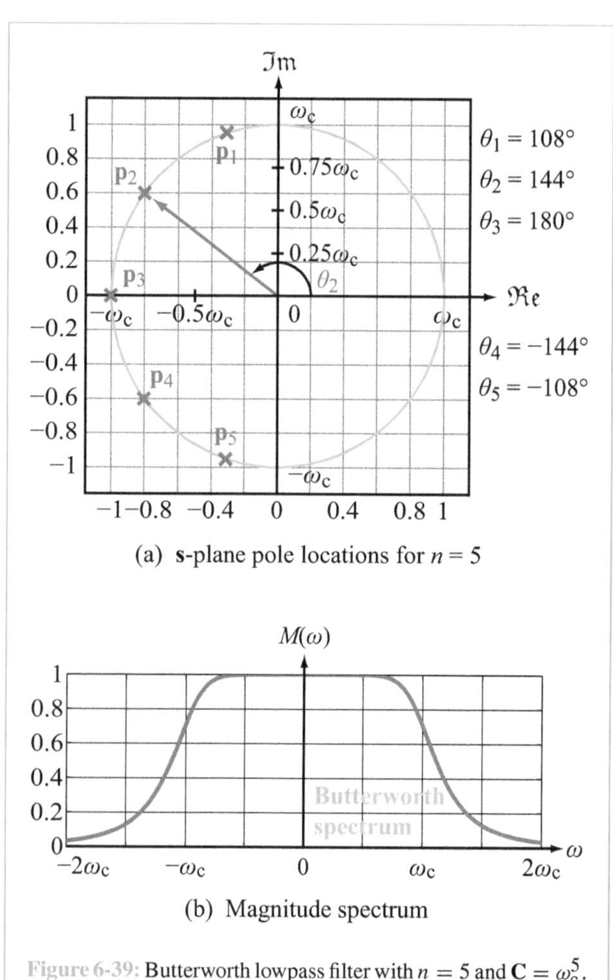

Figure 6-39: Butterworth lowpass filter with $n = 5$ and $\mathbf{C} = \omega_c^5$.

$n =$ odd

Step 1: Around the *full* circle of radius ω_c in the complex **s** plane, distribute $2n$ equally spaced poles around the circle, starting with the first pole at $\mathbf{s} = \omega_c$; that is, at point $(\omega_c + j0)$. The spacing between adjacent poles should be

$$\theta = \frac{2\pi}{2n} = \frac{\pi}{n} \quad \text{(rad)}. \tag{6.83}$$

The placement pattern should be symmetrical with respect to both axes. An example is shown in Fig. 6-40(a) for $n = 7$.

Step 2: Discard all of the poles in the right half-plane.

$n =$ even

Step 1: Around the *full* circle of radius ω_c in the complex **s** plane, distribute $2n$ equally spaced poles around the circle, symmetrically arranged with respect to both axes. When n is even, none of the poles lie on the real axis and the spacing between adjacent poles is π/n radians. An example is shown in Fig. 6-40(b) for $n = 6$.

Step 2: Discard all of the poles in the right half-plane.

Frequency response

To calculate the transfer function, *we use* only *the poles in the OLHP*, and we denote their locations as

$$\mathbf{p}_i = -\alpha_i + jb_i, \qquad \text{for } i = 1, \ldots, n. \tag{6.84}$$

In the OLHP, b_i may be positive, negative, or zero, but α_i is always positive.

The associated Butterworth frequency response is

$$\mathbf{H}_{LP}(\omega) = \mathbf{C} \prod_{i=1}^{n} \left(\frac{1}{j\omega - \mathbf{p}_i} \right). \tag{6.85}$$

(Butterworth lowpass filter)

Computing the magnitude response (gain)

$$M_{LP}(\omega) = |\mathbf{H}_{LP}(\omega)|$$

of an nth-order Butterworth filter is straightforward if we first compute $M_{LP}^2(\omega)$ and then take its square root. The complex conjugate of Eq. (6.85) is

$$\mathbf{H}_{LP}^*(\omega) = \mathbf{C}^* \prod_{i=1}^{n} \frac{1}{-j\omega - \mathbf{p}_i^*} = \frac{\mathbf{C}^*(-1)^n}{\prod_{i=1}^{n}(j\omega + \mathbf{p}_i^*)}. \tag{6.86}$$

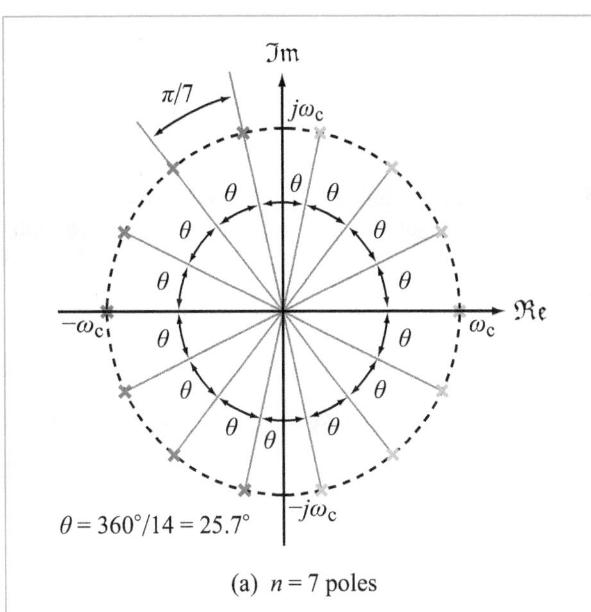

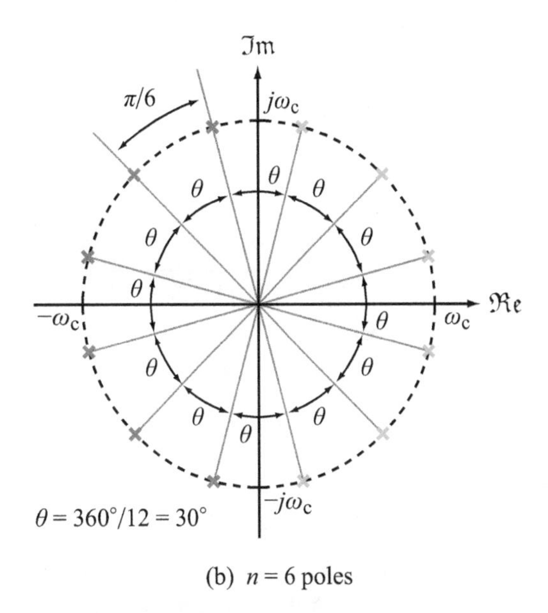

Figure 6-40: Butterworth lowpass filter pole locations for even and odd number of poles. Only the poles in the open left half-plane (red) are used in transfer function expression.

Combining this with Eq. (6.85) gives

$$\begin{aligned} M_{LP}^2(\omega) &= |\mathbf{H}_{LP}(\omega)|^2 \\ &= \mathbf{H}_{LP}(\omega)\, \mathbf{H}_{LP}^*(\omega) \\ &= \frac{\mathbf{C}\mathbf{C}^*(-1)^n}{\prod_{i=1}^{n}(j\omega - \mathbf{p}_i) \prod_{i=1}^{n}(j\omega + \mathbf{p}_i^*)}. \end{aligned} \tag{6.87}$$

From Eq. (6.84), we see that

$$\mathbf{p}_i = -\alpha_i + jb_i, \tag{6.88a}$$

$$-\mathbf{p}_i^* = \alpha_i + jb_i. \tag{6.88b}$$

Each $-\mathbf{p}_i^*$ is the open right half-plane counterpart to an OLHP Butterworth filter pole $\mathbf{p}_i$. Poles $-\mathbf{p}_i^*$ are the open right half-plane poles discarded from the $2n$ poles equally spaced around the circle of radius ω_c.

(a) n = odd

First, consider the case where n is odd. The $2n$ poles are equally spaced around the circle of radius ω_c and have the forms $\{\, \omega_c e^{j2\pi i/(2n)}, i = 0, \ldots, 2n-1 \,\}$, as in Fig. 6-40(a). Each of these poles, when raised to the power $2n$, gives ω_c^{2n}, since

$$(\omega_c e^{j2\pi i/(2n)})^{2n} = \omega_c^{2n} e^{j2\pi i} = \omega_c^{2n}.$$

The denominator of Eq. (6.87) is a polynomial in $j\omega$ and its roots are equally spaced around the circle of radius ω_c. The polynomial simplifies to

$$\prod_{i=1}^{n}(j\omega - \mathbf{p}_i) \prod_{i=1}^{n}(j\omega + \mathbf{p}_i^*) = (j\omega)^{2n} - \omega_c^{2n}. \tag{6.89}$$

Equation (6.87) simplifies to

$$M_{\text{LP}}^2(\omega) = \frac{\mathbf{C}\mathbf{C}^*(-1)^n}{\prod_{i=1}^{n}(j\omega - \mathbf{p}_i) \prod_{i=1}^{n}(j\omega + \mathbf{p}_i^*)} = \frac{|\mathbf{C}|^2(-1)^n}{(j\omega)^{2n} - \omega_c^{2n}} = \frac{|\mathbf{C}|^2}{\omega^{2n} + \omega_c^{2n}}, \tag{6.90}$$

because for odd n, $(-1)^n = -1$, and also $j^{2n} = -1$.

(b) n = even

Next, consider the case where n is even. Now the $2n$ poles are equally spaced around the circle of radius ω_c, but they have the forms $\{\omega_c e^{j2\pi(i+\frac{1}{2})/(2n)}, i = 0, \ldots, 2n-1\}$, as in Fig. 6-40(b). Each of these poles, when raised to the power $2n$, gives $-\omega_c^{2n}$, since

$$(\omega_c e^{j2\pi(i+\frac{1}{2})/(2n)})^{2n} = \omega_c^{2n} e^{j2\pi i} e^{j\pi} = -\omega_c^{2n}.$$

The denominator of Eq. (6.87)) simplifies to

$$\prod_{i=1}^{n}(j\omega - \mathbf{p}_i) \prod_{i=1}^{n}(j\omega + \mathbf{p}_i^*) = (j\omega)^{2n} + \omega_c^{2n}, \tag{6.91}$$

and Eq. (6.87) becomes

$$M_{\text{LP}}^2(\omega) = \frac{\mathbf{C}\mathbf{C}^*(-1)^n}{\prod_{i=1}^{n}(j\omega - \mathbf{p}_i) \prod_{i=1}^{n}(j\omega + \mathbf{p}_i^*)} = \frac{|\mathbf{C}|^2(-1)^n}{(j\omega)^{2n} + \omega_c^{2n}} = \frac{|\mathbf{C}|^2}{\omega^{2n} + \omega_c^{2n}}, \tag{6.92}$$

since for even n we have $(-1)^n = j^{2n} = 1$.

The above derivation shows why the $2n$ poles equally spaced around the circle of radius ω_c must be placed at different, although equally spaced, angles for even and odd n. In either case, taking a square root gives

$$M_{\text{LP}}(\omega) = \frac{|\mathbf{C}|}{\sqrt{\omega^{2n} + \omega_c^{2n}}} = \frac{|\mathbf{C}|}{\omega_c^n} \cdot \frac{1}{\sqrt{1 + (\omega/\omega_c)^{2n}}}. \tag{6.93}$$

(Butterworth lowpass filter)

The family of spectral plots shown in Fig. 6-41, calculated for $\omega_c = 1$ rad/s, demonstrate how the filter's selectivity (sharper roll-off) increases with n. For $\omega \gg \omega_c$, the roll-off rate is

$$S_g = -20n \quad \text{dB/decade}. \tag{6.94}$$

Butterworth filter properties

Butterworth filters are used extensively in many electronic systems, primarily because they possess a number of highly desirable properties. They are causal, BIBO-stable, and realizable using op amps, or inductors and capacitors. Implementation of the Butterworth lowpass filter is fairly straightforward: For any order n and cutoff frequency ω_c, it is easy to determine the locations of its poles, the transfer function $\mathbf{H}_{\text{LP}}(\mathbf{s})$ has a simple form, and the magnitude spectrum is easy to compute. The magnitude spectrum is flat at low frequencies, and the dc gain is $M_0 = |\mathbf{C}|/\omega_c^n$.

6-9.4 Butterworth Lowpass Filter Design

For $j\omega = \mathbf{s}$, Eq. (6.85) becomes

$$\mathbf{H}_{\text{LP}}(\mathbf{s}) = \mathbf{C} \prod_{i=1}^{n} \left(\frac{1}{\mathbf{s} - \mathbf{p}_i} \right). \tag{6.95}$$

All poles are located at a radius ω_c in the OLHP. Pole $\mathbf{p}_i$ can be written as

$$\mathbf{p}_i = -\alpha_i + jb_i = \omega_c e^{j\theta_i}, \tag{6.96}$$

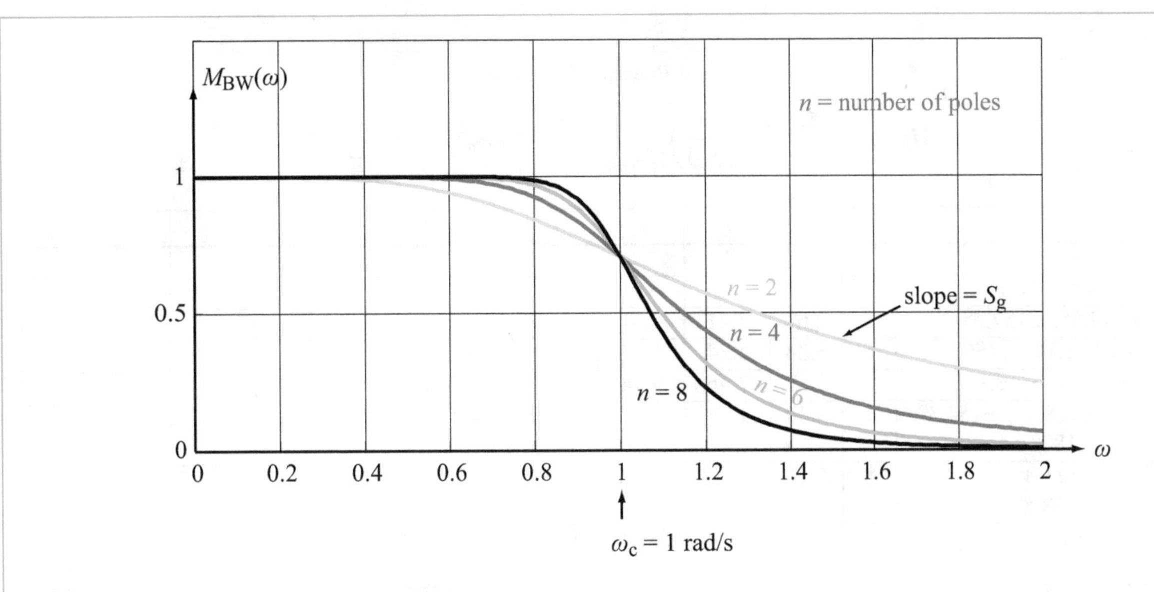

Figure 6-41: Butterworth lowpass filter for $\mathbf{C} = 1$ and $\omega_c = 1$ rad/s; magnitude spectra for various values of the number of poles, n.

with ***pole angle*** θ_i given by

$$\theta_i = 180° - \tan^{-1}\left(\frac{b_i}{|\alpha_i|}\right), \tag{6.97}$$

as shown in Fig. 6-42. Since the magnitudes of all poles $\mathbf{p}_i$ are equal to ω_c, the polynomial expansion of the denominator in Eq. (6.95) will have terms with coefficients proportional to $\omega_c, \omega_c^2, \ldots, \omega_c^n$, which can lead to very large numbers. Consider, for example, a 10th order filter with a cutoff frequency of 1.6 MHz, or, equivalently, $\omega_c = 2\pi f_c \approx 10^7$ rad/s; the largest coefficient in the polynomial will be on the order of $(10^7)^{10} = 10^{70}$! Working with polynomials with such large numbers can be cumbersome. For this reason, Butterworth filters usually are designed in two steps, as follows.

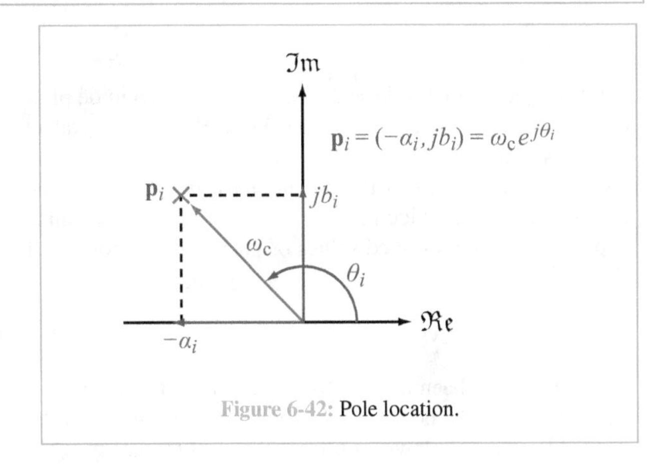

Figure 6-42: Pole location.

Step 1: Normalized transfer function $\mathbf{H}(\mathbf{s}_a)$

Inserting Eq. (6.96) into Eq. (6.95), and then normalizing $\mathbf{s}$ by dividing it by ω_c, gives

$$\begin{aligned}\mathbf{H}_{\text{LP}}(\mathbf{s}) &= \mathbf{C}\prod_{i=1}^{n}\left(\frac{1}{\mathbf{s} - \omega_c e^{j\theta_i}}\right)\\ &= \frac{\mathbf{C}}{\omega_c^n}\prod_{i=1}^{n}\left(\frac{1}{\dfrac{\mathbf{s}}{\omega_c} - e^{j\theta_i}}\right)\\ &= \frac{\mathbf{C}}{\omega_c^n}\prod_{i=1}^{n}\left(\frac{1}{\mathbf{s}_a - e^{j\theta_i}}\right) = \frac{\mathbf{C}}{\omega_c^n}\,\mathbf{H}(\mathbf{s}_a),\end{aligned} \tag{6.98}$$

where $\mathbf{s}_a$ is a normalized complex frequency,

$$\mathbf{s}_a = \frac{\mathbf{s}}{\omega_c}, \tag{6.99}$$

and

$$\mathbf{H}(\mathbf{s}_a) = \prod_{i=1}^{n}\left(\frac{1}{\mathbf{s}_a - e^{j\theta_i}}\right). \tag{6.100}$$

(normalized lowpass transfer function)

Transfer function $\mathbf{H}(\mathbf{s}_a)$ is a ***normalized transfer function***, representing an nth order Butterworth lowpass filter with a

Table 6-3: Butterworth lowpass filter.

$$\mathbf{H}_{\mathrm{LP}}(\mathbf{s}) = \frac{\mathbf{C}}{\omega_c^n} \cdot \left(\frac{1}{\mathbf{D}_n(\mathbf{s}_a)} \right) \Bigg|_{\mathbf{s}_a = \mathbf{s}/\omega_c} ; \qquad \mathbf{D}_n(\mathbf{s}_a) = 1 + \sum_{i=1}^{n} a_i \mathbf{s}^i$$

Pole angles θ_i	n	a_1	a_2	a_3	a_4	a_5	a_6	a_7	a_8	a_9	a_{10}
$\pm 135°$	2	1.41	1								
$\pm 120°$, 180°	3	2	2	1							
$\pm 112.5°$, $\pm 157.5°$	4	2.61	3.41	2.61	1						
$\pm 108°$, $\pm 144°$, 180°	5	3.24	5.24	5.24	3.24	1					
$\pm 105°$, $\pm 135°$, $\pm 165°$	6	3.87	7.46	9.14	7.46	3.87	1				
$\pm 102.9°$, $\pm 128.6°$, $\pm 154.3°$, 180°	7	4.49	10.1	14.59	14.59	10.1	4.49	1			
$\pm 101.3°$, $\pm 123.8°$, $\pm 146.3°$, $\pm 168.8°$	8	5.13	13.14	21.85	25.69	21.85	13.14	5.13	1		
$\pm 100°$, $\pm 120°$, $\pm 140°$, $\pm 160°$, 180°	9	5.76	16.58	31.16	41.99	41.99	31.16	16.58	5.76	1	
$\pm 98°$, $\pm 116°$, $\pm 134°$, $\pm 152°$, $\pm 170°$	10	6.39	20.43	42.8	64.88	74.23	64.88	42.8	20.43	6.39	1

cutoff frequency of 1 rad/s and a dc gain of 1. Magnitude plots of $\mathbf{H}(\mathbf{s}_a)$ are displayed in Fig. 6-41. Since $\mathbf{H}(\mathbf{s}_a)$ is equivalent to $\mathbf{H}_{\mathrm{LP}}(\mathbf{s})$ with $\omega_c = 1$ rad/s, there is no longer a problem with large-size coefficients in the polynomials. For reference and convenience, we provide in Table 6-3 a list of the polynomial expressions and associated values of pole angles θ_i for $n = 1$ to 10.

Step 2: Convert to $\mathbf{H}_{\mathrm{LP}}(\mathbf{s})$

Once $\mathbf{H}(\mathbf{s}_a)$ has been defined for a specified filter of order n, Eq. (6.98) can then be applied to obtain the expression for the desired Butterworth lowpass filter with cutoff frequency ω_c:

$$\mathbf{H}_{\mathrm{LP}}(\mathbf{s}) = \frac{\mathbf{C}}{\omega_c^n} \mathbf{H}(\mathbf{s}_a) \Bigg|_{\mathbf{s}_a = \mathbf{s}/\omega_c} . \qquad (6.101)$$

(denormalization)

Example 6-10: Third-Order LP Filter

Obtain the transfer function for a third-order Butterworth lowpass filter with cutoff frequency of 10^3 rad/s and dc gain of 10.

Solution: From Table 6-3 (or the recipe of Section 6-8.3), we establish that the pole angles of a third-order filter are 180° and $\pm 120°$. Hence, the normalized transfer function is

$$\mathbf{H}(\mathbf{s}_a) = \frac{1}{(\mathbf{s}_a - e^{j180°})(\mathbf{s}_a - e^{j120°})(\mathbf{s}_a - e^{-j120°})} = \frac{1}{\mathbf{s}_a^3 + 2\mathbf{s}_a^2 + 2\mathbf{s}_a + 1} .$$

Next, we convert to $\mathbf{H}_{\mathrm{LP}}(\mathbf{s})$ with $\omega_c = 10^3$ rad/s:

$$\mathbf{H}_{\mathrm{LP}}(\mathbf{s}) = \frac{\mathbf{C}}{\omega_c^3} \mathbf{H}(\mathbf{s}_a) \Bigg|_{\mathbf{s}_a = \mathbf{s}/10^3} = \frac{\mathbf{C}}{10^9} \frac{1}{\left(\frac{\mathbf{s}}{10^3}\right)^3 + 2\left(\frac{\mathbf{s}}{10^3}\right)^2 + 2\left(\frac{\mathbf{s}}{10^3}\right) + 1} = \frac{\mathbf{C}}{\mathbf{s}^3 + (2 \times 10^3)\mathbf{s}^2 + (2 \times 10^6)\mathbf{s} + 10^9} .$$

At dc, $\mathbf{s} = 0$ and $\mathbf{H}_{\mathrm{LP}}(0) = \mathbf{C}/10^9$. To satisfy the condition that the dc gain is 10, it is necessary that $\mathbf{C} = 10^{10}$.

Example 6-11: Op-Amp Realization

The circuit shown in Fig. 6-43 is known as the Sallen-Key op-amp filter.

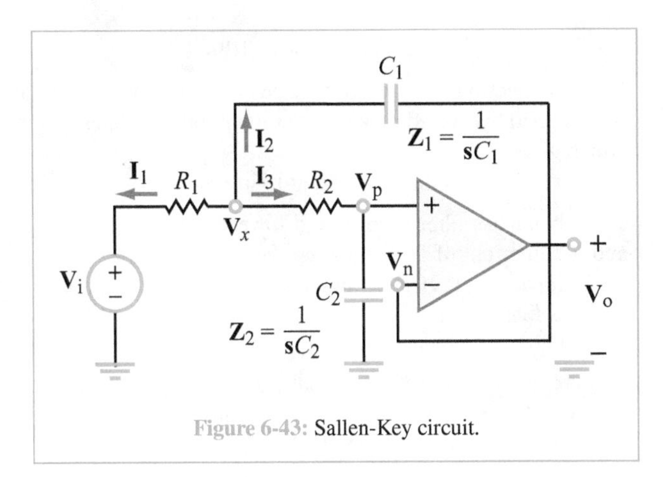

Figure 6-43: Sallen-Key circuit.

(a) Obtain the transfer function $\mathbf{H}(\mathbf{s}) = \mathbf{V}_o(\mathbf{s})/\mathbf{V}_i(\mathbf{s})$.

(b) Obtain the transfer function of a second-order Butterworth filter with cutoff frequency $\omega_c = 10^3$ rad/s and arbitrary dc gain $\mathbf{C}$.

(c) Match the transfer functions in (a) and (b) to specify the values of R_1, R_2, C_1, and C_2.

Solution:

(a) At node $\mathbf{V}_x$, KCL gives

$$\frac{\mathbf{V}_x - \mathbf{V}_i}{R_1} + \frac{\mathbf{V}_x - \mathbf{V}_p}{R_2} + \frac{\mathbf{V}_x - \mathbf{V}_o}{\mathbf{Z}_1} = 0, \tag{6.102}$$

and since $\mathbf{V}_o = \mathbf{V}_n = \mathbf{V}_p$, voltage division gives

$$\mathbf{V}_x = \mathbf{V}_o \left(\frac{R_2 + \mathbf{Z}_2}{\mathbf{Z}_2} \right). \tag{6.103}$$

The combination of the two equations leads to

$$\mathbf{H}(\mathbf{s}) = \frac{\dfrac{1}{R_1R_2C_1C_2}}{\mathbf{s}^2 + \left(\dfrac{1}{R_1C_1} + \dfrac{1}{R_2C_1} \right)\mathbf{s} + \dfrac{1}{R_1R_2C_1C_2}}. \tag{6.104}$$

(b) For a second-order Butterworth lowpass filter, $\theta_i = \pm 135°$ and the normalized transfer function is

$$\mathbf{H}(\mathbf{s}_a) = \frac{1}{(\mathbf{s}_a - e^{j135°})(\mathbf{s}_a - e^{-j135°})} = \frac{1}{\mathbf{s}_a^2 + 1.414\mathbf{s}_a + 1}. \tag{6.105}$$

The corresponding transfer function with $\omega_c = 10^3$ rad/s is

$$\mathbf{H}_{\text{LP}}(\mathbf{s}) = \frac{\mathbf{C}}{\omega_c^2} \mathbf{H}(\mathbf{s}_a)\bigg|_{\mathbf{s}_a = \mathbf{s}/10^3} = \frac{\mathbf{C}}{\mathbf{s}^2 + 1.414 \times 10^3\mathbf{s} + 10^6}. \tag{6.106}$$

(c) To match the expression of the Sallen-Key transfer function given by Eq. (6.104) to the expression for the Butterworth filter given by Eq. (6.106), we need to set

$$\frac{1}{R_1R_2C_1C_2} = \omega_c^2 = 10^6, \qquad \mathbf{C} = 10^6,$$

$$\frac{1}{R_1C_1} + \frac{1}{R_2C_1} = 1.414\omega_c = 1.414 \times 10^3.$$

Since we have four component values to specify (R_1, R_2, C_1, and C_2) and only two constraints to satisfy, there are many possible solutions. To come up with realistic values for resistors and capacitors, we arbitrarily choose

$$R_1 = R_2 = 10 \text{ k}\Omega.$$

The choice leads to

$$C_1 = 0.14\ \mu\text{F} \quad \text{and} \quad C_2 = 70.7 \text{ nF}.$$

Higher-order filters

The procedure used in Example 6-11 for $n = 2$ can be extended to higher even-order filters by building a cascade of Sallen-Key op-amp filters, with each stage designed to match one pair of conjugate poles. That is, if the n poles are located at angles $\pm\theta_1, \pm\theta_2, \ldots,$ then the first Sallen-Key op-amp stage is designed to match angles $\pm\theta_1$, the second stage to match angles $\pm\theta_2$, and so on.

6-9.5 Butterworth Highpass Filter Design

A Butterworth highpass filter with cutoff frequency ω_c can be obtained from a Butterworth lowpass filter with the same cutoff frequency by performing a transformation that maps $\mathbf{s}$ to its reciprocal in the frequency domain. For sinusoidal signals, $\mathbf{s} = j\omega$, so such a transformation would map dc ($\omega = 0$) for $\mathbf{H}_{\text{LP}}(\mathbf{s})$ to ∞ for $\mathbf{H}_{\text{HP}}(\mathbf{s})$, and vice versa. The transformation should be performed on the normalized transfer function $\mathbf{H}(\mathbf{s}_a)$. The process is outlined by the following steps:

Step 1: For a specified value of n, obtain the *lowpass filter* normalized transfer function $\mathbf{H}(\mathbf{s}_a)$, as given by Eq. (6.100).

Step 2: Generate a *highpass filter* normalized transfer function $\mathbf{H}(\mathbf{s}_b)$ by replacing $\mathbf{s}_a$ with $\mathbf{s}_b = 1/\mathbf{s}_a$:

$$\mathbf{H}(\mathbf{s}_b) = \mathbf{H}(\mathbf{s}_a)|_{\mathbf{s}_a = 1/\mathbf{s}_b}. \tag{6.107}$$

(lowpass-to-highpass transformation of normalized transfer functions)

Step 3: Convert $\mathbf{H}(\mathbf{s}_b)$ to the highpass filter transfer function $\mathbf{H}_{\mathrm{HP}}(\mathbf{s})$ with cutoff frequency ω_{c} by applying

$$\mathbf{H}_{\mathrm{HP}}(\mathbf{s}) = \frac{\mathbf{C}}{\omega_{\mathrm{c}}^n}\,\mathbf{H}(\mathbf{s}_b)\bigg|_{\mathbf{s}_b=\mathbf{s}/\omega_{\mathrm{c}}}. \qquad (6.108)$$

(denormalization)

▶ Transfer function $\mathbf{H}_{\mathrm{HP}}(\mathbf{s})$ of a Butterworth highpass filter of order n and cutoff frequency ω_{c} has the same poles as that of $\mathbf{H}_{\mathrm{LP}}(\mathbf{s})$, the transfer function of a Butterworth lowpass filter of the same order and with the same cutoff frequency, but in addition, $\mathbf{H}_{\mathrm{HP}}(\mathbf{s})$, has n zeros, all located at $\mathbf{s} = 0$, whereas $\mathbf{H}_{\mathrm{LP}}(\mathbf{s})$ has none. ◀

Example 6-12: Third-Order HP Filter

Obtain the transfer function for a third-order highpass Butterworth filter with cutoff frequency of 10^3 rad/s and a high-frequency gain of 10.

Solution: From Example 6-10, the normalized transfer function of the third-order lowpass Butterworth filter is given by

$$\mathbf{H}(\mathbf{s}_a) = \frac{1}{\mathbf{s}_a^3 + 2\mathbf{s}_a^2 + 2\mathbf{s}_a + 1}.$$

Transforming from lowpass to highpass entails replacing $\mathbf{s}_a$ with $1/\mathbf{s}_b$. The result is

$$\mathbf{H}(\mathbf{s}_b) = \frac{1}{\dfrac{1}{\mathbf{s}_b^3} + \dfrac{2}{\mathbf{s}_b^2} + \dfrac{2}{\mathbf{s}_b} + 1} = \frac{\mathbf{s}_b^3}{1 + 2\mathbf{s}_b + 2\mathbf{s}_b^2 + \mathbf{s}_b^3}.$$

Finally,

$$\begin{aligned}\mathbf{H}_{\mathrm{HP}}(\mathbf{s}) &= \frac{\mathbf{C}}{\omega_{\mathrm{c}}^3}\,\mathbf{H}(\mathbf{s}_b)\bigg|_{\mathbf{s}_b=\mathbf{s}/\omega_{\mathrm{c}}} \\ &= \frac{\mathbf{C}}{\omega_{\mathrm{c}}^3}\left(\frac{\mathbf{s}^3/\omega_{\mathrm{c}}^3}{1 + \dfrac{2\mathbf{s}}{\omega_{\mathrm{c}}} + \dfrac{2\mathbf{s}^2}{\omega_{\mathrm{c}}^2} + \dfrac{\mathbf{s}^3}{\omega_{\mathrm{c}}^3}}\right) \\ &= \frac{\mathbf{C}}{\omega_{\mathrm{c}}^3}\left(\frac{\mathbf{s}^3}{\mathbf{s}^3 + 2\omega_{\mathrm{c}}\mathbf{s}^2 + 2\omega_{\mathrm{c}}^2\mathbf{s} + \omega_{\mathrm{c}}^3}\right).\end{aligned}$$

As $\mathbf{s} \to \infty$, $\mathbf{H}_{\mathrm{HP}}(\mathbf{s}) \to \mathbf{C}/\omega_{\mathrm{c}}^3$. Hence, to have a high-frequency gain of 10 (with $\omega_{\mathrm{c}} = 10^3$ rad/s), it is necessary that $\mathbf{C} = 10^{10}$. The final expression for the highpass transfer function is

$$\mathbf{H}_{\mathrm{HP}}(\mathbf{s}) = \frac{10\mathbf{s}^3}{\mathbf{s}^3 + (2 \times 10^3)\mathbf{s}^2 + (2 \times 10^6)\mathbf{s} + 10^9}.$$

6-9.6 Butterworth Bandpass Filter Design

Earlier in Section 6-3.3, we illustrated how a bandpass filter can be designed by cascading a lowpass filter and a highpass filter, which gives

$$\mathbf{H}_{\mathrm{BP}}(\mathbf{s}) = \mathbf{H}_{\mathrm{LP}}(\mathbf{s})\,\mathbf{H}_{\mathrm{HP}}(\mathbf{s}). \qquad (6.109)$$

If the bandpass filter is to have a lower cutoff frequency ω_{c_1} and a higher cutoff frequency ω_{c_2}, ω_{c_1} should be assigned to the highpass filter $\mathbf{H}_{\mathrm{HP}}(\mathbf{s})$ and ω_{c_2} should be assigned to the lowpass filter $\mathbf{H}_{\mathrm{LP}}(\mathbf{s})$.

Concept Question 6-13: What statement applies to the poles of a Butterworth filter of any order with a cutoff frequency of 1 rad/s? (See S²)

Concept Question 6-14: Why is it more practical to design a lowpass filter on the basis of poles rather than zeros? (See S²)

Exercise 6-12: Where should the poles of a second-order Butterworth lowpass filter be located, if its cutoff frequency is 3 rad/s?

Answer: $3e^{\pm j135^\circ}$. (See S²)

Exercise 6-13: Where should the poles of a third-order Butterworth lowpass filter be located, if its cutoff frequency is 5 rad/s?

Answer: -5 and $5e^{\pm j120^\circ}$. (See S²)

6-10 Denoising a Trumpet Signal

6-10.1 Noisy Trumpet Signal

Noise may be defined as an unknown and *undesired signal* $x_{\mathrm{n}}(t)$ added to a *desired signal* $x(t)$. One application of signal processing is to reduce the noise component of a noisy signal by filtering out as much of the noise as possible. To demonstrate the filtering process, we will add noise to the trumpet signal, and then explore how to design a filter to remove it. In real filtering applications, the design of the filter depends on knowledge of both the signal and noise spectra. For illustration purposes, we will use a common noise model known as *additive zero-mean white Gaussian noise*. We will examine this name term by term.

- *Additive* means the noise is added to the signal, as opposed to multiplied by the signal. Some noise in images is multiplicative. In this book, we limit our consideration to additive noise.

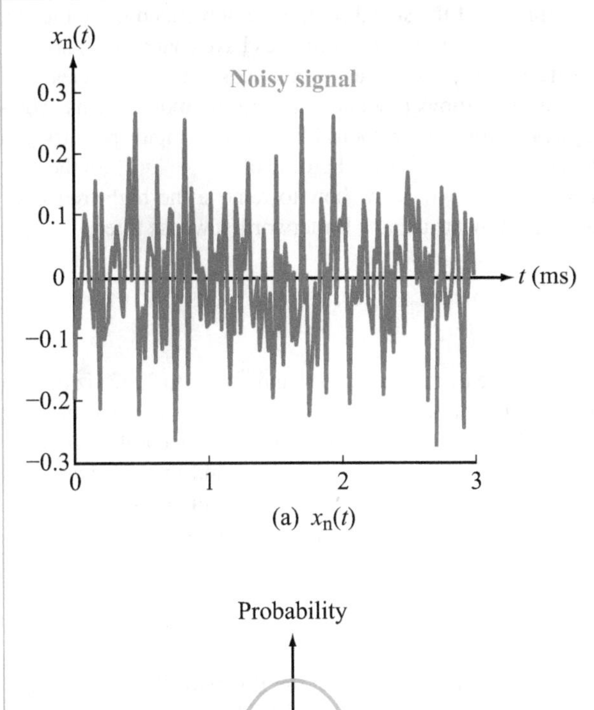

Figure 6-44: Zero-mean white Gaussian noise has a time-average value $(x_n)_{ave} = 0$ and its amplitude is described by a Gaussian probability distribution.

- ***Zero-mean*** means that the *average* value of the noise signal is zero. For a time-varying quantity, such as noise $x_n(t)$, we assume its mean value is

$$(x_n)_{ave} = \lim_{T\to\infty} \frac{1}{T} \int_{-T/2}^{T/2} x_n(t)\,dt = 0. \tag{6.110}$$

If the average value of the noise is non-zero and known, then the mean value can be subtracted from the noisy signal, leaving behind the same signal, plus zero-mean noise.

- ***White*** means that the noise is random; its value at any moment in time is independent of its value at any other moment. That is, the noise signal is completely unpredictable from one instant to another. The spectrum of a white-noise signal extends across all frequencies; hence, the term *white*.

- ***Gaussian*** means that at any instant in time, the amplitude $x_n(t)$ is characterized by a Gaussian probability distribution (Fig. 6-44(b)). The noise is most likely to be small in amplitude (positive or negative), but also has a small chance of being large.

6-10.2 Signal-to-Noise Ratio

Real signals are always accompanied by some amount of noise. The degree to which the noise will interfere with the signal and distort the information it is carrying depends (in part) on the ratio of the signal power to that of the noise power. Recall from Section 1-5 that the power carried by a signal $x(t)$ is $p(t) = |x(t)|^2$. Similarly, noise power is $p_n(t) = |x_n(t)|^2$.

The ***signal-to-noise ratio*** (***SNR***) is defined as the ratio of the time-average value of the signal power to that of the noise power. Accordingly, by Eq. (1.34),

$$\text{SNR} = \lim_{T\to\infty} \left[\frac{\dfrac{1}{T}\displaystyle\int_{-T/2}^{T/2} |x(t)|^2\,dt}{\dfrac{1}{T}\displaystyle\int_{-T/2}^{T/2} |x_n(t)|^2\,dt} \right]. \tag{6.111}$$

A large SNR means that the noise level is low, relative to the signal level, and the interference it causes will be minor. On the other hand, if SNR is low, it means that the noise level is high (relative to that of the signal), in which case it can distort the information content of the signal and reduce its reliability. With a sound signal, noise would change the way we would hear it; and with digital data, noise would introduce errors. In general, for any particular coding scheme, bit error rate varies inversely with SNR.

In practice, *SNR is expressed in power dB:*

$$\text{SNR (dB)} = 10\log[\text{SNR}]. \tag{6.112}$$

Figure 6-45(a) represents a time profile of a noisy trumpet signal given by

$$z(t) = x(t) + x_n(t), \tag{6.113}$$

where $x(t)$ and $x_n(t)$ are the trumpet and noise signals that were introduced earlier in Figs. 6-34(a) and 6-44(a), respectively.

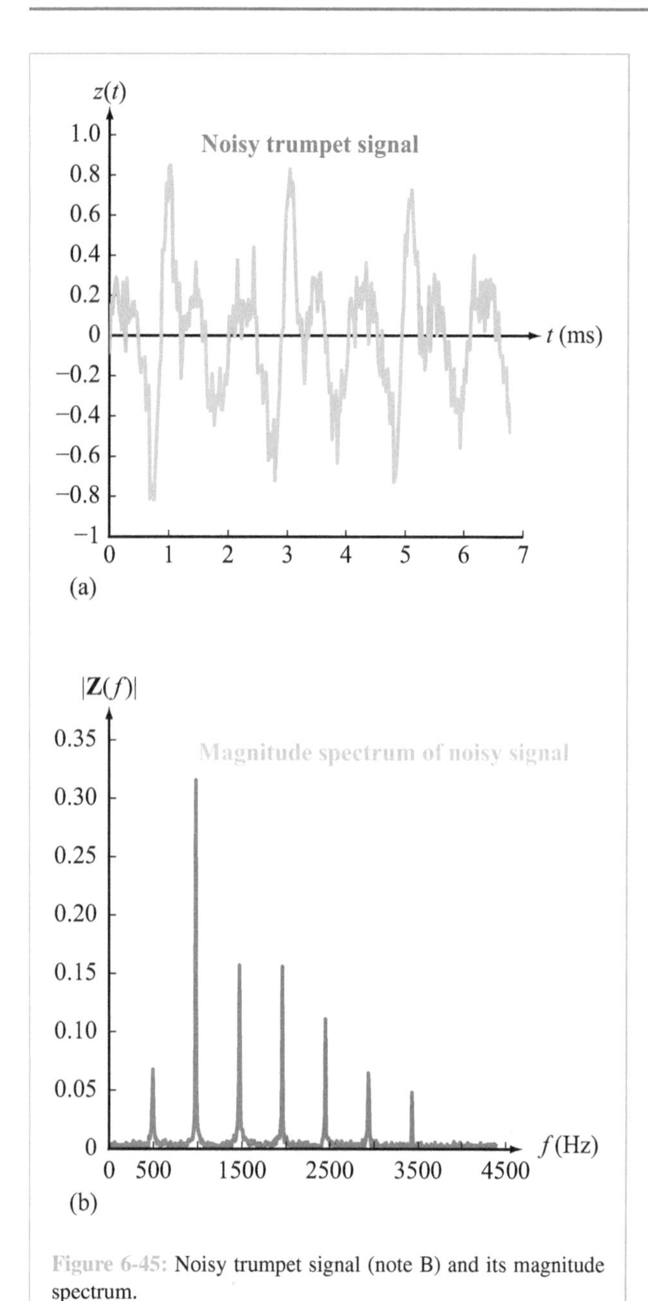

Figure 6-45: Noisy trumpet signal (note B) and its magnitude spectrum.

The signal-to-noise ratio is 13.8, or equivalently, 11.4 dB, which was determined numerically by subjecting the available profiles of $x(t)$ and $x_n(t)$ to Eq. (6.111). This means that, on average, the noise content of the noisy trumpet signal [noise/(signal + noise)] is on the order of 8%. The trumpet harmonics in Fig. 6-45(b) stick up over the noise spectrum like dandelions sticking up over a grassy lawn. However, the valleys (underbrush) of the signal profile are now much more ragged, and even the peaks of the harmonics have changed slightly.

Note that the underbrush noise exists at frequencies well beyond the trumpet sound's eighth harmonic. This *high-frequency noise* is responsible for the rapid perturbation observed in the noisy signal profile of Fig. 6-45(a). In the next subsection, we will show how to remove the high-frequency noise components using a Butterworth lowpass filter.

6-10.3 Lowpass Filtering the Noisy Trumpet Signal

Given that the trumpet signal is bandlimited to 3952 Hz (eighth harmonic of the 494 Hz note B), we shall use a fifth-order ($n = 5$) Butterworth lowpass filter with a cutoff frequency $f_c = 4000$ Hz. The filter's magnitude and impulse responses are displayed in Fig. 6-46. We note that the impulse response becomes negligibly small after about 0.8 ms, which means that the filter's transient response will only affect the initial 0.8 ms of the noisy signal's time profile.

To obtain the lowpass-filtered noisy trumpet signal, we can follow either of two paths.

(1) Multiply the spectrum of the Butterworth filter, $\mathbf{H}_{\text{LP}}(f)$, by the spectrum of the noisy signal, $\mathbf{Z}(f)$, to obtain the spectrum of the filtered signal, $\mathbf{Z}_f(f)$,

$$\mathbf{Z}_f(f) = \mathbf{H}_{\text{LP}}(f)\,\mathbf{Z}(f), \tag{6.114}$$

and then inverse transform $\mathbf{Z}_f(f)$ to get the filtered signal $z_f(t)$.

(2) Convolve the noisy signal $z(t)$ with the filter's impulse response $h_{\text{LP}}(t)$ to get $z_f(t)$,

$$z_f(t) = z(t) * h_{\text{LP}}(t). \tag{6.115}$$

Either approach leads to the filtered noisy trumpet signal shown in Fig. 6-47(a). For comparison, we have included in part (b) of the same figure the original noise-free trumpet signal. We observe that the Butterworth filter has indeed removed the high-frequency part of the noise spectrum, but distortions due to noise below 4000 Hz remain unaffected. That is, the Butterworth filter has improved the quality of the input signal significantly, but since it is a lowpass filter, it imparts no change on either the signal or the noise below 4000 Hz (see S² for more details).

6-11 Resonator Filter

Were we to listen to the noise-free and the filtered noisy signal of the preceding section (which can be done using either the MATLAB code or LabVIEW Module 6.3 on the book website),

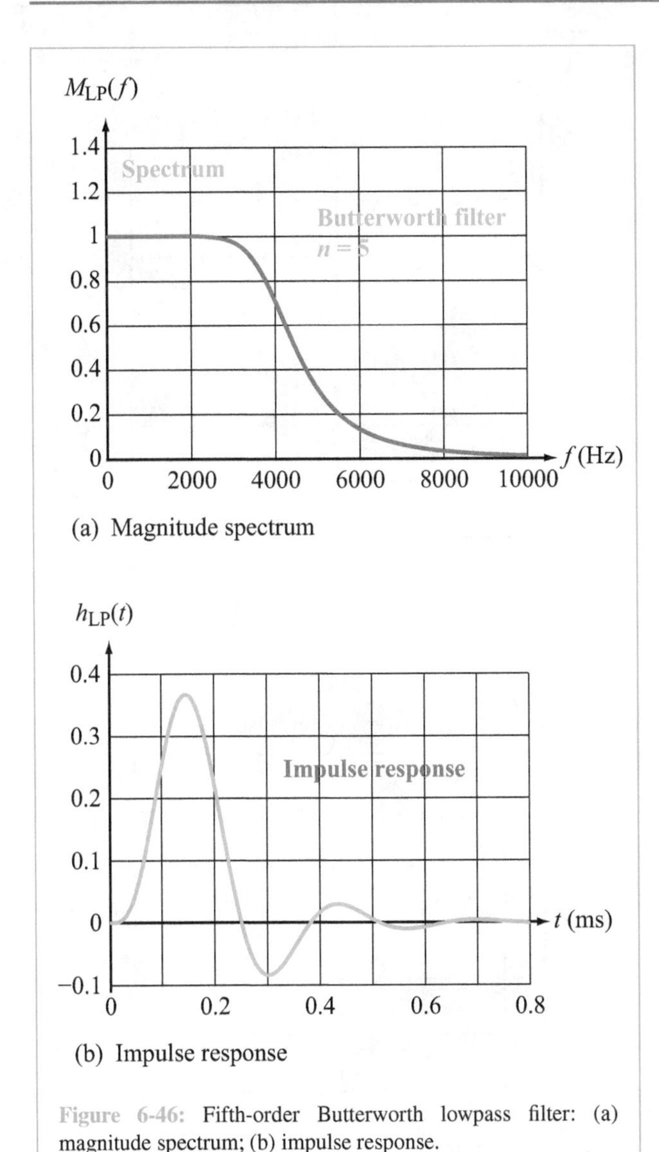

Figure 6-46: Fifth-order Butterworth lowpass filter: (a) magnitude spectrum; (b) impulse response.

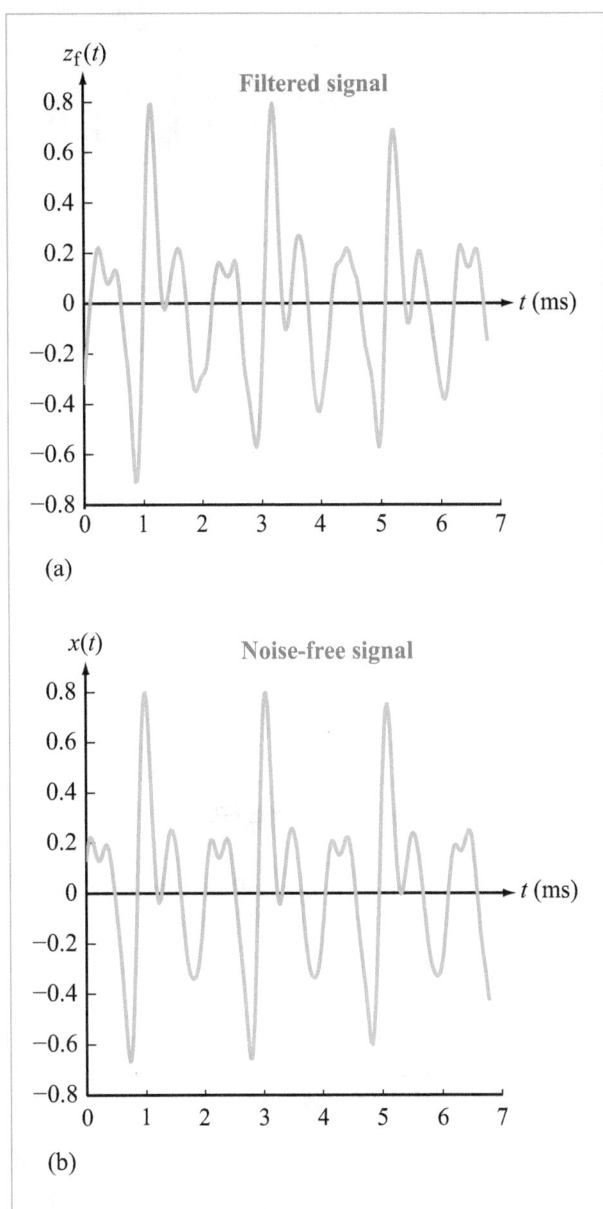

Figure 6-47: Comparison of signal filtered by a fifth-order Butterworth lowpass filter with original noise-free signal. The filtered signal is time-delayed by a fraction of a millisecond.

the two signals would not *sound* the same. This is because the Butterworth lowpass filter did not remove any of the audible part of the noise below 4000 Hz.

Instead of filtering only the high frequencies, we would like to eliminate the noise at all frequencies except the fundamental at 491 Hz and its seven harmonics. [Even though the listed frequency of note B is 494 Hz, this trumpet is playing at 491 Hz.] In other words, we would like to design an ***upside-down comb filter***! Accordingly, its transfer function should be related to that of the comb filter given by Eq. (6.86) as

$$\mathbf{H}_{\text{res}}(\omega) = 1 - \mathbf{H}_{\text{comb}}(\omega) = 1 - \prod_{\substack{k=-n \\ k \neq 0}}^{n} \left[\frac{j\omega - jk\omega_0}{j\omega + \alpha - jk\omega_0} \right]. \tag{6.116}$$

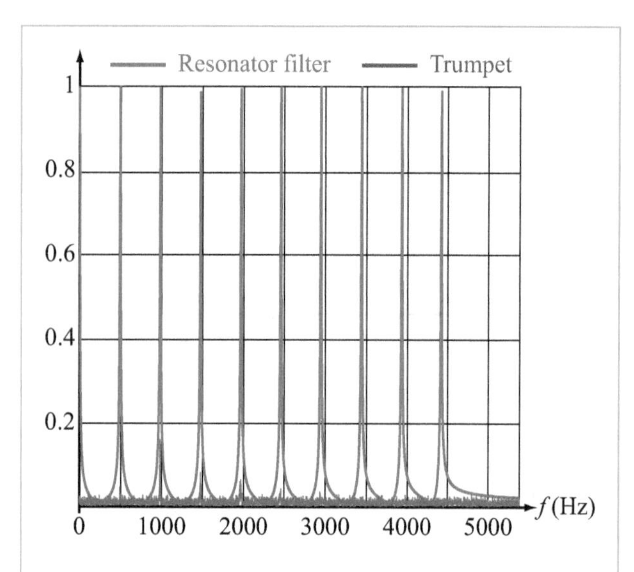

Figure 6-48: Magnitude spectra of resonator filter and noise-free trumpet signal.

▶ We call this a ***resonator filter*** because its poles, being close to the $j\omega$-axis, create resonant peaks in the magnitude spectrum $M_{\text{res}}(\omega)$ at the values of ω corresponding to the parallel pole/zero locations. ◀

It is also known as a ***line enhancement filter*** because it enhances spectral lines, while rejecting the frequencies between them.

Example 6-13: Resonator Trumpet Filter

Design a resonator filter to remove the noise from the noisy trumpet signal of Fig. 6-45(a). Generate plots of the filter's frequency and impulse responses for $\alpha = 25$ s^{-1}, and compare the filtered signal with the original noise-free trumpet signal.

Solution: The fundamental angular frequency is

$$\omega_0 = 2\pi f_0 = 2\pi \times 491 = 3085 \text{ rad/s},$$

and the number of notches is $n = 8$. Hence, Eq. (6.116) becomes

$$\mathbf{H}_{\text{res}}(\omega) = 1 - \prod_{\substack{k=-8 \\ k \neq 0}}^{8} \frac{j\omega - j3085k}{j\omega + 25 - j3085k}.$$

Figure 6-48 displays the magnitude spectra of both the resonator filter and the noise-free trumpet signal. The overlap of

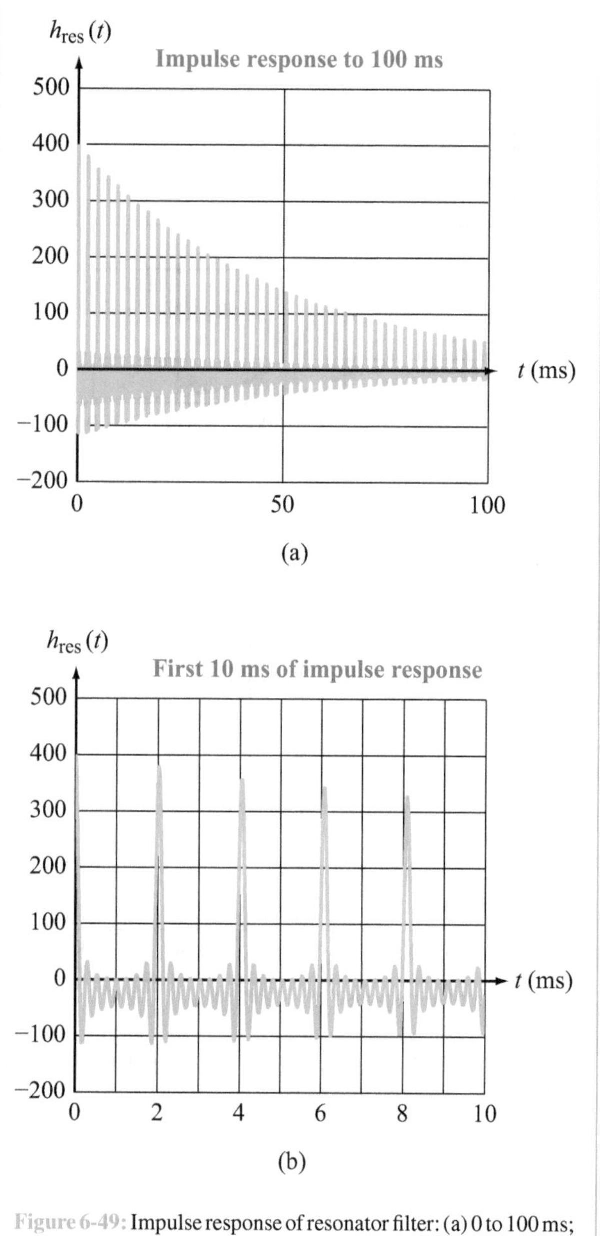

Figure 6-49: Impulse response of resonator filter: (a) 0 to 100 ms; (b) 0 to 10 ms.

the two spectra strongly indicates that the resonator filter should be able to eliminate most of the noise in the noisy trumpet signal.

Following the recipe outlined in Example 6-7, the expressions for $\mathbf{H}_{\text{res}}(\omega)$ can lead us to the impulse response $h_{\text{res}}(t)$, which is plotted in Fig. 6-49 at two different time scales. Part (a) of the figure confirms that $h(t)$ decays rapidly within

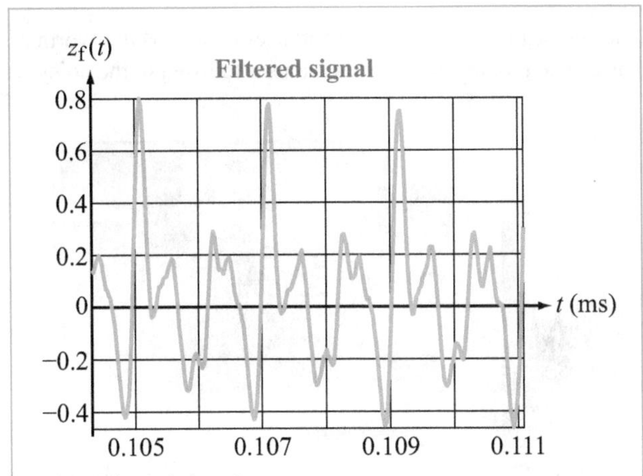

(a) Filtered signal

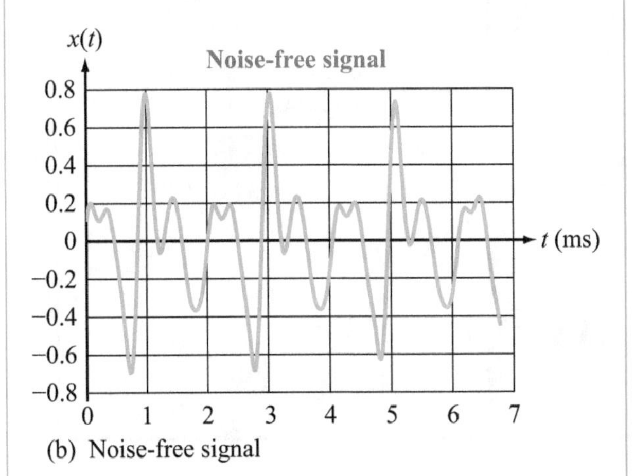

(b) Noise-free signal

Figure 6-50: Comparison of resonator-filtered signal in (a) with original noise-free signal in (b). Even though the two waveforms are not a perfect match, the two signals do sound alike.

0.2 s, and part (b) is an expanded view intended to show the detailed structure of the impulse response during the first 10 ms.

Finally, in Fig. 6-50 we provide a comparison of the resonator-filtered noisy signal with the original noise-free trumpet signal. Even though the two signal waveforms do not look exactly identical, the *sound* of the filtered signal is almost indistinguishable from that of the noise-free signal (see S^2 for more details).

Concept Question 6-15: Why did the resonator filter perform better than the lowpass filter in denoising the trumpet signal? (See S^2)

Concept Question 6-16: Why do the impulse responses of the notch and comb filters include impulses, but the impulse response of the resonator filter does not?
(See S^2)

Exercise 6-14: Obtain the transfer function of a resonator filter designed to enhance 5 Hz sinusoids. Use $\alpha = 2$.

Answer:

$$\mathbf{H}(\mathbf{s}) = \frac{4\mathbf{s}+4}{\mathbf{s}^2+4\mathbf{s}+991}.$$

(See S^2)

Exercise 6-15: Use LabVIEW Module 6.3 to denoise the noisy trumpet signal using a resonator filter, following Example 6-13. Use a noise level of 0.2.

Answer:

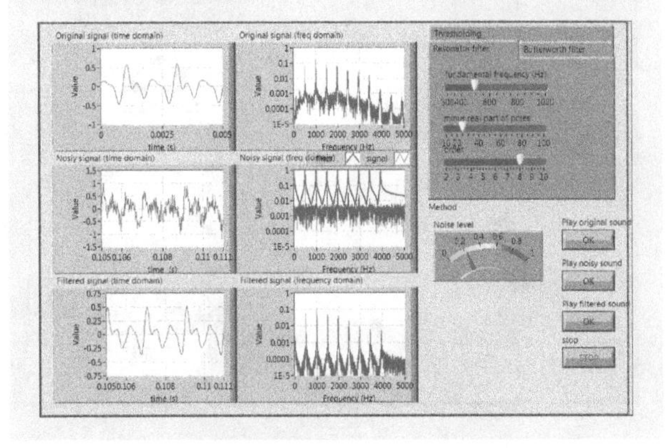

6-12 Modulation

Even though over the past 25 years most analog communication systems have been replaced with digital systems, the technique of ***modulation*** of an analog signal offers great insight into how multiple signals can be *bundled together* and transmitted simultaneously over the same channel (wires, optical fiber, free space, etc.). Accordingly, this section addresses the following topics:

(1) The term ***bandwidth*** has been assigned multiple meanings, not only in common language, but in the technical literature as well. What is the definition of bandwidth in the context of modulation?

(2) ***Double sideband*** (***DSB***) ***modulation*** is a technique used for shifting the spectrum of a signal from being centered at 0 Hz to becoming centered at two new frequencies, $\pm f_c$. How is this done and why?

Module 6.3 Filtering a Noisy Trumpet Signal This module adds noise to a trumpet signal and uses either a Butterworth or resonator lowpass filter, of orders set by the user, to reduce the noise. The module also allows the user to listen to the noisy and filtered signals.

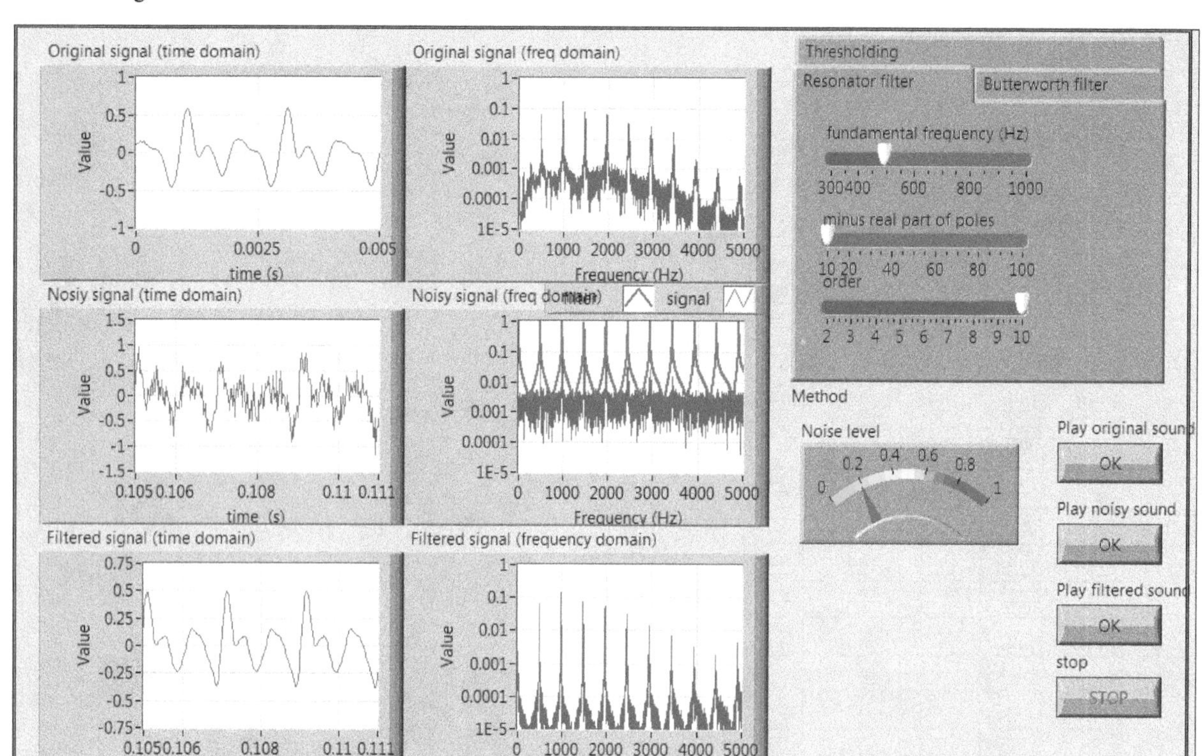

(3) ***Single sideband*** (*SSB*) modulation is similar to DSB, but it requires only half the bandwidth and power that DSB requires. How is this done?

(4) ***Frequency division multiplexing*** (*FDM*) allows us to (a) combine multiple (e.g., telephone) signals—originating from different sources—into a single combined signal, (b) transmit the combined signal to its intended destination, and then (c) unscramble the received combined signal to recover the original individual signals, as if each had traveled alone along the communication channel. How does FDM work?

(5) What is a ***superheterodyne receiver*** and how does it work?

6-12.1 Bandwidth

Men typically can hear audio signals with frequencies from around 20 Hz to around 20 kHz. The bulk of the average power of a man's voice is below 4 kHz. Women typically can hear audio signals with frequencies from around 40 Hz to around 20 kHz, and the average power of a woman's voice extends higher than that of a man's voice, to about 6 kHz. Hence, women's voices tend to be higher-pitched than men's voices. Figure 6-51 displays recorded magnitude spectra of three specific sounds, namely, "oo," "ah," and "ee." The figure also includes a typical spectrum of someone talking, which is plotted (unlike the preceding three figures) using a decibel scale. In all cases, the spectra—which had been generated by Fourier transforming the time records of the spoken sounds—decrease to low levels beyond about 3.5 kHz. Hence, we may regard these spectra as ***bandlimited*** to 3.5 kHz. This could either mean that their spectral contents are actually small (but not necessarily zero) above 3.5 kHz or that the spectra had been subjected to lowpass filtering to remove components above 3.5 kHz. In telephone communication systems, the nominal frequency range allocated

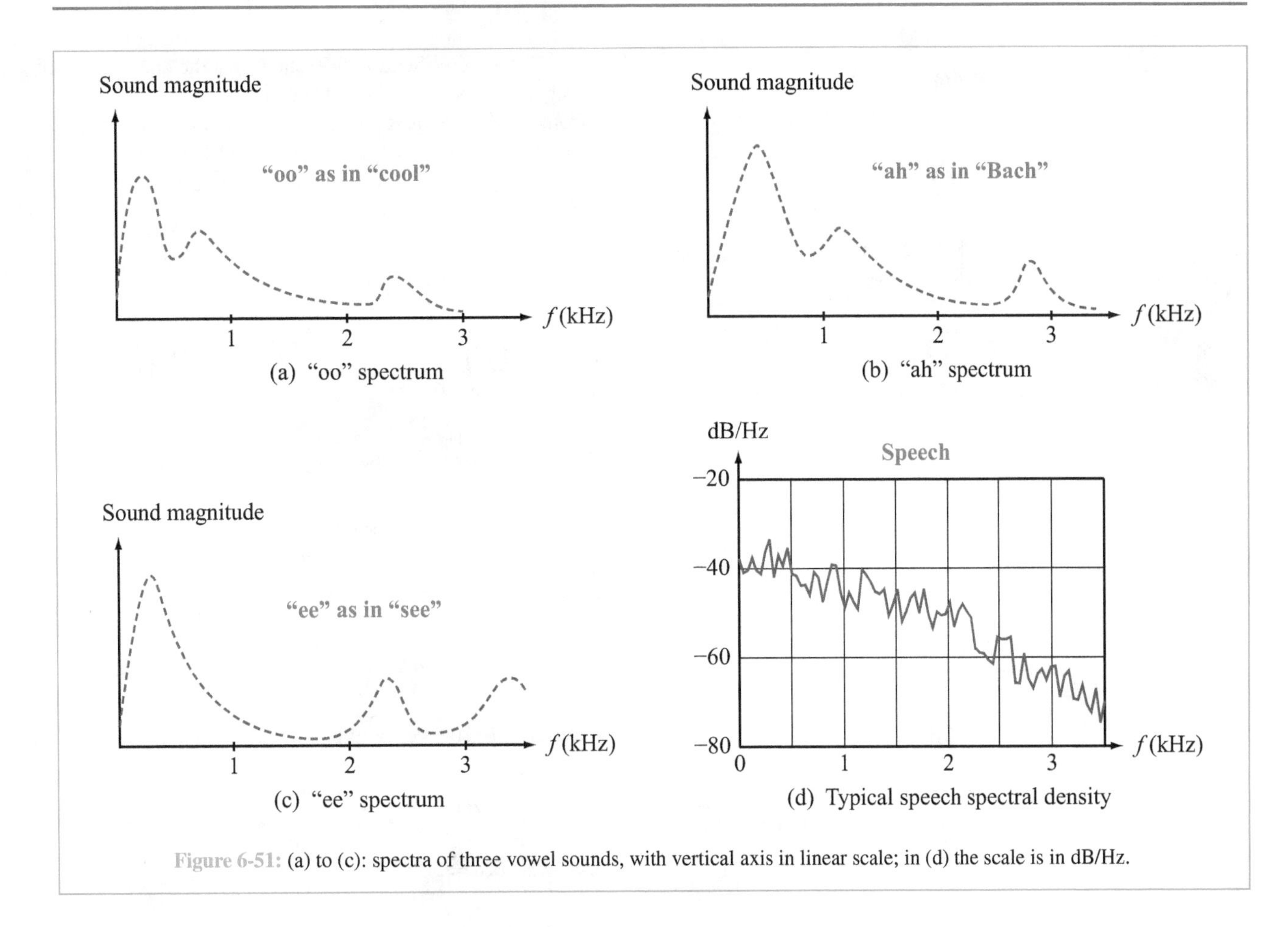

Figure 6-51: (a) to (c): spectra of three vowel sounds, with vertical axis in linear scale; in (d) the scale is in dB/Hz.

to an audio signal is 0.4–3.4 kHz. Because high frequencies are not transmitted through a telephone channel, a person's voice usually sounds differently when transmitted over the phone, as compared with when spoken directly (mouth-to-ear).

At this juncture, we should note the following:

(1) The spectra displayed in Fig. 6-51 are plotted as a function of frequency f in Hz, rather than the angular frequency ω in rad/s. This change is made in deference to the subject matter of the present section in which the terminology is more commonly expressed in terms of f rather than ω.

(2) The recorded spectra are one-sided, plotted as a function of positive values of f only. Since the signals are real-valued, their magnitude spectra are even functions of f, so one-sided plots are quite sufficient to convey all of the information carried by the spectrum.

(3) Figure 6-52 displays bidirectional spectra of two signals. Spectrum $\mathbf{X}_1(f)$ is a *lowpass signal*, non-zero for $|f| < f_{\text{max}}$, and spectrum $\mathbf{X}_2(f)$, which is the result of having passed $x_1(t)$ through an (ideal) highpass filter with cutoff frequency f_ℓ, is a *bandpass signal*. The two spectra are identical between f_ℓ and f_{max}. Based on these spectra, we offer the following bandwidth related definitions:

- Both signals are *bandlimited* to f_{max}, measured in Hz, meaning that their spectra are zero-valued for $f \geq f_{\text{max}}$.

- A signal's *bandwidth* B is the difference between the highest and lowest non-zero frequency components of its one-sided spectrum. For the lowpass signal $x_1(t)$ shown in Fig. 6-52, the bandwidth of its spectrum is $B = f_{\text{max}}$ and its bidirectional bandwidth is $B_b = 2B$, but for bandpass signal $x_2(t)$ its bandwidth is $B = f_{\text{max}} - f_\ell$.

It should be pointed out that there are many other definitions of bandwidth, including half-power bandwidth, rms bandwidth, 99% energy bandwidth, and null-to-null bandwidth. The following example illustrates 99% energy bandwidth, which

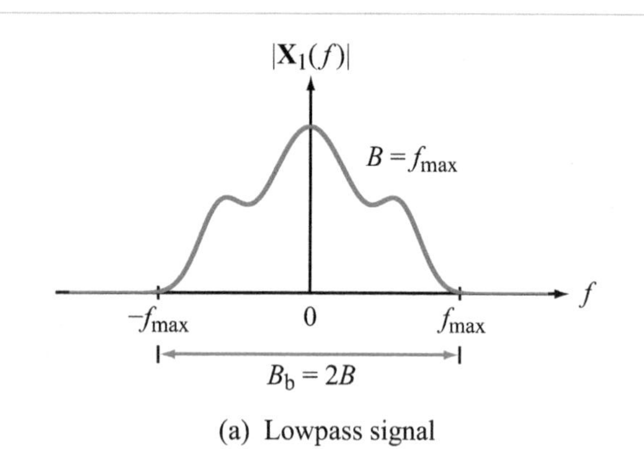

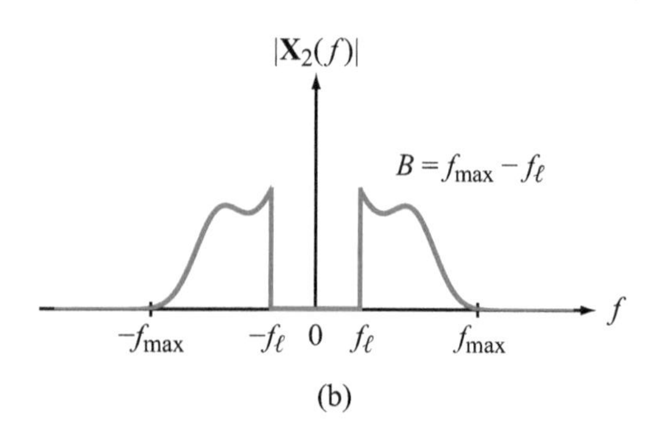

Figure 6-52: (a) Lowpass signal of bandwidth $B = f_{\max}$ and bidirectional bandwidth $B_b = 2B$, (b) bandpass signal's bandwidth is $B = f_{\max} - f_\ell$.

is the definition used by the U.S. Federal Communications Commission (FCC) for radio.

Example 6-14: 99% Bandwidth

Compute the 99% bandwidth of the exponential signal $x(t) = e^{-\alpha t}\, u(t)$, where the 99% bandwidth B_{99} is defined as the (one-sided) frequency bandwidth in Hz containing 99% of the total energy of $x(t)$. Compute B_{99} for $a = 3$.

Solution: From entry #7 in Table 5-6, the Fourier transform of $x(t)$ is

$$e^{-at}\, u(t) \quad \longleftrightarrow \quad \frac{1}{a + j\omega}.$$

The total energy E_T of $x(t)$ is easily computed in the time domain:

$$E_T = \int_{-\infty}^{\infty} |x(t)|^2 \, dt = \int_{0}^{\infty} e^{-2at} \, dt = \frac{1}{2a}.$$

Parseval's theorem (Section 5-9) states that the energy content of a signal can be determined from $|x(t)|^2$ in the time domain or from the square of its magnitude spectrum $|\mathbf{X}(\omega)|^2$ in the frequency domain. The energy contained between $-\omega_0$ and $+\omega_0$ in the spectrum shown in Fig. 6-53 is

$$E_B = \frac{1}{2\pi} \int_{-\omega_0}^{\omega_0} \left| \frac{1}{a + j\omega} \right|^2 d\omega$$

$$= \frac{1}{2\pi} \int_{-\omega_0}^{\omega_0} \frac{1}{a^2 + \omega^2} \, d\omega = \frac{1}{\pi a} \tan^{-1}\left(\frac{\omega_0}{a}\right). \tag{6.117}$$

Setting $E_B = 0.99 E_T$ leads to

$$0.99 = \frac{2}{\pi} \tan^{-1}\left(\frac{\omega_0}{a}\right),$$

which yields

$$\omega_0 = a \tan\left(\frac{0.99\pi}{2}\right) = 63.66a.$$

To convert it to Hz, we should divide by 2π. Hence,

$$B_{99} = \frac{\omega_0}{2\pi} = \frac{63.66a}{2\pi} = 10.13a. \tag{6.118}$$

For $a = 3$, $B_{99} = 30.4$ Hz.

6-12.2 Multiplication of Signals

The simplest way to transmit n different baseband signals over a given channel is to arrange them sequentially. While the process is indeed very simple, it is highly inefficient. Alternatively, we can *bundle* the signals together by attaching each to a separate carrier signal with a different ***carrier frequency***, thereby distributing the n baseband signals along the frequency axis with no overlap between their spectra. The details are covered in Section 6-12.5.

► The processes of attaching a baseband signal to a carrier signal and later detaching it from the same are called ***modulation*** and ***demodulation***, respectively. ◄

As we will see in forthcoming sections, these two processes involve the multiplication of signals. Op-amp circuits can be used to add or subtract two signals, but how do we multiply them together? The answer is the subject of the present subsection.

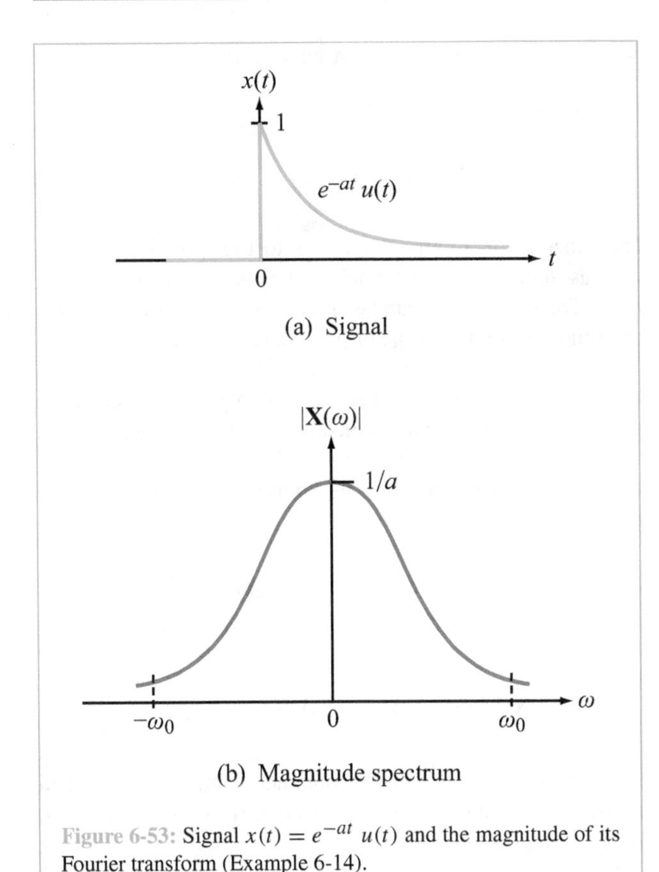

Figure 6-53: Signal $x(t) = e^{-at}\ u(t)$ and the magnitude of its Fourier transform (Example 6-14).

Consider two signals, $x(t)$ and $y(t)$. It is easy to show that their product is equivalent to

$$x(t)\ y(t) = \frac{1}{4}\left\{[x(t) + y(t)]^2 - [x(t) - y(t)]^2\right\}. \quad (6.119)$$

The multiplication operation can be realized by implementing the operations on the right-hand side of Eq. (6.119), which include scaling (by a factor of 1/4), addition and subtraction, and squaring. The recipes outlined in Section 4-5 can be used to construct op-amp circuits to perform the scaling, addition, and subtraction operations. Squaring requires the use of a non-linear ***square-law device*** whose output is proportional to the square of its input. Examples include certain types of diodes and field-effect transistors.

Implementation of the right-hand side of Eq. (6.119) is shown in block-diagram form in Fig. 6-54.

6-12.3 Switching Modulation

Modulating a signal $x(t)$ entails multiplying it by a sinusoidal signal $\cos(2\pi f_c t)$ at frequency f_c. The purpose of modulation is to shift the spectrum of $x(t)$ up and down the frequency axis by f_c (in Hz). ***Switching modulation*** is a simple and commonly used method for multiplying $x(t)$ by $\cos(2\pi f_c t)$. The following derivation of switching modulation shows yet again how useful the concepts of Fourier series, Fourier transform, and their properties can be.

Suppose $x(t)$ is a bandlimited signal with the spectrum shown in Fig. 6-55(a). The actual shape of the spectrum is irrelevant to the present discussion. Switching modulation is illustrated in Fig. 6-55(b). A switch is rapidly opened and closed, with a period $T_c = 1/f_c$. It is not necessary that the switch be closed for the same duration that it is open, although this is commonly the case. The action of opening and closing the switch is equivalent to multiplying $x(t)$ by a square wave of period $T_c = 1/f_c$, as illustrated by Fig. 6-55(c).

Without loss of generality, we choose time $t = 0$ so that the square wave, $x_m(t)$, is symmetric with respect to $t = 0$ (i.e., the square wave is an even function). According to entry #4 in Table 5-4, for $A = 1$, $\tau/T_c = 1/2$, and $T_c = 1/f_c$, $x_m(t)$ is given by the Fourier series

$$x_m(t) = \frac{1}{2} + \frac{2}{\pi}\cos(2\pi f_c t) - \frac{2}{3\pi}\cos(6\pi f_c t) + \cdots .$$

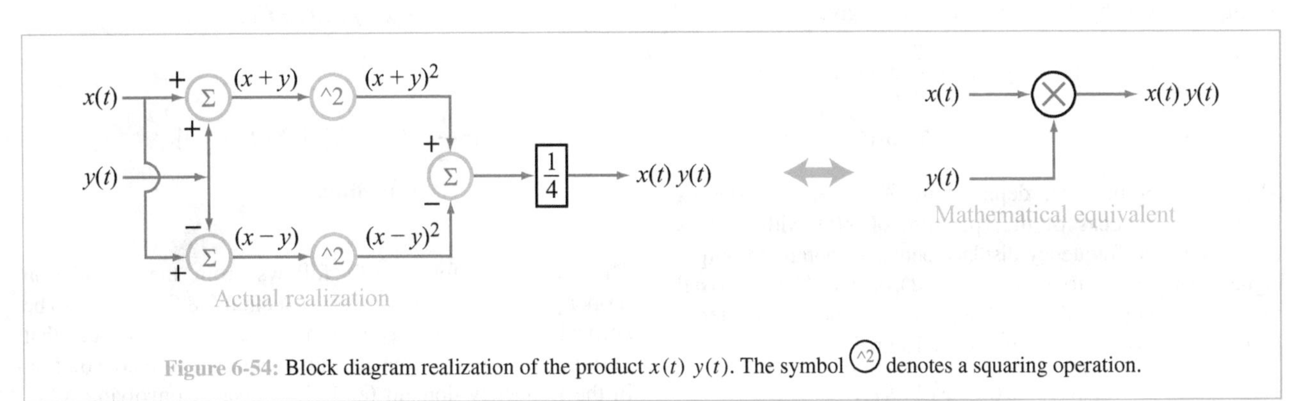

Figure 6-54: Block diagram realization of the product $x(t)\ y(t)$. The symbol (^2) denotes a squaring operation.

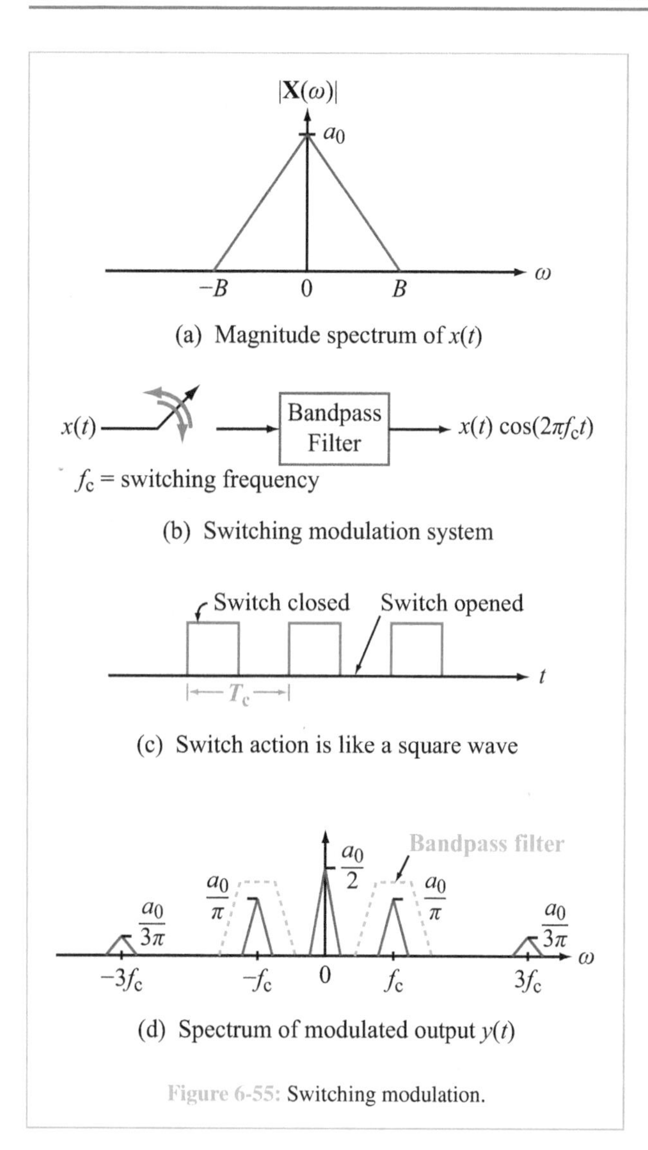

(a) Magnitude spectrum of $x(t)$

(b) Switching modulation system

(c) Switch action is like a square wave

(d) Spectrum of modulated output $y(t)$

Figure 6-55: Switching modulation.

Multiplying $x(t)$ by the square wave $x_{\mathrm{m}}(t)$ gives

$$y(t) = x(t)\, x_{\mathrm{m}}(t) = \frac{1}{2}\, x(t) + \frac{2}{\pi}\, x(t)\, \cos(2\pi f_{\mathrm{c}} t) - \frac{2}{3\pi}\, x(t)\, \cos(6\pi f_{\mathrm{c}} t) + \cdots .$$

The spectrum of $y(t)$, depicted in Fig. 6-55(d), consists of multiple replicas of the spectrum of $x(t)$ with various amplitudes and frequency displacements. To obtain an output signal proportional to $x(t)\cos(2\pi f_{\mathrm{c}} t)$, signal $y(t)$ is passed through a bandpass filter with a gain factor of $\pi/2$, centered at f_{c}, thereby generating a modulated signal

$$y_m(t) = x(t)\, \cos(2\pi f_{\mathrm{c}} t).$$

6-12.4 Double-Sideband Modulation

Consider a ***carrier signal*** of the form

$$x_{\mathrm{c}}(t) = A\cos(2\pi f_{\mathrm{c}} t + \theta_{\mathrm{c}}). \tag{6.120}$$

The function of a carrier signal is to transport information, such as audio, over a channel in the form of a propagating wave. The information can be embedded in the carrier signal's amplitude A, frequency f_{c}, or phase angle θ_{c}. Accordingly, the three processes are known as ***amplitude, frequency,*** and ***phase modulation***. Our current interest is in amplitude modulation, so we will assign f_{c} and θ_{c} constant, non-time varying values. Moreover, for simplicity and without loss of generality, we will set $\theta_{\mathrm{c}} = 0$. Hence, $x_{\mathrm{c}}(t)$ and its Fourier transform $\mathbf{X}_{\mathrm{c}}(f)$ are given by

$$x_{\mathrm{c}}(t) = A\cos(2\pi f_{\mathrm{c}} t) \quad\updownarrow\quad \mathbf{X}_{\mathrm{c}}(f) = \frac{A}{2}[\delta(f - f_{\mathrm{c}}) + \delta(f + f_{\mathrm{c}})]. \tag{6.121}$$

The expression for $\mathbf{X}_{\mathrm{c}}(f)$ was obtained from Table 5-6, then adjusted by a factor of 2π to convert it from ω to f [that is, $\delta(f) = 2\pi\ \delta(\omega)$].

Next, let us assume we have a hypothetical signal $x(t)$ representing a message we wish to transfer between two locations. The magnitude spectrum of $x(t)$ is shown in Fig. 6-56(a). It is bandlimited to B Hz, with $B \ll f_{\mathrm{c}}$. ***Double-sideband (DSB) modulation*** entails replacing A of the carrier signal with $x(t)$, which is equivalent to setting $A = 1$ and multiplying $x(t)$ by $x_{\mathrm{c}}(t)$. The outcome is a modulated signal $y_m(t)$ given by

$$y_m(t) = x(t)\cos(2\pi f_{\mathrm{c}} t) \quad\updownarrow\quad \mathbf{Y}_m(f) = \frac{1}{2}[\mathbf{X}(f - f_{\mathrm{c}}) + \mathbf{X}(f + f_{\mathrm{c}})]. \tag{6.122}$$

(DSB modulation)

The expression for $\mathbf{Y}_m(f)$ follows from the modulation property of the Fourier transform (Section 5-8.8). It can also be derived by applying entry #12 in Table 5-7, which states that multiplication in the time domain is equivalent to convolution in the frequency domain (and vice versa). Convolving $\mathbf{X}(f)$

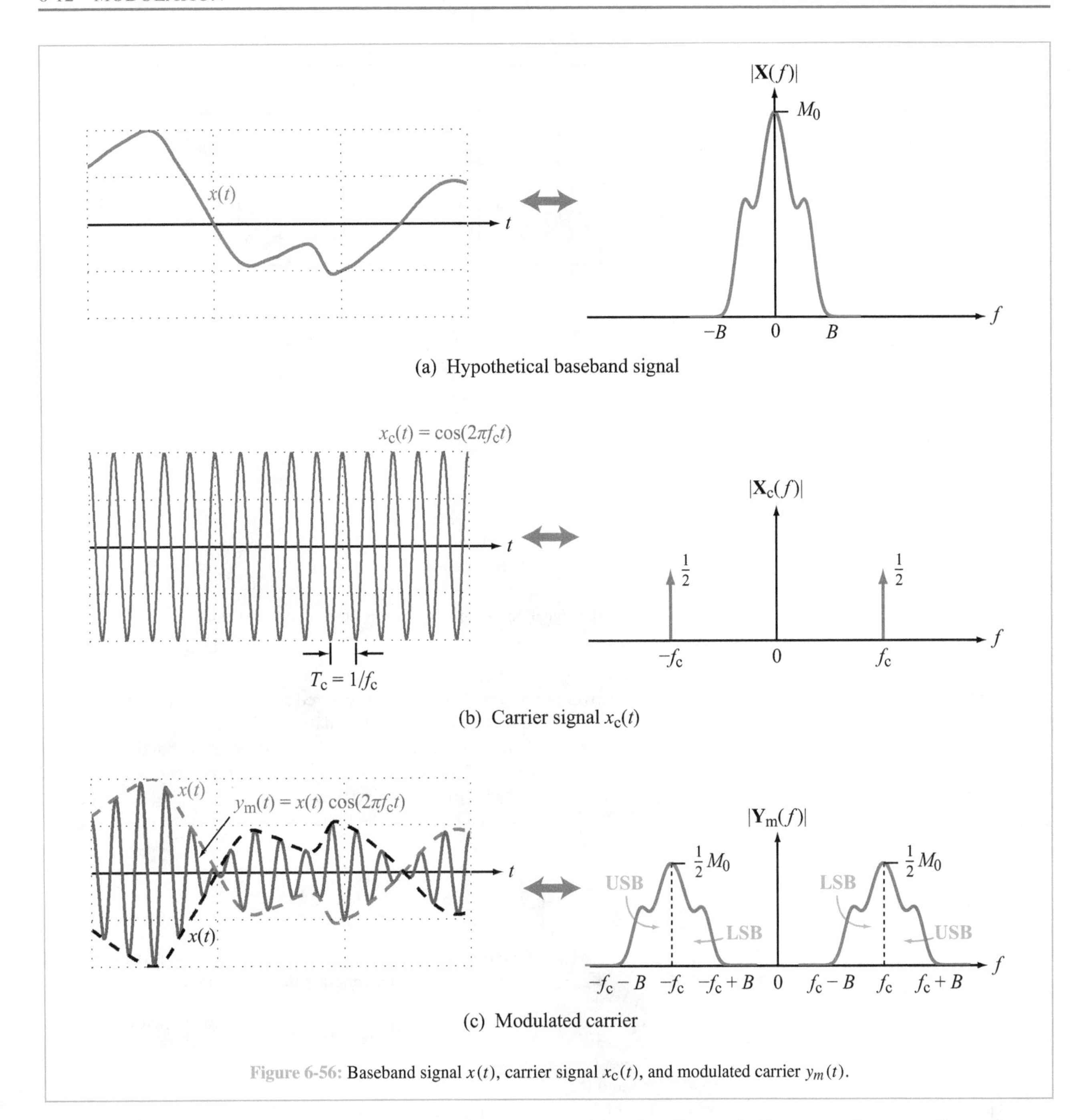

Figure 6-56: Baseband signal $x(t)$, carrier signal $x_c(t)$, and modulated carrier $y_m(t)$.

with two impulses [representing $\cos(2\pi f_c t)$] generates scaled spectra of $\mathbf{X}(f)$, one centered at f_c and another centered at $-f_c$ (Fig. 6-56(c)).

The magnitude spectrum of f_c consists of symmetrical halves, one called the ***upper sideband*** (*USB*), because it is above the carrier frequency f_c, and the other is called the ***lower sideband*** (*LSB*). A similar nomenclature applies to the two half spectra above and below $-f_c$ (Fig. 6-56(c)). Altogether, the bidirectional spectrum has two double-sidebands, each consisting of one LSB and one USB. The information characterizing the real-valued signal $x(t)$ is contained in all four sidebands.

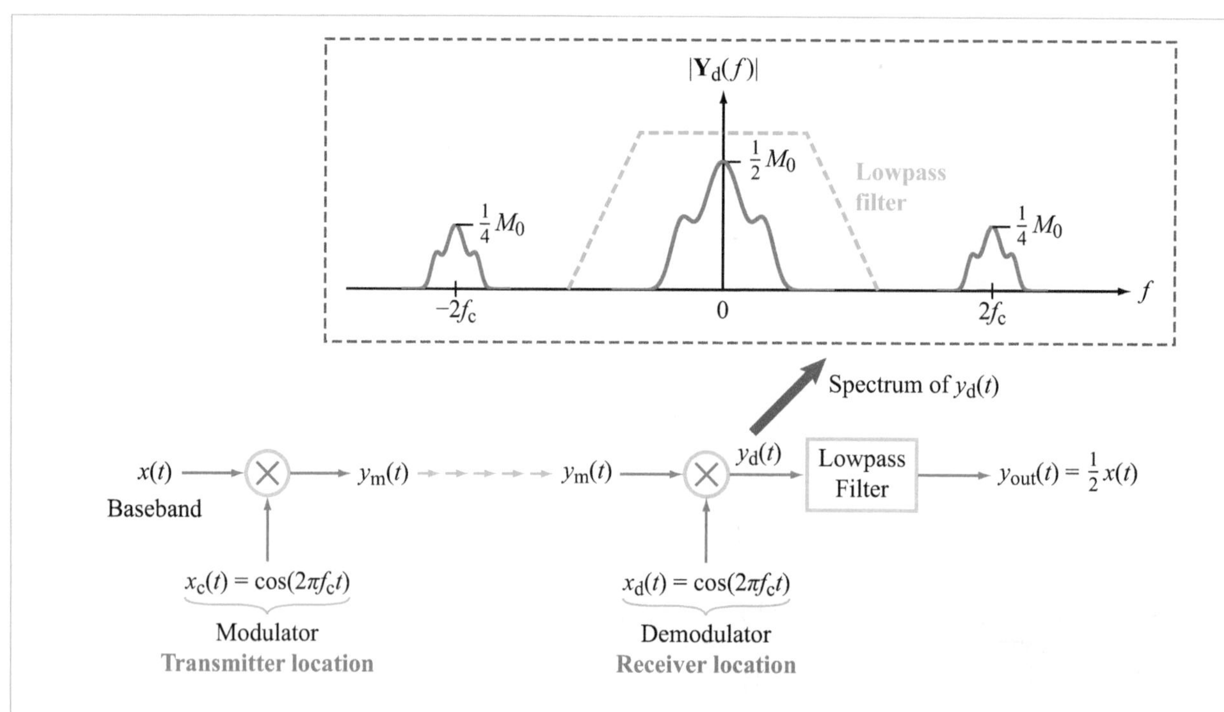

Figure 6-57: The DSB demodulator uses a replica of the carrier signal, followed by a lowpass filter, to recover the baseband signal $x(t)$.

After transmission to its intended destination, the DSB-modulated signal $y_m(t)$ is ***demodulated*** (detected) to recover the original signal $x(t)$. Demodulation consists of two steps (Fig. 6-57).

Step 1: Multiplication of $y_m(t)$ by a replica of the carrier signal generates an intermediate signal $y_d(t)$ given by

$$y_d(t) = y_m(t)\cos(2\pi f_c t) = x(t)\cos^2(2\pi f_c t).$$

Application of the trigonometric relation

$$\cos^2(\theta) = \frac{1}{2}[1 + \cos(2\theta)]$$

leads to

$$y_d(t) = \frac{1}{2}\, x(t) + \frac{1}{2}\, x(t)\cos(4\pi f_c t). \tag{6.123}$$

The first term is a scaled version of the original signal and the second term is a modulated signal at a carrier frequency of $2f_c$. Hence, the spectrum of $y_d(t)$ is

$$\mathbf{Y}_d(f) = \frac{1}{2}\,\mathbf{X}(f) + \frac{1}{4}\,[\mathbf{X}(f - 2f_c) + \mathbf{X}(f + 2f_c)]. \tag{6.124}$$

Step 2: The first term in Eq. (6.124) is centered at dc ($f = 0$), whereas the two components of the second term are centered at $\pm 2f_c$. By passing $y_d(t)$ through a lowpass filter, as shown in Fig. 6-57, the spectra centered at $\pm 2f_c$ can be removed without affecting the original signal $x(t)$. For an audio signal, $B \approx 3$ kHz, compared with an f_c on the order of hundreds of kHz (for AM) or higher, which means that the central spectrum is very far from the other two. Except for a scaling factor of $1/2$, the final output of the DSB demodulator is the original signal $x(t)$.

Example 6-15: DSB Modulation

Given signals $x_1(t) = 4\cos(8\pi t)$, $x_2(t) = 6\cos(6\pi t)$, and $x_3(t) = 4\cos(4\pi t)$, generate the spectrum of

$$y(t) = x_1(t) + x_2(t)\cos(20\pi t) + x_3(t)\cos(40\pi t).$$

Solution:

$$\begin{aligned} y(t) &= x_1(t) + x_2(t)\cos(20\pi t) + x_3(t)\cos(40\pi t) \\ &= 4\cos(8\pi t) \quad \rightarrow \quad @\ \pm 4 \text{ Hz} \\ &+ 6\cos(6\pi t)\cos(20\pi t) \quad \rightarrow \quad @\ \pm [7 \text{ Hz and } 13 \text{ Hz}] \\ &+ 4\cos(4\pi t)\cos(40\pi t) \quad \rightarrow \quad \underbrace{@\ \pm [18 \text{ Hz and } 22 \text{ Hz}]}_{\text{spectral lines}}. \end{aligned}$$

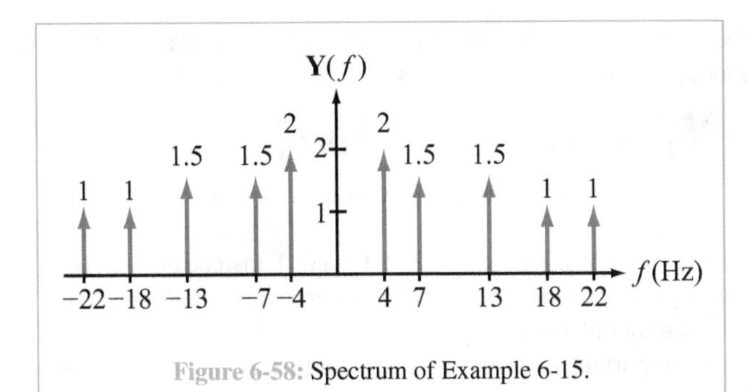

Figure 6-58: Spectrum of Example 6-15.

The spectrum of $y(t)$ is displayed in Fig. 6-58.

In most cases, the transmitter and receiver are located at great distances from one another, which means that the receiver does not have a readily available exact replica of the carrier signal $x_c(t) = \cos(2\pi f_c t)$ to use as demodulator. To recover $x(t)$, the DSB demodulator technique shown in Fig. 6-57 requires that the carrier frequency of the receiver's demodulator be identical with that of the transmitter's modulator, and that the two sinusoids are ***phase coherent*** with respect to one another. As demonstrated shortly in Example 6-16, significant deviation from these requirements can lead to serious distortion of the message embedded in $x(t)$. In view of these requirements, this type of DSB demodulator is known as a ***synchronous*** or ***coherent demodulator***.

Example 6-16: Signal Fading

The term ***fading*** refers to when a received signal "fades in and out," sometimes audibly and sometimes not. Signal fading can occur in DSB systems if the modulation carrier $x_c(t)$ and the demodulation carrier $x_d(t)$ are not in sync with one another. Given

$$x_c(t) = \cos(2\pi f_c t + \theta_c) \tag{6.125a}$$

and

$$x_d(t) = \cos(2\pi f_d t + \theta_d), \tag{6.125b}$$

determine $y_{out}(t)$, the signal at the output of the demodulator in Fig. 6-57, under each of the following conditions: (a) ***frequency offset***: $\theta_c = \theta_d = 0$, but $f_d = f_c + \Delta f$, (b) ***random phase offset***: $f_d = f_c$, but $\theta_c = 0$ and θ_d is random.

Solution:

(a) Frequency offset:
$\theta_c = \theta_d = 0$ and $f_d = f_c + \Delta f$

Using the conditions specified in Eq. (6.125) leads to the following expressions for $y_m(t)$ and $y_d(t)$:

$$\begin{aligned} y_m(t) &= x(t)\cos(2\pi f_c t), \\ y_d(t) &= y_m(t)\, x_d(t) \\ &= x(t)\cos(2\pi f_c t)\cos[(2\pi f_c + 2\pi\ \Delta f)t] \\ &= \frac{1}{2}\, x(t)\cos[2\pi(\Delta f)t] + \frac{1}{2}\, x(t)\cos[(4\pi f_c + 2\pi\ \Delta f)t]. \end{aligned}$$

After lowpass filtering the component whose spectrum is centered at $(2f_c + \Delta f)$, the final output becomes

$$y_{out}(t) = \frac{1}{2}\, x(t)\cos[2\pi(\Delta f)t]. \tag{6.126}$$

The desired baseband signal, $x(t)$, is now multiplied by a time-varying sinusoid which will oscillate between $\pm 1/2$ at a slow frequency of Δf. This is a serious distortion of the message carried by $x(t)$. If $x(t)$ is an audio signal, the Δf-oscillation will cause $x(t)$ to become very loud, then fade down to zero, then repeat the cycle over and over again. Moreover, the distortion will occur even if Δf is only a few hertz.

(b) Random phase offset:
$f_d = f_c$, $\theta_c = 0$, and θ_d is random

If the receiver is able to generate a replica of the carrier signal with identically the same frequency f_c but unable to maintain the phase angle of the receiver constant in time, then by Eq. (6.125b)

$$\begin{aligned} y_m(t) &= x(t)\cos(2\pi f_c t), \\ y_d(t) &= y_m(t)\, x_d(t) \\ &= x(t)\cos(2\pi f_c t)\cos(2\pi f_c t + \theta_d) \\ &= \frac{1}{2}\, x(t)\cos\theta_d + \frac{1}{2}\, x(t)\cos(4\pi f_c t + \theta_d). \end{aligned} \tag{6.127}$$

After removal of the second term by lowpass filtering, the output is

$$y_{out}(t) = \frac{1}{2}\, x(t)\cos\theta_d. \tag{6.128}$$

If θ_d varies randomly in time (because the transmitter and receiver are not synchronized in phase), $y_{out}(t)$ will also, which obviously is a terrible nuisance. And if $\theta_d = 90°$, the signal disappears completely!

The demodulation carrier problem is resolved by transmitting a copy of the modulation carrier along with the modulated signal. This is the basis of amplitude modulation, the subject of Section 6-12.5.

$x(t) \to \Sigma \xrightarrow{A + x(t)} \otimes \to y_m(t) = [A + x(t)] \cos(2\pi f_c t)$

Baseband signal; A dc bias; $x_c(t) = \cos(2\pi f_c t)$ carrier

Figure 6-59: Amplitude modulation includes the addition of a dc bias prior to modulation by the carrier.

6-12.5 Adding a Carrier: Amplitude Modulation

To avoid the problem of fading, we use ***amplitude modulation*** (AM). In amplitude modulation, the baseband signal $x(t)$ is dc-biased by adding a constant A to it prior to modulation by the carrier signal. The process is outlined in Fig. 6-59, and the modulated output in this case is

$$\begin{aligned} y_m(t) &= [A + x(t)] \cos(2\pi f_c t) \\ &= A \cos(2\pi f_c t) + x(t) \cos(2\pi f_c t). \end{aligned} \tag{6.129}$$

The carrier now is modulated by $[A + x(t)]$, instead of just $x(t)$, thereby transmitting a copy of the carrier (of amplitude A) along with the modulated message $x(t)$.

An important parameter in this context is the ***modulation index*** m, defined as

$$m = \frac{|x_{\min}|}{A}, \tag{6.130}$$

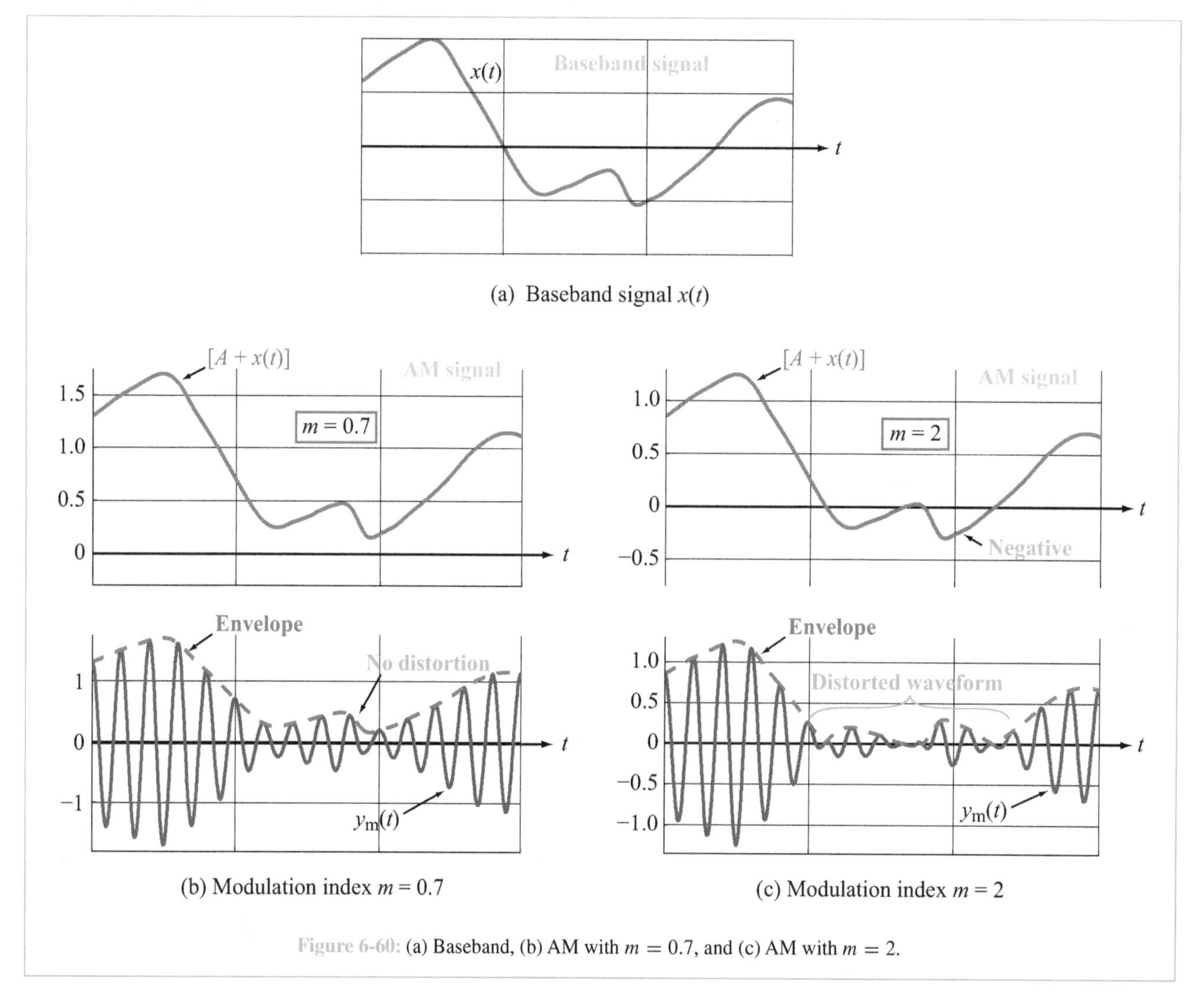

(a) Baseband signal $x(t)$

(b) Modulation index $m = 0.7$

(c) Modulation index $m = 2$

Figure 6-60: (a) Baseband, (b) AM with $m = 0.7$, and (c) AM with $m = 2$.

where $|x_{\min}|$ is the smallest negative amplitude of $x(t)$. If $m < 1$, it means that

$$A + x(t) > 0 \qquad \text{for all } t \qquad (m < 1). \tag{6.131}$$

That is, the dc bias A is large enough to raise the waveform of $x(t)$ to a level that guarantees the condition given by Eq. (6.131). The significance of this condition is illustrated by the waveform displayed in Fig. 6-60 for a hypothetical message $x(t)$. Parts (b) and (c) of the figure display AM signals with $m = 0.7$ and 2, respectively. When $m < 1$, as in Fig. 6-60(b), the upper envelope of $y_m(t)$ is identical with that of $[A + x(t)]$ because the dc bias is greater than the peak negative amplitude. Consequently, if we were to use an ***envelope detector*** (as discussed shortly) at the receiver end of the communication system to recover $[A + x(t)]$, the recovered waveform will be approximately distortion-free. In a subsequent step, the dc bias will be removed to recover the original message $x(t)$.

The envelope detector tracks the positive (upper) envelope of the modulated signal $y_m(t)$. When $m > 1$, which is the case for the modulated signal in Fig. 6-60(c), the envelope becomes ambiguous whenever $[A+x(t)]$ is negative, yielding a distorted waveform. Hence, the use of an envelope detector to recover an approximately distortionless version of the baseband message $x(t)$ is feasible only when:

$$0 < m < 1 \qquad \textbf{(envelope detector condition)}. \tag{6.132}$$

Note that DSB modulation corresponds to AM with $A = 0$ or, equivalently, $m = \infty$, ruling out the applicability of envelope detection entirely.

6-12.6 Envelope Detection

Envelope detection is provided by a circuit (an example of which is shown in Fig. 6-61) with a response fast enough to follow the variations of the envelope of $y_m(t)$, but too slow to respond to the fast oscillations associated with the carrier. If the message signal is bandlimited to $f_{\max}$, the envelope detector circuit should be designed such that its time constant τ_c satisfies the condition

$$f_{\max} \ll \frac{1}{\tau_c} \ll f_c. \tag{6.133}$$

For the simple RC circuit shown in Fig. 6-61(a), $\tau_c = RC$.

By way of an example, let us consider the case of a 1 kHz sinusoid amplitude-modulated by a 10 kHz carrier with $m = 1$.

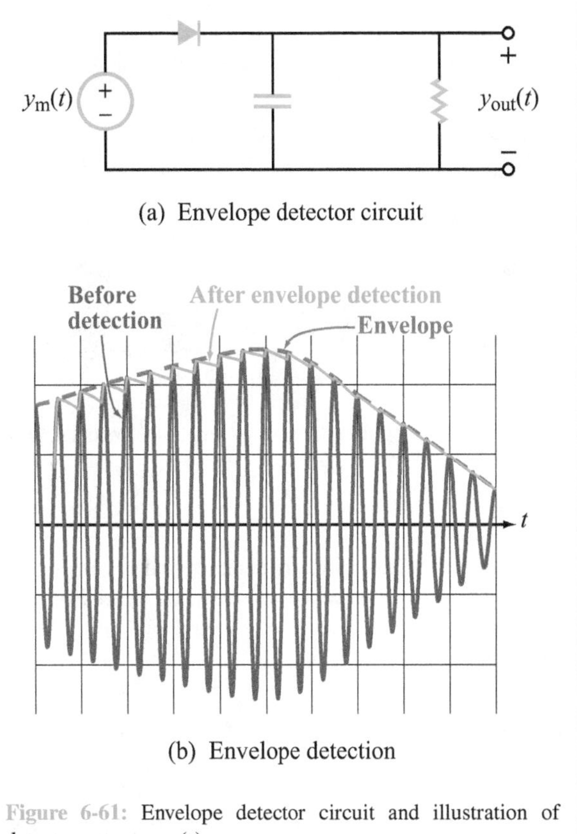

Figure 6-61: Envelope detector circuit and illustration of detector output $y_{out}(t)$.

That is,

$$\begin{aligned} x(t) &= \cos(2 \times 10^3 \pi t), \\ x_c(t) &= \cos(2 \times 10^4 \pi t), \\ y_m(t) &= [1 + x(t)]\, x_c(t) \\ &= [1 + \cos(2\pi \times 10^3 t)] \cos(2\pi \times 10^4 t). \end{aligned}$$

Because the signal is a pure sinusoid, this is called ***tone modulation***. A plot of $y_m(t)$ is displayed in Fig. 6-62; the fast-varying waveform is the 10 kHz carrier, and the envelope is a dc-biased version of $x(t)$. The envelope can be recovered by passing the waveform of $y_m(t)$ through an envelope detector (Fig. 6-61(a)) with $\tau_c \ll 1/f_c = 10^{-4}$ s.

6-12.7 Frequency Translation: Mixing

In communication systems, on both the transmitter and receiver ends, there is often a need to change the frequency of the carrier signal, while preserving the baseband signal it is carrying.

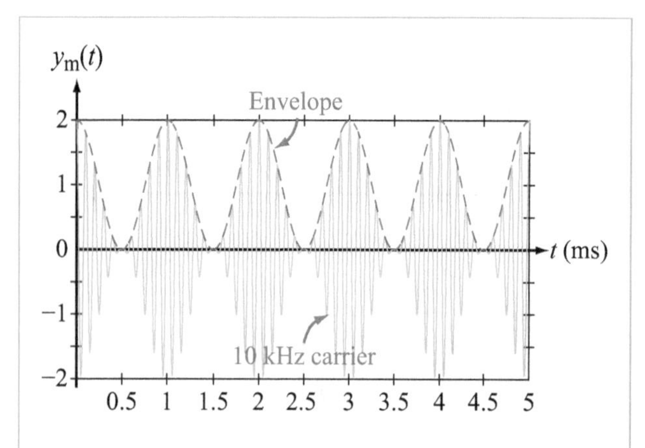

Figure 6-62: 1 kHz tone (sinusoid) amplitude modulating a 10 kHz carrier.

▶ Frequency conversion of a sinusoidal signal from f_1 to f_2 is called ***frequency translation*** or ***mixing***. If $f_2 > f_1$, it is called ***upconversion***, and if $f_2 < f_1$, it is called ***downconversion***. ◀

Changing the frequency of a sinusoidal signal is accomplished by a multiplier (see Section 6-12.2), followed by a bandpass filter. The process is outlined in Fig. 6-63. If the intent is to convert an input signal of carrier frequency f_1 into a new signal of carrier frequency f_2, we use a ***local oscillator*** at a frequency of $(f_1 + f_2)$ or $(f_1 - f_2)$. The intermediate output after the multiplication operation is

$$\begin{aligned} y(t) &= [x(t)\cos(2\pi f_1 t)]\{2\cos[2\pi(f_1 \pm f_2)t]\} \\ &= x(t)\cos(2\pi f_2 t) + x(t)\cos[2\pi(2f_1 \pm f_2)t]. \end{aligned} \quad (6.134)$$

The second term, at an oscillation frequency of $(2f_1 + f_2)$, or $(2f_1 - f_2)$, is far removed from the spectrum of the first

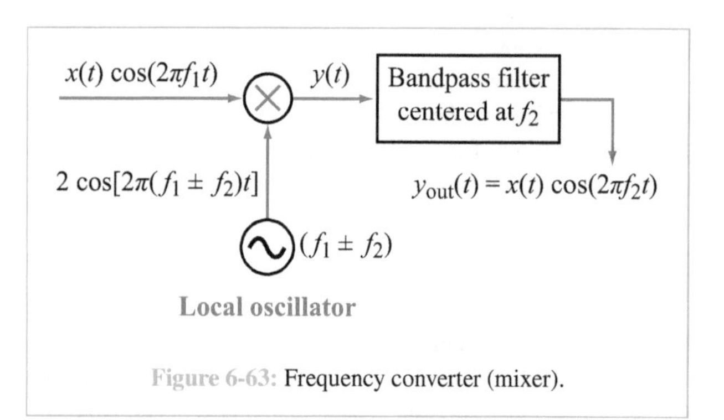

Figure 6-63: Frequency converter (mixer).

term, which is centered at f_2. Hence, if the baseband signal is bandlimited to f_{max}, the second term is removed by passing the signal through a bandpass filter centered at f_2 and of width exceeding $2f_{max}$.

Invented nearly a century ago, the ***superheterodyne receiver*** continues to be the most popular type of receiver used in support of radio communication and radar systems. Figure 6-64 shows a basic block diagram of a superheterodyne receiver. To demonstrate its operation, we will assume the receiver is connected to an antenna and the incoming modulated signal is at a carrier frequency $f_c = 1$ MHz. The ***tuner*** is a bandpass filter whose center frequency can be adjusted so as to allow the intended signal at $f_c = 1$ MHz to pass through, while rejecting signals at other carrier frequencies. After amplification by the ***radio-frequency*** (***RF***) amplifier, the AM signal can either be demodulated directly (which is what receivers did prior to 1918) or it can be converted into an IF signal by ***mixing*** it with another locally generated sinusoidal signal provided by the ***local oscillator***. The frequency of the signal at the mixer's output is

$$f_{IF} = f_{LO} - f_c, \quad (6.135)$$

where f_{LO} is the local-oscillator frequency. The frequency conversion given by Eq. (6.135) assumes that $f_{LO} \geq f_c$; otherwise, $f_{IF} = f_c - f_{LO}$ if $f_{LO} < f_c$. It is important to note that frequency conversion changes the carrier frequency of the AM waveform from f_c to f_{IF}, but the audio signal remains unchanged; it is merely getting carried by a different carrier frequency.

The diagram in Fig. 6-64 indicates that the tuning knob controls the center of the adjustable tuner as well as the local oscillator frequency. By *synchronizing* these two frequencies to each other, the IF frequency always remains constant. This is an important feature of the superheterodyne receiver because it insures that the same ***IF filter/amplifier*** can be used to provide high-selectivity filtering and high-gain amplification, regardless of the carrier frequency of the AM signal. In the ***AM radio band***, the carrier frequency of the audio signals transmitted by an AM radio station may be at any frequency between 530 kHz and 1610 kHz. Because of the built-in synchronization between the tuner and the local oscillator, the IF frequency of an AM receiver is always at 455 kHz, which is the standard IF for AM radio. Similarly, the standard IF for FM radio is 10 MHz, and the standard IF for TV is 45 MHz.

It is impractical to design and manufacture high-performance components at every frequency in the radio spectrum. By designating certain frequencies as IF standards, industry was able to develop devices and systems that operate with very high performance at those frequencies. Consequently, frequency conversion to an IF band is very prevalent not only in radio and

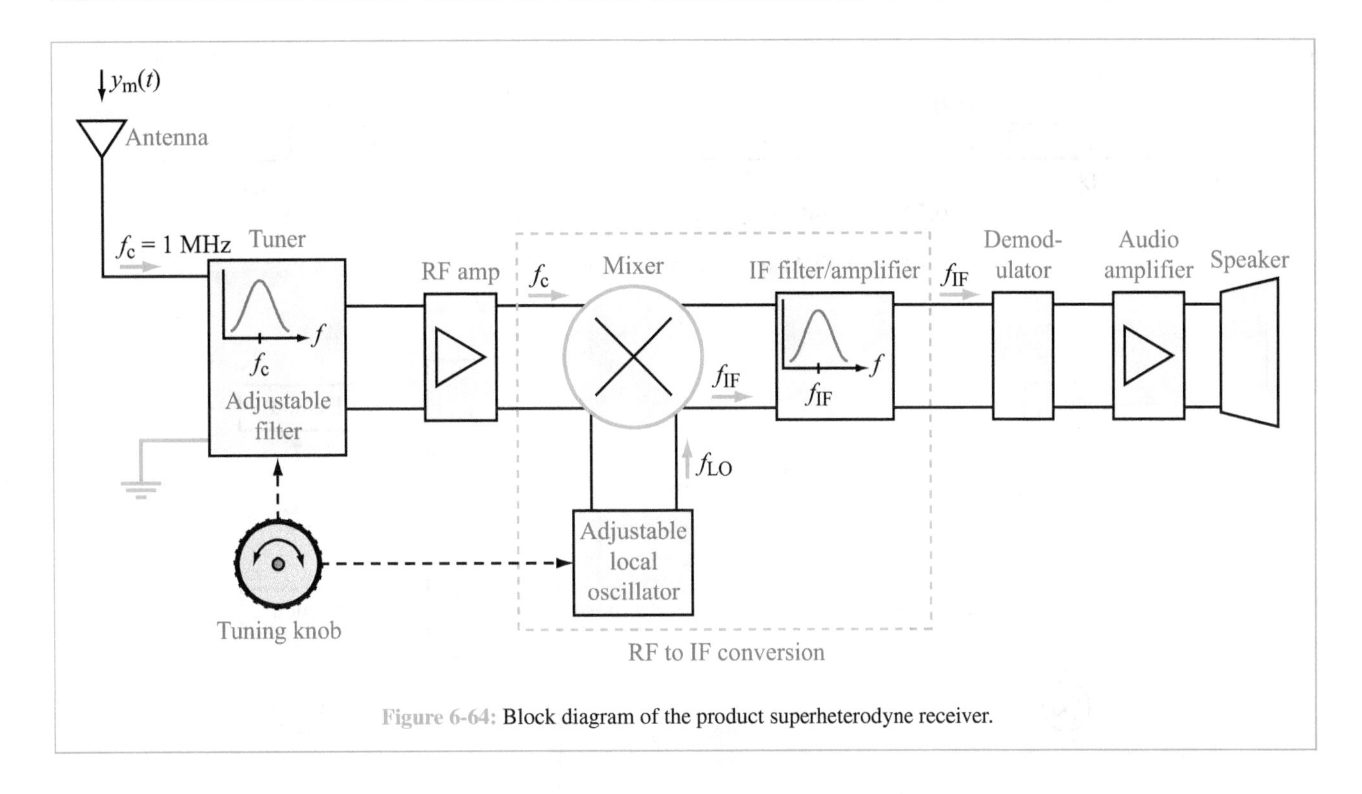

Figure 6-64: Block diagram of the product superheterodyne receiver.

TV receivers, but also in radar sensors, satellite communication systems, and transponders, among others.

After further amplification by the IF amplifier, the modulated signal is demodulated to recover the baseband signal $x(t)$.

6-12.8 Frequency Division Multiplexing (*FDM*)

We noted earlier in this section that multiplication of a baseband message $x(t)$ by a carrier $x_c(t) = \cos(2\pi f_c t)$ leads to duplicate spectra, each of bandwidth $2B$, located at $\pm f_c$ (Fig. 6-56). Each of the two spectra consists of an upper sideband (*USB*) and a lower sideband (*LSB*). Either of the two sidebands, which have even amplitude symmetry and odd phase symmetry, contains all of the information needed to reconstruct $x(t)$. By eliminating one of the sidebands, the bandwidth of the modulated signal is reduced from $2B$ to B, and the signal is then called a ***single sideband*** (*SSB*) signal. The SSB signal is generated from the DSB either by filtering one of the sidebands—which is rather difficult because it requires the use of a brick-wall filter—or by applying a more practical technique called ***phase-shift modulation*** using a ***Hilbert transform***, as shown later in Section 6-12.9.

Suppose we wish to transmit as many phone conversations as possible, each of bandwidth $B = 3.5$ kHz, using radio signals with carrier frequencies in the 1.0 GHz to 1.01 GHz range. Thus, the bandwidth of the available transmission band is 0.01 GHz, or 10 MHz. If we can arrange to modulate each of the phone signals by a different carrier frequency—called ***subcarriers***—so that they occupy separate, 3.5 kHz wide bands across the available transmission bandwidth, we can in principle ***multiplex*** (combine) a large number of signals together. To avoid interference between signals, the spectra of adjacent signals are separated by ***guard bands***. In the present example, if we allow a guard band of 0.5 kHz between adjacent signal spectra (Fig. 6-65(a)), we can potentially multiplex a total of

$$n = \frac{10 \text{ MHz}}{(3.5 + 0.5) \text{ kHz}} = 2500 \text{ signals.}$$

Combining signals along the frequency axis is called ***frequency division multiplexing*** (*FDM*). The FDM process is diagrammed in Fig. 6-65(b) and (c) for the transmitter and receiver ends, respectively. At the transmitter, signal $x_1(t)$ is modulated by subcarrier frequency f_1, $x_2(t)$ by subcarrier frequency f_2, and so on, through $x_n(t)$. Then, the signals are added together and transmitted to their intended destination. At the receiver end, the combined FDM signal is subjected to n parallel bandpass filters yielding n separate modulated signals. Each signal is then demodulated at its respective frequency. The final outcome consists of n separate telephone signals, all

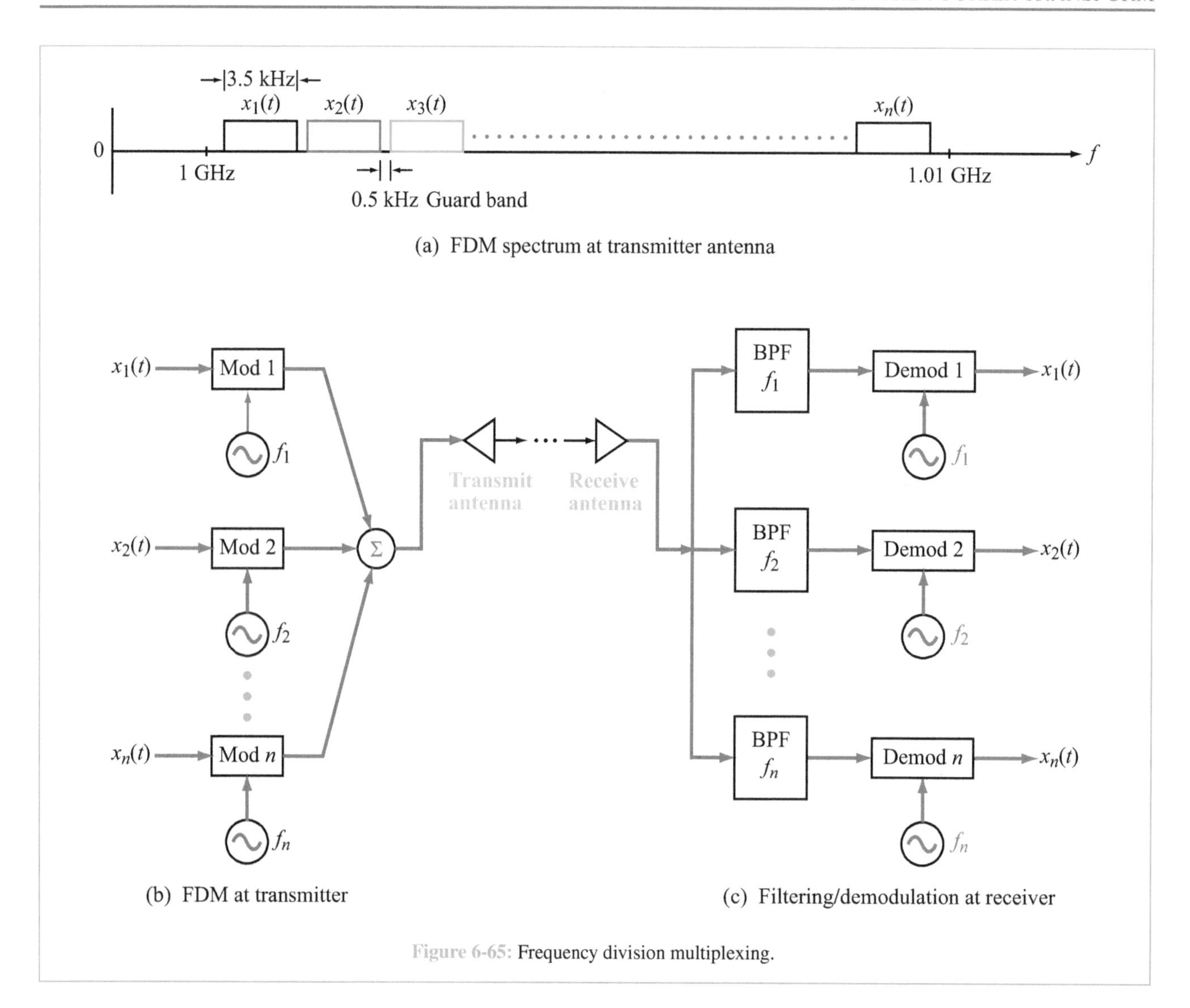

Figure 6-65: Frequency division multiplexing.

of which had traveled together along the same channel at the same time.

Example 6-17: AM versus DSB Spectra

A 1 Hz sinusoid $x(t) = 2\cos(2\pi t)$ is modulated by a 50 Hz carrier. Plot the time waveforms and associated spectra of the modulated sinusoid for (a) DSB modulation and (b) AM with a modulation index of 1.

Solution:

(a) With DSB, the modulated signal is given by

$$y_m(t) = x(t)\cos(2\pi f_c t) = 2\cos(2\pi t)\cos(100\pi t).$$

The waveform of $y_m(t)$ is displayed in Fig. 6-66(a), and the associated spectrum consists of spectral lines at the sum and differences between the frequency of $x(t)$ and that of the carrier. That is, at ± 49 Hz and ± 51 Hz, as shown in Fig. 6-66(b).

(b) With AM, the modulated signal is given by Eq. (6.129). For $m = 1$,

$$y_m(t) = [2 + 2\cos(2\pi t)]\cos(100\pi t).$$

The waveform and its spectrum are displayed in Fig. 6-67. We note that in addition to the spectral lines at ± 49 Hz and ± 51 Hz, the AM spectrum also includes a spectral line at the carrier frequency of 50 Hz.

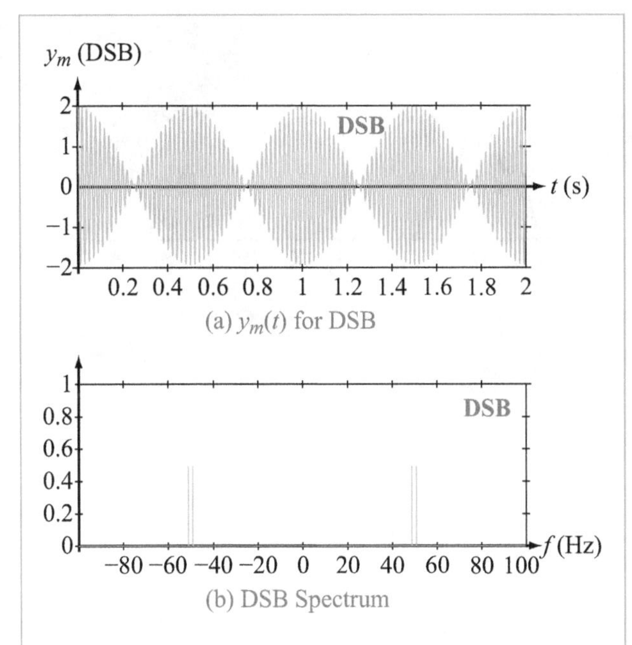

Figure 6-66: Waveform and line spectrum of DSB modulated sinusoid.

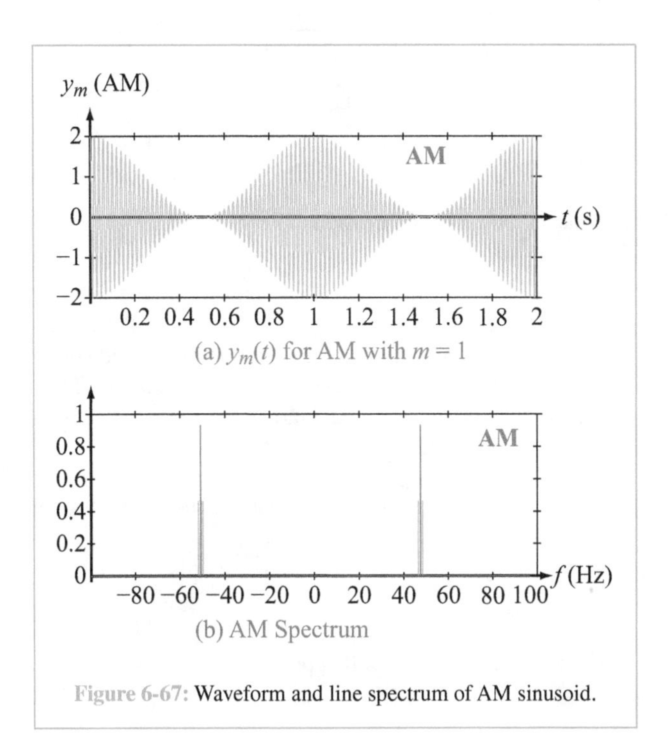

Figure 6-67: Waveform and line spectrum of AM sinusoid.

▶ Note that the envelope of the AM signal is $[1 + \cos(2\pi t)]$, so the tone $x(t)$ can be recovered from $y_m(t)$ using envelope detection. In contrast, the envelope of the DSB-modulated signal is $2|\cos(2\pi t)|$, so only $|x(t)|$, not $x(t)$, can be recovered from $y_m(t)$ using envelope detection. ◀

Concept Question 6-17: Explain how multiple signals can be combined together into a single signal, which, after transmission, can be "un-combined" into the original individual signals? (See S²)

Concept Question 6-18: What is a carrier signal? How is it used? (See S²)

Exercise 6-16: Given 20 signals, each of (two-sided) bandwidth $B_b = 10$ kHz, how much total bandwidth would be needed to combine them using FDM with SSB modulation and no guard bands between adjacent signals?

Answer: 100 kHz. (See S²)

Exercise 6-17: Figure E6-17 depicts the frequency bands allocated by the U.S. Federal Communications Commission (FCC) to four AM radio stations. Each band is 8 kHz in extent. Suppose radio station WJR (with carrier frequency of 760 kHz) were to accidentally transmit a 7 kHz tone, what impact might that have on other stations? (Even though the four stations are separated by long distances, let us assume they are close to one another.)

Answer: After modulation by the 760 kHz carrier, the 7 kHz tone gets transmitted at frequencies equal to the sum and difference of the two frequencies, namely 753 kHz and 767 kHz. Listeners tuned to WSB (750 kHz) and WABC (770 kHz) will hear a tone at 3 kHz.

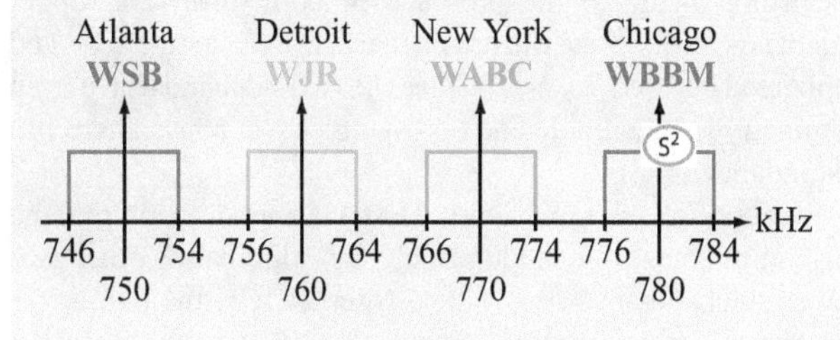

Figure E6-17

(See S²)

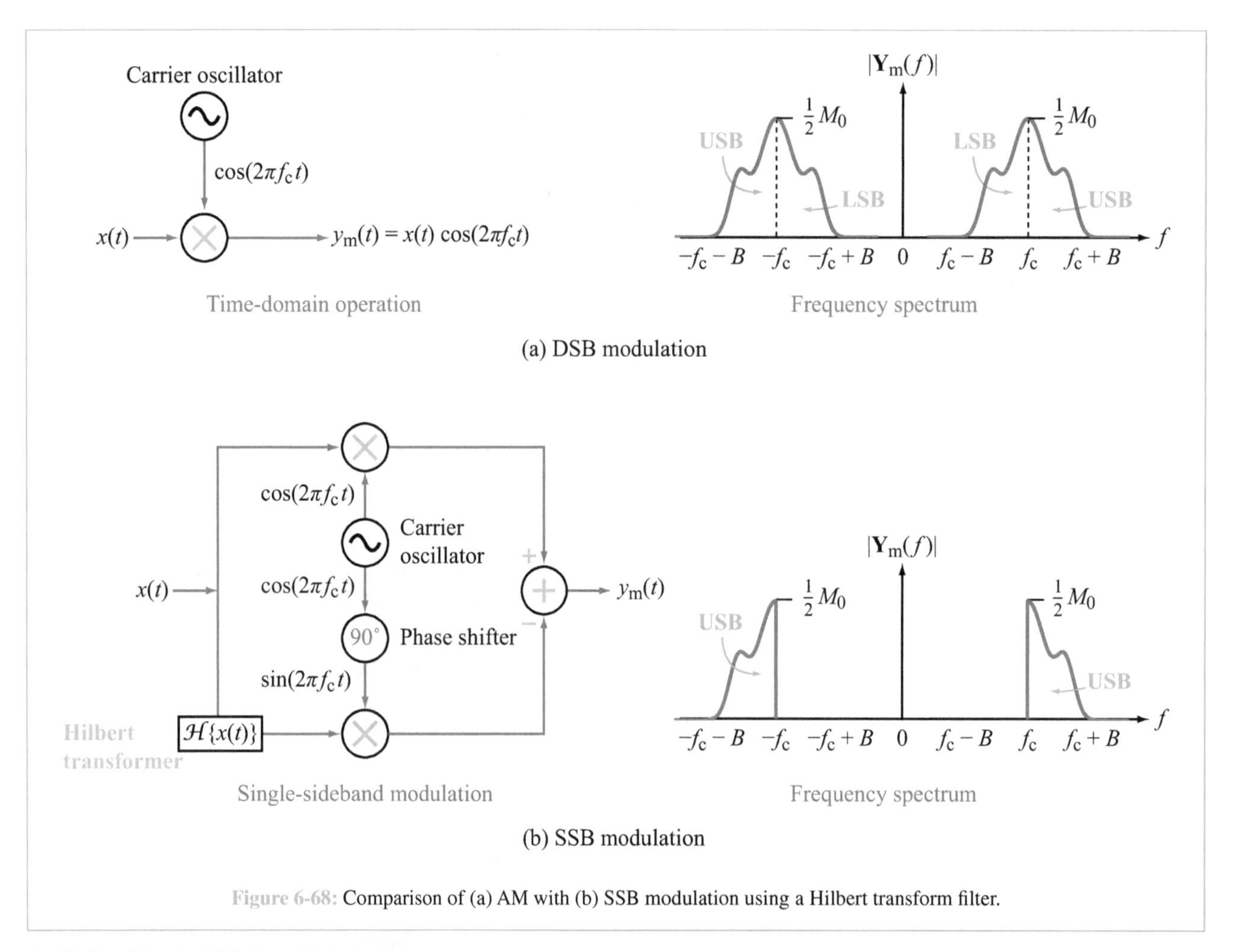

Figure 6-68: Comparison of (a) AM with (b) SSB modulation using a Hilbert transform filter.

6-12.9 Single Sideband Modulation

One problem with all of the modulation schemes (DSB, AM and FDM) presented so far in this section is that they are inefficient in their use of the frequency spectrum. A signal whose maximum frequency is B Hz occupies $2B$ Hz of spectrum, owing to the presence of both lower and upper sidebands. This was illustrated earlier in Fig. 6-56(c) and replicated here as Fig. 6-68. Since the two sidebands have even symmetry in magnitude and odd symmetry in phase, there is a redundancy.

Single sideband modulation (*SSB*) is a modulation scheme that eliminates this redundancy by eliminating the two lower sidebands in Fig. 6-68(a). Alternatively, the two upper sidebands can be eliminated, but we will consider only the case of removing the two lower sidebands. The advantages of SSB modulation over DSB modulation are as follows: (1) The amount of spectrum required to transmit a signal is halved; (2) the amount of power required to transmit a signal is halved; and (3) twice as many signals (of equal bandwidths) can be transmitted in the same amount of spectrum. SSB modulation is used in CB and ham radios, DSL modems, and digital TV.

One way to perform SSB modulation would be to use a brick-wall lowpass filter that passes only the lower sidebands and eliminates the upper sidebands, but brick-wall lowpass filters do not exist in the real world. Instead, we use the system shown in Fig. 6-68(b).

Hilbert transform

It will be convenient later in this presentation to introduce the ***signum function*** $\text{sgn}(x)$ as

$$\text{sgn}(x) = \begin{cases} 1 & \text{for } x > 0, \\ -1 & \text{for } x < 0. \end{cases} \tag{6.136}$$

We also define the ***Hilbert transform*** $\mathcal{H}\{x(t)\}$ of a signal $x(t)$ as the LTI system with frequency response

$$\mathbf{H}(j\omega) = \begin{cases} -j & \text{for } \omega > 0 \\ +j & \text{for } \omega < 0 \end{cases} = -j\,\text{sgn}(\omega). \tag{6.137}$$

The impulse response of the Hilbert transform is the inverse Fourier transform of $\mathbf{H}(j\omega)$, namely

$$\mathcal{H}(t) = \frac{1}{2\pi}\int_{-\infty}^{0} je^{j\omega t}\,d\omega + \frac{1}{2\pi}\int_{0}^{\infty} -je^{j\omega t}\,d\omega. \tag{6.138}$$

$\mathbf{H}(j\omega)$ is a pure imaginary and odd function of ω. Hence, by conjugate symmetry, its inverse Fourier transform $\mathcal{H}(t)$ is a real-valued and odd function of t. Indeed, after a few steps of algebra, $\mathcal{H}(t)$ turns out to be

$$\mathcal{H}(t) = \frac{1}{\pi t}. \tag{6.139}$$

The Hilbert transform is often implemented in discrete time. A discrete-time filter for the Hilbert transform is derived later in Section 9-3.

6-12.10 Implementation

Now consider the modulation scheme shown in Fig. 6-68(b). The signal $x(t)$ is modulated by multiplying it by a carrier $\cos(2\pi f_c t)$, as in DSB and AM modulation. The Hilbert transform $\mathcal{H}\{x(t)\}$ of $x(t)$ is modulated by multiplying it by the phase-shifted carrier $\sin(2\pi f_c t)$. The difference of these two modulated signals gives the SSB-modulated signal $y_m(t)$.

To see why this system eliminates the lower sidebands of the SSB-modulated signal, let $\mathbf{X}(\omega)$ be the Fourier transform of the signal $x(t)$. Then, using Eq. (6.137), we have the following Fourier transform pair:

$$\mathcal{H}\{x(t)\} \quad \longleftrightarrow \quad -j\,\text{sgn}(\omega)\,\mathbf{X}(\omega). \tag{6.140}$$

Recall from Appendix C that

$$\cos(2\pi f_c t) = \frac{e^{j2\pi f_c t}}{2} + \frac{e^{-j2\pi f_c t}}{2} \tag{6.141a}$$

and

$$\sin(2\pi f_c t) = \frac{e^{j2\pi f_c t}}{2j} - \frac{e^{-j2\pi f_c t}}{2j} = \frac{-je^{j2\pi f_c t}}{2} + \frac{je^{-j2\pi f_c t}}{2}. \tag{6.141b}$$

Using the frequency shift property of the Fourier transform in Table 5-7, we have the following Fourier transform pairs:

$$x(t)\cos(2\pi f_c t) \quad \longleftrightarrow \quad \frac{\mathbf{X}(\omega - 2\pi f_c)}{2} + \frac{\mathbf{X}(\omega + 2\pi f_c)}{2}, \tag{6.142a}$$

$$x(t)\sin(2\pi f_c t) \quad \longleftrightarrow \quad \frac{-j\mathbf{X}(\omega - 2\pi f_c)}{2} + \frac{j\mathbf{X}(\omega + 2\pi f_c)}{2}. \tag{6.142b}$$

Applying Eq. (6.142)(b) to Eq. (6.140) requires replacing each ω in Eq. (6.140) with $\omega - 2\pi f_c$, and then with $\omega + 2\pi f_c$. This gives the Fourier transform pair

$$\begin{aligned} &\mathcal{H}\{x(t)\}\sin(2\pi f_c t) \quad \longleftrightarrow \\ &\quad -\frac{\text{sgn}(\omega - 2\pi f_c)\,\mathbf{X}(\omega - 2\pi f_c)}{2} \\ &\quad + \frac{\text{sgn}(\omega + 2\pi f_c)\,\mathbf{X}(\omega + 2\pi f_c)}{2}. \end{aligned} \tag{6.143}$$

Now, note that

$$\frac{1}{2} + \frac{\text{sgn}(\omega - 2\pi f_c)}{2} = \begin{cases} 1 & \text{for } \omega > 2\pi f_c, \\ 0 & \text{for } \omega < 2\pi f_c. \end{cases} \tag{6.144}$$

Subtracting Eq. (6.143) from Eq. (6.142)(a) and using Eq. (6.144) gives

$$\begin{aligned} y_m(t) &= x(t)\cos(2\pi f_c t) - \mathcal{H}\{x(t)\}\sin(2\pi f_c t) \\ &\longleftrightarrow \begin{cases} \mathbf{X}(\omega - 2\pi f_c) & \text{for } \omega > 2\pi f_c, \\ 0 & \text{for } -2\pi f_c < \omega < 2\pi f_c, \\ \mathbf{X}(\omega + 2\pi f_c) & \text{for } \omega < -2\pi f_c. \end{cases} \end{aligned} \tag{6.145}$$

Hence, the lower sidebands of the spectrum of $y_m(t)$ have been eliminated (see Fig. 6-68). The SSB-modulated signal $y_m(t)$ can be demodulated to $x(t)$ using the DSB demodulator shown in Fig. 6-57.

6-13 Sampling Theorem

Today, most processing applied to signals, both natural and artificial, is performed in the digital domain. The audio signal picked up by the built-in microphone in a mobile phone is digitized and coded prior to transmission. Upon reception, the digital bits are converted back into an analog signal, which then drives the speaker of the receiving mobile phone.

Conversion between the analog and digital domains is performed by analog-to-digital converters (ADC) and digital-to-analog converters (DAC). The ADC process consists of two steps (Fig. 6-69):

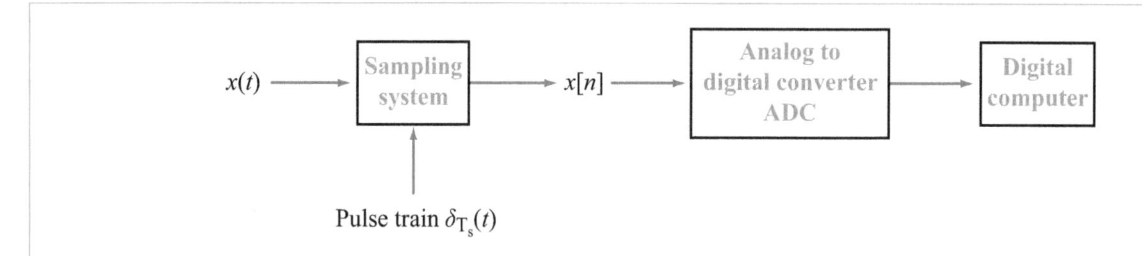

Figure 6-69: The sampling system is like a switch that turns on for an instant every T_s seconds, which is equivalent to multiplying the signal $x(t)$ by a pulse train $\delta_{T_s}(t) = \sum_{n=-\infty}^{\infty} \delta(t - nT_s)$ and then reading off the areas under the impulses to obtain $x[n]$.

(a) sampling the signal at some regular interval T_s and then

(b) converting the analog signal values into digital sequences.

▶ The sampling rate determines how well the variation of the signal with time is captured by the sampling process, whereas the number of bits used to encode the sampled values determines the resolution with which the signal intensity is captured. These are independent processes, one dealing with discretization as a function of time and the other as a function of signal intensity. ◀

An implicit assumption of the sampling process that converts a continuous-time signal $x(t)$ into a ***discrete-time signal*** $x[n]$ is that the reverse process is equally possible [that is, $x(t)$ can be reconstituted from $x[n]$, without error]. Whether or not the reverse process can be realized in practice depends on a sampling-rate condition that involves the sampling rate $f_s = 1/T_s$, relative to the signal bandwidth B. This condition is one of the important attributes of the ***sampling theorem***, the subject of the present section. The sampling theorem is the transition vehicle between continuous-time signals and systems and their counterparts in discrete time (i.e., between the material covered thus far and the material in forthcoming chapters).

The second of the aforementioned steps of the ADC process amounts to storing and processing the analog signal values with finite precision. For the rest of this book, we will assume that all signal values are known to infinite precision. We now focus on the first step: sampling continuous-time signals.

6-13.1 Sampling a Continuous-Time Signal

To ***sample*** a signal $x(t)$, we retain only its values at times t that are integer multiples nT_s of some small time interval T_s. That is, we keep only the values $x(nT_s)$ of $x(t)$ and discard all other values. Sampling converts the continuous-time signal $x(t)$ into a discrete-time signal $x[n]$:

$$x[n] = x(nT_s), \qquad n \in \{\text{integers}\} \tag{6.146}$$

▶ Note that discrete-time notation uses square brackets and index n, instead of curved brackets and time t. The sequence of numbers $x[n]$ is what is stored and processed on a digital device. ◀

The ***sampling interval*** T_s is therefore a discretization length. The ***sampling rate*** f_s is defined as

$$f_s = \frac{1}{T_s}. \tag{6.147}$$

For CDs, the sampling rate is 44,100 samples per second, and the sampling interval is

$$T_s = 1/44100 = 22.7\ \mu\text{s}.$$

The physical act of sampling a continuous-time voltage signal $x(t)$ can be viewed as closing a switch for an instant every T_s seconds, or at a rate of f_s times per second, and storing those values of $x(t)$. By way of an example, consider a sinusoidal signal $x(t) = \cos(2\pi 1000t)$, sampled at $f_s = 8000$ samples per second. The sampling interval is $T_s = 1/8000$ s, and the sampled signal is

$$x[n] = x\left(t = \frac{n}{8000}\right) = \cos\left(2\pi \frac{1000n}{8000}\right) = \cos\left(\frac{\pi}{4}n\right). \tag{6.148}$$

The result is shown in Fig. 6-70. The heights of the circles indicate the values of $x[n]$, and the stems connect them to the times $t = nT_s = n/f_s = n/8000$ at which they are sampled.

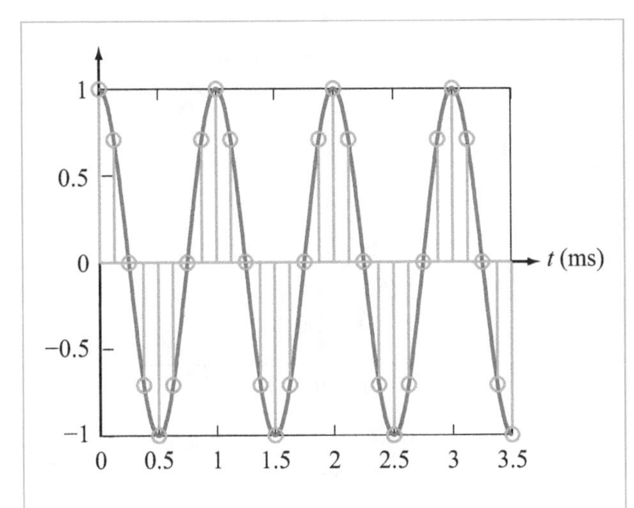

Figure 6-70: Sampling sinusoidal signal $\cos(2000\pi t)$ at 8000 samples per second.

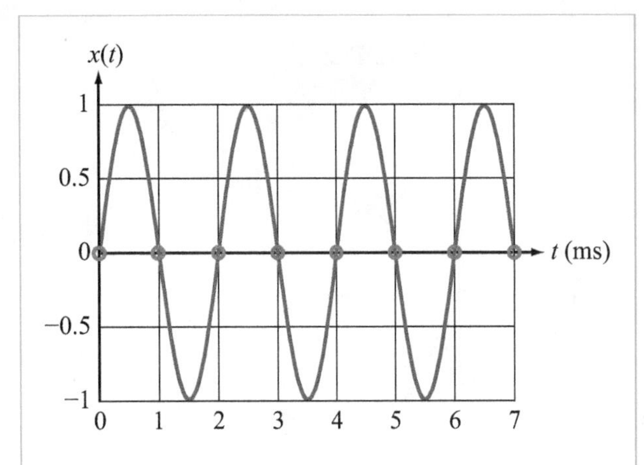

Figure 6-71: A 500 Hz sinusoid, $x(t) = \sin(1000\pi t)$, sampled at the Nyquist rate of 1 kHz.

6-13.2 Statement of the Sampling Theorem

Note that the entire previous section on amplitude modulation was based on a single property of the Fourier transform. An even more remarkable result that also follows from another one of the basic properties of the Fourier transform is the ***sampling theorem.***

Sampling Theorem

- Let $x(t)$ be a real-valued, continuous-time, lowpass signal *bandlimited* to a maximum frequency of B Hz.
- Let $x[n] = x(nT_s)$ be the sequence of numbers obtained by *sampling* $x(t)$ at a sampling rate of f_s samples per second, that is, every $T_s = 1/f_s$ seconds.
- Then $x(t)$ can be *uniquely* reconstructed from its samples $x[n]$ if and only if $f_s > 2B$. The sampling rate must exceed double the bandwidth.
- For a bandpass signal of bandwidth B, a modified constraint applies, as discussed later in Section 6-13.3.

The minimum sampling rate $2B$ is called the ***Nyquist sampling rate***. Although the actual units are $2B$ samples per second, this is usually abbreviated to $2B$ "Hertz," which has the same dimensions as $2B$ samples per second.

Example 6-18: Do We Need $f_s > 2B$ or $f_s \geq 2B$?

A 500 Hz waveform given by $x(t) = \sin(1000\pi t)$ is sampled at the Nyquist sampling rate of 1 kHz (1000 samples per second). Obtain an expression for the sampled signal $x_s[n]$ and determine if the original signal can be reconstructed.

Solution: The sine wave's frequency is $f = 500$ Hz, so a sampling rate of 1 kHz is exactly equal to the Nyquist rate. Figure 6-71 displays the waveform of $x(t)$ and the time locations when sampled at 1 kHz. Every single sample bears a value of zero, because the sampling interval places all sample locations at zero-crossings. Hence,

$$x_s[n] = 0.$$

This is an example of when sampling at exactly the Nyquist rate fails to capture the information in the signal.

6-13.3 Fourier Transform of a Pulse Train

In Fig. 6-72, a lowpass signal $x(t)$ is sampled at an interval T_s by multiplying it by an ideal ***pulse train*** $\delta_{T_s}(t)$ defined as

$$\delta_{T_s}(t) = \sum_{n=-\infty}^{\infty} \delta(t - nT_s). \tag{6.149}$$

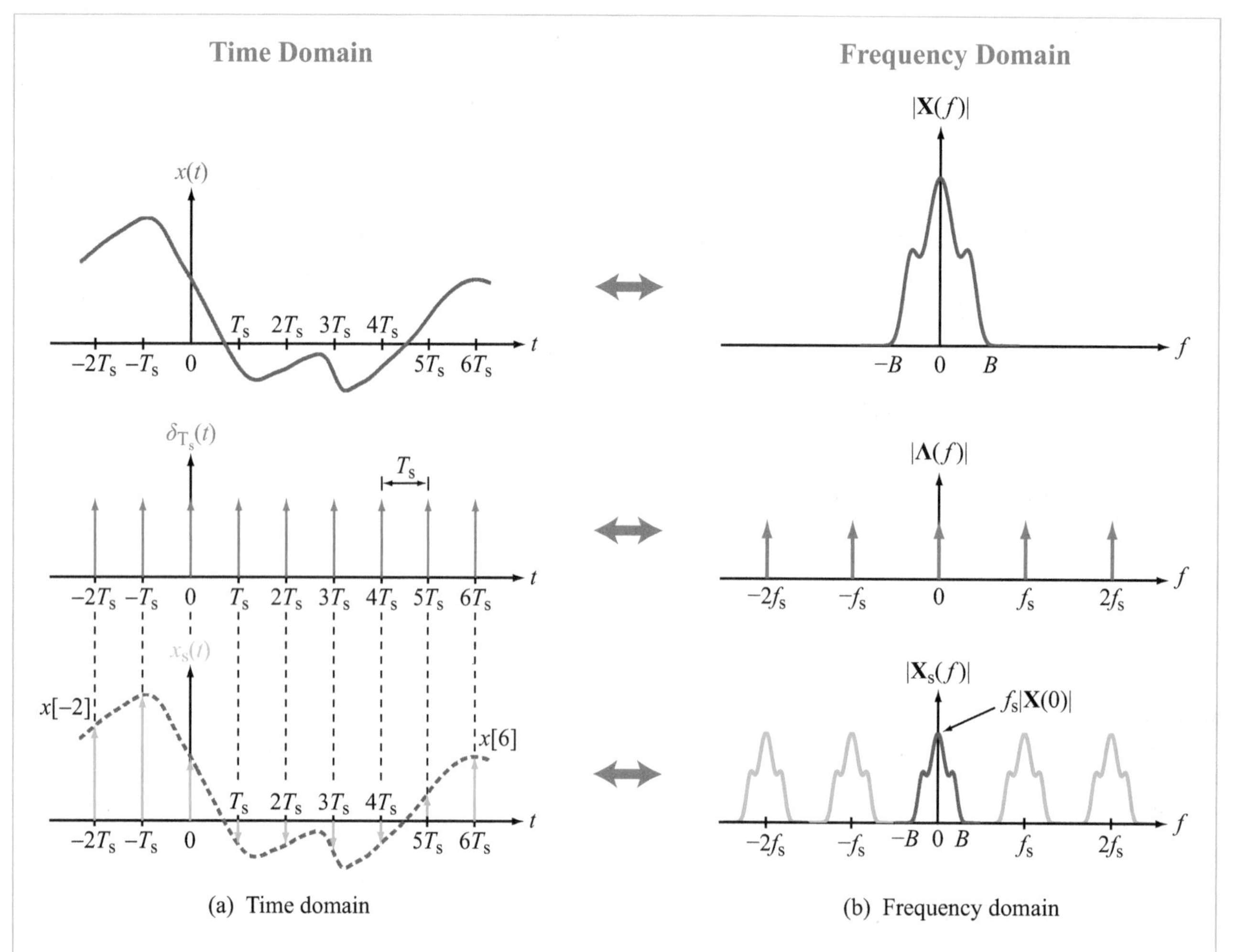

Figure 6-72: In the frequency domain, the spectrum of the sampled signal is a periodic replica of the spectrum of $x(t)$, separated by a spacing $f_s = 1/T_s$.

The pulse train consists of equidistant impulses spanning the full range of t, from $-\infty$ to ∞. The result of the sampling operation is the ***sampled signal*** $x_s(t)$, defined as

$$\begin{aligned} x_s(t) = x(t)\, \delta_{T_s}(t) &= x(t) \sum_{n=-\infty}^{\infty} \delta(t - nT_s) \\ &= \sum_{n=-\infty}^{\infty} x(nT_s)\, \delta(t - nT_s) \\ &= \sum_{n=-\infty}^{\infty} x[n]\, \delta(t - nT_s). \end{aligned} \tag{6.150}$$

By virtue of the impulses, $x_s(t)$ is impulsive at $t = 0, \pm T_s, \pm 2T_s, \ldots,$ and it is zero elsewhere. Thus, $x_s(t)$ and $x[n]$ contain the same information, and each can be obtained from the other, as shown in Fig. 6-72.

Suppose $x(t)$ has a Fourier transform $\mathbf{X}(f)$, bandlimited to B Hz, and suppose we wish to determine the spectrum of the sampled signal $x_s(t)$. When expressed in terms of f as the frequency variable (instead of ω), the frequency-convolution property of the Fourier transform (entry #12 in Table 5-7) takes the form

$$x_1(t)\, x_2(t) \quad \longleftrightarrow \quad \mathbf{X}_1(f) * \mathbf{X}_2(f), \tag{6.151}$$

where $x_1(t)$ and $x_2(t)$ are any two time-varying functions, with corresponding Fourier transforms $\mathbf{X}_1(f)$ and $\mathbf{X}_2(f)$, respectively. In the present case, $x_s(t)$ is the product of $x(t)$ and $\delta_{T_s}(t)$. Hence,

$$x_s(t) = x(t)\,\delta_{T_s}(t) \quad \longleftrightarrow \quad \mathbf{X}_s(f) = \mathbf{X}(f) * \mathbf{\Lambda}(f), \tag{6.152}$$

where $\mathbf{\Lambda}(f)$ is the Fourier transform of $\delta_{T_s}(t)$.

To apply the convolution operation, we first need to establish an expression for $\mathbf{\Lambda}(f)$. Since $\delta_{T_s}(t)$ is a periodic function with period $T_s = 1/f_s$, we can express it in terms of an exponential Fourier series in the form given by Eq. (5.26c) as

$$\delta_{T_s}(t) = \sum_{n=-\infty}^{\infty} \mathbf{d}_n e^{j2\pi n f_s t}, \tag{6.153a}$$

with Fourier coefficients

$$\mathbf{d}_n = \frac{1}{T_s} \int_{-T_s/2}^{T_s/2} \delta_{T_s}(t)\, e^{-j2\pi n f_s t}\, dt. \tag{6.153b}$$

Over the range of integration $[-T_s/2, T_s/2]$, $\delta_{T_s}(t) = \delta(t)$. Hence, the integration leads to

$$\mathbf{d}_n = \frac{1}{T_s} \int_{-T_s/2}^{T_s/2} \delta(t)\, e^{-j2\pi n f_s t}\, dt = \frac{1}{T_s} = f_s, \qquad \text{for all } n. \tag{6.154}$$

Using Eq. (6.154) in Eq. (6.153a) provides the Fourier series representation of the pulse train $\delta_{T_s}(t)$ as

$$\delta_{T_s}(t) = \sum_{n=-\infty}^{\infty} \delta(t - nT_s) = f_s \sum_{n=-\infty}^{\infty} e^{j2\pi n f_s t}. \tag{6.155}$$

6-13.4 Spectrum of the Sampled Signal

Conversion of entry #10 in Table 5-6 from ω to f provides the Fourier transform pair

$$e^{j2\pi n f_s t} \quad \longleftrightarrow \quad \delta(f - nf_s). \tag{6.156}$$

In view of Eq. (6.156), the Fourier transform of $\delta_{T_s}(t)$ is

$$\mathbf{\Lambda}(f) = \mathcal{F}[\delta_{T_s}(t)] = f_s \sum_{n=-\infty}^{\infty} \delta(f - nf_s). \tag{6.157}$$

Using this result in Eq. (6.152), the Fourier transform of the sampled signal becomes

$$\mathbf{X}_s(f) = \mathbf{X}(f) * f_s \sum_{n=-\infty}^{\infty} \delta(f - nf_s) = f_s \sum_{n=-\infty}^{\infty} \mathbf{X}(f - nf_s).$$

In symbolic form, the Fourier transform pair is:

$$x_s(t) = x(t)\,\delta_{T_s}(t) \quad \updownarrow \quad \mathbf{X}_s(f) = f_s \sum_{n=-\infty}^{\infty} \mathbf{X}(f - nf_s). \tag{6.158}$$

6-13.5 Shannon's Sampling Theorem

Part (b) of Fig. 6-72, which displays the spectra of $\mathbf{X}(f)$, $\mathbf{\Lambda}(f)$, and $\mathbf{X}_s(f)$, shows that the spectrum of $\mathbf{X}_s(f)$ consists of a periodic repetition of $\mathbf{X}(f)$, the bandlimited spectrum of the original signal $x(t)$, and the spacing between neighboring spectra is f_s. The spectrum centered at $f = 0$ is called the ***baseband*** spectrum, and the spectra centered at $\pm f_s$, $\pm 2f_s$, etc. are ***image spectra*** that are introduced by the sampling process.

Since the bidirectional bandwidth of each spectrum is $2B$, keeping adjacent spectra from overlapping requires that

$$f_s > 2B \quad \textbf{(Shannon's sampling condition)}. \tag{6.159}$$

The inequality given by Eq. (6.159) is a cornerstone of ***Shannon's sampling theorem***. It states that if a bandlimited signal is sampled at a rate $f_s > 2B$, where B is the signal's one-sided bandwidth, then the signal can be reconstructed without error by passing the sampled signal through an ideal lowpass filter with cutoff frequency B Hz. The filtering process is depicted in Fig. 6-73. The lowpass filter rejects all the spectra of $\mathbf{X}_s$, except the one centered at $f = 0$, which is the spectrum of the original continuous-time signal $x(t)$.

As stated earlier, the condition

$$f_s = 2B \qquad \textbf{(Nyquist rate)} \tag{6.160}$$

is the Nyquist rate, and it represents a lower bound to the rate at which $x(t)$ should be sampled in order to preserve all its information. Sampling at a rate higher than the Nyquist rate does not add information, but sampling at a rate slower than the Nyquist rate is detrimental to recovery. This is because the central spectrum of $\mathbf{X}_s(f)$ will have overlapping sections with neighboring spectra centered at $\pm f_s$, thereby distorting the fidelity of the central spectrum, and in turn, the fidelity of the recovered signal $x(t)$. The three spectra depicted in Fig. 6-74 correspond to undersampled, Nyquist-sampled, and oversampled scenarios.

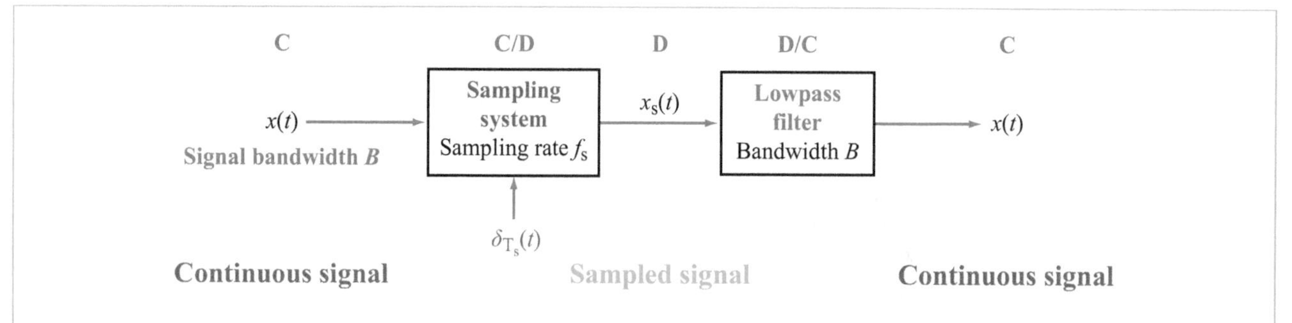

Figure 6-73: A signal $x(t)$, sampled at a sampling frequency $f_s = 1/T_s$ to generate the sampled signal $x_s(t)$, can be reconstructed if $f_s > 2B$. The sampling system is a continuous-time to discrete-time (C/D) converter and the lowpass filter is a D/C converter.

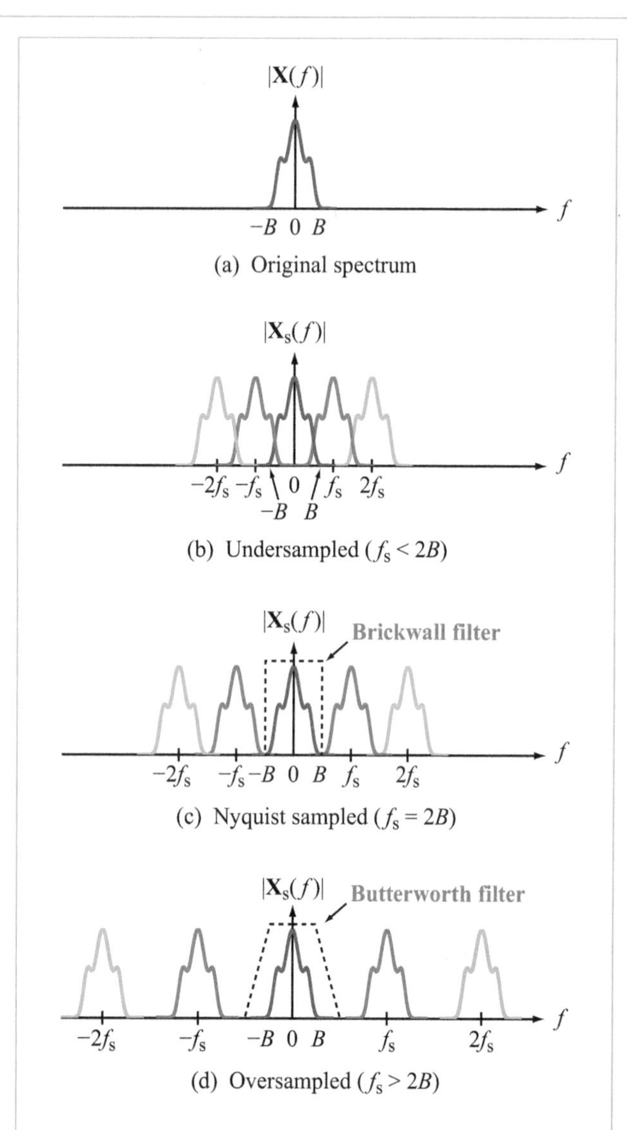

Figure 6-74: Comparison of spectra of $\mathbf{X}_s(f)$, when $x(t)$ is undersampled, Nyquist sampled, and oversampled.

In practice, it is impossible to construct an ideal (brick-wall) lowpass filter with cutoff frequency B, so we must sample $x(t)$ at a rate higher than the Nyquist rate so as to create sufficient separation between adjacent spectra in order to accommodate a realistic lowpass filter around the central spectrum.

6-13.6 Sinc Interpolation Formula

Figure 6-74(c) shows that $x(t)$ can be recovered from $x_s(t)$ using a brick-wall lowpass filter with cutoff frequency $f_s/2$. Equation (6.158) shows that the dc gain of the brick-wall lowpass filter should be $1/f_s$ to offset the multiplication coefficient f_s in Eq. (6.158). The impulse response of this brick-wall lowpass filter is then

$$h(t) = \frac{1}{f_s} \frac{\sin(\pi f_s t)}{\pi t} = \frac{\sin(\pi f_s t)}{\pi f_s t}.$$

According to Eq. (6.150),

$$x_s(t) = \sum_{n=-\infty}^{\infty} x(nT_s)\, \delta(t - nT_s).$$

where $T_s = 1/f_s$. Combining these results leads to

$$\begin{aligned} x(t) = x_s(t) * h(t) &= \sum_{n=-\infty}^{\infty} x(nT_s)\, \delta(t - nT_s) * h(t) \\ &= \sum_{n=-\infty}^{\infty} x(nT_s)\, h(t - nT_s) \\ &= \sum_{n=-\infty}^{\infty} x(nT_s)\, \frac{\sin(\pi f_s(t - nT_s))}{\pi f_s(t - nT_s)}. \end{aligned} \tag{6.161}$$

This is the ***interpolation formula*** for recovering $x(t)$ directly from its samples $x(nT_s)$. To "interpolate" a set of data points means to construct a smooth function that passes through the data points. Interpolation here means to "connect the dots" of the samples $x(nT_s)$ to obtain $x(t)$. While this formula is seldom used directly, it is always nice to have an explicit formula.

Note from Fig. 6-74(d) that if $x(t)$ is oversampled, the cutoff frequency $f_s/2$ of $h(t)$ can be replaced with any number between the maximum frequency B of $x(t)$ and the minimum frequency $(f_s - B)$ of the first copy of the spectrum of $x(t)$ created by sampling. It is still customary to use a cutoff frequency of $f_s/2$, since it lies halfway between B and $f_s - B$.

6-13.7 Summary of Sampling and Reconstruction Operations

Given a continuous-time signal $x(t)$ bandlimited to B Hz, $x(t)$ is sampled by taking its values at times $t = nT_s$ for integers n. The result is the discrete-time signal $x[n] = x(nT_s)$. One way to do this is to multiply $x(t)$ by the impulse train $\delta_{T_s}(t)$. This yields $x_s(t) = x(t)\,\delta_{T_s}(t)$, which is also a train of impulses, only now the area under the impulse at time $t = nT_s$ is $x[n]$.

To reconstruct $x(t)$ from its sampled version $x_s(t)$, the latter is lowpass-filtered (with cutoff frequency B Hz) and divided by f_s to obtain $x(t)$. The process is illustrated in Fig. 6-73.

6-13.8 Aliasing

An ***alias*** is a pseudonym or assumed name. In signal reconstruction, ***aliasing*** refers to reconstructing the wrong signal because of undersampling. The aliasing process is illustrated by Example 6-19.

Example 6-19: Undersampling → Aliasing

A 200 Hz sinusoidal signal given by

$$x(t) = \cos(400\pi t)$$

is sampled at a 500 Hz rate. (a) Generate an expression for the sampled signal $x_s(t)$ and evaluate its five central terms, (b) demonstrate that $x(t)$ can be reconstructed from $x_s(t)$ by passing the latter through an appropriate lowpass filter, and (c) demonstrate that reconstruction would not be possible had the sampling rate been only 333 Hz.

Solution:

(a) With $f_0 = 200$ Hz, $f_s = 500$ Hz (which exceeds the Nyquist rate of 400 Hz), and $T_s = 1/f_s = 2$ ms, the input signal is

$$x(t) = \cos(400\pi t),$$

and its 500 Hz sampled version is

$$\begin{aligned} x_s(t) &= \sum_{n=-\infty}^{\infty} x(t)\,\delta(t - nT_s) \\ &= \sum_{n=-\infty}^{\infty} \cos(400\pi t)\,\delta(t - 2\times 10^{-3}n) \\ &= \sum_{n=-\infty}^{\infty} \cos(0.8\pi n)\,\delta(t - 2\times 10^{-3}n). \end{aligned}$$

The sampled signal is displayed in Fig. 6-75(a). The central five terms correspond to $n = -2$ to $+2$. Thus,

$$\begin{aligned} x_{\text{center }5}(t) &= \cos[400\pi(-4\times 10^{-3})]\,\delta(t+4\text{ ms}) \\ &\quad + \cos[400\pi(-2\times 10^{-3})]\,\delta(t+2\text{ ms}) \\ &\quad + \cos[400\pi(0)]\,\delta(t) \\ &\quad + \cos[400\pi(2\times 10^{-3})]\,\delta(t-2\text{ ms}) \\ &\quad + \cos[400\pi(4\times 10^{-3})]\,\delta(t-4\text{ ms}) \\ &= 0.31\delta(t+4\text{ ms}) - 0.81\delta(t+2\text{ ms}) + \delta(t) \\ &\quad - 0.81\delta(t-2\text{ ms}) + 0.31\delta(t-4\text{ ms}). \end{aligned}$$

(b) Figure 6-75(b) displays the spectrum of $\mathbf{X}_s(f)$. According to Eq. (6.158), the spectrum of the sampled signal, $\mathbf{X}_s(f)$, consists of the spectrum $\mathbf{X}(f)$, plus replicas along the f-axis, shifted by integer multiples of f_s (i.e., by $\pm nf_s$). In the present case, the spectrum of the cosine waveform consists of a pair of impulse functions located at $\pm f_0 = \pm 200$ Hz. Hence, the spectrum of the sampled signal will consist of impulse functions with areas $(f_s/2) = 250$ Hz, at:

$$\begin{aligned} (\pm 200 \pm 500n)\text{ Hz} &= \pm 200\text{ Hz},\ \pm 300\text{ Hz}, \\ &\quad \pm 700\text{ Hz},\ \pm 800\text{ Hz},\ \ldots . \end{aligned}$$

By passing the sampled signal through a lowpass filter with a cutoff frequency of 250 Hz, we recover the spectrum of $\cos(400\pi t)$, while rejecting all other spectral components of $\mathbf{X}_s(f)$, thereby recovering $x(t)$ with no distortion.

(c) Figure 6-76(a) displays the cosine waveform, and its sampled values at $f_s = 333$ Hz, which is below the Nyquist

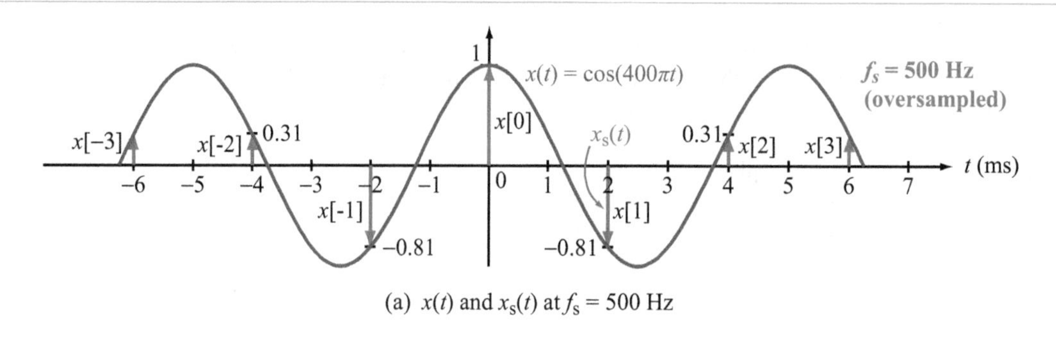

(a) $x(t)$ and $x_s(t)$ at $f_s = 500$ Hz

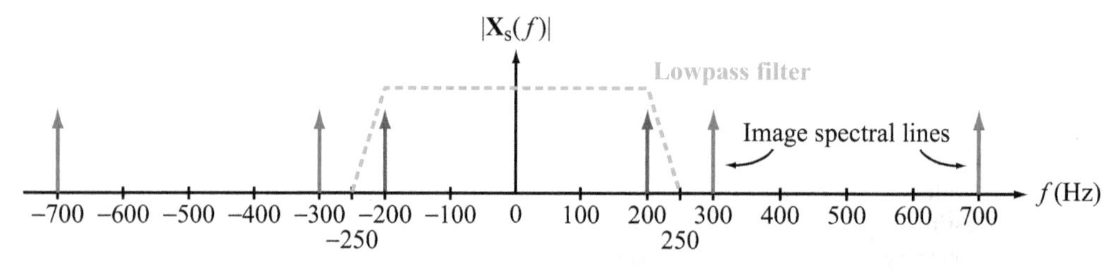

(b) Spectrum of $\mathbf{X}_s(f)$ [blue = spectrum of $x(t)$; red = image spectra]

Figure 6-75: (a) 200 Hz cosine signal sampled at $f_s = 500$ Hz or, equivalently, at a sampling interval $T_s = 1/f_s = 2$ ms; (b) passing the sampled signal through a lowpass filter with a cutoff frequency of 250 Hz would capture the correct spectral lines of the 200 Hz sinusoid. Reconstruction is possible because the sampling rate $f_s > 2B$. For this sinusoid, $B = f_0 = 200$ Hz.

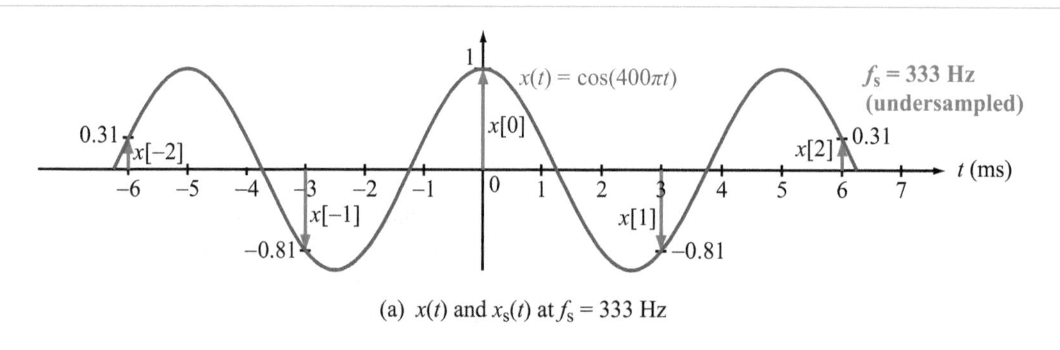

(a) $x(t)$ and $x_s(t)$ at $f_s = 333$ Hz

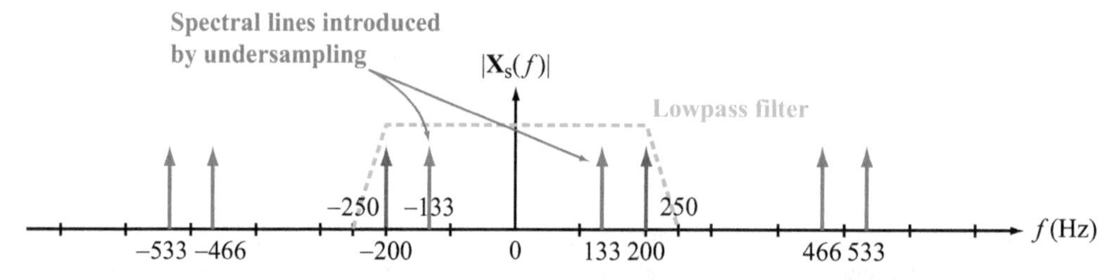

(b) Spectrum of $\mathbf{X}_s(f)$ [blue = spectrum of $x(t)$; red = image spectra]

Figure 6-76: (a) 200 Hz cosine signal sampled at $f_s = 333$ Hz or, equivalently, at a sampling interval $T_s = 1/f_s = 3$ ms; (b) passing the sampled signal through a lowpass filter with a cutoff frequency of 250 Hz would capture the correct spectral lines of the 200 Hz sinusoid, but it will also include the image spectral lines at ± 133 Hz. Perfect reconstruction is not possible because the sampling rate $f_s < 2B$.

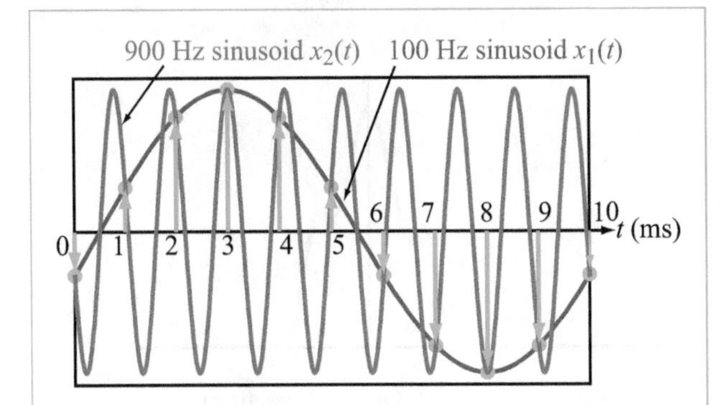

Figure 6-77: Two sinusoids, $x_1(t)$ at $f_1 = 100$ Hz and $x_2(t)$ at $f_2 = 900$ Hz, are sampled at 1 kHz. The two sinusoids have identical sampled values.

rate of 400 Hz. Thus, this is an undersampled waveform. The spectrum consists of impulses at

$$(\pm 200 \pm 333n)\text{ Hz} = \pm 133\text{ Hz}, \ \pm 200\text{ Hz}, \ \pm 466\text{ Hz}, \pm 533\text{ Hz}, \ \ldots .$$

Since the signal's one-sided bandwidth is 200 Hz, the lowpass filter's cutoff frequency has to be greater than that, but such a filter would also encompass the impulse functions at ± 133 Hz, thereby generating a signal that is the sum of two cosines: one at 133 Hz and another at 200 Hz. A narrower, 150 Hz lowpass filter would generate an alias in the form of a 133 Hz cosine waveform instead of the original 200 Hz cosine waveform.

6-13.9 Aliasing in Sinusoids

Let us consider two sinusoids

$$x_1(t) = \cos(2\pi f_1 t) \tag{6.162a}$$

and

$$x_2(t) = \cos(2\pi f_2 t), \tag{6.162b}$$

and let us sample both waveforms at f_s. The nth samples, corresponding to $t = nT_s = n/f_s$, are

$$x_1[n] = x_1(nT_s) = \cos\left(2\pi n \,\frac{f_1}{f_s}\right), \tag{6.163a}$$

$$x_2[n] = x_2(nT_s) = \cos\left(2\pi n \,\frac{f_2}{f_s}\right). \tag{6.163b}$$

Furthermore, let us assume f_1, f_2, and f_s are related by

$$f_2 = \pm(f_1 + m f_s) \qquad (m = \text{integer}). \tag{6.164}$$

Use of Eq. (6.164) in Eq. (6.163b) leads to

$$x_2[n] = \cos\left[2\pi n\left(\frac{f_1}{f_s} + m\right)\right] = \cos\left(2\pi n \frac{f_1}{f_s}\right) = x_1[n].$$

Thus, $x_1(t)$ and $x_2(t)$ will have identical sampled values for all n. This conclusion is illustrated by the two sinusoids in Fig. 6-77, wherein $x_1(t)$ has a frequency $f_1 = 100$ Hz, $x_2(t)$ has a frequency $f_2 = 900$ Hz, and the sampling rate $f_s = 1$ kHz. The three frequencies satisfy Eq. (6.164) for $m = -1$.

The point of this illustration is that the oversampled 100 Hz sinusoid is recoverable, but not so for the undersampled 900 Hz sinusoid. To remove the ambiguity and recover the 900 Hz waveform, it is necessary to sample it at a rate greater than 1800 Hz.

6-13.10 Physical Manifestations of Aliasing

To observe how aliasing distorts a signal, examine the sinusoid shown in Fig. 6-78. The 500 Hz sinusoid is sampled at

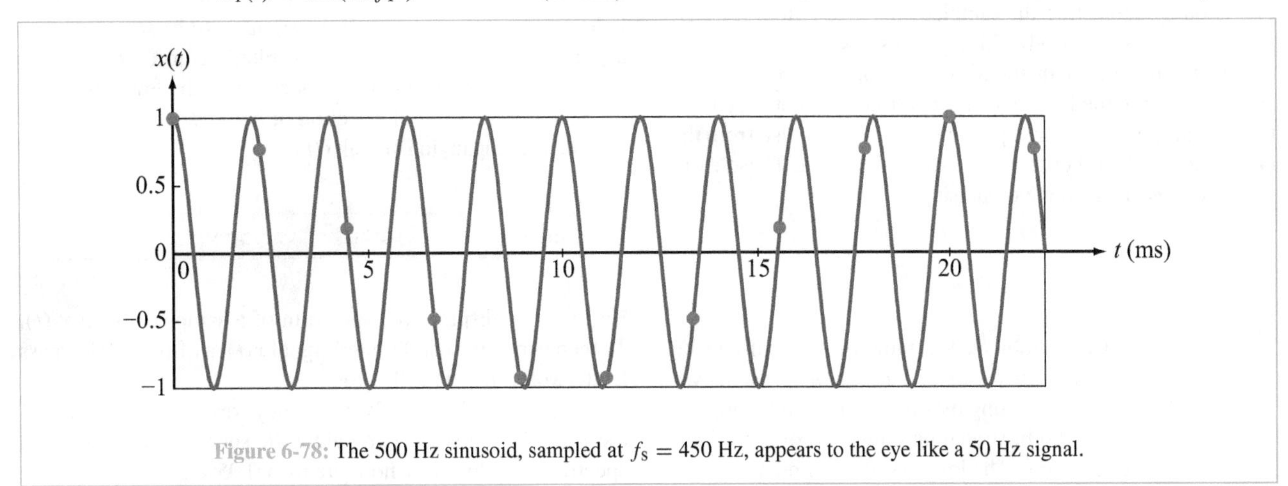

Figure 6-78: The 500 Hz sinusoid, sampled at $f_s = 450$ Hz, appears to the eye like a 50 Hz signal.

450 samples per second. The samples, shown as red dots, appear as if they are part of sampling a 50 Hz sinusoid. Given the choice between interpreting the samples as from a 500 Hz or a 50 Hz signal, the mind chooses the lower frequency (just as the lowpass reconstruction filter would). In this case, the 500 Hz signal masquerades as the 50 Hz sinusoid.

Another visual example of aliasing is found in some old western movies. When a stagecoach wagon wheel with 24 spokes rotating at 1 revolution per second is filmed by a camera that uses 24 frames per second, the wheel *appears stationary*, even though it is moving. And if the wheel has only 23 spokes, it will *appear to rotate backwards slowly!* Filming the rotating wheel at 24 frames per second is equivalent to visually sampling it at a sampling frequency $f_s = 24$ Hz. If the wheel has only 23 spokes and it is rotating at 1 revolution per second, the image of the wheel between successive frames will appear as if the spokes are slightly behind, so the brain interprets it as rotating backwards.

6-13.11 Aliasing in Real Signals

A signal $x(t)$ cannot be time-limited to a finite duration unless its spectrum $\mathbf{X}(f)$ is unbounded (extends in frequency over $(-\infty, \infty)$), and $\mathbf{X}(f)$ cannot be bandlimited to a finite bandwidth unless $x(t)$ is everlasting (exists for all time). No real signal is everlasting, and therefore, no real signal has a perfectly bandlimited spectrum.

Suppose a real signal $x(t)$ with the spectrum $\mathbf{X}(f)$ displayed in Fig. 6-79(a) is sampled at a sampling frequency f_s. The spectrum of the sampled signal, $\mathbf{X}_s(f)$, consists of the sum of replicate spectra of $\mathbf{X}(f)$, which are centered at $f = 0, \pm f_s, \pm 2f_s$, etc. The first overlap occurs at $f_s/2$, which is called the ***folding frequency***. If we were to perform a reconstruction operation by subjecting the sampled signal $x_s(t)$ to a lowpass filter with one-sided bandwidth $f_s/2$, as shown in Fig. 6-79(b), we would recover a distorted version of $x(t)$ because the spectrum within the filter contains contributions not only from the central spectrum [corresponding to $\mathbf{X}(f)$] but also from the tails of all of the other frequency-shifted spectra. These extra contributions are a source of aliasing.

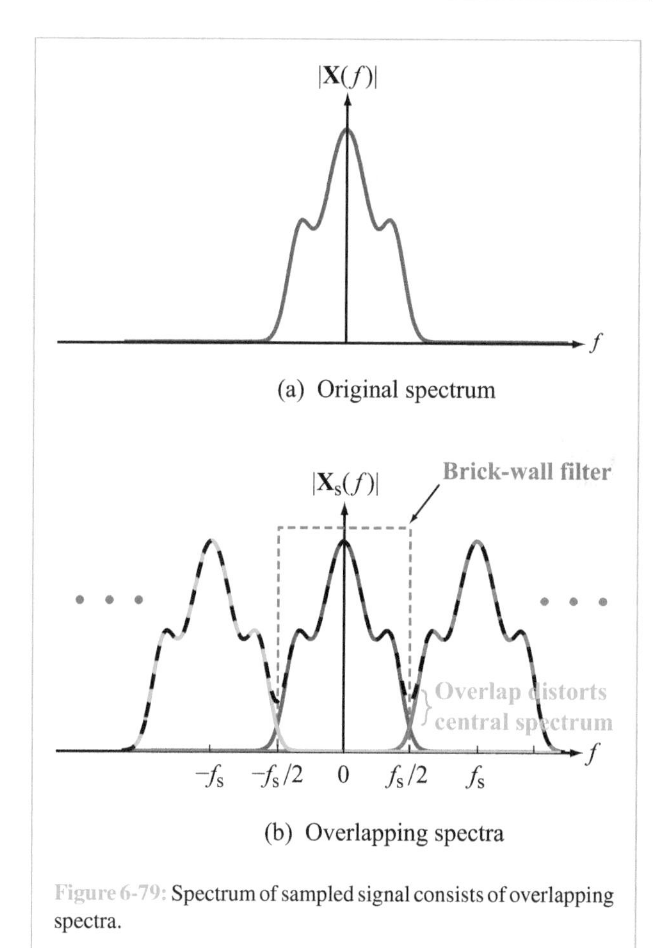

Figure 6-79: Spectrum of sampled signal consists of overlapping spectra.

6-13.12 Antialiasing Filter

To avoid the aliasing problem, the sampling system can be preceded by an analog lowpass ***antialiasing filter***, as shown in Fig. 6-80, thereby converting the spectrum of $x(t)$ from one with unlimited bandwidth, stretching outwards towards $\pm\infty$, to a bandlimited spectrum. The lowpass filter eliminates high-frequency components of the spectrum, while preserving those below its cutoff frequency. If the signal is to be sampled at f_s, a judicious choice is to select the filter's cutoff frequency to be $f_s/2$. An additional bonus is accrued by the lowpass filter's operation in the form of rejection of high frequency noise that may be accompanying signal $x(t)$.

Example 6-20: Importance of Oversampling

Figure 6-81 displays the spectrum of a sampled signal $x_s(t)$. To reconstruct the bandlimited signal $x(t)$, a fifth-order lowpass Butterworth filter is called for.

(a) Specify the filter's frequency response in terms of the signal's one-sided bandwidth B, such that the baseband spectrum is reduced by no more than 0.05%.

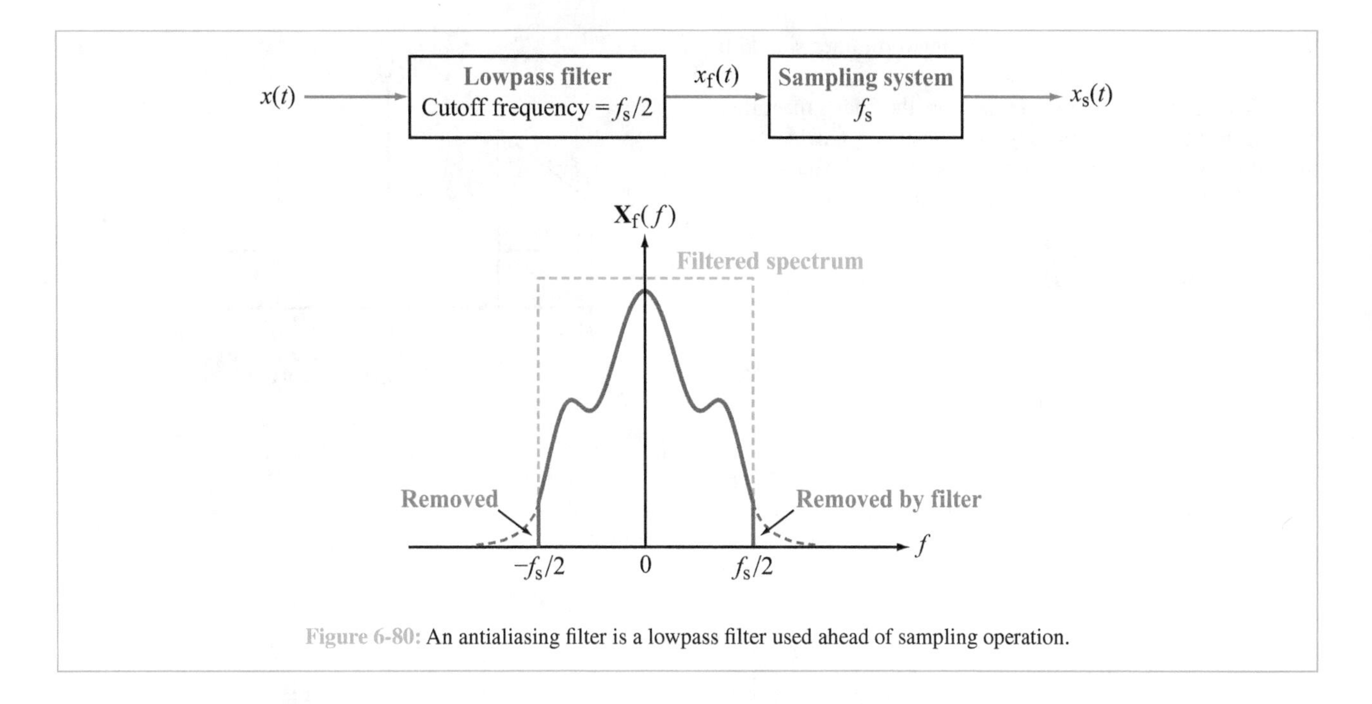

Figure 6-80: An antialiasing filter is a lowpass filter used ahead of sampling operation.

(b) Specify the sampling rate needed so that the image spectra are reduced by the filter down to 0.1% or less of their unfiltered values.

Solution:

(a) From Eq. (6.93), the magnitude response of a Butterworth filter is given by

$$M_{\mathrm{LP}}(\omega) = \frac{|\mathbf{C}|}{\omega_{\mathrm{c}}^n} \cdot \frac{1}{\sqrt{1+(\omega/\omega_{\mathrm{c}})^{2n}}},$$

where ω_{c} is the filter's cutoff angular frequency.

For our present purposes, we will (a) convert ω to $2\pi f$ and ω_{c} to $2\pi f_{\mathrm{c}}$, and (b) choose $\mathbf{C} = \omega_{\mathrm{c}}^n$ so that the dc gain is 1. Hence,

$$M_{\mathrm{LP}}(f) = \frac{1}{\sqrt{1+(f/f_{\mathrm{c}})^{2n}}}.$$

Over the extent of the baseband spectrum, the lowest value of $M_{\mathrm{LP}}(f)$ occurs at $f = B$. Hence, to meet the 0.05% specification with $n = 5$, the response has to satisfy the condition

$$1 - 0.0005 = 0.9995 = \frac{1}{\sqrt{1+(B/f_{\mathrm{c}})^{10}}}.$$

Solving for f_{c} leads to

$$\left(\frac{B}{f_{\mathrm{c}}}\right)^{10} = \frac{1}{(0.9995)^2} - 1 = 0.001,$$

or

$$f_{\mathrm{c}} = (0.001)^{-1/10} B = 2B.$$

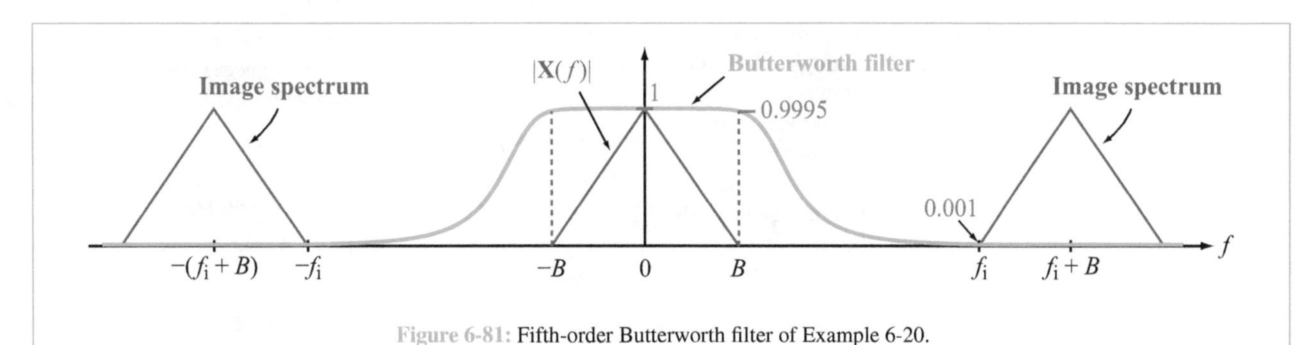

Figure 6-81: Fifth-order Butterworth filter of Example 6-20.

Thus, the cutoff frequency of the Butterworth filter should be twice the one-sided bandwidth of $x(t)$.

(b) The second requirement states that the filter's magnitude at the near-edges of the image spectra (labeled $\pm f_i$ in Fig. 6-81) should be no more than 0.1% of its peak value. That is, with $f_c = 2B$ from part (a), we need to determine f_i such that

$$0.001 = M(f_i) = \frac{1}{\sqrt{1+(f_i/2B)^{10}}}.$$

The solution leads to

$$f_i = 8B,$$

which means that the first image spectra on the two sides of the baseband spectrum are centered at $\pm(8B + B) = \pm 9B$. This corresponds to a sampling rate $f_s = 9B$, or equivalently, an

$$\text{oversampling ratio} = \frac{9B}{2B} = 4.5 \text{ times!}$$

6-13.13 Sampling of Bandpass Signals

Not all signals have spectra that are concentrated around dc. A ***bandpass signal*** is one whose spectrum extends over a bandwidth B between a lower frequency f_ℓ and an upper frequency f_u, as shown in Fig. 6-82(a). According to the foregoing discussion about sampling and reconstruction, it is necessary to sample the signal at more than twice its highest frequency (i.e., at $f_s > 2f_u$) in order to be able to reconstruct it. However, the signal's spectrum contains no information in the spectral range $[-f_\ell, f_\ell]$, so sampling at the traditional Nyquist rate seems excessive. After all, if the signal is mixed with a local-oscillator signal at the appropriate frequency (see Section 6-11.5), its spectrum will get shifted towards dc, becoming a baseband-like signal with bandwidth B, as shown in Fig. 6-82(b). The frequency shifted spectrum continues to contain the same information as the input signal. Hence, a sampling rate that exceeds the Nyquist rate should be quite sufficient to capture the information content of the signal.

Alternatively, instead of frequency translating the bandpass spectrum towards dc, the signal can still be sampled at a rate between $2B$ and $4B$ by selecting the sampling rate f_s such that no overlap occurs between the spectrum $\mathbf{X}(f)$ of the bandpass signal and the image spectra generated by the sampling process. The ***uniform sampling theorem for bandpass signals*** states that a bandpass signal $x(t)$ can be faithfully represented and reconstructed if sampled at a rate f_s such that

$$f_s = \frac{2f_u}{m}, \tag{6.165}$$

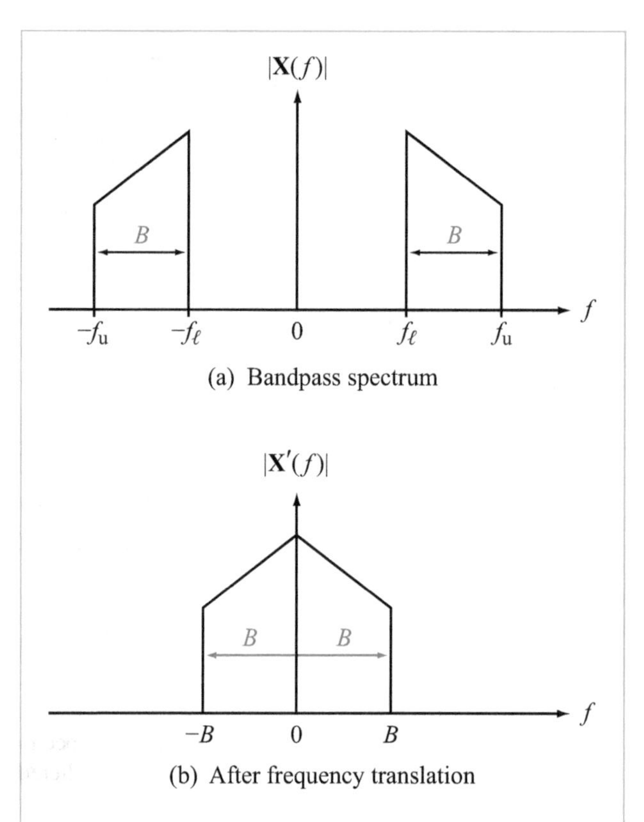

Figure 6-82: A bandpass spectrum can be shifted to create a baseband-like signal by mixing the signal with a local oscillator whose frequency is $f_{LO} = f_\ell$.

where m is the largest integer not exceeding f_u/B. Figure 6-83 is a graphical representation of the stated condition. The baseband-like spectrum corresponds to $f_u/B = 1$.

Example 6-21: Sampling a Bandpass Signal

Bandpass signal $x(t)$ has the spectrum shown in Fig. 6-86(a). Determine the minimum sampling rate necessary to sample the signal so it may be reconstructed with total fidelity.

Solution: From the given spectrum, we obtain

$$f_u = 150 \text{ Hz}, \qquad f_\ell = 50 \text{ Hz},$$

$$B = f_u - f_\ell = 150 - 50 = 100 \text{ Hz},$$

$$\text{and} \quad \frac{f_u}{B} = \frac{150}{100} = 1.5.$$

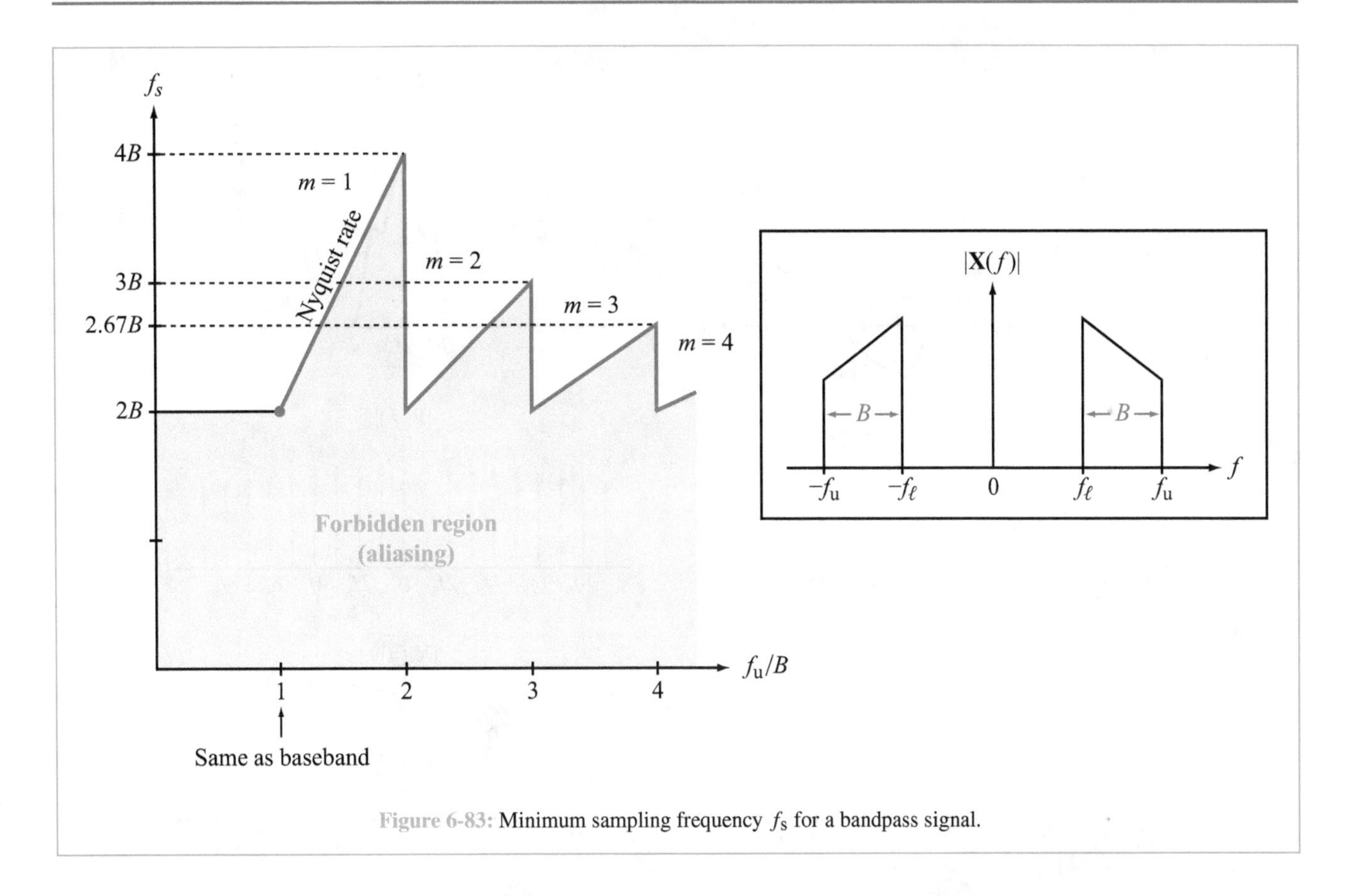

Figure 6-83: Minimum sampling frequency f_s for a bandpass signal.

According to Fig. 6-83, for $f_u/B = 1.5$, the sampling rate should be

$$f_s = 3B = 3 \times 100 = 300 \text{ Hz}.$$

Per Eq. (6.158), the spectrum of the sampled signal is given by

$$\mathbf{X}_s(f) = f_s \sum_{n=-\infty}^{\infty} \mathbf{X}(f - nf_s).$$

Spectrum $\mathbf{X}_s(t)$ consists of the original spectrum, plus an infinite number of duplicates, shifted to the right by nf_s for $n > 0$ and to the left by $|nf_s|$ for $n < 0$. All spectra are scaled by a multiplicative factor of f_s. Figure 6-86(b) displays the spectra for $n = 0$ and ± 1. We note that the spectra do not overlap, which means that the original signal can be reconstructed by passing spectrum $\mathbf{X}_s(f)$ through a lowpass filter with a spectral response that extends between -150 Hz and $+150$ Hz.

6-13.14 Practical Aspects of Sampling

So far, all our discussions of signal sampling assumed the availability of an ideal impulse train composed of impulses. In practice, the sampling is performed by finite-duration pulses that may resemble rectangular or Gaussian waveforms. Figure 6-84 depicts the sampling operation for a signal $x(t)$, with the sampling generated by a train of rectangular pulses given by

$$p_{T_s}(t) = \sum_{n=-\infty}^{\infty} \text{rect}\left(\frac{t - nT_s}{\tau}\right),$$

where T_s is the sampling interval and τ is the pulse width. From Table 5-4, the Fourier series representation of the pulse train is

$$p_{T_s}(t) = \frac{\tau}{T_s} + \sum_{m=1}^{\infty} \frac{2}{m\pi} \sin\left(\frac{m\pi\tau}{T_s}\right) \cos\left(\frac{2m\pi\tau}{T_s}\right). \quad (6.166)$$

The signal sampled by the pulse train is

$$\begin{aligned} x_s(t) &= x(t)\, p_{T_s}(t) \\ &= \frac{\tau}{T_s}\, x(t) + \sum_{m=1}^{\infty} \frac{2}{m\pi}\, x(t) \sin\left(\frac{m\pi\tau}{T_s}\right) \cos\left(\frac{2m\pi t}{T_s}\right), \end{aligned} \quad (6.167)$$

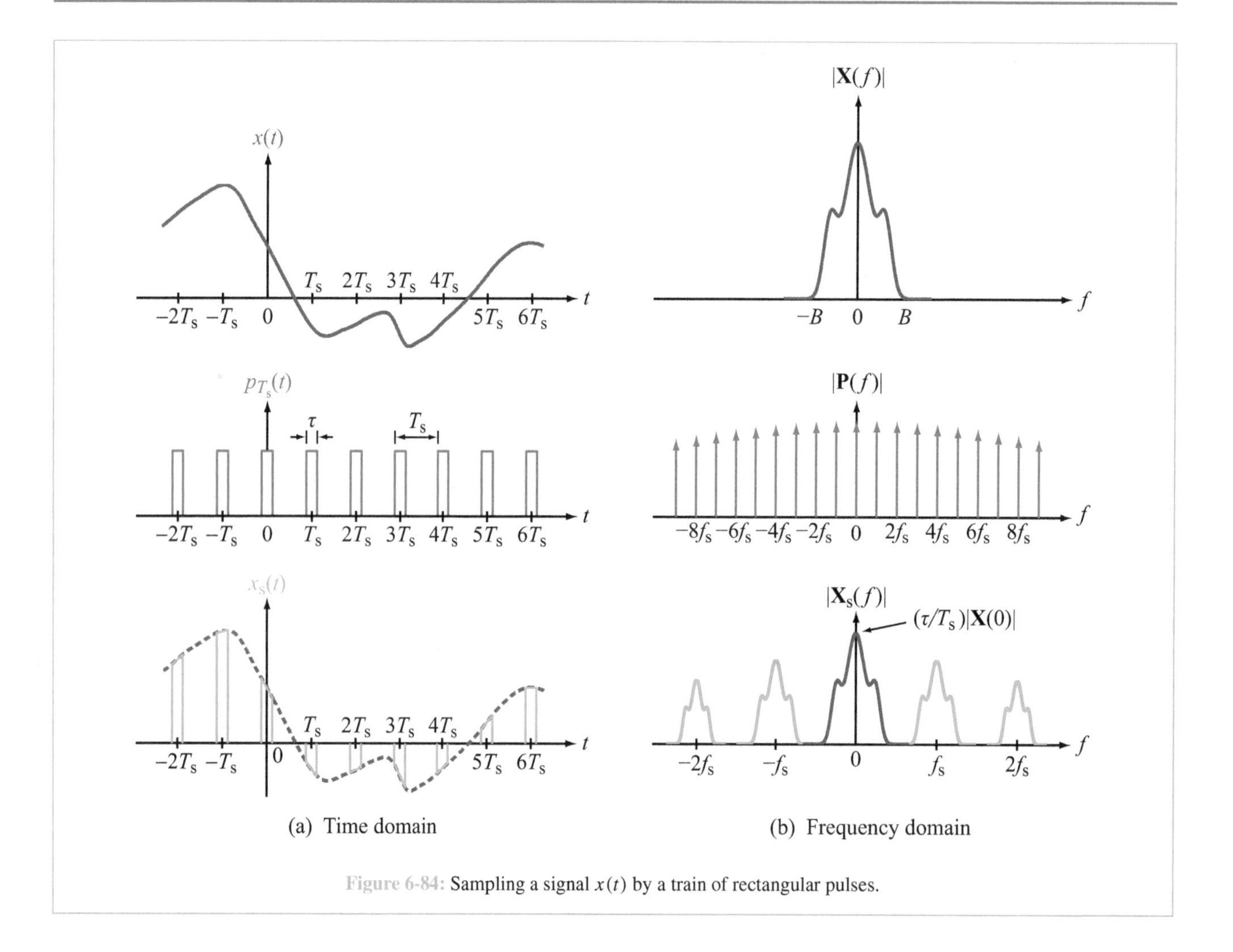

Figure 6-84: Sampling a signal $x(t)$ by a train of rectangular pulses.

which can be cast in the form

$$x_s(t) = A_0\, x(t) + A_1\, x(t)\cos(2\pi f_s t) + A_2\, x(t)\cos(4\pi f_s t) + \cdots, \quad (6.168)$$

with

$$f_s = \frac{1}{T_s}, \qquad A_0 = \frac{\tau}{T_s},$$

$$A_1 = \frac{2}{\pi}\sin(\pi f_s \tau), \qquad A_2 = \frac{1}{\pi}\sin(2\pi f_s \tau), \; \ldots \; .$$

The sequence given by Eq. (6.168) consists of a dc term, $A_0\, x(t)$, and a sum of sinusoids at frequency f_s and its harmonics. The Fourier transform of the A_0 term is $A_0\, \mathbf{X}(f)$, where $\mathbf{X}(f)$ is the transform of $x(t)$. It is represented by the central spectrum of $\mathbf{X}_s(f)$ in Fig. 6-84. The cosine terms in Eq. (6.168) generate image spectra centered at $\pm f_s$ and its harmonics, but their amplitudes are modified by the values of A_1, A_2, etc.

Signal $x(t)$ can be reconstructed from $x_s(t)$ by lowpass filtering it, just as was done earlier with the ideal impulse sampling. The Shannon sampling requirement that f_s should be greater than $2B$ still holds.

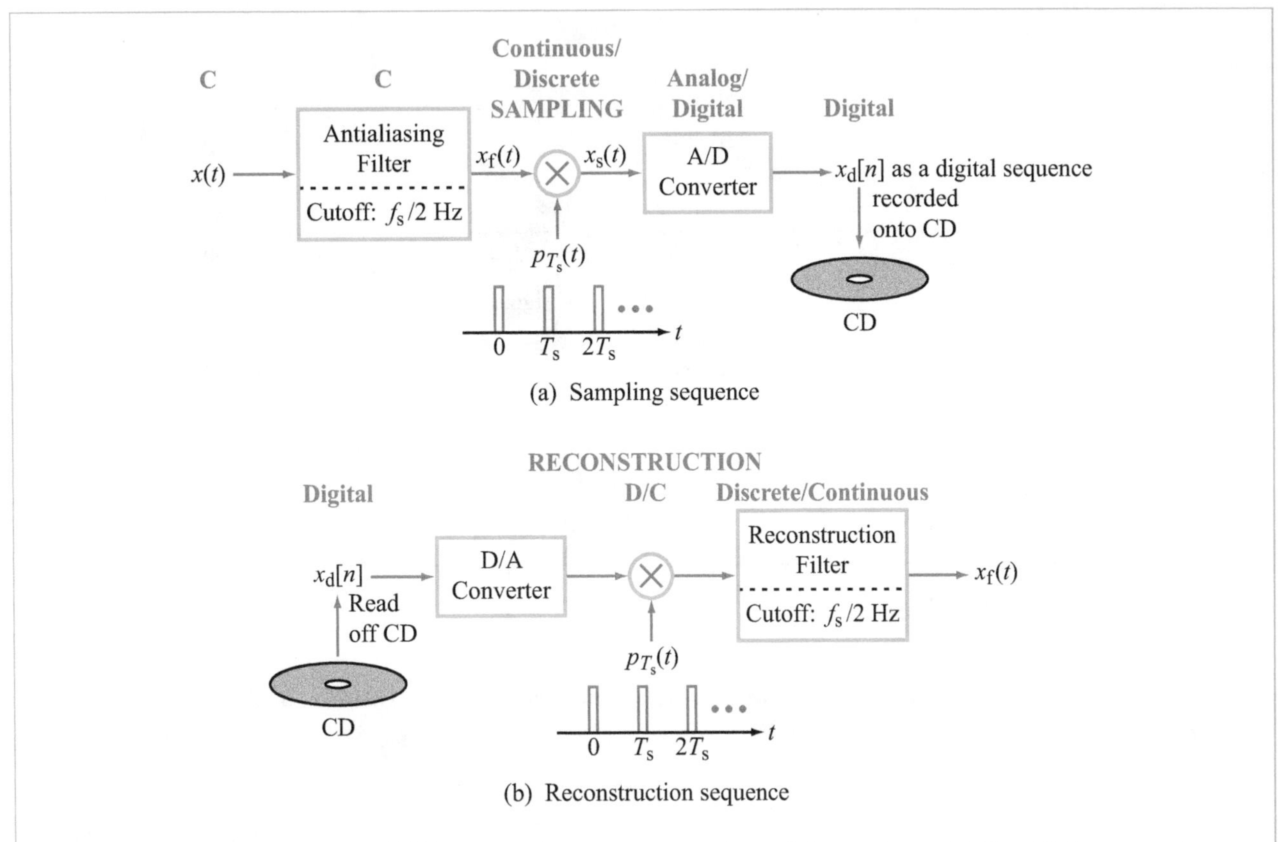

Figure 6-85: DSP system with C/D and D/C to convert between continuous time and discrete time, and A/D and D/A to convert between analog and digital.

By way of an example, we show in Fig. 6-85 a block diagram of a typical DSP system that uses a compact disc (CD) for storage of digital data. The sampling part of the process starts with a continuous-time signal $x(t)$. After bandlimiting the signal to $f_s/2$ by an antialiasing filter, the filtered signal $x_f(t)$ is sampled at a rate f_s and then converted into a digital sequence $x_d[n]$ that gets recorded onto the CD.

Reconstruction performs the reverse sequence, starting with $x_d[n]$ as read off of the CD and concluding in $x_f(t)$, the bandlimited version of the original signal.

Concept Question 6-19: Does sampling a signal at exactly the Nyquist rate guarantee that it can be reconstructed from its discrete samples? (See S²)

Concept Question 6-20: What is signal aliasing? What causes it? How can it be avoided? (See S²)

Concept Question 6-21: If brick-wall lowpass filters are used in connection with a signal bandlimited to f_{max} and sampled at f_s, what should the filter's cutoff frequency be when used as (a) an anti-aliasing filter and (b) a reconstruction filter? (See S²)

Exercise 6-18: What is the Nyquist sampling rate for a signal bandlimited to 5 kHz?

Answer: 10 kHz. (See S²)

Exercise 6-19: A 500 Hz sinusoid is sampled at 900 Hz. No anti-alias filter is used. What is the frequency of the reconstructed sinusoid?

Answer: 400 Hz. (See S²)

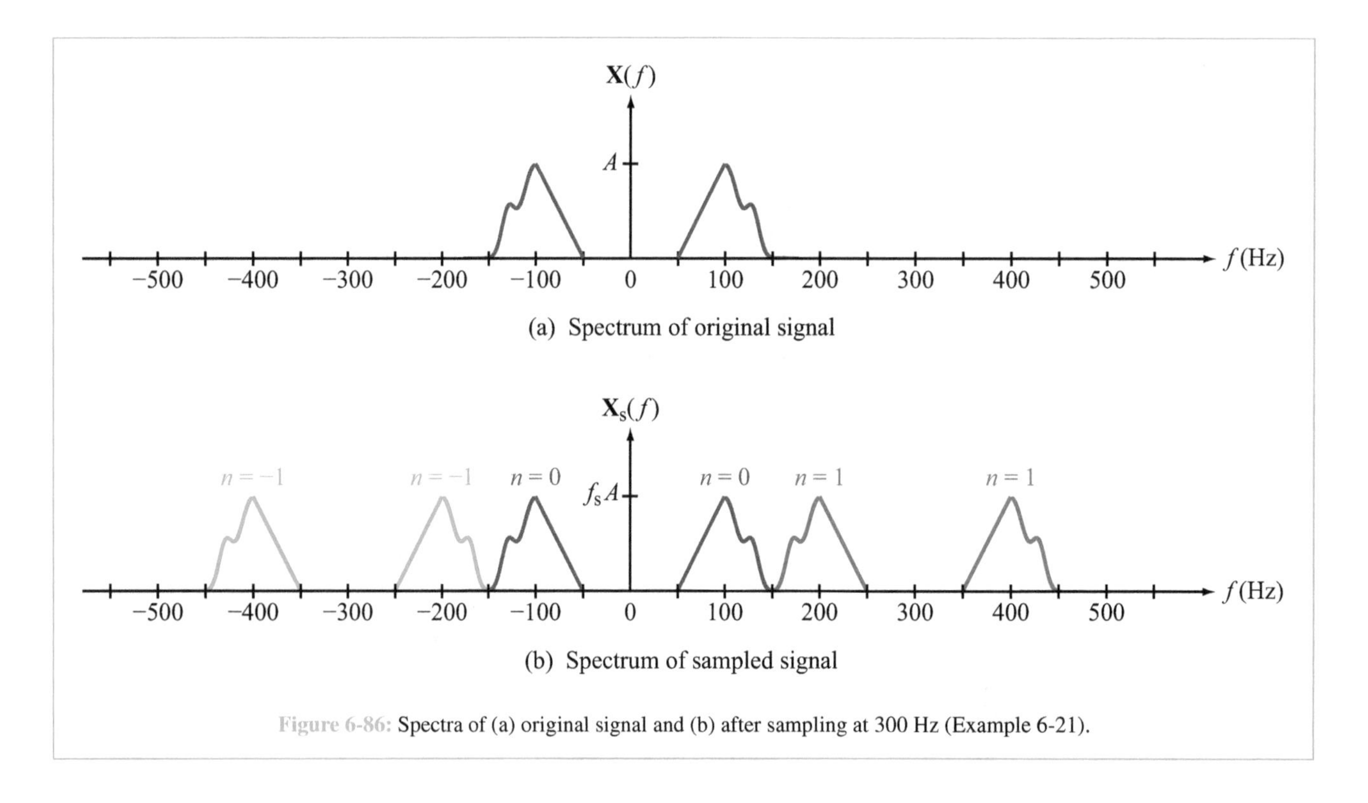

Figure 6-86: Spectra of (a) original signal and (b) after sampling at 300 Hz (Example 6-21).

Summary

Concepts

- Circuits can be designed to serve as lowpass, highpass, bandpass, and bandreject frequency filters.
- The inherent input-output isolation offered by op-amp circuits makes them ideal for cascading multiple stages together to realize the desired spectrum.
- The frequency response of a system is governed by the locations of the poles and zeros of its transfer function $\mathbf{H}(\mathbf{s})$ in the $\mathbf{s}$-plane.
- Notch filters are used to remove sinusoidal interference.
- Comb filters are used to remove periodic interference.
- Butterworth filters can be designed as lowpass, highpass, and bandpass filters with sharp roll-off at their cutoff frequencies.
- Resonator filters are used to reduce noise in periodic signals composed of discrete (line) spectra.
- In amplitude modulation (AM), a signal is dc-biased by adding a constant A to it prior to modulation by the carrier signal.
- In frequency division multiplexing (FDM), a large number of signals can be combined (multiplexed) together, transmitted as a single signal, and then demultiplexed after reception.
- The sampling theorem states that a signal can be recovered from its discrete samples if the sampling rate f_s exceeds $2f_{max}$, where f_{max} is the signal's highest frequency.

Mathematical and Physical Models

Series RLC lowpass and highpass filters

$$\omega_0 = 1/\sqrt{LC} \qquad \text{(resonant frequency)}$$

Quality factor $\quad Q = \omega_0/B$

Notch filter frequency response

$$\mathbf{H}_{\text{notch}}(\omega) = \frac{(j\omega - j\omega_0)(j\omega + j\omega_0)}{(j\omega + \alpha - j\omega_0)(j\omega + \alpha + j\omega_0)}$$

Comb filter frequency response

$$\mathbf{H}_{\text{comb}}(\omega) = \prod_{\substack{k=-n \\ k\neq 0}}^{n} \left[\frac{j\omega - jk\omega_0}{j\omega + \alpha - jk\omega_0}\right]$$

Butterworth lowpass filter response

$$\mathbf{H}_{\text{LP}}(\omega) = \mathbf{C}\prod_{i=1}^{n}\left(\frac{1}{j\omega - \mathbf{p}_i}\right)$$

Resonator filter frequency response

$$\mathbf{H}_{\text{LP}}(\mathbf{s}) = \frac{\mathbf{C}}{\omega_c^n}\,\mathbf{H}(\mathbf{s}_a)\bigg|_{\mathbf{s}_a=\mathbf{s}/\omega_c}$$

Shannon's sampling condition for reconstruction

$$f_s > 2f_{\max}$$

Important Terms

Provide definitions or explain the meaning of the following terms:

active filter
additive zero-mean white noise
aliasing
amplitude modulation (AM)
bandlimited
bandpass filter
bandreject filter
bandwidth B
baseband bandwidth
Bode plot
brick-wall filter
Butterworth filter
carrier frequency
comb filter
corner frequency
cutoff frequency
dc gain
demodulation
double-sideband (DSB) modulation
double-sideband suppressed carrier (DSB-SC)
folding frequency
frequency division multiplexing (FDM)
frequency offset
frequency response $\mathbf{H}(\omega)$
frequency translation (mixing)
gain factor
gain roll-off rate
guard band
half-power frequencies
highpass filter
image spectra
line enhancement filter
local oscillator
lower sideband (LSB)
lowpass filter
magnitude response $M(\omega)$
maximum frequency $f_{\max}$
modulation
modulation index
multiplexing
notch filter
Nyquist rate
oversampling
passband
passive filter
phase coherent
phase offset
phase response $\phi(\omega)$
poles
proper rational function
pulse train
quality factor
resonant frequency
resonator filter
sampling interval
sampling rate f_s
sampling theorem
Shannon's sampling theorem
signal bandwidth
signal-to-noise ratio
single-sideband (SSB) modulation
stopband
subcarrier
superheterodyne receiver
switching modulation
synchronous demodulator
transfer function
tuner
upper sideband (USB)
zeros

PROBLEMS

Section 6-2: Types of Filters

*6.1 Determine the resonant frequency of the circuit shown in Fig. P6.1 given that $R = 100\ \Omega$, $L = 5$ mH, and $C = 1\ \mu$F.

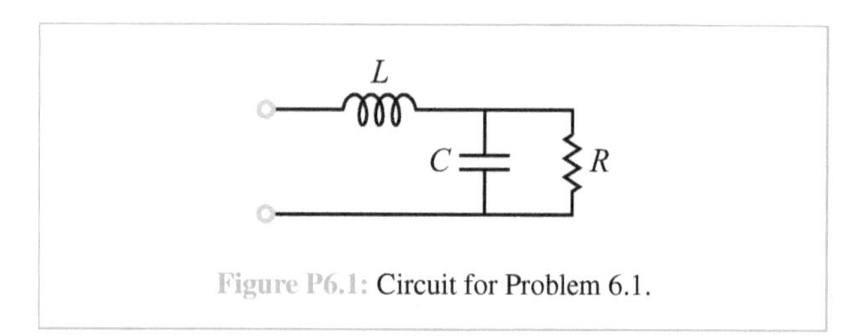

Figure P6.1: Circuit for Problem 6.1.

6.2 Determine the resonant frequency of the circuit shown in Fig. P6.2 given that $R = 100\ \Omega$, $L = 5$ mH, and $C = 1\ \mu$F.

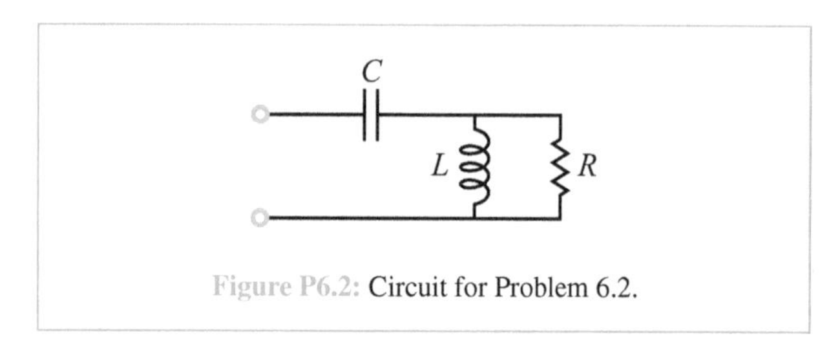

Figure P6.2: Circuit for Problem 6.2.

6.3 For the circuit shown in Fig. P6.3, determine (a) the transfer function $\mathbf{H} = \mathbf{V}_o/\mathbf{V}_i$, and (b) the frequency ω_o at which $\mathbf{H}$ is purely real.

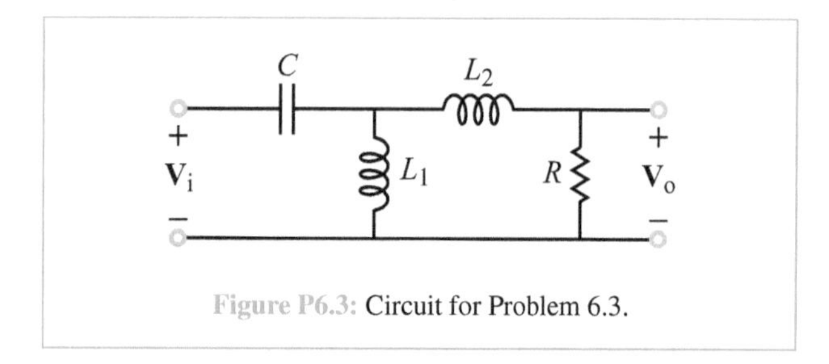

Figure P6.3: Circuit for Problem 6.3.

6.4 For the circuit shown in Fig. P6.4, determine (a) the transfer function $\mathbf{H} = \mathbf{V}_o/\mathbf{V}_i$, and (b) the frequency ω_o at which $\mathbf{H}$ is purely real.

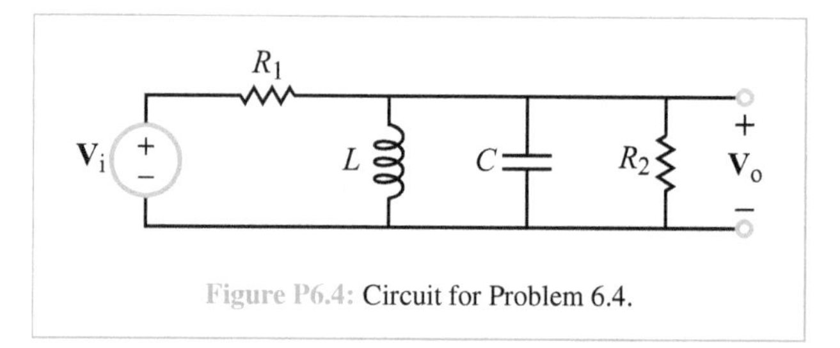

Figure P6.4: Circuit for Problem 6.4.

6.5 Convert the following power ratios to dB.

(a) 3×10^2

*(b) 0.5×10^{-2}

(c) $\sqrt{2000}$

(d) $(360)^{1/4}$

*(e) $6e^3$

(f) $2.3 \times 10^3 + 60$

(g) $24(3 \times 10^7)$

(h) $4/(5 \times 10^3)$

6.6 Convert the following voltage ratios to dB.

(a) 2×10^{-4}

(b) 3000

(c) $\sqrt{30}$

*(d) $6/(5 \times 10^4)$

6.7 Convert the following dB values to voltage ratios.

(a) 46 dB

(b) 0.4 dB

(c) −12 dB

*(d) −66 dB

6.8 Generate magnitude and phase plots for the following voltage transfer functions. Use Fig. 6-7 as a model.

(a) $\mathbf{H}(\omega) = \dfrac{j100\omega}{10 + j\omega}$

(b) $\mathbf{H}(\omega) = \dfrac{0.4(50 + j\omega)^2}{(j\omega)^2}$

(c) $\mathbf{H}(\omega) = \dfrac{(40 + j80\omega)}{(10 + j50\omega)}$

(d) $\mathbf{H}(\omega) = \dfrac{(20 + j5\omega)(20 + j\omega)}{j\omega}$

(e) $\mathbf{H}(\omega) = \dfrac{30(10 + j\omega)}{(200 + j2\omega)(1000 + j2\omega)}$

(f) $\mathbf{H}(\omega) = \dfrac{j100\omega}{(100 + j5\omega)(100 + j\omega)^2}$

(g) $\mathbf{H}(\omega) = \dfrac{(200 + j2\omega)}{(50 + j5\omega)(1000 + j\omega)}$

*Answer(s) in Appendix F.

Section 6-3: Passive Filters

6.9 The element values of a series RLC bandpass filter are $R = 5\ \Omega$, $L = 20$ mH, and $C = 0.5\ \mu$F.

(a) Determine ω_0, Q, B, ω_{c_1}, and ω_{c_2}.

(b) Is it possible to double the magnitude of Q by changing the values of L and/or C, while keeping ω_0 and R unchanged? If yes, propose such values, and if no, why not?

6.10 A series RLC bandpass filter has half-power frequencies at 1 kHz and 10 kHz. If the input impedance at resonance is 6 Ω, what are the values of R, L, and C?

***6.11** A series RLC circuit is driven by an ac source with a phasor voltage $\mathbf{V}_s = 10\angle 30^\circ$ V. If the circuit resonates at 10^3 rad/s and the average power absorbed by the resistor at resonance is 2.5 W, determine the values of R, L, and C given that $Q = 5$.

6.12 The element values of a parallel RLC circuit are $R = 100\ \Omega$, $L = 10$ mH, and $C = 0.4$ mF. Determine ω_0, Q, B, ω_{c_1}, and ω_{c_2}.

6.13 Design a parallel RLC filter with $f_0 = 4$ kHz, $Q = 100$, and an input impedance of 25 kΩ at resonance.

6.14 For the circuit shown in Fig. P6.14 provide the following:

(a) An expression for $\mathbf{H}(\omega) = \mathbf{V}_o/\mathbf{V}_i$ in standard form.

(b) Spectral plots for the magnitude and phase of $\mathbf{H}(\omega)$ given that $R_1 = 1\ \Omega$, $R_2 = 2\ \Omega$, $C_1 = 1\ \mu$F, and $C_2 = 2\ \mu$F.

(c) The cutoff frequency ω_c and the slope of the magnitude (in dB) when $\omega/\omega_c \ll 1$ and when $\omega/\omega_c \gg 1$.

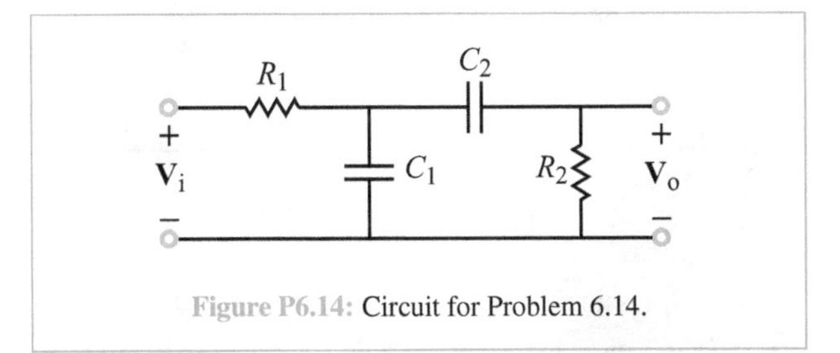

Figure P6.14: Circuit for Problem 6.14.

6.15 For the circuit shown in Fig. P6.15 provide the following:

(a) An expression for $\mathbf{H}(\omega) = \mathbf{V}_o/\mathbf{V}_i$ in standard form.

(b) Spectral plots for the magnitude and phase of $\mathbf{H}(\omega)$ given that $R_1 = 1\ \Omega$, $R_2 = 2\ \Omega$, $L_1 = 1$ mH, and $L_2 = 2$ mH.

(c) The cutoff frequency ω_c and the slope of the magnitude (in dB) when $\omega/\omega_c \ll 1$ and when $\omega/\omega_c \gg 1$.

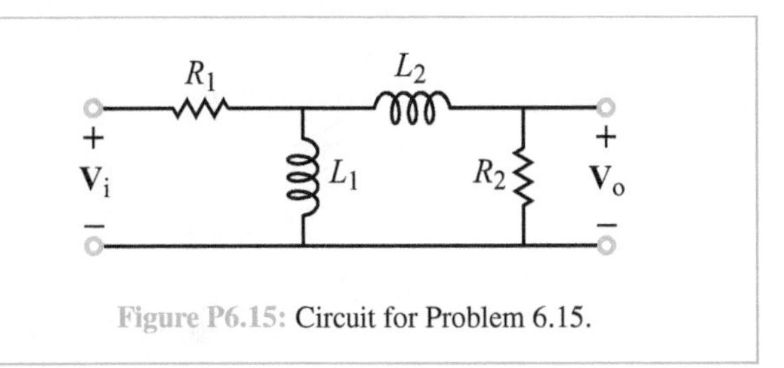

Figure P6.15: Circuit for Problem 6.15.

6.16 For the circuit shown in Fig. P6.16 provide the following:

(a) An expression for $\mathbf{H}(\omega) = \mathbf{V}_o/\mathbf{V}_i$ in standard form.

(b) Spectral plots for the magnitude and phase of $\mathbf{H}(\omega)$ given that $R = 100\ \Omega$, $L = 0.1$ mH, and $C = 1\ \mu$F.

(c) The cutoff frequency ω_c and the slope of the magnitude (in dB) when $\omega/\omega_c \gg 1$.

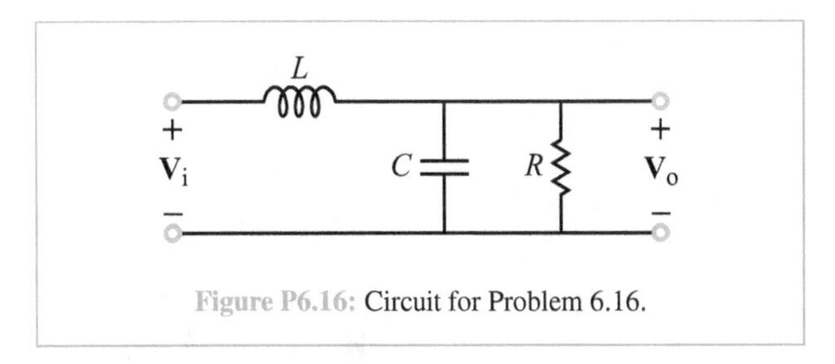

Figure P6.16: Circuit for Problem 6.16.

***6.17** For the circuit shown in Fig. P6.17 provide the following:

(a) An expression for $\mathbf{H}(\omega) = \mathbf{V}_o/\mathbf{V}_i$ in standard form.

(b) Spectral plots for the magnitude and phase of $\mathbf{H}(\omega)$ given that $R = 10\ \Omega$, $L = 1$ mH, and $C = 10\ \mu$F.

(c) The cutoff frequency ω_c and the slope of the magnitude (in dB) when $\omega/\omega_c \ll 1$.

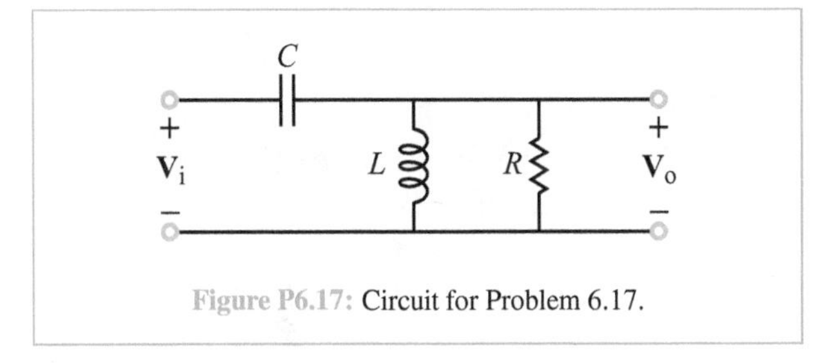

Figure P6.17: Circuit for Problem 6.17.

6.18 For the circuit shown in Fig. P6.18 provide the following:

(a) An expression for $\mathbf{H}(\omega) = \mathbf{V}_o/\mathbf{V}_i$ in standard form.

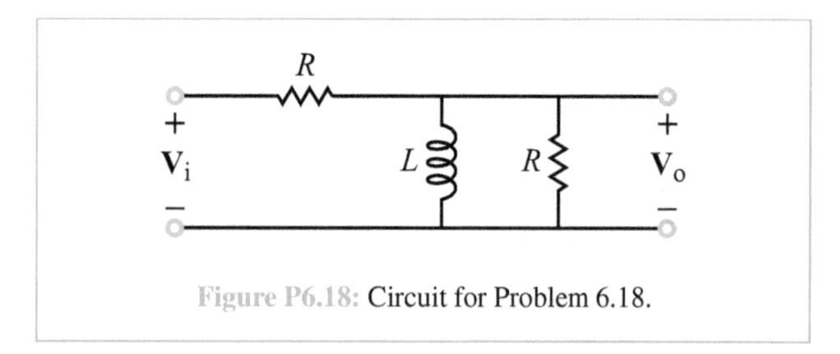

Figure P6.18: Circuit for Problem 6.18.

(b) Spectral plots for the magnitude and phase of $\mathbf{H}(\omega)$ given that $R = 50\ \Omega$ and $L = 2$ mH.

(c) The cutoff frequency ω_c and the slope of the magnitude (in dB) when $\omega/\omega_c \ll 1$.

6.19 For the circuit shown in Fig. P6.19 provide the following:

(a) An expression for $\mathbf{H}(\omega) = \mathbf{V}_o/\mathbf{V}_i$ in standard form.

(b) Spectral plots for the magnitude and phase of $\mathbf{H}(\omega)$ given that $R = 50\ \Omega$ and $L = 2$ mH.

Section 6-4: Active Filters

6.20 For the op-amp circuit of Fig. P6.20 provide the following:

(a) An expression for $\mathbf{H}(\omega) = \mathbf{V}_o/\mathbf{V}_s$ in standard form.

(b) Spectral plots for the magnitude and phase of $\mathbf{H}(\omega)$ given that $R_1 = 1$ kΩ, $R_2 = 4$ kΩ, and $C = 1\ \mu$F.

(c) What type of filter is it? What is its maximum gain?

6.21 For the op-amp circuit of Fig. P6.21 provide the following:

(a) An expression for $\mathbf{H}(\omega) = \mathbf{V}_o/\mathbf{V}_s$ in standard form.

(b) Spectral plots for the magnitude and phase of $\mathbf{H}(\omega)$ given that $R_1 = 99$ kΩ, $R_2 = 1$ kΩ, and $C = 0.1\ \mu$F.

(c) What type of filter is it? What is its maximum gain?

6.22 For the op-amp circuit of Fig. P6.22 provide the following:

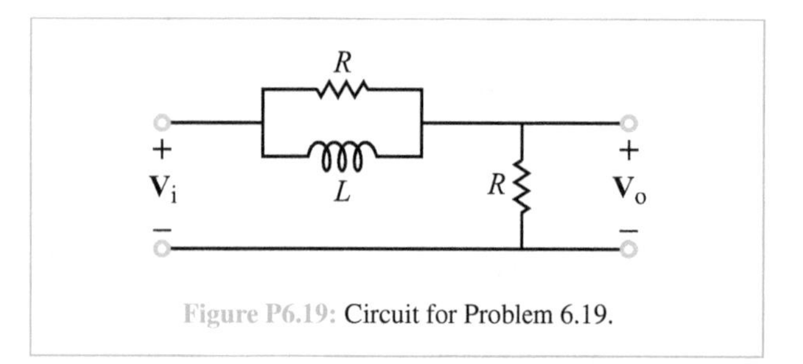

Figure P6.19: Circuit for Problem 6.19.

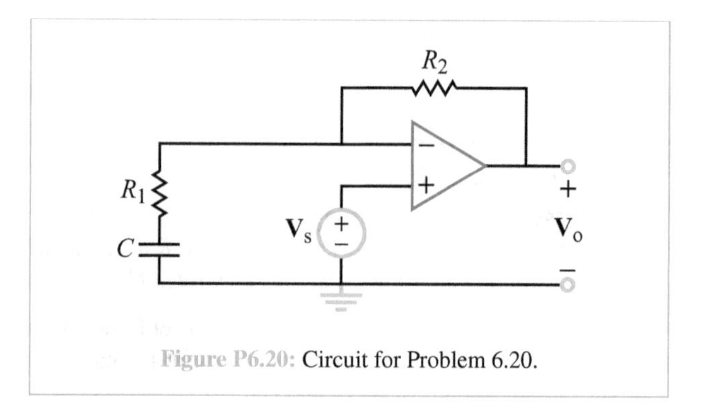

Figure P6.20: Circuit for Problem 6.20.

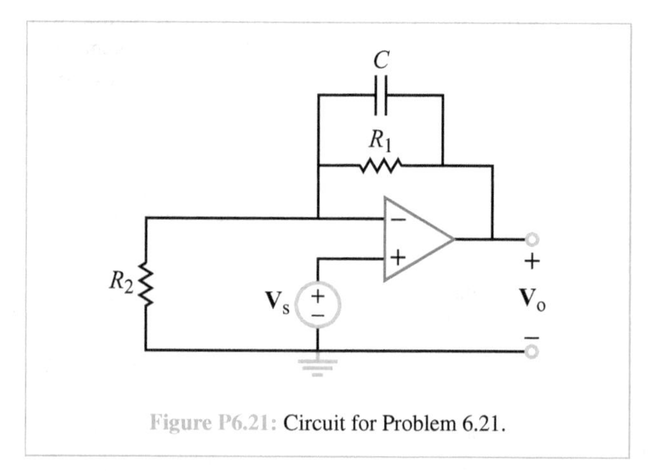

Figure P6.21: Circuit for Problem 6.21.

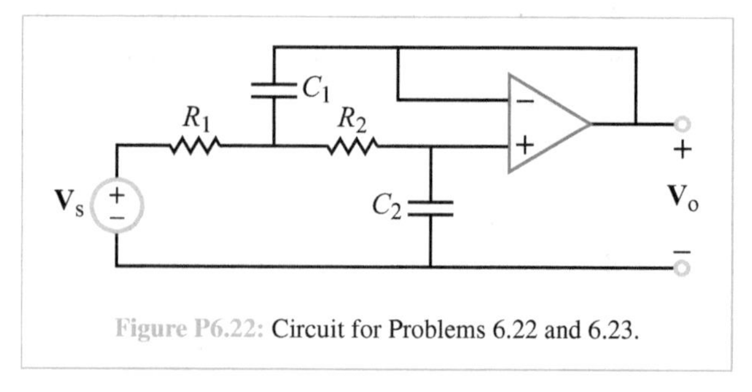

Figure P6.22: Circuit for Problems 6.22 and 6.23.

(a) An expression for $\mathbf{H}(\omega) = \mathbf{V}_o/\mathbf{V}_i$ in standard form.

(b) Spectral plots for the magnitude and phase of $\mathbf{H}(\omega)$ given that $R_1 = R_2 = 100\ \Omega$, $C_1 = 10\ \mu$F, and $C_2 = 0.4\ \mu$F.

(c) What type of filter is it? What is its maximum gain?

6.23 Repeat Problem 6.22 after interchanging the values of C_1 and C_2 to $C_1 = 0.4\ \mu$F and $C_2 = 10\ \mu$F.

***6.24** For the op-amp circuit of Fig. P6.24 provide the following:

(a) An expression for $\mathbf{H}(\omega) = \mathbf{V}_o/\mathbf{V}_s$ in standard form.

(b) Spectral plots for the magnitude and phase of $\mathbf{H}(\omega)$ given that $R_1 = 1$ kΩ, $R_2 = 20\,\Omega$, $C_1 = 5\,\mu$F, and $C_2 = 25$ nF.

(c) What type of filter is it? What is its maximum gain?

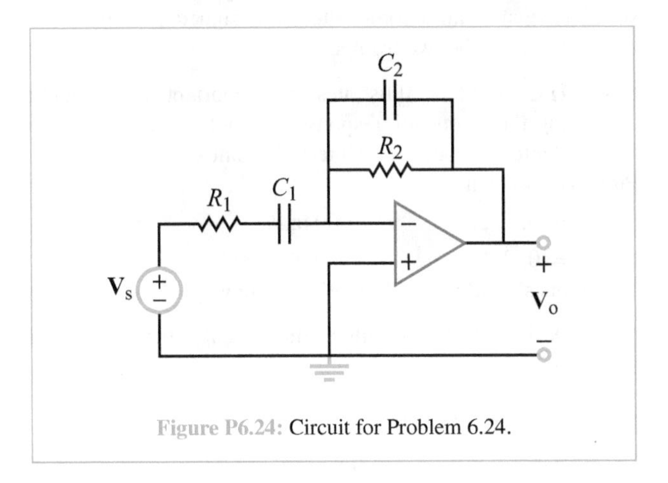

Figure P6.24: Circuit for Problem 6.24.

6.25 Design an active lowpass filter with a gain of 4, a corner frequency of 1 kHz, and a gain roll-off rate of −60 dB/decade.

6.26 Design an active highpass filter with a gain of 10, a corner frequency of 2 kHz, and a gain roll-off rate of 40 dB/decade.

6.27 The element values in the circuit of the second-order bandpass filter shown in Fig. P6.27 are: $R_{f_1} = 100$ kΩ, $R_{s_1} = 10$ kΩ, $R_{f_2} = 100$ kΩ, $R_{s_2} = 10$ kΩ, $C_{f_1} = 3.98 \times 10^{-11}$ F, $C_{s_2} = 7.96 \times 10^{-10}$ F. Generate a spectral plot for the magnitude of $\mathbf{H}(\omega) = \mathbf{V}_o/\mathbf{V}_s$. Determine the frequency locations of the maximum value of M [dB] and its half-power points.

Section 6-5: Ideal Brick-Wall Filters

6.28 Derive the impulse response of a system characterized by the frequency response shown in Fig. P6.28. Express your answer in terms of two sinc functions.

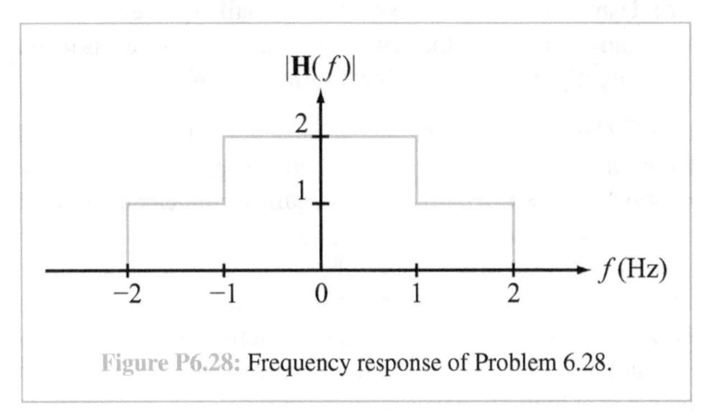

Figure P6.28: Frequency response of Problem 6.28.

6.29 Derive the impulse response of a system characterized by the frequency response shown in Fig. P6.29. Express your answer in terms of three sinc functions.

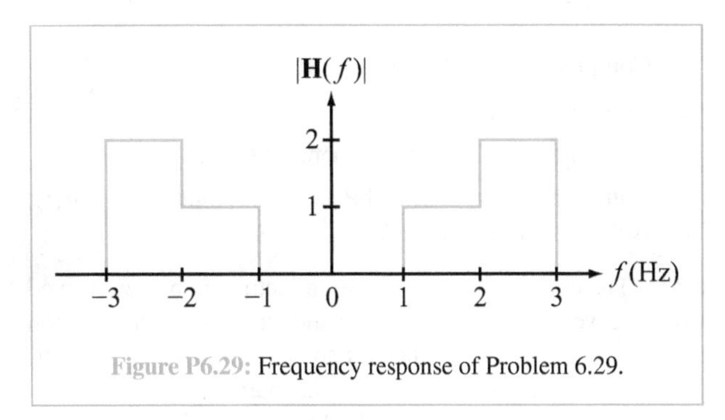

Figure P6.29: Frequency response of Problem 6.29.

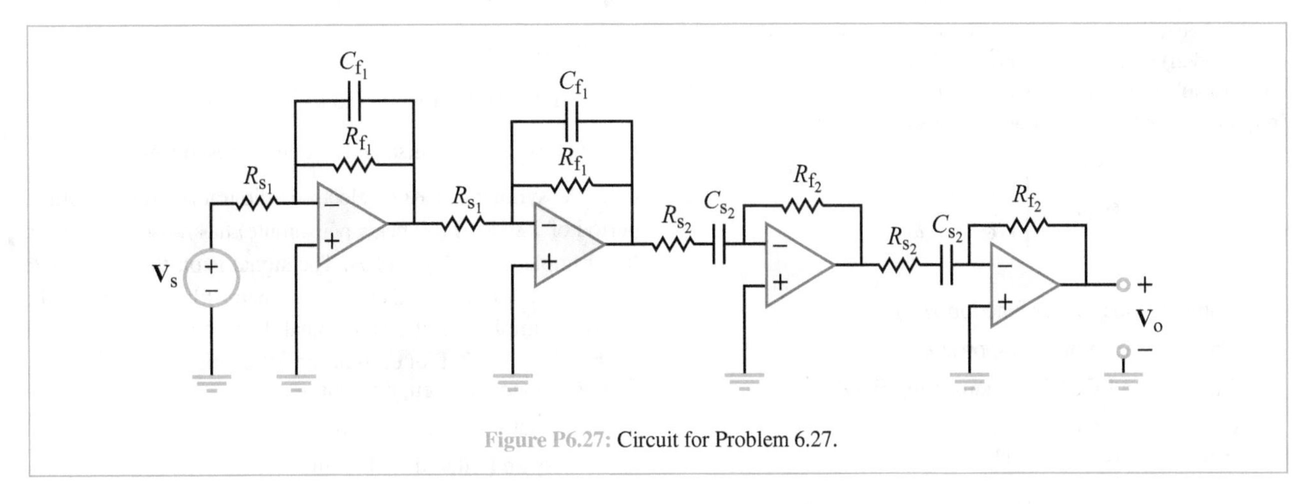

Figure P6.27: Circuit for Problem 6.27.

6.30 A *bandlimited differentiator* system with frequency response $\mathbf{H}(\omega) = j\omega \operatorname{rect}(\omega/\omega_0)$ is used to convert CAT-scan (computed axial tomography x-ray images) signals into medical images.

(a) Obtain the corresponding impulse response $h(t)$.

*(b) Using linear approximations for small arguments of sine and cosine, show that $h(0) = 0$. Can you deduce this result directly from the functional form of $\mathbf{H}(\omega)$?

6.31 The current $i(t)$ through a capacitor can be determined from the voltage $\upsilon(t)$ across the capacitor using the relation $i(t) = C\, d\upsilon/dt$. This requires implementation of the pure differentiator system $y(t) = dx/dt$. However, this system is not BIBO-stable, because a (bounded) input $x(t) = u(t)$ leads to an (unbounded) output $y(t) = \delta(t)$. Using $\mathbf{H}(\omega) = (j\omega a)/(j\omega + a)$ is a useful substitute, since $\mathbf{H}(\omega)$ resembles a differentiator at low frequencies, but it is also bounded at high frequencies:

$$\mathbf{H}(\omega) \approx \begin{cases} j\omega & \omega \ll a, \\ a & \omega \gg a. \end{cases}$$

(a) Compute the transfer function $\mathbf{H}(\mathbf{s})$.

(b) Compute the impulse response $h(t)$.

(c) Compute an LCCDE implementing $\mathbf{H}(\omega)$.

(d) Compute the response to the (bounded) input $x(t) = u(t)$. Is the response bounded?

6.32 The current $i(t)$ through an inductor can be determined from the voltage $\upsilon(t)$ across the inductor using the relation $i(t) = \frac{1}{L}\int_{-\infty}^{t} \upsilon(\tau)\, d\tau$, which is equivalent to $\upsilon(t) = L\, di/dt$. This requires implementation of the pure integrator system $y(t) = \int_{-\infty}^{t} x(\tau)\, d\tau$. However, this system is not BIBO-stable, because a (bounded) input $x(t) = u(t)$ leads to an (unbounded) output $y(t) = r(t)$. Using $\mathbf{H}(\omega) = 1/(j\omega + \epsilon)$ is a useful substitute, since $\mathbf{H}(\omega)$ resembles an integrator at high frequencies, but it is also bounded at low frequencies:

$$\mathbf{H}(\omega) \approx \begin{cases} 1/(j\omega) & \omega \gg \epsilon, \\ 1/\epsilon & \omega \ll \epsilon. \end{cases}$$

(a) Compute the transfer function $\mathbf{H}(\mathbf{s})$.

(b) Compute the impulse response $h(t)$.

(c) Compute an LCCDE implementing $\mathbf{H}(\omega)$.

(d) Compute the response to the (bounded) input $x(t) = u(t)$. Is the response bounded?

Sections 6-7 to 6-11: Filters

6.33 Given a sound recording of two trumpets, one playing note A (440 Hz) and the other playing note E (660 Hz), simultaneously, can a comb filter be designed to remove the sound of one of the two trumpets?

6.34 This problem illustrates an important point about displaying the frequency response of a notch filter. Design a notch filter to remove a 60 Hz interfering sinusoid. Use $\alpha = 0.1$. Plot $|\mathbf{H}(\omega)|$ versus

(a) $\omega = 0, 1, 2, \ldots, 999, 1000$ rad/s,

(b) $\omega = 0, 0.01, 0.02, \ldots, 999.99, 1000$ rad/s.

Your plots should look different! Explain why.

6.35 Design a notch filter that meets the following specifications:

- It eliminates 100 Hz.
- $|h(t)| \le 0.001$ for $t > 0.5$ s.

Plot the resulting impulse and frequency responses.

6.36 Design a comb filter that meets the following specifications:

- It eliminates 5 harmonics of 100 Hz.
- It eliminates dc (zero frequency).
- $|h(t)| \le 0.001$ for $t > 0.5$ s.

Plot the resulting impulse and frequency responses.

6.37 Design a resonator filter that meets the following specifications:

- It passes 5 harmonics of 100 Hz
- It passes dc (zero frequency).
- $|h(t)| \le 0.001$ for $t > 0.5$ s.

Plot the resulting impulse and frequency responses.

6.38 A synthetic EKG (electrocardiogram) signal with a period of 1 second (60 beats per minute) has noise added to it. It is sampled at 256 sample/s. The signal is on the S^2 website as the file `P638.mat`. Design a resonator filter to remove the noise and implement it on the signal. Use poles at $\pm jk - 0.01$ for $k = 1, \ldots, 7$. Plot each of the following:

(a) Resonator frequency response

(b) Resonator impulse response

(c) Noisy and filtered EKG signals

The following three problems implement notch, comb, and resonator filters on simple signals composed of triangles. No MATLAB or MathScript programming is needed.

6.39 The program `P639.m` on the S^2 website generates two triangle waves with periods of 0.02 s and 0.04 s, and generates a comb filter that eliminates the 0.02-s signal and keeps the 0.04-s signal. The user must input the real part a of the poles. Determine the value of a that maximizes the selectivity of the comb filter, while keeping the duration of the impulse response to within ≈ 1 s. Plot each of the following:

(a) The two triangle waves and their spectra

(b) The impulse and frequency responses of the filter

(c) The unfiltered and filtered sums of waves

6.40 The program `P640.m` on the S^2 website generates sinusoid and triangle waves with periods of 0.02 s and 0.04 s, and generates a notch filter that eliminates the sinusoid and keeps the triangle. The user must input the real part a of the poles. Determine the value of a that maximizes the selectivity of the notch filter, while keeping the duration of the impulse response to within ≈ 1 s. Plot each of the following:

(a) The triangle-plus-sinusoid and its spectrum

(b) The impulse and frequency responses of the filter

(c) The filtered signal

*6.41 The program `P641.m` on the S^2 website generates a triangle wave and adds noise to it. It also generates a resonator filter that enhances three of the triangle wave harmonics. The user must input the real part a of the poles. Determine the value of a that maximizes the selectivity of the resonator, while keeping the duration of the impulse response to within ≈ 1 s. Plot each of the following:

(a) The noisy triangle wave and its spectrum

(b) The impulse and frequency responses of the filter

(c) The filtered signal

6.42 For a sixth-order Butterworth lowpass filter with a cutoff frequency of 1 rad/s, compute the following:

(a) The locations of the poles

(b) The transfer function $\mathbf{H}(\mathbf{s})$

(c) The corresponding LCCDE description

6.43 For an eighth-order Butterworth lowpass filter with a cutoff frequency of 1 rad/s, compute the following:

(a) The locations of the poles.

(b) The transfer function $\mathbf{H}(\mathbf{s})$

(c) The corresponding LCCDE description

6.44 Design a Butterworth lowpass filter that meets the following specifications: (1) $|\mathbf{H}(\omega)| = 0.9$ at 10 Hz and (2) $|\mathbf{H}(\omega)| = 0.1$ at 28 Hz. Compute the following:

*(a) The order N and cutoff frequency ω_c

(b) The locations of the poles

(c) The transfer function $\mathbf{H}(\mathbf{s})$

(d) The LCCDE description of the filter

6.45 For a sixth-order Butterworth highpass filter with cutoff frequency 3 rad/s, compute the following:

(a) The locations of the poles

(b) The transfer function $\mathbf{H}(\mathbf{s})$

(c) The LCCDE description of the filter

6.46 For an eighth-order Butterworth highpass filter with cutoff frequency 2 rad/s, compute the following:

(a) The locations of the poles

(b) The transfer function $\mathbf{H}(\mathbf{s})$

(c) The LCCDE description of the filter

6.47 Design a Butterworth highpass filter that meets the following specifications: (1) $|\mathbf{H}(\omega)| = 0.10$ at 6 Hz and (2) $|\mathbf{H}(\omega)| = 0.91$ at 17 Hz. Compute the following:

(a) The order N and cutoff frequency ω_c

(b) The locations of the poles

(c) The transfer function $\mathbf{H}(\mathbf{s})$

(d) The LCCDE description of the filter

6.48 Design a fourth-order Butterworth lowpass filter using two stages of the Sallen-Key circuit shown in Fig. 6-43. In each stage use $R_1 = R_2 = 10$ kΩ and select C_1 and C_2 so that the cutoff frequency is 100 Hz.

6.49 Repeat Problem 6.48 for a sixth-order Butterworth lowpass filter.

6.50 By interchanging the resistors and capacitors in the Sallen-Key circuit of Fig. 6-43, the lowpass circuit becomes a highpass circuit.

(a) Show that the highpass version has the transfer function given by

$$\mathbf{H}(\mathbf{s}) = \frac{\mathbf{s}^2}{\mathbf{s}^2 + \left(\dfrac{1}{R_2C_1} + \dfrac{1}{R_2C_2}\right)\mathbf{s} + \dfrac{1}{R_1R_2C_1C_2}}.$$

(b) Use $R_1 = R_2 = 10$ kΩ and select C_1, and C_2 so that $\mathbf{H}(\mathbf{s})$ matches a second-order Butterworth highpass filter with a cutoff frequency of 1 kHz.

*6.51 In the spring-mass-damper suspension system shown in Fig. P6.51, $b = 4$ N·s/m, $k = 3$ N/m, $m_1 = 6$ kg, and $m_2 = 2$ kg. If $v_{y_1}(t) = dy_1/dt$ is regarded as the input signal and $v_{y_3}(t) = dy_3/dt$ is the output response, show that the suspension system acts like a mechanical third-order Butterworth filter with cutoff frequency of 1 rad/s.

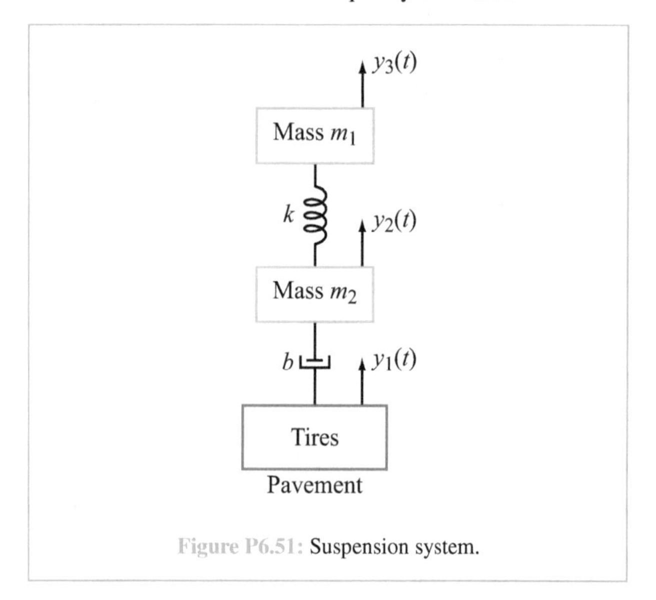

Figure P6.51: Suspension system.

6.52 Suppose in the suspension system of Problem 6.51, mass m_2 is a cart moving at 1.414 m/s along a sinusoidally corrugated roadway whose pavement's vertical displacement is given by $0.3\cos(0.5x)$, where x is the horizontal position in meters. Show that the suspension system keeps the cart (mass m_2) from moving vertically.

6.53 In the suspension system shown in Fig. P6.51, $v_{y_1}(t) = dy_1/dt$ is regarded as the input signal and $v_{y_3}(t) = dy_3/dt$ is the output response. If the associated transfer function takes the form of a third-order Butterworth filter with cutoff frequency ω_c, determine m_2, b, and k in terms of m_1 and ω_c.

Section 6-12: Amplitude Modulation

6.54 More power is required to transmit AM than DSB modulation. For a pure-tone signal $x(t) = A\cos(2\pi f_T t)$ and a carrier $x_c(t) = \cos(2\pi f_c t)$, calculate the average power transmitted when using: (a) DSB modulation, (b) AM with a modulation index of 1.

6.55 A DSB radio station at 760 kHz transmits a 3 kHz tone. A DSB receiver is used, without its final lowpass filter (it has been removed). What would a listener tuned to 770 kHz hear?

6.56 *Quadrature phase* modulation allows transmission of two bandlimited signals within the same frequency band. Let $x(t)$ and $y(t)$ be two voice signals, both bandlimited to 4 kHz. Consider $z(t) = x(t)\cos(2\pi f_c t) + y(t)\sin(2\pi f_c t)$. Show that $z(t)$ occupies 8 kHz of spectrum, but that $x(t)$ and $y(t)$ can both be recovered from modulated signal $z(t)$. [*Hint:* $2\sin(2\pi f_c t)\cos(2\pi f_c t) = \sin(4\pi f_c t)$.]

6.57 *Single sideband* (SSB) modulation is used in ham radio and digital television. One way to implement SSB is to use a sharp filter (here, a brick-wall bandpass filter). Let $x(t)$ be bandlimited to 4 kHz, with the spectrum shown in Fig. P6.57. $x(t)$ is modulated using frequency f_c, then brick-wall bandpass filtered, giving $y(t) = [x(t)\, 2\cos(2\pi f_c t)] * h(t)$, where

$$h(t) = \frac{\sin(2\pi 2000t)}{\pi t}\, 2\cos(2\pi(f_c - 2000)t)$$

is the impulse response of a brick-wall bandpass filter.

(a) Plot the spectrum of $y(t)$.

(b) Show how to recover $x(t)$ from $y(t)$.

(c) Show that SSB requires only half of the power and bandwidth of DSB-SC.

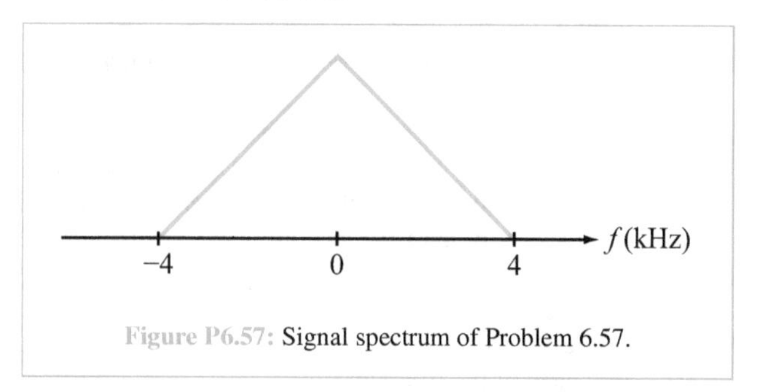

Figure P6.57: Signal spectrum of Problem 6.57.

6.58 In World War II, voice radio scramblers used modulation schemes to distort a signal so that enemy forces could not understand it unless it was demodulated properly. In one scheme, a signal $x(t)$, bandlimited to 4 kHz, is modulated to generate an output signal

$$y(t) = [2x(t)\cos(8000\pi t)] * \left[\frac{\sin(8000\pi t)}{\pi t}\right].$$

(a) Plot the range of the spectrum of $y(t)$.

(b) Describe why the scrambled signal's spectrum is "distorted."

(c) Show that $x(t)$ can be recovered from $y(t)$ using

$$x(t) = [2y(t)\cos(8000\pi t)] * \left[\frac{\sin(8000\pi t)}{\pi t}\right].$$

6.59 FM stereo signals are formed using the system shown in Fig. P6.59, where $L(t)$ is the left speaker signal and $R(t)$ is the right speaker signal. Assume both signals are bandlimited to 15 kHz. Also, signal $C(t)$ is a 38 kHz sinusoidal carrier given by $C(t) = 2\cos(76000\pi t)$. Sketch the spectrum of $z(t)$.

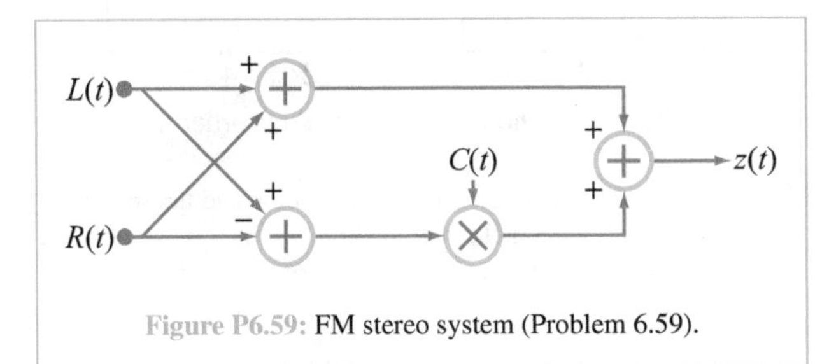

Figure P6.59: FM stereo system (Problem 6.59).

6.60 Design a Butterworth filter that meets the following specifications:

(1) $|\mathbf{H}(\omega)| = 0.995$ at B Hz, and

(2) $|\mathbf{H}(\omega)| = 0.01$ at $2B$ Hz.

Compute the order N and cutoff frequency f_c Hz as a function of B.

6.61 Load the file `P661.mat` from the S^2 website. MATLAB/MathScript variable `S` contains two DSB-modulated signals, modulated with carrier frequencies 10 kHz and 20 kHz, and sampled every 20 μs. Write a short MATLAB/MathScript program that demodulates the two signals from `S`. Listen and describe them both.

Section 6-13: Sampling Theorem

***6.62** The spectrum of the trumpet signal for note G (784 Hz) is negligible above its ninth harmonic. What is the Nyquist sampling rate required for reconstructing the trumpet signal from its samples?

6.63 Compute a Nyquist sampling rate for reconstructing signal

$$x(t) = \frac{\sin(40\pi t)\sin(60\pi t)}{\pi^2 t^2}$$

from its samples.

6.64 Signal

$$x(t) = \frac{\sin(2\pi t)}{\pi t}[1 + 2\cos(4\pi t)]$$

is sampled every 1/6 second. Sketch the spectrum of the sampled signal.

6.65 Signal $x(t) = \cos(14\pi t) - \cos(18\pi t)$ is sampled at 16 sample/s. The result is passed through an ideal brick-wall lowpass filter with a cutoff frequency of 8 Hz. Compute and sketch the spectrum of the output signal.

***6.66** Signal $x(t) = \sin(30\pi t) + \sin(70\pi t)$ is sampled at 50 sample/s. The result is passed through an ideal brick-wall lowpass filter with a cutoff frequency of 25 Hz. Compute and sketch the spectrum of the output signal.

6.67 A signal $x(t)$ has the bandlimited spectrum shown in Fig. P6.67. If $x(t)$ is sampled at 10 samples/s and then passed through an ideal brick-wall lowpass filter with a cutoff frequency of 5 Hz, sketch the spectrum of the output signal.

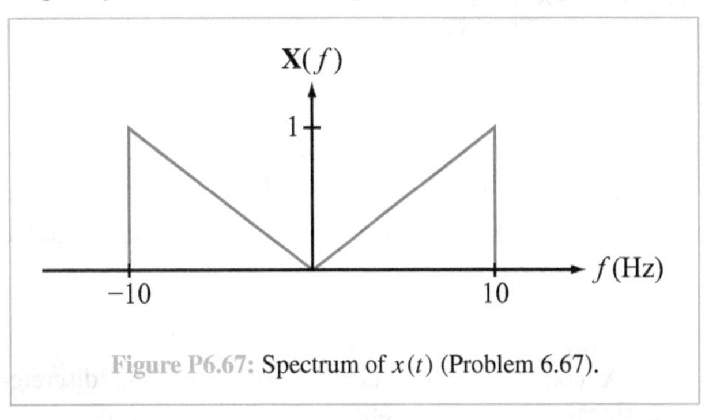

Figure P6.67: Spectrum of $x(t)$ (Problem 6.67).

6.68 A signal $x(t)$, bandlimited to 10 Hz, is sampled at 12 samples/s. what portion of its spectrum can still be recovered from its samples?

6.69 A spoken-word narration recorded on a book-on-CD is bandlimited to 4 kHz. The narration is sampled at the standard CD sampling rate of 44,100 samples/s. To reconstruct it, the sampled narration is passed through a Butterworth lowpass filter with a cutoff frequency of 4 kHz. What is the minimum order required of the Butterworth filter so that all parts of the image spectra are reduced in level to at most 0.001 of their unfiltered values?

6.70 If $x(t)$ is bandlimited to 5 kHz, what is the Nyquist sampling rate for $y(t) = x^2(t)$?

6.71 Compute the Nyquist sampling rates of the following three signals:

(a) $x_1(t) = \dfrac{3t+7}{t}\sin(10\pi t)$

(b) $x_2(t) = \sin(6\pi t)\sin(4\pi t)$

(c) $x_3(t) = \dfrac{\sin(6\pi t)\sin(4\pi t)}{t^2}$

6.72 Show that

$$\sum_{n=-\infty}^{\infty}\left[\frac{\sin(4\pi(t-n))}{4\pi(t-n)} - \frac{\sin(4\pi(t-n-\frac{1}{2}))}{4\pi(t-n-\frac{1}{2})}\right] = \cos(2\pi t).$$

6.73 What does aliasing *sound* like? Load the file `P673.mat` from the S^2 website. This is a speech signal (a single sentence) sampled at 24000 samples/s.

(a) Listen to the signal using `load P673.mat;soundsc(X,24000).` Describe it.

(b) Plot the one-sided magnitude spectrum from 0 to 8 kHz using
```
N=length(X)/3;F=linspace(0,8000,N); FX
=abs(fft(X));plot(F,FX(1:N))
```

(c) Use `Y=X(1:4:end);soundsc(Y,6000).` Describe it. It should sound different.

(d) Plot the one-sided magnitude spectrum of the signal in (c) from 0 to 3 kHz using
```
N=length(Y)/2;F=linspace(0,3000,N); FY
=4*abs(fft(Y));plot(F,FY(1:N))
```

(e) Compare (note differences) answers to (a) and (c), and to (b) and (d).

6.74 A commonly used method for converting discrete-time signals to continuous-time signals is *zero-order hold*, in which $x(t)$ is held constant between its known samples at $t = nT_s$. Zero-order hold interpolation of samples is illustrated in Fig. P6.74(a).

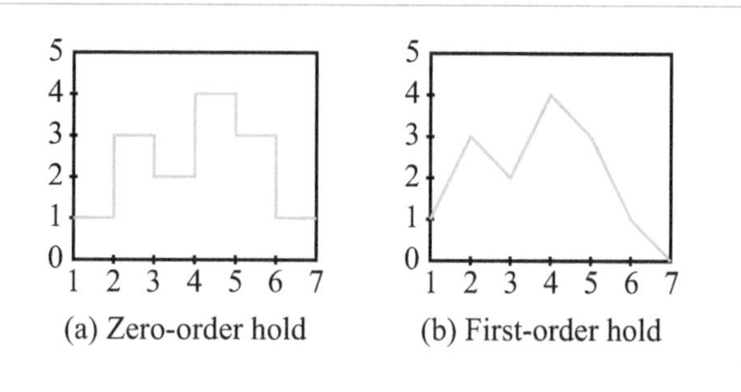

Figure P6.74: Zero-order hold interpolation and first-order hold interpolation.

(a) Zero-order hold is a lowpass filter. Compute its frequency response $\mathbf{H}(\omega)$.

(b) Zero-order hold is used to reconstruct $\cos(0.1t/T_s)$ from its samples at $t = nT_s$. The first copy of the spectrum induced by sampling lies in

$$\frac{2\pi - 0.1}{T_s} \le \omega \le \frac{2\pi + 0.1}{T_s}.$$

Compute the reduction in gain

$$\Delta G = \left| \mathbf{H}\left(\frac{(2\pi - 0.1)/T_s}{\mathbf{H}(0)} \right) \right|$$

between $\omega = 0$ and $\omega = (2\pi - 0.1)/T_s$.

6.75 ***First-order hold*** is commonly used for converting discrete-time signals to continuous-time signals, in which $x(t)$ varies linearly between its known samples at $t = nT_s$. First-order hold interpolation of samples is illustrated in Fig. P6.74(b).

(a) First-order hold is a lowpass filter. Compute its frequency response $\mathbf{H}(\omega)$.

(b) First-order hold is used to reconstruct $\cos(0.1t/T_s)$ from its samples at $t = nT_s$. The first copy of the spectrum induced by sampling lies in

$$\frac{2\pi - 0.1}{T_s} \le \omega \le \frac{2\pi + 0.1}{T_s}.$$

Compute the reduction in gain

$$\Delta G = \left| \mathbf{H}\left(\frac{(2\pi - 0.1)/T_s}{\mathbf{H}(0)} \right) \right|$$

between $\omega = 0$ and $\omega = (2\pi - 0.1)/T_s$.

6.76 This problem relates to computation of spectra $\mathbf{X}(\omega)$ of signals $x(t)$ directly from samples $x(nT_s)$.

(a) Show that the spectrum of the sampled signal

$$x_s(t) = \sum_{n=-\infty}^{\infty} x[n]\,\delta(t - nT_s)$$

is periodic in ω with period $2\pi/T_s$.

(b) Derive a formula for computing $\mathbf{X}(\omega)$ directly from $\{x[n]\}$, if no aliasing exists

LabVIEW Module 6.1

6.77 For an interference frequency of 800 Hz, for what value of the minus real part of the poles does the filtered signal most resemble the original signal?

6.78 For an interference frequency of 1000 Hz, why does the filtered signal look so different from the original signal for any value of the minus real part of the poles?

6.79 For an interference frequency of 1700 Hz, for what value of the minus real part of the poles does the filtered signal most resemble the original signal?

LabVIEW Module 6.2

6.80 Use the comb filter to eliminate the trumpet playing note G. Set the minus real part of poles at 100 and the comb filter order at 9.

6.81 Use a comb filter of order 9 to eliminate the trumpet playing note G. Set the minus real part of poles at 10. Explain why this system does not provide a satisfactory result.

6.82 Use a comb filter of order 9 to eliminate the trumpet playing note A. Set the minus real part of poles at 100.

6.83 Use a comb filter of order 9 to eliminate the trumpet playing note A. Set the minus real part of poles at 10. Explain why this system does not provide a satisfactory result.

LabVIEW Module 6.3

For each of the following problems, display all of the waveforms and spectra, and listen to the original, noisy, and filtered trumpet signals. Use a 10th order filter and a noise level of 0.2.

6.84 For each part, use a Butterworth filter with the specified cutoff frequency. Does the filtered signal waveform resemble the original signal waveform? If not, why not? Has this filter eliminated the audible part of the noise? Does the filtered signal sound like the original signal?

(a) 1000 Hz

(b) 1500 Hz

(c) 2500 Hz

6.85 For each part, use a resonator filter with a fundamental frequency of 488 Hz and the specified minus real part of poles. Has the resonator filter eliminated the audible part of the noise? Does the filtered signal waveform resemble the original signal waveform?

(a) Minus real part of poles is 10.

(b) Minus real part of poles is 100.

(c) Vary minus real part of poles to make the filtered waveform most closely resemble the original signal.

7 Discrete-Time Signals and Systems

Contents

Overview, 347
7-1 Discrete Signal Notation and Properties, 348
7-2 Discrete-Time Signal Functions, 351
7-3 Discrete-Time LTI Systems, 356
7-4 Properties of Discrete-Time LTI Systems, 359
7-5 Discrete-Time Convolution, 363
7-6 The **z**-Transform, 366
7-7 Properties of the **z**-Transform, 369
7-8 Inverse **z**-Transform, 374
7-9 Solving Difference Equations with Initial Conditions, 378
7-10 System Transfer Function, 380
7-11 BIBO Stability of **H**(**z**), 381
7-12 System Frequency Response, 384
7-13 Discrete-Time Fourier Series (DTFS), 389
7-14 Discrete-Time Fourier Transform (DTFT), 394
7-15 Discrete Fourier Transform (DFT), 400
7-16 Fast Fourier Transform (FFT), 407
7-17 Cooley–Tukey FFT, 411
Summary, 414
Problems, 415

Objectives

Learn to:

- Define discrete-time signals and systems.
- Obtain the **z**-transform of a discrete-time signal.
- Characterize discrete-time systems by their **z**-domain transfer functions.
- Use DTFS, DTFT, DFT, and FFT.

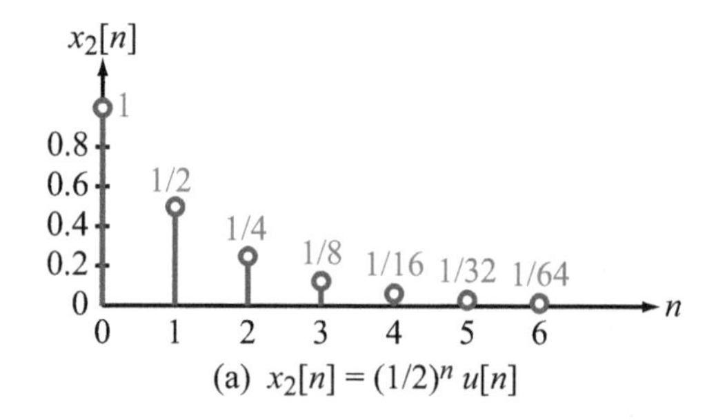

(a) $x_2[n] = (1/2)^n\, u[n]$

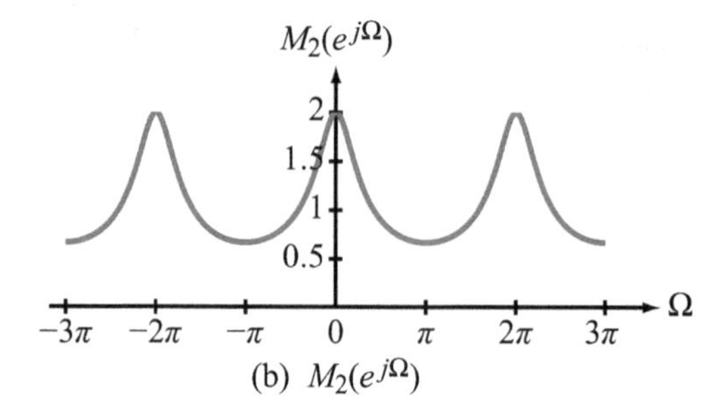

(b) $M_2(e^{j\Omega})$

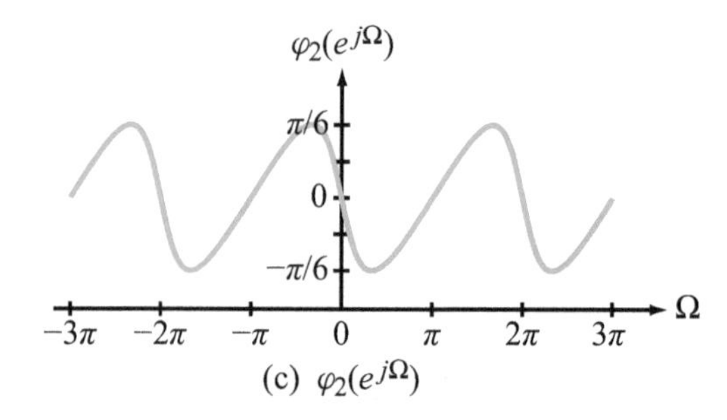

(c) $\varphi_2(e^{j\Omega})$

This chapter introduces ***discrete-time signals and systems***. The ***z-transform*** of a discrete-time signal, which is the analog of the Laplace transform of a continuous-time signal, is examined in great detail in preparation for Chapter 8 where applications of discrete-time signals and systems are highlighted.

Overview

Whereas many types of signals are inherently time- and space-continuous, others are inherently time-discrete. Examples of the latter include demographic data, genetic sequences, the number of flights per day between two specific destinations, and so on. The preceding six chapters dealt with continuous-time signals and systems. This and the following chapter will examine transformation techniques and solution procedures for discrete-time signals and systems.

Recall from Chapter 1 that a ***discrete-time*** signal is defined only at integer-valued times. A ***digital signal*** is a discrete-time signal that has been also quantized to a finite number of values and then stored and processed using bits (0 or 1). *This book will treat only discrete-time signals;* digital signals belong in hardware-oriented computer engineering courses. We provide a bridge to the digital domain, specifically to the study of digital communications systems, digital signal processing (DSP), and related topics.

In the digital domain, a signal is composed of a sequence of binary numbers, 1s and 0s. Conversion of a continuous-time signal $x(t)$ into a binary sequence entails two discretization processes: one involving time and another involving the amplitude of $x(t)$. A digital signal is a coded discrete-time discrete-amplitude signal. Discretization in time is accomplished by sampling $x(t)$ every T_s seconds to generate a discrete-time signal $x[n]$ defined as

$$x[n] = x(nT_s),$$

where as defined previously in Section 6-13, $x(nT_s) = x(t)$ evaluated at $t = nT_s$. The sampling process is illustrated in Fig. 7-1(b).

Discretization in amplitude and conversion into digital format involves the use of an encoder. The binary sequence shown in Fig. 7-1(b) uses a 4-bit encoder, which can convert an analog signal into $2^4 = 16$ binary states ranging from 0 to 15. The analog signal can be scaled and shifted (if necessary) to accommodate its amplitude range within the dynamic range of the encoder. The digital signal corresponding to $x[n]$ is

$$x_d[n] = [a_n b_n c_n d_n],$$

where a_n to d_n assume binary values (0 or 1) to reflect the value of $x[n]$. For example, in Fig. 7-1(b), $x[3] = 12$, and the corresponding digital sequence is $x_d[3] = [1100]$. The correspondence between decimal value and binary sequence for a 4-bit encoder is given in Table 7-1.

Signal $x_d[n]$ becomes the input signal to a digital computer, and after performing the task that it is instructed to do, it generates an output signal, which is denoted $y_d[n]$ in Fig. 7-1. Application of digital-to-analog conversion generates a discrete-time output signal $y[n]$, and when followed by reconstruction, the discrete signal $y[n]$ gets converted into a continuous-time signal $y(t)$, thereby completing the process.

Most signal processing is performed in the digital domain. This includes the application of digital frequency transformations, frequency filtering, coding and decoding, and many other functions. The major advantage of DSP over analog signal processing is the ability to use microprocessors (DSP chips) to perform the sophisticated mathematical operations we will study in this chapter. We will see in Chapters 8–10 that discrete-time signal processing enables filtering with much sharper frequency selectivity than is possible with analog signal processing. Also, operations such as deconvolution and batch signal processing are difficult or impossible to perform using analog signal processing. Furthermore, DSP enables storage on optical media (CDs and DVDs) and computer memory (based on storage of bits).

Table 7-1: Correspondence between decimal value and binary sequence for a 4-bit encoder.

Decimal Value	Binary Sequence	Symbol
0	0000	ıııı
1	0001	ıııl
2	0010	ıılı
3	0011	ııll
4	0100	ılıı
5	0101	ılıl
6	0110	ıllı
7	0111	ılll
8	1000	lııı
9	1001	lııl
10	1010	lılı
11	1011	lıll
12	1100	llıı
13	1101	llıl
14	1110	lllı
15	1111	llll

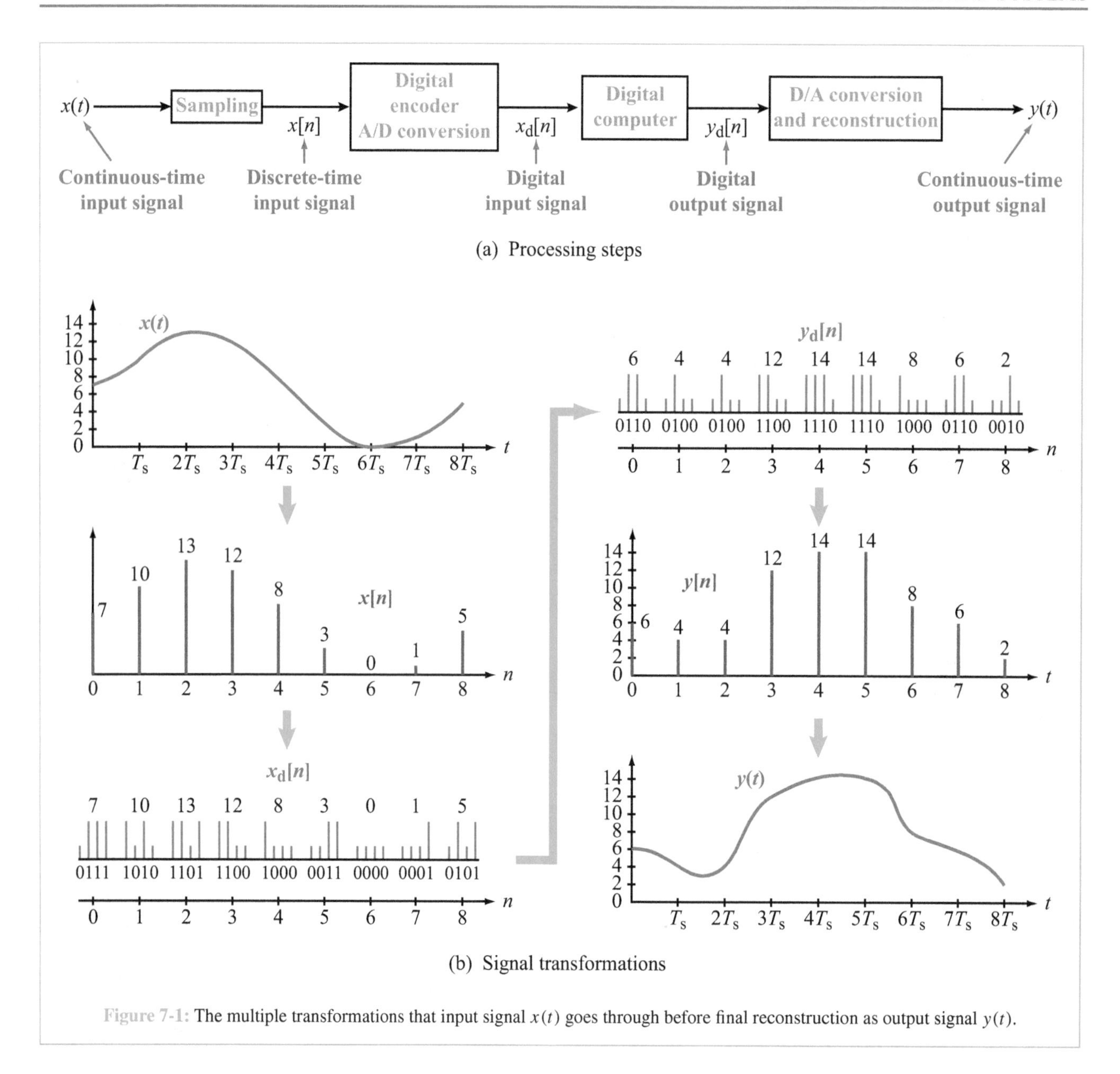

Figure 7-1: The multiple transformations that input signal $x(t)$ goes through before final reconstruction as output signal $y(t)$.

7-1 Discrete Signal Notation and Properties

► Because $x[n]$ is a sampled version of continuous-time signal $x(t)$, integer n is like a discrete surrogate for continuous time t. Hence, n is often referred to as *discrete time* or simply *time* even though (in reality) it is only an integer index. ◄

A discrete time signal $x[n]$ can be specified by listing its values for each integer value of n. For example,

$$x[n] = \begin{cases} 3 & \text{for } n = -1, \\ 2 & \text{for } n = 0, \\ 4 & \text{for } n = 2, \\ 0 & \text{for all other } n. \end{cases} \tag{7.1}$$

Note that because $x[1] = 0$, it need not be listed explicitly, as it is automatically included in the last entry.

7-1.1 Brackets Notation

An alternative format for specifying the values of $x[n]$ uses a ***brackets notation*** in which non-zero values of $x[n]$ are listed sequentially and are separated by commas and the entire sequence is enclosed between two curly brackets. Thus, if $x[n] = 0$ outside the interval $-N \leq n \leq M$ for some positive integers M and N, $x[n]$ can be specified as

$$x[n] = \{x[-N], \ldots, x[-1], \underline{x[0]}, x[1], \ldots, x[M]\}.$$

The underlined value in the sequence designates which value of $x[n]$ is at time $n = 0$. Values not listed explicitly are zero. Some books use an upward-pointing arrow instead of an underline to denote time $n = 0$.

The brackets format is particularly useful for signals of short duration, and for periodic signals. The signal given by Eq. (7.1) assumes the form

$$x[n] = \{3, \underline{2}, 0, 4\}. \tag{7.2}$$

In this format, all sequential values $x[n]$ are listed (including $x[1] = 0$), as long as they are within the range of n bounded by the outer non-zero values of $x[n]$, which in this case are $x[-1] = 3$ and $x[2] = 4$.

When applied to a periodic signal $y[n]$, the brackets format requires that the values of $y[n]$ be specified over two of its periods, and ellipses "..." are added at both ends. For example, the periodic signal $y[n] = \cos\left(\frac{\pi}{2} n\right)$ is written as

$$y[n] = \{\ldots, -1, 0, \underline{1}, 0, -1, 0, 1, 0, \ldots\}. \tag{7.3}$$

Any starting time n may be used, as long as $n = 0$ is included in one of the two periods. In the present example, the two periods start at $n = -2$ and end at $n = 5$.

7-1.2 Stem Plots

Discrete-time signals can also be specified by a graphical format that uses ***stem plots***, so called because they look like a line of dandelions in which the location and height of each dandelion specify the value of n and $x[n]$, respectively. Stem plots emphasize the fact that $x[n]$ is *undefined at non-integer times*. Examples are displayed in Fig. 7-2 for $x[n]$ of Eq. (7.2) and $y[n]$ of Eq. (7.3).

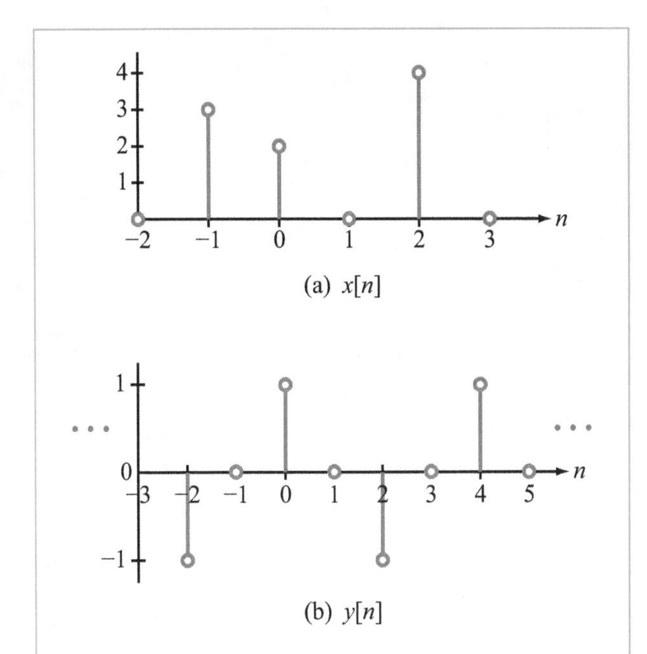

Figure 7-2: Stem plots for $x[n]$ and $y[n]$, as defined by Eqs. (7.2) and (7.3), respectively.

7-1.3 Duration of Discrete-Time Signals

The ***length*** or ***duration*** of a signal in discrete time differs from that in continuous time. Consider:

$$x(t) = 0 \qquad \text{for } t \text{ outside the interval } a \leq t \leq b$$

and

$$x[n] = 0 \qquad \text{for } n \text{ outside the interval } a \leq n \leq b.$$

Boundaries a and b are both integers and $x(a)$, $x(b)$, $x[a]$, and $x[b]$ are all non-zero values. It follows that

$$\text{Duration of } x(t) = b - a, \tag{7.4a}$$

whereas

$$\text{Duration of } x[n] = b - a + 1. \tag{7.4b}$$

The number of integers between a and b (inclusive of the endpoints) is $(b - a + 1)$. Careful counting is needed in discrete time!

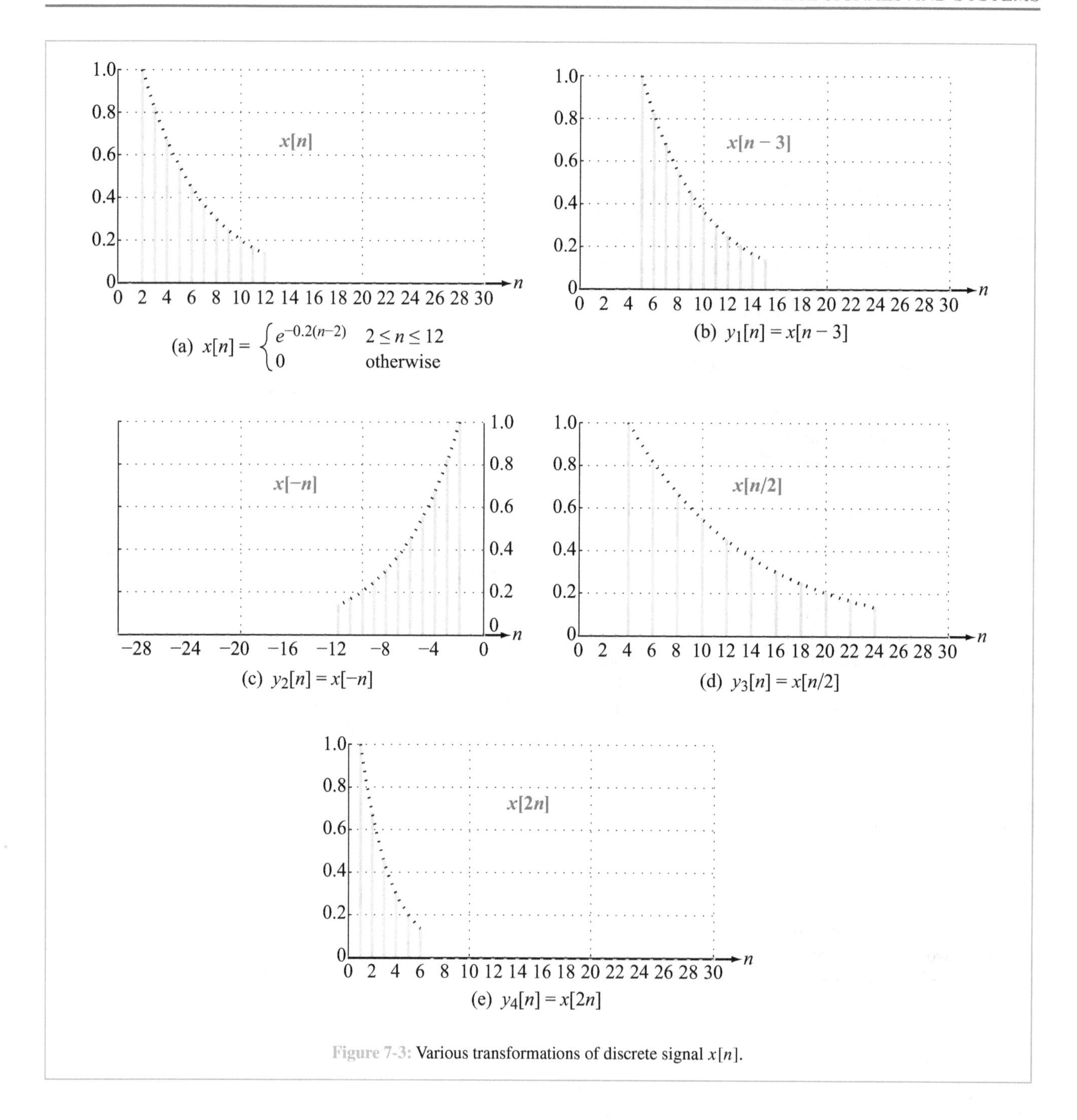

Figure 7-3: Various transformations of discrete signal $x[n]$.

7-1.4 Time Shifting

Figure 7-3(a) displays a discrete-time signal $x[n]$. To time-shift the signal by N units, we replace n with $(n - N)$. The time-shifted version is

$$y[n] = x[n - N] = \begin{cases} \text{right-shifted (delayed)} & \text{if } N > 0, \\ \text{left-shifted (advanced)} & \text{if } N < 0. \end{cases} \tag{7.5}$$

If $N = 3$, for example, $x[n]$ gets shifted to the right by three units. The shifted version is shown is Fig. 7-3(b).

7-1.5 Time Reversal

Replacing n with $-n$ generates a mirror image of $x[n]$, as shown in Fig. 7-3(c). The time-reversed version is

$$y[n] = x[-n]. \quad (7.6)$$

7-1.6 Time Scaling

(a) Downsampling

The discrete-time counterpart for compression, namely $y(t) = x(at)$, where $a > 1$, is

$$y[n] = x[Nn],$$

for some positive integer N. Only every Nth value of $x[n]$, specifically $\{\ldots, x[N], x[2N], x[3N], \ldots\}$ appear in $y[n]$. Hence, this process is called ***decimation***.

Another way of looking at decimation is to regard $x[n]$ as samples of a continuous-time signal $x(t)$ sampled every T_s, so that $x[n] = x(nT_s)$ for integers n. Then, $y[n] = x[Nn] = x(n(NT_s))$, so that $y[n]$ is $x(t)$ sampled every NT_s. This is a lower sampling rate, so the process is also called ***downsampling***.

Figure 7-3(e) displays $y_4[n] = x[2n]$, which is $x[n]$ downsampled by $N = 2$. Note that $x[2n]$ does look like $x[n]$ compressed in time.

(b) Upsampling

The discrete-time counterpart for expansion, namely $y(t) = x(at)$, where $0 < a < 1$, is

$$y[n] = x[n/N],$$

if n is a multiple of N, and $y[n] = 0$ otherwise. If $N = 2$, then $y[n] = \{\ldots, x[0], 0, x[1], 0, x[2], \ldots\}$. This process is called ***zero-stuffing***.

Figure 7-3(d) displays $y_3[n] = x[n/2]$ for even n and $y_3[n] = 0$ for odd n, which is $x[n]$ zero-stuffed with $N = 2$. This does look like $x[n]$ expanded in time, except for the zero values of $y[n]$. In practice, these values are replaced with values interpolated from the non-zero values of $y[n]$. The combined process of zero-stuffing followed by interpolation is equivalent to sampling at a higher sampling rate, so zero-stuffing followed by interpolation is also called ***upsampling***. Sometimes the term "upsampling and interpolation" is also used.

Table 7-2: Discrete-time signal transformations.

Transformation	Outcome $y[n]$
Amplitude scaling	$a\,x[n]$
Amplitude shifting	$x[n] + b$
Time scaling	$x[Nn]$
Time shifting	$x[n-N]$
Time reversal	$x[-n]$
Addition	$x_1[n] + x_2[n]$

7-1.7 Amplitude Transformation

Some applications may call for changing the scale of a discrete-time signal or shifting its average value. The amplitude transformation

$$y[n] = a\,x[n] + b \quad (7.7)$$

can accommodate both operations; coefficient a scales $x[n]$ and coefficient b shifts its dc value.

A summary list of signal transformations is given in Table 7-2.

Concept Question 7-1: What does the underline in $\{3, \underline{1}, 4\}$ mean? (See S²)

Concept Question 7-2: Why are stem plots used in discrete time? (See S²)

Exercise 7-1: Determine the duration of $\{3, \underline{1}, 4, 6\}$.

Answer: $a = -1$ and $b = 2$, so the duration is $b - a + 1 = 4$. (See S²)

Exercise 7-2: If the mean value of $x[n]$ is 3, what transformation results in a zero-mean signal?

Answer: $y[n] = x[n] - 3$. (See S²)

7-2 Discrete-Time Signal Functions

In continuous time, we made extensive use of impulse and step functions, exponential functions, and sinusoids. We now introduce their discrete-time counterparts.

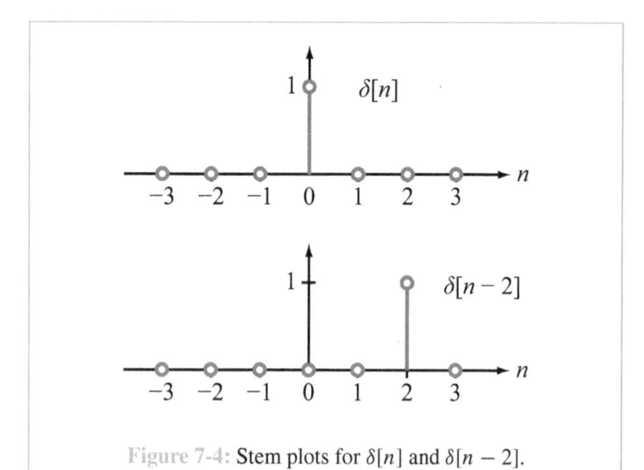

Figure 7-4: Stem plots for $\delta[n]$ and $\delta[n-2]$.

7-2.1 Discrete-Time Impulse $\delta[n]$

In discrete time, impulse $\delta[n]$ is defined as

$$\delta[n] = \{\underline{1}\} = \begin{cases} 1 & \text{for } n = 0, \\ 0 & \text{for } n \neq 0. \end{cases} \tag{7.8}$$

Unlike its continuous-time counterpart $\delta(t)$, $\delta[n]$ does not have to satisfy requirements of zero width, infinite height, and unit area. Figure 7-4 displays stem plots of $\delta[n]$ and its right-shifted version $\delta[n-2]$.

In continuous time, the sampling property of impulses is given in Eq. (1.29) as

$$x(T) = \int_{-\infty}^{\infty} x(t)\, \delta(t-T)\, dt \qquad \textbf{(continuous time)}. \tag{7.9}$$

The ***discrete-time sampling property*** of impulses is

$$x[n] = \sum_{i=-\infty}^{\infty} x[i]\, \delta[i-n] \qquad \textbf{(discrete time)}. \tag{7.10}$$

7-2.2 Discrete-Time Step $u[n]$

The discrete-time step $u[n]$ is defined as

$$u[n] = \{\underline{1}, 1, 1, \ldots\} = \begin{cases} 1 & \text{for } n \geq 0, \\ 0 & \text{for } n < 0, \end{cases} \tag{7.11a}$$

and its shifted version is

$$u[n-N] = \begin{cases} 1 & \text{for } n \geq N, \\ 0 & \text{for } n < N. \end{cases} \tag{7.11b}$$

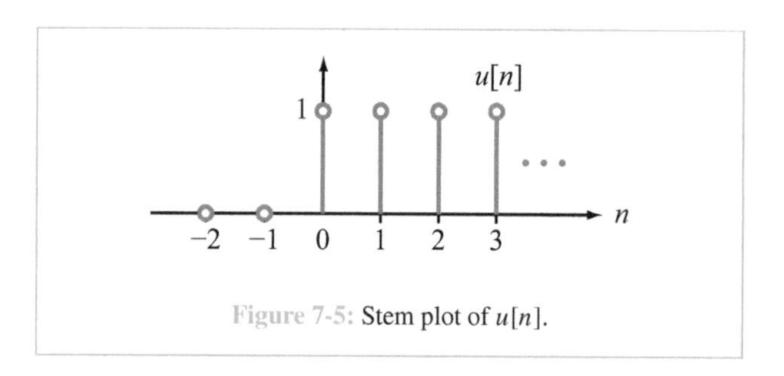

Figure 7-5: Stem plot of $u[n]$.

The relationship between step and impulse is

$$u(t) = \int_{-\infty}^{t} \delta(\tau)\, d\tau \qquad \textbf{(continuous time)} \tag{7.12a}$$

and

$$u[n] = \sum_{i=-\infty}^{n} \delta[i] \qquad \textbf{(discrete time)}. \tag{7.12b}$$

Unlike the continuous-time step, there is no discontinuity at time zero in discrete time; we simply have $u[0] = 1$. *Indeed, the very notion of continuity has no place in discrete time.* A stem-plot representation of $u[n]$ is displayed in Fig. 7-5.

Given that $u[n] = 1$ for all integer values of $n \geq 0$ and $u[n-1] = 1$ for all integer values of $n \geq 1$, it follows that

$$\delta[n] = u[n] - u[n-1]. \tag{7.13}$$

7-2.3 Discrete-Time Geometric Signals

A ***geometric signal*** is defined as

$$x[n] = \mathbf{p}^n\, u[n], \tag{7.14}$$

where $\mathbf{p}$ is a constant that may be real or complex. Geometric signals are the discrete-time counterparts to exponential signals; a geometric signal is an exponential signal that has been sampled at integer times. The geometric signal $(1/2)^n\, u[n]$ and exponential signal $(1/2)^t\, u(t)$ are plotted in Fig. 7-6.

If $\mathbf{p}$ in Eq. (7.14) is a complex number,

$$\mathbf{p} = |\mathbf{p}| e^{j\theta},$$

$x[n]$ will exhibit amplitude and phase plots:

$$|x[n]| = |\mathbf{p}|^n\, u[n] \tag{7.15a}$$

and

$$\underline{/x[n]} = n\theta\, u[n]. \tag{7.15b}$$

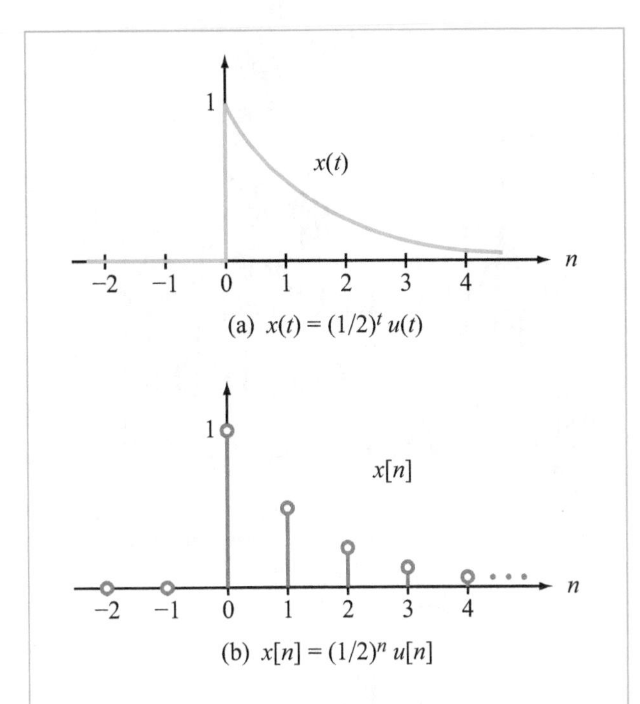

Figure 7-6: Plots of exponential signal $x(t)$ and its geometric counterpart.

7-2.4 Discrete-Time Sinusoids

A discrete-time signal $x[n]$ is periodic with ***fundamental period*** N_0 if, for all integer times n,

$$x[n] = x[n + N_0] \quad \textbf{(any periodic discrete-time signal)}, \tag{7.16}$$

where N_0 is an integer; otherwise, $x[n + N_0]$ would not be defined.

A continuous-time sinusoid of angular frequency ω_0 is given by

$$x(t) = A\cos(\omega_0 t + \theta). \tag{7.17}$$

The corresponding period and frequency are $T_0 = 2\pi/\omega_0$ and $f_0 = 1/T_0$. When sampled every T_s seconds, the continuous-time sinusoid becomes a discrete-time sinusoid given by

$$x[n] = x(nT_s) = A\cos(\omega_0 nT_s + \theta) = A\cos(\Omega n + \theta), \tag{7.18}$$

where

$$\Omega = \omega_0 T_s \qquad \textbf{(rad/sample)} \tag{7.19}$$

is defined as the ***discrete-time angular frequency***. Since θ and Ωn are both in radians and n is a sampling integer, Ω is measured in ***radians per sample***.

(a) Fundamental period of sinusoid

Were we to apply the standard relationship for a continuous-time signal, we would conclude that the discrete-time period of $x[n]$ is

$$N_0 = \frac{2\pi}{\Omega} = \frac{2\pi}{\omega_0 T_s} \qquad \text{(iff } N_0 \text{ is an integer).} \tag{7.20}$$

The expression given by Eq. (7.20) is valid if and only if N_0 is an integer, which is seldom true. To determine the true period N_0, we insert Eq. (7.18) into the periodicity relation given by Eq. (7.16), which yields

$$A\cos(\Omega n + \theta) = A\cos(\Omega n + \theta + \Omega N_0).$$

To satisfy this condition, ΩN_0 has to be equal to an integer multiple of 2π:

$$\Omega N_0 = 2\pi k,$$

where k is an integer. Hence, the ***fundamental period*** of the discrete-time sinusoid is given by

$$N_0 = \frac{2\pi k}{\Omega} = k\,\frac{T_0}{T_s}, \tag{7.21}$$

where T_0 is the period of the continuous-time sinusoid, T_s is the sampling period, and *k has to be selected such that it is the smallest integer value that results in an integer value for N_0.*

Example 7-1: Period of Sinusoid

Determine the discrete-time period of

$$x[n] = \cos(0.3\pi n).$$

A stem plot of $x[n]$ is shown in Fig. 7-7.

Solution: From Eq. (7.21),

$$N_0 = \frac{2\pi k}{\Omega} = \frac{2\pi k}{0.3\pi} = \frac{20k}{3}.$$

Before we select a value for k, we need to make sure that its coefficient is in the form of a non-reducible fraction N/D, which $20/3$ is. Hence, the smallest integer value of k that results in an integer value for N_0 is $k = D = 3$, and that, in turn, leads to

$$N_0 = N = 20 \text{ samples}.$$

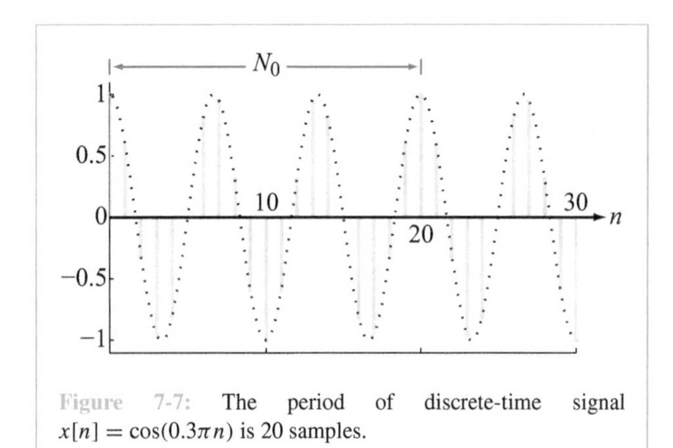

Figure 7-7: The period of discrete-time signal $x[n] = \cos(0.3\pi n)$ is 20 samples.

Thus, the discrete-time sinusoid repeats itself every 20 samples, and the duration of the 20 samples is equal to three periods of the continuous-time sinusoid.

Example 7-2: Sum of Two Sinusoids

Determine the fundamental period of

$$x[n] = \cos(\pi n/3) + \cos(\pi n/4).$$

Solution: The fundamental period of $x[n]$ has to satisfy both sinusoids. For the first sinusoid, Eq. (7.21) gives

$$N_{0_1} = \frac{2\pi k_1}{\Omega_1} = \frac{2\pi k_1}{\pi/3} = 6k_1.$$

The smallest integer value k_1 can assume such that N_{0_1} is an integer is $k_1 = 1$. Hence, $N_{0_1} = 6$.

Similarly, for the second term, we have

$$N_{0_2} = \frac{2\pi k_2}{\Omega_2} = \frac{2\pi k_2}{\pi/4} = 8k_2,$$

which is satisfied by selecting $k_2 = 1$ and $N_{0_2} = 8$.

The fundamental period of $x[n]$ is N_0, selected such that it is simultaneously the smallest integer multiple of both N_{0_1} and N_{0_2}. The value of N_0 that satisfies this condition is:

$$N_0 = 24.$$

Every 24 points, $\cos(\pi n/3)$ repeats four times, $\cos(\pi n/4)$ repeats three times, and their sum repeats once.

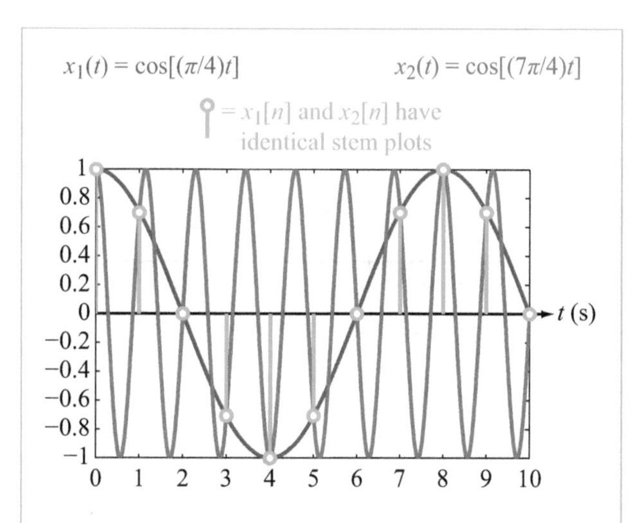

Figure 7-8: Sinusoidal signals at $\omega_1 = \pi/4$ rad/s and $\omega_2 = 7\pi/4$ rad/s are identical at integer times.

(b) Fundamental frequency of sinusoid

Figure 7-8 displays two continuous-time sinusoidal signals: one at angular frequency $\omega_1 = \pi/4$ rad/s and another at $\omega_2 = 7\omega_1 = 7\pi/4$ rad/s. Thus,

$$x_1(t) = \cos\left(\frac{\pi}{4}\,t\right)$$

and

$$x_2(t) = \cos\left(\frac{7\pi}{4}\,t\right).$$

When sampled at $t = nT_s$ with $T_s = 1$ s, their discrete-time equivalents are

$$x_1[n] = \cos\left(\frac{\pi}{4}\,n\right)$$

and

$$x_2[n] = \cos\left(\frac{7\pi}{4}\,n\right).$$

Even though the two discrete-time signals have different angular frequencies, $x_1[n] = x_2[n]$ for all n, as shown in Fig. 7-8. The two signals have identical stem plots, even though they have different angular frequencies. Is the ambiguity related to the fact that the two angular frequencies are integer multiples of one another?

No, the ambiguity is not specific to signals whose angular frequencies are integer multiples of one another. Consider, for example, the two continuous-time sinusoids

$$y_1(t) = \cos\left(\frac{7\pi}{8}\,t\right) \tag{7.22a}$$

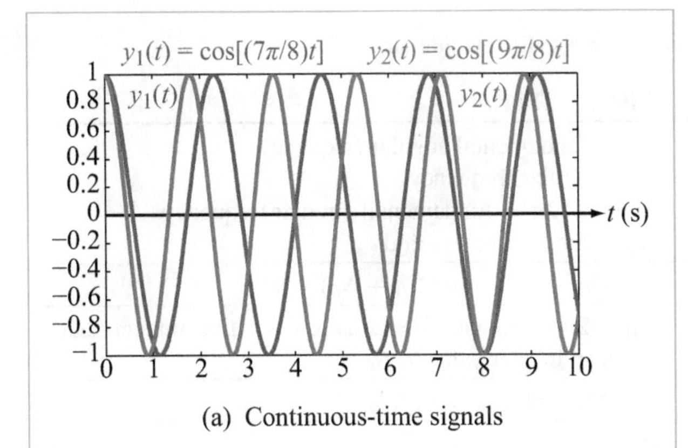

(a) Continuous-time signals

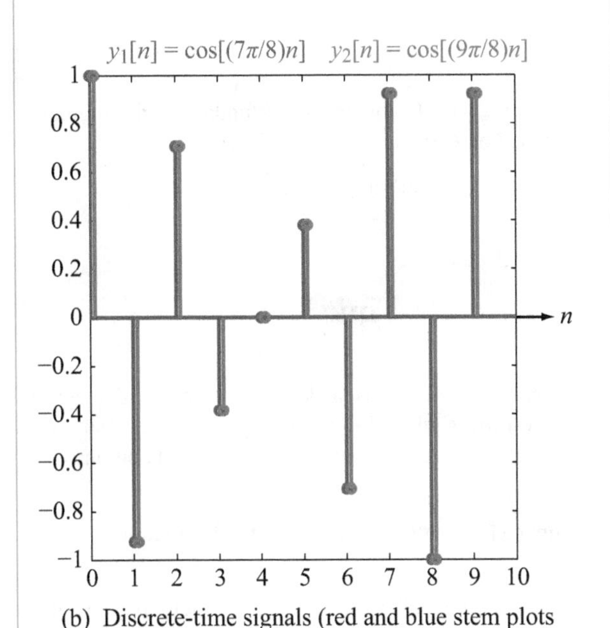

(b) Discrete-time signals (red and blue stem plots are offset slightly for display purposes)

Figure 7-9: Whereas in continuous time, no ambiguity exists between the waveforms of $y_1(t)$ and $y_2(t)$, their discrete-time analogues, $y_1[n]$ and $y_2[n]$, have identical stem plots when sampled at $T_s = 1$ s, even though they have different angular frequencies.

and

$$y_2(t) = \cos\left(\frac{9\pi}{8}\,t\right). \tag{7.22b}$$

Their time plots are displayed in Fig. 7-9(a). Because the two sinusoids have different angular frequencies, their patterns diverge as a function of time, and it is easy to distinguish $y_1(t)$ from $y_2(t)$ and to determine the angular frequency of each from its time plot.

This unambiguous relationship between $\cos(\omega t)$ and its time plot does not carry over into discrete time. When sampled at $t = nT_s$ with $T_s = 1$ s, the discrete-time equivalents of $y_1(t)$ and $y_2(t)$ are

$$y_1[n] = \cos\left(\frac{7\pi}{8}\,n\right) \tag{7.23a}$$

and

$$y_2[n] = \cos\left(\frac{9\pi}{8}\,n\right). \tag{7.23b}$$

Even though $y_1[n]$ and $y_2[n]$ have different discrete-time angular frequencies, they exhibit identical stem plots, as shown in Fig. 7-9(b). This is because for $n = 1$, $\cos\left[\frac{7}{8}\,\pi\right] = \cos\left[\frac{9}{8}\,\pi\right] = -0.92$, and the same is true for other integer values of n.

For a continuous-time sinusoid $x(t) = A\cos(\omega_0 t + \theta)$, there is a one-to-one correspondence between the expression of $x(t)$ and its waveform. If we plot $x(t)$, we get a unique waveform, and conversely, from the waveform, we can deduce the values of A, ω_0, and θ.

The one-to-one correspondence does not apply to discrete-time sinusoids. Sinusoids with different values of Ω can generate the same stem plot, as evidenced by the plots in Figs. 7-8 and 7-9(b). For a specific value of Ω, the stem plot of $\cos(\Omega n)$ is unique, but given a stem plot, there may be multiple values of Ω that would generate the same stem plot.

► Another artifact of discrete-time sinusoids is that a sinusoid may or may not be periodic. ◄

According to Eq. (7.21), a sinusoid $x(t) = \cos(\Omega n + \theta)$ is periodic with period N_0 only if an integer k exists so that $N_0 = (2\pi k/\Omega)$ is an integer. It is seldom possible to satisfy this condition. For example, if $\Omega = 2$ rad/sample, it follows that $N_0 = \pi k$, for which no integer value of k exists that would make N_0 an integer.

If N_0 does exist, the ***fundamental angular frequency*** associated with N_0 is

$$\Omega_0 = \frac{2\pi}{N_0}. \tag{7.24}$$

Since N_0 is a positive integer and the smallest value it can (realistically) assume is 2 samples per cycle, it follows that

$$0 \le \Omega_0 \le \pi. \tag{7.25}$$

Table 7-3: Comparison between continuous-time and discrete-time sinusoids.

Property	Continuous Time	Discrete Time
Nomenclature	ω_0 = angular frequency $f_0 = \omega_0/2\pi$ = circular frequency $T_0 = 1/f_0$ = period	Ω_0 = fundamental angular frequency Ω = angular frequency $\mathbb{F} = \Omega_0/2\pi$ = fundamental circular frequency N_0 = fundamental period
Periodicity Property	$\cos(\omega_0 t + \theta) = \cos(\omega_0(t + T_0) + \theta)$	$\cos(\Omega n + \theta) = \cos(\Omega(n + N_0) + \theta)$
Period	$T_0 = 2\pi/\omega_0$	$N_0 = 2\pi k/\Omega$, with k = smallest positive integer that gives an integer value to N_0
Angular frequency	$0 < \omega_0 < \infty$	$0 \le \Omega_0 \le \pi$; $\Omega_0 = 2\pi/N_0$
Radial frequency	$0 < f_0 < \infty$	$\mathbb{F} = \Omega_0/2\pi,\ 0 \le \mathbb{F} \le \frac{1}{2}$

▶ The principal range of the frequency Ω of a discrete-time sinusoid is limited to 0 to π, inclusive. ◀

Example 7-3: Fundamental Period and Frequency

Compute the fundamental period and fundamental angular frequency of $x[n] = 3\cos(7.2\pi n + 2)$.

Solution: Application of Eq. (7.20) with $\Omega = 7.2\pi$ gives

$$N_0 = \frac{2\pi k}{\Omega} = \frac{2\pi k}{7.2\pi} = \frac{k}{3.6}.$$

The smallest integer value of k that gives an integer value for N_0 is $k = 18$, which yields $N_0 = 5$ samples.

From Eq. (7.24),

$$\Omega_0 = \frac{2\pi}{N_0} = \frac{2\pi}{5} = 0.4\pi \qquad \text{(rad/sample)}.$$

Table 7-3 compares the properties of continuous-time sinusoids with those of discrete-time sinusoids.

Concept Question 7-3: Why is $u[0] = 1$, instead of undefined? (See S²)

Concept Question 7-4: When is a discrete sinusoid not periodic? (See S²)

Exercise 7-3: Determine the fundamental period and fundamental angular frequency of $3\cos(0.56\pi n + 1)$.

Answer: $N_0 = 25$ samples, $\Omega_0 = 2\pi/25$ rad/sample. (See S²)

Exercise 7-4: Compute the fundamental angular frequency of $2\cos(5.1\pi n + 1)$.

Answer: 0.1π rad/sample. (See S²)

7-3 Discrete-Time LTI Systems

A *discrete-time system* is a device or process that accepts a discrete-time signal $x[n]$ as its input, and in response, it produces another discrete-time signal $y[n]$ as its output:

Input $x[n]$ ⟶ System ⟶ Output $y[n]$.

A simple example of a discrete-time system is a weighted moving-average system that generates an output $y[n]$ equal to the weighted average of the three most recent values of the stock market index $x[n]$:

$$y[n] = \frac{6}{11}\,x[n] + \frac{3}{11}\,x[n-1] + \frac{2}{11}\,x[n-2]. \tag{7.26}$$

Output $y[n]$ on day n is the sum of the stock market index on that same day, $x[n]$, plus those on the preceding two days, $x[n-1]$ and $x[n-2]$, with the three input values weighted differently in order to give the largest weight of 6/11 to $x[n]$ and successively smaller weights to $x[n-1]$ and $x[n-2]$. The weights are chosen so that if the stock market index is static (i.e., $x[n] = x[n-1] = x[n-2]$), the sum in Eq. (7.26) would yield $y[n] = x[n]$.

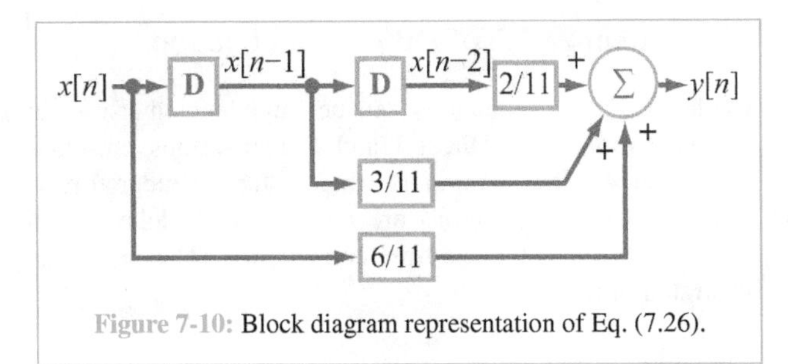

Figure 7-10: Block diagram representation of Eq. (7.26).

7-3.1 System Diagram

The system described by Eq. (7.26) is represented by the block diagram shown in Fig. 7-10. The diagram includes three types of operations.

(a) *Scaling*: Multiplication of $x[n]$, $x[n-1]$, and $x[n-2]$ by $6/11$, $3/11$, and $2/11$, respectively.

(b) *Addition*: Symbolized by the symbol Σ.

(c) *Delay*: Symbolized by the symbol D. Delaying a signal $x[n]$ is an instruction to the system to select its predecessor in the sampled sequence, namely $x[n-1]$.

These are the primary operations carried out by a digital computer to perform addition and subtraction, and to solve difference equations.

7-3.2 Difference Equations

Discrete-time systems work with ***difference equations***, which are the discrete-time counterpart to differential equations in continuous-time. Consider the first-order differential equation

$$\frac{dy}{dt} + c\ y(t) = d\ x(t) \qquad \textbf{(differential equation)}. \qquad (7.27)$$

When sampled at a short time interval T_s, input signal $x(t)$ becomes $x[n] = x(nT_s)$, and similarly, output signal $y(t)$ becomes $y[n] = y(nT_s)$. Also, time derivative dy/dt becomes a difference between $y[n]$ and $y[n-1]$ divided by T_s. Thus,

$$x(t) \rightarrow x[n]$$

$$y(t) \rightarrow y[n]$$

$$\frac{dy}{dt} \rightarrow \frac{y[n]-y[n-1]}{T_s} \qquad (\text{for } T_s \text{ very small, but} \neq 0).$$

The discrete-time equivalent of Eq. (7.27) is then

$$\frac{y[n]-y[n-1]}{T_s} + c\ y[n] = d\ x[n]. \qquad (7.28)$$

Simplification of Eq. (7.28) leads to the ***difference equation***

$$y[n] + a\ y[n-1] = b\ x[n] \qquad \textbf{(difference equation)} \quad (7.29)$$

with

$$a = \frac{-1}{1+cT_s} \quad \text{and} \quad b = \frac{dT_s}{1+cT_s}. \qquad (7.30)$$

Note that coefficients a and b of the difference equation are related to (but different from) coefficients c and d of the differential equation.

In general, the discrete-time counterpart to a continuous-time LCCDE is a difference equation of the same order:

$$\textbf{LCCDE} \qquad \sum_{i=0}^{N} c_{N-i} \frac{d^i y}{dt^i} = \sum_{i=0}^{M} d_{M-i} \frac{d^i x}{dt^i} \qquad (7.31a)$$

and

$$\textbf{Difference Equation} \qquad \sum_{i=0}^{N} a_i\ y[n-i] = \sum_{i=0}^{M} b_i\ x[n-i]. \qquad (7.31b)$$

Coefficients a_i and b_i of the difference equation are related to the coefficient of the LCCDE. Also note that the order of the indices of the coefficients are reversed between continuous and discrete time. More discussion of the latter is given in Chapter 8.

7-3.3 Recursive Equation

By rewriting Eq. (7.29) as

$$y[n] = -a\ y[n-1] + b\ x[n], \qquad (7.32)$$

we obtain a ***recursive form*** in which the value of $y[n]$ is obtained by *updating* the previous value, $y[n-1]$, through a scaling factor $(-a)$ and adding a scaled value of the current input $x[n]$. Expressing a difference equation in recursive form is the format used by DSP chips to perform signal processing.

7-3.4 ARMA Format

The two sides of the difference equation given by Eq. (7.31b) carry special meanings:

(a) **Autoregressive (AR)**

When the input (right-hand side of Eq. (7.31b)) consists of only the input at the present time, namely $x[n]$, Eq. (7.31b) reduces to

$$\sum_{i=0}^{N} a_i \, y[n-i] = x[n] \qquad \textbf{(autoregressive)}. \qquad (7.33)$$

It is called autoregressive because the present output $y[n]$ becomes equal to a linear combination of present input $x[n]$ and previous outputs $y[n-1]$ to $y[n-N]$.

(b) **Moving average (MA)**

Similarly, when the output (left-hand side of Eq. (7.31b)) consists of only the output at the present time, namely $y[n]$, Eq. (7.31b) becomes

$$y[n] = \sum_{i=0}^{M} b_i \, x[n-i] \qquad \textbf{(moving average)}. \qquad (7.34)$$

The right-hand side is the weighted average of the $(M+1)$ most recent values of the input. It is called a moving average because the sequence of input values being averaged is shifted (moved) by one time unit as n is incremented to $(n+1)$, and so on.

For example, the 120-day moving average of a stock index is an MA system of order $M = 119$, with coefficients $b_i = 1/120$ for $i = 0, 1, \dots, 119$. Thus,

$$y[n] = \frac{1}{120} \sum_{i=0}^{119} x[n-i],$$

where $x[n]$ is the present value of the stock index.

Systems described by difference equations that contain multiple terms on both sides are called *ARMA*, denoting the general case where the equation is an amalgamation of both *autoregression and moving average*.

Example 7-4: ARMA Format

Express the following equation in ARMA format:

$$y[n+1] + 2y[n-1] = 3x[n+1] + 4x[n].$$

In ARMA format, $y[n]$ is the latest term on the output side. By changing variables from n to $n-1$, the difference equation becomes

$$y[n] + 2y[n-2] = 3x[n] + 4x[n-1].$$

7-3.5 Realization of Difference Equations

ARMA difference equations can be implemented using the discrete form of the Direct I and II realizations presented in Section 4-7 for continuous-time systems. Integrators in continuous-time realizations are replaced with delays (*shift registers*) in their discrete-time counterparts. The process is illustrated in Example 7-5.

Example 7-5: ARMA Realization

Develop the Direct II realization of the ARMA difference equation

$$y[n] + a_1 \, y[n-1] + a_2 \, y[n-2] = b_0 \, x[n] + b_1 \, x[n-1] + b_2 \, x[n-2].$$

Solution: We define the intermediate variable $z[n]$ as

$$z[n] = b_0 \, x[n] + b_1 \, x[n-1] + b_2 \, x[n-2], \qquad (7.35a)$$

and also as

$$z[n] = y[n] + a_1 \, y[n-1] + a_2 \, y[n-2]. \qquad (7.35b)$$

Figure 7-11(a) displays the realization of Eq. (7.35a) as a linear combination of delayed versions of the input $x[n]$.

Next, we rewrite Eq. (7.35b) so that $y[n]$ alone is on the left-hand side,

$$y[n] = z[n] - a_1 \, y[n-1] - a_2 \, y[n-2]. \qquad (7.35c)$$

Realization of Eq. (7.35c) is shown in Fig. 7-11(b).

We now have two realizations: $x[n] \to z[n]$ and $z[n] \to y[n]$, with each representing a system. The combined realization is equal to the sequential implementation of the two realizations, *in either order*. This is analogous to the Direct Form II implementation described in Section 4-7.2. By reversing the order of the two realizations shown in Fig. 7-11(a) and (b), and replacing the two parallel chains of shift elements with a shared single chain, we obtain the realization shown in Fig. 7-11(c).

Concept Question 7-5: What is a recursive equation? (See S^2)

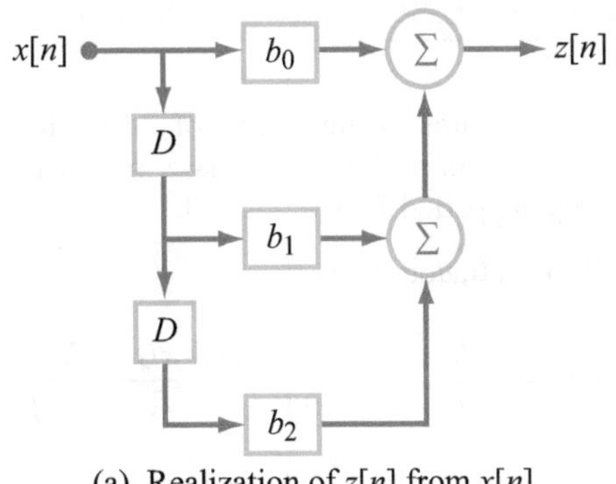

(a) Realization of $z[n]$ from $x[n]$

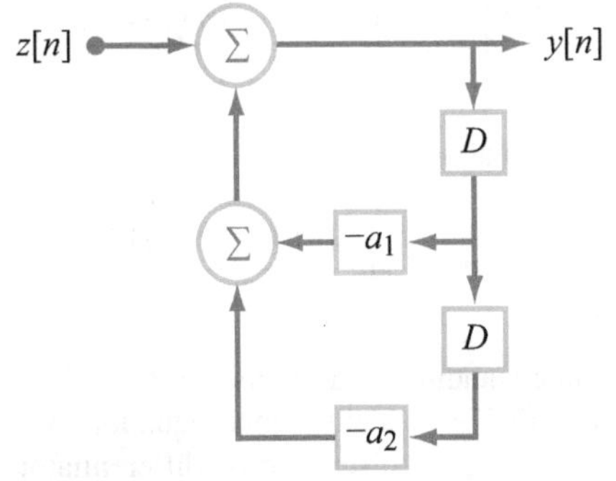

(b) Realization of $y[n]$ from $z[n]$

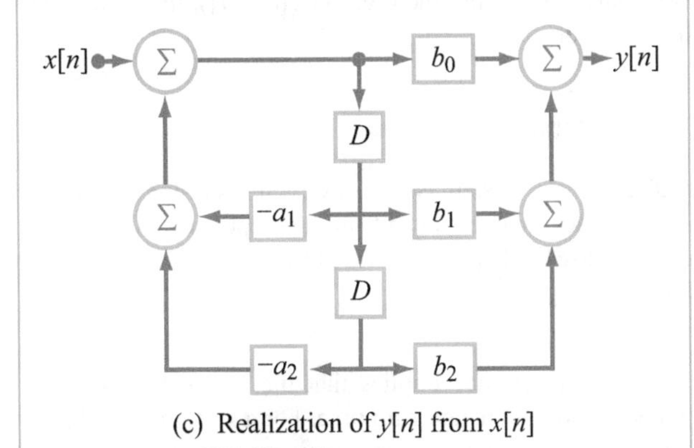

(c) Realization of $y[n]$ from $x[n]$

Figure 7-11: Upon reversing the order of the realizations in (a) and (b) and then combining them by sharing shift registers (delays), the result is the Direct Form II realization shown in (c).

Concept Question 7-6: What is the difference between AR and MA systems? (See S²)

Exercise 7-5: Transform the following equation into the form of an ARMA difference equation:

$$y[n+2]+2y[n]=3x[n+1]+4x[n-1].$$

Answer: $y[n]+2y[n-2]=3x[n-1]+4x[n-3]$ (replace n with $n-2$). (See S²)

7-4 Properties of Discrete-Time LTI Systems

Fundamentally, the properties of scalability, additivity, linearity, and time invariance are exactly the same in discrete time as they are in continuous time. If we were to replace t with n everywhere in Section 2-1, we would obtain the list of properties pertaining to discrete-time LTI systems. Of course, time shifts must be integers in discrete time. A quick review of the aforementioned and other properties of discrete-time LTI systems follows.

7-4.1 Linearity

A system is linear if it has the scaling and additivity properties. For discrete-time systems, we have the following relationships:

Scalability:

Given: $x[n] \to$ System $\to y[n]$,

then $c\,x[n] \to$ System $\to c\,y[n]$.

Additivity:

Given: $x_1[n] \to$ System $\to y_1[n]$

and $x_2[n] \to$ System $\to y_2[n]$,

then $x_1[n]+x_2[n] \to$ System $\to y_1[n]+y_2[n]$.

The ***principle of superposition*** applies to linear systems in discrete time, just as it does to linear systems in continuous time, when initial conditions are zero.

Example 7-6: Difference Equations

Show that all difference equations with constant coefficients are linear.

Solution: Let the system responses to $x_1[n]$ and $x_2[n]$ be $y_1[n]$ and $y_2[n]$, respectively, with the two input-output pairs related by

$$\sum_{i=0}^{N} a_i\ y_1[n-i] = \sum_{i=0}^{M} b_i\ x_1[n-i] \tag{7.36a}$$

and

$$\sum_{i=0}^{N} a_i\ y_2[n-i] = \sum_{i=0}^{M} b_i\ x_2[n-i]. \tag{7.36b}$$

Multiplying Eq. (7.36a) by a constant c gives

$$\sum_{i=1}^{N} a_i(c\ y_1[n-i]) = \sum_{i=1}^{M} b_i(c\ x_1[n-i]),$$

which proves the scaling property.

Adding Eqs. (7.36a and b) yields

$$\sum_{i=1}^{N} a_i(y_1[n-i]+y_2[n-i]) = \sum_{i=1}^{M} b_i(x_1[n-i]+x_2[n-i]),$$

which demonstrates that the additivity property holds.

▶ Any difference equation with constant coefficients describes a linear system. ◀

7-4.2 System Memory

A system may have *memory*, in which case it is called a *dynamic system*, or it may be a *memoryless* or *static system*. A discrete-time system is said to have memory if its output $y[n]$ at time n depends not only on the value of the input $x[n]$ at time n but also on values of $x[n]$ or $y[n]$ at other times. For example,

$$y[n] = 10x[n] \qquad \text{is memoryless,}$$

whereas

$$y[n] = 3x[n] - 2x[n-2] \qquad \text{has memory.}$$

Example 7-7: Discrete-Time Differentiator

Is the discrete-time counterpart of the differentiator $y(t) = 3\ (dx/dt)$ linear? Does it have memory? Assume that the sampling period T_s is very small.

Solution: In continuous time,

$$y(t) = 3\,\frac{dx}{dt} = 3 \lim_{T_s \to 0} \left\{ \frac{x(t) - x(t-T_s)}{T_s} \right\}.$$

Conversion to discrete time is achieved by setting $t = nT_s$:

$$y(nT_s) = 3\left[\frac{x(nT_s) - x(nT_s - T_s)}{T_s}\right],$$

or in discrete signal notation,

$$y[n] = 3\left[\frac{x[n] - x[n-1]}{T_s}\right] \tag{7.37}$$

with the understanding that T_s is very small. The result given by Eq. (7.37) is a difference equation with constant coefficients. Hence, the discrete-time differentiator is an LTI system. Moreover, since $y[n]$ depends on the value of the input at an earlier point in time (i.e., on $x[n-1]$), the system does have memory.

7-4.3 Time Invariance

A discrete-time system is time-invariant if shifting the input by k units leads to an equal shift in the output. That is, if $y[n]$ is the response to $x[n]$, the response to $x[n+k]$ is $y[n+k]$, where k is any integer.

▶ Time invariance implies that the parameters of the system do not change with time. For a system described by a difference equation, its parameters are reflected in the equation's coefficients, so if the coefficients are constant, the system is time invariant. ◀

7-4.4 Impulse Response $h[n]$

The *impulse response* of a discrete-time system is designated $h[n]$. Unlike the continuous-time case, there is no difficulty in *measuring* $h[n]$ for a real-world system, since discrete-time impulses exist physically. For a discrete-time LTI system, its impulse response is the output generated by the system when excited by a discrete-time unit impulse at its input:

$$\underset{\text{Impulse}}{\delta[n]} \longrightarrow \boxed{\text{System}} \longrightarrow \underset{\text{Impulse response}}{h[n]}. \qquad (7.38)$$

In general, the impulse response of an LTI system can be computed from its ARMA difference equation, if it exists, using the **z**-transform, as will be demonstrated in Section 7-6. However, for the special case of a system described by a MA difference equation of the form given by Eq. (7.34), $h[n]$ can be *read off* of its coefficients directly. For

$$y[n] = \sum_{i=0}^{M} b_i\; x[n-i] = b_0\; x[n] + b_1\; x[n-1] + \cdots + b_M\; x[n-M],$$

upon setting $x[n] = \delta[n]$, the output generated by the system is $h[n]$. That is,

$$h[n] = y[n]|_{x[n]=\delta[n]} = \sum_{i=0}^{M} b_i\; \delta[n-i] = \{\underline{b_0}, b_1, \ldots, b_M\}, \qquad (7.39)$$

where in the final step we adopted the brackets notation introduced in Section 7-1.1. Graphically, $h[n]$ is a stem plot with the stem amplitudes being the coefficients of the MA difference equation:

$$y[n] = \sum_{i=0}^{M} b_i\; x[n-i] \longleftrightarrow h[n] = \{\underline{b_0}, b_1, \ldots, b_M\}. \qquad (7.40)$$

7-4.5 Causality and BIBO Stability

The definitions of causality and BIBO stability for discrete LTI systems are identical to those for continuous-time systems.

▶ Specifically, a signal $x[n]$ is ***causal*** if $x[n] = 0$ for all $n < 0$. An LTI system is causal if and only if its impulse response $h[n]$ is causal (i.e., $h[n] = 0$ for all $n < 0$). ◀

ARMA difference equations, like their continuous-time LCCDE counterparts, are inherently causal, since the present value of the output, $y[n]$, can be computed from only present and past values of the input and past values of the output.

▶ A signal $x[n]$ is ***bounded*** if there exists some number L such that $|x[n]| \leq L$. A signal $x[n]$ is ***absolutely summable*** if $\sum_{n=-\infty}^{\infty} |x[n]|$ is finite. ◀

Absolutely summable for a discrete-time signal is the counterpart of absolutely integrable for a continuous-time signal; namely, $\int_{-\infty}^{\infty} |x(t)|\; dt$ is finite.

▶ An LTI system is ***BIBO stable*** if every bounded input $x[n]$ generates a bounded output $y[n]$. Equivalently, an LTI system is BIBO stable if and only if its impulse response $h[n]$ is absolutely summable. ◀

Proof of the preceding statement is almost identical procedurally with that contained in Section 2-6.3 for continuous-time, except that integrals are replaced with summations.

All MA systems are BIBO stable. From Eq. (7.40),

$$h[n] = \left\{\underline{b_0}, b_1, \ldots, b_M\right\},$$

which is absolutely summable because M is finite and so are all of the coefficients.

Example 7-8: MA System Stability

Given the system $y[n] = 2x[n] - 3x[n-1] - 4x[n-2]$, prove that the system is BIBO stable.

Solution: Application of Eq. (7.39) to this MA system gives

$$h[n] = \{\underline{2}, -3, -4\},$$

and its absolute sum is

$$\sum |h[n]| = 2 + 3 + 4 = 9,$$

which is finite. Hence, the system is BIBO stable, as expected.

Example 7-9: BIBO Unstable System

Determine whether or not the system with impulse response

$$h[n] = \frac{(-1)^n}{n+1}\; u[n]$$

is BIBO stable.

Solution: Whereas

$$\sum_{n=-\infty}^{\infty} h[n] = \sum_{n=-\infty}^{\infty} \frac{(-1)^n}{n+1}\, u[n] = \sum_{n=0}^{\infty} \frac{(-1)^n}{n+1} = 1 - \frac{1}{2} + \frac{1}{3} - \frac{1}{4} + \cdots = \ln 2,$$

proving that $h[n]$ is summable, it is not absolutely summable because

$$\sum_{n=-\infty}^{\infty} |h[n]| = 1 + \frac{1}{2} + \frac{1}{3} + \frac{1}{4} + \cdots \to \infty.$$

Since $h[n]$ is not absolutely summable, the system is not BIBO stable.

7-4.6 $h[n]$ as Linear Combination of Geometric Signals

Recall from Section 7-2.3 that a geometric signal given by

$$x[n] = \mathbf{p}^n\, u[n]$$

is the discrete-time equivalent of an exponential signal $e^{\mathbf{p}t}\, u(t)$ in continuous time. Constant $\mathbf{p}$ may be real-valued or complex. Sometimes, the impulse response $h[n]$ of a discrete-time LTI system may be described by a linear combination of geometric signals. That is,

$$h[n] = \sum_{i=1}^{N} \mathbf{C}_i \mathbf{p}_i^n\, u[n]. \tag{7.41}$$

► Such a system is BIBO stable if and only if $|\mathbf{p}_i| < 1$ for all i. ◄

We can prove that the preceding statement is true by proving that $h[n]$ is absolutely summable. By the triangle inequality,

$$\sum_{n=-\infty}^{\infty} |h[n]| = \sum_{n=-\infty}^{\infty} \left| \sum_{i=1}^{N} \mathbf{C}_i \mathbf{p}_i^n\, u[n] \right| \leq \sum_{n=-\infty}^{\infty} \sum_{i=1}^{N} |\mathbf{C}_i \mathbf{p}_i^n|\, u[n].$$

Upon interchanging the order of the two summations and replacing the lower limit on n to zero (as required by $u[n]$), we get

$$\sum_{n=-\infty}^{\infty} |h[n]| \leq \sum_{i=1}^{N} \sum_{n=0}^{\infty} |\mathbf{C}_i \mathbf{p}_i^n| = \sum_{i=1}^{N} |\mathbf{C}_i| \sum_{n=0}^{\infty} |\mathbf{p}_i^n|. \tag{7.42}$$

Recall that the sum of an ***infinite geometric series*** is

$$\sum_{n=0}^{\infty} \mathbf{r}^n = \frac{1}{1-\mathbf{r}} \tag{7.43}$$

if and only if $|\mathbf{r}| < 1$. If $|\mathbf{r}| \geq 1$, the series does not converge. Here, setting $\mathbf{r} = |\mathbf{p}_i|$, we have (if and only if $|\mathbf{p}_i| < 1$)

$$\sum_{n=0}^{\infty} |\mathbf{p}_i^n| = 1 + |\mathbf{p}_i| + |\mathbf{p}_i|^2 + \cdots = \frac{1}{1 - |\mathbf{p}_i|}, \tag{7.44}$$

which is finite. For the summation over i, the sum of N finite quantities is finite. Hence, if $|\mathbf{p}_i| < 1$, $h[n]$ is absolutely summable, and the system is BIBO stable.

By way of comparison, a continuous-time LTI system with an impulse response

$$h(t) = \sum_{i=1}^{N} \mathbf{C}_i e^{\mathbf{p}_i t}\, u(t) \tag{7.45}$$

is BIBO stable if and only if $\mathfrak{Re}\{\mathbf{p}_i\} < 0$ for all i. The continuous-time stability condition of poles $\mathbf{p}_i$ in the open left half-plane in the $\mathbf{s}$-domain becomes the discrete-time stability condition of poles $\mathbf{p}_i$ inside the ***unit circle*** $|\mathbf{p}_i| = 1$.

Concept Question 7-7: Why are all MA systems BIBO stable? (See S²)

Concept Question 7-8: What is the condition for BIBO stability of an LTI system with the impulse response $h[n] = \sum_{i=1}^{N} \mathbf{C}_i \mathbf{p}_i^n\, u[n]$? (See S²)

Exercise 7-6: Is the system with impulse response

$$h[n] = \frac{1}{(n+1)^2}\, u[n]$$

BIBO stable?

Answer: Yes.

$$\sum_{n=-\infty}^{\infty} |h[n]| = \sum_{n=0}^{\infty} \frac{1}{(n+1)^2} = \frac{\pi^2}{6}.$$

This is finite, so the system is BIBO stable. (See S²)

Exercise 7-7: Is the system with $h[n] = \left(\frac{1}{2}\right)^n u[n]$ BIBO stable?

Answer: Yes, since $|\frac{1}{2}| < 1$. (See S²)

Exercise 7-8: A system has an impulse response $h[n] = 0$ for $n < 0$ but $h[0] \neq 0$. Is the system causal?

Answer: Yes. The output of a causal system at time n can depend on the input at the same time n. (See S²)

7-5 Discrete-Time Convolution

Convolution is much simpler to perform in discrete-time than in continuous-time, primarily because there are no integrals to be evaluated. As we will see shortly, computing the convolution of two finite-duration signals is fairly straightforward, and if either signal has infinite duration, the **z**-transform of Section 7-6 should be used.

7-5.1 Derivation of Convolution Sum

The ***convolution sum*** in discrete time is the counterpart of the convolution integral in continuous time. Derivation of the expression defining the convolution sum of two discrete-time signals follows the same basic steps outlined earlier in Section 2-3.1 for the convolution integral.

Step 1: From the definition of the impulse response given by Eq. (7.38),

$$\delta[n] \longrightarrow \boxed{\text{LTI}} \longrightarrow h[n]. \tag{7.46}$$

Step 2: From time-invariance of LTI systems, delaying the input $\delta[n]$ by an integer i delays the output $h[n]$ by the same integer i,

$$\delta[n-i] \longrightarrow \boxed{\text{LTI}} \longrightarrow h[n-i]. \tag{7.47}$$

Step 3: From the scaling property of LTI systems, multiplying the input $\delta[n-i]$ by any *constant* $x[i]$ multiplies the output $h[n-i]$ by the same factor,

$$x[i]\,\delta[n-i] \longrightarrow \boxed{\text{LTI}} \longrightarrow x[i]\,h[n-i]. \tag{7.48}$$

Step 4: From the additivity property of LTI systems, summing the input over i, sums the output over i as well,

$$\sum_{i=-\infty}^{\infty} x[i]\,\delta[n-i] \longrightarrow \boxed{\text{LTI}} \longrightarrow \sum_{i=-\infty}^{\infty} x[i]\,h[n-i]. \tag{7.49}$$

Step 5: From the sampling property of impulses, the input is recognized to be just $x[n]$. Hence,

$$x[n] \longrightarrow \boxed{\text{LTI}} \longrightarrow \sum_{i=-\infty}^{\infty} x[i]\,h[n-i]. \tag{7.50}$$

Using an asterisk $*$ to denote the convolution operation, Eq. (7.50) can be written as

$$y[n] = x[n] * h[n] = \sum_{i=-\infty}^{\infty} x[i]\,h[n-i]. \tag{7.51a}$$

(any pair of functions)

If the two functions represent causal signals or systems, they should be multiplied by step functions (as a reminder) and the limits of the summation should be changed to $i = 0$ to n. That is,

$$\begin{aligned} y[n] &= (x[n]\,u[n]) * (h[n]\,u[n]) \\ &= \sum_{i=-\infty}^{\infty} x[i]\,u[i]\,h[n-i]\,u[n-i] \\ &= u[n] \sum_{i=0}^{n} x[i]\,h[n-i]. \end{aligned} \tag{7.51b}$$

(causal functions)

Example 7-10: Convolution Sum

Given $x[n] = \{\underline{2}, 3, 4\}$ and $h[n] = \{\underline{5}, 6, 7\}$, compute $y[n] = x[n] * h[n]$.

Solution: Both signals have a length of 3 and start at time zero. That is, $x[0] = 2$, $x[1] = 3$, $x[2] = 4$, and $x[i] = 0$ for all other values of i. Similarly, $h[0] = 5$, $h[1] = 6$, $h[2] = 7$, and $h[i] = 0$ for all other values of i.

By Eq. (7.51a), the convolution sum of $x[n]$ and $h[n]$ is

$$y[n] = x[n] * h[n] = \sum_{i=-\infty}^{\infty} x[i]\, h[n-i]. \tag{7.52}$$

Since $h[i] = 0$ for all values of i except $i = 0$, 1, and 2, it follows that $h[n-i] = 0$ for all values of i except for $i = n$, $n-1$, and $n-2$. With this constraint in mind, we can apply Eq. (7.52) at discrete values of n, starting at $n = 0$:

$$y[0] = \sum_{i=0}^{0} x[i]\, h[0-i] = x[0]\, h[0] = 2 \times 5 = 10,$$

$$y[1] = \sum_{i=0}^{1} x[i]\, h[1-i]$$
$$= x[0]\, h[1] + x[1]\, h[0] = 2 \times 6 + 3 \times 5 = 27,$$

$$y[2] = \sum_{i=0}^{2} x[i]\, h[2-i]$$
$$= x[0]\, h[2] + x[1]\, h[1] + x[2]\, h[0]$$
$$= 2 \times 7 + 3 \times 6 + 4 \times 5 = 52,$$

$$y[3] = \sum_{i=1}^{2} x[i]\, h[3-i]$$
$$= x[1]\, h[2] + x[2]\, h[1] = 3 \times 7 + 4 \times 6 = 45,$$

$$y[4] = \sum_{i=2}^{2} x[i]\, h[4-i] = x[2]\, h[2] = 4 \times 7 = 28,$$

$$y[n] = 0, \text{ otherwise.}$$

Hence,

$$y[n] = \{\underline{10}, 27, 52, 45, 28\}.$$

7-5.2 Discrete-Time Convolution Properties

With one notable difference, the properties of the discrete-time convolution are the same as those for continuous time. If (t) is replaced with $[n]$ and integrals are replaced with sums, the convolution properties derived in Chapter 2 lead to those listed in Table 7-4.

The notable difference is associated with property #7. In discrete time, the width (duration) of a signal that is zero-valued outside interval $[a, b]$ is $b - a + 1$, not $b - a$. Consider two signals, $h[n]$ and $x[n]$, defined as follows:

Signal	From	To	Duration
$h[n]$	a	b	$b - a + 1$
$x[n]$	c	d	$d - c + 1$
$y[n]$	$a + c$	$b + d$	$(b + d) - (a + c) + 1$

where $y[n] = h[n] * x[n]$. Note that the duration of $y[n]$ is

$$(b + d) - (a + c) + 1 = (b - a + 1) + (d - c + 1) - 1$$
$$= \text{duration } h[n] + \text{duration } x[n] - 1.$$

7-5.3 Delayed-Impulses Computation Method

For finite-duration signals, computation of the convolution sum can be facilitated by expressing one of the signals as a linear combination of delayed impulses. The process is enabled by the sampling property (#6 in Table 7-4).

Consider, for example, the convolution sum of the two signals defined in Example 7-10, namely

$$y[n] = x[n] * h[n] = \{\underline{2}, 3, 4\} * \{\underline{5}, 6, 7\}. \tag{7.53}$$

The sampling property allows us to express $x[n]$ in terms of impulses,

$$x[n] = 2\delta[n] + 3\delta[n-1] + 4\delta[n-2]. \tag{7.54}$$

Use of Eq. (7.54) in Eq. (7.53) gives

$$y[n] = (2\delta[n] + 3\delta[n-1] + 4\delta[n-2]) * h[n]$$
$$= 2h[n] + 3h[n-1] + 4h[n-2]. \tag{7.55}$$

Given that both $x[n]$ and $h[n]$ are of duration $= 3$, the duration of their sum is $3+3-1 = 5$, and it extends from $n = 0$ to $n = 4$. Computing $y[0]$ using Eq. (7.55) (while keeping in mind that $h[i]$ has a non-zero value for only $i = 0$, 1, and 2) leads to

$$y[0] = 2h[0] + 3h[-1] + 4h[-2]$$
$$= 2 \times 5 + 3 \times 0 + 4 \times 0 = 10.$$

The process can then be repeated to obtain the values of $y[n]$ for $n = 1$, 2, 3, and 4.

Table 7-4: Comparison of convolution properties for continuous-time and discrete-time signals.

Property	Continuous Time	Discrete Time
Definition	$y(t) = h(t) * x(t) = \int_{-\infty}^{\infty} h(\tau)\, x(t-\tau)\, d\tau$	$y[n] = h[n] * x[n] = \sum_{i=-\infty}^{\infty} h[i]\, x[n-i]$
1. Commutative	$x(t) * h(t) = h(t) * x(t)$	$x[n] * h[n] = h[n] * x[n]$
2. Associative	$[g(t) * h(t)] * x(t) = g(t) * [h(t) * x(t)]$	$[g[n] * h[n]] * x[n] = g[n] * [h[n] * x[n]]$
3. Distributive	$x(t) * [h_1(t) + \cdots + h_N(t)] =$ $x(t) * h_1(t) + \cdots + x(t) * h_N(t)$	$x[n] * [h_1[n] + \cdots + h_N[n]] =$ $x[n] * h_1[n] + \cdots + x[n] * h_N[n]$
4. Causal $*$ Causal = Causal	$y(t) = u(t) \int_0^t h(\tau)\, x(t-\tau)\, d\tau$	$y[n] = u[n] \sum_{i=0}^{n} h[i]\, x[n-i]$
5. Time-shift	$h(t-T_1) * x(t-T_2) = y(t-T_1-T_2)$	$h[n-a] * x[n-b] = y[n-a-b]$
6. Convolution with Impulse	$x(t) * \delta(t-T) = x(t-T)$	$x[n] * \delta[n-a] = x[n-a]$
7. Width	width $y(t)$ = width $x(t)$ + width $h(t)$	width $y[n]$ = width $x[n]$ + width $h[n]$ − 1
8. Area	area of $y(t)$ = area of $x(t)$ × area of $h(t)$	$\sum_{n=-\infty}^{\infty} y[n] = \left(\sum_{n=-\infty}^{\infty} h[n]\right)\left(\sum_{n=-\infty}^{\infty} x[n]\right)$
9. Convolution with step	$y(t) = x(t) * u(t) = \int_{-\infty}^{t} x(\tau)\, d\tau$	$x[n] * u[n] = \sum_{i=-\infty}^{n} x[i]$

7-5.4 Graphical Computation Method

The convolution sum can be computed graphically through a four-step process. By way of illustration, we will compute

$$y[n] = x[n] * h[n] = \sum_{i=-\infty}^{\infty} x[i]\, h[n-i], \tag{7.56}$$

with

$$x[n] = \{\underline{2}, 3, 4\}, \qquad h[n] = \{\underline{5}, 6, 7\}. \tag{7.57}$$

Step 1: Replace index n with index i and plot $x[i]$ and $h[-i]$, as shown in Fig. 7-12(a). Signal $h[-i]$ is obtained from $h[i]$ by reflecting it about the vertical axis.

Step 2: Superimpose $x[i]$ and $h[-i]$, as in Fig. 7-12(b), and multiply and sum them. Their product is 10.

Step 3: Shift $h[-i]$ to the right by 1 to obtain $h[1-i]$, as shown in Fig. 7-12(c). Multiplication and summation of $x[i]$ by $h[1-i]$ generates $y[1] = 27$. Shift $h[1-i]$ by one more unit to the right to obtain $h[2-i]$, and then repeat the multiplication and summation process to obtain $y[2]$. Continue the shifting and multiplication and summation processes until the two signals no longer overlap.

Step 4: Use the values of $y[n]$ obtained in step 3 to generate a plot of $y[n]$, as shown in Fig. 7-12(g).

Concept Question 7-9: What does convolution with a step function do to a signal in discrete time? (See S²)

Concept Question 7-10: Why is convolution property #7 in Table 7-4 (duration) different in discrete time? (See S²)

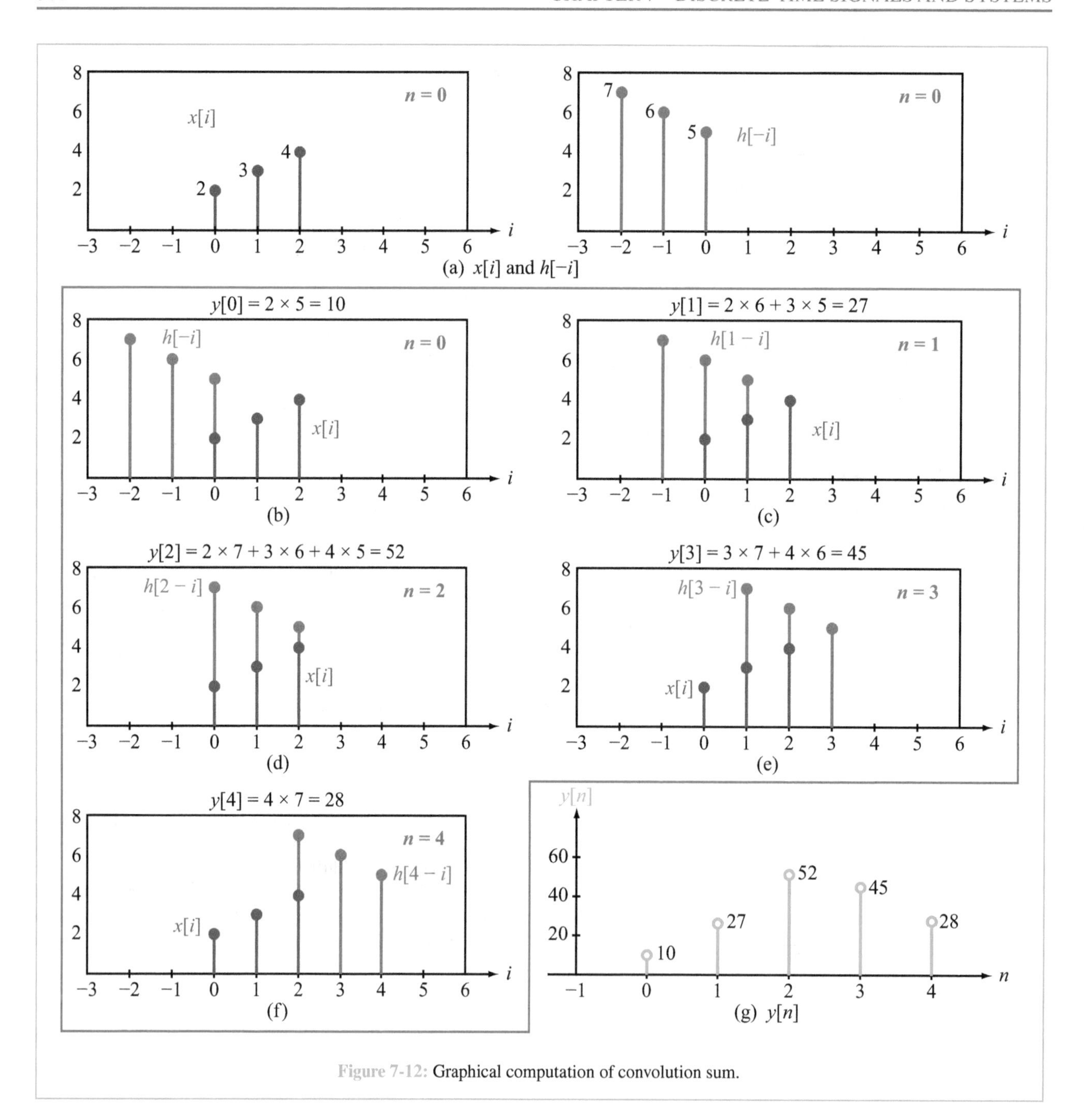

Figure 7-12: Graphical computation of convolution sum.

Exercise 7-9: Compute $y[n] = \{1, \underline{2}\} * \{\underline{0}, 0, 3, 4\}$.

Answer: $y[n] = \{\underline{0}, 3, 10, 8\}$. (See S²)

7-6 The z-Transform

The **z**-transform plays the same role for discrete-time systems that the Laplace transform plays for continuous-time systems. Computations of **z**-transforms and inverse **z**-transforms are performed using procedures and techniques analogous to those used with the Laplace transform.

7-6.1 Definition of the z-Transform

The (unilateral) *z-transform* of a discrete-time signal $x[n]$ is defined as

$$\mathbb{Z}[x[n]] = \mathbf{X}(\mathbf{z}) = \sum_{n=0}^{\infty} x[n]\,\mathbf{z}^{-n}. \tag{7.58a}$$

Note that the summation starts at $n = 0$, so values of $x[n]$ for $n < 0$ (if non-zero valued) are discarded. For the sake of comparison, the Laplace transform of $x(t)$ is

$$\mathcal{L}[x(t)] = \mathbf{X}(\mathbf{s}) = \int_0^{\infty} x(t)\,(e^{\mathbf{s}})^{-t}\,dt, \tag{7.58b}$$

where $\mathbf{z}$ is a complex variable analogous to $e^{\mathbf{s}}$ for the Laplace transform.

The $\mathbf{z}$-transform *transforms* $x[n]$ from the discrete-time domain symbolized by the discrete variable n to the $\mathbf{z}$-domain, where, in general, $\mathbf{z}$ is a complex number. The inverse $\mathbf{z}$-transform performs the reverse operation,

$$x[n] = \mathbb{Z}^{-1}[\mathbf{X}(\mathbf{z})]. \tag{7.59}$$

▶ The uniqueness property of the unilateral $\mathbf{z}$-transform states that a causal signal $x[n]$ has a unique $\mathbf{X}(\mathbf{z})$, and vice versa. ◀

Thus,

$$x[n] \longleftrightarrow \mathbf{X}(\mathbf{z}). \tag{7.60}$$

7-6.2 Examples of z-Transform Pairs

Not all discrete-time signals have $\mathbf{z}$-transforms. Fortunately, however, for the overwhelming majority of signals of practical interest, their $\mathbf{z}$-transforms do exist. Table 7-5 provides a list of $\mathbf{z}$-transform pairs of signals commonly used in science and engineering.

Example 7-11: z-Transforms of $\delta[n]$ and $u[n]$

Verify that the $\mathbf{z}$-transforms of (a) $\delta[n-m]$, (b) $\delta[n]$, and (c) $u[n]$ match those listed in Table 7-5.

Solution:

(a) For $x[n] = \delta[n-m]$, application of the definition of the $\mathbf{z}$-transform given by Eq. (7.58a) gives

$$\mathbf{X}[\mathbf{z}] = \sum_{n=0}^{\infty} x[n]\,\mathbf{z}^{-n} = \sum_{n=0}^{\infty} \delta[n-m]\,\mathbf{z}^{-n} = \mathbf{z}^{-m}.$$

This is because $\delta[n-m] = 1$ for $n = m$, and zero for all other values of n. In shorthand notation,

$$\delta[n-m] \longleftrightarrow \mathbf{z}^{-m}. \tag{7.61}$$

(b) For $m = 0$,

$$\delta[n] \longleftrightarrow 1. \tag{7.62}$$

(c) For $x[n] = u[n]$,

$$\mathbf{X}(\mathbf{z}) = \sum_{n=0}^{\infty} u[n]\,\mathbf{z}^{-n} = 1 + \mathbf{z}^{-1} + \mathbf{z}^{-2} + \cdots. \tag{7.63a}$$

If $|\mathbf{z}^{-1}| < 1$, the infinite geometric series in Eq. (7.63a) can be expressed in closed form as

$$\sum_{n=0}^{\infty} \mathbf{z}^{-n} = \frac{1}{1-\mathbf{z}^{-1}} = \frac{\mathbf{z}}{\mathbf{z}-1} \quad \text{for } |\mathbf{z}^{-1}| < 1. \tag{7.63b}$$

Hence,

$$u[n] \longleftrightarrow \frac{\mathbf{z}}{\mathbf{z}-1} \quad \text{for } |\mathbf{z}| > 1. \tag{7.64}$$

The condition $|\mathbf{z}| > 1$ is called the ***region of convergence (ROC)*** of the $\mathbf{z}$-transform of $u[n]$. The expression $\mathbb{Z}[u[n]] = \mathbf{z}/(\mathbf{z}-1)$ is only valid for $|\mathbf{z}| > 1$. If $|\mathbf{z}| < 1$ the infinite series given by Eq. (7.63a) does not converge, and the expression is not valid. This will be of little practical import in this book. However, books on digital signal processing may use a bilateral $\mathbf{z}$-transform for which ROC will be very important in computing the inverse $\mathbf{z}$-transform.

Table 7-5: Examples of $\mathbf{z}$-transform pairs. ROC stands for region of convergence (validity) in the $\mathbf{z}$-plane.

	z-Transform Pairs			
	$x[n]$		$\mathbf{X}(\mathbf{z}) = \mathbb{Z}[x[n]]$	ROC
1	$\delta[n]$	$\longleftrightarrow$	1	All $\mathbf{z}$
1a	$\delta[n-m]$	$\longleftrightarrow$	$\mathbf{z}^{-m}$, if $m \geq 0$	$\mathbf{z} \neq 0$
2	$u[n]$	$\longleftrightarrow$	$\dfrac{\mathbf{z}}{\mathbf{z}-1}$	$\lvert\mathbf{z}\rvert > 1$
3	$n\,u[n]$	$\longleftrightarrow$	$\dfrac{\mathbf{z}}{(\mathbf{z}-1)^2}$	$\lvert\mathbf{z}\rvert > 1$
3a	$n^2\,u[n]$	$\longleftrightarrow$	$\dfrac{\mathbf{z}(\mathbf{z}+1)}{(\mathbf{z}-1)^3}$	$\lvert\mathbf{z}\rvert > 1$
4	$\mathbf{a}^n\,u[n]$	$\longleftrightarrow$	$\dfrac{\mathbf{z}}{\mathbf{z}-\mathbf{a}}$	$\lvert\mathbf{z}\rvert > \lvert\mathbf{a}\rvert$
4a	$\mathbf{a}^{n-1}\,u[n-1]$	$\longleftrightarrow$	$\dfrac{1}{\mathbf{z}-\mathbf{a}}$	$\lvert\mathbf{z}\rvert > \lvert\mathbf{a}\rvert$
4b	$n\mathbf{a}^n\,u[n]$	$\longleftrightarrow$	$\dfrac{\mathbf{a}\mathbf{z}}{(\mathbf{z}-\mathbf{a})^2}$	$\lvert\mathbf{z}\rvert > \lvert\mathbf{a}\rvert$
4c	$(n-1)\mathbf{a}^{n-2}\,u[n-2]$	$\longleftrightarrow$	$\dfrac{1}{(\mathbf{z}-\mathbf{a})^2}$	$\lvert\mathbf{z}\rvert > \lvert\mathbf{a}\rvert$
4d	$n^2\mathbf{a}^n\,u[n]$	$\longleftrightarrow$	$\dfrac{\mathbf{a}\mathbf{z}(\mathbf{z}+\mathbf{a})}{(\mathbf{z}-\mathbf{a})^3}$	$\lvert\mathbf{z}\rvert > \lvert\mathbf{a}\rvert$
5	$\sin(\Omega n)\,u[n]$	$\longleftrightarrow$	$\dfrac{\mathbf{z}\sin(\Omega)}{\mathbf{z}^2 - 2\mathbf{z}\cos(\Omega)+1}$	$\lvert\mathbf{z}\rvert > 1$
5a	$\mathbf{a}^n \sin(\Omega n)\,u[n]$	$\longleftrightarrow$	$\dfrac{\mathbf{a}\mathbf{z}\sin(\Omega)}{\mathbf{z}^2 - 2\mathbf{a}\mathbf{z}\cos(\Omega)+\mathbf{a}^2}$	$\lvert\mathbf{z}\rvert > \lvert\mathbf{a}\rvert$
6	$\cos(\Omega n)\,u[n]$	$\longleftrightarrow$	$\dfrac{\mathbf{z}^2 - \mathbf{z}\cos(\Omega)}{\mathbf{z}^2 - 2\mathbf{z}\cos(\Omega)+1}$	$\lvert\mathbf{z}\rvert > 1$
6a	$\mathbf{a}^n \cos(\Omega n)\,u[n]$	$\longleftrightarrow$	$\dfrac{\mathbf{z}^2 - \mathbf{a}\mathbf{z}\cos(\Omega)}{\mathbf{z}^2 - 2\mathbf{a}\mathbf{z}\cos(\Omega)+\mathbf{a}^2}$	$\lvert\mathbf{z}\rvert > \lvert\mathbf{a}\rvert$
7	$\cos(\Omega n + \theta)\,u[n]$	$\longleftrightarrow$	$\dfrac{\mathbf{z}^2\cos(\theta) - \mathbf{z}\cos(\Omega-\theta)}{\mathbf{z}^2 - 2\mathbf{z}\cos(\Omega)+1}$	$\lvert\mathbf{z}\rvert > 1$
7a	$2\lvert\mathbf{a}\rvert^n \cos(\Omega n + \theta)\,u[n]$	$\longleftrightarrow$	$\dfrac{\mathbf{z}e^{j\theta}}{\mathbf{z}-\mathbf{a}} + \dfrac{\mathbf{z}e^{-j\theta}}{\mathbf{z}-\mathbf{a}^*}$, $\quad \mathbf{a} = \lvert\mathbf{a}\rvert e^{j\Omega}$	$\lvert\mathbf{z}\rvert > \lvert\mathbf{a}\rvert$

Example 7-12: z-Transform of a Finite Duration Signal

Compute the $\mathbf{z}$-transform of $x[n] = \{\underline{4}, 2, 0, 5\}$.

Solution: Signal $x[n]$ can be written as

$$x[n] = 4\delta[n] + 2\delta[n-1] + 0\delta[n-2] + 5\delta[n-3].$$

Hence,

$$\mathbf{X}(\mathbf{z}) = \sum_{n=0}^{\infty} x[n]\,\mathbf{z}^{-n} = 4 + 2\mathbf{z}^{-1} + 5\mathbf{z}^{-3} = \frac{4\mathbf{z}^3 + 2\mathbf{z}^2 + 5}{\mathbf{z}^3}.$$

For reasons that will become apparent later, $\mathbf{X}(\mathbf{z})$ is expressed in the form of a rational function (ratio of two polynomials).

▶ The relationship between a causal finite-length signal and its $\mathbf{z}$-transform can be cast in the form

$$\{\underline{a_0}, a_1, a_2, \ldots, a_m\} \longleftrightarrow a_0 + \frac{a_1}{\mathbf{z}} + \frac{a_2}{\mathbf{z}^2} + \cdots + \frac{a_m}{\mathbf{z}^m}. \tag{7.65}$$

Example 7-13: z-Transforms of Geometric and Sinusoidal Signals

Find the $\mathbf{z}$-transform of (a) the geometric signal $x_1[n] = \mathbf{a}^n\, u[n]$ and (b) the sinusoidal signal $x_2[n] = A\cos(\Omega n + \theta)\, u[n]$.

Solution:

(a) For $x_1[n]$,

$$\mathbf{X}_1(\mathbf{z}) = \sum_{n=0}^{\infty} \mathbf{a}^n\, u[n]\, \mathbf{z}^{-n} = \sum_{n=0}^{\infty} \left(\frac{\mathbf{a}}{\mathbf{z}}\right)^n.$$

Setting $\mathbf{r} = \mathbf{a}/\mathbf{z}$ in the formula for the infinite geometric series

$$\sum_{n=0}^{\infty} \mathbf{r}^n = \frac{1}{1-\mathbf{r}} \qquad \text{for } |\mathbf{r}| < 1,$$

the expression for $\mathbf{X}_1(\mathbf{z})$ becomes

$$\mathbf{X}_1(\mathbf{z}) = \frac{1}{1-(\mathbf{a}/\mathbf{z})} = \frac{\mathbf{z}}{\mathbf{z}-\mathbf{a}} \qquad \text{for } \left|\frac{\mathbf{a}}{\mathbf{z}}\right| < 1.$$

Hence,

$$\mathbf{a}^n\, u[n] \quad \longleftrightarrow \quad \frac{\mathbf{z}}{\mathbf{z}-\mathbf{a}} \qquad \text{for } |\mathbf{z}| > |\mathbf{a}|. \tag{7.66}$$

(b) For the sinusoidal signal

$$x_2[n] = A\cos(\Omega n + \theta)\, u[n],$$

we start by converting the cosine into complex exponentials using

$$A\cos(\Omega n + \theta) = \frac{A}{2}\, e^{j(\Omega n+\theta)} + \frac{A}{2}\, e^{-j(\Omega n+\theta)} = \frac{A}{2}\, e^{j\theta} e^{j\Omega n} + \frac{A}{2}\, e^{-j\theta} e^{-j\Omega n}.$$

According to Eq. (7.66), if we set $\mathbf{a} = e^{\pm j\Omega}$, then

$$(e^{\pm j\Omega})^n\, u[n] \quad \longleftrightarrow \quad \frac{\mathbf{z}}{\mathbf{z} - e^{\pm j\Omega}},$$

Accordingly,

$$\mathbb{Z}[A\cos(\Omega n + \theta)\, u[n]] = \frac{A}{2}\, e^{j\theta}\, \mathbb{Z}[e^{j\Omega n}\, u[n]] + \frac{A}{2}\, e^{-j\theta}\, \mathbb{Z}[e^{-j\Omega n}\, u[n]] = \frac{A}{2}\, e^{j\theta} \frac{\mathbf{z}}{\mathbf{z}-e^{j\Omega}} + \frac{A}{2}\, e^{-j\theta} \frac{\mathbf{z}}{\mathbf{z}-e^{-j\Omega}}, \tag{7.67}$$

provided $|\mathbf{z}| > |e^{\pm j\Omega}| = 1$. Cross-multiplication leads to the final result

$$A\cos(\Omega n + \theta)\, u[n] \quad \updownarrow \quad A\left[\frac{\mathbf{z}^2\cos(\theta) - \mathbf{z}\cos(\Omega - \theta)}{\mathbf{z}^2 - 2\mathbf{z}\cos(\Omega) + 1}\right]. \tag{7.68}$$

For the two special cases where $A = 1$ and $\theta = 0$ or $-\pi/2$, Eq. (7.68) reduces to, respectively,

$$\cos(\Omega n)\, u[n] \quad \longleftrightarrow \quad \frac{\mathbf{z}^2 - \mathbf{z}\cos(\Omega)}{\mathbf{z}^2 - 2\mathbf{z}\cos(\Omega) + 1}, \tag{7.69a}$$

$$\sin(\Omega n)\, u[n] \quad \longleftrightarrow \quad \frac{\mathbf{z}\sin(\Omega)}{\mathbf{z}^2 - 2\mathbf{z}\cos(\Omega) + 1}. \tag{7.69b}$$

Concept Question 7-11: The $\mathbf{z}$-transform has a unique inverse for what class of signals? (See S²)

Concept Question 7-12: What is the $\mathbf{z}$-transform of $\delta[n]$? $u[n-m]$? $u[n]$? (See S²)

Exercise 7-10: Compute the $\mathbf{z}$-transform of $\{\underline{1}, 2\}$. Put the answer in the form of a rational function.

Answer: $1 + 2\mathbf{z}^{-1} = (\mathbf{z}+2)/\mathbf{z}$. (See S²)

Exercise 7-11: Compute the $\mathbf{z}$-transform of $\{\underline{1}, 1\} + (-1)^n\, u[n]$. Put the answer in the form of a rational function.

Answer:

$$1 + \mathbf{z}^{-1} + \frac{\mathbf{z}}{\mathbf{z}+1} = \frac{\mathbf{z}+1}{\mathbf{z}} + \frac{\mathbf{z}}{\mathbf{z}+1} = \frac{2\mathbf{z}^2 + 2\mathbf{z} + 1}{\mathbf{z}^2 + \mathbf{z}}.$$

(See S²)

7-7 Properties of the z-Transform

Table 7-6 highlights several important properties of the $\mathbf{z}$-transform. Most can be derived directly from the definition given by Eq. (7.58a).

Table 7-6: Properties of the **z**-transform for causal signals for $m > 0$.

Property	$x[n]$		$\mathbf{X}(\mathbf{z})$
1. Linearity	$C_1\, x_1[n] + C_2\, x_2[n]$	$\longleftrightarrow$	$C_1\, \mathbf{X}_1(\mathbf{z}) + C_2\, \mathbf{X}_2(\mathbf{z})$
2. Time delay by 1	$x[n-1]\, u[n]$	$\longleftrightarrow$	$\frac{1}{\mathbf{z}}\, \mathbf{X}(\mathbf{z}) + x[-1]$
2a. Time delay by m	$x[n-m]\, u[n]$	$\longleftrightarrow$	$\frac{1}{\mathbf{z}^m}\, \mathbf{X}(\mathbf{z}) + \frac{1}{\mathbf{z}^m} \sum_{i=1}^{m} x[-i]\, \mathbf{z}^i$
3. Right shift by m	$x[n-m]\, u[n-m]$	$\longleftrightarrow$	$\frac{1}{\mathbf{z}^m}\, \mathbf{X}(\mathbf{z})$
4. Time advance by 1	$x[n+1]\, u[n]$	$\longleftrightarrow$	$\mathbf{z}\, \mathbf{X}(\mathbf{z}) - \mathbf{z}\, x[0]$
4a. Time advance by m	$x[n+m]\, u[n]$	$\longleftrightarrow$	$\mathbf{z}^m\, \mathbf{X}(\mathbf{z}) - \mathbf{z}^m \sum_{i=0}^{m-1} x[i]\, \mathbf{z}^{-i}$
5. Multiplication by a^n	$\mathbf{a}^n\, x[n]\, u[n]$	$\longleftrightarrow$	$\mathbf{X}\left(\frac{\mathbf{z}}{\mathbf{a}}\right)$
6. Multiplication by n	$n\, x[n]\, u[n]$	$\longleftrightarrow$	$-\mathbf{z}\, \frac{d\mathbf{X}(\mathbf{z})}{d\mathbf{z}}$
7. Time scaling	$x[n/k]$	$\longleftrightarrow$	$\mathbf{X}(\mathbf{z}^k)$, k positive integer and n multiple of k
8. Time reversal	$x[-n]$	$\longleftrightarrow$	$\mathbf{X}(1/\mathbf{z})$
9. Summation	$\sum_{k=0}^{n} x[k]\, u[n]$	$\longleftrightarrow$	$\frac{\mathbf{z}}{\mathbf{z}-1}\, \mathbf{X}(\mathbf{z})$
10. Convolution	$x_1[n] * x_2[n]$	$\longleftrightarrow$	$\mathbf{X}_1(\mathbf{z})\, \mathbf{X}_2(\mathbf{z})$
11. Initial value	$x[0]$	$=$	$\lim_{\mathbf{z}\to\infty} \mathbf{X}(\mathbf{z})$
12. Final value	$\lim_{n\to\infty} x[n]$	$=$	$\lim_{\mathbf{z}\to 1} [(\mathbf{z}-1)\, \mathbf{X}(\mathbf{z})]$, if $x[\infty]$ exists

Example 7-14: Linear Combination of Signals

Compute $\mathbb{Z}[\{\underline{1}, 3\} + 4(2^n)\, u[n]]$.

Solution:

$$\mathbb{Z}[\{\underline{1}, 3\} + 4(2^n)\, u[n]] = \mathbb{Z}[\{\underline{1}, 3\}] + 4\mathbb{Z}[2^n\, u[n]]$$

$$= (1 + 3\mathbf{z}^{-1}) + 4\,\frac{\mathbf{z}}{\mathbf{z}-2} = \frac{5\mathbf{z}^2 + \mathbf{z} - 6}{\mathbf{z}^2 - 2\mathbf{z}},$$

where we followed the recipe of Example 7-12 for the finite-length signal (first term), and we applied Eq. (7.66) for the geometric signal.

7-7.1 Right Shifts and Time Delays

In computing the **z**-transform of a shifted signal, we need to distinguish between signals $x[n]$ that are causal ($x[n] = 0$ for $n < 0$) and signals that are noncausal.

► If $x[n]$ is causal, then $x[n] = x[n]\, u[n]$, and right-shifting it by m means replacing $x[n]$ with $x[n-m]\, u[n-m]$. This implies that the signal "starts" at time m, since $x[n-m]\, u[n-m] = 0$ for $n < m$. ◄

But if $x[n]$ is not causal, then right-shifting it means that values of the signal at negative times will now be present at non-negative times. Since the unilateral **z**-transform of a signal uses only values of the signal at non-negative times, computing the **z**-transform of a shifted noncausal signal is more complicated than computing the **z**-transform of a shifted causal signal, as the following example shows.

Example 7-15: z-Transform of Shifted Signals

Let $x_1[n] = \{\underline{3}, 2, 4\}$ and $x_2[n] = \{3, \underline{2}, 4\}$. Compute the **z**-transforms of signals $y_1[n] = x_1[n-1]$ and $y_2[n] = x_2[n-1]$.

Solution: Signal $x_1[n]$ is causal, but signal $x_2[n]$ is not. Signals $y_1[n]$ and $y_2[n]$ are delayed versions of $x_1[n]$ and $x_2[n]$,

$$y_1[n] = x_1[n-1] = \{\underline{0}, 3, 2, 4\},$$
$$y_2[n] = x_2[n-1] = \{\underline{3}, 2, 4\}.$$

Application of Eq. (7.65) leads to

$$\mathbf{X}_1(\mathbf{z}) = 3 + 2\mathbf{z}^{-1} + 4\mathbf{z}^{-2},$$
$$\mathbf{Y}_1(\mathbf{z}) = 3\mathbf{z}^{-1} + 2\mathbf{z}^{-2} + 4\mathbf{z}^{-3},$$
$$\mathbf{X}_2(\mathbf{z}) = 2 + 4\mathbf{z}^{-1},$$
$$\mathbf{Y}_2(\mathbf{z}) = 3 + 2\mathbf{z}^{-1} + 4\mathbf{z}^{-2}.$$

Whereas $\mathbf{Y}_1(\mathbf{z}) = \mathbf{z}^{-1}\,\mathbf{X}_1(\mathbf{z})$, $\mathbf{Y}_2(\mathbf{z})$ is not equal to $\mathbf{z}^{-1}\,\mathbf{X}_2(\mathbf{z})$. This is because $x_2[-1]$ does not appear in $\mathbf{X}_2(\mathbf{z})$, but it does appear in $\mathbf{Y}_2(\mathbf{z})$. Because $x_2[n]$ is not causal, we must use property #2 in Table 7-6 to compute $\mathbf{Y}_2(\mathbf{z})$, whereas property #3 is sufficient to compute $\mathbf{Y}_1(\mathbf{z})$.

We observe from Example 7-15 that even though $x_1[n]$ and $x_2[n]$ consisted of the same sequence of numbers, the fact that one of them is causal and the other is not leads to different expressions for their unilateral **z**-transforms and similarly for the transforms of $y_1[n]$ and $y_2[n]$. To distinguish between signal transformations involving causal and noncausal signals, *we apply "right-shifts" to causal signals and "time delays" to noncausal signals.*

If $x[n]$ is causal, $x[n] = x[n]\,u[n]$, and the right shift of $x[n]$ by m is $x[n-m]\,u[n-m]$. If $x[n]$ is not causal, the time delay of $x[n]$ by m is $x[n-m]\,u[n]$. To delay a noncausal signal by m, we shift it to the right by m and then take only the causal part (the part for $n \geq 0$) of the shifted signal. This distinction is necessary since the unilateral **z**-transform of a signal uses only the causal part of the signal.

The next two subsections show how to compute the **z**-transform of right-shifted (causal) and time-delayed (noncausal) signals.

7-7.2 Right Shift by m Units

Given

$$x[n]\,u[n] \quad \longleftrightarrow \quad \mathbf{X}(\mathbf{z}) = \sum_{n=0}^{\infty} x[n]\,\mathbf{z}^{-n},$$

right-shifting the signal by m units entails not only delaying it by m units, but also starting it m units later. The right-shifted signal is $x[n-m]\,u[n-m]$. Its **z**-transform is

$$\mathbb{Z}[x[n-m]\,u[n-m]] = \sum_{n=0}^{\infty} x[n-m]\,u[n-m]\,\mathbf{z}^{-n}$$
$$= \sum_{n=m}^{\infty} x[n-m]\,\mathbf{z}^{-n}, \qquad (7.70)$$

where we changed the lower limit to m since $u[n-m] = 0$ for $n < m$. Changing variables to $i = n - m$ gives

$$\mathbb{Z}[x[n-m]\,u[n-m]] = \sum_{i=0}^{\infty} x[i]\,\mathbf{z}^{-(i+m)}$$
$$= \frac{1}{\mathbf{z}^m}\sum_{i=0}^{\infty} x[i]\,\mathbf{z}^{-i} = \frac{1}{\mathbf{z}^m}\,\mathbf{X}(\mathbf{z}),$$

which matches property #3 in Table 7-6. In abbreviated notation

$$x[n-m]\,u[n-m] \quad \longleftrightarrow \quad \frac{1}{\mathbf{z}^m}\,\mathbf{X}(\mathbf{z}). \qquad (7.71)$$

(right-shift property)

7-7.3 Time Delay by m Units

Given

$$x[n]\,u[n] \quad \longleftrightarrow \quad \mathbf{X}(\mathbf{z}) = \sum_{n=0}^{\infty} x[n]\,\mathbf{z}^{-n},$$

we wish to relate the **z**-transform of $x[n-m]\,u[n]$—*the original signal delayed by m units*—to $\mathbf{X}(\mathbf{z})$. The **z**-transform of the delayed signal is

$$\mathbb{Z}[x[n-m]\,u[n]] = \sum_{n=0}^{\infty} x[n-m]\,u[n]\,\mathbf{z}^{-n}. \qquad (7.72)$$

By introducing the variable $k = n - m$, Eq. (7.72) becomes

$$\mathbb{Z}[x[n-m]\,u[n]] = \sum_{k=-m}^{\infty} x[k]\,\mathbf{z}^{-(k+m)}$$
$$= \mathbf{z}^{-m}\sum_{k=-m}^{-1} x[k]\,\mathbf{z}^{-k} + \mathbf{z}^{-m}\sum_{k=0}^{\infty} x[k]\,\mathbf{z}^{-k}.$$

Furthermore, we replace index k with $-i$ in the first term, which reverses the order of the summation and leads to

$$\mathbb{Z}[x[n-m]\,u[n]] = \frac{1}{\mathbf{z}^m}\sum_{i=1}^{m} x[-i]\,\mathbf{z}^i + \frac{1}{\mathbf{z}^m}\,\mathbf{X}(\mathbf{z}),$$

which is identical in form to property #2a in Table 7-6. In abbreviated notation, we have

$$x[n-m]\,u[n] \longleftrightarrow \frac{1}{\mathbf{z}^m}\,\mathbf{X}(\mathbf{z}) + \frac{1}{\mathbf{z}^m}\sum_{i=1}^{m} x[-i]\,\mathbf{z}^i. \qquad (7.73)$$

(time delay property)

7-7.4 z-Scaling

► Multiplication of a signal by $\mathbf{a}^n$ leads to scaling the $\mathbf{z}$-variable of its $\mathbf{z}$-transform by a factor $1/\mathbf{a}$. ◄

That is,

$$\text{if} \quad x[n]\,u[n] \longleftrightarrow \mathbf{X}(\mathbf{z}),$$

$$\text{then} \quad \mathbf{a}^n\,x[n]\,u[n] \longleftrightarrow \mathbf{X}\left(\frac{\mathbf{z}}{\mathbf{a}}\right). \qquad (7.74)$$

(z-scaling property)

This property follows from the $\mathbf{z}$-transform definition

$$\mathbb{Z}[\mathbf{a}^n\,x[n]\,u[n]] = \sum_{n=0}^{\infty} x[n]\,\mathbf{a}^n\mathbf{z}^{-n} = \sum_{n=0}^{\infty} x[n]\left(\frac{\mathbf{z}}{\mathbf{a}}\right)^{-n} = \mathbf{X}\left(\frac{\mathbf{z}}{\mathbf{a}}\right).$$

To illustrate the utility of Eq. (7.74), consider the transform pair for the sinusoid given by Eq. (7.69b), namely,

$$\sin(\Omega n)\,u[n] \longleftrightarrow \frac{\mathbf{z}\sin(\Omega)}{\mathbf{z}^2 - 2\mathbf{z}\cos(\Omega) + 1}. \qquad (7.75)$$

If we are working with a signal given by $\mathbf{a}^n \sin(\Omega n)\,u[n]$ and wish to determine its $\mathbf{z}$-transform, we simply apply Eq. (7.74) by replacing $\mathbf{z}$ with $(\mathbf{z}/\mathbf{a})$ everywhere in the $\mathbf{z}$-transform of $\sin(\Omega n)\,u[n]$. Thus,

$$\mathbf{a}^n \sin(\Omega n)\,u[n] \longleftrightarrow \frac{(\mathbf{z}/\mathbf{a})\sin(\Omega)}{(\mathbf{z}/\mathbf{a})^2 - 2\left(\frac{\mathbf{z}}{\mathbf{a}}\right)\cos\Omega + 1} = \frac{\mathbf{az}\sin(\Omega)}{\mathbf{z}^2 - 2\mathbf{az}\cos(\Omega) + \mathbf{a}^2}. \qquad (7.76)$$

7-7.5 z-Derivative

From

$$\mathbf{X}(\mathbf{z}) = \sum_{n=0}^{\infty} x[n]\,\mathbf{z}^{-n}, \qquad (7.77)$$

it follows that

$$\frac{d}{d\mathbf{z}}\,\mathbf{X}(\mathbf{z}) = -\sum_{n=0}^{\infty} n\,x[n]\,\mathbf{z}^{-n-1}. \qquad (7.78)$$

Multiplying both sides by $-\mathbf{z}$ gives

$$-\mathbf{z}\,\frac{d}{d\mathbf{z}}\,\mathbf{X}(\mathbf{z}) = \sum_{n=0}^{\infty} n\,x[n]\,\mathbf{z}^{-n} = \mathbb{Z}[n\,x[n]\,u[n]],$$

or equivalently

$$n\,x[n]\,u[n] \longleftrightarrow -\mathbf{z}\,\frac{d\mathbf{X}(\mathbf{z})}{d\mathbf{z}}. \qquad (7.79)$$

(z-derivative property)

Multiplication of $x[n]$ by n in discrete time is equivalent to taking the negative derivative of the $\mathbf{z}$-transform of $x[n]$ in the $\mathbf{z}$-domain.

7-7.6 Convolution

Recall from Section 7-5 that the convolution of two *causal* signals $x_1[n]$ and $x_2[n]$ is defined as

$$y[n] = x_1[n] * x_2[n] = \sum_{i=0}^{\infty} x_1[i]\,u[i]\,x_2[n-i]\,u[n-i]. \qquad (7.80)$$

The $\mathbf{z}$-transform of $y[n]$ is

$$\mathbb{Z}[y[n]] = \sum_{n=0}^{\infty} \mathbf{z}^{-n} \sum_{i=0}^{\infty} x_1[i]\, u[i]\, x_2[n-i]\, u[n-i]. \quad (7.81)$$

By expressing $\mathbf{z}^{-n} = \mathbf{z}^{-i} \cdot \mathbf{z}^{-(n-i)}$ and interchanging the order of the two summations, we have

$$\mathbb{Z}[y[n]] = \sum_{i=0}^{\infty} x_1[i]\, u[i]\, \mathbf{z}^{-i} \sum_{n=0}^{\infty} x_2[n-i]\, u[n-i]\, \mathbf{z}^{-(n-i)}. \quad (7.82)$$

In the second summation, we change variables by introducing integer $m = n - i$, which leads to

$$\begin{aligned} \mathbb{Z}[y[n]] &= \sum_{i=0}^{\infty} x_1[i]\, u[i]\, \mathbf{z}^{-i} \sum_{m=-i}^{\infty} x_2[m]\, u[m]\, \mathbf{z}^{-m} \\ &= \sum_{i=0}^{\infty} x_1[i]\, u[i]\, \mathbf{z}^{-i} \sum_{m=0}^{\infty} x_2[m]\, \mathbf{z}^{-m} = \mathbf{X}_1(\mathbf{z})\, \mathbf{X}_2(\mathbf{z}). \end{aligned} \quad (7.83)$$

The lower limit on the summation was changed from $m = -i$ to $m = 0$ because $u[m] = 0$ for $m < 0$. The result given by Eq. (7.83) can be encapsulated in the abbreviated form

$$x_1[n] * x_2[n] \longleftrightarrow \mathbf{X}_1(\mathbf{z})\, \mathbf{X}_2(\mathbf{z}). \quad (7.84)$$

(convolution property)

The convolution property states that convolution in discrete time is equivalent to multiplication in the $\mathbf{z}$-domain.

Example 7-16: $\{\underline{1}, 2, -3\} * u[n]$

Compute the convolution $y[n] = \{\underline{1}, 2, -3\} * u[n]$ (a) directly in discrete time and (b) by applying the convolution property of the $\mathbf{z}$-transform.

Solution:

(a) **Discrete-time method**

By application of the sampling property discussed in Section 7-5.3, we have

$$\begin{aligned} y[n] &= \{\underline{1}, 2, -3\} * u[n] \\ &= (\delta[n] + 2\delta[n-1] - 3\delta[n-2]) * u[n] \\ &= u[n] + 2u[n-1] - 3u[n-2] \\ &= \begin{cases} 0 & \text{for } n \leq -1, \\ 1 & \text{for } n = 0, \\ 1+2=3 & \text{for } n = 1, \\ 1+2-3=0 & \text{for } n = 2, \\ 0 & \text{for } n > 2. \end{cases} \end{aligned}$$

Hence,

$$y[n] = \{\underline{1}, 3\}.$$

Note that the convolution of a finite-duration signal and an infinite-duration signal can be of finite duration.

(b) **z-Transform method**

Application of Eq. (7.65) gives

$$x_1[n] = \{\underline{1}, 2, -3\} \longleftrightarrow \mathbf{X}_1(\mathbf{z}) = \left(1 + \frac{2}{\mathbf{z}} - \frac{3}{\mathbf{z}^2}\right).$$

Also, from Eq. (7.64), the $\mathbf{z}$-transform of $u[n]$ is

$$x_2[n] = u[n] \longleftrightarrow \mathbf{X}_2(\mathbf{z}) = \left(\frac{\mathbf{z}}{\mathbf{z}-1}\right).$$

The convolution property given by Eq. (7.84) states that

$$\begin{aligned} \mathbf{Y}(\mathbf{z}) &= \mathbf{X}_1(\mathbf{z})\, \mathbf{X}_2(\mathbf{z}) \\ &= \left(1 + \frac{2}{\mathbf{z}} - \frac{3}{\mathbf{z}^2}\right)\left(\frac{\mathbf{z}}{\mathbf{z}-1}\right) \\ &= \frac{\mathbf{z}^2 + 2\mathbf{z} - 3}{\mathbf{z}(\mathbf{z}-1)} = \frac{(\mathbf{z}+3)(\mathbf{z}-1)}{\mathbf{z}(\mathbf{z}-1)} = \frac{\mathbf{z}+3}{\mathbf{z}} = 1 + \frac{3}{\mathbf{z}}. \end{aligned}$$

The inverse transform of $\mathbf{Y}(\mathbf{z})$ is

$$y[n] = \{\underline{1}, 3\},$$

which is identical with the result obtained in part (a).

7-7.7 Initial and Final Value Theorems

From the definition of the z-transform,

$$\mathbf{X}(\mathbf{z}) = \sum_{n=0}^{\infty} x[n]\, \mathbf{z}^{-n},$$

it is readily apparent that if $\mathbf{z} \to \infty$ all terms in the summation vanish except for $n = 0$. Hence,

$$x[0] = \lim_{\mathbf{z}\to\infty} \mathbf{X}(\mathbf{z}) \quad \textbf{(initial-value theorem)}. \tag{7.85}$$

Recall that a strictly proper rational function is the ratio of two polynomials, with the degree of the numerator's polynomial smaller than the degree of the denominator's polynomial. An immediate consequence of the initial value theorem is that the inverse z-transform $x[n]$ of a strictly proper rational function $\mathbf{X}(\mathbf{z})$ gives $x[0] = 0$. Similarly, a more elaborate derivation leads to

$$\lim_{n\to\infty} x[n] = x[\infty] = \lim_{\mathbf{z}\to 1} [(\mathbf{z}-1)\, \mathbf{X}(\mathbf{z})], \tag{7.86}$$

(final value theorem)

provided $x[\infty]$ exists.

Concept Question 7-13: What is the difference between a time delay and a right shift? (See S²)

Concept Question 7-14: Why does the z-transform map convolutions to products? (See S²)

Exercise 7-12: Compute $\mathbf{Z}[n\, u[n]]$, given that

$$\mathbf{Z}[u[n]] = \frac{\mathbf{z}}{\mathbf{z}-1}.$$

Answer: Using the z-derivative property,

$$\mathbf{Z}[n\, u[n]] = -\mathbf{z}\,\frac{d}{d\mathbf{z}}\left[\frac{\mathbf{z}}{\mathbf{z}-1}\right] = \frac{\mathbf{z}}{(\mathbf{z}-1)^2}.$$

(See S²)

Exercise 7-13: Compute $\mathbf{Z}[n\mathbf{a}^n\, u[n]]$, given that

$$\mathbf{Z}[n\, u[n]] = \frac{\mathbf{z}}{(\mathbf{z}-1)^2}.$$

Answer: Using the z-scaling property,

$$\mathbf{Z}[n\mathbf{a}^n\, u[n]] = \frac{\mathbf{z}/\mathbf{a}}{((\mathbf{z}/\mathbf{a})-1)^2}\,\frac{\mathbf{a}^2}{\mathbf{a}^2} = \frac{\mathbf{a}\mathbf{z}}{(\mathbf{z}-\mathbf{a})^2}.$$

(See S²)

7-8 Inverse z-Transform

The z-transform is used to solve difference equations, in a manner analogous to how the Laplace transform is used to solve differential equations. In the preceding section, we established that the z-transform $\mathbf{X}(\mathbf{z})$ of a given discrete-time signal $x[n]$ can be computed either directly by applying the formal definition given by Eq. (7.58a) or, when more convenient, by taking advantage of the z-transform properties listed in Table 7-6. Now we explore how to perform the inverse process, namely to inverse transform $\mathbf{X}(\mathbf{z})$ to $x[n]$.

The z-transform $\mathbf{X}(\mathbf{z})$ may assume one of two forms:

(1) $\mathbf{X}(\mathbf{z})$ is a polynomial in $\mathbf{z}^{-1}$, or

(2) $\mathbf{X}(\mathbf{z})$ is a rational function (ratio of two polynomials).

We will consider these two cases separately.

7-8.1 $\mathbf{X}(\mathbf{z})$ = Polynomial in $\mathbf{z}^{-1}$

From Eq. (7.65), for a causal signal, we have

$$x[n] = \{\underline{a_0}, a_1, a_2, \ldots, a_m\}$$
$$\Updownarrow \tag{7.87}$$
$$\mathbf{X}(\mathbf{z}) = \left(a_0 + \frac{a_1}{\mathbf{z}} + \frac{a_2}{\mathbf{z}^2} + \cdots + \frac{a_m}{\mathbf{z}^m}\right).$$

Hence, obtaining $x[n]$ from $\mathbf{X}(\mathbf{z})$ when the latter is expressed as a polynomial in $\mathbf{z}^{-1}$ entails no more than reading off the coefficients of $\mathbf{X}(\mathbf{z})$.

Example 7-17: Finite-Length Signals

Obtain the inverse **z**-transforms of

(a) $\mathbf{X}_1(\mathbf{z}) = 2 + 4/\mathbf{z} + 5/\mathbf{z}^3$,

(b) $\mathbf{X}_2(\mathbf{z}) = (7\mathbf{z}^2 + 3\mathbf{z} + 6)/\mathbf{z}^3$, and

(c) $\mathbf{X}_3(\mathbf{z}) = 3/\mathbf{z}^4$.

Solution:

(a) Using Eq. (7.87),

$$x_1[n] = \{\underline{2}, 4, 0, 5\}.$$

Note that $x_1[2] = 0$ because the coefficient of $\mathbf{z}^{-2}$ is zero.

(b) Before we apply Eq. (7.87) to $\mathbf{X}_2(\mathbf{z})$, we need to convert its expression into a polynomial in $\mathbf{z}^{-1}$. That is,

$$\mathbf{X}_2(\mathbf{z}) = \frac{7\mathbf{z}^2 + 3\mathbf{z} + 6}{\mathbf{z}^3} = \frac{7}{\mathbf{z}} + \frac{3}{\mathbf{z}^2} + \frac{6}{\mathbf{z}^3}\,.$$

The corresponding discrete time signal is

$$x_2[n] = \{\underline{0}, 7, 3, 6\}.$$

(c) Since the only non-zero term in $\mathbf{X}_3(\mathbf{z})$ is of power $\mathbf{z}^{-4}$,

$$x_3[n] = \{\underline{0}, 0, 0, 0, 3\}.$$

7-8.2 $\mathbf{X}(\mathbf{z})$ = Rational Function

The procedure for computing the inverse **z**-transform of a rational function is similar to that for the inverse Laplace transform presented in Chapter 3. The only major difference is in the final step of transforming the partial fraction expansion to discrete time. In continuous time, we have

$$e^{\mathbf{a}t}\, u(t) \quad \longleftrightarrow \quad \frac{1}{\mathbf{s} - \mathbf{a}} \qquad \text{(continuous time).}$$

In discrete time, it is the right-shifted version of Eq. (7.66) that has the analogous transform:

$$\mathbf{a}^{n-1}\, u[n-1] \quad \longleftrightarrow \quad \frac{1}{\mathbf{z} - \mathbf{a}} \qquad \text{(discrete time).}$$

Inverse z-Transform Procedure

1. Given $\mathbf{X}(\mathbf{z}) = \dfrac{\mathbf{N}(\mathbf{z})}{\mathbf{D}(\mathbf{z})}$, where $\mathbf{N}(\mathbf{z})$ and $\mathbf{D}(\mathbf{z})$ are polynomials of degrees M and N, respectively, compute poles $\{\mathbf{p}_i = 1, 2, \ldots, N\}$, which are the roots of $\mathbf{D}(\mathbf{z}) = 0$. The poles will be either real or complex, and if complex, they will occur in complex conjugate pairs if the system is real. We assume the poles are distinct: $\mathbf{p}_1 \neq \mathbf{p}_2 \neq \cdots \neq \mathbf{p}_N$.

2. Express $\mathbf{X}(\mathbf{z})$ as

$$\mathbf{X}(\mathbf{z}) = \frac{\mathbf{N}(\mathbf{z})}{D_0(\mathbf{z} - \mathbf{p}_1)(\mathbf{z} - \mathbf{p}_2)\cdots(\mathbf{z} - \mathbf{p}_N)}\,,$$

where D_0 is a constant, and then apply partial fraction expansion to obtain the form

$$\mathbf{X}(\mathbf{z}) = A_0 + \sum_{i=1}^{N} \frac{\mathbf{A}_i}{\mathbf{z} - \mathbf{p}_i}\,. \tag{7.88}$$

Coefficient $A_0 = 0$ if and only if $M < N$; that is, if and only if $\mathbf{X}(\mathbf{z})$ is a strictly proper function.

3. Inverse-transform $\mathbf{X}(\mathbf{z})$ to $x[n]$ by applying transform pair #4a in Table 7-5, namely:

$$\mathbf{a}^{n-1}\, u[n-1] \quad \longleftrightarrow \quad \frac{1}{\mathbf{z} - \mathbf{a}}\,. \tag{7.89}$$

The result is

$$x[n] = A_0\, \delta[n] + \sum_{i=1}^{N} \mathbf{A}_i \mathbf{p}_i^{n-1}\, u[n-1]. \tag{7.90}$$

4. If $\mathbf{A}_i$ and or $\mathbf{p}_i$ are complex quantities, simplify the expression by converting sums of complex conjugate terms to cosine functions. Alternatively, in step 3, use entry #7a in Table 7-5 to convert complex-conjugate terms into cosines directly.

Example 7-18: Rational Functions

Compute

(a) $\mathbb{Z}^{-1}\left(\dfrac{\mathbf{z} - 3}{\mathbf{z}^2 - 3\mathbf{z} + 2}\right)$ and

(b) $\mathbb{Z}^{-1}\left(\dfrac{16\mathbf{z}}{\mathbf{z}^2 - 6\mathbf{z} + 25}\right)$.

Solution:

(a) $\mathbf{X}_1(\mathbf{z}) = \dfrac{\mathbf{z} - 3}{\mathbf{z}^2 - 3\mathbf{z} + 2}$. The roots of $\mathbf{z}^2 - 3\mathbf{z} + 2 = 0$ are the poles $\mathbf{p}_1 = 1$ and $\mathbf{p}_2 = 2$. Hence,

$$\mathbf{X}_1(\mathbf{z}) = \frac{\mathbf{z} - 3}{\mathbf{z}^2 - 3\mathbf{z} + 2} = \frac{\mathbf{z} - 3}{(\mathbf{z} - 1)(\mathbf{z} - 2)} = \frac{A_1}{\mathbf{z} - 1} + \frac{A_2}{\mathbf{z} - 2},$$

with

$$A_1 = (\mathbf{z} - 1)\,\mathbf{X}(\mathbf{z})|_{\mathbf{z}=1} = \left.\frac{\mathbf{z} - 3}{\mathbf{z} - 2}\right|_{\mathbf{z}=1} = 2,$$

$$A_2 = (\mathbf{z} - 2)\,\mathbf{X}(\mathbf{z})|_{\mathbf{z}=2} = \left.\frac{\mathbf{z} - 3}{\mathbf{z} - 1}\right|_{\mathbf{z}=2} = -1.$$

Now that $\mathbf{X}_1(\mathbf{z})$ is given by the standard form

$$\mathbf{X}_1(\mathbf{z}) = \frac{2}{\mathbf{z} - 1} - \frac{1}{\mathbf{z} - 2},$$

it can be inverse $\mathbf{z}$-transformed by applying the relationship given by Eq. (7.90):

$$x_1[n] = 2(1)^{n-1}\,u[n-1] - (2)^{n-1}\,u[n-1]$$
$$= (2 - 2^{n-1})\,u[n-1].$$

(b) $\mathbf{X}_2(\mathbf{z}) = \dfrac{16\mathbf{z}}{\mathbf{z}^2 - 6\mathbf{z} + 25}$

The roots of $\mathbf{z}^2 - 6\mathbf{z} + 25 = 0$ are the poles $\mathbf{p}_1 = 3 + j4$ and $\mathbf{p}_2 = \mathbf{p}_1^* = 3 - j4$. Furthermore,

$$\mathbf{p}_1 = \sqrt{9 + 16}\,e^{j\tan^{-1}(4/3)} = 5e^{j\Omega}$$

and

$$\mathbf{p}_2 = \mathbf{p}_1^* = 5e^{-j\Omega}$$

with $\Omega = 0.93$ rad. Hence,

$$\mathbf{X}_2(\mathbf{z}) = \frac{16\mathbf{z}}{(\mathbf{z} - \mathbf{p}_1)(\mathbf{z} - \mathbf{p}_2)} = \frac{A}{\mathbf{z} - \mathbf{p}_1} + \frac{A^*}{\mathbf{z} - \mathbf{p}_1^*},$$

with

$$A = (\mathbf{z} - \mathbf{p}_1)\,\mathbf{X}_2(\mathbf{z})|_{\mathbf{z}=\mathbf{p}_1}$$
$$= \left.\frac{16\mathbf{z}}{\mathbf{z} - \mathbf{p}_1^*}\right|_{\mathbf{z}=\mathbf{p}_1} = \frac{16(3 + j4)}{(3 + j4) - (3 - j4)} = 10e^{-j36.9^\circ}.$$

Therefore,

$$\mathbf{X}_2(\mathbf{z}) = \frac{10e^{-j36.9^\circ}}{\mathbf{z} - 5e^{j\Omega}} + \frac{10e^{j36.9^\circ}}{\mathbf{z} - 5e^{-j\Omega}}.$$

Use of Eq. (7.89) with $\theta = 36.9^\circ$ gives

$$x_2[n] = 10e^{-j36.9^\circ}(5e^{j\Omega})^{n-1}\,u[n-1]$$
$$+ 10e^{j36.9^\circ}(5e^{-j\Omega})^{n-1}\,u[n-1]$$
$$= 10(5^{n-1})[e^{j[(n-1)\Omega - \theta]} + e^{-j[(n-1)\Omega - \theta]}]\,u[n-1]$$
$$= 20(5^{n-1})\cos((n-1)\Omega - \theta)\,u[n-1]$$
$$= 20(5^{n-1})\cos(0.93(n-1) - 36.9^\circ)\,u[n-1].$$

This can also be done directly using the right-shifted version of entry #7a in Table 7-5.

7-8.3 Alternative Partial Fraction Method

The standard partial-fraction-expansion method generates terms of the form $\mathbf{A}/(\mathbf{z} - \mathbf{p})$. According to Eq. (7.89),

$$\mathbf{A}\mathbf{p}^{n-1}\,u[n-1] \quad \longleftrightarrow \quad \frac{\mathbf{A}}{\mathbf{z} - \mathbf{p}}. \tag{7.91}$$

Hence, all such fraction-expansion terms lead to discrete-time domain terms that include the shifted step $u[n-1]$. *Often, it is more convenient to work with terms involving $u[n]$, rather than $u[n-1]$.* From Table 7-5, transform pair #4 is of the form

$$\mathbf{A}\mathbf{p}^{n}\,u[n] \quad \longleftrightarrow \quad \frac{\mathbf{A}\mathbf{z}}{\mathbf{z} - \mathbf{p}}. \tag{7.92}$$

This means that if we can generate partial fractions that include $\mathbf{z}$ in the numerator, their inverse transformations would yield terms in $u[n]$. To that end, we introduce an alternative partial fraction method consisting of the following simple steps:

Alternative Partial Fraction Method

1. Divide $\mathbf{X}(\mathbf{z})$ by $\mathbf{z}$ to obtain

$$\mathbf{X}'(\mathbf{z}) = \frac{\mathbf{X}(\mathbf{z})}{\mathbf{z}}. \quad (7.93)$$

2. Apply partial fraction expansion to $\mathbf{X}'(\mathbf{z})$ to cast it in the form

$$\mathbf{X}'(\mathbf{z}) = \sum_{i=1}^{N} \frac{\mathbf{A}_i}{\mathbf{z} - \mathbf{p}_i}, \quad (7.94)$$

where $\mathbf{p}_i$ for $i = 1, 2, \ldots, N$ are its poles and $\mathbf{A}_i$ are the associated expansion coefficients (*residues*).

3. Multiply $\mathbf{X}'(\mathbf{z})$ by $\mathbf{z}$ to restore $\mathbf{X}(\mathbf{z})$ in a form compatible with Eq. (7.92):

$$\mathbf{X}(\mathbf{z}) = \sum_{i=1}^{N} \frac{\mathbf{A}_i \mathbf{z}}{\mathbf{z} - \mathbf{p}_i}. \quad (7.95)$$

4. Apply Eq. (7.92) to inverse $\mathbf{z}$-transform $\mathbf{X}(\mathbf{z})$ to $x[n]$.

Example 7-19: Alternative Method

Repeat part (a) of Example 7-18 using the alternative partial fraction expansion method.

Solution: $\mathbf{X}_1(\mathbf{z}) = \dfrac{\mathbf{z} - 3}{\mathbf{z}^2 - 3\mathbf{z} + 2}$

Dividing $\mathbf{X}_1(\mathbf{z})$ by $\mathbf{z}$ gives

$$\mathbf{X}_1'(\mathbf{z}) = \frac{\mathbf{z} - 3}{\mathbf{z}(\mathbf{z}-1)(\mathbf{z}-2)} = \frac{A_1}{\mathbf{z}} + \frac{A_2}{\mathbf{z}-1} + \frac{A_3}{\mathbf{z}-2},$$

with

$$A_1 = \mathbf{z}\, \mathbf{X}_1'(\mathbf{z})\Big|_{\mathbf{z}=0} = \frac{\mathbf{z} - 3}{(\mathbf{z}-1)(\mathbf{z}-2)}\bigg|_{\mathbf{z}=0} = -\frac{3}{2}.$$

Similarly, we determine that $A_2 = 2$ and $A_3 = -1/2$. Hence,

$$\mathbf{X}_1'(\mathbf{z}) = -\frac{3}{2\mathbf{z}} + \frac{2}{\mathbf{z}-1} - \frac{1}{2}\,\frac{1}{\mathbf{z}-2}.$$

Converting back to $\mathbf{X}_1(\mathbf{z})$, we have

$$\mathbf{X}_1(\mathbf{z}) = \mathbf{z}\,\mathbf{X}_1'(\mathbf{z}) = -\frac{3}{2} + \frac{2\mathbf{z}}{\mathbf{z}-1} - \frac{1}{2}\,\frac{\mathbf{z}}{\mathbf{z}-2}.$$

Inverse transforming $\mathbf{X}_1(\mathbf{z})$ leads to

$$x_1[n] = -\frac{3}{2}\,\delta[n] + 2u[n] - \frac{1}{2}\,2^n\,u[n].$$

All terms include $\delta[n]$ or $u[n]$, instead of $u[n-1]$.

At first glance, this expression for $x_1[n]$ seems to differ from the expression for $x_1[n]$ obtained in Example 7-18.

Example 7-18: $x_1[n] = (2 - 2^{n-1})\, u[n-1]$

Example 7-19: $x_1[n] = -\frac{3}{2}\,\delta[n] + 2u[n] - \frac{1}{2}\,2^n\,u[n]$.

However, inserting $n = 0, 1, 2, \ldots$ gives

Example 7-18: $x_1[n] = \{\underline{0}, 1, 0, -2, -6, -14, \ldots\}$,

Example 7-19: $x_1[n] = \{\underline{0}, 1, 0, -2, -6, -14, \ldots\}$.

The two expressions are identical. An important issue in computing inverse $\mathbf{z}$-transform is that different approaches may result in answers that look different but are actually equal. This is easily checked by computing the expressions for several values of n.

Note that by the initial value theorem, we know immediately that $x_1[0] = 0$. The purpose of the impulse in the second expression for $x_1[n]$ is to make $x_1[0] = 0$.

Exercise 7-14: Compute the inverse $\mathbf{z}$-transform of $(\mathbf{z}+3)/(\mathbf{z}+1)$.

Answer:

$$\mathbb{Z}^{-1}\left[\frac{\mathbf{z}+3}{\mathbf{z}+1}\right] = \delta[n] + 2(-1)^{n-1}\,u[n-1].$$

(See S²)

Exercise 7-15: Compute the inverse **z**-transform of $1/[(\mathbf{z}+1)(\mathbf{z}+2)]$.

Answer: $(-1)^{n-1}\, u[n-1] - (-2)^{n-1}\, u[n-1]$. (See S²)

7-9 Solving Difference Equations with Initial Conditions

7-9.1 Solution Procedure

The time-delay property of the **z**-transform can be used to solve difference equations with initial conditions, just as the one-sided Laplace transform can be used to solve LCCDEs with initial conditions. Two important considerations that should be kept in mind when performing the solution procedure are

(a) The time-delay property given by Eq. (7.73) as

$$x[n-m]\, u[n] \quad \longleftrightarrow \quad \frac{1}{\mathbf{z}^m}\, \mathbf{X}(\mathbf{z}) + \frac{1}{\mathbf{z}^m} \sum_{i=1}^{m} x[-i]\, \mathbf{z}^i, \qquad (7.96)$$

which allows relating the **z**-transform of time-delayed signals to that of the undelayed signal.

(b) A standard time reference, which means that unless noted to the contrary, *an input term of the form $x[n-m]$ appearing in the difference equation is implicitly causal and equivalent to $x[n-m]\, u[n]$*. The same notion applies to output-related terms: $y[n-m] = y[n-m]\, u[n]$.

Example 7-20: Difference Equation Solution

Solve for $y[n]$, given the difference equation

$$y[n] - 1.1y[n-1] + 0.3y[n-2] = x[n] + 2x[n-1], \qquad (7.97)$$

with $x[n] = (0.25)^n\, u[n]$ and initial conditions $y[-1] = 3$ and $y[-2] = 4$.

Solution:

1. We start by computing the **z**-transform of all terms in Eq. (7.97), while keeping in mind that because $x[n]$ is causal $x[n] = 0$ for $n < 0$.

$$x[n] = (0.25)^n\, u[n] \quad \longleftrightarrow \quad \mathbf{X}(\mathbf{z}) = \frac{\mathbf{z}}{\mathbf{z} - 0.25}$$

(entry #4, Table 7-5)

$$2x[n-1]\, u[n] \quad \longleftrightarrow \quad \frac{2}{\mathbf{z}}\, \mathbf{X}(\mathbf{z}) + 2x[-1] = \frac{2}{\mathbf{z}-0.25} + 0 = \frac{2}{\mathbf{z}-0.25}$$

(Eq. (7.96) with $m = 1$)

$$y[n] \quad \longleftrightarrow \quad \mathbf{Y}(\mathbf{z})$$

$$y[n-1] \quad \longleftrightarrow \quad \frac{1}{\mathbf{z}}\, \mathbf{Y}(\mathbf{z}) + y[-1] = \frac{1}{\mathbf{z}}\, \mathbf{Y}(\mathbf{z}) \underbrace{+3}_{\text{initial condition}}$$

(Eq. (7.96) with $m = 2$)

$$y[n-2] \quad \longleftrightarrow \quad \frac{1}{\mathbf{z}^2}\, \mathbf{Y}(\mathbf{z}) + \frac{1}{\mathbf{z}}\, y[-1] + y[-2] = \frac{1}{\mathbf{z}^2}\, \mathbf{Y}(\mathbf{z}) \underbrace{+\frac{3}{\mathbf{z}} + 4}_{\text{initial conditions}}.$$

2. We transform Eq. (7.97) to the **z**-domain by replacing each of its terms with the corresponding **z**-transform:

$$\mathbf{Y}(\mathbf{z}) - 1.1\left(\frac{1}{\mathbf{z}}\, \mathbf{Y}(\mathbf{z}) + 3\right) + 0.3\left(\frac{1}{\mathbf{z}^2}\, \mathbf{Y}(\mathbf{z}) + \frac{3}{\mathbf{z}} + 4\right) = \frac{\mathbf{z}}{\mathbf{z}-0.25} + \frac{2}{\mathbf{z}-0.25}. \qquad (7.98)$$

3. Collecting terms involving $\mathbf{Y}(\mathbf{z})$ and then solving for $\mathbf{Y}(\mathbf{z})$ leads to

$$\mathbf{Y}(\mathbf{z}) = \frac{\mathbf{z}(3.1\mathbf{z}^2 + 0.575\mathbf{z} + 0.225)}{(\mathbf{z}-0.25)(\mathbf{z}^2 - 1.1\mathbf{z} + 0.3)} = \frac{\mathbf{z}(3.1\mathbf{z}^2 + 0.575\mathbf{z} + 0.225)}{(\mathbf{z}-0.25)(\mathbf{z}-0.5)(\mathbf{z}-0.6)}.$$

4. Next we apply the alternative partial fraction method of Section 7-8.3 by defining

$$\mathbf{Y}'(\mathbf{z}) = \frac{\mathbf{Y}(\mathbf{z})}{\mathbf{z}} = \frac{3.1\mathbf{z}^2 + 0.575\mathbf{z} + 0.225}{(\mathbf{z}-0.25)(\mathbf{z}-0.5)(\mathbf{z}-0.6)}$$
$$= \frac{A_1}{\mathbf{z}-0.25} + \frac{A_2}{\mathbf{z}-0.5} + \frac{A_3}{\mathbf{z}-0.6}.$$

Evaluation of expansion coefficients A_1 to A_3 leads to

$$\mathbf{Y}'(\mathbf{z}) = \frac{6.43}{\mathbf{z}-0.25} - \frac{51.50}{\mathbf{z}-0.5} + \frac{48.17}{\mathbf{z}-0.6}.$$

5. Finally, we return to $\mathbf{Y}(\mathbf{z})$ by multiplying $\mathbf{Y}'(\mathbf{z})$ by $\mathbf{z}$:

$$\mathbf{Y}(\mathbf{z}) = \frac{6.43\mathbf{z}}{\mathbf{z}-0.25} - \frac{51.50\mathbf{z}}{\mathbf{z}-0.5} + \frac{48.17\mathbf{z}}{\mathbf{z}-0.6}, \quad (7.99a)$$

and with the help of entry #4 in Table 7-5, we inverse $\mathbf{z}$-transform to discrete time:

$$y[n] = [6.43(0.25)^n - 51.50(0.5)^n + 48.17(0.6)^n]\, u[n]. \quad (7.99b)$$

7-9.2 Zero-State/Zero-Input Response

In Section 3-11, we examined several formats (partitions) for how to organize the response of an LTI system, including the zero-state/zero-input partition. In this particular partition, the total response $y(t)$ to an input excitation $x(t)$ is separated into two components:

$$y(t) = y_{\text{ZIR}}(t) + y_{\text{ZSR}}(t),$$

where

(a) $y_{\text{ZIR}}(t)$, called the *zero-input response* (*ZIR*), represents the response of the system to initial conditions alone (i.e., for $x(t) = 0$), and

(b) $y_{\text{ZSR}}(t)$, called the *zero-state response* (*ZSR*), represents the response of the system to the input $x(t)$ alone with all initial conditions set to zero.

An analogous partition can be established for the difference equation. Recall that in Eq. (7.98) the factor 3 inside the first bracketed term actually is the value of $y[-1]$, and the factor $(3/\mathbf{z} + 4)$ inside the second bracketed term is associated with initial conditions $y[-1]$ and $y[-2]$. By rearranging Eq. (7.98) into the form

$$\underbrace{\left(1 - \frac{1.1}{\mathbf{z}} + \frac{0.3}{\mathbf{z}^2}\right)}_{\text{system terms}} \mathbf{Y}(\mathbf{z}) + \underbrace{\left(-3.3 + 1.2 + \frac{0.9}{\mathbf{z}}\right)}_{\text{initial condition terms}}$$
$$= \underbrace{\frac{\mathbf{z}+2}{\mathbf{z}-0.25}}_{\text{input-related terms}}, \quad (7.100)$$

we partition it into one group of terms related to initial conditions and another group related to input $x[n]$. Dividing all terms in Eq. (7.100) by the system terms, moving the second term to the right-hand side of the equation and then simplifying the terms into rational functions leads to

$$\mathbf{Y}(\mathbf{z}) = \underbrace{\frac{\mathbf{z}(2.1\mathbf{z}-0.9)}{(\mathbf{z}-0.5)(\mathbf{z}-0.6)}}_{\text{zero-input response}} + \underbrace{\frac{\mathbf{z}^2(\mathbf{z}+2)}{(\mathbf{z}-0.25)(\mathbf{z}-0.5)(\mathbf{z}-0.6)}}_{\text{zero-state response}}. \quad (7.101)$$

Separate applications of the alternative inverse $\mathbf{z}$-transform partial fraction method to the two terms on the right-hand side of Eq. (7.101) leads to

$$\mathbf{Y}(\mathbf{z}) = \underbrace{\left(\frac{-1.5\mathbf{z}}{\mathbf{z}-0.5} + \frac{3.6\mathbf{z}}{\mathbf{z}-0.6}\right)}_{\text{zero-input response}} + \underbrace{\left(\frac{6.43\mathbf{z}}{\mathbf{z}-0.25} - \frac{50\mathbf{z}}{\mathbf{z}-0.5} + \frac{44.57\mathbf{z}}{\mathbf{z}-0.6}\right)}_{\text{zero-state response}}. \quad (7.102)$$

When simplified, Eq. (7.102) reduces to Eq. (7.99a), as expected. The discrete-time domain counterpart of Eq. (7.102) is

$$y[n] = \underbrace{[-1.5(0.5)^n + 3.6(0.6)^n]\, u[n]}_{\text{zero-input response}} + \underbrace{[6.43(0.25)^n - 50(0.5)^n + 44.57(0.6)^n]\, u[n]}_{\text{zero-state response}}. \quad (7.103)$$

Concept Question 7-15: Why do we solve difference equations using the time-delay property of $\mathbf{z}$-transforms instead of the right-shift property? (See S²)

Exercise 7-16: Use $\mathbf{z}$-transforms to compute the zero-input response of the system

$$y[n] - 2y[n-1] = 3x[n] + 4x[n-1]$$

with initial condition $y[-1] = \frac{1}{2}$.

Answer: $y[n] = 2^n\, u[n]$. (See S²)

7-10 System Transfer Function

In discrete time, the output response $y[n]$ of an LTI system to a causal signal input $x[n]$ is given by the convolution relation

$$y[n] = x[n] * h[n], \qquad (7.104)$$

where $h[n]$ is the impulse response of the system. Since according to Eq. (7.84), convolution in discrete time corresponds to multiplication in the z-domain, it follows that

$$\mathbf{Y(z)} = \mathbf{X(z)}\,\mathbf{H(z)}, \qquad (7.105)$$

where $\mathbf{X(z)}$, $\mathbf{Y(z)}$, and $\mathbf{H(z)}$ are, respectively, the z-transforms of $x[n]$, $y[n]$, and $h[n]$. The ***transfer function*** $\mathbf{H(z)}$ characterizes the LTI system under zero initial conditions:

$$\mathbf{H(z)} = \frac{\mathbf{Y(z)}}{\mathbf{X(z)}} \quad \text{(with zero initial conditions)}. \qquad (7.106)$$

Also,

$$\mathbf{H(z)} = \mathbb{Z}[h[n]]. \qquad (7.107)$$

When $\mathbf{X(z)} = 1$, the z-domain output of an LTI system is its transfer function $\mathbf{H(z)}$. The inverse z-transform of 1 is $\delta[n]$. Hence, the correspondence between the discrete-time domain and the z-domain is described by

$$\text{Discrete time } \delta[n] \longrightarrow \boxed{\text{LTI}} \longrightarrow y[n] = h[n]$$

$$\Updownarrow \qquad\qquad \Updownarrow \qquad\qquad \Updownarrow$$

$$\text{z-domain } 1 \longrightarrow \boxed{\text{LTI}} \longrightarrow \mathbf{Y(z)} = \mathbf{H(z)}.$$

Under zero initial conditions, transformation of a difference equation given by the general form

$$\sum_{i=0}^{N-1} a_i\, y[n-i] = \sum_{i=0}^{M-1} b_i\, x[n-i], \qquad (7.108)$$

(discrete time)

to the z-domain is facilitated by the time-delay property (#2a in Table 7-6):

$$y[n-m] = y[n-m]\,u[n]$$

$$\Updownarrow \qquad (7.109)$$

$$\frac{1}{\mathbf{z}^m}\,\mathbf{Y(z)} + \frac{1}{\mathbf{z}^m}\sum_{i=1}^{m} y[-i]\,\mathbf{z}^i = \frac{1}{\mathbf{z}^m}\,\mathbf{Y(z)} + 0.$$

Note that the summation has been replaced with zero because zero initial conditions means that $y[n] = 0$ for $n < 0$ or, equivalently, $y[-i] = 0$ for $i > 0$.

In view of Eq. (7.109), the z-transform of Eq. (7.108) is

$$\left[\sum_{i=0}^{N-1} a_i \mathbf{z}^{-i}\right] \mathbf{Y(z)} = \left[\sum_{i=0}^{M-1} b_i \mathbf{z}^{-i}\right] \mathbf{X(z)} \qquad (7.110)$$

(z-domain),

which is equivalent to

$$[a_0 + a_1\mathbf{z}^{-1} + \cdots + a_{N-1}\mathbf{z}^{-(N-1)}]\,\mathbf{Y(z)}$$
$$= [b_0 + b_1\mathbf{z}^{-1} + \cdots + b_{M-1}\mathbf{z}^{-(M-1)}]\,\mathbf{X(z)}.$$

The transfer function of the system represented by the difference equation is then given by

$$\mathbf{H(z)} = \frac{\mathbf{Y(z)}}{\mathbf{X(z)}}$$
$$= \mathbf{z}^{N-M}\left[\frac{b_0\mathbf{z}^{M-1} + b_1\mathbf{z}^{M-2} + \cdots + b_{M-1}}{a_0\mathbf{z}^{N-1} + a_1\mathbf{z}^{N-2} + \cdots + a_{N-1}}\right]. \qquad (7.111)$$

▶ The transfer function of a physically realizable LTI system must be a proper rational function. ◀

Example 7-21: Transfer Function Diagram

An LTI system described by the difference equation

$$y[n] - 0.6y[n-1] + 0.08y[n-2] = 4x[n]$$

has zero initial conditions. Obtain the transfer function of the system and its block-diagram realization.

Solution: In the absence of initial conditions, the z-transform of the difference equation is

$$\mathbf{Y(z)} - 0.6\,\frac{1}{\mathbf{z}}\,\mathbf{Y(z)} + 0.08\,\frac{1}{\mathbf{z}^2}\,\mathbf{Y(z)} = 4\mathbf{X(z)}$$

or, equivalently,

$$\left(1 - \frac{0.6}{\mathbf{z}} + \frac{0.08}{\mathbf{z}^2}\right)\mathbf{Y(z)} = 4\mathbf{X(z)}.$$

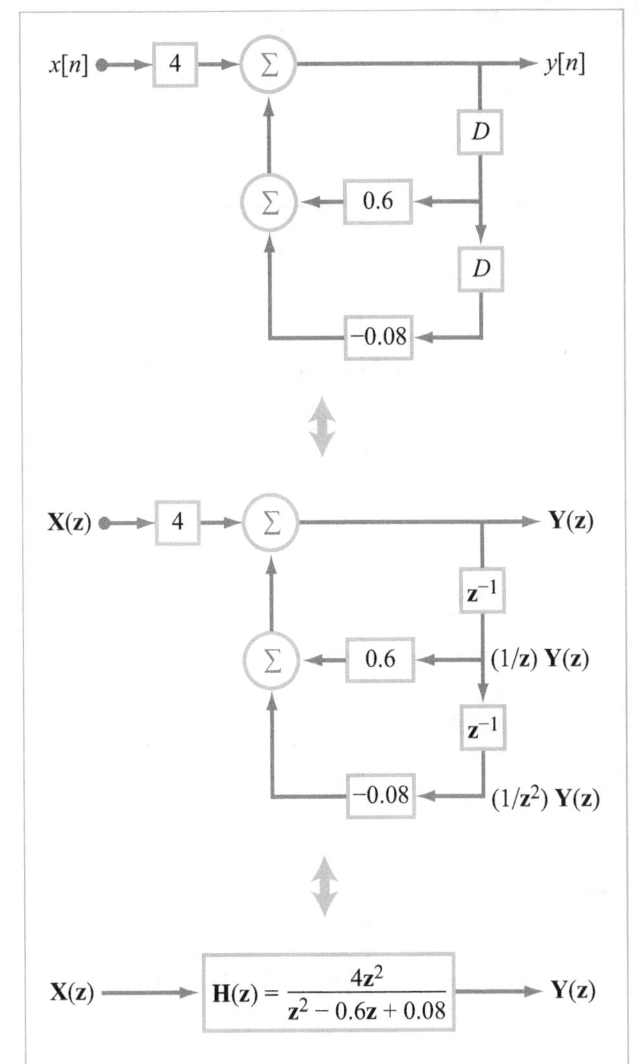

Figure 7-13: Realization of $\mathbf{H}(\mathbf{z})$ in the $\mathbf{z}$-domain, or equivalently $h[n]$ in the time domain, of Example 7-21.

Hence,

$$\mathbf{H}(\mathbf{z}) = \frac{\mathbf{Y}(\mathbf{z})}{\mathbf{X}(\mathbf{z})} = \frac{4\mathbf{z}^2}{\mathbf{z}^2 - 0.6\mathbf{z} + 0.08}, \tag{7.112}$$

and its realization, shown in Fig. 7-13, follows from

$$\mathbf{Y}(\mathbf{z}) = 4\mathbf{X}(\mathbf{z}) + \frac{0.6}{\mathbf{z}}\,\mathbf{Y}(\mathbf{z}) - \frac{0.08}{\mathbf{z}^2}\,\mathbf{Y}(\mathbf{z}).$$

7-11 BIBO Stability of H(z)

According to the BIBO stability criterion, an LTI system is BIBO stable if and only if no bounded input can produce an unbounded output. In discrete time, an input of the form

$$x[n] = (\mathbf{a})^n\, u[n] \tag{7.113}$$

is bounded if and only if $|\mathbf{a}| \le 1$. Otherwise, if $|\mathbf{a}| > 1$, $|\mathbf{a}|^n \to \infty$ as $n \to \infty$. Note that $|\mathbf{a}|$ does not have to be smaller than 1, it can be equal to 1 and $x[n]$ remains bounded. This is an important distinction, as we shall see shortly.

From pair #4 in Table 7-5, the $\mathbf{z}$-transform of $x[n] = \mathbf{a}^n\, u[n]$ is

$$\mathbf{X}(\mathbf{z}) = \frac{\mathbf{z}}{\mathbf{z} - \mathbf{a}}. \tag{7.114}$$

The general form of the transfer function of a discrete-time LTI system with zeros $\{\mathbf{z}_i,\ i = 1, 2, \dots, M\}$ and poles $\{\mathbf{p}_i,\ i = 1, 2, \dots, N\}$ is given by

$$\mathbf{H}(\mathbf{z}) = \frac{\mathbf{C}(\mathbf{z} - \mathbf{z}_1)(\mathbf{z} - \mathbf{z}_2)\dots(\mathbf{z} - \mathbf{z}_M)}{(\mathbf{z} - \mathbf{p}_1)(\mathbf{z} - \mathbf{p}_2)\dots(\mathbf{z} - \mathbf{p}_N)}, \tag{7.115}$$

where $\mathbf{C}$ is a constant.

Assuming that the poles are distinct (i.e., no two poles are identical) and $\mathbf{H}(\mathbf{z})$ is strictly proper, application of partial fraction expansion leads to the general form

$$\mathbf{H}(\mathbf{z}) = \frac{\mathbf{A}_1}{\mathbf{z} - \mathbf{p}_1} + \frac{\mathbf{A}_2}{\mathbf{z} - \mathbf{p}_2} + \cdots \frac{\mathbf{A}_N}{\mathbf{z} - \mathbf{p}_n},$$

where $\mathbf{A}_1$ to $\mathbf{A}_N$ are the expansion coefficients. From the standpoint of our discussion of BIBO stability, let us consider the case of a single-term system given by

$$\mathbf{H}(\mathbf{z}) = \frac{\mathbf{A}_1}{\mathbf{z} - \mathbf{p}}.$$

The stability argument made for a single term can be extended to others. The output of the system with input $\mathbf{X}(\mathbf{z})$ and transfer function $\mathbf{H}(\mathbf{z})$ is

$$\mathbf{Y}(\mathbf{z}) = \mathbf{H}(\mathbf{z})\,\mathbf{X}(\mathbf{z}) = \frac{\mathbf{A}_1\mathbf{z}}{(\mathbf{z} - \mathbf{a})(\mathbf{z} - \mathbf{p})}. \tag{7.116}$$

We shall now consider the special case when $\mathbf{a} = \mathbf{p}$, which yields

$$\mathbf{Y}(\mathbf{z}) = \frac{\mathbf{A}_1\mathbf{z}}{(\mathbf{z} - \mathbf{p})^2}. \tag{7.117a}$$

From entry #4b in Table 7-5, the inverse **z**-transform of $\mathbf{Y}(\mathbf{z})$ is

$$y[n] = \mathbf{A}_1 n \mathbf{p}^n \, u[n]. \tag{7.117b}$$

(a) If $|\mathbf{p}| = |\mathbf{a}| > 1$, $y[n] \to \infty$ as $n \to \infty$. This is not surprising, as the input itself is unbounded in this case.

(b) If $|\mathbf{p}| = |\mathbf{a}| = 1$, $y[n] \to \infty$ as $n \to \infty$.

(c) If $|\mathbf{p}| = |\mathbf{a}| < 1$, $y[n] \to 0$ as $n \to \infty$ because $n\mathbf{p}^n \to 0$ as $n \to \infty$.

Conclusions

▶ BIBO stability for a causal LTI system requires that all poles of $\mathbf{H}(\mathbf{z})$ have magnitudes smaller than 1. Equivalently, all poles $\mathbf{p}_i$ should be located *inside* the *unit circle* defined by $|\mathbf{z}| = 1$. ◀

Location *on* the unit circle (equivalent to $|\mathbf{p}| = 1$) causes the system to be BIBO-unstable.

Figure 7-14 displays, in part (a), the **s**-plane for continuous-time systems, and in part (b), the **z**-plane for discrete-time systems. For continuous-time systems, BIBO stability requires all poles to reside in the open left half-plane. The analogous requirement for discrete-time systems is that the poles of $\mathbf{H}(\mathbf{z})$ should reside inside the unit circle.

▶ For a physically realizable system, complex poles exist in conjugate pairs, and similarly for zeros. ◀

The locations of the poles of $\mathbf{H}(\mathbf{z})$ govern the form of the impulse response $h[n]$. We illustrate the correspondence through three simple examples.

1. A real-valued pole inside the unit circle

Consider a transfer function $\mathbf{H}(\mathbf{z})$ given by

$$\mathbf{H}(\mathbf{z}) = \frac{\mathbf{z}}{\mathbf{z} - \mathbf{p}},$$

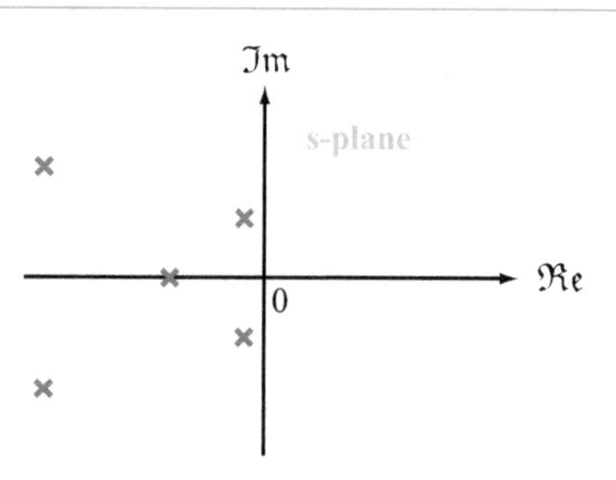

(a) Poles in open left half-plane

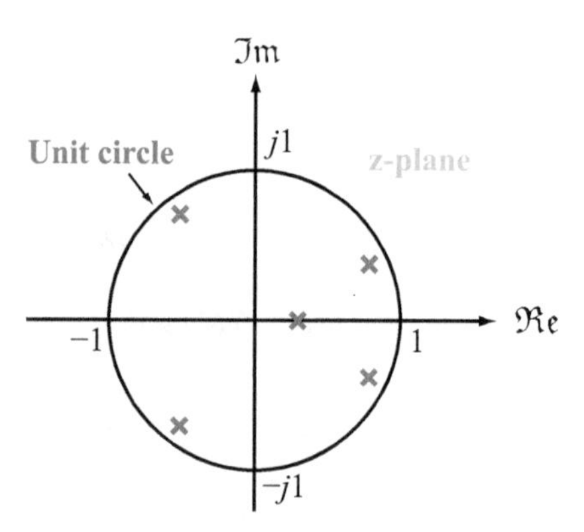

(b) Poles inside unit circle

Figure 7-14: BIBO stability for (a) continuous-time systems requires the poles of the transfer function $\mathbf{H}(\mathbf{s})$ to reside in the open left half-plane, whereas for (b) discrete-time systems, BIBO stability requires the poles of $\mathbf{H}(\mathbf{z})$ to reside inside the unit circle.

with a single zero at the origin and a single pole

$$\mathbf{p} = a + jb,$$

where a and b are real numbers. In the present case, we set $b = 0$ and require that $|a| < 1$ so $\mathbf{p}$ is inside the unit circle.

Case 1: a is positive

From Entry #4 in Table 7-5, it follows that

$$h[n] = a^n \, u[n]. \tag{7.118a}$$

Since a is a positive number smaller than 1, $h[n]$ assumes a decaying geometric pattern, as shown in Fig. 7-15(a) for $a = 0.9$.

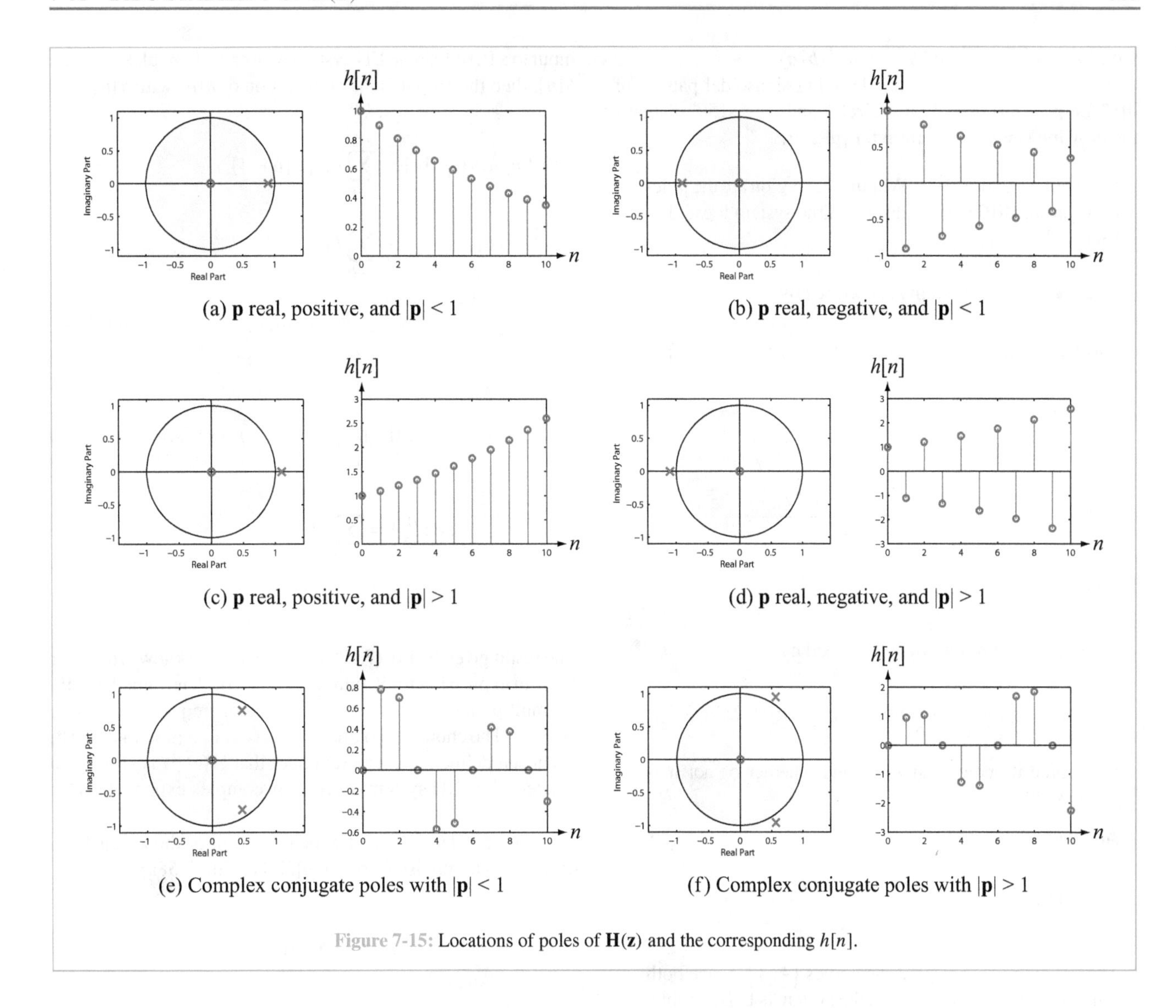

Figure 7-15: Locations of poles of $\mathbf{H(z)}$ and the corresponding $h[n]$.

Case 1: a is negative

For $a = -|a|$,

$$h[n] = a^n\, u[n] = (-|a|)^n\, u[n] = (-1)^n |a|^n\, u[n], \quad (7.118b)$$

which also assumes a decaying geometric pattern, except that the coefficient of $|a|^n$ alternates in sign between $+1$ and -1. The pattern is displayed in Fig. 7-15(b).

2. A real-valued pole outside the unit circle

For $\mathbf{p} = a + jb$, with $b = 0$ and $|a| > 1$ (outside of the unit circle), the expressions given by Eqs. (7.118a and b) remain applicable, but since $|a| > 1$, $h[n]$ has a growing geometric pattern, as shown in Fig. 7-15(c) for $a = 1.1$ and Fig. 7-15(d) for $a = -1.1$.

3. A pair of complex-conjugate poles

Consider the transfer function

$$\mathbf{H(z)} = \frac{\mathbf{z}}{\mathbf{z} - \mathbf{p}} + \frac{\mathbf{z}}{\mathbf{z} - \mathbf{p}^*},$$

with $\mathbf{p} = a + jb$. Both a and b are positive real numbers.

In view of Entry #7a in Table 7-5,

$$h[n] = 2|\mathbf{p}|^n \cos(\Omega n)\, u[n], \quad (7.119)$$

with $|\mathbf{p}| = \sqrt{a^2 + b^2}$ and $\Omega = \tan^{-1}(b/a)$.

Figure 7-15(e) displays the decaying sinusoidal pattern of $h[n]$ for $|\mathbf{p}| < 1$ (inside the unit circle), and Fig. 7-15(f) displays the growing sinusoidal pattern for $|\mathbf{p}| > 1$.

Concept Question 7-16: Within what region are the poles of a causal, BIBO-stable discrete-time system located? (See S²)

Exercise 7-17: A system is described by

$$y[n] - \frac{3}{4}\, y[n-1] + \frac{1}{8}\, y[n-2] = x[n] + 2x[n-1].$$

Compute its transfer function.

Answer:

$$\mathbf{H}(\mathbf{z}) = \frac{\mathbf{Y}(\mathbf{z})}{\mathbf{X}(\mathbf{z})} = \frac{\mathbf{z}^2 + 2\mathbf{z}}{\mathbf{z}^2 - \frac{3}{4}\,\mathbf{z} + \frac{1}{8}}\,.$$

(See S²)

Exercise 7-18: A system is described by

$$y[n] - \frac{3}{4}\, y[n-1] + \frac{1}{8}\, y[n-2] = x[n] + 2x[n-1].$$

Determine its poles and zeros and whether or not it is BIBO stable.

Answer:

$$\mathbf{H}(\mathbf{z}) = \frac{\mathbf{z}^2 + 2\mathbf{z}}{\mathbf{z}^2 - \frac{3}{4}\,\mathbf{z} + \frac{1}{8}} = \frac{\mathbf{z}(\mathbf{z}+2)}{\left(\mathbf{z} - \frac{1}{2}\right)\left(\mathbf{z} - \frac{1}{4}\right)}\,.$$

The system has zeros $\{0, 2\}$ and poles $\left\{\frac{1}{2}, \frac{1}{4}\right\}$. Since both poles are inside the unit circle, the system is BIBO stable. (See S²)

7-12 System Frequency Response

This section examines the response of an LTI system to a specific class of input signals, namely discrete-time complex exponentials and sinusoids.

7-12.1 Complex Exponential Signals

Signal $x[n] = e^{j\Omega n}$ is a discrete-time complex exponential with angular frequency Ω. According to Eq. (7.104), if $x[n]$ is the input to a BIBO-stable LTI system with causal impulse response $h[n]$, then the output is the convolution of $h[n]$ with $x[n]$:

$$\begin{aligned} y[n] = h[n] * x[n] &= \sum_{i=0}^{\infty} h[i]\, x[n-i] \\ &= \sum_{i=0}^{\infty} h[i]\, e^{j\Omega(n-i)} \\ &= e^{j\Omega n} \sum_{i=0}^{\infty} h[i]\, e^{-ji\Omega} = x[n]\, \mathbf{H}(e^{j\Omega}), \end{aligned} \tag{7.120}$$

where we introduce the *frequency response function* defined as

$$\mathbf{H}(e^{j\Omega}) = \sum_{i=0}^{\infty} h[i]\, e^{-ji\Omega}. \tag{7.121}$$

(frequency response)

The result given by Eq. (7.120) states that if we know $\mathbf{H}(e^{j\Omega})$ or have an expression for it, output $y[n]$ can be determined readily by multiplying input $x[n]$ by the system frequency response $\mathbf{H}(e^{j\Omega})$. The choice of notation, namely $\mathbf{H}$ as a function of $e^{j\Omega}$, is chosen deliberately as a reminder that $\mathbf{H}(e^{j\Omega})$ is specific to discrete-time LTI systems excited by complex exponentials (or sinusoids).

Separately, let us consider the definition of the **z**-transform of the causal impulse response $h[i]$. Per Eq. (7.58a),

$$\mathbf{H}(\mathbf{z}) = \sum_{i=0}^{\infty} h[i]\, \mathbf{z}^{-i}, \tag{7.122}$$

where we changed the index from n to i for convenience. Comparison of Eq. (7.121) with Eq. (7.122) reveals that if $\mathbf{z} = e^{j\Omega}$, the two become identical with one another. That is, for an LTI system, $\mathbf{H}(e^{j\Omega})$ can be obtained from $\mathbf{H}(\mathbf{z})$ by setting $\mathbf{z} = e^{j\Omega}$:

$$\underbrace{\mathbf{H}(e^{j\Omega})}_{\text{frequency response}} = \underbrace{\mathbf{H}(\mathbf{z})}_{\text{transfer function}}\Big|_{\mathbf{z}=e^{j\Omega}}. \tag{7.123}$$

In brief, if we know the transfer function $\mathbf{H}(\mathbf{z})$ of an LTI system, we can apply Eq. (7.123) to obtain $\mathbf{H}(e^{j\Omega})$ and then use it to determine the output $y[n]$ due to an input complex exponential $x[n] = e^{j\Omega n}$:

$$e^{j\Omega n} \longrightarrow \boxed{\text{LTI}} \longrightarrow \mathbf{H}(e^{j\Omega})\, e^{j\Omega n} = M(e^{j\Omega})\, e^{j(\Omega n+\theta)}, \tag{7.124}$$

where $M(e^{j\Omega})$ and $\theta(e^{j\Omega})$ are the magnitude (gain) and phase of $\mathbf{H}(e^{j\Omega})$,

$$M(e^{j\Omega}) = |\mathbf{H}(e^{j\Omega})|, \tag{7.125a}$$

$$\theta(e^{j\Omega}) = \angle \mathbf{H}(e^{j\Omega}). \tag{7.125b}$$

The process represented by Eq. (7.124) is entirely analogous to the continuous-time case.

7-12.2 Sinusoidal Signals

An input sinusoidal signal $x(n) = A\cos(\Omega n+\phi)$ can be written as

$$\begin{aligned} x[n] &= A\cos(\Omega n + \phi) \\ &= \frac{A}{2}\,[e^{j(\Omega n+\phi)} + e^{-j(\Omega n+\phi)}] \\ &= \left(\frac{A}{2}\,e^{j\phi}\right) e^{j\Omega n} + \left(\frac{A}{2}\,e^{-j\phi}\right) e^{-j\Omega n}, \end{aligned} \tag{7.126}$$

which now consists of two complex exponential signals: one with angular frequency Ω and another with angular frequency $-\Omega$. Following the same argument made in Section 2-7.4 for the continuous-time sinusoid, we can show that the sum of the corresponding output complex exponentials combine to produce the result

$$A\cos(\Omega n + \phi) \;\longrightarrow\; \boxed{\text{LTI}} \;\longrightarrow\; AM(e^{j\Omega})\cos(\Omega n + \phi + \theta). \tag{7.127}$$

This relation is analogous to Eq. (2.120) for sinusoids in continuous time. An input sinusoid generates an output sinusoid at the same angular frequency. The amplitude and relative phase of the output sinusoid are governed by the frequency response of the system.

Example 7-22: Frequency Response I

The impulse response of an LTI system is

$$h[n] = \left(\frac{1}{2}\right)^n u[n].$$

(a) Determine and plot the frequency response $\mathbf{H}(e^{j\Omega})$ and (b) compute the system response to input $x[n] = \cos\left(\frac{\pi}{3}\, n\right)$.

Solution:

(a) From Eq. (7.121), we have

$$\begin{aligned} \mathbf{H}(e^{j\Omega}) &= \sum_{i=0}^{\infty} h[i]\, e^{-ji\Omega} \\ &= \sum_{i=0}^{\infty} \left(\frac{1}{2}\right)^i e^{-ji\Omega} = \sum_{i=0}^{\infty} \left(\frac{e^{-j\Omega}}{2}\right)^i. \end{aligned} \tag{7.128}$$

In view of

$$\sum_{i=0}^{\infty} \mathbf{r}^i = \frac{1}{1-\mathbf{r}}$$

for any $\mathbf{r}$, so long as $|\mathbf{r}| < 1$, Eq. (7.128) can be written as

$$\mathbf{H}(e^{j\Omega}) = \frac{1}{1 - \frac{1}{2}\, e^{-j\Omega}},$$

because $\left|\frac{1}{2}\, e^{-j\Omega}\right| = \frac{1}{2} < 1$. We can also obtain this result by using Eqs. (7.123) and (7.66). Figure 7-16 displays plots of the magnitude and phase of $\mathbf{H}(e^{j\Omega})$.

► Note that the magnitude spectrum exhibits even symmetry relative to the vertical axis, the phase spectrum exhibits odd symmetry, and both are *continuous* as a function of the angular frequency Ω, and periodic with period 2π. ◄

The fact that the frequency response of a *discrete-time* system is *continuous* in Ω should not be a source of concern. If the input sinusoid is at some angular frequency Ω_0, the output sinusoid will be at the same angular frequency and the connection between the two sinusoids is provided by $\mathbf{H}(e^{j\Omega})$, evaluated at $\Omega = \Omega_0$.

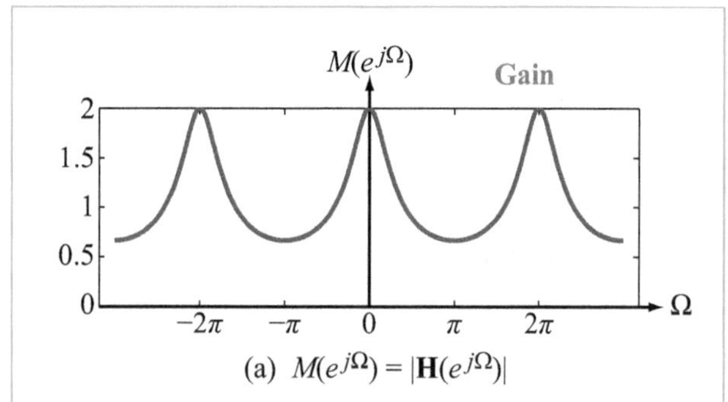

(a) $M(e^{j\Omega}) = |\mathbf{H}(e^{j\Omega})|$

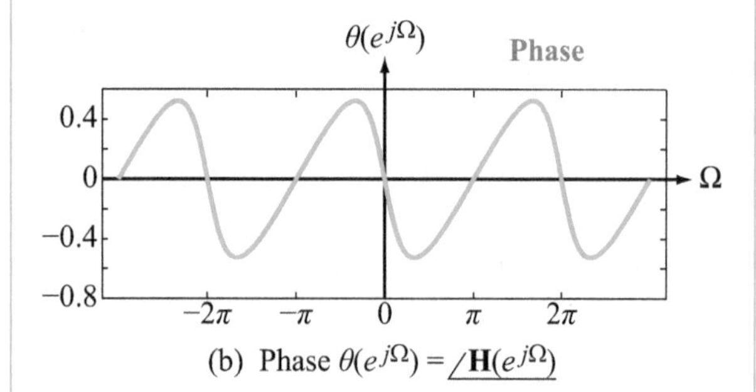

(b) Phase $\theta(e^{j\Omega}) = \underline{/\mathbf{H}(e^{j\Omega})}$

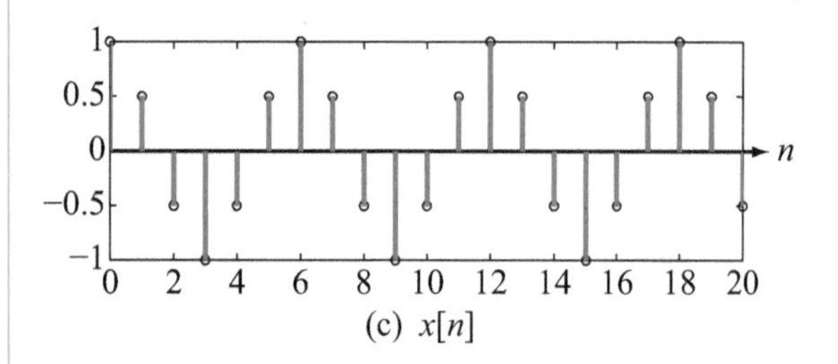

(c) $x[n]$

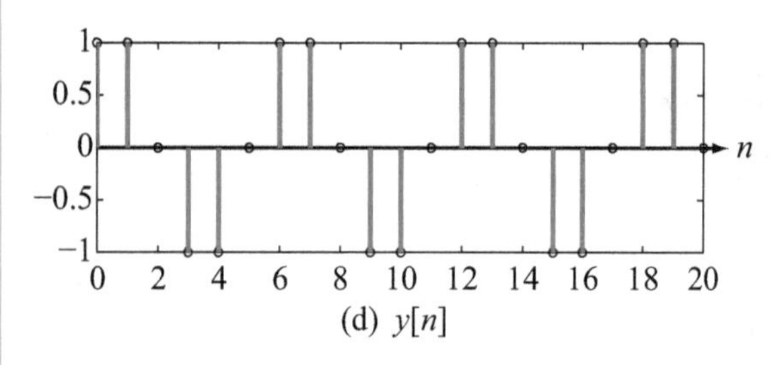

(d) $y[n]$

Figure 7-16: Plots of the magnitude and phase of $\mathbf{H}(e^{j\Omega})$ versus Ω, and $x[n]$ and $y[n]$ versus n (Example 7-22).

(b) For $x[n] = \cos\left(\frac{\pi}{3}\, n\right)$, $\Omega = \frac{\pi}{3}$. At $\Omega = \frac{\pi}{3}$,

$$\mathbf{H}(e^{j\pi/3}) = \frac{1}{1 - \frac{1}{2}\, e^{-j\pi/3}} = 1.155e^{-j\pi/6}.$$

Hence, application of Eq. (7.127) with $\phi = 0$ yields

$$y[n] = |\mathbf{H}(e^{j\pi/3})| \cos\left(\frac{\pi}{3}\, n + \theta\right) = 1.155 \cos\left(\frac{\pi}{3}\, n - \frac{\pi}{6}\right).$$

Plots of $x[n]$ and $y[n]$ are displayed in parts (c) and (d) of Fig. 7-16. The output $y[n]$ may not look like a sinusoid, but it is.

Example 7-23: Two-Point Averager

For the two-point averager system

$$y[n] = \frac{1}{2}(x[n] + x[n-1]),$$

compute $\mathbf{H}(e^{j\Omega})$ and plot it.

Solution: Transforming the difference equation to the $\mathbf{z}$-domain gives

$$\mathbf{Y}(\mathbf{z}) = \frac{1}{2}\, \mathbf{X}(\mathbf{z}) + \frac{1}{2}\, \frac{1}{\mathbf{z}}\, \mathbf{X}(\mathbf{z}),$$

which leads to

$$\mathbf{H}(\mathbf{z}) = \frac{\mathbf{Y}(\mathbf{z})}{\mathbf{X}(\mathbf{z})} = \frac{1}{2}\left(\frac{\mathbf{z}+1}{\mathbf{z}}\right).$$

For sinusoidal input signals, Eq. (7.123) gives

$$\mathbf{H}(e^{j\Omega}) = \mathbf{H}(\mathbf{z})|_{\mathbf{z}=e^{j\Omega}} = \frac{1}{2}\left(\frac{e^{j\Omega}+1}{e^{j\Omega}}\right).$$

Factoring out $e^{j\Omega/2}$ from the numerator provides the ***phase-splitting*** form of $\cos(\Omega/2)$:

$$\mathbf{H}(e^{j\Omega}) = \frac{e^{j\Omega/2}}{e^{j\Omega}}\left(\frac{e^{j\Omega/2} + e^{-j\Omega/2}}{2}\right) = e^{-j\Omega/2}\cos(\Omega/2).$$

Plots of the magnitude and phase of $\mathbf{H}(e^{j\Omega})$ are displayed in Fig. 7-17. The magnitude is even and periodic in Ω with period 2π, as expected. The two-point averager will pass low frequencies ($\Omega \approx 0$) and reject high frequencies ($\Omega \approx \pi$), so it functions as a crude lowpass filter. This agrees with the intuition that averaging smooths out fast variations of a signal, while leaving its slow variations unaffected.

The sawtooth of the phase signal requires some explanation. The phase jumps from $-\pi/2$ to $\pi/2$ at frequency $\Omega = \pi$. This happens because the magnitude is zero at frequency $\Omega = \pi$; the phase can be discontinuous when the magnitude is zero. In fact, the phase is undefined when the magnitude is zero.

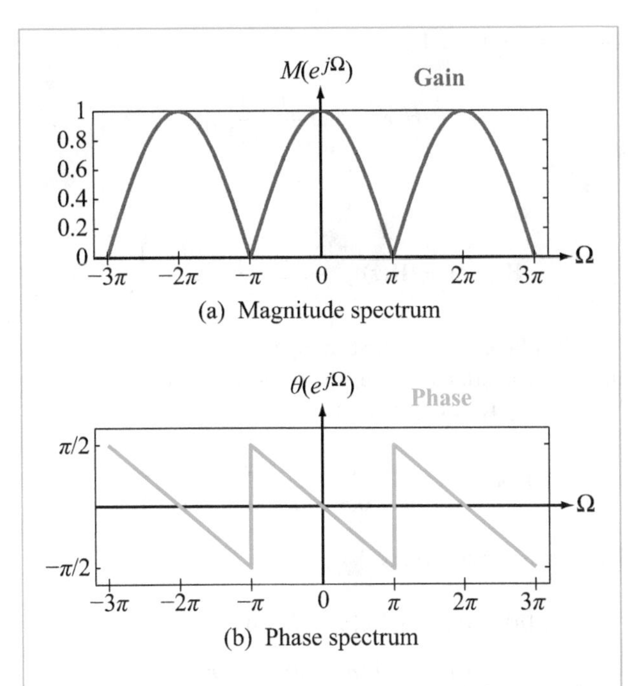

Figure 7-17: Magnitude and phase spectra of the two-point averager (Example 7-23).

The frequency response function $\mathbf{H}(e^{j\Omega}) = e^{-j\Omega/2}\cos(\Omega/2)$ is a continuous function of Ω. The sign of $\cos(\Omega/2)$ changes at frequency $\Omega = \pi$, and this is why the phase jumps by π at frequency $\Omega = \pi$.

A common mistake is to jump to the conclusion that $|\mathbf{H}(e^{j\Omega})| = \cos(\Omega/2)$ and $\angle\mathbf{H}(e^{j\Omega}) = -\Omega/2$, but this is not valid, because $\cos(\Omega/2) < 0$ for some values of Ω.

(a) For $|\Omega| < \pi$,

$$\left.\begin{aligned} |\mathbf{H}(e^{j\Omega})| &= \cos(\Omega/2) \\ \angle\mathbf{H}(e^{j\Omega}) &= -\Omega/2 \end{aligned}\right\} \quad \text{for } |\Omega| < \pi.$$

(b) For $\pi < |\Omega| < 3\pi$, $\cos(\Omega/2) < 0$, which cannot be the magnitude since the magnitude is always non-negative. Instead, the magnitude is $-\cos(\Omega/2)$, which is > 0, and the additional factor of -1 is now included in the phase as an additional phase shift of π. So for $\pi < |\Omega| < 3\pi$, we have

$$\left.\begin{aligned} |\mathbf{H}(e^{j\Omega})| &= -\cos(\Omega/2) \\ \angle\mathbf{H}(e^{j\Omega}) &= \pm\pi - \Omega/2 \end{aligned}\right\} \quad \text{for } \pi < |\Omega| < 3\pi.$$

(c) For $|\Omega| > \pi$, still another issue arises. The phase must be a periodic function of Ω, with period 2π, but the expression $-\Omega/2$ is not periodic. The phase must be reduced mod 2π by subtracting integer multiples of 2π until a value θ such that $|\theta| \leq \pi$ is obtained.

The preceding considerations lead to the plots displayed in Fig. 7-17. Note that the magnitude is an even function of Ω, the phase is an odd function of Ω, and both are periodic with period 2π, as they should be. As a result, once the magnitude and phase have been obtained for $|\Omega| \leq \pi$, their plots can be periodically repeated in Ω so they both have period 2π.

7-12.3 Relating Different Descriptions of LTI Systems

Discrete-time transfer functions play a central role in relating different descriptions of LTI systems, just as they do in continuous time. Transformations between the various descriptions are illustrated through the following series of examples.

Example 7-24: Input-Output Pair to Other Descriptions

The response of an LTI system to input signal

$$x[n] = (-2)^n\, u[n]$$

is

$$y[n] = \frac{2}{3}\,(-2)^n\, u[n] + \frac{1}{3}\,(0.4)^n\, u[n].$$

Determine (a) transfer function $\mathbf{H}(\mathbf{z})$, (b) its poles and zeros, (c) frequency response $\mathbf{H}(e^{j\Omega})$, (d) the impulse response $h[n]$, and (e) the difference equation.

Solution:

(a) In view of Eq. (7.66), the **z**-transform of $x[n]$ is

$$\mathbf{X}(\mathbf{z}) = \frac{\mathbf{z}}{\mathbf{z}+2}.$$

For output $y[n]$, use of Eq. (7.66) leads to

$$\mathbf{Y}(\mathbf{z}) = \frac{2}{3}\,\frac{\mathbf{z}}{\mathbf{z}+2} + \frac{1}{3}\,\frac{\mathbf{z}}{\mathbf{z}-0.4} = \frac{\mathbf{z}(\mathbf{z}+0.4)}{(\mathbf{z}+2)(\mathbf{z}-0.4)}.$$

Hence,

$$\mathbf{H}(\mathbf{z}) = \frac{\mathbf{Y}(\mathbf{z})}{\mathbf{X}(\mathbf{z})} = \frac{\mathbf{z}(\mathbf{z}+0.4)}{(\mathbf{z}+2)(\mathbf{z}-0.4)} \bigg/ \left(\frac{\mathbf{z}}{\mathbf{z}+2}\right) = \frac{\mathbf{z}+0.4}{\mathbf{z}-0.4}.$$

(b) The poles and zeros of $\mathbf{H}(\mathbf{z})$ are: $\mathbf{z} = \{-0.4\}$ and $\mathbf{p} = \{0.4\}$. The system is BIBO stable since $|0.4| < 1$.

(c) From Eq. (7.123), we have

$$\mathbf{H}(e^{j\Omega}) = |\mathbf{H}(\mathbf{z})|_{\mathbf{z}=e^{j\Omega}} = \frac{e^{j\Omega}+0.4}{e^{j\Omega}-0.4}.$$

(d)

$$\begin{aligned} h[n] = \mathbb{Z}^{-1}[\mathbf{H}(\mathbf{z})] &= \mathbb{Z}^{-1}\left[\frac{\mathbf{z}+0.4}{\mathbf{z}-0.4}\right] \\ &= \mathbb{Z}^{-1}\left[\frac{\mathbf{z}}{\mathbf{z}-0.4}\right] + 0.4\,\mathbb{Z}^{-1}\left[\frac{1}{\mathbf{z}-0.4}\right] \\ &= (0.4)^n\, u[n] + 0.4(0.4)^{n-1}\, u[n-1] \\ &= (0.4)^n (u[n] + u[n-1]). \end{aligned}$$

(e) To obtain the difference equation, we arrange $\mathbf{H}(\mathbf{z})$ in power of $\mathbf{z}^{-1}$.

$$\mathbf{H}(\mathbf{z}) = \frac{\mathbf{Y}(\mathbf{z})}{\mathbf{X}(\mathbf{z})} = \frac{\mathbf{z}+0.4}{\mathbf{z}-0.4} = \frac{1+0.4\mathbf{z}^{-1}}{1-0.4\mathbf{z}^{-1}}.$$

Cross-multiplying, we get

$$\mathbf{Y}(\mathbf{z})\,(1-0.4\mathbf{z}^{-1}) = \mathbf{X}(\mathbf{z})\,(1+0.4\mathbf{z}^{-1}),$$

from which we obtain the difference equation

$$y[n] - 0.4y[n-1] = x[n] + 0.4x[n-1].$$

Note that the terms and coefficients of the difference equation can be *read off* almost directly from $\mathbf{H}(\mathbf{z})$.

Example 7-25: Zeros and Poles to Other Descriptions

An LTI system has a zero at 1, a pole at 0.3, and $\mathbf{H}(0) = 1$. Determine (a) $\mathbf{H}(\mathbf{z})$, (b) $\mathbf{H}(e^{j\Omega})$, (c) $h[n]$, and (d) the difference equation describing the system.

Solution:

(a) The poles and zeros determine $\mathbf{H}(\mathbf{z})$ within an unknown scale factor $\mathbf{C}$. Hence,

$$\mathbf{H}(\mathbf{z}) = \mathbf{C}\,\frac{\mathbf{z}-1}{\mathbf{z}-0.3}.$$

The specified value $\mathbf{H}(0) = 1$ allows $\mathbf{C}$ to be determined from

$$1 = \mathbf{H}(0) = \mathbf{C}\,\frac{0-1}{0-0.3} = \frac{\mathbf{C}}{0.3}.$$

Hence, $\mathbf{C} = 0.3$ and

$$\mathbf{H}(\mathbf{z}) = 0.3\,\frac{\mathbf{z}-1}{\mathbf{z}-0.3}.$$

(b)

$$\mathbf{H}(e^{j\Omega}) = |\mathbf{H}(\mathbf{z})|_{\mathbf{z}=e^{j\Omega}} = 0.3\,\frac{e^{j\Omega}-1}{e^{j\Omega}-0.3}.$$

(c) To obtain $h[n]$, we first arrange $\mathbf{H}(\mathbf{z})$ in a form amenable to inverse transformation using the $\mathbf{z}$-transform pairs in Table 7-5. We do so by rewriting $\mathbf{H}(\mathbf{z})$ as

$$\mathbf{H}(\mathbf{z}) = 0.3\,\frac{\mathbf{z}-1}{\mathbf{z}-0.3} = \frac{0.3\mathbf{z}}{\mathbf{z}-0.3} - \frac{0.3}{\mathbf{z}-0.3}.$$

Its inverse $\mathbf{z}$-transform is

$$\begin{aligned} h[n] &= 0.3(0.3)^n\, u[n] - 0.3(0.3)^{n-1}\, u[n-1] \\ &= (0.3)^{n+1}\, u[n] - (0.3)^n\, u[n-1]. \end{aligned}$$

(d) To obtain the difference equation, we rewrite $\mathbf{H}(\mathbf{z})$ in powers of $\mathbf{z}^{-1}$,

$$\mathbf{H}(\mathbf{z}) = \frac{\mathbf{Y}(\mathbf{z})}{\mathbf{X}(\mathbf{z})} = 0.3\,\frac{\mathbf{z}-1}{\mathbf{z}-0.3} = \frac{0.3-0.3\mathbf{z}^{-1}}{1-0.3\mathbf{z}^{-1}}.$$

Cross-multiplication gives

$$\mathbf{Y}(\mathbf{z})\,(1-0.3\mathbf{z}^{-1}) = 0.3\mathbf{X}(\mathbf{z})\,(1-\mathbf{z}^{-1}),$$

whose inverse $\mathbf{z}$-transform is the difference equation

$$y[n] - 0.3y[n-1] = 0.3x[n] - 0.3x[n-1].$$

Example 7-26: Frequency Response to Difference Equation

An LTI system has the frequency response function

$$\mathbf{H}(e^{j\Omega}) = \frac{[\cos(\Omega)-1] + j\sin(\Omega)}{[\cos(2\Omega)-\cos(\Omega)+0.21] + j[\sin(2\Omega)-\sin(\Omega)]}.$$

Obtain (a) $\mathbf{H}(\mathbf{z})$, (b) poles and zeros, (c) $h[n]$, and (d) the difference equation describing the system.

Solution:

(a) To convert $\mathbf{H}(e^{j\Omega})$ to $\mathbf{H}(\mathbf{z})$, the functional variable in $\mathbf{H}(e^{j\Omega})$ should be $e^{j\Omega}$, which calls for replacing cosine and sine functions with the identities

$$\cos(\Omega) = \frac{1}{2}\,(e^{j\Omega} + e^{-j\Omega})$$

and

$$\sin(\Omega) = \frac{1}{2j}\,(e^{j\Omega} - e^{-j\Omega}).$$

After making the substitutions for all sine and cosine functions, a few steps of algebra lead to

$$\mathbf{H}(e^{j\Omega}) = \frac{e^{j\Omega} - 1}{e^{j2\Omega} - e^{j\Omega} + 0.21}\,.$$

Transfer function $\mathbf{H}(\mathbf{z})$ is then given by

$$\mathbf{H}(\mathbf{z}) = \mathbf{H}(e^{j\Omega})\Big|_{e^{j\Omega}=\mathbf{z}} = \frac{\mathbf{z} - 1}{\mathbf{z}^2 - \mathbf{z} + 0.21}\,.$$

(b) The zeros are $\mathbf{z} = \{1\}$, and the poles obtained by setting the denominator $= 0$ are $\mathbf{p} = \{0.7,\ 0.3\}$.

(c) Application of partial fraction expansion to $\mathbf{H}(\mathbf{z})$ leads to

$$\mathbf{H}(\mathbf{z}) = -\frac{3}{4}\,\frac{1}{\mathbf{z} - 0.7} + \frac{7}{4}\,\frac{1}{\mathbf{z} - 0.3}\,.$$

Its inverse $\mathbf{z}$-transform is

$$h[n] = -\frac{3}{4}\,(0.7)^{n-1}\,u[n-1] + \frac{7}{4}\,(0.3)^{n-1}\,u[n-1].$$

(d) To express $\mathbf{H}(\mathbf{z})$ in powers of $\mathbf{z}^{-1}$, we multiply its numerator and denominator by $\mathbf{z}^{-2}$. The new form is

$$\mathbf{H}(\mathbf{z}) = \frac{\mathbf{Y}(\mathbf{z})}{\mathbf{X}(\mathbf{z})} = \frac{\mathbf{z}^{-1} - \mathbf{z}^{-2}}{1 - \mathbf{z}^{-1} - 0.21\mathbf{z}^{-2}}\,.$$

Cross-multiplication gives

$$\mathbf{Y}(\mathbf{z})\,[1 - \mathbf{z}^{-1} + 0.21\mathbf{z}^{-2}] = \mathbf{X}(\mathbf{z})\,[\mathbf{z}^{-1} - \mathbf{z}^{-2}],$$

which leads to the difference equation

$$y[n] - y[n-1] + 0.21y[n-2] = x[n-1] - x[n-2].$$

Concept Question 7-17: What is a fundamental difference between continuous-time and discrete-time frequency response functions? (See S²)

Exercise 7-19: Compute the response of the system $y[n] = x[n] - x[n-2]$ to input $x[n] = \cos(\pi n/4)$.

Answer: $y[n] = 0.707\cos(\pi n/4 + \pi/4)$. (See S²)

Exercise 7-20: An LTI system has

$$\mathbf{H}(e^{j\Omega}) = j\tan(\Omega).$$

Compute the difference equation.

Answer: $y[n] + y[n-2] = x[n] - x[n-2]$. (See S²)

7-13 Discrete-Time Fourier Series (DTFS)

The discrete-time Fourier series (DTFS) is the discrete-time counterpart to Fourier-series expansion of continuous-time periodic signals. Unlike the continuous-time Fourier series (CTFS), the DTFS is finite in length (i.e., it has a finite number of terms).

Computation of its expansion coefficients requires summations rather than integrals. The response of an LTI system to a discrete-time periodic signal can be determined by applying the superposition principle. The process entails the following steps:

Step 1: Compute the DTFS of the input signal.

Step 2: Compute the output response to each DTFS term using the system's frequency response $\mathbf{H}(e^{j\Omega})$, as defined by Eq. (7.121).

Step 3: Sum the results to obtain the total output signal.

7-13.1 Period and Angular Frequency

(a) Continuous-time periodic signal

By way of review, in continuous time, a signal $x(t)$ is periodic with period T_0 if

$$x(t) = x(t + T_0). \tag{7.129}$$

Associated with T_0 is a circular frequency $f_0 = 1/T_0$ and an angular frequency $\omega_0 = 2\pi f_0 = 2\pi/T_0$. The complex-exponential Fourier-series representation of $x(t)$ is given by Eq. (5.57) as

$$x(t) = \sum_{n=-\infty}^{\infty} \mathbf{x}_n e^{jn\omega_0 t} \tag{7.130a}$$

with expansion coefficients

$$\mathbf{x}_n = \frac{1}{T_0}\int_{-T_0/2}^{T_0/2} x(t)\,e^{-jn\omega_0 t}\,dt. \tag{7.130b}$$

(b) Discrete-time periodic signal

Recall from Section 7-2.4 that a discrete-time signal $x[n]$ is periodic with period N_0 if

$$x[n] = x[n + N_0]. \tag{7.131}$$

Associated with the fundamental period N_0 is a ***fundamental angular frequency*** Ω_0 given by

$$\Omega_0 = \frac{2\pi}{N_0}. \tag{7.132}$$

N_0 is an integer and Ω_0 is confined to the range $0 \leq \Omega_0 \leq \pi$.

7-13.2 Orthogonality Property

Before introducing the DTFS representation, we will establish the following ***orthogonality property***:

$$\sum_{n=0}^{N_0-1} e^{j2\pi[(k-m)/N_0]n} = N_0\delta[k-m] = \begin{cases} N_0 & \text{if } k = m, \\ 0 & \text{if } k \neq m, \end{cases} \tag{7.133}$$

where k, m, and n are integers, and none are larger than N_0.

(a) $k = m$:

If $k = m$, Eq. (7.133) reduces to

$$\sum_{n=0}^{N_0-1} e^0 = \sum_{n=0}^{N_0-1} 1 = N_0 \qquad (k = m). \tag{7.134}$$

(b) $k \neq m$:

If in the finite geometric series

$$\sum_{n=0}^{N_0-1} \mathbf{r}^n = \frac{\mathbf{r}^{N_0} - 1}{\mathbf{r} - 1} \qquad \text{for } \mathbf{r} \neq 1,$$

we replace $\mathbf{r}$ with $e^{j2\pi(k-m)/N_0}$, we have

$$\sum_{n=0}^{N_0-1} e^{j2\pi[(k-m)/N_0]n} = \frac{e^{j2\pi[(k-m)/N_0]N_0} - 1}{e^{j2\pi(k-m)/N_0} - 1}.$$

Since $k \neq m$, and k and m cannot be larger than N_0, the exponent $j2\pi(k-m)/N_0$ in the denominator cannot be a multiple of 2π. Hence, the denominator cannot be zero. The exponent in the numerator, on the other hand, is $j2\pi(k-m)$, and therefore, it is always a multiple of 2π. Consequently,

$$\sum_{n=0}^{N_0-1} e^{j2\pi[(k-m)/N_0]n} = 0 \qquad (k \neq m). \tag{7.135}$$

The combination of Eqs. (7.134) and (7.135) constitutes the ***orthogonality property*** given by Eq. (7.133).

7-13.3 DTFS Representation

A discrete-time periodic signal $x[n]$, with fundamental period N_0 and associated fundamental periodic angular frequency $\Omega_0 = 2\pi/N_0$, can be expanded into a ***discrete-time Fourier series*** (***DTFS***) given by

$$x[n] = \sum_{k=0}^{N_0-1} \mathbf{x}_k e^{jk\Omega_0 n}, \qquad n = 0, 1, \ldots, N_0 - 1 \tag{7.136a}$$

with ***expansion coefficients***

$$\mathbf{x}_k = \frac{1}{N_0} \sum_{n=0}^{N_0-1} x[n]\, e^{-jk\Omega_0 n}, \qquad k = 0, 1, \ldots, N_0-1. \tag{7.136b}$$

▶ A periodic signal $x[n]$ of period N_0 is represented in its DTFS by the sum of N_0 complex exponentials at harmonic values of $\Omega_0 = 2\pi/N_0$. ◀

The expression for $\mathbf{x}_k$ is obtained by (1) multiplying both sides of Eq. (7.136a) by $e^{-jm\Omega_0 n}$, (2) summing both sides using $\sum_{n=0}^{N_0-1}$, and then applying the orthogonality property given by Eq. (7.133). That is,

$$\begin{aligned} \sum_{n=0}^{N_0-1} x[n]\, e^{-jm\Omega_0 n} &= \sum_{n=0}^{N_0-1} \sum_{k=0}^{N_0-1} \mathbf{x}_k e^{j(k-m)\Omega_0 n} \\ &= \sum_{k=0}^{N_0-1} \mathbf{x}_k \sum_{n=0}^{N_0-1} e^{j2\pi[(k-m)/N_0]n} \\ &= \sum_{k=0}^{N_0-1} \mathbf{x}_k N_0\, \delta[k-m] = \mathbf{x}_m N_0, \end{aligned} \tag{7.137}$$

because the inner summation is zero except for $k = m$. Changing indices from m to k in the last result leads to Eq. (7.136b).

The Fourier series expansion given by Eq. (7.136a) consists of only N_0 terms, in contrast with the Fourier series for continuous time, which consists of an infinite number of terms. The N_0 terms constitute N_0 harmonics of Ω_0, including the ***mean value*** of $x[n]$ corresponding to $k = 0$, which is given by

$$x_0 = \frac{1}{N_0} \sum_{n=0}^{N_0-1} x[n]. \tag{7.138}$$

In general, the expansion coefficients are complex quantities,

$$\mathbf{x}_k = |\mathbf{x}_k| e^{j\phi_k}. \quad (7.139)$$

▶ Plots of $|\mathbf{x}_k|$ and ϕ_k as a function of k from $k = 0$ to $k = N_0 - 1$ constitute ***magnitude and phase line-spectra***. ◀

Example 7-27: DTFS Computation

Compute the DTFS of the periodic signal

$$x[n] = \{\ldots, \underline{24}, 8, 12, 16, 24, 8, 12, 16, \ldots\}.$$

Solution: From the bracket sequence, the period is $N_0 = 4$, and the fundamental periodic angular frequency is $\Omega_0 = 2\pi/N_0 = \pi/2$. Hence, the series has expansion coefficients x_0 to $\mathbf{x}_3$. From Eq. (7.138), we have

$$x_0 = \frac{1}{N_0}\sum_{n=0}^{N_0-1} x[n] = \tfrac{1}{4}\,(24 + 8 + 12 + 16) = 15. \quad (7.140)$$

For the other coefficients, we need to evaluate Eq. (7.136b) for $\Omega_0 = \pi/2$ and $k = 1$, 2, and 3. Setting $k = 1$ in Eq. (7.136b) and noting that $e^{-j\pi/2} = -j$ gives

$$\begin{aligned}\mathbf{x}_1 &= \frac{1}{4}\sum_{n=0}^{3} x[n]\, e^{-j\Omega_0 n} \\ &= \frac{1}{4}(24e^0 + 8e^{-j\pi/2} + 12e^{-j\pi} + 16e^{-j3\pi/2}) \\ &= \frac{1}{4}(24 - j8 - 12 + j16) = 3 + j2 = 3.6e^{j33.7^\circ}.\end{aligned} \quad (7.141)$$

Similarly, noting that $e^{-j2\pi/2} = e^{-j\pi} = -1$ leads to

$$\mathbf{x}_2 = \tfrac{1}{4}\,(24 - 8 + 12 - 16) = 3 \quad (7.142)$$

and

$$\mathbf{x}_3 = \mathbf{x}_1^* = 3 - j2 = 3.6e^{-j33.7^\circ}. \quad (7.143)$$

Incorporating these coefficients in Eq. (7.136a) leads to

$$\begin{aligned}x[n] &= \sum_{k=0}^{N_0-1} \mathbf{x}_k e^{jk\Omega_0 n} \\ &= x_0 + \mathbf{x}_1 e^{j\pi n/2} + \mathbf{x}_2 e^{j\pi n} + \mathbf{x}_3 e^{j3\pi n/2} \\ &= 15 + 3.6e^{j33.7^\circ} e^{j\pi n/2} + 3e^{j\pi n} + 3.6e^{-j33.7^\circ} e^{j3\pi n/2}.\end{aligned}$$

Recognizing that n is an integer, the middle term reduces to $3\cos(\pi n)$. Also, since $e^{j3\pi n/2} = e^{-j\pi n/2}$ for any integer value of n, the second and last term can be combined into a cosine function. The net result is

$$x[n] = 15 + 7.2\cos\left(\frac{\pi n}{2} + 33.7^\circ\right) + 3\cos(\pi n). \quad (7.144)$$

Inserting $n = 0$ in Eq. (7.144) leads to $x[0] = 24$, as expected. Similar confirmations apply to $n = 1$, 2, and 3.

7-13.4 Spectral Symmetry

(1) In Example 7-27, were we to use Eq. (7.136b) to compute $\mathbf{x}_k$ for negative values of k, the outcome would have been

$$\mathbf{x}_{-k} = \mathbf{x}_k^*, \quad (7.145)$$

which means that the magnitude line spectrum is an even function and the phase line spectrum is an odd function. This result is consistent with similar observations made in Chapter 5 about continuous-time periodic signals.

(2) Were we to compute $\mathbf{x}_k$ for values of k greater than 3, we would discover that the line spectra repeats periodically at a period of four lines. This is very different from the spectra of periodic continuous-time signals.

Figure 7-18 displays line spectra for $|\mathbf{x}_k|$ and $\phi_k = \angle\mathbf{x}_k$ of Example 7-27 as a function of Ω with Ω expressed in multiples of $\Omega_0 = \pi/2$. Thus, line $\mathbf{x}_1$ corresponds to $k = 1$, which in turn corresponds to $\Omega = k\Omega_0 = \pi/2$; $\mathbf{x}_2$ corresponds to $k = 2$ and $\Omega = 2\pi/2 = \pi$; and so on. We note the following observations:

(1) Both spectra repeat every $N_0 = 4$ lines.

(2) Magnitude spectrum $|\mathbf{x}_k|$ is an even function.

(3) Phase spectrum ϕ_k is an odd function.

Because the line spectra are periodic, a periodic signal $x[n]$ can be represented by any contiguous group of harmonics of Ω_0 spanning over a full period N_0. That is, instead of choosing the lower and upper limits of the summation in Eq. (7.136a) to be $k = 0$ and $k = N_0 - 1$, respectively, we may shift the span in

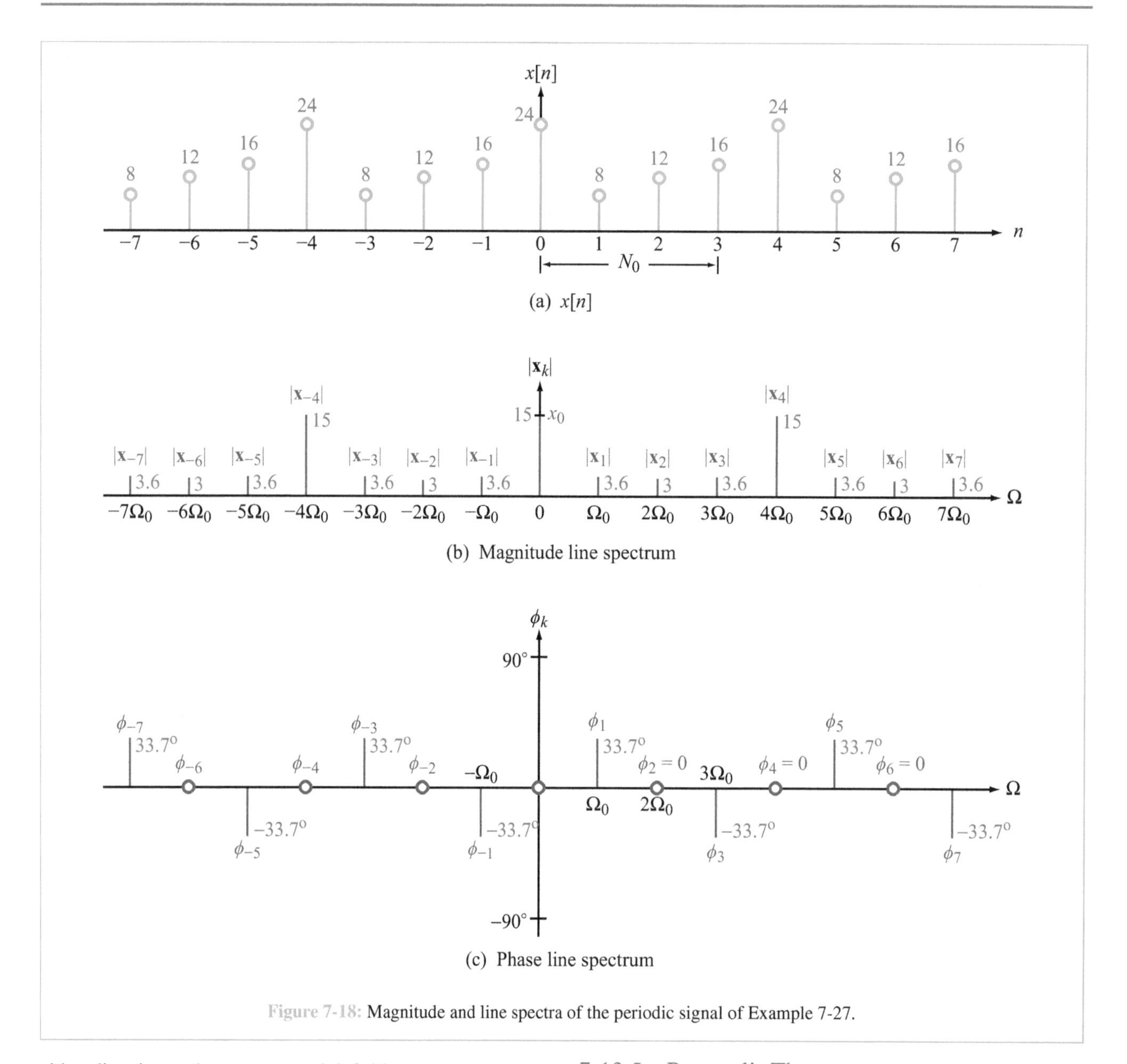

Figure 7-18: Magnitude and line spectra of the periodic signal of Example 7-27.

either direction to the more general definition

$$x[n] = \sum_{k=r}^{N_0-1+r} \mathbf{x}_k e^{jk\Omega_0 n} \tag{7.146}$$

for any integer value of r. If $r = -2$, for example, the span will be from -2 to 1, and if $r = 2$, the span becomes from 2 to 5. This flexibility allows us to reduce the number of expansion coefficients $\mathbf{x}_k$ we need to calculate by taking advantage of the symmetry property $\mathbf{x}_{-k} = \mathbf{x}_k^*$.

7-13.5 Parseval's Theorem

▶ For a discrete-time periodic signal, Parseval's theorem states that its average power is the same when computed in either the time or the frequency domain:

$$\frac{1}{N_0} \sum_{n=0}^{N_0-1} |x[n]|^2 = \sum_{k=0}^{N_0-1} |\mathbf{x}_k|^2. \tag{7.147}$$

◀

Example 7-28: Signal Power

Verify Parseval's theorem for the periodic signal in Example 7-27.

Solution: In Example 7-27, we established that

$$N_0 = 4,$$
$$x[0] = 24,\; x[1] = 8,\; x[2] = 12,\; \text{and } x[3] = 16,$$
$$\mathbf{x}_0 = 15,\; \mathbf{x}_1 = 3.6e^{j33.7^\circ},\; \mathbf{x}_2 = 3,\; \text{and } \mathbf{x}_3 = 3.6e^{-33.7^\circ}.$$

In the time domain, the average power is

$$P_{\text{av}} = \frac{1}{4}\left[(24)^2 + (8)^2 + (12)^2 + (16)^2\right] = 260.$$

In the frequency domain, the average power is

$$P_{\text{av}} = (15)^2 + (3.6)^2 + 3^2 + (3.6)^2 = 260.$$

(Note that 3.6 is actually $\sqrt{13}$ rounded off.)

7-13.6 Output Response Using DTFS

According to Eq. (7.120), the output response $y[n]$ to a complex exponential signal $e^{j\Omega n}$ at the input to an LTI system with frequency response $\mathbf{H}(e^{j\Omega})$ is

$$y[n] = e^{j\Omega n}\,\mathbf{H}(e^{j\Omega}). \tag{7.148}$$

The DTFS representation expresses a periodic signal $x[n]$ in the form of a sum of complex exponentials of complex amplitudes $\mathbf{x}_k$ and corresponding frequencies $k\Omega_0 n$,

$$x[n] = \sum_{k=0}^{N_0-1} \mathbf{x}_k e^{jk\Omega_0 n}. \tag{7.149}$$

By invoking the superposition property of LTI systems, we can write

$$y[n] = \sum_{k=0}^{N_0-1} \mathbf{y}_k e^{jk\Omega_0 n} \tag{7.150a}$$

with

$$\mathbf{y}_k = \mathbf{x}_k\; \mathbf{H}(e^{j\Omega})\Big|_{\Omega=k\Omega_0}. \tag{7.150b}$$

Example 7-29: Averaging a Periodic Signal

The periodic signal given by

$$x[n] = \{\ldots, 4, 0, 1, 0, 1, 0, \underline{4}, 0, 1, 0, 1, 0, \ldots\}$$

is used at the input to the two-point averager of Example 7-23. Determine $y[n]$.

Solution: The period of $x[n]$ is $N_0 = 6$, and from Example 7-23,

$$\mathbf{H}(e^{j\Omega}) = e^{-j\Omega/2}\cos\left(\frac{\Omega}{2}\right). \tag{7.151}$$

Application of Eq. (7.136b) leads to the expansion coefficients:

$$\mathbf{x}_k = \{\underline{1}, 0.5, 0.5, 1, 0.5, 0.5\}.$$

Application of Eq. (7.150b) with $\Omega_0 = 2\pi/N_0 = 2\pi/6 = \pi/3$ gives

$$\mathbf{y}_0 = \mathbf{x}_0\,\mathbf{H}(e^{j0}) = 1\left[e^{-j\Omega/2}\cos\left(\frac{\Omega}{2}\right)\right]\Bigg|_{\Omega=0} = 1,$$

$$\mathbf{y}_1 = \mathbf{x}_1\left[\mathbf{H}(e^{j\Omega})\Big|_{\Omega=\Omega_0=\pi/3}\right]$$
$$= \frac{1}{2}\left[e^{-j\Omega/2}\cos\left(\frac{\Omega}{2}\right)\right]\Bigg|_{\Omega=\Omega_0=\pi/3}$$
$$= \frac{1}{2}\left[e^{-j\pi/6}\cos\left(\frac{\pi}{6}\right)\right] = 0.433e^{-j\pi/6}$$

$$\mathbf{y}_2 = \mathbf{x}_2\left[\mathbf{H}(e^{j\Omega})\Big|_{\Omega=2\Omega_0=2\pi/3}\right]$$
$$= \frac{1}{2}\left[e^{-j2\pi/6}\cos\left(\frac{2\pi}{6}\right)\right] = 0.25e^{-j\pi/3},$$

and similarly,

$$\mathbf{y}_3 = 0,$$
$$\mathbf{y}_4 = \mathbf{y}_2^* = 0.25e^{j\pi/3},$$
$$\mathbf{y}_5 = \mathbf{y}_1^* = 0.433e^{j\pi/6}.$$

Note that when computing y_k, $\mathbf{H}(e^{j\Omega})$ was evaluated at $\Omega = k\Omega_0$, not Ω_0. Thus, there is a one-to-one correspondence between each y_k and $\mathbf{H}(e^{j\Omega})|_{\Omega=k\Omega_0}$.

After combining all six terms of $y[n]$ and recognizing that the complex terms appear in complex conjugates, the expression for $y[n]$ becomes

$$\begin{aligned} y[n] &= 1 + 0.866 \cos\left(\frac{\pi n}{3} - \frac{\pi}{6}\right) + 0.5 \cos\left(\frac{2\pi n}{3} - \frac{\pi}{3}\right) \\ &= 1 + 0.866 \cos\left[\frac{\pi}{3}\left(n - \frac{1}{2}\right)\right] + 0.5 \cos\left[\frac{2\pi}{3}\left(n - \frac{1}{2}\right)\right] \\ &= \{\ldots, 2, 2, 0.5, 0.5, 0.5, 0.5, \underline{2}, 2, 0.5, 0.5, 0.5, 0.5, \ldots\}. \end{aligned} \tag{7.152}$$

The two-point averager has smoothed $x[n]$ by reducing its high-frequency components. Note also that each input component has been "delayed" by 0.5. Of course, a fractional delay in discrete time has no meaning, but the effect of the phase shift is equivalent to what a fractional delay would do to a signal. This is an example of ***linear phase***, where the phase of the frequency response depends linearly on frequency.

Concept Question 7-18: Name one major difference between DTFS and CTFS. (See S²)

Exercise 7-21: Compute the DTFS of $4\cos(0.15\pi n + 1)$.

Answer: $\mathbf{x}_3 = 2e^{j1}$, $\mathbf{x}_{37} = 2e^{-j1}$, and all other $\mathbf{x}_k = 0$ for $k = 0, \ldots, 39$. (See S²)

Exercise 7-22: Confirm Parseval's rule for the above exercise.

Answer: $4^2/2 = 8$ and $|2e^{j1}|^2 + |2e^{-j1}|^2 = 8$. (See S²)

7-14 Discrete-Time Fourier Transform (DTFT)

7-14.1 DTFT Pairs

Recall from Section 5-7 that in continuous time the Fourier transform pair $x(t)$ and $\mathbf{X}(\omega)$ are interrelated by

$$\mathbf{X}(\omega) = \mathcal{F}[x(t)] = \int_{-\infty}^{\infty} x(t)\, e^{-j\omega t}\, dt \tag{7.153a}$$

(continuous time)

and

$$x(t) = \mathcal{F}^{-1}[\mathbf{X}(\omega)] = \frac{1}{2\pi} \int_{-\infty}^{\infty} \mathbf{X}(\omega)\, e^{j\omega t}\, d\omega. \tag{7.153b}$$

The analogous pair of equations for ***nonperiodic discrete-time*** signals are

$$\mathbf{X}(e^{j\Omega}) = \text{DTFT}[x[n]] = \sum_{n=-\infty}^{\infty} x[n]\, e^{-j\Omega n} \tag{7.154a}$$

(discrete time)

and

$$\begin{aligned} x[n] &= \text{DTFT}^{-1}[\mathbf{X}(e^{j\Omega})] \\ &= \frac{1}{2\pi} \int_{\Omega_1}^{\Omega_1+2\pi} \mathbf{X}(e^{j\Omega})\, e^{j\Omega n}\, d\Omega, \end{aligned} \tag{7.154b}$$

where $\mathbf{X}(e^{j\Omega})$ is the ***discrete-time Fourier transform*** (***DTFT***) of $x[n]$, and Ω_1 is any arbitrary value of Ω. As we will see later, $\mathbf{X}(e^{j\Omega})$ is periodic in Ω with period 2π, thereby allowing us to perform the inverse transform operation given by Eq. (7.154b) over any interval of length 2π along the Ω-axis.

The analogy between the DTFT $\mathbf{X}(e^{j\Omega})$ and the continuous-time Fourier transform $\mathbf{X}(\omega)$ can be demonstrated further by expressing the discrete-time signal $x[n]$ in the form of the continuous-time signal $x(t)$ given by

$$x(t) = \sum_{n=-\infty}^{\infty} x[n]\, \delta(t-n). \tag{7.155}$$

The continuous-time Fourier transform of $x(t)$ is

$$\mathbf{X}(\omega) = \sum_{n=-\infty}^{\infty} x[n]\, \mathcal{F}[\delta(t-n)] = \sum_{n=-\infty}^{\infty} x[n]\, e^{-j\omega n} = \mathbf{X}(e^{j\Omega}), \tag{7.156}$$

where in the last step we replaced ω with Ω so we may compare the expression to that given by Eq. (7.154a).

▶ The DTFT is the continuous-time Fourier transform of the continuous-time sum-of-impulses

$$\sum_{n=-\infty}^{\infty} x[n]\, \delta(t-n).$$

◀

As a direct result of Eq. (7.156), most of the properties of the continuous-time Fourier transform also hold for the DTFT. These are summarized in Table 7-7.

Table 7-7: Properties of the DTFT.

Property	$x[n]$		$\mathbf{X}(e^{j\Omega})$
1. Linearity	$k_1\, x_1[n] + k_2\, x_2[n]$	$\longleftrightarrow$	$k_1\, \mathbf{X}_1(e^{j\Omega}) + k_2\, \mathbf{X}_2(e^{j\Omega})$
2. Time shift	$x[n-n_0]$	$\longleftrightarrow$	$\mathbf{X}(e^{j\Omega})\, e^{-jn_0\Omega}$
3. Frequency shift	$x[n]\, e^{j\Omega_0 n}$	$\longleftrightarrow$	$\mathbf{X}(e^{j(\Omega-\Omega_0)})$
4. Multiplication by n (frequency differentiation)	$n\, x[n]$	$\longleftrightarrow$	$j\dfrac{d\mathbf{X}(e^{j\Omega})}{d\Omega}$
5. Time Reversal	$x[-n]$	$\longleftrightarrow$	$\mathbf{X}(e^{-j\Omega})$
6. Time convolution	$x_1[n] * x_2[n]$	$\longleftrightarrow$	$\mathbf{X}_1(e^{j\Omega})\, \mathbf{X}_2(e^{j\Omega})$
7. Frequency convolution	$x_1[n]\, x_2[n]$	$\longleftrightarrow$	$\dfrac{1}{2\pi}\, \mathbf{X}_1(e^{j\Omega}) * \mathbf{X}_2(e^{j\Omega})$
8. Conjugation	$x^*[n]$	$\longleftrightarrow$	$\mathbf{X}^*(e^{-j\Omega})$
9. Parseval's theorem	$\displaystyle\sum_{n=-\infty}^{\infty} \lvert x[n]\rvert^2$	$=$	$\displaystyle\frac{1}{2\pi}\int_{\Omega_1}^{\Omega_1+2\pi} \lvert\mathbf{X}(e^{j\Omega})\rvert^2\, d\Omega$
10. Conjugate symmetry	$\mathbf{X}^*(e^{j\Omega}) = \mathbf{X}(e^{-j\Omega})$		

► In particular, the following conjugate symmetry properties extend directly from the continuous-time Fourier transform to the DTFT:

$x[n]$	$\mathbf{X}(e^{j\Omega})$
Real and even	Real and even
Real and odd	Imaginary and odd
Imaginary and even	Imaginary and even
Imaginary and odd	Real and odd

An important additional property of the DTFT that does not hold for the continuous-time Fourier transform is:

► The DTFT $\mathbf{X}(e^{j\Omega})$ is periodic with period 2π. ◄

This property follows from $e^{j(\Omega+2\pi k)} = e^{j\Omega}$ for any integer k in the definition of the DTFT given by Eq. (7.154a).

► The DTFT can be viewed as a continuous-time Fourier series expansion of the periodic function $\mathbf{X}(e^{j\Omega})$. So the inverse DTFT given by Eq. (7.154b) is simply the formula for computing the continuous-time Fourier series coefficients of a periodic signal with period $T_0 = 2\pi$. ◄

The relation between the DTFT and the **z**-transform is analogous to the relation between the continuous-time Fourier transform and the Laplace transform.

► If, in continuous time, $x(t)$ is causal, then

$$\mathbf{X}(\omega) = \mathbf{X}(\mathbf{s})\big|_{\mathbf{s}=j\omega}. \tag{7.157a}$$

Similarly, in discrete time, if $x[n]$ is causal,

$$\mathbf{X}(e^{j\Omega}) = \mathbf{X}(\mathbf{z})\big|_{\mathbf{z}=e^{j\Omega}}. \tag{7.157b}$$

The relations between periodicity and discreteness in time and frequency can be summarized as follows:

Time Domain	Frequency Domain	Relation
Periodic	Discrete	Fourier Series
Discrete	Periodic	DTFT
Discrete and Periodic	Discrete and Periodic	DTFS

The Discrete-Time Fourier Transform (DTFT)

(a) For a discrete-time signal $x[n]$, its DTFT is its spectrum $\mathbf{X}(e^{j\Omega})$.

(b) For a discrete-time system, the DTFT of its impulse response $h[n]$ is the system's frequency response $\mathbf{H}(e^{j\Omega})$.

Example 7-30: DTFT Computation

Compute the DTFT of

(a) $x_1[n] = \{3, 1, \underline{4}, 2, 5\}$,

(b) $x_2[n] = \left(\frac{1}{2}\right)^n \; u[n]$,

(c) $x_3[n] = 4\sin(0.3n)$,

and plot their magnitude and phase spectra.

Solution:

(a) Signal $x_1[n]$ is noncausal and extends from $n = -2$ to $n = +2$. Hence, application of Eq. (7.154a) gives

$$\mathbf{X}_1(e^{j\Omega}) = 3e^{j2\Omega} + 1e^{j\Omega} + 4 + 2e^{-j\Omega} + 5e^{-j2\Omega}.$$

Its magnitude and phase spectra, given by

$$M_1(e^{j\Omega}) = |\mathbf{X}_1(e^{j\Omega})|$$

and $\phi_1(e^{j\Omega}) = \angle\mathbf{X}_1(e^{j\Omega})$, are displayed in Fig. 7-19. Both are periodic with period 2π. The magnitude spectrum is an even function, and the phase spectrum is an odd function. That is,

$$\mathbf{X}_1(e^{j\Omega}) = \mathbf{X}_1^*(e^{-j\Omega}). \tag{7.158}$$

The periodicity and symmetry properties exhibited by $\mathbf{X}_1(e^{j\Omega})$ are applicable to the DTFT of any discrete-time signal. Note that phase of $+\pi$ is identical to phase of $-\pi$ in Fig. 7-19(c).

(b) For $x_2[n] = \left(\frac{1}{2}\right)^n \; u[n]$, we have

$$\mathbf{X}_2(e^{j\Omega}) = \sum_{n=-\infty}^{\infty} x_2[n]\, e^{-j\Omega n}$$

$$= \sum_{n=0}^{\infty} \left(\frac{1}{2}\right)^n e^{-j\Omega n} = \sum_{n=0}^{\infty} \left(\frac{1}{2e^{j\Omega}}\right)^n. \tag{7.159}$$

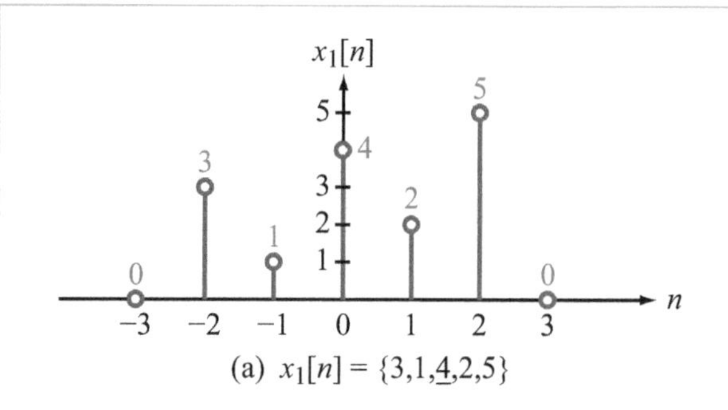

(a) $x_1[n] = \{3,1,\underline{4},2,5\}$

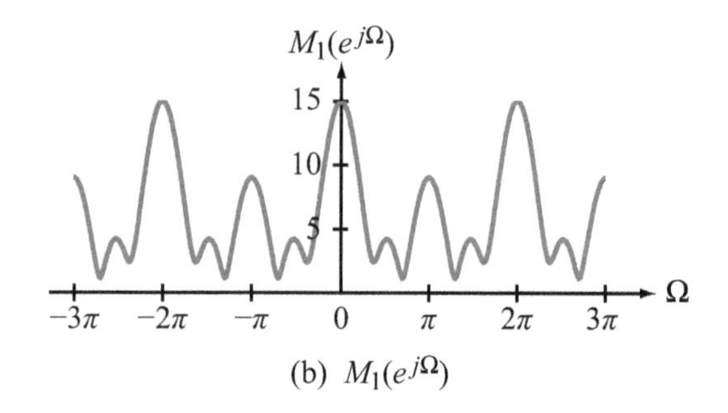

(b) $M_1(e^{j\Omega})$

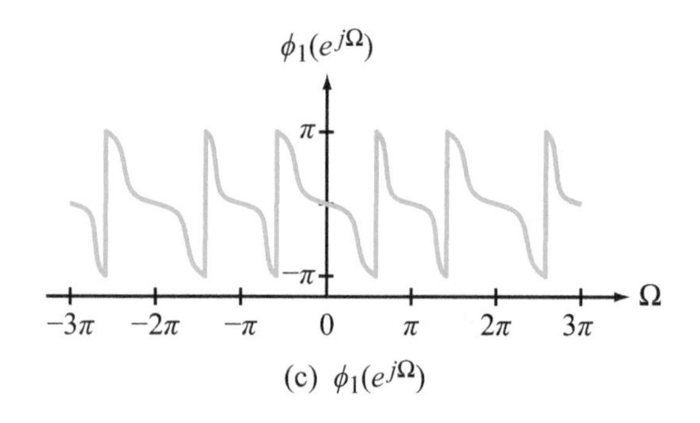

(c) $\phi_1(e^{j\Omega})$

Figure 7-19: Plots of $x_1[n]$ and its DTFT magnitude and phase spectra [part (a) of Example 7-30].

In view of the infinite geometric series, namely,

$$\sum_{n=0}^{\infty} \mathbf{r}^n = \frac{1}{1-\mathbf{r}} \qquad \text{for } |\mathbf{r}| < 1,$$

Eq. (7.159) simplifies to

$$\mathbf{X}_2(e^{j\Omega}) = \frac{1}{1 - \dfrac{1}{2e^{j\Omega}}} = \frac{e^{j\Omega}}{e^{j\Omega} - \dfrac{1}{2}}. \tag{7.160}$$

Plots of $x_2[n]$ and the magnitude and phase spectra of its DTFT are displayed in Fig. 7-20.

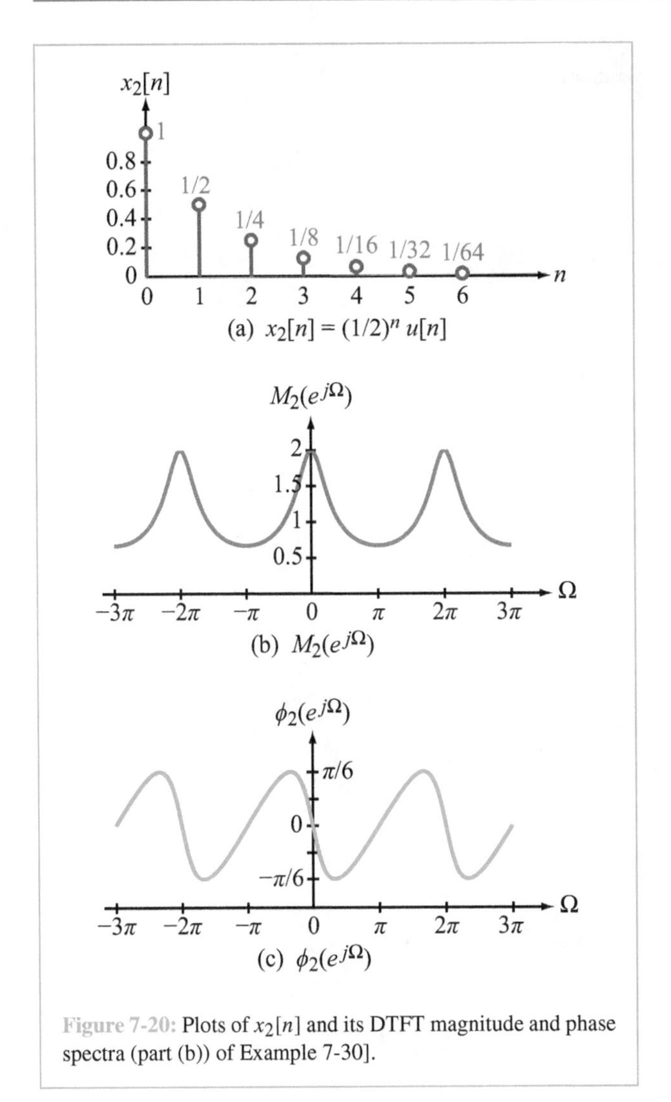

Figure 7-20: Plots of $x_2[n]$ and its DTFT magnitude and phase spectra (part (b)) of Example 7-30].

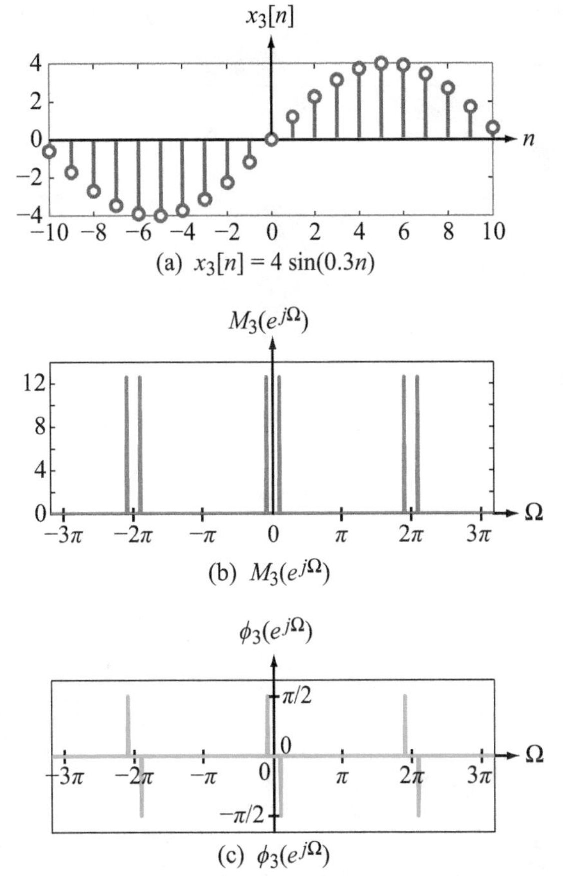

(c) For $x_3[n] = 4\sin(0.3n)$, use of DTFT pair #7 in Table 7-8 gives

$$\mathbf{X}_3(e^{j\Omega}) = \frac{4\pi}{j} \sum_{k=-\infty}^{\infty} [\delta(\Omega - 0.3 - 2\pi k) - \delta(\Omega + 0.3 - 2\pi k)].$$

For the period associated with $k = 0$, we have

$$\mathbf{X}_3(e^{j\Omega}) = j4\pi\, \delta(\Omega + 0.3) - j4\pi\, \delta(\Omega - 0.3), \quad (7.161)$$

which consists of two impulses at $\Omega = \pm 0.3$, as shown in Fig. 7-21. The spectrum repeats every 2π.

Figure 7-21: Plots of $x_3[n]$ and its DTFT magnitude and phase spectra (part (c) of Example 7-30).

7-14.2 Rectangular Pulse

In discrete time, we define the ***rectangular pulse*** $\text{rect}[n/N]$ as

$$\text{rect}\left[\frac{n}{N}\right] = u[n+N] - u[n-1-N] = \begin{cases} 1 & \text{for } |n| \le N, \\ 0 & \text{for } |n| > N. \end{cases} \quad (7.162)$$

The pulse extends from $-N$ to N and is of length $2N+1$. Its DTFT is given by

$$\mathbf{X}(e^{j\Omega}) = \sum_{n=-N}^{N} e^{-j\Omega n}. \quad (7.163)$$

Table 7-8: Discrete-time Fourier transform (DTFT) pairs.

	$x[n]$		$\mathbf{X}(e^{j\Omega})$	Condition
1.	$\delta[n]$	$\longleftrightarrow$	1	
1a.	$\delta[n-m]$	$\longleftrightarrow$	$e^{-jm\Omega}$	m = integer
2.	1	$\longleftrightarrow$	$2\pi \sum_{k=-\infty}^{\infty} \delta(\Omega - 2\pi k)$	
3.	$u[n]$	$\longleftrightarrow$	$\frac{e^{j\Omega}}{e^{j\Omega}-1} + \sum_{k=-\infty}^{\infty} \pi\delta(\Omega - 2\pi k)$	
3a.	$\mathbf{a}^n\, u[n]$	$\longleftrightarrow$	$\frac{e^{j\Omega}}{e^{j\Omega}-\mathbf{a}}$	$\lvert\mathbf{a}\rvert < 1$
3b.	$n\mathbf{a}^n\, u[n]$	$\longleftrightarrow$	$\frac{\mathbf{a}e^{j\Omega}}{(e^{j\Omega}-\mathbf{a})^2}$	$\lvert\mathbf{a}\rvert < 1$
4.	$e^{j\Omega_0 n}$	$\longleftrightarrow$	$2\pi \sum_{k=-\infty}^{\infty} \delta(\Omega - \Omega_0 - 2\pi k)$	
5.	$\mathbf{a}^{-n}\, u[-n-1]$	$\longleftrightarrow$	$\frac{\mathbf{a}e^{j\Omega}}{1-\mathbf{a}e^{j\Omega}}$	$\lvert\mathbf{a}\rvert < 1$
6.	$\cos(\Omega_0 n)$	$\longleftrightarrow$	$\pi \sum_{k=-\infty}^{\infty} [\delta(\Omega - \Omega_0 - 2\pi k) + \delta(\Omega + \Omega_0 - 2\pi k)]$	
7.	$\sin(\Omega_0 n)$	$\longleftrightarrow$	$\frac{\pi}{j} \sum_{k=-\infty}^{\infty} [\delta(\Omega - \Omega_0 - 2\pi k) - \delta(\Omega + \Omega_0 - 2\pi k)]$	
8.	$\mathbf{a}^n \cos(\Omega_0 n + \theta)\, u[n]$	$\longleftrightarrow$	$\frac{e^{j2\Omega}\cos\theta - \mathbf{a}e^{j\Omega}\cos(\Omega_0 - \theta)}{e^{j2\Omega} - 2\mathbf{a}e^{j\Omega}\cos\Omega_0 + \mathbf{a}^2}$	$\lvert\mathbf{a}\rvert < 1$
9.	$\text{rect}\left[\frac{n}{N}\right] = u[n+N] - u[n-1-N]$	$\longleftrightarrow$	$\frac{\sin\left[\Omega\left(N+\frac{1}{2}\right)\right]}{\sin\left(\frac{\Omega}{2}\right)}$	
10.	$\frac{\sin[\Omega_0 n]}{\pi n}$	$\longleftrightarrow$	$\sum_{k=-\infty}^{\infty} [u(\Omega + \Omega_0 - 2\pi k) - u(\Omega - \Omega_0 - 2\pi k)]$	

By changing indices to $m = n + N$, Eq. (7.163) assumes the form

$$\mathbf{X}(e^{j\Omega}) = \sum_{m=0}^{2N} e^{-j\Omega(m-N)} = e^{j\Omega N} \sum_{m=0}^{2N} e^{-j\Omega m}.$$

The summation resembles the finite geometric series

$$\sum_{m=0}^{M-1} \mathbf{r}^m = \frac{1-\mathbf{r}^M}{1-\mathbf{r}} \qquad \text{for } \mathbf{r} \neq 1,$$

so if we set $\mathbf{r} = e^{-j\Omega}$ and $M = 2N+1$, we have

$$\mathbf{X}(e^{j\Omega}) = e^{j\Omega N}\left[\frac{1-(e^{-j\Omega})^{2N+1}}{1-e^{-j\Omega}}\right] = \frac{e^{j\Omega N} - e^{-j\Omega(N+1)}}{1-e^{-j\Omega}}.$$

To convert the numerator and denominator into sine functions, we multiply the numerator and denominator by $e^{j\Omega/2}$. The process leads to

$$\mathbf{X}(e^{j\Omega}) = \frac{\sin\left[\Omega\left(N+\frac{1}{2}\right)\right]}{\sin\left(\frac{\Omega}{2}\right)}. \tag{7.164}$$

This is called a ***discrete sinc function***. Plots of $x[n] = \text{rect}[n/N]$ and its DTFT are displayed in Fig. 7-22 for $N = 10$. We observe that $\mathbf{X}(e^{j\Omega})$ looks like a sinc function that repeats itself every 2π along the Ω-axis.

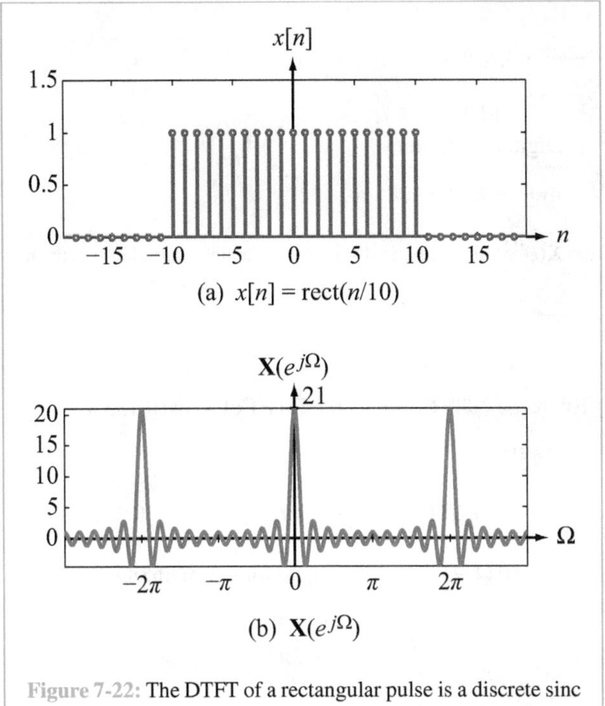

Figure 7-22: The DTFT of a rectangular pulse is a discrete sinc function.

As $\Omega \to 0$, $\sin\Omega \approx \Omega$, which when applied to Eq. (7.164) leads to

$$\lim_{\Omega\to 0} \mathbf{X}(e^{j\Omega}) = \frac{\Omega\left(N+\frac{1}{2}\right)}{(\Omega/2)} = 2N+1.$$

Alternatively, we could have arrived at the same result through

$$\mathbf{X}(e^{j0}) = \sum_{n=-\infty}^{\infty} x[n] = \sum_{n=-N}^{N} 1 = 2N+1.$$

For $N = 10$, $2N+1 = 21$, which matches the peak value of the spectrum in Fig. 7-22(b).

Example 7-31: Brick-Wall Lowpass Filter

A discrete-time lowpass filter has the brick-wall spectrum

$$\mathbf{H}(e^{j\Omega}) = \begin{cases} 1 & 0 \le |\Omega| \le \Omega_0, \\ 0 & \Omega_0 < |\Omega| \le \pi, \end{cases}$$

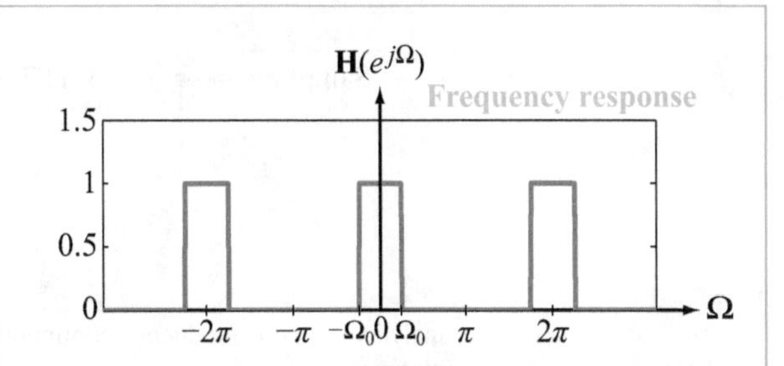

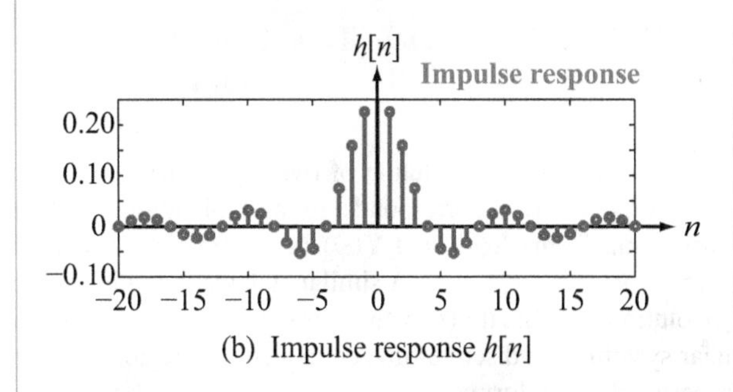

Figure 7-23: The impulse response of a brick-wall lowpass filter is a discrete-time sinc function.

which is defined over $-\pi \le \Omega \le \pi$ and periodic in Ω with period 2π, as shown in Fig. 7-23(a). Determine its corresponding impulse response $h[n]$.

Solution: To obtain $h[n]$, we need to perform the inverse DTFT given by Eq. (7.154b) with $\Omega_1 = -\pi$, namely

$$\begin{aligned} h[n] &= \text{IDTFT}[\mathbf{H}(e^{j\Omega})] \\ &= \frac{1}{2\pi} \int_{-\pi}^{\pi} \mathbf{H}(e^{j\Omega})\, e^{j\Omega n}\, d\Omega \\ &= \frac{1}{2\pi} \int_{-\Omega_0}^{\Omega_0} e^{j\Omega n}\, d\Omega \\ &= \frac{1}{j2\pi n}(e^{j\Omega_0 n} - e^{-j\Omega_0 n}) = \frac{\sin(\Omega_0 n)}{\pi n}, \end{aligned} \tag{7.165}$$

which constitutes a ***discrete-time sinc function***. An example is shown in Fig. 7-23 for $\Omega_0 = \pi/4$. As expected, the transform of a rectangular pulse is a sinc function, and vice-versa.

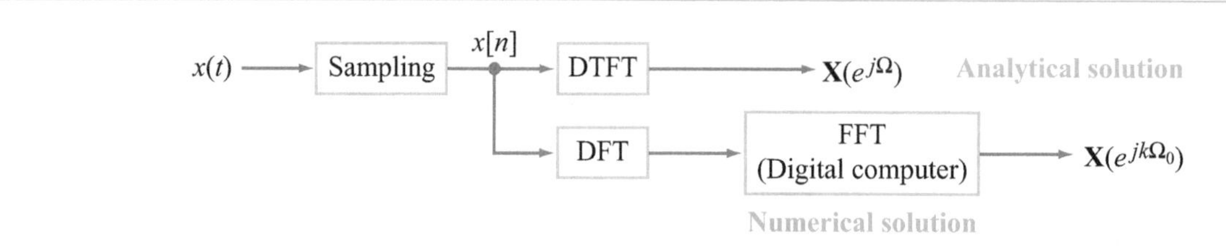

Figure 7-24: Whereas the analytical solution produces a continuous spectrum $\mathbf{X}(e^{j\Omega})$, the numerical solution generates a sampled spectrum at discrete values that are integer multiples of Ω_0.

7-14.3 DTFT of Discrete-Time Convolution

In continuous time, convolution of two signals $x(t)$ and $y(t)$ in the time domain corresponds to multiplication of their Fourier transforms $\mathbf{X}(\omega)$ and $\mathbf{Y}(\omega)$ in the frequency domain (property #11 in Table 5-7). A similar statement applies to the convolution of a signal $x(t)$ with the impulse response $h(t)$ of a linear system. The discrete-time equivalent of the convolution property takes the form

$$x[n] * h[n] \quad \longleftrightarrow \quad \mathbf{X}(e^{j\Omega})\, \mathbf{H}(e^{j\Omega}). \tag{7.166}$$

The convolution of $x[n]$ and $h[n]$ can be computed directly in the discrete-time domain using Eq. (7.51a), or it can be computed indirectly by: (1) computing the DTFTs of $x[n]$ and $h[n]$, (2) multiplying $\mathbf{X}(e^{j\Omega})$ by $\mathbf{H}(e^{j\Omega})$, and then (3) computing the inverse DTFT of the product. Thus,

$$x[n] * h[n] \quad \longleftrightarrow \quad \text{DTFT}^{-1}[\mathbf{X}(e^{j\Omega})\, \mathbf{H}(e^{j\Omega})]. \tag{7.167}$$

The choice between using the direct or indirect approach usually is dictated by overall computational efficiency.

Concept Question 7-19: What is the difference between DTFT and CTFT? (See S²)

Concept Question 7-20: What is the significant difference between discrete-time and continuous-time sinc functions? (See S²)

Exercise 7-23: Compute the DTFT of $4\cos(0.15\pi n + 1)$.

Answer:

$$4\pi e^{j1}\, \delta((\Omega - 0.15)) + 4\pi e^{-j1}\, \delta((\Omega + 0.15)),$$

where $\delta((\Omega - \Omega_0))$ is shorthand for the chain of impulses

$$\sum_{k=-\infty}^{\infty} \delta(\Omega - 2\pi k - \Omega_0).$$

(See S²)

Exercise 7-24: Compute the inverse DTFT of $4\cos(2\Omega) + 6\cos(\Omega) + j8\sin(2\Omega) + j2\sin(\Omega)$.

Answer: $\{6, 4, \underline{0}, 2, -2\}$. (See S²)

7-15 Discrete Fourier Transform (DFT)

The ***discrete Fourier transform*** (*DFT*) is the numerical bridge between the DTFT and the ***fast Fourier transform*** (*FFT*). In Fig. 7-24, we compare two sequences: one leading to an ***analytical solution*** for the DTFT $\mathbf{X}(e^{j\Omega})$ of a discrete-time signal $x[n]$ and another leading to a ***numerical solution*** of the same. The analytical solution uses a DTFT formulation, which includes infinite summations and integrals [as in Eqs. (7.154a and b)]. In practice, discrete-time signal processing is performed numerically by a digital computer using FFTs.

▶ Despite its name, the FFT is *not* a Fourier transform; it is an algorithm for computing the discrete Fourier transform (DFT). ◀

The role of the DFT is to cast the DTFT formulation of an infinite-duration signal in an approximate form composed of finite summations instead of infinite summations and integrals. For finite-duration signals, the DFT formulation is exact.

7-15.1 The DFT Approximation

(a) Nonperiodic signals

For a nonperiodic signal $x[n]$, the DTFT is given by Eq. (7.154) as

$$\mathbf{X}(e^{j\Omega}) = \sum_{n=-\infty}^{\infty} x[n]\, e^{-j\Omega n} \tag{7.168a}$$

and

$$\underbrace{x[n] = \frac{1}{2\pi} \int_{\Omega_1}^{\Omega_1+2\pi} \mathbf{X}(e^{j\Omega})\, e^{j\Omega n}\, d\Omega.}_{\text{Inverse DTFT}} \tag{7.168b}$$

The expression for $\mathbf{X}(e^{j\Omega})$ includes a summation over $-\infty < n < \infty$, and the expression for $x[n]$ involves continuous integration over a segment of length 2π. The discrete Fourier transform (DFT):

(1) limits the span of n of signal $x[n]$ to a finite length N_0, thereby converting the infinite summation into a finite summation (how N_0 is specified is discussed later in Section 7-15.2), and

(2) converts the integral in Eq. (7.168b) into a summation of length N_0.

The DFT formulation is given by

$$\mathbf{X}[k] = \sum_{n=0}^{N_0-1} x[n]\, e^{-jk\Omega_0 n}, \qquad k = 0, 1, \ldots, N_0 - 1 \tag{7.169a}$$

and

$$\underbrace{x[n] = \frac{1}{N_0} \sum_{k=0}^{N_0-1} \mathbf{X}[k] e^{jk\Omega_0 n}, \qquad n = 0, 1, \ldots, N_0 - 1,}_{\text{Inverse DFT}} \tag{7.169b}$$

where $\mathbf{X}[k]$ are called the ***DFT complex coefficients***, and

$$\Omega_0 = \frac{2\pi}{N_0}.$$

This is called an ***N_0-point DFT*** or a DFT of order N_0. Whereas in the DTFT, the Fourier transform $\mathbf{X}(e^{j\Omega})$ is a continuous function of Ω and periodic with period 2π, the Fourier transform of the DFT is represented by N_0 Fourier coefficients $\mathbf{X}[k]$ defined at N_0 harmonic values of Ω_0, where Ω_0 is specified by the choice of N_0. *Thus, N_0 values of $x[n]$ generate N_0 values of $\mathbf{X}[k]$, and vice versa.*

(b) Periodic signals

For a periodic signal $x[n]$, the discrete-time Fourier series (DTFS) of Section 7-13 is given by Eq. (7.136) as

$$\mathbf{x}_k = \frac{1}{N_0} \sum_{n=0}^{N_0-1} x[n]\, e^{-jk\Omega_0 n}, \qquad k = 0, 1, \ldots, N_0 - 1, \tag{7.170a}$$

and

$$\underbrace{x[n] = \sum_{n=0}^{N_0-1} \mathbf{x}_k e^{jk\Omega_0 n}, \qquad n = 0, 1, \ldots, N_0 - 1.}_{\text{DTFS}} \tag{7.170b}$$

Comparison of the DFT formulas in Eq. (7.169) with those in Eq. (7.170) for the DTFS leads to the conclusion that, within a scaling factor, they are identical in every respect. If we define $\mathbf{X}[k] = N_0\mathbf{x}_k$, the pair of expressions in Eq. (7.169) become identical with those in Eq. (7.170).

► The DFT is identical to the DTFS with $\mathbf{X}[k] = N_0\mathbf{x}_k$. ◄

The preceding observations lead to two important conclusions.

(1) The DFT does not offer anything new so far as periodic signals are concerned.

(2) The DFT, in essence, *converts* a nonperiodic signal $x[n]$ into a periodic signal of period N_0 and converts the DTFT into a DTFS (within a scaling constant N_0).

The conversion process is called *windowing* or *truncating*. It entails choosing which span of $x[n]$ to retain and which to delete and how to select the value of N_0. Windowing is the topic of the next subsection.

Example 7-32: DFT of a Periodic Sinusoid

Compute the DFT of $x[n] = A\cos(2\pi(k_0/N_0)n+\theta)$, where k_0 is an integer. Note that this is a *periodic* sinusoid.

Solution: Using the definition of the DFT given by Eq. (7.169a), we obtain

$$\begin{aligned}
\mathbf{X}[k] &= \sum_{n=0}^{N_0-1} A\cos\left(2\pi\,\frac{k_0}{N_0}\,n+\theta\right)\,e^{-j2\pi(k/N_0)n} \\
&= \sum_{n=0}^{N_0-1}\left[\frac{A}{2}\,e^{j\theta}e^{j2\pi(k_0/N_0)n} + \frac{A}{2}\,e^{-j\theta}e^{-j2\pi(k_0/N_0)n}\right] \\
&\qquad \cdot e^{-j2\pi(k/N_0)n} \\
&= \frac{A}{2}\,e^{j\theta}\sum_{n=0}^{N_0-1} e^{j2\pi[(k_0-k)/N_0]n} \\
&\quad + \frac{A}{2}\,e^{-j\theta}\sum_{n=0}^{N_0-1} e^{-j2\pi[(k_0+k)/N_0]n}.
\end{aligned}$$

Using the orthogonality property defined by Eq. (7.133), the Fourier coefficients become

$$\begin{aligned}
\mathbf{X}[k] &= N_0\,\frac{A}{2}\,e^{j\theta} && \text{if } k = k_0, \\
\mathbf{X}[k] &= N_0\,\frac{A}{2}\,e^{-j\theta} && \text{if } k = N_0 - k_0, \\
\mathbf{X}[k] &= 0 && \text{for all other } k.
\end{aligned}$$

As $N_0 \to \infty$, this pair of discrete-time impulses resembles a pair of continuous-time impulses, which is exactly the DTFT of a sinusoid.

7-15.2 Windowing

Consider the nonperiodic signal $x[n]$ shown in Fig. 7-25(a), which represents sampled values measured by a pressure sensor. Even though $x[n]$ has non-zero values over a wide span of n, the bulk of the signal is concentrated over a finite range. If we multiply $x[n]$ by a window function of length N_0, chosen such that it captures the essential variation of $x[n]$, we end up with the truncated signal shown in Fig. 7-25(b). A window function is simply a rectangular function of unit amplitude. Values of $x[n]$ outside the window region are replaced with zeros, so the windowed version of $x[n]$ is

$$x_w[n] = \begin{cases} x[n] & 0 \le n \le N_0 - 1, \\ 0 & \text{otherwise.} \end{cases} \tag{7.171}$$

Next, if we *pretend* that the truncated signal is a single cycle of a periodic signal (Fig. 7-25(c)), we can use the DTFS (or, equivalently, the DFT) to compute its Fourier coefficients.

In the time domain, the windowing action of the DFT limits the number of sampled values of $x[n]$ to N_0. The spectrum of $\mathbf{X}_w(e^{j\Omega})$ is periodic with period 2π. The DFT converts the continuous 2π-period of Ω into N_0 discrete frequencies by sampling the spectrum at multiples of $\Omega_0 = 2\pi/N_0$. This is equivalent to having a spectral resolution $\Delta\Omega = 2\pi/N_0 = \Omega_0$.

In some cases, the available number of samples of $x[n]$ may be limited to some value N_0, but the desired spectral resolution for adequate representation of $x[n]$ requires sampling at a higher rate, N_0'. In that case, the length of signal $x[n]$ is extended from N_0 by appending zeros to the time signal. That is, $x[n]$ is redefined as

$$x'[n] = \begin{cases} x[n] & 0 \le n \le N_0 - 1, \\ 0 & N_0 \le n \le N_0' - 1. \end{cases} \tag{7.172}$$

Extending the length of $x[n]$ by adding zeros is called *zero-padding*. An example of a zero-padded signal is shown is Fig. 7-25(d).

7-15.3 Relation of DFT to z-Transform

Comparison of the expressions defining the **z**-transform [Eq. (7.58a)], the DTFT [Eq. (7.154a)], and the DFT [Eq. (7.169a)], leads to the following conclusion:

$$\underbrace{\mathbf{X}[k]}_{\text{DFT}} = \underbrace{\mathbf{X}(e^{j\Omega})\Big|_{\Omega=k\Omega_0}}_{\text{DTFT}} = \underbrace{\mathbf{X}(\mathbf{z})\big|_{\mathbf{z}=e^{jk\Omega_0}}}_{\text{z-transform}} \tag{7.173}$$

with $\Omega_0 = 2\pi/N_0$.

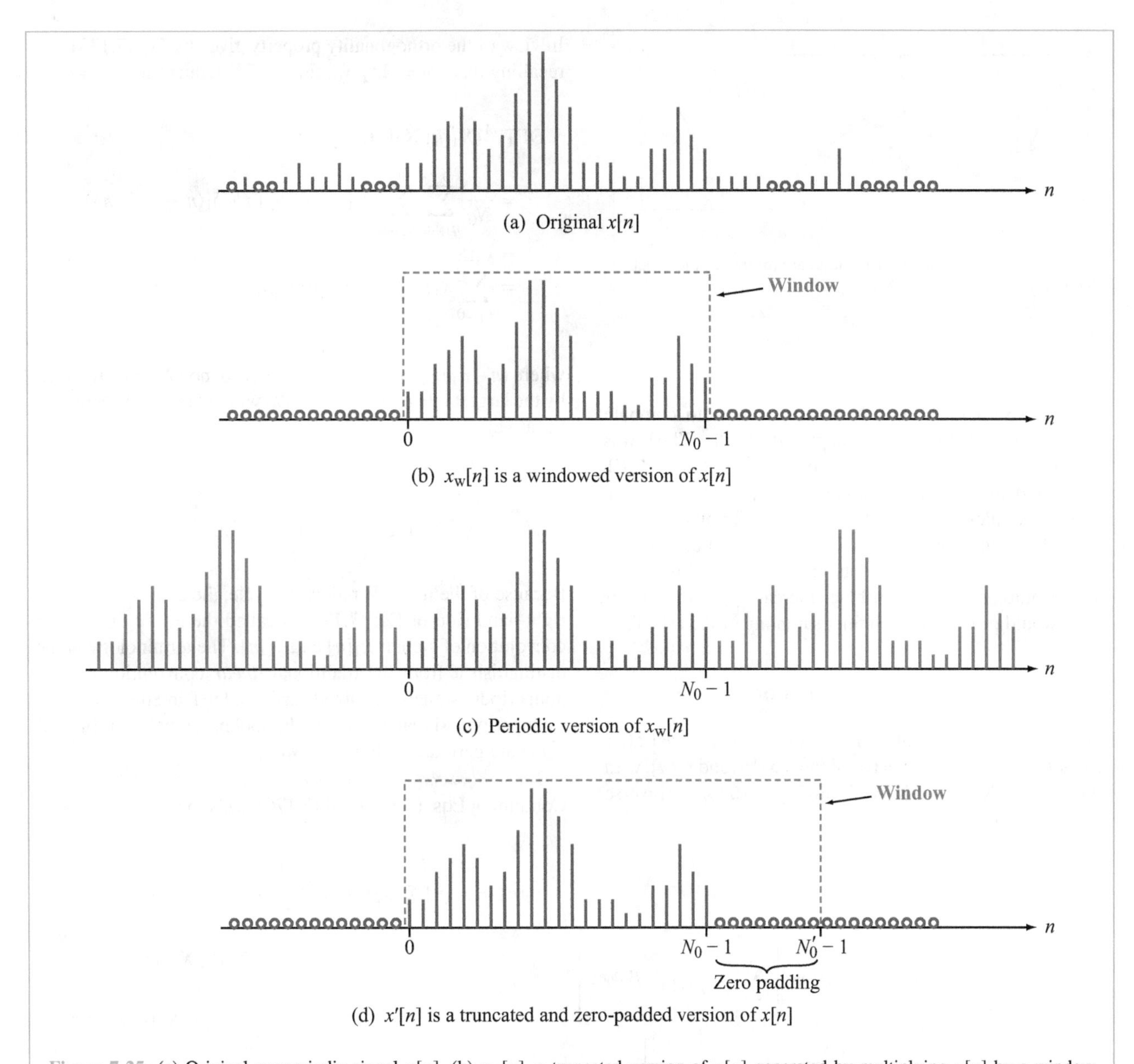

Figure 7-25: (a) Original nonperiodic signal $x[n]$, (b) $x_{\mathrm{w}}[n]$, a truncated version of $x[n]$ generated by multiplying $x[n]$ by a window function, (c) creating an artificial periodic version of the truncated signal $x_{\mathrm{w}}[n]$, and (d) $x'[n]$, a truncated and zero-padded version of $x[n]$.

The DFT is the result of sampling the DTFT at N_0 equally spaced frequencies between 0 and 2π. From the standpoint of the **z**-transform, the DFT is the result of sampling the **z**-transform at N_0 equally spaced points around the unit circle $|\mathbf{z}| = 1$.

Example 7-33: Using DFT to Compute DTFT Numerically

Redo Example 7-30(a) using a 32-point DFT.

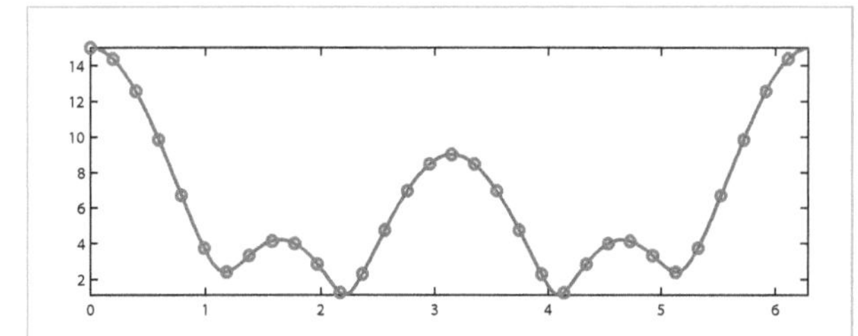

Figure 7-26: A 32-point DFT (red dots) approximates the exact DTFT (blue curve).

Solution: Example 7-30(a) demonstrated how to compute $x_1[n] = \{3, 1, \underline{4}, 2, 5\}$, and the magnitude of the DTFT was plotted in Fig. 7-19(b). Since the magnitude of the DTFT is unaffected by time shifts (property #2 in Table 7-7), we compute and plot the magnitude of the 32-point DFT of $\{\underline{3}, 1, 4, 2, 5\}$, zero-padded with 27 zeros, using Eq. (7.169a). The results are plotted as red dots in Fig. 7-26 and the exact DTFT magnitude is shown in blue. Increasing 32 to a larger number would provide a much finer sampling of the DTFT.

7-15.4 Use of DFT for Convolution

The convolution property of the DTFT extends to the DFT after some modifications. Consider two signals, $x_1[n]$ and $x_2[n]$, with N_0-point DFTs $\mathbf{X}_1[k]$ and $\mathbf{X}_2[k]$. From Eq. (7.169b), the inverse DFT of their product is

$$\begin{aligned}\text{DFT}^{-1}(\mathbf{X}_1[k]\,\mathbf{X}_2[k]) &= \frac{1}{N_0}\sum_{k=0}^{N_0-1}(\mathbf{X}_1[k]\,\mathbf{X}_2[k])e^{jk\Omega_0 n}\\ &= \frac{1}{N_0}\sum_{k=0}^{N_0-1} e^{jk\Omega_0 n}\left[\sum_{n_1=0}^{N_0-1} x_1[n_1]\,e^{-jk\Omega_0 n_1}\right]\\ &\quad\cdot\left[\sum_{n_2=0}^{N_0-1} x_2[n_2]\,e^{-jk\Omega_0 n_2}\right].\end{aligned} \tag{7.174}$$

Rearranging the order of the summations gives

$$\text{DFT}^{-1}(\mathbf{X}_1[k]\,\mathbf{X}_2[k]) = \frac{1}{N_0}\sum_{n_1=0}^{N_0-1}\sum_{n_2=0}^{N_0-1} x_1[n_1]\,x_2[n_2]\sum_{k=0}^{N_0-1} e^{jk\Omega_0(n-n_1-n_2)}. \tag{7.175}$$

In view of the orthogonality property given by Eq. (7.133) and recalling that $\Omega_0 = 2\pi/N_0$, Eq. (7.175) reduces to

$$\begin{aligned}&\text{DFT}^{-1}(\mathbf{X}_1[k]\,\mathbf{X}_2[k])\\ &= \frac{1}{N_0}\sum_{n_1=0}^{N_0-1}\sum_{n_2=0}^{N_0-1} x_1[n_1]\,x_2[n_2]\,N_0\,\delta[(n-n_1-n_2)_{N_0}]\\ &= \sum_{n_1=0}^{N_0-1} x_1[n_1]\,x_2[(n-n_1)_{N_0}],\end{aligned} \tag{7.176}$$

where $(n-n_1)_{N_0}$ means $(n-n_1)$ reduced mod N_0 (i.e., reduced by the largest integer multiple of N_0 without $(n-n_1)$ becoming negative).

7-15.5 DFT and Cyclic Convolution

Because of the mod N_0 reduction cycle, the expression on the right-hand side of Eq. (7.176) is called the ***cyclic*** or circular convolution of signals $x_1[n]$ and $x_2[n]$. The terminology helps distinguish it from the traditional *linear* convolution of two nonperiodic signals. As noted earlier, a DFT in effect converts a nonperiodic signal into a periodic-looking signal, so $x_1[n]$ and $x_2[n]$ are periodic with period N_0.

The symbol commonly used to denote cyclic convolution is ©. Combining Eqs. (7.174) and (7.176) leads to

$$\begin{aligned}y_c[n] = x_1[n] \copyright x_2[n] &= \sum_{n_1=0}^{N_0-1} x_1[n_1]\,x_2[(n-n_1)_{N_0}]\\ &= \text{DFT}^{-1}(\mathbf{X}_1[k]\,\mathbf{X}_2[k])\\ &= \frac{1}{N_0}\sum_{k=0}^{N_0-1}\mathbf{X}_1[k]\,\mathbf{X}_2[k]\,e^{jk\Omega_0 n},\end{aligned} \tag{7.177}$$

where $\Omega_0 = 2\pi/N_0$.

The cyclic convolution $y_c[n]$ can certainly by computed by applying Eq. (7.177), but it can also be computed from the linear convolution $x_1[n] * x_2[n]$ by aliasing the latter. To illustrate, suppose $x_1[n]$ and $x_2[n]$ are both of duration N_0. The linear convolution of the two signals

$$y[n] = x_1[n] * x_2[n] \tag{7.178}$$

is of duration $2N_0 - 1$. Aliasing $y[n]$ means defining $z[n]$, the aliased version of $y[n]$, as

$$\begin{aligned} z[0] &= y[0] + y[0 + N_0] \\ z[1] &= y[1] + y[1 + N_0] \\ &\vdots \\ z[N_0 - 1] &= y[N_0 - 1]. \end{aligned} \tag{7.179}$$

The aliasing process leads to the result that $z[n]$ is the cyclic convolution of $x_1[n]$ and $x_2[n]$:

$$y_c[n] = z[n] = x_1[n] \circledcirc x_2[n]. \tag{7.180}$$

Example 7-34: Cyclic Convolution

Given the two signals

$$\begin{aligned} x_1[n] &= \{\underline{2}, 1, 4, 3\}, \\ x_2[n] &= \{\underline{5}, 3, 2, 1\}, \end{aligned}$$

compute the cyclic convolution of the two signals by

(a) applying the DFT method;

(b) applying the aliasing of the linear convolution method.

Solution:

(a) With $N_0 = 4$ and $\Omega_0 = 2\pi/N_0 = \pi/2$, application of Eq. (7.169a) to $x_1[n]$ and $x_2[n]$ leads to

$$\begin{aligned} \mathbf{X}_1[k] &= \{10,\ -2 + j2,\ 2,\ -2 - j2\}, \\ \mathbf{X}_2[k] &= \{11,\ 3 - j2,\ 3,\ 3 + j2\}. \end{aligned}$$

The point-by-point product of $\mathbf{X}_1[k]$ and $\mathbf{X}_2[k]$ is

$$\begin{aligned} &\mathbf{X}_1[k]\, \mathbf{X}_2[k] \\ &= \{10 \times 11,\ (-2 + j2)(3 - j2),\ 2 \times 3,\ (-2 - j2)(3 + j2)\} \\ &= \{110,\ -2 + j10,\ 6,\ -2 - j10\}. \end{aligned}$$

Application of Eq. (7.177) leads to

$$x_1[n] \circledcirc x_2[n] = \{\underline{28}, 21, 30, 31\}.$$

(b) The linear convolution of $x_1[n]$ and $x_2[n]$ is

$$y[n] = \{\underline{2}, 1, 4, 3\} * \{\underline{5}, 3, 2, 1\} = \{\underline{10}, 11, 27, 31, 18, 10, 3\}.$$

Per Eq. (7.179),

$$\begin{aligned} z[n] &= \{y[0] + y[4],\ \ y[1] + y[5],\ \ y[2] + y[6],\ \ y[3]\} \\ &= \{10 + 18,\ 11 + 10,\ 27 + 3,\ 31\} = \{28, 21, 30, 31\}. \end{aligned}$$

Hence, by Eq. (7.180),

$$x_1[n] \circledcirc x_2[n] = z[n] = \{28, 21, 30, 31\},$$

which is the same answer obtained in part (a).

7-15.6 DFT and Linear Convolution

In the preceding subsection, we examined how the DFT can be used to compute the cyclic convolution of two discrete-time signals (Eq. (7.177)). The same method can be applied to compute the linear convolution of the two signals, provided a preparatory step of ***zero-padding*** the two signals is applied first.

Let us suppose that signal $x_1[n]$ is of duration N_1 and signal $x_2[n]$ is of duration N_2, and we are interested in computing their linear convolution

$$y[n] = x_1[n] * x_2[n].$$

The duration of $y[n]$ is

$$N_c = N_1 + N_2 - 1. \tag{7.181}$$

Next, we zero-pad $x_1[n]$ and $x_2[n]$ so that their durations are equal to or greater than N_c. As we will see in Section 7-16 on how the fast Fourier transform (FFT) is used to compute the DFT, it is advantageous to choose the total length of the zero-padded signals to be M such that $M \geq N_c$, and simultaneously M is a power of 2.

The zero-padded signals are defined as

$$x'_1[n] = \{\underbrace{x_1[n]}_{N_1}, \underbrace{0, \ldots, 0}_{M-N_1}\}, \tag{7.182a}$$

$$x'_2[n] = \{\underbrace{x_2[n]}_{N_2}, \underbrace{0, \ldots, 0}_{M-N_2}\}, \tag{7.182b}$$

and their M-point DFTs are $\mathbf{X}'_1[k]$ and $\mathbf{X}'_2[k]$, respectively. The linear convolution $y[n]$ can now be computed by a modified version of Eq. (7.177), namely

$$y_c[n] = x'_1[n] * x'_2[n] = \text{DFT}^{-1}\{\mathbf{X}'_1[k]\,\mathbf{X}'_2[k]\} = \frac{1}{M}\sum_{k=0}^{M-1} \mathbf{X}'_1[k]\,\mathbf{X}'_2[k]\,e^{j2\pi nk/M}. \quad (7.183)$$

Example 7-35: DFT Convolution

Given signals $x_1[n] = \{\underline{4}, 5\}$ and $x_2[n] = \{\underline{1}, 2, 3\}$, (a) compute their convolution in discrete time, and (b) compare the result with the DFT relation given by Eq. (7.183).

Solution:

(a) Application of Eq. (7.51a) gives

$$x_1[n] * x_2[n] = \sum_{i=0}^{3} x_1[i]\,x_2[n-i] = \{\underline{4}, 13, 22, 15\}.$$

(b) Since $x_1[n]$ is of length $N_1 = 2$ and $x_2[n]$ is of length $N_2 = 3$, their convolution is of length

$$N_c = N_1 + N_2 - 1 = 2 + 3 - 1 = 4.$$

Hence, we need to zero-pad $x_1[n]$ and $x_2[n]$ as

$$x'_1[n] = \{\underline{4}, 5, 0, 0\}$$

and

$$x'_2[n] = \{\underline{1}, 2, 3, 0\}.$$

From Eq. (7.169a) with $\Omega_0 = 2\pi/N_c = 2\pi/4 = \pi/2$, the 4-point DFT of $x'_1[n] = \{\underline{4}, 5, 0, 0\}$ is

$$\mathbf{X}_1[k] = \sum_{n=0}^{3} x'_1[n]\,e^{-jk\pi n/2}, \qquad k = 0, 1, 2, 3,$$

which gives

$$\mathbf{X}'_1[0] = 4(1) + 5(1) + 0(1) + 0(1) = 9,$$
$$\mathbf{X}'_1[1] = 4(1) + 5(-j) + 0(-1) + 0(j) = 4 - j5,$$
$$\mathbf{X}'_1[2] = 4(1) + 5(-1) + 0(1) + 0(-1) = -1,$$

and

$$\mathbf{X}'_1[3] = 4(1) + 5(j) + 0(-1) + 0(j) = 4 + j5.$$

Similarly, the 4-point DFT of $x'_2[n] = \{\underline{1}, 2, 3, 0\}$ gives

$$\mathbf{X}'_2[0] = 6,$$
$$\mathbf{X}'_2[1] = -2 - j2,$$
$$\mathbf{X}'_2[2] = 2,$$

and

$$\mathbf{X}'_2[3] = -2 + j2.$$

Multiplication of corresponding pairs gives

$$\mathbf{X}'_1[0]\,\mathbf{X}'_2[0] = 9 \times 6 = 54,$$
$$\mathbf{X}'_1[1]\,\mathbf{X}'_2[1] = (4 - j5)(-2 - j2) = -18 + j2,$$
$$\mathbf{X}'_1[2]\,\mathbf{X}'_2[2] = -1 \times 2 = -2,$$

and

$$\mathbf{X}'_1[3]\,\mathbf{X}'_2[3] = (4 + j5)(-2 + j2) = -18 - j2.$$

Application of Eq. (7.183) gives

$$y_c[n] = x'_1[n] * x'_2[n] = \frac{1}{N_c}\sum_{k=0}^{N_c-1} \mathbf{X}'_1[k]\,\mathbf{X}'_2[k]\,e^{j2\pi nk/N_c} = \frac{1}{4}\sum_{k=0}^{3} \mathbf{X}'_1[k]\,\mathbf{X}'_2[k]\,e^{jk\pi n/2}.$$

Evaluating the summation for $n = 0, 1, 2$ and 3 leads to

$$y_c[n] = x'_1[n] * x'_2[n] = \{\underline{4}, 13, 22, 15\},$$

which is identical to the answer obtained earlier in part (a). For simple signals like those in this example, the DFT method involves many more steps than does the straightforward convolution method of part (a), but for the type of signals used in practice, the DFT method is computationally superior.

Concept Question 7-21: What is the difference between DFT and DTFS? (See S²)

Concept Question 7-22: How is the DFT related to the DTFT? (See S²)

Exercise 7-25: Compute the 4-point DFT of $\{\underline{4}, 3, 2, 1\}$.

Answer: $\{10, (2 - j2), 2, (2 + j2)\}$. (See S²)

Table 7-9: Comparison of number of complex computations (including multiplication by ± 1 and $\pm j$) required by a standard DFT and an FFT.

N_0	Multiplication		Additions	
	Standard DFT	FFT	Standard DFT	FFT
2	4	1	2	2
4	16	4	12	8
8	64	12	56	24
16	256	32	240	64
⋮	⋮	⋮	⋮	⋮
512	262,144	2,304	261,632	4,608
1,024	1,048,576	5,120	1,047,552	10,240
2,048	4,194,304	11,264	4,192,256	22,528
N_0	N_0^2	$\frac{N_0}{2}\log_2 N_0$	$N_0(N_0-1)$	$N_0 \log_2 N_0$

7-16 Fast Fourier Transform (FFT)

▶ The Fast Fourier Transform (FFT) is a computational algorithm used to compute the Discrete Fourier Transforms (DFT) of discrete signals. Strictly speaking, the FFT is not a transform, but rather an algorithm for computing the transform. ◀

As was mentioned earlier, the fast Fourier transform (FFT) is a highly efficient algorithm for computing the DFT of discrete time signals. An N_0-point DFT performs a linear transformation from an N_0-long discrete-time vector, namely $x[n]$, into an N_0-long frequency domain vector $\mathbf{X}[k]$ for $k = 0, 1, \ldots, N_0 - 1$. Computation of each $\mathbf{X}[k]$ involves N_0 complex multiplications, so the total number of multiplications required to perform the DFT for all $\mathbf{X}[k]$ is N_0^2. This is in addition to $N_0(N_0 - 1)$ complex additions. For $N_0 = 512$, for example, *direct* implementation of the DFT operation requires 262,144 multiplications and 261,632 complex additions.

Contrast these large number of ***multiplications and additions*** (MADs) with the number required using the FFT algorithm: for N_0 large, the number of complex multiplications is reduced from N_0^2 to approximately $(N_0/2)\log_2 N_0$, which is only 2304 complex multiplications for $N_0 = 512$. For complex additions, the number is reduced from $N_0(N_0 - 1)$ to $N_0 \log_2 N_0$ or 4608 for $N_0 = 512$. These reductions, thanks to the efficiency of the FFT algorithm, are on the order of 100 for multiplications and on the order of 50 for addition. The reduction ratios become increasingly more impressive at larger values of N_0 (Table 7-9).

The computational efficiency of the FFT algorithm relies on a "divide and conquer" concept. An N_0-point DFT is ***decomposed*** (divided) into two $(N_0/2)$-point DFTs. Each of the $(N_0/2)$-point DFTs is decomposed further into two $(N_0/4)$-point DFTs. The decomposition process, which is continued until it reaches the 2-point DFT level, is illustrated in the next subsections.

7-16.1 2-Point DFT

For notational efficiency, we introduce the symbols

$$W_{N_0} = e^{-j2\pi/N_0} = e^{-j\Omega_0}, \tag{7.184a}$$

$$W_{N_0}^{nk} = e^{-j2\pi nk/N_0} = e^{-jnk\Omega_0}, \tag{7.184b}$$

and

$$W_{N_0}^{-nk} = e^{j2\pi nk/N_0} = e^{jnk\Omega_0}, \tag{7.184c}$$

where $\Omega_0 = 2\pi/N_0$. Using this shorthand notation, the summations for the DFT, and its inverse given by Eq. (7.169), assume the form

$$\mathbf{X}[k] = \sum_{n=0}^{N_0-1} x[n]\, W_{N_0}^{nk}, \qquad k = 0, 1, \ldots, N_0-1, \tag{7.185a}$$

and

$$x[n] = \frac{1}{N_0} \sum_{k=0}^{N_0-1} \mathbf{X}[k] W_{N_0}^{-nk}, \qquad n = 0, 1, \ldots, N_0 - 1. \tag{7.185b}$$

In this form, the N_0-long vector $\mathbf{X}[k]$ is given in terms of the N_0-long vector $x[n]$, and vice versa, with $W_{N_0}^{nk}$ and $W_{N_0}^{-nk}$ acting as ***weighting coefficients***.

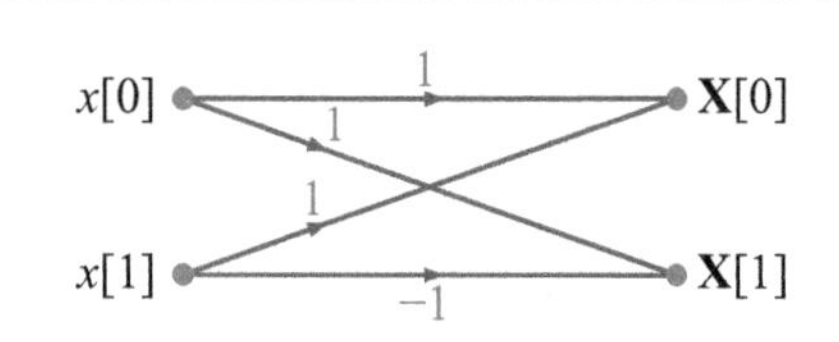

Figure 7-27: Signal flow graph for a 2-point DFT.

For a 2-point DFT,

$$N_0 = 2,$$
$$\Omega_0 = 2\pi/2 = \pi,$$
$$W_2^{0k} = e^{-j0} = 1,$$

and

$$W_2^{1k} = e^{-jk\pi} = (-1)^k.$$

Hence, Eq. (7.185a) yields the following expressions for $\mathbf{X}[0]$ and $\mathbf{X}[1]$:

$$\mathbf{X}[0] = x[0] + x[1] \tag{7.186a}$$

and

$$\mathbf{X}[1] = x[0] - x[1], \tag{7.186b}$$

which can be combined into the compact form

$$\mathbf{X}[k] = x[0] + (-1)^k\, x[1], \qquad k = 0, 1. \tag{7.187}$$

The equations for $\mathbf{X}[0]$ and $\mathbf{X}[1]$ can be represented by the ***signal flow graph*** shown in Fig. 7-27, which is often called a ***butterfly diagram***.

7-16.2 4-Point DFT

For a 4-point DFT, $N_0 = 4$, $\Omega_0 = 2\pi/4 = \pi/2$, and

$$W_{N_0}^{nk} = W_4^{nk} = e^{-jnk\pi/2} = (-j)^{nk}. \tag{7.188}$$

From Eq. (7.185a), we have

$$\begin{aligned} \mathbf{X}[k] &= \sum_{n=0}^{3} x[n]\, W_4^{nk} \\ &= x[0] + x[1]\, W_4^{1k} + x[2]\, W_4^{2k} + x[3]\, W_4^{3k}, \\ &\qquad k = 0, 1, 2, 3. \end{aligned} \tag{7.189}$$

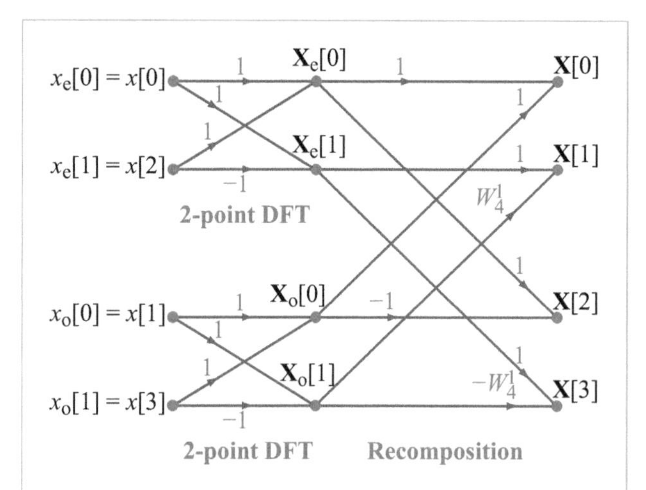

Figure 7-28: Signal flow graph for a 4-point DFT. Weighting coefficient $W_4^1 = -j$. Note that summations occur only at red intersection points.

Upon evaluating W_4^{1k}, W_4^{2k}, and W_4^{3k} and the relationships between them, Eq. (7.189) can be cast in the form

$$\mathbf{X}[k] = \underbrace{[x[0] + (-1)^k\, x[2]]}_{\text{2-point DFT}} + W_4^{1k} \underbrace{[x[1] + (-1)^k\, x[3]]}_{\text{2-point DFT}}, \tag{7.190}$$

which consists of two 2-point DFTs: one that includes values of $x[n]$ for ***even values of*** n, and another for ***odd values of*** n. At this point, it is convenient to define $x_e[n]$ and $x_o[n]$ as $x[n]$ at ***even*** and ***odd*** times:

$$x_e[n] = x[2n], \qquad n = 0, 1, \tag{7.191a}$$
$$x_o[n] = x[2n+1], \qquad n = 0, 1. \tag{7.191b}$$

Thus, $x_e[0] = x[0]$ and $x_e[1] = x[2]$ and, similarly, $x_o[0] = x[1]$ and $x_o[1] = x[3]$. When expressed in terms of $x_e[n]$ and $x_o[n]$, Eq. (7.190) becomes

$$\mathbf{X}[k] = \underbrace{[x_e[0] + (-1)^k\, x_e[1]]}_{\text{2-point DFT of } x_e[n]} + W_4^{1k} \underbrace{[x_o[0] + (-1)^k\, x_o[1]]}_{\text{2-point DFT of } x_o[n]},$$
$$k = 0, 1, 2, 3. \tag{7.192}$$

The FFT computes the 4-point DFT by computing the two 2-point DFTs, followed by a recomposition step that involves multiplying the even 2-point DFT by W_4^{1k} and then adding it to the odd 2-point DFT. The entire process is depicted by the signal flow graph shown in Fig. 7-28. In the graph, Fourier coefficients $\mathbf{X}_e[0]$ and $\mathbf{X}_e[1]$ represent the outputs of the even 2-point DFT, and similarly, $\mathbf{X}_o[0]$ and $\mathbf{X}_o[1]$ represent the outputs of the odd 2-point DFT.

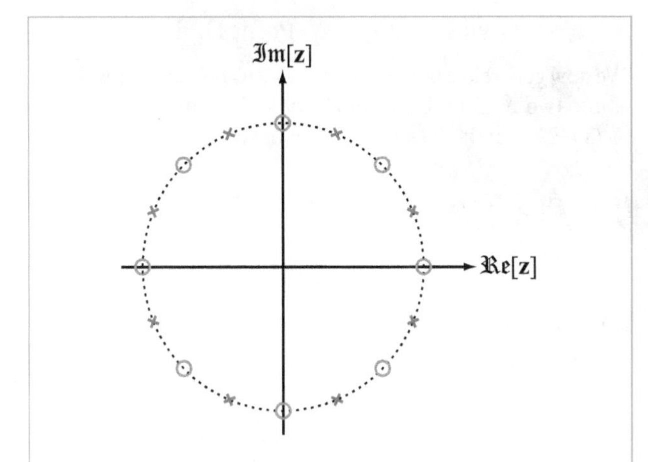

Figure 7-29: A 16-point DFT is $\mathbf{X}(\mathbf{z})$ evaluated at 16 points equally spaced on the unit circle.

7-16.3 16-Point DFT

We now show how to compute a 16-point DFT using two 8-point DFTs and 8 *multiplications and additions* (*MADs*). This *divides* the 16-point DFT into two 8-point DFTs, which in turn can be *divided* into four 4-point DFTs, which are just additions and subtractions. This *conquers* the 16-point DFT by *dividing* it into 4-point DFTs and additional MADs.

Dividing a 16-point DFT

Recall that the 16-point DFT is the $\mathbf{z}$-transform evaluated at 16 equally-spaced points on the unit circle, as shown in Fig. 7-29. Of the 16 points, 8 are depicted by blue circles and the other 8 are depicted by red crosses. Those depicted by circles are points at which $\mathbf{X}(\mathbf{z})$ is evaluated for an 8-point DFT, and the 8 points depicted by crosses are the points at which $\mathbf{X}(\mathbf{z})$ is evaluated for a *modulated 8-point DFT*. Thus, the 16-point DFT can be computed as an 8-point DFT (for even values of k) and as a modulated 8-point DFT (for odd values of k).

Computation at even indices

We consider even and odd indices k separately.

For even values of k, we can write $k = 2k'$ and split the 16-point DFT summation into two summations:

$$\mathbf{X}[2k'] = \sum_{n=0}^{15} x[n]\, e^{-j2\pi(2k'/16)n} = \sum_{n=0}^{7} x[n]\, e^{-j2\pi(2k'/16)n} + \sum_{n=8}^{15} x[n]\, e^{-j2\pi(2k'/16)n}. \tag{7.193}$$

Changing variables from n to $n' = n - 8$ in the second summation, and recognizing $2k'/16 = k'/8$, gives

$$\begin{aligned}\mathbf{X}[2k'] &= \sum_{n=0}^{7} x[n]\, e^{-j2\pi(k'/8)n} + \sum_{n'=0}^{7} x[n'+8]\, e^{-j2\pi(k'/8)(n'+8)} \\ &= \sum_{n=0}^{7} (x[n] + x[n+8]) e^{-j2\pi(k'/8)n} \\ &= \text{DFT}(\{x[n] + x[n+8],\ n = 0, \ldots, 7\}).\end{aligned} \tag{7.194}$$

So for even values of k, the 16-point DFT of $x[n]$ is the 8-point DFT of $\{x[n] + x[n+8],\ n = 0, \ldots, 7\}$. The $\mathbf{z}$-transform is computed at the circles of Fig. 7-29.

Computation at odd indices

For odd values of k, we can write $k = 2k' + 1$ and split the 16-point DFT summation into two summations:

$$\begin{aligned}\mathbf{X}[2k'+1] &= \sum_{n=0}^{15} x[n]\, e^{-j2\pi(2k'+1)/16n} \\ &= \sum_{n=0}^{7} x[n]\, e^{-j2\pi(2k'+1)/16n} \\ &\quad + \sum_{n=8}^{15} x[n]\, e^{-j2\pi(2k'+1)/16n}.\end{aligned} \tag{7.195}$$

Changing variables from n to $n' = n - 8$ in the second summation, and recognizing that $e^{-j2\pi 8/16} = -1$ and

$$\frac{2k'+1}{16} = \frac{k'}{8} + \frac{1}{16},$$

gives

$$\begin{aligned}\mathbf{X}[2k'+1] &= \sum_{n=0}^{7} (x[n]\, e^{-j2\pi(1/16)n}) e^{-j2\pi(k'/8)n} \\ &\quad + \sum_{n'=0}^{7} (x[n'+8]\, e^{-j2\pi(1/16)(n'+8)}) e^{-j2\pi(k'/8)(n'+8)} \\ &= \sum_{n=0}^{7} e^{-j2\pi(1/16)n} (x[n] - x[n+8]) e^{-j2\pi(k'/8)n} \\ &= \text{DFT}(\{e^{-j2\pi(1/16)n}(x[n] - x[n+8]), \\ &\qquad n = 0, \ldots, 7\}).\end{aligned} \tag{7.196}$$

So for odd values of k, the 16-point DFT of $x[n]$ is the 8-point DFT of $\{e^{-j(2\pi/16)n}(x[n]-x[n+8]),\ n=0,\dots,7\}$. The signal $\{x[n]-x[n+8],\ n=0,\dots,7\}$ has been *modulated* through multiplication by $e^{-j(2\pi/16)n}$. The modulation shifts the frequencies, so that the **z**-transform is computed at the crosses of Fig. 7-29. The multiplications by $e^{-j(2\pi/16)n}$ are known as *twiddle multiplications* (mults) by the *twiddle factors* $e^{-j(2\pi/16)n}$.

Example 7-36: Dividing an 8-Point DFT into Two 4-Point DFTs

Divide the 8-point DFT of $\{7,1,4,2,8,5,3,6\}$ into two 4-point DFTs and twiddle mults.

Solution:

- For even values of index k, we have

$$\begin{aligned}\mathbf{X}[0,2,4,6] &= \text{DFT}(\{7+8,\ 1+5,\ 4+3,\ 2+6\})\\ &= \{36,\ 8+j2,\ 8,\ 8-j2\}.\end{aligned}$$

  For odd index values, we need twiddle mults. The twiddle factors are given by $\{e^{-j2\pi n/8}\}$ for $n=0,\ 1,\ 2$, and 3, which reduce to $\{1, \frac{\sqrt{2}}{2}(1-j), -j, \frac{\sqrt{2}}{2}(-1-j)\}$.

- Implementing the twiddle mults gives

$$\begin{aligned}&\{7-8,\ 1-5,\ 4-3,\ 2-6\}\\ &\times\left\{1, \frac{\sqrt{2}}{2}(1-j), -j, \frac{\sqrt{2}}{2}(-1-j)\right\}\\ &=\{-1, 2\sqrt{2}(-1+j), -j, 2\sqrt{2}(1+j)\}.\end{aligned}$$

- For odd values of index k, we have

$$\begin{aligned}\mathbf{X}[1,3,5,7] &= \text{DFT}(\{-1, 2\sqrt{2}(-1+j), -j, 2\sqrt{2}(1+j)\})\\ &= \{-1+j4.66, -1+j6.66, -1-j6.66, -1-j4.66\}.\end{aligned}$$

- Combining these results for even and odd k gives

$$\begin{aligned}&\text{DFT}(\{7,1,4,2,8,5,3,6\})\\ &=\{36,\ -1+j4.7,\ 8+j2,\ -1+j6.7,\ 8,\\ &\quad -1-j6.7,\ 8-j2,\ -1-j4.7\}.\end{aligned}$$

  Note the conjugate symmetry in the second and third lines: $\mathbf{X}[7]=\mathbf{X}^*[1]$, $\mathbf{X}[6]=\mathbf{X}^*[2]$, and $\mathbf{X}[5]=\mathbf{X}^*[3]$.

- This result agrees with direct computation using

```
fft([7 1 4 2 8 5 3 6]).
```

7-16.4 Dividing Up a $2N$-Point DFT

We now generalize the procedure to a $2N$-point DFT by dividing it into two N-point DFTs and N twiddle mults.

(1) For even indices $k=2k'$ we have:

$$\begin{aligned}\mathbf{X}[2k'] &= \sum_{n=0}^{N-1}(x[n]+x[n+N])e^{-j2\pi(k'/N)n}\\ &= \text{DFT}\{x[n]+x[n+N],\ n=0,1,\dots,N-1\}.\end{aligned} \tag{7.197}$$

(2) For odd indices $k=2k'+1$ we have:

$$\begin{aligned}\mathbf{X}[2k'+1] &= \sum_{n=0}^{N-1} e^{-j2\pi(1/(2N))n}(x[n]-x[n+N])e^{-j2\pi(k'/N)n}\\ &= \text{DFT}\{e^{-j2\pi(1/(2N))n}(x[n]-x[n+N])\}.\end{aligned} \tag{7.198}$$

Thus, a $2N$-point DFT can be divided into

- Two N-point DFTs,
- N multiplications by twiddle factors $e^{-j2\pi(1/2N)n}$, and
- $2N$ additions and subtractions.

7-16.5 Dividing and Conquering

Now suppose N *is a power of two;* e.g., $N=1024=2^{10}$. In that case, we can apply the algorithm of the previous subsection recursively to divide an N-point DFT into two $N/2$-point DFTs, then into four $N/4$-point DFTs, then into eight $N/8$-point DFTs, and so on until we reach the following 4-point DFTs:

$$\begin{aligned}\mathbf{X}[0] &= x[0]+x[1]+x[2]+x[3],\\ \mathbf{X}[1] &= x[0]-jx[1]-x[2]+jx[3],\\ \mathbf{X}[2] &= x[0]-x[1]+x[2]-x[3],\\ \mathbf{X}[3] &= x[0]+jx[1]-x[2]-jx[3].\end{aligned} \tag{7.199}$$

At each stage, half of the DFTs are modulated, requiring $N/2$ multiplications. So if N is a power of 2, then an N-point DFT computed using the FFT will require approximately $(N/2)\log_2(N)$ multiplications and $N\log_2(N)$ additions. These can be reduced slightly by recognizing that some multiplications are simply by ± 1 and $\pm j$.

To illustrate the computational significance of the FFT, suppose we wish to compute a 32768-point DFT. Direct computation using Eq. (7.169a) would require $(32768)^2 \approx 1.1\times 10^9$ MADs. In contrast, computation using the FFT would require less than $\frac{32768}{2}\log_2(32768)\approx 250{,}000$ MADs, representing a computational saving of a factor of 4000!

7-17 Cooley-Tukey FFT

The Cooley-Tukey FFT was originally derived by the German mathematician Carl Friedrich Gauss, but its first implementation using computers (with punch cards and tape drives) was carried out by Cooley and Tukey at IBM in 1965. With N defined as the product N_1N_2, the Cooley-Tukey FFT divides an N_1N_2-point DFT into

- N_1 N_2-point DFTs,
- N_2 N_1-point DFTs, and
- $(N_1 - 1)(N_2 - 1)$ twiddle mults.

The goal is to compute the $N = N_1N_2$-point DFT

$$\mathbf{X}[k] = \sum_{n=0}^{N-1} x[n]\, e^{-j2\pi nk/N}, \qquad k = 0, 1, \ldots, N-1. \tag{7.200}$$

N should be chosen to be an integer with a large number of small factors. Then an N-point DFT can be divided into many N_1-point DFTs and N_2 point DFTs. If $N_1 = N_3N_4$ and $N_2 = N_5N_6$, then each N_1-point DFT can be divided into many N_3-point DFTs and N_4-point DFTs, and each N_2-point DFT can be divided into many N_5-point DFTs and N_6-point DFTs, each of which can in turn be further divided. The extreme case of this is when N is a power of two, the case presented in Section 7-16.5.

7-17.1 Coarse and Vernier Indices

The implementation procedure of the Cooley-Tukey FFT starts by defining

(a) the coarse indices n_2 and k_1 and

(b) the vernier indices n_1 and k_2

as the quotients and remainders, respectively, after dividing n by N_1 and k by N_2. That is,

$$\begin{aligned} n &= n_1 + N_1n_2 \text{ for } \begin{cases} n_1 = 0, 1, \ldots, N_1 - 1, \\ n_2 = 0, 1, \ldots, N_2 - 1, \end{cases} \\ k &= k_2 + N_2k_1 \text{ for } \begin{cases} k_1 = 0, 1, \ldots, N_1 - 1, \\ k_2 = 0, 1, \ldots, N_2 - 1. \end{cases} \end{aligned} \tag{7.201}$$

These names come from coarse and fine (vernier) adjustments for gauges and microscopes.

For example, for $N = 6$ with $N_1 = 3$ and $N_2 = 2$, the indices n, n_1, n_2, k, k_1, k_2 assume the values listed in Table 7-10. For example, dividing $n = 5$ by $N_1 = 3$ gives a remainder of $n_1 = 2$ and a quotient of $n_2 = 1$. The DFT $\mathbf{X}[k]$ can then be rewritten as

Table 7-10: Coarse and vernier indices for $N_1 = 3$ and $N_2 = 2$.

n	0	1	2	3	4	5
n_1	0	1	2	0	1	2
n_2	0	0	0	1	1	1
k	0	1	2	3	4	5
k_1	0	0	1	1	2	2
k_2	0	1	0	1	0	1

$$\mathbf{X}[k_2 + N_2k_1] = \sum_{n_1=0}^{N_1-1} \sum_{n_2=0}^{N_2-1} x[n_1 + N_1n_2]\, e^{-j(2\pi/N)nk}. \tag{7.202}$$

The product of indices nk in the DFT exponent can be expressed as

$$\begin{aligned} nk &= (n_1 + N_1n_2)(k_2 + N_2k_1) \\ &= n_1k_2 + N_1n_2k_2 + N_2n_1k_1 + N_1N_2n_2k_1, \end{aligned} \tag{7.203}$$

which, together with replacing N with N_1N_2 in the DFT exponent, leads to

$$\begin{aligned} e^{-j(2\pi/N)nk} &= e^{-j(2\pi/N)n_1k_2} e^{-j(2\pi/N)N_1n_2k_2} \\ &\quad \times e^{-j(2\pi/N)N_2n_1k_1} e^{-j(2\pi/N)N_1N_2n_2k_1} \\ &= e^{-j(2\pi/N)n_1k_2} e^{-j(2\pi/N_2)n_2k_2} e^{-j(2\pi/N_1)n_1k_1}. \end{aligned} \tag{7.204}$$

The double summation in Eq. (7.202) can be rewritten as

$$\begin{aligned} \mathbf{X}[k_2 + N_2k_1] = \underbrace{\sum_{n_1=0}^{N_1-1} e^{-j(2\pi/N_1)n_1k_1}}_{\mathbf{N_1\text{-point DFT}}} \Bigg[\underbrace{e^{-j(2\pi/N)n_1k_2}}_{\textbf{twiddle factor}} \\ \times \underbrace{\sum_{n_2=0}^{N_2-1} e^{-j(2\pi/N_2)n_2k_2} x[n_1 + N_1n_2]}_{\mathbf{N_2\text{-point DFT for each } n_1}} \Bigg]. \end{aligned} \tag{7.205}$$

Hence, the N-point DFT is now composed by N_2 N_1-point DFTs, N_1 N_2-point DFTs, and twiddle mults of twiddle factors $e^{-j(2\pi/N)n_1k_2}$.

7-17.2 Computational Procedure

1. Compute N_1 (for each n_1) N_2-point DFTs of $x[n_1+N_1n_2]$.
2. Multiply the result by *twiddle factors* $e^{-j(2\pi/N)n_1k_2}$.
3. Compute N_2 (for each k_2) N_1-point DFTs of the previous result.

Note that if either $n_1 = 0$ or $k_2 = 0$, then $e^{-j(2\pi/N)n_1k_2} = 1$. Hence, the number of twiddle mults is not N_1N_2 but $(N_1 - 1)(N_2 - 1)$.

7-17.3 2-D Visualization

We can visualize the Cooley-Tukey FFT method by writing both $x[n]$ and X_k as $(N_1 \times N_2)$ arrays, which leads to the following procedure:

1. Map $x[n]$ to x_{n_1,n_2} using $x_{n_1,n_2} = x[n_1 + N_1n_2]$.
2. Determine the N_2-point DFT of each row (for fixed n_1).
3. Multiply this array point-by-point by the array of twiddle factors $e^{-j(2\pi/N)n_1k_2}$.
4. Determine the N_1-point DFT of each column (for fixed k_2). The result of this is $\mathbf{X}[k_1, k_2]$.
5. Map $\mathbf{X}[k_1, k_2] = \mathbf{X}[k_2 + N_2k_1]$.

Note that the mappings used for $x[n]$ and $\mathbf{X}[k]$ are different, but they can be made identical by taking the transpose (exchanging each column for the correspondingly numbered row) of $\mathbf{X}[k_1, k_2]$ before the final mapping. This is called ***index shuffling***. We may exchange "row" and "column" above.

Example 7-37: 4-Point DFT to 2 x 2 Array

Use the 2-D visualization technique to map the general 4-point DFT of $\{x[0], x[1], x[2], x[3]\}$ to 4 2-point DFTs and twiddle mults. Recall that 2-point DFTs involve only additions and subtractions.

Solution:

(1) Map $\{x[0], x[1], x[2], x[3]\}$ to a 2×2 array:

$$\{x[0], x[1], x[2], x[3]\} \rightarrow \begin{bmatrix} x[0] & x[1] \\ x[2] & x[3] \end{bmatrix} = \begin{bmatrix} x[0] \\ x[2] \end{bmatrix} \begin{bmatrix} x[1] \\ x[3] \end{bmatrix}.$$

(2) Compute 2-point DFTs of each column:

$$\begin{bmatrix} x[0]+x[2] \\ x[0]-x[2] \end{bmatrix} \begin{bmatrix} x[1]+x[3] \\ x[1]-x[3] \end{bmatrix} = \begin{bmatrix} x[0]+x[2] & x[1]+x[3] \\ x[0]-x[2] & x[1]-x[3] \end{bmatrix}.$$

(3) Multiply by twiddle factors:

$$\times \begin{bmatrix} 1 & 1 \\ 1 & -j \end{bmatrix} = \begin{bmatrix} x[0]+x[2] & x[1]+x[3] \\ x[0]-x[2] & -j(x[1]-x[3]) \end{bmatrix}.$$

(4) Compute the 2-point DFTs of each row:

$$\begin{bmatrix} (x[0]+x[2])+(x[1]+x[3]) & (x[0]+x[2])-(x[1]+x[3]) \end{bmatrix}$$

$$\begin{bmatrix} (x[0]-x[2])-j(x[1]-x[3]) & (x[0]-x[2])+j(x[1]+x[3]) \end{bmatrix}$$

Read off $\{\mathbf{X}[0], \mathbf{X}[1], \mathbf{X}[2], \mathbf{X}[3]\}$. Note the *index shuffling*: the index locations in the final array are the transpose of the index locations in the first array.

$$\begin{bmatrix} \mathbf{X}[0] & \mathbf{X}[2] \end{bmatrix} \begin{bmatrix} \mathbf{X}[1] & \mathbf{X}[3] \end{bmatrix} = \begin{bmatrix} \mathbf{X}[0] & \mathbf{X}[2] \\ \mathbf{X}[1] & \mathbf{X}[3] \end{bmatrix} \rightarrow \{\mathbf{X}[0], \mathbf{X}[1], \mathbf{X}[2], \mathbf{X}[3]\}$$

7-17.4 Radix-2 Cooley-Tukey FFTs

As noted at the beginning of this section, N should be chosen to be an integer with a large number of small factors. The extreme case of this is when N is a power of two, the case presented in Section 7-16.5. We now specialize the Cooley-Tukey FFT to this case. We will obtain two different forms of the FFT, one of which matches Eq. (7.197) and Eq. (7.198) and one of which is different. If N is an even number, we can write $N = 2(N/2)$. We can then set $N_1 = 2$ and $N_2 = N/2$, or vice versa. This gives two different Radix-2 Cooley-Tukey FFTs, one that decimates (divides up) the computation in the time domain and another that decimates the computation in the frequency domain.

(a) Decimation in time

Inserting $N_1 = 2$ and $N_2 = N/2$ into Eq. (7.205) gives directly

$$\mathbf{X}[k_2] = 1 \underbrace{\sum_{n_2=0}^{N/2-1} e^{-j2\pi(n_2k_2/(N/2))}x[2n_2]}_{\text{N/2-point DFT}}$$
$$+ e^{-j2\pi(1k_2/N)} \underbrace{\sum_{n_2=0}^{N/2-1} e^{-j2\pi(n_2k_2/(N/2))}x[2n_2+1]}_{\text{N/2-point DFT}},$$
$$\mathbf{X}[k_2 + N/2] = 1 \underbrace{\sum_{n_2=0}^{N/2-1} e^{-j2\pi(n_2k_2/(N/2))}x[2n_2]}_{\text{N/2-point DFT}}$$
$$- e^{-j2\pi(1k_2/N)} \underbrace{\sum_{n_2=0}^{N/2-1} e^{-j2\pi(n_2k_2/(N/2))}x[2n_2+1]}_{\text{N/2-point DFT}}.$$

The number of twiddle mults at each stage is $N/2$. This agrees with $(N_1 - 1)(N_2 - 1) = (2 - 1)[(N/2) - 1] \approx N/2$. Since there are $\log_2 N$ stages, the total number of multiplications is $(N/2)\log_2 N$.

(b) Decimation in frequency

Now we reverse the assignments of N_1 and N_2 to $N_1 = \frac{N}{2}$ and $N_2 = 2$. Inserting these into Eq. (7.205) gives directly

$$\mathbf{X}[2k_1] = \underbrace{\sum_{n_1=0}^{N/2-1} e^{-j2\pi(n_1k_1/(N/2))}}_{\text{N/2-point DFT}} \underbrace{[x[n_1] + x[n_1 + N/2]]}_{\text{Half of 2-point DFT}},$$
$$\mathbf{X}[2k_1 + 1] = \underbrace{\sum_{n_1=0}^{N/2-1} e^{-j2\pi(n_1k_1/(N/2))}}_{\text{N/2-point DFT}}$$
$$\cdot \underbrace{[x[n_1] - x[n_1 + N/2]]}_{\text{Half of 2-point DFT}} e^{-j2\pi(1n_1/N)}.$$

We take separate $\frac{N}{2}$-point DFTs of the sums and differences $(x[n] \pm x[n+\frac{N}{2}])$. Only one set of DFTs is multiplied by twiddle factors, so the number of twiddle mults is again $\frac{N}{2}\log_2 N$. For reference, the first FFT derived in this chapter (Section 7-16.3), using **z**-transforms evaluated on the unit circle, is a decimation-in-frequency Radix-2 Cooley-Tukey FFT.

The inverse DFT can be computed using any of these formulas, except that the sign of the exponent in the twiddle mults is changed.

One advantage of decimation-in-frequency over decimation-in-time is that decimation-in-frequency computes $\mathbf{X}[k]$ for even indices k without twiddle mults.

There are many other fast algorithms for computing the DFT besides the Cooley-Tukey FFT, but the Cooley-Tukey FFT is by far the most commonly-used fast algorithm for computing the DFT. The radix-2 Cooley-Tukey FFT is so fast that DFTs of orders N that are not powers of two are often zero-padded, so that N is replaced with the next-highest power of two.

Concept Question 7-23: Why is the FFT so much faster than direct computation of the DFT? (See S²)

Exercise 7-26: How many MADs are needed to compute a 4096-point DFT using the FFT?

Answer: $\frac{4096}{2}\log_2(4096) = 24576.$

Exercise 7-27: Using the decimation-in-frequency FFT, which values of the 8-point DFT of a signal of the form $\{a, b, c, d, e, f, g, h\}$ do not have a factor of $\sqrt{2}$ in them?

Answer: In the decimation-in-frequency FFT, the twiddle multiplications only affect the odd-valued indices. So $\{X_0, X_2, X_4, X_6\}$ do not have a factor of $\sqrt{2}$ in them.

Exercise 7-28: Using the decimation-in-time FFT, show that only two values of the 8-point DFT of a signal of the form $\{a, b, a, b, a, b, a, b\}$ are nonzero.

Answer: In the decimation-in-time FFT, 4-point DFTs of $\{a, a, a, a\}$ and $\{b, b, b, b\}$ are computed. These are both zero except for the dc ($k = 0$) values. So the 8-point DFT has only two nonzero values $X_0 = 4a + 4b$ and $X_4 = 4a - 4b$.

Summary

Concepts

- LTI systems, causality, BIBO stability, and convolution all generalize directly from continuous time to discrete time. Integrals are replaced with summations.
- Transfer functions $\mathbf{H}(\mathbf{z})$ generalize directly and are extremely useful in relating the following descriptions of LTI systems: poles and zeros; impulse response; difference equations; frequency response; specific input-output pairs.
- In discrete time, all functions of Ω are periodic with period 2π.
- The following transforms are directly analogous: **z**-transforms and Laplace transforms; DTFS and Fourier series; DTFT and Fourier transform.
- The FFT is a fast algorithm for computing the DFT.

Mathematical and Physical Models

Convolution

$$y[n] = h[n] * x[n] = \sum_{i=-\infty}^{\infty} h[n-i]\, x[i]$$

Frequency Response

$$\mathbf{H}(e^{j\Omega}) = \textstyle\sum_{n=-\infty}^{\infty} h[n]\, e^{-j\Omega n}$$

$e^{j\Omega n}$ → LTI → $\mathbf{H}(e^{j\Omega})\, e^{j\Omega n}$

$$\mathbf{Z}[x[n]] = \mathbf{X}(\mathbf{z}) = \sum_{n=0}^{\infty} x[n]\, \mathbf{z}^{-n}$$

DTFS

$$x[n] = \sum_{k=0}^{N_0-1} \mathbf{x}_k e^{jk\Omega_0 n}, \qquad n = 0, 1, \ldots, N_0 - 1$$

$$\mathbf{x}_k = \frac{1}{N_0} \sum_{n=0}^{N_0-1} x[n]\, e^{-jk\Omega_0 n}, \qquad k = 0, 1, \ldots, N_0 - 1$$

DTFT

$$\mathbf{X}(e^{j\Omega}) = \sum_{n=-\infty}^{\infty} x[n]\, e^{-j\Omega n}, \quad x[n] = \frac{1}{2\pi} \int_{\Omega_1}^{\Omega_1+2\pi} \mathbf{X}(e^{j\Omega})\, e^{j\Omega n}\, d\Omega$$

DFT

$$\mathbf{X}[k] = \sum_{n=0}^{N_0-1} x[n]\, e^{-jk\Omega_0 n}, \quad k = 0, 1, \ldots, N_0 - 1, \quad \Omega_0 = 2\pi/N_0$$

$A\cos(\Omega n + \phi)$ → LTI → $AM(e^{j\Omega})\cos(\Omega n + \phi + \theta), \quad \theta = \underline{/\mathbf{H}(e^{j\Omega})}$

Important Terms

Provide definitions or explain the meaning of the following terms:

ARMA	DTFT	impulse response	right shift
AR (autoregressive)	difference equation	MA (moving average)	sampling property
brackets notation	discrete time	MAD	signal flow graph
convolution	downsampling	magnitude response	transfer function
decimation	FFT	oversampling	truncating
delay	fundamental angular frequency	Parseval's theorem	windowing
DFT	fundamental period	poles	**z**-transform
DTFS	geometric signal	recursive	zeros

PROBLEMS

Section 7-2: Signals and Functions

7.1 Convert each signal to the form $\{\underline{a}, b, c, d, e\}$.

(a) $u[n] - \delta[n-3] - u[n-4]$

(b) $n\, u[n] - n\, u[n-5]$

*(c) $u[n-1]\, u[4-n]$

(d) $2\delta[n-1] - 4\delta[n-3]$

7.2 Convert each signal to the form $\{\underline{a}, b, c, d, e\}$.

*(a) $u[n] - u[n-4]$

(b) $u[n] - 2u[n-2] + u[n-4]$

(c) $n\, u[n] - 2(n-2)\, u[n-2] + (n-4)\, u[n-4]$

(d) $n\, u[n] - 2(n-1)\, u[n-1] + 2(n-3)\, u[n-3] - (n-4)\, u[n-4]$

7.3 Compute the fundamental periods and fundamental angular frequencies of the following signals:

(a) $7\cos(0.16\pi n + 2)$

(b) $7\cos(0.16n + 2)$

(c) $3\cos(0.16\pi n + 1) + 4\cos(0.15\pi n + 2)$

7.4 Compute the fundamental periods and fundamental angular frequencies of the following signals:

(a) $3\cos(0.075\pi n + 1)$

*(b) $4\cos(0.56\pi n + 0.7)$

(c) $5\cos(\sqrt{2}\,\pi n - 1)$

Section 7-3 and 7-4: Discrete-Time LTI Systems

7.5 Which of the following systems is (i) linear and (ii) time-invariant?

(a) $y[n] = n\, x[n]$

(b) $y[n] = x[n] + 1$

(c) $y[n] + 2y[n-1] = 3x[n] + nx[n-1]$

(d) $y[n] + 2y[n-1] = 3x[n] + 4x[n-1]$

7.6 An ideal digital differentiator is described by the system

$$y[n] = (x[n+1] - x[n-1]) - \tfrac{1}{2}\,(x[n+2] - x[n-2]) + \tfrac{1}{3}\,(x[n+3] - x[n-3]) + \cdots$$

(a) Is the system LTI?

(b) Is it causal?

(c) Prove it is *not* BIBO stable.

(d) Provide a bounded input $x[n]$ that produces an unbounded output $y[n]$.

7.7 The following two input-output pairs are observed for a system known to be linear:

- $\{\underline{1}, 2, 3\} \rightarrow$ Linear $\rightarrow \{\underline{1}, 4, 7, 6\}$
- $\delta[n] \rightarrow$ Linear $\rightarrow \{\underline{1}, 3\}$

Prove the system is *not* time-invariant. (*Hint:* Prove this by contradiction.)

7.8 The following two input-output pairs are observed for a system known to be linear:

- $\{\underline{1}, 1\} \rightarrow$ Linear $\rightarrow \{\underline{5}, 6\}$
- $\delta[n] \rightarrow$ Linear $\rightarrow \{\underline{1}, 3\}$

Prove the system is *not* time-invariant.

Section 7-5: Discrete-Time Convolution

7.9 Compute the following convolutions:

(a) $\{\underline{1}, 2\} * \{\underline{3}, 4, 5\}$

(b) $\{\underline{1}, 2, 3\} * \{\underline{4}, 5, 6\}$

*(c) $\{\underline{2}, 1, 4\} * \{\underline{3}, 6, 5\}$

7.10 Compute the following convolutions:

(a) $\{\underline{3}, 4, 5\} * \{\underline{6}, 7, 8\}$

(b) $\{\underline{1}, 2, -3\} * u[n]$

(c) $\{\underline{3}, 4, 5\} * (u[n] - u[n-3])$

(d) $\{\underline{1}, 2, 4\} * 2\delta[n-2]$

7.11 If $\{\underline{1}, 2, 3\} * x[n] = \{\underline{5}, 16, 34, 32, 21\}$, compute $x[n]$ *without* using **z**-transforms.

7.12 Given that

$\{\underline{1}, 2, 3\} \rightarrow$ System $\rightarrow \{\underline{1}, 4, 7, 6\}$

for a system known to be LTI, compute the system's impulse response $h[n]$ without using **z**-transforms.

7.13 Given the two systems connected in series as

$x[n] \rightarrow$ $h_1[n]$ $\rightarrow w[n] = 3x[n] - 2x[n-1]$,

*Answer(s) in Appendix F.

and

$$w[n] \rightarrow \boxed{h_2[n]} \rightarrow y[n] = 5w[n] - 4w[n-1],$$

compute the overall impulse response.

7.14 The two systems

$$y[n] = 3x[n] - 2x[n-1]$$

and

$$y[n] = 5x[n] - 4x[n-1]$$

are connected in parallel. Compute the overall impulse response.

Sections 7-6 and 7-7: z-Transforms

7.15 Compute the **z**-transforms of the following signals. Cast your answer in the form of a rational fraction.

(a) $(1 + 2^n)\, u[n]$

(b) $2^n u[n] + 3^n u[n]$

(c) $\{\underline{1}, -2\} + (2)^n u[n]$

(d) $2^{n+1} \cos(3n + 4)\, u[n]$

7.16 Compute the **z**-transforms of the following signals. Cast your answer in the form of a rational fraction.

(a) $n\, u[n]$

(b) $(-1)^n 3^{-n}\, u[n]$

(c) $u[n] - u[n-2]$

Section 7-8: Inverse z-Transforms

7.17 Compute the inverse **z**-transforms.

(a) $\dfrac{\mathbf{z}+1}{2\mathbf{z}}$

(b) $\dfrac{\mathbf{z}-1}{\mathbf{z}-2}$

(c) $\dfrac{2\mathbf{z}+3}{\mathbf{z}^2(\mathbf{z}+1)}$

(d) $\dfrac{\mathbf{z}^2+3\mathbf{z}}{\mathbf{z}^2+3\mathbf{z}+2}$

*(e) $\dfrac{\mathbf{z}^2-\mathbf{z}}{\mathbf{z}^2-2\mathbf{z}+2}$

7.18 Compute the inverse **z**-transforms.

(a) $\dfrac{4\mathbf{z}}{\mathbf{z}-1} + \dfrac{5\mathbf{z}}{\mathbf{z}-2} + \dfrac{6\mathbf{z}}{\mathbf{z}-3}$

(b) $\dfrac{(1+j)\mathbf{z}}{\mathbf{z}-(3+4j)} + \dfrac{(1-j)\mathbf{z}}{\mathbf{z}-(3-4j)}$.
Simplify to a geometric-times-sinusoid function.

(c) $\dfrac{(3+j4)\mathbf{z}}{\mathbf{z}-(1+j)} + \dfrac{(3-j4)\mathbf{z}}{\mathbf{z}-(1-j)} + \dfrac{(1+j)\mathbf{z}}{\mathbf{z}-(3+j4)} + \dfrac{(1-j)\mathbf{z}}{\mathbf{z}-(3-j4)}$.
Simplify to a sum of two geometric-times-sinusoids.

(d) $\dfrac{8\mathbf{z}}{\mathbf{z}^2-6\mathbf{z}+25}$

Section 7-9: Solving Difference Equations

7.19 Use the one-sided **z**-transform to solve

$$y[n] - 5y[n-1] + 6y[n-2] = 4u[n]$$

with initial conditions $y[-1] = y[-2] = 1$.

7.20 The sequence of *Fibonacci numbers* $y[n]$ is $\{1, 1, \underline{2}, 3, 5, 8, 13, 21, \ldots\}$. Each Fibonacci number is the sum of the two previous Fibonacci numbers. Use the one-sided **z**-transform to derive an explicit closed-form expression for $y[n]$. Show that for $n > 3$ the ratio $y[n+1]/y[n]$ is equal to $(1+\sqrt{5})/2 \approx 1.618$, which is called the *golden ratio*.

7.21 Given

$$\{\underline{1}, 3, 2\} \rightarrow \boxed{y[n] + 2y[n-1] = 4x[n] + 5x[n-1]} \rightarrow y[n],$$

compute the output $y[n]$.

7.22 Show that the discrete-time sinusoid

$$y[n] = \cos(\Omega_0 n)\, u[n]$$

can be generated using the simple difference equation

$$y[n] - 2\cos(\Omega_0)\, y[n-1] + y[n-2] = 0$$

initialized using

$$y[-1] = \cos(\Omega_0) \quad \text{and} \quad y[-2] = \cos(2\Omega_0).$$

This is useful for simple DSP chips that do not have a built-in cosine function, since the constant $\cos(\Omega_0)$ can be precomputed.

Sections 7-10 and 7-11: Transforms and BIBO Stability

*7.23 The step response (to $u[n]$) of an LTI system is known to be $2u[n] + (-2)^n\, u[n]$. Compute the following:

(a) The transfer function $\mathbf{H}(\mathbf{z})$.

(b) The poles and zeros.

(c) The impulse response $h[n]$.

(d) The difference equation.

7.24 An LTI system has zeros $\{3, 4\}$ and poles $\{1, 2\}$. The transfer function $\mathbf{H}(\mathbf{z})$ has $\mathbf{H}(0) = 6$. Compute the following:

(a) The transfer function $\mathbf{H}(\mathbf{z})$.

(b) The response to $x[n] = \{\underline{1}, -3, 2\}$.

(c) The impulse response $h[n]$.

(d) The difference equation.

7.25 An LTI system has the transfer function

$$\mathbf{H}(\mathbf{z}) = \frac{(\mathbf{z}-1)(\mathbf{z}-6)}{(\mathbf{z}-2)(\mathbf{z}-3)}.$$

Compute the following:

(a) The zeros and poles. Is the system stable?

(b) The difference equation.

(c) The response to input $x[n] = \{\underline{1}, -5, 6\}$.

(d) The impulse response $h[n]$.

7.26 An LTI system has the transfer function

$$\mathbf{H}(\mathbf{z}) = \frac{\mathbf{z}}{(\mathbf{z}-1)(\mathbf{z}-2)}.$$

Compute the following:

(a) The zeros and poles. Is the system stable?

(b) The difference equation.

(c) The response to input $x[n] = \{\underline{2}, -6, 4\}$.

(d) The response to input $x[n] = \{\underline{1}, -1\}$.

(e) The impulse response $h[n]$.

7.27 ***Multichannel blind deconvolution***. We observe the two signals:

$$y_1[n] = \{\underline{1}, -13, 86, -322, 693, -945, 500\},$$

$$y_2[n] = \{\underline{1}, -13, 88, -338, 777, -1105, 750\},$$

where $y_1[n] = h_1[n] * x[n]$, $y_2[n] = h_2[n] * x[n]$, and *all* of $\{h_1[n], h_2[n]$, and $x[n]\}$ are unknown! We know only that all of the signals have finite lengths, are causal, and $x[0] = 1$. Compute $h_1[n]$, $h_2[n]$, and $x[n]$. *Hint:* Use **z**-transforms and zeros.

Section 7-12: Frequency Response

7.28 Compute $\mathbf{H}(e^{j\Omega})$ for the system

$$y[n] + y[n-2] = x[n] - x[n-2].$$

Simplify your answer as much as possible.

7.29 Given

$$\cos\left(\frac{\pi}{2}n\right) \rightarrow \boxed{y[n] = x[n] + 0.5x[n-1] + x[n-2]} \rightarrow y[n$$

*(a) Compute the frequency response $\mathbf{H}(e^{j\Omega})$.

(b) Compute the output $y[n]$.

7.30 Given

$$\cos\left(\frac{\pi}{2}n\right) \rightarrow \boxed{y[n] = 8x[n] + 3x[n-1] + 4x[n-2]} \rightarrow y[n],$$

(a) Compute the frequency response $\mathbf{H}(e^{j\Omega})$.

(b) Compute the output $y[n]$.

7.31 If input $x[n] = \cos(\frac{\pi}{2}n) + \cos(\pi n)$, and

$$x[n] \rightarrow \boxed{y[n] = x[n] + x[n-1] + x[n-2] + x[n-3]} \rightarrow y[n]$$

(a) Compute the frequency response $\mathbf{H}(e^{j\Omega})$.

(b) Compute the output $y[n]$.

7.32 If input $x[n] = 1 + 2\cos(\frac{\pi}{2}n) + 3\cos(\pi n)$, and

$$x[n] \rightarrow \boxed{y[n] = x[n] + 4x[n-1] + 3x[n-3]} \rightarrow y[n],$$

(a) Compute the frequency response $\mathbf{H}(e^{j\Omega})$.

(b) Compute the output $y[n]$.

7.33 If

$$x[n] = 3 + 4\cos\left(\frac{\pi}{2}n + \frac{\pi}{4}\right)$$

$$\downarrow$$

$$\boxed{y[n] + y[n-1] = x[n] - x[n-1]}$$

$$\downarrow$$

$$y[n],$$

(a) Compute the frequency response $\mathbf{H}(e^{j\Omega})$.

(b) Compute the output $y[n]$.

7.34 If input $x[n] = 9 + 2\cos(\frac{\pi}{2}n) + 3\cos(\pi n)$, and

$$x[n] \longrightarrow \boxed{\begin{aligned} &5y[n] + 3y[n-1] + y[n-2] \\ &= 7x[n] + 6x[n-1] - x[n-2] \end{aligned}} \longrightarrow y[n].$$

(a) Compute the frequency response $\mathbf{H}(e^{j\Omega})$.

(b) Compute the output $y[n]$.

7.35 If $x[n] = 4 + 3\cos(\frac{\pi}{3}n) + 2\cos(\frac{\pi}{2}n)$, and

$$x[n] \longrightarrow \boxed{\begin{aligned} &y[n] + 2y[n-1] + y[n-2] \\ &= x[n] + x[n-1] + x[n-2] \end{aligned}} \longrightarrow y[n],$$

(a) Compute the frequency response $\mathbf{H}(e^{j\Omega})$. Simplify your answer as much as possible.

(b) Compute the output $y[n]$.

***7.36** If $x[n] = \cos(2\frac{\pi}{3}n) + \cos(\frac{\pi}{2}n)$, and

$$x[n] \longrightarrow \boxed{\begin{aligned} &y[n] + ay[n-1] = \\ &x[n] + bx[n-1] + x[n-2] \end{aligned}} \longrightarrow y[n],$$

for what constants a and b is $y[n] = \sin(\frac{\pi}{2}n)$?

7.37 Given the system:

$$\begin{aligned} y[n] = {} & 0.5x[n] + 0.29(x[n+1] + x[n-1]) \\ & - 0.042(x[n+3] + x[n-3]) \\ & + 0.005(x[n+5] + x[n-5]), \end{aligned}$$

(a) Compute an expression for $\mathbf{H}(e^{j\Omega})$. Express it as a sum of 3 cosines and a constant.

(b) Plot $|\mathbf{H}(e^{j\Omega})|$ at $\Omega = \frac{2\pi k}{200}$, $0 \le k \le 99$.

(c) Describe in words what function the system performs on $x[n]$.

7.38 A system has $2N$ poles $\{\mathbf{p}_i, \mathbf{p}_i^*, i = 1, \ldots, N\}$ in complex conjugate pairs, and $2N$ zeros $\{1/\mathbf{p}_i, 1/\mathbf{p}_i^*, i = 1, \ldots, N\}$ in complex conjugate pairs at reciprocals of $\{\mathbf{p}_i, \mathbf{p}_i^*\}$. Show that the gain $|\mathbf{H}(e^{j\Omega})|$ of the system is constant for all Ω. This is an *all-pass system*. All-pass systems are used to alter a filter's phase response without affecting its gain. *Hint:* For any complex number $\mathbf{z} = |\mathbf{z}|e^{j\theta}$, the ratio

$$\frac{\mathbf{z}}{\mathbf{z}^*} = \frac{|\mathbf{z}|e^{j\theta}}{|\mathbf{z}|e^{-j\theta}} = e^{j2\theta}$$

has magnitude 1.

Section 7-13: DTFS

7.39 Given:

$$x[n] = \{\ldots, \underline{5}, 3, 1, 3, 5, 3, 1, 3, 5, 3, 1, 3, \ldots\},$$

compute its DTFS expansion.

7.40 Given:

$$x[n] = \{\ldots, \underline{3}, -1, -1, -1, 3, -1, -1, -1, \ldots\},$$

compute its DTFS expansion.

7.41 Given:

$$x[n] = \{\ldots, 18, 12, 6, 0, 6, 12, \underline{18}, 12, 6, 0, 6, 12, 18, \ldots\},$$

(a) Compute its DTFS expansion. (*Hint:* $x[n]$ has period $= 6$, is real, and is an even function.)

(b) Compute its average power in the time domain.

(c) Compute its average power in the frequency domain.

7.42 The goal of this problem is to show how the conjugate symmetry relation $\mathbf{x}_{-k} = \mathbf{x}_k^*$ can greatly simplify the computation of DTFS from its coefficients.

Let $x[n]$ have a DTFS with coefficients

$$\mathbf{x}_k = \cos(\pi k/4) + \sin(3\pi k/4).$$

Note that $\mathbf{x}_k$ are real-valued, but not a symmetric function of index k.

(a) Explain why $x[n]$ has period $N_0 = 8$.

(b) Compute the real part of $x[n]$ from the symmetric component of $\mathbf{x}_k$.

(c) Compute the imaginary part of $x[n]$ from the antisymmetric component of $\mathbf{x}_k$.

(d) Compute the average powers in the time and frequency domains. Confirm that they agree.

7.43 Given input:

$$x[n] = \{\ldots, 4, -2, 0, -2, \underline{4}, -2, 0, -2, 4, -2, 0, -2, \ldots\},$$

and system impulse response $h[n] = \sin(\pi n/3)/(\pi n)$, compute output $y[n]$.

***7.44** Given input $x[n]$ and output $y[n]$:

$$\begin{aligned} x[n] &= \{\ldots, 4, 2, 1, 0, \underline{4}, 2, 1, 0, 4, 2, 1, 0, \ldots\}, \\ y[n] &= \{\ldots, 10, 4, \underline{10}, 4, 10, 4, \ldots\}, \end{aligned}$$

and system impulse response $h[n] = \{\underline{a}, b, c\}$, determine the values of a, b, and c.

Section 7-14: DTFT

7.45 Compute the DTFTs of the following signals (simplify answers to sums of sines and cosines).

(a) $\{1, 1, \underline{1}, 1, 1\}$

(b) $\{3, \underline{2}, 1\}$

7.46 Compute the inverse DTFT of

$$\mathbf{X}(e^{j\Omega}) = [3+2\cos(\Omega)+4\cos(2\Omega)]+j[6\sin(\Omega)+8\sin(2\Omega)].$$

7.47 Compute the inverse DTFT of

$$\mathbf{X}(e^{j\Omega}) = [7+5\cos(\Omega)+3\cos(2\Omega)]+j[\sin(\Omega)+\sin(2\Omega)].$$

7.48 Use inverse DTFTs to evaluate these integrals.

(a) $\displaystyle \frac{1}{2\pi}\int_{-\pi}^{\pi} \frac{e^{j\Omega}e^{j3\Omega}}{e^{j\Omega}-\frac{1}{2}}\,d\Omega$

(b) $\displaystyle \frac{1}{2\pi}\int_{-\pi}^{\pi} e^{j3\Omega}2\cos(3\Omega)\,d\Omega$

7.49 Given $x[n] = \{1, 4, 3, 2, 5, 7, \underline{-45}, 7, 5, 2, 3, 4, 1\}$, compute the following:

(a) $\mathbf{X}(e^{j\pi})$

(b) $\angle\mathbf{X}(e^{j\Omega})$

(c) $\int_{-\pi}^{\pi} \mathbf{X}(e^{j\Omega})\,d\Omega$

(d) $\int_{-\pi}^{\pi} |\mathbf{X}(e^{j\Omega})|^2\,d\Omega$

Section 7-15: DFT

7.50 Compute the DFTs of each of the following signals:

*(a) $\{\underline{12}, 8, 4, 8\}$

(b) $\{\underline{16}, 8, 12, 4\}$

7.51 Determine the DFT of a single period of each of the following signals:

(a) $\cos(\frac{\pi}{4}n)$

(b) $\frac{1}{4}\sin(\frac{3\pi}{4}n)$

7.52 Compute the inverse DFTs of the following:

(a) $\{0, 0, 3, 0, 4, 0, 3, 0\}$

(b) $\{0, 3+j4, 0, 0, 0, 0, 0, 3-j4\}$

7.53 Use DFTs to compute the convolution

$$\{\underline{1}, 3, 5\} * \{\underline{7}, 9\}.$$

7.54 Let $x[n]$ have even duration N and ***half-wave antisymmetry*** $x[n+\frac{N}{2}] = -x[n]$. An example of half-wave antisymmetry is $x[n] = \{\underline{3}, 1, 4, 2, -3, -1, -4, -2\}$. Show that the DFT $\mathbf{X}[k]$ of $x[n]$ is 0 for even integers k using:

(a) The definition of the DFT.

(b) The decimation-in-frequency FFT.

7.55 An important application of the FFT is to compute the cyclic convolution $y[n] = h[n] \circledcirc x[n]$, because the linear convolution $y[n] = h[n]*x[n]$ can be computed from the cyclic convolution of zero-padded $h[n]$ and $x[n]$. Then, $y[n]$ can be computed from

$$y[n] = \text{DFT}^{-1}\{\text{DFT}\{h[n]\}\cdot\text{DFT}\{x[n]\}\},$$

which requires computation of two DFTs and one inverse DFT. Show that the DFTs of two real-valued signals $h[n]$ and $x[n]$ can be determined from the single DFT of the complex-valued signal $z[n] = h[n] + jx[n]$, using conjugate symmetry.

8 Applications of Discrete-Time Signals and Systems

Contents

Overview, 421
8-1 Discrete-Time Filters, 421
8-2 Notch Filters, 427
8-3 Comb Filters, 434
8-4 Deconvolution and Dereverberation, 439
8-5 Bilateral **z**-Transforms, 445
8-6 Inverse Bilateral **z**-Transforms, 447
8-7 ROC, Stability, and Causality, 449
8-8 Deconvolution and Filtering Using the DFT, 450
8-9 Computing Spectra of Periodic Signals, 457
8-10 Computing Spectra of Nonperiodic Signals, 462
Summary, 469
Problems, 469

Objectives

Learn to:

- Design discrete-time filters, including notch and comb filters.
- Perform discrete-time filtering and deconvolution using the discrete Fourier transform.
- Compute the spectra of both periodic and nonperiodic discrete-time signals.
- Perform deconvolution using real-time signal processing.

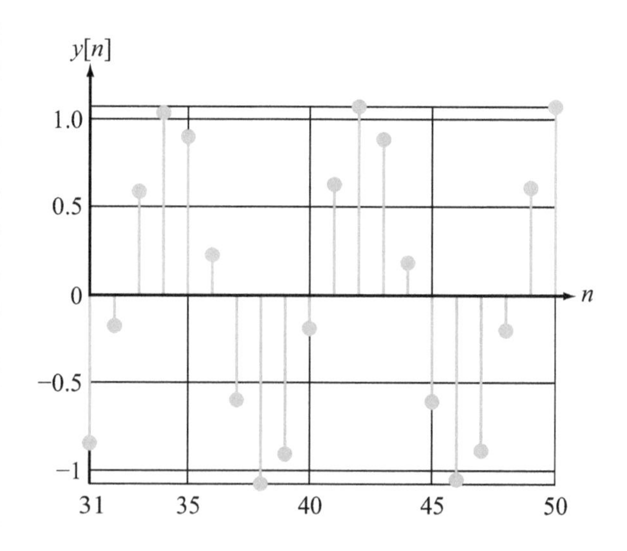

In ***real-time signal processing*** the output signal is generated from the input signal as it arrives, using an ARMA difference equation. In ***batch signal processing***, a previously digitized and stored signal can be processed in its entirety, all at once. This chapter presents examples of how ***noise filtering*** and other applications are performed in ***discrete time***.

Overview

Chapters 4, 6, and 8 share a common feature: Each covers applications that demonstrate the utility of the theoretical concepts introduced in the chapters that preceded it. Chapter 4 focused on applications of the Laplace transform, and Chapter 6 did likewise for the Fourier transform. Both dealt exclusively with continuous-time signals. Chapter 8 is the analogue to the combination of Chapters 4 and 6, but for discrete-time signals.

The material covered in Chapter 8 consists of four core topics.

(1) Discrete-time filtering: This topic includes the design and implementation of general highpass, lowpass, bandpass, and bandreject filters, as well as specialized filters that perform ***deconvolution*** ("undoing" convolution) and ***dereverberation*** (removing echoes).

(2) Batch signal processing: By storing the entire signal digitally prior to signal processing, it is possible to process the entire signal all at once, and to do so to both causal and noncausal signals.

(3) Bilateral z-transforms: By allowing noncausal systems with noncausal, but one-sided impulse responses, deconvolution of non-minimum-phase systems can be performed.

(4) Computing spectra of continuous-time signals: The discrete Fourier transform (DFT) and the fast Fourier transform (FFT) algorithm are used as computational tools to compute the spectra of periodic and nonperiodic continuous-time signals.

► Unless stated to the contrary, all systems considered in this chapter are assumed to be causal. ◄

8-1 Discrete-Time Filters

In our examination of continuous-time filters in Chapter 6, we noted that the filter's frequency response is governed by the proximity of the locations of the poles and zeros of its transfer function $\mathbf{H}(\mathbf{s})$ to the vertical axis in the $\mathbf{s}$-plane. Also, BIBO stability requires that the poles reside in the open left half-plane. A similar notion applies to discrete-time filters, except that the reference boundary is the ***unit circle*** defined by $|\mathbf{z}| = 1$. The poles of $\mathbf{H}(\mathbf{z})$ must reside inside the unit circle, and the shape of the frequency response $\mathbf{H}(e^{j\Omega})$ is dictated by the locations of the poles and zeros of $\mathbf{H}(\mathbf{z})$ relative to the unit circle.

8-1.1 Roles of Poles and Zeros of H(z)

The transfer function of an LTI system with zeros $\{\mathbf{z}_i,\ i = 1, 2, \ldots, M\}$ and poles $\{\mathbf{p}_i,\ i = 1, 2, \ldots, N\}$ is given by Eq. (7.115) as

$$\mathbf{H}(\mathbf{z}) = \frac{\mathbf{C}(\mathbf{z} - \mathbf{z}_1)(\mathbf{z} - \mathbf{z}_2)\cdots(\mathbf{z} - \mathbf{z}_M)}{(\mathbf{z} - \mathbf{p}_1)(\mathbf{z} - \mathbf{p}_2)\cdots(\mathbf{z} - \mathbf{p}_N)}, \tag{8.1}$$

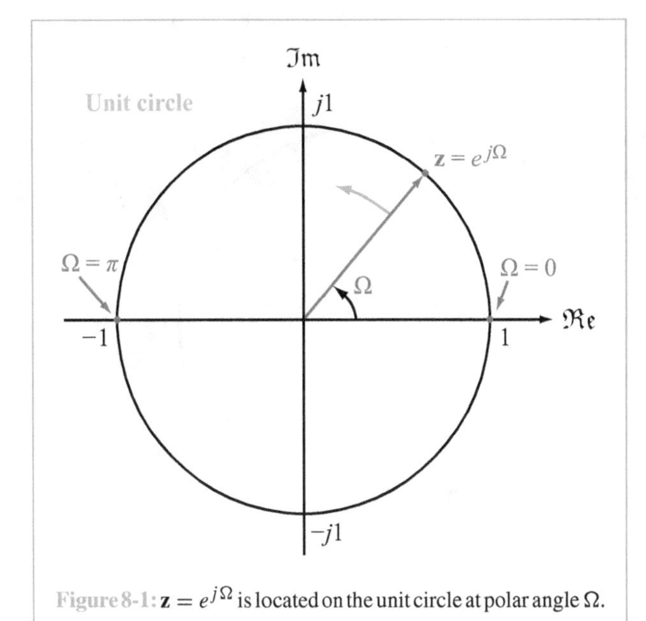

Figure 8-1: $\mathbf{z} = e^{j\Omega}$ is located on the unit circle at polar angle Ω.

where $\mathbf{C}$ is a constant. Replacing $\mathbf{z}$ with $e^{j\Omega}$ gives the frequency response

$$\mathbf{H}(e^{j\Omega}) = \frac{\mathbf{C}(e^{j\Omega} - \mathbf{z}_1)(e^{j\Omega} - \mathbf{z}_2)\cdots(e^{j\Omega} - \mathbf{z}_M)}{(e^{j\Omega} - \mathbf{p}_1)(e^{j\Omega} - \mathbf{p}_2)\cdots(e^{j\Omega} - \mathbf{p}_N)}. \tag{8.2}$$

Figure 8-1 displays the unit circle in the $\mathbf{z}$-plane. Because

$$|e^{j\Omega}| = 1 \qquad \text{and} \qquad \angle e^{j\Omega} = \Omega, \tag{8.3}$$

setting $\mathbf{z} = e^{j\Omega}$ implicitly defines a point in the polar $\mathbf{z}$-plane whose radius is 1 and polar angle is Ω. Thus, replacing $\mathbf{z}$ in $\mathbf{H}(\mathbf{z})$ with $e^{j\Omega}$ is equivalent to placing $\mathbf{z}$ on the unit circle at angle Ω. Point $(1 + j0)$ in the $\mathbf{z}$-plane corresponds to angular frequency $\Omega = 0$, and point $(-1 + j0)$ corresponds to $\Omega = \pi$. Increasing the angular frequency of the frequency response $\mathbf{H}(e^{j\Omega})$ is equivalent to moving point $e^{j\Omega}$ in Fig. 8-1 counterclockwise along the unit circle.

To develop a qualitative understanding of the expression given by Eq. (8.2), let us consider a simple LTI system characterized by a frequency response that has one zero at $(-1 + j0)$ and two poles, as shown in Fig. 8-2. Unless they reside along the real axis, the poles and zeros of a physically realizable system always occur in complex conjugate pairs. Hence, in the present case, $\mathbf{p}_2 = \mathbf{p}_1^*$. Also, let us assume constant $\mathbf{C} = 1$. From Fig. 8-2,

$$e^{j\Omega} - \mathbf{z}_1 = l_1 e^{j\psi_1},$$

$$e^{j\Omega} - \mathbf{p}_1 = d_1 e^{j\delta_1},$$

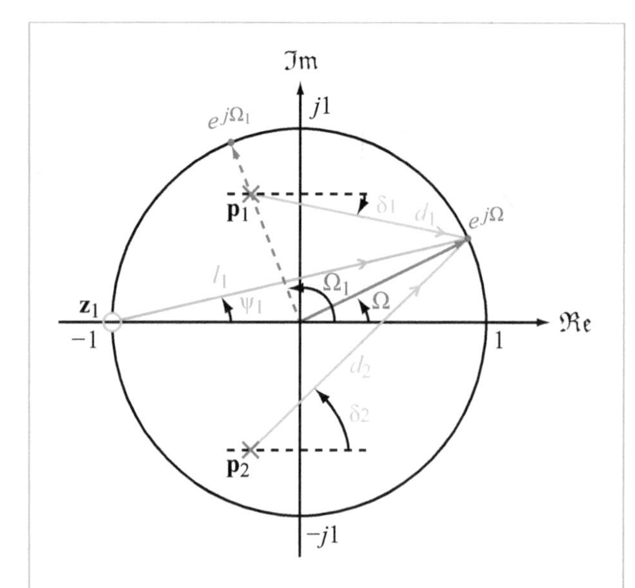

Figure 8-2: Locations of zero $\mathbf{z}_1$ and conjugate poles $\mathbf{p}_1$ and $\mathbf{p}_2$, relative to $e^{j\Omega}$.

and

$$e^{j\Omega} - \mathbf{p}_2 = d_2 e^{j\delta_2},$$

where l_1, d_1, and d_2 are the distances between the "observation point" $e^{j\Omega}$ on the unit circle and the locations of the zero and two poles, and ψ_1, δ_1, and δ_2, are the associated angles. Hence, $\mathbf{H}(e^{j\Omega})$ is given by

$$\mathbf{H}(e^{j\Omega}) = \frac{e^{j\Omega} - \mathbf{z}_1}{(e^{j\Omega} - \mathbf{p}_1)(e^{j\Omega} - \mathbf{p}_2)} = \frac{l_1}{d_1 d_2}\, e^{j(\psi_1 - \delta_1 - \delta_2)} = M(e^{j\Omega})\, e^{j\theta(e^{j\Omega})} \tag{8.4}$$

with

$$M(e^{j\Omega}) = \frac{l_1}{d_1 d_2}, \tag{8.5a}$$

$$\theta(e^{j\Omega}) = \psi_1 - \delta_1 - \delta_2. \tag{8.5b}$$

The magnitude and phase spectra of $\mathbf{H}(e^{j\Omega})$ are generated by moving $e^{j\Omega}$ from $(1 + j0)$, corresponding to $\Omega = 0$, all the way around the unit circle, all the while computing $M(e^{j\Omega})$ and $\theta(e^{j\Omega})$ as functions of Ω. Since for a physically realizable system the poles and zeros either lie on the real axis or exist in conjugate pairs, their locations create an inherently symmetrical arrangement with respect to $e^{j\Omega}$ and $e^{-j\Omega}$. Consequently,

▶ • $\mathbf{H}(e^{j\Omega})$ is periodic in Ω with period 2π.

• Its magnitude $M(e^{j\Omega})$ is an even function: $M(e^{j\Omega}) = M(e^{-j\Omega})$.

• Its phase $\theta(e^{j\Omega})$ is an odd function: $\theta(e^{j\Omega}) = -\theta(e^{-j\Omega})$. ◀

In Fig. 8-2, zero $\mathbf{z}_1$ is located on the unit circle at $\Omega = \pi$. This means that $l_1 = 0$ when $\Omega = \pi$, creating a null in the spectrum of $\mathbf{H}(e^{j\Omega})$ at that frequency. Similarly, had the zero been located on the unit circle at an angle Ω_1 instead, $\mathbf{H}(e^{j\Omega})$ would have exhibited a null at Ω_1.

Whereas zeros generate nulls in the spectrum, poles generate peaks. A pole located close to the unit circle along a direction Ω_1, for example, would cause a peak in the spectrum at $\Omega = \Omega_1$. In summary:

▶ • A zero of $\mathbf{H}(\mathbf{z})$ located near (or on) the unit circle at $\mathbf{z} = e^{j\Omega_1}$ leads to a dip (or null) in the magnitude spectrum $M(e^{j\Omega})$ at $\Omega = \Omega_1$.

• A pole of $\mathbf{H}(\mathbf{z})$ located near the unit circle at $\mathbf{z} = e^{j\Omega_2}$ leads to a peak in the magnitude spectrum $M(e^{j\Omega})$ at $\Omega = \Omega_2$.

• Multiple poles or multiple zeros at the same location generate magnitude spectra with steeper slopes. ◀

We will now explore the utility of these pole-zero properties for several types of filters.

8-1.2 Lowpass Filter

Let us consider a simple lowpass filter whose transfer function has one conjugate pair of poles and one conjugate pair of zeros. To emphasize low frequencies, which lie on the right half of the unit circle, we place poles at $0.5e^{\pm j60^\circ}$, as shown in Fig. 8-3(a). By the same logic, to dampen the magnitude of the frequency response at high frequencies, we place zeros in the left half of the circle, at $1e^{\pm j139^\circ}$. Inserting these values in Eq. (8.2) with $\mathbf{C} = 1$ gives

$$\mathbf{H}(e^{j\Omega}) = \frac{(e^{j\Omega} - e^{j139^\circ})(e^{j\Omega} - e^{-j139^\circ})}{(e^{j\Omega} - 0.5e^{j60^\circ})(e^{j\Omega} - 0.5e^{-j60^\circ})}. \tag{8.6}$$

The magnitude spectrum displayed in Fig. 8-3(b) was generated by computing $M(e^{j\Omega}) = |\mathbf{H}(e^{j\Omega})|$ at multiple values of Ω between 0 and π. Since $M(e^{j\Omega})$ is an even function, it is sufficient to display it over $0 \le \Omega \le \pi$. The general shape of the magnitude spectrum does indeed resemble that of a lowpass

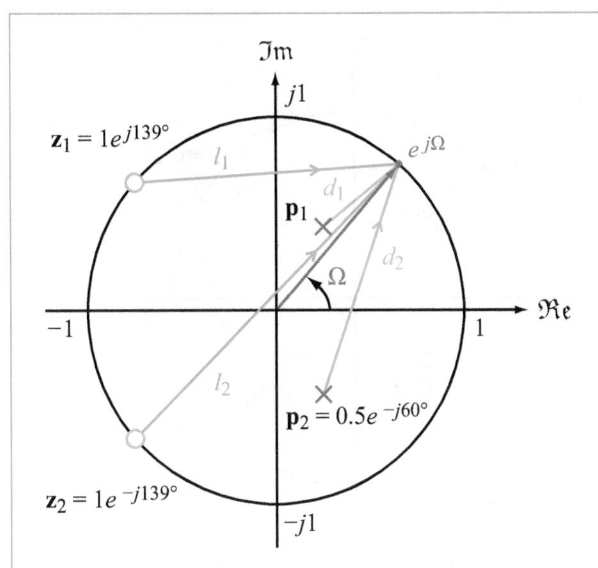

(a) Poles @ $0.5e^{\pm j60°}$ and zeros @ $1e^{\pm j139°}$

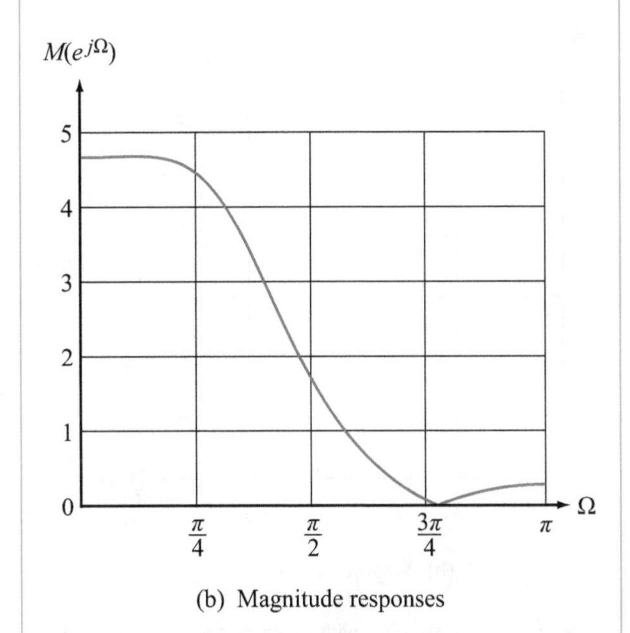

(b) Magnitude responses

Figure 8-3: Lowpass filter with conjugate poles at $0.5e^{\pm j60°}$ and zeros at $1e^{\pm j139°}$.

filter, albeit not a steep one. A spectrum with a steeper slope requires more poles and/or zeros.

Note that in Fig. 8-3(b), the magnitude at $\Omega = 0$ is 4.7, corresponding to setting $\Omega = 0$ in Eq. (8.6) and then computing the absolute value of $\mathbf{H}(e^{j\Omega})$. The scaling constant $\mathbf{C}$ had been set to 1. By choosing a different value for $\mathbf{C}$, we can scale the vertical axis in Fig. 8-3(b) to suit the intended application. For example, setting $\mathbf{C} = 1/4.7$ leads to a magnitude spectrum with a dc value of 1.

Example 8-1: Half-Band Lowpass Filter

A ***half-band lowpass filter*** passes discrete-time signals with angular frequencies below $\pi/2$ and rejects signals with angular frequencies above $\pi/2$. For such a filter with:

- Zeros: $\{e^{\pm j\pi/2},\ e^{\pm j3\pi/4},\ e^{j\pi}\}$,
- Poles: $\{0.6,\ 0.8e^{\pm j\pi/4},\ 0.8e^{\pm j\pi/2}\}$,

Obtain (a) its $\mathbf{H}(\mathbf{z})$, (b) its difference equation, and (c) generate a plot of its magnitude spectrum $M(e^{j\Omega})$. Assume $\mathbf{C} = 1$.

Solution:

(a) The transfer function has five equally spaced zeros on the left half of the unit circle, and five poles at their mirror angles near the right half of the circle (but with varying radii). The locations of all ten poles and zeros are mapped in Fig. 8-4(a).

The filter transfer function $\mathbf{H}(\mathbf{z})$ is obtained by inserting the specified values for the poles and zeros in Eq. (8.1) with $\mathbf{C} = 1$. The process leads to

$$\begin{aligned}\mathbf{H}(\mathbf{z}) = \big[&(\mathbf{z} - e^{j\pi/2})(\mathbf{z} - e^{-j\pi/2})(\mathbf{z} - e^{j3\pi/4})\\ &\cdot (\mathbf{z} - e^{-j3\pi/4})(\mathbf{z} - e^{j\pi})\big]\\ &\cdot \big[(\mathbf{z} - 0.6)(\mathbf{z} - 0.8e^{j\pi/4})(\mathbf{z} - 0.8e^{-j\pi/4})(\mathbf{z} - 0.8e^{j\pi/2})\\ &\cdot (\mathbf{z} - 0.8e^{-j\pi/2})\big]^{-1}.\end{aligned}$$

After a few simple steps of algebra, the expression reduces to

$$\mathbf{H}(\mathbf{z}) = \frac{\mathbf{z}^5 + 2.414\mathbf{z}^4 + 3.414\mathbf{z}^3 + 3.414\mathbf{z}^2 + 2.414\mathbf{z} + 1}{\mathbf{z}^5 - 1.73\mathbf{z}^4 + 1.96\mathbf{z}^3 - 1.49\mathbf{z}^2 + 0.84\mathbf{z} - 0.25}. \tag{8.7}$$

(b) Noting that $\mathbf{H}(\mathbf{z}) = \mathbf{Y}(\mathbf{z})/\mathbf{X}(\mathbf{z})$, the difference equation can be obtained formally by rearranging Eq. (8.7) in powers of $\mathbf{z}^{-1}$, cross multiplying, and then inverse $\mathbf{z}$-transforming to discrete time, or it can be obtained informally by simply reading off the coefficients of the numerator and denominator. Either approach leads to

$$\begin{aligned}y[n] &- 1.73y[n-1] + 1.96y[n-2] - 1.49y[n-3]\\ &+ 0.84y[n-4] - 0.25y[n-5]\\ &= x[n] + 2.414x[n-1] + 3.414x[n-2]\\ &+ 3.414x[n-3] + 2.414x[n-4] + x[n-5].\end{aligned} \tag{8.8}$$

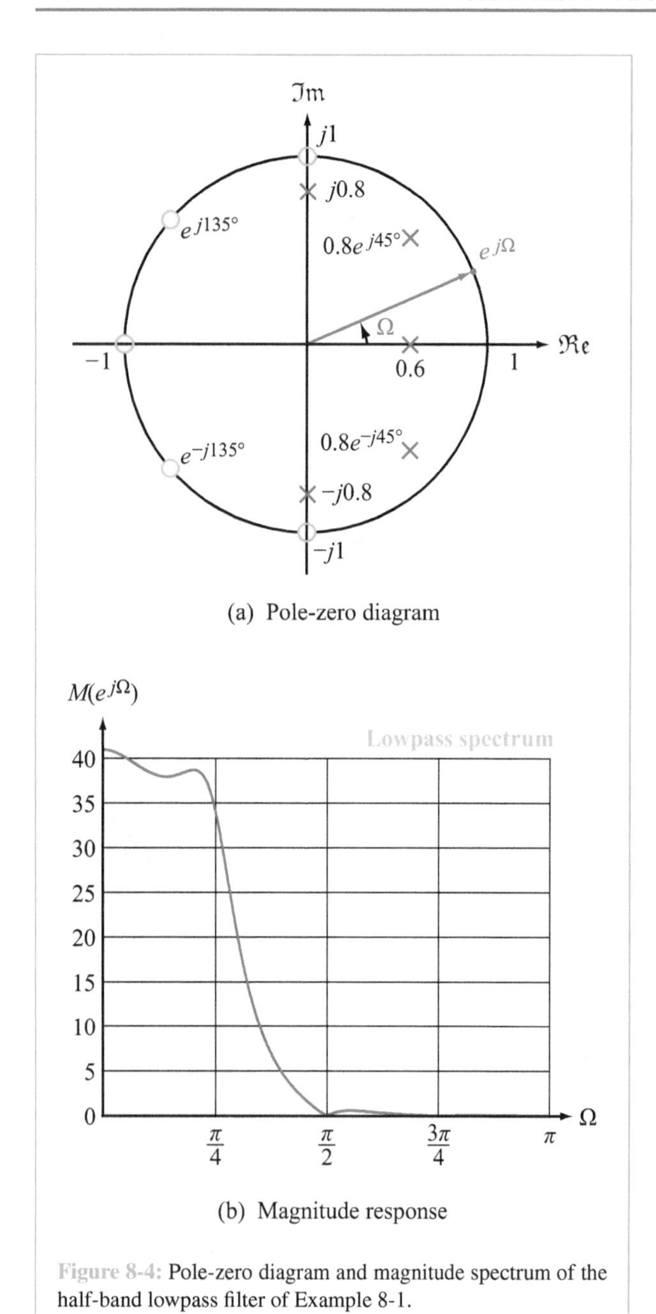

Figure 8-4: Pole-zero diagram and magnitude spectrum of the half-band lowpass filter of Example 8-1.

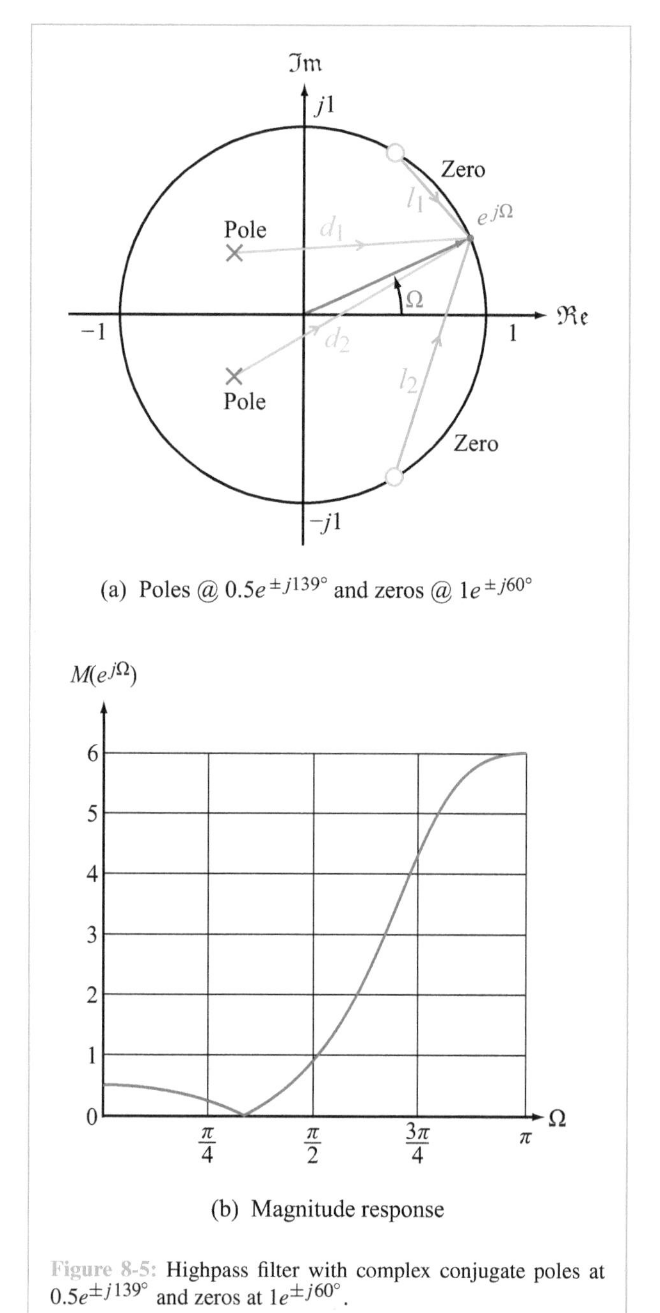

Figure 8-5: Highpass filter with complex conjugate poles at $0.5e^{\pm j139°}$ and zeros at $1e^{\pm j60°}$.

The ARMA difference equation is initialized with $y[n] = 0$ for $n < 5$. The first recursion computes $y[5]$ from inputs $\{x[0], \ldots, x[5]\}$.

(c) Figure 8-4(b) displays the magnitude spectrum $M(e^{j\Omega})$, obtained by replacing $\mathbf{z}$ with $e^{j\Omega}$ in Eq. (8.7) and then computing $M(e^{j\Omega})$ for various values of Ω over the range $0 \leq \Omega \leq \pi$. (See S² for more details.)

8-1.3 Highpass Filter

Reversing the angular locations of the poles and zeros of the lowpass filter configuration shown earlier in Fig. 8-3 leads to a highpass filter with the magnitude spectrum shown in Fig. 8-5. By the same token, if the poles and zeros of the half-band lowpass filter shown in Fig. 8-4(a) were to be moved to their mirror image locations relative to the imaginary axis, the lowpass-filter spectrum would become a ***half-band highpass-filter*** spectrum.

8-1.4 Bandpass Filter

Placing conjugate poles along the vertical axis, close to the unit circle, generates a simple bandpass filter, as shown in Fig. 8-6. The location of the peak corresponds to the angles of the conjugate poles.

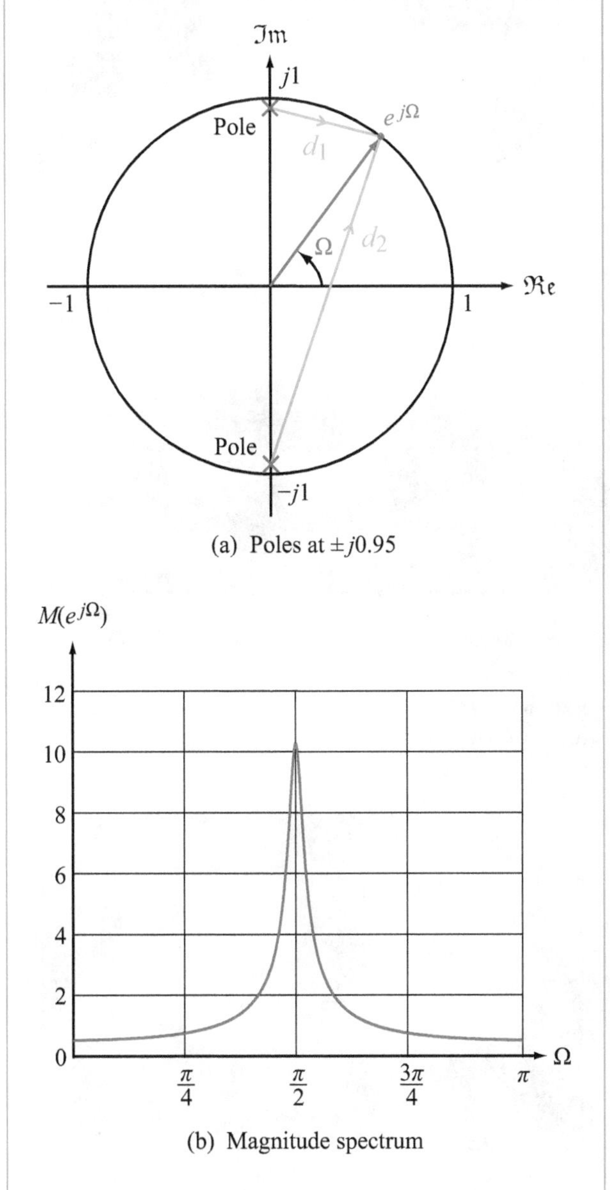

Figure 8-6: Bandpass filter consisting of complex conjugate poles at $\pm j0.95$.

8-1.5 Bandreject Filter

To create a bandreject filter, we need a configuration with zeros placed in the part of the spectrum where the rejection band is intended to be. In Fig. 8-7(a), four zeros are used, a pair at $1e^{\pm j72^\circ}$ and another at $1e^{\pm j108^\circ}$. The result is a

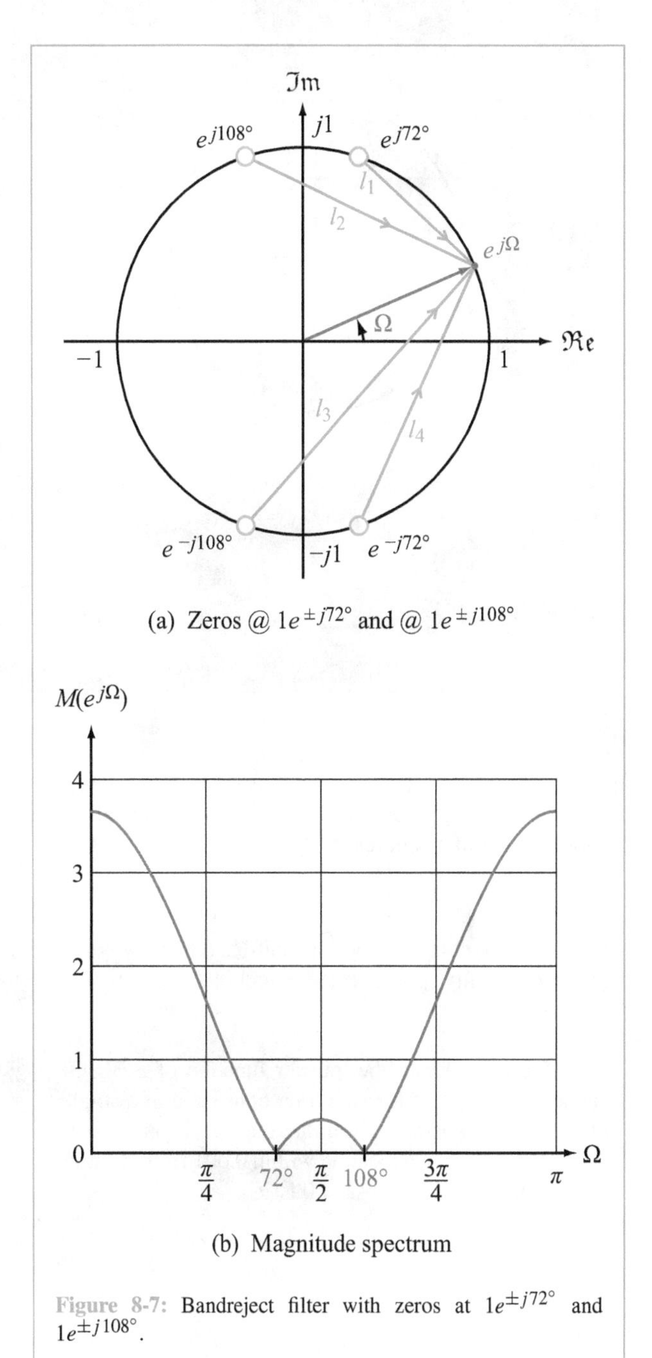

Figure 8-7: Bandreject filter with zeros at $1e^{\pm j72^\circ}$ and $1e^{\pm j108^\circ}$.

Module 8.1 Discrete-Time Frequency Response from Poles and Zeros This module allows the user to specify locations of up to 6 complex conjugate pairs of poles and zeros, and computes the discrete-time frequency response of the resulting system. This is useful for discrete-time filter design.

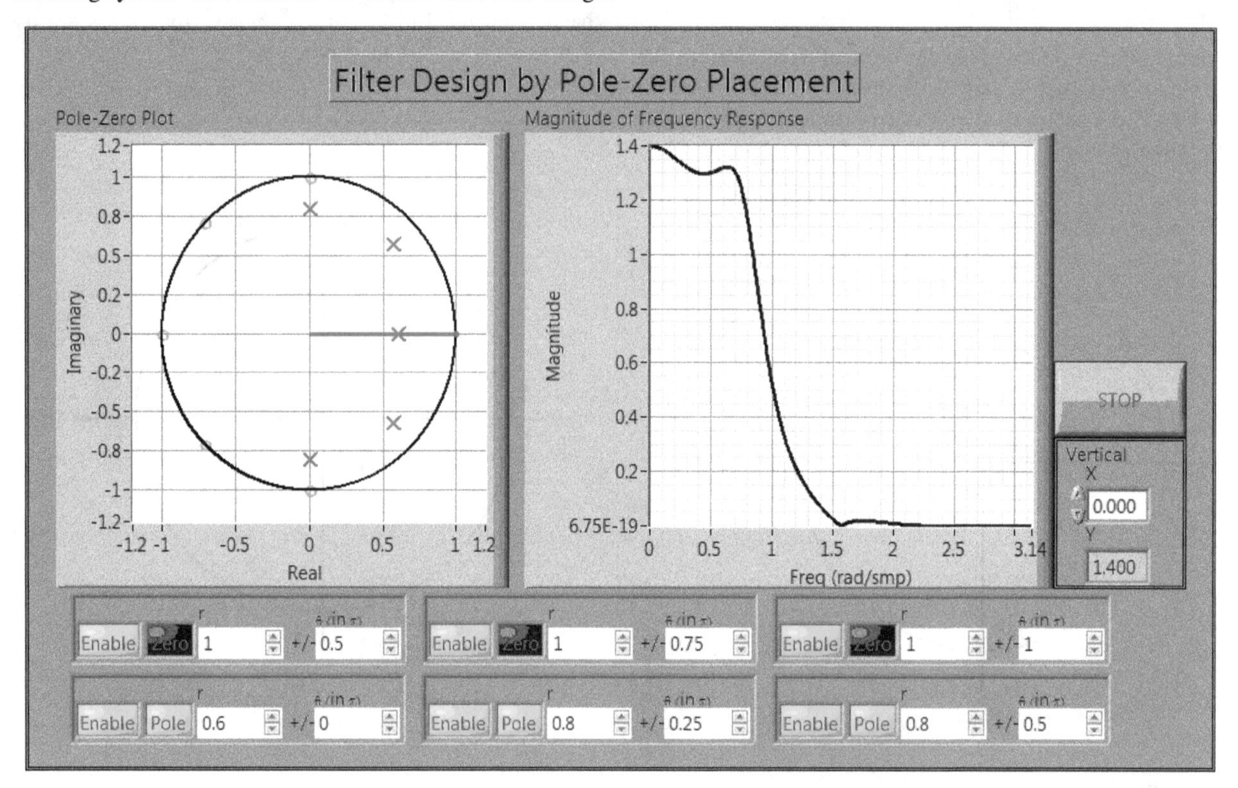

magnitude spectrum (Fig. 8-7(b)) that has nulls at values of Ω corresponding to $\pm 72°$ and $\pm 108°$, and an overall shape resembling that of a bandreject filter.

Concept Question 8-1: For the half-band filter, why are the zeros equally spaced on the unit circle? (See S²)

Exercise 8-1: Obtain the transfer function of a BIBO-stable, discrete-time lowpass filter consisting of a single pole and a single zero, given that the zero is on the unit circle, the pole is at a location within 0.001 from the unit circle, and the dc gain at $\Omega = 0$ is 1.

Answer: $\mathbf{H}(\mathbf{z}) = 0.0005(\mathbf{z}+1)/(\mathbf{z}-0.999)$. (See S²)

Exercise 8-2: Use LabVIEW Module 8.1 to replicate the result of Section 8-1.2 and produce Fig. 8-3.

Answer:

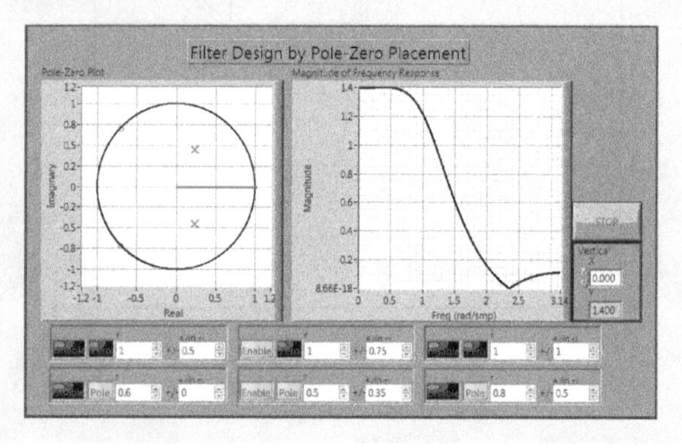

Exercise 8-3: Use LabVIEW Module 8.1 to replicate Example 8-1 and produce Fig. 8-4.

Answer:

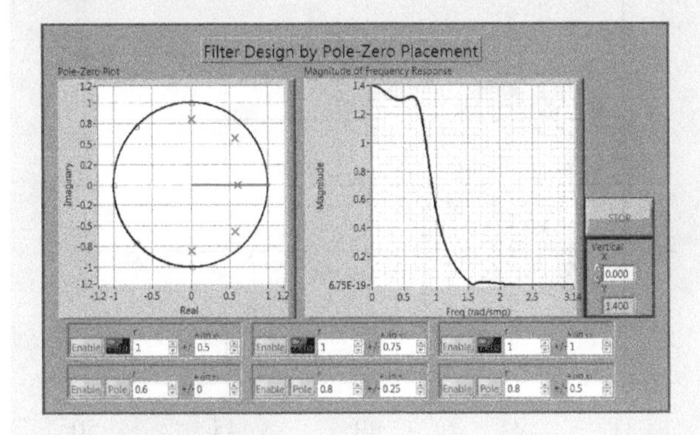

8-2 Notch Filters

Continuous-time notch filters were discussed at length in Chapter 6. The present section provides a parallel treatment for discrete-time notch filters. In both cases, the goal is to design a filter that can eliminate an interfering sinusoid of frequency f_i Hz from a signal $x(t)$, while leaving other frequency components of $x(t)$ (almost) unaffected.

In the discrete-time case, signal $x(t)$ is sampled every T_s seconds (at a sampling rate $f_s = 1/T_s$ samples per second) to generate $x[n] = x(nT_s)$. The same process holds for the interfering sinusoid $x_i(t)$. Thus, if

$$x_i(t) = A\cos(2\pi f_i t + \phi) \tag{8.9}$$

is sampled at $t = nT_s$, we obtain the discrete form

$$x_i[n] = A\cos(2\pi f_i(nT_s) + \phi) = A\cos(\Omega_i n + \phi), \tag{8.10}$$

where Ω_i is the ***discrete-time angular frequency*** of the interfering sinusoid, defined as

$$\Omega_i = 2\pi f_i T_s = \frac{2\pi f_i}{f_s}. \tag{8.11}$$

To remove a sinusoid at angular frequency Ω_i, the filter's transfer function $\mathbf{H}(\mathbf{z})$ should have zeros at $e^{\pm j\Omega_i}$. Furthermore, to minimize the impact of the two zeros on other frequencies, the transfer function should have two poles at $ae^{\pm j\Omega_i}$, with a chosen to be slightly smaller than 1. Closely placed pole/zero pairs effectively cancel each other at frequencies not close to Ω_i. By choosing $a < 1$, we ensure BIBO stability of the filter.

With these poles and zeros, $\mathbf{H}(\mathbf{z})$ is given by

$$\mathbf{H}(\mathbf{z}) = \frac{(\mathbf{z}-e^{j\Omega_i})(\mathbf{z}-e^{-j\Omega_i})}{(\mathbf{z}-ae^{j\Omega_i})(\mathbf{z}-ae^{-j\Omega_i})} = \frac{\mathbf{z}^2 - 2\mathbf{z}\cos\Omega_i + 1}{\mathbf{z}^2 - 2a\mathbf{z}\cos\Omega_i + a^2}. \tag{8.12}$$

Application of partial fraction expansion leads to the impulse response

$$h[n] = \mathbf{Z}^{-1}[\mathbf{H}(\mathbf{z})] = \delta[n] + Ka^n\cos(\Omega_i n + \phi')\,u[n-1], \tag{8.13}$$

where constants K and ϕ' are functions of a and Ω_i. As in the continuous case, a is a trade-off parameter between resolution (sharpness of the dip of $\mathbf{H}(e^{j\Omega})$ at Ω_i) and the duration of $h[n]$.

Implementation of the notch filter is realized using a difference equation, obtained by: (1) rearranging the numerator and denominator of $\mathbf{H}(\mathbf{z})$ in powers of $\mathbf{z}^{-1}$,

$$\frac{\mathbf{Y}(\mathbf{z})}{\mathbf{X}(\mathbf{z})} = \frac{\mathbf{z}^2 - 2\mathbf{z}\cos\Omega_i + 1}{\mathbf{z}^2 - 2a\mathbf{z}\cos\Omega_i + a^2} = \frac{1 - 2\mathbf{z}^{-1}\cos\Omega_i + \mathbf{z}^{-2}}{1 - 2a\mathbf{z}^{-1}\cos\Omega_i + a^2\mathbf{z}^{-2}},$$

(2) cross-multiplying to get

$$\mathbf{Y}(\mathbf{z})[1 - 2a\mathbf{z}^{-1}\cos\Omega_i + a^2\mathbf{z}^{-2}] = \mathbf{X}(\mathbf{z})[1 - 2\mathbf{z}^{-1}\cos\Omega_i + \mathbf{z}^{-2}],$$

and (3) inverse $\mathbf{z}$-transforming to discrete time,

$$\begin{aligned} y[n] &- 2a\cos\Omega_i\, y[n-1] + a^2\, y[n-2] \\ &= x[n] - 2\cos\Omega_i\, x[n-1] + x[n-2]. \end{aligned} \tag{8.14}$$

According to Eq. (8.14), the notch filter is implemented as a linear combination of the present input and the two most recent inputs and outputs. Note that the coefficients of the ARMA difference equation can be read off directly from the expression of $\mathbf{H}(\mathbf{z})$.

Example 8-2: Notch-Filter Design I

Signal $x(t) = \sin(250\pi t) + \sin(400\pi t)$ is the sum of two sinusoids, one at $f_1 = 125$ Hz and another at $f_2 = 200$ Hz. The signal is sampled by a digital signal processing (DSP) system at 1000 samples per second. Design a notch filter with $a = 0.9$ to reject the 200 Hz sinusoid, compare the input and output signals, and plot the filter's magnitude frequency response.

Solution: At $f_s = 1000$ samples/s, $T_s = 1/f_s = 10^{-3}$ s, and the angular frequency of the sinusoid to be rejected is

$$\Omega_2 = 2\pi f_2 T_s = 2\pi \times 200 \times 10^{-3} = 0.4\pi.$$

With $\Omega_i = \Omega_2 = 0.4\pi$ and $a = 0.9$, Eq. (8.12) becomes

$$\begin{aligned}\mathbf{H}(\mathbf{z}) &= \frac{\mathbf{z}^2 - 2\mathbf{z}\cos\Omega_2 + 1}{\mathbf{z}^2 - 2a\mathbf{z}\cos\Omega_2 + a^2} \\ &= \frac{\mathbf{z}^2 - 2\mathbf{z}\cos(0.4\pi) + 1}{\mathbf{z}^2 - 2\times 0.9\mathbf{z}\cos(0.4\pi) + (0.9)^2} \\ &= \frac{\mathbf{z}^2 - 0.62\mathbf{z} + 1}{\mathbf{z}^2 - 0.56\mathbf{z} + 0.81}. \end{aligned} \tag{8.15}$$

By reading off the coefficients of the numerator and denominator, we obtain the ARMA difference equation

$$\begin{aligned} y[n] - 0.56y[n-1] + 0.81y[n-2] \\ = x[n] - 0.62x[n-1] + x[n-2]. \end{aligned}$$

The input, $x[n]$, is the sampled version of signal $x(t)$, namely,

$$x[n] = \sin(\Omega_1 n) + \sin(\Omega_2 n),$$

with $\Omega_1 = 0.25\pi$ and $\Omega_2 = 0.4\pi$. The corresponding filtered output is $y[n]$. The process can be initiated by setting $y[-2] = y[-1] = x[-2] = x[-1] = 0$. From Eq. (8.13), we see that the non-impulse term decays as $a^n = 0.9^n$. At $n = 30$ (which occurs at 30 ms because the sampling interval is 1 ms), $(0.9)^{30} \approx 0.04$, which means that by then, the transient component of the output signal has essentially concluded. Figure 8-8 displays plots of $x[n]$ and $y[n]$, starting at $n = 31$, thereby avoiding the transient part of the output signal. We observe that $y[n]$ is clearly a single sinusoid; the notch filter has indeed filtered out the 200 Hz sinusoid.

The frequency response of the notch filter, $\mathbf{H}(e^{j\Omega})$, is obtained from Eq. (8.15) by setting $\mathbf{z} = e^{j\Omega}$,

$$\mathbf{H}(e^{j\Omega}) = \frac{e^{j2\Omega} - 0.62e^{j\Omega} + 1}{e^{j2\Omega} - 0.56e^{j\Omega} + 0.81}.$$

At the two ends of the spectrum, namely at $\Omega = 0$ and π,

$$\mathbf{H}(e^{j0}) = \frac{1 - 0.62 + 1}{1 - 0.56 + 0.81} = 1.104$$

and

$$\mathbf{H}(e^{j\pi}) = \frac{1 + 0.62 + 1}{1 + 0.56 + 0.81} = 1.105.$$

The magnitude of an ideal notch filter should be 1 at frequencies away from the notch frequency $\Omega_2 = 0.4\pi$. Hence, we can modify $\mathbf{H}(e^{j\Omega})$ by multiplying it by 1/1.104. The magnitude of the unmodified response is shown in Fig. 8-9. See S² for details.

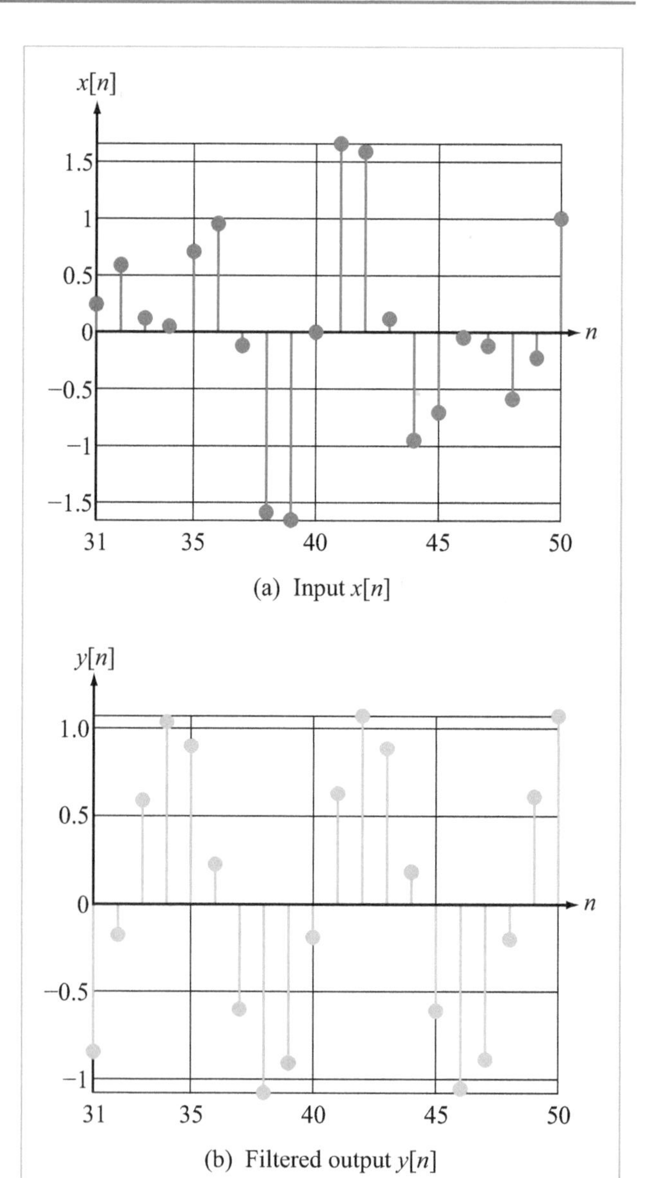

Figure 8-8: Time waveforms of (a) the original and (b) filtered signals of Example 8-2.

Example 8-3: Notch Filter Design II

Design a discrete-time filter that fulfills the following specifications:

(1) It rejects 1000 Hz.

(2) Its magnitude response is 0.9 at both 900 Hz and 1100 Hz.

(3) Its magnitude response is greater than 0.9 outside the range 900 Hz $< f <$ 1100 Hz.

(4) The magnitude at 0 Hz (dc) is 1.

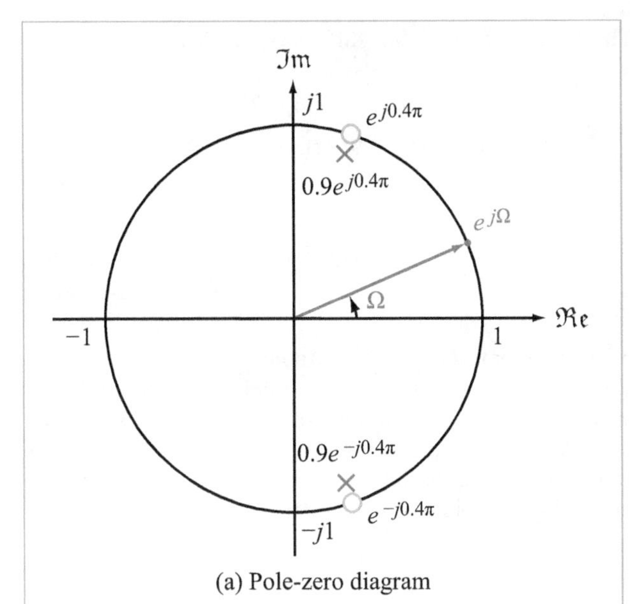

(a) Pole-zero diagram

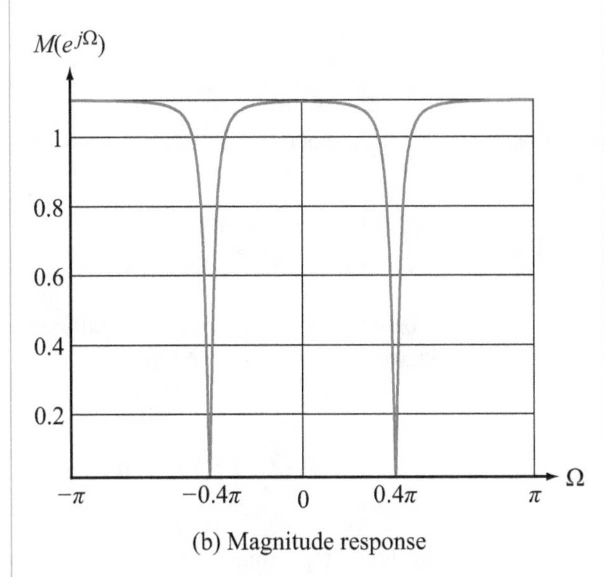

(b) Magnitude response

Figure 8-9: Pole-zero diagram and magnitude response of notch filter of Example 8-2.

The filter (intended for continuous-time signals that had been sampled at 6000 samples/s) should use two poles and two zeros.

Solution: When sampled at a rate $f_s = 6000$ samples/s, a continuous-time sinusoidal signal of angular frequency ω becomes a discrete-time sinusoid of angular frequency

$$\Omega = \frac{\omega}{f_s} = \frac{2\pi f}{f_s} . \tag{8.16}$$

Accordingly, the notch frequency $f_i = 1000$ Hz corresponds to

$$\Omega_i = \frac{2\pi \times 1000}{6000} = \frac{\pi}{3} .$$

To reject a discrete-time sinusoid at Ω_i, the filter's transfer function should have zeros at $e^{\pm j\Omega_i}$ and neighboring poles at $ae^{\pm j\Omega_i}$ (Fig. 8-10(a)), with a close to but smaller than 1 to ensure BIBO stability.

The frequency response of the notch filter, $\mathbf{H}(e^{j\Omega})$, is obtained from Eq. (8.12) by setting $\mathbf{z} = e^{j\Omega}$. Thus,

$$\mathbf{H}(e^{j\Omega}) = \mathbf{C}\, \frac{e^{j2\Omega} - e^{j\Omega} + 1}{e^{j2\Omega} - ae^{j\Omega} + a^2} , \tag{8.17}$$

where $\mathbf{C}$, an as yet to be determined constant, has been added as a multiplicative constant so as to satisfy the problem specifications. One of the filter specifications states that $\mathbf{H}(e^{j0}) = 1$, which requires

$$1 = \mathbf{C}\, \frac{1 - 1 + 1}{1 - a + a^2} ,$$

or

$$\mathbf{C} = 1 - a + a^2. \tag{8.18}$$

The constant a is determined from the specification that $M(e^{j\Omega}) = 0.9$ at $f = 900$ Hz and 1100 Hz or, equivalently, at

$$\Omega_L = 2\pi\, \frac{900}{6000} = 0.3\pi \tag{8.19a}$$

and

$$\Omega_H = 2\pi\, \frac{1100}{6000} = 0.367\pi. \tag{8.19b}$$

Thus, the specifications require

$$0.9 = (1 - a + a^2) \left| \frac{e^{j0.6\pi} - e^{j0.3\pi} + 1}{e^{j0.6\pi} - ae^{j0.3\pi} + a^2} \right| \tag{8.20a}$$

and

$$0.9 = (1 - a + a^2) \left| \frac{e^{j0.734\pi} - e^{j0.367\pi} + 1}{e^{j0.734\pi} - ae^{j0.367\pi} + a^2} \right| . \tag{8.20b}$$

By trial and error, we find that $a = 0.95$ provides a satisfactory solution. The corresponding magnitude response of the filter is shown in Fig. 8-10(b). The horizontal axis was converted from Ω to f using Eq. (8.16).

In conclusion, with $a = 0.95$ and

$$\mathbf{C} = 1 - a + a^2 = 1 - 0.95 + (0.95)^2 = 0.953,$$

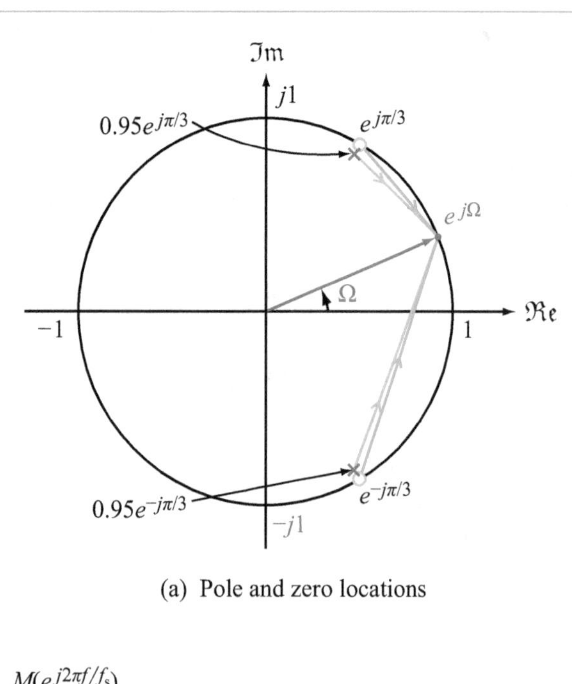

(a) Pole and zero locations

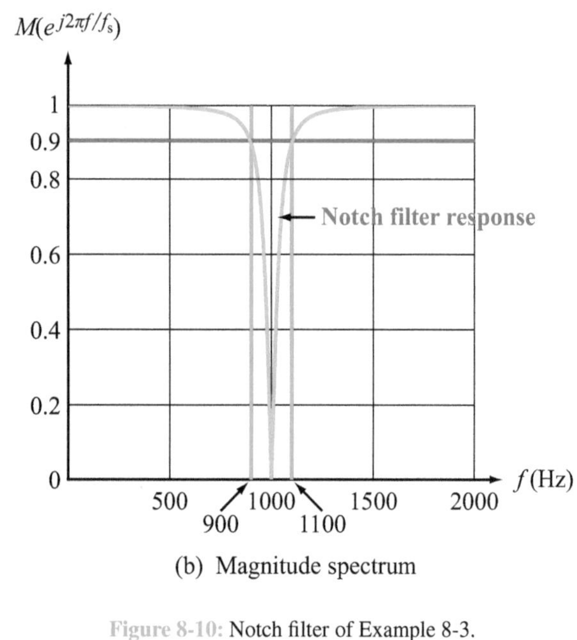

(b) Magnitude spectrum

Figure 8-10: Notch filter of Example 8-3.

replacing $e^{j\Omega}$ with $\mathbf{z}$ in Eq. (8.17) leads to the notch filter transfer function

$$\mathbf{H}(\mathbf{z}) = 0.953\left(\frac{\mathbf{z}^2 - \mathbf{z} + 1}{\mathbf{z}^2 - 0.95\mathbf{z} + 0.9}\right).$$

The corresponding ARMA difference equation is

$$\begin{aligned} y[n] - 0.95y[n-1] + 0.9y[n-2] \\ = 0.953x[n] - 0.953x[n-1] + 0.953x[n-2]. \end{aligned}$$

Example 8-4: Trumpet Notch Filter

A signal generated by an actual trumpet playing note B (which has frequency harmonics of 491 Hz) had an interfering 1200 Hz sinusoid added to it. The trumpet-plus-sinusoid was sampled at the standard CD sampling rate of 44,100 samples/s. Design and implement a discrete-time notch filter to eliminate the interfering sinusoid. Use $a = 0.99$.

Solution: At $f_s = 44,100$ samples/s, the angular frequency of the interfering sinusoid is

$$\Omega_i = \frac{2\pi f_i}{f_s} = \frac{2\pi \times 1200}{44,100} = 0.171.$$

With zeros at $e^{\pm j\Omega_i}$ and poles at $0.99e^{\pm j\Omega_i}$, $\mathbf{H}(\mathbf{z})$ assumes the form

$$\mathbf{H}(\mathbf{z}) = \frac{\mathbf{z}^2 - 2\mathbf{z}\cos(0.171) + 1}{\mathbf{z}^2 - 1.98\mathbf{z}\cos(0.171) + (0.99)^2}. \tag{8.21}$$

The notch filter can be implemented by an ARMA difference equation with coefficients read off $\mathbf{H}(\mathbf{z})$, namely

$$\begin{aligned} y[n] - 1.95y[n-1] + 0.98y[n-2] \\ = x[n] - 1.97x[n-1] + x[n-2]. \end{aligned}$$

The harmonics of the trumpet are at $491k$ (Hz), with $k = 1, 2, \ldots,$ and their corresponding discrete-time angular frequencies are

$$\Omega_t = \frac{2\pi \times 491k}{44,100} = 0.07k, \qquad k = 1, 2, \ldots .$$

Figure 8-11 displays the magnitude spectra of (a) the trumpet signal, (b) the interfering 1200 Hz sinusoid (with $\Omega_i = 0.171$), and (c) the notch filter.

Comparison of the trumpet-plus-sinusoid waveforms, before and after passing through the notch filter, is available in Fig. 8-12. The filtering was performed in discrete time, then the signal was reconstructed to continuous time. Also shown is the

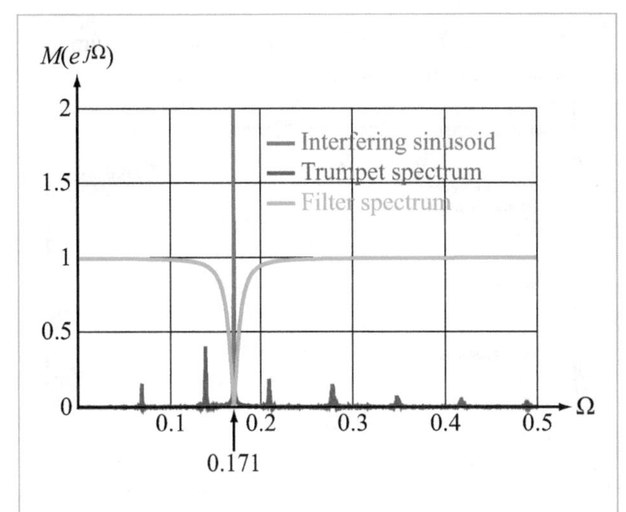

Figure 8-11: Spectra of trumpet signal, interfering sinusoid and notch filter (Example 8-4).

original trumpet signal before adding the interfering sinusoid to it. We observe that the signals in (b) and (c) of Fig. 8-12 are very similar, yet both are very different in both scale and waveform shape from the trumpet-plus-sinusoid signal in part (a) of the figure. Thus, the notch filter did indeed recover the original trumpet signal almost exactly.

As a final note, we should examine how long it takes the transient component of the filter's impulse response to decay to a negligible level. The factor $a^n = (0.99)^n$ decays to ≈ 0.04 if $n = 315$. In continuous time, this corresponds to $t = 315/(44,100) = 0.007$ s. Hence, the filter provides its steady state response within 7 ms after the appearance of the signal at its input terminals. (See S² for more details.)

Concept Question 8-2: Where are the poles and zeros of a discrete-time notch filter located? (See S²)

Exercise 8-4: Determine the ARMA difference equation for the notch filter that rejects a 250 Hz sinusoid. The sampling rate is 1000 samples per second. Use $a = 0.99$.

Answer: $y[n] + 0.98y[n-2] = x[n] + x[n-2]$. (See S²)

Exercise 8-5: Use LabVIEW Module 8.1 to replicate Example 8-2 and produce the pole-zero and gain plots of Fig. 8-9.

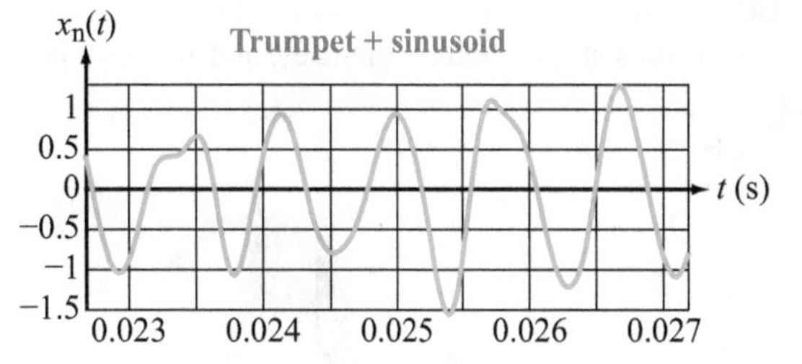

(a) Waveform of trumpet-plus sinusoid

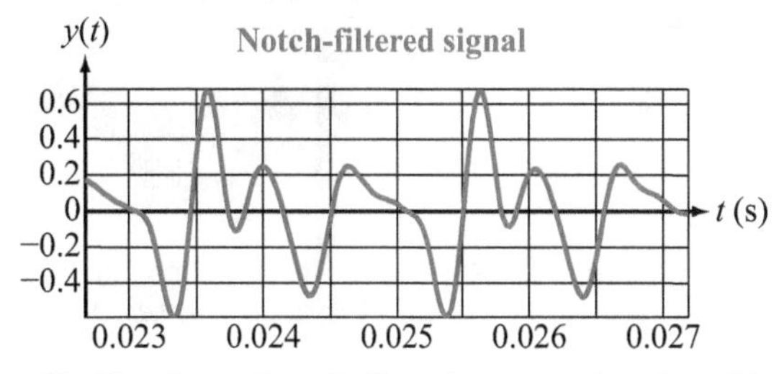

(b) Waveform of notch-filtered trumpet-plus-sinusoid

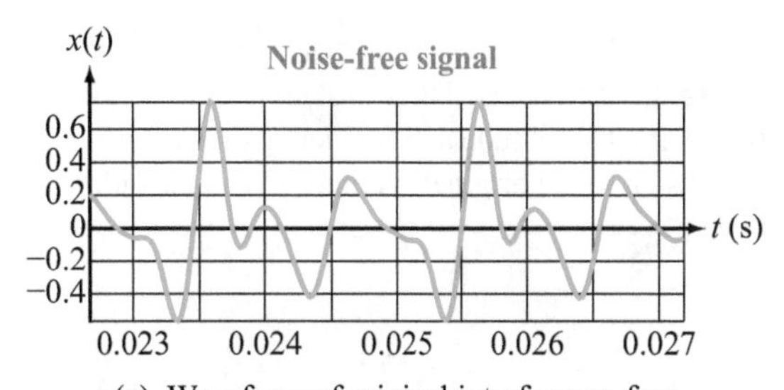

(c) Waveform of original interference-free trumpet signal.

Figure 8-12: Comparison of signal waveforms: (a) trumpet-plus-sinusoid $x_n(t)$, (b) $y(t)$, the notch-filtered version of $x_n(t)$, and (c) original interference-free trumpet signal $x(t)$.

Answer:

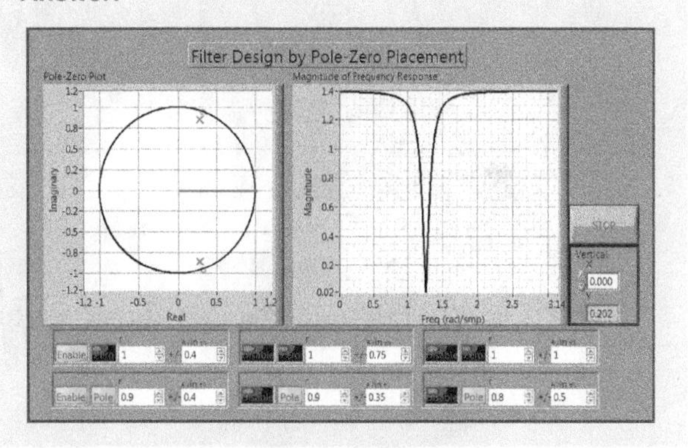

Module 8.2 Discrete-Time Notch Filter to Eliminate One of Two Sinusoids This module computes the sum of two sinusoids at specified frequencies, and applies a notch filter to eliminate one of the sinusoids.

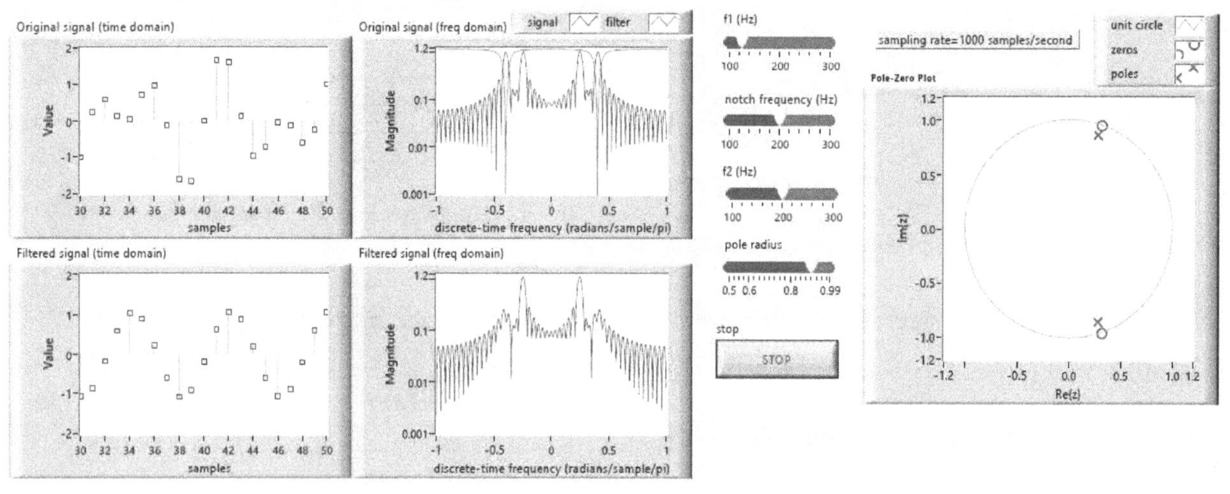

Module 8.3 Discrete-Time Notch Filter to Eliminate Sinusoid from Trumpet Signal This module adds a sinusoid with specified amplitude and frequency to a trumpet signal and uses a notch filter to eliminate the sinusoid. It also allows the user to listen to the original, trumpet-plus-sinusoid, and filtered signals.

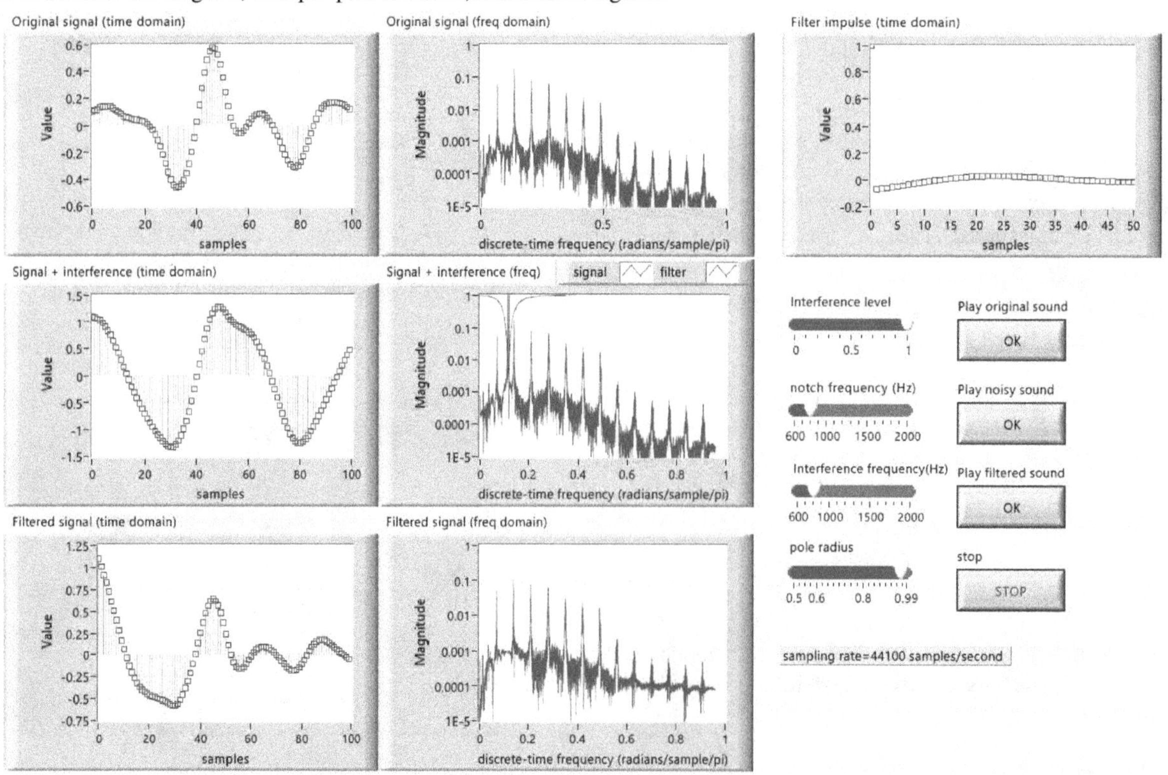

Exercise 8-6: Use LabVIEW Module 8.2 to replicate Example 8-2 and produce the time waveforms of Fig. 8-8.

Answer:

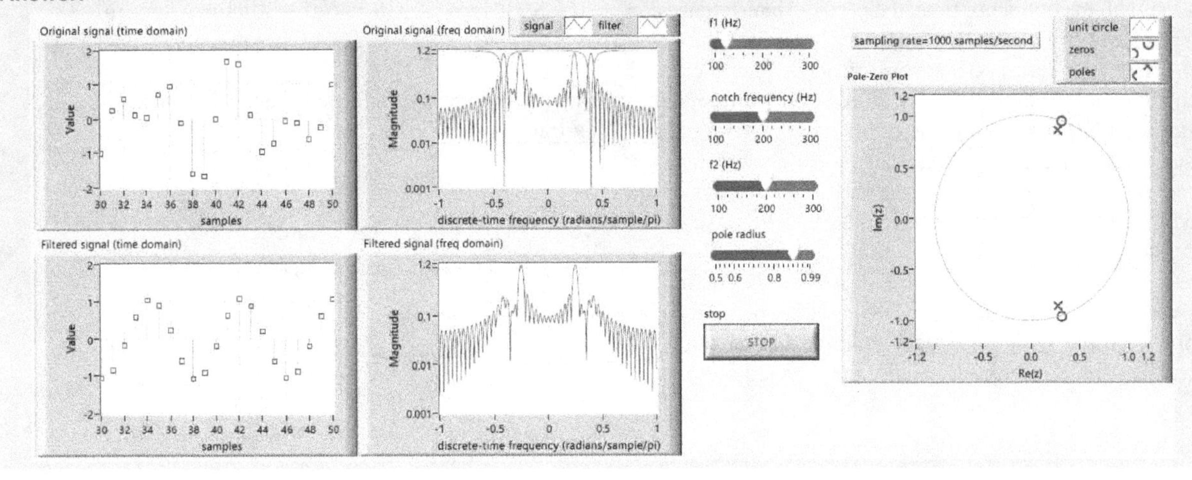

Exercise 8-7: Use LabVIEW Module 8.3 to replicate Example 8-4 and produce the time waveforms of Fig. 8-12 (as stem plots).

Answer:

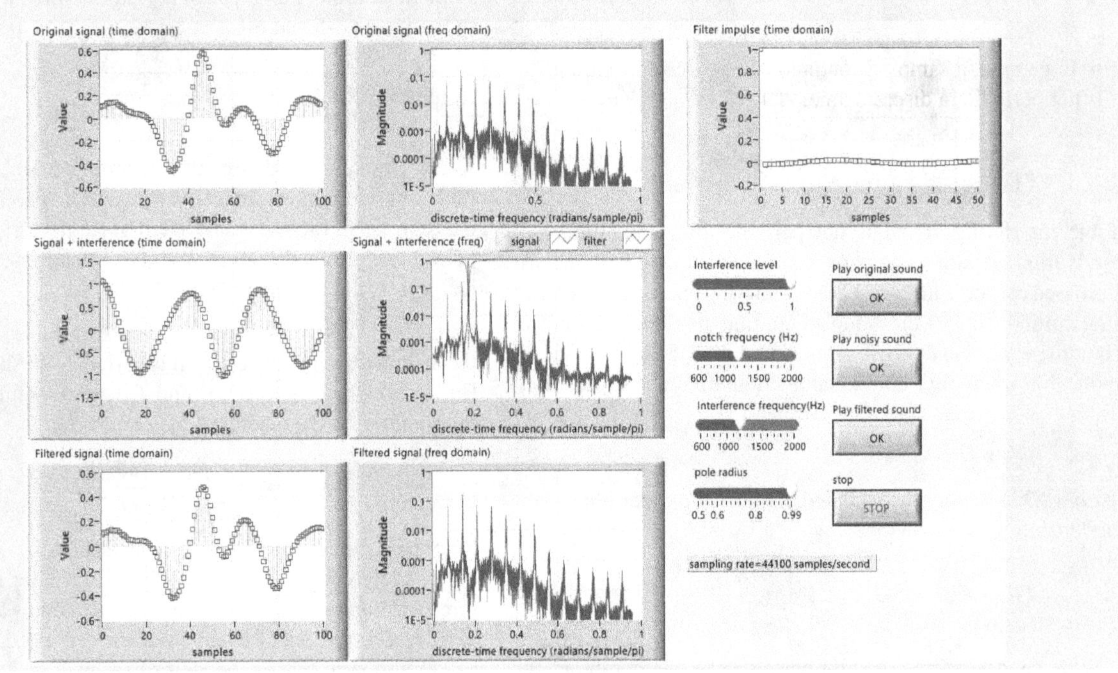

8-3 Comb Filters

The purpose of a comb filter is to remove *periodic* interference from a signal. In continuous time, the periodic interference can be expanded in a Fourier series. Hence, it can be eliminated using a cascade (series connection) of notch filters, each of which eliminates a single harmonic of the interfering signal.

After sampling, each harmonic of the interference becomes a discrete-time sinusoid, which can be eliminated using a discrete-time notch filter. Hence, the various harmonics of the periodic interference again can be removed using a cascade of notch filters—just as in continuous time.

We should note that a signal that is periodic in continuous time with period T_0 *will not necessarily* be periodic after sampling, unless T_0 is an integer multiple of the sampling interval T_s. That is, periodicity in discrete time requires that

$$\frac{T_0}{T_s} = k, \qquad k = \text{integer} \tag{8.22a}$$

or, equivalently,

$$\frac{f_s}{f_0} = k, \qquad k = \text{integer}, \tag{8.22b}$$

where $f_s = 1/T_s$ is the sampling rate and $f_0 = 1/T_0$ is the frequency of the continuous-time periodic signal.

► In general, a sampled continuous-time periodic signal is not periodic in discrete time. ◄

8-3.1 Cascaded Comb Filter

If the interfering signal is not periodic after sampling, the comb filter should consist of a cascade of notch filters, as discussed earlier, with each aimed at filtering one of the discrete-time components of the sampled continuous-time harmonics. The ***cascaded comb filter***, demonstrated in Example 8-5, is applicable whether or not the sampled interference is periodic.

8-3.2 Periodic Comb Filter

If Eq. (8.22) is true, the sampled signal will be periodic with fundamental period N_0 given by

$$N_0 = \frac{f_s}{f_0} = k. \tag{8.23}$$

Whereas designing a comb filter as a cascade combination of multiple notch filters is a perfectly acceptable approach, the periodic nature of the interfering original offers an additional approach characterized by the simple ARMA difference equation given by

$$y[n] - a^{N_0}\, y[n - N_0] = x[n] - x[n - N_0], \tag{8.24}$$

(periodic comb filter)

where a is a selectable constant such that $0 < a < 1$. The **z**-transform of Eq. (8.24) is

$$\mathbf{Y(z)} - a^{N_0}\mathbf{z}^{-N_0}\,\mathbf{Y(z)} = \mathbf{X(z)} - \mathbf{z}^{-N_0}\,\mathbf{X(z)}, \tag{8.25}$$

which leads to the transfer function

$$\mathbf{H(z)} = \frac{\mathbf{Y(z)}}{\mathbf{X(z)}} = \frac{1 - \mathbf{z}^{-N_0}}{1 - a^{N_0}\mathbf{z}^{-N_0}} = \frac{\mathbf{z}^{N_0} - 1}{\mathbf{z}^{N_0} - a^{N_0}}\,. \tag{8.26}$$

The zeros of $\mathbf{H(z)}$ are

$$\mathbf{z}_k = \{e^{j2\pi k/N_0},\ k = 0, 1, \ldots, N_0 - 1\},$$

and its poles are

$$\mathbf{p}_k = \{a e^{j2\pi k/N_0},\ k = 0, 1, \ldots, N_0 - 1\}.$$

The corresponding angular frequencies are

$$\Omega_k = \frac{2\pi k}{N_0}, \qquad k = 0, 1, \ldots, N_0 - 1. \tag{8.27}$$

Implementation of this type of periodic comb filter is illustrated in Examples 8-5 and 8-6.

Example 8-5: Comb Filter Design

A 30 Hz sinusoidal signal is corrupted by a zero-mean, 60 Hz periodic interference from a motor. The interference is bandlimited to 180 Hz. Using a DSP system with a sampling rate of 480 samples/s, design a comb filter that passes the sinusoid and rejects the interference. Use $a = 0.95$.

Solution: Our first step should be to determine whether or not the sampled interference signal is periodic. We do so by testing Eq. (8.23). Since $f_0 = 60$ Hz and $f_s = 480$ samples/s, the ratio, if an integer, is the discrete-time period N_0:

$$N_0 = \frac{f_s}{f_0} = \frac{480}{60} = 8.$$

Since the sampled interference is indeed periodic, we can pursue either of two paths: (a) designing a notch filter to remove the three harmonics of the interfering signal, namely at 60, 120, and 180 Hz ("zero-mean" means that there is no dc component) or (2) applying the model described by Eq. (8.24), which is applicable to periodic discrete-time signals only.

We will explore both options.

(a) Cascaded comb filter

The transfer function of the comb filter consists of a series connection of three notch filters with frequencies

$$\Omega_{01} = \frac{2\pi \times 60}{480} = \frac{\pi}{4},$$
$$\Omega_{02} = 2\Omega_0 = \frac{\pi}{2},$$

and

$$\Omega_{03} = 3\Omega_0 = \frac{3\pi}{4}.$$

Hence, $\mathbf{H(z)}$ is given by

$$\mathbf{H(z)} = \prod_{k=1}^{3} \frac{(\mathbf{z} - e^{jk\pi/4})(\mathbf{z} - e^{-jk\pi/4})}{(\mathbf{z} - 0.95e^{jk\pi/4})(\mathbf{z} - 0.95e^{-jk\pi/4})},$$

where we selected $a = 0.95$ for all three notch filters. After several steps of algebra, the expression for $\mathbf{H(z)}$ simplifies to

$$\mathbf{H(z)} = \frac{\mathbf{z}^6 + \mathbf{z}^4 + \mathbf{z}^2 + 1}{\mathbf{z}^6 + (0.95)^2\mathbf{z}^4 + (0.95)^4\mathbf{z}^2 + (0.95)^6}. \tag{8.28}$$

The pole-zero diagram and magnitude spectrum of $\mathbf{H}(e^{j\Omega})$ are displayed in Fig. 8-13.

The comb filter is implemented by an ARMA difference equation, which can be readily read off $\mathbf{H(z)}$ as

$$\begin{aligned} y[n] + (0.95)^2 y[n-2] + (0.95)^4 y[n-4] \\ + (0.95)^6 y[n-6] \\ = x[n] + x[n-2] + x[n-4] + x[n-6]. \end{aligned} \tag{8.29}$$

To initialize the computation of $y[n]$, we start by setting $y[n] = 0$ for $n < 6$. The first recursion computes $y[6]$ in terms of $\{x[0], \ldots, x[6]\}$. The waveforms of the original and filtered signals are shown in Fig. 8-14. It is clear that after passing through the comb filter, only the desired 30 Hz sinusoid remains. As with the notch filter of the preceding section, the dc magnitude $\mathbf{H}(e^{j0})$ is slightly greater than 1. This can again easily be corrected by multiplying $\mathbf{H(z)}$ by $1/\mathbf{H}(e^{j0})$.

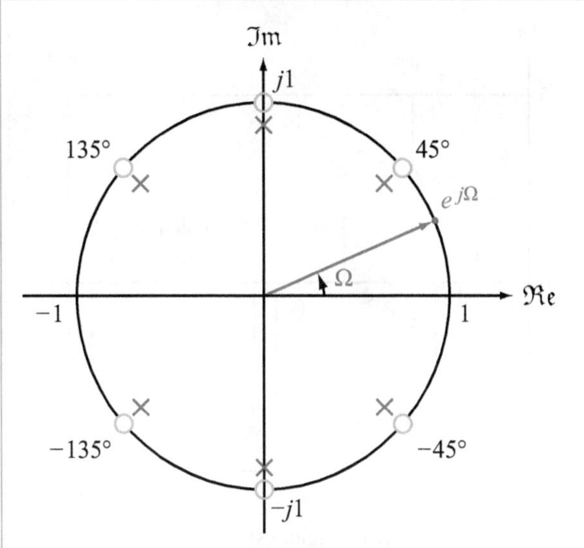

(a) Pole-zero diagram

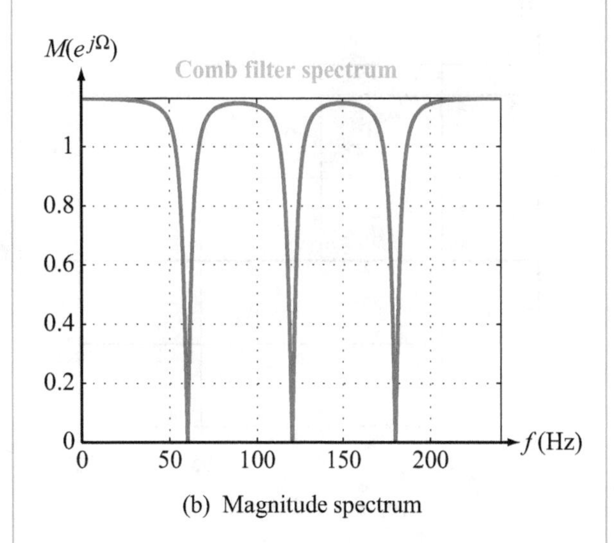

(b) Magnitude spectrum

Figure 8-13: Pole-zero diagram and magnitude spectrum of comb filter (Example 8-5a).

(b) Periodic comb filter

We implement the filter given by Eq. (8.24) with $N_0 = 8$ and $a = 0.95$:

$$y[n] - 0.95^8 y[n-8] = x[n] - x[n-8]. \tag{8.30}$$

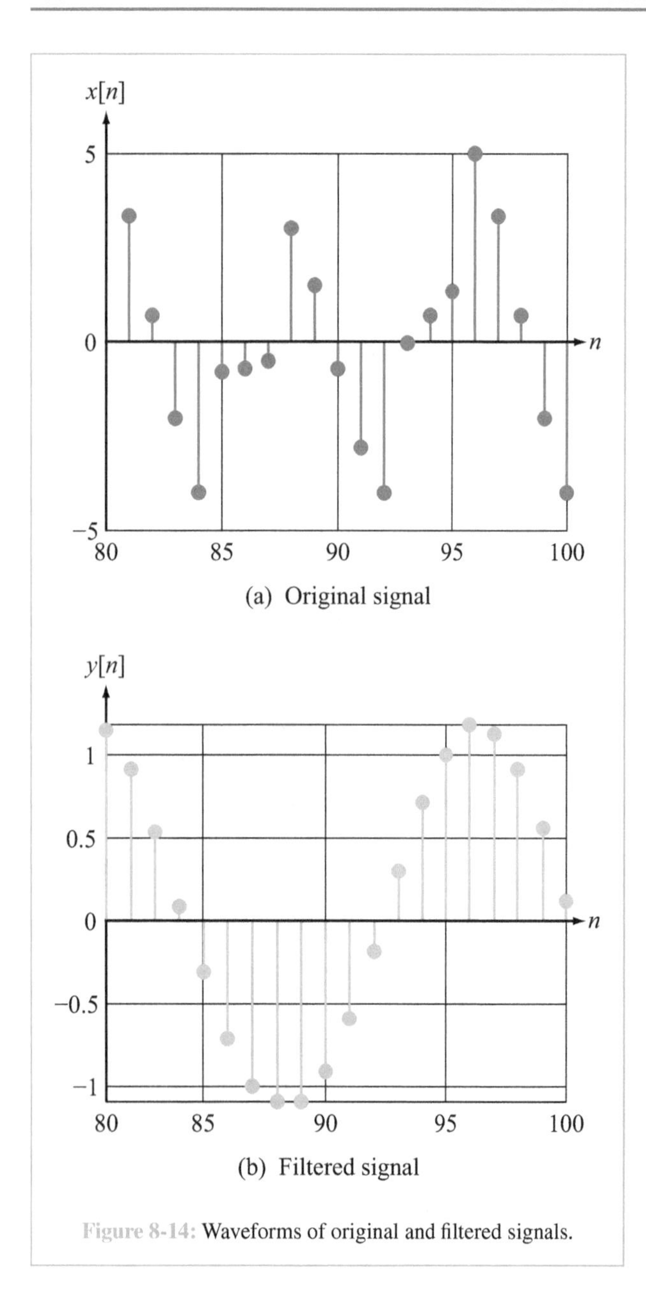

Figure 8-14: Waveforms of original and filtered signals.

The filter has eight zeros and eight poles, at angles Ω_k given by Eq. (8.27) as

$$\Omega_k = \frac{2\pi k}{8} = \frac{\pi k}{4}, \qquad k = 0, 1, \ldots, 7.$$

Six of these pole/zero pairs are identical to those of the transfer function of the filter in part (a). The periodic filter has two additional notches that would reject signals with $\Omega = 0$ and $\Omega = \pi$, but neither is associated with any components of the interfering periodic signal. The main advantage of the periodic notch filter described by Eq. (8.30) over the cascaded notch filter given by Eq. (8.29) is implementation; far fewer addition and multiplication operations are required. (See S² for details.)

Example 8-6: Comb-Filtering Trumpet Signal

Consider a signal consisting of the superposition of two actual trumpets: one playing note A at a fundamental frequency of 440 Hz and another playing note B at a frequency of 491 Hz. The signal was sampled at the standard CD sampling rate of 44,100 sample/s. Design and implement a discrete-time comb filter to eliminate the signal of the trumpet playing note A, while keeping the signal of note B.

Solution: The angular frequencies corresponding to notes A and B are

$$\Omega_{A_k} = \frac{2\pi f_A k}{f_s} = \frac{2\pi \times 440k}{44,100} = 0.0627k, \qquad k = 1, 2, \ldots$$

and

$$\Omega_{B_k} = \frac{2\pi f_B k}{44,100} = \frac{2\pi \times 491k}{44,100} = 0.07k, \qquad k = 1, 2, \ldots.$$

The goal is to design a comb filter that passes all harmonics of Ω_B and rejects all harmonics of Ω_A. For the sampled signal to be exactly periodic at a period corresponding to Ω_A, the ratio f_s/f_A has to be an integer. In the present case, $f_s/f_A = 44,100/440 = 100.227$, which is not an integer, but we may approximate the ratio as 100 or 101 so as to use the periodic comb filter given by Eq. (8.24) with $a = 0.99$:

$$y[n] - 0.99^{N_0} y[n - N_0] = x[n] - x[n - N_0]. \qquad (8.31)$$

The filter was evaluated for $N_0 = 100$ and $N_0 = 101$, and the latter value was found to provide better performance.

Figure 8-15(a) displays spectra of note A, note B, and the filter's magnitude response. For comparison, we show in Fig. 8-15(b) and (c) the waveforms of the combined trumpet signals (notes A and B together) before and after getting filtered

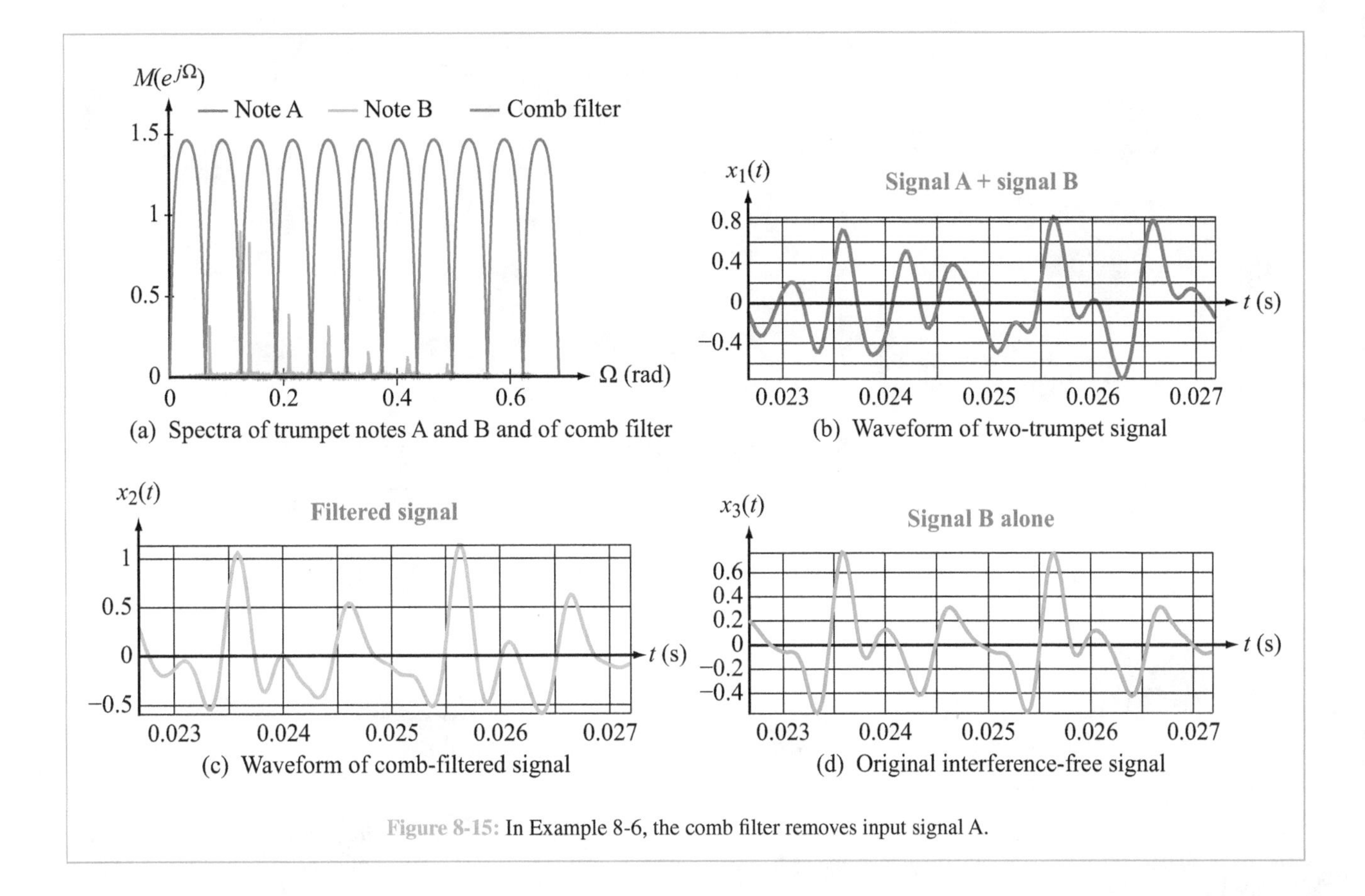

Figure 8-15: In Example 8-6, the comb filter removes input signal A.

by the periodic comb filter. The filtered waveform is a close, but not perfect, reproduction of the original waveform of note B alone (which is shown in part (d) of the figure).

The filter succeeds in removing all of the harmonics of note A, but it also removes those harmonics of note B that happen to coincide with one or more of the filter's notch frequencies. When a harmonic of note A overlaps with a harmonic of note B, it is difficult for any filter to distinguish one from the other. For example, the angular frequency of the seventh harmonic of note A is $\Omega_{A_7} = 0.4388$ rad, which is close to the sixth harmonic of note B, $\Omega_{B_6} = 0.4197$ rad. To distinguish between two closely spaced harmonics, the notches in the filter's frequency response have to be much narrower than the spacing between the harmonics.

The subject of frequency filtering a periodic interference will be revisited in Section 8-8, where we discuss ***batch filtering***, a technique that can outperform both the periodic and cascaded comb filters. (See S² for details.)

Concept Question 8-3: What is the relation between discrete-time notch and comb filters? (See S²)

Exercise 8-8: Determine the ARMA difference equation for a comb filter that rejects periodic interference that has period = 0.01 s and is bandlimited to 200 Hz. The sampling rate is 600 samples per second. Use $a = 0.99$.

Answer: Harmonics at 100 and 200 Hz;

$$y[n] + 0.98y[n-2] + 0.96y[n-4]$$
$$= x[n] + x[n-2] + x[n+4].$$

(See S²)

Exercise 8-9: Use LabVIEW Module 8.4 to replicate Example 8-5 and produce the pole-zero and gain plots of Fig. 8-13 and time waveforms of Fig. 8-14.

Answer:

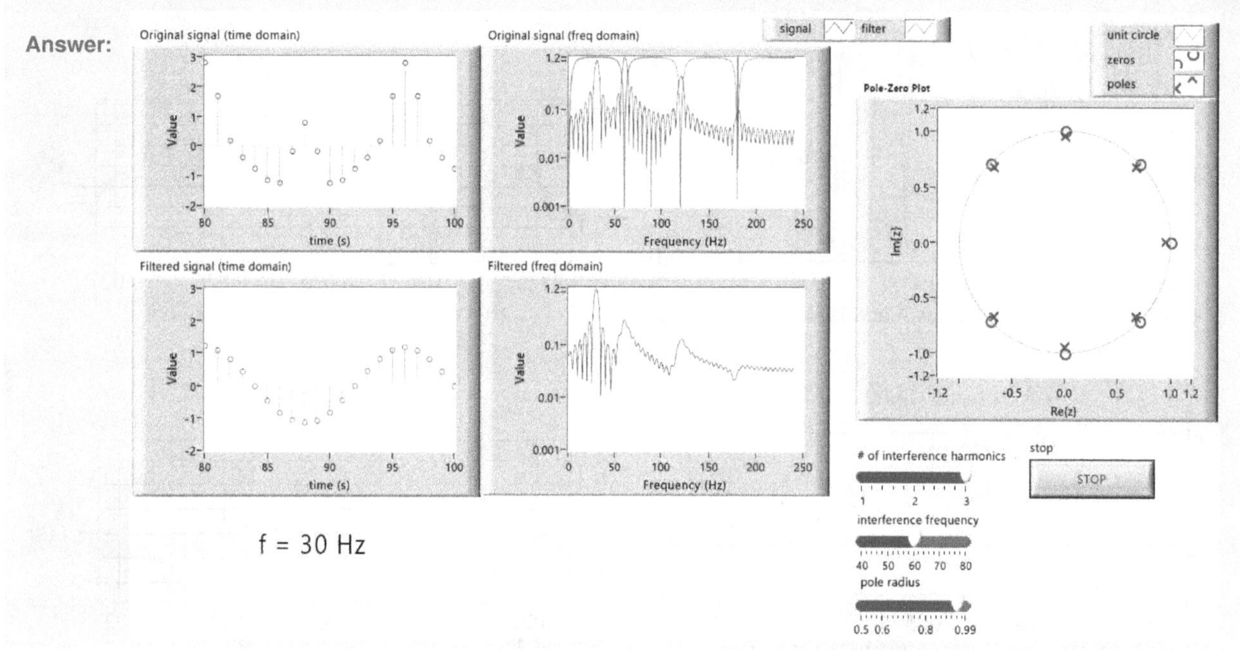

Module 8.4 Discrete-Time Comb Filter to Eliminate Periodic Signal from Sinusoid This module generates a periodic signal with a fundamental frequency and up to 3 harmonics. It then uses a comb filter to eliminate the periodic signal, while retaining the sinusoid.

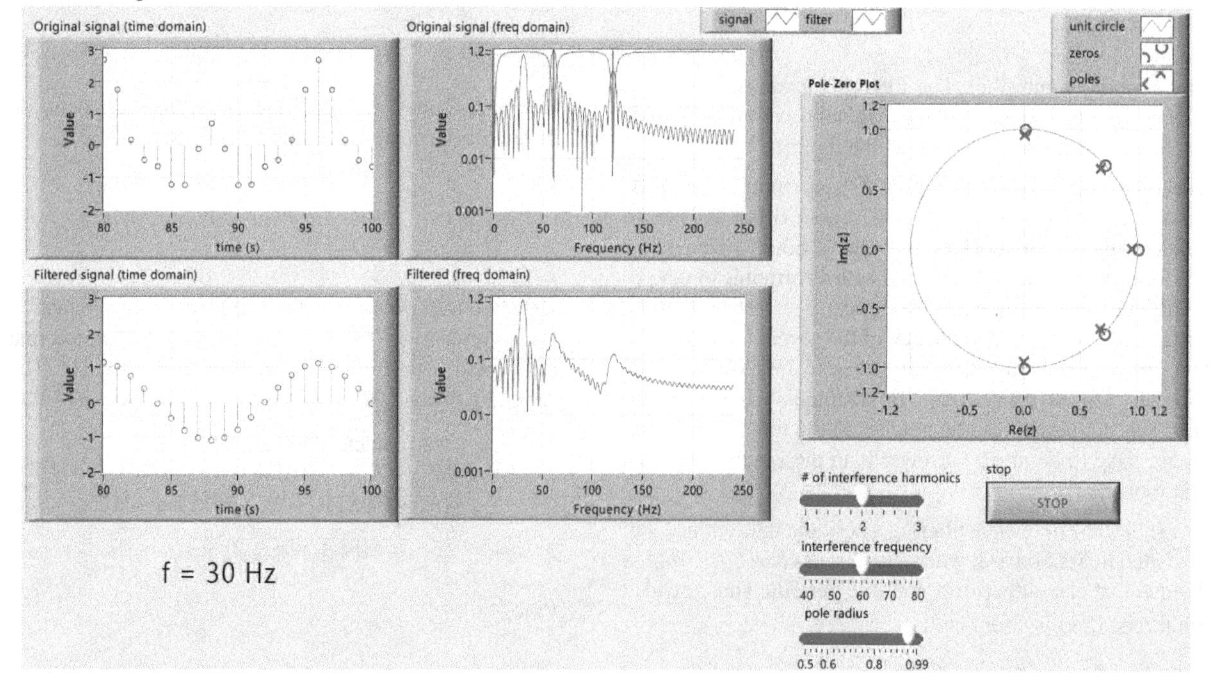

Module 8.5 **Discrete-Time Comb Filter to Separate Two Trumpet Signals** This module adds the waveforms of trumpets playing notes A and B, and uses a comb filter to separate them. The module also allows the user to listen to the original (two trumpets) and filtered (one trumpet) signals.

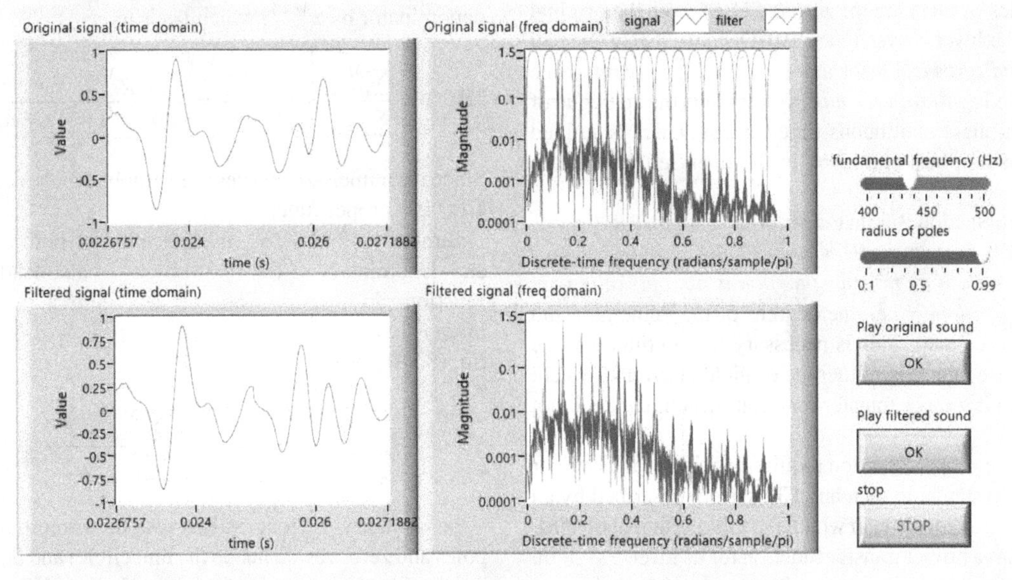

8-4 Deconvolution and Dereverberation

In some system applications, the discrete-time output $y[n]$ is measured and recorded, the system has been fully characterized (which means that its impulse response $h[n]$ is known), and the quantity we wish to determine is the input signal $x[n]$. An example might be a sensor that "measures" a physical quantity such as temperature or pressure, but the measurement process is not just a direct proportional relationship; rather, it is a convolution involving $x[n]$ and the sensor's impulse response $h[n]$:

$$y[n] = x[n] * h[n]. \tag{8.32}$$

To recover the quantity of interest (namely, $x[n]$) we need to perform a ***deconvolution*** operation, wherein we *undo* the convolution given by Eq. (8.32) to determine $x[n]$ from $y[n]$ and $h[n]$.

One possible approach to solving the deconvolution problem is to determine the ***impulse response of the inverse system***, $g[n]$, or equivalently, the ***transfer function of the inverse system***, $\mathbf{G(z)}$. When **z**-transformed, the convolution given by Eq. (8.32) becomes a product, namely,

$$\mathbf{Y(z)} = \mathbf{H(z)}\,\mathbf{X(z)}. \tag{8.33}$$

The input is then given by

$$\mathbf{X(z)} = \frac{\mathbf{Y(z)}}{\mathbf{H(z)}} = \mathbf{G(z)}\,\mathbf{Y(z)}, \tag{8.34}$$

where

$$\mathbf{G(z)} = \frac{1}{\mathbf{H(z)}}. \tag{8.35}$$

In symbolic form, we have

$$x[n] \longrightarrow \boxed{\begin{matrix}\text{System}\\ \mathbf{H(z)}\end{matrix}} \longrightarrow y[n] \tag{8.36a}$$

and

$$y[n] \longrightarrow \boxed{\begin{matrix}\text{Inverse system}\\ \mathbf{G(z)}\end{matrix}} \longrightarrow x[n]. \tag{8.36b}$$

Conceptually, the deconvolution solution is straightforward: (1) **z**-transform $h[n]$ and $y[n]$ to obtain $\mathbf{H(z)}$ and $\mathbf{Y(z)}$, (2) obtain the transfer function of the inverse system, $\mathbf{G(z)} = 1/\mathbf{H(z)}$, (3) apply Eq. (8.34) to find $\mathbf{X(z)}$, and (4) inverse **z**-transform $\mathbf{X(z)}$ to obtain the sought-after input $x[n]$. *In practice, however, we should ensure that* $\mathbf{G(z)}$ *is both BIBO stable and causal.*

8-4.1 BIBO Stability and Causality

The system described by $\mathbf{H}(\mathbf{z})$ is BIBO stable if and only if all of its poles reside inside the unit circle defined by $|\mathbf{z}| = 1$. Since the poles of $\mathbf{G}(\mathbf{z})$ are the zeros of $\mathbf{H}(\mathbf{z})$, both the original system and the inverse system are BIBO stable if $\mathbf{H}(\mathbf{z})$ has all of its poles *and* zeros inside the unit circle. Such a discrete-time system is called a ***minimum-phase*** system. It is the analogue of the minimum-phase continuous-time system, which is defined to have all of its poles and zeros in the open left half of the **s**-plane.

Recall from Section 3-8 that *a continuous-time BIBO stable and causal LTI system has a BIBO stable and causal LTI inverse system if and only if its transfer function is not only minimum phase, but also proper;* i.e., the degrees of the numerator and denominator are equal. This is necessary for continuous-time systems because otherwise either the original or inverse system would have an improper transfer function, which renders it not BIBO stable.

However, the proper-function requirement is not an issue for discrete-time systems. In fact, any LTI system described by an ARMA difference equation in which b_0, the coefficient of $x[n]$, is non-zero, has a proper transfer function for its inverse system. Moreover, even if $b_0 = 0$, we can still develop a viable inverse system by expressing $\mathbf{G}(\mathbf{z})$ in terms of an auxiliary minimum-phase transfer function $\widetilde{\mathbf{G}}(\mathbf{z})$.

8-4.2 ARMA Difference Equation for Inverse System

From Eq. (7.108), the general form of the ARMA difference equation describing an LTI system is given by

$$\sum_{i=0}^{N-1} a_i \, y[n-i] = \sum_{i=0}^{M-1} b_i \, x[n-i], \tag{8.37}$$

and the corresponding transfer function is

$$\mathbf{H}(\mathbf{z}) = \frac{\sum_{i=0}^{M-1} b_i \mathbf{z}^{-i}}{\sum_{i=0}^{N-1} a_i \mathbf{z}^{-i}}. \tag{8.38}$$

Coefficient a_0 belongs to $y[n]$, the output at "present time" n, so it is by definition non-zero. Moreover, it is customary to normalize the values of all of the other coefficients so that $a_0 = 1$. On the input side, if the coefficient of $x[n]$ (namely, b_0) is not zero, it means $y[n]$ does depend on $x[n]$ and possibly earlier values of the input as well. But if $b_0 = 0$, it means that $y[n]$ depends on $x[n-1]$ and earlier values but not on the current input $x[n]$. Similarly, if $b_0 = b_1 = 0$, it means that $y[n]$ depends on $x[n-2]$ and possibly on earlier inputs. We will consider the cases $b_0 \neq 0$ and $b_0 = 0$ (but $b_1 \neq 0$) separately and then extend our conclusions to other cases.

(a) Case 1: $b_0 \neq 0$

If $M = N$ in Eq. (8.38), then $\mathbf{H}(\mathbf{z})$ is automatically a proper function. If $M > N$, we multiply both the numerator and denominator by $\mathbf{z}^{M-1}$, which leads to

$$\mathbf{H}(\mathbf{z}) = \frac{\sum_{i=0}^{M-1} b_i \mathbf{z}^{M-1-i}}{\sum_{i=0}^{N-1} a_i \mathbf{z}^{M-1-i}} = \frac{b_0 \mathbf{z}^{M-1} + \cdots + b_{M-1}}{a_0 \mathbf{z}^{M-1} + \cdots + a_{N-1} \mathbf{z}^{M-N}}. \tag{8.39}$$

Since the numerator and denominator both have degree $(M-1)$, $\mathbf{H}(\mathbf{z})$ is a proper function.

Similarly, if $M < N$, multiplication of both the numerator and denominator of Eq. (8.38) by $\mathbf{z}^{N-1}$ also leads to a proper function. So *no matter how M compares with N, $\mathbf{H}(\mathbf{z})$ is a proper function, and so will be its reciprocal $\mathbf{G}(\mathbf{z})$.* Inverting Eq. (8.38) gives

$$\mathbf{G}(\mathbf{z}) = \frac{1}{\mathbf{H}(\mathbf{z})} = \frac{\sum_{i=0}^{N-1} a_i \mathbf{z}^{-i}}{\sum_{i=0}^{M-1} b_i \mathbf{z}^{-i}}. \tag{8.40}$$

In summary, as long as the system is minimum-phase (its poles and zeros reside inside the unit circle) and b_0 in Eq. (8.39) is not zero, its transfer function $\mathbf{H}(\mathbf{z})$ and inverse transfer function $\mathbf{G}(\mathbf{z})$ are both proper functions and the inverse system is causal and BIBO stable.

(a) Case 2: $b_0 = 0$ but $b_1 \neq 0$

If $b_0 = 0$ but $b_1 \neq 0$, the transfer function of the original system should be rewritten as

$$\mathbf{H}(\mathbf{z}) = \frac{\sum_{i=1}^{M-1} b_i \mathbf{z}^{-i}}{\sum_{i=0}^{N-1} a_i \mathbf{z}^{-i}} = \frac{\sum_{j=0}^{M-2} b_{j+1} \mathbf{z}^{-j} \mathbf{z}^{-1}}{\sum_{i=0}^{N-1} a_i \mathbf{z}^{-i}} = \mathbf{z}^{-1} \, \widetilde{\mathbf{H}}(\mathbf{z}), \tag{8.41}$$

where in the numerator we introduce (a) the index $j = i - 1$ and (b) an ***auxiliary transfer function*** $\widetilde{\mathbf{H}}(\mathbf{z})$ defined as

$$\widetilde{\mathbf{H}}(\mathbf{z}) = \frac{\sum_{j=0}^{M-2} b_{j+1} \mathbf{z}^{-j}}{\sum_{i=0}^{N-1} a_i \mathbf{z}^{-i}}. \tag{8.42}$$

If $M-2 = N-1$, then $M-N = 1$ and $\widetilde{\mathbf{H}}(\mathbf{z})$ is automatically a proper function. If $M-N > 1$, multiplication of the numerator and denominator by $\mathbf{z}^{M-2}$ renders $\widetilde{\mathbf{H}}(\mathbf{z})$ a proper function (the numerator and denominator both have degrees $M-2$). Also, if $M-N < 1$, multiplication of the numerator and denominator by $\mathbf{z}^{N-1}$ also renders $\widetilde{\mathbf{H}}(\mathbf{z})$ a proper function. Hence, while the original transfer function $\mathbf{H}(\mathbf{z})$ is not a proper function, the auxiliary transfer function $\widetilde{\mathbf{H}}(\mathbf{z})$ is, as is the auxiliary inverse transfer function $\widetilde{\mathbf{G}}(\mathbf{z}) = 1/\widetilde{\mathbf{H}}(\mathbf{z})$.

According to property #2 in Table 7-6, in the absence of initial conditions, multiplication of the $\mathbf{z}$-transform by $\mathbf{z}^{-1}$ is (in the discrete-time domain) a unit time delay. Hence, the original system described by $\mathbf{H}(\mathbf{z})$ and corresponding impulse response $h[n]$ can be regarded as a system described by an invertible auxiliary transfer function $\widetilde{\mathbf{H}}(\mathbf{z})$ in series with a unit time delay. Functionally, this means that

$$x[n] \rightarrow \boxed{\text{System } \mathbf{H}(\mathbf{z})} \rightarrow y[n], \tag{8.43a}$$

$$y[n] \rightarrow \boxed{\text{Inverse system } \widetilde{\mathbf{G}}(\mathbf{z})} \rightarrow x'[n] = x[n-1]. \tag{8.43b}$$

where $\widetilde{\mathbf{G}}(\mathbf{z})$ is an *auxiliary transfer function of the inverse system:*

$$\widetilde{\mathbf{G}}(\mathbf{z}) = \frac{1}{\widetilde{\mathbf{H}}(\mathbf{z})} = \frac{1}{\mathbf{z}\,\mathbf{H}(\mathbf{z})}. \tag{8.44}$$

The operation described by Eq. (8.43b) accomplishes the deconvolution operation, *except for a time delay*. This is as expected; the condition $b_0 = 0$ and $b_1 \neq 0$ in Eq. (8.37) implies that output $y[n]$ at present discrete time n is not a function of input $x[n]$ at that same time, but it is a function of $x[n-1]$ at time $(n-1)$, and possibly at earlier times, so there is no way to recover $x[n]$ from $y[n]$.

Because $b_0 = 0$, the system described by $\mathbf{G}(\mathbf{z})$ is not causal, so it is not possible to recover $x[n]$. However, it is possible to recover a delayed version, $x[n-1]$, by using the inverse system described by Eq. (8.43b).

By extension, if $b_0 = b_1 = 0$ and $b_2 \neq 0$, the output of the inverse system will be $x[n-2]$, and so on.

Example 8-7: Deconvolution of Minimum-Phase System

Given the system

$$y[n] = x[n] - \frac{5}{6}\,x[n-1] + \frac{1}{6}\,x[n-2], \tag{8.45}$$

(a) determine if it is minimum-phase and (b) solve for $x[n]$ if $y[n] = \{\underline{12}, 8, -7, -2, 1\}$.

Solution:

(a) The impulse response can be read off directly by evaluating $y[n]$ with $x[n] = \delta[n]$, while keeping in mind that

$$\delta[n-m] = \begin{cases} 1 & n = m, \\ 0 & n \neq m. \end{cases} \tag{8.46}$$

The result is

$$h[n] = \left\{\underline{1}, -\tfrac{5}{6}, \tfrac{1}{6}\right\}, \tag{8.47}$$

and the corresponding transfer function is

$$\mathbf{H}(\mathbf{z}) = 1 - \frac{5}{6}\,\mathbf{z}^{-1} + \frac{1}{6}\,\mathbf{z}^{-2} = \frac{\mathbf{z}^2 - (5/6)\mathbf{z} + (1/6)}{\mathbf{z}^2}. \tag{8.48}$$

The zeros of $\mathbf{H}(\mathbf{z})$ are $\mathbf{z}_k = \left\{\frac{1}{2}, \frac{1}{3}\right\}$ and its poles are $\mathbf{p}_k = \{0, 0\}$. Because its poles and zeros reside inside the unit circle, the system is minimum phase, and since $b_0 = 1 \neq 0$, it has a BIBO stable and causal inverse system.

(b) We can solve for $x[n]$ by implementing either of the following two approaches:

(1) Apply the convolution property of the $\mathbf{z}$-transform, which entails inverting $\mathbf{H}(\mathbf{z})$ to obtain its inverse $\mathbf{G}(\mathbf{z})$, obtaining $\mathbf{Y}(\mathbf{z})$ from $y[n]$, solving for $\mathbf{X}(\mathbf{z}) = \mathbf{G}(\mathbf{z})\,\mathbf{Y}(\mathbf{z})$, and then inverse $\mathbf{z}$-transforming to get $x[n]$.

(2) Simply rewrite Eq. (8.45) so that $x[n]$ is on one side all by itself, and then evaluate it for $n = 0$, 1, and 2. While doing so, we have to keep in mind that $x[n]$ is causal, and therefore $x[i] = 0$ for $i < 0$.

To demonstrate the second approach, we rewrite Eq. (8.45) as

$$x[n] = y[n] + \frac{5}{6}\,x[n-1] - \frac{1}{6}\,x[n-2]. \tag{8.49}$$

For $y[n] = \{\underline{12}, 8, -7, -2, 1\}$, we obtain

$$x[0] = y[0] + \frac{5}{6}\,x[-1] - \frac{1}{6}\,x[-2] = 12 + 0 + 0 = 12,$$

$$x[1] = y[1] + \frac{5}{6}\,x[0] - \frac{1}{6}\,x[-1] = 8 + \frac{5}{6}\,(12) - 0 = 18,$$

$$x[2] = y[2] + \frac{5}{6}\,x[1] - \frac{1}{6}\,x[0] = -7 + \frac{5}{6}\,(18) - \frac{1}{6}\,(12) = 6.$$

Further recursions show that $x[n] = 0$ for $n \geq 3$. Hence,

$$x[n] = \{\underline{12}, 18, 6\}.$$

Example 8-8: Deconvolving a System with $b_0 = 0$

Find the inverse system of

$$y[n] - 0.5y[n-1] = 0.5x[n-1] \tag{8.50}$$

to obtain $x[n]$ from $y[n]$. Note that $b_0 = 0$.

Solution: The $\mathbf{z}$-transform of Eq. (8.50) gives

$$\mathbf{Y}(\mathbf{z}) - 0.5\mathbf{z}^{-1}\,\mathbf{Y}(\mathbf{z}) = 0.5\mathbf{z}^{-1}\,\mathbf{X}(\mathbf{z}), \tag{8.51}$$

which leads to

$$\mathbf{H}(\mathbf{z}) = \frac{\mathbf{Y}(\mathbf{z})}{\mathbf{X}(\mathbf{z})} = \frac{0.5\mathbf{z}^{-1}}{1 - 0.5\mathbf{z}^{-1}} = \frac{1}{2\mathbf{z} - 1}. \tag{8.52}$$

The absence of a term of power $\mathbf{z}^0$ in the numerator means that $b_0 = 0$, so the system is not invertible. To appreciate the significance of this fact, we will pursue the deconvolution problem without taking this fact into account and then repeat it again using an auxiliary inverse system.

(a) Direct inverse deconvolution

The transfer function of the inverse system is

$$\mathbf{G}(\mathbf{z}) = \frac{1}{\mathbf{H}(\mathbf{z})} = 2\mathbf{z} - 1. \tag{8.53}$$

The corresponding impulse response is

$$g[n] = \{2, \underline{-1}\}, \tag{8.54}$$

which is noncausal because it is non-zero at time $n = -1$. The difference equation for the inverse system is

$$x[n] = 2y[n+1] - y[n], \tag{8.55}$$

which is noncausal because $x[n]$ at time n depends on $y[n+1]$ at later time $(n+1)$. This expression could have also been obtained directly from Eq. (8.50) upon replacing n with $n+1$ everywhere.

(b) Auxiliary inverse deconvolution

The noncausal deconvolution result given by Eq. (8.55) is a consequence of the fact that the inverse system is not causal. To obtain a causal inverse system, we implement the recipe of Section 8-4.2, which calls for establishing an auxiliary transfer function of the inverse system given by

$$\widetilde{\mathbf{G}}(\mathbf{z}) = \frac{1}{\mathbf{z}\,\mathbf{H}(\mathbf{z})} = \frac{2\mathbf{z} - 1}{\mathbf{z}} = 2 - \mathbf{z}^{-1}. \tag{8.56}$$

If we define $\widetilde{\mathbf{G}}(\mathbf{z}) = \mathbf{X}'(\mathbf{z})/\mathbf{Y}(\mathbf{z})$, we get

$$\frac{\mathbf{X}'(\mathbf{z})}{\mathbf{Y}(\mathbf{z})} = 2 - \mathbf{z}^{-1},$$

which after cross multiplication and inverse $\mathbf{z}$-transformation leads to

$$x'[n] = 2y[n] - y[n-1]. \tag{8.57}$$

From Eq. (8.43b), $x'[n] = x[n-1]$. Hence,

$$x[n-1] = 2y[n] - y[n-1]. \tag{8.58}$$

This expression could have also been obtained from Eq. (8.55) by replacing n with $(n-1)$ everywhere.

As stated earlier, the auxiliary transfer function approach does not recover $x[n]$, but it recovers a delayed version of it, and it does so through a BIBO stable and causal auxiliary transfer function $\widetilde{\mathbf{G}}(\mathbf{z})$.

Example 8-9: Deconvolution of Minimum-Phase System

Given

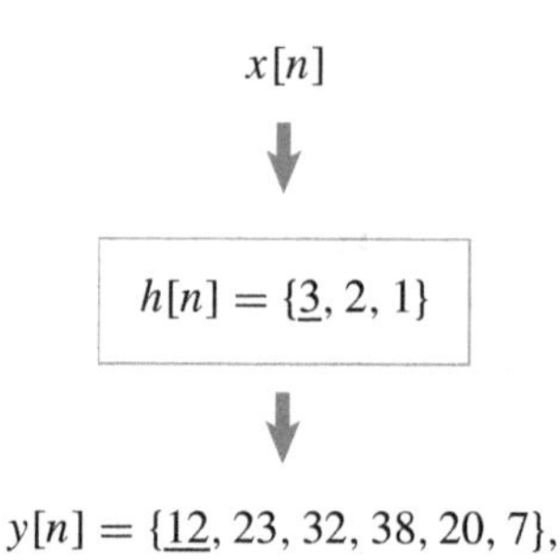

$$y[n] = \{\underline{12}, 23, 32, 38, 20, 7\},$$

determine $x[n]$.

Solution: We know the system is causal because $h[n]$ starts at $n = 0$, but we also need to determine if it is minimum phase. The transfer function corresponding to $h[n] = \{\underline{3}, 2, 1\}$ is

$$\mathbf{H}(\mathbf{z}) = 3 + 2\mathbf{z}^{-1} + \mathbf{z}^{-2} = \frac{3\mathbf{z}^2 + 2\mathbf{z} + 1}{\mathbf{z}^2}. \tag{8.59}$$

The degrees of the numerator and denominator polynomials are the same, so $\mathbf{H}(\mathbf{z})$ is a proper function.

The roots of the numerator $(3\mathbf{z}^2 + 2\mathbf{z} + 1) = 0$ are

$$\mathbf{z}_k = \{-0.33 \pm j0.47\} = \{0.58e^{\pm j125^\circ}\},$$

which reside inside the unit circle. The poles are $\mathbf{p}_k = \{0, 0\}$. Since all of the poles and zeros of $\mathbf{H}(\mathbf{z})$ reside inside the unit circle $|\mathbf{z}| = 1$, the system is minimum phase. Hence, both $\mathbf{H}(\mathbf{z})$ and its inverse $\mathbf{G}(\mathbf{z}) = 1/\mathbf{H}(\mathbf{z})$ are causal and BIBO stable.

Next, we use $h[n] = \{\underline{3}, 2, 1\}$ to write down the ARMA difference equation of the original system:

$$y[n] = 3x[n] + 2x[n-1] + x[n-2]. \tag{8.60}$$

Having established that both the system and its inverse are stable and causal, we can rewrite Eq. (8.60) in the form

$$x[n] = \frac{1}{3}\, y[n] - \frac{2}{3}\, x[n-1] - \frac{1}{3}\, x[n-2]. \tag{8.61}$$

The given output is of duration $N_0 = 6$. Hence, we recursively evaluate $x[n]$ for $n = 0, 1, \ldots, 5$, keeping in mind all along that $x[i] = 0$ for $i < 0$. For $y[n] = \{\underline{12}, 23, 32, 38, 20, 7\}$, the outcome is

$$x[0] = \frac{1}{3}(12) - \frac{2}{3}(0) - \frac{1}{3}(0) = 4,$$

$$x[1] = \frac{1}{3}(23) - \frac{2}{3}(4) - \frac{1}{3}(0) = 5,$$

$$x[2] = \frac{1}{3}(32) - \frac{2}{3}(5) - \frac{1}{3}(4) = 6,$$

$$x[3] = \frac{1}{3}(38) - \frac{2}{3}(6) - \frac{1}{3}(5) = 7,$$

$$x[4] = \frac{1}{3}(20) - \frac{2}{3}(7) - \frac{1}{3}(6) = 0,$$

and

$$x[5] = \frac{1}{3}(7) - \frac{2}{3}(0) - \frac{1}{3}(7) = 0.$$

Hence, the solution of the deconvolution problem is

$$x[n] = \{\underline{4}, 5, 6, 7\}.$$

8-4.3 Dereverberation

Reverberation of a signal is the act of adding delayed and geometrically weighted copies of the signal to itself. Each copy of the signal is equal to the previous copy multiplied by a ***reflection coefficient*** r and delayed by D seconds. The reverberated version $y(t)$ of a signal $x(t)$ is

$$y(t) = x(t) + r x(t-D) + r^2 x(t-2D) + \cdots = \sum_{i=0}^{\infty} r^i x(t - iD). \tag{8.62}$$

Since reflection cannot physically gain strength, $|r| < 1$.

Reverberation of a voice can make it sound richer and fuller if the delay is a few hundredths of a second. This is why a singing voice sounds much better in a confined cavity with reflecting walls, such as a shower stall. In music recording, reverbing is called ***overdubbing***.

Reverberation occurs in many natural and artificial systems, including multiple reflections of sonar signals by the sea floor and air-sea boundary, reflections by mirrors at the two ends of a longitudinal, gas-laser cavity, back-and-forth echoes of signals on mismatched transmission lines, among many others. In many of these cases, ***dereverberation*** of the signal is desired.

Suppose a digital signal processing system with a sampling rate of f_s samples/s is used to convert the original signal $x(t)$ and reverberated version $y(t)$ into $x[n]$ and $y[n]$, respectively. A delay of D seconds is a delay of Df_s samples. The sampling rate f_s is chosen such that the discrete-time delay M is an integer,

$$M = Df_s. \tag{8.63}$$

After sampling, the reverberating system is

$$\begin{aligned} y[n] &= x[n] + r\, x[n-M] + r^2\, x[n-2M] + \cdots \\ &= \sum_{i=0}^{\infty} r^i\, x[n - iM]. \end{aligned} \tag{8.64}$$

The goal of dereverberation is to recover $x[n]$ from $y[n]$.

The **z**-transform of Eq. (8.64) is

$$\mathbf{Y}(\mathbf{z}) = \sum_{i=0}^{\infty} r^i \mathbf{z}^{-iM}\, \mathbf{X}(\mathbf{z}) = \mathbf{X}(\mathbf{z}) \sum_{i=0}^{\infty} (r\mathbf{z}^{-M})^i. \tag{8.65}$$

In view of the geometric series,

$$\sum_{i=0}^{\infty} \mathbf{a}^i = \frac{1}{1-\mathbf{a}} \quad \text{if} \quad |\mathbf{a}| < 1, \tag{8.66}$$

Eq. (8.65) can be recast in the compact form

$$\mathbf{Y}(\mathbf{z}) = \frac{\mathbf{X}(\mathbf{z})}{1 - r\mathbf{z}^{-M}}, \tag{8.67}$$

from which we obtain the transfer function

$$\mathbf{H}(\mathbf{z}) = \frac{\mathbf{Y}(\mathbf{z})}{\mathbf{X}(\mathbf{z})} = \frac{1}{1 - r\mathbf{z}^{-M}} = \frac{\mathbf{z}^M}{\mathbf{z}^M - r}. \tag{8.68}$$

Transfer function $\mathbf{H}(\mathbf{z})$ has M zeros at the origin ($\mathbf{z} = 0$) and M poles inside the unit circle at $\mathbf{p}_k = \{r^{1/M} e^{j2\pi k/M}, k = 0, 1, \ldots, M-1\}$. Hence, $\mathbf{H}(\mathbf{z})$ is a minimum-phase system, and it has a BIBO stable and causal inverse system

$$\mathbf{G}(\mathbf{z}) = \frac{\mathbf{X}(\mathbf{z})}{\mathbf{Y}(\mathbf{z})} = 1 - r\mathbf{z}^{-M}. \tag{8.69}$$

Cross multiplication, followed by an inverse **z**-transformation, leads to the original input signal

$$x[n] = y[n] - r\, y[n-M]. \tag{8.70}$$

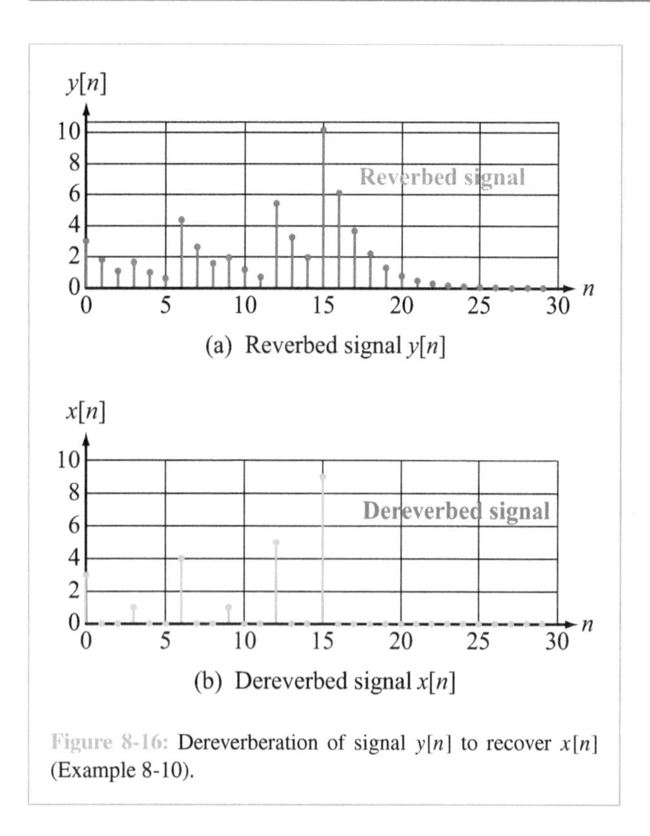

Figure 8-16: Dereverberation of signal $y[n]$ to recover $x[n]$ (Example 8-10).

In the spirit of completeness, we note that the impulse response $g[n]$ of the inverse system is

$$g[n] = \{1, \underbrace{0, \ldots, 0}_{M-1 \text{ zeros}}, -r\}. \tag{8.71}$$

Example 8-10: Dereverberation of a Signal

The signal shown in Fig. 8-16(a) was produced by reverbing a short-duration signal with reflection coefficient $r = 0.6$ and time delay $M = 1$. Compute the original signal.

Solution: Inserting $r = 0.6$ and $M = 1$ in Eq. (8.70) gives

$$x[n] = y[n] - 0.6y[n-1].$$

Computing $x[n]$ recursively for each of the given values of $y[n]$, while recognizing that $x[-1] = 0$, leads to the plot shown in Fig. 8-16(b). (See S² for details.)

Concept Question 8-4: What specific property of the reverberation problem permits its solution using an inverse system? (See S²)

Module 8.6 Dereverberation of a Simple Signal This module reverbs the short signal shown for specified values of the reflection coefficient and time delay between copies. It then dereverbs it to recover the original signal.

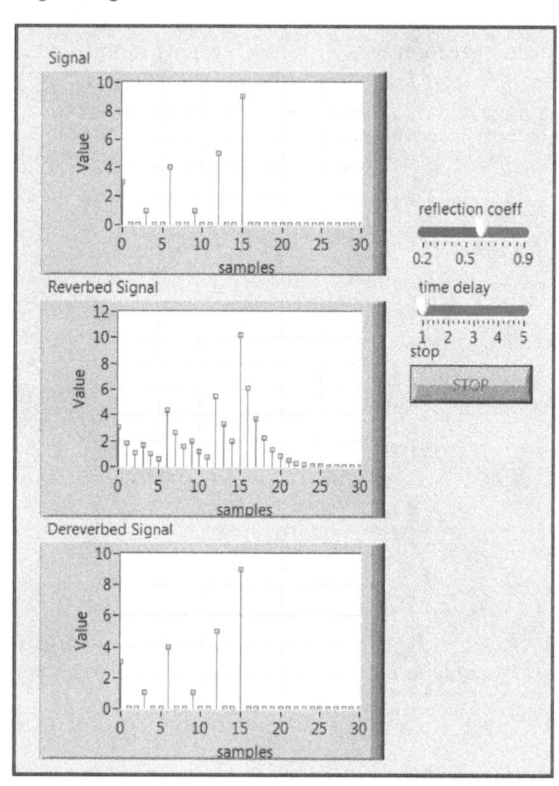

Concept Question 8-5: What is the definition of a discrete-time minimum phase system? (See S²)

Exercise 8-10: Is deconvolution using real-time signal processing possible for the system $y[n] = x[n] - 2x[n-1]$?

Answer: No. The system has a zero at 2, so it is not minimum phase. (See S²)

Exercise 8-11: A system is given by

$$y[n] = x[n] - 0.5x[n-1] + 0.4x[n-2].$$

What is the difference equation of its inverse system?

Answer: The system is minimum phase, so its inverse system is

$$x[n] = y[n] + 0.5x[n-1] - 0.4x[n-2].$$

(See S²)

8-5 Bilateral z-Transforms

Section 8-4 seems to indicate that it is impossible to find an inverse system for a non-minimum-phase system. This is not quite correct. It is impossible to find a stable and causal inverse system for a non-minimum-phase system. But it is possible to find a stable and noncausal inverse system for a non-minimum-phase system with no zeros on the unit circle. While a noncausal system cannot be implemented as is, it may be possible to implement a delayed version of a noncausal system. To explore this possibility, we need to extend the unilateral **z**-transform defined in Chapter 7 to a bilateral **z**-transform. We perform this extension in this section.

8-5.1 Definitions

(a) Bilateral z-Transforms

The bilateral or two-sided **z**-transform is defined as

$$\mathbf{X}(\mathbf{z}) = \sum_{n=-\infty}^{\infty} x[n]\, \mathbf{z}^{-n}. \tag{8.72}$$

This differs from the unilateral **z**-transform defined in Eq. (7.58a) in that the lower limit of the summation is changed from $n = 0$ to $n = -\infty$. The relation between the unilateral and bilateral **z**-transforms is analogous to the relation between the unilateral and bilateral Laplace transforms, defined in Eq. (3.117) and Eq. (3.119). The unilateral **z**-transforms listed in Table 7-5, and the unilateral **z**-transform properties listed in Table 7-6, also apply to the bilateral **z**-transform, except time delays and right shifts are now identical, and time advances and left shifts are also identical.

For example, the unilateral **z**-transform of $x[n] = \{3, \underline{1}, 4\}$ is $(1 + 4\mathbf{z}^{-1})$, while the bilateral **z**-transform of $x[n]$ is $(3\mathbf{z} + 1 + 4\mathbf{z}^{-1})$.

(b) Left-, Right-, and Two-Sided Signals

We offer the following definitions for a signal $x[n]$. The signal $x[n]$ is:

- *Causal* if $x[n] = 0$ for $n < 0$.
- *Anticausal* if $x[n] = 0$ for $n \geq 0$.
- *Two-sided* if $x[n]$ is neither causal nor anticausal.
- *Right-sided* if $x[n] = 0$ for $n < N < 0$ for some N. For example, $\{3, \underline{1}, 4 \dots\}$ is right-sided.
- *Left-sided* if $x[n] = 0$ for $n > N \geq 0$ for some N. For example, $\{\dots 3, \underline{1}, 4, 1\}$ is left-sided.

8-5.2 Geometric Signals

(a) Causal Geometric Signals

Consider the causal geometric signal $x_c[n]$ defined as

$$x_c[n] = \mathbf{a}^n\, u[n]. \tag{8.73}$$

Causality is assumed by the presence of $u[n]$. The bilateral **z**-transform $\mathbf{X}_c(\mathbf{z})$ of $x_c[n]$ is computed by applying Eq. (8.72), which gives

$$\mathbf{X}_c(\mathbf{z}) = \sum_{n=-\infty}^{\infty} \mathbf{a}^n\, u[n]\mathbf{z}^{-n} = \sum_{n=0}^{\infty} \left(\frac{\mathbf{a}}{\mathbf{z}}\right)^n. \tag{8.74}$$

In view of the infinite geometric series

$$\sum_{n=0}^{\infty} \mathbf{r}^n = \frac{1}{1-\mathbf{r}} \text{ if and only if } |\mathbf{r}| < 1, \tag{8.75}$$

we can rewrite Eq. (8.74) as

$$\mathbf{X}_c(\mathbf{z}) = \frac{1}{1 - \left(\frac{\mathbf{a}}{\mathbf{z}}\right)} = \frac{\mathbf{z}}{\mathbf{z} - \mathbf{a}}, \tag{8.76}$$

so long as $|\mathbf{a}/\mathbf{z}| < 1$, which requires that $|\mathbf{z}| > |\mathbf{a}|$.

Since $x_c[n]$ is causal, the result given by Eq. (8.76) agrees with the unilateral **z**-transform of $x[n]$ given in Eq. (7.66), and listed as item #4 in Table 7-5.

This result is only true if $|\mathbf{z}| > |\mathbf{a}|$. This condition is listed in the third column of Table 7-5, under the heading "ROC," but it had no significance in Chapter 7. The ***region of convergence*** (*ROC*) is the set of **z** values for which the infinite geometric series used to compute an expression for $\mathbf{X}(\mathbf{z})$ converges. This expression for $\mathbf{X}_c(\mathbf{z})$ given by Eq. (8.76) is only true for $\mathbf{z} \in$ ROC.

(b) Anticausal Geometric Signals

Now we define the anticausal geometric signal $x_a[n]$ as (note the signs)

$$x_a[n] = -\mathbf{a}^n\, u[-n-1], \tag{8.77}$$

which is equal to zero for $n \geq 0$. The bilateral **z**-transform $\mathbf{X}_a(\mathbf{z})$ of $x_a[n]$ is then

$$\mathbf{X}_a(\mathbf{z}) = -\sum_{n=-\infty}^{\infty} \mathbf{a}^n\, u[-n-1]\mathbf{z}^{-n} = -\sum_{n=-\infty}^{-1} \left(\frac{\mathbf{a}}{\mathbf{z}}\right)^n. \tag{8.78}$$

Changing variables from n to $n' = -n$ gives

$$\mathbf{X}_a(\mathbf{z}) = -\sum_{n'=1}^{\infty} \left(\frac{\mathbf{a}}{\mathbf{z}}\right)^{-n'} = -\sum_{n'=1}^{\infty} \left(\frac{\mathbf{z}}{\mathbf{a}}\right)^{n'}. \tag{8.79}$$

The summation begins at $n' = 1$. To take advantage of the geometric series given by Eq. (8.75), we add and subtract the $n' = 0$ element, namely $-(\mathbf{z}/\mathbf{a})^0 = -1$. Thus, Eq. (8.79) is rewritten as

$$\mathbf{X}_a(\mathbf{z}) = -\sum_{n'=0}^{\infty} \left(\frac{\mathbf{z}}{\mathbf{a}}\right)^{n'} + 1. \tag{8.80}$$

Use of Eq. (8.75) leads to

$$\mathbf{X}_a(\mathbf{z}) = -\frac{1}{1-\left(\frac{\mathbf{z}}{\mathbf{a}}\right)} + 1 = \frac{\mathbf{z}}{\mathbf{z}-\mathbf{a}} = \mathbf{X}_c(\mathbf{z}), \tag{8.81}$$

provided $|\mathbf{z}/\mathbf{a}| < 1$, which requires that $|\mathbf{z}| < |\mathbf{a}|$. The two different signals $x_c[n]$ and $x_a[n]$ both seem to have the same bilateral **z**-transform $\mathbf{X}_c(\mathbf{z}) = \mathbf{X}_a(\mathbf{z})$! However, the ROCs of the **z**-transforms are different. The bilateral **z**-transform of $x_c[n]$ is $\mathbf{X}_c(\mathbf{z})$ only for the ROC $\{\mathbf{z} : |\mathbf{z}| > |\mathbf{a}|\}$ and the bilateral **z**-transform of $x_a[n]$ is $\mathbf{X}_a(\mathbf{z})$ only for the ROC $\{\mathbf{z} : |\mathbf{z}| < |\mathbf{a}|\}$.

In the sequel, we abbreviate $\{\mathbf{z} : |\mathbf{z}| > |\mathbf{a}|\}$ to $\{|\mathbf{z}| > |\mathbf{a}|\}$, and "**z**-transform" will mean the bilateral **z**-transform.

(c) Two-Sided Geometric Signals

The **z**-transform of a sum of geometric signals, where each signal is either causal or anticausal, is the sum of the **z**-transforms of each of the individual signals. The ROC of the sum of the signals is the intersection of the ROCs of the individual signals. If this is an empty set, the **z**-transform is undefined.

Consider the two-sided signal given by

$$x[n] = \mathbf{a}^n\, u[n] + \mathbf{b}^n\, u[-n-1]. \tag{8.82}$$

Using the combination of Eqs. (8.76) and (8.81) leads to

$$\mathbf{X}(\mathbf{z}) = \frac{\mathbf{z}}{\mathbf{z}-\mathbf{a}} - \frac{\mathbf{z}}{\mathbf{z}-\mathbf{b}}. \tag{8.83}$$

Case 1: $|\mathbf{b}| > |\mathbf{a}|$

The ROC of $\mathbf{X}(\mathbf{z})$ is

$$\{|\mathbf{z}| > |\mathbf{a}|\} \bigcap \{|\mathbf{z}| < |\mathbf{b}|\} = \{|\mathbf{a}| < |\mathbf{z}| < |\mathbf{b}|\}.$$

Case 2: $|\mathbf{b}| < |\mathbf{a}|$

The ROC of $\mathbf{X}(\mathbf{z})$ is

$$\{|\mathbf{z}| > |\mathbf{a}|\} \bigcap \{|\mathbf{z}| < |\mathbf{b}| < |\mathbf{a}|\} = \emptyset.$$

So the **z**-transform is undefined.

For the general case where $x[n]$ is the sum of many causal and anticausal geometric signals, we have

$$x[n] = \underbrace{\sum_i A_i \mathbf{p}_i^n\, u[n]}_{\text{causal}} + \underbrace{\sum_j B_j \mathbf{q}_j^n\, u[-n-1]}_{\text{anticausal}}, \tag{8.84a}$$

$$\mathbf{X}(\mathbf{z}) = \sum_i A_i \frac{\mathbf{z}}{\mathbf{z}-\mathbf{p}_i} - \sum_j B_j \frac{\mathbf{z}}{\mathbf{z}-\mathbf{q}_j}, \tag{8.84b}$$

and

$$\begin{aligned} \text{ROC} &= \left[\bigcap_i \{|\mathbf{z}| > |\mathbf{p}_i|\}\right] \bigcap \left[\bigcap_j \{|\mathbf{z}| < |\mathbf{q}_j|\}\right] \\ &= [\max\{|\mathbf{p}_i|\}] < |\mathbf{z}| < [\min\{|\mathbf{q}_j|\}]. \end{aligned} \tag{8.84c}$$

From Eq. (8.84c), we see that the ROC of the **z**-transform of a sum $x[n]$ of causal and anticausal geometric signals has the following properties:

▶ The ROC is an *annulus* (ring) whose inner radius is the largest-magnitude pole of the causal part of $x[n]$, and whose outer radius is the smallest-magnitude pole of the anticausal part of $x[n]$. Unless all of the poles of the causal part of $x[n]$ have smaller magnitudes than all of the poles of the anticausal part of $x[n]$, the **z**-transform of $x[n]$ is undefined. If $x[n]$ is causal, the outer radius is ∞, and if $x[n]$ is anticausal, the inner radius is 0. ◀

Example 8-11: z-Transform of Two-Sided Signal

Compute the **z**-transform of

$$9(2)^n\, u[n] + 8(3)^n\, u[n] - 7(4)^n\, u[-n-1] + 6(5)^n\, u[-n-1].$$

Solution:

$$\mathbf{X(z)} = \frac{9\mathbf{z}}{\mathbf{z}-2} + \frac{8\mathbf{z}}{\mathbf{z}-3} + \frac{7\mathbf{z}}{\mathbf{z}-4} - \frac{6\mathbf{z}}{\mathbf{z}-5}.$$

The ROC is

$$\{|\mathbf{z}| > 2\} \cap \{|\mathbf{z}| > 3\} \cap \{|\mathbf{z}| < 4\} \cap \{|\mathbf{z}| < 5\} = \{3 < |\mathbf{z}| < 4\}.$$

Example 8-12: Another Two-Sided Signal

Determine the ROC of the **z**-transform of

$$x[n] = (-2)^n\, u[n] + (\tfrac{1}{2})^n\, u[-n-1].$$

Solution:

$$\text{ROC} = \{|\mathbf{z}| > |-2| = 2\} \bigcap \{|\mathbf{z}| < \tfrac{1}{2}\} = \emptyset.$$

The **z**-transform is undefined. There is no **z** for which both of the two geometric series converge.

(d) **ROCs for Finite-Duration Left-Sided and Right-Sided Signals**

The above rules for ROC must be modified slightly for signals that are not sums of geometric signals:

Signal Type and Length	ROC
Causal & finite duration	$\{0 < \lvert\mathbf{z}\rvert \leq \infty\}$
Anticausal & finite length	$\{0 \leq \lvert\mathbf{z}\rvert < \infty\}$
Two-sided & finite length	$\{0 < \lvert\mathbf{z}\rvert < \infty\}$
Right-sided signal	$\{0 < \lvert\mathbf{z}\rvert < \infty\}$
Left-sided signal	$\{0 < \lvert\mathbf{z}\rvert < \infty\}$
Impulse $\delta[n]$	$\{0 \leq \lvert\mathbf{z}\rvert \leq \infty\}$

These rules are easily established by considering whether

$$\mathbf{X(z)} = \cdots + x[-1]\,\mathbf{z} + x[0] + x[1]\,\mathbf{z}^{-1} + \cdots \quad (8.85)$$

will blow up at $\mathbf{z} = 0$ or $|\mathbf{z}| \to \infty$. In particular, if $x[n]$ is causal, $\mathbf{X(z)}$ has the form

$$\mathbf{X(z)} = x[0] + x[1]\,\mathbf{z}^{-1} + \cdots, \quad (8.86)$$

which blows up at $\mathbf{z} = 0$ but not as $|\mathbf{z}| \to \infty$. It is standard to abuse notation slightly and write $|\mathbf{z}| = \infty$ to emphasize that $|\mathbf{z}| = \infty$ is included in the ROC.

8-6 Inverse Bilateral z-Transforms

The inverse **z**-transform of $\frac{\mathbf{z}}{\mathbf{z}-2}$ could be either $(2)^n\, u[n]$ or $-(2)^n\, u[-n-1]$! Clearly, this is an unacceptable ambiguity, but the ambiguity can be resolved by specifying the ROC of $\mathbf{X(z)}$. Specifically, the inverse **z**-transforms of

- $\frac{\mathbf{z}}{\mathbf{z}-2}$ with ROC $\{\mathbf{z} : |\mathbf{z}| > 2\}$ is $(2)^n\, u[n]$.
- $\frac{\mathbf{z}}{\mathbf{z}-2}$ with ROC $\{\mathbf{z} : |\mathbf{z}| < 2\}$ is $-(2)^n\, u[-n-1]$.

In fact, we will soon realize that it is the ROC, *not* $\mathbf{X(z)}$, that determines whether the inverse **z**-transform is stable, causal, anticausal, or two-sided.

8-6.1 Multiple Inverse z-Transforms

Consider a **z**-transform $\mathbf{X(z)}$ for which the partial fraction expansion of $\frac{\mathbf{X(z)}}{\mathbf{z}}$ is

$$\frac{\mathbf{X(z)}}{\mathbf{z}} = A_0\, \frac{1}{\mathbf{z}-0} + \sum_{i=1}^{N} \mathbf{A}_i\, \frac{1}{\mathbf{z}-\mathbf{p}_i}. \quad (8.87)$$

Multiplying both side by **z** gives:

$$\mathbf{X(z)} = A_0 + \sum_{i=1}^{N} \mathbf{A}_i\, \frac{\mathbf{z}}{\mathbf{z}-\mathbf{p}_i}, \quad (8.88)$$

where $\{\mathbf{p}_i\}$ are the poles of $\mathbf{X(z)}$ and $\{\mathbf{A}_i\}$ are their associated residues. The (causal) inverse *unilateral* **z**-transform of $\mathbf{X(z)}$ is, from Eq. (7.92), given by

$$x[n] = A_0\, \delta[n] + \sum_{i=1}^{N} \mathbf{A}_i \mathbf{p}_i^n\, u[n]. \quad (8.89)$$

In contrast, the inverse *bilateral* **z**-transform is given by

$$x[n] = A_0\, \delta[n] + \sum_{i=1}^{N} \mathbf{A}_i \begin{cases} \mathbf{p}_i^n\, u[n], & \text{if } \{|\mathbf{z}| > |\mathbf{p}_i|\}, \\ -\mathbf{p}_i^n\, u[-n-1], & \text{if } \{|\mathbf{z}| < |\mathbf{p}_i|\}, \end{cases} \quad (8.90)$$

because there are two choices (causal or anticausal) for each term in the partial fraction expansion, depending on the choice of ROC for that term.

This suggests that there are 2^N possible inverse **z**-transforms, since the choice for each term can be made independently. But there are actually only $N+1$ possible inverse **z**-transforms, since the other choices lead to ROCs that are empty sets. To see why, consider the following two examples.

Example 8-13: Inverse z-Transform

Compute all of the possible inverse **z**-transforms of

$$\mathbf{X}(\mathbf{z}) = \frac{2\mathbf{z}^2 - 2.5\mathbf{z}}{\mathbf{z}^2 - 2.5\mathbf{z} + 1}.$$

Solution: The partial fraction expansion of $\frac{\mathbf{X}(\mathbf{z})}{\mathbf{z}}$ is

$$\frac{\mathbf{X}(\mathbf{z})}{\mathbf{z}} = \frac{1}{\mathbf{z} - (1/2)} + \frac{1}{\mathbf{z} - 2}. \tag{8.91}$$

Multiplying through by **z** gives

$$\mathbf{X}(\mathbf{z}) = \frac{\mathbf{z}}{\mathbf{z} - (1/2)} + \frac{\mathbf{z}}{\mathbf{z} - 2}. \tag{8.92}$$

Each of the two terms has two possible inverse **z**-transforms, one causal and one anticausal. So it would appear that there are $2^2 = 4$ possible inverse two-sided **z**-transforms of $\mathbf{X}(\mathbf{z})$:

- $x_1[n] = (\frac{1}{2})^n\, u[n] + (2)^n\, u[n]$,

  with ROC

  $$\{|\mathbf{z}| > \tfrac{1}{2}\} \bigcap \{|\mathbf{z}| > 2\} = \{|\mathbf{z}| > 2\} \text{ (causal)}.$$

- $x_2[n] = -(\frac{1}{2})^n\, u[-n-1] - (2)^n\, u[-n-1]$

  with ROC

  $$\{|\mathbf{z}| < \tfrac{1}{2}\} \bigcap \{|\mathbf{z}| < 2\} = \{|\mathbf{z}| < \tfrac{1}{2}\} \text{ (anticausal)}.$$

- $x_3[n] = (\frac{1}{2})^n\, u[n] - (2)^n\, u[-n-1]$

  with ROC

  $$\{|\mathbf{z}| > \tfrac{1}{2}\} \bigcap \{|\mathbf{z}| < 2\} = \{\tfrac{1}{2} < |\mathbf{z}| < 2\} \text{ (stable)}.$$

- $x_4[n] = -(\frac{1}{2})^n\, u[-n-1] + (2)^n\, u[n]$

  with ROC

  $$\{|\mathbf{z}| < \tfrac{1}{2}\} \bigcap \{|\mathbf{z}| > 2\} = \emptyset \text{ (not valid for any z)}.$$

So there are actually only 3, not 4, possible inverse two-sided **z**-transforms of $\mathbf{X}(\mathbf{z})$.

Example 8-14: Inverse z-Transform II

Compute all of the possible inverse **z**-transforms of

$$\mathbf{X}(\mathbf{z}) = \frac{\mathbf{z}}{\mathbf{z}-2} + \frac{\mathbf{z}}{\mathbf{z}-3} + \frac{\mathbf{z}}{\mathbf{z}-4}.$$

Solution: Given that $\mathbf{X}(\mathbf{z})$ has three poles, the maximum possible number of **z**-transforms is $2^3 = 8$, but not all might by viable. The bilateral **z**-transform is

$$x[n] = \begin{cases} (2)^n\, u[n] \\ -(2)^n\, u[-n-1] \end{cases} + \begin{cases} (3)^n\, u[n] \\ -(3)^n\, u[-n-1] \end{cases} + \begin{cases} (4)^n\, u[n] \\ -(4)^n\, u[-n-1], \end{cases}$$

and the associated ROCs for the various combinations are

- $\{|\mathbf{z}| < 2\} \cap \{|\mathbf{z}| < 3\} \cap \{|\mathbf{z}| < 4\} = \{|\mathbf{z}| < 2\}$.
- $\{|\mathbf{z}| > 2\} \cap \{|\mathbf{z}| < 3\} \cap \{|\mathbf{z}| < 4\} = \{2 < |\mathbf{z}| < 3\}$.
- $\{|\mathbf{z}| < 2\} \cap \{|\mathbf{z}| > 3\} \cap \{|\mathbf{z}| < 4\} = \emptyset$.
- $\{|\mathbf{z}| > 2\} \cap \{|\mathbf{z}| > 3\} \cap \{|\mathbf{z}| < 4\} = \{3 < |\mathbf{z}| < 4\}$.
- $\{|\mathbf{z}| < 2\} \cap \{|\mathbf{z}| < 3\} \cap \{|\mathbf{z}| > 4\} = \emptyset$.
- $\{|\mathbf{z}| > 2\} \cap \{|\mathbf{z}| < 3\} \cap \{|\mathbf{z}| > 4\} = \emptyset$.
- $\{|\mathbf{z}| < 2\} \cap \{|\mathbf{z}| > 3\} \cap \{|\mathbf{z}| > 4\} = \emptyset$.
- $\{|\mathbf{z}| > 2\} \cap \{|\mathbf{z}| > 3\} \cap \{|\mathbf{z}| > 4\} = \{|\mathbf{z}| > 4\}$.

Hence, excluding the four with null sets, we have only 4 possible inverse **z**-transforms:

$\{|\mathbf{z}| < 2\}$: $-(2)^n\, u[-n-1] - (3)^n\, u[-n-1] - (4)^n\, u[-n-1]$,

$\{2 < |\mathbf{z}| < 3\}$: $(2)^n\, u[n] - (3)^n\, u[-n-1] - (4)^n\, u[-n-1]$,

$\{3 < |\mathbf{z}| < 4\}$: $(2)^n\, u[n] + (3)^n\, u[n] - (4)^n\, u[-n-1]$,

$\{|\mathbf{z}| > 4\}$: $(2)^n\, u[n] + (3)^n\, u[n] + (4)^n\, u[n]$ (causal).

For pairs of complex poles, the same ROC must be used for each pole and its complex conjugate.

8-6.2 Stable Inverse z-Transform

In general, there is always one causal and one anticausal inverse **z**-transform. If there are no poles on the unit circle $\{|\mathbf{z}| = 1\}$, then there is also one BIBO stable inverse **z**-transform. Here BIBO stable means that the inverse **z**-transform is absolutely summable, so it is the impulse response of a BIBO stable system (see Section 7-4.5).

To compute the BIBO stable inverse **z**-transform of $\mathbf{X}(\mathbf{z})$, proceed as follows:

- Order the poles $\{\mathbf{p}_1 \ldots \mathbf{p}_N\}$ of $\mathbf{X}(\mathbf{z})$ in increasing order of magnitudes (group conjugate poles):

$$|\mathbf{p}_1| < \cdots < |\mathbf{p}_{M-1}| < 1 < |\mathbf{p}_M| < \cdots < |\mathbf{p}_N|.$$

- Compute the partial fraction expansion of $\mathbf{X}(\mathbf{z})/\mathbf{z}$:

$$\frac{\mathbf{X}(\mathbf{z})}{\mathbf{z}} = A_0 \frac{1}{\mathbf{z} - 0} + \sum_{i=1}^{N} \mathbf{A}_i \frac{1}{\mathbf{z} - \mathbf{p}_i}.$$

- Multiply this by $\mathbf{z}$:

$$\mathbf{X}(\mathbf{z}) = A_0 + \sum_{i=1}^{N} \mathbf{A}_i \frac{\mathbf{z}}{\mathbf{z} - \mathbf{p}_i}.$$

- The stable inverse **z**-transform is then

$$A_0\, \delta[n] + \sum_{i=1}^{M-1} \mathbf{A}_i \mathbf{p}_i^n\, u[n] - \sum_{i=M}^{N} \mathbf{A}_i \mathbf{p}_i^n\, u[-n-1].$$

8-7 ROC, Stability, and Causality

We have established that it is the ROC, not $\mathbf{X}(\mathbf{z})$, that determines whether the inverse **z**-transform is stable or causal. If the ROC is defined by the form $|\mathbf{a}| < |\mathbf{z}| < |\mathbf{b}|$, then:

Signal Type	ROC
BIBO Stable	$|\mathbf{a}| < 1 < |\mathbf{b}|$
Causal	$|\mathbf{b}| \to \infty$
Anticausal	$\{\mathbf{a} = 0\} \subset$ ROC
Two-sided	$0 < |\mathbf{a}|$ & $|\mathbf{b}| < \infty$
Stable & causal	$|\mathbf{a}| < 1$ & $|\mathbf{b}| \to \infty$

Note that a stable and causal system has an ROC of the form $\{|\mathbf{z}| > |\mathbf{a}| < 1\}$, so all poles are inside the unit circle.

Example 8-15: ROC, Stability, and Causality

For each of the following ROCs, describe the causality and stability of the associated **z**-transforms:

(1) $\{\mathbf{z} : |\mathbf{z}| > a > 1\}$
(2) $\{\mathbf{z} : |\mathbf{z}| > a < 1\}$
(3) $\{\mathbf{z} : |\mathbf{z}| < a < 1\}$
(4) $\{\mathbf{z} : |\mathbf{z}| < a > 1\}$
(5) $\{\mathbf{z} : 1 < a < |\mathbf{z}| < b\}$
(6) $\{\mathbf{z} : a < |\mathbf{z}| < b < 1\}$
(7) $\{\mathbf{z} : 1 > a < |\mathbf{z}| < b > 1\}$

Solution:

(1) $\{\mathbf{z} : |\mathbf{z}| > a > 1\}$ → causal & unstable.
(2) $\{\mathbf{z} : |\mathbf{z}| > a < 1\}$ → causal & stable.
(3) $\{\mathbf{z} : |\mathbf{z}| < a < 1\}$ → anticausal & unstable.
(4) $\{\mathbf{z} : |\mathbf{z}| < a > 1\}$ → anticausal & stable.
(5) $\{\mathbf{z} : 1 < a < |\mathbf{z}| < b\}$ → 2-sided & unstable.
(6) $\{\mathbf{z} : a < |\mathbf{z}| < b < 1\}$ → 2-sided & unstable.
(7) $\{\mathbf{z} : 1 > a < |\mathbf{z}| < b > 1\}$ → 2-sided & stable.

8-7.1 Deconvolving Non-Minimum-Phase Multipath Systems

We noted in Section 8-4.1 that a system is called a ***minimum-phase*** system if the transfer function $\mathbf{H}(\mathbf{z})$ of the system and the transfer function $\mathbf{G}(\mathbf{z})$ of its inverse ($\mathbf{G}(\mathbf{z}) = 1/\mathbf{H}(\mathbf{z})$) are both BIBO stable. To realize BIBO stability, all of the poles and zeros of $\mathbf{H}(\mathbf{z})$ have to reside inside the unit circle. Non-minimum phase systems arise in:

- Manipulation of flexible links in robotics.
- Rudders in ships and ailerons in planes.
- Driving a car backwards.
- Some multipath systems.

In continuous time, if the system is not minimum phase, there is no way to reconstruct the input from the output in real time. But, in discrete-time signal processing, we can reconstruct a *delayed* version of the input from the output, if the original system has no zeros *on* the unit circle. We can do this by taking the inverse **z**-transform and choosing the ROC so that the resulting inverse system $g[n]$ is stable but non-causal. The non-causal part of $g[n]$ decays rapidly as $n \to -\infty$, so it can be *truncated* to a finite number N of non-causal terms. Then $g[n-N]$ is causal, and the output is $x[n-N]$, a *delayed* version of the original signal $x[n]$.

The following example shows how to deconvolve non-minimum-phase multipath systems.

Example 8-16: Deconvolution of Non-Minimum-Phase Multipath System

We are given the system

$$y[n] = x[n] - \frac{3}{5}\,x[n-1] - \frac{18}{25}\,x[n-2]. \qquad (8.93)$$

The goal is to solve the deconvolution problem to reconstruct $x[n]$ from $y[n]$.

Solution: The impulse response is $h[n] = \{\underline{1}, -\frac{3}{5}, -\frac{18}{25}\}$. The transfer function is the **z**-transform of $h[n]$, or

$$\mathbf{H}(\mathbf{z}) = 1 - \frac{3}{5}\,\mathbf{z}^{-1} - \frac{18}{25}\,\mathbf{z}^{-2} = \frac{\mathbf{z}^2 - \frac{3}{5}\mathbf{z} - \frac{18}{25}}{\mathbf{z}^2}. \qquad (8.94)$$

The zeros of $\mathbf{H}(\mathbf{z})$ are $\{\frac{6}{5}, -\frac{3}{5}\}$ and the poles are $\{0, 0\}$. Because one of the zeros $(6/5)$ lies outside the unit circle, the system is not minimum phase, and therefore it does not have a stable and causal inverse system. However, we can obtain an approximate inverse system that is stable and causal. To do so, we start with the transfer function of the inverse system,

$$\mathbf{G}(\mathbf{z}) = \frac{1}{\mathbf{H}(\mathbf{z})} = \frac{\mathbf{z}^2}{\mathbf{z}^2 - \frac{3}{5}\mathbf{z} - \frac{18}{25}}. \qquad (8.95)$$

The partial fraction expansion of $\mathbf{G}(\mathbf{z})/\mathbf{z}$ is

$$\frac{\mathbf{G}(\mathbf{z})}{\mathbf{z}} = \frac{1/3}{\mathbf{z}+\frac{3}{5}} + \frac{2/3}{\mathbf{z}-\frac{6}{5}}. \qquad (8.96)$$

Multiplication by **z** leads to

$$\mathbf{G}(\mathbf{z}) = \frac{1}{3}\,\frac{\mathbf{z}}{\mathbf{z}+\frac{3}{5}} + \frac{2}{3}\,\frac{\mathbf{z}}{\mathbf{z}-\frac{6}{5}}. \qquad (8.97)$$

The corresponding *stable* inverse **z**-transform consists of a causal component associated with the pole $p_1 = -3/5$ and an anticausal component associated with pole $p_2 = 6/5$:

$$g[n] = \underbrace{\frac{1}{3}\left(\frac{-3}{5}\right)^n u[n]}_{\text{causal}} - \underbrace{\frac{2}{3}\left(\frac{6}{5}\right)^n u[-n-1]}_{\text{anticausal}}. \qquad (8.98)$$

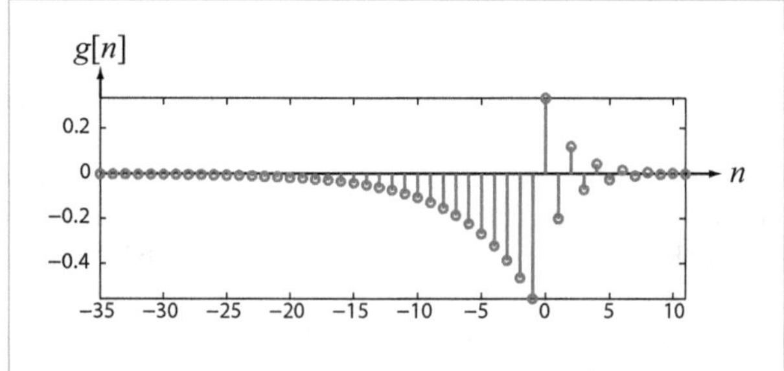

Figure 8-17: Stable but noncausal inverse filter $g[n]$.

While $g[n]$ is noncausal, $|g[n]| < 0.001$ for $n < -35$ (Fig. 8-17). By defining values of $|g[n]| < 0.001$ as negligible, $g[n-35]$ becomes causal, because $|g[n-35]| < 0.001$ for $n < 0$. Furthermore, $|g[n]| < 0.001$ for $n > 11$. So we can define the *approximate inverse filter* $\tilde{g}[n]$ as

$$\tilde{g}[n] = \begin{cases} g[n-35] & 0 \le n \le 46, \\ 0 & \text{otherwise.} \end{cases} \qquad (8.99)$$

Hence, $\tilde{g}[n]$ is a *delayed* causal moving average (MA) of length $11 - (-35) + 1 = 47$. We note that

$$h[n] * \tilde{g}[n] \approx \delta[n-35]. \qquad (8.100)$$

This is because by time invariance,

$$y[n] \longrightarrow \boxed{g[n]} \longrightarrow x[n]$$

implies that

$$y[n] \longrightarrow \boxed{g[n-35]} \longrightarrow x[n-35].$$

We can, to a very good approximation (neglecting $|g[n]| < 0.001$), recover $x[n]$ *delayed by* 35. At the standard CD sampling rate of 44100 samples/s, this is less than 1 millisecond, which may be quite acceptable.

8-8 Deconvolution and Filtering Using the DFT

So far, in this book, all signal processing has been performed, or simulated, in ***real time***, wherein a signal is processed by passing it through a causal LTI system. In continuous time, the system might be an electronic circuit or a physical device. In discrete time, the system is a difference equation, implemented on a computer.

For real-time signal processing, a stable and causal system is required. In many applications, however, the processing need not be performed in real time, which allows us to use ***batch signal processing*** instead. In batch processing, the entire signal is recorded and stored (and therefore known for all past and future times) before it is processed, and when it is processed, it is processed in its entirety.

► The advantage of batch signal processing is that the LTI system no longer needs to be causal. ◄

This flexibility greatly expands the class of systems that can be used for signal processing, including non-minimum-phase systems. Batch signal processing usually is performed using the FFT algorithm of Section 7-17.

► All of the examples to follow in this chapter use batch signal processing, which requires that the entire signal be known and available before any signal processing is performed. ◄

8-8.1 Deconvolution of Non-Minimum-Phase Systems

Recall that the objective of deconvolution is to reconstruct the input $x[n]$ of a system from measurements of its output $y[n]$ and knowledge of its impulse response $h[n]$. That is, we seek to solve $y[n] = h[n] * x[n]$ for $x[n]$, given $y[n]$ and $h[n]$.

Deconvolution using DFT

If $x[n]$ has duration M and $h[n]$ has duration L, then $y[n]$ has duration $N_0 = L + M - 1$. Let us define the ***zero-padded functions***

$$\tilde{h}[n] = \{h[n], \underbrace{0, \ldots, 0}_{N_0 - L \text{ zeros}}\}, \tag{8.101a}$$

$$\tilde{x}[n] = \{x[n], \underbrace{0, \ldots, 0}_{N_0 - M \text{ zeros}}\}. \tag{8.101b}$$

With $\tilde{x}[n]$, $\tilde{h}[n]$, and $y[n]$ all now of duration N_0, we can obtain their respective N_0-point DFTs, $\tilde{\mathbf{X}}[k]$, $\tilde{\mathbf{H}}[k]$, and $\mathbf{Y}[k]$, which are interrelated by

$$\mathbf{Y}[k] = \tilde{\mathbf{H}}[k]\tilde{\mathbf{X}}[k]. \tag{8.102}$$

Upon dividing by $\tilde{\mathbf{H}}[k]$ and taking an N_0-point inverse DFT, we have

$$\begin{aligned}\tilde{x}[n] &= \text{DFT}^{-1}\{\tilde{\mathbf{X}}[k]\} \\ &= \text{DFT}^{-1}\left\{\frac{\mathbf{Y}[k]}{\tilde{\mathbf{H}}[k]}\right\} \\ &= \text{DFT}^{-1}\left\{\text{DFT}\{y[n]\}\Big/\text{DFT}\{\tilde{h}[n]\}\right\}.\end{aligned} \tag{8.103}$$

Discarding the $(N_0 - M)$ final zeros in $\tilde{x}[n]$ gives $x[n]$. The zero-padding and unpadding processes allow us to perform the deconvolution problem for any system, whether minimum phase or not.

FFT implementation issues

(a) To use the FFT algorithm (Section 7-17) to compute the three DFTs, N_0 should be rounded up to the next power of 2 because the FFT can be computed more rapidly.

(b) In some cases, some of the values of $\tilde{\mathbf{H}}[k]$ may be zero, which is problematic because the computation of $\mathbf{Y}[k]/\tilde{\mathbf{H}}[k]$ would involve dividing by zero. A possible solution to the division-by-zero problem is to change the value of N_0. Suppose $\tilde{\mathbf{H}}[k] = 0$ for some value of index k, such as $k = 3$. This corresponds to $\tilde{\mathbf{H}}(\mathbf{z})$ having a zero at $e^{j2\pi k/N_0}$ for $k = 3$, because by definition, the DFT is the $\mathbf{z}$-transform sampled on the unit circle at $\mathbf{z} = e^{j2\pi k/N_0}$ for integers k. Changing N_0 to, say, $N_0 + 1$ (or some other suitable integer) means that the DFT is now the $\mathbf{z}$-transform $\tilde{\mathbf{H}}(\mathbf{z})$ sampled at $\mathbf{z} = e^{j2\pi k/(N_0+1)}$, so the zero at $k = 3$ when the period was N_0 may now get missed with the sampling at the new period of $N_0 + 1$. Changing N_0 to $N_0 + 1$ may avoid one or more zeros in $\tilde{\mathbf{H}}[k]$, but it may also introduce new ones. It may be necessary to try multiple values of N_0 to satisfy the condition that $\tilde{\mathbf{H}}[k] \neq 0$ for all k.

Example 8-17: DFT Deconvolution

In response to an input $x[n]$, an LTI system with an impulse response $h[n] = \{\underline{1}, 2, 3\}$ generated an output $y[n] = \{\underline{6}, 19, 32, 21\}$. Determine $x[n]$, given that it is of finite duration.

Solution: The output is of duration $N_0 = 4$, so we should zero-pad $h[n]$ to the same duration by defining

$$\tilde{h}[n] = \{\underline{1}, 2, 3, 0\}. \tag{8.104}$$

From Eq. (7.169a), the 4-point DFT of $\tilde{h}[n]$ is

$$\tilde{\mathbf{H}}[k] = \sum_{n=0}^{3} \tilde{h}[n]\, e^{-j2\pi kn/4}, \qquad k = 0, 1, 2, 3, \tag{8.105}$$

which yields

$$\widetilde{\mathbf{H}}[0] = 1(1) + 2(1) + 3(1) + 0(1) = 6,$$

$$\widetilde{\mathbf{H}}[1] = 1(1) + 2(-j) + 3(-1) + 0(j) = -2 - j2,$$

$$\widetilde{\mathbf{H}}[2] = 1(1) + 2(-1) + 3(1) + 0(-1) = 2,$$

and

$$\widetilde{\mathbf{H}}[3] = 1(1) + 2(j) + 3(-1) + 0(-j) = -2 + j2.$$

Similarly, the 4-point DFT of $y[n] = \{\underline{6}, 19, 32, 21\}$ is

$$\mathbf{Y}[k] = \sum_{n=0}^{3} y[n]\, e^{-j2\pi kn/4}, \qquad k = 0, 1, 2, 3,$$

which yields

$$\mathbf{Y}[0] = 6(1) + 19(1) + 32(1) + 21(1) = 78,$$

$$\mathbf{Y}[1] = 6(1) + 19(-j) + 32(-1) + 21(j) = -26 + j2,$$

$$\mathbf{Y}[2] = 6(1) + 19(-1) + 32(1) + 21(-1) = -2,$$

and

$$\mathbf{Y}[3] = 6(1) + 19(j) + 32(-1) + 21(-j) = -26 - j2.$$

The 4-point DFT of $x[\tilde{n}]$ is, therefore,

$$\widetilde{\mathbf{X}}[0] = \frac{\mathbf{Y}[0]}{\widetilde{\mathbf{H}}[0]} = \frac{78}{6} = 13,$$

$$\widetilde{\mathbf{X}}[1] = \frac{\mathbf{Y}[1]}{\widetilde{\mathbf{H}}[1]} = \frac{-26 + j2}{-2 - j2} = 6 - j7,$$

$$\widetilde{\mathbf{X}}[2] = \frac{\mathbf{Y}[2]}{\widetilde{\mathbf{H}}[2]} = \frac{-2}{2} = -1,$$

and

$$\widetilde{\mathbf{X}}[3] = \frac{\mathbf{Y}[3]}{\widetilde{\mathbf{H}}[3]} = \frac{-26 - j2}{-2 + j2} = 6 + j7.$$

By Eq. (7.169b), the inverse DFT of $\widetilde{\mathbf{X}}[k]$ is

$$\tilde{x}[n] = \frac{1}{4} \sum_{k=0}^{3} \widetilde{\mathbf{X}}[k] e^{j2\pi kn/4}, \qquad n = 0, 1, 2, 3, \tag{8.106}$$

which yields

$$\tilde{x}[n] = \{\underline{6}, 7, 0, 0\}.$$

Given that $y[n]$ is of duration $N_0 = 4$ and $h[n]$ is of duration $L = 3$, it follows that $x[n]$ must be of duration $M = N_0 - L + 1 = 4 - 3 + 1 = 2$, if its duration is finite. Deletion of the zero-pads from $\tilde{x}[n]$ leads to

$$x[n] = \{\underline{6}, 7\}, \tag{8.107}$$

whose duration is indeed 2.

8-8.2 Filtering Noisy Signals by Thresholding

In both continuous and discrete time, perfect brick-wall filters with abrupt boundaries are physically impossible to realize. To generate approximate brick-wall-like frequency responses, the filters' transfer functions must have many (appropriately placed) poles and zeros.

In contrast, implementing a brick-wall filter in batch signal processing is a trivial task. All that is necessary is to set the coefficients of undesired frequency components of $\mathbf{X}[k]$ to zero. Ideal brick-wall filters can be constructed with any desired frequency response, including lowpass, highpass, bandpass, and bandreject.

However, even though brick-wall filters can be implemented easily using the DFT, a filter with a tapered frequency response may be preferable. The impulse response of a brick-wall lowpass filter is a discrete-time sinc function, and while its noncausality is no longer an issue, its slow decay with time means that the duration of the impulse response is longer than the duration of the signal. Consequently, the convolution of the oscillatory impulse response with the signal exhibits oscillations, called "ringing," in the filtered signal. The ringing pattern can be avoided by using a filter with a tapered frequency response.

If noise with a broad spectrum has been added to a periodic signal with a distinctive line spectrum, the bulk of the noise can be removed by setting all frequency components to zero, except for those associated with the signal's line spectrum. Only the noise added to the spectral lines themselves remains. The process is illustrated by the following two examples.

MATLAB/MathScript Recipe for Brick-Wall Lowpass Filter

To eliminate all frequency components above F Hz in a signal sampled at S samples per second and stored in MATLAB/MathScript as vector **X**, apply the following code:

```
N=length(X);
FX=fft(X);
K=ceil(N*F/S)+1;
FX(K:N+2-K)=0;
Y=real(ifft(FX));
```

Note that 1 should be added to NF/S because MATLAB/MathScript indexing starts at 1, not 0.

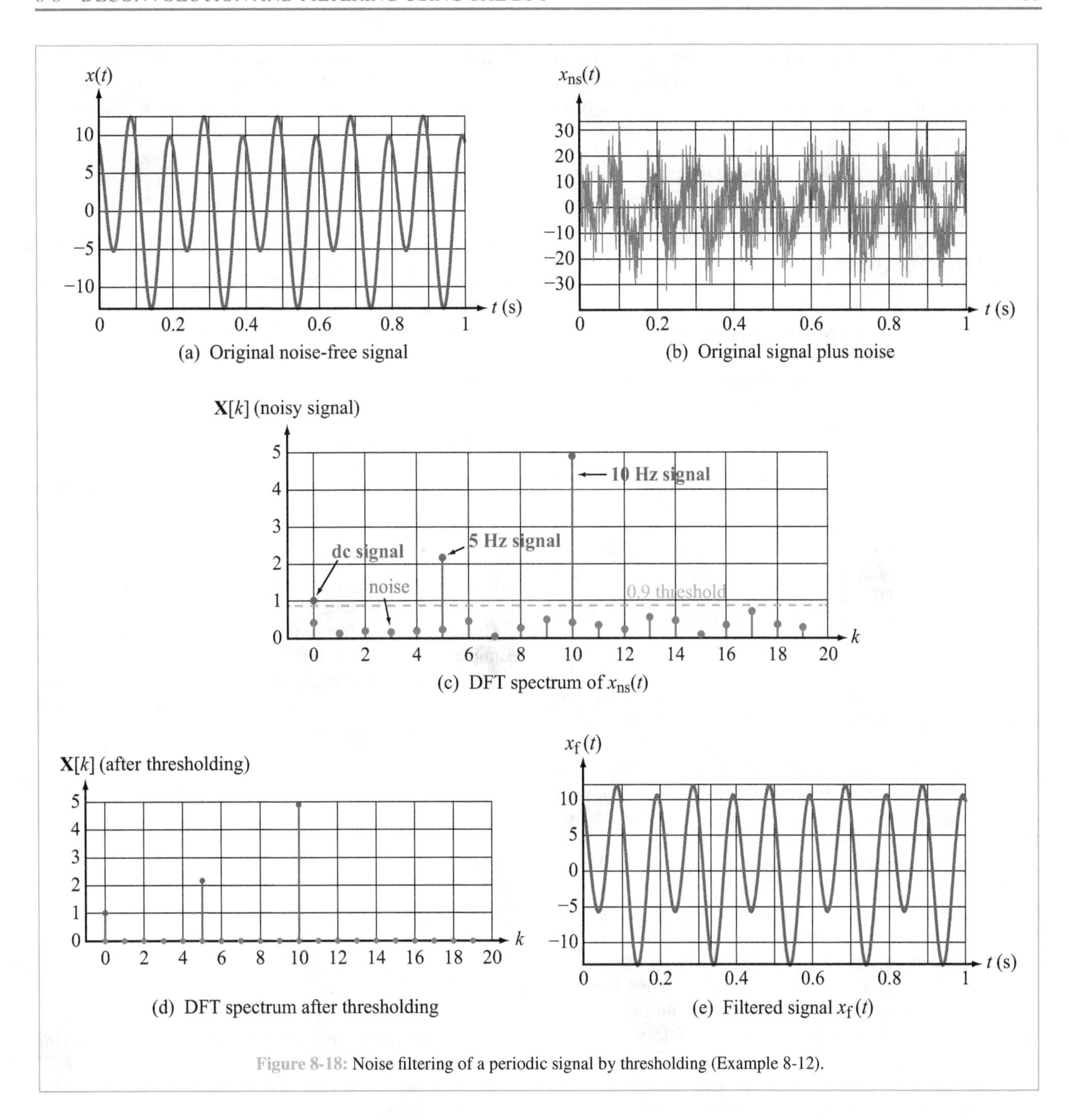

Figure 8-18: Noise filtering of a periodic signal by thresholding (Example 8-12).

Example 8-18: Noise Filtering by Thresholding

Figures 8-18(a) and (b) display a signal given by

$$x(t) = 1 + 4\sin(10\pi t) + 10\cos(20\pi t + 0.6435),$$

before and after broad-spectrum noise was added to it. The goal is to retrieve the original signal $x(t)$ in Fig. 8-18(a) from the noisy signal $x_{\text{ns}}(t)$ in Fig. 8-18(b), *without* knowledge of the original signal. The signal-to-noise ratio is −2.56 dB. Develop a simple technique to retrieve $x(t)$ from the noisy signal.

Solution: A negative value (in dB) for the signal-to-noise ratio means that the total noise power is *greater* than the total signal power. Inspection of the noisy signal in Fig. 8-18(b) suggests that:

(1) the signal is a 10 Hz sinusoid (10 cycles over the waveform's 1-second recording), plus possibly other harmonic sinusoids and a dc component, and

(2) the noise has a broad spectrum.

After sampling the continuous-time noisy signal $x_{ns}(t)$, at 1000 samples/s, a 1000-point DFT was computed to generate the spectrum of $x_{ns}(t)$. Figure 8-18(c) displays only the first 20 spectral components, as higher frequency components are judged to contain only noise and no signal. The prominent components are at $k = 0$, 5, and 10. These are the components associated with the periodic signal.

To clarify the example, the signal and noise components at indices $k = 0, 5, 10$ are shown separately, although in practice only their sums are known. Note that since the phases of the signal and noise components differ, the magnitude of the sum is not the sum of the magnitudes, but if $|\mathbf{X}[k]| \gg |\mathbf{N}[k]|$, then $|\mathbf{X}[k] + \mathbf{N}[k]| \approx |\mathbf{X}[k]|$.

An easy solution to eliminating most of the noise is to choose an appropriate ***threshold*** and then assign zero values to all spectral components whose amplitudes are below the selected threshold. Since the amplitude of the dc component is approximately 1.0, we set the threshold level at 0.9.

The thresholded spectrum is shown in Fig. 8-18(d). Upon performing a 1000-point inverse DFT on the thresholded spectrum, we obtain a filtered discrete-time signal $x_f[n]$, with $n = 0, 1, \dots, 1000$. Reconstruction of $x_f[n]$ to continuous time leads to signal $x_f(t)$ shown in Fig. 8-18(e), which bears a very close resemblance to the original noise-free waveform shown in Fig. 8-18(a).

The most remarkable aspect of this example is that we did not use any knowledge of the signal or noise strengths. The only assumption we made was that the harmonics of the signal were larger than any components of the noise, so that a simple thresholding strategy would eliminate most of the noise while leaving the signal intact. The threshold value of 0.9 was chosen in part by trial and error. An important clue is that if the signal is periodic, the harmonics occur at integer multiples of some fundamental frequency; the threshold should be chosen to preserve this structure in the filtered signal. Statistics of the signal and noise, if known, can be used to compute a value of the threshold, but this is far beyond the scope of this book. (See S² for details.)

Noise filtering by thresholding can be applied not only to sinusoids, as was demonstrated in Example 8-18, but to other periodic and nonperiodic signals as well. This is illustrated in Example 8-19.

Example 8-19: Noise Filtering Trumpet Signal by Thresholding

The signal of an actual trumpet playing note B has broad-spectrum noise added to it. The goal is to retrieve the original signal, shown in Fig. 8-19(a), from the noisy signal, shown in Fig. 8-19(b). The trumpet signal is sampled at 44,100 samples/s. The signal-to-noise ratio is 8.54 dB.

Solution: The spectrum of the noisy signal is shown in Fig. 8-19(c). The threshold is set (by trial and error) at 0.0015. The thresholded spectrum is shown in Fig. 8-19(d). The inverse DFT of this spectrum, followed by reconstruction to continuous time, is shown in Fig. 8-19(e).

Comparing Fig. 8-19(a) and Fig. 8-19(e) shows that the noise has been virtually eliminated. Listening to the noisy and filtered trumpet signals would confirm this. Note that the additive white noise sounds like hissing.

This example reaffirms the conclusions of the previous example, namely that the thresholding technique is an effective filtering tool so long as the signal we wish to preserve is periodic and its spectral components are distinctly larger than those of the noise. (See S² for details.)

8-8.3 Removal of Periodic Interference

Periodic interference can be removed easily by simply setting the harmonics of the interference to zero. No thresholding is necessary if the period of the interfering signal is known. This is illustrated in Example 8-20.

Example 8-20: Two-Trumpets Signal, Revisited

We are given the signal of two actual trumpets playing simultaneously notes A and B. The goal is to use the DFT to eliminate the trumpet playing note A, while preserving the trumpet playing note B. We need only know that note B is at a higher frequency than note A.

Solution: The two-trumpets signal time-waveform is shown in Fig. 8-20(a), and the corresponding spectrum is shown in Fig. 8-20(b). We note that the spectral lines occur in pairs of harmonics with the lower harmonic of each pair associated with

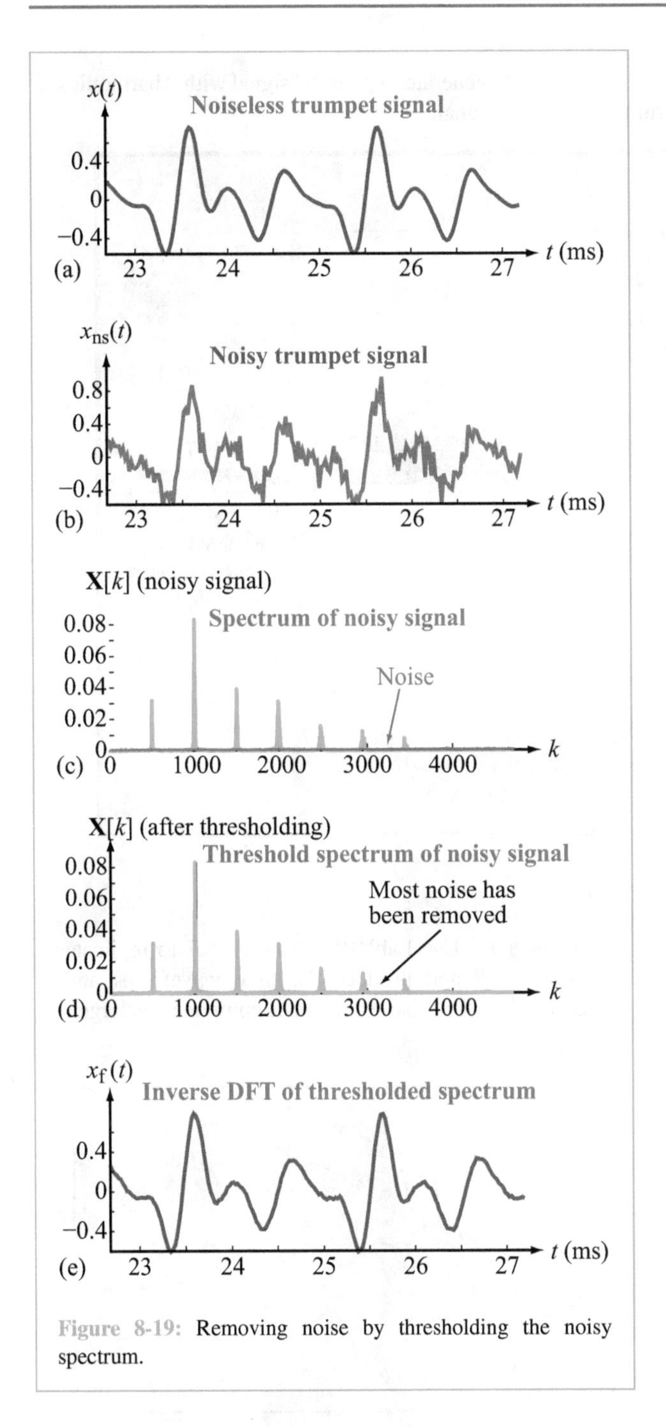

Figure 8-19: Removing noise by thresholding the noisy spectrum.

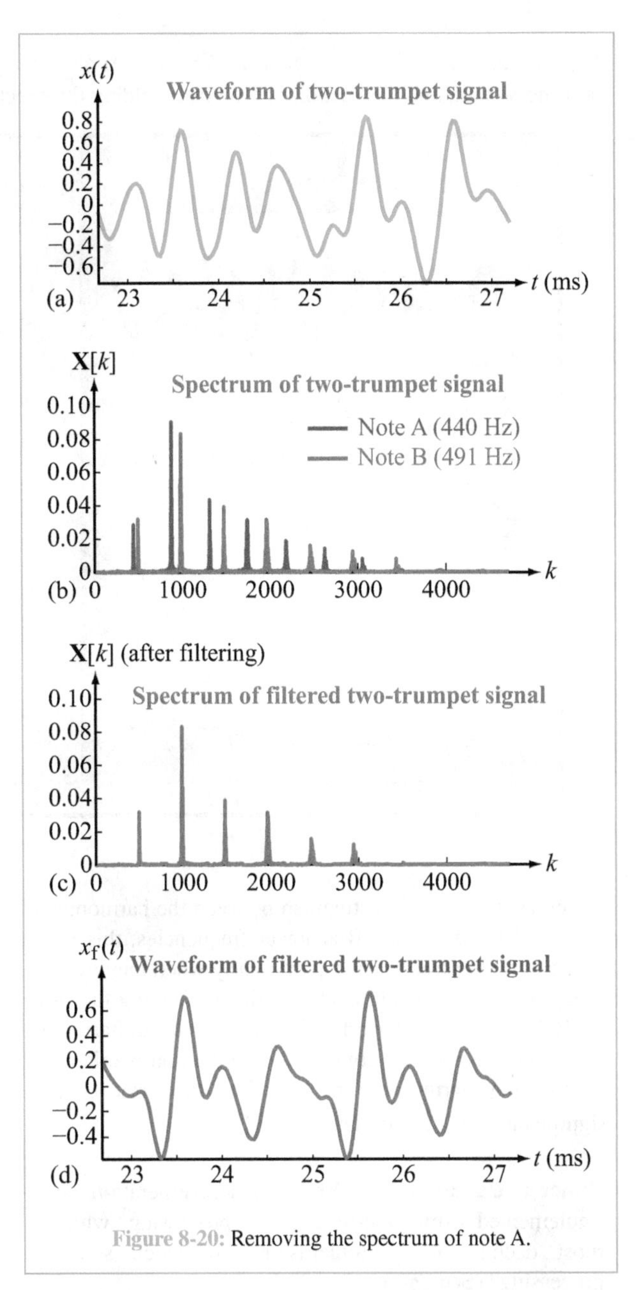

Figure 8-20: Removing the spectrum of note A.

note A and the higher harmonic of each pair associated with note B.

Since we wish to eliminate note A, we set the lower component of each pair of spectral lines to zero. The modified spectrum is shown in Fig. 8-20(c). The inverse DFT of this spectrum, followed by reconstruction to continuous time, is shown in Fig. 8-20(d).

The filtering process eliminated the signal due to the trumpet playing note A, while preserving the signal due to note B, almost completely. This can be confirmed by listening to the signals before and after filtering.

Module 8.7 Denoising a Periodic Signal by Thresholding This module generates a periodic signal with 2 harmonics, adds noise to it, and then denoises it by thresholding the spectrum of the noisy signal.

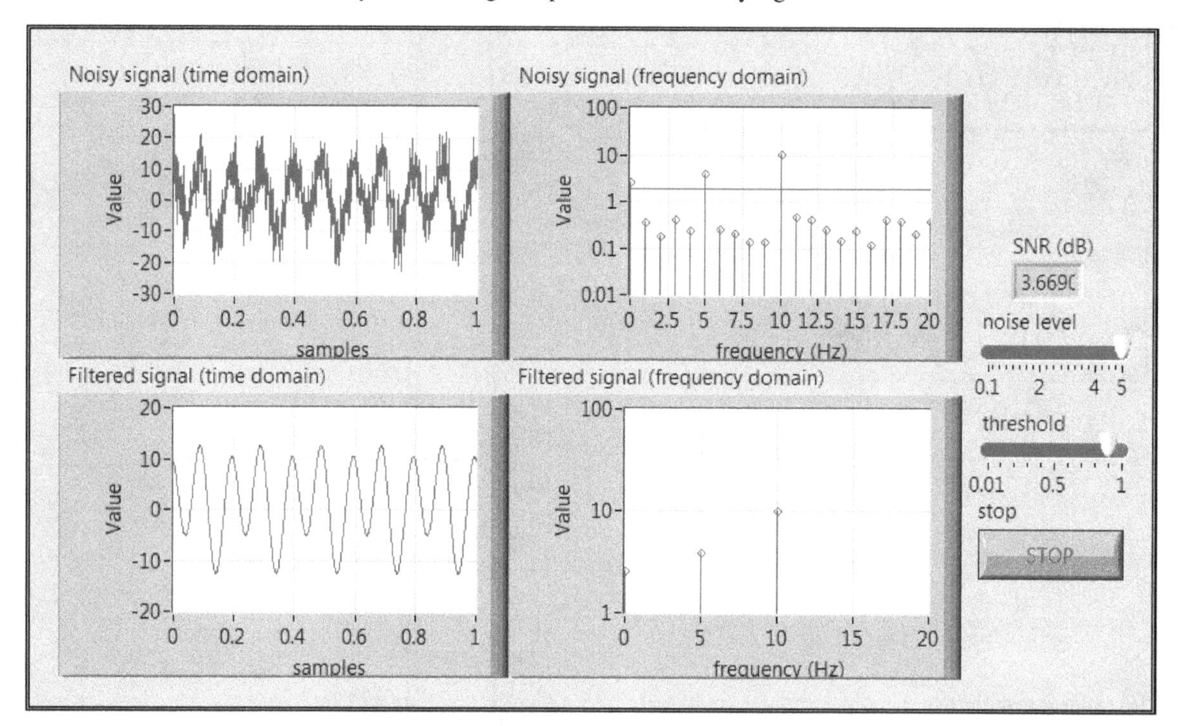

Whereas it is easy to distinguish between the harmonics of note A and those of note B at lower frequencies, this is not the case at higher frequencies, particularly when they overlap. Hence, one cannot be eliminated without affecting the other slightly. Fortunately, the overlapping high-frequency harmonics contain very little power compared with the non-overlapping, low-frequency harmonics, and therefore, their role is quite insignificant. (See S^2 for details.)

Concept Question 8-6: Why can dereverberation be implemented using real-time signal processing, while most deconvolution problems require batch signal processing? (See S^2)

Concept Question 8-7: In DFT computation, why do we zero-pad the *output* to the next highest power of 2? (See S^2)

Concept Question 8-8: For what type of signal and noise spectra is filtering by thresholding effective at removing the noise? (See S^2)

Exercise 8-12: Use LabVIEW Module 8.7 to replicate Example 8-18 and produce the time waveforms and spectra of Fig. 8-18. Note that the dc component is larger.

Answer:

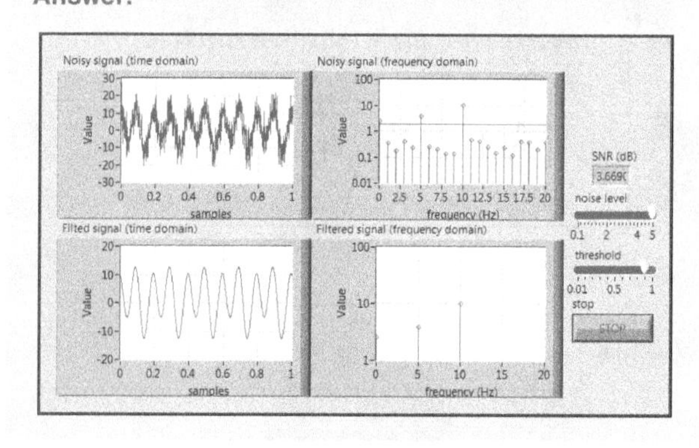

Module 8.8 Separating Two Trumpet Signals Using the DFT This module adds the waveforms of trumpets playing notes A and B and uses the DFT to separate them by setting a band of frequencies around each harmonic of one of the trumpets to zero. The module also allows the user to listen to the original (two trumpets) and filtered (one trumpet) signals.

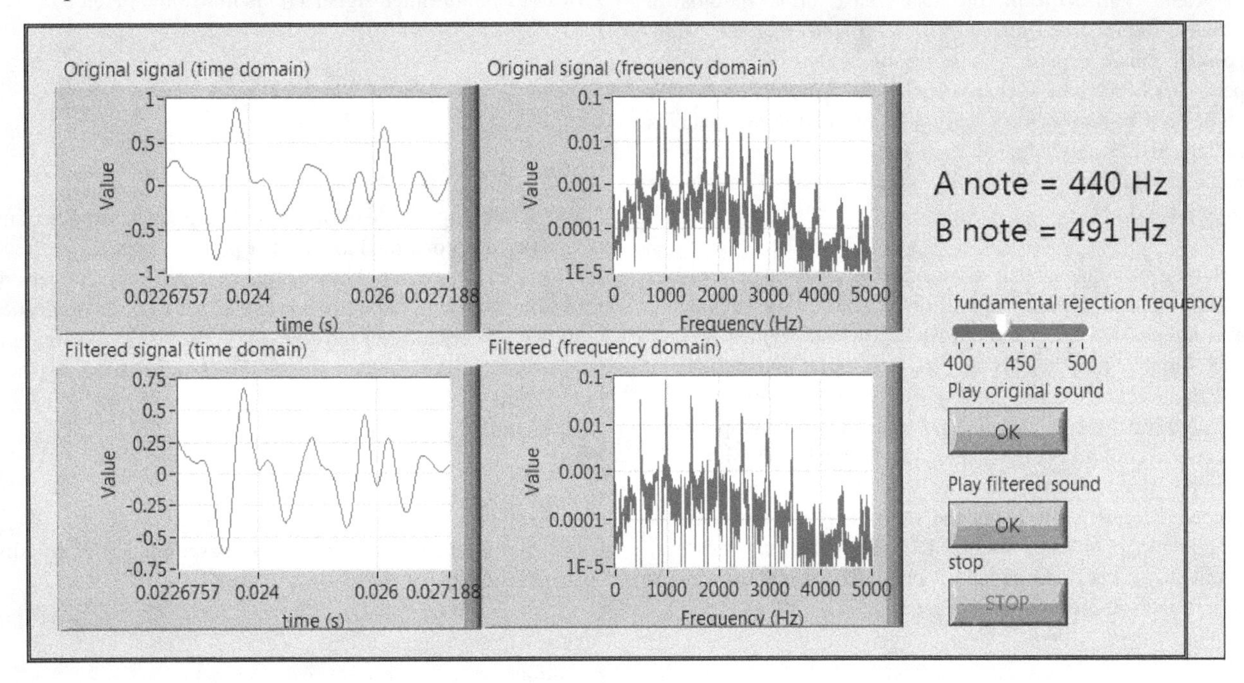

Exercise 8-13: Use LabVIEW Module 8.8 to replicate Example 8-20 and produce the time waveforms and spectra of Fig. 8-20. The time waveforms are different.

Answer:

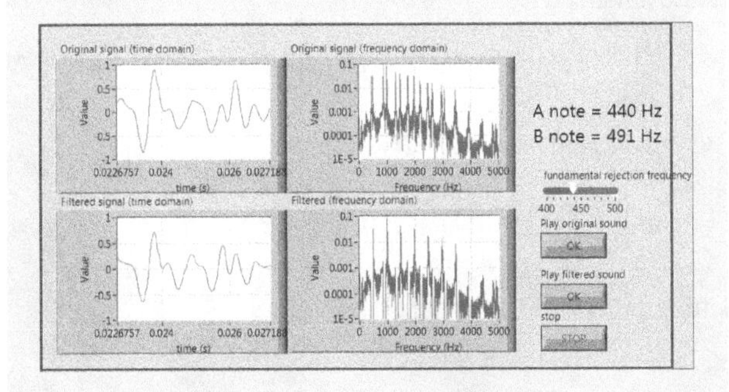

8-9 Computing Spectra of Periodic Signals

In many system applications (including signal storage and transmission, sensing and actuating, and adaptive control of physical systems) knowledge of the spectra of the relevant signals is key to the effective implementation of those applications. We cite three simple examples.

(a) CD Capacity: Suppose we wish to compile a set of spoken-word CDs of a famous speaker or to record audio readings of certain books. The common CD sampling rate is 44,100 samples/s, so a CD with a capacity of 670 megabytes can hold about 75 minutes of material. But if we compute the spectrum of the speaker's voice and determine that it is limited to *less than* 5 kHz, we can use a sampling rate of only 2×5 kHz $= 10,000$ sample/s. Not only does such a rate exceed the Nyquist rate, but it also means that we can increase the capacity of a CD from a little over one hour to about five hours of recorded material.

(b) Noise Filtering: If we know a signal is bandlimited to, say, 5 kHz, we can use a lowpass filter to eliminate noise at frequencies above 5 kHz. But if the signal's spectrum extends over the range from 5 kHz to 10 kHz, for example, we use a bandpass filter instead. If the signal is periodic with a line spectrum, we can eliminate additive noise at all frequencies except at those pertaining to the signal. Knowledge of the signal's spectrum allows us to design a noise filter to match it.

(c) Music Transcription: In music transcription, the goal is to transcribe recorded music to musical staff notation. Many types of music consist of a series of notes or chords. The spectrum of each note or chord consists of harmonics of a single frequency (for a note) or of harmonics of multiple frequencies (for chords). Knowledge of their fundamental frequency or frequencies allows us to identify which notes and chords are being played.

Having established that knowledge of a signal's spectrum is useful and important, we will now explore how in practice to compute that spectrum. Most real-world signals do not have simple analytic expressions, making it impractical to obtain their spectra by directly applying the Fourier-series (for periodic signals) and Fourier-transform (for nonperiodic signals) techniques of Chapter 5. Instead, the signal is sampled at an appropriate rate, and then its spectrum is computed numerically using the DFT. This section presents the mathematical basis for the DFT computational process, specifically for periodic signals. A parallel presentation is given in Section 8-10 for nonperiodic signals.

8-9.1 Fourier Series of a Sampled Signal

Suppose $x(t)$ is a continuous-time periodic signal with ***period*** T_0, so that $x(t) = x(t + nT_0)$, where n is any integer. Also, $x(t)$ is ***bandlimited*** to $f_{\max}$ (Hz); i.e., *at and above* $f_{\max}$, the signal energy is zero. Strictly speaking, a signal cannot be bandlimited to a finite-value frequency $f_{\max}$ unless it is everlasting, and real signals are not everlasting. When characterizing a real signal as bandlimited to $f_{\max}$, we mean that ignoring components of the signal's spectrum at frequencies higher than $f_{\max}$ results in negligible distortion of the signal's waveform and information content.

Our goal is to compute the ***complex Fourier coefficients*** $\mathbf{x}_k$ of the Fourier series expansion of $x(t)$. The sequence $\{\mathbf{x}_k\}$ constitutes the ***spectrum*** of periodic signal $x(t)$.

(a) Finite Fourier series

Since $x(t)$ is periodic with ***frequency*** $f_0 = 1/T_0$, its spectrum consists of harmonics at frequencies kf_0 for integer values of k. Being bandlimited to $f_{\max}$, the highest-frequency harmonic of $x(t)$ is at $f_{\max}$, and the corresponding ***maximum value of index*** k is

$$K = \frac{f_{\max}}{f_0} = f_{\max} T_0. \tag{8.108}$$

Hence, the summation index in the Fourier-series expansion given by Eq. (5.57) is limited to the range from $-K$ to K [or $(2K+1)$ terms], not from $-\infty$ to $+\infty$. Accordingly, the Fourier series of a bandlimited signal $x(t)$ is finite, and given by

$$x(t) = \sum_{k=-K}^{K} \mathbf{x}_k e^{j2\pi kt/T_0}. \tag{8.109}$$

(b) Sampling

For a periodic signal, the information contained in any one of its periods is also contained in all other periods. Hence, when we sample $x(t)$, we need only to sample a single complete period. From the sampling theorem, we know that in order to ensure unique reconstruction of the signal it is necessary to sample it at a ***sampling rate*** f_s that exceeds the Nyquist sampling rate. For a signal bandlimited to $f_{\max}$, the ***Nyquist rate*** is $2f_{\max}$. Consequently, the sampling condition is

$$f_s > 2f_{\max}. \tag{8.110}$$

If one period of $x(t)$, of duration T_0, is sampled at N equally spaced times,

$$f_s = \frac{N}{T_0}. \tag{8.111}$$

Combining Eqs. (8.108), (8.110), and (8.111) leads to

$$N > 2K, \tag{8.112}$$

which states that the number of sampled times should be larger than twice the value of index K in the Fourier series summation given by Eq. (8.109). The condition described by Eq. (8.112) can be amended to

$$N \geq (2K + 1). \tag{8.113}$$

The N-sampled values of $x(t)$ occur at

$$t = \left(\frac{n}{N}\right) T_0, \qquad n = 0, 1, \ldots, N-1. \tag{8.114}$$

Setting $t = (n/N)T_0$ in Eq. (8.109) gives

$$x\left(\frac{n}{N} T_0\right) = \sum_{k=-K}^{K} \mathbf{x}_k \, e^{j2\pi k[(n/N)T_0]/T_0} = \sum_{k=-K}^{K} \mathbf{x}_k \, e^{j2\pi kn/N}, \qquad n = 0, 1, \ldots, N-1. \tag{8.115}$$

Note the resemblance of Eq. (8.115) to the DTFS. It is relatively straightforward to show that the Fourier coefficient $\mathbf{x}_k$ can be computed from the sampled values $x(nT_0/N)$ through the summation

$$\mathbf{x}_k = \frac{1}{N} \sum_{n=0}^{N-1} x\left(\frac{n}{N}\, T_0\right) e^{-j2\pi kn/N}, \qquad k = -K, \dots, K. \tag{8.116}$$

Index k extends from $-K$ to $+K$, which includes $k = 0$. Hence, we have $(2K + 1)$ Fourier coefficients. To compute $(2K + 1)$ unknowns, we need at least that many known values of $x(t)$. Hence, the number of samples, N should equal or exceed $(2K + 1)$. This is the condition defined by Eq. (8.113).

(c) Computing $\mathbf{x}_k$ by DFT

When we introduced the DFT in Section 7-15, we defined the DFT complex coefficients $\mathbf{X}[k]$ in terms of a summation involving discrete-time signal $x[n]$, specifically

$$\mathbf{X}[k] = \sum_{n=0}^{N_0-1} x[n]\, e^{-j2\pi kn/N_0}, \qquad k = 0, 1, \dots, N_0 - 1. \tag{8.117}$$

Because of the close similarity between the forms of Eqs. (8.116) and (8.117), we can implement the following sequence:

Step 1: Define $x'\left(\frac{n}{N}\, T_0\right) = \frac{1}{N}\, x\left(\frac{n}{N}\, T_0\right)$.

Step 2: Apply an N-point DFT to samples $x'\left(\frac{n}{N}\, T_0\right)$, using the FFT algorithm, to compute $\mathbf{x}_k$ for $k = 0, 1, \dots, K$.

Step 3: Use the relationship $\mathbf{x}_{-k} = \mathbf{x}_k^*$ to compute $\mathbf{x}_k$ for negative values of k.

Example 8-21: DFT Computation of Fourier Series

At a sampling rate of 50 samples/s, the number of samples generated by a periodic signal with period $T_0 = 0.2$ s and $f_{\max} = 25$ Hz is $N = f_s T_0 = 50 \times 0.2 = 10$ samples. Compute the Fourier-series coefficients of $x(t)$, given its sampled values: {<u>9</u>, 0.117, −5.195, 1.859, 11.53, 9, −4.585, −12.8, −5.75, 6.827}.

► Actually, the signal is given by

$$x(t) = 1 + 4\sin(10\pi t) + 10\cos(20\pi t + 0.6435),$$

but presumably that information is not available to us. The actual signal is bandlimited to 10 Hz, so the specified information that $f_{\max} = 25$ Hz is conservative. ◄

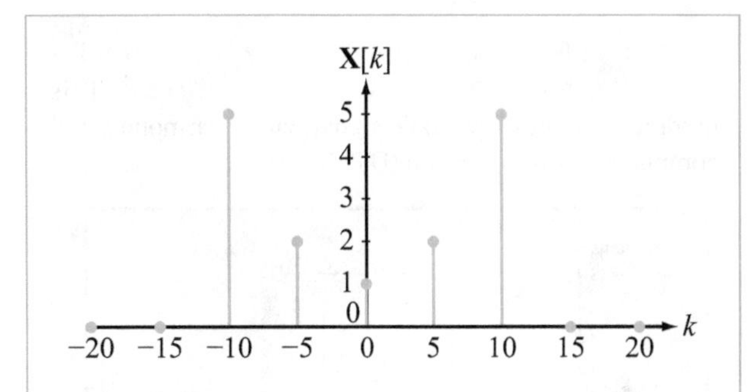

Figure 8-21: DFT spectrum generated from ten sampled values of $x(t)$.

Solution: A 10-point DFT of the 10 samples is computed using the command `fft` in MATLAB or MathScript. We can either normalize the sampled values by dividing them by 10 prior to computing the DFT, or the DFT output can be divided by 10 after computing it with the unnormalized samples. The output is then shifted using the command `fftshift` so that dc ($f = 0$) is at the middle of the spectrum instead of at the beginning. The magnitude spectrum is plotted in Fig. 8-21 as a function of $k/T_0 = k/0.2 = 5k$, with $k = -4, \dots, 4$. (See S² for details.)

(d) Interpretation of DFT output

Suppose we are given the output `fft(X)`, where `X` is a vector of samples of a signal $x(t)$ sampled at S samples/s (MATLAB and MathScript use symbol S for sampling rate). The number of samples is `N=length(X)`. How do we interpret the output `fft(X)`?

The DFT/FFT computation treats the signal as if it is periodic, whether it is or not. The computation assumes that the signal has a period T_0 and the N samples are equally spaced over a time duration T_0. Thus, it assumes the signal is periodic with period $T_0 = N/S$ seconds. It then follows that the signal has a line spectrum of harmonics at $f = k/T_0 = kS/N$ (Hz). The magnitudes and phases of the harmonics are

`abs(fft(X))/N` and `angle(fft(X))`. (8.118)

Only the first half of the output of `fft` should be used to plot the one-sided spectrum. The second half of the output is the mirror image of the first half (by conjugate symmetry), so it need not be plotted.

Module 8.9 Computing Spectra of Discrete-Time Periodic Signals Using the DTFS This module generates a periodic signal with 2 harmonics and computes its line spectrum (DTFS).

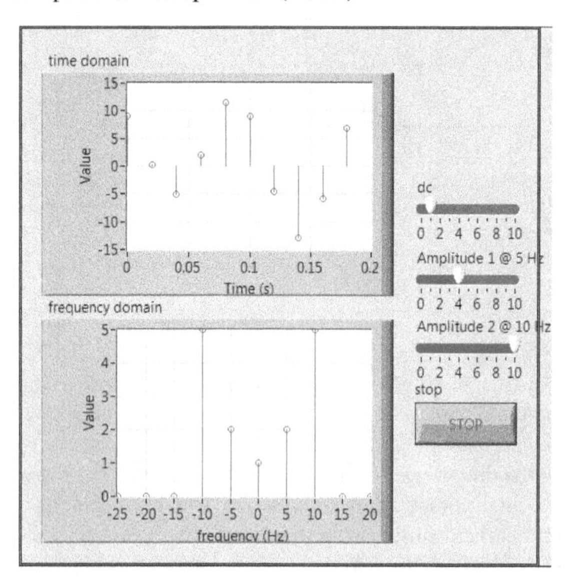

Example 8-22: Interpreting DFT Output

A signal $x(t)$ is sampled at 1024 samples per second and stored in vector X. Then `fft(X)/4096` gives

$$[8, \underbrace{0\ldots0}_{31}, 2{+}2i, \underbrace{0\ldots0}_{95}, 6i, \underbrace{0\ldots0}_{3839}, -6i, \underbrace{0\ldots0}_{95}, 2{-}2i, \underbrace{0\ldots0}_{31}.]$$

Assuming no aliasing has occurred, what is $x(t)$?

Solution: The length is `N=length(X)=4096` samples. The duration of the signal is $T_0 = 4096/1024 = 4$ s. There are five peaks, located at indices $K = [1, 33, 129, 3969, 4065]$. Note that the `fft` indexing starts at $K = 1$, but DFT indexing starts at $k = 0$. Hence we must subtract 1 from indices K to get DFT indices k. Based on the output, $\mathbf{x}[k] = 0$ for all k except the following:

- $\mathbf{x}_0 = 8$
- $\mathbf{x}_{32} = \mathbf{x}^*_{4064} = 2 + j2 = 2\sqrt{2}\, e^{j\pi/4}$
- $\mathbf{x}_{128} = \mathbf{x}^*_{3968} = j6 = 6e^{j\pi/2}$

The frequency associated with $\mathbf{x}_{32}$ is

$$\frac{k}{T_0} = k\,\frac{S}{N} = \frac{32 \times 1024}{4096} = 8 \text{ Hz},$$

and similarly, the frequency associated with $\mathbf{x}_{128}$ is

$$\frac{k}{T_0} = k\,\frac{S}{N} = \frac{128 \times 1024}{4096} = 32 \text{ Hz}.$$

Components at indices above $4096/2 = 2048$ are discarded for the one-sided line spectrum.

The signal $x(t)$ is therefore

$$x(t) = 8 + 4\sqrt{2}\cos(16\pi t + \pi/4) + 12\cos(64\pi t + \pi/2).$$

8-9.2 Spectral Leakage

Suppose we are given two signals, namely,

$$x(t) = 2\cos(880\pi t), \qquad 0 \le t < 1, \tag{8.119a}$$

and

$$y(t) = 2\cos(881\pi t), \qquad 0 \le t < 1. \tag{8.119b}$$

The two signals are identical in every respect, except that $f_0 =$ 440 Hz for $x(t)$ and 440.5 Hz for $y(t)$. Both sinusoids are 1 s long ***segments***, and both are sampled at $f_s = 1024$ samples/s.

Recall from Chapter 5 that the spectrum of a sinusoid of frequency f_0 consists of two spectral lines located at $\pm f_0$. Given a ***signal's duration*** T_d (which should not be confused with the sinusoid's period T_0) and sampling rate f_s, the number of samples is

$$N = f_s T_d.$$

In the present case, for both signals $x(t)$ and $y(t)$, $f_s = 1024$ samples/s, $T_d = 1$ s, and $N = 1024$ samples. When we perform a 1024-point DFT on sampled signal $x(t)$, we obtain the spectrum shown in Fig. 8-22(a), consistent with our expectations that the spectrum should consist of lines at ± 440 Hz, and their heights should be 1.

In contrast, application of the same steps to signal $y(t)$, whose frequency is 440.5 Hz, leads to two lines that are spread out at their bases, as shown in Fig. 8-22(b), and their heights are significantly lower than 1.

Why do we get different spectra for the 440 Hz and 440.5 Hz sinusoids when computed using the DFT? The simple answer has to do with the value of the product $f_0 T_d$:

$$\begin{aligned} x(t)&: \quad f_0 T_d = 440 \times 1 = 440 \text{ (which is an integer)}, \\ y(t)&: \quad f_0 T_d = 440.5 \times 1 = 440.5 \text{ (not an integer)}. \end{aligned}$$

When $f_0 T_d$ is an integer, the DFT generates an exact Fourier transform, but when $f_0 T_d$ is not an integer, it generates a spectrum with ***spectral leakage*** (the name given to spectral lines that are spread out at their bases). Close examination of the base spread shows that it is a discrete sinc function.

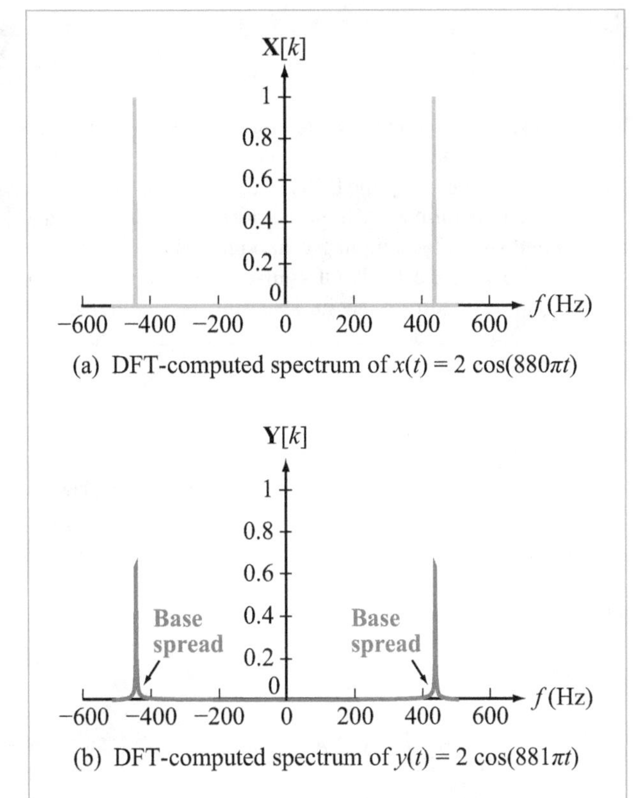

Figure 8-22: Spectral leakage (base spread of spectral lines) occurs when $f_0 T_d$ is not an integer, where f_0 is the sinusoid's frequency and T_d is the segment duration. For $x(t)$ at 440 Hz and $T_d = 1$ s, $f_0 T_d = 440$, whereas for $y(t)$ at 440.5 Hz and $T_d = 1$ s, $f_0 T_d = 440.5$ is not an integer.

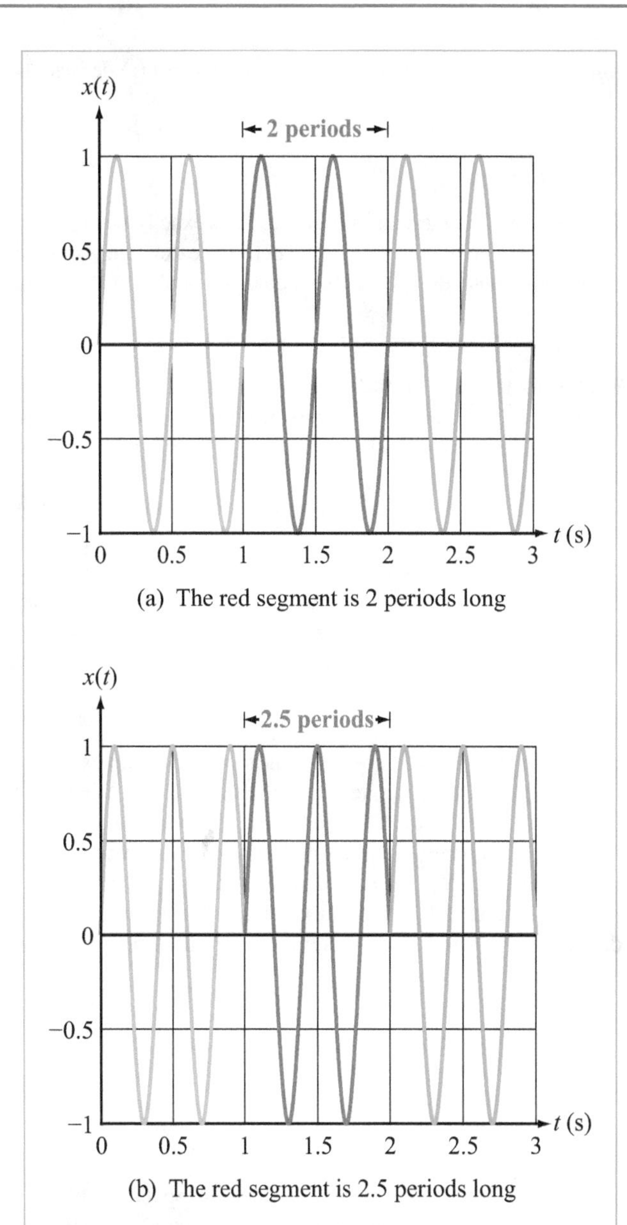

Figure 8-23: Periodic extension generates the correct parent sinusoid from the available signal segment when the segment is exactly an integer multiple of the sinusoid's period, as in (a), but it fails to do so otherwise.

The more elaborate explanation for what causes spectral leakage when the DFT is used to compute the spectrum of a periodic signal is related to an implicit assumption of the DFT that the sampled signal is a single period of a periodic signal. The spectral-leakage problem arises because in most cases the sampled signal is *not* a single period. Instead, it consists of several periods and a fraction of a period of the periodic signal.

To appreciate the implications of the DFT assumption, we should distinguish between "sinusoid" and "a segment of a sinusoid":

$$\begin{aligned} &\text{Sinusoid:} && 2\cos(2\pi f_0 t) && -\infty < t < \infty, \\ &T_d\text{-long segment:} && 2\cos(2\pi f_0 t) && 0 \le t < T_d. \end{aligned}$$

A sinusoid is everlasting, whereas a finite-duration segment is not. The DFT is formulated to compute the Fourier series of a periodic signal extending from $-\infty$ to ∞ on the basis of sampled values taken across one period or multiple periods (but not fractions of periods).

If the input signal segment is the 1 s long sinusoid shown in red in Fig. 8-23(a), the DFT assumes that it is adjoined (preceded and followed) by identical copies over and over in time. Thus, $x(t)$ is part of a ***periodic extension*** $\tilde{x}(t)$ defined as

$$\tilde{x}(t) = \{\ldots, x(t+T_d),\ x(t),\ x(t-T_d),\ \ldots\}. \quad (8.120)$$

(a) T_d/T_0 = integer

The red sinusoid segment in Fig. 8-23(a) is exactly two periods long ($T_d = 2T_0$). Consequently, its periodic extension is indeed the parent sinusoid of which the signal is a part.

(b) $T_d/T_0 \neq$ integer

That is not the case for the red signal segment in Fig. 8-23(b). The segment is 2.5 periods long, so the replication and adjoining process generates a periodic extension that is *not* the true parent sinusoid.

The DFT computes the Fourier series of a sampled version of $\tilde{x}(t)$. If $\tilde{x}(t)$ is a true parent of the periodic signal, which occurs when the segment length T_d is an integer multiple of the periodic signal's period T_0, no spectral leakage occurs. On the other hand, if T_d/T_0 is not an integer, $\tilde{x}(t)$ is not the true parent of $x(t)$ and spectral leakage will occur.

(c) Summary

For a sinusoid $x(t) = A\cos(\omega_0 t)$ sampled at f_s samples/s over a duration T_d, the discrete-time sinusoid $x[n] = A\cos(\Omega n)$ has angular frequency $\Omega = (\omega_0/f_s)$ and fundamental period $N_0 = 2\pi k/\Omega$, where k is the smallest integer that results in an integer value for N_0. The total number of samples is $N = f_s T_d$. When a N-point DFT is used to compute the spectrum of the discrete-time sinusoid, no spectral leakage occurs if N/N_0 is an integer.

Concept Question 8-9: Why is it possible to compute, using the DFT, the exact spectrum of a signal from its samples? (See S²)

Concept Question 8-10: When does the spectrum of a pure sinusoid computed using the DFT contain spectral leakage? (See S²)

Exercise 8-14: The spectrum of

$$\{\cos(0.3\pi n),\ n = 0, \ldots, N-1\}$$

is to be computed using the DFT. For what values of N will there be no spectral leakage?

Answer: N = integer multiple of 20. (See S²)

8-10 Computing Spectra of Nonperiodic Signals

In the preceding section, we computed the Fourier series expansion of a periodic and bandlimited signal by sampling it in time and then applying the DFT. In this section, we compute the Fourier transform of a finite-duration (***time-limited***) and bandlimited signal by sampling it in both time and frequency, and also applying the DFT. Of course, a time-limited signal has an infinite spectrum, so its spectrum is bandlimited by artificially applying a finite-width window that captures the most significant part of its spectral energy.

8-10.1 Timelimited-Bandlimited Signals

Our goal is to develop a formulation that takes advantage of the DFT to compute the ***continuous-time Fourier transform*** (CTFT) of a nonperiodic signal $x(t)$ and its inverse:

$$\mathbf{X}(f) = \int_{-\infty}^{\infty} x(t)\, e^{-j2\pi f t}\, dt, \quad (8.121a)$$

$$x(t) = \int_{-\infty}^{\infty} \mathbf{X}(f)\, e^{j2\pi f t}\, df. \quad (8.121b)$$

The frequency variable is represented by f in Hz, instead of ω in radians/s. Using f in the material that follows makes the results somewhat easier to follow.

Signal $x(t)$ and its spectrum $\mathbf{X}(f)$ are defined to be

(1) Time-limited to T_d :

$$x(t) = 0, \text{ unless } -\frac{T_d}{2} < t < \frac{T_d}{2}, \quad (8.122a)$$

(2) Bandlimited to $f_{max} = \frac{B_b}{2}$:

$$\mathbf{X}(f) = 0, \text{ unless } -\frac{B_b}{2} < f < \frac{B_b}{2}. \quad (8.122b)$$

In Chapter 6, we defined the bandwidth B of a lowpass signal whose positive spectrum extends to frequency f_{max} as $B = f_{max}$. We also defined the ***bidirectional bandwidth*** extending from $-f_{max}$ to $+f_{max}$ as

$$B_b = 2B = 2f_{max}. \quad (8.123)$$

To establish symmetry between the time dimension (over which the signal is time-limited to T_d; i.e., it is zero outside the range $-T_d/2$ to $+T_d/2$) and the frequency dimension, we defined

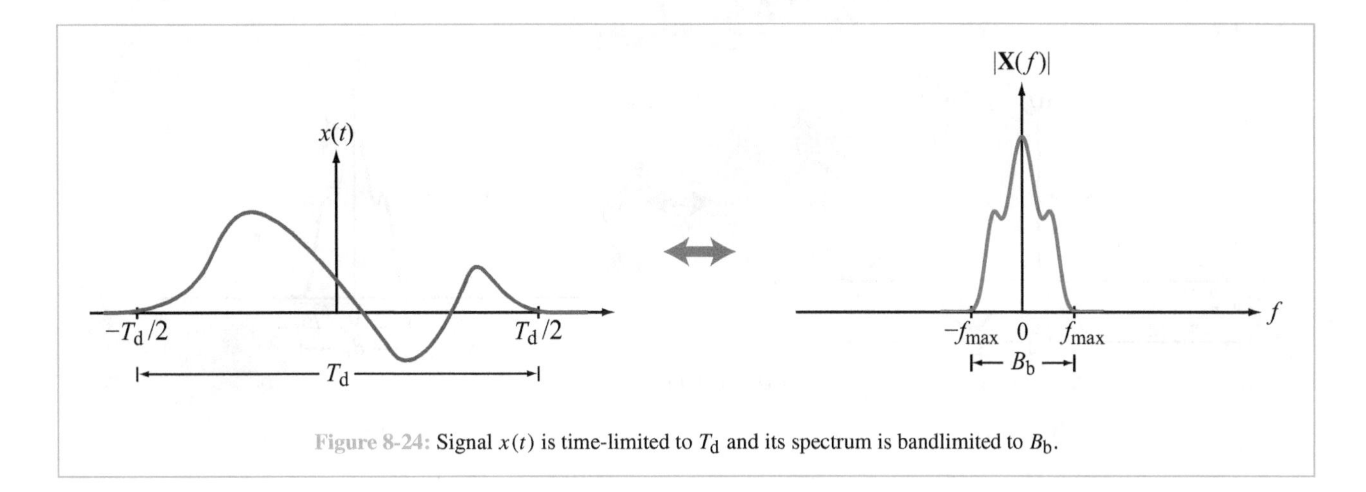

Figure 8-24: Signal $x(t)$ is time-limited to T_d and its spectrum is bandlimited to B_b.

spectrum $\mathbf{X}(f)$ in Eq. (8.122b) as bandlimited to $B_b/2$, rather than to B. The notational symmetry creates duality between the time and frequency dimensions, which is useful when transforming results obtained in either dimension to the other.

Note that $x(t) = 0$ at the edges of its duration $(\pm T_d/2)$, which implies that its non-zero duration is slightly shorter than T_d (Fig. 8-24). Similarly, $\mathbf{X}(f)$ is defined to be zero at $\pm B_b/2$, and therefore the signal's bandwidth is slightly narrower than B_b. These may seem like subtle distinctions, but they will prove significant when we discuss sampling rates in the next subsection.

For computational reasons, T_d and/or B_b are rounded up so that their dimensionless product

$$N_0 = T_d B_b, \tag{8.124}$$

is an *odd integer*. A related integer we will use shortly is

$$M = \frac{N_0 - 1}{2}. \tag{8.125}$$

The fact that N_0 is an odd integer guarantees M to be always an integer.

In preparation for the material in the next subsection, signal $x(t)$ is defined in Eq. (8.122a) such that its duration T_d is centered at $t = 0$. Should that not be the case, an additional preparatory step is needed. For example, if a signal $y(t)$ is defined over the time span $a < t < b$, we introduce the parameters

$$t_c = \frac{b+a}{2}, \qquad T_d = b - a, \tag{8.126}$$

so as to define $y(t)$ relative to a time-centered signal $x(t)$,

$$x(t) = y(t + t_c), \qquad -\frac{T_d}{2} < t < \frac{T_d}{2}. \tag{8.127}$$

After computing $\mathbf{X}(f)$, the Fourier transform of $x(t)$, using the procedure outlined in the next subsection, we apply the time-shift property of the Fourier transform (#4 in Table 5-7) to obtain the Fourier transform $\mathbf{Y}(f)$ of the original signal $y(t)$,

$$\mathbf{Y}(f) = \mathbf{X}(f)\, e^{-j2\pi f t_c}. \tag{8.128}$$

8-10.2 Sampling in Time

By sampling a nonperiodic signal $x(t)$ of duration T_d and bandwidth B_b (Fig. 8-25(a) at a sampling rate $f_s = B_b$, we avoid aliasing issues and preserve the signal's information content completely. The rate $f_s = B_b = 2f_{max}$ exceeds the Nyquist rate because in Eq. (8.122b) we defined the spectrum such that $\mathbf{X}(f) = 0$ at $|f| \geq f_{max}$, the implication being that the signal bandwidth is slightly narrower than $B_b = 2f_{max}$. For a signal of duration T_d, sampling it at $f_s = B_b$ generates a total of $N_0 = f_s T_d = T_d B_b$ samples. The sampling times are $t = n/B_b$, with n spanning the range between $-M$ and $+M$, where $M = (N_0 - 1)/2$. From Eq. (6.150), the ***sampled continuous-time signal*** $x_s(t)$ for $T_s = 1/f_s = 1/B_b$ is

$$x_s(t) = \sum_{n=-\infty}^{\infty} x(t)\,\delta\left(t - \frac{n}{B_b}\right) = \sum_{n=-\infty}^{\infty} x[n]\,\delta\left(t - \frac{n}{B_b}\right), \tag{8.129}$$

where $x[n] = x(n/B_b)$ is the sampled version of $x(t)$. The sampled signal is shown in Fig. 8-25(b), along with its spectrum $\mathbf{X}_s(f)$. From Eq. (6.158), the spectrum of the sampled signal, $\mathbf{X}_s(f)$, is related to the spectrum of the continuous-time signal, $\mathbf{X}(t)$, by

$$\mathbf{X}_s(t) = f_s \sum_{m=-\infty}^{\infty} \mathbf{X}(f - mf_s) = B_b \sum_{m=-\infty}^{\infty} \mathbf{X}(f - mB_b), \tag{8.130}$$

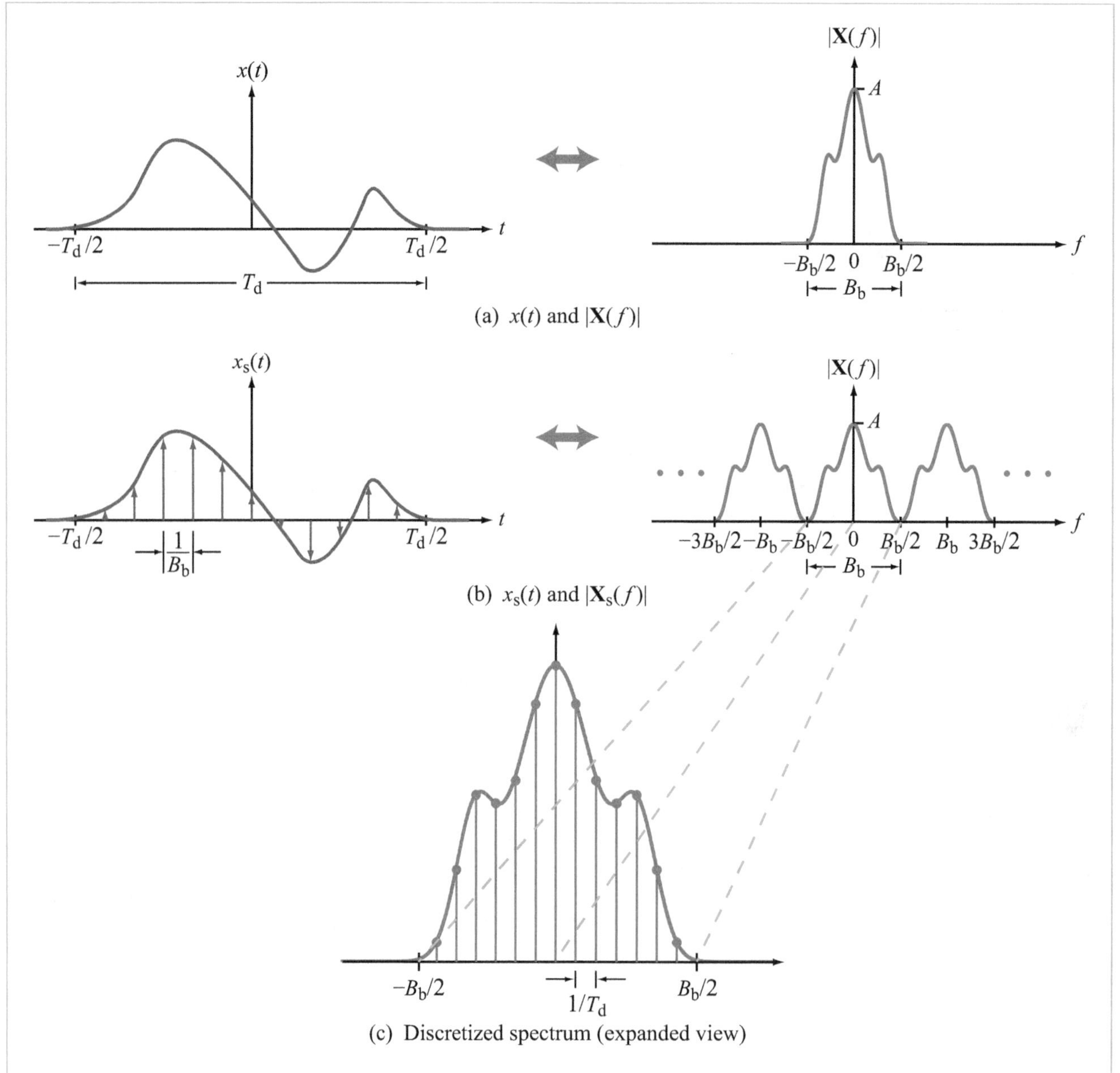

Figure 8-25: Continuous-time signal $x(t)$ and its spectrum $\mathbf{X}(f)$; sampling $x(t)$ in time leads to a periodic spectrum of period B_b in frequency; and discretizing the spectrum at T_d samples/Hz generates a line spectrum with an inter-line spacing of $1/T_d$.

where, to avoid confusion, the summation index has been changed to m. Sampling in the time domain leads to spectrum replication in the frequency domain. That is, $\mathbf{X}_s(f)$ is periodic with period B_b. Also, the spectra $\mathbf{X}(f)$ of the unsampled signal and the central segment of spectrum $\mathbf{X}_s(f)$ of the sampled signal are related by the factor B_b,

$$\mathbf{X}_s(f) = B_b\, \mathbf{X}(f), \qquad -\frac{B_b}{2} < f < \frac{B_b}{2}. \tag{8.131}$$

From the basic definition of the Fourier transform, $\mathbf{X}_s(f)$ of signal $x_s(t)$ is

$$\mathbf{X}_s(f) = \int_{-\infty}^{\infty} x_s(t)\, e^{-j2\pi ft}\, dt$$
$$= \int_{-\infty}^{\infty} \sum_{n=-\infty}^{\infty} x[n]\, \delta\left(t - \frac{n}{B_b}\right) e^{-j2\pi ft}\, dt. \quad (8.132)$$

Exchanging the order of the integration and summation gives

$$\mathbf{X}_s(f) = \sum_{n=-\infty}^{\infty} x[n] \int_{-\infty}^{\infty} \delta\left(t - \frac{n}{B_b}\right) e^{-j2\pi ft}\, dt$$
$$= \sum_{n=-M}^{M} x[n]\, e^{-j2\pi nf/B_b}, \quad (8.133)$$

where in the summation n is confined to the range $-M$ to $+M$ since $x[n] = 0$ outside that range. In view of Eq. (8.131), it follows that the spectrum $\mathbf{X}(f)$ of the unsampled signal is

$$\mathbf{X}(f) = \frac{1}{B_b}\mathbf{X}_s(f) = \frac{1}{B_b}\sum_{n=-M}^{M} x[n]\, e^{-j2\pi nf/B_b}. \quad (8.134)$$

8-10.3 Sampling in Frequency

Numerical computation of the Fourier transform and its inverse requires not only that $x(t)$ be sampled in time, but also that its spectrum $\mathbf{X}(f)$ be sampled in frequency. To avoid aliasing, $\mathbf{X}(f)$ should be sampled at a rate T_d samples per Hz, analogous to sampling $x(t)$ at B_b samples per second. Sampling a bandwidth B_b at T_d samples per Hz generates $N_0 = T_d B_b$ samples. Thus, both $x(t)$ and its spectrum $\mathbf{X}(f)$ consist of N_0 samples each.

Spectrum $\mathbf{X}(f)$ is sampled at frequencies $f = k/T_d$, with k spanning the range $-M$ to M. The sampled ***Fourier spectral components*** are labeled $\mathbf{X}_k$. Thus,

$$\mathbf{X}_k = \mathbf{X}(f)\big|_{f=k/T_d}$$
$$= \frac{1}{B_b}\sum_{n=-M}^{M} x[n]\, e^{-j2\pi nf/B_b}\Big|_{f=k/T_d}$$
$$= \frac{1}{B_b}\sum_{n=-M}^{M} x[n]\, e^{-j2\pi nk/N_0}, \qquad |k| \le M, \quad (8.135)$$

where $N_0 = T_d B_b$. By comparing Eq. (8.135) with Eq. (7.168a), we recognize the summation as an N_0-point DFT of $x[n]$. That is,

$$\underbrace{\mathbf{X}_k}_{\text{Fourier transform}} = \frac{1}{B_b}\underbrace{\mathbf{X}[k]}_{\text{DFT}}, \quad (8.136)$$

where $\mathbf{X}[k]$ is the kth DFT complex coefficient. In the formal definition of the DFT [Eq. (7.168)], the summation index n spans the range 0 to N_0, but in Eq. (8.135) its range is from $-M$ to M. However, since $N_0 = 2M+1$, both contain the same number of samples. Furthermore, because the spectrum $\mathbf{X}_s(f)$ is periodic, the DFT can be computed on the basis of any N_0-consecutive samples of $x[n]$.

Use of the Fourier duality allows us to repeat the preceding analysis that led to Eqs. (8.134) and (8.135) by exchanging t and f, $x(t)$ and $\mathbf{X}(f)$, T_d and B_b. The process leads to the inverse Fourier transform

$$x[n] = \frac{1}{T_d}\sum_{k=-M}^{M} \mathbf{X}_k e^{j2\pi nk/N_0}$$
$$= \frac{1}{T_d B_b}\sum_{k=-M}^{M} \mathbf{X}[k] e^{j2\pi nk/N_0}$$
$$= \frac{1}{N_0}\sum_{k=-M}^{M} \mathbf{X}[k] e^{j2\pi nk/N_0}, \qquad |n| \le M. \quad (8.137)$$

The final result is the N_0-point inverse DFT, expressed in terms of the DFT Fourier coefficients $\mathbf{X}[k]$ (not to be confused with the Fourier transform component $\mathbf{X}_k$).

8-10.4 Summary

In conclusion, to take advantage of the computational process of the DFT and its associated FFT algorithm, we note the following:

(a) A signal $x(t)$ time-limited to T_d and bandlimited to B_b should be sampled at B_b samples/s and its spectra $\mathbf{X}(f)$ can be represented by a discrete spectrum with spacing of $1/T_d$ between adjacent spectral components $\mathbf{X}_k$. Both $x(t)$ and $\mathbf{X}(f)$ consist of $N_0 = T_d B_b$ samples.

(b) Application of an N_0-point DFT on $x[n]$ generates DFT components $\mathbf{X}[k]$ for $|k| \le M = (N_0 - 1)/2$. The spectral components of the continuous signal $x(t)$ are given by $\mathbf{X}_k = \mathbf{X}[k]/B_b$.

(c) Applications of an N_0-point inverse DFT on DFT components $\mathbf{X}[k]$ generates $x[n]$ for $|n| \le M$.

8-10.5 Comparison with Discretization

Returning to the definitions for the Fourier transform and its inverse given by Eq. (8.121), if we discretize the integrals by converting them into summations with

$$t = n\,\Delta t, \qquad f = k\,\Delta f,$$
$$\Delta t = \frac{1}{B_b}, \qquad \Delta f = \frac{1}{T_d},$$

we obtain the results

$$\mathbf{X}_k = \frac{1}{B_b} \sum_{n=-M}^{M} x\left(\frac{n}{B_b}\right) e^{-j2\pi nk/N_0}, \qquad |k| \le M \tag{8.138a}$$

and

$$x\left(\frac{n}{B_b}\right) = \frac{1}{T_d} \sum_{k=-M}^{M} \mathbf{X}_k\, e^{j2\pi nk/N_0}, \qquad |n| \le M \tag{8.138b}$$

with $N_0 = T_d B_b$. The fact that these results, based on discretizing the integrals, are identical with those given by Eqs. (8.135) and (8.137) is not surprising, because the discretization spacings $\Delta t = 1/B_b$ and $\Delta f = 1/T_d$, correspond exactly to the sampling rates of B_b samples/s and T_d samples/Hz used in the denominators of Eqs. (8.135) and (8.137).

Example 8-23: Computing CTFT by DFT

Use the DFT to compute the Fourier transform of the continuous signals:

(a) $x_1(t) = e^{-|t|}$,

(b) $x_2 = \frac{1}{\sqrt{2\pi}}\, e^{-t^2/2}$.

Solution:
(a) Our first task is to assign realistic values for the signal duration T_d and the width of its spectrum B_b. It is an "educated" trial-and-error process. At $t = 6$, $e^{-6} = 0.0025$, so we will assume that $x_1(t) \approx 0$ for $|t| > 6$. Since $x_1(t) = e^{-|t|}$ is symmetrical with respect to the vertical axis,

$$T_d = 2 \times 6 = 12 \text{ s}.$$

To specify a value for B_b, we need to know $\mathbf{X}(f)$, which is the quantity we are trying to compute. Because we usually do not know $\mathbf{X}(f)$, we have to select a value for B_b, perform the computation, increase the value of B_b, perform the computation again, and then compare the spectrum obtained with the higher value of B_b with that obtained with the smaller value. If there is a discernible difference between the two spectra, B_b should be increased again and the process repeated until we reach a point of no discernible change in the computed spectrum of $\mathbf{X}(f)$. In the present case, that point is reached when $B_b = 16$. To generate a symmetrical spectrum, $N_0 = T_d B_b$ should be an odd integer. Hence, we increase B_b to 16.083 so that

$$N_0 = T_0 B_b = 12 \times 16.083 = 193.$$

The results, based on a 193-point DFT computation, are shown in Fig. 8-26(a). For the sake of comparison, we also show exact values of the Fourier transform of $x_1(t) = e^{-|t|}$ at each of the 193 spectral points. The exact values were calculated from knowledge of the analytical expression for the Fourier transform, namely

$$\mathbf{X}_1(f) = \frac{2}{4\pi^2 f^2 + 1}.$$

Despite the relatively coarse discretization, the DFT has done a excellent job of computing $\mathbf{X}_1(f)$.

(b) For $x_2(t) = \frac{1}{\sqrt{2\pi}}\, e^{-t^2/2}$,

$$x_2(t) < 0.00013 \qquad \text{for } |t| \ge 4.$$

Hence, we assign

$$T_d = 2 \times 4 = 8 \text{ s}.$$

By trial and error, we determine that $B_b = 1.2$ Hz is sufficient to characterize $\mathbf{X}_2(f)$. The combination gives

$$N_0 = T_d B_b = 8 \times 1.2 = 9.6.$$

To increase the value of N_0 to an odd integer, we increase B_b to 1.375 Hz, which results in $N_0 = 11$. In Fig. 8-26(b) computed values of the discretized spectrum of $x_2(t)$ are compared with exact values based on evaluating the analytical expression for $\mathbf{X}_2(f) = e^{-2\pi^2 f^2}$. The comparison provides an excellent demonstration of the power of the sampling theorem; representing $x_2(t)$ by only 11 equally spaced samples is sufficient to capture its information content and generate its Fourier transform with high fidelity.

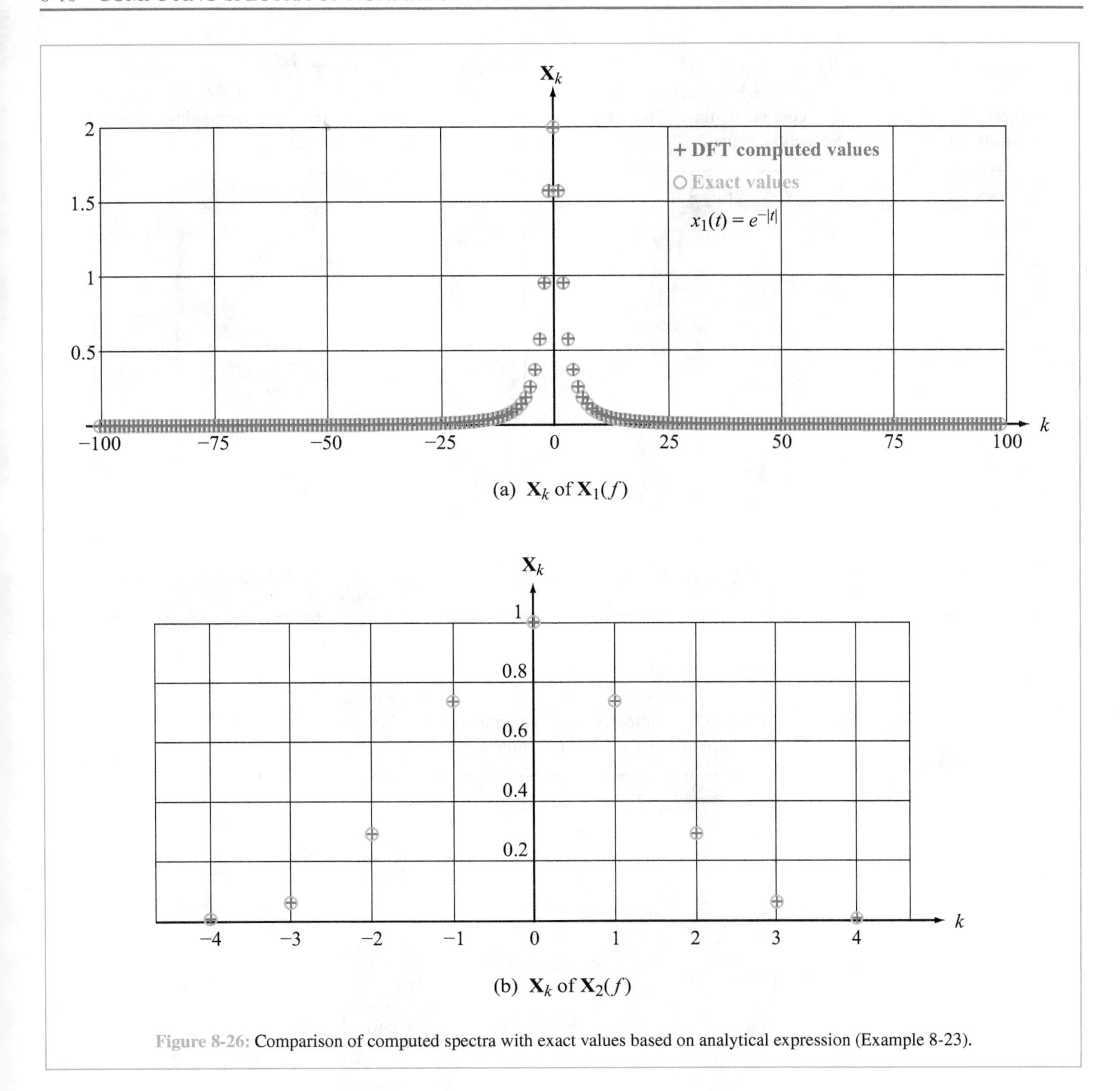

Figure 8-26: Comparison of computed spectra with exact values based on analytical expression (Example 8-23).

Concept Question 8-11: Why can Fourier transforms be computed so accurately with the DFT? (See S²)

Concept Question 8-12: Why can some Fourier transforms be computed using as few as 11 discretization points? (See S²)

Module 8.10a Computing Continuous-Time Fourier Transforms Using the DFT This module numerically computes the continuous-time Fourier transform of a two-sided exponential signal.

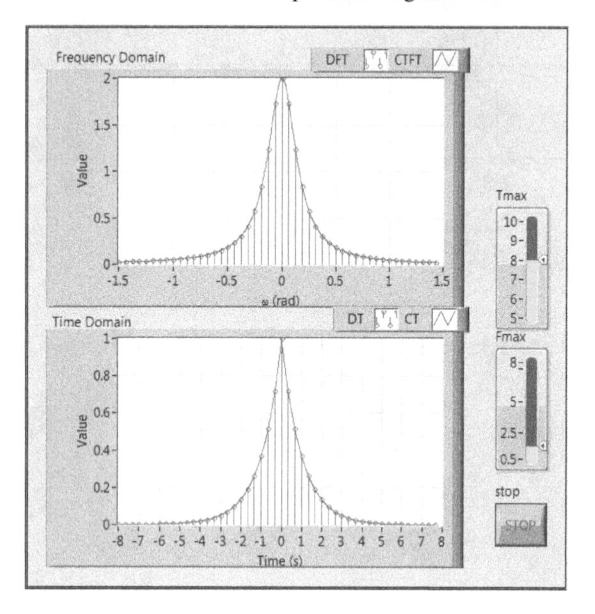

Module 8.10b Computing Continuous-Time Fourier Transforms Using the DFT This module numerically computes the continuous-time Fourier transform of a Gaussian signal.

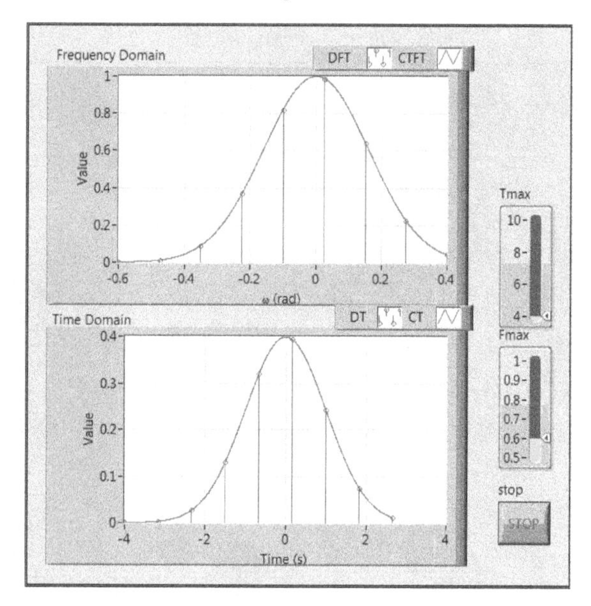

Module 8.10c Computing Continuous-Time Fourier Transforms Using the DFT This module numerically computes the continuous-time Fourier transform of a one-sided exponential signal.

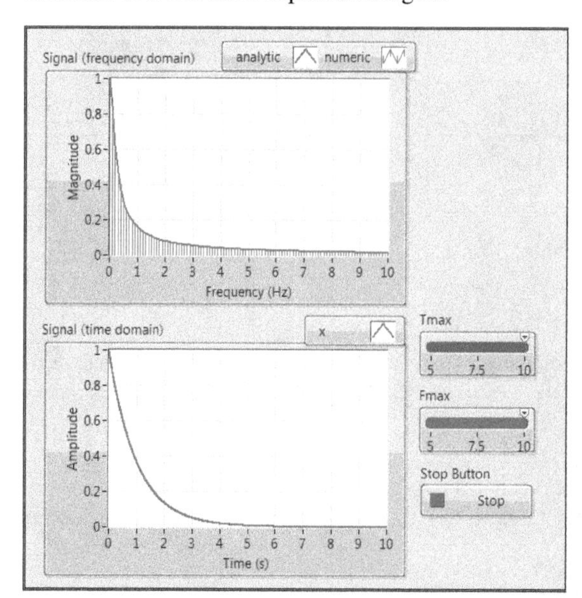

Summary

Concepts

- Single-frequency signals can be rejected by notch filters.
- Periodic signals can be rejected by comb filters.
- For minimum-phase systems, deconvolution can be performed in real time, using inverse systems.
- For non-minimum-phase systems, deconvolution can be performed using batch processing and the DFT or by using the bilateral **z**-transform.
- The Fourier series of periodic signals and Fourier transforms of nonperiodic signals can be computed using the DFT.

Mathematical and Physical Models

Transfer Function of Notch Filter

$$\mathbf{H}(\mathbf{z}) = \frac{\mathbf{z}^2 - 2\mathbf{z}\cos\Omega_\mathrm{i} + 1}{\mathbf{z}^2 - 2a\mathbf{z}\cos\Omega_\mathrm{i} + a^2}$$

Ω_i = angular frequency of interfering sinusoid

Transfer Function of Comb Filter

Series connection of notch filters

Reverberation

Reverberated signal: $y[n] = \sum_{i=0}^{\infty} r^i \, x[n - iM]$

Dereverberated signal: $x[n] = y[n] - r\, y[n - M]$

Computation of Fourier Series

Given: $x(t)$ periodic with period T_0 and bandlimited to $f_{\max}$

$$x(t) = \sum_{k=-K}^{K} \mathbf{x}_k e^{j2\pi kt/T_0}$$

with $K = f_{\max} T_0$, $\mathbf{x}_k = \frac{1}{N}\,\mathrm{DFT}[\{x[n]\}]$

Computation of Fourier Transform

Given: $x(t)$ time-limited to $T_\mathrm{d}/2$ and its spectrum bandlimited to $B_\mathrm{b}/2$

$$\mathbf{X}_k = \frac{1}{B_\mathrm{b}} \sum_{n=-M}^{M} x\left(\frac{n}{B_\mathrm{b}}\right) e^{-j2\pi nk/N_0} \qquad |k| \le M$$

$$x[n] = \frac{1}{T_\mathrm{d}} \sum_{k=-M}^{M} \mathbf{X}_k e^{j2\pi nk/N_0} \qquad N_0 = B_\mathrm{b} T_\mathrm{d}$$

Important Terms

Provide definitions or explain the meaning of the following terms:

batch processing	dereverberation	notch filter	signal duration
bilateral **z**-transform	half-band filter	periodic extension	spectral leakage
comb filter	minimum-phase system	region of convergence (ROC)	thresholding
deconvolution	noncausal inverse **z**-transform	reverberation	

PROBLEMS

Section 8-2: Notch Filters

*8.1 Design an ARMA difference equation that rejects a 500 Hz sinusoid sampled at 2000 samples per second. Use $a = 0.99$.

8.2 Design an ARMA difference equation that rejects a 250 Hz sinusoid sampled at 1500 samples per second. Use $a = 0.99$.

8.3 Design an ARMA difference equation that rejects a 250 Hz sinusoid sampled at 2000 samples per second. Use $a = 0.99$.

8.4 Design an ARMA difference equation that rejects a 750 Hz sinusoid sampled at 2000 samples per second. Use $a = 0.99$.

8.5 Design an ARMA difference equation that eliminates both sinusoids $\cos(\frac{\pi}{3}\,n)$ and $\cos(\frac{2\pi}{3}\,n)$. Use $a = 0.99$.

*Answer(s) in Appendix F.

*8.6 Design an ARMA difference equation that rejects 100 Hz and 500 Hz sinusoids sampled 1200 samples per second. Use $a = 0.99$.

8.7 Design an ARMA difference equation that rejects 500 Hz and 1000 Hz sinusoids sampled 3000 samples per second. Use $a = 0.99$.

8.8 Write a MATLAB or MathScript program that adds a 750 Hz sinusoid to the trumpet signal (note B), designs a notch filter that eliminates the sinusoid, and then implements the notch filter. The trumpet signal is sampled at 44,100 samples/s. Attempt to listen to the trumpet-plus-sinusoid signal before and after filtering by the notch filter. Use $a = 0.99$.

Section 8-3: Comb Filters

8.9 Design an ARMA difference equation that eliminates periodic interference of period = 0.001 s, sampled at 50000 samples per second. Use $a = 0.99$.

8.10 A triangle wave with period 1/750 s has the truncated Fourier series expansion

$$\cos(2\pi 750t) + \frac{1}{9}\cos(2\pi(3)750t) + \frac{1}{25}\cos(2\pi(5)750t).$$

Write a MATLAB or MathScript program that adds this triangle wave to the trumpet signal, designs a comb filter that eliminates the triangle wave, and implements the comb filter. The trumpet signal is sampled at 44,100 samples/s. Attempt to listen to the trumpet-plus-triangle signal before and after filtering by the comb filter. Use $a = 0.99$.

8.11 If $x[n]$ is any real-valued, even, and periodic signal with period $= 8$, show that the output of a system described by $y[n] = x[n] - x[n-2] + x[n-4] - x[n-6]$ is $y[n] = A\cos(\pi n/2)$ for some constant A.

8.12 If $x[n]$ is any real-valued, zero-mean, and periodic signal with period $= 8$, show that the output of a system described by $y[n] = x[n] + x[n-1] + \cdots + x[n-7]$ is $y[n] = 0$.

Section 8-4: Deconvolution and Dereverberation

8.13 Show that a reverbed sinusoid is just another sinusoid at the same frequency.

*8.14 Design a dereverbing system for a signal that has been reverbed with reflection coefficient 0.8 and time delay 0.001 seconds between echoes, sampled 3000 samples/s.

8.15 Solve the following deconvolution problem for $x[n]$ *without* using MATLAB or MathScript:

$$x[n] * \{\underline{1}, 1, \tfrac{1}{2}\} = \{\underline{4}, 6, 12, 15, 10, 3\}.$$

(*Hint:* The system is minimum phase.)

8.16 Solve the following deconvolution problem for $x[n]$ *without* using MATLAB or MathScript:

$$x[n] * \{\underline{1}, \tfrac{1}{2}, \tfrac{1}{2}\} = \{\underline{8}, 6, 11, 8, 5, 2\}.$$

(*Hint:* The system is minimum phase.)

Section 8-8: Deconvolution and Filtering Using the DFT

8.17 Solve each of the following deconvolution problems for input $x[n]$. Use MATLAB or MathScript.

(a) $x[n] * \{\underline{1}, 2, 3\} = \{\underline{7}, 15, 27, 13, 24, 27, 34, 15\}$.

(b) $x[n] * \{\underline{1}, 3, 5\} = \{\underline{3}, 10, 22, 18, 28, 29, 52, 45\}$.

(c) $x[n] * \{\underline{1}, 4, 2, 6, 5, 3\} =$ $\{\underline{2}, 9, 11, 31, 48, 67, 76, 78, 69, 38, 12\}$.

8.18 Solve each of the following deconvolution problems for input $x[n]$. Use MATLAB or MathScript.

(a) $x[n] * \{\underline{3}, 1, 4, 2\} = \{\underline{6}, 23, 18, 57, 35, 37, 28, 6\}$.

(b) $x[n] * \{\underline{1}, 7, 3, 2\} = \{\underline{2}, 20, 53, 60, 53, 54, 21, 10\}$.

(c) $x[n] * \{\underline{2}, 2, 3, 6\} =$ $\{\underline{12}, 30, 42, 71, 73, 43, 32, 45, 42\}$.

*8.19 Solve the following deconvolution problem for input $x[n]$. Use MATLAB or MathScript. Note that the N-point DFT is zero at $k = N/2$.

$$x[n] * \{\underline{1}, 3, 2\} = \{\underline{3}, 10, 13, 15, 16, 26, 37, 18\}.$$

8.20 Solve the following deconvolution problem for input $x[n]$. Use MATLAB or MathScript. Note that the DFT at $k = 0$ is zero for *any* DFT order! (*Hint:* Modulate all three signals by $\{e^{j2\pi n/16}\}$.)

$$x[n] * \{\underline{1}, 2, -3\} = \{\underline{2}, 11, 9, -11, 15, -12, 10, -24\}.$$

8.21 Write a MATLAB or MathScript program that adds a 750 Hz sinusoid to the trumpet signal, and then uses the DFT to eliminate it. The trumpet signal is sampled at 44,100 samples per second. Attempt to listen to the trumpet-plus-sinusoid signal before and after filtering using the DFT.

8.22 A triangle wave with period 1/750 s has the truncated Fourier series expansion

$$\cos(2\pi 750t) + \frac{1}{9}\cos(2\pi(3)750t) + \frac{1}{25}\cos(2\pi(5)750t).$$

Write a MATLAB or MathScript program that adds this triangle wave to the trumpet signal, and then uses the DFT to eliminate it. The trumpet signal is sampled at 44,100 samples per second. Attempt to listen to the trumpet-plus-triangle signal before and after filtering using the DFT.

8.23 A triangle wave with period 1/20 s has the truncated Fourier series expansion

$$\cos(2\pi 20t) + \frac{1}{9}\cos(2\pi(3)20t) + \frac{1}{25}\cos(2\pi(5)20t).$$

Write a MATLAB or MathScript program that adds noise `randn(1,N)` to this triangle wave. Use a sampling rate of 44,100 samples per second and signal length $N = 11025$. Use thresholding and the DFT to eliminate the noise in the signal. Plot the noisy and filtered waveforms for $0 \leq t < 0.25$ s.

8.24 A triangle wave with period 1/20 s has the truncated Fourier series expansion

$$\cos(2\pi 20t) + \frac{1}{9}\cos(2\pi(3)20t) + \frac{1}{25}\cos(2\pi(5)20t).$$

Write a MATLAB or MathScript program that adds noise `randn(1,N)` to this triangle wave. Use a sampling rate of 44,100 samples per second and signal length $N = 11025$. Knowing only that: (1) the signal is periodic with period 1/20 s, (2) the signal is bandlimited to 100 Hz, use the DFT to eliminate the noise in the signal. Plot the noisy and filtered waveforms for $0 \leq t < 0.25$ s.

Section 8-9: Computing Spectra of Periodic Signals

8.25 A continuous-time signal $x(t)$ is sampled at 40,000 samples per second, resulting in signal $x[n]$. The length of $x[n]$ is 8000 samples. MATLAB or MathScript's `fft` function is applied to $x[n]$, and the output is zero except at the indices $\{201, 401, 601, 801, 7201, 7401, 7601, 7801\}$.

(a) What is the duration of $x(t)$?

(b) $x(t)$ consists of sinusoids at what specific frequencies, in Hz?

(c) What is the period of $x(t)$?

*8.26 A continuous-time signal $x(t)$ is sampled at 44,100 samples per second, resulting in signal $x[n]$. The length of $x[n]$ is 11,025 samples. MATLAB or MathScript's `fft` function is applied to $x[n]$. The output is zero except at indices 111 and 10,916.

(a) What is the duration of $x(t)$?

(b) What is the frequency in Hz of $x(t)$?

(c) What pure tonal note is being played?

8.27 A continuous-time signal $x(t)$ is sampled at 44,100 samples per second, resulting in signal $x[n]$. The length of $x[n]$ is 1000 samples. MATLAB or MathScript's `fft` function is applied to $x[n]$, and the output is zero except for these indices:

- At index 51, `fft` gives $1500 + j2000$.
- At index 101, `fft` gives $2500 + j6000$.
- At index 901, `fft` gives $2500 - j6000$.
- At index 951, `fft` gives $1500 - j2000$.

Determine a formula for the signal $x(t)$.

8.28 A continuous-time signal $x(t)$ is sampled at 44,100 samples per second, resulting in signal $x[n]$. The length of $x[n]$ is 100 samples. MATLAB or MathScript's `fft` function is applied to $x[n]$, and the output is zero except for these indices:

- At index 2, `fft` gives $600 + j800$.
- At index 3, `fft` gives $700 + j2400$.
- At index 99, `fft` gives $700 - j2400$.
- At index 100, `fft` gives $600 - j800$.

Determine a formula for the signal $x(t)$.

8.29 We are given the discrete-time sinusoidal segment $\{3\cos(0.4\pi n + 0.2), n = 0, \ldots, (N-1)\}$. For what values of N will the spectrum computed using the DFT have no spectral leakage?

8.30 We are given the discrete-time sinusoidal segment $\{6.2\cos(0.75\pi n + 1), n = 0, \ldots, (N-1)\}$. For what values of N will the spectrum computed using the DFT have no spectral leakage?

8.31 File `P831.mat` on the S² website contains signals for the 12 keys on a touch-tone phone keypad, in the order {1, 2, 3, 4, 5, 6, 7, 8, 9, ∗, 0, #}. The signal is sampled at 8192 samples per second. Segment the signal into 12 parts (one per key), then use the DFT to determine the frequencies of the sinusoids present in the signal for each key. Make a diagram that summarizes the frequencies.

Section 8-10: Computing Spectra of Nonperiodic Signals

8.32 Use a 40-point DFT to compute the inverse Fourier transform of

$$\mathbf{X}(\omega) = \left(\frac{\sin(\omega/2)}{\omega/2}\right)^2.$$

Assume that $\mathbf{X}(\omega) \approx 0$ for $|f| > 10$ Hz and $x(t) \approx 0$ for $|t| > 1$ s. Plot the actual and computed inverse Fourier transforms on the same plot to show the close agreement between them.

8.33 Use an 80-point DFT to compute the inverse Fourier transform of

$$\mathbf{X}(\omega)\,\mathbf{H}(\omega) = \left(\frac{\sin(\omega/2)}{\omega/2}\right)^2 (1 + e^{-j\omega}).$$

Assume that $\mathbf{X}(\omega) \approx 0$ for $|f| > 10$ Hz and $x(t) \approx 0$ for $|t| > 2$ s. Plot the actual and computed inverse Fourier transforms on the same plot to show the close agreement between them.

LabVIEW Module 8.1

8.34 Replicate the design of the half-band lowpass filter in Example 8-1.

8.35 In Example 8-1, why must the pole on the positive real axis have magnitude 0.6, when all of the other poles had magnitudes 0.8? Redo Problem 8.34, but change the pole at 0.6 to 0.8.

8.36 Design a comb filter that eliminates periodic non-zero-mean interference with discrete-time period 8 and 2 harmonics. Use a magnitude of 0.9 for all poles.

8.37 Replicate the design of the notch filter in Example 8-3, with the following changes:

- The rejected frequency is 1200 Hz.
- The dc gain is close to 1.4.
- The gain is 1.0 at 960 Hz and 1440 Hz.
- The gain exceeds 1.0 outside the range

$$960 \text{ Hz} < f < 1440 \text{ Hz}.$$

LabVIEW Module 8.2

For Problems 8.38 and 8.39, set the notch frequency to the frequency we wish to to eliminate, and display all stem plots and spectra.

8.38 For each part, set $f_1 = 100$ Hz and $f_2 = 200$ Hz, and use the notch filter with the specified pole magnitude to eliminate the 200 Hz sinusoid. Confirm that the filtered signal is a sinusoid.

(a) Use a pole of magnitude 0.8.

(b) Use a pole of magnitude 0.5. Explain why the phases of the filtered signals in (a) and (b) are different.

(c) Use a pole of magnitude 0.99. Explain why the notch filter does not work as intended.

8.39 Set $f_1 = 100$ Hz and $f_2 = 250$ Hz. For what range of pole magnitudes can you eliminate the 200 Hz sinusoid? Explain your answer.

LabVIEW Module 8.3

In Problems 8.40 to 8.43, the interference level is set to 1, and the interference and notch frequencies are equal. Display all stem plots and spectra. Listen to all signals.

8.40 For an interference frequency of 800 Hz, for what pole radius does the filtered signal most resemble, and sound like, the original signal?

8.41 For an interference frequency of 1000 Hz, which is very close to the second harmonic of the trumpet signal, why is it possible to still eliminate it?

8.42 Repeat Problem 8.41 for an interference frequency of 1500 Hz.

8.43 For an interference frequency of 1500 Hz, what is the smallest value of pole radius that makes the filtered stem plot resemble the original stem plot?

LabVIEW Module 8.4

For Problems 8.44 to 8.47, the original signal is a 30 Hz sinusoid plus periodic interference consisting of 1, 2 or 3 harmonics, with a specified fundamental frequency (labeled as "interference frequency").

8.44 For interference with period $1/60$ s and 3 harmonics, use the comb filter with pole magnitude 0.8 to eliminate interference, leaving just the 30 Hz sinusoid.

8.45 Repeat Problem 8.44, but use pole radius 0.99. Why does this not work?

8.46 For interference with period $1/40$ s and 3 harmonics, use the comb filter with pole magnitude 0.8 to eliminate interference, leaving just the 30 Hz sinusoid.

8.47 Repeat Problem 8.46, but use pole radius 0.99. Why does this not work?

LabVIEW Module 8.5

Display all plots and listen to original and filtered sounds.

8.48 Use the comb filter to eliminate the trumpet playing note A. Use a pole magnitude of 0.99. Confirm that the comb filter has eliminated one of the two trumpets.

8.49 Repeat Problem 8.48, but eliminate the trumpet playing note B.

8.50 Repeat Problem 8.49, but use a pole magnitude of 0.8. Why does this still work? What advantage is there in using a smaller pole radius?

8.51 Use the comb filter with fundamental frequency of 460 Hz and a pole magnitude of 0.99. Confirm that both trumpets are still present in the filtered signal.

LabVIEW Module 8.6

8.52 Plot the reverbed signal for reflection coefficient of 0.9 and time delay of 1. The result is a sequence of geometrically decaying signals. Explain why you cannot just read off the peak values of each geometrically decaying sequence to reconstruct the original signal from the reverbed signal.

8.53 Plot the reverbed signal for reflection coefficient of 0.2 and time delay of 5. Why can you (roughly) reconstruct the original from the reverbed signal?

LabVIEW Module 8.8

Display all plots and spectra and listen to the original and filtered signals.

8.54 Eliminate the trumpet playing note A. Use a fundamental rejection frequency of 435 Hz. Confirm that only one trumpet signal is left.

8.55 Eliminate the trumpet playing note B. Use a fundamental rejection frequency of 490 Hz. Confirm that only one trumpet signal is left.

8.56 Use a fundamental rejection frequency of 460 Hz. Confirm that both trumpet signals remain, except for the fundamentals of both trumpet signals.

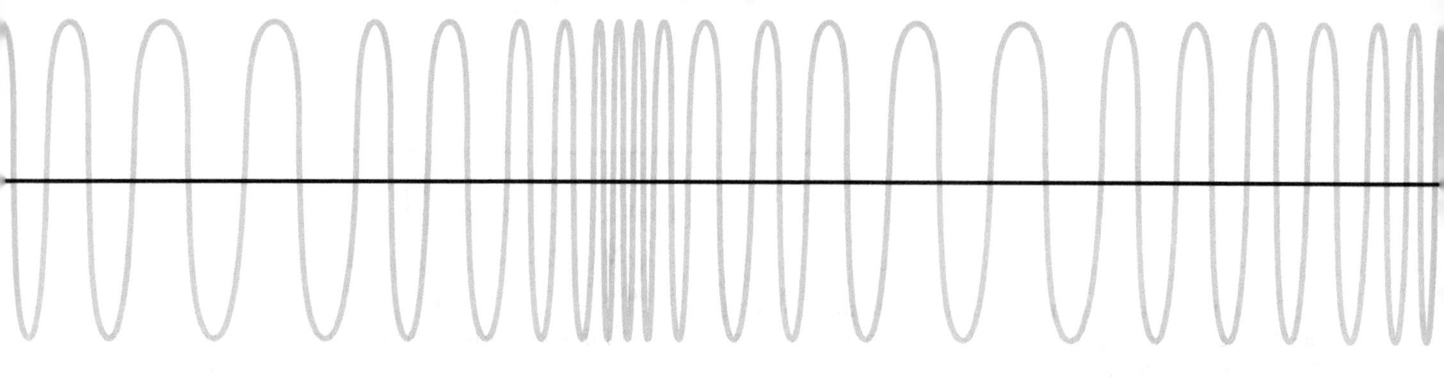

9 Filter Design, Multirate, and Correlation

Contents

Overview, 475
9-1 Data Windows, 475
9-2 Spectrograms, 485
9-3 Finite Impulse Response (FIR) Filter Design, 492
9-4 Infinite Impulse Response (IIR) Filter Design, 503
9-5 Multirate Signal Processing, 512
9-6 Downsampling, 513
9-7 Upsampling, 516
9-8 Interpolation, 517
9-9 Multirate Signal Processing Examples, 518
9-10 Oversampling by Upsampling, 520
9-11 Audio Signal Processing, 525
9-12 Correlation, 527
9-13 Biomedical Applications, 534
Summary, 538
Problems, 539

Objectives

Learn to:

- Apply discrete-time rectangular, Bartlett, Hanning, Hamming, and Blackman data windows.
- Compute spectrograms.
- Design finite-impulse-response (FIR) and infinite-impulse-response (IIR) filters.
- Use multirate signal processing, including downsampling, upsampling, and interpolation.
- Use autocorrelation and cross-correlation in practical applications.

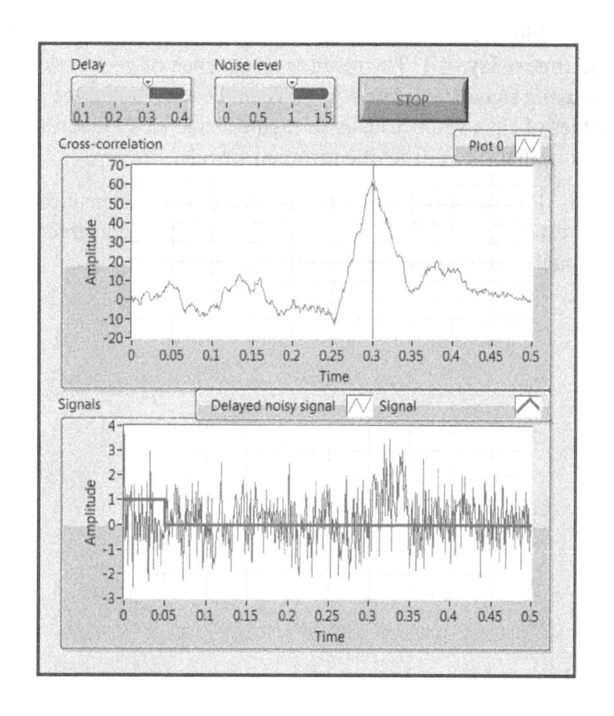

After learning about data windows and how they are used to limit the extent of discrete-time signals prior to computing their spectra, this chapter explores multiple approaches to discrete-time filter design, multirate signal processing, and correlation, and illustrates these techniques through examples of biomedical applications.

Overview

The first six chapters of this book dealt with continuous-time signals and systems. Chapter 7 introduced their discrete-time analogues and Chapter 8 provided application examples of discrete-time signals and systems. Now, in Chapter 9, we present additional applications in discrete-time signal processing that have no obvious continuous-time counterparts.

These applications include: ***spectral leakage*** and the consequent need for the imposition of ***data windows***; ***spectrograms***, which are time-varying spectra computed using a series of data windows centered at different times; discrete-time ***filter design*** to remove undesired signal components; ***downsampling*** and ***upsampling***, which require discrete-time filters and which constitute an important ingredient of ***wavelet transforms*** (Chapter 10); and various forms of ***correlation***, an important tool for computing time delay, and the period of a periodic signal, and to classify a signal as one of several possible types, each with its own set of properties. Additionally, the utility of correlation is demonstrated through three biomedical signal processing examples.

9-1 Data Windows

A critical first step in the computation of the spectrum of a discrete-time signal is to assign an appropriate ***data window*** that limits the extent of the signal and shapes its amplitude profile. The length of the window and its shape determine the spectral ***resolution*** of the processed signal and the ***degree of interference*** between adjacent sinusoids. This section examines the spectra of data windows, and the trade-off between the two aforementioned properties for each of five types of commonly used windows.

The preceding chapter included several computed spectra of an actual trumpet signal, for which the spectral lines had broad bases. This is due in part to the trumpet signal not being perfectly periodic, but it is also due in part to ***spectral leakage***, so called because the spectral lines seem to be "leaking."

Spectral leakage occurs because the spectrum is computed using only a finite number of samples of the signal. While spectral leakage cannot be eliminated entirely (except in special cases), its effects can be mitigated using ***data windows***. Multiplying the finite number of samples by a data window reduces the ***sidelobes***, making it easier to interpret and to spot small spectral peaks.

The concept of spectral leakage was introduced in Section 8-10.2, where it was demonstrated through Fig. 8-22 that the shape of the spectrum computed for a single sinusoid (of frequency f_0 and sampled over a duration T_d) depends on the product $f_0 T_d$. More specifically, leakage does not occur only if $f_0 T_d$ is an integer value. In the present section we examine the underlying reasons for this conclusion.

9-1.1 Spectrum of a Rectangular Pulse

We start by examining the spectrum of a discrete-time rectangular pulse of length L given by

$$w_R[n] = \begin{cases} 1 & n = 0, 1, \ldots, L-1, \\ 0 & \text{otherwise}, \end{cases} \tag{9.1}$$

which is equivalent to a constant signal sampled across a ***rectangular window*** with a uniform amplitude of unity. Per Eq. (7.154a), the DTFT $\mathbf{W}_R(e^{j\Omega})$ of $w_R[n]$ is

$$\mathbf{W}_R(e^{j\Omega}) = \sum_{n=-\infty}^{\infty} w_R[n]\, e^{-j\Omega n} = \sum_{n=0}^{L-1} 1 e^{-j\Omega n}. \tag{9.2}$$

By setting $\mathbf{r} = e^{-j\Omega}$ in Eq. (9.2) and taking advantage of the geometric series relationship

$$\sum_{n=0}^{L-1} \mathbf{r}^n = \frac{1-\mathbf{r}^L}{1-\mathbf{r}}, \tag{9.3}$$

the expression for $\mathbf{W}_R(e^{j\Omega})$ can be written as

$$\begin{aligned} \mathbf{W}_R(e^{j\Omega}) &= \frac{1-e^{-j\Omega L}}{1-e^{-j\Omega}} = \left(\frac{e^{j\Omega L/2} - e^{-j\Omega L/2}}{e^{j\Omega/2} - e^{-j\Omega/2}}\right) \frac{e^{-j\Omega L/2}}{e^{-j\Omega/2}} \\ &= \frac{\sin(\Omega L/2)}{\sin(\Omega/2)}\, e^{-j\Omega(L-1)/2}. \end{aligned} \tag{9.4}$$

This is a discrete sinc function multiplied by the factor $e^{-j\Omega(L-1)/2}$, which is due to the pulse being delayed by $(L-1)/2$ to make it causal. Recall that delaying a signal by D multiplies its DTFT by $e^{-j\Omega D}$.

Ignoring the factor $e^{-j\Omega(L-1)/2}$, $\mathbf{W}_R(e^{j\Omega})$ is plotted in Fig. 9-1 for $L = 21$. We observe that $\mathbf{W}_R(e^{j\Omega})$ does indeed look like a continuous-time ***sinc function*** (multiplied by L). However, unlike the continuous-time sinc function, $\mathbf{W}_R(e^{j\Omega})$ is also periodic with period 2π.

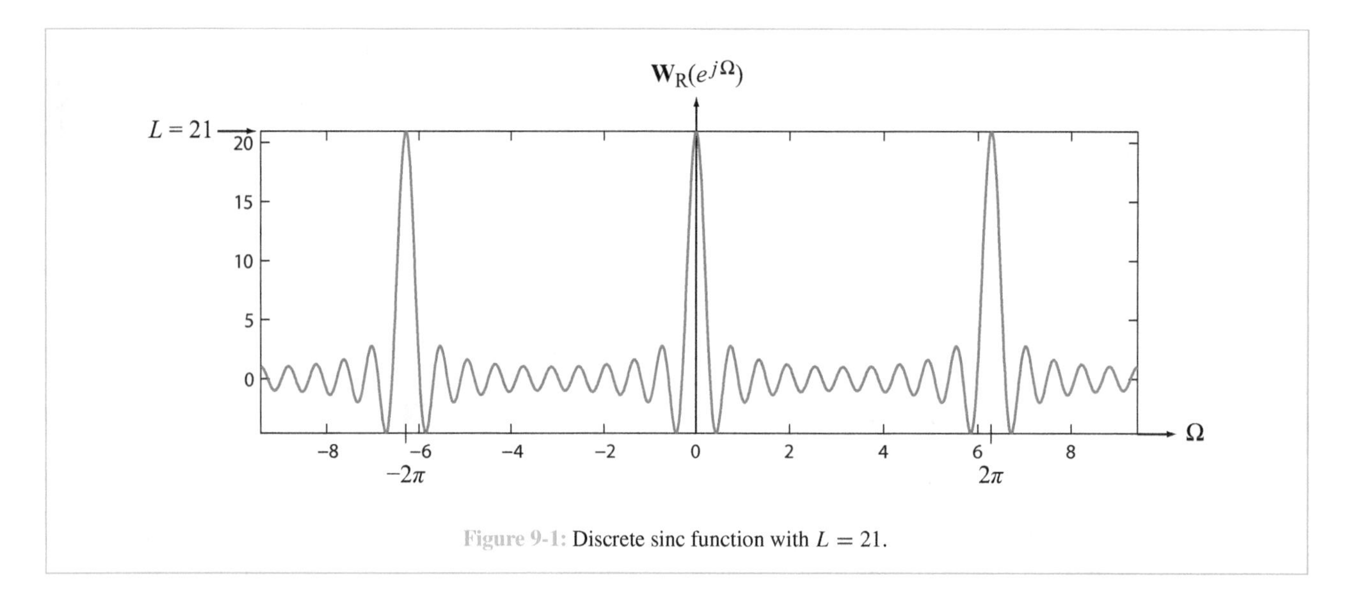

Figure 9-1: Discrete sinc function with $L = 21$.

9-1.2 Spectrum of a Sinusoidal Segment

Now consider the discrete-time sinusoid

$$x[n] = 2\cos(\Omega_0 n), \qquad n = 0, \pm 1, \dots \tag{9.5}$$

defined at all integer values of n. When limited to a causal segment of finite length L, we obtain the function

$$\begin{aligned} y[n] &= \begin{cases} 2\cos(\Omega_0 n) & n = 0, 1, \dots, L-1 \\ 0 & \text{otherwise} \end{cases} \\ &= 2\cos(\Omega_0 n)\; w_\mathrm{R}[n] = e^{j\Omega_0 n}\; w_\mathrm{R}[n] + e^{-j\Omega_0 n}\; w_\mathrm{R}[n], \end{aligned} \tag{9.6}$$

with $w_\mathrm{R}[n]$ as defined by Eq. (9.1). Using the modulation property of the DTFT, the DTFT of $y[n]$ is given by

$$\mathbf{Y}(e^{j\Omega}) = \mathbf{W}_\mathrm{R}(e^{j(\Omega-\Omega_0)}) + \mathbf{W}_\mathrm{R}(e^{j(\Omega+\Omega_0)}). \tag{9.7}$$

Upon shifting the expression for $\mathbf{W}_\mathrm{R}(e^{j\Omega})$ given by Eq. (9.4) by $\pm\Omega_0$, Eq. (9.7) becomes

$$\begin{aligned} \mathbf{Y}(e^{j\Omega}) &= \frac{\sin((\Omega-\Omega_0)L/2)}{\sin((\Omega-\Omega_0)/2)} e^{-j(\Omega-\Omega_0)(L-1)/2} \\ &\quad + \frac{\sin((\Omega+\Omega_0)L/2)}{\sin((\Omega+\Omega_0)/2)} e^{-j(\Omega+\Omega_0)(L-1)/2}. \end{aligned} \tag{9.8}$$

The spectrum of the sinusoidal segment $y[n]$, shown in Fig. 9-2, consists of two discrete sinc functions, one centered at Ω_0 and another centered at $-\Omega_0$. Each line in a line spectrum gets replaced with a discrete sinc function.

9-1.3 Use of Nonrectangular Functions

In Eq. (9.6) the discrete-time sinusoid $x[n]$ was multiplied by a rectangular pulse $w_\mathrm{R}[n]$ to produce the sinusoidal segment $y[n]$. Sometimes, it may be more useful to multiply $x[n]$ by a window function $w[n]$ that may have a different amplitude distribution within the segment $0 \le n \le L-1$, but is zero outside that range, the same as $w_\mathrm{R}[n]$. Thus, $w[n]$ has the general form

$$w[n] \begin{cases} \text{specified distribution} & \text{for } 0 \le n \le L-1, \\ = 0 & \text{otherwise.} \end{cases} \tag{9.9}$$

The shapes and properties of several specified amplitude distributions are discussed in later subsections. Accordingly, for the discrete time sinusoid $x[n]$ defined by Eq. (9.5), we define $y_w[n]$ as

$$\begin{aligned} y_w[n] &= \begin{cases} 2\cos(\Omega_0 n)\; w[n] & \text{for } 0 \le n \le L-1, \\ 0 & \text{otherwise,} \end{cases} \\ &= 2\cos(\Omega_0 n)\; w[n]. \end{aligned} \tag{9.10}$$

By analogy with Eq. (9.7), the DTFT $\mathbf{Y}_\mathrm{W}(e^{j\Omega})$ of $y_w[n]$ is given by

$$\mathbf{Y}_\mathrm{W}(e^{j\Omega}) = \mathbf{W}(e^{j(\Omega-\Omega_0)}) + \mathbf{W}(e^{j(\Omega+\Omega_0)}), \tag{9.11}$$

where $\mathbf{W}(e^{j\Omega})$ is the DTFT of $w[n]$. The spectrum of $y_w[n]$ is a pair of DTFTs $\mathbf{W}(e^{j\Omega})$, one centered at Ω_0 and the other centered at $-\Omega_0$.

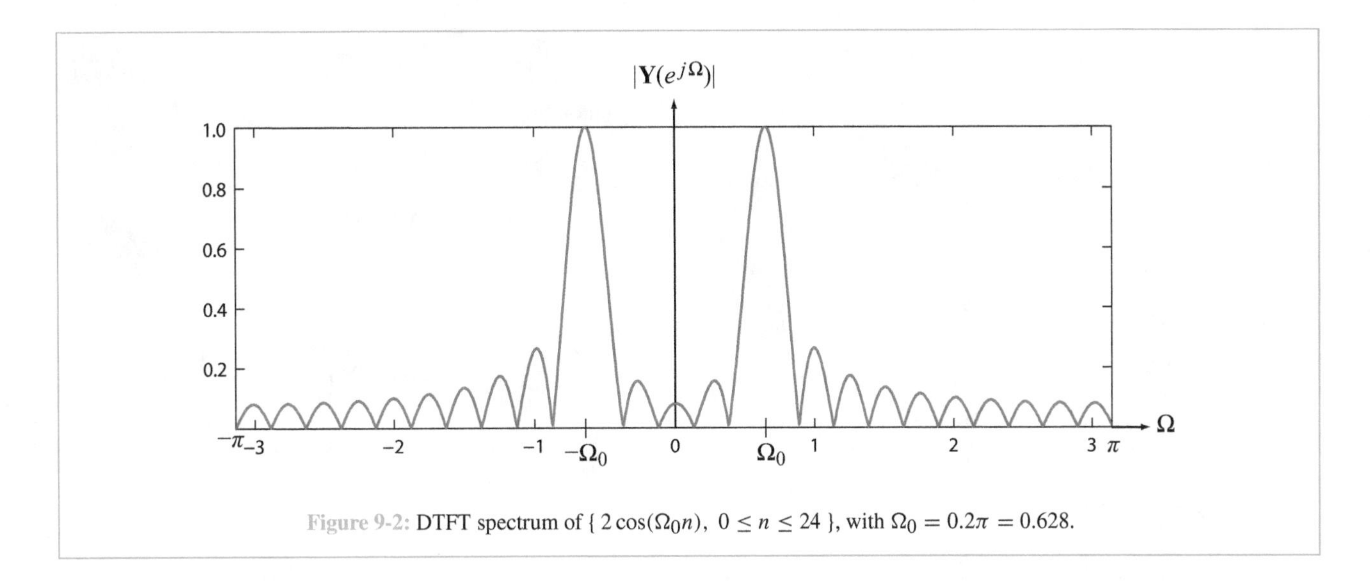

Figure 9-2: DTFT spectrum of $\{\, 2\cos(\Omega_0 n),\ 0 \leq n \leq 24 \,\}$, with $\Omega_0 = 0.2\pi = 0.628$.

For the rectangular window $w_R[n]$, its DTFT is a sinc function (Fig. 9-1), but for the general case, $\mathbf{W}(e^{j\Omega})$ need not be a discrete sinc function. It can be any function whose inverse DTFT $w[n]$ is zero outside the range $0 \leq n \leq L-1$.

The function $w[n]$ is called a *data window*. Multiplying the given data by $w[n]$ is called *windowing the data*. So "window" is both a noun and a verb. It is important to remember that the data are *multiplied by* $w[n]$, not convolved with it.

► To use the window $w[n]$ on data segment $x[n]$, compute the DFT of $w[n]\, x[n]$, **not** of $w[n] * x[n]$. ◄

We now introduce two parameters that will guide our choices for possible data windows.

9-1.4 Parameters of Data Windows

As noted earlier in Section 7-15.3, for a data segment of length L, increasing the order of the DFT, N, is equivalent to discretizing the DTFT to more points in the range $|\Omega| \leq \pi$, but it does not change the shape of the spectrum. Increasing N is also equivalent to applying zero-padding to extend the data segment of length L to create a segment of length $N > L$.

We now introduce two important parameters of the window spectrum $\mathbf{W}(e^{j\Omega})$, namely its *mainlobe width* Ω_M and its *sidelobe attenuation* Γ.

(a) Mainlobe width Ω_M

Figure 9-3 shows a typical spectrum, with a mainlobe centered at $\Omega = 0$, and several sidelobes on both sides. The *mainlobe width* extends between the first minima on the two sides of the peak of the spectrum. This is an important parameter because it is a measure of the resolution of the window—*the wider the mainlobe, the poorer the resolution.*

(b) Sidelobe attenuation Γ

The sidelobe attenuation Γ is defined as the ratio of the height of the first sidelobe in the amplitude spectrum to the height of the mainlobe. It is common practice to express Γ in dB:

$$\Gamma = -20 \log_{10} \frac{|\mathbf{W}_1|}{|\mathbf{W}_0|} \ \text{[dB]}, \tag{9.12a}$$

where $|\mathbf{W}_0|$ is the peak magnitude of the window spectrum (usually at $\Omega = 0$) and $|\mathbf{W}_1|$ is the peak magnitude of the first sidelobe (at $\Omega = \Omega_1$ in Fig. 9-3).

Alternatively, we can remove the minus sign and interchange the two magnitudes, to obtain

$$\Gamma = 20 \log_{10} \frac{|\mathbf{W}_0|}{|\mathbf{W}_1|} \ \text{[dB]}. \tag{9.12b}$$

Sidelobes are a source of interference to neighboring sinusoids, so it is highly desirable to reduce their levels as much as possible. As we will see shortly, the sidelobes can be reduced by tapering the amplitude distribution across the window, but there is often a trade-off between Ω_M and Γ.

► For a data segment of length L, lowering the sidelobe levels may lead to broadening of the mainlobe. ◄

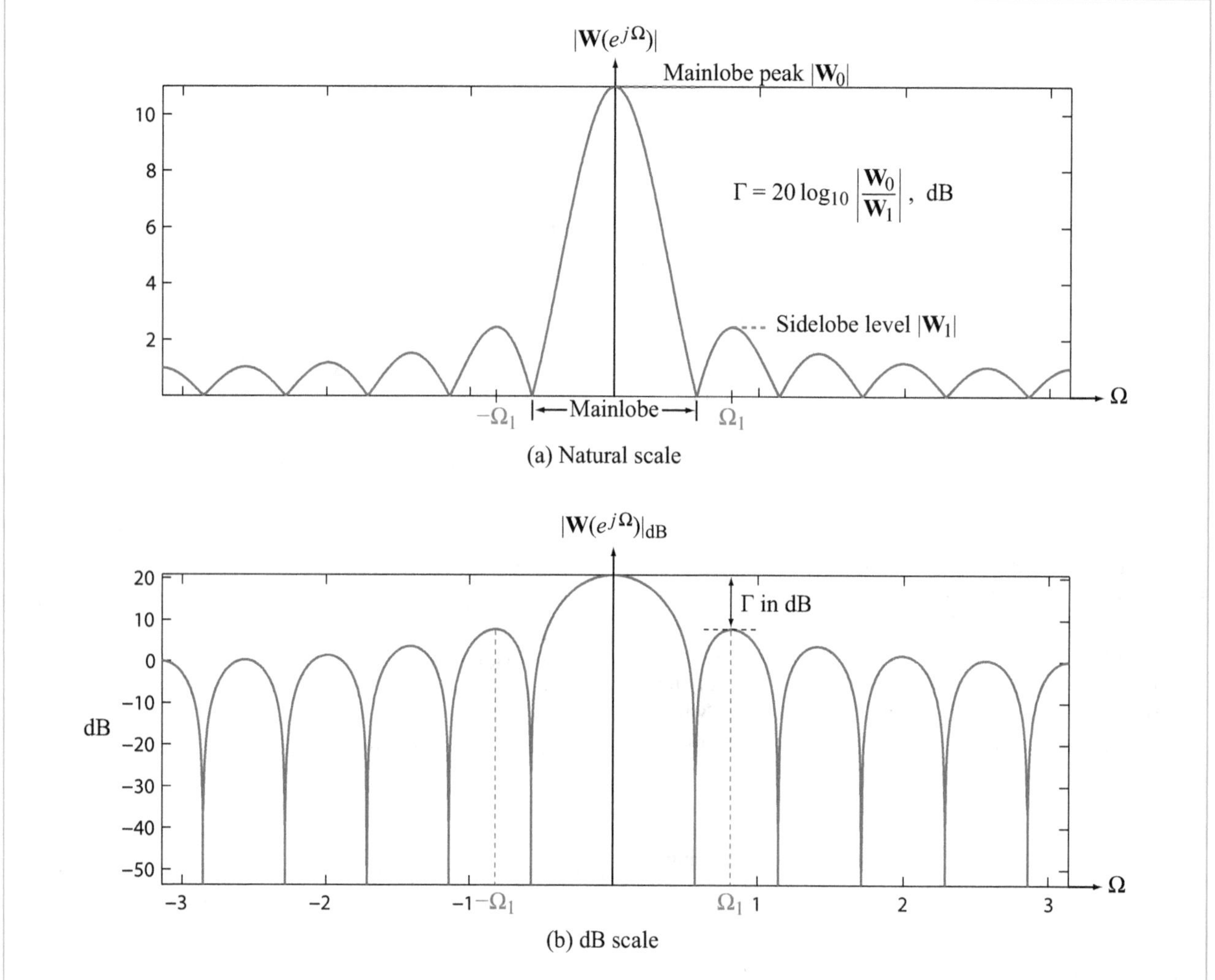

Figure 9-3: Magnitude of window spectrum displayed in (a) natural units and in (b) dB, with $|\mathbf{W}(e^{j\Omega})|_{\text{dB}} = 20\log_{10}|\mathbf{W}(e^{j\Omega})|$. The zero nulls on the natural scale transform to $-\infty$ on the dB scale.

9-1.5 Rectangular Window

For a rectangular window with constant amplitude, its DTFT $\mathbf{W}_{\text{R}}(e^{j\Omega})$ is given by Eq. (9.4) as

$$\mathbf{W}_{\text{R}}(e^{j\Omega}) = \frac{\sin(\Omega L/2)}{\sin(\Omega/2)}\, e^{-j\Omega(L-1)/2}. \tag{9.13}$$

The first nulls of the spectrum of the magnitude $|\mathbf{W}_{\text{R}}(e^{j\Omega})|$ occur at values of Ω corresponding to $\sin(\Omega L/2) = 0$, or equivalently

$$\frac{\Omega L}{2} = \pm\pi. \tag{9.14}$$

Thus, the first nulls are at

$$\Omega_{\text{null}} = \pm\frac{2\pi}{L}, \tag{9.15}$$

and the width of the mainlobe between them is

$$\Omega_{\text{M}} = 2|\Omega_{\text{null}}| = \frac{4\pi}{L} \qquad \textbf{(rectangular window)}. \tag{9.16}$$

▶ The width of the mainlobe is inversely proportional to L, so a longer window in the time domain generates a narrower spectrum in the frequency domain. ◀

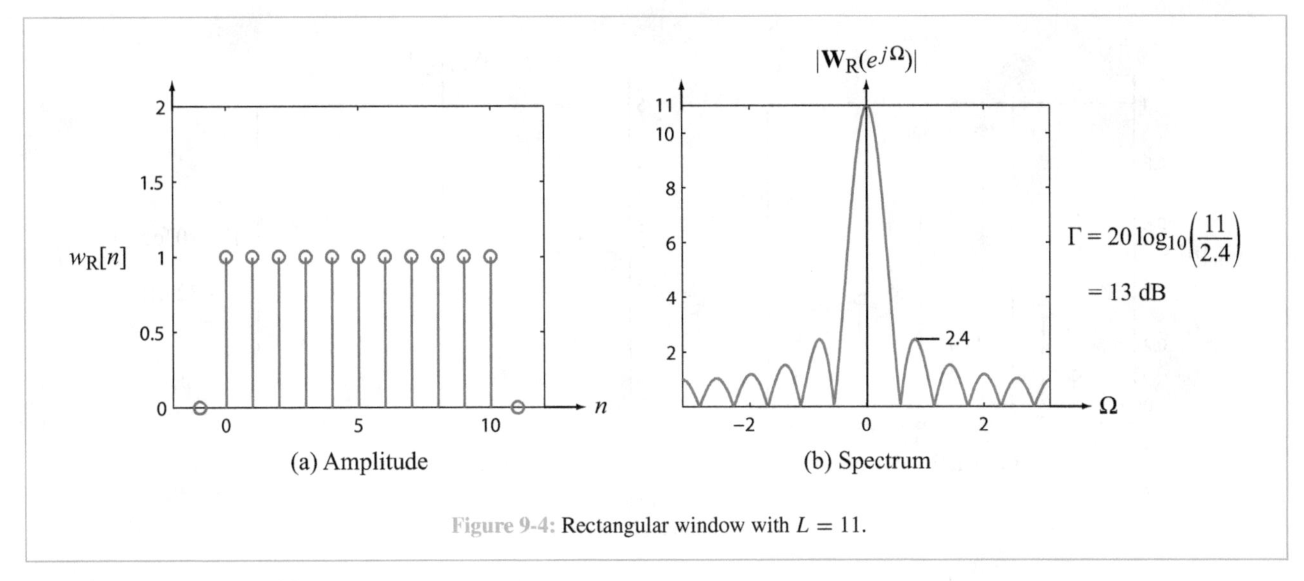

Figure 9-4: Rectangular window with $L = 11$.

The peak value of $|\mathbf{W}_R(e^{j\Omega})|$ occurs at $\Omega = 0$ and has a value of L. The peak values of the sidelobes can be established by (1) determining the locations of the sidelobe peaks and then (2) evaluating the expression for $|\mathbf{W}_R(e^{j\Omega})|$ at those locations. The process can be accomplished analytically or numerically. Either approach leads to the conclusion that the sidelobe attenuation of the rectangular window is

$$\Gamma = 13 \text{ dB} \qquad \textbf{(rectangular window)}.$$

The amplitude across a rectangular window of length $L = 11$ and the corresponding spectrum are displayed in Fig. 9-4(b). The rectangular window has the narrowest mainlobe, with a peak value $|\mathbf{W}_0| = L = 11$. The peak value of its first sidelobe is $|\mathbf{W}_1| = 2.4$ (or $2.4/11 = 0.22$, relative to the peak value, which is equivalent to $\Gamma = 20 \log_{10}(11/2.4) = 13$ dB).

9-1.6 Bartlett Triangular Window

A data window in essence modulates the amplitude of the signal contained within it. In a rectangular window, all signal samples are assigned equal weight, but we can use windows with different shapes that contain a ***tapered*** profile. A triangular taper assigns a maximum weight to the central sample, as shown in Fig. 9-5(a), and progressively smaller weights to neighboring

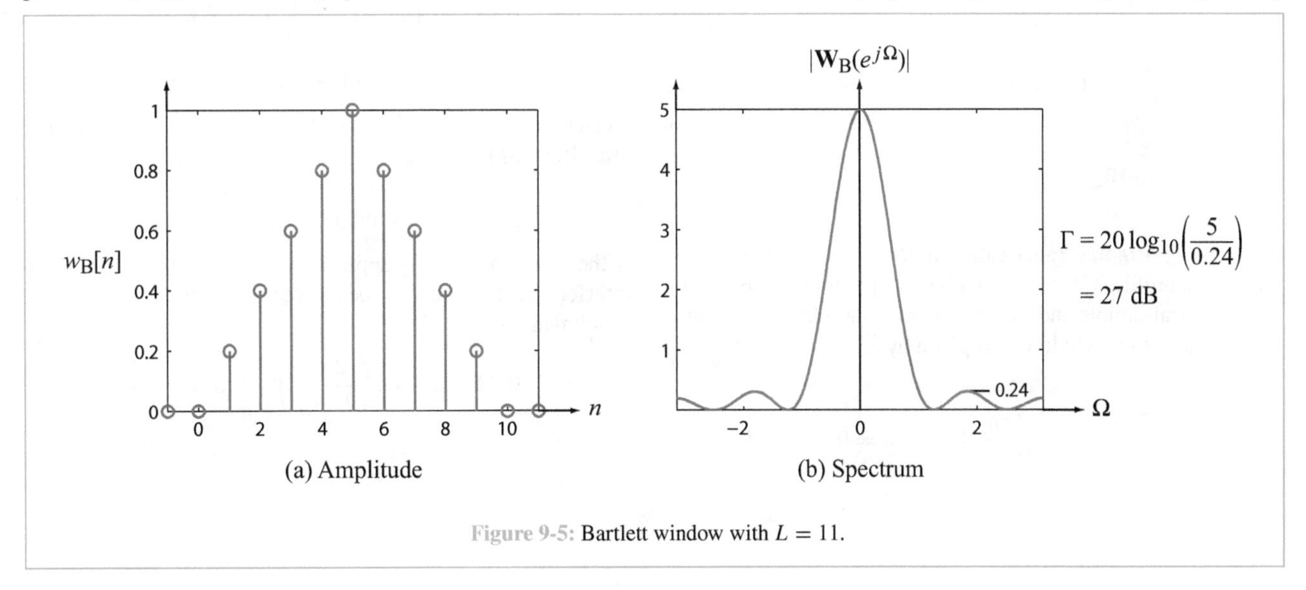

Figure 9-5: Bartlett window with $L = 11$.

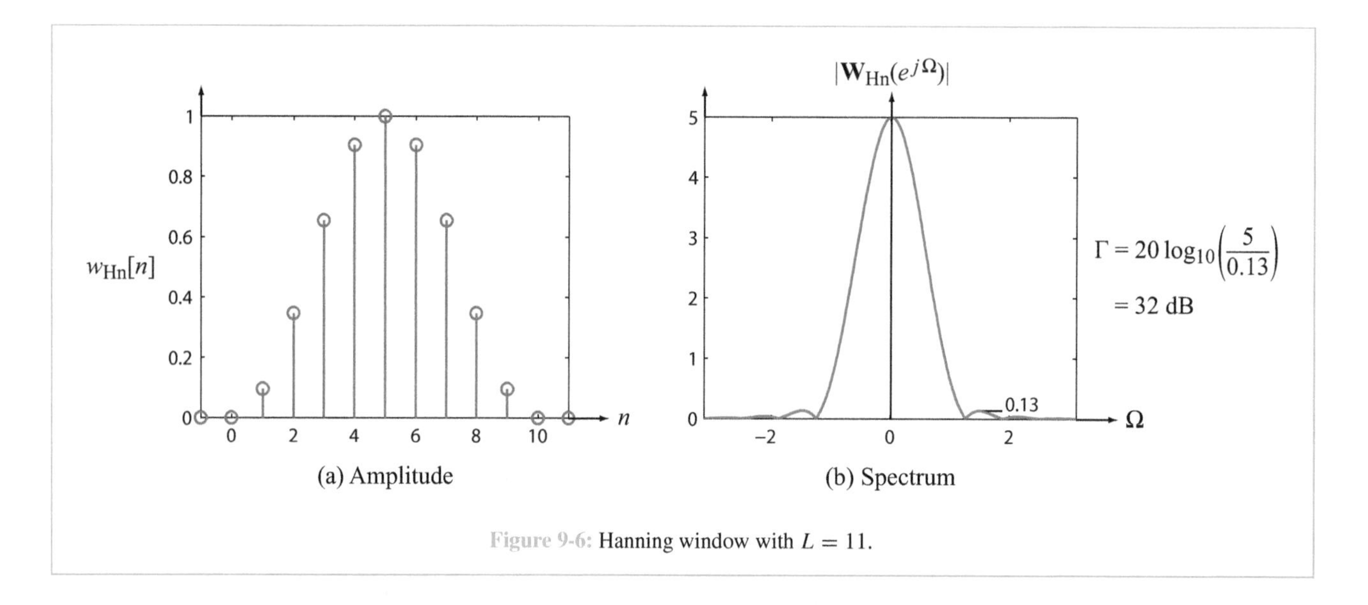

Figure 9-6: Hanning window with $L = 11$.

samples, with the weights assuming a triangular shape given by

$$w_{\mathrm{B}}[n] = 1 - \frac{2}{L-1}\left|n - \frac{L-1}{2}\right|, \qquad n = 0, 1, \ldots, L-1. \tag{9.17a}$$

(Bartlett triangular window)

The corresponding spectrum (Fig. 9-5(b)) has a sidelobe attenuation of 27 dB (compared with only 13 dB for the rectangular window), but the mainlobe width is twice as wide as that of the rectangular window,

$$\Omega_{\mathrm{M}} = \frac{8\pi}{L} \qquad \textbf{(Bartlett triangular window).} \tag{9.17b}$$

9-1.7 Hanning Window

A ***Hanning window*** (also called a ***Hann window***) offers a sinusoidal profile (Fig. 9-6(a)) with less taper for samples close to the central sample and steeper taper for samples close to the outer edges of the window. It is given by

$$w_{\mathrm{Hn}}[n] = \frac{1}{2} - \frac{1}{2}\cos\left(\frac{2\pi n}{L-1}\right), \qquad n = 0, 1, \ldots, L-1. \tag{9.18}$$

(Hanning window)

The corresponding spectrum (Fig. 9-6(b)) has a mainlobe level with the same width as that of the triangular window, but the sidelobe attenuation is increased further to 32 dB. The Hann window is named after Julius von Hann.

9-1.8 Hamming Window

By adjusting the values of the two constant coefficients in Eq. (9.18), it is possible to increase the sidelobe attenuation further to 43 dB, without sacrificing any additional increase in the width of the mainlobe. This optimized window is called a ***Hamming window*** and it is given by

$$w_{\mathrm{Hm}}[n] = 0.54 - 0.46\cos\left(\frac{2\pi n}{L-1}\right), \qquad n = 0, 1, \ldots, L-1, \tag{9.19}$$

(Hamming window)

and it is displayed in (Fig. 9-7). The Hamming window is named after Richard Hamming.

9-1.9 Blackman Window

If the signal-processing application calls for still better noise-interference performance (i.e., lower sidelobes), we can use the Blackman window given by

$$w[n] = 0.42 - 0.5\cos\left(\frac{2\pi n}{L-1}\right) + 0.08\cos\left(\frac{4\pi n}{L-1}\right), \tag{9.20}$$

$$n = 0, 1, \ldots, L-1.$$

(Blackman window)

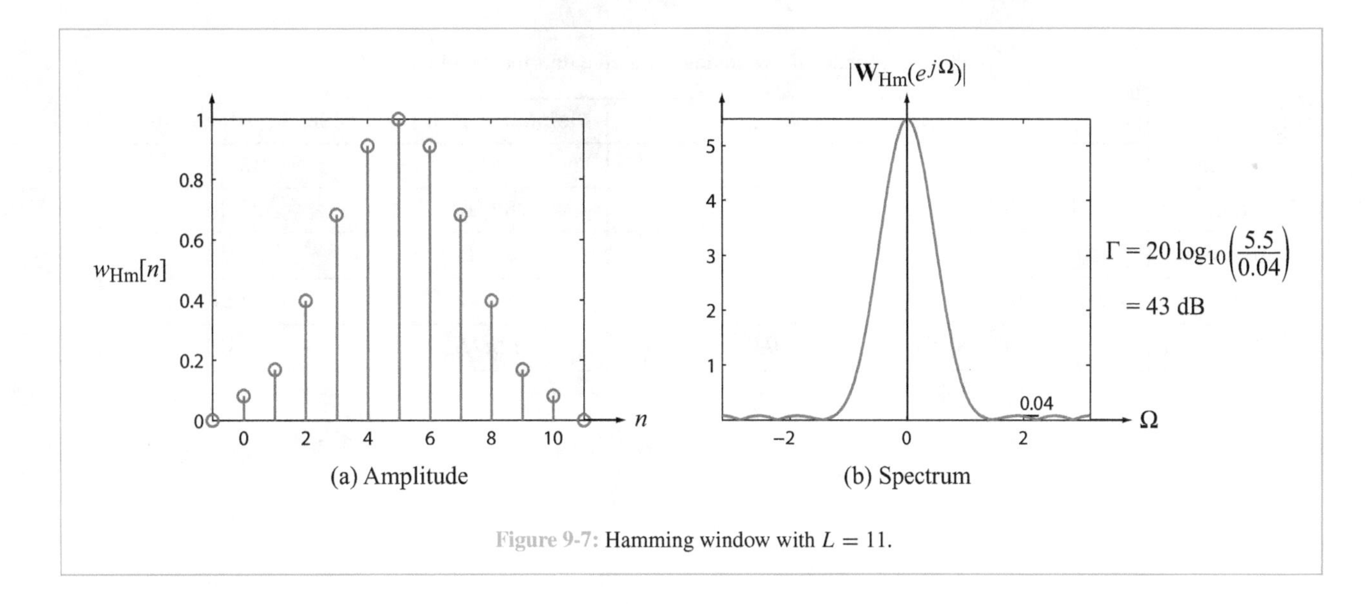

Figure 9-7: Hamming window with $L = 11$.

The sidelobe attenuation is 58 dB (Fig. 9-8), but it is attained at the expense of the mainlobe width, which is three times wider than that of the rectangular filter and 50% wider than that of the other three types of filters. Table 9-1 provides a summary of the properties of all five types of data windows.

Example 9-1: Using a Data Window

Use a 256-point DFT to compute the one-sided spectrum of the segment $\{\cos(0.2\pi n), 0 \leq n \leq 24\}$ using (a) a rectangular window and (b) a Hamming window, both of length 25.

Solution: Using MATLAB code Ex91.m (on the book website), we obtain the results shown in Fig. 9-9. Comparison of the two spectra shows that the Hamming window generates a spectrum with a mainlobe that is twice as wide as that generated by the rectangular window, but the sidelobes in the Hamming window spectrum are barely apparent.

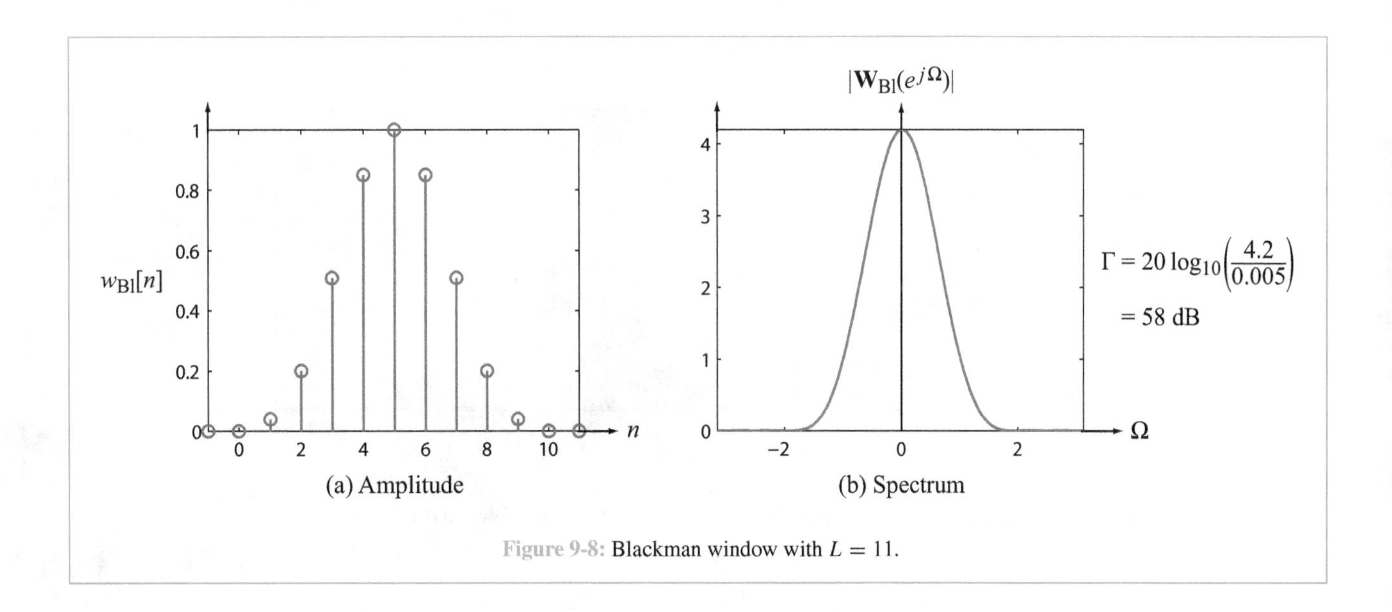

Figure 9-8: Blackman window with $L = 11$.

Table 9-1: **Properties of five commonly used data windows of length L.**

Name	$w[n],\ 0 \le n \le L-1$	Mainlobe width Ω_{M}	Sidelobe attenuation Γ
Rectangle	$w_{\mathrm{R}}[n] = 1$	$4\pi/L$	13 dB
Bartlett	$w_{\mathrm{B}}[n] = 1 - 2\left\|n - \frac{L-1}{2}\right\|/(L-1)$	$8\pi/L$	27 dB
Hanning	$w_{\mathrm{Hn}}[n] = 0.50 - 0.50\cos\left(\frac{2\pi n}{L-1}\right)$	$8\pi/L$	32 dB
Hamming	$w_{\mathrm{Hm}}[n] = 0.54 - 0.46\cos\left(\frac{2\pi n}{L-1}\right)$	$8\pi/L$	43 dB
Blackman	$w_{\mathrm{Bl}}[n] = 0.42 - 0.50\cos\left(\frac{2\pi n}{L-1}\right) + 0.08\cos\left(\frac{4\pi n}{L-1}\right)$	$12\pi/L$	58 dB

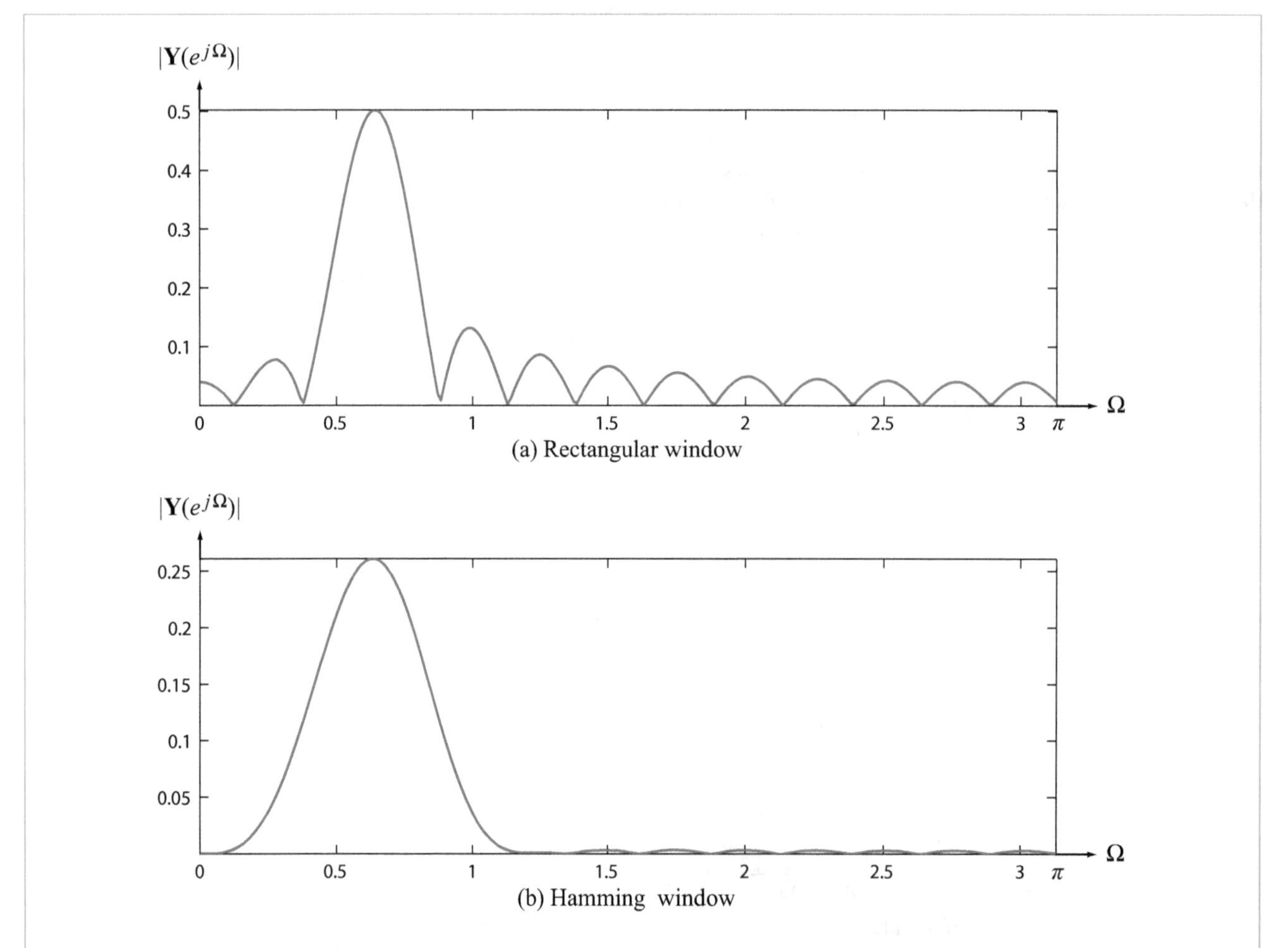

Figure 9-9: Computed spectrum for $\{\cos(0.2\pi n), 0 \le n \le 24\}$ using (a) a rectangular window and (b) a Hamming window (Example 9-1).

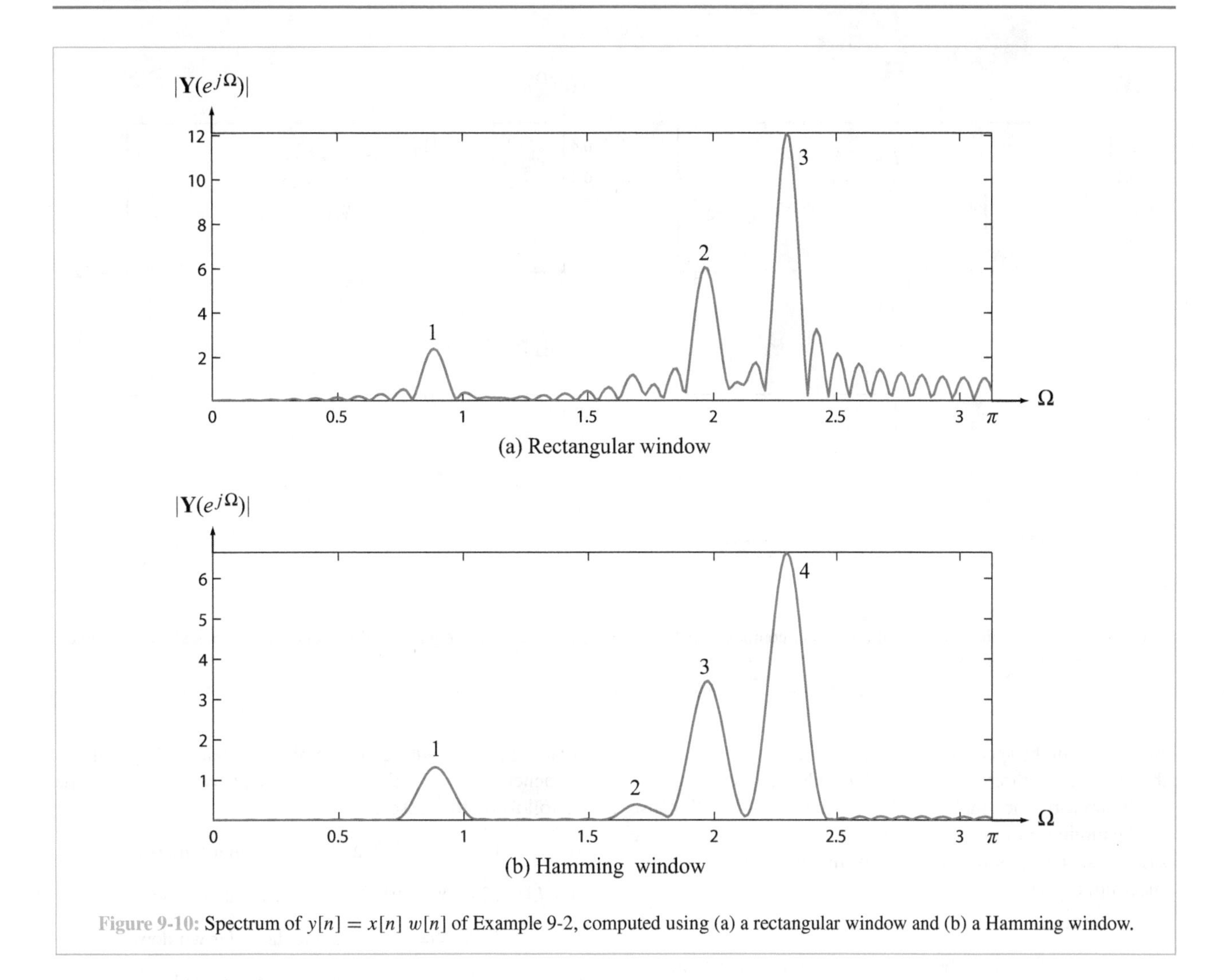

Figure 9-10: Spectrum of $y[n] = x[n]\, w[n]$ of Example 9-2, computed using (a) a rectangular window and (b) a Hamming window.

9-1.10 The Significance of Data Windows

We established in the preceding subsection that the length and shape (taper profile) of the data window determine the width of the mainlobe Ω_M of the associated spectrum and the sidelobe attenuation Γ. The obvious question is: *which type of window should one use?* The simple answer is: *it depends on the intended application*, as illustrated next through the following two examples.

Example 9-2: Detecting Different Sinusoids

We are given a signal $x[n]$ of length $L = 75$ composed of the sum of 4 sinusoids with different amplitudes and frequencies, namely

$$\begin{aligned} x[n] = {} & 5\cos(0.282\pi n + 53^\circ) \\ & + \sqrt{2}\,\cos(0.542\pi n + 45^\circ) \\ & + 13\cos(0.628\pi n + 67^\circ) \\ & + 25\,\cos(0.730\pi n + 73^\circ). \end{aligned} \tag{9.21}$$

Using MATLAB code Ex92.m (on the book website), a 256-point DFT was used to compute the one-sided spectrum of $y[n] = x[n]\, w[n]$, once with a rectangular window ($w[n] = w_R[n]$) and a second time with a Hamming window ($w[n] = w_{Hm}[n]$). Both windows are of length 75. The results are displayed in Fig. 9-10. How many distinct sinusoids can be discerned from each of the two spectra?

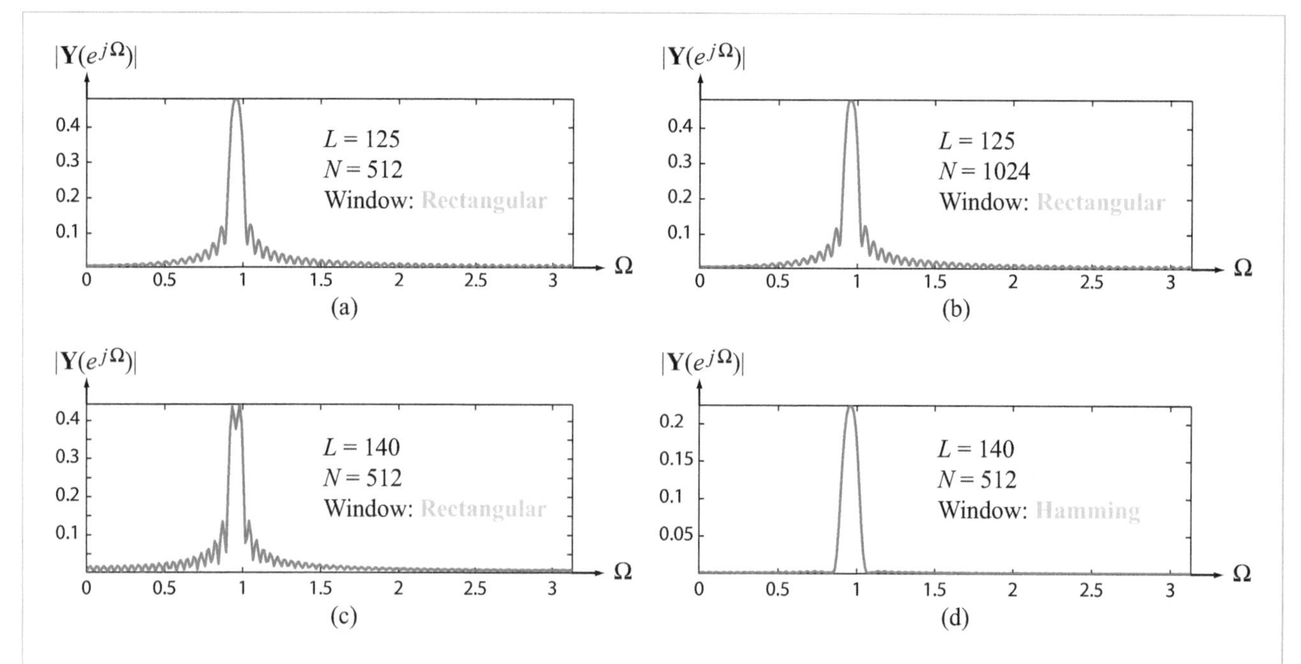

Figure 9-11: Computed spectra of a segment composed of the sum of two sinusoids of equal amplitude and frequencies $\Omega_1 = 0.3\pi$ and $\Omega_2 = 0.31\pi$ (Example 9-3).

Solution: In the spectrum of Fig. 9-10(a), it is easy to discern the presence of three of the four sinusoids, but the fourth one is obscured by the sidelobes of the other sinusoids. Because the Hamming window offers much better suppression of the sidelobes, it is possible to discern all four sinusoids in the spectrum of Fig. 9-10(b).

We note that because the frequencies of the four sinusoids in Example 9-2 are spaced sufficiently apart, the broader mainlobe of the Hamming-window spectrum (by a factor of 2 compared with that of the rectangular window) did not cause overlap between the mainlobes of the spectra of the different sinusoids. Hence, the Hamming window offered a distinct advantage over the rectangular window, and no disadvantage. However, we could have reached a very different conclusion had the sinusoids been spaced more closely together.

Example 9-3: Resolving Two Closely Spaced Sinusoids

Use an N-point DFT to compute the spectrum of $y[n] = x[n]\, w[n]$, where $w[n]$ is a data window of length L and $x[n]$ is a segment, also of length L, composed of the sum of two sinusoids with equal amplitudes, but slightly different frequencies, namely $\Omega_1 = 0.3\pi$ and $\Omega_2 = 0.31\pi$. Consider the following four cases:

(a) $L = 125$, $N = 512$, and a rectangular window.

(b) $L = 125$, $N = 1024$, and a rectangular window.

(c) $L = 140$, $N = 512$, and a rectangular window.

(d) $L = 140$, $N = 512$, and a Hamming window.

How resolvable are the two sinusoids in each case?

Solution: Using MATLAB code Ex93.m (on book website), we obtain the four one-sided spectra displayed in Fig. 9-11.

(a) The two closely spaced sinusoids are not resolvable.

(b) Increasing N by a factor of 2 translates into a finer discretization of the DTFT, but the two sinusoids remain unresolvable. Hence, resolution is not a discretization issue.

(c) Increasing L (the length of the data window) from 125 to 140 results in a proportionately narrower mainlobe; the width of the mainlobe of the rectangular window is given by $\Omega_M = 4\pi/L$ (Table 9-1). Consequently, because the mainlobes of the spectra of the two sinusoids are now slightly narrower, it is possible to discern from the spectrum shown in Fig. 9-11(c) that the mainlobe consists of two partially overlapping peaks.

(d) Repeating the conditions in (c) but replacing the rectangular window with a Hamming window causes the mainlobes of the peaks representing the two sinusoids to double in width, thereby losing the ability to resolve the two sinusoids.

Data Windows

- To apply a data window to a discrete-time signal $x[n]$ defined over $0 \leq n \leq L-1$, multiply $x[n]$ by the window function $w[n]$, point by point, to obtain $y[n] = x[n]\, w[n]$.
- On the DFT spectrum of a sampled signal containing multiple sinusoids, two adjacent sinusoids are distinguishable if their separation exceeds the spectral resolution of the data window, which is commonly defined as the width of the mainlobe.
- Among the various types of commonly used data windows, the rectangular window has the narrowest mainlobe.
- For any type of data window, increasing the window length L results in a proportionate reduction in mainlobe width, and hence better spectral resolution.
- So long as the DTFT has been properly discretized, increasing the DFT order N does not improve the spectral resolution.
- Data windows with a tapered amplitude distribution have sidelobe levels smaller than those exhibited by the spectrum of a rectangular window, which makes it easier to discern the presence of weak (small-amplitude) sinusoids, so long as their frequencies are separated from others by at least the width of one mainlobe of the data window's spectrum.

Concept Question 9-1: Why do we multiply a signal by a data window, instead of convolving it with the data window? (See S²)

Concept Question 9-2: Why do most data windows look alike? (See S²)

Exercise 9-1: Compute the coefficients of a 5-point (a) Bartlett window and (b) Hamming window.

Answer: (a) $\{0, \frac{1}{2}, 1, \frac{1}{2}, 0\}$,
(b) $\{0.08, 0.54, 1.00, 0.54, 0.08\}$.

Module 9.1 Discrete-Time Lowpass Filter Design Using Windowing This module designs a lowpass filter of duration $2L+1$ using a Hamming window on the inverse DTFT (a sinc function) of a brickwall lowpass filter frequency response. Cutoff frequency and filter duration are selectable variables.

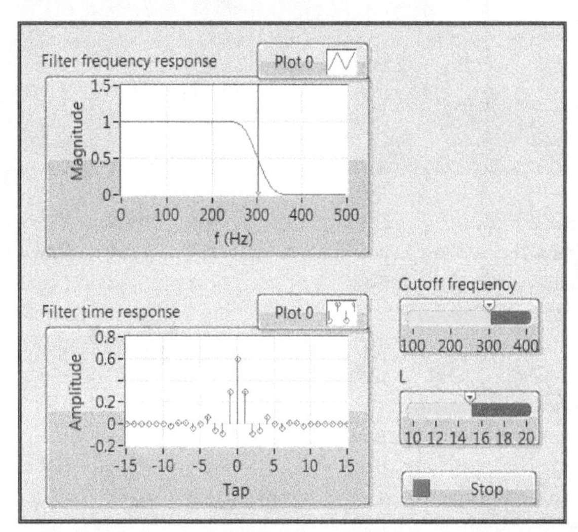

9-2 Spectrograms

So far we have considered the spectra of simple signals consisting of one or a few sinusoids. The spectra of most signals of practical interest are composed of numerous sinusoids, and since the signals are time-varying, so are their spectra. Examples of signals with time-varying spectra include:

- **Musical Signals**: The note played by, say, a trumpet changes at specific times, so its line spectrum shifts frequencies abruptly at specific times.
- **Radar and Sonar Signals**: The frequency of the transmitted signal gets Doppler shifted due to the motion of the illuminated target with respect to the radar antenna or sonar transducer, and the Doppler shift may vary with time depending on the illumination configuration.
- **Chirp Signals**: These are sinusoidal signals whose frequency changes linearly with time. Bird chirps and dolphin clicks can be modeled well as chirp signals.

Other examples include non-repetitive speech and many forms of audio and video signals.

By way of an example, let us consider the musical signal $x(t)$ given by

$$x(t) = \begin{cases} \sum_{k=1}^{\infty} A_{k,1} \cos\left(2\pi f_{k,1} t + \theta_{k,1}\right), & T_0 < t < T_1 \\ \sum_{k=1}^{\infty} A_{k,2} \cos\left(2\pi f_{k,2} t + \theta_{k,2}\right), & T_1 < t < T_2 \\ \sum_{k=1}^{\infty} A_{k,3} \cos\left(2\pi f_{k,3} t + \theta_{k,3}\right), & T_2 < t < T_3 \\ \vdots & \vdots \end{cases} \tag{9.22}$$

where duration T_0 to T_1 defines the first time window, T_1 to T_2 defines the second time window, and so on. In some but not all cases, durations T_i to T_{i+1} of each note are known because whole notes have the same duration and half notes have half of that duration, etc. So, segmenting the signal into multiple intervals is rather straightforward. Also, in each octave, the frequencies f_k can only take on twelve different values, all known (but this is not true in the general case).

Spectrograms are a useful tool for analyzing signals with time-varying spectra. A spectrogram is generated by computing spectra of segments of the signal using a series of data windows, with varying delays, and then plotting the computed spectra alongside each other and viewing the result from above as an image.

To generate a spectrogram of the musical signal given by Eq. (9.22), for example, spectra are computed for each of the individual time durations, and then combined to form an image. Multiple examples will be presented later in this section.

9-2.1 Continuous-Time Spectrogram

The spectrogram computes the spectral density (squared magnitude of the Fourier transform) $|\mathbf{X}(\omega)|^2$ of each segment $\{x(t), T_i < t < T_{i+1}\}$ of the signal $x(t)$. The signal is sampled, and the spectrum of each segment is computed using the DFT with a data window over that interval. Since the T_i are usually unknown, intervals $\{t : \tau - T/2 < t < \tau + T/2\}$ of equal lengths T are used, centered at times τ. In continuous time, the ***spectrogram*** $\mathcal{S}(\omega, \tau)$ is defined as

$$\mathcal{S}(\omega, \tau) = \left| \int_{\tau - T/2}^{\tau + T/2} w(t - \tau + T/2)\, x(t)\, e^{-j\omega t}\, dt \right|^2 . \tag{9.23}$$

The squared Fourier transform magnitude is used so that $\mathcal{S}(\omega, \tau)$ has units of energy. When $\mathcal{S}(\omega, \tau)$ is displayed as an image with ω and τ as axes, peaks appear as bright pixels.

9-2.2 Discrete-Time Spectrogram

In discrete time, the signal is sampled to $x[n]$, a data window $w[n]$ (such as one of those described in Section 9-1) is used and the intervals are defined as $\{N - L/2 \le n \le N + L/2\}$, centered at N and of length $L+1$, with L even. In discrete time, the spectrogram $\mathcal{S}(e^{j\Omega}, N)$ is defined as

$$\mathcal{S}(e^{j\Omega}, N) = \left| \sum_{n=N-L/2}^{N+L/2} w[n - N + L/2]\, x[n]\, e^{-j\Omega n} \right|^2 . \tag{9.24}$$

Here, we consider only non-overlapping intervals and, for simplicity, we use a rectangular window. Choosing a non-rectangular window does not affect much the process of generating and displaying $\mathcal{S}(e^{j\Omega}, N)$ as an image.

9-2.3 MATLAB/MathScript Recipe for Spectrograms

For a signal sampled at S samples per second and stored in MATLAB/MathScript as vector X, we can apply the following code to generate a spectrogram and display it as an image. The interval length is denoted by L.

MATLAB/MathScript Spectrogram Recipe

```
LX=length(X);%L=interval length
XX=reshape(X,L,length(X)/L);
FXX=abs(fft(XX)).*abs(fft(XX));
imagesc(FXX)
```

The length of X must be a multiple of the interval length L, so that an integer number of intervals can be used. The horizontal axis of the image is the time of the left end (not the center) of the interval over which the spectrum is computed. The horizontal axis should be multiplied by $1/S$ seconds per sample to be interpreted as seconds.

The vertical axis of the image is the frequency axis for the spectrum displayed in each vertical slice of the spectrogram. The vertical axis should be multiplied by S/L Hz. MATLAB/MathScript displays images with the origin at the upper left corner, but since $\mathcal{S}(e^{j\Omega}, N)$ is conjugate symmetric in the vertical axis, the lower left corner can also be treated as the origin.

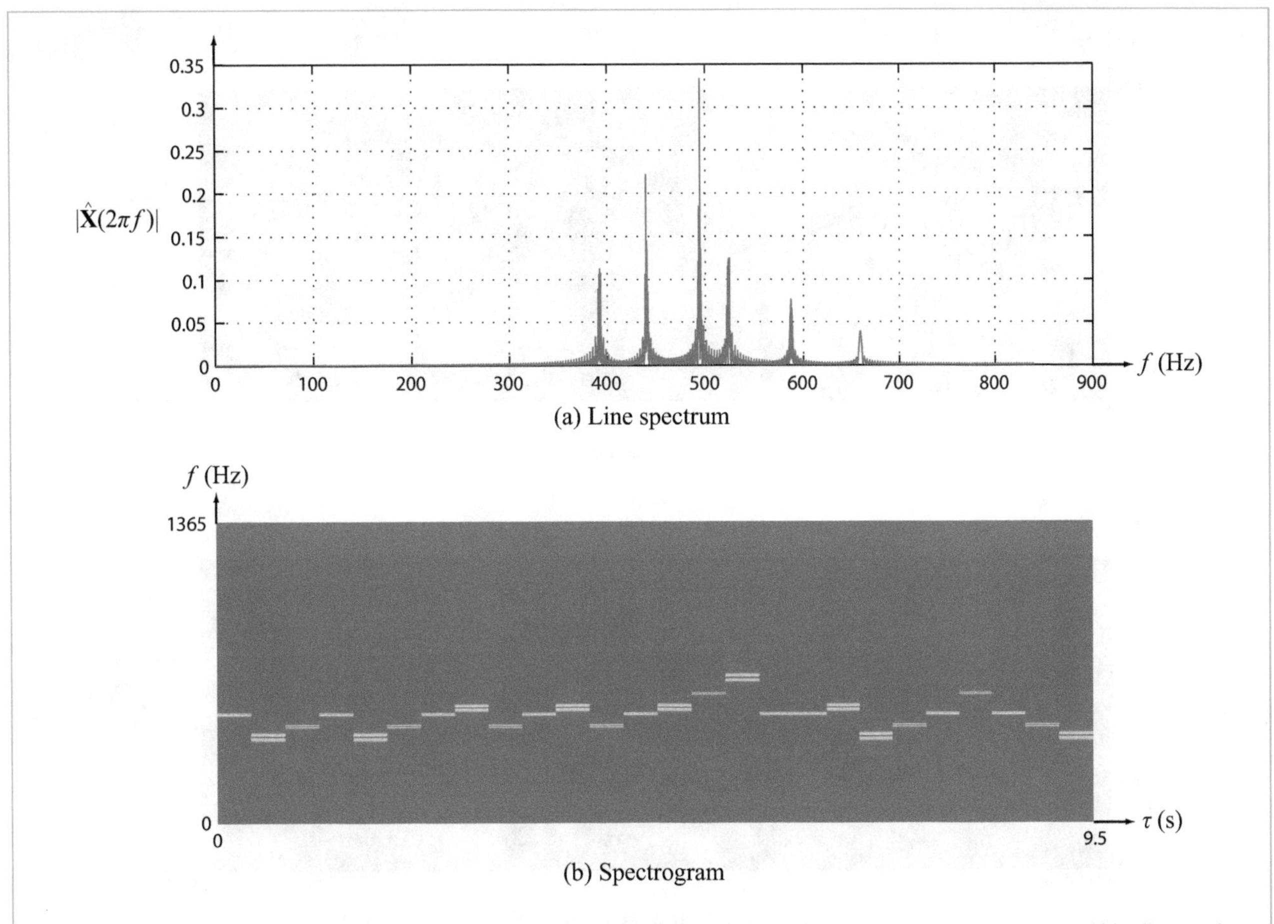

Figure 9-12: Line spectrum and spectrogram of music signal of Example 9-4. Even though spectrograms are computed in discrete time using N and Ω (see Eq. (9.24)), usually they are labeled in continuous time in terms of f (Hz) and τ (time).

Example 9-4: Spectrogram of a Musical Signal

Interpret a music signal sampled at 8192 samples per second and stored in MATLAB Code Ex94.mat on the book website. The signal consists entirely of whole notes of duration 0.3662 seconds each.

Solution: The line spectrum shown in Fig. 9-12(a) was computed from the recorded signal contained on the referenced website. The spectrum is a histogram of the musical notes, indicating *how often* each note was played, but not *when* it was played. The spectrogram of the signal, using 26 segments of lengths 3000 each, is depicted in Fig. 9-12(b). The duration of each segment is $\frac{L}{S} = \frac{3000}{8192} = 0.3662$ s. The spectrum of each segment is one sinusoid, and the vertical position of the line indicates the frequency of that sinusoid. To obtain the frequency in Hz, we should multiply the vertical index by $\frac{S}{L} = \frac{8192}{3000}$ Hz.

The spectrogram shows that the signal is a piece of music. It is a pure tonal version of the chorus of "The Victors," the fight song of the University of Michigan. An extension of this example is provided in Problem 9.13, in which the spectrogram is used to identify two fight songs, making it possible to eliminate one of them.

9-2.4 Trade-Off between Time and Frequency Resolution

Consider a sampled signal $x[n]$ of duration N, divided into M non-overlapping segments each of length L. That is,

$$N = L\,M. \tag{9.25}$$

Since M is an integer (number of segments), it can assume only certain values, namely those for which N is an integer multiple of L. The smallest value of M is $M = 1$ segment,

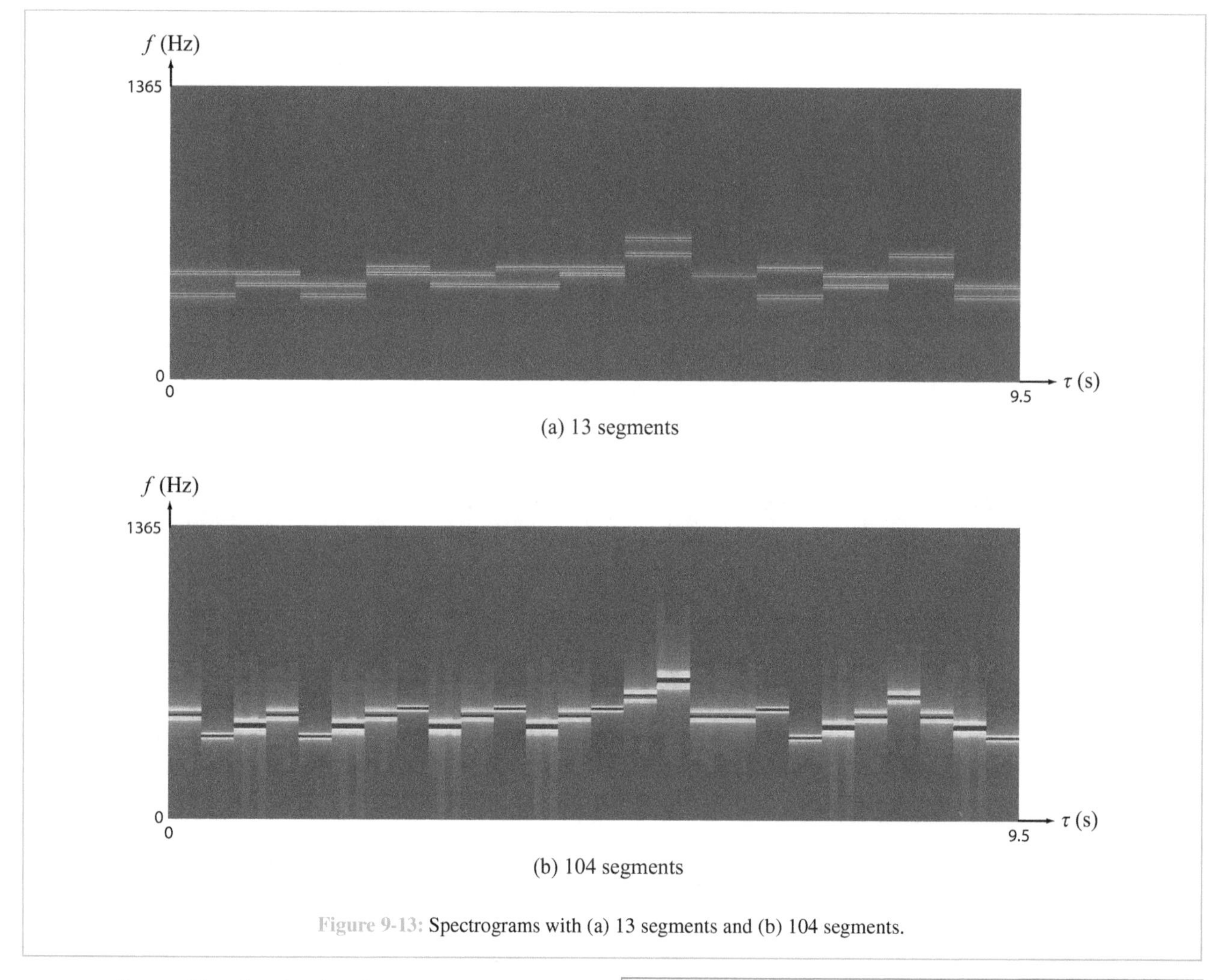

Figure 9-13: Spectrograms with (a) 13 segments and (b) 104 segments.

corresponding to $N = L$, wherein all N samples are used in computing the DFT. At the other extreme, the largest possible value of M is $M = N$, corresponding to segments of length $L = 1$ sample each.

Resolution in time is measured by the length of the processed segment, L, whereas resolution in the frequency domain is inversely proportional to L (longer segments generate narrower mainlobes).

▶ Hence, for a fixed duration N, increasing L leads to poorer resolution in time, but better resolution in frequency (easier discrimination of closely spaced sinusoids); and conversely, decreasing L allows more rapid tracking of the spectrum, but at a poorer spectral resolution. ◀

Example 9-5: Resolution Trade-Off

The music signal recorded in MATLAB Code Ex94.mat (of Example 9-4) consists of 78,000 samples. Subdivide the record (a) into 13 segments of lengths 6000 samples each, and also (b) into 104 segments of lengths 750 samples each. Generate spectrograms for the two cases and compare them.

Solution: The spectrogram using only 13 segments is shown in part (a) of Fig. 9-13, and that using 104 segments is in part (b). A longer segment generates a finer-resolution spectrum, so the spectral lines in Fig. 9-13(a) are distinct, but two spectral lines appear in every segment, even though the two notes were not played together. In contrast, when the number of segments is increased to 104, different notes appear in different segments, but the line spectrum representing each sinusoid is smeared out, because the mainlobe is wider. Thus, the frequencies of the notes are not indicated as sharply as was the case for the longer (but fewer) segments. The intent of the example is to illustrate the trade-off between the time and frequency resolutions.

9-2.5 Chirp Signals

A ***chirp*** signal is a sinusoid whose frequency increases or decreases linearly with time over a specific duration τ. Some birds and animals can produce chirped sounds, but chirping also is used as a form of modulation in some high-resolution radar systems. In general, the frequency of the sinusoid varies linearly over a bandwidth B extending between a start frequency f_1 and an end frequency f_2. For simplicity, we limit our consideration to the case where the chirp signal starts with 0 Hz at time $t = 0$ and increases to an upper frequency limit $f_{\max}$ at $t = \tau$. For a single sinusoid of amplitude A and angular frequency ω, the signal is given by

$$x(t) = A\cos(\omega t),$$

but for a chirp, ω increases linearly as

$$\omega = \alpha t, \tag{9.26a}$$

where α is the ***chirp rate*** given by

$$\alpha = \frac{2\pi f_{\max}}{\tau}. \tag{9.26b}$$

Hence,

$$x(t) = A\cos(\alpha t^2), \qquad 0 \le t \le \tau. \tag{9.27}$$

As will be demonstrated through Example 9-6, the spectrogram of a chirp signal is a line with a slope proportional to $2f_{\max}$.

Example 9-6: Spectrogram of a Chirp Signal

Compute and plot the spectrogram of a chirp signal with $A = 1$, $\alpha = 1$, and $\tau = 81.91$ s, using a sampling rate of 100 samples/second, and then divide the sampled signal into 32 non-overlapping segments. Also, determine the maximum frequency $f_{\max}$.

Solution: Figure 9-14(a) displays the first 20 s of the chirp; early on the variation is slow as a function of time, but the oscillations become more rapid as time goes on.

For $A = 1$, $\alpha = 1$, and $\tau = 81.91$ s,

$$x(t) = \cos(t^2), \qquad 0 \le t \le 81.91 \text{ s}.$$

A signal $x(t)$ of 81.91 s in duration and sampled at 100 samples/second yields a sampled signal

$$x[n] = \left\{ \cos\left(\frac{n^2}{10{,}000}\right), \quad 0 \le n \le 8191 \right\},$$

where t was replaced with $n/100$.

We have a total of $N = 8192$ samples, which when divided among $M = 32$ segments, gives

$$L = \frac{N}{M} = \frac{8192}{32} = 256.$$

The spectrogram shown in Fig. 9-14(b) was computed, using a 256-point DFT, for 32 segments of $x[n]$. The details are contained in MATLAB Code Ex96.m.

The spectrogram of the chirp signal is a staircase, depicting a sinusoid whose frequency increases linearly with time. Based on Eq. (9.26b) with $\alpha = 1$ and $\tau = 81.91$ s, we would conclude that the final instantaneous frequency of the chirp signal is

$$f_{\max} = \frac{\alpha\tau}{2\pi} = \frac{1 \times 81.91}{2\pi} = 13 \text{ Hz}.$$

This conclusion is erroneous, because the exact definition of the instantaneous frequency of a signal $x(t) = A\cos(\phi(t))$ is

$$f = \frac{1}{2\pi}\frac{d\phi}{dt}. \tag{9.28}$$

For a single sinusoid at frequency f_0

$$\phi = 2\pi f_0 t \qquad \text{(single sinusoid)},$$

and application of Eq. (9.28) leads to $f = f_0$. But for a chirp signal

$$\phi = \alpha t^2 \qquad \text{(chirp)},$$

and consequently,

$$f = \frac{1}{2\pi}\frac{d}{dt}(\alpha t^2) = \frac{2\alpha t}{2\pi}.$$

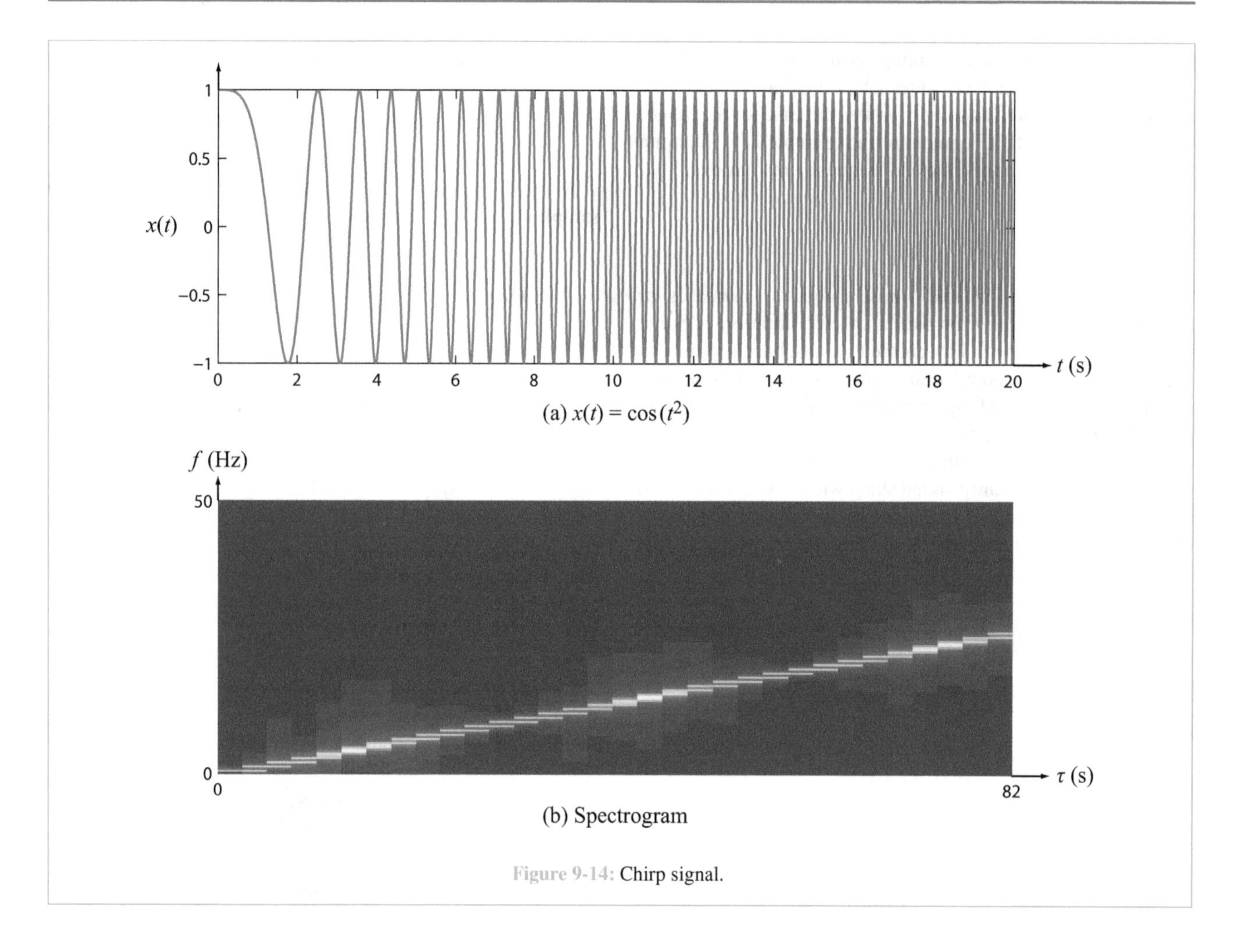

Figure 9-14: Chirp signal.

The chirp reaches the true maximum frequency $f'_{\max}$ at $t = \tau$:

$$f'_{\max} = \frac{2\alpha\tau}{2\pi} = \frac{\alpha\tau}{\pi} = \left(\frac{2\pi f_{\max}}{\tau}\right)\frac{\tau}{\pi} = 2f_{\max},$$

where we used Eq. (9.26b) for α. Hence, the final instantaneous frequency of the chirp is $2f_{\max}$, not $f_{\max}$. So, for the present case, the maximum frequency is $2 \times 13 = 26$ Hz. This conclusion can be confirmed from the spectrogram in Fig. 9-14(b).

We conclude this section with a spectrogram example of a spoken phrase. Figure 9-15(a) is a time-domain record of a 3 s long phrase, sampled at 24,000 samples per second, and in part (b) of the same figure, we show the corresponding spectrogram. The spectrogram is based on 288 segments, each 250 samples in length (for a total of $288 \times 250 = 72{,}000$ samples).

Concept Question 9-3: A time-varying spectrum is a contradiction in terms. What does a spectrogram actually compute? (See S²)

Exercise 9-2: What would the spectrogram of $\cos(t^3)$ look like?

Answer: A parabola, since the instantaneous frequency is

$$f = \frac{1}{2\pi}\frac{dt^3}{dt} = \frac{3}{2\pi}\,t^2.$$

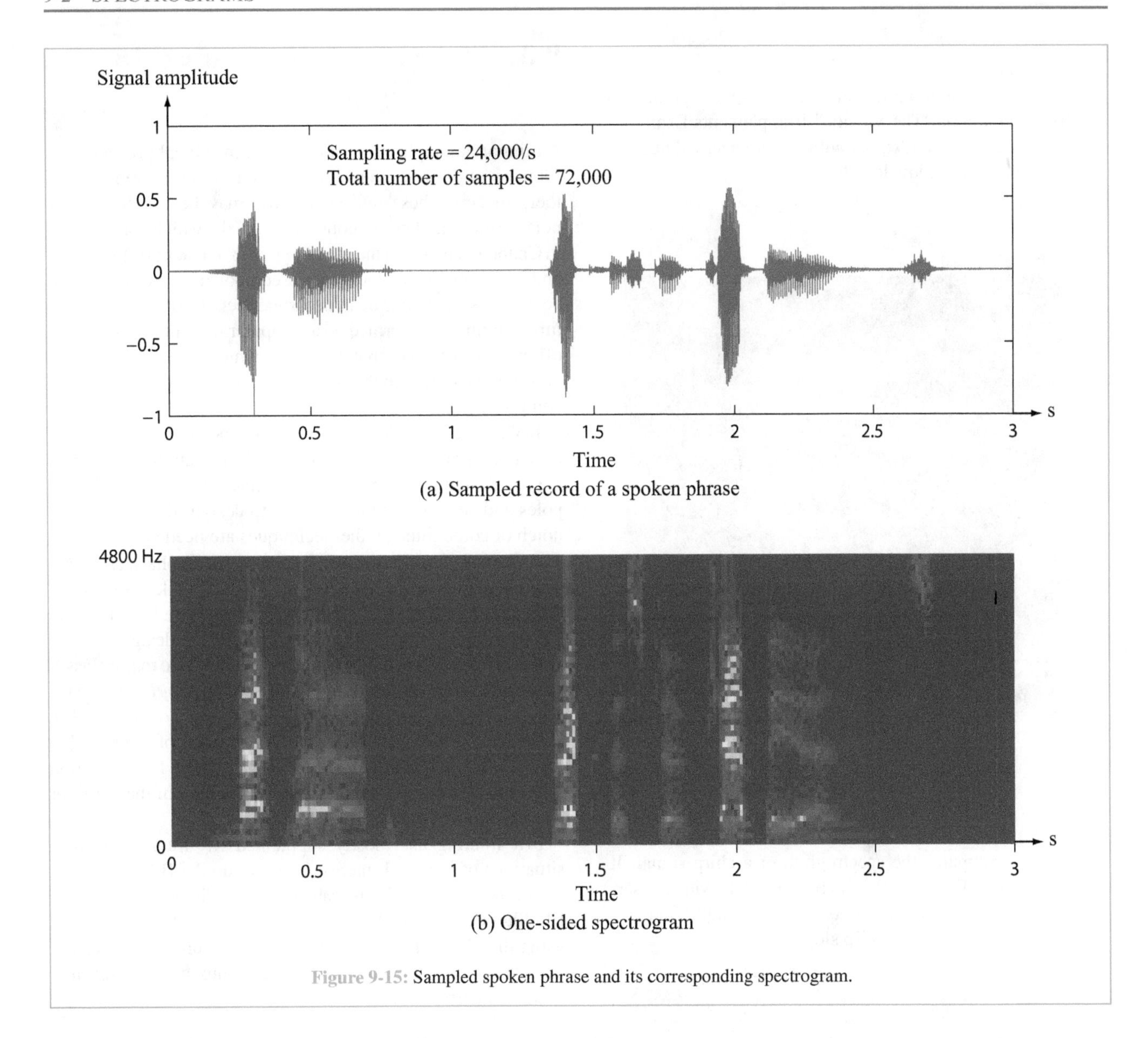

(a) Sampled record of a spoken phrase

(b) One-sided spectrogram

Figure 9-15: Sampled spoken phrase and its corresponding spectrogram.

Exercise 9-3: Use LabVIEW Module 9.2 to display the spectrogram of "The Victors." Choose the window length so that the notes do not overlap in time.

Answer: (See Module 9.2.)

Exercise 9-4: Use LabVIEW Module 9.3 to display the spectrogram of a chirp with slope 1.0 using window length 32.

Answer: (See Module 9.3.)

Module 9.2 Spectrogram of Tonal Version of "The Victors" This module computes the spectrogram of a tonal version of "The Victors." It implements Examples 9-4 and 9-5 in the text, but with a user-specified number of segments (window length).

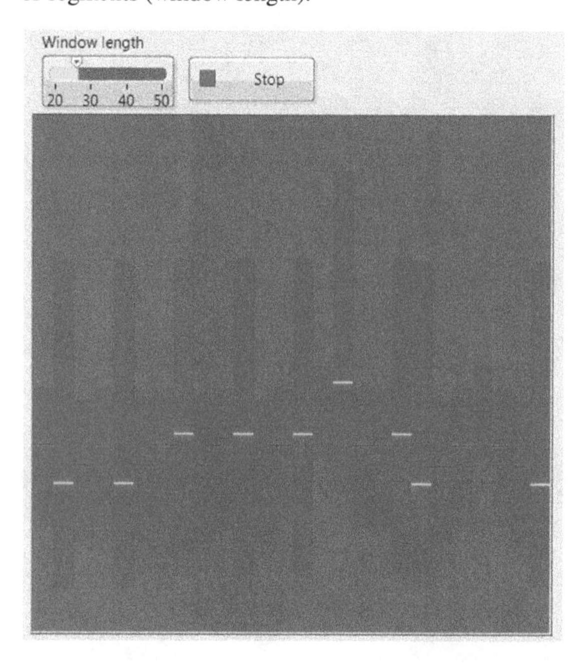

Module 9.3 Spectrogram of a Chirp Signal This module computes the spectrogram of a chirp signal. It implements Example 9-6 in the text, but with a user-specified number of segments (window length) and rate of frequency increase (chirp slope).

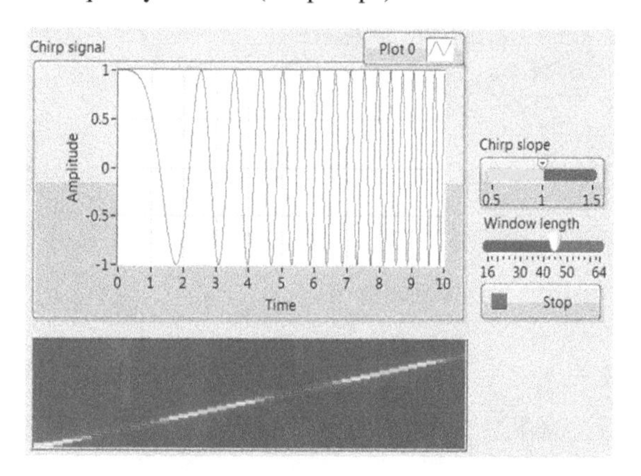

9-3 Finite Impulse Response (FIR) Filter Design

An important application of discrete-time signal processing is *filtering* of discrete-time signals to remove undesired parts of their spectra. These undesired parts may be interference or noise. Filtering of continuous-time signals, which was covered in Chapter 6, involves the use of op-amps, capacitors, inductors, and resistors, and since these components are imperfect, continuous-time filtering is also imperfect. In contrast, discrete-time filtering is performed on a computer with almost no round-off error, or on a computer chip, in which case round-off can be significant, but much less so than imperfect physical circuit components.

In Chapter 8, we showed how placing poles and zeros can be used to design notch and comb filters. But an attempt (Example 8-1) to design a half-band lowpass filter showed that placing poles and zeros is not the best way to design filters other than notch or comb filters. Other techniques are needed.

A *discrete-time filter* is an LTI system that emphasizes certain desired frequency components of a signal while reducing or eliminating others. Notch and comb filters were introduced in Chapter 8, including examples to illustrate the design process. In this and the succeeding section we examine two major classes of discrete-time filters, namely the *finite-impulse response* (*FIR*) filter and the *infinite-impulse response* (*IIR*) filter. The fundamental difference between the two types of filters is that for the FIR filter its impulse response $h[n]$ has a finite duration (and is therefore more practical), whereas $h[n]$ of the IIR filter has an infinite duration.

FIR filtering has mostly supplanted IIR filtering in most situations of practical interest because, unlike IIR filters, FIR filters are always stable and cause no phase distortion. However, IIR filters can be made more selective than FIR filters while using the same number of memory storage units. By covering both types of filters, we gain insight into filter design and implementation.

9-3.1 FIR Filtering Configuration

In the frequency domain, the role of an FIR filter is to emphasize certain frequency components over others. The filtering operation can be realized in the frequency domain by multiplying each frequency component by a specified weight or, equivalently, by convolving the discrete-time signal $x[n]$ with the impulse response of the filter, $h[n]$, to obtain an output signal $y[n]$:

$$y[n] = h[n] * x[n]. \tag{9.29}$$

If $h[n]$ is causal and given by

$$h[n] = b_n \qquad \text{for } 0 \leq n \leq M, \tag{9.30}$$

and $x[n]$ is of duration N, then

$$y[n]=h[n]*x[n]=b_0\, x[n]+b_1\, x[n-1]+\cdots+b_M\, x[n-M], \quad \text{for } 0 \leq n \leq N+M. \tag{9.31}$$

Note that the filter $h[n]$ is of *order* M and its *duration* is $M+1$.

The FIR filter is similar to the moving average (MA) filter described earlier in Section 7-3.4 in that the output $y[n]$ at each discrete time n is equal to the weighted sum of prior nonzero values of $x[n]$, up to and including the present-time value $x[n]$.

Sometimes, the FIR filter impulse response $h[n]$ may not be causal; that is, it may be defined over a discrete-time interval that starts at negative values of n. Consider, for example, the case where $h[n] \neq 0$ for $\{-L \leq n \leq L\}$. To perform the filtering operation, we need to convert $h[n]$ into a causal equivalent, which is easily realized by simply delaying $h[n]$ by L to obtain the delayed version $\tilde{h}[n]$:

$$\tilde{h}[n] = h[n-L], \qquad \text{for } 0 \leq n \leq 2L. \tag{9.32}$$

By invoking the time-shift property of convolution (property #5 in Table 7-4), the original output response $y[n]$ given by

$$y[n] = h[n] * x[n] = \sum_{i=-L}^{L} h[i]\, x[n-i], \tag{9.33a}$$

can be rewritten as a delayed response

$$y[n-L] = h[n-L]*x[n] = \tilde{h}[n]*x[n] = \sum_{i=0}^{2L} \tilde{h}[i]\, x[n-i]. \tag{9.33b}$$

The original noncausal FIR filter $h[n]$ has been converted into a causal MA system of order $2L$. The only consequence is that the output is now delayed by L time samples, which is seldom an issue in most practical situations. For example, at the standard CD sampling rate of 44100 samples/s, a filter of length $L = 100$ samples leads to a delay in the filtered output signal by only $100/44100 = 2.2$ ms. Also, in the case of image processing applications, noncausal filters can be implemented as is (without a shift) because the entire image is available before the start of the filtering process.

The effect of the time delay $\tilde{h}[n] = h[n-L]$ is equivalent to the addition of *linear phase*. The time-delay property of the DTFT (Table 7-7) states that if $\mathbf{H}(e^{j\Omega})$ is the frequency response (DTFT) of $h[n]$, then the frequency response of $\tilde{h}[n]$ is

$$\tilde{\mathbf{H}}(e^{j\Omega}) = \mathbf{H}(e^{j\Omega})e^{-j\Omega L}, \tag{9.34}$$

so that the phase of $\tilde{\mathbf{H}}(e^{j\Omega})$ is the phase of $\mathbf{H}(e^{j\Omega})$ minus ΩL, which is proportional to frequency Ω. Hence, the term "linear phase."

▶ Linear phase in the frequency domain (DTFT) is equivalent to time delay in the time domain. ◀

An example of linear phase is given by Eq. (9.4) for the rectangular window $w_R[n]$.

There are 3 major approaches to FIR filter design:

- *Windowing* the ideal FIR filter, computed as the inverse DTFT of the desired frequency response function, usually using a Hamming window.
- *Frequency sampling*, in which the desired frequency response is attained exactly, but only at a finite number of frequencies, usually equally spaced.
- *Minimax*, in which the iterative Parks-McClellan algorithm is used to *minimize the maximum* (minimax) absolute weighted error.

Before presenting these approaches (in upcoming subsections), we present some common forms of the desired (ideal) filter frequency response function.

9-3.2 Desired Frequency Response Functions

Let $\mathbf{H}_D(e^{j\Omega})$ be the *desired* (ideal) frequency response function of the FIR filter. Recall that $\mathbf{H}_D(e^{j\Omega})$ is periodic in Ω with period 2π, and conjugate symmetric:

$$\mathbf{H}_D(e^{-j\Omega}) = \mathbf{H}_D^*(e^{j\Omega}). \tag{9.35}$$

A. Forms of frequency response

- For a *lowpass filter* with cutoff frequency Ω_0:

$$\mathbf{H}_D(e^{j\Omega}) = \begin{cases} 1 & \text{for } 0 \leq |\Omega| < \Omega_0, \\ 0 & \text{for } \Omega_0 < |\Omega| \leq \pi. \end{cases} \tag{9.36}$$

For a *bandpass filter* with cutoff frequencies Ω_L and Ω_H:

$$\mathbf{H}_D(e^{j\Omega}) = \begin{cases} 0 & \text{for } 0 \leq |\Omega| < \Omega_L, \\ 1 & \text{for } \Omega_L < |\Omega| < \Omega_H, \\ 0 & \text{for } \Omega_H < |\Omega| \leq \pi. \end{cases} \tag{9.37}$$

- For an *ideal differentiator* (used in digital speedometers and process control):

$$\mathbf{H}_\mathrm{D}(e^{j\Omega}) = j\Omega \text{ for } |\Omega| < \pi. \tag{9.38}$$

For a *Hilbert transform* (Section 6-12.9):

$$\mathbf{H}_\mathrm{D}(e^{j\Omega}) = \begin{cases} -j & \text{for } 0 < \Omega < \pi, \\ j & \text{for } -\pi < \Omega < 0. \end{cases} \tag{9.39}$$

Note that for the differentiator and Hilbert transform, $\mathbf{H}_\mathrm{D}(e^{j\Omega})$ is discontinuous at $\Omega = \pm\pi$.

B. **Forms of impulse response**

From conjugate symmetry of $\mathbf{H}_\mathrm{D}(e^{j\Omega})$, its inverse DTFT $h_\mathrm{D}[n]$ should have the following forms:

- $\mathbf{H}_\mathrm{D}(e^{j\Omega})$ for the lowpass and bandpass filters is real and even, so $h_\mathrm{D}[n]$ should be real and even (see Section 7-14).
- $\mathbf{H}_\mathrm{D}(e^{j\Omega})$ for the differentiator and Hilbert transform is pure imaginary and odd, so $h_\mathrm{D}[n]$ should be real and odd.
- We design $h_\mathrm{D}[n]$ to be real, even or odd, and noncausal. Then we delay it by half its length to make it causal, as in Eq. (9.33b).

9-3.3 FIR Filter Design by Windowing

The desired frequency response $\mathbf{H}_\mathrm{D}(e^{j\Omega})$ of each of the four filters described by Eqs. (9.36) to (9.39) is an ideal function with one or more discontinuities. Consequently, its corresponding impulse responses $h_\mathrm{D}[n]$ (i.e., the inverse DTFT of $\mathbf{H}_\mathrm{D}(e^{j\Omega})$) is infinite in duration and is not BIBO stable. To implement the filtering procedure described in the preceding subsection, we can truncate $h_\mathrm{D}[n]$ and delay it appropriately so as to generate a modified impulse response $h[n]$, such that it is both causal and BIBO stable.

The modification process is described by the diagram in Fig. 9-16, using a lowpass filter as an example.

Step 1: The desired frequency response of the filter, $\mathbf{H}_\mathrm{D}(e^{j\Omega})$, with a cutoff frequency $\Omega_0 = 0.5\pi$, is given by Eq. (9.36) with $\Omega_0 = 0.5\pi$. That is,

$$\mathbf{H}_\mathrm{D}(e^{j\Omega}) = \begin{cases} 1 & \text{for } 0 \le |\Omega| < 0.5\pi, \\ 0 & \text{for } 0.5\pi < |\Omega| \le \pi. \end{cases} \tag{9.40}$$

Step 2: The corresponding impulse response $h_\mathrm{D}[n]$ is obtained by computing the inverse DTFT of $\mathbf{H}_\mathrm{D}(e^{j\Omega})$:

$$h_\mathrm{D}[n] = \frac{1}{2\pi} \int_{-0.5\pi}^{0.5\pi} 1 e^{j\Omega}\, d\Omega = \frac{\sin(0.5\pi n)}{\pi n}. \tag{9.41}$$

Step 3: The impulse response $h_\mathrm{D}[n]$ is noncausal and not BIBO stable, so it is not yet suitable as a lowpass filter. It needs to be truncated and delayed. Truncation is realized by multiplying $h_\mathrm{D}[n]$ by an appropriate window. In this example, we choose a Hamming window of length 11. Since the Hamming window now extends over the range $-5 \le n \le 5$ instead of $0 \le n \le 10$, we must replace n with $n+5$ (a time advance by 5). Hence,

$$\begin{aligned} w[n] &= w_\mathrm{Hm}[n+5] \\ &= 0.54 - 0.46 \cos\left(\frac{2\pi(n+5)}{10}\right) \\ &= 0.54 + 0.46 \cos\left(\frac{2\pi n}{10}\right), \qquad -5 \le n \le 5. \end{aligned} \tag{9.42}$$

The multiplication yields

$$\begin{aligned} h[n] &= w[n]\; h_\mathrm{D}[n] \\ &= \left[0.54 + 0.46 \cos\left(\frac{2\pi n}{10}\right)\right] \frac{\sin(0.5\pi n)}{\pi n}, \\ &\quad -5 \le n \le 5. \end{aligned} \tag{9.43}$$

Step 4: Delaying $h[n]$ by 5 leads to

$$\begin{aligned} \widetilde{h}[n] &= h[n-5] \\ &= 0.54 + 0.46 \cos\left(\frac{2\pi(n-5)}{10}\right) \frac{\sin(0.5\pi(n-5))}{\pi(n-5)}, \\ &\quad 0 \le n \le 10. \end{aligned} \tag{9.44}$$

Step 5: In this demonstration example, the input signal $x[n]$ is the sum of two sinusoids with frequencies $\Omega_1 = 0.2\pi$ and $\Omega_2 = 0.7\pi$:

$$x[n] = \cos(0.2\pi n) + \cos(0.7\pi n). \tag{9.45}$$

The delayed output $y[n-L]$ is obtained by convolving $x[n]$ with $\widetilde{h}[n]$:

$$y[n-L] = \widetilde{h}[n] * x[n]. \tag{9.46}$$

The output, displayed in Fig. 9-16, shows that the sinusoid with $\Omega_2 = 0.7\pi$ in $x[n]$ has been almost (but not completely)

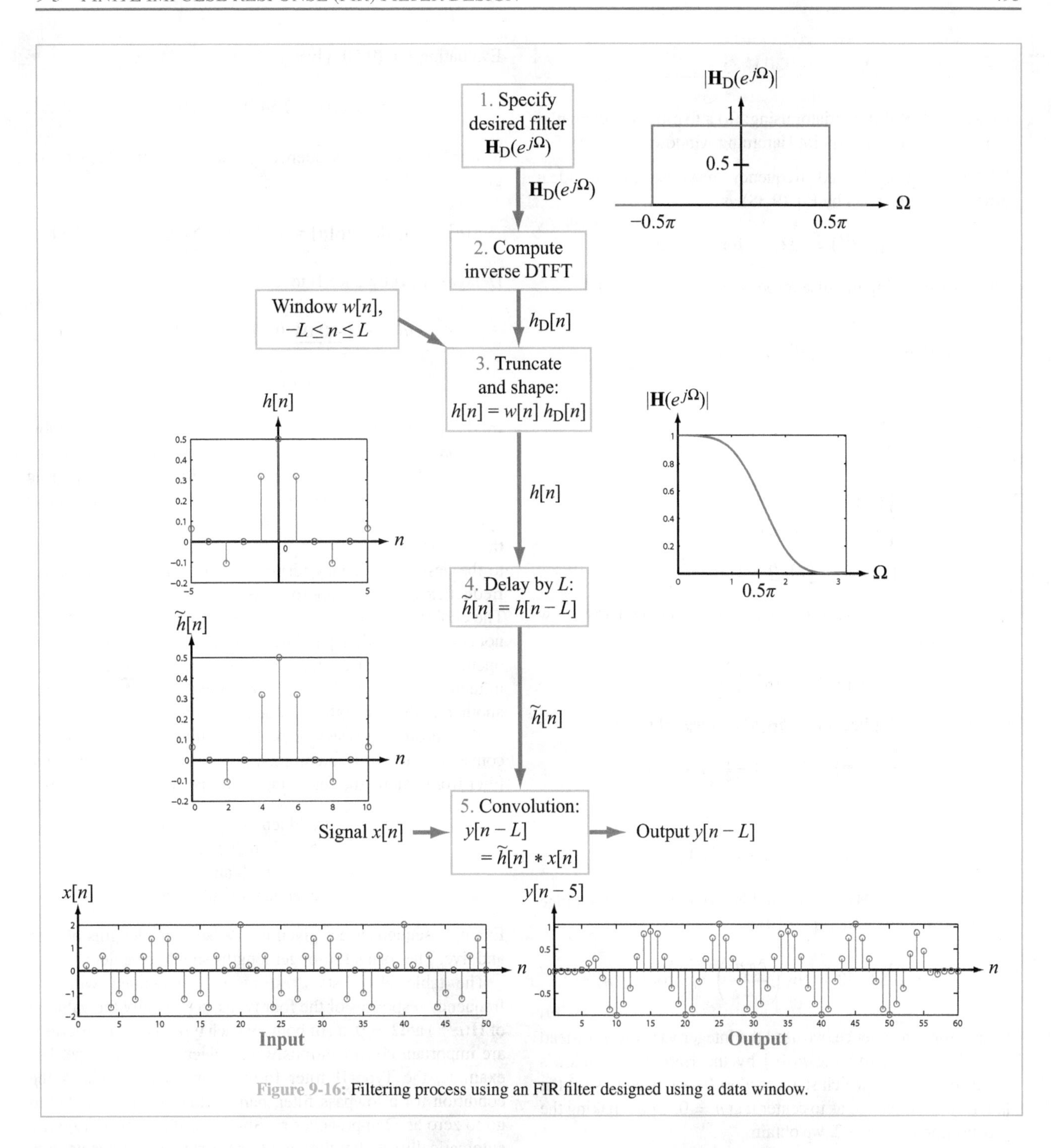

Figure 9-16: Filtering process using an FIR filter designed using a data window.

filtered out. The lower-frequency sinusoid with $\Omega_1 = 0.2\pi$ is essentially the same as in the original $x[n]$. Note the transients at the beginning and end of the output. These are caused by the finite length of $x[n]$. In practice, the input is so much longer than the filter that these transients become irrelevant.

Example 9-7: Window Design of FIR Differentiator

Design an FIR differentiator using: (a) a five-point rectangular window and (b) a five-point Hamming window.

Solution: The desired frequency response of the ideal differentiator is given by Eq. (9.38) as

$$\mathbf{H}_\mathrm{D}(e^{j\Omega}) = j\Omega \qquad \text{for } |\Omega| < \pi.$$

The corresponding impulse response is its inverse DTFT:

$$\begin{aligned} h_\mathrm{D}[n] &= \frac{1}{2\pi}\int_{-\pi}^{\pi} \mathbf{H}_\mathrm{D}(e^{j\Omega})\, e^{j\Omega n}\, d\Omega \\ &= \frac{1}{2\pi}\int_{-\pi}^{\pi} j\Omega e^{j\Omega n}\, d\Omega \\ &= \begin{cases} (-1)^n/n & n \neq 0 \\ 0 & n = 0 \end{cases} \\ &= \{\ldots, \tfrac{1}{3}, -\tfrac{1}{2}, 1, \underline{0}, -1, \tfrac{1}{2}, -\tfrac{1}{3}, \ldots\}. \end{aligned} \tag{9.47}$$

(a) The impulse response of a five-point rectangular window centered at $n = 0$ is

$$w_\mathrm{R}[n] = \{1, 1, \underline{1}, 1, 1\}. \tag{9.48}$$

One-to-one multiplication of $h_\mathrm{D}[n]$ and $w_\mathrm{R}[n]$ gives

$$h[n] = w_\mathrm{R}[n]\, h_\mathrm{D}[n] = \{-\tfrac{1}{2}, 1, \underline{0}, -1, \tfrac{1}{2}\},$$

and the delayed version is

$$\widetilde{h}_\mathrm{R}[n] = h[n-2] = \{\underline{-\tfrac{1}{2}}, 1, 0, -1, \tfrac{1}{2}\}.$$

(b) A five-point Hamming window extending over the range $0 \leq n' \leq 4$ is given by Eq. (9.19) as

$$w_\mathrm{Hm}[n'] = 0.54 - 0.46\cos\left(\frac{2\pi n'}{4}\right) \qquad \text{for } 0 \leq n' \leq 4, \tag{9.49}$$

where, for convenience, we used the integer variable n' instead of n. Before multiplying $h_\mathrm{D}[n]$ by the Hamming window's impulse response (which starts at $n' = 0$), we need to shift the latter by 2 samples so as to center it at $n = 0$. Upon making the substitution $n' = n + 2$, we obtain

$$w_\mathrm{Hm}[n] = 0.54 - 0.46\cos\left(\frac{2\pi}{4}(n+2)\right), \qquad -2 \leq n \leq 2. \tag{9.50}$$

Evaluating Eq. (9.50) gives

$$w_\mathrm{Hm}[n] = \{0.08, 0.54, \underline{1}, 0.54, 0.08\}. \tag{9.51}$$

Multiplication of the sequences given by Eqs. (9.47) and (9.51) gives

$$h[n] = w_\mathrm{Hm}[n]\, h_\mathrm{D}[n] = \{-0.04, 0.54, \underline{0}, -0.54, 0.04\}.$$

Delaying $h[n]$ by 2 leads to

$$\widetilde{h}[n] = \{\underline{-0.04}, 0.54, 0, -0.54, 0.04\}.$$

9-3.4 FIR Filter Design by Frequency Sampling

In the preceding subsection, we examined how a window function of length $2L+1$ can be used to truncate the extent of the ideal impulse response $h_\mathrm{D}[n]$, as well as to shape the weighting pattern across the extent of the impulse response. An alternative approach is to choose an FIR filter $\widetilde{h}[n]$ of length $2L+1$ such that the DTFT of $\widetilde{h}[n]$, namely $\widetilde{\mathbf{H}}(e^{j\Omega})$, is identically equal to the desired frequency response $\widetilde{\mathbf{H}}_\mathrm{D}(e^{j\Omega})$ at $2L+1$ choice frequencies. Such an approach is called ***frequency sampling***. The choice frequencies can be chosen to be equally spaced or not equally spaced, depending on the objectives of the filtering operation. In either case, the specified choice frequencies will influence how close (or not) $\widetilde{\mathbf{H}}(e^{j\Omega})$ and $\widetilde{\mathbf{H}}_\mathrm{D}(e^{j\Omega})$ are to one another at all of the other frequencies.

The design process can be facilitated by symmetry considerations, so it is common practice to use an applicable filter from among the following four types of filters:

Type I:	odd length/symmetric
Type II:	even length/symmetric
Type III:	odd length/antisymmetric
Type IV:	even length/antisymmetric

Example sequences are given in Table 9-2 for lengths of four and five. The forms for longer lengths should be evident.

The table also lists "***restrictions***" associated with the frequency responses of the four types, specifically the value(s) of $\widetilde{\mathbf{H}}(e^{j\Omega})$ at $\Omega = 0, \pi$, or both for each type. These restrictions are important considerations when selecting a filter type. For example, the Type II filter form automatically satisfies the condition for a lowpass filter, namely that its $\widetilde{\mathbf{H}}(e^{j\Omega})$ should go to zero as Ω approaches π. Similarly, the filter of Type IV automatically satisfies the condition for a highpass filter, namely that its $\widetilde{\mathbf{H}}(e^{j\Omega})$ is zero at dc ($\Omega = 0$).

FIR Type I filters offer no restrictions, and the restrictions listed in Table 9-2 for the other types follow from the forms

Table 9-2: **Forms of FIR filters for lengths of 4 or 5 elements.**

Type	Form of $\widetilde{h}[n]$	Restriction(s)
I	$\{\underline{b}, a, c, a, b\}$	None
II	$\{\underline{b}, a, a, b\}$	$\widetilde{\mathbf{H}}(e^{j\pi}) = 0$
III	$\{\underline{b}, a, 0, -a, -b\}$	$\widetilde{\mathbf{H}}(e^{j0,j\pi}) = 0$
IV	$\{\underline{b}, a, -a, -b\}$	$\widetilde{\mathbf{H}}(e^{j0}) = 0$

of their $h[n]$. Recall from Eq. (7.65) that the **z**-transform of a causal finite-length signal is given by

$$\{\underline{a_0}, a_1, a_2, \ldots, a_m\} \quad \Longleftrightarrow \quad a_0 + \frac{a_1}{\mathbf{z}} + \frac{a_2}{\mathbf{z}^2} + \cdots + \frac{a_m}{\mathbf{z}^m}. \tag{9.52}$$

Applying this correspondence to the functional forms given in Table 9-2 for filter types II–IV, and then evaluating them at $\mathbf{z} = e^{j\pi}$ or $\mathbf{z} = e^{j0}$, leads to

- Type II: $\widetilde{h}[n] = \{\underline{b}, a, a, b\}$

$$\longrightarrow \quad \widetilde{\mathbf{H}}(e^{j\pi}) = b + \frac{a}{e^{j\pi}} + \frac{a}{e^{j2\pi}} + \frac{b}{e^{j3\pi}}$$
$$= b - a + a - b = 0.$$

- Type III: $\widetilde{h}[n] = \{\underline{b}, a, 0, -a, -b\}$

$$\longrightarrow \quad \widetilde{\mathbf{H}}(e^{j\pi}) = b + \frac{a}{e^{j\pi}} + 0 - \frac{a}{e^{j3\pi}} - \frac{b}{e^{j4\pi}}$$
$$= b - a + a - b = 0$$

and

$$\widetilde{\mathbf{H}}(e^{j0}) = b + \frac{a}{e^{j0}} + 0 - \frac{a}{e^{j0}} - \frac{b}{e^{j0}}$$
$$= b + a - a - b = 0.$$

- Type IV: $\widetilde{h}[n] = \{\underline{b}, a, -a, -b\}$

$$\longrightarrow \quad \widetilde{\mathbf{H}}(e^{j0}) = b + \frac{a}{e^{j0}} + 0 - \frac{a}{e^{j0}} - \frac{b}{e^{j0}} = 0.$$

► Alternatively, we can compute the DTFT of $\{\underline{a_0}, a_1, \ldots, a_m\}$ to obtain $\mathbf{H}(e^{j\Omega})$ directly. ◄

Types II and III FIR filters automatically satisfy the lowpass filter criterion of rejecting the signal's spectrum at $\Omega = \pi$, and types III and IV automatically satisfy the highpass filter criterion of rejecting the signal's spectrum at $\Omega = 0$ (dc). With their odd-numbered lengths, types I and III are easier to design with than types II and IV, so we will henceforth limit our consideration to symmetric and antisymmetric FIR filters with odd durations.

9-3.5 FIR Filter Design by Solving a Linear System of Equations

As noted earlier, the impulse response $h[n]$ of the FIR filter is of length $2L+1$, extending between $-L$ and L. In practice, the length is a user-specified parameter.

► The idea behind the frequency sampling approach is to select the $2L+1$ values of $h[n]$ such that its DTFT $\mathbf{H}(e^{j\Omega})$ matches the desired frequency response $H_{\mathrm{D}}(e^{j\Omega})$ at exactly $2L+1$ values of Ω. ◄

These select values of Ω can be chosen to be (a) equally spaced across the range of Ω between $-\pi$ and π, (b) clustered across a narrow range of particular interest, or (c) distributed so as to emphasize specific frequency components of interest.

For the equally spaced case, the choice frequencies are given by

$$\Omega_k = \frac{2\pi k}{2L+1}, \qquad k = -L, \ldots, L, \tag{9.53}$$

and requiring $\mathbf{H}(e^{j\Omega})$ of the FIR filter to be equal to $\mathbf{H}_{\mathrm{D}}(e^{j\Omega})$ of the desired filter at those choice frequencies is given mathematically by

$$\mathbf{H}(e^{j\Omega})\Big|_{\Omega=\Omega_k} = \mathbf{H}_{\mathrm{D}}(e^{j\Omega})\Big|_{\Omega=\Omega_k},$$

or equivalently,

$$\sum_{n=-L}^{L} h[n]\, e^{-j\Omega n}\Big|_{\Omega=\Omega_k} = \mathbf{H}_{\mathrm{D}}(e^{j\Omega})\Big|_{\Omega=\Omega_k}. \tag{9.54}$$

Inserting Eq. (9.53) into Eq. (9.54) leads to

$$\sum_{n=-L}^{L} h[n]\, e^{-j2\pi nk/(2L+1)} = \mathbf{H}_{\mathrm{D}}(e^{j2\pi k/(2L+1)}), \tag{9.55}$$
$$k = -L, \ldots, L,$$

which represents a linear system of $2L+1$ equations (one for each value of k between $-L$ and L) in $2L+1$ unknowns. Solution of the system of equations provides the values of the unknowns, namely $h[n]$ for $-L \le n \le L$.

Example 9-8: Design of Lowpass Filter Using Frequency Sampling

Design a Type I FIR lowpass filter of length 5 using frequency sampling (Table 9-2). The filter should satisfy the following conditions: $\mathbf{H}(e^{j\Omega}) = 1$ at $\Omega = 0$, $\mathbf{H}(e^{j\Omega}) = 0$ at $\Omega = \pi$, and $\mathbf{H}(e^{j\Omega}) = 0.75$ at $\Omega = \pi/2$.

Solution: A Type I filter of length 5 and centered at $n = 0$ is given by

$$h[n] = \{a, b, \underline{c}, b, a\}. \tag{9.56}$$

The corresponding frequency response is

$$\begin{aligned}\mathbf{H}(e^{j\Omega n}) &= \sum_{n=-L}^{L} h[n]\, e^{-j\Omega n} \\ &= ae^{-j2\Omega} + be^{-j\Omega} + c + be^{j\Omega} + ae^{j2\Omega} \\ &= c + 2b\cos(\Omega) + 2a\cos(2\Omega).\end{aligned} \tag{9.57}$$

Application of the three specified conditions leads to:

$$\mathbf{H}(e^{j\Omega}) = 1\Big|_{\Omega=0} \quad \longrightarrow \quad 1 = c + 2b + 2a, \tag{9.58a}$$

$$\mathbf{H}(e^{j\Omega}) = 0\Big|_{\Omega=\pi} \quad \longrightarrow \quad 0 = c - 2b + 2a, \tag{9.58b}$$

$$\mathbf{H}(e^{j\Omega}) = 0.75\Big|_{\Omega=\pi/2} \quad \longrightarrow \quad 0.75 = c - 2a. \tag{9.58c}$$

Simultaneous solution of the three equations of Eq. (9.58) gives

$$a = -\frac{1}{16}, \qquad b = \frac{1}{4}, \qquad c = \frac{5}{8}.$$

So the filter is

$$h[n] = \left\{ -\tfrac{1}{16}, \tfrac{1}{4}, \underline{\tfrac{5}{8}}, \tfrac{1}{4}, -\tfrac{1}{16} \right\}, \tag{9.59a}$$

and the delayed version is

$$\widetilde{h}[n] = \left\{ \underline{-\tfrac{1}{16}}, \tfrac{1}{4}, \tfrac{5}{8}, \tfrac{1}{4}, -\tfrac{1}{16} \right\}. \tag{9.59b}$$

9-3.6 FIR Filter Design Parameters

The magnitude spectrum $|\mathbf{H}_D(e^{j\Omega})|$ of an ideal brickwall lowpass filter looks like a perfect rectangle with sharp edges. An example was shown earlier in Fig. 9-16. To generate such a frequency response, the corresponding time-domain impulse response $h_D[n]$ would have to be infinite in duration. In practice, $h_D[n]$ is multiplied by a window function that truncates its length, and possibly shapes its amplitude profile. The simplest such window is the rectangular window $w_R[n]$ given by Eq. (9.1). The finite-length filter $h[n] = w_R[n]\, h_D[n]$ exhibits a lowpass spectrum $\mathbf{H}(e^{j\Omega})$ that differs from that of the ideal filter, $\mathbf{H}_D(e^{j\Omega})$, in a number of ways. Figure 9-17 displays the impulse response $h[n]$ and its corresponding spectrum $|\mathbf{H}(e^{j\Omega})|$ for rectangular windows of lengths $N = 201$ and 21. Both are for a filter with a cutoff frequency $\Omega_0 = \pi/2$. In the transition regions between the passband and the two stopbands, the slope is much steeper for the spectrum of $h[n]$ with $N = 201$ than for the spectrum of the shorter impulse response. Also, both spectra exhibit ripples, particularly near the edges of the transition regions, but the character of the ripples is different for the two cases. Moreover, the ripple effect would look different had we used a window different from the rectangular function $w_R[n]$.

In general, the deviation of a filter's spectrum from the desired ideal spectrum is characterized by the following attributes:

- Passband ripple R_p: Relative to the ideal spectrum with $|\mathbf{H}_D(e^{j\Omega})| = 1$ in the passband, the spectrum of a real filter exhibits fluctuations that range between a minimum $1 - R_p$ and a maximum $1 + R_p$ (Fig. 9-17(b)).

- Stopband ripple R_s: This is the peak value of the sidelobes in the stopband.

- Rolloff rate S_g: This is the slope at the middle of the transition region between the passband and stopband.

The FIR filter design approaches considered thus far—namely, the windowing and frequency sampling methods of Sections 9-3.3 and 9-3.4—can be implemented relatively easily and neither method requires excessive computation. Their spectra, however, are not optimized to meet specific design criteria, such as the maximum acceptable passband ripple or the minimum acceptable sidelobe attenuation. Consequently, a filter designer would have to pursue a "trial and error" approach in the hope of approaching an impulse response $h[n]$ with an acceptable spectrum. An alternative, and more systematic, approach is to use an error criterion to iteratively arrive at a "quasi optimal" design of $h[n]$. This is the idea behind the Parks and McClellan algorithm described in the next subsection.

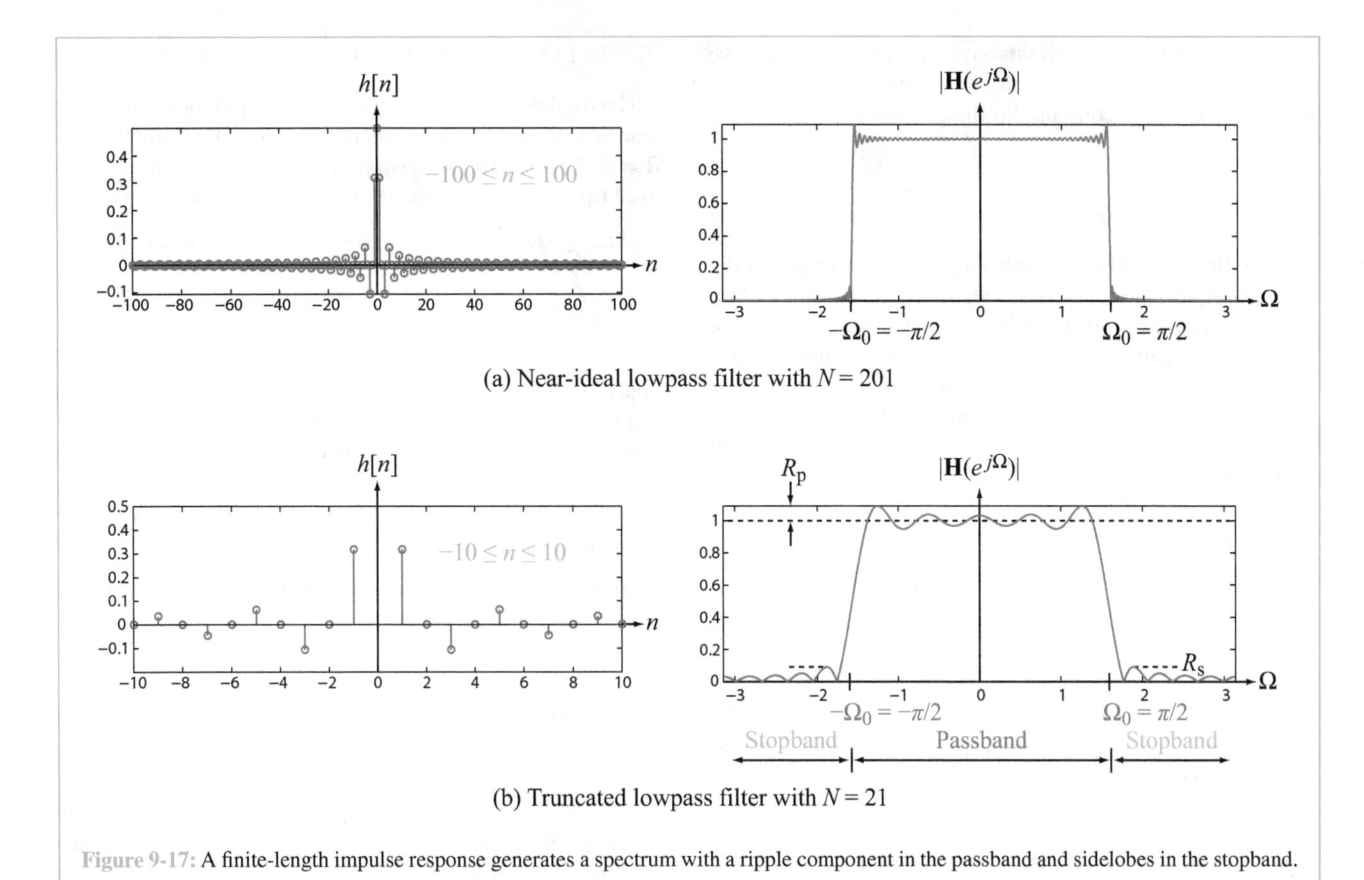

Figure 9-17: A finite-length impulse response generates a spectrum with a ripple component in the passband and sidelobes in the stopband.

9-3.7 FIR Filter Design Using the Minimax Criterion

This subsection presents an approach to FIR filter design that allows the filter designer to prioritize the relative importance of the order of the FIR filter, the size of the ripples in its passband(s) and stopband(s), and the rolloff rates in the transition regions between the passband(s) and stopband(s). It uses an iterative algorithm, called the ***Parks-McClellan algorithm***, to minimize the weighted maximum magnitude of the difference between the ideal desired frequency response $\mathbf{H}_\mathrm{D}(e^{j\Omega})$ and the designed frequency response $\mathbf{H}(e^{j\Omega})$. The ***weighted error*** $E(e^{j\Omega})$ is defined as

$$E(e^{j\Omega}) = W(e^{j\Omega})\,|\mathbf{H}_\mathrm{D}(e^{j\Omega}) - \mathbf{H}(e^{j\Omega})|, \tag{9.60}$$

where $W(e^{j\Omega})$ is the ***weight*** assigned to penalize the error at Ω. The weighting function is selected by the designer to emphasize or de-emphasize the importance of the deviation between the desired spectrum and the spectrum of the filter under design, at specific frequencies. Since $\mathbf{H}(e^{j\Omega})$ is related to the impulse response $h[n]$ by

$$\mathbf{H}(e^{j\Omega}) = \sum_{n=0}^{N-1} h[n]\, e^{-j\Omega n}, \tag{9.61}$$

the selectable parameters of $h[n]$ are its length N and the amplitudes of its components $\{\, h[0],\ h[1], \ldots, h[N-1]\,\}$. The Parks-McClellan algorithm is configured to minimize the maximum value of $E(e^{j\Omega})$ across the range of Ω, which explains the meaning of the term "minimax." It is an iterative algorithm that typically converges in 3–5 iterations for simple filters, such as lowpass and highpass filters. More iterations may be required for more complicated filters.

Notationally, the minimax criterion is expressed as

$$\underset{h[n]}{\mathrm{MIN}}\ \underset{\Omega}{\mathrm{MAX}}\ \{\, E(e^{j\Omega})\,\},$$

where the L amplitudes of $h[n]$ are the selectable variables, $0 \le \Omega \le \pi$, and $E(e^{j\Omega})$ is the error defined by Eq. (9.60).

Filters designed using the minimax criterion tend to have gains that oscillate ("ripple") around the desired gain, but the size of the ripple is constant. So these filters are often called ***equiripple*** filters.

9-3.8 Parks-McClellan Algorithm Implementation

A proof that the Parks-McClellan algorithm converges to the solution of the minimax criterion, as stated by Eq. (9.60), requires many pages of complicated mathematics, including such topics as ***Chebyschev polynomials***, ***Chebyschev approximation theory***, ***Remez exchange theorem***, and ***alternation theorem***. Instead, we present a summary of how to implement the Parks-McClellan algorithm in MATLAB, followed by two examples to illustrate its operation.

N, F, and G are input parameters, defined as follows:

- N is the order of the FIR filter. The filter is of length $N+1$.
- F is a vector of pairs of normalized frequencies: $F = \Omega/\pi$, extending over the range from 0 to 1. F must be even in length and include both 0 and 1.
- G is a vector of gains representing an initial $|\mathbf{H}(e^{j\Omega})|$ at normalized frequencies F.

The MATLAB Signal Processing Toolbox command `firpm` designs a filter with output $h[n]$. [The command is not available in MathScript.] Note that $\mathbf{H}(e^{j\Omega})$ is the frequency response used to *initialize* the iterative process, whereas $h[n]$ is the impulse response corresponding to the *final* frequency response after convergence has been achieved. That is, $h[n]$ *is not* the impulse response corresponding to $\mathbf{H}(e^{j\Omega})$.

The default design is a symmetric $h[n]$, suitable for lowpass, highpass, and bandpass filters. To obtain an antisymmetric $h[n]$, suitable for an ideal Hilbert transform given by

$$\mathbf{H}_{\mathrm{D}}(e^{j\Omega}) = \begin{cases} -j & \text{for } 0 < \Omega < \pi, \\ j & \text{for } -\pi < \Omega < 0, \end{cases} \tag{9.62}$$

it is necessary to use the command

```
h=firpm(N,[0 1],[1 1],'hilbert');
```

The frequency response of the designed filter will then be pure imaginary.

A differentiator given by

$$\mathbf{H}_{\mathrm{D}}(e^{j\Omega}) = j\Omega, \qquad |\Omega| < \pi, \tag{9.63}$$

can be designed by the command

```
h=firpm(N,[0 1],[0 pi],'hilbert');
```

The ripples in $E(e^{j\Omega})$ have the same amplitudes, but ripples near $\Omega = 0$ create a larger *percentage* error than ripples near $\Omega = \pi$. To make the *percentage* errors $E(e^{j\Omega})/|\Omega|$ have same-sized ripples at all frequencies Ω, use instead the command

```
h=firpm(N,[0 1],[0 pi],'differentiator');
```

Example 9-9: Parks-McClellan Lowpass Filter

Apply the Parks-McClellan algorithm to design a half-band (cutoff frequency $\Omega_0 = 0.5\pi$) lowpass filter of order 21 (which means that the duration of its impulse response is $N+1 = 21+1 = 22$). Specify the transition region to be between $\Omega = 0.4\pi$ and $\Omega = 0.6\pi$. Display the gain of the frequency response after 1, 2, and 3 iterations.

Solution: Given the specified information, we begin by defining the initial frequency response as

$$|\mathbf{H}(e^{j\Omega})| = \begin{cases} 1 & \text{for } 0 \le \Omega < 0.4\pi, \\ \text{dc} & \text{for } 0.4\pi < |\Omega| < 0.6\pi, \\ 0 & \text{for } 0.6\pi < |\Omega| \le \pi, \end{cases}$$

where "dc" stands for "don't care." We have specified a unity gain in the passband (0 to 0.4π) and zero gain in the stopband ($\geq 0.6\pi$), but assigned no specifications across the transition region. Hence, N, F, and G are defined as

$$N = 21, \qquad F = [0.0 \quad 0.4 \quad 0.6 \quad 1.0],$$
$$G = [1.0 \quad 1.0 \quad 0.0 \quad 0.0].$$

Application of the MATLAB command

```
h=firpm(N,F,G)
```

generates the three plots shown in Fig. 9-18. The algorithm was stopped after 1, 2, and 3 iterations, and then the DTFT was computed for each case.

- The gain plot for $|\mathbf{H}(e^{j\Omega})|$ realized after 1 iteration is shown in red. It exhibits approximately flat responses in the passband and stopband, and an approximately linear profile in the transition region (with $|\mathbf{H}(e^{j\Omega})| = 0.5$ at the cutoff frequency $\Omega_0 = 0.5\pi$).
- The output of the algorithm after 2 iterations is displayed in green. The frequency response includes ripples in the passband and stopband, but the rolloff rate (slope) in the transition region is much greater.

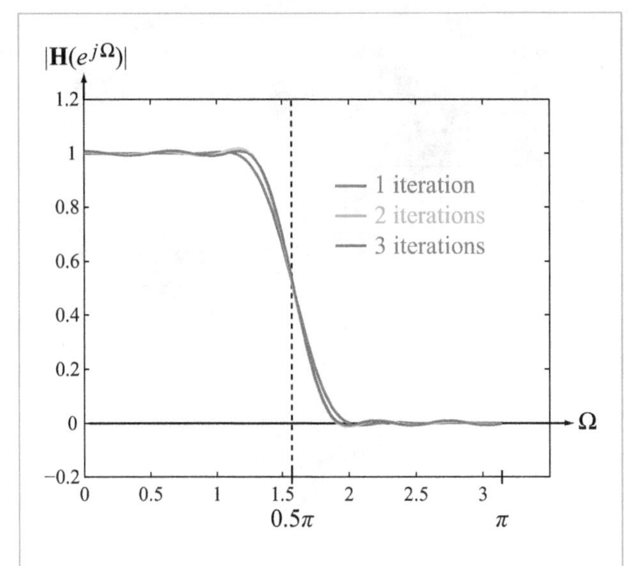

Figure 9-18: Results of Parks-McClellan algorithm after: 1 iteration (red); 2 iterations (green); 3 iterations (blue).

- The blue plot corresponds to three iterations. It is similar to the two-iteration response, but the slope is slightly steeper in the transition region, and the ripples are now smaller and have equal amplitudes across both the passband and the stopband. The frequency response remains essentially unchanged with the application of additional iterations.

If the objective of the filter designer is to avoid ripples, then $h[n]$ corresponding to the red frequency response would be the preferred option, but if small ripples can be tolerated so as to achieve a steep response in the transition region, $h[n]$ corresponding to the blue response would be the better choice. In the present example, the transition region extended from 0.4π to 0.6π. Making it narrower or wider can impact the size of the ripples. Also, increasing the filter order would decrease the ripple amplitude, but increase their number.

Example 9-10: Equiripple Bandpass Filter

Implement the Parks-McClellan algorithm for a bandpass filter defined by the initial frequency response

$$|\mathbf{H}(e^{j\Omega})| = \begin{cases} 0 & \text{for } 0 \le |\Omega| < 0.2\pi, \\ \text{dc} & \text{for } 0.2\pi < |\Omega| < 0.3\pi, \\ 1 & \text{for } 0.3\pi < |\Omega| < 0.7\pi, \\ \text{dc} & \text{for } 0.7\pi < |\Omega| < 0.8\pi, \\ 0 & \text{for } 0.8\pi < |\Omega| \le \pi. \end{cases}$$

Compare filters of order: (a) $N = 40$ and (b) $N = 60$. Allow the algorithm to converge.

Solution:

(a) $N = 40$,

$$F = [0.0 \quad 0.2 \quad 0.3 \quad 0.7 \quad 0.8 \quad 1.0],$$
$$G = [0.0 \quad 0.0 \quad 1.0 \quad 1.0 \quad 0.0 \quad 0.0],$$

and

```
h=firpm(N,F,G),
```

which leads to the two-sided plots shown in part (a) of Fig. 9-19.

The frequency response includes noticeable-size ripples in the passbands and stopbands, but the rolloffs are fairly sharp.

(b) Upon changing N to 60, implementation of the MATLAB code with the same values for F and G leads to the plot for $|\mathbf{H}(e^{j\Omega})|$ shown in Fig. 9-19(b), which exhibits smaller-size ripples and sharper rolloffs.

Concept Question 9-4: Why are FIR filter impulse responses almost always even or odd? (See S²)

Concept Question 9-5: What is the difference between an FIR filter and an MA system? (See S²)

Concept Question 9-6: What does the frequency sampling design procedure do? (See S²)

Concept Question 9-7: What does the Parks-McClellan design procedure do? (See S²)

Exercise 9-5: Design a differentiator of length 3 using a rectangular data window. Interpret your answer.

Answer: $h[n] = \{1, 0, -1\}$ becomes

$$y[n] = x[n+1] - x[n-1],$$

which is a difference operator.

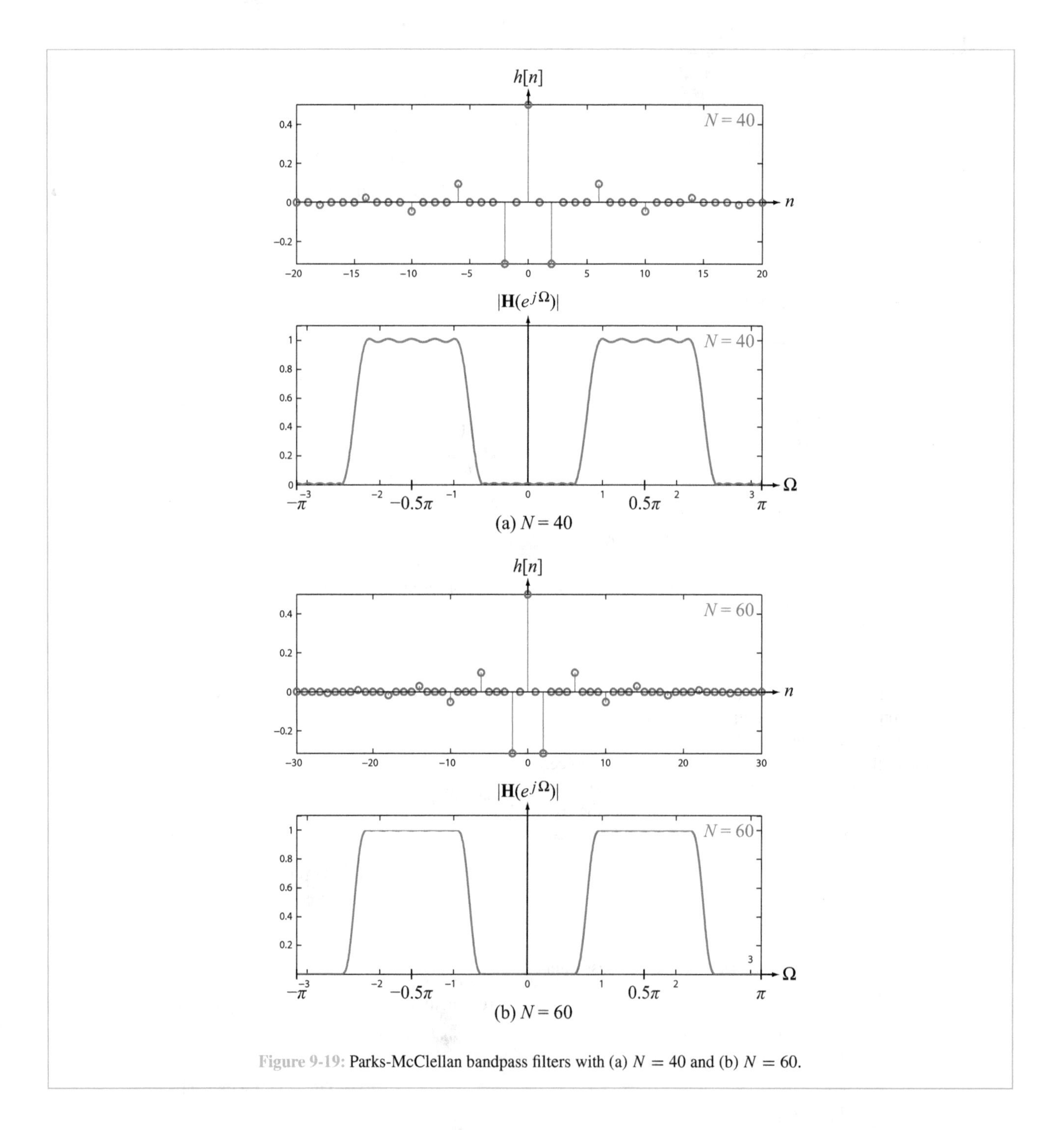

Figure 9-19: Parks-McClellan bandpass filters with (a) $N = 40$ and (b) $N = 60$.

9-4 Infinite Impulse Response (IIR) Filter Design

9-4.1 Overview of IIR Filter Design

An IIR filter has the form of an ARMA difference equation (see Section 7-3):

$$\begin{aligned} y[n] &+ a_1\, y[n-1] + \cdots + a_N\, y[n-N] \\ &= b_0\, x[n] + b_1\, x[n-1] + \cdots + b_M\, x[n-M], \end{aligned} \tag{9.64}$$

for a set of constant coefficients $\{a_n,\ 1 \leq n \leq N\}$ and $\{b_n,\ 0 \leq n \leq M\}$, and constant orders (N, M). An IIR filter can also be specified by its transfer function

$$\mathbf{H}(\mathbf{z}) = \frac{\mathbf{Y}(\mathbf{z})}{\mathbf{X}(\mathbf{z})} = \frac{b_0 + b_1\mathbf{z}^{-1} + \cdots + b_M\mathbf{z}^{-M}}{1 + a_1\mathbf{z}^{-1} + \cdots + a_M\mathbf{z}^{-N}}\,, \tag{9.65}$$

or its impulse response

$$h[n] = \sum_{i=1}^{N} \mathbf{C}_i \mathbf{p}_i^n\; u[n], \tag{9.66}$$

where $\{\mathbf{p}_i\}$ are the poles of the transfer function $\mathbf{H}(\mathbf{z})$ and $\{\mathbf{C}_i\}$ are constants determined by partial fraction expansion (Section 7-8). Since the discrete-time index n is unbounded, $h[n]$ is infinite in duration, and therefore it is called an ***infinite impulse response*** (IIR) filter.

9-4.2 Notch Filters: FIR versus IIR

An IIR filter can be much more selective in frequency than an FIR filter with the same number of coefficients. To illustrate with an example, let us design FIR and IIR notch filters that reject a 125 Hz sinusoid, and let us use a sampling rate $f_s = 1000$ samples/s. The discrete-time frequency to be rejected is

$$\Omega_0 = 2\pi\,\frac{f_0}{f_s} = 2\pi\,\frac{125}{1000} = \frac{\pi}{4}\,. \tag{9.67}$$

(a) FIR notch filter

To reject $\Omega_0 = \pi/4$, the notch filter's transfer function $\mathbf{H}(\mathbf{z})$ must have two conjugate zeros at $e^{\pm j\pi/4}$. Also, causality requires $\mathbf{H}(\mathbf{z})$ to have at least as many poles. By placing two poles at the origin in the complex plane, we generate a causal impulse response of finite duration. Thus,

$$\begin{aligned} \mathbf{H}(\mathbf{z}) &= \frac{(\mathbf{z} - e^{j\pi/4})(\mathbf{z} - e^{-j\pi/4})}{\mathbf{z}^2} \\ &= 1 - \frac{2}{\mathbf{z}}\cos\left(\frac{\pi}{4}\right) + \frac{1}{\mathbf{z}^2} = 1 - \frac{1.414}{\mathbf{z}} + \frac{1}{\mathbf{z}^2}\,. \end{aligned} \tag{9.68}$$

The corresponding impulse response is

$$h[n] = \{\underline{1}, -1.414, 1\}, \tag{9.69}$$

and by application of the convolution property (#4 in Table 7-4), the output $y[n]$ is given by

$$\begin{aligned} y[n] = x[n] * h[n] &= \sum_{i=0}^{2} x[n-i]\; h[i] \\ &= x[n]\; h[0] + x[n-1]\; h[1] + x[n-2]\; h[2] \\ &= x[n] - 1.414x[n-1] + x[n-2]. \end{aligned} \tag{9.70}$$

The magnitude of the frequency response $\mathbf{H}(e^{j\Omega})$, obtained by setting $\mathbf{z} = e^{j\Omega}$ in Eq. (9.68), is shown graphically in Fig. 9-20(a). The response does have a null at $\Omega_0 = \pi/4$, but it is not very selective, as it also eliminates nearby frequencies.

(b) IIR notch filter

The transfer function of the IIR notch filter must also have zeros at $e^{\pm j\pi/4}$, but its two poles can be placed very close to these zeros, just inside the unit circle so as to insure BIBO stability (Section 8-2). We will select poles at $0.99e^{\pm j\pi/4}$. Consequently, its transfer function is given by

$$\begin{aligned} \mathbf{H}(\mathbf{z}) &= \frac{(\mathbf{z} - e^{j\pi/4})(\mathbf{z} - e^{-j\pi/4})}{(\mathbf{z} - 0.99e^{j\pi/4})(\mathbf{z} - 0.99e^{-j\pi/4})} \\ &= \frac{\mathbf{z}^2 - 2\cos(\pi/4)\;\mathbf{z} + 1}{\mathbf{z}^2 - 2(0.99)\cos(\pi/4)\;\mathbf{z} + 0.99^2} \\ &= \frac{\mathbf{z}^2 - 1.414\mathbf{z} + 1}{\mathbf{z}^2 - 1.40\mathbf{z} + 0.98}\,. \end{aligned} \tag{9.71}$$

The corresponding ARMA difference equation is given by

$$\begin{aligned} y[n] &- 1.40y[n-1] + 0.98y[n-2] \\ &= x[n] - 1.414x[n-1] + x[n-2]. \end{aligned} \tag{9.72}$$

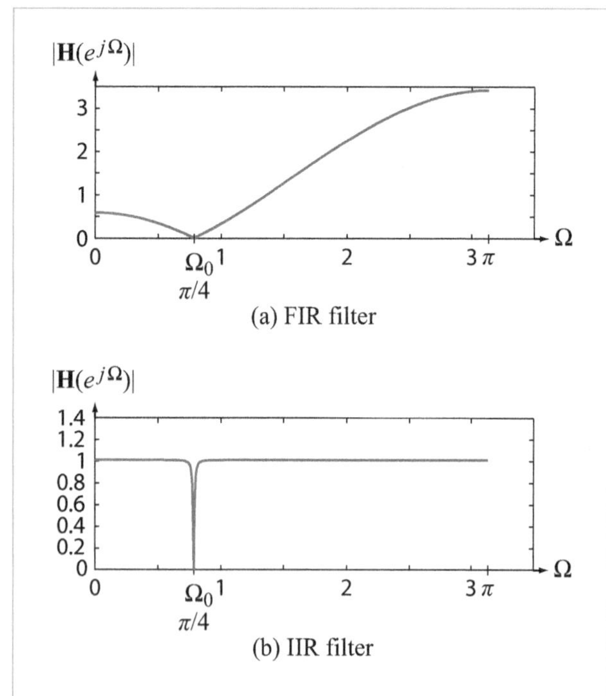

Figure 9-20: (a) FIR and (b) IIR notch filters, both designed to eliminate a sinusoid with $\Omega_0 = \pi/4$.

Upon replacing $\mathbf{z}$ with $e^{j\Omega}$ and then computing $|\mathbf{H}(e^{j\Omega})|$ as a function of Ω over the range 0 to π, we obtain the plot displayed in Fig. 9-20(b). We observe that the IIR notch filter is highly selective; it only eliminates $\Omega_0 = \pi/4$, but leaves nearby frequencies unaffected.

The FIR notch filter requires one multiplication and two additions per recursion, and two memory registers to store the two most recent input values $x[n-1]$ and $x[n-2]$. The IIR notch filter requires three multiplications and four additions per recursion, and four memory registers to store the two most recent input values $x[n-1]$ and $x[n-2]$ and the two most recent output values $y[n-1]$ and $y[n-2]$. However, the IIR notch filter is far more selective.

The impulse response $h[n]$ of the IIR filter is the inverse $\mathbf{z}$-transform of $\mathbf{H}(\mathbf{z})$. Using the methods of Section 7-8, $h[n]$ is found to be

$$h[n] = \delta[n] - 0.0202(0.99)^{n-1} \cdot \cos\left(\frac{\pi}{4}(n-1) - 2.3512\right) u[n-1], \quad (9.73)$$

which has infinite duration but is absolutely summable, so the IIR notch filter is BIBO stable. We should note, however, that if the poles are very close to the unit circle, roundoff error due to finite precision in a DSP chip can make a mathematically stable filter unstable. Also, the transient response is very long, because $(0.99)^n$ decays very slowly as a function of n.

The philosophy behind IIR filter design is to take a continuous-time filter, such as the Butterworth filter of Section 6-8, and transform it into a discrete-time filter. The process has led to two major approaches to IIR filter design:

- *Impulse invariance*: the continuous-time impulse response of a particular filter of interest, denoted here as $h_a(t)$, is sampled to a discrete-time impulse response $h[n]$.
- *Bilinear transform*: the continuous-time transfer function $\mathbf{H}_a(\mathbf{s})$ of a particular filter of interest is mapped to a discrete-time transfer function $\mathbf{H}(\mathbf{z})$.

Throughout this section, the subscript "a" designates a continuous-time (analog) LTI system.

9-4.3 IIR Filter Design Using Impulse Invariance

The idea behind designing an IIR filter using impulse invariance is to sample the impulse response $h_a(t)$ of a suitable continuous-time filter using a sampling interval T_s. This gives

$$h[n] = T_s\, h_a(nT_s). \quad (9.74)$$

Multiplication by T_s is required dimensionally. To see why this is necessary, consider the RC circuit first presented in Fig. 2-5 and which reappears throughout Chapters 2 and 3. The input and output of this circuit are both voltages, so its transfer function should be dimensionless. Indeed, from Eq. (4.43), its transfer function is

$$\mathbf{H}_a(\mathbf{s}) = \frac{1/RC}{\mathbf{s} + 1/RC},$$

which is dimensionless. But the impulse response of this RC circuit is, from Eq. (2.17),

$$h_a(t) = \frac{1}{RC}\, e^{-t/RC}\, u(t),$$

which has units of 1/time. In discrete time, however, $\mathbf{H}(\mathbf{z})$, $h[n]$, $\mathbf{z}$, and n are all dimensionless. So to make the units match, it is necessary to scale $h_a(t)$ as in Eq. (9.74).

The recipe for IIR filter design using impulse invariance is as follows:

(1) Select a suitable continuous-time filter, such as a Butterworth filter, with transfer function $\mathbf{H}_a(\mathbf{s})$.

(2) Compute its continuous-time impulse response $h_a(t) = \mathcal{L}^{-1}\{\mathbf{H}_a(\mathbf{s})\}$ using the inverse Laplace transform, if $h_a(t)$ is not already known.

(3) Select the sampling interval T_s so that the sampling frequency $1/T_s$ exceeds the Nyquist frequency (double the maximum frequency in the spectrum) of $h_a(t)$.

(4) Sample $h_a(t)$ to generate $h[n] = T_s\, h_a(nT_s)$.

► Impulse invariance generates stable filters. ◄

FIR filters are always BIBO stable, regardless of the method used to generate them. This is because the impulse response of an FIR filter is necessarily absolutely summable (see Section 7-4.5). However, an IIR filter is only stable if all of its poles lie inside the unit circle (see Section 7-11). So an issue that must be addressed in any IIR filter design procedure is whether the transformation from continuous time to discrete time preserves stability; does stability of the continuous-time filter guarantee stability of the discrete-time filter designed from it?

Recall from Section 3-7 that a continuous-time system is BIBO stable if and only if all of its poles lie in the OLHP. The OLHP is the open left half of the complex $\mathbf{s}$ plane, i.e., $\mathfrak{Re}\{\mathbf{s}\} < 0$. We now show that the transfer function $\mathbf{H}(\mathbf{z})$ designed using impulse invariance of $\mathbf{H}_a(\mathbf{s})$ has all of its poles inside the unit circle if $\mathbf{H}_a(\mathbf{s})$ has all of its poles in the OLHP. That is, impulse invariance preserves stability.

Let the poles of $\mathbf{H}_a(\mathbf{s})$ be $\{\mathbf{p}_i\}$, and let $\{\mathbf{A}_i\}$ be the residues of the partial fraction expansion of $\mathbf{H}_a(\mathbf{s})$. Then the impulse response $h_a(t) = \mathcal{L}^{-1}\{\mathbf{H}_a(\mathbf{s})\}$ is

$$h_a(t) = \sum_{i=1}^{N} \mathbf{A}_i e^{\mathbf{p}_i t}\, u(t). \tag{9.75}$$

Applying the continuous-to-discrete-time mapping defined by Eq. (9.74) leads to

$$h[n] = T_s\, h_a(nT_s) = \sum_{i=1}^{N} (\mathbf{A}_i T_s)(e^{\mathbf{p}_i T_s})^n\, u[n]. \tag{9.76}$$

Note that we used the property

$$u(nT_s) = \begin{cases} 1 & \text{for } n > 0, \\ 0 & \text{for } n < 0, \end{cases} \tag{9.77}$$

which is the same as $u[n]$ if we define $u(0T_s) = 1$.

The poles of the transfer function $\mathbf{H}(\mathbf{z})$ designed using impulse invariance are $\{e^{\mathbf{p}_i T_s}\}$. Poles $\{\mathbf{p}_i = -a_i + jb_i\}$ are in the OLHP if and only if $a_i > 0$, in which case the discrete-time poles $\{e^{\mathbf{p}_i T_s}\}$ have magnitudes

$$|e^{\mathbf{p}_i T_s}| = |e^{-a_i T_s}| \cdot |e^{jb_i T_s}| = e^{-a_i T_s} < 1. \tag{9.78}$$

The discrete-time poles lie inside the unit circle, and therefore the discrete-time system is stable.

Example 9-11: Impulse Invariance IIR Filter

The continuous-time filter

$$\mathbf{H}_a(\mathbf{s}) = \frac{\mathbf{s} + 0.1}{(\mathbf{s} + 0.1)^2 + 16} \tag{9.79}$$

has a sharp resonant peak at $\omega_0 = 4$ rad/s in its frequency response $\mathbf{H}_a(j\omega)$. This is because its poles $\{-0.1 \pm j4\}$ are close to the imaginary axis (see Section 6-5). Use impulse invariance to design an IIR filter that also has a sharp resonant peak in its frequency response. Keep T_s unspecified.

Solution: Using entry #14 in Table 3-2, we obtain the continuous-time impulse response:

$$h_a(t) = \mathcal{L}^{-1}\{\mathbf{H}_a(\mathbf{s})\} = e^{-0.1t}\cos(4t)\, u(t). \tag{9.80}$$

Per Eq. (9.74), the discrete-time impulse response designed using impulse invariance is given by

$$h[n] = T_s\, h_a(nT_s) = T_s\, e^{(-0.1T_s)n}\cos((4T_s)n)\, u[n]. \tag{9.81}$$

By comparison with entry #6 in Table 7-5, the corresponding discrete-time transfer function is

$$\mathbf{H}(\mathbf{z}) = \mathbf{Z}\{h[n]\} = T_s\, \frac{\mathbf{z}^2 - \mathbf{z}e^{-0.1T_s}\cos(4T_s)}{\mathbf{z}^2 - \mathbf{z}2e^{-0.1T_s}\cos(4T_s) + e^{-0.2T_s}}. \tag{9.82}$$

The poles of $\mathbf{H}(\mathbf{z})$ are the roots of the denominator when set to zero. By construction, these are $\{e^{\mathbf{p}_i T_s}\} = \{e^{-0.1T_s}e^{\pm j4T_s}\}$. For small T_s, these poles are very close to the unit circle, and they produce a sharp resonant peak in the frequency response $\mathbf{H}(e^{j\Omega})$ at $\Omega_0 = 4T_s$.

A plot of the frequency response $\mathbf{H}(e^{j\Omega})$, obtained by replacing $\mathbf{z}$ with $e^{j\Omega}$ in Eq. (9.82), is displayed in Fig. 9-21.

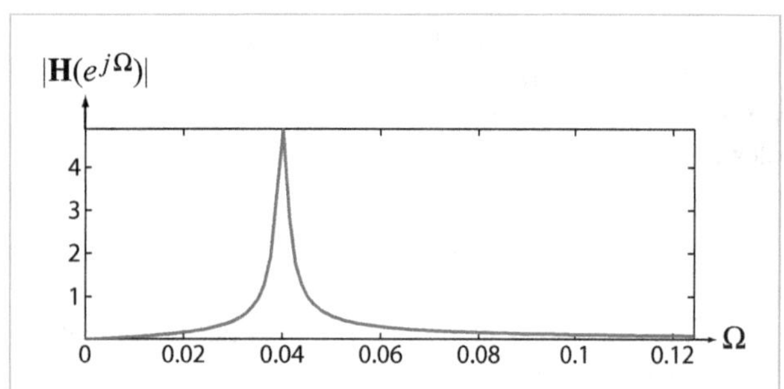

Figure 9-21: One-sided frequency response for the IIR design in Example 9-11, with $T_s = 0.01$. As expected, there is a resonant peak at $\Omega_0 = 4T_s = 0.04$.

9-4.4 IIR Filter Design Using Bilinear Transformation

The idea behind designing an IIR filter using bilinear transformation is to map the given continuous-time transfer function $\mathbf{H}_a(\mathbf{s})$ to the discrete-time transfer function $\mathbf{H}(\mathbf{z})$ using the ***bilinear transformation***

$$\mathbf{s} = \frac{2}{T}\left(\frac{\mathbf{z}-1}{\mathbf{z}+1}\right), \tag{9.83}$$

where T is a positive ***transformation factor*** ($T > 0$). The bilinear transformation gets its name from the functional form of Eq. (9.83), being the ratio of two linear functions. Here, T is a constant to be specified, and should not be confused with the sampling interval T_s used in impulse invariance.

The resulting discrete-time transfer function $\mathbf{H}(\mathbf{z})$ is then

$$\mathbf{H}(\mathbf{z}) = \mathbf{H}_a(\mathbf{s})\Big|_{\mathbf{s}=\frac{2}{T}\left(\frac{\mathbf{z}-1}{\mathbf{z}+1}\right)}. \tag{9.84}$$

The recipe for IIR filter design using the bilinear transformation is as follows:

(1) Select a suitable continuous-time filter, such as a Butterworth filter, with transfer function $\mathbf{H}_a(\mathbf{s})$.

(2) Denote the degree of the denominator polynomial of $\mathbf{H}_a(\mathbf{s})$ as N.

(3) Choose an appropriate value for T, which usually depends on the filter specifications.

(4) Replace $\mathbf{s}$ with $\mathbf{s} = \frac{2}{T}\left(\frac{\mathbf{z}-1}{\mathbf{z}+1}\right)$ everywhere in the expression for $\mathbf{H}_a(\mathbf{s})$. Call the result $\mathbf{H}(\mathbf{z})$.

(5) Simplify $\mathbf{H}(\mathbf{z})$ by multiplying it by $(\mathbf{z}+1)^N/(\mathbf{z}+1)^N$. This converts $\mathbf{H}(\mathbf{z})$ into a ratio of two polynomials.

A simple example of IIR filter design by bilinear transformation follows.

Example 9-12: Bilinear Transformation IIR Filter

Use the bilinear transformation to design a discrete-time lowpass filter from a second-order continuous-time Butterworth lowpass filter with cutoff frequency $\omega_0 = 1$ rad/s. Use $T = 2$ s to make the algebra easier to follow.

Solution: From Table 6-3, the second-order continuous-time lowpass Butterworth filter with $\omega_0 = 1$ rad/s has transfer function

$$\mathbf{H}_a(\mathbf{s}) = \frac{1}{\mathbf{s}^2 + \sqrt{2}\mathbf{s} + 1}. \tag{9.85}$$

The bilinear transformation with $T = 2$ is

$$\mathbf{s} = \frac{\mathbf{z}-1}{\mathbf{z}+1}. \tag{9.86}$$

Substituting Eq. (9.86) into Eq. (9.85) gives the discrete-time transfer function

$$\mathbf{H}(\mathbf{z}) = \frac{1}{\left(\frac{\mathbf{z}-1}{\mathbf{z}+1}\right)^2 + \sqrt{2}\left(\frac{\mathbf{z}-1}{\mathbf{z}+1}\right) + 1}. \tag{9.87}$$

Multiplying Eq. (9.87) by $\left(\frac{\mathbf{z}+1}{\mathbf{z}+1}\right)^2$ leads to

$$\mathbf{H}(\mathbf{z}) = \frac{(\mathbf{z}+1)^2}{(\mathbf{z}-1)^2 + \sqrt{2}(\mathbf{z}-1)(\mathbf{z}+1) + (\mathbf{z}+1)^2} = \frac{\mathbf{z}^2 + 2\mathbf{z} + 1}{(2+\sqrt{2})\mathbf{z}^2 + (2-\sqrt{2})}. \tag{9.88}$$

Transfer function $\mathbf{H}(\mathbf{z})$ has poles at

$$\left\{\pm j\,\frac{\sqrt{2-\sqrt{2}}}{\sqrt{2+\sqrt{2}}}\right\},$$

both of which lie inside the unit circle, so the discrete-time filter is BIBO stable. $\mathbf{H}(\mathbf{z})$ has a double zero at $\{-1\}$, so it rejects $\Omega = \pi$ (see Section 8-1) and is therefore a (crude) lowpass filter. It may be implemented using the ARMA difference equation

$$y[n] + \frac{2-\sqrt{2}}{2+\sqrt{2}}\,y[n-2] = \frac{1}{2+\sqrt{2}}\,(x[n] + 2x[n-1] + x[n-2]), \tag{9.89}$$

which requires only 4 additions and two multiplications per recursion, since multiplying $y[n-1]$ by 2 is equivalent to adding it to itself.

9-4.5 Bilinear Transformation and BIBO Stability

► The bilinear transformation generates stable filters. ◄

Now, we show that the bilinear transformation maps the OLHP in the $\mathbf{s}$ plane for the continuous-time filter to the interior of the unit circle in the $\mathbf{z}$ plane for the discrete-time filter, thereby preserving stability. Next we show that the reverse process also is true, confirming that the bilinear transformation is an invertible transformation. Finally, we show that the bilinear transformation maps the imaginary axis $\mathbf{s} = j\omega$ to the unit circle $\mathbf{z} = e^{j\Omega}$ in a one-to-one mapping. This allow a discrete-time filter designer to map a specific continuous-time frequency ω to a specific discrete-time frequency Ω.

(a) OLHP to interior of unit circle

The poles $\{\mathbf{p}_i = -a_i + jb_i\}$ of $\mathbf{H}_a(\mathbf{s})$ are all in the OLHP if $a_i > 0$. We now show that the bilinear transformation maps the entire left OLHP, $\mathfrak{Re}[\mathbf{s}] < 0$, to the interior $|\mathbf{z}| < 1$ of the unit circle.

Solving the bilinear transformation given by Eq. (9.83) for $\mathbf{z}$ in terms of $\mathbf{s}$ gives

$$\mathbf{s} = \frac{2}{T}\left(\frac{\mathbf{z}-1}{\mathbf{z}+1}\right) \quad \rightarrow \quad \mathbf{z} = \frac{1+\mathbf{s}T/2}{1-\mathbf{s}T/2}. \tag{9.90}$$

For a pole at $\mathbf{s} = -a + jb$,

$$\mathbf{z} = \frac{(1-aT/2)+j(bT/2)}{(1+aT/2)-j(bT/2)}. \tag{9.91}$$

The squared magnitude $|\mathbf{z}|^2$ of $\mathbf{z}$ is

$$|\mathbf{z}|^2 = \frac{(1-aT/2)^2+(bT/2)^2}{(1+aT/2)^2+(bT/2)^2}. \tag{9.92}$$

Since all poles $\{\mathbf{p}_i = -a_i + jb_i\}$ are located in the OLHP, it follows that $a_i > 0$ for all i. Hence, for the pole under consideration, $-aT < aT$ since $a > 0$ and $T > 0$. Adding $1 + (aT/2)^2$ to both sides of this inequality leads to

$$\left(1-\frac{aT}{2}\right)^2 < \left(1+\frac{aT}{2}\right)^2. \tag{9.93}$$

Consequently, the numerator of Eq. (9.92) is smaller than the denominator, which implies that $|\mathbf{z}|^2 < 1$. Hence, the OLHP of $\mathbf{s}$ is mapped to the interior $|\mathbf{z}| < 1$ of the unit circle, thereby demonstrating that the bilinear transformation, like impulse invariance, preserves stability.

(b) Interior of unit circle to OLHP

Conversely, we now show that the bilinear transformation also maps the interior $|\mathbf{z}| < 1$ of the unit circle to the OLHP of $\mathbf{s}$.

Recall that for any complex number $\mathbf{z}$ we have

$$\mathbf{z}\mathbf{z}^* = |\mathbf{z}|^2, \qquad \mathbf{z} - \mathbf{z}^* = j2 \cdot \mathfrak{Im}[\mathbf{z}], \tag{9.94}$$

Upon multiplying both the numerator and denominator on the right-hand side of Eq. (9.83) by $(\mathbf{z}^* + 1)$, we have

$$\mathbf{s} = \frac{2}{T}\frac{\mathbf{z}-1}{\mathbf{z}+1}\left[\frac{\mathbf{z}^*+1}{\mathbf{z}^*+1}\right] = \frac{2}{T}\frac{\mathbf{z}\mathbf{z}^*+\mathbf{z}-\mathbf{z}^*-1}{(\mathbf{z}+1)(\mathbf{z}^*+1)} = \frac{2}{T}\frac{|\mathbf{z}|^2-1}{|\mathbf{z}+1|^2} + j\,\frac{4}{T}\frac{\mathfrak{Im}[\mathbf{z}]}{|\mathbf{z}+1|^2}. \tag{9.95}$$

The real part of this expression is

$$\mathfrak{Re}(\mathbf{s}) = \frac{2}{T}\frac{|\mathbf{z}|^2-1}{|\mathbf{z}+1|^2} < 0, \tag{9.96}$$

if and only if $|\mathbf{z}| < 1$. So, the bilinear transformation also maps the inside of the unit circle $|\mathbf{z}| = 1$ to the OLHP of $\mathbf{s}$.

9-4.6 Frequency Warping

The significance of the bilinear transformation as a filter design technique is that it maps the imaginary axis defined by $\mathfrak{Re}[\mathbf{s}] = 0$ to the unit circle $|\mathbf{z}| = 1$. This follows from the previous subsection by setting $a = 0$. Equally significant is the fact that the frequency response $\mathbf{H}(e^{j\Omega})$ of the discrete-time filter has the same shape as the frequency response $\mathbf{H}_a(j\omega)$ of the continuous-time filter. The difference is that the frequency axis has been warped, to compress the interval $\{0 \le \omega \le \infty\}$ to $\{0 \le \Omega \le \pi\}$. We now prove this assertion.

Upon setting $\mathbf{z} = e^{j\Omega}$ in Eq. (9.83), the bilinear transformation becomes

$$\mathbf{s} = \frac{2}{T}\frac{e^{j\Omega}-1}{e^{j\Omega}+1} = \frac{2}{T}\frac{e^{j\Omega/2}-e^{-j\Omega/2}}{e^{j\Omega/2}+e^{-j\Omega/2}}\left[\frac{e^{j\Omega/2}}{e^{j\Omega/2}}\right] = \frac{2}{T}\frac{2j\sin(\Omega/2)}{2\cos(\Omega/2)} = j\,\frac{2}{T}\tan\left(\frac{\Omega}{2}\right). \tag{9.97}$$

By definition, from Eq. (3.2), $\mathbf{s} = \sigma + j\omega$. Since the expression for $\mathbf{s}$ given by Eq. (9.97) is purely imaginary, it follows that

$$\omega = \left(\frac{2}{T}\right)\tan\left(\frac{\Omega}{2}\right), \tag{9.98}$$

which is known as the *prewarping formula*. Continuous-time frequency ω maps to discrete-time frequency Ω.

The tangent function maps the interval $\{0 \leq \Omega \leq \pi\}$ to the interval $\{0 \leq \omega \leq \infty\}$.

The discrete-time frequency response $\mathbf{H}(e^{j\Omega})$ is the continuous-time frequency response $\mathbf{H}_a(j\omega)$ *nonlinearly compressed in frequency*:

$$\mathbf{H}(e^{j\Omega}) = \mathbf{H}_a(\mathbf{s})\Big|_{\mathbf{s}=j\frac{2}{T}\tan\left(\frac{\Omega}{2}\right)} \tag{9.99}$$

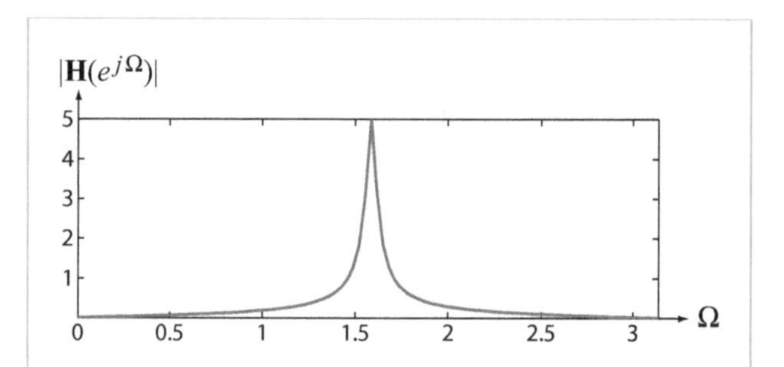

Figure 9-22: IIR resonant filter designed using the bilateral transformation with a peak at $\Omega_0 = \frac{\pi}{2}$.

Example 9-13: IIR Filter Design Using Prewarping

Repeat Example 9-11 using the bilinear transformation instead of impulse invariance. Design the filter such that the peak of the continuous-time filter at $\omega_0 = 4$ rad/s maps to $\Omega_0 = \pi/2$ for the discrete-time filter.

Solution: First, we choose T to map $\omega = 4$ rad/s to $\Omega = \pi/2$ using the prewarping formula given by Eq. (9.98):

$$\omega = \left(\frac{2}{T}\right)\tan\left(\frac{\Omega}{2}\right)$$

$$\rightarrow \quad 4 = \left(\frac{2}{T}\right)\tan\frac{\pi/2}{2} = \frac{2}{T},$$

which has the solution $T = \frac{1}{2}$.

Second, we use this value of T in the bilinear transformation formula given by Eq. (9.83):

$$\mathbf{s} = \frac{2}{1/2}\frac{\mathbf{z}-1}{\mathbf{z}+1} = 4\,\frac{\mathbf{z}-1}{\mathbf{z}+1}. \tag{9.100}$$

Third, we apply the bilinear transformation given by Eq. (9.84) to the expression for $\mathbf{H}_a(\mathbf{s})$ given by Eq. (9.79):

$$\mathbf{H}(\mathbf{z}) = \mathbf{H}_a(\mathbf{s})\big|_{\mathbf{s}=4\left(\frac{\mathbf{z}-1}{\mathbf{z}+1}\right)} = \frac{4\,\frac{\mathbf{z}-1}{\mathbf{z}+1} + 0.1}{\left(4\,\frac{\mathbf{z}-1}{\mathbf{z}+1} + 0.1\right)^2 + 16}. \tag{9.101}$$

Simplification of Eq. (9.101) leads to

$$\mathbf{H}(\mathbf{z}) = \frac{\frac{1}{8}\mathbf{z}^2 + 0.0061\mathbf{z} - 0.119}{\mathbf{z}^2 + 0.0006\mathbf{z} + 0.9512}. \tag{9.102}$$

By setting the numerator of Eq. (9.102) equal to zero and solving for the roots, we obtain the zeros of $\mathbf{H}(\mathbf{z})$. A similar procedure applied to the denominator yields the value of the poles. The process yields the following poles and zeros of $\mathbf{H}(\mathbf{z})$:

- **Poles at** $\{-0.0003 \pm j0.9753\} \approx \{0.9753e^{\pm j\pi/2}\}$
- **Zeros at** $\{-1, 0.951\}$.

The zero at -1 confirms that the frequency response $\mathbf{H}(e^{j\Omega})$ is zero at $\Omega = \pi$ (see Section 8-1). The zero at 0.951 confirms that the frequency response is close to zero at $\Omega = 0$. The poles at $0.9753e^{\pm j\pi/2}$ confirm that there will be a large peak at $\Omega = \frac{\pi}{2}$, as desired.

To perform the bilinear transformation in MATLAB, we first expand the denominator in Eq. (9.79) to obtain:

$$\mathbf{H}_a(\mathbf{s}) = \frac{\mathbf{s}+0.1}{(\mathbf{s}+0.1)^2+16} = \frac{\mathbf{s}+0.1}{\mathbf{s}+0.2\mathbf{s}+16.01}. \tag{9.103}$$

Next, we use the coefficients of the polynomials in the numerator and denominator in the command

```
[B A]=bilinear([1 0.1], [1 0.2 16.01], 2),
```

with the last entry representing the factor $1/T = 1/(1/2) = 2$. The output is

```
B=[0.125 0.0061 -0.119]
A=[1.000 0.0006 0.9512],
```

which are the coefficients of the numerator and denominator in the expression for $\mathbf{H}(\mathbf{z})$ in Eq. (9.102).

The frequency response $\mathbf{H}(e^{j\Omega})$ is obtained from Eq. (9.102) by replacing $\mathbf{z}$ with $e^{j\Omega}$:

$$\mathbf{H}(e^{j\Omega}) = \frac{\frac{1}{8}e^{j2\Omega} + 0.0061e^{j\Omega} - 0.119}{e^{j2\Omega} + 0.0006e^{j\Omega} + 0.9512}. \tag{9.104}$$

A plot of the magnitude of $\mathbf{H}(e^{j\Omega})$ is displayed in Fig. 9-22. We note that the filter's frequency response exhibits a large peak at $\Omega_0 = \pi/2$, as desired.

Example 9-14: Lowpass Filter Design

Use the bilinear transformation to design a discrete-time lowpass filter with cutoff frequency $\Omega_0 = \frac{\pi}{3}$ from a third-order continuous-time Butterworth lowpass filter with cutoff frequency $\omega_0 = 1$ rad/s.

Solution: From Table 6-3 the third-order continuous-time lowpass Butterworth filter with $\omega_0 = 1$ rad/s has transfer function

$$\mathbf{H}_a(\mathbf{s}) = \frac{1}{\mathbf{s}^3 + 2\mathbf{s}^2 + 2\mathbf{s} + 1}. \quad (9.105)$$

We must choose T in the bilinear transformation so that the continuous-time cutoff frequency $\omega_0 = 1$ rad/s is mapped to the discrete-time cutoff frequency $\Omega_0 = \frac{\pi}{3}$. This is performed using the prewarping formula given by Eq. (9.98) as follows:

$$\omega = \frac{2}{T} \tan \frac{\Omega}{2}$$

$$\rightarrow \quad 1 = \left(\frac{2}{T}\right) \tan \frac{\pi/3}{2} = \frac{2}{\sqrt{3}\, T}, \quad (9.106)$$

which has the solution $T = 2/\sqrt{3}$.

We use this value of T in the bilinear transformation formula given by Eq. (9.83):

$$\mathbf{s} = \frac{2}{T} \frac{\mathbf{z} - 1}{\mathbf{z} + 1} = \sqrt{3}\, \frac{\mathbf{z} - 1}{\mathbf{z} + 1}. \quad (9.107)$$

Substituting Eq. (9.107) into Eq. (9.105) gives the discrete-time transfer function

$$\mathbf{H}(\mathbf{z}) = \frac{1}{\left(\sqrt{3}\,\frac{\mathbf{z}-1}{\mathbf{z}+1}\right)^3 + 2\left(\sqrt{3}\,\frac{\mathbf{z}-1}{\mathbf{z}+1}\right)^2 + 2\left(\sqrt{3}\,\frac{\mathbf{z}-1}{\mathbf{z}+1}\right) + 1}. \quad (9.108)$$

Multiplying Eq. (9.108) by $\left(\frac{\mathbf{z}+1}{\mathbf{z}+1}\right)^3$ gives, after much algebra,

$$\mathbf{H}(\mathbf{z}) = \frac{(\mathbf{z}+1)^3}{(7+5\sqrt{3})\mathbf{z}^3 - (3+7\sqrt{3})\mathbf{z}^2 + (7\sqrt{3}-3)\mathbf{z} + (7-5\sqrt{3})}. \quad (9.109)$$

Transfer function $\mathbf{H}(\mathbf{z})$ has a triple zero at $\{-1\}$, so it rejects $\Omega = \pi$ (see Section 8-1), and is therefore a lowpass filter. The pole-zero diagram is shown in Fig. 9-23(a). Note the triple zero at -1 and the arc of poles in the right half of the $\mathbf{z}$ plane.

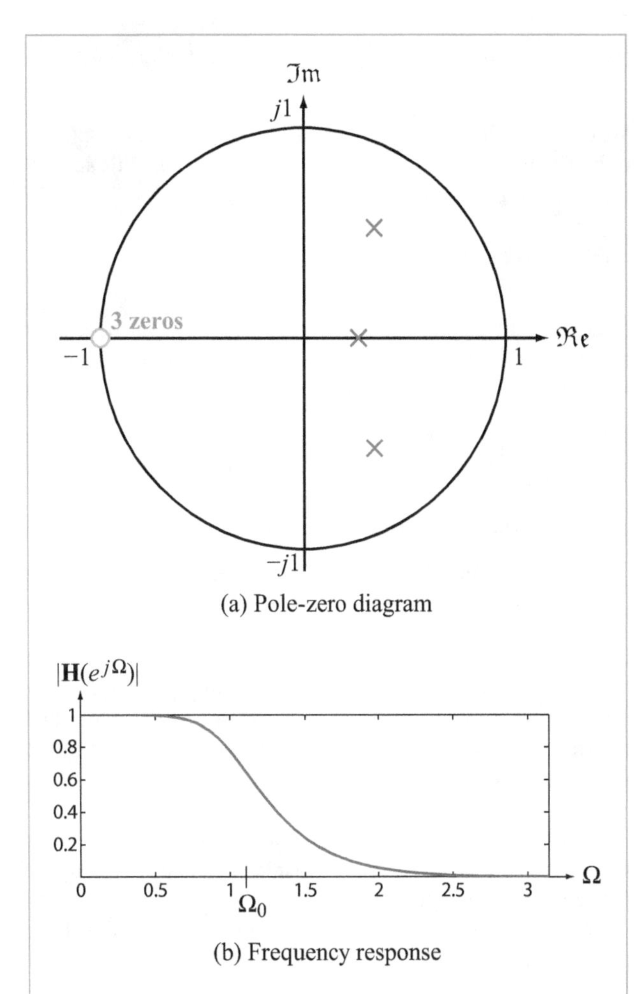

Figure 9-23: (a) Pole-zero diagram for the IIR Butterworth lowpass filter with cutoff frequency $\Omega_0 = \frac{\pi}{3}$, and (b) the corresponding frequency response.

The frequency response $\mathbf{H}(e^{j\Omega})$, obtained by setting $\mathbf{z} = e^{j\Omega}$ in Eq. (9.109), is shown in Fig. 9-23(b). The filter is indeed a lowpass filter with cutoff frequency $\Omega_0 = \frac{\pi}{3}$. Like its continuous-time namesake, the filter gain is strictly decreasing, with absolutely no ripple at all. The rolloff is gradual, but it could be made sharper by using a higher-order Butterworth filter. The discrete-time lowpass filter can be implemented as an ARMA difference equation of order (3,3).

The MATLAB code for this example is on the book website (Ex914.m).

9-4.7 Comparison of Different Approaches to Filter Design

We conclude this section by comparing the frequency responses of half-band lowpass filters designed using various FIR and IIR filter design techniques.

Example 9-15: Lowpass Filter Designs

Design a half-band (cutoff frequency $\Omega_0 = \frac{\pi}{2}$) lowpass filter using 20 coefficients in an MA or ARMA difference equation, with the following specifications:

(a) Hamming window.

(b) Frequency sampling.

(c) Minimax with transition 0.4π to 0.6π.

(d) Discrete-time 10th-order Butterworth filter using the bilinear transformation.

Solution: The results are shown in Fig. 9-24. For each of the first three design techniques, the impulse response is represented by the stem plot in the left half of the figure and the frequency response is given by the continuous plot in the right half. They are all quite similar. The minimax design has a sharper transition, but also has ripples in the passband and stopband. For the Butterworth filter, which exhibits the steepest slope in the transition region, we display the pole-zero diagram instead of the impulse response.

The MATLAB code for this example is on the book website (Ex915.m).

Concept Question 9-8: Why would you ever use an IIR filter, when FIR filters have no phase distortion and are guaranteed to be stable? (See S^2)

Concept Question 9-9: Why is the bilinear transformation of a piecewise-constant gain also piecewise constant? (See S^2)

Exercise 9-6: Using the continuous-time filter

$$h_a(t) = \delta(t) - 3e^{-3t}\ u(t)$$

and $T_s = 2$, design a discrete-time filter using impulse invariance.

Answer: The impulse is just feedthrough.

$$h[n] = \delta[n] - T_s\ h_a(nT_s) = \delta[n] - 6e^{-6n}\ u[n].$$

Exercise 9-7: Using the continuous-time filter $\mathbf{H}_a(\mathbf{s}) = \mathbf{s}/(\mathbf{s}+1)$ and $T = 2$, design a discrete-time filter using bilinear transformation.

Answer: Setting $\mathbf{s} = \frac{2}{2}\frac{\mathbf{z}-1}{\mathbf{z}+1}$ in $\mathbf{H}_a(\mathbf{s})$ gives

$$\mathbf{H}(\mathbf{z}) = \frac{(\mathbf{z}-1)/(\mathbf{z}+1)}{1+(\mathbf{z}-1)/(\mathbf{z}+1)} = \frac{\mathbf{z}-1}{(\mathbf{z}+1)+(\mathbf{z}-1)} = \frac{1}{2}(1-\mathbf{z}^{-1}).$$

So $h[n] = \{\underline{\frac{1}{2}}, -\frac{1}{2}\}$ is actually FIR here!

Exercise 9-8: We wish to design an IIR discrete-time lowpass filter with cutoff frequency $\Omega_0 = \frac{\pi}{2}$ using bilinear transformation with $T = 0.001$. Determine the continuous-time lowpass filter cutoff frequency ω.

Answer:

$$\omega = \frac{2}{T}\tan\left(\frac{\Omega_0}{2}\right) = \frac{2}{0.001}\tan\left(\frac{\pi/2}{2}\right) = 2000\tan\left(\frac{\pi}{4}\right) = 2000 \text{ rad/s}.$$

Exercise 9-9: Using bilinear transformation with $T = 0.1$, the continuous-time frequency $\omega = 20$ rad/s maps to what discrete-time frequency?

Answer:

$$20 = \omega = \frac{2}{0.1}\tan\left(\frac{\Omega}{2}\right) \rightarrow 1 = \tan\left(\frac{\Omega}{2}\right) \rightarrow \Omega = \frac{\pi}{2}.$$

Exercise 9-10: Use bilinear transformation with $T = 2$ to design an IIR ideal differentiator.

Answer: From Chapter 3, $\mathbf{H}_a(\mathbf{s})=\mathbf{s}$, $\mathbf{s} = \frac{2}{2}\frac{\mathbf{z}-1}{\mathbf{z}+1}$. So

$$\mathbf{H}(\mathbf{z}) = \frac{\mathbf{z}-1}{\mathbf{z}+1} = \frac{\mathbf{Y}(\mathbf{z})}{\mathbf{X}(\mathbf{z})} \rightarrow y[n] + y[n-1] = x[n] - x[n-1].$$

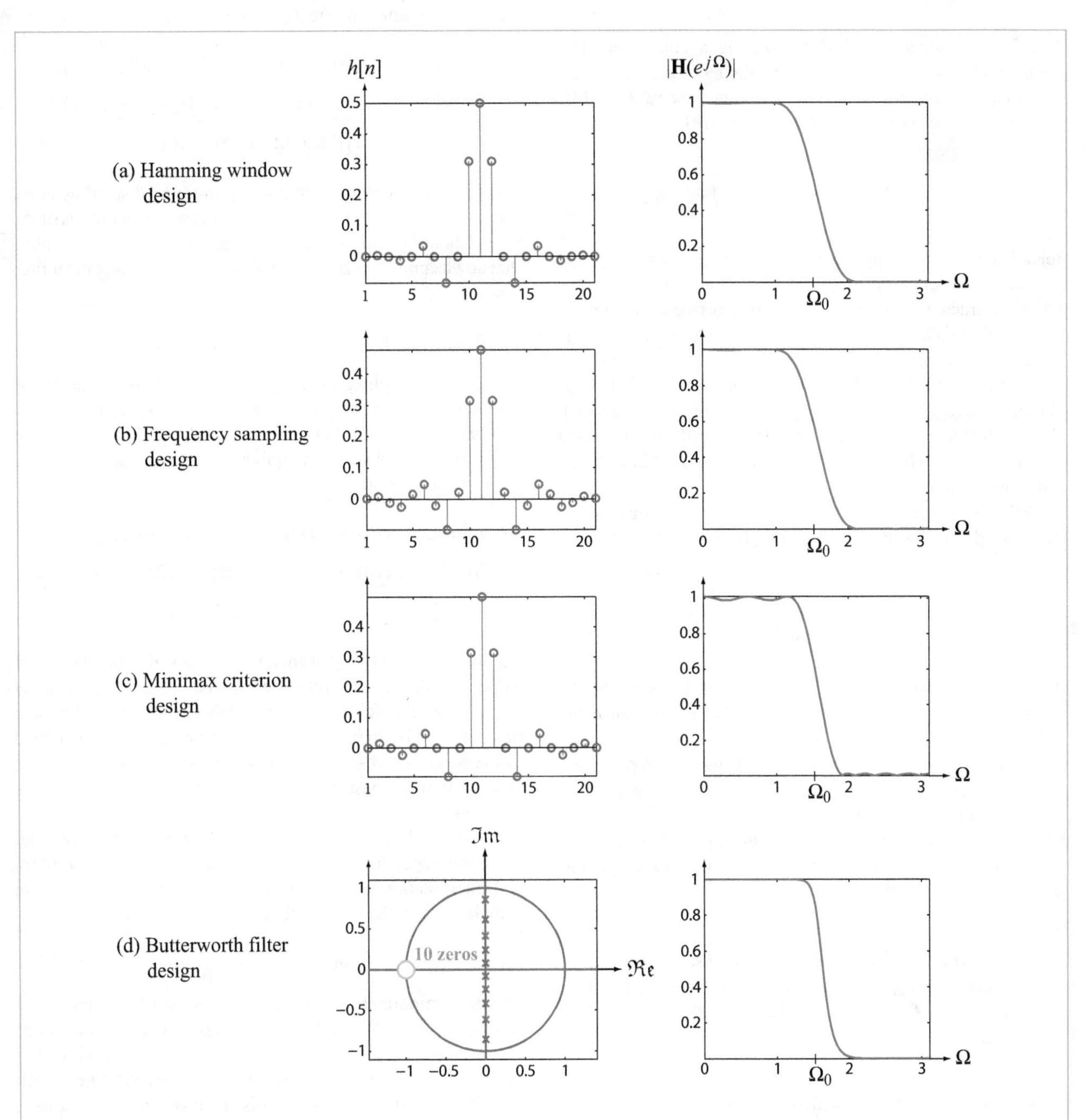

Figure 9-24: Half-band lowpass filter designed using: (a) Hamming window, (b) frequency sampling, (c) minimax criterion, and (d) Butterworth filter, all for a cutoff frequency $\Omega_0 = \pi/2$.

9-5 Multirate Signal Processing

Recall from Section 6-13 that sampling a continuous-time signal $x(t)$ at a ***sampling rate*** f_s entails recording the values of $x(t)$ at integer multiples of the ***sampling interval*** $T_s = 1/f_s$ and storing them as a discrete-time signal $x[n]$:

$$x[n] = x(nT_s) = x\left(\frac{n}{f_s}\right),$$

for all integers n. Sampling $x(t)$ at a different sampling rate produces a different discrete-time signal, but $x(t)$ can always be reconstructed from any sampled discrete-time version, so long as f_s satisfies the ***Nyquist criterion***, which states that f_s should exceed the ***highest-frequency component*** f_{max} present in the spectrum of $x(t)$ by at least a factor of 2. That is, f_s should exceed $2f_{max}$. If the Nyquist criterion is not satisfied, the reconstructed continuous-time signal will be an aliased version of $x(t)$, which differs from the true $x(t)$. More details are available in Section 6-13.

Let $x_1[n]$ be the samples resulting from a sampling at a rate f_{s_1} and $x_2[n]$ be the samples resulting from sampling at f_{s_2}:

$$x_1[n] = x(n/f_{s_1}),$$
$$x_2[n] = x(n/f_{s_2}). \tag{9.110}$$

The goal of multirate signal processing is to compute the samples $x_2[n]$ from the samples $x_1[n]$, that is, to change the sampling rate from f_{s_1} to f_{s_2} after the fact.

One obvious way to accomplish the change in sampling rate would be to reconstruct $x(t)$ from $x_1[n]$ and then resample $x(t)$ at sampling rate f_{s_2} to obtain $x_2[n]$. But if f_{s_2}/f_{s_1} is a rational number (ratio of two integers), this can be performed directly in the discrete-time (n) domain. The goal of this section is to demonstrate the process.

▶ We assume throughout that the *original* sampling rate, f_{s_1}, exceeds $2f_{max}$. Furthermore, to avoid generating an aliased signal, f_{s_2} should exceed $2f_{max}$ as well. ◀

9-5.1 Operations of Multirate Signal Processing

(a) f_{s_1}/f_{s_2} = integer

If the original sampling rate f_{s_1} is an integer multiple of the new sampling rate f_{s_2}, then obtaining $x_2[n]$ from $x_1[n]$ is trivial. For example, if $f_{s_1} = 1000$ sample/s and $f_{s_2} = 500$ sample/s, then the sampling intervals are $T_{s_1} = 1$ ms and $T_{s_2} = 2$ ms. Thus,

$$x_1[n] = \{\dots, \underline{x(0)}, x(.001), x(.002), x(.003), \dots\},$$
$$x_2[n] = \{\dots, \underline{x(0)}, x(.002), x(.004), x(.006), \dots\}$$
$$= \{\dots, \underline{x_1[0]}, x_1[2], x_1[4], x_1[6], \dots\}. \tag{9.111}$$

So $x_2[n]$ can be obtained from $x_1[n]$ simply by discarding every other value of $x_1[n]$. The extension to larger integer ratios of f_{s_2} to f_{s_1} should be apparent. This is called ***downsampling***. Note that downsampling may induce aliasing if f_{s_2} is not greater than $2f_{max}$.

(b) f_{s_1}/f_{s_2} = 1/integer

If the new sampling rate, f_{s_2}, is an integer multiple of the original sampling rate, f_{s_1}, then obtaining $x_2[n]$ from $x_1[n]$ is more complicated. For example, if $f_{s_1} = 500$ sample/s and $f_{s_2} = 1000$ sample/s, the sampling intervals are $T_{s_1} = 2$ ms and $T_{s_2} = 1$ ms, in which case

$$x_1[n] = \{\dots, \underline{x(0)}, x(.002), x(.004), x(.006), \dots\}$$
$$x_2[n] = \{\dots, \underline{x(0)}, x(.001), x(.002), x(.003), \dots\}$$
$$= \{\dots, \underline{x_1[0]}, ?, x_1[1], ?, x_1[2], ?, x_1[3], \dots\}. \tag{9.112}$$

Half of the values of $x_2[n]$ are known values of $x_1[n]$. The other values of $x_2[n]$ are not yet known, and these are designated by the symbol "?". We show below that these unknown values can be computed by first replacing them with zeros (this is called ***upsampling*** or ***zero-stuffing***), and then filtering the result with a discrete-time lowpass filter with cutoff frequency $\Omega_0 = \frac{\pi}{2}$ (this is called ***interpolation***).

If we wish to change the sampling rate by a rational but non-integer factor, we use a combination of upsampling and interpolation and downsampling. The collection of these techniques is called ***multirate signal processing***.

(c) Multirate signal processing applications

The most important application of multirate signal processing is ***oversampling***. Recall from Section 6-13 that the final stage of reconstructing a signal $x(t)$ from its samples $x(nT_s)$ involves the use of a continuous-time (analog) lowpass filter, such as a Butterworth filter. Since this filter is a physical device, a fast rolloff requires many physical components (op-amps, resistors, capacitors, wires). If the sampling rate $f_s = 1/T_s$ greatly exceeds double the maximum frequency in the spectrum of $x(t)$, this filter will be small. But a huge sampling rate means more numbers need to be stored and processed. Instead, using upsampling and interpolation, we can compute the samples that would have resulted from a huge sampling rate, without actually using a huge sampling rate. In effect, we will replace a fast-rolloff *continuous*-time filter with a fast-rolloff *discrete*-time filter, which can be implemented using a computer.

We present in a later section an application that involves the use of the musical ***circle of fifths*** to synthesize the sounds of a musical instrument playing all musical notes from the sound of a musical instrument playing a single note. If a snippet of a musical instrument playing a single note can be found, a music synthesizer for that instrument playing all possible notes can be programmed on a computer. Multirate signal processing will also be used in the next chapter for the discrete-time wavelet transform.

The next few sections address the following operations of multirate signal processing:

- Downsampling (also known as ***decimating***).
- Upsampling (also known as ***zero-stuffing***).
- Interpolation (low-pass filtering).
- Multirate processing (combining multiple operations).

► Throughout, we should keep in mind that spectra of discrete-time signals *must* repeat in Ω every 2π (and possibly more often). ◄

9-6 Downsampling

9-6.1 Downsampling in the Time Domain

Downsampling, also called ***decimation***, reduces the sampling rate by an integer factor. This is very easy to do. To change the sampling rate from 1200 samples/s to 600 samples/s, we simply omit every other sample. To change the rate from 1200 sample/s to 400 samples/s, we simply omit two out of three samples, keeping only every third sample.

Downsampling by an integer factor L is depicted using the notation

$$x[n] \rightarrow \boxed{\downarrow L} \rightarrow y[n] = x[Ln]. \tag{9.113}$$

The MATLAB/MathScript recipe for downsampling by L is:

```
Y=X[1:L:end];
```

For example, for $L = 3$, we have:

$$y[0] = x[0], \quad y[1] = x[3], \quad y[2] = x[6], \quad y[3] = x[9], \dots,$$

so if $x[n]$ is given by

$$x[n] = \{\dots, \underline{3}, 1, 4, 1, 5, 9, 2, 6, 5, 8, \dots\},$$

then downsampling it by 3 gives

$$y[n] = \{\dots, \underline{3}, 1, 2, 8, \dots\}.$$

We could also retain $x[n]$ at times $n = 1, 4, 7, \dots$. This would be a different ***polyphase*** component of $x[n]$. Usually, we assume that downsampling retains $x[0]$ specifically.

9-6.2 Downsampling in the z-Transform Domain

By extending the unilateral **z**-transform defined by Eq. (7.58a) to the bilateral **z**-transform as defined in Eq. (8.72), we have

$$\mathbf{X}(\mathbf{z}) = \mathbb{Z}\{x[n]\} = \sum_{n=-\infty}^{\infty} x[n]\, \mathbf{z}^{-n}. \tag{9.114}$$

This is analogous to the extension of the unilateral Laplace transform to the bilateral Laplace transform in Section 3-9.

Given that $\mathbf{X}(\mathbf{z})$ is the bilateral **z**-transform of $x[n]$, we now seek to find an expression for $\mathbf{Y}(\mathbf{z})$ in terms of $\mathbf{X}(\mathbf{z})$, where $\mathbf{Y}(\mathbf{z})$ is the bilateral **z**-transform of

$$y[n] = x[Ln]. \tag{9.115}$$

That is, $y[n]$ is a downsampled version of $x[n]$ and L is its downsampling integer factor.

(a) $L = 2$

As a first step towards relating $\mathbf{Y}(\mathbf{z})$ to $\mathbf{X}(\mathbf{z})$, we consider the simple case where $L = 2$. The bilateral **z**-transform $\mathbf{Y}(\mathbf{z})$ is then given by

$$\begin{aligned} \mathbf{Y}(\mathbf{z}) &= \sum_{n=-\infty}^{\infty} y[n]\, \mathbf{z}^{-n} \\ &= \sum_{n=-\infty}^{\infty} x[2n]\, \mathbf{z}^{-n} \\ &= \cdots + x[-2]\, \mathbf{z} + x[0] + x[2]\, \mathbf{z}^{-1} + \cdots. \end{aligned} \tag{9.116}$$

Next, we write the sums for $\mathbf{X}(\mathbf{z})$ and $\mathbf{X}(-\mathbf{z})$:

$$\mathbf{X}(\mathbf{z}) = \cdots + x[-2]\,\mathbf{z}^2 + x[-1]\,\mathbf{z} + x[0] + x[1]\,\mathbf{z}^{-1} + x[2]\,\mathbf{z}^{-2} + \cdots, \quad (9.117a)$$

$$\mathbf{X}(-\mathbf{z}) = \cdots + x[-2]\,\mathbf{z}^2 - x[-1]\,\mathbf{z} + x[0] - x[1]\,\mathbf{z}^{-1} + x[2]\,\mathbf{z}^{-2} - \cdots. \quad (9.117b)$$

The sum of the expressions is

$$\mathbf{X}(\mathbf{z}) + \mathbf{X}(-\mathbf{z}) = \cdots + 2x[-2]\,\mathbf{z}^2 + 2x[0] + 2x[2]\,\mathbf{z}^{-2} + \cdots. \quad (9.118)$$

All odd powers of $\mathbf{z}$, as well as all $x[n]$ for odd times n, have been eliminated. If we replace $\mathbf{z}$ with $\mathbf{z}^{1/2}$ everywhere in Eq. (9.118), the new sum becomes equal to twice the sum in Eq. (9.116), from which we conclude that

$$\mathbf{Y}(\mathbf{z}) = \frac{1}{2}\,[\mathbf{X}(\mathbf{z}^{1/2}) + \mathbf{X}(-\mathbf{z}^{1/2})] \qquad (L = 2). \quad (9.119)$$

Hence, given $x[n]$ and its corresponding $\mathbf{z}$-transform $\mathbf{X}(\mathbf{z})$, we can apply Eq. (9.119) to obtain the $\mathbf{z}$-transform of $\mathbf{Y}(\mathbf{z})$ directly from $\mathbf{X}(\mathbf{z})$.

(b) $L \geq 2$

The relationship given by Eq. (9.119) applies to $L = 2$. For $L \geq 2$, the relationship takes the form

$$\mathbf{Y}(\mathbf{z}) = \frac{1}{L}\sum_{k=0}^{L-1} \mathbf{X}(e^{-j2\pi k/L}\,\mathbf{z}^{1/L}) \qquad (L \geq 2). \quad (9.120)$$

To derive Eq. (9.120), we start by introducing the ***switching function*** $s[n]$ defined as

$$s[n] = \frac{1}{L}\sum_{k=0}^{L-1} e^{j2\pi nk/L}. \quad (9.121)$$

For $n =$ multiple of L, $2\pi nk/L$ is either zero or a multiple of 2π. Hence,

$$s[n] = 1, \qquad \text{for } n = \text{multiple of } L. \quad (9.122)$$

For $n \neq$ multiple of L, we can use the finite geometric series relationship

$$\sum_{k=0}^{L-1} \mathbf{r}^k = \frac{\mathbf{r}^L - 1}{\mathbf{r} - 1} \qquad \text{for } \mathbf{r} \neq 1, \quad (9.123)$$

to express $s[n]$ as

$$s[n] = \frac{1}{L}\left[\frac{(e^{j2\pi n/L})^L - 1}{e^{j2\pi n/L} - 1}\right] = \frac{1}{L}\left[\frac{1 - 1}{e^{j2\pi n/L} - 1}\right] = 0, \qquad \text{for } n \neq \text{multiple of } L. \quad (9.124)$$

Combining the results given by Eqs. (9.122) and (9.124) with Eq. (9.121) leads to

$$s[n] = \frac{1}{L}\sum_{k=0}^{L-1} e^{j2\pi nk/L} \quad (9.125a)$$

$$= \begin{cases} 1 & \text{for } n = \text{multiple of } L, \\ 0 & \text{otherwise.} \end{cases} \quad (9.125b)$$

The $\mathbf{z}$-transform of the product $s[n]\,x[n]$ is given by

$$\mathbb{Z}\{s[n]\,x[n]\} = \frac{1}{L}\sum_{k=0}^{L-1} \mathbb{Z}\{e^{j2\pi nk/L}\,x[n]\}. \quad (9.126)$$

The $\mathbf{z}$-scaling property of the $\mathbf{z}$-transform (property #5 in Table 7-6) states that

$$\mathbf{a}^n\,x[n] \longleftrightarrow \mathbf{X}\left(\frac{\mathbf{z}}{\mathbf{a}}\right).$$

If we set $\mathbf{a} = e^{j2\pi k/L}$, the scaling property allows us to rewrite Eq. (9.126) in the form:

$$\mathbb{Z}\{s[n]\,x[n]\} = \frac{1}{L}\sum_{k=0}^{L-1} \mathbf{X}(e^{-j2\pi k/L}\,\mathbf{z}). \quad (9.127)$$

By application of Eq. (9.125b), the product $s[n]\,x[n]$ is given by

$$s[n]\,x[n] = \{\ldots, \underline{x[0]}, \underbrace{0, \ldots, 0}_{L-1}, x[L], \underbrace{0, \ldots, 0}_{L-1}, x[2L], \ldots\}, \quad (9.128)$$

and its $\mathbf{z}$-transform is

$$\mathbb{Z}\{s[n]\,x[n]\} = \{\cdots + x[0] + \underbrace{0 + \cdots + 0}_{L-1} + x[L]\,\mathbf{z}^{-L} + \underbrace{0 + \cdots + 0}_{L-1} + x[2L]\,\mathbf{z}^{-2L} + \cdots\}$$
$$= \{\cdots + x[0] + x[L]\,\mathbf{z}^{-L} + x[2L]\,\mathbf{z}^{-2L} + \cdots\}. \quad (9.129)$$

For the downsampled signal

$$y[n] = x[Ln] = \{\dots, x[0], x[L], x[2L], \dots\}, \quad (9.130)$$

its **z**-transform is

$$\mathbf{Y}(\mathbf{z}) = \{\dots + x[0] + x[L]\,\mathbf{z}^{-1} + x[2L]\,\mathbf{z}^{-2} + \dots\}. \quad (9.131)$$

If we replace **z** with $\mathbf{z}^L$ everywhere in Eq. (9.131), we end up with the same expression we have on the right-hand side of Eq. (9.129). Furthermore, the right-hand sides of Eqs. (9.127) and (9.129) are both equal to $\mathbf{Z}\{s[n]\,x[n]\}$. The net result is that

$$\mathbf{Y}(\mathbf{z}^L) = \mathbf{Z}\{s[n]\,x[n]\} = \frac{1}{L}\sum_{k=0}^{L-1} \mathbf{X}(e^{-j2\pi k/L}\,\mathbf{z}). \quad (9.132)$$

Replacing **z** with $\mathbf{z}^{1/L}$ gives

$$\mathbf{Y}(\mathbf{z}) = \frac{1}{L}\sum_{k=0}^{L-1} \mathbf{X}(e^{-j2\pi k/L}\,\mathbf{z}^{1/L}), \quad (9.133)$$

which is identical with Eq. (9.120), thereby affirming the relationship between the **z**-transform of the original signal, $\mathbf{X}(\mathbf{z})$, and the transform of the downsampled signal, $\mathbf{Y}(\mathbf{z})$.

9-6.3 Downsampling in the Frequency Domain

The DTFT $\mathbf{Y}(e^{j\Omega})$ of $y[n]$ can be obtained from $\mathbf{Y}(\mathbf{z})$ upon setting $\mathbf{z} = e^{j\Omega}$. Application of the recipe to Eq. (9.133) leads to

$$\mathbf{Y}(e^{j\Omega}) = \frac{1}{L}\sum_{k=0}^{L-1} \mathbf{X}(e^{-j2\pi k/L}e^{j\Omega/L}) = \frac{1}{L}\sum_{k=0}^{L-1} \mathbf{X}(e^{j(\Omega-2\pi k)/L}). \quad (9.134)$$

The downsampled discrete-time signal $y[n] = x[Ln]$, in effect, compresses time by a factor of L, which causes the frequency response $\mathbf{X}(e^{j\Omega})$ to expand in the frequency domain by the same factor. Consequently, the period increases from 2π for $\mathbf{X}(e^{j\Omega})$ to $2\pi L$ for $\mathbf{X}(e^{j\Omega/L})$. The spectrum of any discrete-time signal, including $\mathbf{Y}(e^{j\Omega})$, must be periodic with period 2π. This condition is accommodated by the transformation given by Eq. (9.133) through the addition of $L-1$ copies of the expanded spectrum $\mathbf{X}(e^{j\Omega/L})$, shifted in frequency by $2\pi k$ from the original copy. The net consequence is that $\mathbf{Y}(e^{j\Omega})$ is indeed periodic with period 2π, as required.

▶ Downsampling by L stretches the spectrum by L. ◀

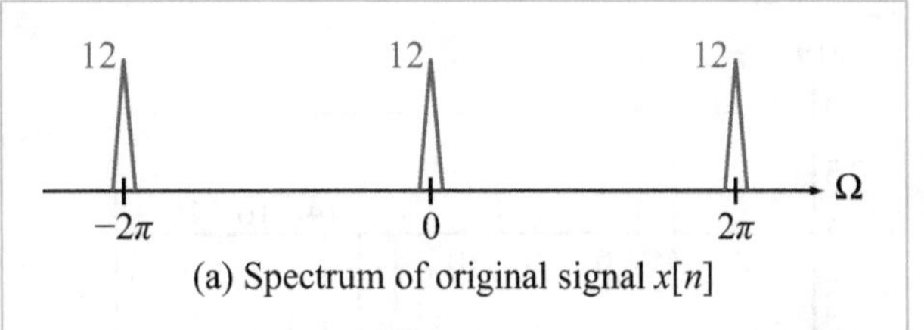

(a) Spectrum of original signal $x[n]$

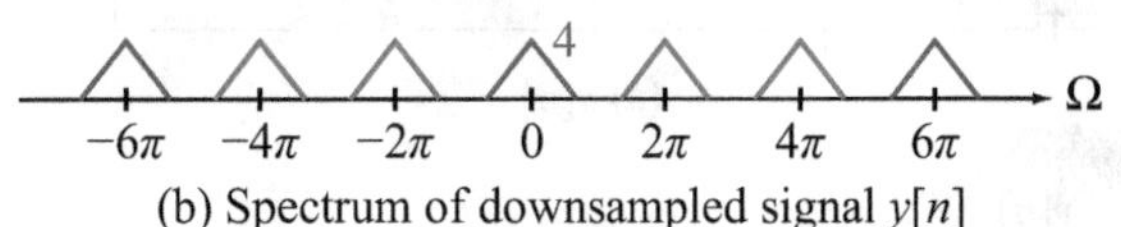

(b) Spectrum of downsampled signal $y[n]$

Figure 9-25: Comparison of spectra of the original signal and the signal downsampled by 3. The red spectra represent the new copies introduced as a result of the downsampling operation.

Example 9-16: Downsampling in the Frequency Domain

A signal has the spectrum shown in Fig. 9-25(a). Plot the spectrum of the signal after downsampling it by a factor of 3.

Solution: Downsampling by a factor of 3 causes the spectrum to stretch out by the same factor and causes the amplitude to decrease, also by the same factor. Additionally, new copies of the spectrum are introduced as shown in Fig. 9-25(b).

9-6.4 Downsampling a Sinusoid

Downsampling a discrete-time sinusoid by an integer factor L can be expressed symbolically as

$$\sin(\Omega_0 n) \rightarrow \boxed{\downarrow L} \rightarrow \sin(\Omega_0 L n) = \sin((\Omega_0 L)n). \quad (9.135)$$

The frequency of the downsampled sinusoid is higher by a factor L. Figure 9-26 displays plots for

$$x[n] = \sin(0.1\pi n)$$

and for

$$y[n] = x[2n] = \sin(0.2\pi n).$$

Even though the amplitudes of both $x[n]$ and $y[n]$ are the same, the amplitudes of their corresponding spectra differ by a factor of 2. This is because from Eq. (9.120), the spectrum of the downsampled signal is multiplied by the factor $1/L$.

Concept Question 9-10: Does downsampling always cause aliasing? (See S[2])

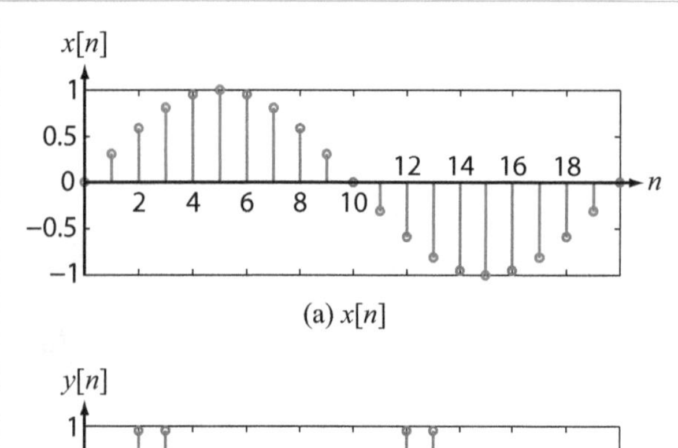

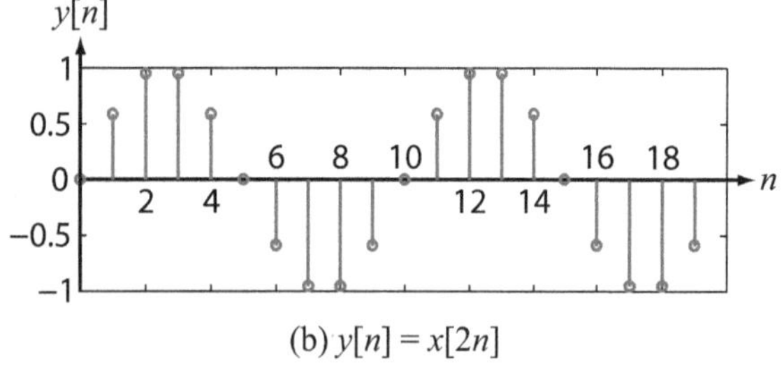

Figure 9-26: Effect of downsampling on a sinusoid: (a) $x[n] = \sin(0.1\pi n)$; (b) $y[n] = x[2n] = \sin(0.2\pi n)$.

Exercise 9-11: $\cos(0.6\pi n) \rightarrow \boxed{\downarrow 3} \rightarrow ?$

Answer: $\Omega = 0.6\pi$ becomes $\Omega = 1.8\pi$, which aliases to $\Omega = 0.2\pi$, since $\cos(1.8\pi) \equiv \cos(-0.2\pi) \equiv \cos(0.2\pi)$.

Exercise 9-12: $\cos(0.8\pi n) \rightarrow \boxed{\downarrow 4} \rightarrow ?$

Answer: $\Omega = 0.8\pi$ becomes $\Omega = 3.2\pi$, which aliases to $\Omega = 0.8\pi$, since $\cos(3.2\pi) \equiv \cos(-0.8\pi) \equiv \cos(0.8\pi)$.

9-7 Upsampling

9-7.1 Upsampling in the Time Domain

Sometimes, the term ***upsampling*** is used to denote the combination of two sequential operations: *zero stuffing* and ***interpolation***. To maintain notational symmetry in this presentation, we will use upsampling to mean zero stuffing only. Interpolation follows later in Section 9-8.

To upsample a discrete-time signal $x[n]$ by an integer factor L, we simply insert $L-1$ zeros between each pair of adjacent samples of $x[n]$. The resultant upsampled signal $y[n]$ is then given by

$$y[n] = \{\ldots, \underline{x[0]}, \underbrace{0, \ldots, 0}_{L-1}, x[1], \underbrace{0, \ldots, 0}_{L-1}, x[2], \ldots\}.$$

Upsampling by a factor L is depicted notationally as

$$x[n] \rightarrow \boxed{\uparrow L} \rightarrow$$

$$y[n] = \begin{cases} x[n/L] & \text{for } n/L = \text{integer}, \\ 0 & \text{for } n/L \neq \text{integer}. \end{cases} \qquad (9.136)$$

For example, if $L = 3$,

$$y[0] = x[0], \qquad y[3] = x[1], \qquad y[6] = x[2], \ldots$$

$$y[1] = y[2] = y[4] = y[5] = y[7] = y[8] = \cdots = 0.$$

The MATLAB/MathScript recipe for upsampling by L is:

```
Z=[X;zeros(L-1,length(X))];Y=Z(:)';
```

Example 9-17: Upsampling in the Time Domain

Upsample the signal

$$x[n] = \{\ldots, \underline{3}, 1, 4, 1, 5, 9, \ldots\}$$

by a factor of 2.

Solution: Inserting $2-1=1$ zero between samples gives

$$y[n] = \{\ldots, \underline{3}, 0, 1, 0, 4, 0, 1, 0, 5, 0, 9, \ldots\}$$

9-7.2 Upsampling in the z-Transform Domain

From Eq. (9.114), the bilateral **z**-transform of the discrete-time signal $x[n]$ is given by the sum

$$\mathbf{X}(\mathbf{z}) = \{\cdots + x[-2]\,\mathbf{z}^2 + x[-1]\,\mathbf{z} + x[0] + x[1]\,\mathbf{z}^{-1} + x[2]\,\mathbf{z}^{-2} + \cdots\}. \qquad (9.137)$$

Upsampling $x[n]$ by a factor of 2 yields

$$y[n] = \{\ldots, x[-2], 0, x[-1], 0, \underline{x[0]}, 0, x[1], 0, x[2], \ldots\}, \qquad (9.138)$$

and its **z**-transform is

$$\mathbf{Y}(\mathbf{z}) = \{\ldots, x[-2]\,\mathbf{z}^4 + 0 + x[-1]\,\mathbf{z}^2 + 0 + x[0] + 0 + x[1]\,\mathbf{z}^{-2} + 0 + x[2]\,\mathbf{z}^{-4} + \cdots\}. \qquad (9.139)$$

Comparison of Eq. (9.137) with Eq. (9.139) leads to the conclusion that

$$\mathbf{Y}(\mathbf{z}) = \mathbf{X}(\mathbf{z}^2) \qquad (\text{for } L = 2). \qquad (9.140)$$

For the general case,

$$\mathbf{Y}(\mathbf{z}) = \mathbf{X}(\mathbf{z}^L) \qquad (L \geq 2). \qquad (9.141)$$

9-7.3 Upsampling in the Frequency Domain

The DTFT $\mathbf{Y}(e^{j\Omega})$ of $y[n]$ is found by setting $\mathbf{z} = e^{j\Omega}$ in Eq. (9.141):

$$\mathbf{Y}(e^{j\Omega}) = \mathbf{X}(e^{j\Omega L}). \tag{9.142}$$

Since upsampling stretches discrete time by a factor of L, frequency is compressed by L in the spectrum of $\mathbf{X}(e^{j\Omega L})$, relative to the spectrum of $\mathbf{X}(e^{j\Omega})$. The period of $\mathbf{X}(e^{j\Omega L})$ is $2\pi/L$ and the spectrum repeats every 2π, as required for any DTFT.

▶ Upsampling by L compresses the spectrum by L. ◀

Example 9-18: Upsampling in Frequency Domain by 2

Given the spectrum of signal $x[n]$ shown in Fig. 9-27(a), plot the spectrum of the signal after upsampling it by a factor of 2.

Solution: We denote the upsampled signal $y[n]$ and its frequency response

$$\mathbf{Y}(e^{j\Omega}) = \mathbf{X}(e^{j2\Omega}).$$

The spectrum $\mathbf{Y}(e^{j\Omega})$ shown in Fig. 9-27(b) has the same amplitude as that of the original signal, but its width is narrower by a factor of 2, and repeats every π.

Concept Question 9-11: Can upsampling ever cause aliasing? (See S²)

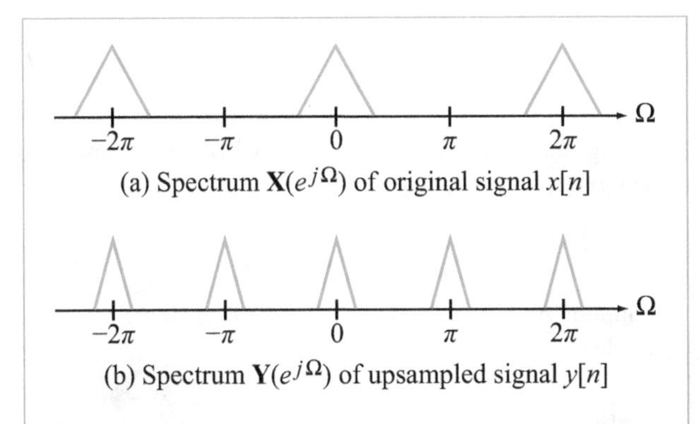

Figure 9-27: Upsampling by a factor of 2 causes the spectrum to narrow by the same factor and to repeat at twice the rate (Example 9-18).

Exercise 9-13: $\cos(0.4\pi n)$ → [↑4] → ?

Answer: $\Omega = \{0.4\pi, (2-0.4)\pi, (2+0.4)\pi, (4-0.4)\pi\}$ become $\Omega = \{0.1\pi, 0.4\pi, 0.6\pi, 0.9\pi\}$. The input was a single sinusoid, but the output is four sinusoids.

Exercise 9-14: $\cos(0.8\pi n)$ → [↑4] → ?

Answer: $\Omega = \{0.8\pi, (2-0.8)\pi, (2+0.8)\pi, (4-0.8)\pi\}$ become $\Omega = \{0.2\pi, 0.3\pi, 0.7\pi, 0.8\pi\}$. The input was a single sinusoid, but the output is four sinusoids.

9-8 Interpolation

Interpolating a signal $x(t)$ with specified values at times $t = nT$, where $n = 0, 1, \ldots$, entails assigning values to $x(t)$ at times $t \neq nT$, based on the given values $\{x(nT)\}$. A simple example is linear interpolation wherein over every interval between nT and $(n+1)T$, the newly assigned values of $x(t)$ vary linearly between $x(nT)$ and $x((n+1)T)$. In essence, interpolation *connects the dots* between given values of $\{x(nT)\}$.

Interpolation can also be implemented entirely in discrete time. Let us consider a discrete-time signal $x[n]$ that gets upsampled by a factor L to generate an upsampled version $x_u[n]$ given by:

$$\begin{aligned} x_u[n] &= \{\ldots,\ x[0],\ \ldots\ldots,\ x[1],\ \ldots\ldots,\ x[2], \ldots\} \\ &= \{\ldots, x_u[0], \underbrace{0, \ldots, 0}_{L-1}, x_u[L], \underbrace{0, \ldots, 0}_{L-1}, x_u[2L], \ldots\}, \end{aligned} \tag{9.143}$$

where $x_u[nL] = x[n]$. The upsampled signal can be interpolated by replacing the zero values with the values that make the interpolated $x_u[n]$ have a maximum frequency of π/L. The interpolation task can be accomplished by filtering $x_u[n]$ using a brickwall lowpass filter with a cutoff frequency π/L. As noted earlier, when a signal is upsampled by a factor L, its spectrum shrinks by the same factor. For any discrete-time signal, its one-sided spectrum extends between zero and π, so the spectrum of the upsampled $x_u[n]$ extends to π/L, and repeats afterward. Accordingly, the frequency response of the ideal brickwall lowpass filter should be

$$\mathbf{H}(e^{j\Omega}) = \begin{cases} L & \text{for } 0 \leq |\Omega| < \pi/L, \\ 0 & \text{for } \pi/L < |\Omega| < \pi. \end{cases} \tag{9.144a}$$

The corresponding impulse response is

$$h[n] = L\,\frac{\sin(\pi n/L)}{\pi n}. \qquad (9.144b)$$

In the frequency domain, the spectrum $\mathbf{Y}(e^{j\Omega})$ of the lowpass-filtered signal $y[n]$ is

$$\mathbf{Y}(e^{j\Omega}) = \mathbf{H}(e^{j\Omega})\,\mathbf{X}_{\mathrm{u}}(e^{j\Omega}), \qquad (9.145)$$

where $\mathbf{X}_{\mathrm{u}}(e^{j\Omega})$ is the spectrum of $x_{\mathrm{u}}[n]$. Multiplication in the frequency domain is equivalent to convolution in the discrete-time domain. Hence,

$$y[m] = \sum_{n=-\infty}^{\infty} x_{\mathrm{u}}[n]\,h[m-n] = \sum_{n=-\infty}^{\infty} x_{\mathrm{u}}[n]\,L\,\frac{\sin\left(\frac{\pi}{L}\,(m-n)\right)}{\pi(m-n)}. \qquad (9.146)$$

Hence, $x_{\mathrm{u}}[n]$ is the upsampled signal given by Eq. (9.143), and $y[m]$ is the interpolated version. Signal $y[m]$ preserves the values of $x_{\mathrm{u}}[n]$ specified at $n = \{\ldots, -2L, -L, 0, L, 2L, \ldots\}$, but also assigns values at $L-1$ locations between values of n that are multiples of L, for every n. Each interpolated value of $y[m]$ is obtained by applying a weighted average to $x_{\mathrm{u}}[n]$, with the weighting provided by the sinc function contained in Eq. (9.146).

We asserted earlier that $y[m]$ preserves the values of $x_{\mathrm{u}}[n]$ for n equal to a multiple of L. To demonstrate the validity of this statement, we introduce two new variables: $n' = n/L$ and $m' = m/L$. Since $x_{\mathrm{u}}[n] = 0$ except for n a multiple of L, index $n' = \{\ldots, -2, -1, 0, 1, 2, \ldots\}$ denotes the times of the nonzero specified values of $x[n] = x[n'L]$. The uninterpolated values of $y[m] = y[m'L]$ are those at $m' = \{\ldots, -2, -1, 0, 1, 2, \ldots\}$.

Upon replacing n with $n'L$ and m with $m'L$ in Eq. (9.146), we have

$$\begin{aligned} y[m'L] &= \sum_{n'=-\infty}^{\infty} x_{\mathrm{u}}[n'L]\,L\,\frac{\sin((\pi/L)(m'L-n'L))}{\pi(m'L-n'L)} \\ &= \sum_{n'=-\infty}^{\infty} x_{\mathrm{u}}[n'L]\,\frac{\sin(\pi(m'-n'))}{\pi(m'-n')} \\ &= \sum_{n'=-\infty}^{\infty} x_{\mathrm{u}}[n'L]\,\delta[m'-n'] = x_{\mathrm{u}}[m'L]. \end{aligned} \qquad (9.147)$$

Hence, $y[m]$ does indeed preserve $x_{\mathrm{u}}[n]$ at the latter's nonzero specified values.

► Following upsampling with interpolation removes duplicate copies of the spectrum. ◄

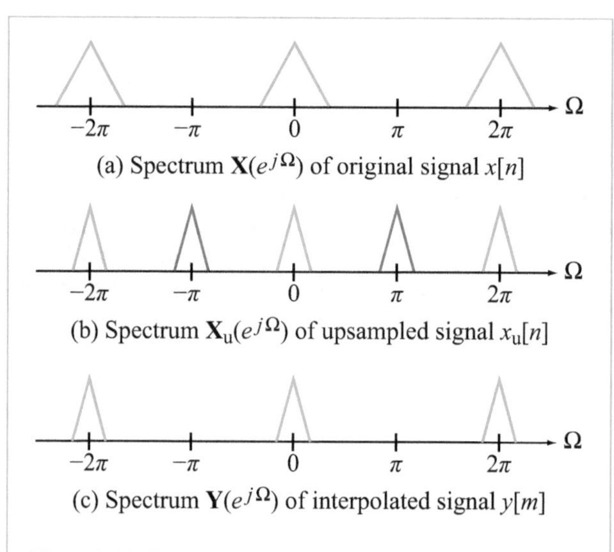

Figure 9-28: The combination of upsampling and interpolation.

Upsampling a signal $x[n]$ by a factor L not only compresses the spectrum of $x[n]$ in frequency by a factor of L, but it also introduces additional copies of the compressed spectrum centered at frequencies $\Omega = 2\pi k/L$ for $k = 0, 1, \ldots, L-1$. An example is shown in Fig. 9-28(b) for $L = 2$. The additional copies of the spectrum get removed by the lowpass filter during the interpolation operation. The combined operations of upsampling and interpolation compress the spectrum without introducing additional copies of the spectrum (Fig. 9-28(c)).

Concept Question 9-12: Can upsampling and interpolation ever cause aliasing? (See S²)

9-9 Multirate Signal Processing Examples

We now examine various sequences of downsampling, upsampling, and interpolation.

9-9.1 Upsampling Followed by Downsampling

Upsampling by L, followed by downsampling by L, yields the original signal:

$$x[n] \rightarrow \boxed{\uparrow L} \rightarrow \boxed{\downarrow L} \rightarrow x[n]. \qquad (9.148)$$

This is because upsampling inserts zeros and downsampling promptly removes the zeros.

9-9.2 Downsampling Followed by Upsampling

Reversing the order of the process given by Eq. (9.148) leads to

$$x[n] \rightarrow \boxed{\downarrow L} \rightarrow \boxed{\uparrow L} \rightarrow \{\dots, x[0], \underbrace{0, \dots, 0}_{L-1}, x[L], \underbrace{0, \dots, 0}_{L-1}, x[2L], \dots\}. \tag{9.149}$$

The process changes all values to zero, except for those at times n that are multiples of L.

9-9.3 Downsampling Followed by Upsampling and Interpolation

If there is no aliasing in the downsampling step in the sequence represented by Eq. (9.149), the addition of an interpolation step leads to restoration of $x[n]$:

$$x[n] \rightarrow \boxed{\downarrow L} \rightarrow \boxed{\uparrow L} \rightarrow \boxed{h[n] = L\,\frac{\sin\left(\frac{\pi}{L}\,n\right)}{\pi n}} \rightarrow x[n]. \tag{9.150}$$

9-9.4 Upsampling Followed by Interpolation and Downsampling

To multiply the effective sampling rate by a rational number M/N, the signal can be upsampled by M, interpolated, then downsampled by N:

$$x[n] \rightarrow \boxed{\uparrow M} \rightarrow \boxed{h[n] = M\,\frac{\sin\left(\frac{\pi}{M}\,n\right)}{\pi n}} \rightarrow \boxed{\downarrow N} \rightarrow y[n]. \tag{9.151}$$

The MATLAB/MathScript recipe for multirate signal processing a sampling with rate M/N is

MATLAB/MathScript

```
L=length(X);Z=[X;zeros(M-1,L)];Y=Z(:)';
F=fft(Y);F(L/2:L*M+2-L/2)=0;Y=ifft(F);
Y=Y(1:N:end);
```

Performing upsampling before downsampling avoids aliasing, provided the final sampling rate (after the downsampling step) exceeds twice the maximum frequency in the spectrum of the original signal $x[n]$.

Example 9-19: Multirate Signal Processing Combinations

Consider the following three systems:

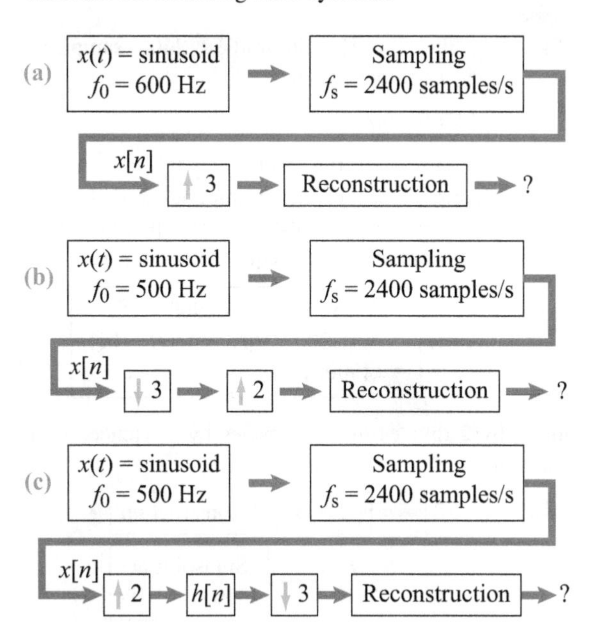

In each case, the input is a sinusoid $x(t)$ at the designated frequency f_0, which gets sampled at $f_s = 2400$ samples/s. The reconstruction step converts discrete-time sinusoids into continuous-time sinusoids. Determine the frequencies of the output sinusoids, in Hz. Also, in part (c), specify the cutoff frequency of the brickwall lowpass filter $h[n]$. For simplicity, ignore the back-and-forth conversion between continuous-time frequency in Hz and discrete-time frequency $\Omega = 2\pi f/f_s$.

Solution:

(a) Per Section 6-13.4, when a continuous-time sinusoid $x(t)$ of frequency f_0 is sampled at a rate f_s to produce $x[n]$, the spectrum of $x[n]$ contains components at

$$\pm\{f_0, f_0 \pm f_s, f_0 \pm 2f_s, \dots\}.$$

In the present case, $f_0 = 600$ Hz and $f_s = 2400$ samples/s. Hence, the frequency components of the sampled signal are

$$\pm\{600, 1800, 3000, \dots\}.$$

Upsampling $x[n]$ by 3 shrinks its spectrum by the same factor. So, the components of the output signal after reconstruction are those indicated at the output of the following diagram:

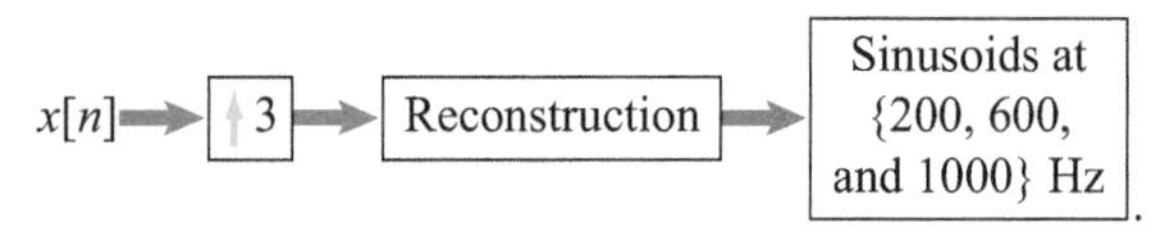

Since $f_0 = 600$ Hz, the Nyquist rate is 1200 Hz. Consequently, higher-frequency components will not survive the reconstruction process.

(b) Sampling the 500 Hz sinusoid at 2400 samples/s generates frequency components at

$$\pm\{500, 1900, 2900, 4300, \ldots\}.$$

Downsampling by 3 expands the spectrum by the same factor. Also, another consequence of downsampling is that extra components get added to make the spectrum periodic with period 2400 Hz (in continuous-time equivalent). Hence,

$$x[n] \rightarrow \boxed{\downarrow 3} \rightarrow \begin{array}{l} \pm\{3 \times 500 - f_s, 3 \times 1900 - 2f_s, \\ \quad 3 \times 2900 - 3f_s, 3 \times 4300 - 4f_s\} \\ = \pm\{900, 1500, 3300, \ldots\} \end{array}.$$

Upsampling by 2 divides the frequencies by 2. Hence, after reconstruction,

$$x[n] \rightarrow \boxed{\downarrow 3} \rightarrow \boxed{\uparrow 2} \rightarrow \boxed{\text{Reconstruction}} \rightarrow \boxed{\text{Sinusoids at } \{450, 750\} \text{ Hz}}.$$

The $3300/2 = 1650$ Hz component is above the Nyquist rate, so it gets removed in the reconstruction process.

(c) Sampling the 500 Hz sinusoid generates the same frequency components as in part (b). Upsampling by 2 shrinks the spectrum by half. Hence

$$x[n] \rightarrow \boxed{\uparrow 2} \rightarrow \boxed{\pm\{250, 950, 1450, \ldots\}}.$$

To perform the proper interpolation after upsampling by a factor $L = 2$, the lowpass filter should have a cutoff frequency

$$\Omega_0 = \frac{\pi}{L} = \frac{\pi}{2}.$$

The corresponding continuous-time cutoff frequency is

$$f_0 = \frac{\Omega_0 f_s}{2\pi} = \frac{\pi}{2} \times \frac{2400}{2\pi} = 600 \text{ Hz}.$$

The lowpass filter removes all components of the upsampled signal except for ±250 Hz. The final step of downsampling by 3 increases the remaining frequency component by a factor of 3 to 750 Hz. Hence,

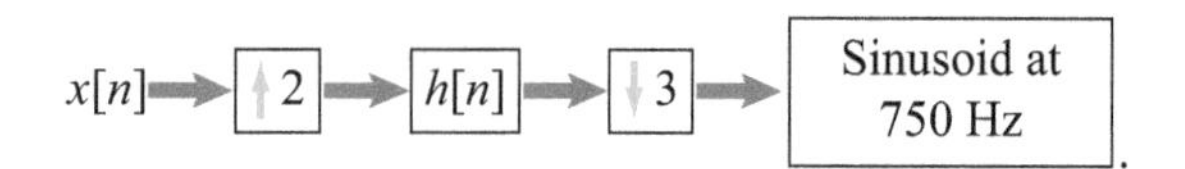

Concept Question 9-13: Why would we ever want to downsample? (See S²)

9-10 Oversampling by Upsampling

As noted earlier in Section 6-13, sampling a continuous-time signal at f_s samples/s creates a spectrum with extra copies of the original signal spectrum, repeated every f_s Hz along the frequency axis (Fig. 9-29(b)). Ultimately, most discrete-time signals must be converted back (*reconstructed*) to continuous time so they may be physically heard or displayed. To perform the reconstruction process with minimal distortion, it is necessary to eliminate the extra spectral copies and retain only the original spectrum of the signal. This calls for the use of an appropriate lowpass filter such as the Butterworth filter presented in Section 6-9 and depicted pictorially in Fig. 6-39. The filtering process is performed by a physical filter composed of op amps, resistors, and capacitors. As we will explain shortly, the degree of complexity of the filter (number of stages and components) is governed (in practice) by the ratio of the sampling rate f_s to the Nyquist rate of the signal, $2f_{\max}$.

For a fixed-duration continuous-time signal, the sampling rate determines the total number of discrete samples used to represent the signal. A high sampling rate translates into more computer storage space to store the samples, as well as more computation when the samples are used to perform signal processing tasks. Hence, from the standpoint of signal processing, the sampling rate should be selected to be as low as possible. To avoid loss of information, however, f_s should exceed the Nyquist rate of $2f_{\max}$. This means that the signal should be sampled at a rate only slightly greater than the Nyquist rate. Such a choice, however, places a stringent constraint on the roll-off rate S_g (Fig. 6-3(a)) of the lowpass filter in the transition band between the passband and rejection band. A steeper slope (higher roll-off rate) requires more op-amp stages, and therefore more hardware (Fig. 6-19(c)).

To reduce the cost and complexity of the lowpass filter, we can sample the original signal at a rate many times greater than the Nyquist rate, thereby creating greater spacing along the frequency axis between the central spectrum and the copies generated by the sampling process. Hence, far fewer op-amp stages are needed because the roll-off rate need not very high, but the higher sampling rate means that we have many more samples to deal with.

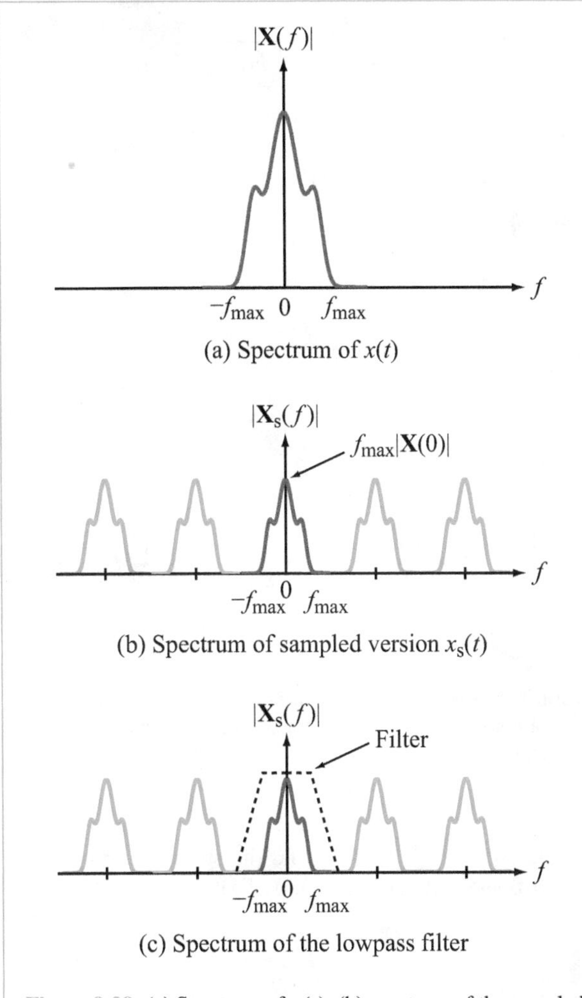

Figure 9-29: (a) Spectrum of $x(t)$, (b) spectrum of the sampled version $x_s(t)$, and (c) spectrum of the lowpass filter used to remove the extra spectral copies in the reconstruction process.

Instead of having to make the difficult trade-off between hardware complexity and data volume, we can circumvent the issue by adding one step prior to the reconstruction step. We can start by sampling the original signal at a rate only slightly higher than the Nyquist rate, but before we apply the lowpass filtering, we apply upsampling and interpolation to the discrete signal so as to widen the gap between the central spectrum and its neighboring copies. Upsampling by a factor L causes the central spectrum and all of its copies to shrink by the same factor, and interpolation then removes the extra copies, leaving the (narrower) copies, thereby increasing the frequency space between them. The process is called ***oversampling***. For example, "32X oversampling" means that the implemented oversampling rate is 32 times the actual sampling rate f_s. The signal is sampled at f_s, goes through all of the intended signal processing steps, and then just before reconstructing it into a continuous-time signal, it is upsampled and interpolated. This next-to-final step relaxes the constraints on the structure and complexity of the hardware of the lowpass filter, and yet the high data rate occurs only at the very end of the process.

9-10.1 Case Study: Reconstruction with and without Oversampling

To illustrate the utility of oversampling, we consider a specific scenario twice, once without the use of oversampling, and again with oversampling. We will demonstrate that for the specific system described shortly:

▶ With $\times 1$ oversampling (i.e., no oversampling), a 10th-order Butterworth filter is required to meet the specified criteria. Using the Sallen-Key circuit of Fig. 6-43, a 10th-order filter requires 5 op amps, 10 capacitors, and 10 resistors.

In contrast, with $11\times$ oversampling, only a 2nd-order filter, composed of 1 op amp, 2 capacitors, and 2 resistors, is needed, representing a saving of 80% in hardware components! ◀

9-10.2 Without Oversampling

Consider a continuous-time signal $x(t)$ with a corresponding frequency spectrum (Fourier transform)

$$\mathbf{X}(f) = \int_{-\infty}^{\infty} x(t)\, e^{-j2\pi f t}\, dt. \tag{9.152}$$

The spectrum has bandwidth B, with $\mathbf{X}(f) = 0$ for $|f| > B$ Hz. The total bidirectional bandwidth is $B_b = 2B$, and $f_{max} = B$ (see Fig. 6-52(a)).

The Nyquist rate, which is the minimum sampling rate that permits perfect reconstruction of $x(t)$ from its samples, is $2B$ samples/s. To be on the safe side, $x(t)$ was sampled at 1.5 times the Nyquist rate; that is, at $f_s = 3B$. The equivalent sampling interval is $T_s = 1/f_s = 1/(3B)$, and the sampled signal is then given by Eq. (6.150) as

$$x_s(t) = \sum_{n=-\infty}^{\infty} x[n]\, \delta(t - nT_s) = \sum_{n=-\infty}^{\infty} x[n]\, \delta\left(t - \frac{n}{3B}\right). \tag{9.153}$$

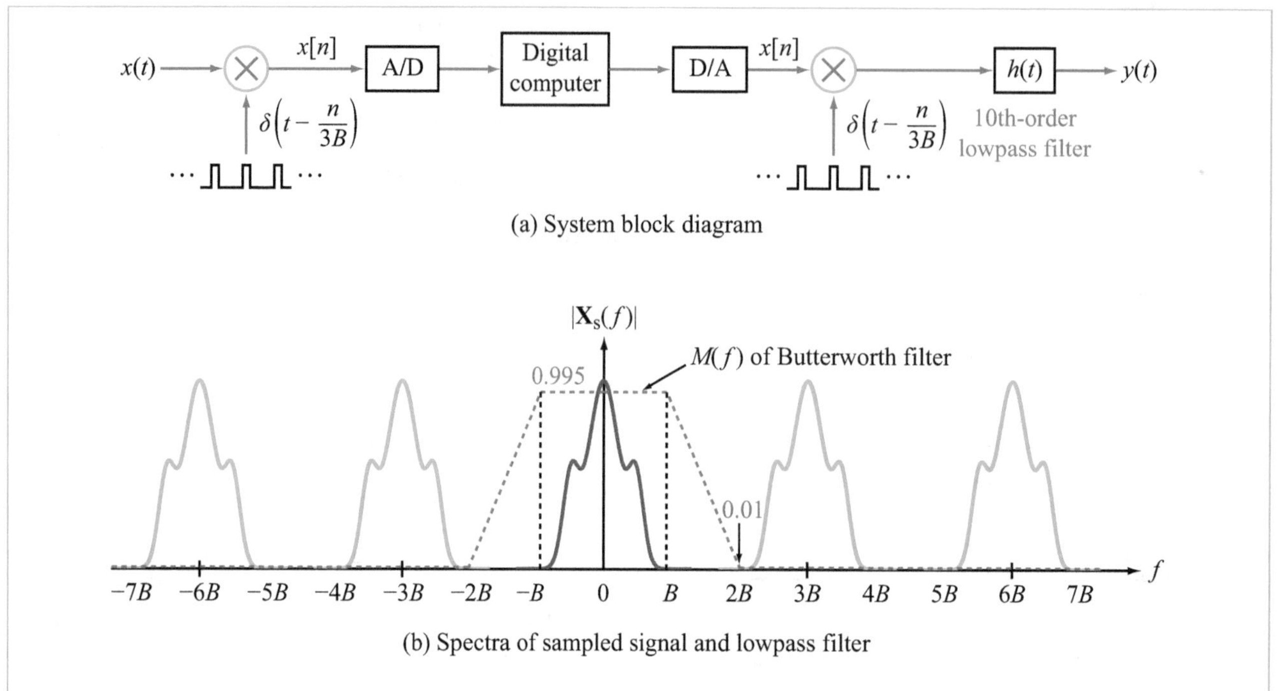

Figure 9-30: Signal $x(t)$ is sampled at $3B$ samples/s, 1.5 times the Nyquist rate to generate $x[n]$, which is then lowpass-filtered by a 10th-order Butterworth filter $h(t)$. The A/D converter changes the sampled signal into a binary sequence for computer storage and signal modification, and the D/A converts the digital signal back into analog discrete-time format.

A system block diagram is shown in Fig. 9-30(a). As noted earlier, the sampling process generates copies of the spectrum at a spacing of $f_s = 3B$, as depicted in Fig. 9-30(b). Because the central spectrum extends to B and the first copy starts at $2B$, it is necessary to use a highly selective continuous-time lowpass filter in order to reconstruct $x(t)$ from $x[n]$. The desired filter performance is specified in terms of its gain function $M(f)$:

$$\begin{aligned} 0.995 \le M(f) \le 1 \qquad & \text{for } 0 \le |f| \le B, \\ 0 \le M(f) \le 0.01 \qquad & \text{for } 2B \le |f| \le \infty. \end{aligned} \tag{9.154}$$

The specifications call for a gain between 0.995 and 1 in the passband, and no greater than 0.01 in the stop band. The steep change in gain level has to occur between $f = B$ and $f = 2B$.

The filter performance is to be realized using a Butterworth lowpass filter. From Example 6-20, the gain function of a Butterworth lowpass filter with cutoff frequency f_c, gain at dc (0 Hz) of 1, and order N is given by

$$M(f) = \frac{1}{\sqrt{1 + (f/f_c)^{2N}}} . \tag{9.155}$$

Our unknowns are f_c and N, and the specifications call for $M(B) = 0.995$ and $M(2B) = 0.01$. That is,

$$0.995 = \frac{1}{\sqrt{1 + (B/f_c)^{2N}}} , \tag{9.156a}$$

$$0.01 = \frac{1}{\sqrt{1 + (2B/f_c)^{2N}}} . \tag{9.156b}$$

Upon squaring the two equations and then combining them together to solve for the unknown quantities, we obtain:

$$N = 10, \qquad f_c = \sqrt[3]{2}\, B = 1.259B.$$

Hence, a 10th-order filter is required.

9-10.3 With 11× Oversampling

The system shown is Fig. 9-31(a) is similar to the one in Fig. 9-30(a), except for the addition of two operations, namely upsampling by 11 and interpolation by a discrete-time Butterworth filter with impulse response $h_1[n]$. The upsampling and interpolation increases the sampling rate of $x(t)$ from $3B$ to

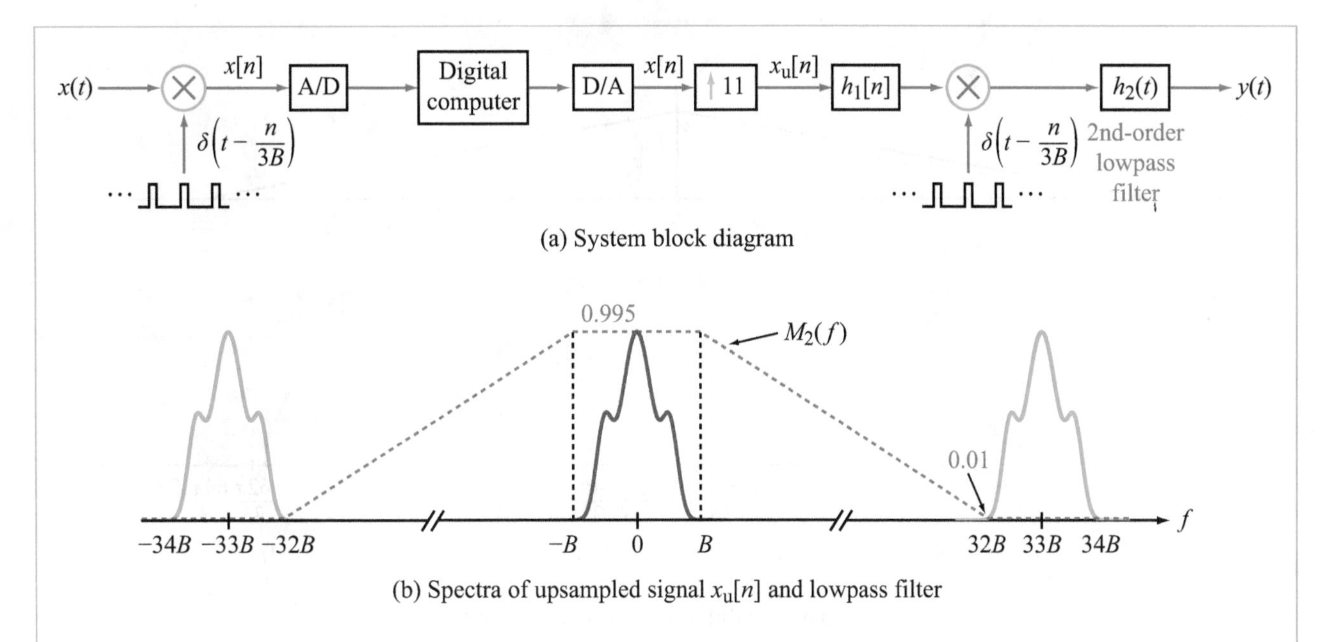

Figure 9-31: Modified version of the system and associated spectrum shown in Fig. 9-30; the modified system includes upsampling by 11 followed by interpolation performed by a discrete-time filter with impulse response $h_1[n]$.

an apparent rate of $33B$. Consequently, the separation between the central spectrum and the neighboring copy increases from B (as in Fig. 9-30(b)) to $31B$ (Fig. 9-31(b)). Accordingly, the specifications on the gain function of the analog Butterworth filter $h_2(t)$ are now given by

$$\begin{aligned} 0.995 \leq M_2(f) &\leq 1 && \text{for } 0 \leq |f| \leq B, \\ 0 \leq M_2(f) &\leq 0.01 && \text{for } 32B \leq |f| \leq \infty. \end{aligned} \tag{9.157}$$

The equations analogous to Eq. (9.156) are

$$0.995 = \frac{1}{\sqrt{1 + (B/f_c)^{2N}}}, \tag{9.158a}$$

$$0.01 = \frac{1}{\sqrt{1 + (32B/f_c)^{2N}}}, \tag{9.158b}$$

which together lead to the solution:

$$N = 2, \qquad f_c = 3.162B.$$

The upsampling/interpolation operations allowed us to realize the same final result using only a 2nd-order filter instead of a 10th-order filter.

9-10.4 Discrete-Time Interpolation Filter

The block diagram shown in Fig. 9-31(a) includes an interpolation filter with impulse response $h_1[n]$. The spectrum of the original signal $x(t)$ extends between $-B$ and $+B$. When sampled at a rate $f_s = 3B$, the discrete-time signal $x[n]$ extends over Ω between

$$\Omega_{\min} = -\frac{2\pi f_{\max}}{f_s} = -\frac{2\pi B}{3B} = -\frac{2\pi}{3},$$

and $\Omega_{\max} = |\Omega_{\min}|$. The spectrum of $x[n]$ is depicted in black in Fig. 9-32(a).

Upsampling $x[n]$ by 11 shrinks the spectrum by the same factor, and adds copies at f_s, as depicted by the blue spectrum shown in Fig. 9-32(b). The function of the interpolation filter $h_1[n]$ is to remove the higher frequency copies. Its proposed frequency response $\mathbf{H}_1(e^{j\Omega})$, which is shown in red, obeys the following specifications:

$$\begin{aligned} 0.995 \leq |\mathbf{H}_1(e^{j\Omega})| &\leq 1 && \text{for } 0 \leq |\Omega| \leq \frac{2\pi}{33}, \\ 0 \leq |\mathbf{H}_1(e^{j\Omega})| &\leq 0.01 && \text{for } \frac{4\pi}{33} \leq |\Omega| \leq \frac{62\pi}{33}. \end{aligned} \tag{9.159}$$

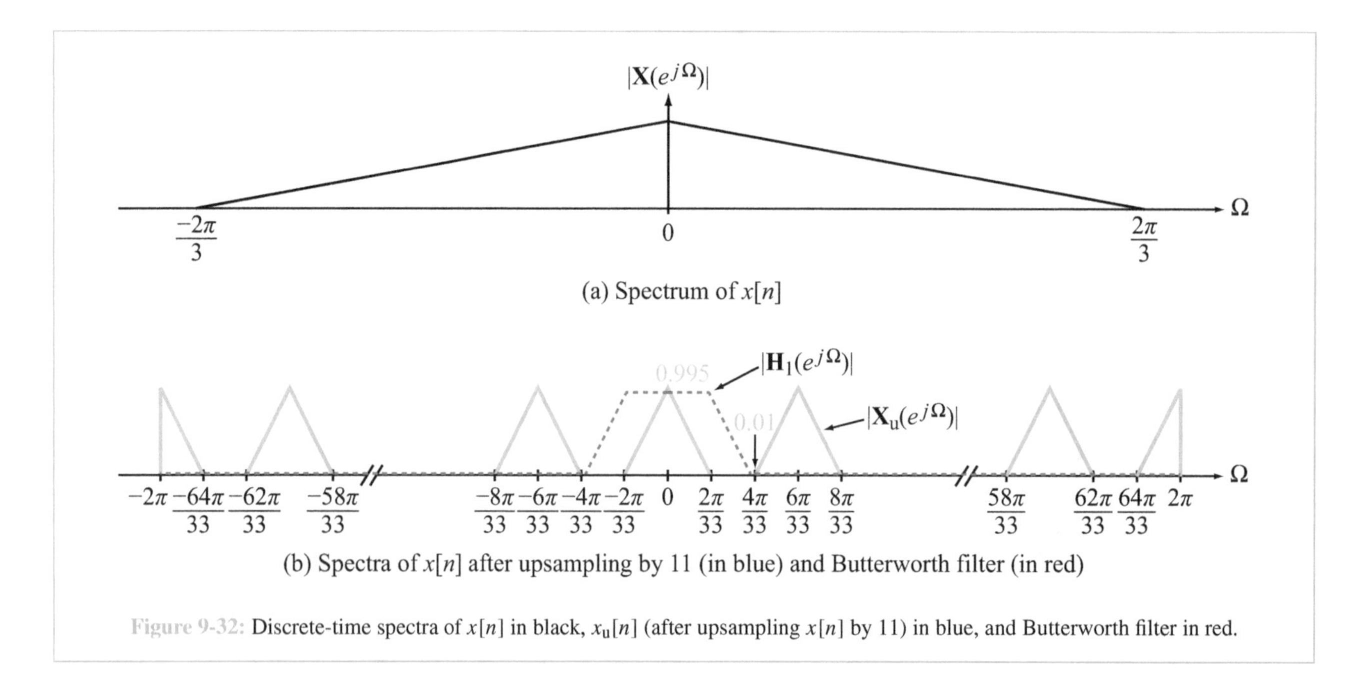

Figure 9-32: Discrete-time spectra of $x[n]$ in black, $x_u[n]$ (after upsampling $x[n]$ by 11) in blue, and Butterworth filter in red.

▶ This is a highly selective filter, except now it is a discrete-time filter that can be implemented computationally. ◀

The filter specifications are identical in form to those given earlier by Eq. (9.154), except for scaling the frequency by $\Omega = 2\pi f/(33B)$. That is, for an equivalent continuous-time Butterworth filter, the order should be $N = 10$ and the cutoff frequency should be $f_c = \sqrt[3]{2}\, B$. The equivalent discrete-time cutoff frequency is

$$\Omega_c = \sqrt[3]{2} \times \frac{2\pi B}{33B} = \sqrt[3]{2} \times \frac{2\pi}{33}. \tag{9.160}$$

The filter $h_1[n]$ can be designed using the bilinear transformation introduced in Section 9-4.4. The prewarping formula relating continuous-time angular frequency $\omega = 2\pi f$ to discrete-time angular frequency Ω is given by Eq. (9.98) as

$$\omega = \frac{2}{T} \tan\left(\frac{\Omega}{2}\right). \tag{9.161}$$

We choose the prewarping factor T by matching the cutoff frequency ω_c in continuous time to the cutoff frequency Ω_c in discrete time. The latter is given by Eq. (9.160) and the former is

$$\omega_c = 2\pi f_c = 2\pi \sqrt[3]{2}\, B. \tag{9.162}$$

Upon replacing ω with ω_c and Ω with Ω_c in Eq. (9.161), we have

$$\omega_c = \frac{2}{T} \tan\left(\frac{\Omega_c}{2}\right),$$

or

$$2\pi \sqrt[3]{2}\, B = \frac{2}{T} \tan\left(\sqrt[3]{2}\, \frac{2\pi}{2 \times 33}\right).$$

The angle of the tangent function is very small, which allows us to use the approximation $\tan\theta \approx \theta$. Consequently, we determine that

$$T = \frac{1}{33B}.$$

Despite the fact that the prewarping formula is basically nonlinear in character, the tangent approximation makes the relationship between ω and Ω essentially linear, thereby satisfying the conditions specified by Eq. (9.159).

The actual discrete-time filter $\mathbf{H(z)}$ is designed from the continuous-time Butterworth filter $\mathbf{H(s)}$ as follows:

(1) $\mathbf{H(s)}$ is a 10th-order Butterworth filter with cutoff frequency $\omega_c = 2\pi \sqrt[3]{2}\, B$. From Table 6-3 and Eq. (6.96), the 10 poles of a 10th-order Butterworth filter are:

$$\{\omega_c e^{\pm j98^\circ}, \omega_c e^{\pm j116^\circ}, \omega_c e^{\pm j134^\circ}, \omega_c e^{\pm j152^\circ}, \omega_c e^{\pm j170^\circ}\}.$$

These 10 poles are in 5 complex conjugate pairs. We label the 5 pole angles as

$$\theta_1 = 98^\circ,\ \theta_2 = 116^\circ,\ \theta_3 = 134^\circ,\ \theta_4 = 152^\circ,\ \theta_5 = 170^\circ.$$

For the first pair of conjugate poles:

$$(\mathbf{s} - \omega_c e^{j98°})(\mathbf{s} - \omega_c e^{-j98°}) = \mathbf{s}^2 - 2\omega_c \cos(98°)\, \mathbf{s} + \omega_c^2, \tag{9.163}$$

and similarly for the other 4 complex conjugate pairs of poles.

(2) The transfer function $\mathbf{H}_1(\mathbf{s})$ of the 10th-order Butterworth lowpass filter with cutoff frequency ω_c is

$$\mathbf{H}_1(\mathbf{s}) = \prod_{k=1}^{5} \frac{\omega_c^2}{\mathbf{s}^2 - 2\omega_c \cos(\theta_k)\mathbf{s} + \omega_c^2}. \tag{9.164}$$

At dc($\mathbf{s} = j\omega = 0$), the gain of $\mathbf{H}_1(j\omega)$ is 1.

(3) Bilinear transformation is performed by substituting for $\mathbf{s}$ in $\mathbf{H}_1(\mathbf{s})$ the bilinear transform

$$\mathbf{s} = \frac{2}{T}\frac{\mathbf{z}-1}{\mathbf{z}+1} = 66B\,\frac{\mathbf{z}-1}{\mathbf{z}+1}. \tag{9.165}$$

The transfer function for the discrete-time Butterworth filter is then

$$\mathbf{H}_1(\mathbf{z}) = \mathbf{H}_1(\mathbf{s})\Big|_{\mathbf{s}=\left(66B\,\frac{\mathbf{z}-1}{\mathbf{z}+1}\right)} \tag{9.166}$$

$$= \prod_{k=1}^{5} \frac{\omega_c^2}{\left[66B\,\frac{\mathbf{z}-1}{\mathbf{z}+1}\right]^2 - 2\omega_c\cos(\theta_k)\left[66B\,\frac{\mathbf{z}-1}{\mathbf{z}+1}\right] + \omega_c^2}. \tag{9.167}$$

As presented in Section 7-10, the corresponding impulse response $h_1[n]$, the poles and zeros, and the ARMA difference equation can all be obtained from $\mathbf{H}_1(\mathbf{z})$. The ARMA difference equation requires 18 storage registers to implement it, and each recursion requires 18 multiplications and 18 additions. The 18 multiplications can be reduced to 13 by taking advantage of the even symmetry of the right-hand side of the ARMA difference equation. The upsampled $x[n]$ is 91% zeros, so the computation of filtering with $h[n]$ is reduced to 9%.

Concept Question 9-14: Why can upsampling and interpolation obviate the need for a continuous-time lowpass filter with many circuit elements? (See S²)

9-11 Audio Signal Processing

Suppose we are given a snippet of a trumpet playing a single note. The goal is to generate snippets of the trumpet playing all of the other musical notes from the single note. We can do this using multirate filtering and the "circle of fifths" from music theory.

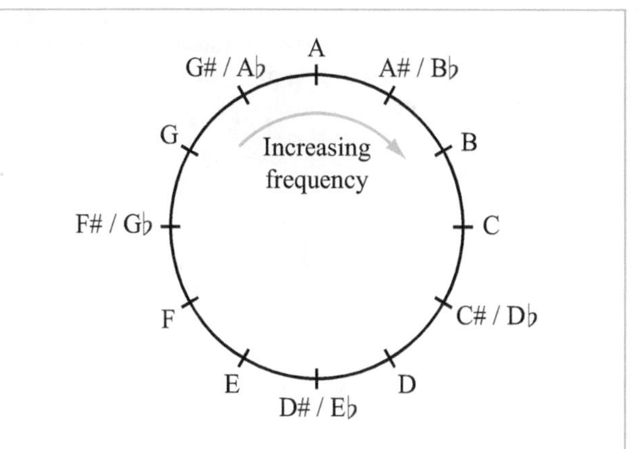

Figure 9-33: Letter names of the pitches (notes) in Western music, arranged by frequency. The notes are arranged in a circle because the pattern repeats itself every octave.

9-11.1 Music Notation and Frequencies

An *octave* is the distance or *interval* between two musical pitches whose frequency ratio is 1 : 2. Pitches, or musical notes, are assigned letter names according to their frequency, and notes an octave apart are assigned the same letter. In most Western music, the octave is partitioned into 12 distinct pitches, as shown in Fig. 9-33. The symbols "♯" and "♭" are read as "sharp" and "flat," respectively; A♯ and B♭ are alternative names for the same note. The white keys of the piano keyboard produce A, B, C, D, E, F, and G, while the black keys produce the other five notes.

For at least 200 years, most keyboard instruments have been tuned according to the system of *equal temperament*, in which the ratio of the frequencies of any pair of adjacent notes is $2^{1/12} \approx 1.0595$. This system was adopted to ensure that the frequency ratios of intervals and chords would remain unchanged in music in all possible keys. The most common pitch standard used for tuning instruments today is A = 440 Hz, where A is the A above middle C on the piano. Table 9-3 shows the approximate frequencies of the twelve notes from this A up to the next G♯.

The interval between adjacent notes is called a *semitone*. We note from Table 9-3 that the ratio between the frequencies of notes that are 7 semitones apart, such as A (440 Hz) and E (659 Hz), is approximately 2 : 3. This is not a coincidence: this frequency ratio is basic to Western music as well as the musics of many other cultures, and one of the advantages of the equal-temperament system is that it incorporates a close approximation to the 2 : 3 ratio.

Table 9-3: Frequencies of the twelve notes from A = 440 Hz to G♯ in order of increasing frequency, in the equal-temperament system.

Note	Frequency (Hz)
A	440
A♯	$440 \times 2^{1/12} \approx 466$
B	$440 \times 2^{2/12} \approx 494$
C	$440 \times 2^{3/12} \approx 523$
C♯	$440 \times 2^{4/12} \approx 554$
D	$440 \times 2^{5/12} \approx 587$
D♯	$440 \times 2^{6/12} \approx 622$
E	$440 \times 2^{7/12} \approx 659$
F	$440 \times 2^{8/12} \approx 698$
F♯	$440 \times 2^{9/12} \approx 740$
G	$440 \times 2^{10/12} \approx 784$
G♯	$440 \times 2^{11/12} \approx 830$

Table 9-4: Frequencies of 12 notes from A = 440 Hz up to the next G♯, derived by multiplying 440 by successive powers of 3/2 and then dividing by powers of 2 as needed to bring the result within the octave range 440–880 Hz.

Note	Frequency (Hz)
A	440
E	$440 \times (3/2)^1 \approx 660$
B	$440 \times (3/2)^2/2 \approx 495$
F♯	$440 \times (3/2)^3/2 \approx 742$
C♯	$440 \times (3/2)^4/2^2 \approx 557$
G♯	$440 \times (3/2)^5/2^2 \approx 835$
D♯	$440 \times (3/2)^6/2^3 \approx 626$
A♯	$440 \times (3/2)^7/2^4 \approx 470$
F	$440 \times (3/2)^8/2^4 \approx 705$
C	$440 \times (3/2)^9/2^5 \approx 529$
G	$440 \times (3/2)^{10}/2^5 \approx 793$
D	$440 \times (3/2)^{11}/2^6 \approx 595$

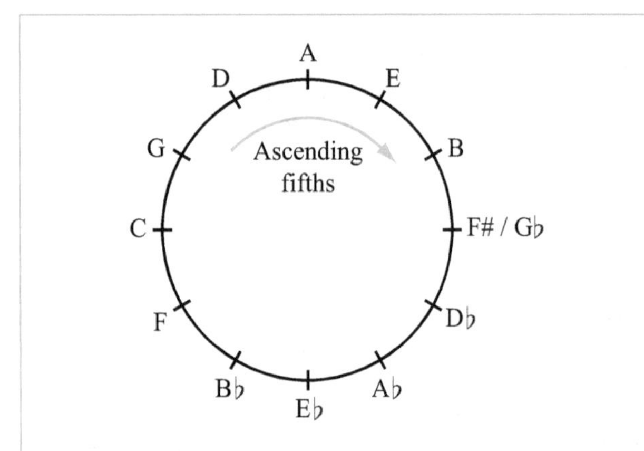

Figure 9-34: The circle of fifths. The circle is usually portrayed rotated clockwise by 90° in music theory, so that C is at the top.

If we proceed clockwise around the circle of Fig. 9-33 by taking repeated jumps of 7 "hours" (semitones) each, we will visit every note, since 7 and 12 are relatively prime. By taking the notes in this order, we can create another circle known as the ***circle of fifths*** (Fig. 9-34), so called because adjacent notes in this circle are five letter names apart; for example, D and A, which are adjacent on the circle, are spanned by D, E, F, G, and A.

9-11.2 Application of Multirate Signal Processing

By repeatedly upsampling by 2 and downsampling by 3, we can start with the snippet of a trumpet playing note A and multiply that frequency (440 Hz) by powers of $\frac{3}{2}$. This sweeps out all of the 12 notes, in the order shown in Table 9-4. Extra upsamplings by 2 are frequently necessary to keep the frequencies within this octave. We can then obtain the 12 notes in all other octaves by upsampling and downsampling by 2 to synthesize the same note in a different octave.

Ironically, this would be easier to implement in continuous time. A variable speed tape recorder or record turntable could implement multirate processing by speeding up or slowing down the tape or turntable, using a variable-speed motor. The music "group" Alvin and the Chipmunks (the vocals were actually sung by one person) was recorded in this way in 1958 by speeding up the playback by a factor of two on a variable-speed tape recorder. It was then necessary to (literally!) cut and paste snippets of tape. The original Chipmunks record won the first Grammy award for best-engineered nonclassical album in 1959.

Concept Question 9-15: Why does the circle of fifths make it simple to use multirate filtering to generate all notes from a single note? (See S^2)

9-12 Correlation

The term *correlation* encompasses three different types of relationships between two real-valued signals $x[n]$ and $y[n]$:

(a) Autocorrelation

$$r_x[n] = x[n] * x[-n], \qquad (9.168a)$$

and the duration of $r_x[n]$ is $2N-1$, where N is the duration of $x[n]$.

(b) Cross-Correlation

$$r_{xy}[n] = x[n] * y[-n], \qquad (9.168b)$$

and the duration of $r_{xy}[n]$ is $N_x + N_y - 1$, where N_x and N_y are the durations of $x[n]$ and $y[n]$, respectively.

(c) Correlation

$$r_{xy}[0],$$

which denotes the simultaneous presence or absence of signals $x[n]$ and $y[n]$.

Whereas the autocorrelation and cross-correlation operations generate new signals, correlation generates a single number.

9-12.1 Autocorrelation

The convolution operation of two discrete-time signals, as given by Eq. (7.51a), involves the multiplication of one of the signals with a *reflected and delayed* image of the other. The correlation functions defined in this section involve a similar form of convolution, but without the reflection.

The autocorrelation $r_x[n]$ of $x[n]$ is given by

$$r_x[n] = x[n] * x[-n] = \sum_{i=-\infty}^{\infty} x[i]\, x[i+n]. \qquad (9.169)$$

The *lag* n of $r_x[n]$ has units of time, but strictly speaking, it is an index. Lag is the delay of $x[n]$ relative to a copy of itself.

Based on the definition for $r_x[n]$, the expressions for lags $n = 0, \pm 1$, and ± 2 are

$$r_x[0] = \cdots + (x[-1])^2 + (x[0])^2 + (x[1])^2 + \cdots$$

$$r_x[\pm 1] = \cdots + x[-1]\, x[0] + x[0]\, x[1] + x[1]\, x[2] + \cdots$$

$$r_x[\pm 2] = \cdots + x[-1]\, x[1] + x[0]\, x[2] + x[1]\, x[3] + \cdots .$$

Properties of $r_x[n]$

Property 1: $r_x[n]$ is an even function: $r_x[n] = r_x[-n]$.

Property 2: $r_x[0] = \sum_{i=-\infty}^{\infty} (x[i])^2 \geq 0 =$ energy of $x[n]$.

Property 3: $r_x[0] \geq r_x[n]$ for any lag n.

Property 4: The DTFT $\mathbf{R}_x(e^{j\Omega})$ of $r_x[n]$ is related to the DTFT $\mathbf{X}(e^{j\Omega})$ of $x[n]$ as follows:

$$\mathbf{R}_x(e^{j\Omega}) = \mathbf{X}(e^{j\Omega})\, \mathbf{X}(e^{-j\Omega}) = \mathbf{X}(e^{j\Omega})\, \mathbf{X}^*(e^{j\Omega}) = |\mathbf{X}(e^{j\Omega})|^2. \qquad (9.170)$$

The final step was obtained by using the time-reversal and conjugate symmetry properties of the DTFT. Note that $\mathbf{R}_x(e^{j\Omega})$ is a real and even function.

Property 5: Delaying $x[n]$ by D does not affect $r_x[n]$.

This property can be demonstrated by applying the time-delay property of the DTFT, which states that the DTFT of $x'[n] = x[n-D]$ is given by

$$\mathbf{X}'(e^{j\Omega}) = \mathbf{X}(e^{j\Omega})\, e^{-j\Omega D}. \qquad (9.171)$$

The DTFT of $x'[n]$ is

$$\mathbf{R}_{x'}(e^{j\Omega}) = |\mathbf{X}'(e^{j\Omega})|^2 = |\mathbf{X}(e^{j\Omega})\, e^{-j\Omega D}|^2$$
$$= |\mathbf{X}(e^{j\Omega})|^2 = \mathbf{R}_x(e^{j\Omega}). \qquad (9.172)$$

Thus, the DTFTs of the autocorrelation $r_x[n]$ of $x[n]$ and of $r_{x'}[n]$ of the delayed version $x'[n] = x[n-D]$ are the same. Since the DTFTs are the same, it follows that the autocorrelations themselves are the same.

MATLAB/MathScript Recipe

Computing the Autocorrelation of X

```
N=length(X);M=nextpow2(2*N-1);
RX=real(ifft(abs(fft(X,M)).^2));
RX=fftshift(RX) puts rx(0) in the middle.
```

Example 9-20: Autocorrelation of $x[n] = \{3, 1, 4\}$

Compute $r_x[n]$ of $x[n]$.

Solution:

$$r_x[0] = 3^2 + 1^2 + 4^2 = 26,$$

$$r_x[\pm 1] = (3)(1) + (1)(4) = 7,$$

$$r_x[\pm 2] = (3)(4) = 12.$$

Hence,

$$r_x[n] = \{12, 7, \underline{26}, 7, 12\}.$$

$r_x[n]$ is even and $r_x[0] \geq r_x[n]$, as expected.

9-12.2 Using Autocorrelation to Compute Period of Noisy Signal

Suppose $x[n]$ is a signal with unknown period N. The goal is to estimate N from the noisy observations

$$y[n] = x[n] + w[n], \tag{9.173}$$

where $w[n]$ is white Gaussian noise (Section 6-10.1) with zero mean and variance σ^2.

The autocorrelation $r_y[n]$ of $y[n]$ is

$$\begin{aligned} r_y[n] &= (x[n] + w[n]) * (x[-n] + w[-n]) \\ &= x[n] * x[-n] + w[n] * w[-n] \\ &\quad + \underbrace{x[n] * w[-n]}_{r_{xw}[n]\approx 0} + \underbrace{x[-n] * w[n]}_{r_{xw}[-n]\approx 0} \\ &\approx r_x[n] + r_w[n] \approx r_x[n] + \sigma^2 \delta[n]. \end{aligned} \tag{9.174}$$

The final expression of Eq. (9.174) ignores the terms involving correlations between $x[n]$ and $w[n]$ because $w[n]$ is zero-mean uncorrelated noise.

Since the signal period is N,

$$x[n] = x[n+N],$$

and the autocorrelations at lags N, $2N$, etc., are

$$\begin{aligned} r_x[N] &= \sum x[i]\, x[i+N] = \sum x[i]\, x[i] = r_x[0], \\ r_x[2N] &= \sum x[i]\, x[i+2N] = \sum x[i]\, x[i] = r_x[0], \\ &\vdots \quad \vdots \end{aligned} \tag{9.175}$$

which leads to the conclusion that $r_x[n]$ is periodic with period N. So, to a good approximation,

$$r_y[n] \approx \begin{cases} \sigma^2 + \sum x^2[i] & \text{for } n = 0, \\ \sum x^2[i] & \text{for } n = N, \\ \sum x^2[i] & \text{for } n = 2N, \\ \vdots & \vdots \\ 0 & \text{otherwise.} \end{cases} \tag{9.176}$$

So the period N is found by searching for large peaks in $r_y[n]$, which will be at $n = 0, N, 2N, \ldots$. The peak at $n = 0$ is much larger, because σ^2 is added to it.

To illustrate with an example, we show in parts (a) and (b) of Fig. 9-35 a trumpet signal before and after the addition of random noise to it. The autocorrelation of the noisy signal is displayed in Fig. 9-35(c). Ignoring the peak at $n = 0$, the next highest peak is at $n = 90$ and multiples thereof. The original signal had been sampled at 44100 samples/s. Hence, the frequency of the signal corresponding to the peak at $n = 90$ is given by

$$f = \frac{f_s}{90} = \frac{44100}{90} = 490 \text{ Hz}.$$

9-12.3 Cross-Correlation

The ***cross-correlations*** $r_{xy}[n]$ and $r_{yx}[n]$ of signals $x[n]$ and $y[n]$ are defined as

$$r_{xy}[n] = x[n] * y[-n] = \sum_{i=-\infty}^{\infty} x[i]\, y[i-n], \tag{9.177a}$$

$$r_{yx}[n] = y[n] * x[-n] = \sum_{i=-\infty}^{\infty} y[i]\, x[i-n]. \tag{9.177b}$$

Upon replacing i with $(i+n)$ in the infinite summation, we have

$$r_{yx}[n] = \sum_{i=-\infty}^{\infty} x[i]\, y[i+n] = r_{xy}[-n]. \tag{9.177c}$$

The ***lag*** n of $r_{xy}[n]$ and $r_{yx}[n]$ represents the delay of one signal relative to the other.

Writing out the definition of $r_{xy}[n]$ for $n = 0, 1, 2$ gives

$$\begin{aligned} r_{xy}[0] &= \ldots x[-1]\, y[-1] + x[0]\, y[0] + x[1]\, y[1] + \cdots, \\ r_{xy}[1] &= \ldots x[0]\, y[-1] + x[1]\, y[0] + x[2]\, y[1] + \cdots, \\ r_{xy}[2] &= \ldots x[0]\, y[-2] + x[1]\, y[-1] + x[2]\, y[0] + \cdots. \end{aligned}$$

Cross-correlation, unlike autocorrelation, is not an even function. The DTFT $\mathbf{R_{xy}}(e^{j\Omega})$ of $r_{xy}[n]$ is related to $\mathbf{X}(e^{j\Omega})$ and $\mathbf{Y}(e^{j\Omega})$ by

$$\mathbf{R_{xy}}(e^{j\Omega}) = \mathbf{X}(e^{j\Omega})\, \mathbf{Y}^*(e^{j\Omega}). \tag{9.178}$$

The MATLAB/MathScript recipe for computing cross-correlation is

```
L=length([X Y]);M=nextpow2(L-1);
RXY=real(ifft(fft(X,M).*conj(fft(Y,M))));
RXY=fftshift(RXY)
```

Example 9-21: Compute Cross-Correlation

Compute the cross-correlation of $x[n] = \{3, 1, 4\}$ and $y[n] = \{2, 7, 1\}$.

Solution: The duration of r_{xy} is $N_x + N_y - 1 = 2 + 2 - 1 = 5$. The elements of $r_{xy}[n]$ are

$$\begin{aligned} r_{xy}[-2] &= (3)(1) = 3, \\ r_{xy}[-1] &= (3)(7) + (1)(1) = 22, \\ r_{xy}[0] &= (3)(2) + (1)(7) + (4)(1) = 17, \\ r_{xy}[1] &= (1)(2) + (4)(7) = 30, \\ r_{xy}[2] &= (4)(2) = 8. \end{aligned}$$

Hence,

$$r_{xy}[n] = \{3, 22, \underline{17}, 30, 8\}.$$

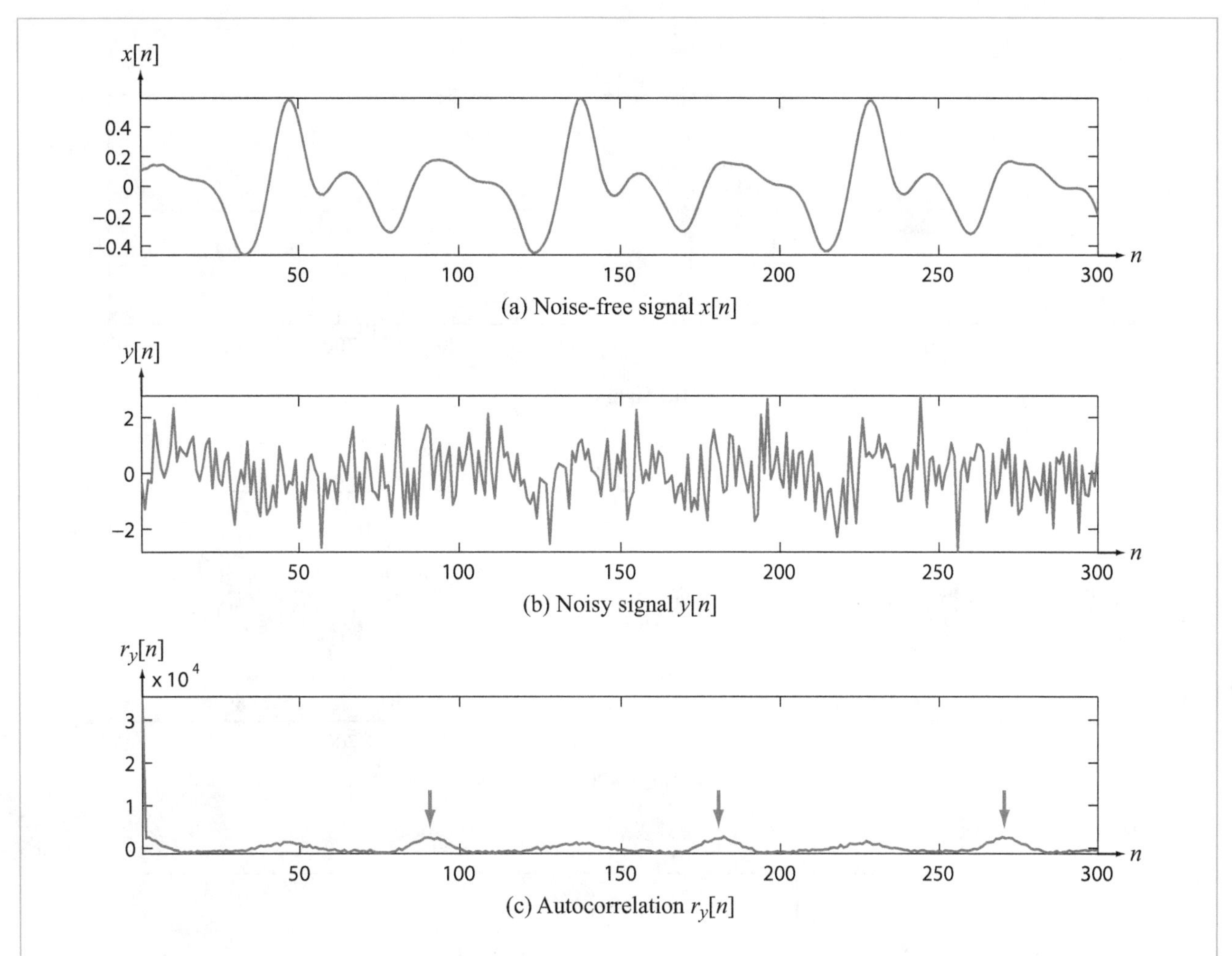

Figure 9-35: Estimating the period of a trumpet signal. (a) Noise-free signal, (b) noisy signal, (c) autocorrelation of noisy signal. The period is 90, corresponding to a frequency $f = 490$ Hz.

9-12.4 Using Cross-Correlation to Compute Time Delay

Let $x[n]$ be a known signal with unknown time delay Δ. Without loss of generality, let the energy of $x[n]$ be one. The goal is to estimate Δ from the noisy observations

$$y[n] = x[n - \Delta] + w[n], \tag{9.179}$$

where $w[n]$ is white Gaussian noise. The cross-correlation $r_{yx}[n]$ is

$$\begin{aligned} r_{yx}[n] &= (x[n-\Delta] + w[n]) * x[-n] \\ &= \underbrace{x[n-\Delta] * x[-n]}_{r_x[n-\Delta]} + \underbrace{w[n] * x[-n]}_{r_{wx}[n]\approx 0} \\ &\approx r_x[n-\Delta] \approx \delta[n-\Delta]. \end{aligned} \tag{9.180}$$

Even though the final approximation $r_x[n - \Delta] \approx \delta[n - \Delta]$ is rather tenuous, the cross-correlation of the observations with the known signal is expected to have a peak at $n = \Delta$.

In part (a) of Fig. 9-36, we display a short random pulse $x[n]$ delayed by an unknown Δ, relative to a reference discrete-

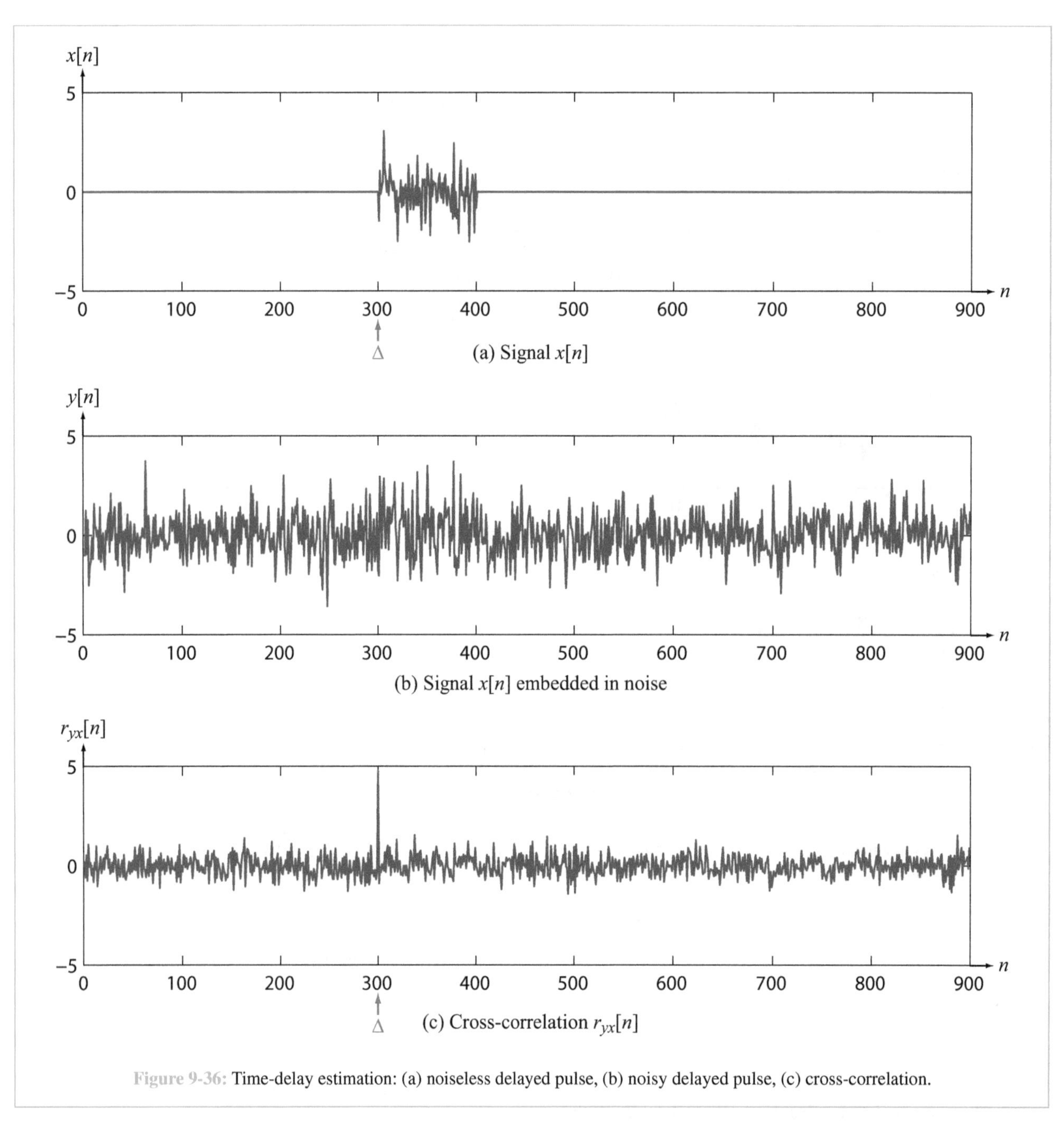

Figure 9-36: Time-delay estimation: (a) noiseless delayed pulse, (b) noisy delayed pulse, (c) cross-correlation.

time scale, and in part (b) we show the signal after adding it to white noise. Because (a) the amplitude variations of the signal and the noise are comparable to one another, and (b) both the signal and the noise are random in nature, it is difficult to discern the presence of the signal in the pattern shown in Fig. 9-36(b). Computing the cross-correlation between the known signal $x[n]$ and the recorded noisy signal $y[n]$ entails multiplying $y[n]$ with many shifted versions of $x[n]$. The peak in the record of $r_{yx}[n]$, displayed in Fig. 9-36(c), corresponds to the shift that generates the highest correlation, from which we deduce that the unknown delay is $\Delta = 300$.

9-12.5 Correlation

The ***correlation*** $r_{xy}[0]$ between signals $x[n]$ and $y[n]$, both of duration N, is given by

$$r_{xy}[0] = \sum_{i=0}^{N-1} x[i]\, y[i]. \tag{9.181}$$

If the two signals are initially not of the same duration, then *zero-padding* (Section 7-15.2) should be applied to make their lengths the same.

The sum in Eq. (9.181) is the inner product of vectors $\mathbf{x}$ and $\mathbf{y}$ with components $x[i]$ and $y[i]$, namely

$$\mathbf{x}^T \mathbf{y} = ||\mathbf{x}|| \cdot ||\mathbf{y}|| \cos\theta, \tag{9.182}$$

where $\mathbf{x}^T$ is the transpose of $\mathbf{x}$ and θ is the angle between the two vectors. To compute the inner product in MATLAB/MathScript, we use

$$X' * Y.$$

The correlation $r_{xy}[0]$ is a measure of ***signal similarity*** between signals $x[n]$ and $y[n]$. To quantify the degree of similarity, we consider the following inequality:

$$0 \le \sum_{n=0}^{N-1} \left[\frac{x[n]}{\sqrt{r_x[0]}} \pm \frac{y[n]}{\sqrt{r_y[0]}} \right]^2 . \tag{9.183}$$

Since the quantity on the right-hand side is raised to the second power, the inequality always applies. Expanding Eq. (9.183) gives

$$0 \le \sum_{n=0}^{N-1} \frac{x^2[n]}{r_x[0]} + \sum_{n=0}^{N-1} \frac{y^2[n]}{r_y[0]} \pm 2 \sum_{n=0}^{N-1} \frac{x[n]\, y[n]}{\sqrt{r_x[0]\, r_y[0]}} . \tag{9.184}$$

By property #2 of the autocorrelation function, Eq. (9.184) simplifies to

$$0 \le 1 + 1 \pm 2\, \frac{r_{xy}[0]}{\sqrt{r_x[0]\, r_y[0]}} . \tag{9.185}$$

The equality part of the inequality in Eq. (9.185) should apply when $x[n] = y[n]$, which requires that

$$\pm \frac{r_{xy}[0]}{\sqrt{r_x[0]\, r_y[0]}} \le 1 . \tag{9.186}$$

Equivalently, we choose the negative sign in Eq. (9.185) and rewrite the expression as

$$0 \le 2 \left[1 - \frac{|r_{xy}[0]|}{\sqrt{r_x[0]\, r_y[0]}} \right], \tag{9.187}$$

which leads to the ***Cauchy-Schwartz inequality***

$$|r_{xy}[0]| \le \sqrt{r_x[0]\, r_y[0]} . \tag{9.188}$$

The ratio of the correlations represents the angle θ defined in connection with Eq. (9.182). That is,

$$\cos\theta = \frac{r_{xy}[0]}{\sqrt{r_x[0]\, r_y[0]}} = \frac{r_{xy}[0]}{\sqrt{E_x E_y}} , \tag{9.189}$$

where $E_x = r_x[0]$ is the energy of $x[n]$, and a similar definition applies to E_y.

The factor $\cos\theta$ is called the ***correlation coefficient*** of $x[n]$ and $y[n]$, and represents an energy-normalized version of correlation. The closer θ is to zero, the more similar are $x[n]$ and $y[n]$, excluding a scale factor.

9-12.6 Classification of Signals

Suppose $y[n]$ is a noisy signal consisting of white noise $w[n]$ added to one of L possible signals:

$$y[n] = w[n] + \begin{cases} x_1[n], \text{ or} \\ x_2[n], \text{ or} \\ \vdots \\ x_L[n]. \end{cases} \tag{9.190}$$

The identity of the specific signal contained in $y[n]$ is unknown, so to identify it, we choose the signal that minimizes the quantity

$$\sum_{n=0}^{N-1} \left[\frac{y[n]}{\sqrt{r_y[0]}} - \frac{x_i[n]}{\sqrt{r_{x_i}[0]}} \right]^2 , \tag{9.191}$$

which represents the *squared-distance* between the energy-normalized data $y[n]$ and each of the energy-normalized signals $x_i[n]$. By applying the algebra steps that led earlier to the Cauchy-Schwartz inequality of Eq. (9.188), it is easy to show that minimizing Eq. (9.191) is equivalent to computing the correlation $r_{yx_i}[0]$ for $\{x_1[n], x_2[n], \ldots, x_L[n]\}$ and then choosing the signal $x_i[n]$ that generates the largest (closest to 1.0) correlation coefficient:

$$\cos\theta_i = \frac{r_{yx_i}[0]}{\sqrt{r_y[0]\, r_{x_i}[0]}} = \frac{\sum_{n=0}^{N-1} y[n]\, x_i[n]}{\sqrt{\sum_{n=0}^{N-1} y^2[n] \sum_{n=0}^{N-1} x_i^2[n]}} . \tag{9.192}$$

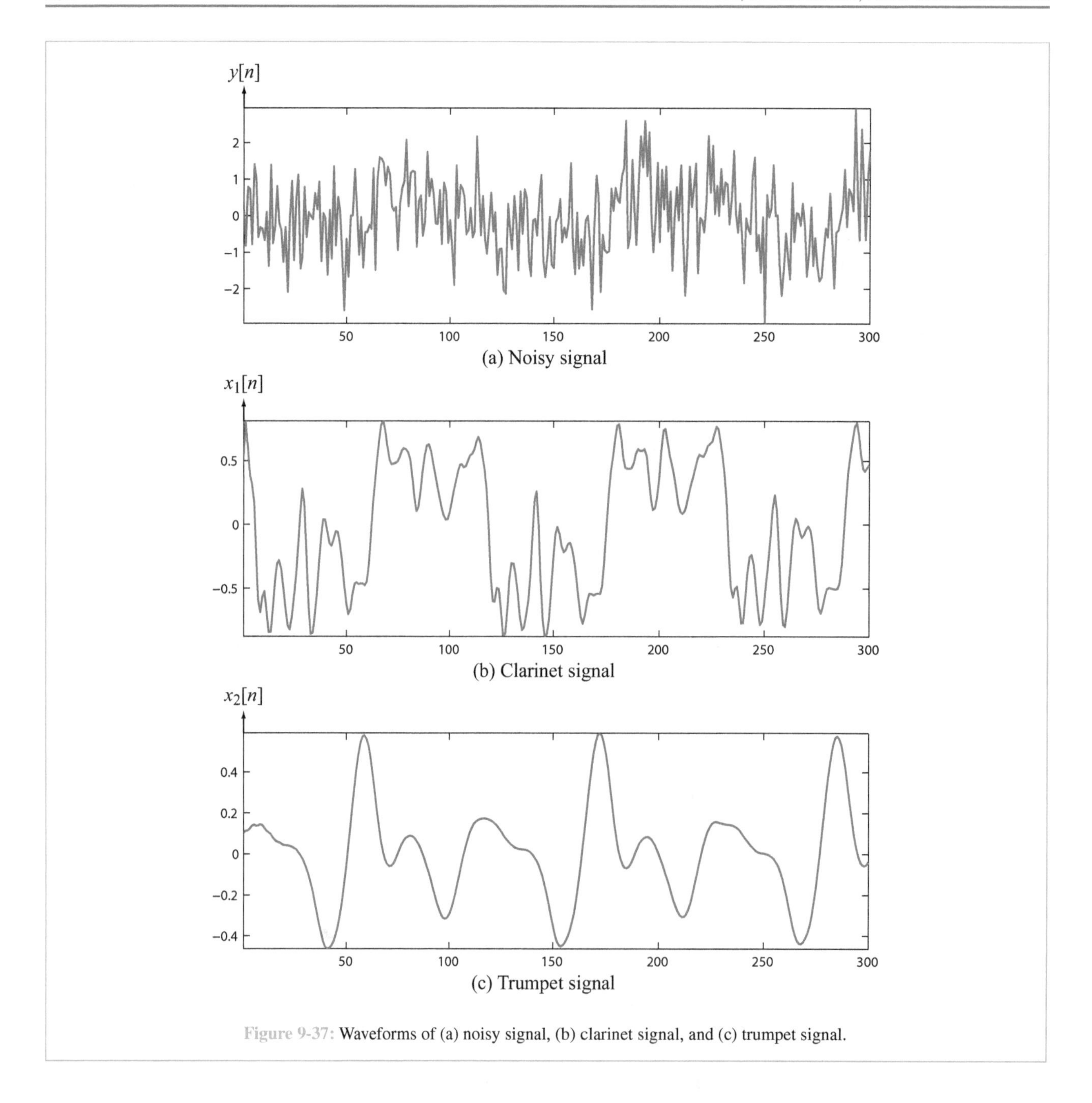

Figure 9-37: Waveforms of (a) noisy signal, (b) clarinet signal, and (c) trumpet signal.

By way of illustration, Fig. 9-37(a) displays a noisy signal with a signal-to-noise ratio of −5.95 dB. The negative sign signifies that, on average, the noise is much larger than the signal. Suppose we know that the signal contained in the noisy signal is the waveform of either a clarinet ($x_1[n]$) or a trumpet ($x_2[n]$). To determine which one of the two musical signals is contained in the noisy waveform, we compute the correlation coefficient between the noisy signal $y[n]$ and the musical signals $x_1[n]$ and $x_2[n]$. Using MATLAB, we obtain the results:

$$\cos\theta_1 = 0.441 \quad \text{(clarinet)},$$
$$\cos\theta_2 = -0.055 \quad \text{(trumpet)}.$$

Module 9.4 Use of Autocorrelation to Estimate Period This module generates a periodic signal, adds noise to it, and computes its autocorrelation. The first peak of the autocorrelation indicates the period. Period and noise level are both set by sliders.

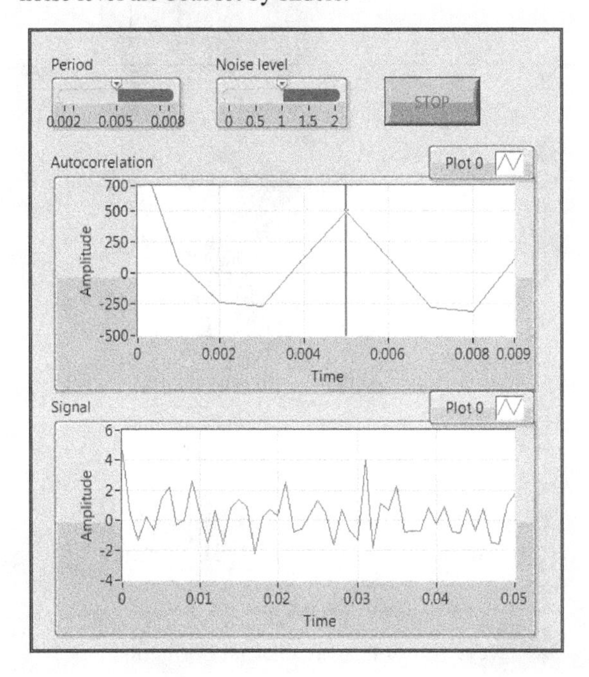

Module 9.5 Use of Cross-Correlation to Estimate Time Delay This module generates a signal, delays it, adds noise to the delayed signal, and computes the cross-correlation between original and noisy delayed signals. The peak in the cross-correlation indicates the delay. Delay and noise level are both set by sliders.

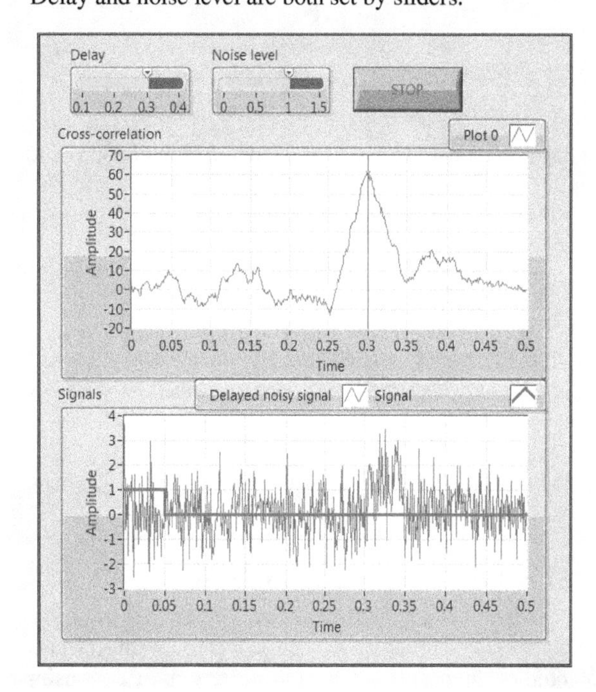

The distinction between the magnitudes of the two correlation coefficients is sufficiently large for us to conclude that the unknown signal embedded in the noisy signal is that of the clarinet. The signal-to-noise ratio of -5.95 dB means that the energy content of the signal is only about 25% of that of the noise. Had the signal been larger, correlation coefficient $\cos\theta_1$ would have been closer to 1.0, but $\cos\theta_2$ would remain close to 0.

Using correlation to classify signals can also be used as a tool to determine the time delay of a signal. In the example presented earlier in Fig. 9-36, signal $x[n]$ was delayed by an unknown delay Δ, and cross-correlation was performed to determine Δ. An analogous procedure that would lead to the same result is to formulate the time-delay problem as

$$y[n] = w[n] + \begin{cases} x[n], \text{ or} \\ x[n-1], \text{ or} \\ \vdots \\ x_L[n-L], \end{cases} \tag{9.193}$$

where the unknown delay Δ may assume any value between 0 and L. Because all of the signals $x[n-i]$ have the same energy, no energy normalization is necessary, so it is sufficient to compute

$$\sqrt{E_y E_{x_i}} \cos\theta_i = \sum_{n=0}^{N-1} y[n]\, x_i[n] = \sum_{n=0}^{N-1} y[n]\, x[n-i] = r_{yx}[i]. \tag{9.194}$$

The unknown delay Δ is the index i for which $\cos\theta_i$ is the closest to 1.0. This is because E_{x_i} are all equal.

Concept Question 9-16: What are the applications of each of the three forms of correlation in this section? (See S^2)

Exercise 9-15: Use LabVIEW Module 9.4 to estimate the period of the waveform with period 0.005 and noise level 1.

Answer: (See Module 9.4.)

Exercise 9-16: Use LabVIEW Module 9.5 to estimate the time delay of the signal when its actual delay is 0.3 and the noise level is 1.

Answer: (See Module 9.5.)

9-13 Biomedical Applications

We now demonstrate how the signal correlation properties presented in the preceding section can be used in support of three biomedical applications:

- Computation of heartbeat rate from an electrocardiogram (EKG) using autocorrelation.
- Classification of an EKG record into one of three possible types of waveforms, only one of which is considered "normal."
- Computation of time delay in ultrasound.

9-13.1 Electrocardiograms (EKGs)

An ***electrocardiogram*** (***EKG***) is a record of the electrical potential of the heart, measured as a function of time using a number of wire leads taped to the surface of the chest. To remove possible 60 Hz interference picked up by the wire leads, a notch filter (Section 8-2) is used before recording the EKG.

Under normal conditions, the EKG waveform exhibits a periodic pattern with a period of about 1 s, corresponding to 60 beats per minute. The rate may change, of course, with physical activity and/or emotional stress.

EKG waveforms are used for diagnosing heart malfunctions. Three synthetic EKG waveforms (with good resemblance to actual EKG waveforms) are displayed in Fig. 9-38. The first is an example of a waveform belonging to a heart operating normally, whereas the waveforms in parts (b) and (c) are associated with hearts characterized by ***atrial flutter*** and ***atrial fibrillation***, respectively. The three waveforms are each 10 s in duration, sampled at 100 samples/s, thereby generating records of 1000 samples each. Strictly speaking, the horizontal axes in the three plots shown in Fig. 9-38 should be discrete time n, but we opted to display t in seconds instead.

Whereas the "normal heart" waveform in Fig. 9-38(a) consists of regular pulses at 1 pulse per second, separated by minor fluctuations, the flutter waveform (Fig. 9-38(b)) contains three additional bumps (pulses) in the waveform. This means that the spectrum of the flutter waveform will be dominated by a component at 240 beats per minute or, equivalently, 4 Hz.

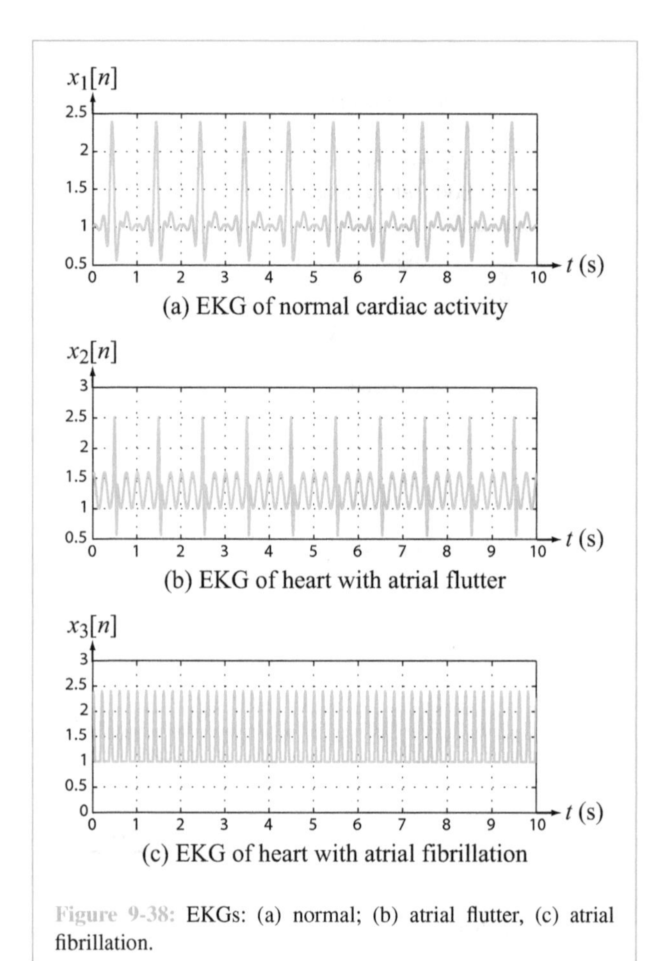

Figure 9-38: EKGs: (a) normal; (b) atrial flutter, (c) atrial fibrillation.

The fibrillation waveform is Fig. 9-38(c) consists entirely of equal-amplitude bumps at 300 beats per minute, corresponding to a spectral line at 5 Hz. Atrial fibrillation is a serious heart condition requiring a pacemaker to correct its behavior into normal cardiac action.

9-13.2 Measuring Heart Rate by Autocorrelation

To simulate the effects of additive noise, zero-mean white Gaussian noise with a standard deviation $\sigma = 0.5$ was added to each of the three signal waveforms of Fig. 9-38. Only 2 s long

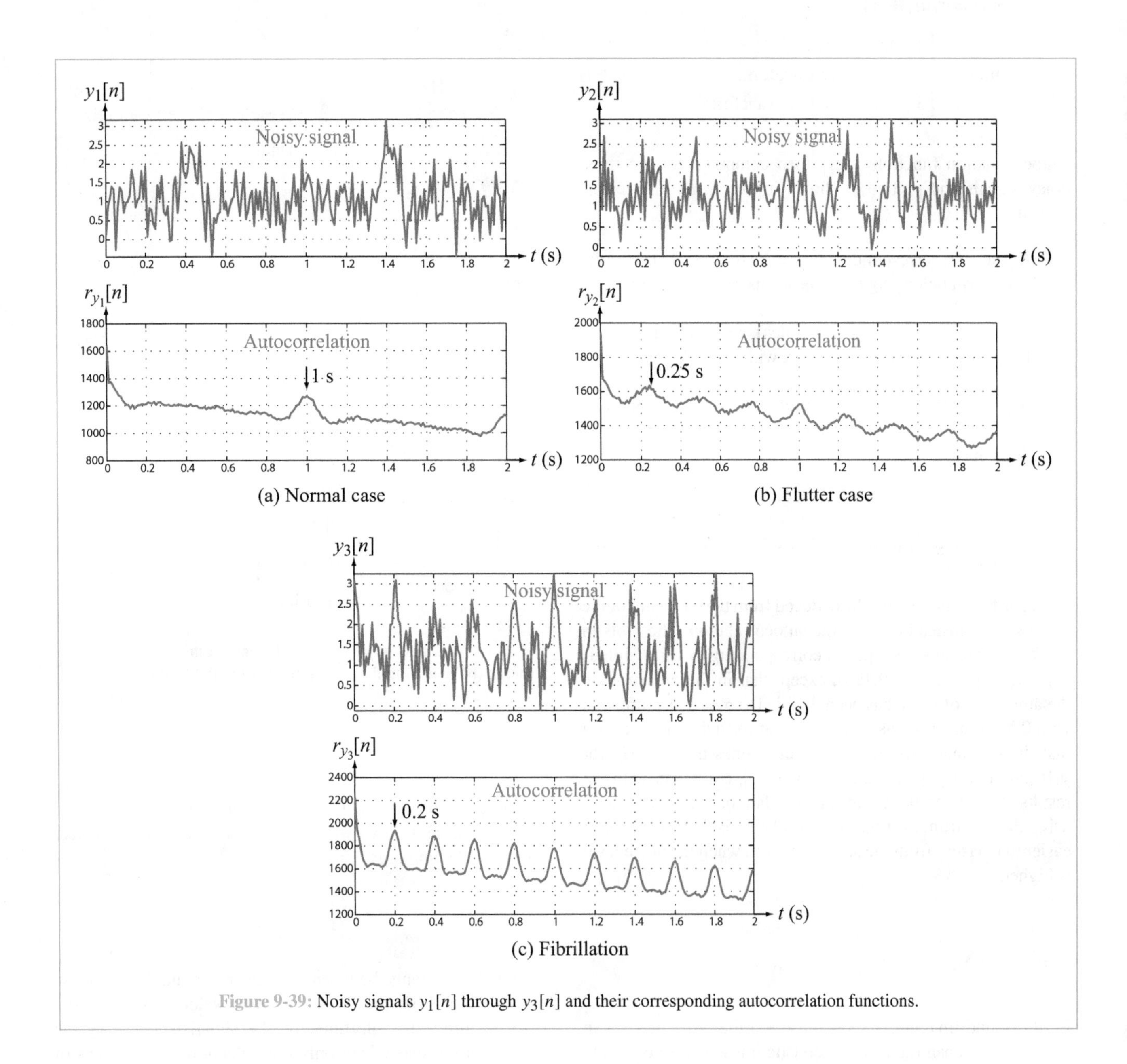

Figure 9-39: Noisy signals $y_1[n]$ through $y_3[n]$ and their corresponding autocorrelation functions.

segments are used, and labeled

$$y_i[n] = x_i[n] + w[n]$$

$$\text{with } i = \begin{cases} 1 & \text{for normal,} \\ 2 & \text{for flutter waveform,} \\ 3 & \text{for fibrillation waveform.} \end{cases} \tag{9.195}$$

Autocorrelation functions were then computed for the three noisy signals and plotted as shown in Fig. 9-39. The results indicate:

(a) A peak at n corresponding to $t = 1$ s (or $f = 1$ Hz) for the waveform belonging to the heart operating normally.

(b) A peak at 0.25 s (and its multiples), with a corresponding frequency of 4 Hz, for the heart with atrial flutter.

(c) A peak at 0.2 s (and its multiples) with a corresponding frequency of 5 Hz, for the heart afflicted with atrial fibrillation.

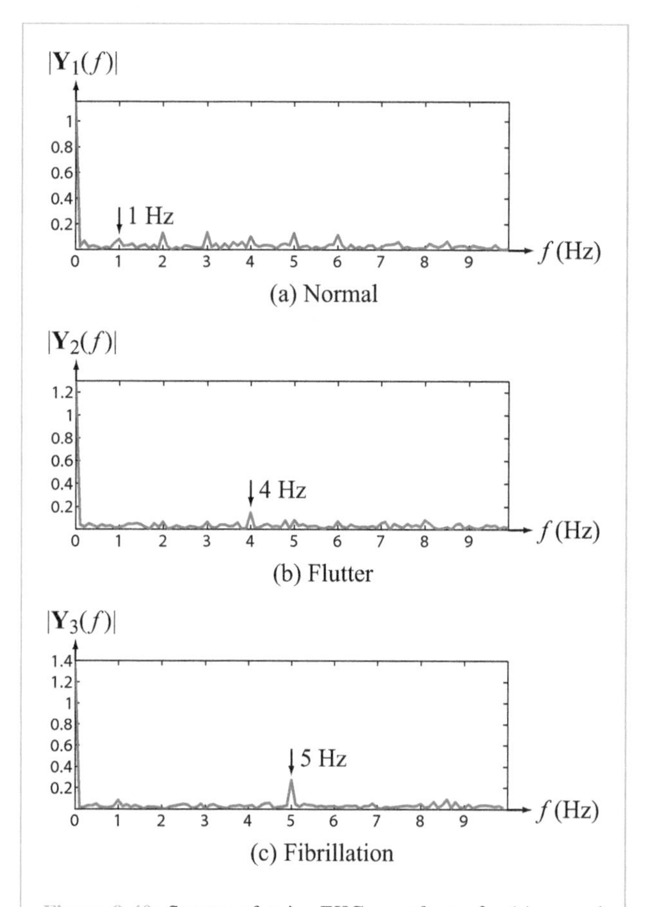

Figure 9-40: Spectra of noisy EKG waveforms for (a) normal heart, (b) heart with flutter, and (c) heart with fibrillation.

9-13.3 Measuring Heart Rate from the Spectrum

The heartbeat rate can also be deduced from the spectrum of the noisy signal, instead of from the autocorrelation. The plots in Fig. 9-40 are computed spectra corresponding to noisy signals $y_1[n]$ to $y_3[n]$ of Eq. (9.195), except that in the present case the amplitude of noise has been doubled to $\sigma = 1.0$, compared with 0.5 for the previous case. It is clear from the three spectra that the dominant nonzero Hz spectral lines are at 1 Hz for $y_1[n]$, 4 Hz for $y_2[n]$, and 5 Hz for $y_3[n]$, consistent with the results obtained earlier using autocorrelation. The advantage of using the spectrum over the autocorrelation is that the period is easier to measure from the spectrum, even when the noise level is higher.

9-13.4 Ultrasound Time Delay

In ultrasound imaging, a short pulse (on the order of 1 μs in duration) is transmitted by an acoustic transducer into a body part. The pulse propagates into the body through soft tissue, but is reflected by body material with distinctly different acoustic indices of refraction, such as bones, organs, and inclusions. Part of the reflected energy is received and recorded by the transducer. The time delay t_d between the transmitted and received pulse is the two-way travel time between the transducer and the reflecting organ or inclusion:

$$t_d = 2\,\frac{R}{v}\,, \tag{9.196}$$

where R is the range between the transducer and the reflecting organ or inclusion, and v is the acoustic velocity in soft tissue (≈ 1.54 mm/μs). Unfortunately, the signal received by the transducer consists of not only the reflections from organs of interest, but also reflections of lower-level intensity from the soft-body tissue surrounding those organs. This second group of reflections manifest themselves as the equivalent of additive random noise.

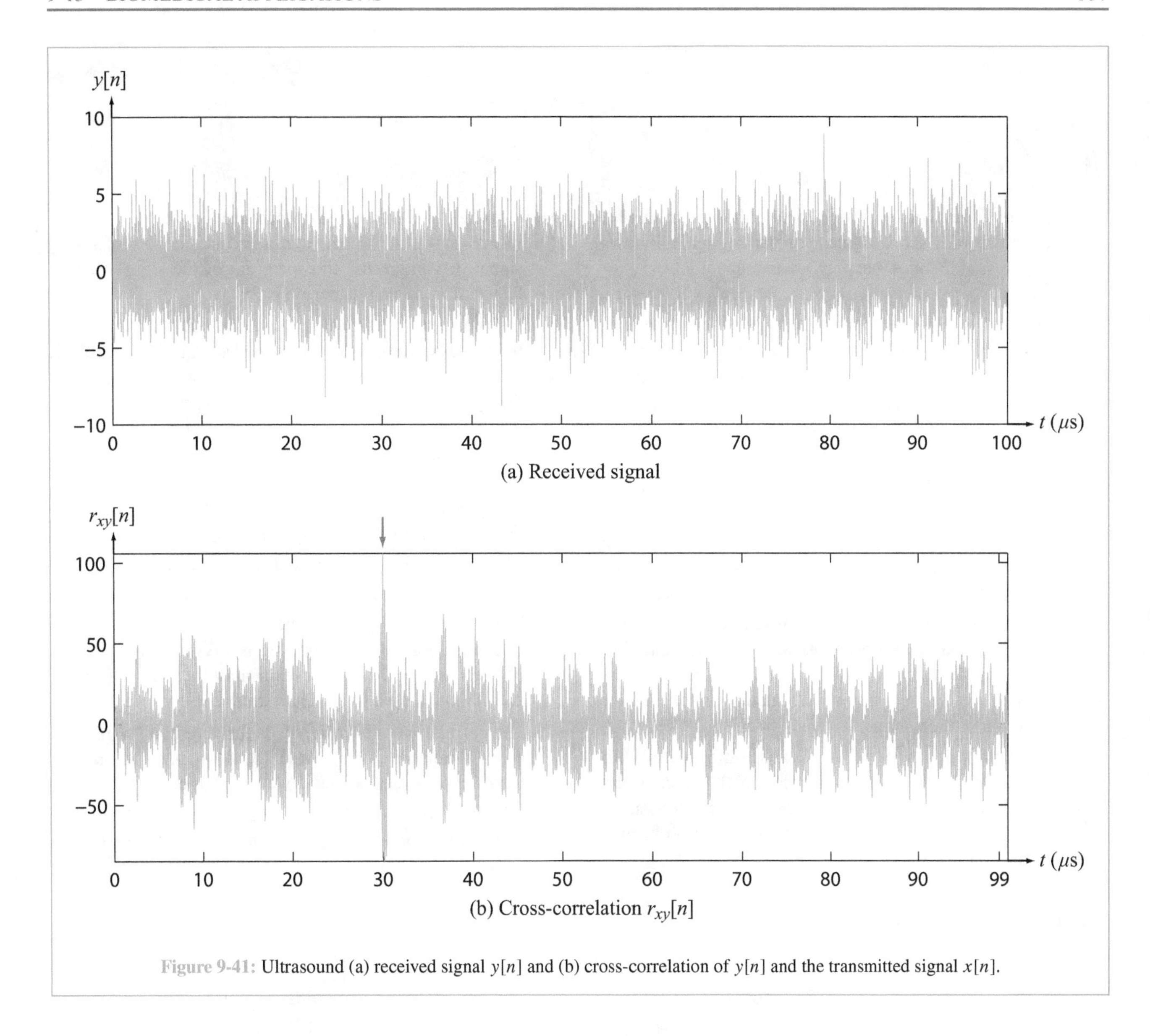

Figure 9-41: Ultrasound (a) received signal $y[n]$ and (b) cross-correlation of $y[n]$ and the transmitted signal $x[n]$.

The range to an organ of interest can be discerned by computing the cross-correlation between the received signal and a copy of the transmitted signal. Because of the extensive amount of noise accompanying the signal reflected by a certain organ, it is almost impossible to discern the reflection from the signal waveform of the received signal displayed in Fig. 9-41(a), but the delay time is clearly evident in the plot of the cross-correlation function $r_{xy}[n]$, where $x[n]$ is a copy of the transmitted signal pulse and $y[n]$ is the noisy received signal. The duration of the transmitted pulse is only 1 μs, while the duration of the received pulse is 100 μs. From Fig. 9-41(b), $r_{xy}[n]$ has a peak at $t = 30$ μs. Hence, by Eq. (9.196), the range to the reflecting organ is

$$R = \frac{vt_{\rm d}}{2} = \frac{1.54 \text{ mm} \times 30}{2} = 2.31 \text{ cm}.$$

Summary

Concepts

- A data window is used to compute the spectrum of a signal from a short segment of that signal. Windows suppress side lobes, but may broaden main lobes.
- Spectrograms depict a time-varying signal spectrum by segmenting the signal using data windows, computing the spectrum of each segment, and displaying the result as an image.
- Discrete-time filters have impulse responses that have either finite durations (FIR) or infinite durations (IIR).
- FIR filters can be designed by: (1) using a data window on the inverse DTFT of the desired frequency response, (2) computing the impulse response of a function whose frequency response matches samples of the desired frequency response, or (3) using a minimax criterion, which generates an equiripple frequency response.
- IIR filters are designed from a given continuous-time filter, such as a Butterworth filter, by: (1) sampling the impulse response of the continuous-time filter, or (2) performing a bilinear transformation on the transfer function of the continuous-time filter.
- Downsampling a signal by L keeps only every Lth sample and deletes all other values. It stretches the spectrum of the signal by a factor of L, and introduces $L-1$ copies of one period of the spectrum to keep it periodic with period 2π.
- Upsampling a signal by L, also known as zero-stuffing, inserts $L-1$ zeros between adjacent samples of the signal. It compresses the spectrum of the signal by a factor of L, so that its period is $2\pi/L$.
- Interpolation of a signal upsampled by L is performed by lowpass filtering the upsampled signal, using cutoff frequency $2\pi/L$. This maintains the nonzero values of the upsampled signal, and changes its zero values to the values that make the interpolated upsampled signal have a maximum frequency of $2\pi/L$.
- Multirate signal processing uses upsampling by L, interpolation, and downsampling by M, in that order, to alter the sampling rate by a factor of L/M.
- The autocorrelation of a signal is the convolution of the signal with its time reversal. Autocorrelation is used to compute the period of a periodic signal.
- The cross-correlation of two signals is the convolution of one signal with the time reversal of the other. Cross-correlation is used to compute time delay.
- The correlation of two signals is the zeroth lag of their cross-correlation, or equivalently, their inner product. Correlation is used to classify a signal as one of several possible signals.
- Autocorrelation, cross-correlation, and correlation have applications to EKG and ultrasound imaging.

Mathematical and Physical Models

$x_{\text{windowed}}[n] = x[n]\, w[n]$

Hamming window $w[n] = 0.54 - 0.46\cos(2\pi n/(L-1))$

Spectrogram

$$\mathcal{S}(e^{j\Omega}, N) = \left| \sum_{n=N-L/2}^{N+L/2} w[n-N+L/2]\, x[n]\, e^{-j\Omega n} \right|^2$$

Bilinear transformation $\mathbf{s} = \dfrac{2}{T}\dfrac{\mathbf{z}-1}{\mathbf{z}+1}$

Frequency (pre)warping $\omega = \dfrac{2}{T}\tan\left(\dfrac{\Omega}{2}\right)$

Upsampling

$x[n] \rightarrow \boxed{\uparrow L} \rightarrow y[n] \leftrightarrow \mathbf{Y}(e^{j\Omega}) = \mathbf{X}(e^{j\Omega L})$

Downsampling

$x[n] \rightarrow \boxed{\downarrow L} \rightarrow y[n] \leftrightarrow y[n] = x[Ln]$

Autocorrelation $r_x[n] = x[n] * x[-n]$

Cross-correlation $r_{xy}[n] = x[n] * y[-n]$

Correlation $r_{xy}[0] = \sum x[n]\, y[n]$

Important Terms

Provide definitions or explain the meaning of the following terms:

autocorrelation	EKG	impulse invariance	resolution
bilinear transformation	equiripple design	interpolation	sidelobe
chirp	FIR filter	mainlobe	spectral leakage
correlation	frequency sampling	minimax criterion	spectrogram
cross-correlation	frequency (pre)warping	multirate	upsampling
data window	Hamming window	Parks-McClellan algorithm	windowing inverse DTFT
downsampling	IIR filter	rectangular window	zero-stuffing

PROBLEMS

Section 9-1: Data Windows

9.1 This problem demonstrates the utility of data windows. Signal $x(t) = \cos(200\pi t) + 0.02\cos(400\pi t)$ is observed for only $0 \leq t < 0.1$. Then, $x(t)$ is sampled at a sampling rate of 1000 samples/second.

(a) Compute and plot the spectrum of $x(t)$ from its samples.

(b) Repeat (a) using a Hamming window instead of a rectangular window.

(c) What does the plot in (b) show that the plot in (a) does not?

9.2 Signal $x[n] = \sin(0.3\pi n) + \sin(0.4\pi n)$ is observed for $1 \leq n \leq L$ for some L. The goal is to determine the smallest value of L that resolves the two peaks. Plot the spectrum using

```
plot(abs(fft(sin(0.3*pi*[1:L])
+sin(0.4*pi*[1:L]),N))).
```

*(a) Estimate the minimum value of L needed to split (resolve) the two peaks using the optics resolution formula $|\omega_2 - \omega_1| \geq 2\pi/L$.

*(b) Using $N = 256$, find the smallest value of L that resolves the two peaks. "Resolves" means there is a dip (but not all the way to zero) between two peaks. Provide two plots with consecutive values of L with unsplit and split peaks.

(c) Repeat (b) with N doubled to 512. How does this affect resolution?

(d) Repeat (b) using a Hamming window. How does this affect resolution?

9.3 Load MATLAB file `P93.mat`.
`X1` is 75 samples of a sum of sinusoids.

(a) Plot its spectrum: `plot(abs(fft(X1,256)))`. How many sinusoids are there?

(b) Repeat (a) using a Hamming window. Now how many sinusoids are there?

(c) Estimate the frequencies of the sinusoids using the plot from part (b). The sampling rate was 1000 samples/second.

*__9.4__ The four plots shown in Fig. P9.4 are actual plots of signal spectra using either a length $L = 10$ or 20, either a DFT order $N = 64$ or 256, and either a Hamming or a rectangular window. For each plot, choose which values of L and N and which window was used.

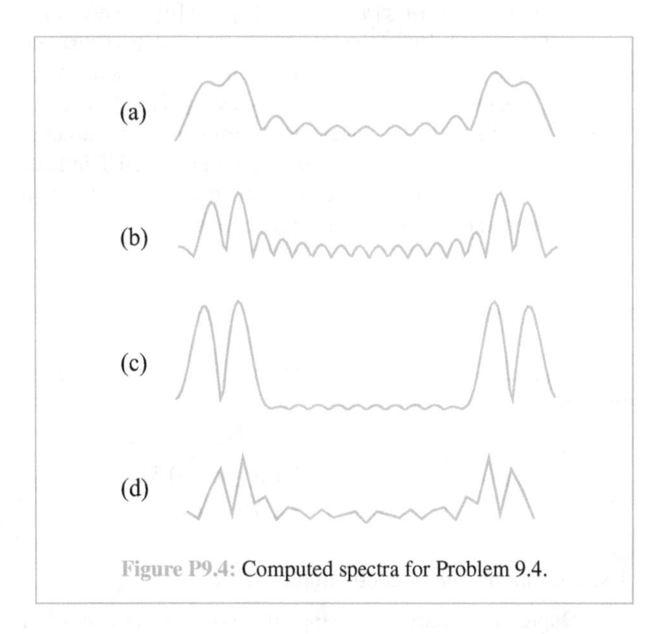

Figure P9.4: Computed spectra for Problem 9.4.

9.5 For a window of duration L, where $L \gg 3\pi$ (which is true in practice), show that the ratio of the height of the first sidelobe to the height of the mainlobe is

(a) -13 dB for a rectangular window.

(b) -27 dB for a Bartlett window.

---
*Answer(s) in Appendix F.

9.6 This problem shows how data windows suppress sidelobes. Recall that a Hanning window is given by

$$w_{\text{Hn}}[n] = \frac{1}{2} - \frac{1}{2}\cos\left(\frac{2\pi n}{L-1}\right)$$

for $0 \leq n \leq L-1$. Advance in time by $(L-1)/2$ and change the denominator of the cosine from $L-1$ to L and define $L' = L-1$ to obtain

$$w'[n] = \begin{cases} \frac{1}{2} + \frac{1}{2}\cos(2\pi n/(L'+1)) & \text{for } |n| \leq L'/2, \\ 0 & \text{for } |n| > L'/2. \end{cases}$$

(a) Compute the DTFT $\mathbf{W}(e^{j\Omega})$ of $w'[n]$. Your answer should have three terms. *Hint:* Use the modulation property of the DTFT.

(b) Plot each term, and their sum, separately using different colors.

(c) Using the plots, explain how using this window helps suppress sidelobes.

9.7 The spectrum of $x[n] = 2\cos(\Omega_0 n)\, w[n]$, where $w[n]$ is a rectangular pulse of length N, consists of two discrete sinc functions, centered at $\pm\Omega_0$. These two sinc functions are spectral leakage. But Section 8-9.2 showed that if N is an integer multiple of $2\pi/\Omega_0$, then the spectrum of $x[n]$ computed using a DFT of order N is zero except at the two DFT indices corresponding to $\pm\Omega_0$, so there is no spectral leakage. Explain what happened to the discrete sinc functions.

Section 9-2: Spectrograms

9.8 This problem shows how a spectrogram tracks a sudden frequency change. Let

$$x(t) = \begin{cases} \cos(200\pi t) & \text{for } 0 \leq t < 0.5, \\ \cos(400\pi t) & \text{for } 0.5 \leq t < 1.0. \end{cases}$$

Use a sampling rate of 2000 samples/second:

(a) Depict the spectrogram using 20 bins of 100 samples each using
```
imagesc(abs(fft(reshape(X',100,20))))
```

(b) Depict the spectrogram using 25 bins of 80 samples each using
```
imagesc(abs(fft(reshape(X',80,25))))
```

(c) Explain why the spectrograms look different.

9.9 Repeat part (b) of Problem 9.8 using rectangular and Hamming windows.

9.10 Load MATLAB file `P910.mat`. It contains two variables `X1` and `X2`.

(a) Listen to `X1` using `soundsc(X1,10000)`. Describe it.

(b) Depict its spectrogram using
```
imagesc(abs(fft(reshape(X1',100,100)))).
```
Describe the signal using its spectrogram.

9.11 Load MATLAB file `P910.mat`. It contains two variables `X1` and `X2`.

(a) Listen to `X2` using `soundsc(X2,10000)`. Describe it.

(b) Depict its spectrogram using
```
imagesc(abs(fft(reshape(X2',100,100)))).
```
Describe the signal using its spectrogram.

9.12 Load MATLAB file `P912.mat`. `Y` is part of the Michigan fight song.

(a) Listen to `Y` using `soundsc(Y,44100)`. Describe it.

(b) Depict only the lowest frequencies of its spectrogram using
```
YY=abs(fft(reshape(Y,32768,15)));
imagesc(YY(30769:32768,:))
```
Explain what this spectrogram is depicting.

9.13 Load MATLAB file `P913.mat`.
`X2` is the sum of tonal versions of two fight songs.

(a) Listen to `X2` using `soundsc(X2,8192)`. Describe it.

(b) Depict spectrogram using
```
imagesc(abs(fft(reshape(X2',3000,26)))).
```
This should make it apparent that the two songs are in different octaves.

(c) Eliminate one of the songs by setting some values of `fft(X2)` to zero. Plot the spectrogram of result `Y` using `imagesc(abs(fft(reshape(Y',3000,26))))`. Listen to `Y` using `soundsc(Y,8192)`. Describe it.

Section 9-3: FIR Filter Design

9.14 Let $h[n]$ be a lowpass filter with cutoff frequency Ω_0.

(a) Show that $2h[n]\cos(\Omega_c n)$ is a bandpass filter with cutoff frequencies $\Omega_c - \Omega_0$ and $\Omega_c + \Omega_0$.

(b) Show that $h[n]\cos(\pi n)$ is a highpass filter with cutoff frequency $\pi - \Omega_0$.

(c) Show that $\mathbf{H}(-\mathbf{z})$ is a highpass filter with cutoff frequency $\pi - \Omega_0$.

9.15 Design a discrete-time bandpass filter with cutoff frequencies $\pi/4$ and $3\pi/4$ using:

(a) A 5-point rectangular window.

(b) Frequency sampling at $\Omega = \{0,\ \pi/3,\ 2\pi/3,\ \pi\}$. Use a filter of duration five.

(c) Compare your answers to (a) and (b).

9.16 A *Hilbert transformer* has the frequency response

$$\mathbf{H}(e^{j\Omega}) = \begin{cases} -j & \text{for } 0 < \Omega < +\pi, \\ +j & \text{for } -\pi < \Omega < 0, \end{cases}$$

and a corresponding impulse response

$$h[n] = \frac{1-(-1)^n}{\pi n},$$

with $h[0] = 0$. Design a discrete-time Hilbert transformer using

*(a) A 5-point rectangular window.

*(b) Frequency sampling at $\Omega = \pi/3$ and $2\pi/3$.

(c) Compare your answers with (a) and (b).

9.17 Design a 31-point half-band (i.e., with cutoff frequency at $\pi/2$) lowpass, discrete-time filter by windowing the ideal impulse response using a Hamming window. Plot the impulse response and plot the gain of the resulting filter for $0 \le \Omega \le \pi$.

9.18 Design a 31-point bandlimited (to π) ideal differentiator by windowing the ideal impulse response using a Hamming window. Plot the impulse response and plot the gain of the resulting filter for $0 \le \Omega \le \pi$.

9.19 Design a 31-point bandpass filter with cutoff frequencies $\pi/4$ and $3\pi/4$ by windowing the ideal impulse response using a Hamming window. Plot the impulse response and plot the gain of the resulting filter for $0 \le \Omega \le \pi$.

9.20 This problem is about a frequency sampling FIR filter design that uses an inverse DFT. We wish to design a symmetric noncausal FIR filter

$$h[n] = \{h[-L] \ldots \underline{h[0]} \ldots h[L]\}$$

of odd length $2L+1$ using the frequencies

$$\Omega = \left\{0, \pm\frac{2\pi}{2L+1}, \pm\frac{4\pi}{2L+1}, \pm\frac{6\pi}{2L+1}, \ldots, \pm\frac{2\pi L}{2L+1}\right\}.$$

(a) Show that $h[n]$ can be computed without solving a linear system of equations using only an inverse DFT of order $2L+1$ and some reordering of the gains. *Hint:* The DFT and DTFS differ by only a scale factor.

(b) Design a noncausal lowpass filter $h[n] = \{a, b, \underline{c}, b, a\}$ with the gains specified in the following table:

Ω	0	$\pm\frac{2\pi}{5}$	$\pm\frac{4\pi}{5}$
$\mathbf{H}(e^{j\Omega})$	1	$\frac{1}{2}$	0

by solving a linear system of equations.

(c) Design $h[n]$ by computing an inverse DFT.

9.21 Repeat Problem 9.20 to design a lowpass filter with

Ω	0	$\pm\frac{2\pi}{5}$	$\pm\frac{4\pi}{5}$
$\mathbf{H}(e^{j\Omega})$	1	1	0

9.22 This problem requires MATLAB's Signal Processing Toolbox. Use the Parks-McClellan algorithm to design a half-band (cutoff frequency $\Omega_0 = \pi/2$) lowpass filter of order 21 (the impulse response duration is 22). The gain is to be

$$|\mathbf{H}(e^{j\Omega})| = \begin{cases} 1 & \text{for } 0 \le |\Omega| \le 0.45\pi, \\ \text{dc} & \text{for } 0.45\pi < |\Omega| < 0.55\pi, \\ 0 & \text{for } 0.55\pi \le |\Omega| \le \pi. \end{cases}$$

Optional: Display the gains computed after 1, 2, 3, and 4 iterations by setting the maximum number of iterations to be 1, 2, 3, 4 in the mfile.

9.23 Determine the FIR filter $h[n]$ of length L that minimizes the mean square error

$$e = \frac{1}{2\pi}\int_{-\pi}^{\pi} |\mathbf{H}_\mathrm{D}(e^{j\Omega}) - \sum_{n=0}^{L-1} h[n]\, e^{-j\Omega n}|^2\, d\Omega,$$

where $\mathbf{H}_\mathrm{D}(e^{j\Omega})$ is the desired frequency response. *Hint:* Use Parseval's theorem for the DTFT.

Section 9-4: IIR Filter Design

9.24 The impulse response of a brickwall lowpass filter with cutoff frequency $\Omega_0 = \pi/2$ is

$$h[n] = \frac{\sin(\frac{\pi}{2}\, n)}{\pi n}.$$

This $h[n]$ is clearly noncausal. Show that $h[n]$ is also not BIBO stable.

*9.25 Using the analog system $\mathbf{H}_a(\mathbf{s}) = 1000/(\mathbf{s}+1000)$ and $T = 0.001$, design a discrete-time filter using:

(a) Impulse invariance (find the impulse response).

(b) Bilinear transformation (find the transfer function).

9.26 Using the analog filter

$$\mathbf{H}_a(\mathbf{s}) = \frac{1}{(\mathbf{s}+1)^2}$$

and $T = 2$, design a digital filter using:

(a) Impulse invariance (find the impulse response).

(b) Bilinear transformation (find the impulse response).

9.27 Using the analog filter

$$\mathbf{H}_a(\mathbf{s}) = \frac{1}{\mathbf{s}^2+\mathbf{s}+1}$$

and $T = 2$, design a digital filter using

(a) Impulse invariance (find the impulse response).

(b) Bilinear transformation (find the transfer function).

*9.28 Use bilinear transformation with $T = 2$ to design an IIR double differentiator.

9.29 Find the gain of an IIR filter designed using the bilinear transformation with $T = 2$ for a continuous-time Butterworth filter of order N and cutoff frequency $\Omega_0 = 1$.

9.30 We are given an analog filter whose gain (frequency response magnitude) is as given in Fig. P9.30.

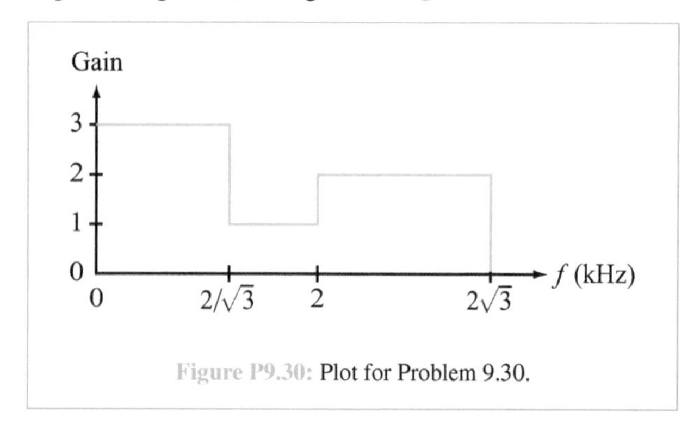

Figure P9.30: Plot for Problem 9.30.

Draw the gain of an IIR filter design using the bilinear transformation with $T = 0.001/2\pi$ s.

9.31 A continuous-time filter has the frequency response $\mathbf{H}(j\omega)$ shown in Fig. P9.31. The bilinear transformation is used to design a discrete-time filter from this filter. If $T = 1.02$, plot the frequency response $\mathbf{H}(e^{j\Omega})$ of the resulting discrete-time filter and the frequency response $\mathbf{H}(j\omega)$ of the continuous-time filter on the same plot.

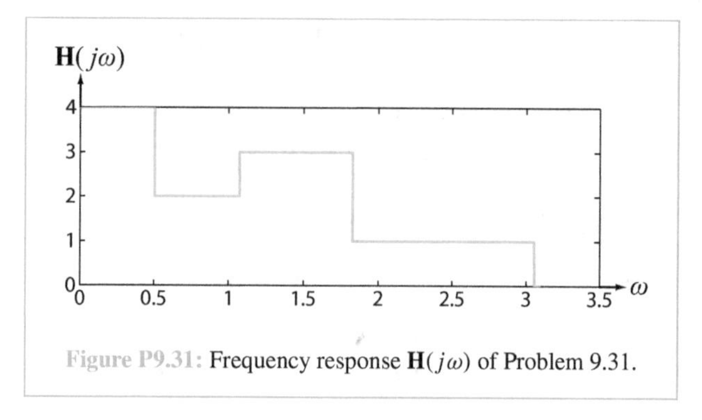

Figure P9.31: Frequency response $\mathbf{H}(j\omega)$ of Problem 9.31.

9.32 Prove that the bilinear transformation maps:

(a) The imaginary axis $\mathfrak{Re}[\mathbf{s}] = 0$ to the unit circle $|\mathbf{z}| = 1$.

(b) The left half-plane $\mathfrak{Re}[\mathbf{s}] < 0$ to the interior of the unit circle $|\mathbf{z}| < 1$.

(c) The right half-plane $\mathfrak{Re}[\mathbf{s}] > 0$ to the exterior of the unit circle $|\mathbf{z}| > 1$.

Hence, a discrete-time filter designed from a stable continuous-time filter is also stable.

9.33 Use the bilinear transformation to design a discrete-time Butterworth filter with cutoff frequency $\Omega_0 = \pi/2$ from a continuous-time Butterworth filter of order three. Compute the transfer function, poles, and zeros, and difference equation. *Note:* This problem does not require MATLAB/MathScript.

9.34 Use the bilinear transformation to design a discrete-time Butterworth filter with cutoff frequency $\Omega_0 = 2\pi/3$ from a continuous-time Butterworth filter of order three. Compute the transfer functions, poles, and zeros, and difference equation. *Note:* This problem does not require MATLAB/MathScript.

9.35 Show that if the bilinear tranformation is used to design a discrete-time Butterworth filter with cutoff frequency $\Omega_0 = \pi/2$ from a continuous-time Butterworth filter of *any* order, then the zeros are all at -1 and the poles are all on the imaginary axis.

Section 9-5: Multirate Signal Processing

9.36 Design a multirate system that converts a 300 Hz sinusoid sampled at 1000 samples/s to 450 Hz sinusoid without any aliasing.

9.37 A spoken-word recording is bandlimited to a maximum frequency of 4410 Hz. It is sampled at 11025 samples/s and burned onto a CD to be sold in stores. But CD players operate a 44100 samples/s. Design a discrete-time system that upsamples the samples at 11025 samples/s to 44100 samples/s. Using a Hamming window on the impulse response of an ideal lowpass filter, design the discrete-time lowpass filter in this system to meet these specifications:

(a) All portions of the spoken-word baseband spectrum are passed with gain > 0.9899.

(b) All portions of the first image spectrum are rejected with gain < 0.0101.

9.38 Prove that these systems are equivalent if and only if L and M are relatively prime:

$$\{x[n] \rightarrow \boxed{\downarrow M} \rightarrow \boxed{\uparrow L} \rightarrow y[n]\} \equiv$$

$$\{x[n] \rightarrow \boxed{\uparrow L} \rightarrow \boxed{\downarrow M} \rightarrow y[n]\}.$$

Hint: Use time domain.

9.39 Prove the so-called *noble identities*:

(a)

$$\{x[n] \rightarrow \boxed{\mathbf{H(z)}} \rightarrow \boxed{\uparrow L} \rightarrow y[n]\} \equiv$$

$$\{x[n] \rightarrow \boxed{\uparrow L} \rightarrow \boxed{\mathbf{H(z^L)}} \rightarrow y[n]\}.$$

Hint: Use **z**-transforms.

(b)

$$\{x[n] \rightarrow \boxed{\downarrow L} \rightarrow \boxed{\mathbf{H(z)}} \rightarrow y[n]\} \equiv$$

$$\{x[n] \rightarrow \boxed{\mathbf{H(z^L)}} \rightarrow \boxed{\downarrow L} \rightarrow y[n]\}.$$

Hint: Use time domain.

9.40 Let $x[n]$ be bandlimited to $\pi/2$, so $\mathbf{X}(e^{j\Omega}) = 0$ for $\pi/2 < |\Omega| \leq \pi$. Filter $x[n]$ using

$$x[n] \rightarrow \boxed{\downarrow 2} \rightarrow \boxed{\uparrow 2} \rightarrow \boxed{h[n] = 2\,\frac{\sin(\pi n/2)}{\pi n}} \rightarrow y[n].$$

Show by direct computation (perform the convolution) that $y[n] = x[n]$ for even n values.

*9.41 For the system

$$\cos(0.6\pi n) \rightarrow \boxed{\downarrow L} \rightarrow y[n],$$

for which of these values of L: $\{2, 3, 4, 5, 6\}$, does $y[n]$ include a component $A\cos(0.6\pi n)$?

9.42 A signal $x(t)$ bandlimited to 499 Hz is sampled at 1000 samples/s, giving $x[n]$. We have only a poor *analog* lowpass filter to perform digital-to-analog conversion. Design a DSP system that will allow good reconstruction of $x(t)$ from its samples $x[n]$. You may use a discrete-time sinc function as a digital filter.

9.43 A more realistic version of the previous problem is as follows: A signal with maximum frequency B Hz is sampled at double its Nyquist rate. Design an oversampling DSP system with upsampling by 10 and interpolation. Give the specifications for the discrete-time and continuous-time filters. The copies of the signal spectrum induced by sampling are considered to have been eliminated if the filter gain for them is 0.001 or less.

9.44 You are given the signal of a trumpet playing a note. Use multirate filtering and the circle of fifths to generate the signal of the trumpet playing all musical notes.

Section 9-12: Correlation

9.45 This problem investigates *fractional time delay* (time delay by a non-integer). Consider a system

$$y[n] = x\left[n - \frac{1}{2}\right],$$

which makes no sense since n is an integer. But we can implement a time delay of $\frac{1}{2}$ as follows:

(a) Determine the frequency response of a system that delays the input by $\frac{1}{2}$.

(b) Compute the impulse response of a system that delays the input by $\frac{1}{2}$.

(c) Interpret your answer to (b) in terms of upsampling and interpolation.

9.46 Signals $x[n]$ and $y[n]$ are given by

$$x[n] = A\cos\left(2\pi\,\frac{M}{N}\,n + \theta_1\right)$$

and

$$y[n] = B\cos\left(2\pi\,\frac{M}{N}\,n + \theta_2\right),$$

and both have period N.

(a) Show that the correlation coefficient between $x[n]$ and $y[n]$ is $\cos(\theta_1 - \theta_2)$.

(b) Show that the correlation coefficient between $A\cos(2\pi\,\frac{M}{N}\,n)$ and $B\sin(2\pi\,\frac{M}{N}\,n)$ is zero.

For pairs of periodic signals, correlation is defined over a single period of the signals.

9.47 Signals $x[n]$ and $y[n]$ are given by

$$x[n] = A\cos\left(2\pi \frac{L}{N} n\right)$$

and

$$y[n] = B\cos\left(2\pi \frac{M}{N}\right),$$

where $0 \leq L, M \leq N$. Show that unless $L = M$, the correlation between $x[n]$ and $y[n]$ is zero. For pairs of periodic signals, correlation is defined over a single period of the signals. *Note:* This result is still true even if $x[n]$ and $y[n]$ have nonzero phase shifts.

9.48 Show that if the spectra of two signals do not overlap, their cross-correlation is zero.

9.49 For each of the following two signals, show that its autocorrelation is just the signal itself.

(a)

$$x_1[n] = \frac{\sin(\Omega_0 n)}{\pi n}$$

for any constant $|\Omega_0| < \pi$.

(b)

$$x_2[n] = \frac{\sin(\Omega_0 n)}{\pi n} 2\cos(\Omega_1 n)$$

for any constants $|\Omega_0| < \pi$ and $|\Omega_1| < \frac{\pi}{2}$.

9.50 A causal signal $x[n]$ of duration 3 has

$$r_x[n] = \{6, 14, \underline{41}, 14, 6\}.$$

Determine $x[n]$. There are four solutions, all closely related to each other.

9.51 A causal signal $x[n]$ of length $2N+1$ has a known $r_x[n]$ of length $4N+1$. The $2N$ zeros of $\mathbf{X(z)}$ are in N complex conjugate pairs, none on the unit circle. Show that if $x[0] \neq 0$, then the solution of $x[n] * x[-n] = r_x[n]$ can be satisfied by 2^{N+1} different signals $x[n]$.

LabVIEW Module 9.1

9.52 Design a lowpass filter with the specified duration and cutoff frequency. Specify whether the impulse response looks like a windowed sinc function.

(a) Duration 41 and cutoff 200 Hz.

(b) Duration 21 and cutoff 100 Hz.

(c) Duration 41 and cutoff 400 Hz.

9.53 Although the sampling rate is not specified, explain how you can determine that the sampling rate is 1000 sample/s.

LabVIEW Module 9.2

9.54 In each of these problems, display the spectrogram for the specified length.

(a) Length 26. The display is the notes of the chorus of what famous college fight song? *Hint:* The authors' affiliation.

(b) Length 20. How is the display related to the display of the previous problem? What are the additional features that appear in this spectrogram?

(c) Repeat (b) for window length 50.

9.55 What simple action could be taken to reduce the additional features appearing in the spectrograms in (b) and (c) of Problem 9.54?

LabVIEW Module 9.3

9.56 In each of these problems, display the spectrogram for the specified window length and chirp slope.

(a) Slope 1.0 and window length 45.

(b) Slope 1.5 and window length 58.

(c) Slope 1.5 and window length 16.

9.57 Why is each spectrogram segment thicker in (c) of Problem 9.56?

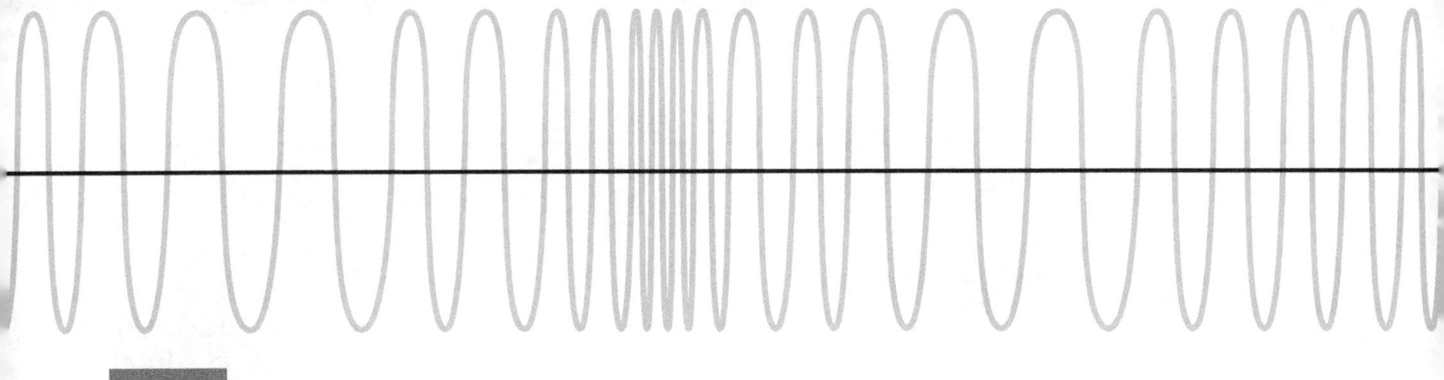

10 Image Processing, Wavelets, and Compressed Sensing

Contents

Overview, 546
10-1 Image Processing Basics, 546
10-2 Discrete-Space Fourier Transform, 549
10-3 2-D DFT, 553
10-4 Downsampling and Upsampling of Images, 554
10-5 Image Denoising, 555
10-6 Edge Detection, 559
10-7 Image Deconvolution, 565
10-8 Overview of the Discrete-Time Wavelet Transform, 569
10-9 Haar Wavelet Transform, 572
10-10 The Family of Wavelet Transforms, 577
10-11 Non-Haar Single-Stage Perfect Reconstruction, 581
10-12 Daubechies Scaling and Wavelet Functions, 584
10-13 2-D Wavelet Transform, 590
10-14 Denoising by Thresholding and Shrinking, 595
10-15 Compressed Sensing, 599
10-16 Computing Solutions to Underdetermined Equations, 601
10-17 Landweber Algorithm, 604
10-18 Compressed Sensing Examples, 606
Summary, 614
Problems, 614

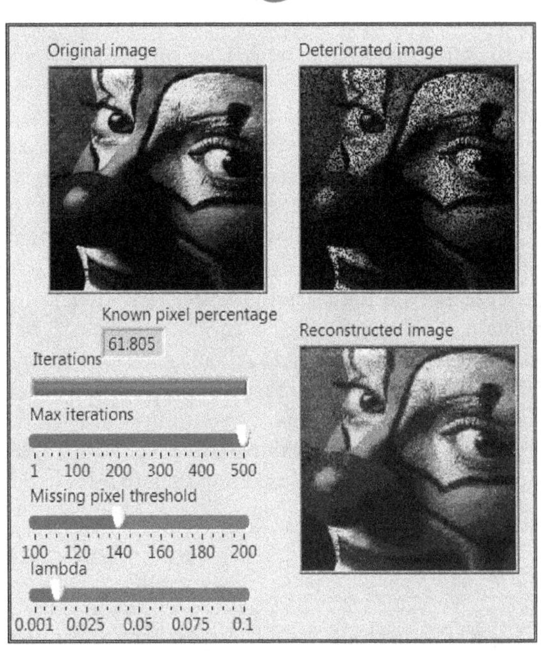

The one-dimensional (1-D) discrete-time signals and systems tools are extended in this chapter to 2-D spatial images, which then are used to perform image-processing enhancements, including denoising, edge detection, and deconvolution. This is followed by a treatment of the discrete-space wavelet transform and examples of its many applications, including inpainting and compressed sensing.

Objectives

Learn to:

- Compute the 2-D discrete-space Fourier transform of an image.
- Denoise an image.
- Apply edge detection and deconvolution.
- Use the wavelet transform in image denoising, inpainting, and compressed sensing.

Overview

With the exception of the image examples presented in Sections 5-13 and 6-1, the preceding nine chapters dealt exclusively with one-dimensional (1-D) signals and systems, and the dimension under consideration was either continuous time t or discrete time n. An image is a two-dimensional (2-D) configuration, and the two dimensions are spatial rather than temporal. A discrete-space image consists of a 2-D array of numbers, often referred to as *pixels* (short for picture elements). Many, but not all, of the transformations and techniques developed for 1-D signals are *extendable* to 2-D images. The first part of this chapter provides the 1-D to 2-D signal-processing extension tools and associated nomenclature and notation. These tools are then used to perform Fourier-based ***image-processing enhancements***, such as denoising, edge detection, and deconvolution.

The second major topic treated in this chapter is the discrete-space ***wavelet transform*** and its applications in image compression and denoising. In recent years, the wavelet transform has become an important tool in image processing because it can offer a performance superior to that provided by Fourier-based techniques for denoising and compression.

Compressed sensing, the third topic of this chapter, makes use of the fact that many signals and images of interest can be represented using a sparse (mostly zero-valued) linear combination of wavelet functions. This means that a signal or image can be reconstructed from a set of linear combinations of itself that is much smaller in size than the signal or image itself. This is useful for: deconvolution, reconstruction of a signal or image from a relatively small number of 2-D DFT values, and restoring missing pixels. We present an introduction to this topic, showing the reader how to make use of it without getting into the details of why it works so well.

10-1 Image Processing Basics

10-1.1 Extending 1-D to 2-D

An image is an array of numbers, called ***pixels***, while a signal is a vector of numbers. We denote an image as $\{x[m,n], 0 \leq m \leq M, 0 \leq n \leq N\}$, analogous to the representation $x[n]$ of a 1-D signal. Often, but not always, the upper left corner of an image is designated to be location $[0,0]$ (the origin), with indices m and n increasing downward and rightward, respectively. So $x[m,n]$ is the intensity (brightness) of the pixel in the $(m+1)$th row and $(n+1)$th column of the image (Fig. 10-1). This follows MATLAB/Mathscript matrix notation, except that in MATLAB/Mathscript matrix notation the origin has coordinates $[1,1]$ because in MATLAB, indexing does not include zero. Note that this is not the same as Cartesian coordinates, in which the origin is the lower left corner of the image.

A discrete-time linear time-invariant system is characterized by an impulse response $h[n]$. The analogous 2-D system response is called the ***point-spread function*** $h[m,n]$, which will be introduced shortly. Unlike the common image format in which the origin is located at the upper left corner, the format used with $h[m,n]$ often defines the origin at the center of the image. Hence, when applying a multistep process to an image, it is critically important to keep track of the coordinate systems associated with each step, and to make the necessary spatial shifts to align them when necessary.

► Image $x[m-M,\ n-N]$ represents image $x[m,n]$ shifted down by M and to the right by N. ◄

An example is shown in Fig. 10-1.

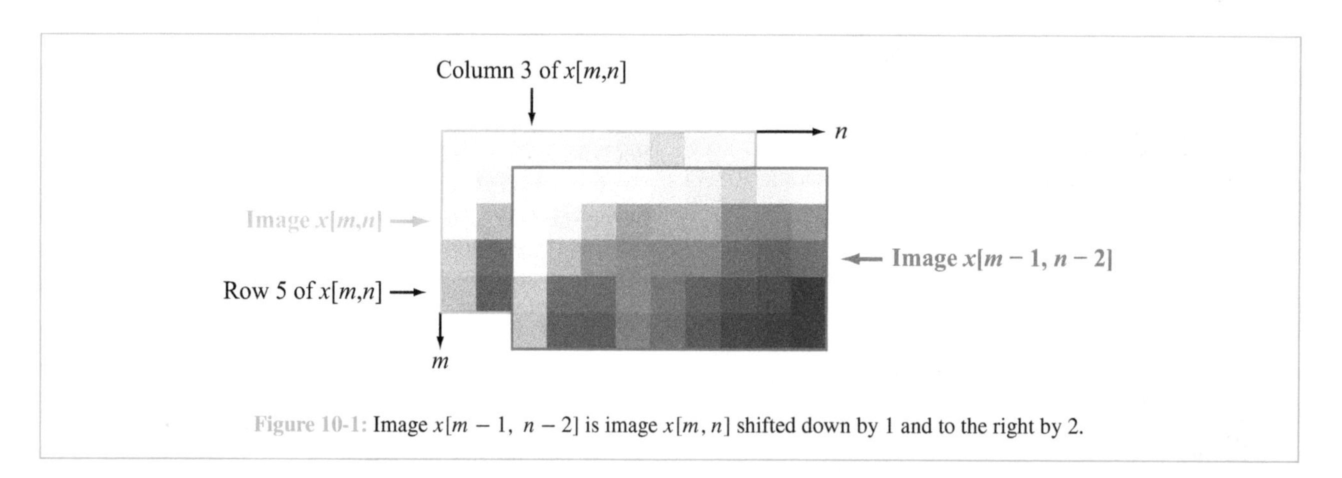

Figure 10-1: Image $x[m-1,\ n-2]$ is image $x[m,n]$ shifted down by 1 and to the right by 2.

Color images consist of three separate images, each representing a different color. Usually these colors are red, green, and blue. A color image is a triplet of pixel values at each location $[m, n]$:

$$\{x_{\text{red}}[m, n], x_{\text{green}}[m, n], x_{\text{blue}}[m, n].\}$$

In this book, we limit our attention to ***grayscale*** (black and white) images, described by a single pixel value $x[m, n]$ at each location $[m, n]$. In grayscale images, the brightness at location $[m, n]$ is proportional to $x[m, n]$, scaled so that the maximum value of $x[m, n]$ is depicted as white, and the minimum value of $x[m, n]$ is depicted as black. Since most images are non-negative ($x[m, n] \geq 0$), black usually depicts $x[m, n] = 0$.

A grayscale image with pixels $x[m, n]$ stored in the $M \times N$ MATLAB/Mathscript array `X` can be displayed using `imagesc(X),colormap(gray)`. Omitting `colormap(gray)` results in a depiction of the image using colors to represent different values. This should not be confused with a color image.

As we will see shortly, the following concepts generalize in a straightforward manner from 1-D to 2-D:

- LTI, impulses, convolution, impulse response (now called ***point-spread function***).
- Frequency (now called ***wavenumber***) response.
- Fourier transforms and the sampling theorem.
- DTFT (now called DSFT), DFT, FFT, **z**-transforms.

In contrast, the following concepts do *not* generalize in a useful manner from 1-D to 2-D:

- Laplace transforms and difference equations.
- Transfer functions, poles and zeros.
- Partial fraction expansions.

10-1.2 2-D Sampling

Usually, a discrete-space image is obtained by sampling a continuous-space image signal $x(\xi, \eta)$ at spatial locations $\xi = mT_s$ and $\eta = nT_s$ for some small ***discretization length*** T_s. The discrete form of $x(\xi, \eta)$ is denoted $x[m, n]$. In some cases, however, the image is generated in discrete format directly, as part of the sensing process. An example is a camera that acquires images via an array of sensors composed of charge-coupled devices (CCDs).

10-1.3 2-D Reconstruction

The 2-D Fourier transform of a continuous-space image signal was introduced in Chapter 5 in the form of Eq. (5.143a):

$$\mathbf{X}(\omega_1, \omega_2) = \mathcal{F}[x(\xi, \eta)]$$
$$= \int_{-\infty}^{\infty}\int_{-\infty}^{\infty} x(\xi, \eta)\, e^{-j\omega_1\xi} e^{-j\omega_2\eta}\, d\xi\, d\eta, \tag{10.1}$$

where (ξ, η) are the (horizontal, vertical) image coordinates, $x(\xi, \eta)$ is the image intensity, and (ω_1, ω_2) are spatial frequencies—called ***wavenumbers***—along the vertical and horizontal directions in the frequency domain. The 2-D Fourier transform consists of two transformations, one from ξ in the spatial domain to ω_1 in the frequency domain, and another from η to ω_2.

If the spectrum $\mathbf{X}(\omega_1, \omega_2)$ is bandlimited to a maximum wavenumber $\omega_{\max}$ along both wavenumber dimensions, then

$$\omega_1(\max) = \omega_2(\max) = \omega_{\max}. \tag{10.2}$$

To satisfy the Nyquist sampling criterion, the sampling length T_s (which is proportional to the reciprocal of the sampling wavenumber ω_s) should be short enough to guarantee that

$$\omega_s = \frac{2\pi}{T_s} > 2\omega_{\max}. \tag{10.3}$$

In Section 6-13.6, we explained how the sinc interpolation formula given by Eq. (6.161) can be applied to reconstruct a continuous-time signal $x(t)$ from its samples $x[n] = x(nT_s)$. The process is equally applicable to discretized images, except that the interpolation process has to be performed both horizontally and vertically (or vice versa). To obtain $x(\xi, \eta)$ from $x[m, n] = x(mT_s, nT_s)$, the 2-D sinc interpolation formula assumes the form

$$x(\xi, \eta) = \sum_{m=-\infty}^{\infty}\sum_{n=-\infty}^{\infty} x(mT_s, nT_s)$$
$$\times \frac{\sin\left(\frac{\pi}{T_s}(\xi - mT_s)\right) \cdot \sin\left(\frac{\pi}{T_s}(\eta - nT_s)\right)}{\left(\frac{\pi}{T_s}(\xi - mT_s)\right) \cdot \left(\frac{\pi}{T_s}(\eta - nT_s)\right)}. \tag{10.4}$$

(reconstruction from discrete to continuous)

The interpolation process is equally applicable to other manifestations of of $x[m, n]$, such as after getting denoised or filtered to enhance certain features. If the available image $x[m, n]$ was generated in discrete space to start with, we can still apply the interpolation formula by choosing a value of T_s that is sufficiently short as to produce a visually acceptable continuous-space image.

10-1.4 LSI Systems

A 2-D system with input $x[m, n]$ and output $y[m, n]$ is denoted as

$$x[m, n] \rightarrow \boxed{\text{SYSTEM}} \rightarrow y[m, n].$$

The system is considered ***linear-shift invariant*** (***LSI***) if it satisfies the following properties:

(a) **Shift Invariant**

If

$$x[m, n] \rightarrow \boxed{\text{LSI}} \rightarrow y[m, n],$$

it then follows that, for any integers M and N,

$$x[m-M,\ n-N] \rightarrow \boxed{\text{LSI}} \rightarrow y[m-M,\ n-N]. \tag{10.5}$$

(b) **Scalable**

For any constant c,

$$cx[m, n] \rightarrow \boxed{\text{LSI}} \rightarrow cy[m, n]. \tag{10.6}$$

(c) **Superposition**

If

$$x_1[m, n] \rightarrow \boxed{\text{LSI}} \rightarrow y_1[m, n]$$

$$\text{and } x_2[m, n] \rightarrow \boxed{\text{LSI}} \rightarrow y_2[m, n],$$

it then follows that, for any constants c_1 and c_2,

$$c_1x_1[m, n]+c_2x_2[m, n] \rightarrow \boxed{\text{LSI}} \rightarrow c_1y_1[m, n]+c_2y_2[m, n]. \tag{10.7}$$

10-1.5 Point-Spread Function

In 2-D, impulse $\delta[m, n]$ is defined as

$$\delta[m, n] = \begin{cases} 1 & \text{for } m = n = 0, \\ 0 & \text{otherwise.} \end{cases} \tag{10.8}$$

The impulse response of a 2-D LSI system is called the ***point-spread function*** (***PSF***) of the system and is denoted $h[m, n]$. By analogy with Eq. (7.46),

$$\delta[m, n] \rightarrow \boxed{\text{LSI}} \rightarrow h[m, n]. \tag{10.9}$$

In medical imaging systems, the PSF $h[m, n]$ is sometimes *measured* directly by imaging a small bead, which acts like a 2-D impulse $\delta[m, n]$. The *spread* of the bead by the *imperfect* imaging system is the ***point-spread function*** of that imaging system. A similar procedure is used in astronomy, wherein $h[m, n]$ of the imaging telescope is determined by measuring the image of a tiny star acting like an impulse.

> ▶ The PSF of an imaging system is the *identity* image of that system. ◀

10-1.6 2-D Convolution

By analogy with Eq. (7.51a), the response of a 2-D LSI system to any image $x[m, n]$ is given by the 2-D convolution of $x[m, n]$ with the system's PSF $h[m, n]$:

$$\begin{aligned} y[m, n] &= x[m, n] ** h[m, n] \\ &= \sum_{i=-\infty}^{\infty} \sum_{j=-\infty}^{\infty} x[i, j]\, h[m-i,\ n-j]. \end{aligned} \tag{10.10}$$

Note that the symbol for 2-D convolution consists of 2 stars: $**$. Also note that convolving the PSF of an imaging system with an image is no different mathematically from convolving two images together.

Example 10-1: 2-D Convolution of 2 × 2 Images

Compute the 2-D convolution

$$\begin{bmatrix} 1 & 2 \\ 3 & 4 \end{bmatrix} ** \begin{bmatrix} 5 & 6 \\ 7 & 8 \end{bmatrix}.$$

Solution: The definition given by Eq. (10.10) entails multiplication of the first matrix by a shifted version of the second matrix, which translates into

$$\begin{bmatrix} 1 & 2 \\ 3 & 4 \end{bmatrix} ** \begin{bmatrix} 5 & 6 \\ 7 & 8 \end{bmatrix}$$

$$= 1 \begin{bmatrix} 5 & 6 & 0 \\ 7 & 8 & 0 \\ 0 & 0 & 0 \end{bmatrix} + 2 \begin{bmatrix} 0 & 5 & 6 \\ 0 & 7 & 8 \\ 0 & 0 & 0 \end{bmatrix} + 3 \begin{bmatrix} 0 & 0 & 0 \\ 5 & 6 & 0 \\ 7 & 8 & 0 \end{bmatrix} + 4 \begin{bmatrix} 0 & 0 & 0 \\ 0 & 5 & 6 \\ 0 & 7 & 8 \end{bmatrix}$$

$$= \begin{bmatrix} 5 & 16 & 12 \\ 22 & 60 & 40 \\ 21 & 52 & 32 \end{bmatrix}.$$

Note how each member of the first matrix is multiplied by a shifted version of the second matrix, and then all four matrices are added together. The 2-D convolution can be implemented in MATLAB/Mathscript by the command

```
conv2([1 2; 3 4], [5 6; 7 8]).
```

Concept Question 10-1: What 1-D concepts generalize usefully to 2-D? (See S^2)

Concept Question 10-2: What 1-D concepts do not generalize usefully to 2-D? (See S^2)

10-2 Discrete-Space Fourier Transform

For a nonperiodic discrete-time signal $x[n]$, we defined its discrete-time Fourier transform (DTFT) in Eq. (7.154a) as

$$\mathbf{X}(e^{j\Omega}) = \sum_{n=-\infty}^{\infty} x[n]\, e^{-j\Omega n},$$

where Ω is an angular frequency in discrete time. The 2-D analogue of the DTFT is the ***discrete-space Fourier transform*** (***DSFT***), computed by applying a DTFT to the rows and then a second DTFT to the columns (or vice versa). That is,

$$\mathbf{X}(e^{j\Omega_1}, e^{j\Omega_2}) = \text{DSFT}(x[m,n]) = \sum_{m=-\infty}^{\infty} \sum_{n=-\infty}^{\infty} x[m,n]\, e^{-j(\Omega_1 m + \Omega_2 n)}. \tag{10.11}$$

The DSFT is periodic in Ω_1 and Ω_2, both with periods 2π. The inverse DSFT is computed by applying the inverse DTFT twice, once for rows and then for columns (or vice versa):

$$x[m,n] = \text{DSFT}^{-1}[\mathbf{X}(e^{j\Omega_1}, e^{j\Omega_2})] = \frac{1}{4\pi^2} \int_{-\pi}^{\pi} \int_{-\pi}^{\pi} \mathbf{X}(e^{j\Omega_1}, e^{j\Omega_2}) \cdot e^{j(\Omega_1 m + \Omega_2 n)}\, d\Omega_1\, d\Omega_2. \tag{10.12}$$

10-2.1 Conjugate Symmetry Properties of the DSFT

(1) If the image $x[m,n]$ is real-valued, then its DSFT has conjugate symmetry:

$$\mathbf{X}^*(e^{j\Omega_1}, e^{j\Omega_2}) = \mathbf{X}(e^{-j\Omega_1}, e^{-j\Omega_2}), \tag{10.13a}$$

$$|\mathbf{X}(e^{j\Omega_1}, e^{j\Omega_2})| = |\mathbf{X}(e^{-j\Omega_1}, e^{-j\Omega_2})|, \tag{10.13b}$$

$$-\angle\mathbf{X}(e^{j\Omega_1}, e^{j\Omega_2}) = \angle\mathbf{X}(e^{-j\Omega_1}, e^{-j\Omega_2}), \tag{10.13c}$$

(2) If $x[m,n]$ is real and an even function:

$$x[m,n] = x[-m,-n],$$

then $\mathbf{X}(e^{j\Omega_1}, e^{j\Omega_2})$ is real and an even function also:

$$\mathbf{X}(e^{j\Omega_1}, e^{j\Omega_2}) = \mathbf{X}(e^{-j\Omega_1}, e^{-j\Omega_2}).$$

(3) If $x[m,n]$ is real and an odd function:

$$x[m,n] = -x[-m,-n],$$

then $\mathbf{X}(e^{j\Omega_1}, e^{j\Omega_2})$ is pure imaginary and an odd function:

$$\mathbf{X}(e^{j\Omega_1}, e^{j\Omega_2}) = -\mathbf{X}(e^{-j\Omega_1}, e^{-j\Omega_2}).$$

The second and third properties require that $x[m,n]$ be defined for negative values of indices m and n. If $[0,0]$ is in the center of the image, property 2 implies that the image has diagonal symmetry: pixel $[2,3]$, for example, has the same brightness as pixel $[-2,-3]$, and pixel $[2,-3]$ has the same brightness as pixel $[-2,3]$. Property 3 is applicable only if the image brightness is scaled so that "zero brightness" is defined to be at some intermediate level between black and white, thereby allowing the brightness to be both positive and (artificially) negative in intensity.

10-2.2 Wavenumber Response

► The DSFT of the point-spread function $h[m,n]$ is the ***wavenumber response*** $\mathbf{H}(e^{j\Omega_1}, e^{j\Omega_2})$:

$$\mathbf{H}(e^{j\Omega_1}, e^{j\Omega_2}) = \text{DSFT}(h[m,n]) = \sum_{m=-\infty}^{\infty} \sum_{n=-\infty}^{\infty} h[m,n]\, e^{-j(\Omega_1 m+\Omega_2 n)}. \quad (10.14)$$

Related properties include

$$e^{j(\Omega_1 m+\Omega_2 n)} \rightarrow \boxed{\mathbf{H}(e^{j\Omega_1}, e^{j\Omega_2})} \rightarrow \mathbf{H}(e^{j\Omega_1}, e^{j\Omega_2})\, e^{j(\Omega_1 m+\Omega_2 n)} \quad (10.15a)$$

and

$$\cos(\Omega_1 m+\Omega_2 n) \rightarrow \boxed{\mathbf{H}(e^{j\Omega_1}, e^{j\Omega_2})} \rightarrow |\mathbf{H}(e^{j\Omega_1}, e^{j\Omega_2})| \cos(\Omega_1 m+\Omega_2 n+\theta), \quad (10.15b)$$

where θ is the phase angle of $\mathbf{H}(e^{j\Omega_1}, e^{j\Omega_2})$. These two properties are extensions of their 1-D counterparts given by Eqs. (7.124) and (7.127). Equation (10.15b) can be derived from Eq. (10.15a) in the same way that Eq. (7.127) can be derived from Eq. (7.124).

10-2.3 2-D Spectrum

Since the DSFT is doubly periodic in Ω_1 and Ω_2, it is customary to depict the 2-D spectrum over the ranges $-\pi < \Omega_1, \Omega_2 < \pi$, with the origin $(\Omega_1, \Omega_2) = (0,0)$ at the center of the frequency plane. This is analogous to two-sided 1-D spectra. In MATLAB/ Mathscript, depicting the ***2-D spectrum*** with the origin at the center of an image stored in array X of size $M \times N$ is performed using

```
imagesc(fftshift(abs(fft2(X,M,N)))).
```
(10.16)

An important difference from 1-D signals is that most images have all non-negative pixel values. The appearance of the 2-D spectrum of a non-negative image gets dominated by a single bright dot at the origin $(\Omega_1, \Omega_2) = (0,0)$. This is the dc value of the spectrum, and is given by

$$\mathbf{X}(e^{j0}, e^{j0}) = \sum_{m=0}^{M-1} \sum_{n=0}^{N-1} x[m,n]. \quad (10.17)$$

The dc value is the sum of MN non-negative pixel values, making it much larger than other values in the spectrum.

Instead of displaying an image of the 2-D spectrum $|\mathbf{X}(e^{j\Omega_1}, e^{j\Omega_2})|$, it is common practice to display an image of its logarithm:

$$\log_{10} |\mathbf{X}(e^{j\Omega_1}, e^{j\Omega_2})|.$$

An example is shown in Fig. 10-2. The original image, displayed in part (a), is a group of numbers, and its linear and logarithmic spectra are displayed in parts (b) and (c), respectively. The ***logarithmic format*** compresses the range of the spectrum, so it reduces the dominance by the central pixel. Consider, for example, a central pixel with a magnitude of 100, another pixel with a magnitude of 20, and a third one with a magnitude of 10. On a linear scale, the latter two pixels appear almost black, compared to the white pixel at the center. The logarithmic values of 10, 20, and 100 are 1, 1.3, and 2, so the range 10 to 100 gets compressed to 1 to 2, thereby allowing a viewer to "see" the lower intensity pixels. The logarithmic scale also is used in the ***Richter scale*** for earthquakes, stellar magnitudes in astronomy, and in displaying quantitative data that extend over multiple orders of magnitude.

Note the appearance of multiple bright lines in the spectrum shown in Fig. 10-2(c). These lines are associated with the geometrical shapes of the letters in the original image. For example, the horizontal segments of the letters, such as the top and bottom parts of the letter E, are responsible for generating the vertical line in the spectrum. The direction of the spectral line is orthogonal to the direction of the segments responsible for generating that line. Similar associations apply to the other spectral lines. This is explained in Problem 10.20.

An alternative to displaying the spectrum in black and white using a logarithmic scale is to display it in color without the logarithmic compression. The available range of colors is *mapped* onto the available range of values of $|\mathbf{X}(e^{j\Omega_1}, e^{j\Omega_2})|$. The result is displayed in Fig. 10-2(d).

Example 10-2: 2-D Two-Point Average

Image $\{x[m,n],\ 0 \le m \le 9,\ 0 \le n \le 9\}$ is displayed in Fig. 10-3(a) using a colormap display of pixel values. The actual pixel values are shown in Fig. 10-3(b). The image consists of $10^2 = 100$ pixels.

Image $x[m,n]$ is defined by the analytical expression

$$x[m,n] = x_1[m]\, x_2[n],$$

where

$$x_1[m] = 9 - 2(4.5 - m), \qquad 0 \le m \le 9,$$

$$x_2[n] = 10(1 - e^{-0.14n}), \qquad 0 \le n \le 9.$$

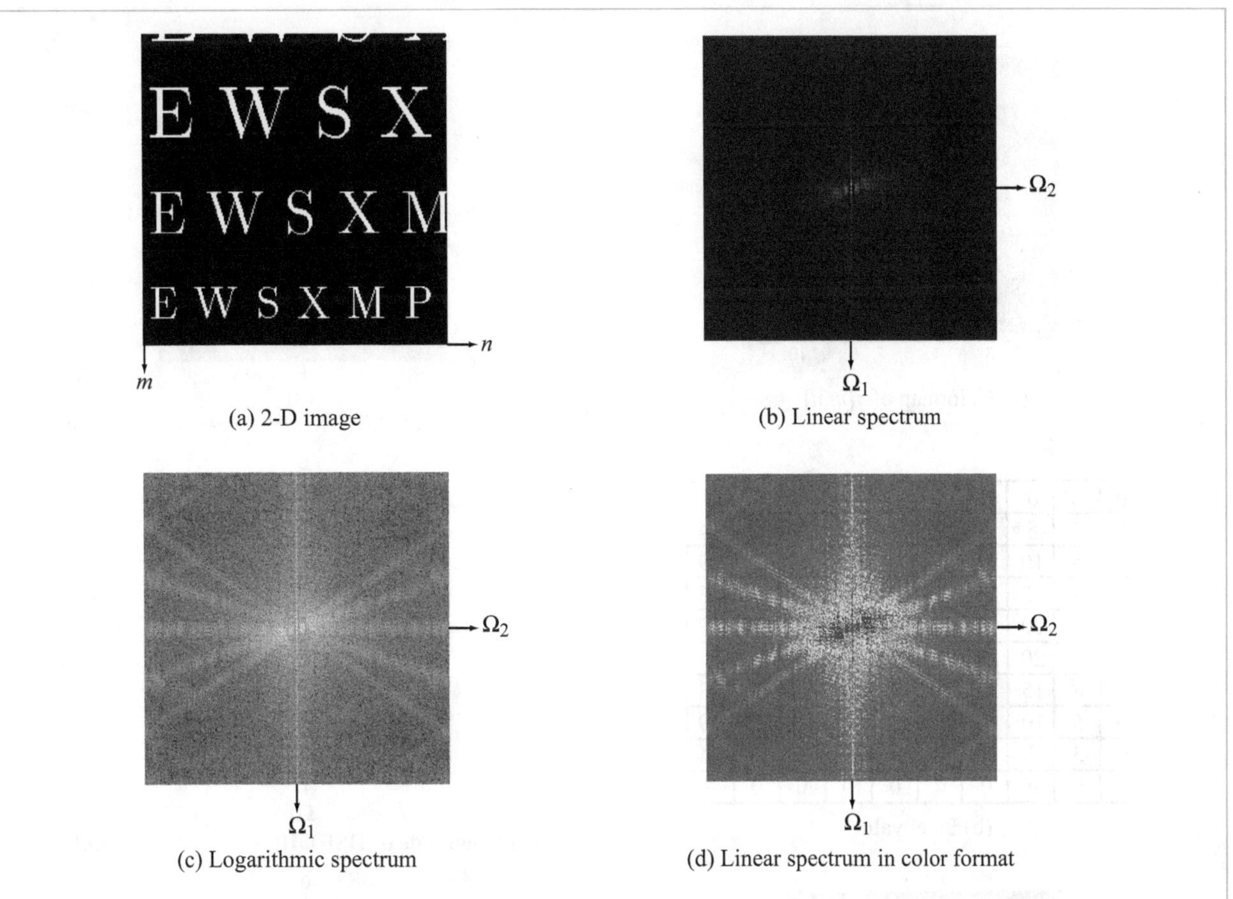

(a) 2-D image (b) Linear spectrum

(c) Logarithmic spectrum (d) Linear spectrum in color format

Figure 10-2: (a) 2-D image $x[m,n]$, (b) spectrum $|\mathbf{X}(e^{j\Omega_1}, e^{j\Omega_2})|$ in linear scale, (c) logarithmic spectrum $\log_{10} |\mathbf{X}(e^{j\Omega_1}, e^{j\Omega_2})|$, and (d) linear spectrum in color format, with different colors representing different values of $|\mathbf{X}(e^{j\Omega_1}, e^{j\Omega_2})|$.

A 2-D version of the 1-D two-point averager is a system that averages the pixel values in a 2×2 block of the image. Its point-spread function is given by

$$h[m,n] = \frac{1}{4}\begin{bmatrix} \underline{1} & 1 \\ 1 & 1 \end{bmatrix}.$$

The pixel at the origin (upper left corner) is underlined, just as it is in a 1-D bracket notation. It is understood, by definition, that $h[m,n] = 0$ outside the 2×2 block. That is,

$$h[m,n] = \begin{cases} \frac{1}{4} & \text{for } 0 \leq m,n \leq 1, \\ 0 & \text{otherwise.} \end{cases}$$

Apply the 2-D two-point averager to the image in Fig. 10-3(b), and display the output image $y[m,n]$, by:

(a) performing the convolution directly in the discrete-space domain, and

(b) computing the DSFTs of $x[m,n]$ and $h[m,n]$, multiplying them, and then computing the inverse DSFT of the product to obtain $y[m,n]$.

Solution:

(a) Application of the convolution operation given by Eq. (10.10) leads to the smoothed image $y[m,n]$ displayed in Fig. 10-3(c). Since image $x[m,n]$ is $N_1 \times N_1$ with $N_1 = 10$, and $h[m,n]$ is $N_2 \times N_2$ with $N_2 = 2$, the size of the convolved image $y[m,n]$ is $N_3 \times N_3$ with

$$N_3 = N_1 + N_2 - 1 = 10 + 2 - 1 = 11.$$

The smoothing effect of the averager is quite noticeable, particularly when we compare the right sides of images $x[m,n]$ in Fig. 10-3(a) and $y[m,n]$ in Fig. 10-3(c).

(b) Using MATLAB/Mathscript (Eq. (10.16)), the DSFTs $\mathbf{X}(e^{j\Omega_1}, e^{j\Omega_2})$ of $x[m,n]$ and $\mathbf{H}(e^{j\Omega_1}, e^{j\Omega_2})$ of $h[m,n]$ were

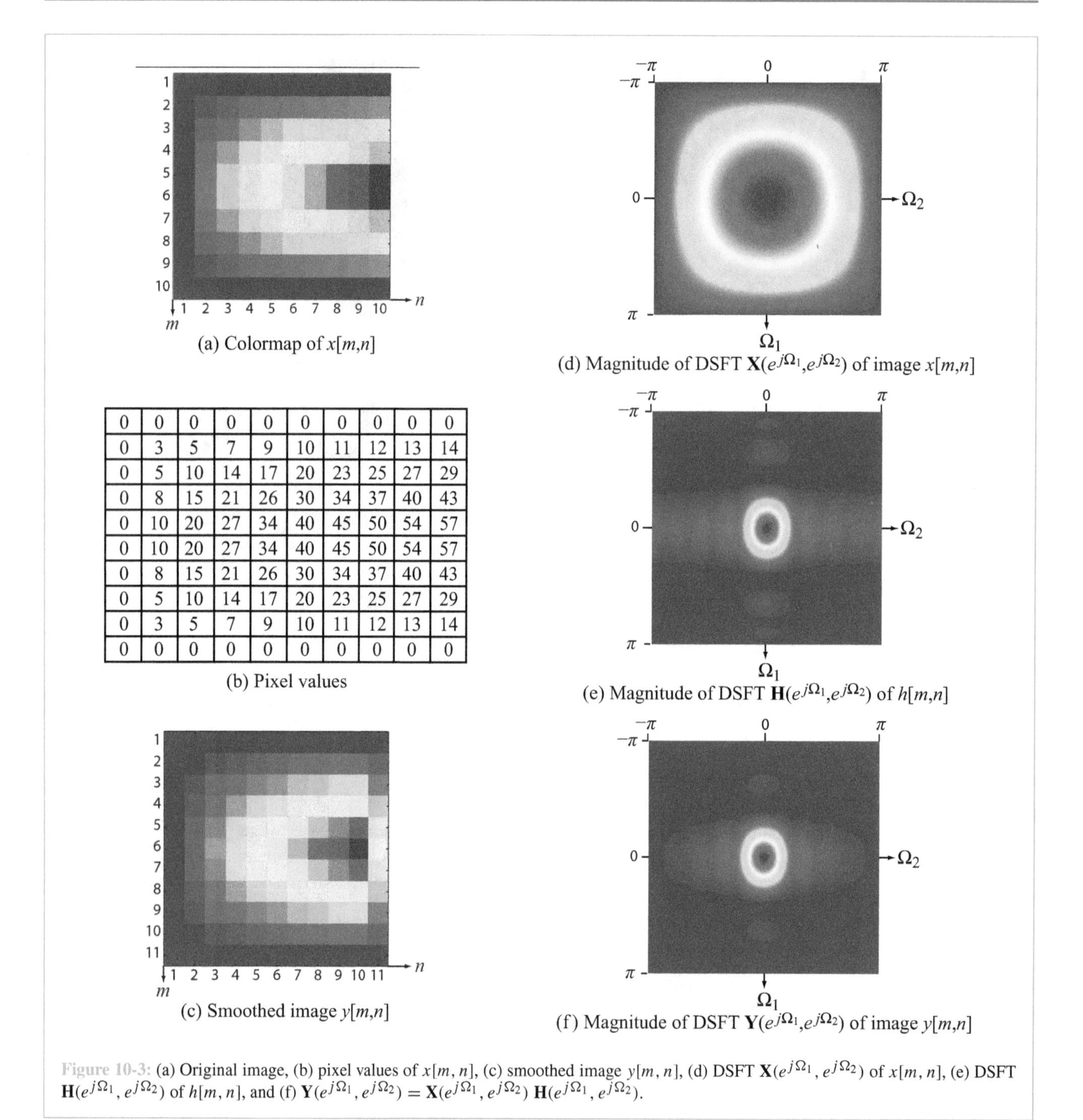

0	0	0	0	0	0	0	0	0	0
0	3	5	7	9	10	11	12	13	14
0	5	10	14	17	20	23	25	27	29
0	8	15	21	26	30	34	37	40	43
0	10	20	27	34	40	45	50	54	57
0	10	20	27	34	40	45	50	54	57
0	8	15	21	26	30	34	37	40	43
0	5	10	14	17	20	23	25	27	29
0	3	5	7	9	10	11	12	13	14
0	0	0	0	0	0	0	0	0	0

Figure 10-3: (a) Original image, (b) pixel values of $x[m,n]$, (c) smoothed image $y[m,n]$, (d) DSFT $\mathbf{X}(e^{j\Omega_1},e^{j\Omega_2})$ of $x[m,n]$, (e) DSFT $\mathbf{H}(e^{j\Omega_1},e^{j\Omega_2})$ of $h[m,n]$, and (f) $\mathbf{Y}(e^{j\Omega_1},e^{j\Omega_2}) = \mathbf{X}(e^{j\Omega_1},e^{j\Omega_2})\,\mathbf{H}(e^{j\Omega_1},e^{j\Omega_2})$.

computed and their magnitudes were then displayed in parts (d) and (e) of Fig. 10-3. The product

$$\mathbf{Y}(e^{j\Omega_1},e^{j\Omega_2}) = \mathbf{H}(e^{j\Omega_1},e^{j\Omega_2})\,\mathbf{X}(e^{j\Omega_1},e^{j\Omega_2}) \qquad (10.18)$$

was computed and its magnitude is displayed in Fig. 10-3(f). Finally, the inverse DSFT of $\mathbf{Y}(e^{j\Omega_1},e^{j\Omega_2})$ was computed using Eq. (10.12). As expected, the result is identical with the smoothed image obtained in part (a) of the solution, and displayed in Fig. 10-3(c).

Concept Question 10-3: Why is it difficult to display 2-D spectra of images? (See S^2)

Exercise 10-1: Use LabVIEW Module 10.1 to show the effect of drastic lowpass filtering on the letters image. Set both sliders to their minimum values.

Answer:

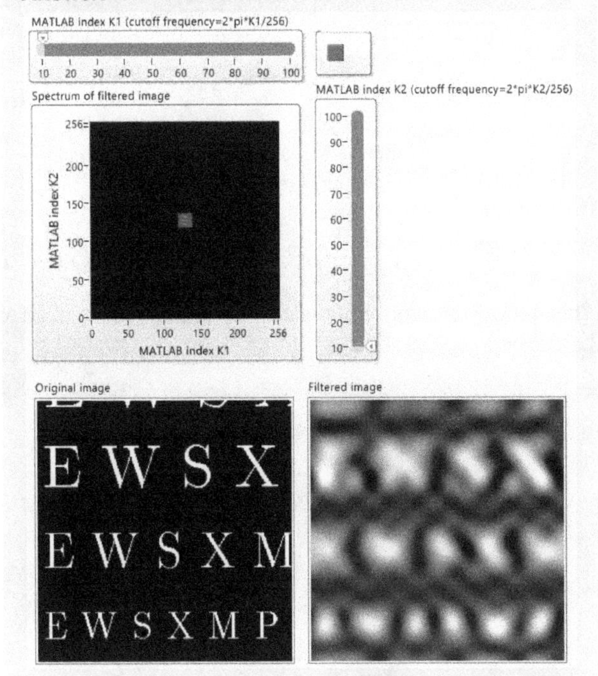

10-3 2-D DFT

Whereas image signal $x[m, n]$ is defined in discrete space, its DSFT $\mathbf{X}(e^{j\Omega_1}, e^{j\Omega_2})$ is defined in continuous 2-D wavenumber space (Ω_1, Ω_2). The same correspondence is true for the point spread function $h[m, n]$ and its transform $\mathbf{H}(e^{j\Omega_1}, e^{j\Omega_2})$. Recall from Section 7-15 that the discrete Fourier transform (DFT) is a numerical recipe for representing the spectrum of a nonperiodic signal in terms of a finite set of discrete frequency components. The DFT representation provides the platform for applying the highly efficient fast Fourier transform (FFT) to compute the spectra of signals of interest.

Module 10.1 Effect of Lowpass Filtering an Image This module applies a 2-D brickwall lowpass filter to the "letters" image using the 2-D DFT. The cutoff wavenumbers are selectable indices.

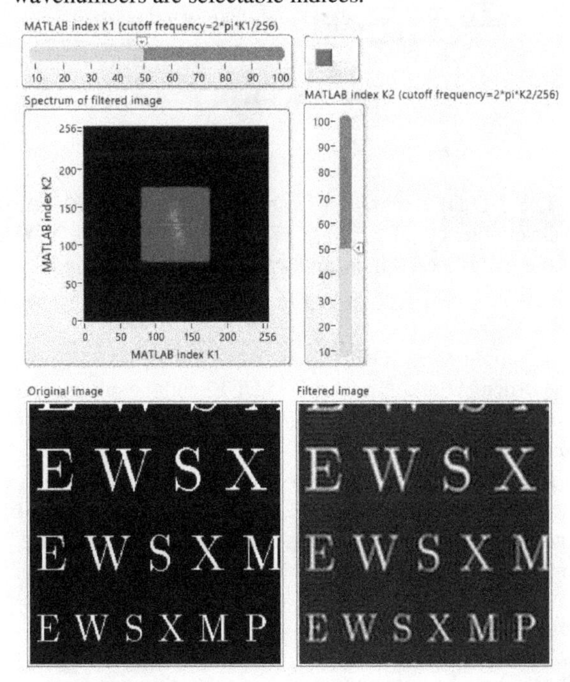

The DFT is equally applicable to 2-D discrete-space signals. For an $(M \times N)$ image $x[m, n]$, the wavenumber space (Ω_1, Ω_2) is discretized into

$$\Omega_1 = 2\pi \, \frac{k_1}{M}, \qquad k_1 = 0, 1, \ldots, M-1, \tag{10.19a}$$

$$\Omega_2 = 2\pi \, \frac{k_2}{N}, \qquad k_2 = 0, 1, \ldots, N-1, \tag{10.19b}$$

and the summations in Eq. (10.11) are limited to finite ranges extending between 0 and $M-1$ for m and between 0 and $N-1$ for n. The two modifications convert the definitions for the DSFT and its inverse given by Eqs. (10.11) and (10.12) into

$$\mathbf{X}[k_1, k_2] = \sum_{m=0}^{M-1} \sum_{n=0}^{N-1} x[m, n] \, e^{-j2\pi(k_1(m/M)+k_2(n/N))} \tag{10.20a}$$

2-D DFT

and its inverse 2-D DFT,

$$x[m,n] = \frac{1}{MN} \sum_{k_1=0}^{M-1} \sum_{k_2=0}^{N-1} \mathbf{X}[k_1,k_2]\, e^{j2\pi(k_1(m/M)+k_2(n/N))}.$$

Inverse 2-D DFT

(10.20b)

The 2-D DFT of $x[m,n]$ can be computed rapidly using a 2-D version of the FFT. In MATLAB/Mathscript, the 2-D DFT of the image stored in array X can be computed using

```
fft2(X)=fft(fft(X).'),
```

which applies the 1-D FFT to each row, and then to each column. The orders M and N of the 2-D DFT can be made larger than the size of the image by zero-padding the image, just as in 1-D (see Section 8-8).

Image processing is usually performed using batch processing, often using the 2-D DFT, computed using the 2-D FFT. The FFT is even more important in 2-D than in 1-D, since 2-D images often contain more samples (pixels) than 1-D signals.

10-3.1 2-D Cyclic Convolution

The 1-D cyclic convolution of two finite-duration signals was presented in Section 7-15.5. The *cyclic* aspect is a consequence of converting the nonperiodic signals into periodic signals in preparation for applying the DFT to compute their spectra.

For two images of equal size, with M rows and N columns, the ***2-D cyclic convolution*** consists of a 1-D cyclic convolution applied to each row, and then to each column, or vice versa. The 2-D extension of Eq. (7.184) is

$$\begin{aligned} y_c[m,n] &= x_1[m,n] \circledcirc \circledcirc x_2[m,n] \\ &= \text{DFT}^{-1}(\mathbf{X}_1[k_1,k_2]\,\mathbf{X}_2[k_1,k_2]) \\ &= \frac{1}{MN} \sum_{k_1=0}^{M-1} \sum_{k_2=0}^{N-1} \mathbf{X}_1[k_1,k_2]\,\mathbf{X}_2[k_1,k_2] \\ &\quad \cdot e^{j2\pi(k_1(m/M)+k_2(n/N))}. \end{aligned} \quad (10.21)$$

Example 10-3: Cyclic Convolution of Two 2 x 2 Images

This example is intended to demonstrate the procedure involved in the application of the DFT method to compute the cyclic convolution of two images. Hence, image size is limited to 2×2.

Apply the DFT method to obtain the cyclic convolution of

$$x_1[m,n] = \begin{bmatrix} 1 & 2 \\ 3 & 4 \end{bmatrix}, \qquad x_2[m,n] = \begin{bmatrix} 5 & 6 \\ 7 & 8 \end{bmatrix}.$$

Solution: With $M = N = 2$, application of Eq. (10.20a) yields

$$\mathbf{X}_1[k_1,k_2] = \begin{bmatrix} 10 & -2 \\ -4 & 0 \end{bmatrix}, \qquad \mathbf{X}_2[k_1,k_2] = \begin{bmatrix} 26 & -2 \\ -4 & 0 \end{bmatrix}.$$

Their point-by-point product is

$$\mathbf{X}_1[k_1,k_2]\,\mathbf{X}_2[k_1,k_2] = \begin{bmatrix} 10\times26 & (-2)\times(-2) \\ (-4)\times(-4) & 0\times0 \end{bmatrix} = \begin{bmatrix} 260 & 4 \\ 16 & 0 \end{bmatrix}.$$

From Eq. (10.21), the cyclic convolution of $x_1[m,n]$ and $x_2[m,n]$ can be obtained from the inverse 2-D DFT of $\mathbf{X}_1[k_1,k_2]\,\mathbf{X}_2[k_1,k_2]$, which yields

$$y_c[m,n] = \frac{1}{2^2} \begin{bmatrix} 260+4+16 & 260+16-4 \\ 260+4-16 & 260-4-16 \end{bmatrix} = \begin{bmatrix} 70 & 68 \\ 62 & 60 \end{bmatrix}.$$

This 2-D cyclic convolution can be implemented in MATLAB/Mathscript as

```
FX=fft2([1 2;3 4]);FY=fft2([5 6;7 8]);
        Y=real(ifft2(FX.*FY));
```

Concept Question 10-4: Is there a 2-D version of the FFT? (See S^2)

10-4 Downsampling and Upsampling of Images

The concepts of downsampling (decimation) and upsampling (zero-padding) and interpolation generalize directly from 1-D signals to 2-D images. Downsampling in 2-D involves downsampling in both directions. For example:

$$x[m,n] \rightarrow \boxed{\downarrow (2,3)} \rightarrow x[2m,3n].$$

The downsampling factor in this case is 2 along the vertical direction and 3 along the horizontal direction. To illustrate the process, we apply it to a 5×7 image:

$$\begin{bmatrix} 1 & 2 & 3 & 4 & 5 & 6 & 7 \\ 8 & 9 & 10 & 11 & 12 & 13 & 14 \\ 15 & 16 & 17 & 18 & 19 & 20 & 21 \\ 22 & 23 & 24 & 25 & 26 & 27 & 28 \\ 29 & 30 & 31 & 32 & 33 & 34 & 35 \end{bmatrix} \rightarrow \boxed{\downarrow (2,3)} \rightarrow \begin{bmatrix} 1 & 4 & 7 \\ 15 & 18 & 21 \\ 29 & 32 & 35 \end{bmatrix}.$$

Upsampling in 2-D involves upsampling in both directions. Upsampling a signal $x[m,n]$ by a factor 2 along the vertical and by a factor 3 along the horizontal, for example, is denoted symbolically as

$$x[m,n] \rightarrow \boxed{\uparrow(2,3)} \rightarrow \begin{cases} x\left[\frac{m}{2}, \frac{n}{3}\right] & \text{for } m = \text{integer multiple of 2} \\ & \text{and } n = \text{integer multiple of 3,} \\ 0 & \text{otherwise.} \end{cases}$$

Applying this upsampling operation to a simple 2×2 image yields

$$\begin{bmatrix} 1 & 2 \\ 3 & 4 \end{bmatrix} \rightarrow \boxed{\uparrow(2,3)} \rightarrow \begin{bmatrix} 1 & 0 & 0 & 2 \\ 0 & 0 & 0 & 0 \\ 3 & 0 & 0 & 4 \\ 0 & 0 & 0 & 0 \end{bmatrix}.$$

Downsampling and upsampling are used extensively in image processing operations. A few examples follow.

10-4.1 Thumbnail Images

Thumbnail images are miniature-size versions of full-size images, designed to contain sufficient detail so as to resemble their full-size parents, but with only a fraction of the number of pixels contained in the parent images. The size reduction allows for the simultaneous use of many thumbnail images on computer screens and in printed matter.

Reducing the size of an image is accomplished by downsampling it, but the downsampling step should be preceded with a lowpass-filtering step to avoid introducing aliasing into the downsampled image. Specifically, if $x[m,n]$ is the original full-size image, $x_{\text{th}}[m,n]$ is the thumbnail image to be created from $x[m,n]$, and the size reduction factor is $L \times L$, the two-step process is:

Step 1: Apply lowpass filtering to $x[m,n]$ using a brickwall lowpass filter with cutoff wavenumber $\Omega_0 = \pi/L$ in both directions. The process reduces the maximum wavenumber of $\mathbf{X}(e^{j\Omega_1}, e^{j\Omega_2})$ from π to π/L. The ***lowpass-filtered image*** is denoted $\tilde{x}[m,n]$.

Step 2: Downsample the filtered image $\tilde{x}[m,n]$ by a factor L in both directions. The result is the thumbnail image $x_{\text{th}}[m,n]$, which satisfies the Nyquist criterion, and therefore is immune to aliasing. The filtering step reduces the wavenumber range by a factor of L and the downsampling increases it by the same factor. Hence, the maximum wavenumber of the thumbnail image is again $\Omega = \pi$. The lowpass filter eliminates aliasing.

To illustrate the process with an example, let us consider the clown image shown in Fig. 10-4(a), which is composed of 200×200 pixels. Our goal is to downsample by $L = 4$ in both directions. Figure 10-4(b) displays a downsampled version of the original image $x[m,n]$, without the a priori application of lowpass filtering. For comparison, the image shown in Fig. 10-4(c) had undergone lowpass filtering before downsampling. The latter image is a better thumbnail representation than the former, because the absence of the lowpass filtering step allowed aliasing to occur in the downsampling process.

10-4.2 Upsampling and Interpolating Small Images

Sometimes we may need to create a large image from a small one; i.e., to perform the reverse of the thumbnail image process. Enlarging an $M \times M$ image by an integer factor L to $LM \times LM$ requires upsampling and interpolation. The upsampling involves multiplying the number of rows and columns by L and inserting many zeros, and the interpolation entails subjecting the upsampled image to a lowpass filter with a cutoff wavenumber $\Omega_0 = \pi/L$ and gain L^2. Application of the two-step process to the thumbnail image in Fig. 10-4(c) led to the image in Fig. 10-5, which should be identical with the lowpass-filtered version of the original clown image of Fig. 10-4(a).

Concept Question 10-5: Why is lowpass filtering before downsampling needed to produce a thumbnail image? (See S²)

10-5 Image Denoising

The spectral content of an image carries information about the spatial variability contained in the image. Low wavenumbers in the spectrum are associated with slowly varying spatial tones across the image, and high wavenumbers are associated with fast (sudden) spatial variations, such as edges between high (white) and low (dark) brightness levels. Consequently, the distribution of energy in the wavenumber spectrum depends on the shapes and sizes of the objects contained in the image, as well as on the degree of tonal contrast between the different objects or between the objects and the background.

Noise also has a spectrum and the shape of the noise spectrum is related to the mechanism that generates the noise, but in most circumstances the noise behaves like a Gaussian random process. In communication systems, ***white noise*** refers to a zero-mean random process with a Gaussian amplitude distribution (Section 6-10.1). Its corresponding spectrum has a ("flat") uniform distribution in the frequency domain, akin to ***white light*** containing spectral components of all colors.

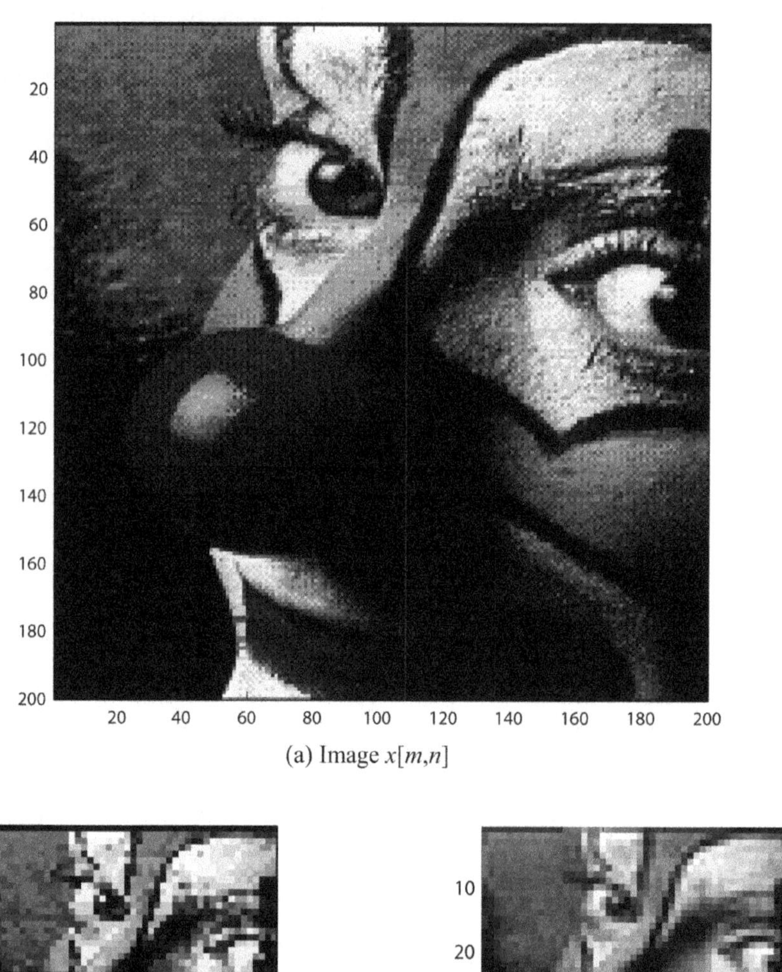

(a) Image $x[m,n]$

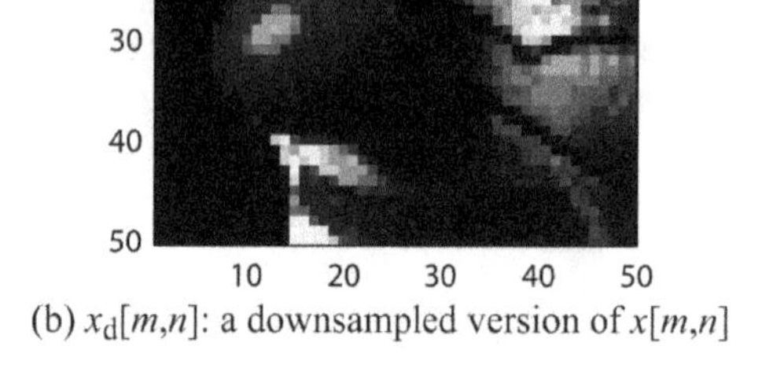

(b) $x_{\mathrm{d}}[m,n]$: a downsampled version of $x[m,n]$

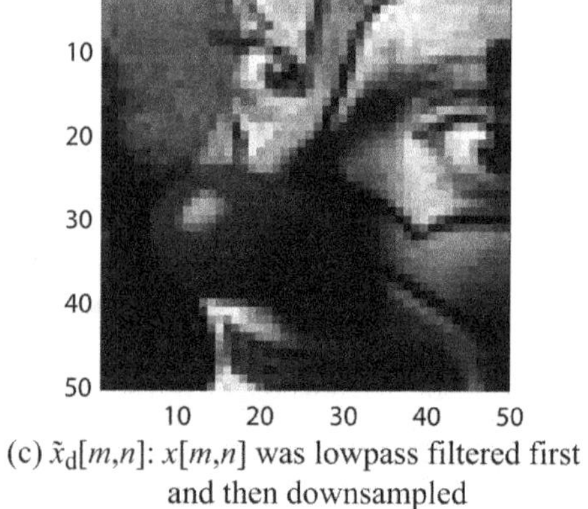

(c) $\tilde{x}_{\mathrm{d}}[m,n]$: $x[m,n]$ was lowpass filtered first and then downsampled

Figure 10-4: (a) Clown image $x[m,n]$, (b) image $x_{\mathrm{d}}[m,n]$, obtained by downsampling $x[m,n]$ by $L = 4$ in both directions, (c) $\tilde{x}_{\mathrm{d}}[m,n]$, obtained by lowpass filtering $x[m,n]$ with a lowpass filter with cutoff wavenumber $\Omega_0 = \pi/4$, followed by downsampling by a factor of 4.

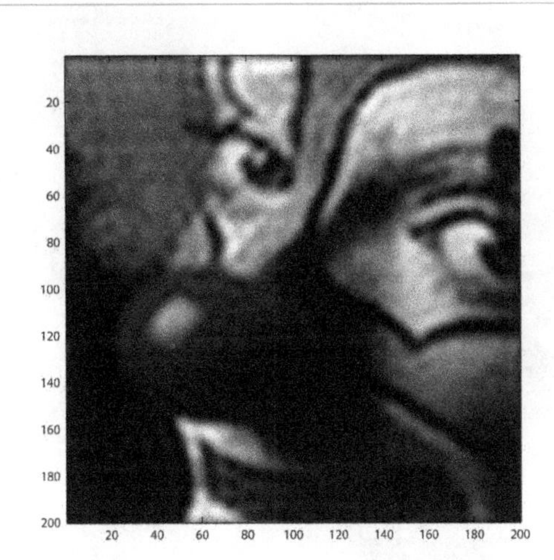

Figure 10-5: Image $\tilde{x}[m,n]$, generated by upsampling and interpolating image $\tilde{x}_{\rm d}[m,n]$ in Fig. 10-4(c).

In an image, the pixel intensity is often non-negative, so the zero reference is some level intermediate between the tones of black and very bright white.

Image denoising refers to the (partial) removal of additive noise from an image. A relatively simple way to denoise an image is to set the high-wavenumber components of the noisy image to zero, which can be implemented using a 2-D brickwall lowpass filter. An important consideration is the cutoff wavenumber $\Omega_{\rm c}$ selected for the filter, relative to the image spectrum. Consider, for example, the 1-D image spectrum shown in Fig. 10-6, corresponding to a single row of a particular image.

Also shown is the white noise spectrum, and lowpass spectra for two candidate filters. Filter #1 preserves most of the image spectrum (thereby preserving edges of objects), but does not remove much of the noise. In contrast, Filter #2 removes much more of the noise, but by filtering out high-wavenumber components, it also blurs the edges of objects in the image.

To illustrate with a real image example, we started with the image shown earlier in Fig. 10-2(a). The image size is 256×256 and the (signal) pixel amplitude $x_{\rm s}[m,n]$ was set at 0 for black pixels and at 255 for white pixels. Next, noise $x_{\rm n}[m,n]$ was added to each pixel of the image $x_{\rm s}[m,n]$. Each value of $x_{\rm n}[m,n]$ had a random value between 0 and 500. Consequently, the amplitudes of the noisy image,

$$y[m,n] = x_{\rm s}[m,n] + x_{\rm n}[m,n],$$

have a range from 0 to a maximum of 755. The noisy image is displayed in Fig. 10-7(a).

Recall from Section 6-10.2 that the ***signal-to-noise ratio*** (***SNR***) is a measure of how significant (or insignificant) the presence of noise is and the degree to which it is likely to distort the information carried by the signal. For a 2-D image, SNR in dB is defined as

$$\text{SNR} = 10\log_{10}\left(\frac{\sum\sum x_{\rm s}^2[m,n]}{\sum\sum x_{\rm n}^2[m,n]}\right), \tag{10.22}$$

where $x_{\rm s}[m,n]$ is the amplitude of the noise-free image pixel (m,n), $x_{\rm n}[m,n]$ is the amplitude of the noise added to pixel (m,n), and the summations are performed over all image

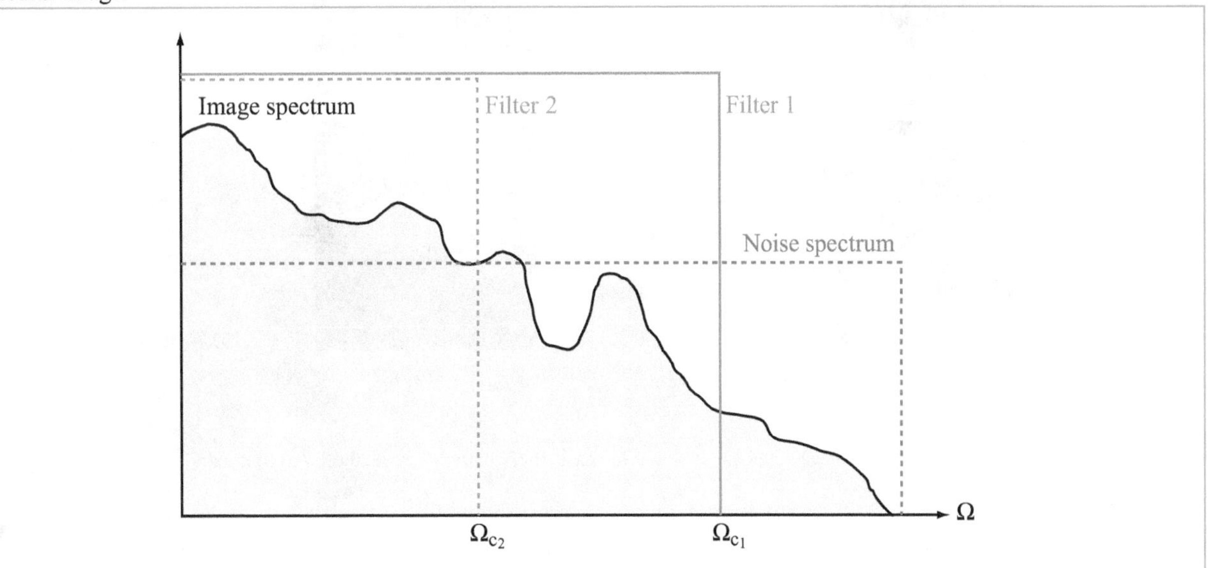

Figure 10-6: 1-D spectrum of a single image row (black), noise spectrum (red), and spectra of two candidate brickwall lowpass filters (blue).

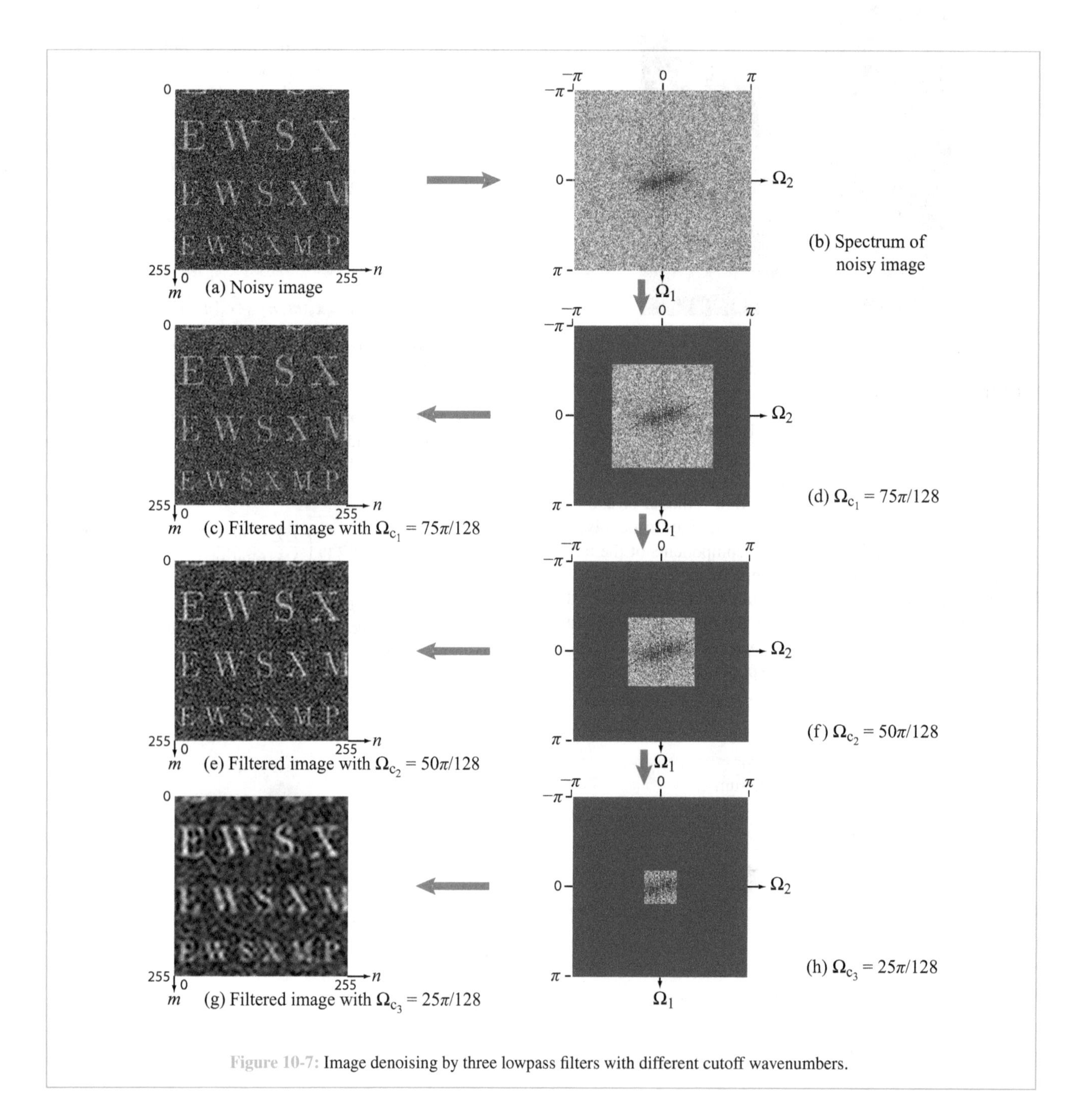

Figure 10-7: Image denoising by three lowpass filters with different cutoff wavenumbers.

pixels. For the synthesized image shown in Fig. 10-7(a), SNR $= -12.8$ dB, which means that the total signal power across the image is only 5% of the total noise added to the image. In other words, the image is very noisy!

A colormap depiction of the spectrum of the noisy image is shown in Fig. 10-7(b), which also is 256×256 pixels. In Ω space, *spectrum pixel* $(128, 128)$ *corresponds to the dc value* $(\Omega_1, \Omega_2) = (0, 0)$ *of the spectrum*. For the purpose of comparison, three new images were generated, all displayed along the left-hand side of Fig. 10-7, and their corresponding spectra are displayed in the right-hand column. To generate the filtered image shown in Fig. 10-7(c), we applied a brickwall lowpass filter with cutoff wavenumber $\Omega_{c_1} = 75\pi/128$ to the spectrum of the noisy image, and then transformed the filtered spectrum back to the spatial domain. Similar processes with narrower filters were applied to the spectra in parts (f) and (h) of Fig. 10-7.

Comparison of the filtered images provides a qualitative sense of what happens to the image as the cutoff wavenumber is moved progressively towards the center of the spectrum.

(a) The spectrum of the original noisy image in Fig. 10-7(a) extends to $\pm\pi$.

(b) The spectrum of the image in Fig. 10-7(c) extends to $\pm 75\pi/128$, or approximately $(75/128)^2 = 34\%$ of the central spectrum of the original noisy image. Some of the noise has been removed, without distorting the shapes of the letters.

(c) Narrowing the spectrum to $(50/128)^2 \approx 15\%$ of the spectrum of the original noisy image removes more noise, and the shapes of the letters are only slightly distorted (Fig. 10-7(e)).

(d) The narrowest filter removes all but $(25/128)^2 \approx 4\%$ of the image spectrum. Consequently, the edges of the letters appear fuzzy (Fig. 10-7(g)).

These images illustrate the trade-off inherent in lowpass filtering; we can reduce noise, but at the expense of distorting the image.

Concept Question 10-6: What is the trade-off inherent in denoising using the 2-D DFT? (See S^2)

Exercise 10-2: Use LabVIEW Module 10.2 to denoise the letters image using a lowpass filter. Set "K" to 0.5, L to 5, and noise level to 100.

Answer:

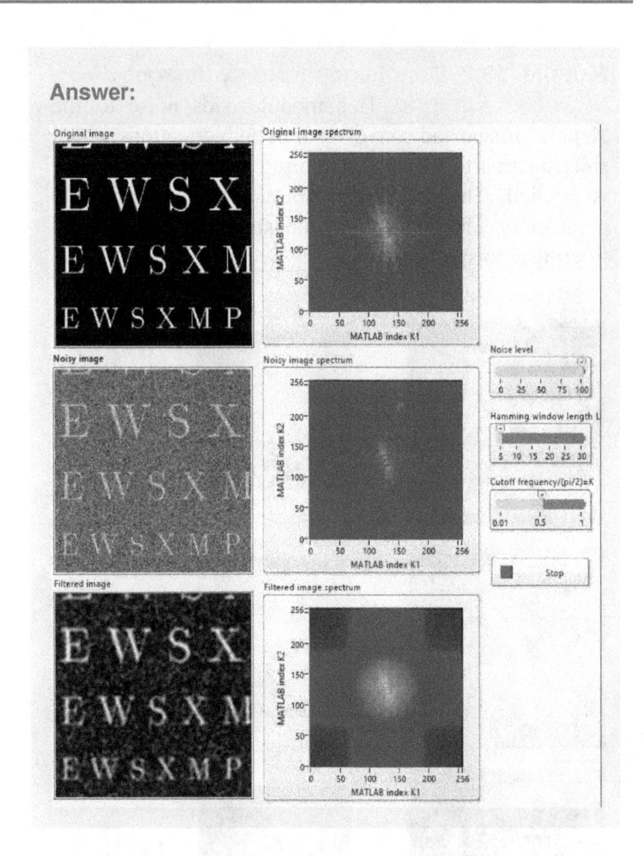

10-6 Edge Detection

An ***edge*** in an image is a sharp boundary between two different regions of an image. Here "sharp" means a width of at most a few pixels, and a "boundary" means that significant differences exist in the pixel values between the two sides of the edge. "Significant" is not clearly defined; its definition depends on the characteristics of the image and the reason why edges are of interest. This is a nebulous definition, but there is no uniform definition of an edge.

The goal of ***edge detection*** is to determine the locations $[m, n]$ of edges in an image $x[m, n]$. Edge detection is used to ***segment*** an image into different regions, or to determine the boundaries of a region of interest. For example, a medical image may consist of different human organs. Interpretation of the image is easier if (say) the region of the image corresponding to the pancreas is identified separately from the rest of the image. Identification of a face is easier if the eyes in an image of the face are identified as a region separate from the rest of the image of the face. Ideally, an edge is a contour that encloses a region of the image whose values differ significantly from the values around it. Edge detection also is important in computer vision.

Module 10.2 Denoising a Noisy Image Using Lowpass Filtering This module adds noise to the "letters" image and convolves it in both directions with a sinc function to which a Hamming window of length L is applied. The cutoff and noise level are selectable parameters. The goal is to demonstrate the trade-offs in filtering a noisy image.

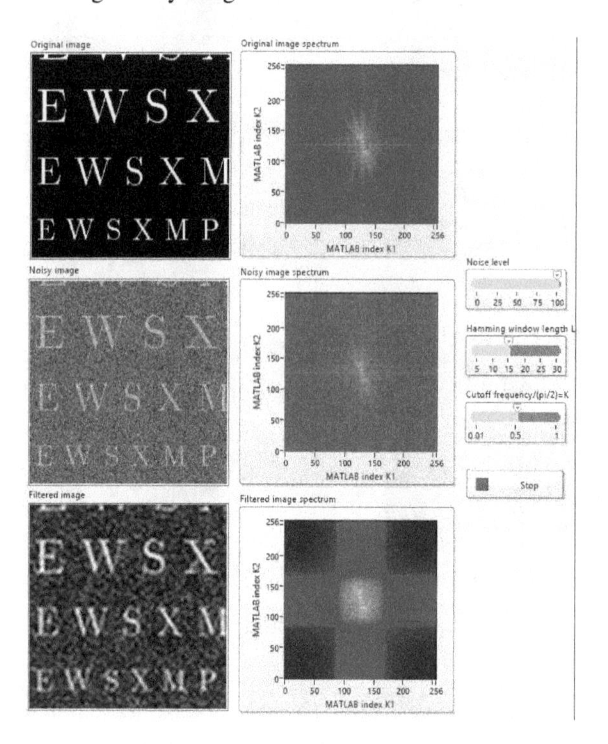

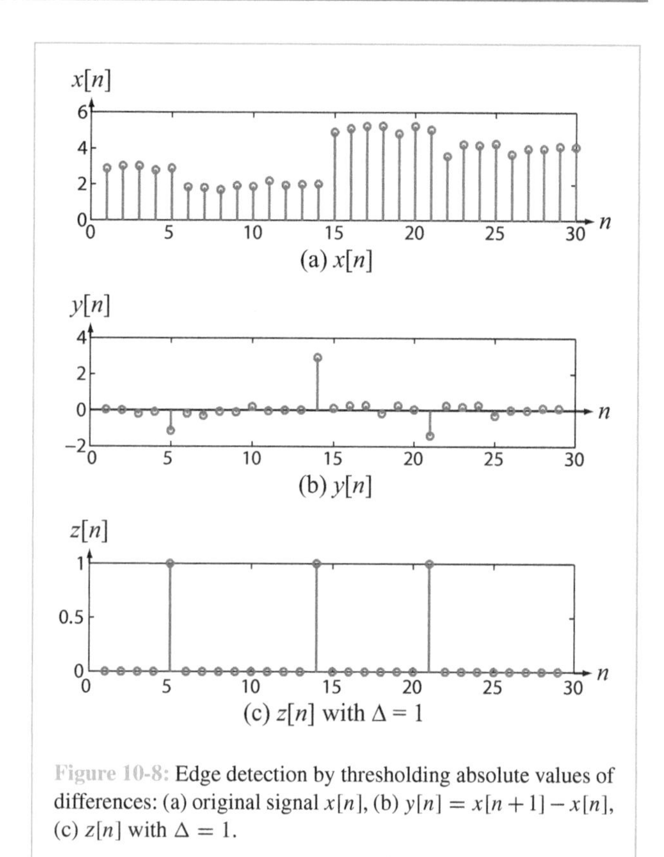

Figure 10-8: Edge detection by thresholding absolute values of differences: (a) original signal $x[n]$, (b) $y[n] = x[n+1] - x[n]$, (c) $z[n]$ with $\Delta = 1$.

10-6.1 1-D Edge Detection

We start by examining a simple 1-D edge-detection method, as it forms the basis for a commonly used 2-D edge-detection algorithm.

An obvious approach to detecting the locations of sharp changes in a 1-D signal $x(t)$ is to compute its derivative $x'(t) = dx/dt$. Rapid changes of $x(t)$ with t generate derivatives with large magnitudes, and slow changes generate derivatives with small magnitudes. The times t_0 at which $|x'(t_0)|$ is large represent *potential edges* of $x(t)$. The threshold for "large" has to be defined in the context of the signal $x(t)$ itself.

For a 1-D discrete-time signal $x[n]$, the discrete-time counterpart to the derivative is the ***difference operator***

$$d[n] = x[n+1] - x[n]. \tag{10.23}$$

The difference $d[n]$ is large when $x[n]$ changes rapidly with n, making it possible to easily pinpoint the time n_0 of an edge. As simple as it is, computing the difference $d[n]$ and thresholding $|d[n]|$ is a very effective method for detecting 1-D edges. If the threshold is set at a value Δ, the edge-detection algorithm can be cast as

$$z[n] = \begin{cases} 1 & \text{for } |d[n]| > \Delta, \\ 0 & \text{for } |d[n]| < \Delta. \end{cases} \tag{10.24}$$

The times n_i at which $z[n_i] = 1$ denote the edges of $x[n]$. Specification of the threshold level Δ depends on the *character* of $x[n]$. In practice, for a particular class of signals, the algorithm is tested for several values of Δ so as to determine the value that provides the best results for the intended application.

For the signal $x[n]$ displayed in Fig. 10-8(a), the difference operator $d[n]$ was computed using Eq. (10.24) and then plotted in Fig. 10-8(b). It is evident that $|d[n]|$ exhibits significant values at $n = 5, 14$, and 21. Setting $\Delta = 1$ would detect all three edges, as shown in part (c) of the figure, but had we chosen Δ to be 2, for example, only the edge at $n = 15$ would have been detected. The choice depends on the intended application.

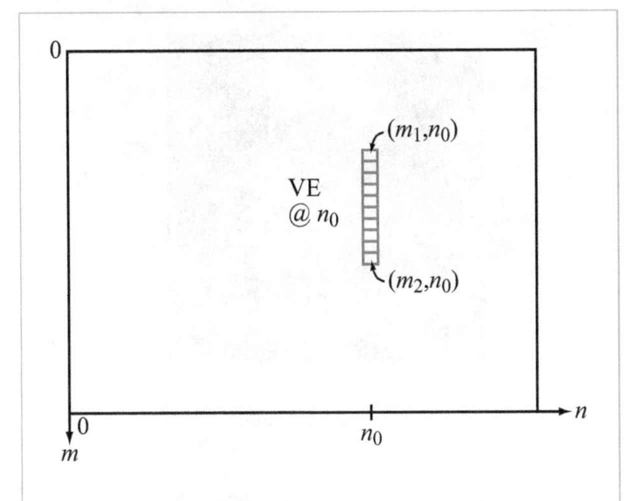

Figure 10-9: Vertical edge VE at $n = n_0$ extends from $m = m_1$ to $m = m_2$.

10-6.2 2-D Edge Detection

The 1-D edge detection method can be extended to edge detection in 2-D images. Let us define a ***vertical edge*** (VE) as a vertical line at $n = n_0$, extending from $m = m_1$ to $m = m_2$, as shown in Fig. 10-9. That is,

$$\text{VE} = \{[m,n]\colon m_1 \le m \le m_2;\ n = n_0\}. \tag{10.25}$$

The total length of VE is $(m_2 - m_1 + 1)$.

One way to detect a vertical edge is to apply the difference operator given by Eq. (10.23) to each row of the image. In 2-D, the difference operator for row m is given by

$$d[m,n] = x[m,\ n+1] - x[m,n]. \tag{10.26}$$

If $d[m,n]$ satisfies a specified threshold for a group of continuous pixels (all at $n = n_0$) extending between $m = m_1$ and m_2, then we call the group a vertical edge. In real images, we may encounter situations where $d[m,n]$ may exhibit a large magnitude, but it is associated with a local variation in tone, not an edge. A vertical edge at $n = n_0$ requires that not only $d[m,n_0]$ at row m be large, but also that $d[m+1,\ n_0]$ at the row above m and $d[m-1,\ n_0]$ at the row below row m be large as well. All three differences should be large and of the same polarity in order for the three pixels to qualify as a vertical edge.

This requirement suggests that a vertical edge detector should not only compute horizontal differences, but also vertical sums of the differences. The magnitude of a vertical sum becomes an indicator of the presence of a true vertical edge. A relatively simple edge operator is illustrated in Fig. 10-10 for both

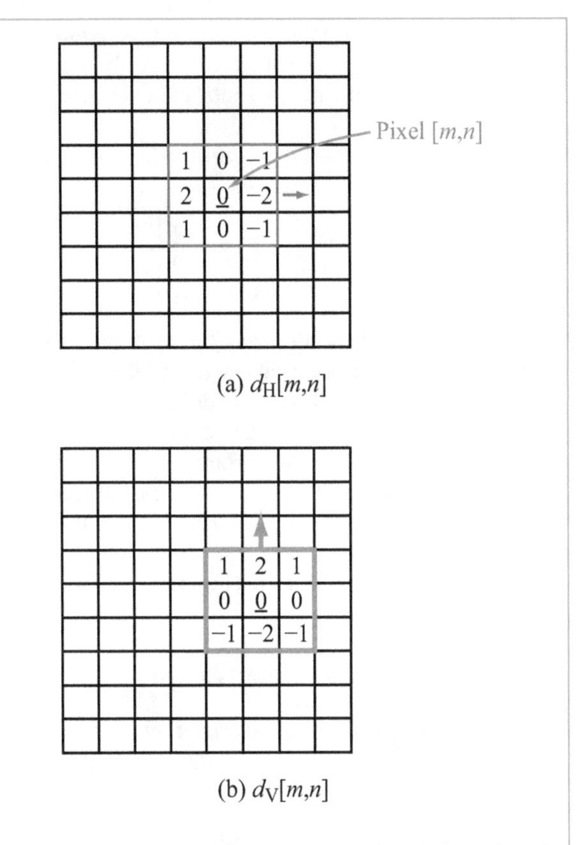

Figure 10-10: Point spread functions $d_H[m,n]$ (in red) and $d_V[m,n]$ (in blue).

a ***horizontal-direction vertical-edge detector*** $d_H[m,n]$ and a ***vertical-direction horizontal-edge detector*** $d_V[m,n]$. Each detector consists of a 3×3 window centered at the pixel of interest. Detector $d_H[m,n]$ computes the difference between the values of pixel $[m,\ n+1]$ and pixel $[m,\ n-1]$, whose positions are to the left and right of pixel $[m,n]$, respectively. Similar differences are performed for the row above and the row below row m. Then, the three differences are added up together, with the middle difference assigned twice the weight of the two other differences. The net result is

$$\begin{aligned} d_H[m,n] = {} & x[m+1,\ n+1] - x[m+1,\ n-1] \\ & + 2x[m,\ n+1] - 2x[m,\ n-1] \\ & + x[m-1,\ n+1] - x[m-1,\ n-1]. \end{aligned} \tag{10.27}$$

The coefficients of the six terms of Eq. (10.27) are the nonzero weights shown in Fig. 10-10(a).

Computing $d_{\mathrm{H}}[m,n]$ for every pixel is equivalent to convolving image $x[m,n]$ with the window's point spread function $h_{\mathrm{H}}[m,n]$ along the horizontal direction. That is,

$$d_{\mathrm{H}}[m,n] = x[m,n] ** h_{\mathrm{H}}[m,n], \tag{10.28}$$

with

$$h_{\mathrm{H}}[m,n] = \begin{bmatrix} 1 & 0 & -1 \\ 2 & \underline{0} & -2 \\ 1 & 0 & -1 \end{bmatrix}. \tag{10.29}$$

To compute $d_{\mathrm{H}}[m,n]$ for all pixels $[m,n]$ in the image, it is necessary to add an extra row above of and identical with the top row, and a similar add-on is needed at the bottom end of the image. The decision as to whether or not a given pixel is part of a vertical edge is made by comparing the magnitude of $d_{\mathrm{H}}[m,n]$ with a predefined threshold Δ whose value is selected heuristically (based on practical experience for the class of images under consideration).

Horizontal edges can be detected by a vertical-direction edge detector $d_{\mathrm{V}}[m,n]$ given by

$$d_{\mathrm{V}}[m,n] = x[m,n] ** h_{\mathrm{V}}[m,n], \tag{10.30}$$

where $h_{\mathrm{V}}[m,n]$ is the ***point spread function*** (***PSF***) for a pixel (m,n). By exchanging the roles of the rows and column in Eq. (10.29), we have

$$h_{\mathrm{V}}[m,n] = \begin{bmatrix} 1 & 2 & 1 \\ 0 & \underline{0} & 0 \\ -1 & -2 & -1 \end{bmatrix}. \tag{10.31}$$

Of course, most edges are neither purely horizontal nor purely vertical, so an edge at an angle different from 0° or 90° (with 0° denoting the horizontal dimension of the image) should have edge components along both the horizontal and vertical directions. Hence, the following ***edge-detection gradient*** is often used:

$$g[m,n] = \sqrt{d_{\mathrm{H}}^2[m,n] + d_{\mathrm{V}}^2[m,n]}\,. \tag{10.32}$$

For each pixel $[m,n]$, we define the ***edge indicator*** $z[m,n]$ as

$$z[m,n] = \begin{cases} 1 & \text{if } g[m,n] > \Delta, \\ 0 & \text{if } g[m,n] < \Delta, \end{cases} \tag{10.33}$$

where Δ is a prescribed ***gradient threshold***. In the image, pixels for which $z[m,n] = 1$ are shown in white, and those with $z[m,n] = 0$ are shown in black. Usually, the value of Δ is selected empirically by examining a histogram of $g[m,n]$ or through repeated trials.

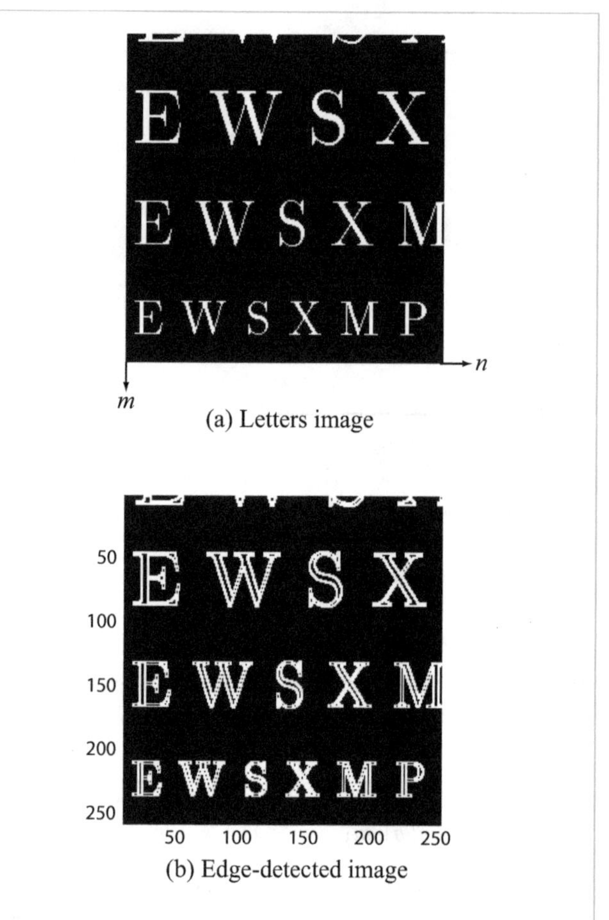

(a) Letters image

(b) Edge-detected image

Figure 10-11: Application of the Sobel edge detector to the image in (a) with $\Delta = 200$ led to the image in (b).

Sobel edge detector examples

The gradient algorithm given by Eq. (10.33) is known as the ***Sobel edge detector***, named after Irwin Sobel, who developed it in 1968, when computer-based image processing was in its infancy and only simple algorithms could be used. Application of the Sobel edge detector to the letters image in part (a) of Fig. 10-11 leads to the image in part (b). Through repeated applications using different values of Δ, it was determined that $\Delta = 200$ provided an image with clear edges, including diagonal and curved edges. The value specified for Δ depends in part on the values assigned to black and white tones in the image.

The Sobel edge detector does not always capture all of the major edges contained in an image. When applied to the clown image of Fig. 10-12(a), the edge detector identified

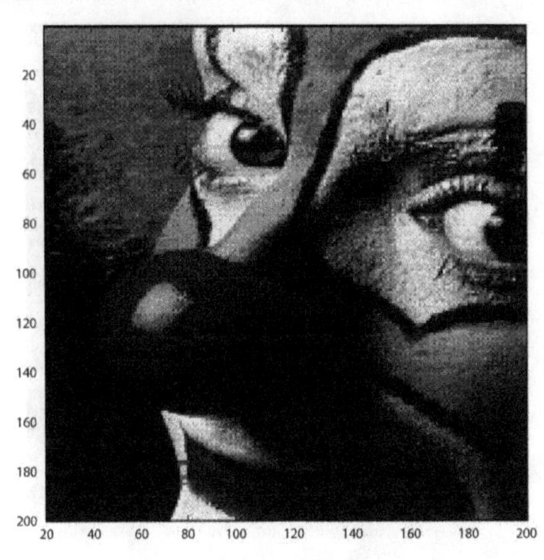

(a) Clown image

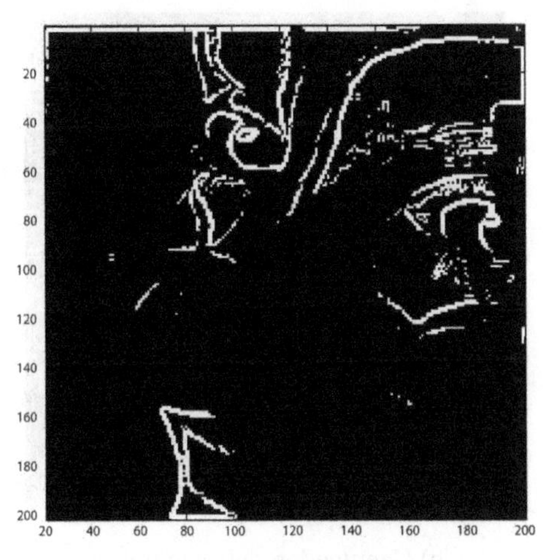

(b) Sobel edge-detected image

Figure 10-12: Application of the Sobel edge detector to the image in (a) captures some of the edges in the image, but also misses others.

some parts of continuous edges, but failed to identify others, which suggests the need for a detector that can track edges and complete edge contours as needed. Such a capability is provided by the Canny edge detector, the subject of the next subsection.

10-6.3 Canny Edge Detector

The ***Canny edge detector*** is a commonly used algorithm that extends the capabilities of the Sobel detector by applying preprocessing and postprocessing steps. The preprocessing step involves the use of a 2-D Gaussian PSF to reduce image noise and to filter out isolated image features that are not edges. After computing the Sobel operator given by Eq. (10.32), the Canny algorithm performs an ***edge thinning*** step, separating detected edges into different candidate categories, and then applies certain criteria to decide whether or not the candidate edges should be connected together. The five-step process of the Canny detection algorithm are:

Step 1: Image $x[m,n]$ is *blurred* (filtered) by convolving it with a truncated Gaussian point spread function. An example of a practical function that can perform the desired operation is the 5×5 PSF given by

$$h_G[m,n] = \frac{1}{159}\begin{bmatrix} 2 & 4 & 5 & 4 & 2 \\ 4 & 9 & 12 & 9 & 4 \\ 5 & 12 & 15 & 12 & 5 \\ 4 & 9 & 12 & 9 & 4 \\ 2 & 4 & 5 & 4 & 2 \end{bmatrix}. \tag{10.34}$$

The standard deviation of the truncated Gaussian function is 1.4.

Application of $h_G[m,n]$ to image $x[m,n]$ generates a filtered image $x_1[m,n]$ given by

$$x_1[m,n] = h_G[m,n] ** x[m,n]. \tag{10.35}$$

Step 2: For image $x_1[m,n]$, compute the horizontal and vertical edge detectors given by Eqs. (10.28) and (10.30).

Step 3: Compute the gradient magnitude and orientation:

$$g[m,n] = \sqrt{d_H^2[m,n] + d_V^2[m,n]} \tag{10.36a}$$

and

$$\theta[m,n] = \tan^{-1}\left(\frac{d_V[m,n]}{d_H[m,n]}\right). \tag{10.36b}$$

For a vertical edge, $d_V[m,n] = 0$, and therefore $\theta[m,n] = 0$. Similarly, for a horizontal edge, $d_H[m,n] = 0$ and $\theta = 90°$.

Step 4: At each pixel $[m,n]$, round $\theta[m,n]$ to the nearest of $\{0°, 45°, 90°, 135°\}$. Next, determine whether to keep the value of $g[m,n]$ of pixel $[m,n]$ as is or to replace it with zero. The decision logic is as follows:

(a) For a pixel $[m,n]$ with $\theta[m,n] = 0°$, compare the value of $g[m,n]$ to the values of $g[m,\ n+1]$ and $g[m,\ n-1]$, corresponding to the pixels at the immediate right and left of pixel $[m,n]$. If $g[m,n]$ is the largest of the three gradients, keep its value as is; otherwise, set it to zero.

(b) For a pixel $[m, n]$ with $\theta = 45°$, compare the value of $g[m, n]$ to the values of $g[m+1,\ n-1]$ and $g[m-1,\ n+1]$, corresponding to the pixel neighbors along the 45° diagonal. If $g[m, n]$ is the largest of the three gradients, keep its value as is; otherwise, set it to zero.

(c) For a pixel $[m, n]$ with $\theta = 90°$, compare the value of $g[m, n]$ to the values of $g[m-1,\ n]$ and $g[m+1,\ n]$, corresponding to pixels immediately above and below pixel $[m, n]$. If $g[m, n]$ is the largest of the three gradients, keep its value as is; otherwise, set it to zero.

(d) For a pixel $[m, n]$ with $\theta = 135°$, compare the value of $g[m, n]$ to the values of $g[m-1,\ n-1]$ and $g[m+1,\ n+1]$. If $g[m, n]$ is the largest of the three gradients, keep its value as is; otherwise, set it to zero.

The foregoing operation is called ***edge thinning***, as it avoids making an edge wider than necessary in order to indicate its presence.

Step 5: Replace the edge indicator algorithm given by Eq. (10.33) with a double-threshold algorithm given by

$$z[m,n] = \begin{cases} 2 & \text{if } g[m,n] > \Delta_2, \\ 1 & \text{if } \Delta_1 < g[m,n] < \Delta_2, \\ 0 & \text{if } g[m,n] < \Delta_1. \end{cases} \tag{10.37}$$

The edge indicator $z[m, n]$ may assume one of three values, indicating the *presence* of an edge ($z[m, n] = 2$), the *possible presence* of an edge ($z[m, n] = 1$), and the *absence* of an edge ($z[m, n] = 0$). The middle category requires resolution into one of the other two categories. This is accomplished by converting pixel $[m, n]$ with $z[m, n] = 1$ into an edge if any one of its nearest 8 neighbors is a confirmed edge. That is, pixel $[m, n]$ is an edge location only if it adjoins another edge location.

The values assigned to thresholds Δ_1 and Δ_2 are selected through multiple trials. For example, the clown edge-image shown in Fig. 10-13 was obtained by applying the Canny algorithm to the clown image with $\Delta_1 = 0.05$ and $\Delta_2 = 0.125$. This particular combination provides an edge-image that successfully captures the contours that segment the clown image.

Edge detection can be implemented in MATLAB's Image Processing Toolbox using the commands

```
E=edge(X,'sobel',T1) for Sobel and
E=edge(X,'canny',T1,T2) for Canny.
```

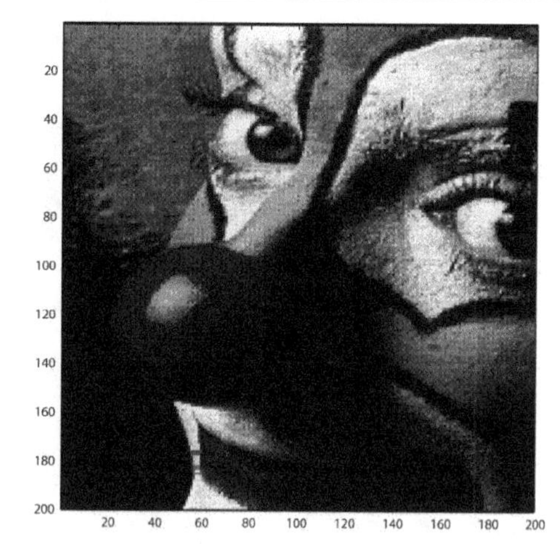

(a) Clown image

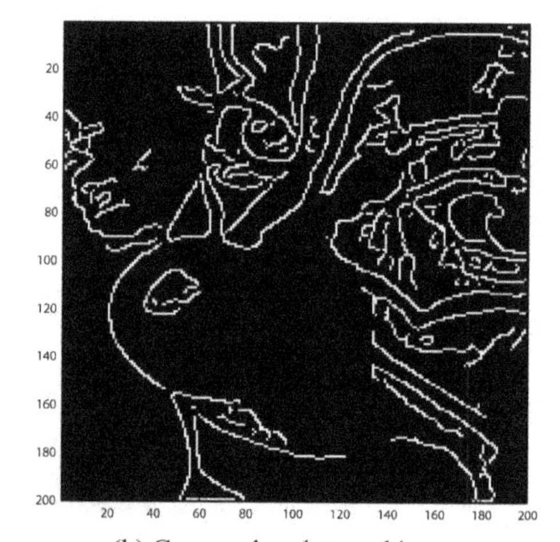

(b) Canny edge-detected image

Figure 10-13: The Canny edge detector provides better edge-detection performance than the Sobel detector in Fig. 10-12.

The image is stored in array X, the edge image is stored in array E and T1 and T2 are the thresholds. MATLAB assigns default values to the thresholds, computed from the image, if they are not specified.

Concept Question 10-7: What is the general concept behind edge detection? (See S²)

10-7 Image Deconvolution

When our eyes view a scene, they form an approximate image of the scene, because the optical imaging process performed by the eyes *distorts* the true scene, with the degree of distortion being dependent on the imaging properties of the eyes' lenses. The same is true when imaging with a camera, a medical imaging system, and an optical or radio telescope. Distortion can also be caused by the intervening medium between the imaged scene and the imaging sensor. Examples include the atmosphere when a telescope is used to image a distant object, or body tissue when a medical ultrasound sensor is used to image body organs. In all cases, the imaging process involves the convolution of a *true image scene* $x[m,n]$ with a *point spread function* $h[m,n]$ representing the imaging sensor (and possibly the intervening medium). The *recorded* (sensed) image $y[m,n]$ is, therefore, given by

$$y[m,n] = h[m,n] * *x[m,n]. \qquad (10.38)$$

The goal of ***image deconvolution*** is to *deconvolve* the recorded image so as to extract the true image $x[m,n]$, or a close approximation of it. Doing so requires knowledge of the PSF $h[m,n]$. In terms of size:

- $x[m,n]$ is the *unknown true image*, with size $(M \times M)$.
- $h[m,n]$ is the *known PSF*, with size $(L \times L)$.
- $y[m,n]$ is the *known recorded image*, with size $(L+M-1) \times (L+M-1)$.

As noted earlier, the PSF of the imaging sensor can be established by imaging a small object representing an impulse.

Since convolution in the discrete-time domain translates into multiplication in the frequency domain, the DSFT-equivalent of Eq. (10.38) is given by

$$\mathbf{Y}(e^{j\Omega_1}, e^{j\Omega_2}) = \mathbf{H}(e^{j\Omega_1}, e^{j\Omega_2})\, \mathbf{X}(e^{j\Omega_1}, e^{j\Omega_2}). \qquad (10.39)$$

Before we perform the frequency transformation of $y[m,n]$, we should round up $(L+M-1)$ to N, where N is the smallest power of 2 greater than $(L+M-1)$. The rounding-up step allows us to use the fast radix-2 2-D FFT to compute 2-D DFTs of order $(N \times N)$. Alternatively, the Cooley-Tukey FFT (Section 7-17) can be used, in which case N should be an integer with a large number of small factors.

Sampling the DSFT at $\Omega_1 = 2\pi k_1/N$ and $\Omega_2 = 2\pi k_2/N$ for $k_1 = 0, 1, \ldots, N-1$ and $k_2 = 0, 1, \ldots, N-1$ provides the DFT complex coefficients $\mathbf{Y}[k_1, k_2]$.

A similar procedure can be applied to $h[m,n]$ to obtain coefficients $\mathbf{H}[k_1, k_2]$, after zero-padding $h[m,n]$ so that it also is of size $(N \times N)$. The DFT equivalent of Eq. (10.39) is then given by

$$\mathbf{Y}[k_1, k_2] = \mathbf{H}[k_1, k_2]\, \mathbf{X}[k_1, k_2]. \qquad (10.40)$$

The objective of deconvolution is to compute the DFT coefficients $\mathbf{X}[k_1, k_2]$, given the DFT coefficients $\mathbf{Y}[k_1, k_2]$ and $\mathbf{H}[k_1, k_2]$.

10-7.1 Nonzero $\mathbf{H}[k_1, k_2]$ Coefficients

In the ideal case where none of the DFT coefficients $\mathbf{H}[k_1, k_2]$ are zero, the DFT coefficients $\mathbf{X}[k_1, k_2]$ of the unknown image can be obtained through simple division,

$$\mathbf{X}[k_1, k_2] = \frac{\mathbf{Y}[k_1, k_2]}{\mathbf{H}[k_1, k_2]}. \qquad (10.41)$$

Exercising the process for all possible values of k_1 and k_2 leads to an $(N \times N)$ 2-D DFT for $\mathbf{X}[k_1, k_2]$, whereupon application of an inverse 2-D DFT process yields a zero-padded version of $x[m,n]$. Upon discarding the zeros, we obtain the true image $x[m,n]$. The deconvolution procedure is straightforward, but it hinges on a critical assumption, namely that none of the DFT coefficients of the imaging system's transfer function is zero. Otherwise, division by zero in Eq. (10.41) would lead to undeterminable values for $\mathbf{X}[k_1, k_2]$.

10-7.2 Image Deconvolution Example

To demonstrate the performance of the deconvolution process, we used a noise-free version of the letters image, shown in Fig. 10-14(a), which we denote $x[m,n]$. Then, we convolved it with a truncated 2-D Gaussian PSF (a common test PSF) given by

$$h[m,n] = e^{-(m^2+n^2)/20}, \qquad -10 \leq m, n \leq 10. \qquad (10.42)$$

The PSF represents the imaging system. The convolution process generated the blurred image shown in Fig. 10-14(b), which we label $y[m,n]$. The image sizes are:

- Original letters image $x[m,n]$: 256×256
- Gaussian PSF $h[m,n]$: 21×21
- Blurred image $y[m,n]$: $256 + 21 - 1 = 276 \times 276$

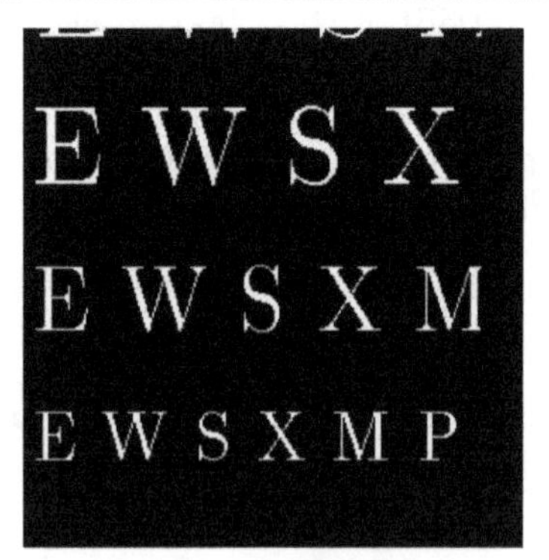

(a) Letters image $x[m,n]$

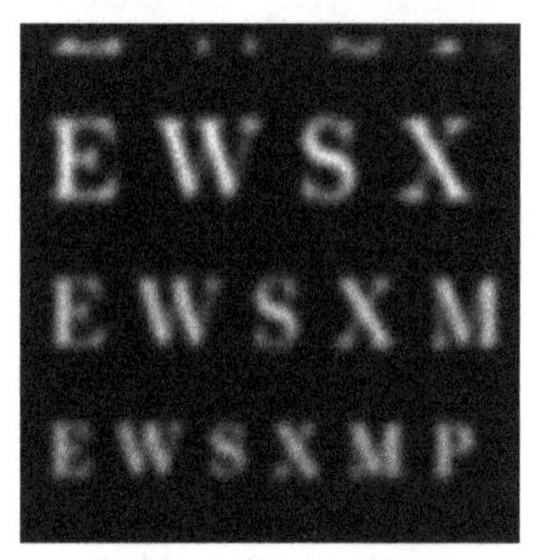

(b) Blurred image $y[m,n]$

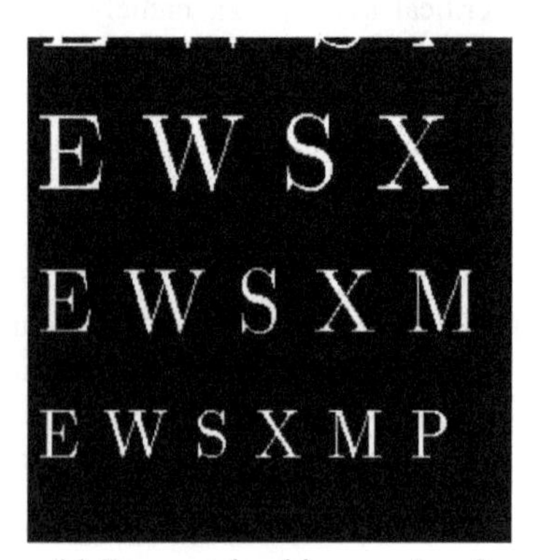

(c) Deconvolved image $x[m,n]$

Figure 10-14: The blurred image in (b) was generated by convolving $x[m,n]$ with a Gaussian PSF, and the image in (c) was recovered through deconvolution of $y[m,n]$.

After zero-padding all three images to 280 × 280, the Cooley-Tukey 2-D FFT was applied to all three images using the recipe in Section 7-17. Then, Eq. (10.41) was applied to find coefficients $\mathbf{X}[k_1,k_2]$, which ultimately led to the deconvolved image $x[m,n]$ displayed in Fig. 10-14(c). The deconvolved image matches the original image shown in Fig. 10-14(a). The process was successful because none of the $\mathbf{H}[k_1,k_2]$ coefficients had zero (or near-zero) values and $\mathbf{Y}[k_1,k_2]$ was noise-free.

10-7.3 Tikhonov Image Regularization

All electronic imaging systems generate some noise of their own. The same is true for the eye-brain system. Hence, Eq. (10.40) should be modified to

$$\mathbf{Y}[k_1,k_2] = \mathbf{H}[k_1,k_2]\,\mathbf{X}[k_1,k_2] + \mathbf{V}[k_1,k_2], \tag{10.43}$$

where $\mathbf{V}[k_1,k_2]$ represents the spectrum of the *additive noise* contributed by the imaging system. The known quantities are the measured image $\mathbf{Y}[k_1,k_2]$ and the PSF of the system, $\mathbf{H}[k_1,k_2]$, and the sought-out quantity is the true image $\mathbf{X}[k_1,k_2]$. Dividing both sides of Eq. (10.43) by $\mathbf{H}[k_1,k_2]$ and solving for $\mathbf{X}[k_1,k_2]$ gives

$$\mathbf{X}[k_1,k_2] = \frac{\mathbf{Y}[k_1,k_2]}{\mathbf{H}[k_1,k_2]} - \frac{\mathbf{V}[k_1,k_2]}{\mathbf{H}[k_1,k_2]}\,. \tag{10.44}$$

In many practical applications, $\mathbf{H}[k_1,k_2]$ may assume very small values for large values of (k_1,k_2). Consequently, the second term in Eq. (10.44) may end up *amplifying* the noise component and may *drown out* the first term. To avoid the noise-amplification problem, the deconvolution can be converted into a ***regularized*** estimation process. Regularization involves the use of a ***cost function*** that trades off ***estimation accuracy*** (of $x[m,n]$) against ***measurement precision***. The process generates an estimate $\hat{x}[m,n]$ of the true image $x[m,n]$. Accuracy refers to a bias associated with all pixel values of the reconstructed image $\hat{x}[m,n]$ relative to $x[m,n]$. Precision refers to the ± uncertainty associated with each individual pixel value due to noise.

A commonly used regularization model is the ***Tikhonov regularization***, which seeks to minimize the cost function

$$e = \sum_{m=0}^{N-1}\sum_{n=0}^{N-1}[(y[m,n] - h[m,n]**\hat{x}[m,n])^2 + (\lambda\,\hat{x}[m,n])^2], \tag{10.45}$$

where zero-padding to size $N \times N$ has been implemented so that all quantities in Eq. (10.45), except for λ, are of the same order. The parameter λ is non-negative and it is called a ***regularization*** *parameter*. The first term on the right-hand side of Eq. (10.45) represents the bias error associated with $\hat{x}[m,n]$ and the second term represents the variance. Setting $\lambda = 0$ reduces Eq. (10.45) to the unregularized state we dealt with earlier in Section 10-7.1, wherein the measurement process was assumed to be noise-free. For realistic imaging processes, λ should be greater than zero, but there is no simple method for specifying its value, so usually its value is selected heuristically (trial and error).

The estimation process may be performed iteratively in the discrete-time domain by selecting an initial estimate $\hat{x}[m,n]$ and then recursively iterating the estimate until the error e approaches a minimum level. Alternatively, the process can be performed in the frequency domain using a Wiener filter, as discussed next.

10-7.4 Wiener Filter

The frequency domain DFT equivalent of the Tikhonov cost function given by Eq. (10.45) is

$$E = \frac{1}{MN} \sum_{k_1=0}^{N-1} \sum_{k_2=0}^{N-1} [|\mathbf{Y}[k_1,k_2] - \mathbf{H}[k_1,k_2]\,\hat{\mathbf{X}}[k_1,k_2]|^2 + \lambda^2 |\mathbf{X}[k_1,k_2]|^2]. \tag{10.46}$$

The error can be minimized separately for each (k_1, k_2) combination. The process can be shown (Problem 10.5) to lead to the solution

$$\hat{\mathbf{X}}[k_1,k_2] = \mathbf{Y}[k_1,k_2]\, \frac{\mathbf{H}^*[k_1,k_2]}{|\mathbf{H}[k_1,k_2]|^2 + \lambda^2}, \tag{10.47}$$

where $\mathbf{H}^*[k_1,k_2]$ is the complex conjugate of $\mathbf{H}[k_1,k_2]$. The quantity multiplying $\mathbf{Y}[k_1,k_2]$ is called a *Wiener filter* $\mathbf{G}[k_1,k_2]$. That is,

$$\hat{\mathbf{X}}[k_1,k_2] = \mathbf{Y}[k_1,k_2]\,\mathbf{G}[k_1,k_2], \tag{10.48a}$$

with

$$\mathbf{G}[k_1,k_2] = \frac{\mathbf{H}^*[k_1,k_2]}{|\mathbf{H}[k_1,k_2]|^2 + \lambda^2}. \tag{10.48b}$$

The operation of the Wiener filter is summarized as follows:

(a) For values of (k_1,k_2) such that $|\mathbf{H}[k_1,k_2]| \gg \lambda$, the Wiener filter implementation leads to

$$\hat{\mathbf{X}}[k_1,k_2] \approx \mathbf{Y}[k_1,k_2]\, \frac{\mathbf{H}^*[k_1,k_2]}{|\mathbf{H}[k_1,k_2]|^2} = \frac{\mathbf{Y}[k_1,k_2]}{\mathbf{H}[k_1,k_2]}, \tag{10.49a}$$

which is the same as Eq. (10.41).

(b) For values of (k_1,k_2) such that $|\mathbf{H}[k_1,k_2]| \ll \lambda$, the Wiener filter implementation leads to

$$\hat{\mathbf{X}}[k_1,k_2] \approx \mathbf{Y}[k_1,k_2]\, \frac{\mathbf{H}^*[k_1,k_2]}{\lambda^2}. \tag{10.49b}$$

In this case, the Wiener filter may underestimate the value of $\hat{\mathbf{X}}[k_1,k_2]$, but it avoids the noise amplification problem that would have occurred with the use of the unregularized deconvolution given by Eq. (10.41).

Wiener filter deconvolution example

To demonstrate the capabilities of the Wiener filter, we compare image deconvolution performed with and without regularization. The demonstration process involves images at various stages, namely:

- $x[m,n]$: true letters image (Fig. 10-15(a)).
- $y[m,n] = h[m,n] ** x[m,n] + v[m,n]$: the imaging process not only distorts the image (through the PSF), but also adds random noise $v[m,n]$. The result, displayed in Fig. 10-15(b), is an image with signal-to-noise ratio of 10.8 dB, which means that the random noise energy is only about 8% of that of the signal.
- $\hat{x}_1[m,n]$: estimate of $x[m,n]$ obtained without regularization (i.e., using Eq. (10.41)). Image $\hat{x}_1[m,n]$, displayed in Fig. 10-15(c), does not show any of the letters present in the original image, despite the fact that the noise level is small relative to the signal.
- $\hat{x}_2[m,n]$: estimate of $x[m,n]$ obtained using the Wiener filter of Eq. (10.47) with $\lambda^2 = 5$. The deconvolved image (Fig. 10-15(d)) displays all of the letters contained in the original image, but some high wavenumber noise also is present (see S² for more details).

10-7.5 Median Filtering

Median filtering is used to remove *salt-and-pepper* noise, often due to bit errors or *shot noise* associated with electronic devices. The concept of median filtering is very straightforward:

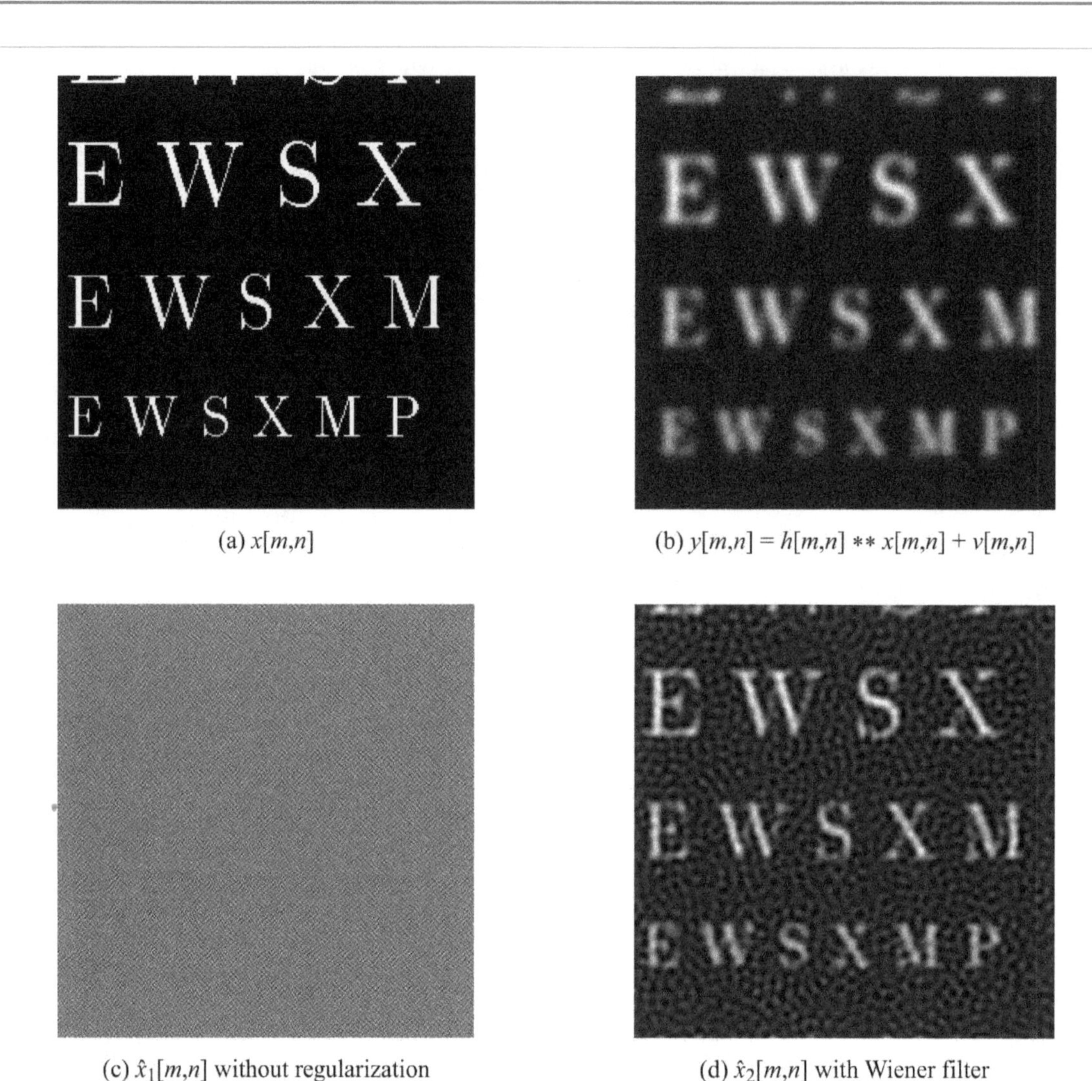

Figure 10-15: (a) Original noise-free undistorted letters image $x[m, n]$, (b) blurred image due to imaging system PSF and addition of random noise $v[m, n]$, (c) deconvolution using Eq. (10.41), and (d) deconvolution using Eq. (10.47) with $\lambda^2 = 5$.

► A median filter of order L replaces each pixel with the median value of the L^2 pixels in the $L \times L$ block centered on that pixel. ◄

For example, a median filter of order $L = 3$ replaces each pixel $[m, n]$ with the median value of the $3 \times 3 = 9$ pixels centered at $[m, n]$. Figure 10-16(a) shows an image corrupted with salt-and-pepper noise, and part (b) of the same figure shows the image after the application of a median filter of order $L = 5$.

Concept Question 10-8: Why is Tikhonov regularization needed? (See S^2)

Exercise 10-3: Use LabVIEW Module 10.3 to deconvolve the letters image from a noisy blurred version of it. Set the noise level to 1000 and L to 1.

Answer: (See Module 10.3.)

(a) Noisy image (b) After median filtering

Figure 10-16: Median filtering example: (a) letters image corrupted by salt-and-pepper noise, and (b) image after application of median filtering using a 5×5 window.

Module 10.3 Deconvolution from a Noisy Blurred Image Using Wiener Filter This module blurs the "letters" image with the 2-D Gaussian PSF used in Section 10-7.2, adds noise to the blurred image, and then deconvolves the image using a Wiener filter. The noise level and Tikhonov regularization parameter λ (L) are selectable parameters.

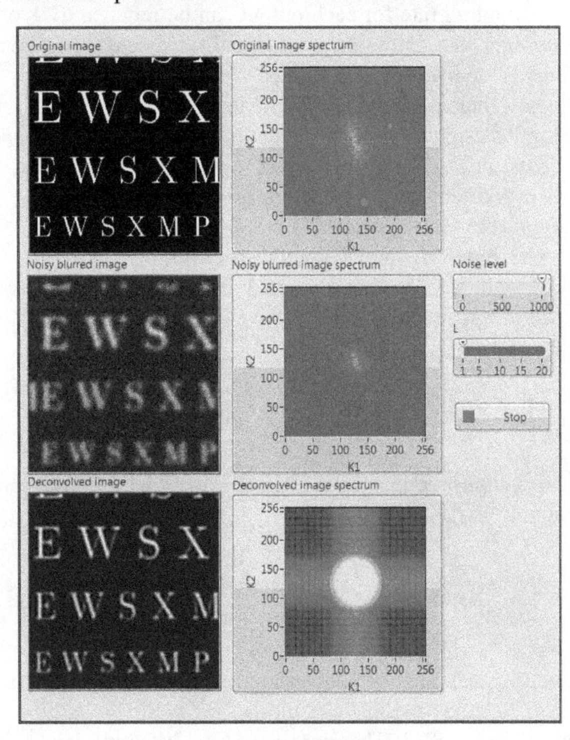

10-8 Overview of the Discrete-Time Wavelet Transform

The ***wavelet transform*** is an important signal processing tool for representing signals or images that consist mostly of slowly varying regions, with a few fast-varying regions. Like the discrete Fourier transform (DFT), it represents signals or images as a linear combination of ***basis functions***. Because basis functions play a vital role in this section, we begin with a review of what they are and how they relate to signal processing.

10-8.1 Basis Functions

The general form of the decomposition of a signal $x[n]$ into basis functions $\{\phi_k[n]\}$ with coefficients $\{x_k\}$ is

$$x[n] = \sum_{k=1}^{\infty} x_k \ \phi_k[n]. \tag{10.50}$$

The coefficients $\{x_k\}$ are the transform of $x[n]$ using the basis functions $\{\phi_k[n]\}$. For the DFT, for example, the basis functions are the complex exponential functions $\phi_k[n] = e^{j2\pi(k/N)n}$.

Basis functions $\{\phi_k[n]\}$ should be ***orthogonal***, which means that, for some constant C,

$$\sum_{n=-\infty}^{\infty} \phi_i[n] \ \phi_j^*[n] = C\delta[i-j]. \tag{10.51}$$

If the basis functions $\{\phi_k[n]\}$ are orthogonal, the coefficients x_k can be computed from $x[n]$ using

$$x_k = \frac{1}{C} \sum_{n=-\infty}^{\infty} x[n] \ \phi_k^*[n]. \tag{10.52}$$

If $C = 1$, the basis functions are ***orthonormal***, in which case Parseval's theorem holds:

$$\sum_{n=-\infty}^{\infty} x^2[n] = \sum_{k=-\infty}^{\infty} |x_k|^2. \tag{10.53}$$

Similar relationships apply to the exponential form of the continuous-time Fourier series, the DTFS, and the DFT. Our current interest is in the wavelet transform.

A note on nomenclature: we use the term ***discrete time wavelets*** to represent wavelet decompositions of discrete-time signals and images. In the literature, the "discrete wavelet transform" represents continuous-time signals and images using discrete-indexed continuous-time basis wavelet functions. The "continuous wavelet transform" represents continuous-time signals using continuous-indexed continuous-time basis wavelet functions. In this chapter we consider only discrete-time signals and images.

10-8.2 Advantage of Wavelets over DFT

One characteristic of the DFT is that signals and images, even with only a few fast-varying segments such as edges, require high-wavenumber complex exponentials to represent them. Hence, most or all of the DFT values $\mathbf{X}[k_1, k_2]$ are nonzero. In contrast, the basis functions used in wavelet transforms are localized in time and frequency. This means that a signal that is mostly slowly-varying but has a few localized fast-varying regions requires only a few low-resolution basis functions to represent the slowly-varying regions, and a few high-resolution basis functions to represent just the localized fast-varying regions. Many of the wavelet transform values are thus zero (or near zero). This feature leads to the following three major applications of wavelet transforms:

- *Compression* of signals and images: they are represented in the wavelet transform domain by many fewer numbers than in the original signal or image. The JPEG-2000 image compression standard uses the wavelet transform.
- *Compressed sensing* of signals and images: since the signal or image in the wavelet transform domain requires many fewer numbers to represent it, it can be reconstructed from many fewer observations than would be required to reconstruct the original signal or image. An introduction to compressed sensing is presented later in this chapter.
- *Filtering of signals and images*: since the signal or image in the wavelet transform domain requires many fewer numbers to represent it, thresholding small values of the wavelet transform of a noisy signal or image to zero reduces the noise in the original signal or image. We will show that the combination of *thresholding* and *shrinkage* gives results far superior to using the 2-D DFT for noise reduction.

After this Overview section, we present the *Haar* wavelet transform, which is the simplest wavelet transform, and yet illustrates many features of the family of wavelet transforms. We then present *quadrature-mirror filters* (*QMFs*) and derive the *Smith-Barnwell condition* for perfect reconstruction of the original signal from its wavelet transform. We conclude our treatment of wavelets by deriving the *Daubechies wavelet function*, which is the most commonly used wavelet function because it has multiple zeros at $\mathbf{z} = 1$, and so it sparsifies many real-world signals. Finally, examples of signal and image compression and denoising are provided.

10-8.3 Historical Overview

Wavelet transforms can be viewed as a generalization of *filter banks* and *subband coding*. A filter bank is a set of bandpass filters connected in parallel; each bandpass filter passes a different range of frequencies. So the signal input into the filter bank is separated into different components, each of which consists of a different part of the spectrum of the input signal. The right half of Fig. 6-65, which is a receiver for frequency-domain-multiplexing of signals, is a filter bank that separates the received signal into different frequency bands (before modulating them back down to baseband). The DFT can be viewed as an extreme case of a filter bank; the DFT separates $x[n]$ into individual complex exponentials at frequencies $\Omega_k = 2\pi k/N$.

Filter banks are used in audio signal processing. In human hearing, some frequencies cannot be heard as well as others. Also a large component at one frequency can mask a component at another frequency. So it makes sense to keep only the frequency bands that humans can hear. This is the basic idea behind coding of signals: omit the frequency bands that contribute little to the perception of the signal. This is called *subband coding*. The mp3 coding of music uses this idea (and many others).

For sampled signals, and especially for images that are inherently sampled into pixels, it was recognized that separating frequency bands using a parallel bank of filters is inefficient. Since the bandwidth of the output of each filter in a bank of L filters is only $\frac{1}{L}$ of the bandwidth of the original signal, the sampling rate for each output can be reduced to $\frac{1}{L}$ of the sampling rate of the original signal, as discussed in Section 6-13.13 on bandpass sampling.

An efficient tree-like filter structure for separating a 1-D sampled signal into different frequency bands (*subband decomposition*) is shown in Fig. 10-17, in which $g[n]$ is a lowpass filter and $h[n]$ is a highpass filter. The concept is easily extendable to 2-D images.

The input signal $x[n]$ with spectrum (DTFT) $\mathbf{X}(e^{j\Omega})$ is separated into a low-frequency-band signal $x_{\text{L}}[n]$ whose spectrum is roughly

$$\mathbf{X}_{\text{L}}(e^{j\Omega}) = \begin{cases} \mathbf{X}(e^{j\Omega}) & \text{for } 0 \le |\Omega| < \pi/2, \\ 0 & \text{for } \pi/2 < |\Omega| \le \pi, \end{cases}$$

and a high-frequency-band signal $x_{\text{H}}[n]$ whose spectrum is roughly

$$\mathbf{X}_{\text{H}}(e^{j\Omega}) = \begin{cases} 0 & \text{for } 0 \le |\Omega| < \pi/2, \\ \mathbf{X}(e^{j\Omega}) & \text{for } \pi/2 < |\Omega| \le \pi. \end{cases}$$

Each signal can be downsampled by 2 without aliasing (see Section 6-13.3), resulting in $x_{\text{LD}}[n] = x_{\text{L}}[2n]$ and $x_{\text{HD}}[n] = x_{\text{H}}[2n]$. There are now two different signals, each

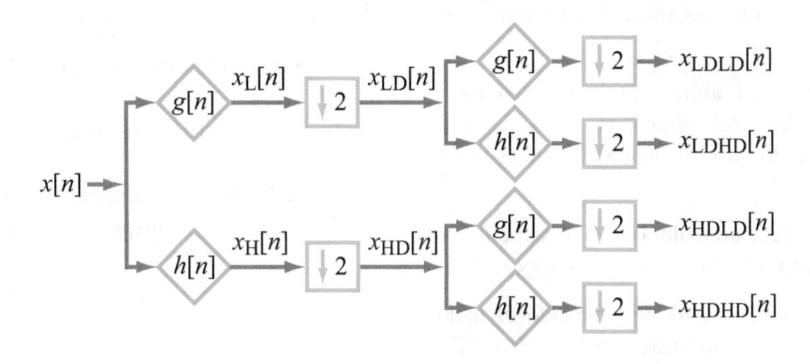

Figure 10-17: Tree-like filter structure for subband decomposition. The diamond-shaped operations denote lowpass and highpass frequency filters, realized through cyclic convolution.

of which is sampled only half as often as $x[n]$, so the total number of samples is unaltered, and each represents a different frequency band of the original signal.

This same decomposition can then be applied to each of the two downsampled signals x_{LD} and x_{HD}. This results in four signals, each of which is sampled only one fourth as often as $x[n]$, so the total number of samples is the same, and each represents a different frequency band of bandwidth $\pi/4$. Repeating this decomposition N times, $x[n]$ can be decomposed into 2^N signals, each of which represents a different frequency band of bandwidth $\pi/2^N$ and is sampled only $1/2^N$ as often as $x[n]$. Use of $N = 5$, resulting in $2^5 = 32$ subbands, is a common choice.

10-8.4 Significance of Wavelets

The wavelet transform differs from this subband coding in that $x[n]$ is not recursively decomposed explicitly into lower and higher frequency bands, although the decomposition is still roughly into lower and higher frequency bands. Instead, the lower frequency bands are replaced with signals that represent slowly-varying parts of $x[n]$, and the higher frequency bands are replaced with signals that represent fast-varying parts of $x[n]$. The latter signals are mostly zero-valued if $x[n]$ is slowly-varying most of the time. Real-world signals and images do tend to consist of mostly slowly-varying regions, containing a few localized regions in which they are fast-varying. Wavelets are good at representing such signals with wavelet transforms that are mostly zero-valued.

To see why representing a signal using only a few wavelet transform components is useful, consider periodic signals. A periodic signal $x(t)$ with period T_0 and maximum frequency B Hz can be represented in the frequency domain using only BT_0 frequencies, since its spectrum consists of harmonics at frequencies k/T_0 Hz for integers k. The maximum frequency B Hz must equal N/T_0 for some integer N, or equivalently, $N = BT_0$ frequencies. Instead of storing $x(t)$, we can generate it using BT_0 sinusoidal generators, each of which requires only an amplitude and phase. So $x(t)$ can be ***compressed*** into $2BT_0$ (plus a dc term, if present) numbers.

If noise had been added to $x(t)$, most of the noise can be eliminated because any part of the spectrum of the noisy $x(t)$ that is not at a harmonic k/T_0 Hz is noise and can be filtered out, as was done in Sections 6-10 and 8-8. We will perform similar actions on signals and images that are not periodic, but which have wavelet transforms that are mostly zero-valued. This includes many real-world signals and images.

10-8.5 Review of Cyclic Convolution

Cyclic convolutions will be used throughout forthcoming sections to compute the wavelet transforms of signals and images, so a short review of cyclic convolutions is in order. Cyclic convolution was covered in Section 7-15.5.

The cyclic or circular convolution of two signals $h[n]$ and $x[n]$, each having duration N, was defined earlier by Eq. (7.177) as

$$y_c[n] = h[n] \circledcirc x[n] = \sum_{i=0}^{N-1} h[i]\, x[(n-i)_N], \qquad (10.54)$$

where, without loss of generality, N_0 has been replaced with N, n_1 replaced with i, $x_1[n]$ replaced with $h[n]$, and $x_2[n]$ replaced with $x[n]$. Using cyclic convolutions instead of linear convolutions amounts to sampling all DTFTs at $\Omega = 2\pi\,\frac{k}{N}$, which is equivalent to replacing all DTFTs with Nth-order DFTs (see Section 7-15).

For applications involving wavelets, the following conditions apply:

- Batch processing is used almost exclusively. This is because the entire original signal is known before processing begins, so the use of noncausal filters is not a problem.
- The order N of the cyclic convolution is the same as the duration of the signal $x[n]$, which usually is very large.
- ***Filtering*** $x[n]$ with a filter $h[n]$ will henceforth mean computing the cyclic convolution $h[n] \copyright x[n]$. The duration L of filter $h[n]$ is much smaller than N $(L \ll N)$, so $h[n]$ gets zero-padded (see Section 8-8.1) with $(N-L)$ zeros. The result of the cyclic convolution is the same as $h[n] * x[n]$, except for the first $L-1$ values, which are aliased, and the final $L-1$ values, which are no longer present, but added to the first $L-1$ values.
- Filtering $x[n]$ with the noncausal filter $h[-n]$ gives the same result as the linear convolution $h[-n] * x[n]$, except that the noncausal part of the latter will alias the final $L-1$ places of the cyclic convolution.
- Zero-padding does not increase the computation, since multiplication by zero is known to give zero, so it need not be computed.
- For two filters $g[n]$ and $h[n]$, both of length L, $g[n] \copyright h[n]$ consists of $g[n] * h[n]$ followed by $N-(2L+1)$ zeros.
- As long as the final result has length N, linear convolutions may be replaced with cyclic convolutions and the final result will be the same.

To illustrate some of these properties, let $h[n] = \{\underline{1}, 2\}$ and $x[n] = \{\underline{3}, 4, 5, 6, 7, 8\}$. Then $N = 6$, $L = 2$, and

$$
\begin{aligned}
h[n] * x[n] &= \{\underline{3}, 10, 13, 16, 19, 22, 16\} \\
h[n] \copyright x[n] &= \{16+3, 10, 13, 16, 19, 22\} \\
h[-n] * x[n] &= \{6, \underline{11}, 14, 17, 20, 23, 8\} \\
h[-n] \copyright x[n] &= \{11, 14, 17, 20, 23, 6+8\}
\end{aligned}
$$

Note that $h[n] \copyright x[n]$ is determined from $h[n]*x[n]$ by adding the final value of $h[n] * x[n]$ to its initial value, and then eliminating that final value from the sequence altogether.

10-9 Haar Wavelet Transform

The Haar transform is by far the simplest wavelet transform, and yet it illustrates many of the concepts of how the wavelet transform works.

10-9.1 Single-Stage Decomposition

Consider the finite-duration signal $x[n]$

$$x[n] = \{\underline{a}, b, c, d, e, f, g, h\}. \tag{10.55}$$

Define the lowpass and highpass filters with impulse responses $g_{\text{haar}}[n]$ and $h_{\text{haar}}[n]$, respectively, as

$$g_{\text{haar}}[n] = \{\underline{1}, 1\}, \tag{10.56a}$$

$$h_{\text{haar}}[n] = \{\underline{1}, -1\}. \tag{10.56b}$$

Recall from Example 7-23 that the frequency responses of these filters are the DTFTs given by

$$\mathbf{G}_{\text{haar}}(e^{j\Omega}) = 1 + e^{-j\Omega} = 2\cos(\Omega/2)e^{-j\Omega/2}, \tag{10.57a}$$

$$\mathbf{H}_{\text{haar}}(e^{j\Omega}) = 1 - e^{-j\Omega} = 2\sin(\Omega/2)je^{-j\Omega/2}, \tag{10.57b}$$

which have lowpass and highpass frequency responses, respectively.

Define the ***average*** (lowpass) signal $x_{\text{L}}[n]$ as

$$
\begin{aligned}
x_{\text{L}}[n] &= x[n] \copyright g_{\text{haar}}[n] \\
&= \{\underline{a+h},\ b+a,\ c+b,\ d+c,\ e+d \ldots\}
\end{aligned} \tag{10.58a}
$$

and the ***detail*** (highpass) signal $x_{\text{H}}[n]$

$$
\begin{aligned}
x_{\text{H}}[n] &= x[n] \copyright h_{\text{haar}}[n] \\
&= \{\underline{a-h},\ b-a,\ c-b,\ d-c,\ e-d \ldots\}.
\end{aligned} \tag{10.58b}
$$

Next, define the ***downsampled average signal*** $x_{\text{LD}}[n]$ as

$$x_{\text{LD}}[n] = x_{\text{L}}[2n] = \{\underline{a+h},\ c+b,\ e+d,\ g+f\} \tag{10.59a}$$

and the ***downsampled detail signal*** $x_{\text{HD}}[n]$ as

$$x_{\text{HD}}[n] = x_{\text{H}}[2n] = \{\underline{a-h},\ c-b,\ e-d,\ g-f\}. \tag{10.59b}$$

The signal $x[n]$ of duration 8 has been replaced by the two signals $x_{\text{LD}}[n]$ and $x_{\text{HD}}[n]$, each of durations 4, so no information about $x[n]$ has been lost. We use cyclic convolutions instead of linear convolutions so that the cumulative length of the downsampled signals equals the length of the original signal. Using linear convolutions, each convolution with $g_{\text{haar}}[n]$ or $h_{\text{haar}}[n]$ would lengthen the signal unnecessarily. As we shall see, using cyclic convolutions instead of linear convolutions is sufficient to recover the original signal from its Haar wavelet transform.

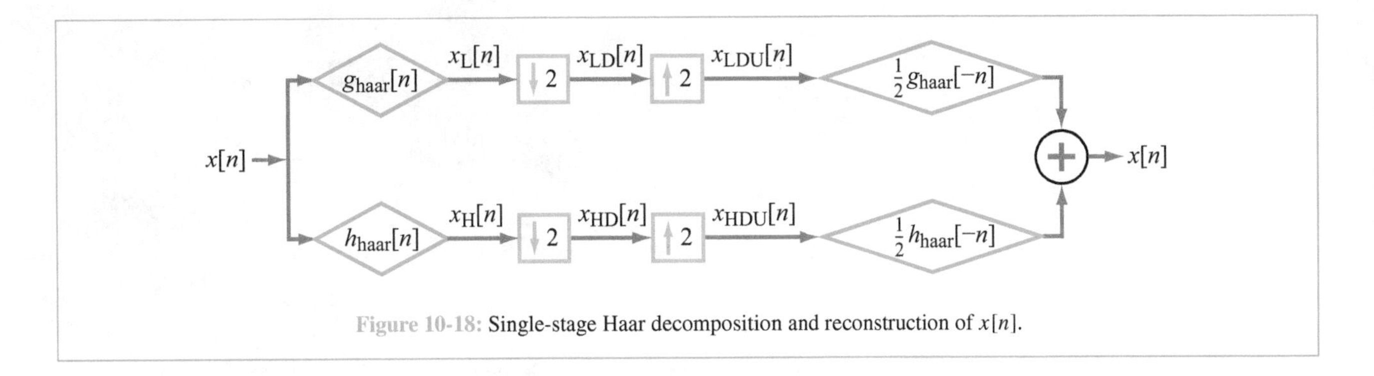

Figure 10-18: Single-stage Haar decomposition and reconstruction of $x[n]$.

10-9.2 Single-Stage Reconstruction

The signal $x[n]$ can be reconstructed from $x_{LD}[n]$ and $x_{HD}[n]$ as follows:

1. Define the upsampled (zero-stuffed) signal $x_{LDU}[n]$ as

$$x_{LDU}[n] = \begin{cases} x_{LD}[n/2] & \text{for } n \text{ even} \\ 0 & \text{for } n \text{ odd} \end{cases}$$
$$= \{\underline{a+h},\ 0,\ c+b,\ 0,\ e+d,\ 0,\ g+f,\ 0\}, \quad (10.60a)$$

and the upsampled (zero-stuffed) signal $x_{HDU}[n]$ as

$$x_{HDU}[n] = \begin{cases} x_{HD}[n/2] & \text{for } n \text{ even} \\ 0 & \text{for } n \text{ odd} \end{cases}$$
$$= \{\underline{a-h},\ 0,\ c-b,\ 0,\ e-d,\ 0,\ g-f,\ 0\}. \quad (10.60b)$$

Note that downsampling by 2 followed by upsampling by 2 replaces values of $x[n]$ with zeros for odd times n.

2. Next, filter $x_{LDU}[n]$ and $x_{HDU}[n]$ with filters $g_{haar}[-n]$ and $h_{haar}[-n]$, respectively.

As noted earlier in Section 10-8.4, the term "filter" in the context of the wavelet transform means "cyclic convolution." Filters $g_{haar}[n]$ and $h_{haar}[n]$ are called ***analysis filters***, because they are used to compute the Haar wavelet transform. Their time reversals $g_{haar}[-n]$ and $h_{haar}[-n]$ are called ***synthesis filters***, because they are used to compute the inverse Haar wavelet transform (that is, to reconstruct the signal from its Haar wavelet transform). The reason for using time reversals here is explained below.

The cyclic convolutions of $x_{LDU}[n]$ with $g_{haar}[-n]$ and $x_{HDU}[n]$ with $h_{haar}[-n]$ yield:

$$x_{LDU}[n] \copyright\ g_{haar}[-n] = \{\underline{a+h},\ c+b,\ c+b, \ldots, a+h\}, \quad (10.61a)$$

$$x_{HDU}[n] \copyright\ h_{haar}[-n] = \{\underline{a-h},\ b-c,\ c-b, \ldots, h-a\}. \quad (10.61b)$$

3. Adding the outcomes of the two cyclic convolutions and dividing the sum by 2 gives $x[n]$:

$$x[n] = \tfrac{1}{2}\, x_{LDU}[n] \copyright\ g_{haar}[-n] + \tfrac{1}{2}\, x_{HDU}[n] \copyright\ h_{haar}[-n]. \quad (10.62)$$

The single-stage Haar decomposition and reconstruction is depicted in Fig. 10-18.

It is still not evident why this is worth doing. The following example provides a partial answer.

Consider the finite-duration ($N = 16$) signal $x[n]$:

$$x[n] = \begin{cases} 4 & \text{for } 0 \le n \le 4 \\ 1 & \text{for } 5 \le n \le 9 \\ 3 & \text{for } 10 \le n \le 14 \\ 4 & \text{for } n = 15. \end{cases} \quad (10.63)$$

The Haar-transformed signals are

$$x_{LD}[n] = x_L[2n] = \{\underline{8}, 8, 8, 2, 2, 4, 6, 6\},$$
$$x_{HD}[n] = x_H[2n] = \{\underline{0}, 0, 0, 0, 0, 2, 0, 0\}. \quad (10.64)$$

These can be derived as follows. We have

$$x[n] = \{4, 4, 4, 4, 4, 1, 1, 1, 1, 1, 3, 3, 3, 3, 3, 4\},$$
$$x_L[n] = x[n] \copyright\ \{1, 1\}$$
$$= \{8, 8, 8, 8, 8, 5, 2, 2, 2, 2, 4, 6, 6, 6, 6, 7\},$$
$$x_H[n] = x[n] \copyright\ \{1, -1\}$$
$$= \{0, 0, 0, 0, 0, 3, 0, 0, 0, 0, 2, 0, 0, 0, 0, 1\},$$
$$x_{LD}[n] = x_L[2n] = \{8, 8, 8, 2, 2, 4, 6, 6\},$$
$$x_{HD}[n] = x_H[2n] = \{0, 0, 0, 0, 0, 2, 0, 0\}.$$

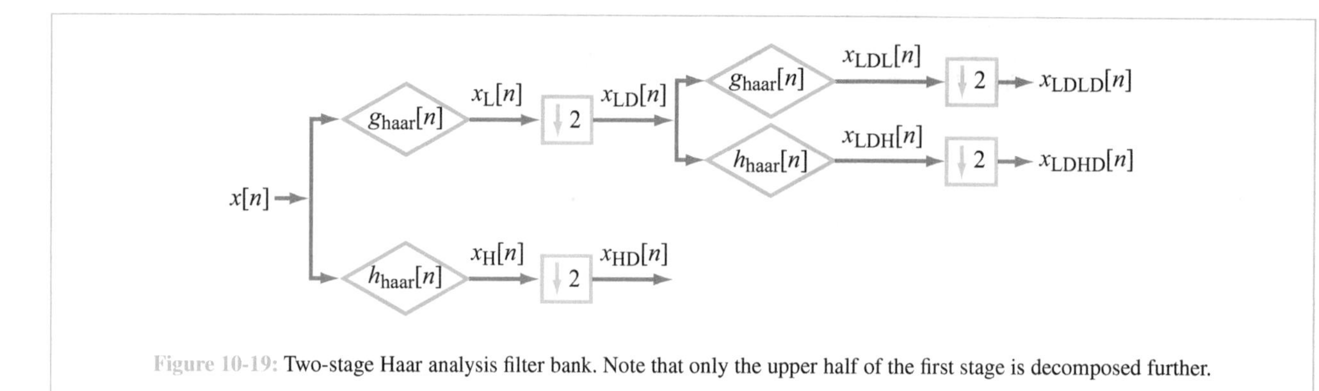

Figure 10-19: Two-stage Haar analysis filter bank. Note that only the upper half of the first stage is decomposed further.

The original signal $x[n]$ can be recovered from $x_{LD}[n]$ and $x_{HD}[n]$ by

$$x_{LDU}[n] = \{8, 0, 8, 0, 8, 0, 2, 0, 2, 0, 4, 0, 6, 0, 6, 0\},$$

$$x_{HDU}[n] = \{0, 0, 0, 0, 0, 0, 0, 0, 0, 0, 2, 0, 0, 0, 0, 0\},$$

$$x_{LDU}[n] \circledcirc \{1, \underline{1}\} = \{8, 8, 8, 8, 8, 2, 2, 2, 2, 4, 4, 6, 6, 6, 6, 8\},$$

$$x_{HDU}[n] \circledcirc \{-1, \underline{1}\} = \{0, 0, 0, 0, 0, 0, 0, 0, 0, -2, 2, 0, 0, 0, 0, 0\},$$

$$\frac{x_{LDU}[n] \circledcirc \{1, \underline{1}\} + x_{HDU}[n] \circledcirc \{-1, \underline{1}\}}{2} = \{4, 4, 4, 4, 4, 1, 1, 1, 1, 1, 3, 3, 3, 3, 3, 4\} = x[n].$$

We observe that the outcome of the second cyclic convolution is *sparse* (mostly zero-valued). The Haar transform allows $x[n]$, which has duration 16, to be represented using the eight values of $x_{LD}[n]$ and the single nonzero value (and its location $n = 5$) of $x_{HD}[n]$. This saves almost half of the storage required for $x[n]$. Hence, $x[n]$ has been *compressed* by 43%.

Even though $x[n]$ is not sparse, it was transformed, using the Haar transform, into a sparse representation with the same number of samples, meaning that most of the values of the Haar-transformed signal are zero-valued. This reduces the amount of memory required to store $x[n]$, because only the times at which nonzero values occur (as well as the values themselves) need be stored. The few bits (0 or 1) required to store locations of nonzero values are considered to be negligible in number compared with the many bits required to store the actual nonzero values. Since the Haar transform is orthogonal, $x[n]$ can be recovered perfectly from its Haar-transformed values.

10-9.3 Multistage Decomposition and Reconstruction

In the simple example used in the preceding subsection, only 1 element of the Haar-transformed signal $x_{HD}[n]$ is nonzero, but all 8 elements of $x_{LD}[n]$ are nonzero. We can reduce the number of nonzero elements of $x_{LD}[n]$ by applying a second Haar transform stage to it. That is, $x_{LD}[n]$ can be transformed into the two signals $x_{LDLD}[n]$ and $x_{LDHD}[n]$ by applying the steps outlined in Fig. 10-19. Thus,

$$\begin{aligned}
x_{LD}[n] &= \{8, 8, 8, 2, 2, 4, 6, 6\}, \\
x_{LDL}[n] &= x_{LD}[n] \circledcirc \{1, 1\} \\
&= \{14, 16, 16, 10, 4, 6, 10, 12\}, \\
x_{LDH}[n] &= x_{LD}[n] \circledcirc \{1, -1\} \\
&= \{2, 0, 0, -6, 0, 2, 2, 0\}, \\
x_{LDLD}[n] &= x_{LDL}[2n] = \{14, 16, 4, 10\}, \\
x_{LDHD}[n] &= x_{LDH}[2n] = \{2, 0, 0, 2\}.
\end{aligned} \tag{10.65}$$

Signal $x_{LDHD}[n]$ is again sparse: only two of its four values are nonzero. So $x[n]$ can now be represented by the four values of $x_{LDLD}[n]$, the two nonzero values of $x_{LDHD}[n]$, and the one nonzero value of $x_{HD}[n]$. This reduces the storage required for $x[n]$ by 57%.

The average signal $x_{LDLD}[n]$ can in turn be decomposed even further. The result is an ***analysis filter bank*** that computes the Haar wavelet transform of $x[n]$. This analysis filter bank consists of a series of sections like the left half of Fig. 10-18, connected as in Fig. 10-19, except that each average signal $x_{LL}[n]$ is decomposed further. The signals computed at the right end of this analysis filter bank constitute the Haar wavelet transform of $x[n]$. Reconstruction of $x[n]$ is shown in Fig. 10-20.

10-9.4 Haar Wavelet Transform Filter Banks

Decomposition

A signal $x[n]$ of duration $N = 2^K$ (with K an integer) can be represented by the Haar wavelet transform through a

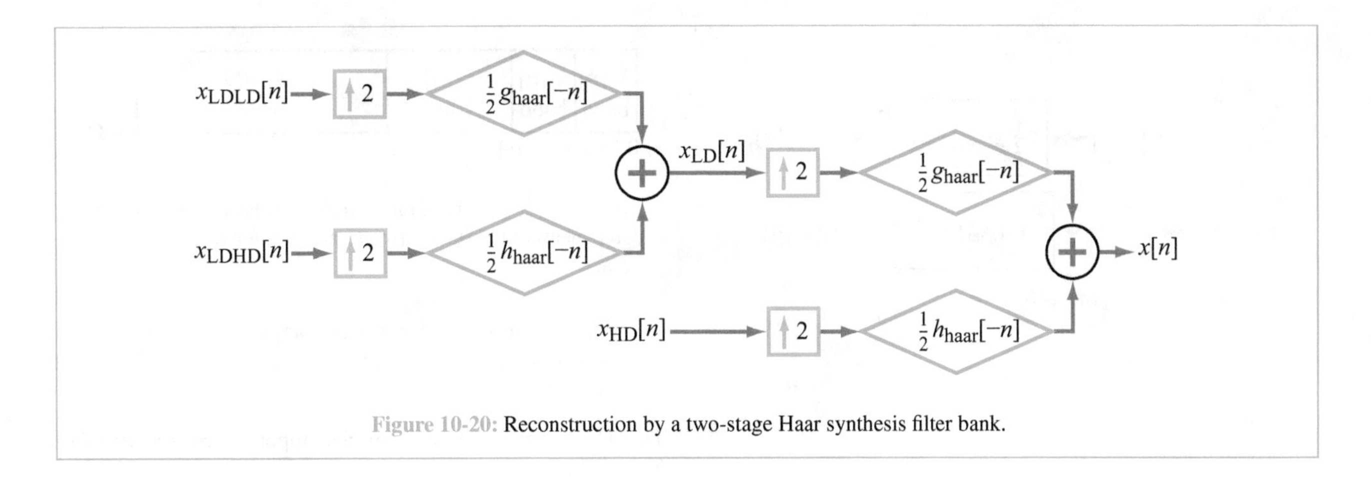

Figure 10-20: Reconstruction by a two-stage Haar synthesis filter bank.

K-stage decomposition process involving cyclic convolutions with filters $g_{\text{haar}}[n]$ and $h_{\text{haar}}[n]$, as defined by Eq. (10.56). The signal $x[n]$ can be zero-padded so that its length is a power of 2, if that is not already the case, just as is done for the FFT. The sequential process is:

Stage 1:

$$x[n] \rightarrow \boxed{h_{\text{haar}}[n]} \rightarrow \boxed{\downarrow 2} \rightarrow \tilde{x}_1[n] = x_{\text{HD}}[n],$$

$$x[n] \rightarrow \boxed{g_{\text{haar}}[n]} \rightarrow \boxed{\downarrow 2} \rightarrow \widetilde{X}_1[n] = x_{\text{LD}}[n].$$

Stage 2:

$$\widetilde{X}_1[n] \rightarrow \boxed{h_{\text{haar}}[n]} \rightarrow \boxed{\downarrow 2} \rightarrow \tilde{x}_2[n] = x_{\text{LDHD}}[n],$$

$$\widetilde{X}_1[n] \rightarrow \boxed{g_{\text{haar}}[n]} \rightarrow \boxed{\downarrow 2} \rightarrow \widetilde{X}_2[n] = x_{\text{LDLD}}[n].$$

$$\vdots$$

Stage K:

$$\widetilde{X}_{K-1}[n] \rightarrow \boxed{h_{\text{haar}}[n]} \rightarrow \boxed{\downarrow 2} \rightarrow \tilde{x}_K[n],$$

$$\widetilde{X}_{K-1}[n] \rightarrow \boxed{g_{\text{haar}}[n]} \rightarrow \boxed{\downarrow 2} \rightarrow \widetilde{X}_K[n].$$

The Haar transform of $x[n]$ consists of the combination of $K+1$ signals:

$$\text{Duration} \Rightarrow \{\underbrace{\tilde{x}_1[n]}_{N/2}, \underbrace{\tilde{x}_2[n]}_{N/4}, \underbrace{\tilde{x}_3[n]}_{N/8}, \ldots, \underbrace{\tilde{x}_K[n]}_{N/2^K}, \underbrace{\widetilde{X}_K[n]}_{N/2^K}\}. \tag{10.66}$$

To represent $x[n]$, we need to retain the "high-frequency" outputs of all K stages (i.e., $\{\tilde{x}_1[n], \tilde{x}_2[n], \ldots, \tilde{x}_K[n]\}$), but only the final output of the "low-frequency" sequence, namely $\widetilde{X}_K[n]$. The total duration of all of the $K+1$ Haar transform signals is

$$\frac{N}{2} + \frac{N}{4} + \frac{N}{8} + \cdots + \frac{N}{2^K} + \frac{N}{2^K} = N, \tag{10.67}$$

which equals the duration N of $x[n]$. We use cyclic convolutions instead of linear convolutions so that the total lengths of the downsampled signals equals the length of the original signal. Were we to use linear convolutions, each convolution with $g_{\text{haar}}[n]$ or $h_{\text{haar}}[n]$ would lengthen the signal unnecessarily.

- The $\tilde{x}_k[n]$ for $k = 1, 2, \ldots, K$ are called ***detail*** signals.
- The $\widetilde{X}_K[n]$ is called the ***average*** signal.

Reconstruction

The inverse Haar wavelet transform can be computed in reverse order, starting with $\{\widetilde{X}_K[n], \tilde{x}_K[n], \tilde{x}_{K-1}[n], \ldots\}$.

Stage 1:

$$\widetilde{X}_K[n] \rightarrow \boxed{\uparrow 2} \rightarrow \boxed{\tfrac{1}{2}\, g_{\text{haar}}[-n]} \rightarrow A_{K-1}[n],$$

$$\tilde{x}_K[n] \rightarrow \boxed{\uparrow 2} \rightarrow \boxed{\tfrac{1}{2}\, h_{\text{haar}}[-n]} \rightarrow B_{K-1}[n],$$

$$\widetilde{X}_{K-1}[n] = A_{K-1}[n] + B_{K-1}[n].$$

Stage 2:

$\widetilde{X}_{K-1}[n] \rightarrow \boxed{\uparrow 2} \rightarrow \boxed{\tfrac{1}{2}\, g_{\text{haar}}[-n]} \rightarrow A_{K-2}[n]$,

$\widetilde{x}_{K-1}[n] \rightarrow \boxed{\uparrow 2} \rightarrow \boxed{\tfrac{1}{2}\, h_{\text{haar}}[-n]} \rightarrow B_{K-2}[n]$,

$\widetilde{X}_{K-2}[n] = A_{K-2}[n] + B_{K-2}[n]$.

⋮

Stage K:

$\widetilde{X}_{1}[n] \rightarrow \boxed{\uparrow 2} \rightarrow \boxed{\tfrac{1}{2}\, g_{\text{haar}}[-n]} \rightarrow A_{0}[n]$,

$\widetilde{x}_{1}[n] \rightarrow \boxed{\uparrow 2} \rightarrow \boxed{\tfrac{1}{2}\, h_{\text{haar}}[-n]} \rightarrow B_{0}[n]$,

$x[n] = A_0[n] + B_0[n]$.

10-9.5 Haar Wavelet Transform in the Frequency Domain

As noted in Section 10-8, the original objective of subband coding was to decompose a signal into different frequency bands. Wavelets are more general than simple subbanding in that decomposition of the spectrum of the input into different bands is not the explicit purpose of the wavelet transform. Nevertheless, the decomposition does approximately allocate the signal into different frequency bands, so it is helpful to track and understand the approximate decomposition.

Recall from Eq. (10.57) that $\mathbf{G}_{\text{haar}}(e^{j\Omega})$ is approximately a lowpass filter and $\mathbf{H}_{\text{haar}}(e^{j\Omega})$ is approximately a highpass filter. Hence, at the output of the first stage of the Haar wavelet transform:

- $\widetilde{X}_1[n]$ is the lowpass-frequency part of $x[n]$, covering the approximate range $0 \le |\Omega| \le \pi/2$, and

- $\widetilde{x}_1[n]$ is the highpass-frequency part of $x[n]$, covering the approximate range $\pi/2 \le |\Omega| \le \pi$.

At each stage, downsampling expands the spectrum of each signal to the full range $0 \le |\Omega| \le \pi$. Hence, at the output of the second stage,

- $\widetilde{X}_2[n]$ covers the frequency range $0 \le |\Omega| \le \pi/2$ of $\widetilde{X}_1[n]$, which corresponds to the range $0 \le |\Omega| \le \pi/4$ of $x[n]$.

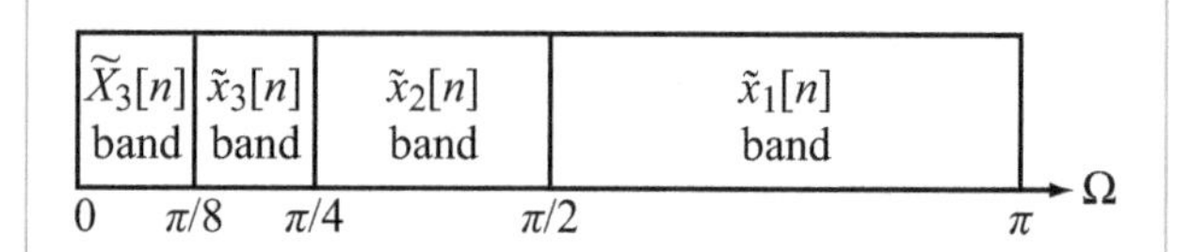

Figure 10-21: Approximate frequency-band coverage by components of the Haar wavelet transform for $K = 3$.

- $\widetilde{x}_2[n]$ covers the frequency range $\pi/2 \le |\Omega| \le \pi$ of $\widetilde{X}_1[n]$, which corresponds to the range $\pi/4 \le |\Omega| \le \pi/2$ of $x[n]$.

The Haar wavelet transform decomposes the spectrum of $\mathbf{X}(e^{j\Omega})$ into *octaves*, with

$$\text{Octave } \frac{\pi}{2^k} \le |\Omega| \le \frac{\pi}{2^{k-1}} \text{ represented by } \widetilde{x}_k[n].$$

Since the width of this band is $\pi/2^k$, the sampling rate can be reduced by a factor of 2^k using downsampling.

For a signal of duration $N = 2^K$, the Haar wavelet transform decomposes $x[n]$ into $K+1$ components, namely $\widetilde{X}_K[n]$ and $\{\widetilde{x}_1[n], \widetilde{x}_2[n], \ldots, \widetilde{x}_K[n]\}$, with each component representing a frequency octave. The spectrum decomposition is illustrated in Fig. 10-21 for $K = 3$. Because the different components cover different octaves, they can be sampled at different rates. Furthermore, if the signal or image consists of slowly-varying segments with occasional fast-varying features, as many real-world signals and images do, then $\widetilde{x}_k[n]$ for small k will be sparse, requiring few samples to represent it.

10-9.6 Normalized Haar Functions

Each stage in the Haar reconstruction process, such as in the 2-stage diagram shown in Fig. 10-20, includes the summation of 2 signals, both of which need to be divided by a factor of 2. To avoid the division step, we can replace the Haar functions defined by Eq. (10.56) with normalized versions given by

$$\begin{aligned} \tilde{g}_{\text{haar}}[n] &= \frac{1}{\sqrt{2}}\,\{\underline{1}, 1\}, \\ \tilde{h}_{\text{haar}}[n] &= \frac{1}{\sqrt{2}}\,\{\underline{1}, -1\}. \end{aligned} \tag{10.68}$$

These are called the ***normalized Haar scaling*** (lowpass) and ***wavelet*** (highpass) functions, respectively. They are also energy-normalized:

$$\sum_{n=0}^{1} \tilde{g}_{\text{haar}}[n]^2 = \sum_{n=0}^{1} \tilde{h}_{\text{haar}}[n]^2 = 1. \tag{10.69}$$

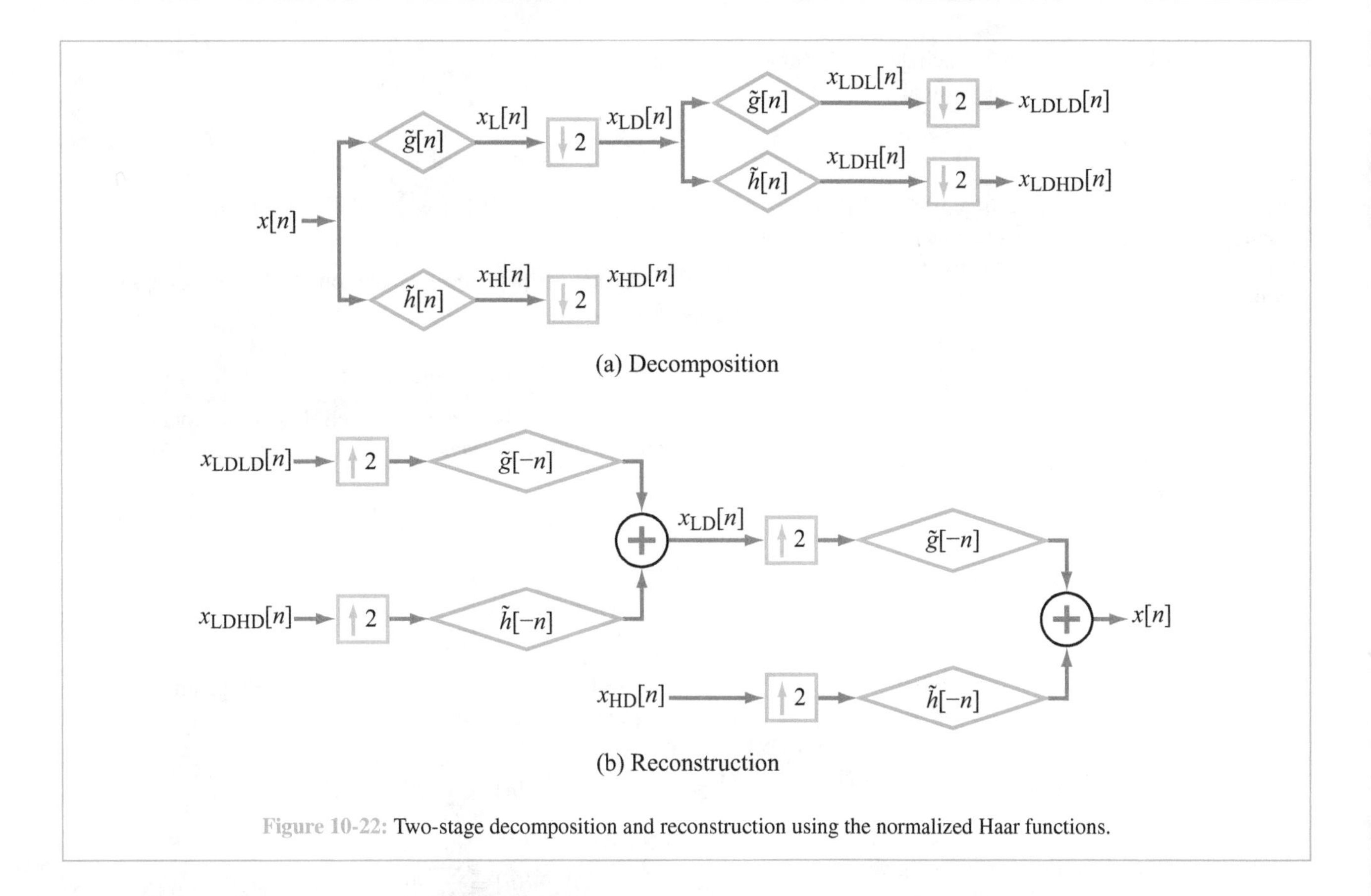

Figure 10-22: Two-stage decomposition and reconstruction using the normalized Haar functions.

In terms of these normalized Haar functions, the two-stage decomposition and reconstruction diagrams shown earlier in Figs. 10-19 and 10-20 assume the forms displayed in Fig. 10-22.

Concept Question 10-9: What is the difference between using the Haar wavelet transform and a set of bandpass filters? (See S²)

Concept Question 10-10: How is it possible that the wavelet transform requires less computation than the FFT? (See S²)

10-10 The Family of Wavelet Transforms

The preceding section focused on the Haar wavelet transform, which is only one member of an extended family of wavelet transforms. A few other members are treated in future sections. Common among all of these members of the wavelet transform family are a set of formulations and orthogonality properties, so we devote the present section to a presentation that applies to all wavelet transforms. For the sake of easier readability, the presentation will adhere to the following notational guidelines:

- $x[n]$ is of duration N with $N = 2^K$ and K is a positive integer.
- $g[n]$ is a *scaling* function of duration 2 or longer. Its specific form depends on the specific wavelet transform. For the Haar wavelet transform, $g_{\text{haar}}[n]$ is of duration 2 and is defined by Eq. (10.56a).
- $h[n]$ is a *wavelet* function of duration 2 or longer. Its specific form depends on the specific wavelet transform. For the Haar wavelet transform, $h_{\text{haar}}[n]$ is of duration 2 and is defined by Eq. (10.56b).
- All convolutions are cyclic, and the "mod N" notation is suppressed for easy readability.

- Most limits on summations (which quickly become very complicated expressions) are omitted.

10-10.1 First Recursion Step

When introduced earlier in Section 10-9.4, $\widetilde{X}_1[n]$ and $\widetilde{x}_1[n]$ were defined as the cyclic convolutions of $x[n]$ with $g[n]$ and $h[n]$, respectively, followed by downsampling, which replaces n with $2n$. That is,

$$\widetilde{X}_1[n] = \sum_{i=0}^{N-1} x[i]\, g[2n-i]. \tag{10.70}$$

In stage 2, $\widetilde{X}_2[n]$ is the cyclic convolution of $\widetilde{X}_1[n]$ and $g[n]$, followed by downsampling:

$$\begin{aligned}\widetilde{X}_2[n] &= \sum_{i=0}^{N-1} \widetilde{X}_1[i]\, g[2n-i] \\ &= \sum_{i=0}^{N-1}\sum_{j=0}^{N-1} x[j]\, g[2i-j]\, g[2n-i] \\ &= \sum_{j=0}^{N-1} x[j] \sum_{i=0}^{N-1} g[2i-j]\, g[2n-i]. \end{aligned} \tag{10.71}$$

Changing variables from i to $i' = 2n - i$ gives

$$\begin{aligned}\widetilde{X}_2[n] &= \sum_{j} x[j] \sum_{i'} g[2(2n-i')-j]\, g[i'] \\ &= \sum_{j} x[j] \sum_{i'} g[2^2 n - 2i' - j]\, g[i'] \\ &= \sum_{j} x[j]\, g^{(2)}[2^2 n - j], \end{aligned} \tag{10.72}$$

where $g^{(2)}[n]$ is defined as

$$g^{(2)}[n] = \sum_{i'} g[n-2i']\, g[i']. \tag{10.73}$$

Note that $g^{(2)}[n]$ is the convolution of $g[n]$ with an upsampled $g[n]$.

Similarly, $\widetilde{x}_2[n]$ is the cyclic convolution of $\widetilde{X}_1[n]$ and $h[n]$, followed by downsampling. Repeating the earlier argument gives

$$\widetilde{x}_1[n] = \sum_{i=0}^{N-1} x[i]\, h[2n-i] \tag{10.74a}$$

and

$$\widetilde{x}_2[n] = \sum_{j} x[j]\, h^{(2)}[2^2 n - j], \tag{10.74b}$$

where $h^{(2)}[n]$ is defined as

$$h^{(2)}[n] = \sum_{i'} h[n-2i']\, g[i']. \tag{10.75}$$

Note that $h^{(2)}[n]$ is the convolution of $g[n]$ and an upsampled $h[n]$.

10-10.2 Continuing Recursion Steps

For stage k, where k is between 1 and K, continuing the recursion process leads to

$$\widetilde{X}_k[n] = \sum_{i=0}^{N-1} x[i]\, g^{(k)}[(2^k n - i)], \tag{10.76a}$$

$$\widetilde{x}_k[n] = \sum_{i=0}^{N-1} x[i]\, h^{(k)}[(2^k n - i)], \tag{10.76b}$$

where filters $g^{(k)}[n]$ and $h^{(k)}[n]$ can be computed recursively ahead of time using the two formulas

$$g^{(k+1)}[n] = \sum_{i=0}^{L} g[i]\, g^{(k)}[(n - 2^k i)], \tag{10.77a}$$

$$h^{(k+1)}[n] = \sum_{i=0}^{L} h[i]\, g^{(k)}[(n - 2^k i)], \tag{10.77b}$$

and initialized using

$$g^{(1)}[n] = g[n], \tag{10.78a}$$

$$h^{(1)}[n] = h[n]. \tag{10.78b}$$

These formulas are simpler than they appear—each recursion is a cyclic convolution of the preceding filter with $g[n]$ or $h[n]$, upsampled by 2, k times.

As an aside, an alternative method for computing $g^{(k)}[n]$ and $h^{(k)}[n]$ is to use the **z**-transform. Since convolution in the time domain corresponds to multiplication in the frequency domain, we can transform $g[n]$ and $h[n]$ to the **z**-domain to obtain $\mathbf{G(z)}$ and $\mathbf{H(z)}$, and then use the following formulas to compute $\mathbf{G}^{(k)}\mathbf{(z)}$ and $\mathbf{H}^{(k)}\mathbf{(z)}$:

$$\mathbf{G}^{(k)}(\mathbf{z}) = \mathbf{G(z)}\, \mathbf{G(z^2)}\, \mathbf{G(z^4)} \ldots \mathbf{G}(\mathbf{z}^{2^{k-1}}), \tag{10.79a}$$

$$\mathbf{H}^{(k)}(\mathbf{z}) = \mathbf{G(z)}\, \mathbf{G(z^2)}\, \mathbf{G(z^4)} \ldots \mathbf{H}(\mathbf{z}^{2^{k-1}}). \tag{10.79b}$$

Inverse transformation to the discrete-time domain yields $g^{(k)}[n]$ and $h^{(k)}[n]$. An example of this operation is given later in Section 10-10.5.

10-10.3 Reconstruction

Given $\widetilde{X}_K[n]$ and $\{\widetilde{x}_1[n], \widetilde{x}_2[n], \ldots, \widetilde{x}_K[n]\}$, reconstruction uses similar cyclic convolutions, except that the synthesis functions are $g[-n]$ and $h[-n]$, instead of $g[n]$ and $h[n]$. The reconstructed $x[n]$ is given by

$$x[n] = \sum_{k=1}^{K} \sum_{i=0}^{N-1} \widetilde{x}_k[i]\, h^{(k)}[(2^k i - n)] + \sum_{i=0}^{N-1} \widetilde{X}_K[i]\, g^{(K)}[(2^K i - n)]. \tag{10.80}$$

10-10.4 Expansion in Basis Functions

The preceding expressions may seem different from the expansion in basis functions given by Eq. (10.50), but in fact they can be cast in that form. If in Eq. (10.76), we replace n with m, and then i with n, we have

$$\widetilde{X}_k[m] = \sum_{n=0}^{N-1} x[n]\, g^{(k)}[(2^k m - n)], \tag{10.81a}$$

$$\widetilde{x}_k[m] = \sum_{n=0}^{N-1} x[n]\, h^{(k)}[(2^k m - n)]. \tag{10.81b}$$

Next, replacing i with m in Eq. (10.80) gives

$$x[n] = \sum_{k=1}^{K} \sum_{m=0}^{N-1} \widetilde{x}_k[m]\, h^{(k)}[(2^k m - n)] + \sum_{m=0}^{N-1} \widetilde{X}_K[m]\, g^{(K)}[(2^K m - n)]. \tag{10.82}$$

The formulations in Eqs. (10.82) and (10.81) have, respectively, the same form as Eqs. (10.50) and (10.52),

$$\phi_{m,k,g}[n] = g^{(k)}[(2^k m - n)], \tag{10.83a}$$

$$\phi_{m,k,h}[n] = h^{(k)}[(2^k m - n)]. \tag{10.83b}$$

Note that we have not shown that the basis functions are orthonormal, as defined by Eq. (10.51). However, since the output of the decomposition matches the input of the reconstruction, the basis functions must be orthogonal.

10-10.5 Haar Transform Functions

The multistage decomposition and reconstruction expressions developed in the preceding subsections apply to any member of the wavelet transform family. By way of an example, we now repeat them for the specific case of the Haar wavelet transform. We start by defining $g[n]$ and $h[n]$ as the normalized Haar functions given by Eq. (10.68),

$$g[n] = \tilde{g}_{\text{haar}}[n] = \frac{1}{\sqrt{2}}\, \{\underline{1}, 1\}, \tag{10.84a}$$

$$h[n] = \tilde{h}_{\text{haar}}[n] = \frac{1}{\sqrt{2}}\, \{\underline{1}, -1\}. \tag{10.84b}$$

The recursion process is initialized by

$$g^{(1)}[n] = \tilde{g}_{\text{haar}}[n] = \frac{1}{\sqrt{2}}\, \{\underline{1}, 1\}, \tag{10.85a}$$

$$h^{(1)}[n] = \tilde{h}_{\text{haar}}[n] = \frac{1}{\sqrt{2}}\, \{\underline{1}, -1\}. \tag{10.85b}$$

Then, Eq. (10.77) with $k = 1$ gives

$$\begin{aligned} g^{(2)}[n] &= \underbrace{g[0]\, g^{(1)}[n-0]}_{i=0} + \underbrace{g[1]\, g^{(1)}[n-2^1]}_{i=1} \\ &= \frac{1}{\sqrt{2}}\frac{1}{\sqrt{2}}\, \{\underline{1}, 1\} + \frac{1}{\sqrt{2}}\frac{1}{\sqrt{2}}\, \{\underline{0}, 0, 1, 1\} \\ &= \frac{1}{2}\, \{\underline{1}, 1, 1, 1\}, \end{aligned} \tag{10.86a}$$

and

$$\begin{aligned} h^{(2)}[n] &= \underbrace{h[0]\, g^{(1)}[n-0]}_{i=0} + \underbrace{h[1]\, g^{(1)}[n-2^1]}_{i=1} \\ &= \frac{1}{\sqrt{2}}\frac{1}{\sqrt{2}}\, \{\underline{1}, 1\} - \frac{1}{\sqrt{2}}\frac{1}{\sqrt{2}}\, \{\underline{0}, 0, 1, 1\} \\ &= \frac{1}{2}\, \{\underline{1}, 1, -1, -1\}. \end{aligned} \tag{10.86b}$$

Similarly, Eq. (10.77) with $k = 2$ gives

$$\begin{aligned} g^{(3)}[n] &= \underbrace{g[0]\, g^{(2)}[n-0]}_{i=0} + \underbrace{g[1]\, g^{(2)}[n-2^2]}_{i=1} \\ &= \frac{1}{\sqrt{2}}\frac{1}{2}\, \{\underline{1}, 1, 1, 1\} + \frac{1}{\sqrt{2}}\frac{1}{2}\, \{\underline{0}, 0, 0, 0, 1, 1, 1, 1\} \\ &= \frac{1}{2\sqrt{2}}\, \{\underline{1}, 1, 1, 1, 1, 1, 1, 1\}, \end{aligned} \tag{10.87a}$$

and

$$h^{(3)}[n] = \underbrace{h[0]\,g^{(2)}[n-0]}_{i=0} + \underbrace{h[1]\,g^{(2)}[n-2^2]}_{i=1}$$
$$= \frac{1}{\sqrt{2}}\frac{1}{2}\{\underline{1},1,1,1\} - \frac{1}{\sqrt{2}}\frac{1}{2}\{\underline{0},0,0,0,1,1,1,1\}$$
$$= \frac{1}{2\sqrt{2}}\{\underline{1},1,1,1,-1,-1,-1,-1\}. \qquad (10.87b)$$

Further recursions are straightforward to compute.

10-10.6 Haar Wavelet Transform by Matrix-Vector Product

For a signal $x[n]$ of duration $N = 8$, $K = \log_2 N = 3$. The completely decomposed normalized Haar wavelet transform can be computed using the matrix-vector product:

$$\begin{bmatrix} \tilde{x}_1[0] \\ \tilde{x}_1[1] \\ \tilde{x}_1[2] \\ \tilde{x}_1[3] \\ \tilde{x}_2[0] \\ \tilde{x}_2[1] \\ \tilde{x}_3[0] \\ \tilde{X}_3[0] \end{bmatrix} = \mathcal{H} \begin{bmatrix} x[7] \\ x[0] \\ x[1] \\ x[2] \\ x[3] \\ x[4] \\ x[5] \\ x[6] \end{bmatrix}, \qquad (10.88)$$

where the matrix $\mathcal{H}$ is defined as

$$\mathcal{H} = \frac{1}{2\sqrt{2}} \times \begin{bmatrix} -2 & 2 & 0 & 0 & 0 & 0 & 0 & 0 \\ 0 & 0 & -2 & 2 & 0 & 0 & 0 & 0 \\ 0 & 0 & 0 & 0 & -2 & 2 & 0 & 0 \\ 0 & 0 & 0 & 0 & 0 & 0 & -2 & 2 \\ -\sqrt{2} & -\sqrt{2} & \sqrt{2} & \sqrt{2} & 0 & 0 & 0 & 0 \\ 0 & 0 & 0 & 0 & -\sqrt{2} & -\sqrt{2} & \sqrt{2} & \sqrt{2} \\ -1 & -1 & -1 & -1 & 1 & 1 & 1 & 1 \\ 1 & 1 & 1 & 1 & 1 & 1 & 1 & 1 \end{bmatrix}. \qquad (10.89)$$

The elements of matrix $\mathcal{H}$ are obtained from the computed values of functions $g^{(k)}[n]$ and $h^{(k)}[n]$ for $k = 1$, 2, and 3. Specifically:

- $\tilde{x}_1[0]$: The top row (Row 1) represents a zero-padded version of $h^{(0)}[n]$, with its elements arranged in reverse order (because the cyclic convolution multiplies $x[i]$ by $h^{(1)}[2n-i] = h[2n-i]$.
- $\tilde{x}_1[1]$, $\tilde{x}_1[2]$, and $\tilde{x}_1[3]$: Rows 2 through 4 from the top represent Row 1, shifted successively by 2 to the right.
- $\tilde{x}_2[0]$: Row 5 represents $h^{(2)}[n]$ with zero padding and reversal.
- $\tilde{x}_2[1]$: Row 6 represents Row 5 with $2 \times 2 = 4$ shifts to the right.
- $\tilde{x}_2[2]$ and $\tilde{x}_2[3]$ are zero; hence, they are not listed in the column on the left-hand side of Eq. (10.88).
- $\tilde{x}_3[0]$: Row 7 represents $h^{(3)}[n]$ with reversal.
- $\tilde{x}_3[1] = \tilde{x}_3[2] = \tilde{x}_3[3] = 0$. Hence, they are not included.
- $\tilde{X}_3[0]$: Row 8 represents $g^{(3)}[n]$ with reversal (which is irrelevant in the present case because all elements of $g^{(3)}[n]$ are the same).

The inverse length-8 Haar wavelet transform can be computed using the matrix-vector product

$$\begin{bmatrix} x[7] \\ x[0] \\ x[1] \\ x[2] \\ x[3] \\ x[4] \\ x[5] \\ x[6] \end{bmatrix} = \mathcal{H}^T \begin{bmatrix} \tilde{x}_1[0] \\ \tilde{x}_1[1] \\ \tilde{x}_1[2] \\ \tilde{x}_1[3] \\ \tilde{x}_2[0] \\ \tilde{x}_2[1] \\ \tilde{x}_3[0] \\ \tilde{X}_3[0] \end{bmatrix}. \qquad (10.90)$$

Note that $\mathcal{H}^{-1} = \mathcal{H}^T$, so $\mathcal{H}$ is an *orthogonal* matrix. This means that if $x[n]$ is perturbed to $x[n] + \delta x[n]$—due to noise or other errors—then the wavelet transform $\{\tilde{x}_k[n]\}$ of $x[n]$ (including $\tilde{X}_0[n]$) is perturbed by $\delta\tilde{x}_k[n]$, and the energies of the two perturbations are equal:

$$\sum_{n=0}^{7} (\delta x[n])^2 = \sum_{k=1}^{3}\sum_{n} (\delta\tilde{x}_k[n])^2. \qquad (10.91)$$

This implies that a small change in any of the $\tilde{x}_k[n]$ will result in a small change in the reconstructed $x[n]$. So thresholding small values of $\tilde{x}_k[n]$ to zero will have only a small effect on the reconstructed $x[n]$. This is useful for compression, as illustrated later.

Concept Question 10-11: What is the significance of the orthonormality of the wavelet transform? (See S^2)

Exercise 10-4: Show that $\tilde{g}_{\text{haar}}[n]$ and $\tilde{h}_{\text{haar}}[n]$ are energy-normalized functions.

Answer: From Eq. (10.68) we have

$$(1/\sqrt{2})^2 + (1/\sqrt{2})^2 = (1/\sqrt{2})^2 + (-1/\sqrt{2})^2 = 1.$$

Exercise 10-5: Show that $\mathcal{H}^{-1} = \mathcal{H}^T$ for Eq. (10.89).

Answer: The product $\mathcal{H}\mathcal{H}^T = I$; hence $\mathcal{H}^T = \mathcal{H}^{-1}$.

10-11 Non-Haar Single-Stage Perfect Reconstruction

We demonstrated in earlier sections that the Haar wavelet transform is capable of *perfect reconstruction* of $x[n]$. Now, we explore the *conditions* that need to be satisfied so as to perfectly reconstruct $x[n]$ by any non-Haar wavelet transform that obeys the basis-functions formulation given in Section 10-10. These conditions pertain to non-Haar $g[n]$ and $h[n]$ functions.

We consider only FIR filters $g[n]$ and $h[n]$ of lengths $L+1$, with L being an odd integer. Even though the time-reversed impulse responses $g[-n]$ and $h[-n]$ used in the synthesis (reconstruction) filter bank are noncausal, their delayed versions $g[-(n-L)]$ and $h[-(n-L)]$ are causal:

$$g[-(n-L)] = g[L-n] = \{\underline{g[L]}, g[L-1], \dots, g[0]\}.$$

As a precursor to the derivation of the conditions required for perfect reconstruction, it will prove useful to provide a review of relevant downsampling and upsampling relationships.

10-11.1 Review of Downsampling and Upsampling

The topics of downsampling and upsampling were covered in Sections 9-6 and 9-7. We restate the key results here.

Downsampling a signal $x[n]$ by 2 means deleting every other value of $x[n]$. The result of downsampling by 2 is

$$x[n] \rightarrow \boxed{\downarrow 2} \rightarrow y[n] = x[2n] = \{\dots, \underline{x[0]}, x[2], x[4], \dots\}. \tag{10.92}$$

The $\mathbf{z}$-transforms $\mathbf{X}(\mathbf{z})$ of $x[n]$ and $\mathbf{Y}(\mathbf{z})$ of $y[n]$ are related by

$$\mathbf{Y}(\mathbf{z}) = \frac{1}{2}\,\mathbf{X}(\mathbf{z}^{1/2}) + \frac{1}{2}\,\mathbf{X}(-\mathbf{z}^{1/2}). \tag{10.93}$$

Downsampling by 2

Upsampling a signal $x[n]$ by 2 means inserting zeros between successive values of $x[n]$. The result of upsampling by 2 is

$$x[n] \rightarrow \boxed{\uparrow 2} \rightarrow y[n] = \begin{cases} x[n/2] & \text{for } n \text{ even} \\ 0 & \text{for } n \text{ odd}, \end{cases} \tag{10.94a}$$

$$y[n] = \{\dots, \underline{x[0]}, 0, x[1], 0, x[2], 0, x[3], \dots\}. \tag{10.94b}$$

The $\mathbf{z}$-transforms $\mathbf{X}(\mathbf{z})$ of $x[n]$ and $\mathbf{Y}(\mathbf{z})$ of $y[n]$ are related by

$$\mathbf{Y}(\mathbf{z}) = \mathbf{X}(\mathbf{z}^2). \tag{10.95}$$

Upsampling by 2

10-11.2 Derivation of Perfect Reconstruction Condition Using z-Transforms

The diagram in Fig. 10-23 represents one-stage decomposition and reconstruction processes involving scaling and wavelet functions $g[n]$ and $h[n]$ belonging to any member of the wavelet transform family. In view of the downsampling property given by Eq. (10.93), the $\mathbf{z}$-transforms $\mathbf{X}_{\text{HD}}(\mathbf{z})$ of $x_{\text{HD}}[n]$ and $\mathbf{X}_{\text{LD}}(\mathbf{z})$ of $x_{\text{LD}}[n]$ are given by

$$\mathbf{X}_{\text{HD}}(\mathbf{z}) = \frac{1}{2}\,[\mathbf{H}(\mathbf{z}^{1/2})\,\mathbf{X}_{\text{in}}(\mathbf{z}^{1/2}) + \mathbf{H}(-\mathbf{z}^{1/2})\,\mathbf{X}_{\text{in}}(-\mathbf{z}^{1/2})], \tag{10.96a}$$

$$\mathbf{X}_{\text{LD}}(\mathbf{z}) = \frac{1}{2}\,[\mathbf{G}(\mathbf{z}^{1/2})\,\mathbf{X}_{\text{in}}(\mathbf{z}^{1/2}) + \mathbf{G}(-\mathbf{z}^{1/2})\,\mathbf{X}_{\text{in}}(-\mathbf{z}^{1/2})]. \tag{10.96b}$$

Here, $\mathbf{X}_{\text{in}}(\mathbf{z})$ is the $\mathbf{z}$-transform of the input $x_{\text{in}}[n]$, and $\mathbf{X}_{\text{HD}}(\mathbf{z})$ and $\mathbf{X}_{\text{LD}}(\mathbf{z})$ are the outputs of the decomposition process. For the reconstruction process of Fig. 10-23, use of the upsampling relationship given by Eq. (10.95), in combination with the sampling property of the $\mathbf{z}$-transform (property #8 in Table 7-6), leads to:

$$\mathbf{X}_{\text{out}}(\mathbf{z}) = \mathbf{H}(1/\mathbf{z})\,\mathbf{X}_{\text{HD}}(\mathbf{z}^2) + \mathbf{G}(1/\mathbf{z})\,\mathbf{X}_{\text{LD}}(\mathbf{z}^2). \tag{10.97}$$

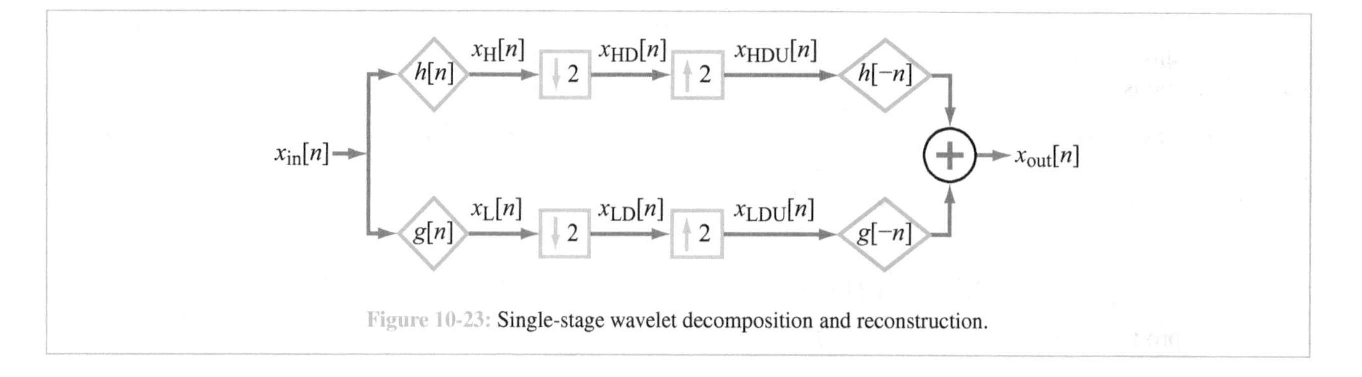

Figure 10-23: Single-stage wavelet decomposition and reconstruction.

Substituting Eq. (10.96) into Eq. (10.97) gives

$$\mathbf{X}_{\text{out}}(\mathbf{z}) = \frac{1}{2}\,\mathbf{H}(1/\mathbf{z})\,[\mathbf{H}(\mathbf{z})\,\mathbf{X}_{\text{in}}(\mathbf{z}) + \mathbf{H}(-\mathbf{z})\,\mathbf{X}_{\text{in}}(-\mathbf{z})] + \frac{1}{2}\,\mathbf{G}(1/\mathbf{z})\,[\mathbf{G}(\mathbf{z})\,\mathbf{X}_{\text{in}}(\mathbf{z}) + \mathbf{G}(-\mathbf{z})\,\mathbf{X}_{\text{in}}(-\mathbf{z})]. \tag{10.98}$$

Perfect reconstruction means that $\mathbf{X}_{\text{out}}(\mathbf{z}) = \mathbf{X}_{\text{in}}(\mathbf{z})$. Setting $\mathbf{X}_{\text{in}} = \mathbf{X}_{\text{out}} = \mathbf{X}$ in Eq. (10.98) and then rearranging terms leads to

$$\left\{1 - \tfrac{1}{2}\,[\mathbf{H}(1/\mathbf{z})\,\mathbf{H}(\mathbf{z}) + \mathbf{G}(1/\mathbf{z})\,\mathbf{G}(\mathbf{z})]\right\}\,\mathbf{X}(\mathbf{z}) + \tfrac{1}{2}\,[\mathbf{H}(1/\mathbf{z})\,\mathbf{H}(-\mathbf{z}) + \mathbf{G}(1/\mathbf{z})\,\mathbf{G}(-\mathbf{z})]\,\mathbf{X}(-\mathbf{z}) = 0. \tag{10.99}$$

To satisfy this equation for any $\mathbf{X}(\mathbf{z})$, it is necessary to satisfy two conditions:

(a) The coefficient of $\mathbf{X}(\mathbf{z})$ must be zero, which requires

$$\mathbf{H}(1/\mathbf{z})\,\mathbf{H}(\mathbf{z}) + \mathbf{G}(1/\mathbf{z})\,\mathbf{G}(\mathbf{z}) = 2. \tag{10.100}$$

(b) The coefficient of $\mathbf{X}(-\mathbf{z})$ in Eq. (10.98) must also be zero. Thus, we need to satisfy the condition

$$\mathbf{H}(1/\mathbf{z})\,\mathbf{H}(-\mathbf{z}) + \mathbf{G}(1/\mathbf{z})\,\mathbf{G}(-\mathbf{z}) = 0. \tag{10.101}$$

To guarantee perfect reconstruction, both Eq. (10.100) and Eq. (10.101) must be satisfied by $\mathbf{G}(\mathbf{z})$ and $\mathbf{H}(\mathbf{z})$. The two conditions can be converted into a specific condition on $\mathbf{G}(\mathbf{z})$ alone, and a second condition on $\mathbf{H}(\mathbf{z})$ in terms of $\mathbf{G}(\mathbf{z})$. Several solution methods are available, two of which will be pursued in the present and succeeding subsections.

The condition given by Eq. (10.101) can be satisfied by setting

$$\mathbf{H}(\mathbf{z}) = -\mathbf{G}(-1/\mathbf{z})\,\mathbf{z}^{-L}, \tag{10.102}$$

where L is any odd integer. Replacing $\mathbf{z}$ with $1/\mathbf{z}$ gives

$$\mathbf{H}(1/\mathbf{z}) = -\mathbf{G}(-\mathbf{z})\,\mathbf{z}^{L}, \tag{10.103}$$

and replacing $\mathbf{z}$ with $-\mathbf{z}$ in Eq. (10.102) gives

$$\mathbf{H}(-\mathbf{z}) = -\mathbf{G}(1/\mathbf{z})\,(-\mathbf{z})^{-L}. \tag{10.104}$$

Since L is odd, $(-\mathbf{z})^{-L} = -\mathbf{z}^{-L}$. Hence,

$$\mathbf{H}(-\mathbf{z}) = \mathbf{G}(1/\mathbf{z})\,\mathbf{z}^{-L}. \tag{10.105}$$

Using Eqs. (10.103) and (10.105), the product constituting the first term in Eq. (10.101) becomes

$$\mathbf{H}(1/\mathbf{z})\,\mathbf{H}(-\mathbf{z}) = [-\mathbf{G}(-\mathbf{z})\,\mathbf{z}^{L} \cdot \mathbf{G}(1/\mathbf{z})\,\mathbf{z}^{-L}] = -\mathbf{G}(-\mathbf{z})\,\mathbf{G}(1/\mathbf{z}), \tag{10.106}$$

which satisfies Eq. (10.101).

Having demonstrated that the relationship between $\mathbf{H}(\mathbf{z})$ and $\mathbf{G}(\mathbf{z})$ given by Eq. (10.102) does indeed satisfy the second of the two conditions, we now use Eqs. (10.102) and (10.103) in Eq. (10.100) in order to obtain an expression for $\mathbf{G}(\mathbf{z})$ alone:

$$[-\mathbf{G}(-\mathbf{z})\,\mathbf{z}^{L}][-\mathbf{G}(-1/\mathbf{z})\,\mathbf{z}^{-L}] + \mathbf{G}(1/\mathbf{z})\,\mathbf{G}(\mathbf{z}) = 2. \tag{10.107}$$

Simplification of Eq. (10.107) reduces to

$$\mathbf{G}(-\mathbf{z})\,\mathbf{G}(-1/\mathbf{z}) + \mathbf{G}(1/\mathbf{z})\,\mathbf{G}(\mathbf{z}) = 2, \tag{10.108}$$

which is known as the ***Smith-Barnwell condition*** for perfect reconstruction. Any $g[n]$ with a $\mathbf{z}$-transform $\mathbf{G}(\mathbf{z})$ that satisfies Eq. (10.108) will result in perfect reconstruction of $x[n]$ at the output of Fig. 10-23 if $\mathbf{H}(\mathbf{z})$ is determined from $\mathbf{G}(\mathbf{z})$ using Eq. (10.102).

Once $g[n]$ has been specified, $h[n]$ can be obtained by implementing the inverse $\mathbf{z}$-transform to Eq. (10.102). Using properties #3, 5, and 8 in Table 7-6, we can easily show that

$$h[n] = (-1)^n\,g[L-n]. \tag{10.109}$$

By selecting the odd integer L in Eq. (10.102) to be the same as the order of the scaling function $g[n]$—which means that $g[n]$ is of duration $L+1$—then $g[L-n]$ is causal, and so is $h[n]$.

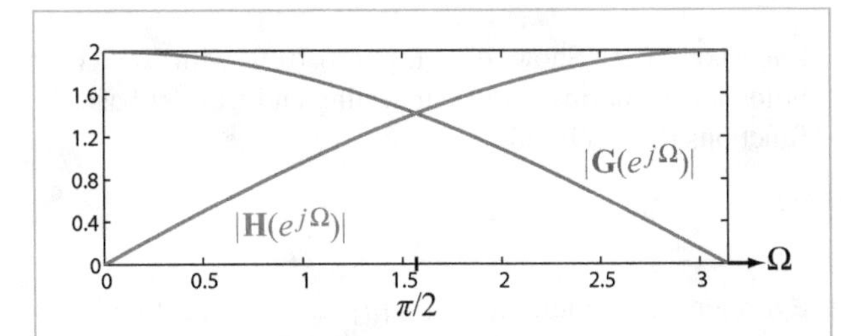

Figure 10-24: $|\mathbf{G}(e^{j\Omega})|$ and $|\mathbf{H}(e^{j\Omega})|$ for the Haar transform. This QMF Pair has symmetry about the $\Omega = \pi/2$ axis.

10-11.3 Perfect Reconstruction Condition Using the DTFT

The conditions for perfect reconstruction, given by Eqs. (10.102) and (10.108), can also be derived in the Ω-domain by setting $\mathbf{z} = e^{j\Omega}$ and repeating all of the steps given in the preceding subsection. Alternatively, we can convert Eqs. (10.102) and (10.108) into the Ω-domain by noting that

$$-\mathbf{z} = -e^{j\Omega} = e^{\pm j\pi} e^{j\Omega} = e^{j(\Omega \pm \pi)}, \tag{10.110a}$$

$$\frac{1}{\mathbf{z}} = e^{-j\Omega}, \tag{10.110b}$$

and

$$-\frac{1}{\mathbf{z}} = e^{-j(\Omega \pm \pi)}. \tag{10.110c}$$

Transformation of Eqs. (10.102) and (10.108) to the Ω-domain gives

$$\mathbf{H}(e^{j\Omega}) = -\mathbf{G}(e^{-j(\Omega \pm \pi)})\, e^{-j\Omega L} \tag{10.111}$$

and

$$\mathbf{G}(e^{j(\Omega \pm \pi)})\, \mathbf{G}(e^{-j(\Omega \pm \pi)}) + \mathbf{G}(e^{-j\Omega})\, \mathbf{G}(e^{j\Omega}) = 2, \tag{10.112}$$

which is equivalent to

$$|\mathbf{G}(e^{j(\Omega \pm \pi)})|^2 + |\mathbf{G}(e^{j\Omega})|^2 = 2. \tag{10.113}$$

Filters $\mathbf{G}(e^{j\Omega})$ and $\mathbf{H}(e^{j\Omega})$ are like complementary filters; if $\mathbf{G}(e^{j\Omega})$ is a lowpass filter, then $\mathbf{H}(e^{j\Omega})$ is a highpass filter, because its frequency response is shifted by π and the DTFT is periodic in Ω with period 2π. To illustrate, we show in Fig. 10-24 the behavior of $|\mathbf{G}_{\text{haar}}(e^{j\Omega})|$ and $|\mathbf{H}_{\text{haar}}(e^{j\Omega})|$ of the Haar wavelet transform given by Eq. (10.57). The two spectra are mirror images of one another with respect to the $\Omega = \pi/2$ axis. Hence, $\mathbf{G}(e^{j\Omega})$ and $\mathbf{H}(e^{j\Omega})$ are called a pair of ***quadrature mirror filters*** (***QMF***).

The relationship between $\mathbf{G}(e^{j\Omega})$ and $\mathbf{H}(e^{j\Omega})$ given by Eq. (10.111) in the Ω-domain is equivalent to the relationship between $h[n]$ and $g[n]$ given by Eq. (10.109) in the discrete-time domain.

10-11.4 Orthogonality of Single-Stage Basis Functions

In the preceding subsection,, we showed that if perfect reconstruction occurs, then the wavelet transform can be written as an expansion in basis functions, as in Eq. (10.82), with the wavelet transform computed using Eq. (10.81). The basis functions were given by Eq. (10.83) as

$$\begin{aligned} \phi_{m,k,g}[n] &= g^{(k)}[2^k m - n], \\ \phi_{m,k,h}[n] &= h^{(k)}[2^k m - n]. \end{aligned} \tag{10.114}$$

We did not show formally that the set of basis functions $\{\phi_{m,k,g}[n], \phi_{m,k,h}[n]\}$ are orthonormal. Now that we have derived the Smith-Barnwell condition for perfect reconstruction, we can quickly show the orthonormality of these basis functions for $k = 1$. The extension to larger values of k can be performed using induction, but it is too lengthy to give here.

For $k = 1$, Eq. (10.114) becomes

$$\begin{aligned} \phi_{m,1,g}[n] &= g[2m - n], \\ \phi_{m,1,h}[n] &= h[2m - n]. \end{aligned} \tag{10.115}$$

The basis functions $\{\phi_{m,1,g}[n]\}$ are orthonormal if, for any m_1 and m_2,

$$\begin{aligned} &\sum_{n=-\infty}^{\infty} \phi_{m_1,1,g}[n]\, \phi_{m_2,1,g}[n] = \\ &\sum_{n=-\infty}^{\infty} g[2m_1 - n]\, g[2m_2 - n] = \delta[m_1 - m_2]. \end{aligned} \tag{10.116}$$

Changing variables from n to $n' = 2m_2 - n$ gives

$$\begin{aligned} &\sum_{n=-\infty}^{\infty} g[2m_1 - n]\, g[2m_2 - n] \\ &= \sum_{n'=-\infty}^{\infty} g[2m_1 - 2m_2 + n']\, g[n'] \\ &= r_g[2(m_1 - m_2)], \end{aligned} \tag{10.117}$$

where $r_g[n]$ is the autocorrelation of $g[n]$, which was defined in Section 9-12.1 as

$$r_g[n] = g[n] * g[-n] = \sum_{i=-\infty}^{\infty} g[n+i]\, g[i]. \tag{10.118}$$

Hence, the basis functions $\{\phi_{m,1,g}[n]\}$ are orthonormal if $r_g[2(m_1 - m_2)] = \delta[m_1 - m_2]$.

Using the time-reversal property of the **z**-transform (entry #8 of Table 7-6), the **z**-transform of $r_g[n]$ is

$$\mathbf{R_g}(\mathbf{z}) = \mathbf{G}(\mathbf{z})\, \mathbf{G}(1/\mathbf{z}). \tag{10.119}$$

Consequently, the Smith-Barnwell condition for perfect reconstruction, Eq. (10.108), which we repeat here as

$$\mathbf{G}(-\mathbf{z})\, \mathbf{G}(-1/\mathbf{z}) + \mathbf{G}(\mathbf{z})\, \mathbf{G}(1/\mathbf{z}) = 2. \tag{10.120}$$

is actually equivalent to

$$\mathbf{R_g}(-\mathbf{z}) + \mathbf{R_g}(\mathbf{z}) = 2. \tag{10.121}$$

In the derivation of the expression given by Eq. (10.93) for the **z**-transform of a downsampled signal, we noted that for any polynomial $\mathbf{R_g}(\mathbf{z})$, the sum $\mathbf{R_g}(\mathbf{z}) + \mathbf{R_g}(-\mathbf{z})$ is a polynomial whose coefficients of odd powers of **z** are zero. Consequently, in view of Eq. (10.94b), the inverse **z**-transform of the Smith-Barnwell condition of Eq. (10.108) is that $r_g[n]$ be equal to 0 for even and nonzero n.

However, the basis functions $\{\phi_{m,1,g}[n]\}$ are orthonormal if $r_g[2(m_1 - m_2)] = \delta[m_1 - m_2]$; i.e., $r_g[n] = 0$ for even and nonzero n. This is precisely the Smith-Barnwell condition, so the basis functions $\{\phi_{m,1,g}[n]\}$ are orthonormal if the Smith-Barnwell condition is satisfied.

Upon repeating the above argument after replacing $g[n]$ with $h[n]$, we can show that the basis functions $\{\phi_{m,1,h}[n]\}$ are orthonormal if $h[n]$ is determined from $g[n]$ using Eq. (10.109) and $g[n]$ satisfies the Smith-Barnwell condition (Exercise 10-8).

Concept Question 10-12: What is the Smith-Barnwell condition for? (See S^2)

Exercise 10-6: Show that the normalized Haar scaling function $\tilde{g}_{\text{haar}}[n]$ in Eq. (10.68) satisfies the Smith-Barnwell condition given by Eq. (10.113).

Answer: See S^2.

Exercise 10-7: Show that Eq. (10.109) with $L = 1$ holds for the normalized Haar scaling and wavelet basis functions in Eq. (10.68).

Answer: See S^2.

Exercise 10-8: Show that if $g[n]$ satisfies the Smith-Barnwell condition and $h[n]$ is determined from $g[n]$ using Eq. (10.109), then $h[n]$ satisfies the Smith-Barnwell condition.

Answer: See S^2.

10-12 Daubechies Scaling and Wavelet Functions

A commonly used wavelet transform relies on a pair of functions known as the ***dbL Daubechies scaling and wavelet functions***, $g_{\text{dbL}}[n]$, and $h_{\text{dbL}}[n]$, respectively, attributed to Ingrid Daubechies, who also made many other wavelet-related contributions. The scaling function $g_{\text{dbL}}[n]$ is characterized by a **z**-transform $\mathbf{G}_{\text{dbL}}(\mathbf{z})$ that has L zeros at $\mathbf{z} = -1$, in addition to satisfying the Smith-Barnwell condition given by Eq. (10.108). The corresponding wavelet function $h_{\text{dbL}}[n]$, obtained through Eq. (10.109), is characterized by a **z**-transform $\mathbf{H}_{\text{dbL}}(\mathbf{z})$ that has L zeros at $\mathbf{z} = 1$. This latter property is responsible for the success of the *dbL* Daubechies wavelet functions in ***sparsifying*** many real-world signals and images. Sparsifying a signal means making most of its members zero-valued. More details are forthcoming in the next subsection.

10-12.1 H(z) with Multiple Zeros at z = 1

(a) Definition of piecewise-polynomial signals

A signal $x[n]$ is defined to be ***piecewise-L^{th}-degree polynomial*** if it has the form

$$x[n] = \begin{cases} \sum_{k=0}^{L} a_{0,k}\, n^k & -\infty < n \le N_0, \\ \sum_{k=0}^{L} a_{1,k}\, n^k & N_0 < n \le N_1, \\ \sum_{k=0}^{L} a_{2,k}\, n^k & N_1 < n \le N_2, \\ \vdots \end{cases} \tag{10.122}$$

Signal $x[n]$ is segmented into intervals, and in each interval $x[n]$ is a polynomial in time n and of degree L. The times N_i at which the coefficients $\{a_{i,k}\}$ change values are sparse, meaning that they are scattered over time n. In continuous time, such a signal would be a ***spline***, except that for a spline the derivatives of the signal must match at the *knots* (the times where the coefficients $\{a_{i,k}\}$ change values). The idea here is that the coefficients $\{a_{i,k}\}$ can change completely at the times N_i; there is no *smoothness* requirement. Indeed, these times N_i are the edges of $x[n]$.

We first consider piecewise-constant, then piecewise-linear, then piecewise-polynomial signals, and in each case we examine the effect of zeros of $\mathbf{H}(\mathbf{z})$ at $\mathbf{z} = 1$.

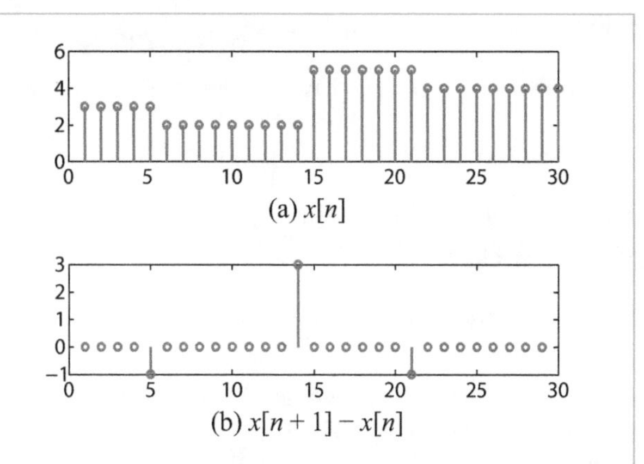

Figure 10-25: A system with a zero at $\mathbf{z} = 1$ compresses piecewise-constant signals: (a) A piecewise-constant signal, (b) signal differences.

(b) Piecewise-constant signals

If a transfer function $\mathbf{H}(\mathbf{z})$ has a single zero at $\mathbf{z} = 1$, it can be expressed in the form

$$\mathbf{H}(\mathbf{z}) = (\mathbf{z} - 1)\, \mathbf{Q}(\mathbf{z}), \tag{10.123}$$

where $\mathbf{Q}(\mathbf{z})$ is a rational function. The inverse $\mathbf{z}$-transform of $(\mathbf{z} - 1)$ is $\{1, \underline{-1}\}$. The overall impulse response of LTI systems connected in series is the convolution of their impulse responses (see Section 2-5.2). Hence, the system with transfer function $\mathbf{H}(\mathbf{z})$ can be implemented by two systems connected in series:

$$x[n] \to \boxed{w_1[n] = x[n+1] - x[n]} \to w_1[n] \to \boxed{q[n]} \to y[n], \tag{10.124}$$

where $q[n]$ is the inverse $\mathbf{z}$-transform of $\mathbf{Q}(\mathbf{z})$.

Now let $x[n]$ be piecewise constant, meaning that $x[n]$ has the form

$$x[n] = \begin{cases} a_0 & -\infty < n \leq N_0, \\ a_1 & N_0 < n \leq N_1, \\ a_2 & N_1 < n \leq N_2, \\ \vdots & \end{cases} \tag{10.125}$$

The value of $x[n]$ changes only at a few scattered times. The amount by which $x[n]$ changes between $n = N_i$ and $n = N_i + 1$ is the jump $a_{i+1} - a_i$:

$$x[n+1] - x[n] = \begin{cases} 0 & \text{for } n \neq N_i, \\ a_{i+1} - a_i & \text{for } n = N_i. \end{cases} \tag{10.126}$$

This can be restated as

$$x[n+1] - x[n] = \sum_i (a_{i+1} - a_i)\, \delta[n - N_i], \tag{10.127}$$

and is illustrated in Fig. 10-25, where we observe that $x[n+1] - x[n]$ is zero-valued, except at junctions between adjacent segments. This property causes the wavelet transform of $x[n]$ to be sparsified as well. Consider the convolution

$$\begin{aligned} x[n] * h[n] &= x[n] * \{1, \underline{-1}\} * q[n] \\ &= (x[n+1] - x[n]) * q[n] \\ &= \sum_i (a_{i+1} - a_i)\, \delta[n - N_i] * q[n] \\ &= \sum_i (a_{i+1} - a_i)\, q[n - N_i]. \end{aligned} \tag{10.128}$$

In the final step, we used the time-shift property of convolution (entry #5 in Table 7-4). In practice, $q[n]$ is a very short (length 3 or 4) FIR filter, so $x[n] * h[n]$ is still mostly zero-valued.

The normalized Haar wavelet function

$$\tilde{h}_{\text{haar}}[n] = \frac{1}{\sqrt{2}}\, \{\underline{1}, -1\}$$

has transfer function

$$\tilde{H}_{\text{haar}}(\mathbf{z}) = \frac{\mathbf{z} - 1}{\sqrt{2}\,\mathbf{z}},$$

which has a zero at $\mathbf{z} = 1$. So the Haar wavelet transform compresses piecewise constant signals. For the Haar wavelet function, $\mathbf{Q}(\mathbf{z}) = 1/(\sqrt{2}\,\mathbf{z})$.

(c) Piecewise-linear signals

A transfer function $\mathbf{H}(\mathbf{z})$ with two zeros at $\mathbf{z} = 1$ can be written as:

$$\mathbf{H}(\mathbf{z}) = (\mathbf{z} - 1)^2\, \mathbf{Q}(\mathbf{z}) \tag{10.129}$$

for some rational function $\mathbf{Q}(\mathbf{z})$. The inverse $\mathbf{z}$-transform of $(\mathbf{z} - 1)$ is $\{1, \underline{-1}\}$, so the inverse $\mathbf{z}$-transform of $(\mathbf{z} - 1)^2$ is

$$\mathbb{Z}^{-1}\{(\mathbf{z} - 1)^2\} = \{1, \underline{-1}\} * \{1, \underline{-1}\}. \tag{10.130}$$

The overall impulse response of LTI systems connected in series is the convolution of their impulse responses (see Section 2-5.2). So the system with transfer function $\mathbf{H}(\mathbf{z})$ can be implemented by three systems connected in series:

$x[n] \longrightarrow$ [$w_1[n] = x[n+1] - x[n]$] $\longrightarrow w_1[n]$
$\longrightarrow$ [$w_2[n] = w_1[n+1] - w_1[n]$] $\longrightarrow w_2[n]$
$\longrightarrow$ [$q[n]$] $\longrightarrow y[n]$,

(10.131)

where $q[n]$ is the inverse $\mathbf{z}$-transform of $\mathbf{Q}(\mathbf{z})$.

A piecewise linear signal $x[n]$ has the form

$$x[n] = \begin{cases} a_{0,1}\, n + a_{0,0} & -\infty < n \le N_0, \\ a_{1,1}\, n + a_{1,0} & N_0 < n \le N_1, \\ a_{2,1}\, n + a_{2,0} & N_1 < n \le N_2, \\ \vdots & \end{cases} \tag{10.132}$$

Proceeding as we did earlier for the case of a piecewise-constant $x[n]$, taking differences, and then taking differences of the differences, ends up sparsifying the piecewise-linear signal $x[n]$. The process is illustrated in Fig. 10-26. The bottom signal in Fig. 10-26 is in turn convolved with $q[n]$, resulting in a series of scaled and delayed values of $q[n]$. As noted earlier, $q[n]$ usually is a very short FIR filter, so $x[n] * h[n]$ is still mostly zero-valued.

(d) Piecewise-polynomial signals

In the preceding parts of this subsection, we used a transfer function $\mathbf{H}(\mathbf{z})$ with a single zero at $\mathbf{z} = 1$ to compress a piecewise-constant signal, and another with 2 zeros to compress a piecewise-linear signal. In both cases, the objective is to compress the signal to a highly sparsified function $y[n]$. Now, we extend our examination to the general case of a piecewise-polynomial signal.

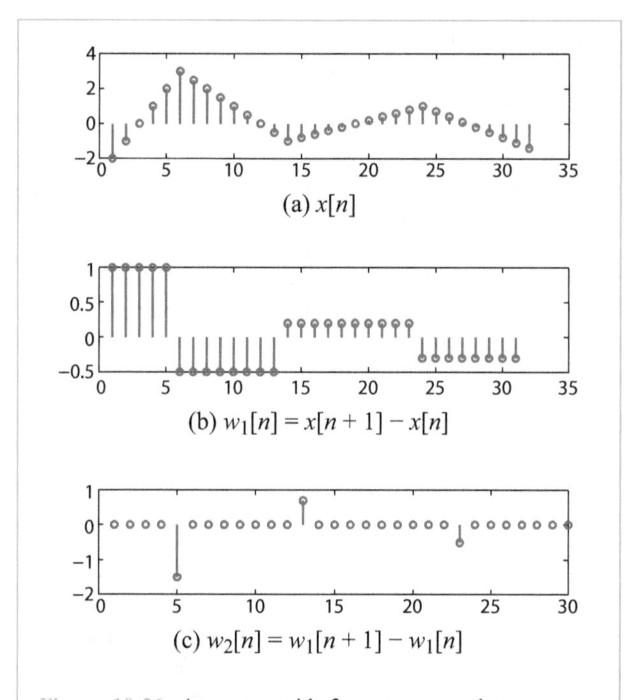

Figure 10-26: A system with 2 zeros at $\mathbf{z} = 1$ compresses piecewise-linear signals. (a) Signal $x[n]$, (b) differences of signal $x[n]$, (c) differences of the middle signal.

Let us consider a signal $x[n]$ given by a polynomial of degree $L - 1$:

$$x[n] = a_0\, n^{L-1} + a_1\, n^{L-2} + \cdots + a_{L-1}, \tag{10.133}$$

where a_0 through a_{L-1} are constant coefficients. To sparsify the signal, we need to use a transfer function $\mathbf{H}(\mathbf{z})$ that has L zeros at $\mathbf{z} = 1$, which means that we can express it in the form

$$\mathbf{H}(\mathbf{z}) = (\mathbf{z} - 1)^L\, \mathbf{Q}(\mathbf{z}) \tag{10.134}$$

for some rational function $\mathbf{Q}(\mathbf{z})$. The system can be regarded as a series connection of the $L + 1$ systems:

$x[n] \longrightarrow$ [$w_1[n] = x[n+1] - x[n]$] $\longrightarrow w_1[n]$
$\longrightarrow$ [$w_2[n] = w_1[n+1] - w_1[n]$] $\longrightarrow w_2[n]$
$\longrightarrow$ [$w_3[n] = w_2[n+1] - w_2[n]$] $\longrightarrow w_3[n] \ldots w_L[n]$
$\longrightarrow$ [$q[n]$] $\longrightarrow y[n]$,

(10.135)

where $q[n]$ is the inverse $\mathbf{z}$-transform of $\mathbf{Q}(\mathbf{z})$.

The process in Eq. (10.135) includes L *forward difference operators*, labeled $w_1[n]$ through $w_L[n]$. The first difference operation yields

$$w_1[n] = x[n+1] - x[n]. \tag{10.136}$$

To compute $w_1[n]$, we need $x[n]$—which is given by Eq. (10.133)—and $x[n+1]$, which can be obtained from Eq. (10.133) by replacing n with $n+1$:

$$x[n+1] = a_0\ (n+1)^{L-1} + a_1\ (n+1)^{L-2} + \cdots + a_{L-1}. \tag{10.137}$$

Application of the binomial expansion to $(n+1)^{L-1}$ yields

$$(n+1)^{L-1} = n^{L-1} + (L-1)\ n^{L-2} + \text{terms of degree} < (L-2). \tag{10.138}$$

Use of the expressions given by Eqs. (10.133), (10.137), and (10.138) in Eq. (10.136) leads to

$$w_1[n] = (L-1)\ n^{L-2} + \text{terms of degree} < (L-2).$$

Hence, the difference operator $w_1[n]$ results in a polynomial of degree $L-2$, compared to $x[n]$ which is of degree $L-1$. Each additional difference-operator step reduces the degree of the output by 1, so after L such steps, the final output $w_L[n]$ will be a polynomial of degree zero. Consequently, signals that are piecewise-polynomial functions of time n, also of degree $L-1$, will be compressed to zero, except in the vicinity of the times at which the polynomial coefficients change. At these times, scaled and delayed versions of the impulse response $q[n] = \mathbb{Z}^{-1}\{\mathbf{Q(z)}\}$ will appear. Since in practice $q[n]$ is a short FIR filter of length 3 or 4, $y[n]$ will be mostly zero-valued.

The continuous-time analogue of the forward difference operator is the time derivative:

$$w(t) = \frac{d^L x}{dt^L}$$

for any $(L-1)$th degree piecewise-polynomial $x(t)$ defined over 1 or more time segments. Each derivative step reduces the degree of the polynomial by 1, so at the end of L derivative steps, the polynomial is reduced to zero, except at the boundary between different time segments.

10-12.2 Computation of the dbL Daubechies Scaling Function

We now derive the dbL Daubechies scaling function $g[n]$, whose $\mathbf{z}$-transform has L zeros at $\mathbf{z} = -1$ and satisfies the Smith-Barnwell condition given by Eq. (10.108) for perfect reconstruction. The dbL Daubechies wavelet function compresses piecewise-polynomial signals of degree $L-1$ to zero, except near the times n at which the polynomial coefficients change. Some authors use the nomenclature that the $D2L$ Daubechies wavelet function has L zeros, so that the "DL Daubechies wavelet" function is undefined for L odd. This seems wasteful, so we use dbL to denote L zeros.

From Eq. (10.102), the $\mathbf{z}$-transform $\mathbf{H(z)}$ of the wavelet function $h[n]$ has L zeros at $\mathbf{z} = 1$ if the $\mathbf{z}$-transform $\mathbf{G(z)}$ of the scaling function $g[n]$ has L zeros at $\mathbf{z} = -1$.

Let $g[n]$ be the dbL Daubechies scaling function. For its $\mathbf{z}$-transform $\mathbf{G(z)}$ to have L zeros at $\mathbf{z} = -1$, and for $g[n]$ to be causal, $\mathbf{G(z)}$ must have the form

$$\mathbf{G(z)} = \frac{(\mathbf{z}+1)^L}{\mathbf{z}^L}\ \mathbf{Q(z)} = (1+\mathbf{z}^{-1})^L\ \mathbf{Q(z)}, \tag{10.139}$$

for some polynomial $\mathbf{Q(z)}$ in $\mathbf{z}^{-1}$ (so that $q[n]$ and $g[n]$ are causal) of degree $L-1$. After specifying the scaling function design procedure, it will become evident why $\mathbf{Q(z)}$ should have degree $L-1$.

The $\mathbf{z}$-transform $\mathbf{G(z)}$ of the scaling function must also satisfy the Smith-Barnwell condition given by Eq. (10.108):

$$\mathbf{G(-z)}\ \mathbf{G(-z^{-1})} + \mathbf{G(z)}\ \mathbf{G(z^{-1})} = 2. \tag{10.140}$$

Let $r[n]$ be the autocorrelation of $q[n]$:

$$r[n] = r[-n] = q[n] * q[-n]. \tag{10.141}$$

The $\mathbf{z}$-transform $\mathbf{R(z)}$ of $r[n]$ is then

$$\mathbf{R(z)} = \mathbf{R(1/z)} = \mathbf{Q(z)}\ \mathbf{Q(1/z)}. \tag{10.142}$$

Substituting Eq. (10.139) and Eq. (10.142) in the Smith-Barnwell condition given by Eq. (10.140) gives

$$(1-\mathbf{z})^L(1-\mathbf{z}^{-1})^L\ \mathbf{R(-z)} + (1+\mathbf{z})^L(1+\mathbf{z}^{-1})^L\ \mathbf{R(z)} = 2. \tag{10.143}$$

This is a linear system of equations of size L in the $\mathbf{z}$-transform of the unknown function $r[n]$. Its solution determines $\mathbf{R(z)}$.

Now that the complete procedure has been specified, we examine why $\mathbf{Q(z)}$ should have degree $L-1$. We start by supposing that $\mathbf{Q(z)}$ has degree $L-1$, in which case:

- From Eq. (10.139), $\mathbf{G(z)}$ has degree $2L-1$.

- From Eq. (10.142), $\mathbf{R}(\mathbf{z})$ has degree $2L-2$.
- In Eq. (10.143), each term has degree $4L-2$.
- In Eq. (10.143), coefficients of $(-\mathbf{z})^m$ are identical to those for $\mathbf{z}^m$. These terms cancel for odd m. So the coefficients of odd powers of $\mathbf{z}$ cancel.
- In Eq. (10.143), equations for negative powers of $\mathbf{z}$ are identical to those for positive powers of $\mathbf{z}$. This reduces by half the number of independent equations.
- Equation (10.143) is L equations in L unknowns $r[n]$. Hence, if $\mathbf{Q}(\mathbf{z})$ is of degree $L-1$, as supposed earlier, then there are just enough unknowns to satisfy the linear equations in the inverse $\mathbf{z}$-transform of Eq. (10.143), as illustrated further in the next subsection.

Next, we show how to compute $\mathbf{Q}(\mathbf{z})$ from $\mathbf{R}(\mathbf{z})$. From Eq. (10.142), $\mathbf{R}(\mathbf{z})=\mathbf{R}(1/\mathbf{z})$, so if $\mathbf{z}_0$ is a zero of $\mathbf{R}(\mathbf{z})$ then $1/\mathbf{z}_0$ is also a zero of $\mathbf{R}(\mathbf{z})$. Since $r[n]$ is real-valued, the zeros also occur in complex conjugate pairs: $\mathbf{z}_0^*$ is also a zero. In conclusion, the zeros of $\mathbf{R}(\mathbf{z})$ occur in *conjugate reciprocal quadruples*, each of which has the form, for some zero $\mathbf{z}_0$,

$$\{\,\mathbf{z}_0, \mathbf{z}_0^*, 1/\mathbf{z}_0, 1/\mathbf{z}_0^*\,\}.$$

$\mathbf{Q}(\mathbf{z})$ is then computed from $\mathbf{R}(\mathbf{z})$ by performing a *spectral factorization* of $\mathbf{R}(\mathbf{z})$, as follows:

- Zeros of $\mathbf{R}(\mathbf{z})$ inside the unit circle ($|\mathbf{z}_0|<1$) are assigned to be the zeros of $\mathbf{Q}(\mathbf{z})$.
- Zeros of $\mathbf{R}(\mathbf{z})$ outside the unit circle ($|\mathbf{z}_0|>1$) are assigned to be the zeros of $\mathbf{Q}(1/\mathbf{z})$.
- In practice, there are no zeros of $\mathbf{R}(\mathbf{z})$ on the unit circle ($|\mathbf{z}_0|=1$).

This determines the minimum-phase $\mathbf{Q}(\mathbf{z})$, and hence $\mathbf{G}(\mathbf{z})$, to a scale factor determined by Eq. (10.143). The procedure is illustrated in Section 10-12.3.

10-12.3 Computation of the $D2$ Daubechies Scaling Function for $L=2$

From Eq. (10.141), $r[n]=r[-n]$. Hence, for $L=2$, $\mathbf{R}(\mathbf{z})$ can be written as

$$\mathbf{R}(\mathbf{z})=r[1]\,\mathbf{z}+r[0]+r[1]\,\mathbf{z}^{-1}, \tag{10.144}$$

in which case Eq. (10.143) becomes

$$\begin{aligned}&(1+\mathbf{z})^2(1+\mathbf{z}^{-1})^2(r[1]\,\mathbf{z}+r[0]+r[1]\,\mathbf{z}^{-1})\\&+(1-\mathbf{z})^2(1-\mathbf{z}^{-1})^2(-r[1]\,\mathbf{z}+r[0]-r[1]\,\mathbf{z}^{-1})=2.\end{aligned} \tag{10.145}$$

Expanding Eq. (10.145) gives

$$\begin{aligned}&(\mathbf{z}^2+4\mathbf{z}+6+4\mathbf{z}^{-1}+\mathbf{z}^{-2})\\&\quad\times(r[1]\mathbf{z}+r[0]+r[1]\,\mathbf{z}^{-1})\\&\quad+(\mathbf{z}^2-4\mathbf{z}+6-4\mathbf{z}^{-1}+\mathbf{z}^{-2})\\&\quad\times(-r[1]\mathbf{z}+r[0]-r[1]\,\mathbf{z}^{-1})=2.\end{aligned} \tag{10.146}$$

The two polynomials being added in Eq. (10.146) both have degree 6, since the powers of $\mathbf{z}$ in Eq. (10.146) vary from -3 to $+3$. However, the coefficients of odd powers of $\mathbf{z}$ cancel, and the coefficients of negative powers of $\mathbf{z}$ are identical to the coefficients of the corresponding positive powers of $\mathbf{z}$. So equating coefficients of powers of $\mathbf{z}$ in Eq. (10.146) yields two equations (the coefficients of $\mathbf{z}^1$ and $\mathbf{z}^3$) in two unknowns ($r[0]$ and $r[1]$). Equating coefficients of $\mathbf{z}^2$ and $\mathbf{z}^0$, respectively, gives the two equations

$$\begin{aligned}2r[0]+8r[1]&=0,\\12r[0]+16r[1]&=2,\end{aligned} \tag{10.147}$$

which jointly have the solution

$$r[0]=1/4,\qquad r[1]=-1/16. \tag{10.148}$$

Hence $\mathbf{R}(\mathbf{z})$ of Eq. (10.144) becomes

$$\mathbf{R}(\mathbf{z})=-(1/16)\mathbf{z}^{-1}+1/4-(1/16)\mathbf{z}. \tag{10.149}$$

The two zeros of $\mathbf{R}(\mathbf{z})$ (the roots of $\mathbf{R}(\mathbf{z})=0$) are

$$\mathbf{z}_0=\mathbf{z}_0^*=2-\sqrt{3}=0.2679, \tag{10.150a}$$

$$\frac{1}{\mathbf{z}_0}=\frac{1}{\mathbf{z}_0^*}=2+\sqrt{3}=3.7321. \tag{10.150b}$$

The complex conjugate quadruple condenses to just a pair, since both zeros are real-valued.

The spectral factorization of $\mathbf{R}(\mathbf{z})$ is performed by choosing $\mathbf{z}_0$ to be the zero of $\mathbf{Q}(\mathbf{z})$, since $|\mathbf{z}_0|<1$. Then, Eq. (10.139) becomes, for some constant C,

$$\begin{aligned}\mathbf{G}(\mathbf{z})&=C(1+\mathbf{z}^{-1})^2(1-0.2679\mathbf{z}^{-1})\\&=C(1+1.7321\mathbf{z}^{-1}+0.4642\mathbf{z}^{-2}-0.2679\mathbf{z}^{-3}).\end{aligned} \tag{10.151}$$

Table 10-1: **Coefficients of Daubechies scaling functions.**

$g[n]$	*db*1	*db*2	*db*3	*db*4
$g[0]$	0.7071	0.4830	0.3327	0.2304
$g[1]$	0.7071	0.8365	0.8069	0.7148
$g[2]$	0	0.2241	0.4599	0.6309
$g[3]$	0	−0.1294	−0.1350	−0.0280
$g[4]$	0	0	−0.0854	−0.1870
$g[5]$	0	0	0.0352	0.0308
$g[6]$	0	0	0	0.0329
$g[7]$	0	0	0	−0.0106

The **z**-transform of the scaling function $\mathbf{G(z)}$ has been determined to a scale factor C. The constant C is computed by inserting Eq. (10.151) into Eq. (10.140). This gives $C = 0.4830$. With this newly found value of C, inverse **z**-transforming Eq. (10.151) leads to the $D2$ Daubechies scaling function $g[n]$:

$$g[n] = \{\underline{0.4830}, 0.8365, 0.2242, -0.1294\}, \qquad (10.152)$$

whose duration is 4. From Eq. (10.109), the causal *db*2 Daubechies wavelet function $h[n]$ is

$$\begin{aligned} h[n] &= (-1)^n \, g[3-n] \\ &= \{\underline{-0.1294}, -0.2242, 0.8365, -0.4830\}. \qquad (10.153) \end{aligned}$$

Note that $g[n]$ and $h[n]$ are energy-normalized:

$$\sum_{n=0}^{3} g[n]^2 = \sum_{n=0}^{3} h[n]^2 = 1. \qquad (10.154)$$

This concludes the procedure for calculating $g[n]$ and $h[n]$ for L =2. Repeating the procedure for L =1, 3, and 4 leads to the coefficients of the *db*1, *db*2, *db*3, and *db*4 Daubechies scaling functions listed in Table 10-1. The corresponding wavelet functions can be obtained from these scaling functions using Eq. (10.109).

10-12.4 Amount of Computation

The total amount of computation required to compute the *dbL* Daubechies wavelet transform of a signal $x[n]$ of duration N can be determined as follows. The duration of the *dbL* Daubechies wavelet and scaling functions is $2L$. Convolving both of these with $x[n]$ requires $2(2L)N = 4LN$ ***multiplications-and-additions*** (***MADs***). But since the results will be downsampled by two, only half of the convolution outputs must be computed, reducing the number to $2LN$.

At each successive decomposition, these functions are convolved with the average signal from the previous stage. So these functions are convolved with signals with respective durations,

$$\{\underbrace{\widetilde{X}_1[n]}_{N/2}, \underbrace{\widetilde{X}_2[n]}_{N/4}, \dots, \underbrace{\widetilde{X}_K[n]}_{N/2^K}\},$$

where K is the number of decomposition stages. The total number of MADs required is therefore

$$2L\left(N + \frac{N}{2} + \frac{N}{4} + \cdots + \frac{N}{2^K}\right) < 4LN. \qquad (10.155)$$

The additional computation for computing more decompositions (i.e., increasing K) is minimal.

Since L is small, this is comparable to the amount of computation $\frac{N}{2}\log_2(N)$ required to compute the DFT, using the FFT, of a signal $x[n]$ of duration N. But the DFT requires complex additions and multiplications, while the wavelet transform uses only real additions and multiplications. So the savings are even greater than they first appear.

In summary, the *dbL* Daubechies scaling $g[n]$ and wavelet $h[n]$ functions have the following properties:

- Both $g[n]$ and $h[n]$ have durations $2L$.
- Lowpass $\mathbf{G(z)}$ has $L-1$ zeros at $\mathbf{z} = -1$.
- Highpass $\mathbf{H(z)}$ has $L-1$ zeros at $\mathbf{z} = 1$.
- Convolution with $h[n]$ sparsifies signals that are piecewise-$(L-1)$th-degree polynomial signals.
- Computation of the complete wavelet transform of $x[n]$ of duration N requires $4LN$ real MADs.

Given $g_{\text{dbL}}[n]$ and $h_{\text{dbL}}[n]$ for $L = 2$, namely Eqs. (10.152) and (10.153), let us apply them to compute the *db*2 Daubechies wavelet transform of the piecewise-linear signal shown in Fig. 10-27(a). The two-stage procedure involves the computation of average signals $\widetilde{X}_1[n]$ and $\widetilde{X}_2[n]$, and detail signals $\widetilde{x}_1[n]$ and $\widetilde{x}_2[n]$, using Eqs. (10.70), (10.72), and (10.74). The results are displayed in parts (b) to (e) of Fig. 10-27. We note that

- Average signals $\widetilde{X}_1[n]$ and $\widetilde{X}_2[n]$ are low-resolution versions of $x[n]$.
- Detail signals $\widetilde{x}_1[n]$ and $\widetilde{x}_2[n]$ are sparse (mostly zero), and their nonzero values are small in magnitude.

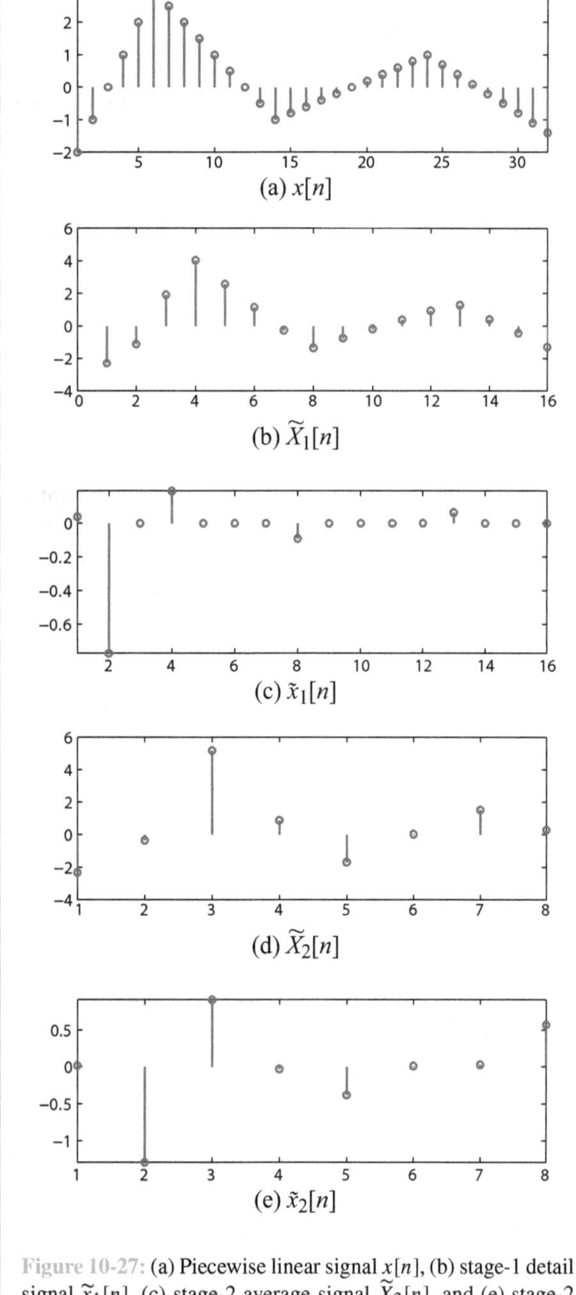

Figure 10-27: (a) Piecewise linear signal $x[n]$, (b) stage-1 detail signal $\tilde{x}_1[n]$, (c) stage-2 average signal $\widetilde{X}_2[n]$, and (e) stage-2 detail signal $\tilde{x}_2[n]$.

- The $db2$ wavelet transform of the given $x[n]$ consists of $\tilde{x}_1[n]$, $\tilde{x}_2[n]$, and $\widetilde{X}_2[n]$.

These patterns explain the terms "average" and "detail." The $db2$ wavelet transform of the given $x[n]$ consists of $\tilde{x}_1[n]$, $\tilde{x}_2[n]$, and $\widetilde{X}_2[n]$.

Concept Question 10-13: Why does the wavelet transform sparsify real-world signals and images? (See S²)

Concept Question 10-14: What attribute of Daubechies wavelet functions makes them sparsify piecewise-polynomial functions? (See S²)

Exercise 10-9: Show that $db1$ Daubechies scaling function $g[n]$ is the normalized Haar scaling function $\tilde{g}_{\text{haar}}[n]$.

Answer: See S².

Exercise 10-10: Show that $db1$ Daubechies scaling function $g[n]$ is orthogonal to even-valued translations of $g[n]$.

Answer: See S².

Exercise 10-11: Show that a system with two zeros at $\mathbf{z} = 1$ compresses signals linear in time n to zero.

Answer: See S².

10-13 2-D Wavelet Transform

The real power of the wavelet transform becomes apparent when it is applied to 2-D images. A 512×512 image has more than a quarter-million pixel values. Storing a sparse representation of an image, rather than the image itself, saves a huge amount of memory. Compressed sensing (covered later in Section 10-15) becomes very powerful when applied to images.

10-13.1 Image Analysis Filter Bank

The generalization of the wavelet transform from signals to images is straightforward if ***separable*** scaling and wavelet functions are used. In this book we restrict our attention to separable functions; i.e., the 2-D filter $g_2[m,n] = g[m]\, g[n]$. With $g[n]$ and $h[n]$ denoting scaling and wavelet functions, such as the Haar or Daubechies functions, the image analysis filter bank performs the following operations:

(1) **Stage-1 decomposition**

$$x[m,n] \rightarrow \boxed{g[m]\,g[n]} \rightarrow \boxed{\downarrow (2,2)} \rightarrow \tilde{x}_{\text{LL}}^{(1)}[m,n] \tag{10.156a}$$

$$x[m,n] \rightarrow \boxed{g[m]\,h[n]} \rightarrow \boxed{\downarrow (2,2)} \rightarrow \tilde{x}_{\text{LH}}^{(1)}[m,n] \tag{10.156b}$$

$$x[m,n] \rightarrow \boxed{h[m]\,g[n]} \rightarrow \boxed{\downarrow (2,2)} \rightarrow \tilde{x}_{\text{HL}}^{(1)}[m,n] \tag{10.156c}$$

$$x[m,n] \rightarrow \boxed{h[m]\,h[n]} \rightarrow \boxed{\downarrow (2,2)} \rightarrow \tilde{x}_{\text{HH}}^{(1)}[m,n] \tag{10.156d}$$

(2) **Stage-2 to stage-K decomposition**

$$\tilde{x}_{\text{LL}}^{(1)}[m,n] \rightarrow \boxed{g[m]\,g[n]} \rightarrow \boxed{\downarrow (2,2)} \rightarrow \tilde{x}_{\text{LL}}^{(2)}[m,n] \tag{10.157a}$$

$$\tilde{x}_{\text{LL}}^{(1)}[m,n] \rightarrow \boxed{g[m]\,h[n]} \rightarrow \boxed{\downarrow (2,2)} \rightarrow \tilde{x}_{\text{LH}}^{(2)}[m,n] \tag{10.157b}$$

$$\tilde{x}_{\text{LL}}^{(1)}[m,n] \rightarrow \boxed{h[m]\,g[n]} \rightarrow \boxed{\downarrow (2,2)} \rightarrow \tilde{x}_{\text{HL}}^{(2)}[m,n] \tag{10.157c}$$

$$\tilde{x}_{\text{LL}}^{(1)}[m,n] \rightarrow \boxed{h[m]\,h[n]} \rightarrow \boxed{\downarrow (2,2)} \rightarrow \tilde{x}_{\text{HH}}^{(2)}[m,n] \tag{10.157d}$$

The decomposition process is continued as above, until output $\tilde{x}_{\text{LL}}^{K}[m,n]$ is reached, where K is the total number of stages.

(3) **Final wavelet transform**

At the conclusion of the decomposition process, $x[m,n]$ consists of:

(a) The coarsest *average image* $\tilde{x}_{\text{LL}}^{(K)}[m,n]$.

(b) Three *detail images* at each stage:

$$\{\tilde{x}_{\text{LH}}^{(K-1)}[m,n],\ \tilde{x}_{\text{HL}}^{(K-1)}[m,n],\ \tilde{x}_{\text{HH}}^{(K-1)}[m,n]\}$$
$$\{\tilde{x}_{\text{LH}}^{(K-2)}[m,n],\ \tilde{x}_{\text{HL}}^{(K-2)}[m,n],\ \tilde{x}_{\text{HH}}^{(K-2)}[m,n]\},$$

up to the largest (in size) three detail images:

$$\{\tilde{x}_{\text{LH}}^{(1)}[m,n],\ \tilde{x}_{\text{HL}}^{(1)}[m,n],\ \tilde{x}_{\text{HH}}^{(1)}[m,n]\}.$$

The average images $\tilde{x}_{\text{LL}}^{(k)}[m,n]$ are analogous to the average signals $\tilde{X}_k[n]$ of a signal, except that they are low-resolution versions of a 2-D image $x[m,n]$ instead of a 1-D signal $x[n]$. In 2-D, there are now three detail images, while in 1-D there is only one detail signal.

In 1-D, the detail signals are zero except near edges, representing abrupt changes in the signal or in its slope. In 2-D, the three detail images play the following roles:

(a) $\tilde{x}_{\text{LH}}^{(k)}[m,n]$ picks up vertical edges,

(b) $\tilde{x}_{\text{HL}}^{(k)}[m,n]$ picks up horizontal edges,

(c) $\tilde{x}_{\text{HH}}^{(k)}[m,n]$ picks up diagonal edges.

10-13.2 Image Synthesis Filter Bank

The image synthesis filter bank combines all of the detail images, and the coarsest average image, $\tilde{x}_{\text{LL}}^{(K)}[m,n]$, into the original image $x[m,n]$, as follows:

$$\tilde{x}_{\text{LL}}^{(K)}[m,n] \rightarrow \boxed{\uparrow (2,2)} \rightarrow \boxed{g[-m]\,g[-n]} \rightarrow A_{\text{LL}}^{(K)}[m,n]$$

$$\tilde{x}_{\text{LH}}^{(K)}[m,n] \rightarrow \boxed{\uparrow (2,2)} \rightarrow \boxed{g[-m]\,h[-n]} \rightarrow A_{\text{LH}}^{(K)}[m,n]$$

$$\tilde{x}_{\text{HL}}^{(K)}[m,n] \rightarrow \boxed{\uparrow (2,2)} \rightarrow \boxed{h[-m]\,g[-n]} \rightarrow A_{\text{HL}}^{(K)}[m,n]$$

$$\tilde{x}_{\text{HH}}^{(K)}[m,n] \rightarrow \boxed{\uparrow (2,2)} \rightarrow \boxed{h[-m]\,h[-n]} \rightarrow A_{\text{HH}}^{(K)}[m,n]$$

Average signal $\tilde{x}_{\text{LL}}^{(K-1)}[m,n]$ is the sum of the above four outputs:

$$\tilde{x}_{\text{LL}}^{(K-1)}[m,n] = A_{\text{LL}}^{(K)}[m,n] + A_{\text{LH}}^{(K)}[m,n] + A_{\text{HL}}^{(K)}[m,n] + A_{\text{HH}}^{(K)}[m,n]. \tag{10.158}$$

Here $\{A_{\text{LL}}^{(K)}[m,n],\ A_{\text{LH}}^{(K)}[m,n],\ A_{\text{HL}}^{(K)}[m,n],\ A_{\text{HH}}^{(K)}[m,n]\}$ are just four temporary quantities to be added. The analogy to signal analysis and synthesis filter banks is evident, except that at each stage there are three detail images instead of one detail signal.

Repetition of the reconstruction step represented by Eq. (10.158) through an additional $K-1$ stages leads to

$$x[m,n] = \tilde{x}_{\text{LL}}^{(0)}[m,n].$$

The condition for perfect reconstruction is the 2-D version of the Smith-Barnwell condition defined by Eq. (10.113), namely

$$|\mathbf{G}_2(e^{j\Omega_1}, e^{j\Omega_2})|^2 + |\mathbf{G}_2(e^{j(\Omega_1+\pi)}, e^{j(\Omega_2+\pi)})|^2 + |\mathbf{G}_2(e^{j(\Omega_1+\pi)}, e^{j\Omega_2})|^2 + |\mathbf{G}_2(e^{j\Omega_1}, e^{j(\Omega_2+\pi)})|^2 = 4, \quad (10.159)$$

where $\mathbf{G}_2(e^{j\Omega_1}, e^{j\Omega_2})$ is the DSFT of the 2-D scaling function $g_2[m,n] = g[m]\,g[n]$. Since the 2-D scaling function $g_2[m,n]$ is separable, its DSFT is also separable:

$$\mathbf{G}_2(e^{j\Omega_1}, e^{j\Omega_2}) = \mathbf{G}(e^{j\Omega_1})\,\mathbf{G}(e^{j\Omega_2}), \quad (10.160)$$

where $\mathbf{G}(e^{j\Omega_1})$ is the DTFT of $g[n]$. The 2-D Smith-Barnwell condition is satisfied if the 1-D Smith-Barnwell condition given by Eq. (10.113) is satisfied (see Exercise 10-12).

10-13.3 2-D Haar Wavelet Transform of Shepp-Logan Phantom

The ***Shepp-Logan phantom*** is a piecewise-constant image that has been a test image for tomography algorithms since the 1970s. A 256 × 256 image of the Shepp-Logan phantom is displayed in Fig. 10-28(a). To illustrate what the coarse and detail images generated by the application of the 2-D wavelet transform look like, we applied a 4-stage 2-D Haar wavelet transform to the image in Fig. 10-28(a). The results are displayed per the pattern in Fig. 10-28(b), with:

(1) Stage-4 images: **16 x 16**

The coarsest average image, $x_{\text{LL}}^{(4)}[m,n]$, is used as a thumbnail image, and placed at the upper left-hand corner in Fig. 10-28(c). The three stage-4 detail images are arranged clockwise around image $x_{\text{LL}}^{(4)}[m,n]$.

(2) Stage-3 images: **32 x 32**

The three stage-3 detail images are four times as large, and are arranged clockwise around the stage-4 images.

(3) Stage-2 images: **64 x 64**

(4) Stage-1 images: **128 x 128**

The largest images in Fig. 10-28(c) are the three stage-1 images. The number of pixels in the 2-D Haar transform can

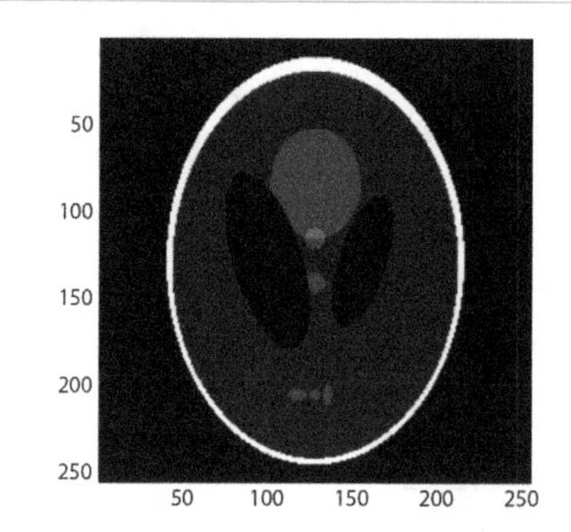

(a) 256 × 256 Shepp-Logan phantom image

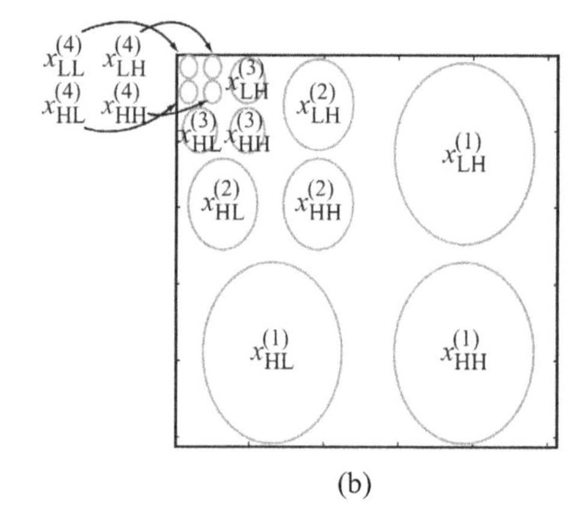

(b)

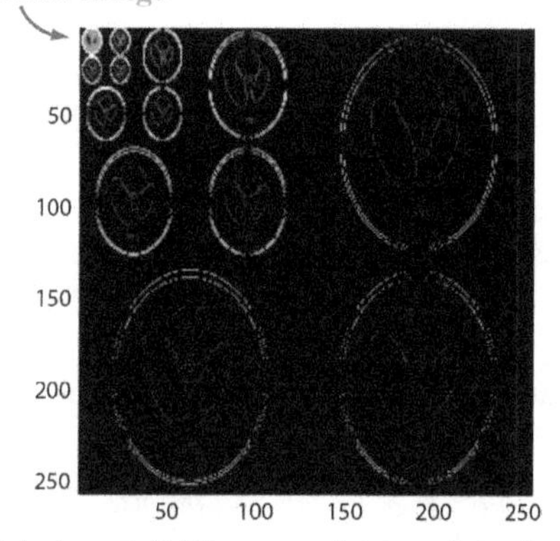

(c) 4-stage 2-D Haar wavelet transform images

Figure 10-28: (a) 256 × 256 test image, (b) arrangement of images generated by a 3-stage Haar wavelet transform, and (c) the images represented in (b). A logarithmic scale is used to display the values.

be computed similarly to the 1-D Haar transform durations in Eq. (10.66) and Eq. (10.69):

Stage 4 images: The 4 (16×16) images (1 average and 3 detail) contain a total of $4(16)^2 = 1024$ pixels.

Stage 3 images: The 3 (32×32) detail images contain a total of $3(32)^2 = 3072$ pixels. The fourth 32×32 image is the average image, which is decomposed into the stage 4 images.

Stage 2 images: The 3 (64×64) detail images contain a total of $3(64)^2 = 12288$ pixels. The fourth 64×64 image is the average image, which is decomposed into the stage 3 images.

Stage 1 images: The 3 (128×128) detail images contain a total of $3(128)^2 = 49152$ pixels. The fourth 128×128 image is the average image, which is decomposed into the stage 2 images.

The total number of pixels in the wavelet transform of the Shepp-Logan phantom is then:

$$1024 + 3072 + 12288 + 49152 = 65536.$$

This equals the number of pixels in the Shepp-Logan phantom, which is $256^2 = 65536$.

Even though the coarsest average image is only 16×16, it contains almost all of the large numbers in the 2-D wavelet transform of the original image, and captures most of its primary features. That is why it is used as a ***thumbnail image***. The pixel values of the stage-4 detail images are almost entirely zeros. The 3-stage composition process preserves the information content of the original image—composed of $256^2 = 65536$ pixels—while compressing it down to 4 images containing only 3619 nonzero pixels. This is a 94.5% reduction in the number of nonzero pixels (see S² for more details).

10-13.4 2-D $D3$ Daubechies Wavelet Transform of Clown Image

For a second example, we repeated the steps outlined in the preceding subsection, but this time we used a 2-D $D3$ Daubechies wavelet transform on the 200×200 clown image shown in Fig. 10-29(a). The images generated by the 3-stage decomposition process are displayed in part (b) of the figure, using the same arrangement as shown earlier in Fig. 10-29(b) (see S² for more details).

Module 10.4a Wavelet Transform of Shepp-Logan Phantom Using Haar Wavelets This module computes the Haar wavelet transform of the Shepp-Logan phantom. The number of levels of decomposition is a selectable parameter. This module is an interactive version of Fig. 10-28.

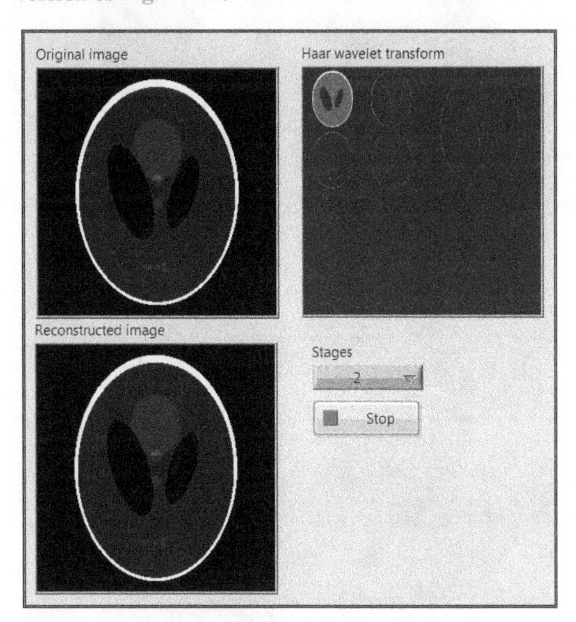

10-13.5 Image Compression by Thresholding Its Wavelet Transform

Image compression is an important feature of the wavelet transform. Not only is the original image represented by fewer pixels, but also many of the pixels of the wavelet-transform images are zero-valued. The compression ratio can be improved further by thresholding the output images of the final stage of the wavelet-transform decomposition process. Thresholding a pixel means replacing it with zero if its absolute value is below a given ***threshold level*** λ. As noted earlier in connection with 1-D signals and 2-D images, most of the wavelet-transform detail signals and images have very small values, so little information is lost by setting their values to zero, which means that they no longer need to be stored, thereby reducing the storage capacity needed to store the wavelet transform of the image. Furthermore, since the wavelet transform is composed of orthogonal basis functions, a small change in the wavelet transform of an image will produce only a small change in the image reconstructed from these values.

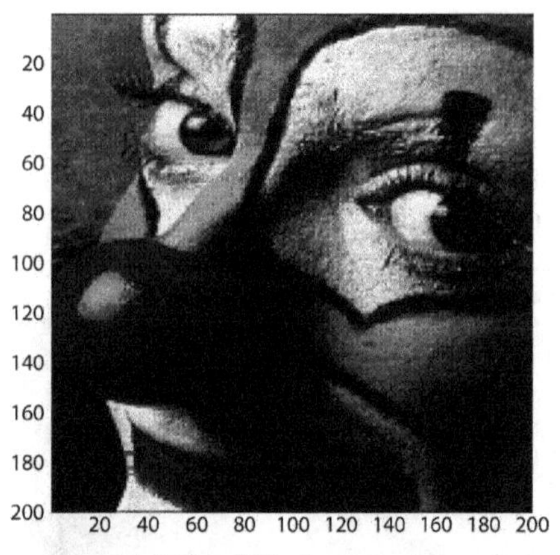

(a) 200 × 200 clown image

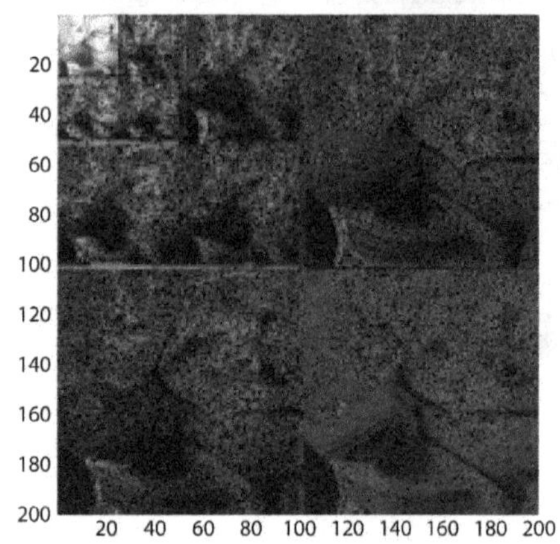

(b) 2-D D3 Daubechies wavelet transform of the clown image

Figure 10-29: (a) Clown image and (b) its 3-stage wavelet-transform images. A logarithmic scale is used to display the values.

To illustrate with an example, we compare in Fig. 10-30 two images:

(a) In part (a), we show the original 200 × 200 clown image, and

(b) in part (b) we show a reconstructed clown image, generated from the $D3$ Daubechies wavelet-transform images after thresholding the images with $\lambda = 0.11$.

The reconstructed image looks almost identical to the original image, even though only 6% of the pixels in the wavelet-transform images are nonzero.

Module 10.4b Wavelet Transform of Clown Image Using Daubechies Wavelets This module computes the $D3$ Daubechies wavelet transform of the clown image. The number of levels of decomposition is a selectable parameter. This module is an interactive version of Fig. 10-29.

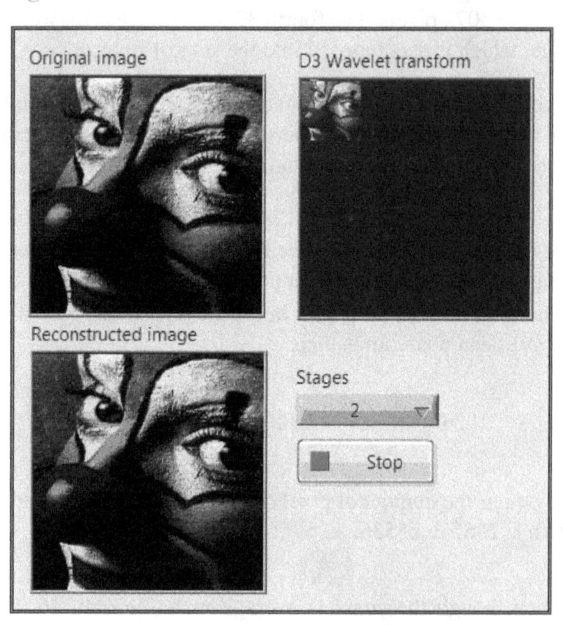

Concept Question 10-15: Why is the 2-D wavelet transform depicted as a thumbnail image and other near-zero images? (See S^2)

Exercise 10-12: Show that for separable 2-D scaling and wavelet functions, the 2-D Smith-Barnwell condition Eq. (10.159) is satisfied if the 1-D Smith-Barnwell condition given by Eq. (10.113) is satisfied.

Answer: See S^2.

Exercise 10-13: Use LabVIEW Module 10.5 to compress and then decompress the clown image. Use a threshold of 0.5. What compression ratio does this produce?

Answer: 52.1512.

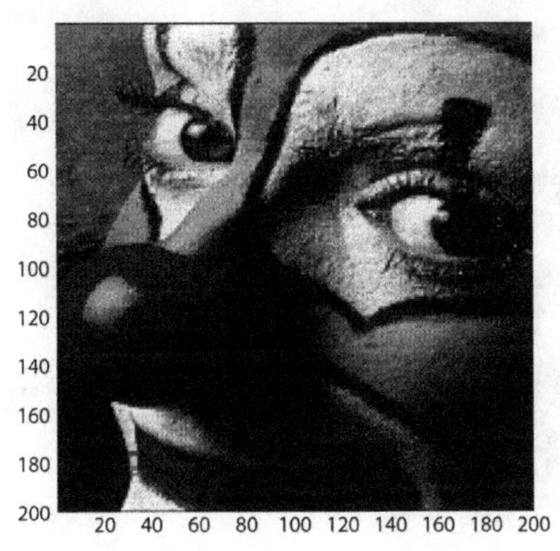
(a) 200 × 200 clown image

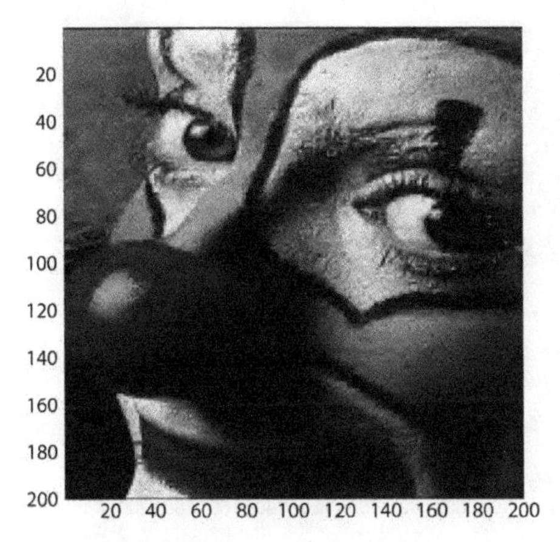
(b) Reconstructed clown image

Figure 10-30: (a) Original clown image, and (b) image reconstructed from thresholded $D3$ Daubechies wavelet transform images, requiring only 6% as much storage capacity as the original image.

Module 10.5 Compression of Clown Image Using Daubechies Wavelets This module computes the $D3$ Daubechies wavelet transform of the clown image, sets to zero values below a threshold, and reconstructs the clown image from its thresholded wavelet transform. The threshold is set by a slider, and the resulting compression ratio (number of pixels divided by number of nonzero wavelet coefficients) is computed. This module is an interactive version of Fig. 10-30.

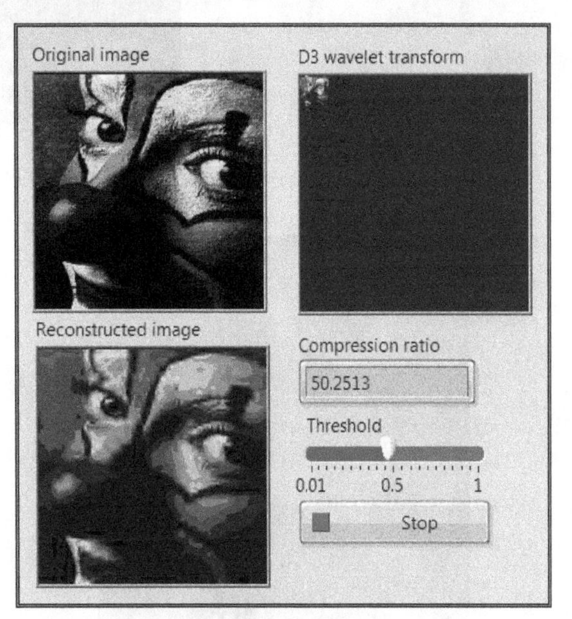

10-14 Denoising by Thresholding and Shrinking

A noisy image is given by

$$y[m,n] = x[m,n] + v[m,n], \tag{10.161}$$

where $x[m,n]$ is the desired image and $v[m,n]$ is the noise that had been added to it. The goal of denoising is to recover the original image $x[m,n]$, or a close approximation thereof, from the noisy image $y[m,n]$. We now show that the obvious approach of simply thresholding the wavelet transform of the image does not work well. Then we show that the combination of thresholding and ***shrinking*** the wavelet transform of the image does work well.

10-14.1 Denoising by Thresholding Alone

One approach to denoising is to threshold the wavelet transform of $y[m,n]$. The idea is that for small wavelet transform values, the signal-to-noise ratio is low, so little of value is lost by thresholding these small values to zero. For large wavelet transform values, the signal-to-noise ratio is large, so these large values should be kept. This approach worked quite well in Example 8-18, but it works poorly on wavelet transforms of noisy images, as the following example shows.

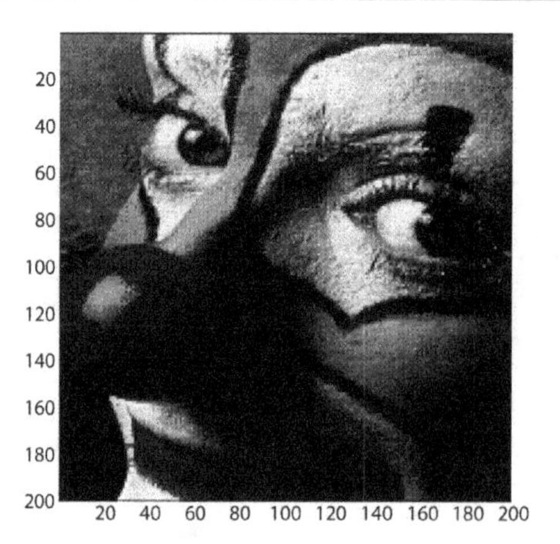

(a) Original noise-free image

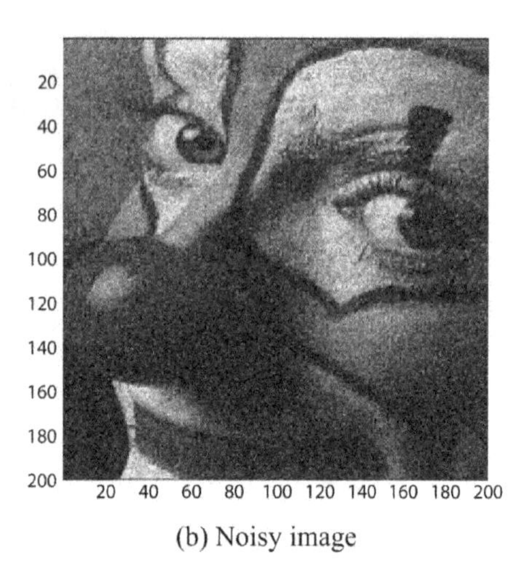

(b) Noisy image

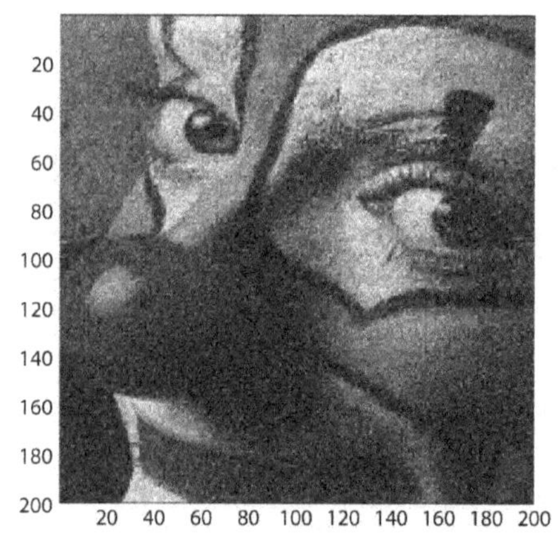

(c) Image denoised by thresholding its wavelet transform

Figure 10-31: (a) Noise-free clown image, (b) noisy image with SNR = 11.5, and (c) image reconstructed from thresholded wavelet transform.

Zero-mean 2-D white Gaussian noise with standard deviation $\sigma = 0.1$ was added to the clown image of Fig. 10-31(a). The noisy image, shown in Fig. 10-31(b), has a signal-to-noise ratio (SNR) of 11.5, which means that the noise level is, on average, only about 8.7% of that of the signal.

The $D3$ Daubechies wavelet transform was computed for the noisy image, then thresholded with $\lambda = 0.11$, which appeared to provide the best results. Finally, the image was reconstructed from the thresholded wavelet transform, and it now appears in Fig. 10-31(c). Upon comparing the images in parts (b) and (c) of the figure, we conclude that the thresholding operation failed to reduce the noise by any appreciable amount.

10-14.2 Denoising by Thresholding and Shrinkage

We now show that a combination of ***thresholding*** small wavelet-transform values to zero and ***shrinking*** other wavelet transform values by a small number λ performs much better in denoising images. First we show that shrinkage comes from minimizing a cost functional, just as Wiener filtering given by Eq. (10.46) came from minimizing the Tikhonov cost functional given by Eq. (10.45).

In the material that follows, we limit our treatment to 1-D signals. The results are readily extendable to 2-D images.

Module 10.6 Wavelet Denoising of Clown Image Using Daubechies Wavelets This module adds noise to the clown image, computes the $D3$ Daubechies wavelet transform of the noisy clown image, sets to zero values below a threshold, and reconstructs the clown image from its thresholded wavelet transform. The noise level and threshold are selectable parameters. This module is an interactive version of Fig. 10-31.

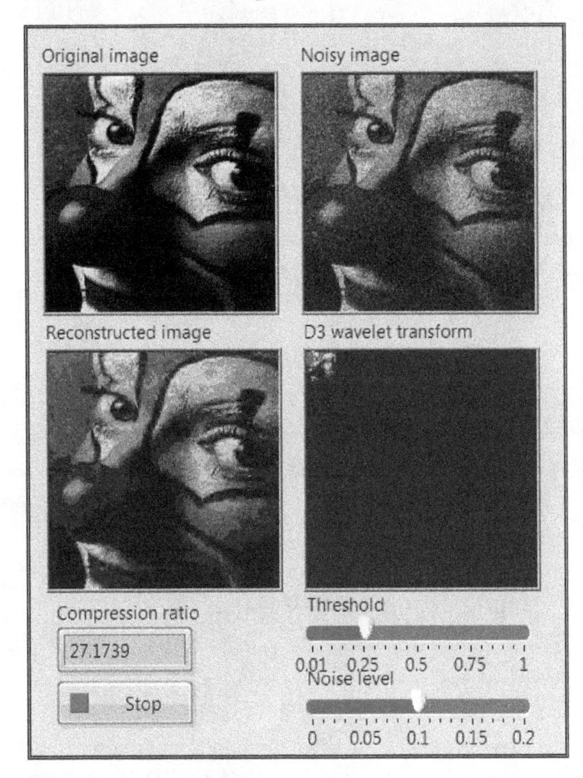

Suppose we are given noisy observations $y[n]$ of a signal $x[n]$ that is known to be sparse (mostly zero),

$$y[n] = x[n] + v[n]. \tag{10.162}$$

All three signals have durations N and the noise $v[n]$ is known to have zero-mean. The goal is to estimate $x[n]$ from the noisy observations $y[n]$; i.e., to denoise $y[n]$ with the additional information that $x[n]$ is mostly zero-valued.

The prior knowledge that $x[n]$ is sparse can be incorporated by estimating $x[n]$ from $y[n]$, not by simply using $y[n]$ as an estimator for $x[n]$, but by minimizing over $x[n]$ the ***LASSO cost functional***

$$\Lambda = \underbrace{\frac{1}{2}\sum_{n=0}^{N-1}(y[n]-x[n])^2}_{\text{fidelity to data } y[n]} + \lambda \underbrace{\sum_{n=0}^{N-1}|x[n]|}_{\text{sparsity}}. \tag{10.163}$$

Readers familiar with basic estimation theory will note that Λ is the negative log-likelihood function for zero-mean white Gaussian noise $v[n]$ with independent Laplacian *a priori* distributions for each $x[n]$. The coefficient λ is a ***trade-off parameter*** between fidelity to the data $y[n]$ and imposition of sparsity. If $\lambda = 0$, then the estimator $\hat{x}[n]$ of $x[n]$ is just $\hat{x}[n] = y[n]$. Nonzero λ emphasizes sparsity, while allowing some difference between $\hat{x}[n]$ and $y[n]$, which takes into account the noise $v[n]$.

LASSO is an acronym for ***least absolute shrinkage and selection operator***.

10-14.3 Minimization of LASSO Cost Functional

The minimization of Λ decouples in time n, so each term can be minimized separately. The goal then is to find an ***estimate*** of $x[n]$, which we denote $\hat{x}[n]$, that minimizes over $x[n]$ the nth term Λ_n of Eq. (10.163), namely

$$\Lambda_n = \frac{1}{2}(y[n]-x[n])^2 + \lambda|x[n]|. \tag{10.164}$$

The expression given by Eq. (10.164) is the sum of two terms. The value of λ is selected to suit the specific application; if fidelity to the measured observations $y[n]$ is highly prized, then λ is assigned a small value, but if data storage is an important attribute, then λ may be assigned a large value.

Keeping in mind that $\lambda \geq 0$, the minimization problem consists of four possible scenarios:

(1) $y[n] \geq 0$ and $y[n] \leq \lambda$
(2) $y[n] \geq 0$ and $y[n] \geq \lambda$
(3) $y[n] \leq 0$ and $|y[n]| \leq \lambda$
(4) $y[n] \leq 0$ and $|y[n]| \geq \lambda$

Case 1: $y[n] \geq 0$ and $y[n] \leq \lambda$

Let us consider the following example for a particular value of n:

Measurement $y[n] = 1$
Trade-off parameter $\lambda = 2$
Signal $x[n]$: unknown

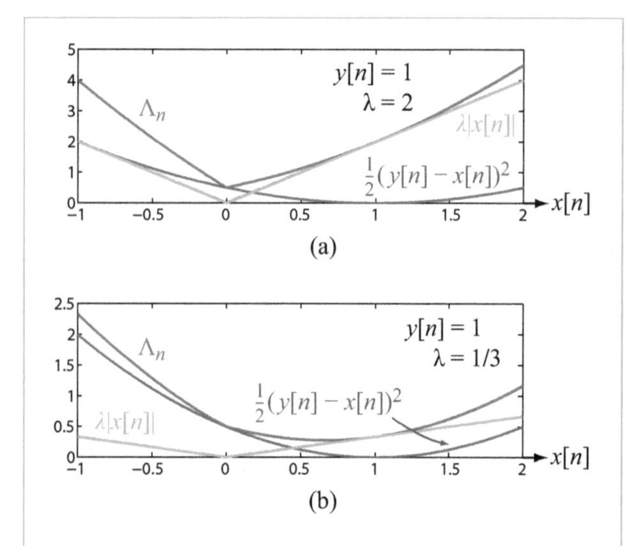

Figure 10-32: Plots of the first and second terms of Λ_n, and their sum for: (a) $y[n] = 1$ and $\lambda = 2$, and (b) $y[n] = 1$ and $\lambda = 1/3$.

The estimated value $\hat{x}[n]$ of $x[n]$ is found by minimizing Λ_n. Figure 10-32(a) displays three plots, corresponding to the first and second terms in Eq. (10.164), and their sum. It is evident from the plot of Λ_n that Λ_n is minimized at $x[n] = 0$. Hence,

$$\hat{x}[n] = 0 \qquad \text{for } y[n] \geq 0 \text{ and } y[n] \leq \lambda. \tag{10.165a}$$

Case 2: $y[n] \geq 0$ and $y[n] \geq \lambda$

Repetition of the scenario described by case 1, but with λ changed from 2 to 1/3, leads to the plots shown in Fig. 10-32(b). In this case, Λ_n is parabolic-like in shape, and its minimum occurs at a positive value of $x[n]$ at which the slope of Λ_n is zero. That is,

$$\frac{d\Lambda_n}{dx[n]} = \frac{d}{dx[n]}\left(\frac{1}{2}\,(y[n] - x[n])^2 + \lambda\, x[n]\right) = 0,$$

which leads to

$$\hat{x}[n] = y[n] - \lambda, \quad \text{for } y[n] \geq 0 \text{ and } y[n] \geq \lambda. \tag{10.165b}$$

Case 3: $y[n] \leq 0$ and $|y[n]| \leq \lambda$

This case, which is identical to case 1 except that now $y[n]$ is negative, leads to the same result, namely

$$\hat{x}[n] = 0, \qquad \text{for } y[n] \leq 0 \text{ and } |y[n]| \leq \lambda. \tag{10.165c}$$

Case 4: $y[n] \leq 0$ and $|y[n]| \geq \lambda$

Repetition of the analysis of case 2, but with $y[n]$ negative, leads to

$$\hat{x}[n] = y[n] + \lambda, \qquad \text{for } y[n] \leq 0 \text{ and } |y[n]| \geq \lambda. \tag{10.165d}$$

The four cases can be combined into

$$\hat{x}[n] = \begin{cases} y[n] - \lambda & \text{for } y[n] > +\lambda, \\ y[n] + \lambda & \text{for } y[n] < -\lambda, \\ 0 & \text{for } |y[n]| < \lambda. \end{cases} \tag{10.166}$$

Values of $y[n]$ smaller in absolute value than the threshold λ are ***thresholded*** (set) to zero. Values of $y[n]$ larger in absolute value than the threshold λ are ***shrunk*** by λ, making their absolute values smaller. So $\hat{x}[n]$ is computed by ***thresholding and shrinking*** $y[n]$. This is usually called (ungrammatically) "thresholding and shrinkage."

The next example shows that denoising images works much better with thresholding and shrinkage than with thresholding alone.

When we applied thresholding alone to the noisy image of Fig. 10-31(b), we obtained the image shown in Fig. 10-31(c), which we repeat here in Fig. 10-33(a). Application of thresholding and shrinkage in combination, with $\lambda = 0.11$, leads to the image in Fig. 10-33(b), which provides superior rendition of the clown image by filtering much more of the noise, while preserving the real features of the image (see S^2 for more details).

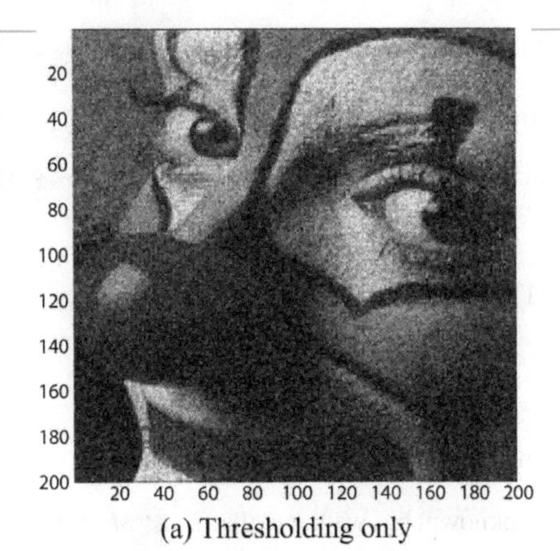

(a) Thresholding only

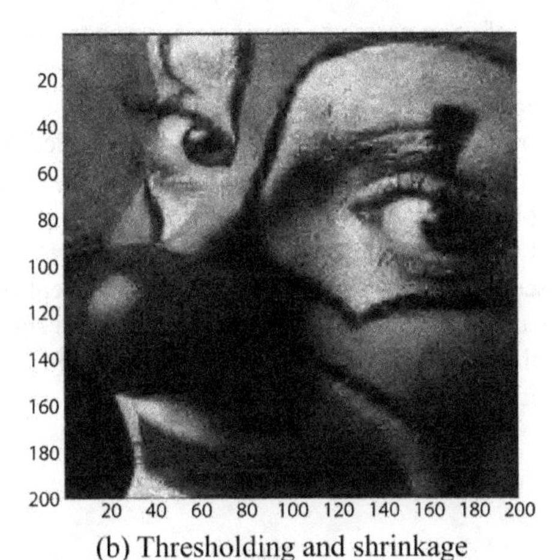

(b) Thresholding and shrinkage

Figure 10-33: Denoising the clown image: (a) denoising by thresholding alone, (b) denoising by thresholding and shrinkage in combination.

Concept Question 10-16: Why is shrinkage, in addition to thresholding, needed for noise reduction? (See S^2)

Concept Question 10-17: Why does wavelet-based denoising work so much better than lowpass filtering? (See S^2)

Exercise 10-14: Use LabVIEW Module 10.6 to denoise the clown image. Use a noise level of 0.2 and threshold of 1. Discuss the result.

Answer:

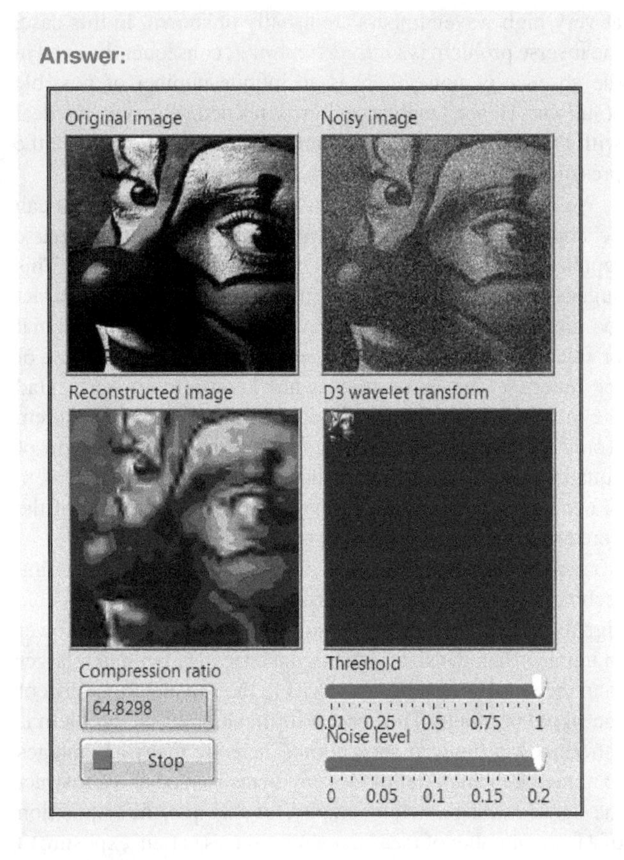

10-15 Compressed Sensing

The solution of an ***inverse problem*** in signal and image processing is the reconstruction of an unknown signal or image from measurements (known linear combinations) of the values of the signal or image. Such inverse problems arise in medical imaging, radar imaging, optics, and many other fields. For example, in tomography and ***magnetic resonance imaging*** (***MRI***), the inverse problem is to reconstruct an image from measurements of some (but not all) of its 2-D Fourier transform values.

If the number of measurements equals or exceeds the size (duration in 1-D, number of pixels in 2-D) of the unknown signal or image, solution of the inverse problem in the absence of noise becomes a solution of a linear system of equations. In practice, there is always noise in the measurements, so some sort of regularization is required. In Section 10-7.3, the deconvolution problem required Tikhonov regularization to produce a recognizable solution when noise was added to the data. Furthermore, often the number of observations is less than the size of the unknown signal or image. For example, in tomography, the 2-D Fourier transform values of the image at very high wavenumbers are usually unknown. In this case, the inverse problem is ***underdetermined***; consequently, even in the absence of noise there is an infinite number of possible solutions. Hence, regularization is needed, not only to deal with the underdetermined formulation, but also to manage the presence of noise in the measurements.

We have seen that many real-world signals and images can be compressed, using the wavelet transform, into a sparse representation in which most of the values are zero. This suggests that the number of measurements needed to reconstruct the signal or image can be less than the size of the signal or image, because in the wavelet-transform domain, most of the values to be reconstructed are known to be zero. Had the locations of the nonzero values been known, the problem would have been reduced to a solution of a linear system of equations smaller in size than that of the original linear system of equations. In practice, however, neither the locations of the nonzero values nor their values are known.

Compressed sensing refers to a set of signal processing techniques used for reconstructing wavelet-compressible signals and images from measurements that are much fewer in number than the size of the signal or image, but much larger than the number of nonzero values in the wavelet transform of the signal or image. The general formulation of the problem is introduced in the next subsection. There are many advantages to reducing the number of measurements needed to reconstruct the signal or image. In tomography, for example, the acquisition of a fewer number of measurements reduces patient exposure to radiation. In MRI, this reduces acquisition time inside the MRI machine, and in smartphone cameras, it reduces the exposure time and energy required to acquire an image.

This section presents the basic concepts behind compressed sensing and applies these concepts to a few signal and image inverse problems. Compressed sensing is an active area of research and development, and will experience significant growth in applications in the future.

10-15.1 Problem Formulation

To cast the compressed sensing problem into an appropriate form, we define the following quantities:

(a) $\{x[n], n = 0 \ldots N-1\}$ is an *unknown signal* of length N

(b) The corresponding (unknown) wavelet transform of $x[n]$ is

$$\{\tilde{x}_1[n], \tilde{x}_2[n], \ldots, \tilde{x}_L[n], \tilde{X}_L[n]\},$$

and the wavelet transform of $x[n]$ is *sparse:* only K values of all of the $\{\tilde{x}_k[n]\}$ are nonzero, with $K \ll N$.

(c) $\{y[n], n = 0, 1, \ldots, M-1\}$ are M *known measurements*

$$y[0] = a_{0,0}\ x[0] + a_{0,1}\ x[1] + \cdots + a_{0,N-1}\ x[N-1],$$
$$y[1] = a_{1,0}\ x[0] + a_{1,1}\ x[1] + \cdots + a_{1,N-1}\ x[N-1],$$
$$\vdots$$
$$y[M-1] = a_{M-1,0}\ x[0] + a_{M-1,1}\ x[1] + \cdots + a_{M-1,N-1}\ x[N-1],$$

where $\{a_{n,i},\ n=0, 1, \ldots, M-1, \text{and } i=0, 1, \ldots, N-1\}$ are known

(d) K is unknown, but we know that $K \ll M < N$.

The goal of compressed sensing is to compute signal $\{x[n], n = 0, 1, \ldots, N-1\}$ from the M known measurements $\{y[n], n = 0, 1, \ldots, M-1\}$.

The compressed sensing problem can be divided into two components, a ***direct problem*** and an ***inverse problem***. In the direct problem, the independent variable (input) is $x[n]$ and the dependent variable (output) is the measurement $y[n]$. The roles are reversed in the inverse problem: the measurements become the independent variables (input) and the unknown signal $x[n]$ becomes the output. The relationships between $x[n]$ and $y[n]$ involve vectors and matrices:

Signal vector

$$\underline{x} = [x[0], x[1], \ldots, x[N-1]]^T, \tag{10.167}$$

where T denotes the transpose operator, which converts a row vector into a column vector.

$$\underline{y} = [y[0], y[1], \ldots, y[M-1]]^T. \tag{10.168}$$

Wavelet transform vector

$$\underline{z} = \begin{bmatrix} z_1 \\ z_2 \\ \vdots \\ \vdots \\ z_N \end{bmatrix} = \begin{bmatrix} \tilde{x}_1[n] \\ \tilde{x}_2[n] \\ \vdots \\ \tilde{x}_L[n] \\ \tilde{X}_L[n] \end{bmatrix}, \tag{10.169}$$

with $\underline{z}$ of length N, and $\tilde{x}_1[n]$ to $\tilde{x}_L[n]$ are the detail signals of the wavelet transform and $\tilde{X}_L[n]$ is the coarse signal. This notation appeared previously in Eq. (10.90).

Wavelet transform matrix

$$\underline{z} = W\,\underline{x}, \tag{10.170}$$

where W is a known $N \times N$ ***wavelet transform matrix*** that implements the wavelet transform of $x[n]$ to obtain $\underline{z}$. For example, for the Haar wavelet transform of a signal $x[n]$ of duration 8, W is given by $\mathcal{H}$ of Eq. (10.89).

Direct-problem formulation

$$\underline{y} = A\,\underline{x}, \tag{10.171}$$

where A is an $M \times N$ matrix. Usually, A is a known matrix based on a physical model or direct measurement of $\underline{y}$ for a known $\underline{x}$. Combining Eqs. (10.170) and (10.171) gives

$$\underline{y} = A\,W^{-1}\,\underline{z} = A_{\mathrm{w}}\,\underline{z}, \tag{10.172a}$$

where

$$A_{\mathrm{w}} = A\,W^{-1}. \tag{10.172b}$$

Since A and W are both known matrices, A_{w} also is known.

Inverse-problem formulation

If, somehow, $\underline{z}$ can be determined from the measurement vector $\underline{y}$, then $\underline{x}$ can be computed by inverting Eq. (10.170):

$$\underline{x} = W^{-1}\,\underline{z}. \tag{10.173}$$

For the orthogonal wavelet transforms, such as the Haar and Daubechies transforms covered in Sections 10-9 and 10-12, $W^{-1} = W^T$, so the inverse wavelet transform can be computed as easily as the wavelet transform. In practice, both are computed using analysis and synthesis filter banks, as discussed in earlier sections.

The crux of the compressed sensing problem reduces to finding $\underline{z}$, given $\underline{y}$. An additional factor to keep in mind is that only K values of the elements of $\underline{z}$ are nonzero, with $K \ll M$. Algorithms for computing $\underline{z}$ from $\underline{y}$ rely on iterative approaches, as discussed in future sections.

Because $\underline{z}$ is of length N, $\underline{y}$ of length M, and $M < N$ (fewer measurements than unknowns), Eq. (10.172) represents an ***underdetermined system*** of linear equations, whose solution is commonly called an ***ill-posed problem***.

10-15.2 Inducing Sparsity into Solutions

In seismic signal processing, explosions are set off on the Earth's surface, and echoes of the seismic waves created by this explosion are measured by seismometers. In the 1960s, sedimentary media (such as the bottom of the Gulf of Mexico) were modeled as a stack of layers, so the seismometers would record occasional sharp pulses reflected off of the interfaces between the layers. The amplitudes and times of the pulses would allow the layered medium to be reconstructed. However, the occasional pulses had to be deconvolved from the source pulse created by the explosions. The deconvolution problem was modeled as an underdetermined linear system of equations.

A common approach to finding a sparse solution to a system of equations is to choose the solution that minimizes the sum of absolute values of the solution. This is known as the ***minimum ℓ_1 norm*** solution. The ℓ_1 norm is denoted by the symbol $||\underline{z}||_1$ and defined as

$$||\underline{z}||_1 = \sum_{i=1}^{N} |z_i|. \tag{10.174a}$$

The goal is to find the solution to the system of equations that minimizes $||\underline{z}||_1$.

A second approach,which does not provide a sparse solution, called the ***squared ℓ_2 norm*** finds the solution that minimizes the sum of squares of the solution. The squared ℓ_2 norm is denoted by the symbol $||\underline{z}||_2^2$ and is defined as

$$||\underline{z}||_2^2 = \sum_{i=1}^{N} |z_i|^2. \tag{10.174b}$$

10-16 Computing Solutions to Underdetermined Equations

10-16.1 Basis Pursuit

As noted earlier, the unknown signal vector $\underline{x}$ can be determined from Eq. (10.173), provided we have a viable solution for $\underline{z}$ (because W is a known matrix). One possible approach to finding $\underline{z}$ is to implement the minimum ℓ_1 norm given by Eq. (10.174a), subject to the constraint given by Eq. (10.172a). That is, the goal is to find vector $\underline{z}$ from measurement vector $\underline{y}$ such that

$$\sum_{i=1}^{N} |z_i| \text{ is minimum},$$

and

$$\underline{y} = A_{\mathrm{w}}\,\underline{z}.$$

The solution method is known as ***basis pursuit***, and it can be formulated as a ***linear programming*** problem by defining the positive $\underline{z}^+$ and negative $\underline{z}^-$ parts of $\underline{z}$ as:

$$z_i^+ = \begin{cases} +z_i & \text{if } z_i \geq 0, \\ 0 & \text{if } z_i < 0, \end{cases} \tag{10.175a}$$

$$z_i^- = \begin{cases} -z_i & \text{if } z_i \leq 0, \\ 0 & \text{if } z_i > 0. \end{cases} \tag{10.175b}$$

Vector $\underline{z}$ is then given by

$$\underline{z} = \underline{z}^+ - \underline{z}^-, \tag{10.176}$$

and its ℓ_1 norm is

$$||\underline{z}||_1 = \sum_{i=1}^{N} (z_i^+ + z_i^-). \tag{10.177}$$

In terms of $\underline{z}^+$ and $\underline{z}^-$, the basis pursuit problem becomes

$$\text{Minimize } \sum_{i=1}^{N} (z_i^+ + z_i^-) \quad \text{subject to } \underline{y} = A_w\, \underline{z}^+ - A_w\, \underline{z}^-, \tag{10.178}$$

$z_i^+ \geq 0$ and $z_i^- \geq 0$, which is a straightforward linear programming problem that can be solved using *linprog* in MATLAB's Optimization Toolbox. The basis pursuit method is limited to noise-free signal and image problems, so in the general case, more sophisticated approaches are called for.

10-16.2 LASSO Cost Functional

A serious shortcoming of the basis pursuit solution method is that it does not account for the presence of noise in the observations $\underline{y}$. As noted earlier in Section 10-14.2, the ***least absolute shrinkage and selection operator*** (*LASSO*) provides an effective approach for estimating the true signal from noisy measurements. In the present context, the LASSO functional is defined as

$$\Lambda = \underbrace{\tfrac{1}{2}\, ||\underline{y} - A_w\, \underline{z}||_2^2}_{\text{fidelity}} + \underbrace{\lambda ||\underline{z}||_1}_{\text{sparsity}}, \tag{10.179}$$

where λ is a trade-off parameter between sparsity of $\underline{z}$ and fidelity to the measurement $\underline{y}$. Choosing the solution $\underline{z}$ that minimizes Eq. (10.179), for a specified value of λ, is called ***basis pursuit denoising***. The solution requires the use of an iterative algorithm. Two such algorithms are presented in later subsections, preceded by a short review of ***pseudo inverses***.

10-16.3 Review of Pseudo-Inverses

(a) Overdetermined system

Consider the linear system of equations given by

$$\underline{y} = A_w\, \underline{z}, \tag{10.180}$$

with $\underline{z}$ of length N, $\underline{y}$ of length M, and A_w of size $M \times N$ with full rank. If $M > N$, then the system is ***overdetermined*** (more measurements than unknowns) and, in general, it has no solution. The vector $\hat{\underline{z}}$ that minimizes $||\underline{y} - A_w\, \underline{z}||_2^2$ is called the ***pseudo-inverse solution***, and is given by the *estimate*

$$\hat{\underline{z}} = (A_w^T\, A_w)^{-1}\, A_w^T\, \underline{y}. \tag{10.181a}$$

Note that $A_w^T\, A_w$ is an $N \times N$ matrix with full rank.

To avoid matrix-inversion problems, $\hat{\underline{z}}$ should be computed not by inverting $A_w^T\, A_w$, but by solving the linear system of equations

$$(A_w^T A_w)\, \hat{\underline{z}} = A_w^T\, \underline{y} \tag{10.181b}$$

using the ***LU decomposition*** method or similar techniques. Here, LU stands for *lower upper*, in reference to the lower triangular submatrix and the upper triangular submatrix multiplying the unknown vector $\hat{\underline{z}}$.

(b) Underdetermined system

Now consider the underdetermined system characterized by $M < N$ (fewer measurements than unknowns). In this case, there is an infinite number of possible solutions. The vector $\hat{\underline{z}}$ that minimizes $||\underline{z}||_2^2$ among this infinite number of solutions also is called the pseudo-inverse solution, and is given by the estimate

$$\hat{\underline{z}} = A_w^T\, (A_w\, A_w^T)^{-1}\, \underline{y}. \tag{10.182}$$

In the present case, $A_w\, A_w^T$ is an $M \times M$ matrix with full rank. Solution $\hat{\underline{z}}$ should be computed not by inverting $A_w\, A_w^T$, but by initially solving the linear system

$$(A_w\, A_w^T)\, \hat{\underline{r}} = \underline{y}, \tag{10.183a}$$

to compute an intermediate estimate $\hat{\underline{r}}$, and then computing $\hat{\underline{z}}$ by applying

$$\hat{\underline{z}} = A_w^T\, \hat{\underline{r}}. \tag{10.183b}$$

10-16.4 Iterative Reweighted Least Squares (IRLS) Algorithm

According to Eq. (10.172a), measurement vector $\underline{y}$ and wavelet transform vector $\underline{z}$ are related by

$$\underline{y} = A_w\, \underline{z}. \tag{10.184}$$

The ***iterative reweighted least squares*** (*IRLS*) algorithm uses the Tikhonov regularization functional given by Eq. (10.45), together with a diagonal weighting matrix D to minimize the cost function

$$\mathcal{T} = \underbrace{\tfrac{1}{2}\, ||\underline{y} - A_w\, \underline{z}||_2^2}_{\text{fidelity}} + \underbrace{\lambda ||D\, \underline{z}||_2^2}_{\text{size}}, \tag{10.185}$$

where λ is the trade-off parameter between the size of $\underline{z}$ and the fidelity to the data $\underline{y}$. The goal is to trade off small differences between $\underline{y}$ and $A_w\,\underline{z}$ so as to keep $\underline{z}$ small. To compute $\underline{z}$, we first rewrite Eq. (10.185) in the expanded form

$$\mathcal{T} = \frac{1}{2}\left\| \underbrace{\begin{bmatrix} y[0] \\ y[1] \\ \vdots \\ y[M-1] \end{bmatrix}}_{M\times 1} - \underbrace{\begin{bmatrix} a_{0,0} & a_{0,1} & \cdots & a_{0,N-1} \\ a_{1,0} & a_{1,1} & \cdots & a_{1,N-1} \\ \vdots \\ a_{M-1,0} & a_{M-1,1} & \cdots & a_{M-1,\,N-1} \end{bmatrix}}_{M\times N} \underbrace{\begin{bmatrix} z_1 \\ z_2 \\ \vdots \\ \vdots \\ \vdots \\ z_N \end{bmatrix}}_{N\times 1} \right\|_2^2 + \lambda \left\| \underbrace{\begin{bmatrix} D_{11} & & & & \\ & \ddots & & 0 & \\ & & \ddots & & \\ & 0 & & \ddots & \\ & & & & D_{NN} \end{bmatrix}}_{N\times N} \underbrace{\begin{bmatrix} z_1 \\ z_2 \\ \vdots \\ \vdots \\ \vdots \\ z_N \end{bmatrix}}_{N\times 1} \right\|_2^2, \tag{10.186}$$

where $a_{i,j}$ is the (i,j)th element of $\mathbf{A}_w$.

Both the unknown vector $\underline{x}$ and its wavelet vector $\underline{z}$ are of size $N \times 1$. This is in contrast with the much shorter measurement vector $\underline{y}$, which is of size $M \times 1$, with $M < N$.

We now introduce vector $\mathbf{y}$ and matrix $\mathbf{A}_w$ as

$$[\mathbf{y}] = \underbrace{\begin{bmatrix} \underline{y} \\ \underline{0} \end{bmatrix}}_{(M+N)\times 1} \begin{matrix} \}M\times 1 \\ \}N\times 1 \end{matrix}, \tag{10.187a}$$

$$[\mathbf{A}_w] = \underbrace{\begin{bmatrix} A_w \\ \sqrt{2\lambda}\, D \end{bmatrix}}_{(M+N)\times N} \begin{matrix} \}M\times N \\ \}N\times N \end{matrix}. \tag{10.187b}$$

Vector $\mathbf{y}$ is vector $\underline{y}$ of length M stacked on top of vector $\underline{0}$, which is a column vector of zeros of length N. Similarly, matrix $\mathbf{A}_w$ is matrix A_w (of size $M \times N$) stacked on top of matrix D (of size $N \times N$), multiplied by the scalar factor $\sqrt{2\lambda}$.

Next, we introduce the new cost function $\mathcal{T}_1$ as

$$\begin{aligned} \mathcal{T}_1 &= \frac{1}{2}\,\|\mathbf{y} - \mathbf{A}_w\,\underline{z}\|_2^2 \\ &= \frac{1}{2}\left\| \begin{bmatrix} \underline{y} \\ \underline{0} \end{bmatrix} - \begin{bmatrix} A_w \\ \sqrt{2\lambda}\, D \end{bmatrix} \underline{z} \right\|_2^2 \\ &= \frac{1}{2}\,\|\underline{y} - A_w\,\underline{z}\|_2^2 + \lambda\,\|0 - D\,\underline{z}\|_2^2 \\ &= \frac{1}{2}\,\|\underline{y} - A_w\,\underline{z}\|_2^2 + \lambda\,\|D\,\underline{z}\|_2^2 = \mathcal{T}. \end{aligned} \tag{10.188}$$

Hence, Eq. (10.185) can be rewritten in the form

$$\mathcal{T} = \tfrac{1}{2}\,\|\mathbf{y} - \mathbf{A}_w\,\underline{z}\|_2^2. \tag{10.189}$$

The vector $\underline{z}$ minimizing $\mathcal{T}$ is the pseudo-inverse given by

$$\begin{aligned} \hat{\underline{z}} &= (\mathbf{A}_w^T\mathbf{A}_w)^{-1}\mathbf{A}_w^T\,\mathbf{y} \\ &= \left([A_w^T \;\; \sqrt{2\lambda}\, D^T] \begin{bmatrix} A_w \\ \sqrt{2\lambda}\, D \end{bmatrix} \right)^{-1} [A_w^T \;\; \sqrt{2\lambda}\, D^T] \begin{bmatrix} \underline{y} \\ \underline{0} \end{bmatrix} \\ &= (A_w^T\,A_w + 2\lambda\, D^T\, D)^{-1}\, A_w^T\,\underline{y}. \end{aligned} \tag{10.190}$$

As always, instead of performing matrix inversion (which is susceptible to noise amplification), vector $\hat{\underline{z}}$ should be computed by solving

$$(A_w^T\,A_w + 2\lambda\, D^T\, D)\,\hat{\underline{z}} = A_w^T\,\underline{y}. \tag{10.191}$$

Once $\hat{\underline{z}}$ has been determined, the unknown vector $\underline{x}$ can be computed by solving Eq. (10.170).

To solve Eq. (10.191) for $\hat{\underline{z}}$, however, we need to know A_w, λ, D, and $\underline{y}$. From Eq. (10.172b), $A_w = A\,W^{-1}$, where A is a known matrix based on a physical model or calibration data, and W is a known wavelet transform matrix. The parameter λ is specified by the user to adjust the intended balance between data fidelity and storage size (as noted in connection with Eq. (10.185)), and $\underline{y}$ is the measurement vector. The only remaining quantity is the diagonal matrix D, whose function is to assign weights to z_1 through z_N so as to minimize storage size by having many elements of $\underline{z} \to 0$. Initially, D is unknown, but it is possible to propose an initial function for D and then iterate to obtain a solution for $\underline{z}$ that minimizes the number of nonzero elements, while still satisfying Eq. (10.191).

The Tikhonov function given by Eq. (10.185) reduces to the LASSO functional given by Eq. (10.179) if

$$D = \text{diag}\left[\frac{1}{\sqrt{|z_n|}}\right]. \tag{10.192}$$

This is because the second terms in the two equations become identical:

$$||D\,\underline{z}||_2^2 = \sum_{n=1}^{N} \frac{z_n^2}{|z_n|} = \sum_{n=1}^{N} |z_n| = ||\underline{z}||_1. \quad (10.193)$$

Given this correspondence between the two cost functionals, the IRLS algorithm uses the following iterative procedure to find $\underline{z}$:

(a) Initial solution: Set $D = I$ and then compute $\underline{z}^{(1)}$, the initial iteration of $\underline{z}$, by solving Eq. (10.191).

(b) Initial D: Use $\underline{z}^{(1)}$ to compute $D^{(1)}$, the initial iteration of D:

$$D^{(1)} = \text{diag}\left[\frac{1}{\sqrt{|z_n^{(1)}| + \epsilon}}\right] \quad (10.194)$$

(c) Second iteration: Use $D^{(1)}$ to compute $\underline{z}^{(2)}$ by solving Eq. (10.191) again.

(d) Recursion: Continue to iterate by computing $D^{(k)}$ from $\underline{z}^{(k)}$ using

$$D^{(k)} = \text{diag}\left[\frac{1}{\sqrt{|z_n^{(k)}| + \epsilon}}\right] \quad (10.195)$$

for a small deviation ϵ inserted in the expression to keep $D^{(k)}$ finite when elements of $\underline{z}^{(k)} \to 0$.

The iterative process ends when no significant change occurs between successive iterations. The algorithm, also called ***focal underdetermined system solver*** (FOCUSS), is guaranteed to converge under mild assumptions. However, because the method requires a solution of a large system of equations at each iteration, the algorithm is considered unsuitable for most signal and image processing applications. Superior-performance algorithms are introduced in succeeding sections.

10-17 Landweber Algorithm

The ***Landweber algorithm*** is a recursive algorithm for solving linear systems of equations $\underline{y} = Ax$. The ***iterative shrinkage and thresholding algorithm*** (ISTA) consists of the Landweber algorithm,with thresholding and shrinkage applied at each recursion. Thresholding and shrinkage were used in Section 10-14 to minimize the LASSO functional.

10-17.1 Underdetermined System

For an underdetermined system $\underline{y} = A\underline{x}$ with $M < N$, the solution $\hat{\underline{x}}$ that minimizes the sum of squares of the elements of $\underline{x}$ is, by analogy with Eq. (10.182), given by

$$\hat{\underline{x}} = A^T (A\,A^T)^{-1}\,\underline{y}. \quad (10.196)$$

A useful relationship in matrix algebra states that if all of the eigenvalues λ_i of $(A\,A^T)$ lie in the interval $0 < \lambda_i < 2$, then the coefficient of $\underline{y}$ in Eq. (10.196) can be written as

$$A^T(AA^T)^{-1} = \sum_{k=0}^{\infty}(I - A^T A)^k\,A^T. \quad (10.197)$$

The symbol λ_i for eigenvalue is unrelated to the trade-off parameter λ in Eq. (10.185).

Using Eq. (10.197) in Eq. (10.196) leads to

$$\hat{\underline{x}} = \sum_{k=0}^{\infty}(I - A^T A)^k\,A^T\,\underline{y}. \quad (10.198)$$

A recursive implementation of Eq. (10.198) assumes the form

$$\underline{x}^{(K+1)} = \sum_{k=0}^{K}(I - A^T A)^k\,A^T\,\underline{y}, \quad (10.199)$$

where the upper limit in the summation is now K (instead of ∞). For $K = 0$ and $K = 1$, we obtain the expressions

$$\underline{x}^{(1)} = A^T\,\underline{y}, \quad (10.200a)$$

$$\underline{x}^{(2)} = A^T\,\underline{y} + (I - A^T A)A^T\,\underline{y} = (I - A^T A)\,\underline{x}^{(1)} + A^T\,\underline{y}. \quad (10.200b)$$

Extending the process to $K = 2$ gives

$$\underline{x}^{(3)} = \underbrace{A^T\,\underline{y}}_{k=0} + \underbrace{(I - A^T A)A^T\,\underline{y}}_{k=1} + \underbrace{(I - A^T A)^2\,A^T\,\underline{y}}_{k=2}$$
$$= \underline{x}^{(2)} + A^T(\underline{y} - A\,\underline{x}^{(2)}). \quad (10.200c)$$

Continuing the pattern leads to

$$\underline{x}^{(k+1)} = \underline{x}^{(k)} + A^T(\underline{y} - A\,\underline{x}^{(k)}). \quad (10.201)$$

The process can be initialized by $\underline{x}^{(0)} = \underline{0}$, which makes $\underline{x}^{(1)} = A^T\,\underline{y}$, as it should.

The recursion process is called the ***Landweber iteration***, which in optics is known as the ***van Cittert iteration***. It is guaranteed to converge to the solution of $\underline{y} = A\underline{x}$ that minimizes the sum of squares of the elements of $\underline{x}$, provided that the eigenvalues λ_i of AA^T are within the range $0 < \lambda_i < 2$. If this condition is not satisfied, the formulation may be scaled to

$$\frac{\underline{y}}{c} = \frac{A}{c}\,\underline{x}$$

or, equivalently,

$$\underline{u} = B\underline{x}, \tag{10.202}$$

where $\underline{u} = \underline{y}/c$ and $B = A/c$. The constant c is chosen so that the eigenvalue condition is satisfied. For example, if c is chosen to be equal to the sum of the squares of the magnitudes of all of the elements of A, then the eigenvalues λ_i' of BB^T will be in the range $0 < \lambda_i' < 1$.

10-17.2 Overdetermined System

In analogy with Eq. (10.181a), the solution for an overdetermined system $\underline{y} = A\underline{x}$ is given by

$$\hat{\underline{x}} = (A^T A)^{-1} A^T \underline{y}. \tag{10.203}$$

Using the equality

$$(A^T A)^{-1} = \sum_{k=0}^{\infty} (I - A^T A)^k \tag{10.204}$$

we can rewrite Eq. (10.203) in the same form as Eq. (10.198), namely

$$\hat{\underline{x}} = \sum_{k=0}^{\infty} (I - A^T A)^k \, A^T \, \underline{y}. \tag{10.205}$$

Hence, the Landweber algorithm is equally applicable to solving overdetermined systems of linear equations.

10-17.3 Iterative Shrinkage and Thresholding Algorithm (ISTA)

For a linear system given by

$$\underline{y} = A\,\underline{x}, \tag{10.206}$$

where $\underline{x}$ is the unknown signal of length N and $\underline{y}$ is the (possibly noisy) observation of length M, the LASSO cost functional is

$$\Lambda = \frac{1}{2} \sum_{n=0}^{N-1} (y[n] - (A\underline{x})[n])^2 + \lambda \sum_{n=0}^{N-1} |x[n]|, \tag{10.207}$$

where matrix A is $M \times N$ and $(A\,\underline{x})[n]$ is the nth element of $A\,\underline{x}$.

► In the system described by Eq. (10.206), $\underline{x}$ and $\underline{y}$ are generic input and output vectors. The Landweber algorithm provides a good estimate of $\underline{x}$, given $\underline{y}$. The estimation algorithm is equally applicable to any other linear system, including the system $\underline{y} = A_w \underline{z}$, where A_w is the matrix given by Eq. (10.172b) and $\underline{z}$ is the wavelet transform vector. ◄

The ISTA algorithm combines the Landweber algorithm with the thresholding and shrinkage operation outlined earlier in Section 10-14.3, and summarized by Eq. (10.166). After each iteration, elements $x^{(k)}[n]$, of vector $\underline{x}^{(k)}$, whose absolute values are smaller than the trade-off parameter λ are thresholded to zero, and those whose absolute values are larger than λ are shrunk by λ. Hence, the ISTA algorithm combines Eq. (10.201) with Eq. (10.166):

$$\underline{x}^{(0)} = \underline{0}, \tag{10.208a}$$

$$\underline{x}^{(k+1)} = \underline{x}^{(k)} + A^T(\underline{y} - A\,\underline{x}^{(k)}), \tag{10.208b}$$

with

$$x_i^{(k+1)} = \begin{cases} x_i^{(k+1)} - \lambda & \text{if } x_i^{(k+1)} > \lambda, \\ x_i^{(k+1)} + \lambda & \text{if } x_i^{(k+1)} < -\lambda, \\ 0 & \text{if } |x_i^{(k+1)}| < \lambda, \end{cases} \tag{10.208c}$$

where $x_i^{(k+1)}$ is the ith component of $\underline{x}^{(k+1)}$.

The ISTA algorithm converges to the value of $\underline{x}$ that minimizes the LASSO functional given by Eq. (10.207), provided all of the eigenvalues λ_i of AA^T obey $|\lambda_i| < 1$.

The combination of thresholding and shrinking is often called ***soft thresholding***, while thresholding small values to zero without shrinking is called ***hard thresholding***. There are many variations on ISTA, with names like SPARSA (***sparse reconstruction by separable approximation***), FISTA (***fast iterative shrinkage and thresholding algorithm***), and TWISTA (***two-step iterative shrinkage and thresholding algorithm***).

Concept Question 10-18: What is the difference between Landweber and ISTA? (See S²)

Concept Question 10-19: What condition guarantees that the Landweber iteration converges to the solution? (See S²)

10-18 Compressed Sensing Examples

To illustrate the utility of the ISTA described in the preceding section, we present four examples of compressed sensing:

- **Reconstruction** of an image from *some*, but not all, of its 2-D DFT values.
- **Image inpainting**, which entails filling in holes (missing pixel values) in an image.
- **Valid deconvolution** of an image from only part of its convolution with a known point spread function (PSF).
- **Tomography**, which involves reconstruction of a 3-D image from slices of its 2-D DFT.

These are only a few of many more possible types of applications of compressed sensing.

The ISTA was used in all four cases, and the maximum number of possible iterations was set at 1000, or fewer if the algorithm converges to where no apparent change is observed in the reconstructed images. The LASSO functional parameter was set at $\lambda = 0.01$, as this value seemed to provide the best results (see S² for more details on all four applications).

10-18.1 Image Reconstruction from Subset of DFT Values

Suppose that after computing the 2-D DFT of an image $x[m,n]$, some of the DFT values $\mathbf{X}[k_1,k_2]$ were lost or no longer available. The goal is to reconstruct image $x[m,n]$ from the partial subset of its DFTs. Since the available DFT values are fewer than those of the unknown signal, the system is underdetermined and the application is a good illustration of compressed sensing.

For a 1-D signal $\{x[n], n = 0, 1, \ldots, N-1\}$, its DFT can be implemented by the matrix-vector product

$$\underline{y} = A\,\underline{x}, \tag{10.209}$$

where the (k,n)th element of A is $A_{k,n} = e^{-j2\pi nk/N}$, the nth element of vector $\underline{x}$ is $x[n]$, and the kth element of vector $\underline{y}$ is $\mathbf{X}[k]$. Multiplication of both sides by A^H implements an inverse 1-D DFT within a factor $1/N$.

The 2-D DFT of an $N \times N$ image can be implemented by multiplication by an $N^2 \times N^2$ block matrix B whose (k_2,n_2)th block is the $N \times N$ matrix A multiplied by the scalar $e^{-j2\pi n_2k_2/N}$. So the element $B_{k_1+Nk_2,n_1+Nn_2}$ of B is $e^{-j2\pi n_1k_1/N}e^{-j2\pi n_2k_2/N}$, where $0 \le n_1, n_2, k_1, k_2 \le N-1$.

The absence of some of the DFT values is equivalent to deleting some of the rows of B and $\underline{y}$, thereby establishing an underdetermined linear system of equations. To illustrate the reconstruction process for this underdetermined system, we consider two different scenarios applied to the same set of data.

► For convenience, we call the values $\mathbf{X}[k_1,k_2]$ of the 2-D DFT the ***pixels of the DFT image***. ◄

(a) Least-squares reconstruction

Starting with the 256×256 image shown in Fig. 10-34(a), we compute its 2-D DFT and then we randomly select a subset of the pixels in the DFT image and label them as *unknown*. The complete 2-D DFT image consists of $256^2 = 65536$ pixel values of $\mathbf{X}[k_1,k_2]$. Of those, 35755 are unaltered (and therefore have known values), and the other 29781 pixels have unknown values. Figure 10-34(b) displays the locations of pixels with known DFT values as white dots and those with unknown values as black dots.

In the least-squares reconstruction method, all of the pixels with unknown values are set to zero, and then the inverse 2-D DFT is computed. The resulting image, shown in Fig. 10-34(c), is a poor rendition of the original image in part (a) of the figure.

(b) ISTA reconstruction

The ISTA reconstruction process consists of two steps:

(1) Measurement vector $\underline{y}$, representing the 35755 DFT pixels with known values, is used to estimate the (entire 65536) wavelet transform vector $\underline{z}$ by applying the recipe outlined in Section 10-17.3 with $\lambda = 0.01$ and 1000 iterations. The relationship between $\underline{y}$ and $\underline{z}$ is given by $\underline{y} = A_w\underline{z}$, with $A_w = BW$ and some rows of B deleted.

(2) Vector $\underline{z}$ is then used to reconstruct $\underline{x}$ by applying the relation $\underline{x} = W^{-1}\underline{z}$. The Haar transform was used in this step.

The reconstructed image, displayed in Fig. 10-34(d), is an excellent rendition of the original image.

It is important to note that while B and W^{-1} are each $N^2 \times N^2$, with $N = 65536$, neither matrix is ever computed or stored during the implementation of the reconstruction process. Multiplication by W^{-1} is implemented by a 2-D filter bank, and multiplication by B is implemented by a 2-D FFT. Consequently, ISTA is a very fast algorithm.

10-18.2 Image Inpainting

In an image inpainting problem, some of the pixel values of an image are unknown, either because those pixel values have been corrupted, or because they represent some unwanted feature of the image that we wish to remove. The goal is to restore the image to its original version, in which the unknown pixel

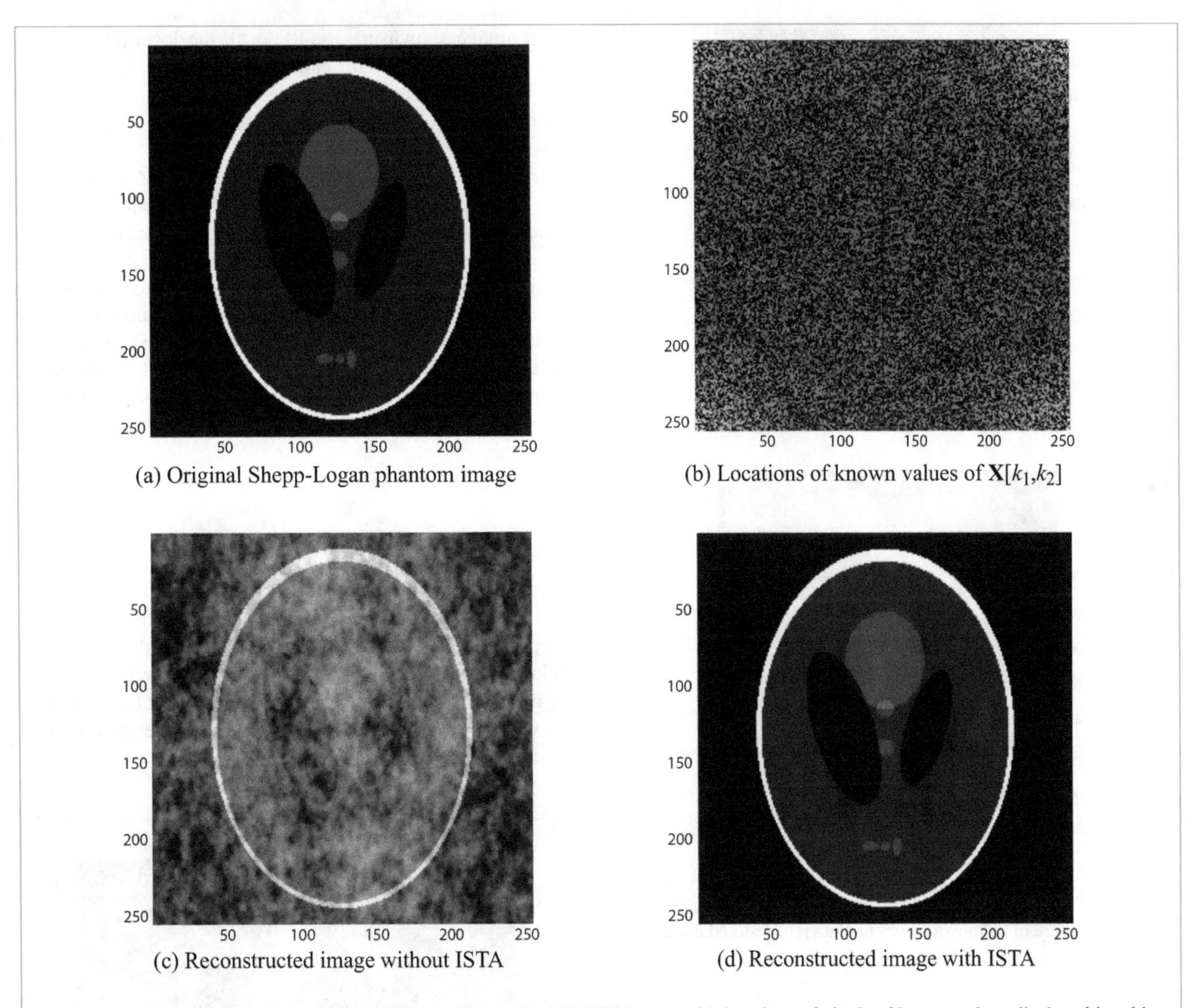

Figure 10-34: (a) Original Shepp-Logan phantom image, (b) 2-D DFT image with locations of pixels of known values displayed in white and those of unknown values displayed in black, (c) reconstructed image using available DFT pixels and (d) reconstructed image after filling in missing DFT pixel values with estimates provided by ISTA.

values are replaced with the, hitherto, unknown pixel values of the original image. This can be viewed as a kind of interpolation problem.

It is not at all evident that this can be done at all—how can we restore unknown pixel values? But under the assumption that the Daubechies wavelet transform of the image is sparse (mostly zero-valued), image inpainting can be formulated as a compressed sensing problem. Let $\underline{y}$ be the vector of the known pixel values and $\underline{x}$ be the vector of the wavelet transform of the image. Note that all elements of $\underline{x}$ are unknown, even though some of the pixel values are actually known. Then the problem can be formulated as an underdetermined linear system $\underline{y} = A\underline{x}$.

For example, if $\underline{x}$ is a column vector of length five, and only the first, third, and fourth elements of $\underline{x}$ are known, the problem can be formulated as

$$\begin{bmatrix} y_1 \\ y_2 \\ y_3 \end{bmatrix} = \begin{bmatrix} x_1 \\ x_3 \\ x_4 \end{bmatrix} = \begin{bmatrix} 1 & 0 & 0 & 0 & 0 \\ 0 & 0 & 1 & 0 & 0 \\ 0 & 0 & 0 & 1 & 0 \end{bmatrix} \begin{bmatrix} x_1 \\ x_2 \\ x_3 \\ x_4 \\ x_5 \end{bmatrix}.$$

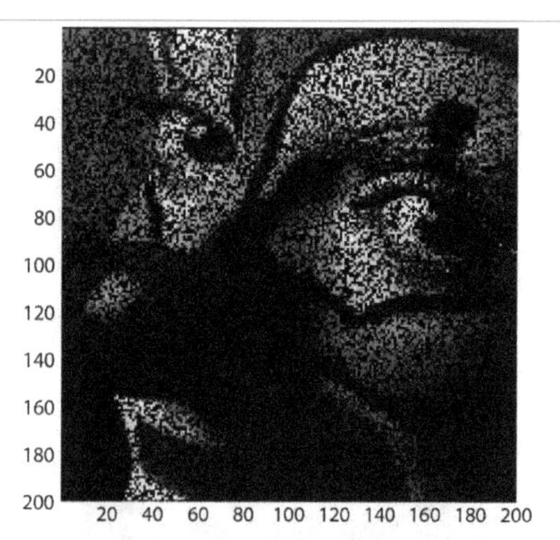

(a) Locations of known values of image

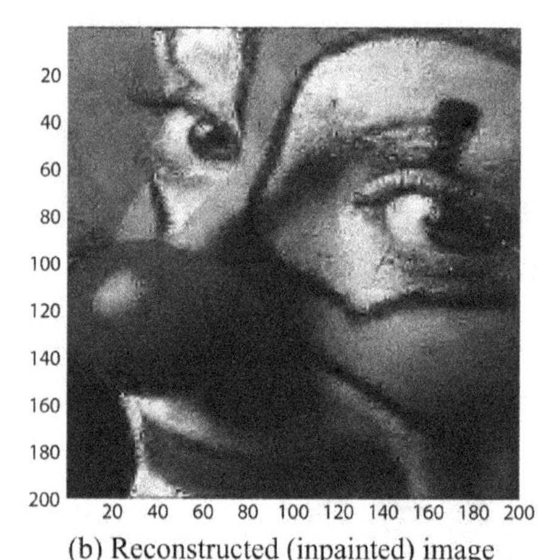

(b) Reconstructed (inpainted) image

Figure 10-35: (a) Locations of known values of clown image in white and those of unknown values in black; (b) restored image.

One application of image inpainting is to restore a painting in which the paint in some regions of the painting has been chipped off, scraped off, damaged by water or simply faded, but most of the painting is unaffected. Another application is to remove unwanted letters or numbers from an image. Still another application is "wire removal" in movies, the elimination of wires used to suspend actors or objects used for an action stunt in a movie scene.

In all of these cases, damage to the painting, or presence of unwanted objects in the image, has made some small regions of the painting or image unknown. The goal is to fill in the unknown values to restore the (digitized) painting or image to its original version.

Using a $200 \times 200 = 40000$-pixel clown image, 19723 pixels were randomly selected and their true values were deleted. In the image shown in Fig. 10-35(a), the locations of pixels with unknown values are painted black, while the remaining half (approximately) have their correct values. The goal is to reconstruct the clown image from the remaining half. The db3 wavelet transform was used.

In terms of the formulation $\underline{y} = AW^T\underline{z}$, $M = 20277$ and $N = 40000$, so that just over half of the clown image pixel values are known. The ISTA is a good algorithm to solve this compressed sensing problem, since the matrix vector multiplication $\underline{y} = AW^T\underline{z}$ can be implemented quickly by taking the inverse wavelet transform of the current iteration (multiplication by W^T), and then selecting a subset of the pixel values (multiplication by A). The result, after 1000 iterations, is shown in Fig. 10-35(b). The image has been reconstructed quite well, but not perfectly.

Module 10.7 Inpainting of Clown Image Using Daubechies Wavelets and IST Algorithm This module deletes pixels of the clown image at randomly selected locations, and uses the IST algorithm to reconstruct the db3 Daubechies wavelet transform of the clown image from the known clown pixel values. The "missing pixel threshold" slider controls what fraction of pixels is deleted. This module is an interactive version of Fig. 10-35. *Note:* The module must be rerun every time any slider is changed.

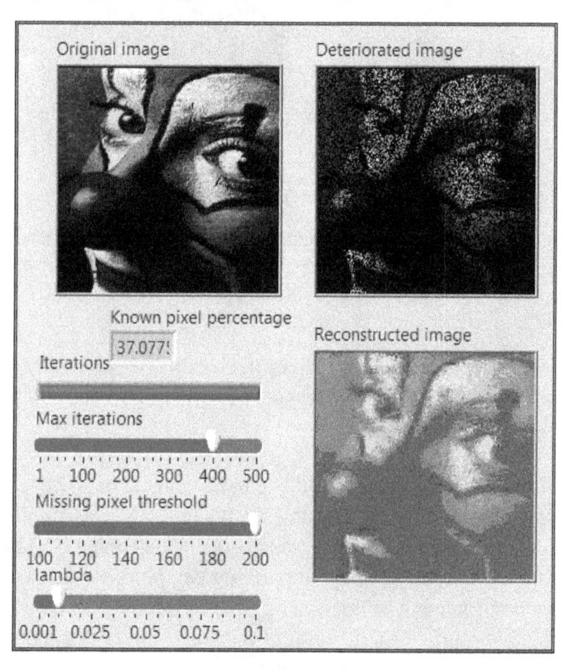

10-18.3 Valid 2-D Deconvolution

(a) **Definition of valid convolution**

Given an $M \times M$ image $x[m,n]$ and an $L \times L$ point spread function (PSF) $h[m,n]$, their 2-D convolution generates an $(M+L-1)\times(M+L-1)$ blurred image $y[m,n]$. The process is reversible: the blurred image can be deconvolved to reconstruct $x[m,n]$ by subjecting $y[m,n]$ to a Wiener filter, as described earlier in Section 10-7. To do so, however, requires that all of $y[m,n]$ be known.

Often, we encounter deconvolution applications where only a fraction of $y[m,n]$ is known, specifically, the part of $y[m,n]$ called the ***valid convolution***. This is the part whose convolution computation does not require the image $x[m,n]$ to be zero-valued outside the square $0 \leq m,n \leq M-1$.

For $L < M$ (image larger than PSF), the valid 2-D convolution of $h[m,n]$ and $x[m,n]$ is defined as

$$\begin{aligned} y_V[m,n] &= \sum_{i=0}^{M-1}\sum_{j=0}^{M-1} x[i,j]\, h[m-i,\, n-j] \\ &= h[m,n] * x[m,n], \text{ restricted to} \\ &\quad \{L-1 \leq m,n \leq M-1\}. \end{aligned} \tag{10.210}$$

A valid convolution omits all end effects in 1-D convolution and all edge effects in 2-D convolution. Consequently, the size of $y_V[m,n]$ is $(M-L+1)\times(M-L+1)$, instead of $(M+L-1)\times(M+L-1)$ for the complete convolution $y[m,n]$.

To further illustrate the difference between $y[m,n]$ and $y_V[m,n]$, let us consider the following example:

$$x[m,n] = \begin{bmatrix} 1 & 2 & 3 \\ 4 & 5 & 6 \\ 7 & 8 & 9 \end{bmatrix}$$

and

$$h[m,n] = \begin{bmatrix} 11 & 12 \\ 13 & 14 \end{bmatrix}.$$

Since $M = 3$ and $L = 2$, the 2-D convolution is $(M+L-1)\times(M+L-1) = 4\times 4$, and $y[m,n]$ is given by

$$y[m,n] = \begin{bmatrix} 11 & 34 & 57 & 36 \\ 57 & 143 & 193 & 114 \\ 129 & 293 & 343 & 192 \\ 91 & 202 & 229 & 126 \end{bmatrix}.$$

In contrast, the size of the valid 2-D convolution is $(M-L+1)\times(M-L+1) = 2\times 2$, and $y_V[m,n]$ is given by

$$y_V[m,n] = \begin{bmatrix} 143 & 193 \\ 293 & 343 \end{bmatrix}.$$

The valid convolution $y_V[m,n]$ is the central part of $y[m,n]$, obtained by deleting the edge rows and columns from $y[m,n]$. In MATLAB, the valid 2-D convolution of X and H can be computed using the command

```
Y=conv2(X,H,'valid').
```

(b) Reconstruction from $y_V[m,n]$

The valid 2-D deconvolution problem is to reconstruct an unknown image from its valid 2-D convolution with a known PSF. The 2-D DFT and Wiener filter cannot be used here, since not all of the blurred image $y[m,n]$ is known. It may seem that we may simply ignore, or set to zero, the unknown parts of $y[m,n]$ and still obtain a decent reconstructed image using a Wiener filter, but as we will demonstrate with an example, such an approach does not yield fruitful results.

The valid 2-D deconvolution problem is clearly underdetermined, since the $(M-L+1)\times(M-L+1)$ portion of the blurred image is smaller than the $M\times M$ unknown image. But if $x[m,n]$ is sparsifiable, then valid 2-D deconvolution can be formulated as a compressed sensing problem and solved using the ISTA. The matrix A turns out to be a ***block Toeplitz with Toeplitz blocks*** matrix, but multiplication by A is implemented as a valid 2-D convolution. Multiplication by A^T is implemented as a valid 2-D convolution.

The valid 2-D convolution can be implemented as $\underline{y}_V = A\underline{x}$ where

$$\underline{x} = [1\ 2\ 3\ 4\ 5\ 6\ 7\ 8\ 9]^T,$$

$$\underline{y}_V = [143\ 193\ 293\ 343]^T,$$

and the matrix A is composed of the elements of $h[m,n]$ as follows:

$$A = \begin{bmatrix} 14 & 13 & 0 & 12 & 11 & 0 & 0 & 0 & 0 \\ 0 & 14 & 13 & 0 & 12 & 11 & 0 & 0 & 0 \\ 0 & 0 & 0 & 14 & 13 & 0 & 12 & 11 & 0 \\ 0 & 0 & 0 & 0 & 14 & 13 & 0 & 12 & 11 \end{bmatrix}.$$

Note that A is a 2×3 block matrix of 2×3 blocks. Each block is constant along its diagonals, and the blocks are constant along block diagonals. This is the block Toeplitz with Toeplitz blocks structure. Also note that images $x[m,n]$ and $y_V[m,n]$ have been unwrapped row by row, starting with the top row, and the transposes of the rows are stacked into a column vector. Finally, note that multiplication by A^T can be implemented as a valid 2-D convolution with the doubly reversed version of $h[m,n]$. For example, if

$$z[m,n] = \begin{bmatrix} 1 & 2 & 3 & 4 \\ 5 & 6 & 7 & 8 \\ 9 & 10 & 11 & 12 \\ 13 & 14 & 15 & 16 \end{bmatrix}$$

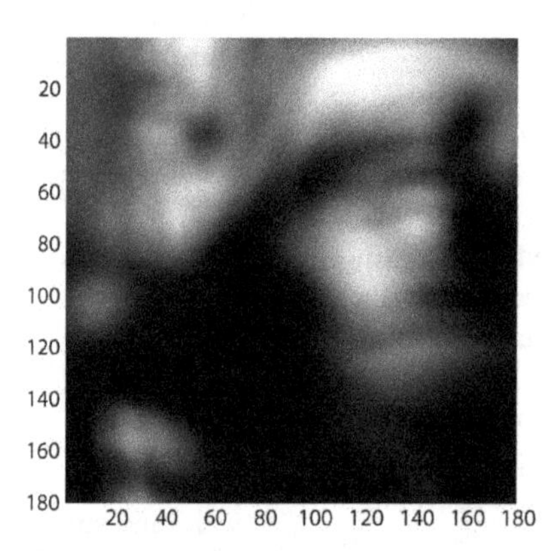

(a) Valid 2-D convolution $y_{\text{V}}[m,n]$ of clown image

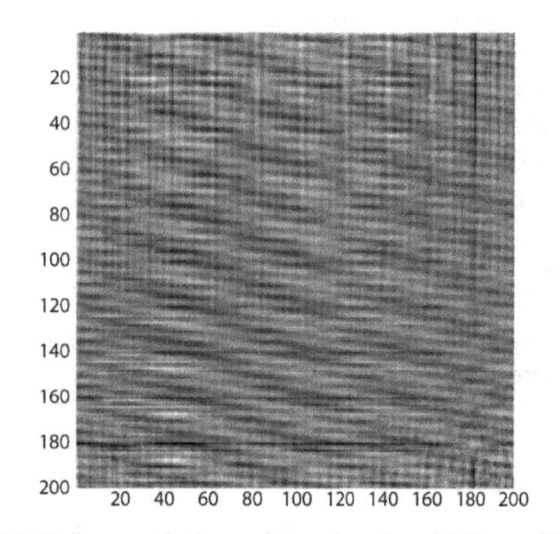

(b) 2-D deconvolution using $y_{\text{V}}[m,n]$ and Wiener filter

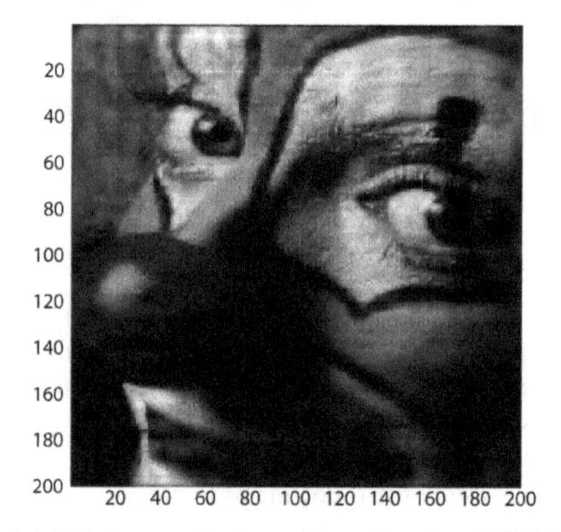

(c) 2-D deconvolution using $y_{\text{V}}[m,n]$ and ISTA

Figure 10-36: (a) Valid 2-D convolution $y_{\text{V}}[m,n]$ of clown image, (b) deconvolution using Wiener filter, and (c) deconvolution using ISTA.

and

$$g[m,n] = \begin{bmatrix} 14 & 13 \\ 12 & 11 \end{bmatrix} = h[1-m,\ 1-n], \quad \text{with } m,n = 0,1,$$

then the valid 2-D convolution of $z[m,n]$ and $g[m,n]$ is

$$w_{\text{v}}[m,n] = \begin{bmatrix} 184 & 234 & 284 \\ 384 & 434 & 484 \\ 584 & 634 & 684 \end{bmatrix}.$$

This valid 2-D convolution can also be implemented as $\underline{w} = A^T \underline{z}$ where

$$\underline{z} = [1\ 2\ 3\ 4\ 5\ 6\ 7\ 8\ 9\ 10\ 11\ 12\ 13\ 14\ 15\ 16]^T$$

and

$$\underline{w} = [184\ 234\ 284\ 384\ 434\ 484\ 584\ 634\ 684]^T.$$

To illustrate the process with an image, we computed the valid 2-D convolution of the 200×200 clown image with a 20×20 PSF. The goal is to reconstruct the clown image from the 181×181 blurred image shown in Fig. 10-36(a). The $db3$ Daubechies wavelet function was used to sparsify the image. Here, $M = 200$ and $L = 20$, so the valid 2-D convolution has size $(M-L+1) \times (M-L+1) = 181 \times 181$. In terms of $\underline{y}_{\text{V}} = A\underline{x}$, A is $181^2 \times 200^2 = 32761 \times 40000$.

Parts (b) and (c) of Fig. 10-36 show reconstructed versions of the clown image, using a Wiener filter and ISTA, respectively. Both images involve deconvolution using the restricted valid convolution data $y_{\text{V}}[m,n]$. In the Wiener-image approach, the unknown parts of the blurred image (beyond the edges of $y_{\text{V}}[m,n]$) were ignored, and the resultant image bears no real resemblance to the original clown image. In contrast, the ISTA approach provides excellent reconstruction of the original image. This is because ISTA is perfectly suited for solving underdetermined systems of linear equations with sparse solutions.

10-18.4 Computed Axial Tomography (CAT)

Computed axial tomography, also known as *CAT scan*, is a technique capable of generating 3-D images of the X-ray attenuation (absorption) properties of an object, such as the human body. The X-ray absorption coefficient of a material is strongly dependent on the density of that material. CAT has the sensitivity necessary to image body parts across a wide range of densities, from soft tissue to blood vessels and bones.

As depicted in Fig. 10-37(a), a CAT scanner uses an X-ray source, with a narrow slit to generate a fan-beam, wide enough to encompass the extent of the body, but only a few millimeters in thickness. The attenuated X-ray beam is captured by an array of ~ 700 detectors. The X-ray source and the detector array are mounted on a circular frame that rotates in steps of a fraction of a degree over a full 360° circle around the object or patient, each time recording an X-ray attenuation profile from a different angular direction. Typically, on the order of 1000 such profiles are recorded, each composed of measurements by 700 detectors. For each horizontal slice of the body, the process is completed in less than 1 second. CAT performs a deconvolution to generate a 2-D image of the absorption coefficient of that horizontal slice. To image an entire part of the body, such as the chest or head, the process is repeated over multiple slices (layers). Our current interest is in the deconvolution process, so we limit our treatment to the 2-D case.

For each anatomical slice, the CAT scanner generates on the order of 7×10^5 measurements (1000 angular orientations $\times 700$ detectors). In terms of the coordinate system shown in Fig. 10-37(b), we define $\alpha(\xi, \eta)$ as the absorption coefficient of the object under test at location (ξ, η). The X-ray beam is directed along the ξ direction at $\eta = \eta_0$. The X-ray intensity received by the detector located at $\xi = \xi_0$ and $\eta = \eta_0$ is given by

$$I(\xi_0, \eta_0) = I_0 \exp\left[-\int_0^{\xi_0} \alpha(\xi, \eta_0)\, d\xi \right], \qquad (10.211)$$

where I_0 is the X-ray intensity radiated by the source. Outside the body, $\alpha(\xi, \eta) = 0$. The corresponding logarithmic ***path attenuation*** $p(\xi_0, \eta_0)$ is defined as

$$p(\xi_0, \eta_0) = -\log \frac{I(\xi_0, \eta_0)}{I_0} = \int_0^{\xi_0} \alpha(\xi, \eta_0)\, d\xi. \qquad (10.212)$$

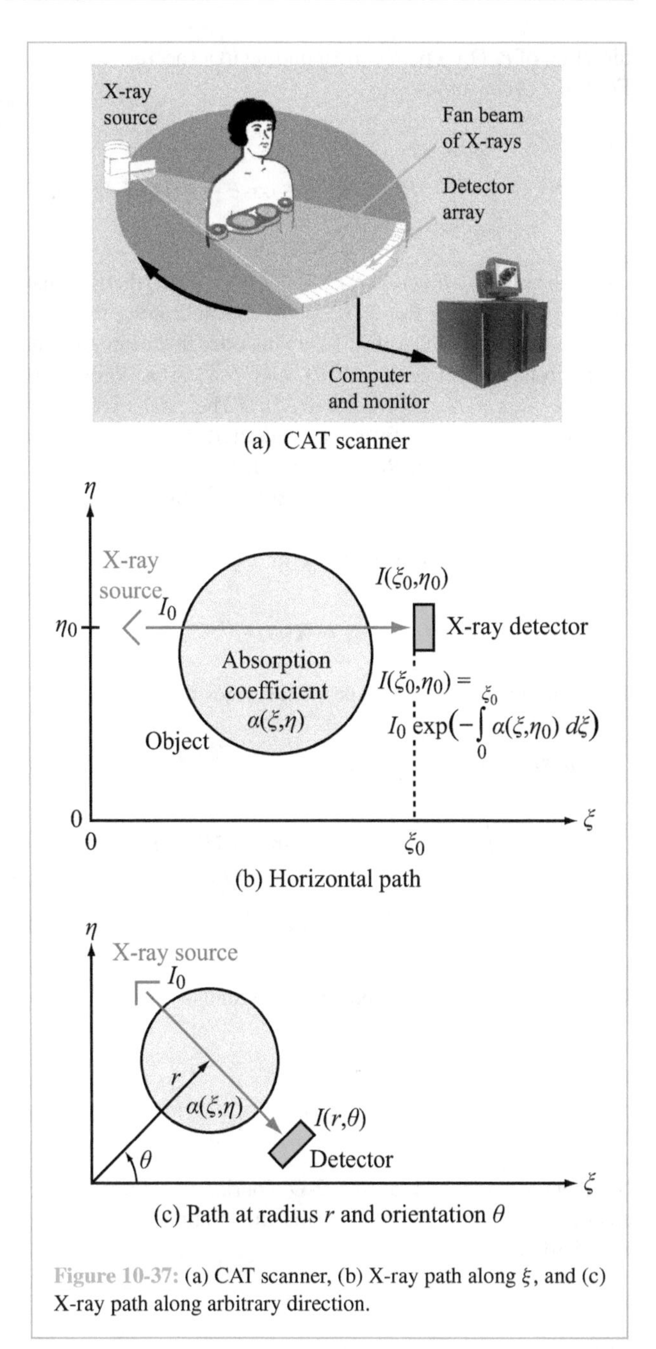

Figure 10-37: (a) CAT scanner, (b) X-ray path along ξ, and (c) X-ray path along arbitrary direction.

The path attenuation $p(\xi_0, \eta_0)$ is the integrated absorption coefficient across the X-ray path.

In the general case, the path traversed by the X-ray source is at a range r and angle θ in a polar coordinate system, as depicted in Fig. 10-37(c). The direction of the path is orthogonal to the direction of r. For a path corresponding to a specific set (r, θ), Eq. (10.212) becomes

$$p(r,\theta) = \int_{-\infty}^{\infty}\int_{-\infty}^{\infty} \alpha(\xi,\eta)\,\delta(r - \xi\cos\theta - \eta\sin\theta)\,d\xi\,d\eta, \tag{10.213}$$

where the impulse function $\delta(r - \xi\cos\theta - \eta\sin\theta)$ dictates that only those points in the (ξ, η) plane that fall along the path specified by fixed values of (r, θ) are included in the integration.

The relation between $p(r, \theta)$ and $\alpha(\xi, \eta)$ is known as the 2-D ***Radon transform*** of $\alpha(\xi, \eta)$. The goal of CAT is to reconstruct $\alpha(\xi, \eta)$ from the measured path attenuations $p(r, \theta)$, by inverting the Radon transform given by Eq. (10.213). We do so with the help of the Fourier transform.

Recall from entry #1 in Table 5-6 that for variable r, $\mathcal{F}\{\delta(r)\} = 1$, and from entry #4 in Table 5-7 that the shift property is

$$\mathcal{F}\{x(r - r_0)\} = \mathbf{X}(\omega)\,e^{-j\omega r_0}.$$

The combination of the two properties leads to

$$\begin{aligned}\mathcal{F}\{\delta(r - \xi\cos\theta - \eta\sin\theta)\} &= \int_0^{\infty} \delta(r - \xi\cos\theta - \eta\sin\theta)\,e^{-j\omega r}\,dr \\ &= e^{-j\omega(\xi\cos\theta + \eta\sin\theta)} = e^{-j(\omega_1\xi + \omega_2\eta)},\end{aligned} \tag{10.214}$$

where we define angular frequencies ω_1 and ω_2 as

$$\omega_1 = \omega\cos\theta, \tag{10.215a}$$

$$\omega_2 = \omega\sin\theta. \tag{10.215b}$$

Next, let us define $\mathbf{A}$ as the 2-D Fourier transform of the absorption coefficient $\alpha(\xi, \eta)$ using the relationship given by Eq. (5.143a):

$$\mathbf{A}(\omega_1,\omega_2) = \int_{-\infty}^{\infty}\int_{-\infty}^{\infty} \alpha(\xi,\eta)\,e^{-j\omega_1\xi}e^{-j\omega_2\eta}\,d\xi\,d\eta. \tag{10.216}$$

If we know $\mathbf{A}(\omega_1, \omega_2)$, we can perform an inverse 2-D Fourier transform to retrieve $\alpha(\xi, \eta)$. To do so, we need to relate $\mathbf{A}(\omega_1, \omega_2)$ to the measured path attenuation profiles $p(r, \theta)$. To that end, we use Eq. (10.213) to compute $\mathbf{P}(\omega, \theta)$, the 1-D Fourier transform of $p(r, \theta)$:

$$\begin{aligned}\mathbf{P}(\omega,\theta) &= \int_0^{\infty} p(r,\theta)\,e^{-j\omega r}\,dr \\ &= \int_0^{\infty}\Bigg[\int_{-\infty}^{\infty}\int_{-\infty}^{\infty} \alpha(\xi,\eta) \\ &\qquad \cdot\,\delta(r - \xi\cos\theta - \eta\sin\theta)\,d\xi\,d\eta\Bigg]e^{-j\omega r}\,dr.\end{aligned} \tag{10.217}$$

By reversing the order of integration, we have

$$\begin{aligned}\mathbf{P}(\omega,\theta) = \int_{-\infty}^{\infty}\int_{-\infty}^{\infty} &\alpha(\xi,\eta) \\ &\cdot\left[\int_0^{\infty} \delta(r - \xi\cos\theta - \eta\sin\theta)\,e^{-j\omega r}\,dr\right]d\xi\,d\eta.\end{aligned} \tag{10.218}$$

We recognize the integral inside the square bracket as the Fourier transform of the shifted impulse function, as given by Eq. (10.214). Hence, Eq. (10.218) simplifies to

$$\mathbf{P}(\omega,\theta) = \int_{-\infty}^{\infty}\int_{-\infty}^{\infty} \alpha(\xi,\eta)\,e^{-j(\omega_1\xi + \omega_2\eta)}\,d\xi\,d\eta, \tag{10.219}$$

which is identical to Eq. (10.216). Hence,

$$\mathbf{A}(\omega_1,\omega_2) = \mathbf{P}(\omega,\theta), \tag{10.220}$$

where $\mathbf{A}(\omega_1, \omega_2)$ is the 2-D Fourier transform of $\alpha(\xi, \eta)$, and $\mathbf{P}$ is the 1-D Fourier transform (with respect to r) of $p(r, \theta)$. The variables (ω_1, ω_2) and (ω, θ) and related by Eq. (10.215).

If $p(r, \theta)$ is measured for all r across the body of interest and for all directions θ, then its 1-D Fourier transform $\mathbf{P}(\omega, \theta)$ can be computed, and then converted to $\mathbf{A}(\omega_1, \omega_2)$ using Eq. (10.215). The conversion is called the ***projection-slice theorem***. In practice, however, $p(r, \theta)$ is measured for only a finite number of angles θ, so $\mathbf{A}(\omega_1, \omega_2)$ is known only along radial slices in the 2-D wavenumber domain (ω_1, ω_2). Reconstruction to find $\alpha(\xi, \eta)$ from a subset of its 2-D Fourier transform values is a perfect example of compressed sensing.

Image reconstruction from partial radial slices

To demonstrate the reconstruction process, we computed the 2-D DFT $\mathbf{X}[k_1, k_2]$ of a 256×256 Shepp-Logan phantom image, and then retained the data values corresponding to only 12 radial slices, as shown in Fig. 10-38(a). These radial slices simulate $\mathbf{P}(\omega, \theta)$, corresponding to 12 radial measurements $p(r, \theta)$. In terms of $\underline{y} = A\underline{x}$, the number of pixels in the frequency domain image is $N = 65536$, and the number of values contained in the 12 radial slices is $M = 11177$.

(a) Least-squares reconstruction: Unknown values of $\mathbf{X}[k_1, k_2]$ were set to zero, and then the inverse 2-D DFT was computed. The resulting image is displayed in Fig. 10-38(b).

(b) ISTA reconstruction: Application of ISTA with $\lambda = 0.01$ for 1000 iterations led to the image in Fig. 10-38(c), which bears very good resemblance to the original image.

Concept Question 10-20: Why is it possible to reconstruct a real-world image almost perfectly from only a subset of its 2-D DFT values, or a subset of its pixel values? (See S^2)

Exercise 10-15: Use LabVIEW Module 10.7 to inpaint the clown image. Use lambda = 0.01, missing pixel threshold = 140, and max iterations = 500.

Answer:

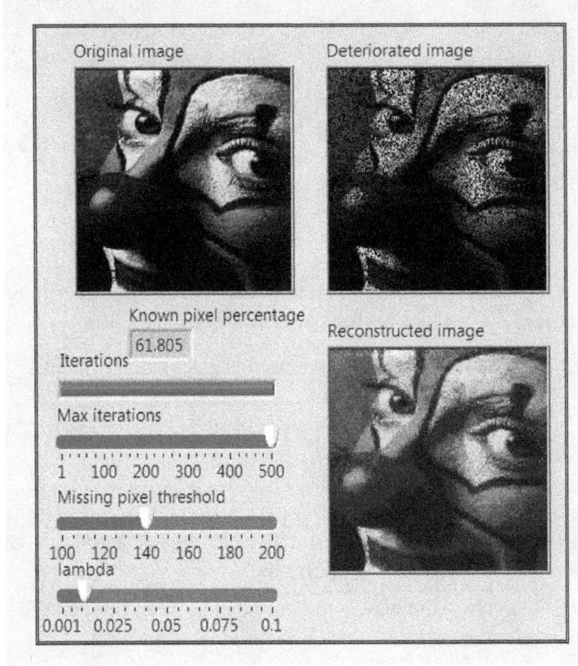

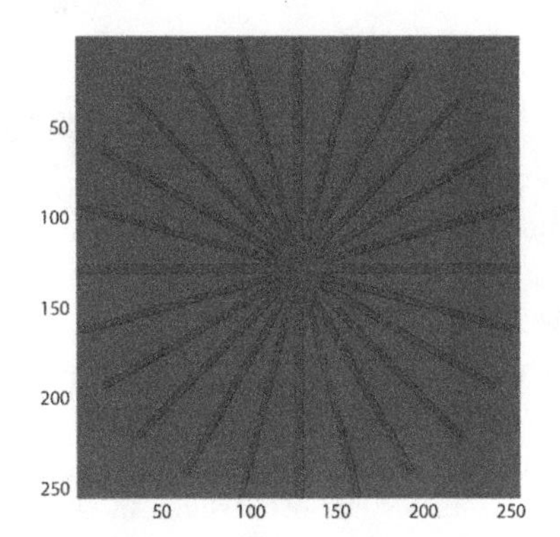

(a) Locations of known values of $\mathbf{X}[k_1,k_2]$

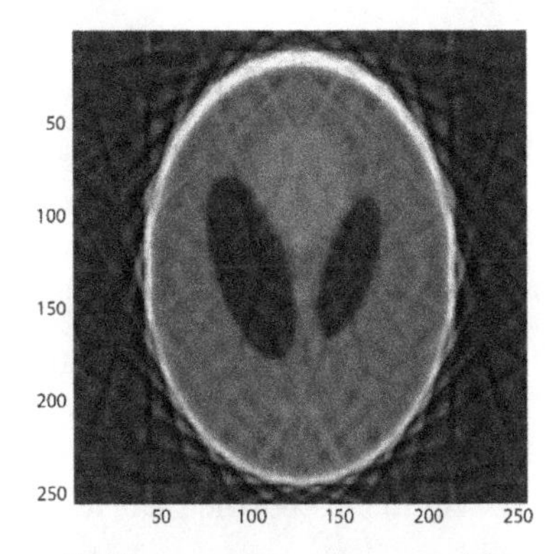

(b) Least-squares reconstruction

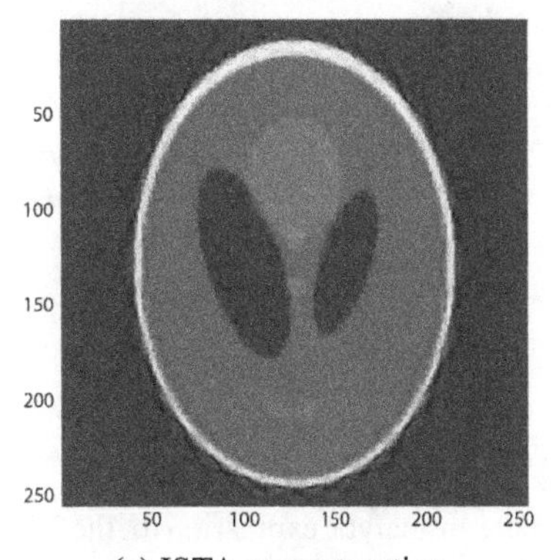

(c) ISTA reconstruction

Figure 10-38: Shepp-Logan phantom image reconstruction from partial radial slices of its 2-D DFT: (a) radial slices of $\mathbf{X}[k_1, k_2]$, (b) least-squares reconstruction, and (c) ISTA reconstruction.

Summary

Concepts

- LTI, convolution, impulse response (now PSF), DTFT (now DSFT), frequency (now wavenumber) response, DFT, FFT all generalize directly from 1-D (signal) to 2-D (image) processing.
- Deconvolution from noisy data requires Tikhonov regularization, which is performed by a Wiener filter.
- The discrete-time wavelet transform splits signals into average signals and detail signals, whose total length matches that of the original signal. But the detail signals are sparse (mostly zero). This also applies to images. It is very fast.
- The Daubechies DL wavelet transform models signals as piecewise-$(L-1)$th polynomials. The Haar transform is a $D1$ wavelet transform.
- Wavelet transforms are useful for compressing, denoising, deconvolving, reconstructing, and inpainting images.
- Compressed sensing allows reconstruction of sparsifiable signals or images using many fewer linear combinations of the signal or image values than the number of values.
- IRLS and IST are iterative algorithms for compressed sensing problems.

Mathematical and Physical Models

Wiener Filter $\mathbf{G}[k_1,k_2]=\mathbf{H}[k_1,k_2]^*/[|\mathbf{H}[k_1,k_2]|^2+\lambda^2]$

ℓ_1 Norm $||x[n]||_1=\sum|x[n]|$

LASSO functional $\Lambda=\frac{1}{2}||\underline{y}-A\,W^{-1}\underline{z}||_2^2+\lambda||\underline{z}||_1$

Landweber $x^{(k+1)}=x^{(k)}+A^T(y-Ax^{(k)})$

Shrinkage $y[n]=x[n]-\lambda\cdot\text{sign}(x[n])$

Threshold $y[n]=x[n]$ if $|x[n]|>\lambda$; $y[n]=0$ if $|x[n]|<\lambda$

IST algorithm Landweber with thresholding and shrinkage

Important Terms

Provide definitions or explain the meaning of the following terms:

analysis filter bank	image deconvolution	QMF filter pair	Tikhonov regularization
average signal	image denoising	scaling function	wavelet function
compressed sensing	IST algorithm	Shepp-Logan phantom	wavelet transform
Daubechies wavelets	ℓ_1 norm	Smith-Barnwell condition	wavenumber response
detail signal	perfect reconstruction	sparse signal or image	Wiener filter
DSFT (2-D DTFT)	piecewise polynomial	synthesis filter bank	
Haar wavelets	point-spread function	threshold and shrink	

PROBLEMS

Sections 10-1 to 10-7: Image Processing

*10.1 Compute an analytic expression for the 2-D wavenumber response of the 2-D LSI system

$$y[m,n]=\frac{1}{4}\,x[m,n]+\frac{1}{4}\,x[m-1,\;n-1]+\frac{1}{4}\,x[m-1,\;n]+\frac{1}{4}\,x[m,\;n-1].$$

This system averages the pixels in a 2×2 block of the image, so it is a 2-D version of the two-point averager.

10.2 Show that the 2-D wavenumber response of the PSF

$$h[m,n]=\begin{bmatrix}1&2&1\\2&\underline{4}&2\\1&2&1\end{bmatrix}$$

is very close to circularly symmetric, making it a good lowpass filter. Demonstrate this property by:

(a) Displaying the wavenumber response as an image with dc at the center.

*Answer(s) in Appendix F.

(b) Using the expansion

$$\cos(\Omega) = 1 - \frac{\Omega^2}{2!} + \frac{\Omega^4}{4!} - \cdots$$

and neglecting all terms of degree four or higher.

10.3 The 2-D discrete Laplacian operator has PSF

$$h[m,n] = \begin{bmatrix} 0 & 1 & 0 \\ 1 & \underline{-4} & 1 \\ 0 & 1 & 0 \end{bmatrix}.$$

(a) Show that for small (Ω_1, Ω_2), the 2-D wavenumber response of the 2-D Laplacian

$$\nabla^2 x = \frac{\partial^2 x}{\partial \xi^2} + \frac{\partial^2 x}{\partial \eta^2}$$

has a continuous-space 2-D Fourier transform given by $\mathbf{X}(\omega_1, \omega_2) = -\omega_1^2 - \omega_2^2$. Use

$$\cos(\omega) = 1 - \frac{\omega^2}{2!} + \frac{\omega^4}{4!} - \cdots$$

and neglect all terms of degree four or higher.

(b) A common *edge detection* technique is to threshold the 2-D Laplacian. Apply this to the image in the file `letters.mat` using

```
load letters.mat;
Y=conv2(X,[0 1 0;1 -4 1;0 1 0]);
Y(abs(Y)<1)=0;
subplot(221),imagesc(X),colormap(gray)
subplot(222),imagesc(abs(Y)),
colormap(gray)
```

(The thresholding actually has no effect in this case.)

10.4 This problem can be solved entirely using only techniques from Chapter 2! *Motion blur* occurs when a photo is taken of an object in constant linear motion, relative to the camera. The object moves a horizontal distance T while the camera lens shutter is open. (If the motion is not in the horizontal direction, we can rotate the coordinate system.) If $u(\xi, \eta)$ is the image of the scene under static conditions, the blurred image is then

$$v(\xi, \eta) = \int_0^T u(\xi - \tau,\ \eta)\, d\tau.$$

(a) Show that $v(\xi, \eta) = u(\xi, \eta) * h(\xi)$ (convolution in ξ for each η). What is $h(\xi)$?

(b) Show that $w(\xi, \eta) = \partial v / \partial \xi = u(\xi, \eta) - u(\xi - 2T,\ \eta)$. *Hint:* Use Table 2-1. This looks like a double exposure of the static image.

(c) Show that $w(\xi, \eta) + w(\xi - T,\ y) = u(\xi - 2T,\ \eta)$. The two images are farther apart. Repeating step (c) a few more times will separate the images completely.

(d) Apply the procedure to deblur the blurred image `V` in the file `P104.mat.` $T = 50$ pixels. Approximate derivatives by differences:

```
W=[V zeros(225,1)]-[zeros(225,1) V];
```

Display the original and deblurred images, and compare them.

10.5 Derive the Wiener filter by showing that the $x[m,n]$ minimizing the Tikhonov functional

$$e = \sum_{m=0}^{N-1} \sum_{n=0}^{N-1} [(y[m,n] - h[m,n] ** \hat{x}[m,n])^2 + (\lambda\ \hat{x}[m,n])^2]$$

has the 2-D DFT

$$\mathbf{X}[k_1,k_2] = \mathbf{Y}[k_1,k_2]\, \frac{\mathbf{H}[k_1,k_2]}{|\mathbf{H}[k_1,k_2]|^2 + \lambda^2}.$$

Hints: Use Parseval's theorem and

$$|a+b|^2 = aa^* + ab^* + ba^* + bb^*.$$

Add and subtract $|\mathbf{Y}\mathbf{H}^*|^2$, divide by $(\mathbf{H}\mathbf{H}^* + \lambda^2)$, and complete the square.

10.6 Deblurring due to an out-of-focus camera can be modeled crudely as a 2-D convolution with a disk-shaped point-spread function

$$h[m,n] = \begin{cases} 1 & \text{for } m^2 + n^2 < R^2, \\ 0 & \text{for } m^2 + n^2 > R^2, \end{cases}$$

where R is the radius of a circle within which pixels are unblurred, and outside of which pixels are totally blurred. This problem deblurs an out-of-focus image in the (unrealistic) absence of noise.

(a) Blur the letter image with an (approximate) disk PSF using

```
H(25,25)=0;for I=1:25;for J=1:25;
if((I-13)*(I-13)+(J-13)*(J-13)<145);
H(I,J)=1;end;end;end;
load letters;Y=conv2(X,H);
subplot(221),imagesc(Y),colormap(gray)
```

(b) Deblur this out-of-focus image using the command

```
Z=real(ifft2(fft2(Y)./fft2(H,280,280)));
subplot(222),imagesc(Z),colormap(gray)
```

Note that the size of the blurred image is

$$256 + 25 - 1 = 280.$$

(c) Explain why this approach will not work in the real world (i.e., in the presence of noise).

10.7 Repeat Problem 10.6, but now add noise to the blurred image:

(a) Add noise to the blurred image using

```
Y=Y+100*randn(280,280).
```

(b) Deblur the image using the command

```
Z=real(ifft2(fft2(Y)./fft2(H,280,280)));
subplot(222),imagesc(Z),colormap(gray)
```

(c) Deblur the image using a Wiener filter, using

```
FH=fft2(H,280,280);
W=real(ifft2(fft2(Y).*conj(FH)./
(abs(FH).*abs(FH)+10)));
subplot(221),imagesc(Z),colormap(gray)
subplot(222),imagesc(W),colormap(gray)
```

10.8 This problem shows how *thumbnail* images are created from larger images.

(a) Downsample the "letters" image $\downarrow$ (4 × 4) and display it:

```
load letters;subplot(221),
imagesc(X(1:4:end,1:4:end)),
colormap(gray)
```

How many of the letters can you read?

(b) Now lowpass-filter the image with a quarter-band 2-D lowpass filter that passes only $\mathbf{X}(e^{j\Omega_1}, e^{j\Omega_1})$ for $0 \le |\Omega_1, \Omega_2| < \pi/4$:

```
FX=fft2(x);
FY=FX;FY(32:258-32,32:258-32)=0;
Y=real(ifft2(FY));
subplot(222),imagesc(Y(1:4:end,1:4:end)),
colormap(gray)
```

Now how many of the letters can you read?

10.9 This problem investigates how to denoise images by 2-D brickwall lowpass filtering. The program adds noise to the clown image, then 2-D brickwall lowpass filters it:

```
load clown.mat;
Y=X+0.2*randn(200,200);FY=fft2(Y);
FZ=FY;L=??;FZ(L:202-L,L:202-L)=0;
Z=real(ifft2(FZ));imagesc(Z),colormap(gray)
```

(a) Run the program for $L = 101$, 20, and 10. Display the filtered images.

(b) Discuss the trade-offs involved in varying the cutoff frequency.

10.10 This problem denoises images by 2-D lowpass filtering with a separable 2-D lowpass filter $h[m, n] = h[m]\, h[n]$, where $h[n]$ and $h[m]$ are each an FIR lowpass filter designed by windowing the impulse response of a brickwall lowpass filter, which suppresses "ringing."

(a) Design a 1-D lowpass filter $h[n]$ of duration 31 by using a Hamming window on the impulse response of a brickwall lowpass filter with cutoff frequency $\Omega = \pi/3$. Design a similar filter $h[m]$.

(b) Filter the "letters" image by 2-D convolution with $h[m]\, h[n]$. Display the result.

(c) Try varying the filter duration and cutoff frequency to see if you can improve the image quality.

Sections 10-8 to 10-14: Wavelets

10.11 Show that the **z**-transform of Eq. (10.77) is

$$\mathbf{G}^{(k)}(\mathbf{z}) = \mathbf{G}(\mathbf{z})\, \mathbf{G}(\mathbf{z}^2)\, \mathbf{G}(\mathbf{z}^4) \;\ldots\; \mathbf{G}(\mathbf{z}^{2^{k-1}}),$$

$$\mathbf{H}^{(k)}(\mathbf{z}) = \mathbf{G}(\mathbf{z})\, \mathbf{G}(\mathbf{z}^2)\, \mathbf{G}(\mathbf{z}^4) \;\ldots\; \mathbf{H}(\mathbf{z}^{2^{k-1}}).$$

Use these relationships to generate the rows of Eq. (10.89).

10.12 Why are the time-reversals necessary in the synthesis filters? Show that using $-g[n]$ and $h[n]$ instead of $g[-n]$ and $h[-n]$ leads to versions of Eq. (10.100) and Eq. (10.101) to which there is no solution.

10.13 Compute the *db*3 Daubechies scaling function. Confirm that your answer matches the coefficients listed in Table 10-1.

10.14 Take the inverse DTFT of the Smith-Barnwell condition to show that $g[n]$ must be orthonormal to its even-valued translations.

10.15 Use the result of Problem 10.14 to derive the *db*2 scaling function.

10.16 This problem investigates denoising images by thresholding and shrinking the 2-D Haar wavelet transform of the noisy image.

(a) Run the program `P1016.m` on the book website. This adds noise to the "letters" image, computes its 2-D Haar transform, thresholds and shrinks the wavelet transform, computes the inverse 2-D Haar wavelet transform of the result, and displays images.

(b) Why does this provide better results than the 2-D DFT or convolution with a lowpass filter?

Section 10-15: Compressed Sensing

10.17 Even if a compressed sensing problem is only slightly underdetermined, and it has a mostly sparse solution, there is no guarantee that the sparse solution is unique. The worst case for compressed sensing is as follows:

Let $a_{m,n} = e^{-j2\pi mn/N}$ for $n = 0, 1, \ldots, N-1$ and for $m = 0, 1, \ldots, N-1$, but skipping every multiple of N/L in m. For example, if $N = 12$ and $L = 4$, $m = 1, 2, 4, 5, 7, 8, 10, 11$ and A is an 8×12 matrix. Let

$$\underline{z}_n = \begin{cases} 1 & \text{for } n \text{ a multiple of } L, \\ 0 & \text{for } n \text{ not a multiple of } L. \end{cases}$$

Continuing the example, $\underline{z}_n = 1$ for $n = 0, 4, 8$ and 0 for other values of n, so $\underline{z}_n$ is sparse.

Now consider the compressed sensing problem of reconstructing an unknown and (N/L)-sparse $\underline{x}$ from observations $\underline{y} = A\underline{x}$. The unknown $\underline{x}$ is zero unless n is a multiple of L, so it is indeed (N/L)-sparse.

Show that $\underline{x} + c\underline{z}$, where c is any constant, is another (N/L)-sparse solution to the compressed sensing problem. This shows that the (N/L)-sparse solution is not unique, so compressed sensing will not work for this problem.

10.18 Free the clown from his cage. Run the program `P1018.m`. This sets horizontal and vertical bands of the clown image to zero, making it appear that the clown is confined to a cage. *Free the clown:* The program then uses inpainting to replace the bands of zeros with pixels by regarding the bands of zeros as unknown pixel values of the clown. Change the widths of the bands of zeros and see how this affects the reconstruction.

10.19 De-square the clown image. Run the program `P1019.m`. This sets 81 small squares of the clown image to zero, decimating it. The program then uses inpainting to replace the small squares with pixels by regarding the 81 small squares as unknown pixel values of the clown. Change the sizes of the squares and see how this affects the reconstruction.

10.20 In Section 10-2 we saw that the 2-D spectrum of the letters image was dominated by four lines. Explain this using the Radon transform. *Hint:* Rotating an image rotates its 2-D Fourier transform by the same angle.

LabVIEW Module 10.1

10.21 Set both slides to 10. What effects did this produce in the blurred image?

10.22 Set both slides to 25. What features of the filtered image are still noticeably blurred? Why are they still blurred?

10.23 Set both slides to 50. How does the filtered image compare to the original image? What does this say about the image sampling rate?

LabVIEW Module 10.2

10.24 Set noise level to 100, L to 30, and K to 1. What effects did this produce in the filtered image?

10.25 Repeat Problem 10.24 with K set to 0.1.

10.26 For noise level at 100 and $L = 30$, what value of K seems to give the best performance in readability of the letters?

10.27 Repeat Problem 10.26, but reduce L to 5. Does the change help?

LabVIEW Module 10.3

10.28 Describe the reconstructed image for each of the following scenarios:

(a) Noise level = 0 and $L = 1$.

(b) Noise level = 1000 and $L = 1$.

(c) Noise level = 1000 and $L = 10$.

10.29 If noise level = 1000 and $L = 0$ (out of the slide range), what would happen to the reconstructed image?

LabVIEW Module 10.5

Roughly how high can the threshold for setting the wavelet transform values to zero be set without the specified visual effect on the reconstructed image? What is the associated compression ratio?

10.30 A noticeable effect.

10.31 A significant effect.

LabVIEW Module 10.6

10.32 Using a threshold of 0.25, what is the maximum noise level for which much of the noise can be eliminated without significantly affecting the denoised image?

10.33 Using a noise level of 0.2, do the best you can to denoise the image. What could you do that is not in the module to do a better job?

LabVIEW Module 10.7

For the following problems, set max iterations to 500, use the specified ***missing pixel threshold*** (*MPT*) and λ (L) values, inpaint the deteriorated image, and display the results.

10.34 MPT of 200 and L of 0.003.

10.35 MPT of 140 and L of 0.003.

10.36 MPT of 180 and L of 0.011.

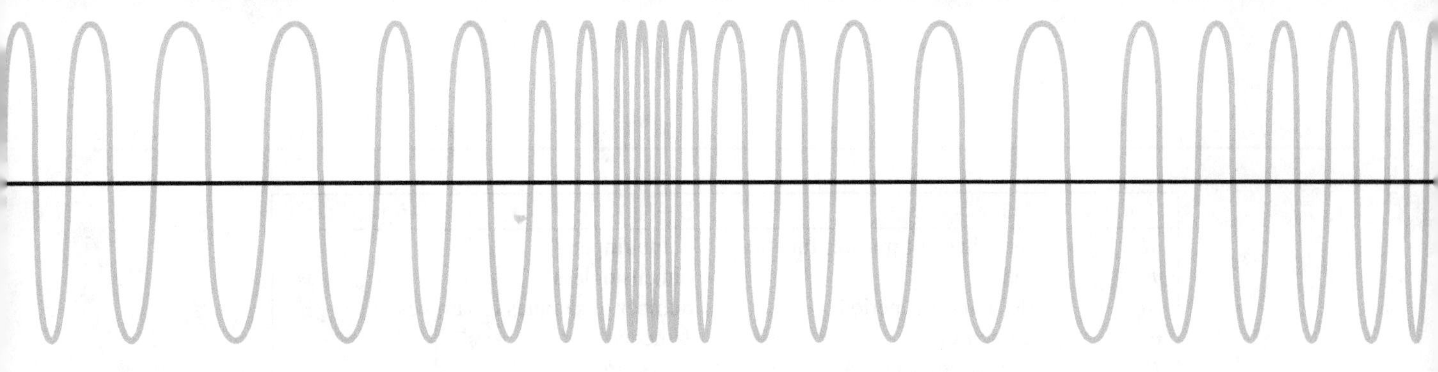

A Symbols, Quantities, and Units

Symbol	Quantity	SI Unit	Unit Abbreviation
a_n	Fourier coefficient	same as function	
b	damping coefficient	newton·seconds/meter	N·s/m
B	bandwidth	hertz	Hz
b_n	Fourier coefficient	same as function	
C	capacitance	farad	F
C	heat capacity	joules/°C	J/°C
c_n	Fourier coefficient	same as function	
E	energy	signal-specific	
f	circular frequency	hertz	Hz
$\mathbf{F}$	force	newton	N
$\mathcal{F}\{\ \}$	Fourier transform		
G	gain	output/input	
$\mathbf{G(s)}$	feedback transfer function	system-specific	
$\mathbf{G(z)}$	inverse transfer function	input/output	
$\mathbf{H(s)}$	transfer function	output/input	
$\mathbf{H(z)}$	discrete-time transfer function	output/input	
$h[n]$	discrete-time impulse response	output/input	
$h(t)$	impulse response	output/input	
$\mathbf{H}(e^{j\Omega})$	discrete-time frequency response	output/input	
$\mathbf{H}(\omega)$	frequency response	output/input	
i	current	amp	A
i	index	dimensionless	—
$\mathbf{I(s)}$	**s**-domain current	amp	A
k	index	dimensionless	—
k	spring constant	newtons/meter	N/m
L	inductance	henry	H
L	length	meters	m
$\mathcal{L}\{\ \}$	Laplace transform		
m	mass	kilogram	kg
m	modulation index	dimensionless	—
m	index	dimensionless	—

Symbol	Quantity	SI Unit	Unit Abbreviation
M	magnitude of transfer function	output/input	
n	index	dimensionless	—
N_0	discrete-time period	number of samples	samples
$\mathbf{p}$	complex pole	1/second	s^{-1}
P	time-average power	signal-specific	
q	rate of heat flow	joules/s	J/s
Q	enthalpy	joules	J
Q	quality factor	dimensionless	—
$\mathbf{Q}(\mathbf{s})$	transfer function with feedback	system-specific	
$r(t)$	unit ramp function	second	s
r_{xy}	correlation	signals specific	
R	resistance	ohm	Ω
R	thermal resistance	°C/watt	°/W
$\text{rect}(t)$	rectangle function	dimensionless	—
$\mathbf{s}$	complex frequency	1/second	s^{-1}
$\mathcal{S}$	spectrogram	joules	J
SNR	signal-to-noise ratio	dimensionless	—
T_0	period	second	s
T	temperature	degrees Celsius	°C
$\mathcal{T}$	relative temperature	degrees centigrade	°C
$u(t)$	unit step function	dimensionless	—
v	velocity	meters/second	m/s
υ	voltage	volt	V
$\mathbf{V}(\mathbf{s})$	**s**-domain voltage	volt	V
$w[n]$	discrete-time window		
$\mathbf{x}_n$	complex Fourier coefficient	same as function	
$\mathbf{X}(\omega_1, \omega_2)$	2-D Fourier transform	meter^2	m^2
$\mathbf{z}$	complex zero	1/second	s^{-1}
$\mathbf{z}$	complex variable		
$\mathbb{Z}\{\ \}$	**z**-transform		
$\mathbf{Z}(\mathbf{s})$	**s**-domain impedance	ohm	Ω
α	attenuation coefficient	nepers/s	Np/s
α	real part of pole or zero	1/second	s^{-1}
Γ	sidelobe attenuation		
$\delta[n]$	discrete-time impulse	1/second	s^{-1}
$\delta(t)$	impulse (delta) function	1/second	s^{-1}
θ	rotation or phase angle	degrees or radians	° or rad
$\boldsymbol{\theta}(\mathbf{s})$	**s**-domain rotation angle	degrees or radians	° or rad
ξ	damping coefficient	dimensionless	—
τ	time constant	second	s
ϕ	phase angle	degrees or radians	° or rad
ω	angular frequency	radians/second	rad/s
Ω	discrete-time angular frequency	radians/sample	rad/sample
Ω_M	mainlobe width	radians	rad

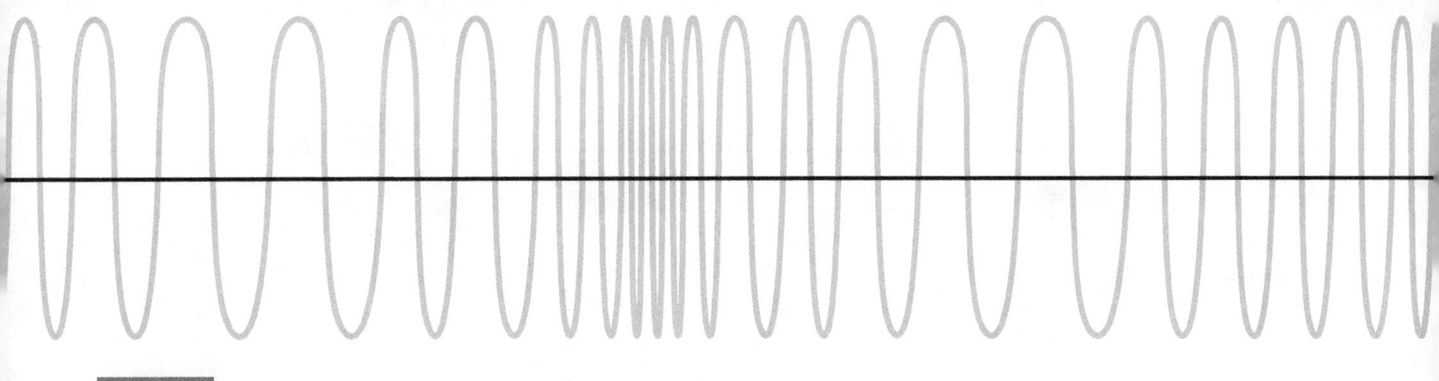

B Review of Complex Numbers

A *complex number* $\mathbf{z}$ may be written in the *rectangular form*

$$\mathbf{z} = x + jy, \qquad \text{(B.1)}$$

where x and y are the *real* ($\mathfrak{Re}$) and *imaginary* ($\mathfrak{Im}$) parts of $\mathbf{z}$, respectively, and $j = \sqrt{-1}$. That is,

$$x = \mathfrak{Re}(\mathbf{z}), \qquad y = \mathfrak{Im}(\mathbf{z}). \qquad \text{(B.2)}$$

Note that $\mathfrak{Im}(3 + j4) = 4$, not $j4$.

Alternatively, $\mathbf{z}$ may be written in *polar form* as

$$\mathbf{z} = |\mathbf{z}|e^{j\theta} = |\mathbf{z}|\underline{/\theta} \qquad \text{(B.3)}$$

where $|\mathbf{z}|$ is the magnitude of $\mathbf{z}$, θ is its phase angle, and the form $\underline{/\theta}$ is a useful shorthand representation commonly used in numerical calculations. By applying *Euler's identity*,

$$e^{j\theta} = \cos\theta + j\sin\theta, \qquad \text{(B.4)}$$

we can convert $\mathbf{z}$ from polar form, as in Eq. (B.3), into rectangular form, as in Eq. (B.1),

$$\mathbf{z} = |\mathbf{z}|e^{j\theta} = |\mathbf{z}|\cos\theta + j|\mathbf{z}|\sin\theta, \qquad \text{(B.5)}$$

which leads to the relations

$$x = |\mathbf{z}|\cos\theta, \qquad y = |\mathbf{z}|\sin\theta, \qquad \text{(B.6)}$$

$$|\mathbf{z}| = \sqrt{x^2 + y^2}, \qquad \theta = \tan^{-1}(y/x). \qquad \text{(B.7)}$$

The two forms of $\mathbf{z}$ are illustrated graphically in Fig. B-1. Because in the complex plane, a complex number assumes the form of a vector, it is represented by a bold letter.

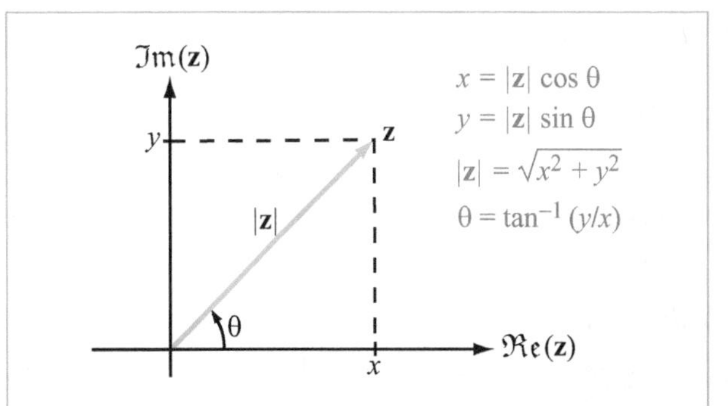

Figure B-1: Relation between rectangular and polar representations of a complex number $\mathbf{z} = x + jy = |\mathbf{z}|e^{j\theta}$.

When using Eq. (B.7), care should be taken to ensure that θ is in the proper quadrant by noting the signs of x and y individually, as illustrated in Fig. B-2. Specifically,

$$\theta = \begin{cases} \tan^{-1}(y/x) & \text{if } x > 0, \\ \tan^{-1}(y/x) \pm \pi & \text{if } x < 0, \\ \pi/2 & \text{if } x = 0 \text{ and } y > 0, \\ -\pi/2 & \text{if } x = 0 \text{ and } y < 0. \end{cases}$$

Complex numbers $\mathbf{z}_2$ and $\mathbf{z}_4$ point in opposite directions and their phase angles θ_2 and θ_4 differ by 180°, despite the fact that (y/x) has the same value in both cases.

The *complex conjugate* of $\mathbf{z}$, denoted with a star superscript (or asterisk), is obtained by replacing j (wherever it appears) with $-j$, so that

$$\mathbf{z}^* = (x + jy)^* = x - jy = |\mathbf{z}|e^{-j\theta} = |\mathbf{z}|\underline{/-\theta}. \qquad \text{(B.8)}$$

The magnitude $|\mathbf{z}|$ is equal to the positive square root of the product of $\mathbf{z}$ and its complex conjugate:

$$|\mathbf{z}| = \sqrt{\mathbf{z}\mathbf{z}^*}. \qquad \text{(B.9)}$$

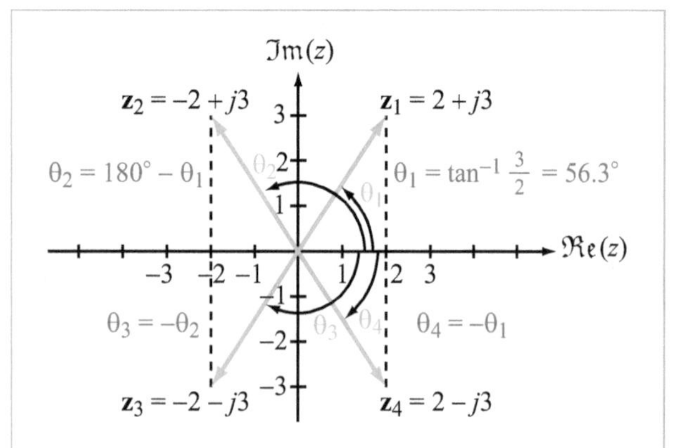

Figure B-2: Complex numbers $\mathbf{z}_1$ to $\mathbf{z}_4$ have the same magnitude $|\mathbf{z}| = \sqrt{2^2 + 3^2} = 3.61$, but their polar angles depend on the polarities of their real and imaginary components.

We now highlight some of the salient properties of complex algebra.

Equality: If two complex numbers $\mathbf{z}_1$ and $\mathbf{z}_2$ are given by

$$\mathbf{z}_1 = x_1 + jy_1 = |\mathbf{z}_1|e^{j\theta_1}, \qquad \text{(B.10a)}$$

$$\mathbf{z}_2 = x_2 + jy_2 = |\mathbf{z}_2|e^{j\theta_2}, \qquad \text{(B.10b)}$$

then $\mathbf{z}_1 = \mathbf{z}_2$ if and only if (*iff*) $x_1 = x_2$ and $y_1 = y_2$ or, equivalently, $|\mathbf{z}_1| = |\mathbf{z}_2|$ and $\theta_1 = \theta_2$.

Addition:

$$\mathbf{z}_1 + \mathbf{z}_2 = (x_1 + x_2) + j(y_1 + y_2). \qquad \text{(B.11)}$$

Multiplication:

$$\begin{aligned}\mathbf{z}_1\mathbf{z}_2 &= (x_1 + jy_1)(x_2 + jy_2)\\ &= (x_1x_2 - y_1y_2) + j(x_1y_2 + x_2y_1),\end{aligned} \qquad \text{(B.12a)}$$

or

$$\begin{aligned}\mathbf{z}_1\mathbf{z}_2 &= |\mathbf{z}_1|e^{j\theta_1} \cdot |\mathbf{z}_2|e^{j\theta_2}\\ &= |\mathbf{z}_1||\mathbf{z}_2|e^{j(\theta_1+\theta_2)}\\ &= |\mathbf{z}_1||\mathbf{z}_2|[\cos(\theta_1 + \theta_2) + j\sin(\theta_1 + \theta_2)].\end{aligned} \qquad \text{(B.12b)}$$

Division: For $\mathbf{z}_2 \neq 0$,

$$\begin{aligned}\frac{\mathbf{z}_1}{\mathbf{z}_2} &= \frac{x_1 + jy_1}{x_2 + jy_2}\\ &= \frac{(x_1 + jy_1)}{(x_2 + jy_2)} \cdot \frac{(x_2 - jy_2)}{(x_2 - jy_2)}\\ &= \frac{(x_1x_2 + y_1y_2) + j(x_2y_1 - x_1y_2)}{x_2^2 + y_2^2},\end{aligned} \qquad \text{(B.13a)}$$

or

$$\begin{aligned}\frac{\mathbf{z}_1}{\mathbf{z}_2} &= \frac{|\mathbf{z}_1|e^{j\theta_1}}{|\mathbf{z}_2|e^{j\theta_2}}\\ &= \frac{|\mathbf{z}_1|}{|\mathbf{z}_2|}e^{j(\theta_1-\theta_2)}\\ &= \frac{|\mathbf{z}_1|}{|\mathbf{z}_2|}[\cos(\theta_1 - \theta_2) + j\sin(\theta_1 - \theta_2)].\end{aligned} \qquad \text{(B.13b)}$$

Powers: For any positive integer n,

$$\begin{aligned}\mathbf{z}^n &= (|\mathbf{z}|e^{j\theta})^n\\ &= |\mathbf{z}|^n e^{jn\theta} = |\mathbf{z}|^n(\cos n\theta + j\sin n\theta),\end{aligned} \qquad \text{(B.14)}$$

$$\begin{aligned}\mathbf{z}^{1/2} &= \pm|\mathbf{z}|^{1/2}e^{j\theta/2}\\ &= \pm|\mathbf{z}|^{1/2}[\cos(\theta/2) + j\sin(\theta/2)].\end{aligned} \qquad \text{(B.15)}$$

Useful relations:

$$-1 = e^{j\pi} = e^{-j\pi} = 1\angle 180°, \qquad \text{(B.16a)}$$

$$j = e^{j\pi/2} = 1\angle 90°, \qquad \text{(B.16b)}$$

$$-j = -e^{j\pi/2} = e^{-j\pi/2} = 1\angle -90°, \qquad \text{(B.16c)}$$

$$\sqrt{j} = (e^{j\pi/2})^{1/2} = \pm e^{j\pi/4} = \frac{\pm(1 + j)}{\sqrt{2}}, \qquad \text{(B.16d)}$$

$$\sqrt{-j} = \pm e^{-j\pi/4} = \frac{\pm(1 - j)}{\sqrt{2}}. \qquad \text{(B.16e)}$$

For quick reference, the preceding properties of complex numbers are summarized in Table B-1. Note that if a complex number is given by $(a + jb)$ and $b = 1$, it can be written either as $(a + j1)$ or simply as $(a + j)$. Thus, j is synonymous with $j1$.

Example B-1: Working with Complex Numbers

Given two complex numbers

$$\mathbf{V} = 3 - j4,$$

$$\mathbf{I} = -(2 + j3),$$

(a) express **V** and **I** in polar form, and find (b) **VI**, (c) **VI***, (d) **V**/**I**, and (e) $\sqrt{\mathbf{I}}$.

Table B-1: Properties of complex numbers.

Euler's Identity: $e^{j\theta} = \cos\theta + j\sin\theta$	
$\sin\theta = \dfrac{e^{j\theta} - e^{-j\theta}}{2j}$	$\cos\theta = \dfrac{e^{j\theta} + e^{-j\theta}}{2}$
$\mathbf{z} = x + jy = \lvert\mathbf{z}\rvert e^{j\theta}$	$\mathbf{z}^* = x - jy = \lvert\mathbf{z}\rvert e^{-j\theta}$
$x = \mathfrak{Re}(\mathbf{z}) = \lvert\mathbf{z}\rvert\cos\theta$	$\lvert\mathbf{z}\rvert = \sqrt{\mathbf{z}\mathbf{z}^*} = \sqrt{x^2 + y^2}$
$y = \mathfrak{Im}(\mathbf{z}) = \lvert\mathbf{z}\rvert\sin\theta$	$\theta = \tan^{-1}(y/x)$
$\mathbf{z}^n = \lvert\mathbf{z}\rvert^n e^{jn\theta}$	$\mathbf{z}^{1/2} = \pm\lvert\mathbf{z}\rvert^{1/2} e^{j\theta/2}$
$\mathbf{z}_1 = x_1 + jy_1$	$\mathbf{z}_2 = x_2 + jy_2$
$\mathbf{z}_1 = \mathbf{z}_2$ iff $x_1 = x_2$ and $y_1 = y_2$	$\mathbf{z}_1 + \mathbf{z}_2 = (x_1 + x_2) + j(y_1 + y_2)$
$\mathbf{z}_1\mathbf{z}_2 = \lvert\mathbf{z}_1\rvert\lvert\mathbf{z}_2\rvert e^{j(\theta_1+\theta_2)}$	$\dfrac{\mathbf{z}_1}{\mathbf{z}_2} = \dfrac{\lvert\mathbf{z}_1\rvert}{\lvert\mathbf{z}_2\rvert} e^{j(\theta_1-\theta_2)}$
$-1 = e^{j\pi} = e^{-j\pi} = 1\angle\pm 180^\circ$	
$j = e^{j\pi/2} = 1\angle 90^\circ$	$-j = e^{-j\pi/2} = 1\angle -90^\circ$
$\sqrt{j} = \pm e^{j\pi/4} = \pm\dfrac{(1+j)}{\sqrt{2}}$	$\sqrt{-j} = \pm e^{-j\pi/4} = \pm\dfrac{(1-j)}{\sqrt{2}}$

Solution:

(a)

$$
\begin{aligned}
\lvert\mathbf{V}\rvert &= \sqrt{\mathbf{V}\mathbf{V}^*} \\
&= \sqrt{(3 - j4)(3 + j4)} = \sqrt{9 + 16} = 5, \\
\theta_V &= \tan^{-1}(-4/3) = -53.1^\circ, \\
\mathbf{V} &= \lvert\mathbf{V}\rvert e^{j\theta_V} = 5e^{-j53.1^\circ} = 5\angle -53.1^\circ,
\end{aligned}
$$

$$\lvert\mathbf{I}\rvert = \sqrt{2^2 + 3^2} = \sqrt{13} = 3.61.$$

Since $\mathbf{I} = (-2 - j3)$ is in the third quadrant in the complex plane (Fig. B-3),

$$
\begin{aligned}
\theta_\mathbf{I} &= -180^\circ + \tan^{-1}\left(\tfrac{3}{2}\right) = -123.7^\circ, \\
\mathbf{I} &= 3.61\angle -123.7^\circ.
\end{aligned}
$$

Alternatively, whenever the real part of a complex number is negative, we can factor out a (-1) multiplier and then use Eq. (B.16a) to replace it with a phase angle of either $+180^\circ$ or -180°, as needed. In the case of $\mathbf{I}$, the process is as follows:

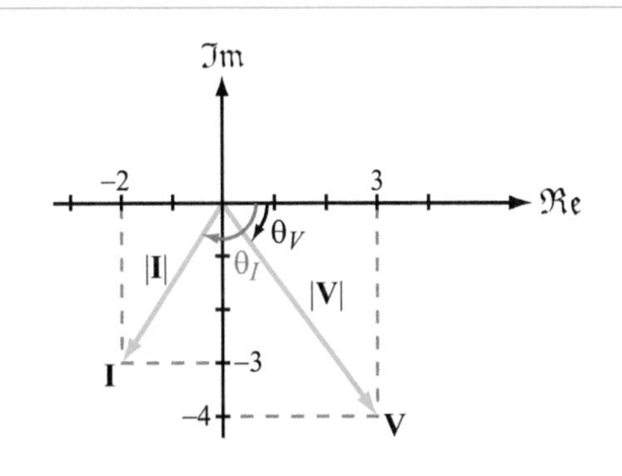

Figure B-3: Complex numbers $\mathbf{V}$ and $\mathbf{I}$ in the complex plane (Example B-1).

$$
\begin{aligned}
\mathbf{I} &= -2 - j3 = -(2 + j3) \\
&= e^{\pm j180^\circ} \cdot \sqrt{2^2 + 3^2}\, e^{j\tan^{-1}(3/2)} \\
&= 3.61 e^{j57.3^\circ} e^{\pm j180^\circ}.
\end{aligned}
$$

Since our preference is to end up with a phase angle within the range between $-180°$ and $+180°$, we will choose $-180°$. Hence,

$$\mathbf{I} = 3.61e^{-j123.7°}.$$

(b)

$$\begin{aligned}\mathbf{VI} &= (5\angle{-53.1°})(3.61\angle{-123.7°}) \\ &= (5 \times 3.61)\angle(-53.1° - 123.7°) = 18.05\angle{-176.8°}.\end{aligned}$$

(c)

$$\mathbf{VI}^* = 5e^{-j53.1°} \times 3.61e^{j123.7°} = 18.05e^{j70.6°}.$$

(d)

$$\frac{\mathbf{V}}{\mathbf{I}} = \frac{5e^{-j53.1°}}{3.61e^{-j123.7°}} = 1.39e^{j70.6°}.$$

(e)

$$\begin{aligned}\sqrt{\mathbf{I}} &= \sqrt{3.61e^{-j123.7°}} \\ &= \pm\sqrt{3.61}\, e^{-j123.7°/2} = \pm 1.90e^{-j61.85°}.\end{aligned}$$

Exercise B-1: Express the following complex functions in polar form:

$$\mathbf{z}_1 = (4 - j3)^2,$$

$$\mathbf{z}_2 = (4 - j3)^{1/2}.$$

Answer: $\mathbf{z}_1 = 25\angle{-73.7°}$, $\mathbf{z}_2 = \pm\sqrt{5}\,\angle{-18.4°}$. (See S²)

Exercise B-2: Show that $\sqrt{2j} = \pm(1 + j)$. (See S²)

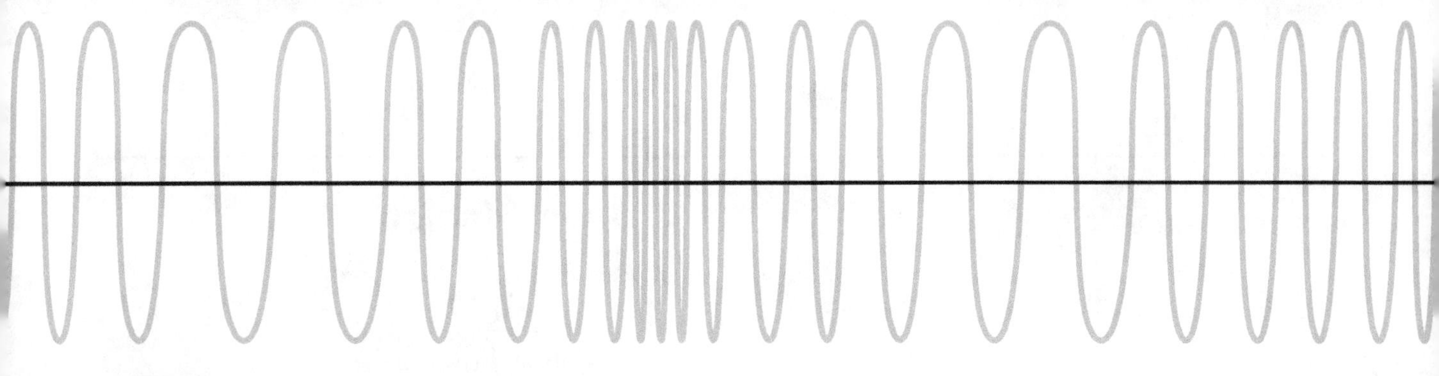

C Mathematical Formulas

C-1 Trigonometric Relations

$$\sin x = \pm\cos(x \mp 90°)$$
$$\cos x = \pm\sin(x \pm 90°)$$
$$\sin x = -\sin(x \pm 180°)$$
$$\cos x = -\cos(x \pm 180°)$$
$$\sin(-x) = -\sin x$$
$$\cos(-x) = \cos x$$
$$\sin^2 x = \frac{1}{2}(1 - \cos 2x)$$
$$\cos^2 x = \frac{1}{2}(1 + \cos 2x)$$
$$\sin(x \pm y) = \sin x \cos y \pm \cos x \sin y$$
$$\cos(x \pm y) = \cos x \cos y \mp \sin x \sin y$$
$$2\sin x \sin y = \cos(x - y) - \cos(x + y)$$
$$2\sin x \cos y = \sin(x + y) + \sin(x - y)$$
$$2\cos x \cos y = \cos(x + y) + \cos(x - y)$$
$$\sin 2x = 2\sin x \cos x$$
$$\cos 2x = 1 - 2\sin^2 x$$
$$\sin x + \sin y = 2\sin\left(\frac{x+y}{2}\right)\cos\left(\frac{x-y}{2}\right)$$
$$\sin x - \sin y = 2\cos\left(\frac{x+y}{2}\right)\sin\left(\frac{x-y}{2}\right)$$
$$\cos x + \cos y = 2\cos\left(\frac{x+y}{2}\right)\cos\left(\frac{x-y}{2}\right)$$
$$\cos x - \cos y = -2\sin\left(\frac{x+y}{2}\right)\sin\left(\frac{x-y}{2}\right)$$
$$e^{jx} = \cos x + j\sin x \qquad \text{(Euler's identity)}$$
$$\sin x = \frac{e^{jx} - e^{-jx}}{2j}$$
$$\cos x = \frac{e^{jx} + e^{-jx}}{2}$$
$$\cos^2 x + \sin^2 x = 1$$
$$2\pi \text{ rad} = 360°$$
$$1 \text{ rad} = 57.30°$$

C-2 Indefinite Integrals

(a and b are constants)

$$\int \sin ax\, dx = -\frac{1}{a}\cos ax$$
$$\int \cos ax\, dx = \frac{1}{a}\sin ax$$
$$\int e^{ax}\, dx = \frac{1}{a}e^{ax}$$
$$\int \ln x\, dx = x\ln x - x$$
$$\int xe^{ax}\, dx = \frac{e^{ax}}{a^2}(ax - 1)$$
$$\int x^2 e^{ax}\, dx = \frac{e^{ax}}{a^3}(a^2x^2 - 2ax + 2)$$
$$\int x\sin ax\, dx = \frac{1}{a^2}\sin ax - \frac{x}{a}\cos ax$$
$$\int x\cos ax\, dx = \frac{1}{a^2}\cos ax + \frac{x}{a}\sin ax$$
$$\int x^2\sin ax\, dx = \frac{2x}{a^2}\sin ax - \frac{a^2x^2 - 2}{a^3}\cos ax$$
$$\int x^2\cos ax\, dx = \frac{2x}{a^2}\cos ax + \frac{a^2x^2 - 2}{a^3}\sin ax$$

$$\int e^{ax} \sin bx\, dx = \frac{e^{ax}}{a^2+b^2}(a \sin bx - b \cos bx)$$

$$\int e^{ax} \cos bx\, dx = \frac{e^{ax}}{a^2+b^2}(a \cos bx + b \sin bx)$$

$$\int e^{ax} \sin^2 bx\, dx =$$

$$\frac{e^{ax}}{a^2+4b^2}\left[(a \sin bx - 2b \cos bx) \sin bx + \frac{2b^2}{a}\right]$$

$$\int e^{ax} \cos^2 bx\, dx =$$

$$\frac{e^{ax}}{a^2+4b^2}\left[(a \cos bx + 2b \sin bx) \cos bx + \frac{2b^2}{a}\right]$$

$$\int \sin ax \sin bx\, dx =$$

$$\frac{\sin(a-b)x}{2(a-b)} - \frac{\sin(a+b)x}{2(a+b)}, \quad a^2 \neq b^2$$

$$\int \cos ax \cos bx\, dx =$$

$$\frac{\sin(a-b)x}{2(a-b)} + \frac{\sin(a+b)x}{2(a+b)}, \quad a^2 \neq b^2$$

$$\int \sin ax \cos bx\, dx =$$

$$-\frac{\cos(a-b)x}{2(a-b)} - \frac{\cos(a+b)x}{2(a+b)}, \quad a^2 \neq b^2$$

$$\int \sin^2 ax\, dx = \frac{x}{2} - \frac{\sin 2ax}{4a}$$

$$\int \cos^2 ax\, dx = \frac{x}{2} + \frac{\sin 2ax}{4a}$$

$$\int \frac{dx}{x^2+a^2} = \frac{1}{a} \tan^{-1}\frac{x}{a}$$

$$\int \frac{dx}{(x^2+a^2)^2} = \frac{1}{2a^2}\left(\frac{x}{x^2+a^2} + \frac{1}{a}\tan^{-1}\frac{x}{a}\right)$$

$$\int \frac{x^2\, dx}{a^2+x^2} = x - a \tan^{-1}\frac{x}{a}$$

C-3 Definite Integrals

(m and n are integers)

$$\int_0^{2\pi} \sin nx\, dx = \int_0^{2\pi} \cos nx\, dx = 0$$

$$\int_0^{\pi} \sin^2 nx\, dx = \int_0^{\pi} \cos^2 nx\, dx = \frac{\pi}{2}$$

$$\int_0^{\pi} \sin nx \sin mx\, dx = 0, \quad n \neq m$$

$$\int_0^{\pi} \cos nx \cos mx\, dx = 0, \quad n \neq m$$

$$\int_0^{\pi} \sin nx \cos nx\, dx = 0$$

$$\int_0^{\pi} \sin nx \cos mx\, dx = 0$$

$$\int_0^{2\pi} \sin nx \cos mx\, dx = 0$$

$$\int_0^{\infty} \frac{\sin ax}{ax}\, dx = \frac{\pi}{2a}$$

C-4 Approximations for Small Quantities

For $|x| \ll 1$,

$$(1 \pm x)^n \approx 1 \pm nx$$

$$(1 \pm x)^2 \approx 1 \pm 2x$$

$$\sqrt{1 \pm x} \approx 1 \pm \frac{x}{2}$$

$$\frac{1}{\sqrt{1 \pm x}} \approx 1 \mp \frac{x}{2}$$

$$e^x = 1 + x + \frac{x^2}{2!} + \cdots \approx 1 + x$$

$$\ln(1 + x) \approx x$$

$$\sin x = x - \frac{x^3}{3!} + \frac{x^5}{5!} + \cdots \approx x$$

$$\cos x = 1 - \frac{x^2}{2!} + \frac{x^4}{4!} + \cdots \approx 1 - \frac{x^2}{2}$$

$$\lim_{x \to 0} \frac{\sin x}{x} = 1$$

$$\sum_{n=0}^{N} \mathbf{r}^n = \frac{1 - \mathbf{r}^{N+1}}{1 - \mathbf{r}}, \qquad \text{for } \mathbf{r} \neq 1$$

$$\sum_{n=0}^{\infty} \mathbf{r}^n = \frac{1}{1 - \mathbf{r}}, \qquad \text{for } |\mathbf{r}| < 1$$

C-5 Polar-Rectangular Forms

$$e^{j0} = 1, \qquad e^{j\pi/2} = j$$

$$e^{j\pi} = -1, \qquad e^{j3\pi/2} = -j$$

$$e^{j\pi/4} = \frac{1}{\sqrt{2}}(1 + j), \qquad e^{-j\pi/4} = \frac{1}{\sqrt{2}}(1 - j)$$

$$ae^{j\theta} = a\cos\theta + ja\sin\theta$$

The N solutions to $z^N = 1$:

$$z = e^{j2\pi k/N}; \quad \{k = 0, 1, \ldots, N - 1\}$$

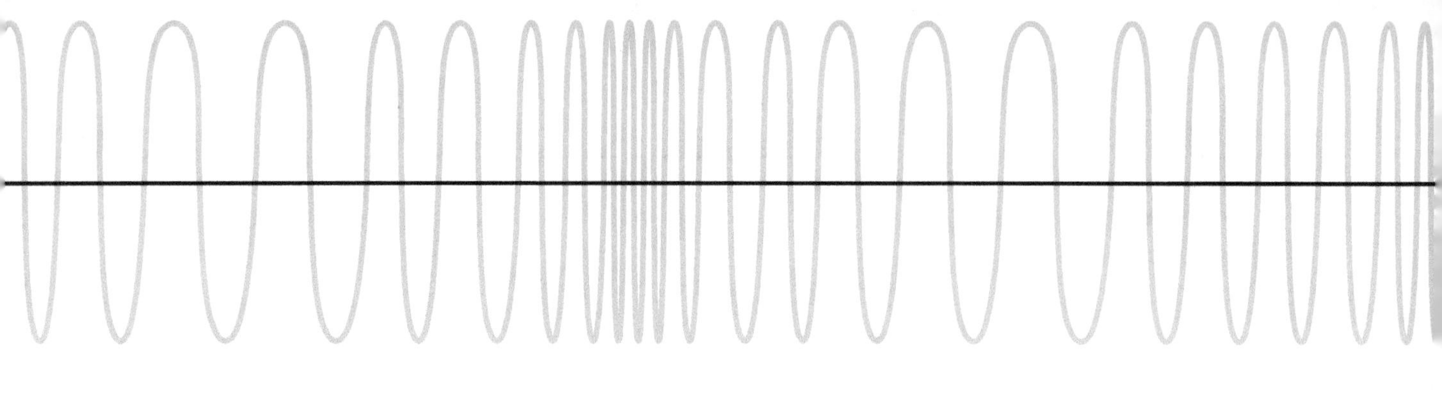

MATLAB® and MathScript

A Short Introduction for Use in Signals and Systems

D-1 Background

"A computer will always do exactly what you tell it to do. But that may not be what you had in mind"—a quote from the 1950s.

This Appendix is a short introduction to MATLAB and MathScript *for this book*. It is not comprehensive; only commands directly applicable to signals and systems are covered. No commands in any of MATLAB's Toolboxes are included, since these commands are not included in basic MATLAB or MathScript. Programming concepts and techniques are not included, since they are not used anywhere in this book.

MATLAB

MATLAB is a computer program developed and sold by The Mathworks, Inc. It is the most commonly used program in signal processing, but it is also used in all fields of engineering.

MATLAB (**mat**rix **lab**oratory) was originally based on a set of numerical linear algebra programs, written in FORTRAN, called LINPACK. So MATLAB tends to formulate problems in terms of vectors and arrays of numbers, and often solves problems by formulating them as linear algebra problems.

MathScript

MathScript is a computer program developed and sold by National Instruments, as a module in LabVIEW. The basic commands used by MATLAB also work in MathScript, but higher-level MATLAB commands, and those in Toolboxes, usually do not work in MathScript. Unless otherwise noted, all MATLAB commands used in this book and website also work in MathScript.

A student version of MathScript is included on the website accompanying the book. *Access to MATLAB is not required to use this book.* In this sequel, we use "M/M" to designate "MATLAB or MathScript."

Getting started

To install the student version of MathScript included on this website, follow the instructions.

When you run M/M, a **prompt** `>>` will appear when it is ready. Then you can type commands. Your first command should be `>>cd mydirectory`, to change directory to your working directory, which we call "mydirectory" here.

We will use `this font` to represent typed commands and generated output. You can get help for any command, such as `plot`, by typing at the prompt `help plot`.

Some basic things to know about M/M:

- Inserting a **semicolon** ";" at the end of a command suppresses the output; without it M/M will type the results of the computation. This is harmless, but it is irritating to have numbers flying by on your screen.
- Inserting **ellipses** "`...`" at the end of a command means it is continued on the next line. This is useful for long commands.
- Inserting "%" at the beginning of a line makes the line a **comment**; it will not be executed. Comments are used to explain what the program is doing at that point.
- `clear` eliminates all present variables. Programs should start with a `clear`.
- `whos` shows all variables and their sizes.
- M/M variables are case-sensitive: `t` and `T` are different variables.
- `save myfile X,Y` saves the variables `X` and `Y` in the file myfile.mat for use in another session of M/M at another time.
- `load myfile` loads all variables saved in myfile.mat, so they can now be used in the present session of M/M.
- `quit` ends the present session of M/M.

.m Files

An M/M program is a list of commands executed in succession. Programs are called "m-files" since their extension is ".m," or "scripts."

To write an .m file, at the upper left, click:
File → New → m-file
This opens a window with a text editor.
Type in your commands and then type:
File → Save as → myname.m
Make sure you save it with an .m extension. Then you can run the file by typing its name at the prompt: `>> myname`. Make sure the file name is not the same as a MATLAB command! Using your own name is a good idea.

You can access previously-typed commands using uparrow and downarrow on your keyboard.

To *download a file* from a website, right-click on it, select **save target as**, and use the menu to select the proper file type (specified by its file extension).

D-2 Basic Computation

D-2.1 Basic Arithmetic

- Addition: `3+2` gives `ans=5`
- Subtraction: `3-2` gives `ans=1`
- Multiplication: `2*3` gives `ans=6`
- Division: `6/2` gives `ans=3`
- Powers: `2^3` gives `ans=8`
- Others: `sin,cos,tan,exp,log,log10`
- Square root: `sqrt(49)` gives `ans=7`
- Conjugate: `conj(3+2j)` gives `ans=3-2i`

Both `i` or `j` represent $\sqrt{-1}$; answers use `i`. `pi` represents π. `e` does not represent 2.71828.

D-2.2 Entering Vectors and Arrays

To enter *row vector* [1 2 3] and store it in `A` type at the prompt `A=[1 2 3];` or `A=[1,2,3];`

To enter the same numbers as a *column vector* and store it in `A`, type at the prompt *either* `A=[1;2;3];` *or* `A=[1 2 3];A=A';` Note `A=A'` replaces `A` with its transpose. "Transpose" means "convert rows to columns, and vice-versa."

To enter a vector of consecutive or equally-spaced numbers, follow these examples:

- `[2:6]` gives `ans=2 3 4 5 6`
- `[3:2:9]` gives `ans=3 5 7 9`
- `[4:-1:1]` gives `ans=4 3 2 1`

To enter an *array* or matrix of numbers, type, for example, `B=[3 1 4;1 5 9;2 6 5];` This gives the array `B` and its transpose `B'`

$$B = \begin{bmatrix} 3\,1\,4 \\ 1\,5\,9 \\ 2\,6\,5 \end{bmatrix} \qquad B' = \begin{bmatrix} 3\,1\,2 \\ 1\,5\,6 \\ 4\,9\,5 \end{bmatrix}$$

Other basics of arrays:

- `ones(M,N)` is an $M \times N$ array of "1"
- `zeros(M,N)` is an $M \times N$ array of "0"
- `length(X)` gives the length of vector `X`
- `size(X)` gives the size of array `X`
  For `B` above, `size(B)` gives `ans=3 3`
- `A(I,J)` gives the (I,J)th element of `A`. For `B` above, `B(2,3)` gives `ans=9`

D-2.3 Array Operations

Arrays add and subtract point-by-point:
`X=[3 1 4];Y=[2 7 3];X+Y` gives `ans=5 8 7`
But `X*Y` generates an error message.
To compute various types of vector products:

- To multiply element-by-element, use `X.*Y` This gives `ans=6 7 12`. To divide element-by-element, type `X./Y`
- To find the inner product of `X` and `Y` (3)(2)+(1)(7)+(4)(3)=25, use `X*Y'` This gives `ans=25`
- To find the outer product of `X` and `Y`

$$\begin{bmatrix} (3)(2) & (3)(7) & (3)(3) \\ (1)(2) & (1)(7) & (1)(3) \\ (4)(2) & (4)(7) & (4)(3) \end{bmatrix} \quad \text{use } X'{*}Y$$

This gives the above matrix.

A common problem is when you think you have a row vector when in fact you have a column vector. Check by using `size(X)`; in the present example, the command gives `ans=1,3` which tells you that `X` is a 1×3 (row) vector.

- The following functions operate on each element of an array separately, giving another array: `sin,cos,tan,exp,log,log10,sqrt` `cos([0:3]*pi)` gives `ans=1 -1 1 -1`
- To compute n^2 for $n = 0, 1 \ldots 5$, use `[0:5].^2` which gives `ans=0 1 4 9 16 25`
- To compute 2^n for $n = 0, 1 \ldots 5$, use `2.^[0:5]` which gives `ans=1 2 4 8 16 32`

Other array operations include:

- `A=[1 2 3;4 5 6];(A(:))'` Stacks A by columns into a column vector and transposes the result to a row vector. In the present example, the command gives `ans=1 4 2 5 3 6`
- `reshape(A(:),2,3)` Unstacks the column vector to a 2×3 array which, in this case, is the original array `A`.
- `X=[1 4 1 5 9 2 6 5];C=X(2:8)-X(1:7)` Takes differences of successive values of `X`. In the present example, the command gives `C=3 -3 4 4 -7 4 -1`
- `D=[1 2 3]; E=[4 5 6]; F=[D E]` This *concatenates* the vectors `D` and `E` (i.e., it appends `E` after `D` to get vector `F`) In the present example, the command gives `F=1 2 3 4 5 6`
- `I=find(A>2)` stores in `I` locations (indices) elements of vector `A` that exceed 2. `find([3 1 4 1 5]<2)` gives `ans=2 4`
- `A(A>2)=0` sets to 0 all values of elements of vector `A` exceeding 2. `A=[3 1 4 1 5]; A(A<2)=0` gives `A=3 0 4 0 5`

M/M indexing of arrays starts with 1, while signals and systems indexing starts with 0. For example, the DFT is defined using index $n = 0, 1 \ldots N-1$, for $k = 0, 1 \ldots N-1$. `fft(X)`, which computes the DFT of `X`, performs

```
fft(X)=X*exp(-j*2*pi*[0:N-1]'*[0:N-1]/N);
```

D-2.4 Solving Systems of Equations

To solve the linear system of equations

$$\begin{bmatrix} 1\,2 \\ 3\,4 \end{bmatrix} \begin{bmatrix} x \\ y \end{bmatrix} = \begin{bmatrix} 17 \\ 39 \end{bmatrix}$$

using

```
A=[1 2;3 4];Y=[17;39];X=A\Y;X'
```

gives `ans=5.000 6.000`, which is the solution $[x\ y]'$.

To solve the complex system of equations

$$\begin{bmatrix} 1+2j & 3+4j \\ 5+6j & 7+8j \end{bmatrix} \begin{bmatrix} x \\ y \end{bmatrix} = \begin{bmatrix} 16+32j \\ 48+64j \end{bmatrix}$$

`[1+2j 3+4j;5+6j 7+8j]\[16+32j;48+64j]` gives

$$\texttt{ans=}\begin{matrix} 2-2i \\ 6+2i \end{matrix}\text{'}$$

which is the solution.

These systems can also be solved using `inv(A)*Y`, but we do not recommend it because computing the matrix inverse of `A` takes much more computation than just solving the system of equations. Computing a matrix inverse can lead to numerical difficulties for large matrices.

D-3 Plotting

D-3.1 Plotting Basics

To plot a function $x(t)$ for $a \leq t \leq b$:

- Generate, say, 100 values of t in $a \leq t \leq b$ using `T=linspace(a,b,100);`
- Generate and store 100 values of $x(t)$ in `X`
- Plot each computed value of `X` against its corresponding value of `T` using `plot(T,X)`
- If you are making several different plots, put them all on one page using `subplot`. `subplot(324),plot(T,X)` divides a figure into a 3-by-2 array of plots, and puts the `X` vs. `T` plot into the 4th place in the array (the middle of the right-most column).

Print out the current figure (the one in the foreground; click on a figure to bring it to the foreground) by typing `print`

Print the current figure to a encapsulated postscript file myname.eps by typing `print -deps2 myname.eps`. Type `help print` for a list of printing options for your computer. For example, use `-depsc2` to save a figure in color.

To make separate plots of $\cos(4t)$ and $\sin(4t)$ for $0 \leq t \leq 5$ in a single figure, use the following:

```
T=linspace(0,5,100);X=cos(4*T);Y=sin(4*T);
subplot(211),plot(T,X)
subplot(212),plot(T,Y)
```

These commands produce the following figure:

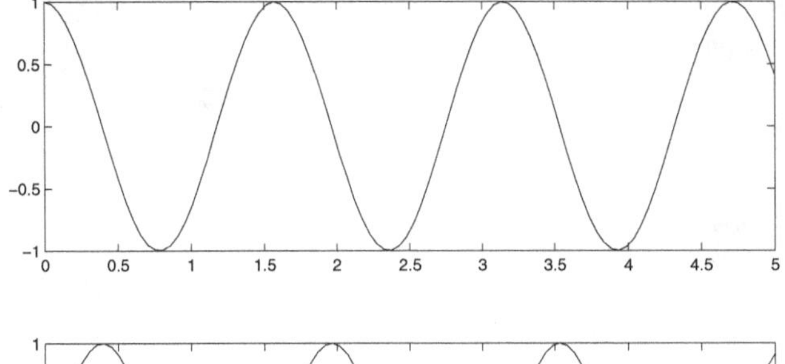

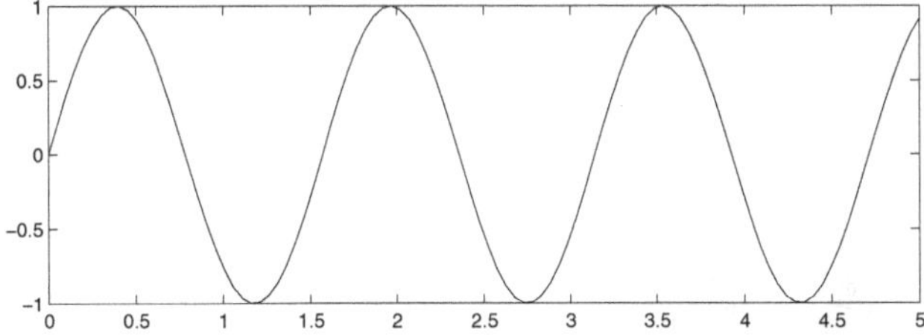

The default is that `plot(X,Y)` plots each of the 100 ordered pairs `(X(I),Y(I))` for `I` $= 1 \ldots 100$, and connects the points with straight lines. If there are only a few data points to be plotted, they should be plotted as individual ordered pairs, not connected by lines. This can be done using `plot(X,Y,'o')` (see Fig. 6-78 below).

D-3.2 Plotting Problems

Common problems encountered using `plot`:
`T` and `X` must have the same lengths; and
Neither `T` nor `X` should be complex; use
`plot(T,real(X))` if necessary.
The above `linspace` command generates 100 equally-spaced numbers between a and b, *including a and b*. This is *not* the same as sampling $x(t)$ with a sampling interval of $(b-a)/100$. To see why:

- `linspace(0,1,10)` gives 10 numbers between 0 and 1 inclusive, spaced by 0.111
- `[0:.1:1]` gives 11 numbers spaced by 0.1

Try the following yourself on M/M:

- `T=[0:10];X=3*cos(T);plot(T,X)`
  This should be a very jagged-looking plot, since it is only sampled at 11 integers, and the samples are connected by lines.
- `T=[0:0.1:10];X=3*cos(T);plot(T,X)`
  This should be a much smoother plot, since there are now 101 (not 100) samples.
- `T=[1:4000];X=cos(2*pi*440*T/8192); sound(X,8192)` This is musical note "A." `sound(X,Fs)` plays `X` as sound, at a sampling rate of `Fs` samples/second.
- `plot(X)`. This should be a blue smear! It's about 200 cycles squished together.
- `plot(X(1:100))` This "zooms in" on the first 100 samples of X to see the sinusoid.

D-3.3 More Advanced Plotting

Plots should be labelled and annotated:

- `title('Myplot')` adds the title "Myplot"
- `xlabel('t')` labels the x-axis with "t"
- `ylabel('x')` labels the y-axis with "x"
- `\omega` produces ω in `title,xlabel` and `ylabel`. Similarly for other Greek letters. Note ' (not ') should be used everywhere.
- `axis tight` contracts the plot borders to the limits of the plot itself
- `axis([a b c d])` changes the horizontal axis limits to $a \leq x \leq b$ and the vertical axis limits to $c \leq y \leq d$.
- `grid on` adds grid lines to the plot
- `plot(T,X,'g',T,Y,'r')` plots on the same plot (`T,X,Y` must all have the same lengths) `X` vs. `T` in green and `Y` vs. `T` in red.

D-3.4 Plotting Examples

A good way to learn how to plot is to study specific examples. The figures in this book were redrawn from figures generat using MATLAB. Four specific illustrative examples follo

(a) Figure 4-32(b):

This example shows how to plot two different functions in a single plot, using different colors, and insert a title. The following .m file

```
clear;T=linspace(0,600,1000);
A=0.01;K=0.04;B=A+K;
SA=3*(1-exp(-A*T));
SB=3*(1-exp(-B*T));
plot(T,SA,'b',T,SB,'r')
title('STEP RESPONSE WITH
AND WITHOUT FEEDBACK')
grid on,print -depsc2 m1.eps
```

generates the following figure:

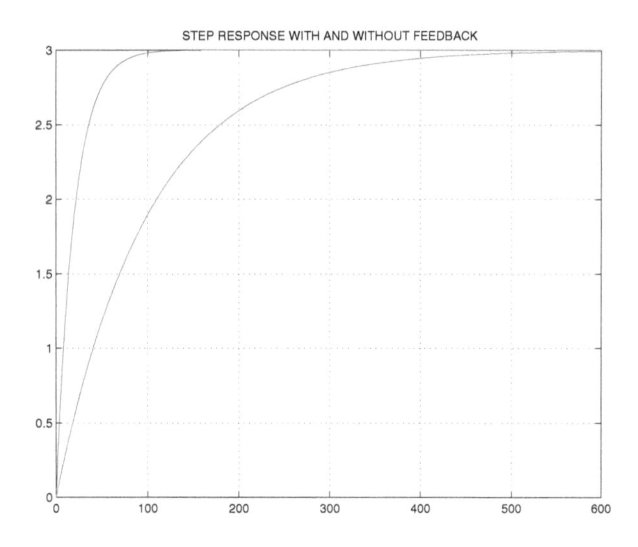

(b) Figure 6-25(b):

This example demonstrates ./ and .*
The following .m file

```
clear;W=linspace(-8,8,1000);V=j*W;
N=(V+2j).*(V-2j).*(V-.1+4j).*(V-.1-4j);
D1=(V+.5+1j).*(V+.5-1j).*(V+.5+3j);
D2=(V+.5-3j).*(V+.5+5j).*(V+.5-5j);
plot(W,abs(N./D1./D2),'r'),grid on
```

generates the following figure:

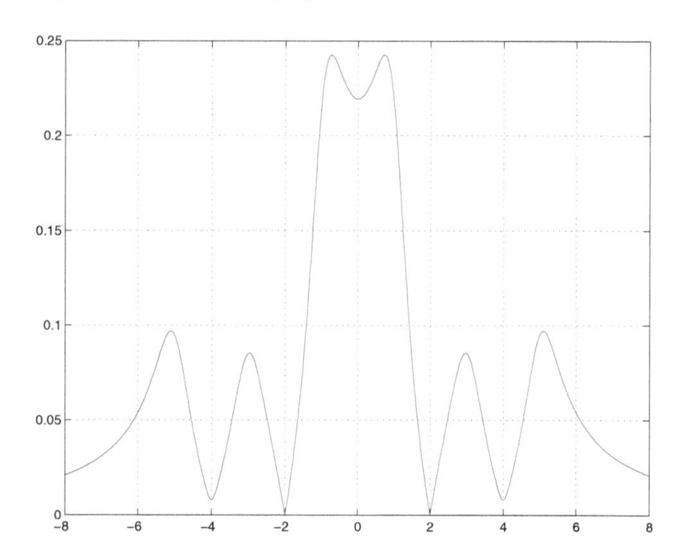

(c) Figure 6-39(b):

This example shows how to use a loop: `H=H./(V-P(I));` is executed for `I=1,2,3,4,5` in succession. The following .m file

```
clear;W=linspace(-2,2,1000);V=j*W;
P=exp(j*2*pi*[3:7]/10);
H=ones(1,1000);
for I=1:5;H=H./(V-P(I));end
subplot(211),plot(W,abs(H)),grid on
```

generates the following figure:

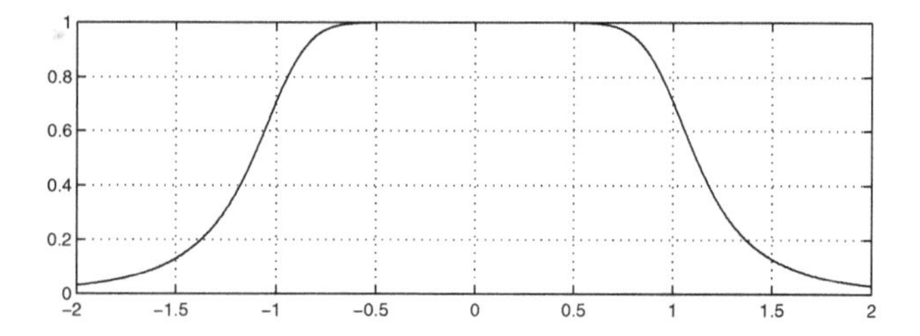

(d) Figure 6-78:

This example shows how to use `hold` to superpose two plots, how to plot individual points, use `subplot` to change the aspect ratio of a figure, and `axis tight` to tighten it. It also uses `[a:0.001:b]`, not `linspace(a,b,1000)`, to sample every 0.001. The following .m file

```
T1=[0:1/45000:1/45];
X1=cos(2*pi*500*T1);
T2=[0:1/450:1/45];
X2=cos(2*pi*500*T2);
subplot(211),plot(T2,X2,'or'),hold,
subplot(211),plot(T1,X1),axis tight
```

generates the following figure:

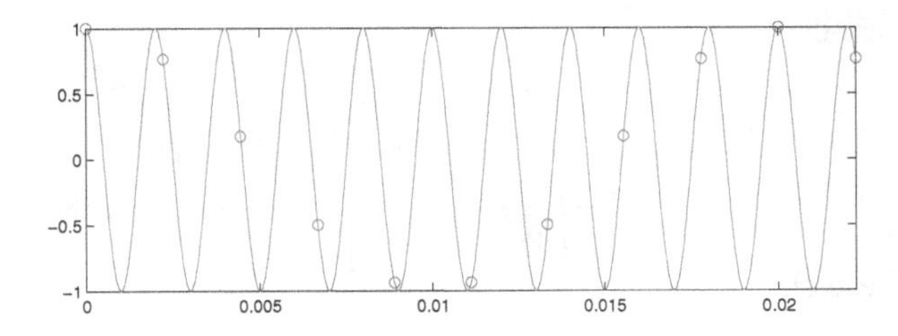

These do not include the computer examples, whose programs are listed elsewhere on this website.

D-4 Partial Fractions

D-4.1 Rectangular-to-Polar Complex Conversion

If an M/M result is a complex number, then it is presented in its rectangular form `a+bj`. M/M recognizes both `i` and `j` as $\sqrt{-1}$, so that complex numbers can be entered as `3+2j` or `3+2i`.

To convert a complex number `X` to polar form, use `abs(X),angle(X)` to get its magnitude and phase (in radians), respectively. To get its phase in degrees, use `angle(X)*180/pi`

Note `atan(imag(X)/real(X))` will **not** give the correct phase, since this formula is only valid if the real part is positive. `angle` corrects this.

The real and imaginary parts of `X` are found using `real(X)` and `imag(X)`, respectively.

D-4.2 Polynomial Zeros

To compute the zeros of a polynomial, enter its coefficients as a *row* vector `P` and use `R=roots(P)`. For example, to find the zeros of $3x^3 - 21x + 18$ (the roots of $3x^3 - 21x + 18 = 0$) use `P=[3 0 -21 18];R=roots(P);R'`, giving `ans= -3.0000 2.0000 1.0000`, which are the roots.

To find the monic (leading coefficient is one) polynomial from the values of its zeros, enter the numbers as a *column* vector `R` and use `P=poly(R)`. For example, to find the polynomial having $\{1, 3, 5\}$ as its zeros, use `R=[1;3;5];P=poly(R)`, giving `P=1 -9 23 -15`. The polynomial is therefore $x^3 - 9x^2 + 23x - 15$.

Note that polynomial are stored as row vectors, and roots are stored as column vectors.

Pole-zero diagrams are made using `zplane`. To produce the pole-zero diagram of

$$\mathbf{H(z)} = \frac{\mathbf{z}^2 + 3\mathbf{z} + 2}{\mathbf{z}^2 + 5\mathbf{z} + 6},$$

type `zplane([1 3 2],[1 5 6])`. The unit circle $|\mathbf{z}|=1$ is also plotted, as a dotted line.

D-4.3 Partial Fraction Expansions

Partial fraction expansions are a vital part of signals and systems, and their computation is onerous (see Chapter 3). M/M computes partial fraction expansions using `residue`. Specifically,

$$\mathbf{H(s)} = \frac{b_0\mathbf{s}^M + b_1\mathbf{s}^{M-1} + \cdots + b_M}{a_0\mathbf{s}^N + a_1\mathbf{s}^{N-1} + \cdots + a_N}$$

has the partial fraction expansion (if $M \leq N$)

$$\mathbf{H(s)} = K + \frac{\mathbf{R_1}}{\mathbf{s} - \mathbf{p_1}} + \cdots + \frac{\mathbf{R_N}}{\mathbf{s} - \mathbf{p_N}}$$

The poles $\{\mathbf{p_i}\}$ and residues $\{\mathbf{R_i}\}$ can be computed from coefficients $\{a_i\}$ and $\{b_i\}$ using

```
B=[b0 b1 ... bM];A=[a0 a1 ... aN]
[R P]=residue(B,A);[R P]
```

The residues $\{R_i\}$ are given in column vector `R`, and poles $\{p_i\}$ are given in column vector `P`.

To compute the partial fraction expansion of

$$\mathbf{H(s)} = \frac{3\mathbf{s} + 6}{\mathbf{s}^2 + 5\mathbf{s} + 4},$$

use the command

```
[R P]=residue([3 6],[1 5 4]);[R P]
```

This gives $\begin{bmatrix}2 & -4\\ 1 & -1\end{bmatrix}$, so R= $\begin{bmatrix}2\\1\end{bmatrix}$ and P= $\begin{bmatrix}-4\\-1\end{bmatrix}$, from which we read off

$$\mathbf{H(s)} = \frac{2}{\mathbf{s}+4} + \frac{1}{\mathbf{s}+1}.$$

In practice, the poles and residues both often occur in complex conjugate pairs. Then use

$$\mathbf{R}e^{\mathbf{p}t} + \mathbf{R}^* e^{\mathbf{p}^* t} = 2|\mathbf{R}|e^{at}\cos(\omega t + \theta),$$

$\mathbf{R} = |\mathbf{R}|e^{j\theta}$ and $\mathbf{p} = a + j\omega$, to simplify the result.

To compute the partial fraction expansion of

$$\mathbf{H(s)} = \frac{\mathbf{s}+7}{\mathbf{s}^2 + 8\mathbf{s} + 25},$$

use the command

```
[R P]=residue([1 7],[1 8 25]);[R P]
```

This gives

$$\begin{bmatrix}0.5000 - 0.5000i & -4.000 + 3.000i\\ 0.5000 + 0.5000i & -4.000 - 3.000i\end{bmatrix}$$

from which we have

$$\mathbf{H(s)} = \frac{0.5 - j0.5}{\mathbf{s}+4-j3} + \frac{0.5 + j0.5}{\mathbf{s}+4+j3},$$

which has the inverse Laplace transform

$$h(t) = (0.5 - j0.5)e^{(-4+j3)t} + (0.5 + j0.5)e^{(-4-j3)t}$$

From `abs(0.5-0.5j),angle(0.5-0.5j)`,

$$h(t) = 2\frac{\sqrt{2}}{2}e^{-4t}\cos\left(3t - \frac{\pi}{4}\right) = \sqrt{2}e^{-4t}\cos\left(3t - \frac{\pi}{4}\right).$$

Both $h(t)$ expressions are valid for $t > 0$.

If $\mathbf{H(s)}$ is proper but not strictly proper, the constant K is nonzero. It is computed using

```
[R P K]=residue(B,A);[R P],K
```

since K has size different from R and P.

To find the partial fraction expansion of

$$\mathbf{H(s)} = \frac{\mathbf{s}^2 + 8\mathbf{s} + 9}{\mathbf{s}^2 + 3\mathbf{s} + 2},$$

use the command

```
[R P K]=residue([1 8 9],[1 3 2]);[R P] K
```

gives $\begin{bmatrix}3 & -2\\ 2 & -1\end{bmatrix}$, K=1 so R= $\begin{bmatrix}3\\2\end{bmatrix}$, P= $\begin{bmatrix}-2\\-1\end{bmatrix}$, from which we read off

$$\mathbf{H(s)} = 1 + \frac{3}{\mathbf{s}+2} + \frac{2}{\mathbf{s}+1}.$$

Double poles are handled as follows:

To find the partial fraction expansion of

$$\mathbf{H(s)} = \frac{8\mathbf{s}^2 + 33\mathbf{s} + 30}{\mathbf{s}^3 + 5\mathbf{s}^2 + 8\mathbf{s} + 4},$$

use the command

```
[R P]=residue([8 33 30],[1 5 8 4]);
```

[R P] gives $\begin{bmatrix}3 & -2\\ 4 & -2\\ 5 & -1\end{bmatrix}$, so R= $\begin{bmatrix}3\\4\\5\end{bmatrix}$, P= $\begin{bmatrix}-2\\-2\\-1\end{bmatrix}$. We then read off

$$\mathbf{H(s)} = \frac{3}{\mathbf{s}+2} + \frac{4}{(\mathbf{s}+2)^2} + \frac{5}{\mathbf{s}+1}.$$

In practice, we are interested not in an analytic expression for $h(t)$, but in computing $h(t)$ sampled every T_s seconds. These samples can be computed directly from R and P, for $0 \leq t \leq T$:

```
t=[0:Ts:T];H=real(R.'*exp(P*t));
```

Since R and P are column vectors, and t is a row vector, H is the inner products of R with each column of the array `exp(P*t)`. `R.'` transposes R without also taking complex conjugates of its elements. `real` is necessary since roundoff error creates a tiny (incorrect) imaginary part in H.

D-4.4 Frequency Response

`polyval(P,W)` evaluates the polynomial whose coefficents are stored in row vector P at the elements of vector W. For example, to evaluate the polynomial $x^2 - 3x+2$ at $x = 4$, `polyval([1 -3 2],4)` gives `ans=6`

The continuous-time **frequency response** of

$$\mathbf{H(s)} = \frac{b_0\mathbf{s}^M + b_1\mathbf{s}^{M-1} + \cdots + b_M}{a_0\mathbf{s}^N + a_1\mathbf{s}^{N-1} + \cdots + a_N}$$

can be plotted for $0 \leq \omega \leq W$ using

```
B=[b0 b1 ... bM];A=[a0 a1 ... aN]
w=linspace(0,W,1000);
H=polyval(B,j*w)./polyval(A,j*w);
subplot(211),plot(w,abs(H))
```

D-4.5 Discrete-Time Commands

- `stem(X)` produces a stem plot of `X`
- `conv(X,Y)` convolves `X` and `Y`
- `fft(X,N)` computes the N-point DFT of `X`
- `ifft(F)` computes the inverse DFT of `F`. Due to roundoff error, use `real(ifft(F))`.
- `sinc(X)` compute $\frac{\sin(\pi x)}{\pi x}$ for each element.

D-4.6 Figure 6-31 Example

We combine many of the above commands to show how Fig. 6-31 in the book was produced.

Example 6-8 plots the impulse and magnitude frequency responses of a comb filter that eliminates 1-kHz and 2-kHz sinusoids, using poles with a real part of -100. Both plots are given in Fig. 6-31, which was redrafted from plots generated by: (% denotes a comment statement)

```
F=linspace(0,3000,100000);
W=j*2*pi*F;A=100;
%Compute frequency response:
Z=j*2*pi*1000*[-2 -1 1 2];
N=poly(Z);D=poly(Z-A);
FH=polyval(N,W)./polyval(D,W);
subplot(211),plot(F,abs(FH))
%Compute impulse response:
T=linspace(0,0.05,10000);
[R P K]=residue(N,D);
H=real(R.'*exp(P*T));
subplot(212),plot(T,H)
```

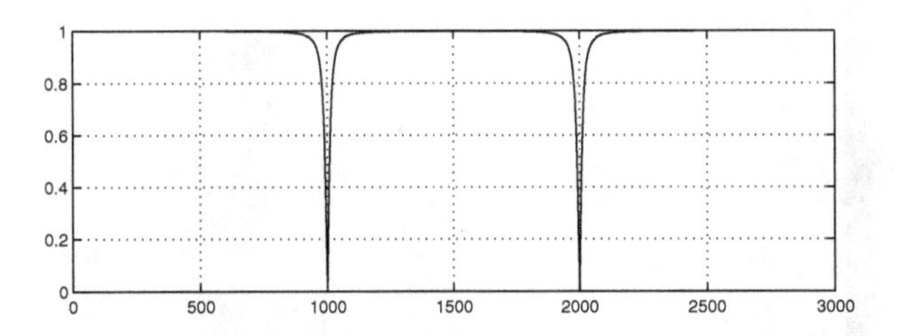

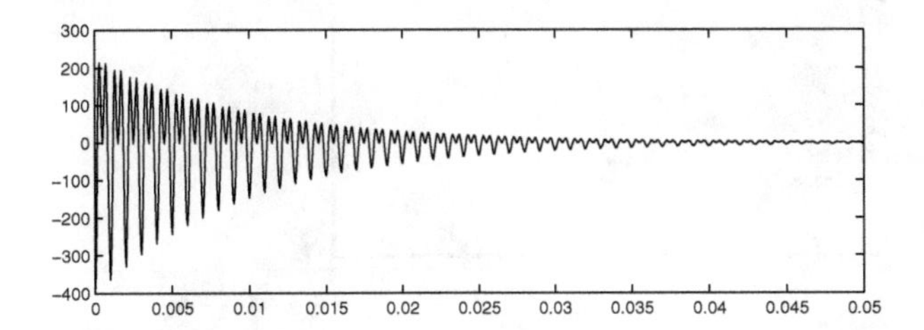

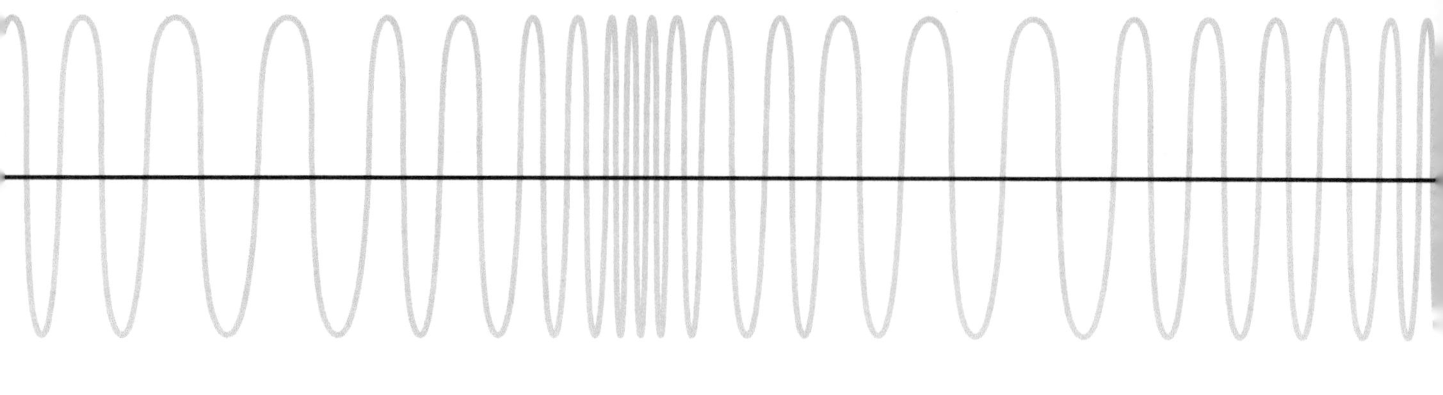

A Guide to Using LabVIEW Modules

On the s^2 website:

1. Open the LabVIEW program.
2. Click on **Open Existing** in the right panel, and select a LabVIEW module (which has extension `.vi`) from the list. Selecting Module LV 2.1 (file `LV2.1.vi`), for example, opens Fig. E-1.
3. On the top row, click on **Operate** and select **Run**. Use the sliders to select parameter values. Move a slider and watch the response change in real time!
4. **To print the window**: On the top row, click on **File**, select **Print**, and choose a printer. LabVIEW can print to an .rtf or .html file, but not to a .pdf or .eps file.
5. To see the **block diagram** of the `.vi` file: On the top row, click on **Window** and select **Show Block Diagram**.
6. To run a different LabVIEW module, click on **File**, select **Open**, and click on a different `.vi` file.

LabVIEW, like MATLAB and other computer programs, can give erroneous results. See, for example, Problems 2.55 and 6.34.

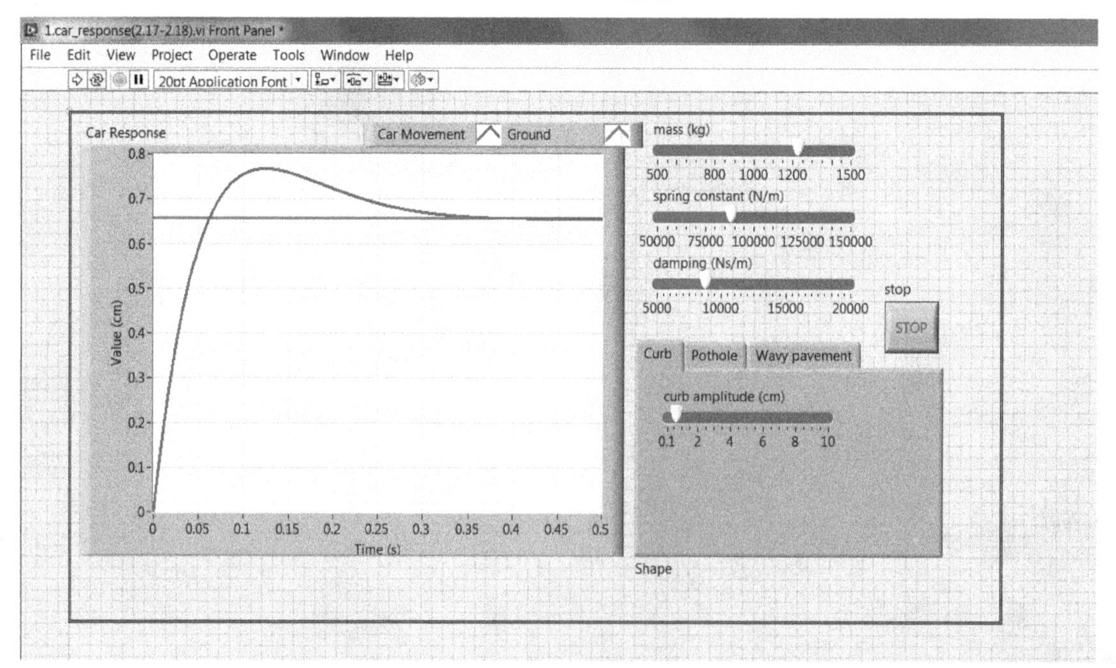

Figure E-1: Module LV 2.1.

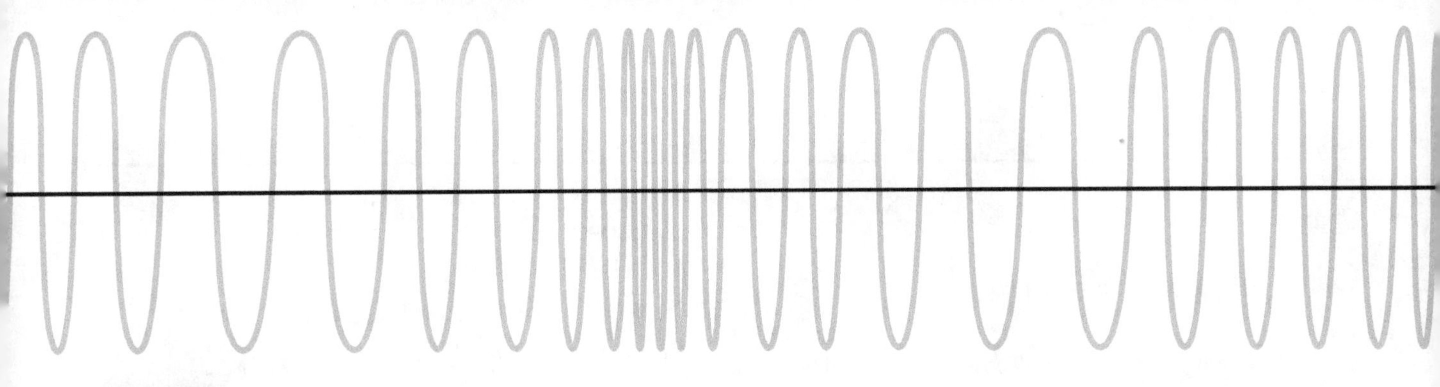

F Answers to Selected Problems

Chapter 1

1.3(b) Analog and continuous in space and discrete in time

1.6(a) See plot on (S²).

1.11(b) See plot on (S²).

1.16(b) Odd symmetry

1.20(b) See plot on (S²).

1.22(c) See plot on (S²).

1.26(d) 0.5

1.29(b) 1/6

1.34(b) Neither

1.39(c) 1

Chapter 2

2.1(b) Linear, but not time invariant

2.5(b) $y(t) = y_{\text{ramp}}(t) - 2y_{\text{ramp}}(t-1) + y_{\text{ramp}}(t-2)$

2.10(c) See plot on (S²).

2.15(a) $u(t) - 3u(t-1) + 2u(t-2)$

2.17(a) $e^{-t}\,u(t) - e^{-2t}\,u(t)$

2.22(d) Causal but not BIBO stable

2.23(d) BIBO stable but not causal

2.29(c) $\cos(4t - 53^\circ)$

2.38 $4\cos(3t - 81.9^\circ) + 3\cos(4t - 98.1^\circ)$

2.43(c) $B \leq 0$

Chapter 3

3.1(a) $\mathbf{X}_1(\mathbf{s}) = \frac{1}{\mathbf{s}}(4 - 2e^{-\mathbf{s}} - 2e^{-2\mathbf{s}})$

3.4(c) $\mathbf{X}_3(\mathbf{s}) = 60/(\mathbf{s}+2)^4$

3.7(a) $\mathbf{X}_1(\mathbf{s}) = 21.65$

3.11 $x(0^+) = 0,\ x(\infty) = 2$

3.15(c) $x_3(t) = 2e^{-3t}\cos(2t + 45^\circ)\,u(t)$

3.18(b) $x_2(t) = [e^{-2(t-6)} - 2\sin(4(t-6))]\,u(t-6)$

3.24(a) $y_1(t) = [2 + 2e^{-t} - 4e^{-5t}]\,u(t)$

3.32 $\mathbf{H}(\mathbf{s}) = 10^{10}/(\mathbf{s}+10^5)^2$; $h(t) = 10^{10}te^{-10^5 t}\,u(t)$

3.37 $g(t) = \delta(t) + 2e^{-t}\,u(t)$

3.41(b) $\mathbf{p} = \{-1, -2\},\ \mathbf{z} = \{-3, -4\}$

3.43(d) $y(t) = e^{-3t}\,u(t) - e^{-4t}\,u(t)$

3.46(c) $h(t) = 30e^{-5t}\,u(t) - 15e^{-3t}\,u(t)$

3.49(f) $y_{\text{FORCED}}(t) = 20\cos(3t - 36.9^\circ)\,u(t)$

3.50 $y(0) = 16$ V

Chapter 4

4.1 $\upsilon(t) = [1.5 - 1.572e^{-4t} + 0.072e^{-12t}]\,u(t)$

4.7 $\upsilon_{C_2}(t) = 50e^{-6t}\,u(t)$

4.13 $\upsilon_{\text{C}}(t) = 10(1 + e^{-t})\,u(t)$

4.19 $\upsilon_{\text{out}}(t) = \left[\frac{1}{3} + \frac{8}{75}e^{-3t} + \frac{1}{10}\cos(4t - 53^\circ)\right] u(t)$ V

4.23(b) $\mathbf{H}(\mathbf{s}) = 10/(\mathbf{s}+10)$

4.24(c) $h(t) = 3\sin(3t)\,u(t)$

4.28(b) $\mathbf{H}(\mathbf{s}) = 2/(\mathbf{s}^2 + 7\mathbf{s} + 6)$

4.32 $h(t) = te^{-10^4 t}\,u(t)$

4.35 See diagram on (S²).

4.42(a) $K = 10$

4.52 $T_\infty = 1785.3^\circ$C

4.57 $K_1 = 300;\ K_2 = 19$

Chapter 5

5.1(a) $y(t) = 2\cos(400t - 83.13^\circ)$

5.5
$y(t) = \cos(1000t - 53.13^\circ) - 1.17\cos(2000t - 69.44^\circ)$

5.9 (See S^2).

5.18 (See S^2).

5.25 (See S^2).

5.31 (See S^2).

5.37 $P_{av} = 147\ \mu W$

5.47 $\mathbf{F}(\omega) = [20\pi \cos(\omega/2)/(\pi^2 - \omega^2)$

5.50(a)
$\mathbf{F}(\omega) = 3/[(0.5 + j\omega)^2 + 36] + 1/[(0.5 + j\omega)^2 + 4]$

5.57 $y(t) = 3\sin(t) + 0.2\sin(3t)$

Chapter 6

6.1 $\omega_0 = 10^4$ rad/s

6.5(b) -23 dB

(e) 20.81 dB

6.6(d) -78.4 dB

6.7(d) 5×10^{-4}

6.11 $R = 20\ \Omega,\ L = 0.1\ \text{H},\ C = 10\ \mu\text{F}$

6.17 (See S^2).

6.24 (See S^2).

6.30(b) $h(t) \approx 0$ for small t

6.41 (See S^2).

6.44(a) $N = 3,\ \omega_c = 26\pi$ rad/s

6.51 $\mathbf{H}(\mathbf{s}) = 1/(\mathbf{s}^3 + 2\mathbf{s}^2 + 2\mathbf{s} + 1)$

6.62 14,112 samples/s

6.66 spectrum of output signal = 0

Chapter 7

7.1(c) $\{\underline{0}, 1, 1, 1, 1\}$

7.2(a) $\{\underline{1}, 1, 1, 1, 0\}$

7.4(b) $N_0 = 25$ samples

7.9(c) $\{\underline{6}, 15, 28, 29, 20\}$

7.17(e) $(\sqrt{2})^n \cos\left(\frac{\pi n}{4}\right)\ u[n]$

7.23(a) $\mathbf{H}(\mathbf{z}) = 3(\mathbf{z} + 1)/(\mathbf{z} + 2)$

(b) $\mathbf{p} = \{-2\},\ \mathbf{z} = \{-1\}$

(c) $h[n] = 3(-2)^n\ u[n] + 3(-2)^{n-1}\ u[n-1]$

(d) $y[n] + 2y[n-1] = 3x[n] + 3x[n-1]$

7.29(a) $\mathbf{H}(e^{j\Omega}) = 1 + 0.5e^{-j\Omega} + e^{-j2\Omega}$

7.36 $a = 0$ and $b = 1$

7.44 $h[n] = \{\underline{2}, 0, 2\}$

7.50(a) $\mathbf{X}_k = \{\underline{32}, 8, 0, 8\}$

Chapter 8

8.1 $y[n] - 0.98y[n-2] = x[n] + x[n-2]$

8.6
$y[n] - 0.98y[n-2] + 0.96y[n-4] = x[n] - x[n-2] + x[n-4]$

8.14 $x[n] = y[n] - 0.8y[n-3]$

8.19 $\{\underline{3}, 1, 4, 1, 5, 9\}$

8.26(a) 0.25 s

(b) 440 Hz

(c) Note A

Chapter 9

9.2(a) 20

(b) 13

9.4(a) $L = 10,\ N = 256$, rectangular

9.16(a) $\{0, -\frac{2}{\pi}, \underline{0}, \frac{2}{\pi}, 0\}$

(b) $\{0, -\frac{1}{\sqrt{3}}, \underline{0}, \frac{1}{\sqrt{3}}, 0\}$

9.25(a) $h[n] = e^{-n}\ u[n]$

(b) $\mathbf{H}(\mathbf{z}) = \dfrac{\mathbf{z} + 1}{3\mathbf{z} - 1}$

9.28
$y[n] + 2y[n-1] + y[n-2] = x[n] - 2x[n-1] + x[n-2]$

9.41 None

Chapter 10

10.1 $\mathbf{H}(e^{j\Omega_1}, e^{j\Omega_2}) = \cos(\Omega_1/2)\cos(\Omega_2/2)\ e^{-j(\Omega_1 + \Omega_2)/2}$

Index

A

Absolutely integrable systems, 59–60
Acoustic pressure waveform, 3–4
Active filters, 263, 272–275
Additive zero-mean white Gaussian noise, 298–299
Additivity property, 32, 41
Affine system, 35
Aliasing, 325–328
AM (amplitude modulation), 312–313, 316–317
Amplifier gain-bandwidth product, 171–174
Amplitude modulation (AM), 312–313, 316–317
Amplitude/phase representation, Fourier series, 197, 201–204
Amplitude spectrum, 202–203, 206–207
Amplitude transformation, discrete-time signals, 351
Analog signals, 4–6
Angular frequency, 195, 197, 355, 384–387, 389–390
  discrete-time Fourier series (DTFS), 389–390
  discrete-time signals, 338, 384–387
  fundamental, 195, 197, 355, 390
Antialiasing filter, 328–330
Anticausal systems, 4
Anticipatory systems, 58
Aperiodic waveforms, 10. *See also* Nonperiodic waveforms
Approximation formulas for small quantities, 627
Area property, LTI convolution, 54–55
ARMA, 358, 440–441
Associative property, LTI convolution, 50–51
Attenuation coefficient, 67
Audio signal processing, 525–526
Autocorrelation, 527–528, 534–536
Autoregression and moving average (ARMA), 358, 440–441

B

Bandlimited signals, 458, 462–463
Bandpass filters, 256–258, 261–267, 270–271, 277, 298, 305–306, 425
  brick-wall, 277
  Butterworth, 298
  discrete-time, 425
  frequency responses, 256–258
  order of, 270–271
  passive filter configuration, 263–267, 270–271
  RL circuit, 261–263
  transfer functions of, 263–267
Bandpass signals, 265, 305–306, 330–333
  sampling, 330–333
  bandwidth, 265, 305–306
Bandreject filters, 256–258, 269–270, 277–278, 425–426
  brick-wall filter, 277–278
  discrete-time, 425–426
  frequency responses, 256–258
  passive filter configuration, 269–270
  transfer functions of, 269–270

Bandwidth, 171–174, 258, 265, 304–306
  amplifier gain-bandwidth product, 171–174
  bandlimited spectra, 304–307
  filter signals, 258, 265, 305–306
  modulation, 304–306
  open-loop mode, 171–172
  rejection, 258
Bartlett triangular window, 479–480
Basis functions, wavelet transforms, 569, 579, 583–584
Basis pursuit solution method, 601–602
Batch filtering, 437
Batch signal processing, 420–421, 451
BIBO, 58–61, 67, 111–112, 165, 175–176, 361–362, 381–384, 440, 449–450, 507
Bilateral Laplace transforms, 86, 113–114
Bilateral **z**-transforms, 445–450
  deconvolution of non-minimum-phase systems, 449–450
  geometric signals, 445–446
  inverse, 447–450
  region of convergence (ROC), 445–446, 449
Bilinear transformation, IIR filters, 504, 506–507
Biomechanical models, 146–149
Biomedical imaging, 534–537
Blackman window, 480–481
Bode plots, 261–263
Bounded signals, 58
Bounded-input/bounded-output (BIBO) stability, 58–61, 67, 111–112, 165, 175–176, 361–362, 381–384, 440, 449–450, 507
Brick-wall (ideal) filters, 258, 275–278
  bandpass filters, 277
  bandreject filters, 277–278
  conjugate symmetry, 276
  cutoff frequency, 258, 276
  frequency responses of, 275–278
  highpass filters, 277
  lowpass filters, 258, 276–277
  modulation property, 277
Butterworth filters, 289–298
  bandpass, 298
  design of, 294–298
  frequency response, 293–294
  highpass, 297–298
  lowpass, 289–296
  normalized transfer functions, 295–296
  pole placement, 292–294
  third-order, 296, 298

C

Canny edge detector, 563–564
Capacitors in **s**-domain, 132–133
Car suspension system, 72–77
Carrier signal, 308
Cascade connections, 50, 66, 154–155, 273–274, 434–435
  active filters, 273–274
  comb filters, 434–435
  LTI systems, 50, 66
  transfer functions, 154–155
CAT, 611–613
Causal systems, 4, 42, 52, 58, 60, 235, 361–362, 381–382, 440
Central nervous system (CNS), 2–3
Characteristic equations, 65
Chirp signals, 489–492
Circle of fifths, 513, 526
Circuit analysis, 36–40, 42–46, 56–57, 62, 98–99, 119–120, 131–191, 213–216, 236–237, 258–271
Circuit element models in **s**-domain, 132–134
Closed-loop systems, 163–166, 172–174, 176–177
Comb filters, 254, 285–287, 291, 301–303, 434–439
Combined signal transformations, 7–8
Commutative property, LTI convolution, 42, 50
Completeness property, 204
Complex frequency domain, **s**, 86. *See also* **s**-domain
Complex numbers, 621–624
Complex poles, 104–106
Compressed sensing, 546, 570, 599–604, 606–614
Compression, wavelet transforms, 570
Computed axial tomography (CAT), 611–613
Conjugate symmetry, 62–63, 204, 233, 276, 549
Continuous-time Fourier series (CTFS), 389
Continuous-time Fourier transforms (CTFT), 394–395
Continuous-time spectrograms, 486
Control systems, 160, 162–172, 178–182. *See also* Feedback
Convergence, 87, 234, 222–223
Convolution, 31, 40–58, 363–366, 372–374, 400, 404–406, 548–549, 554, 571–572
Cooley–Tukey FFTs, 411–414
Corner frequency, 256–258, 269
Correlation, 526–533
Cosine waveform, Fourier transforms, 233
Coupled first-order equations, 65–66

Course and Vernier indices, 411
Critically damped LTI system response, 67, 69–70
Cross-correlation, 528–530
Current, electrical systems, 142
Current constraints, op amps, 150
Cutoff frequency, 258, 276
Cyclic convolution, 404–405, 554, 571–572

D

Damped natural frequency, 68–69
Damper force, 72–73, 142
Damping coefficient, 67
Data windows, 475–485
  Bartlett triangular window, 479–480
  Blackman window, 480–481
  Hamming window, 480–481
  Hanning window, 480
  rectangular window, 478–479
Daubechies scaling and wavelet functions, 584–590, 593–594
Decaying exponentials, BIBO stability with, 60–61
Decimation, 351, 512–516, 518–519
Deconvolution, 439–443, 449–452, 565–569, 609–611
  discrete-time systems, 439–443, 449–452
  image processing, 565–569, 609–611
Demodulation, 310
Denoising, 298–301, 555, 557–560, 595–599
  additive zero-mean white Gaussian, 298–299
  images, 555, 557–560, 595–599
  instrument (trumpet) signals, 298–301
  lowpass filtering, 300–301, 597–560
  shrinking, 596–597
  signal-to-noise ratio (SNR), 299–300, 557–559
  thresholding, 595–597
Dereverberation, 443–444
Derivative property, **z**-transforms, 372
DFT, 403–404
Difference equations, discrete-time LTI systems, 357–359, 378–381
Differentiation, 55–58, 193–194
Differentiator, op amps, 151
Digital signal processing (DSP), 5–6, 347–348
Digital signals, analog signals compared with, 4–6
Dirac (delta) function, $\delta(t)$, 16–19. *See also* Impulse functions
Direct form I (DFI) and II (DFII) realization, 157–161
Dirichlet conditions, 197, 234
Discrete Fourier transforms (DFT), 400–410, 450–468, 553–554, 570
Discrete-space Fourier transform (DSFT), 549–553
  conjugate symmetry of, 549
  two-dimensional spectrum, 550–553
  wavenumber response, 550
Discrete-space signals, 4
Discrete-time filters, 421–439
  bandpass, 425
  bandreject, 425–426
  cascaded, 434–435
  comb, 434–439
  highpass, 424–425
  lowpass, 422–424
  notch, 427–433
  poles and zeros, 421–422
  transfer functions, 421–422
Discrete-time Fourier series (DTFS), 389–394, 458–460
  fundamental period, 390
  orthogonality property, 390
  Parseval's theorem, 392–393
  periodic signals, 389–390, 458–460
  spectral symmetry, 391–392
  spectrum computations, 458–460
Discrete-time Fourier transforms (DTFT), 394–400, 403–404
  DFT numerical computation, 403–404
  discrete-time convolution, 400
  pairs, 394–398
Discrete-time LTI systems, 356–362, 378–384
Discrete-time signals, 3–4, 5–6, 346–419, 420–473
Discrete-time spectrograms, 486
Distinct complex poles, 104–105
Distinct poles, 101–102
Distributive property, LTI convolution, 51–52
Domain transformation, 86, 117–122
Double-sideband (DSB) modulation, 303, 308–313, 316–317
Downsampling (decimation), 351, 512–516, 518–519, 554–555
DSFT, *see* Discrete-space Fourier transform
DTFS, *see* Discrete-time Fourier series
DTFT, *see* Discrete-time Fourier transform
Duality of frequency and time domains, 225–226
Dynamic (physical) systems, 35–36

E

Edge detection, images, 559–564
Electrocardiograms (EKG), 534
Electromechanical analogues, 132, 140–149
Energy, signal power, 21–23
Energy spectral density, Fourier transforms, 231–232
Enthalpy, 167
Envelope detection, 313
Equal temperament system, 525–526
Error signal, 163
Everlasting signals, 86, 113, 235
Expansion coefficients (residues), 101, 104, 390
Exponential waveforms, 19–21

F

Fast Fourier transform (FFT), 400–401, 407–413, 451
FDM, 230, 304, 315–316
Feedback, 154–155, 160, 162–171, 177–182
- BIBO stability, 165
- closed-loop transfer function, 163–166
- inverse system construction, 164
- inverted pendulum control, 178–182
- multiple system configurations, 154–155
- negative, 154–155, 160, 162–171
- positive, 160
- proportional, 164–166, 179–180
- proportional-plus-derivative (PD), 166, 177, 180–181
- proportional-plus-integral (PI), 178, 181
- system stabilization, 164–166, 179–182
- temperature control using, 162–163, 167–171

Filter banks, wavelet transforms, 570, 574–576, 590–592
Filter design, 278–281, 285–286, 294–298, 427–430, 492–511
- Butterworth filters, 294–298
- comb filters, 285–286
- finite impulse response (FIR) filters, 492–502
- infinite impulse response (IIR) filters, 492, 503–511
- notch filters, 427–430
- poles and zeros, 278–281
- single-pole/single-zero transfer functions, 278–280

Filters, 254–287, 289–298, 300–303, 421–439, 567–568
- active, 263, 272–275
- bandpass, 256–258, 261–267, 270–271, 277, 298, 425
- bandreject, 256–258, 269–270, 277–278, 425–426
- Bode plots, 261–263
- brick-wall (ideal), 258, 275–278
- Butterworth, 289–298
- cascaded, 273–274, 434–435
- comb, 254, 285–286, 434–439
- discrete-time, 421–439
- frequency rejection, 281–287
- highpass, 255–256, 256–258, 260–263, 267–268, 273, 277, 297–298, 424–425
- image deconvolution, 567–568
- line enhancement, 302
- lowpass, 254–255, 256–258, 259–260, 269–277, 289–297, 300–301, 305–306, 328–329, 422–424
- median filtering, 567–568
- notch, 254, 281–285, 290, 427–433
- resonator, 254, 300–303
- stopband, 258
- Wiener, 567–568

Final-value theorem, 94–95, 374
Finite duration signals, 368
Finite impulse response (FIR) filters, 492–502, 503
FIR filters, 492–502
Folding frequency, 328
Forced response, 119
Forces systems, 72–73, 140–149
Fourier analysis, 192–252, 253–345
Fourier integrals, 198–199, 209, 216–217, 222–223
Fourier series, 195–218, 235–236. *See also* Discrete-time Fourier series (DTFS)
Fourier transforms, 218–238, 253–345. *See also* Discrete Fourier transforms (DFT); Discrete-time Fourier transforms (DTFT)
Frequency differentiation property, 95
Frequency division multiplexing (FDM), 230, 304, 315–316
Frequency filters, 254. *See also* Filters
Frequency integration property, 95
Frequency rejection filters, 281–287
Frequency response, 62–64, 116, 235–236, 256–260, 263, 275–281, 293–294, 384–389, 493–494
- brick-wall filters, 275–278
- Butterworth filters, 293–294
- conjugate symmetry of, 62–63
- filter design, 278–281

Frequency sampling, 493, 496–497
Frequency-shift property, 92
Frequency warping, 507–508
Fundamental period, 10, 353–356, 390
- discrete-time Fourier series (DTFS), 390
- discrete-time signals, 353–356
- waveforms, 10

G

Gain-bandwidth product (GBP), 172–174
Gaussian model for impulse functions, 18
Gaussian probability distribution, 299
Geometric signals, 352–353, 362
Gibbs phenomenon, 212, 234–235
Graphical convolution technique, 46–50, 365–366

H

Haar wavelet transform, 569, 572–577, 579–581, 592–593
Hamming window, 480–481
Hanning window, 480
Heart rate measurement, 534–536
Heat capacity, 167–168
Heat transfer model, 167–168, 170
Highpass filters, 255–256, 256–258, 260–263, 267–268, 273–274, 277, 297–298, 424–425
Hilbert transform, 318–319

I

IIR filters, 492, 503–511
Image inpainting, 606–608
Image processing, 238–243, 254–256, 545–617
  compressed sensing, 546, 570, 599–601, 606–603
  convolution, 548–549
  deconvolution, 565–569, 609–611
  denoising, 555, 557–559, 595–599
  discrete Fourier transform (DFT), 553–554
  discrete-space Fourier transform (DSFT), 549–553
  downsampling and upsampling, 554–557
  edge detection, 559–564
  filtering, 254–256, 567–568, 570–571
  Fourier transforms, 238–243, 254–256, 569–570
  highpass filtering, 255–256
  LandWeber algorithm, 604–605
  lowpass filtering, 254–255
  MATLAB software, 241–243
  point-spread function (PSF), 546, 548, 562
  sampling, 547
  spatial Fourier transforms, 238–239
  thumbnails, 555–556
  wavelet transforms, 546, 569–595
Image reconstruction, 241–243, 547, 573–576, 581–584, 579, 581–584, 606–607, 613
  DTF value subsets, 606
  ISTA, 604–607, 613
  least-squares, 613
  MATLAB software, 241–243
  multistage, 574–576
  perfect, 581–584
  single-stage, 573–574
  wavelet transforms, 579, 581–584
Impedances in **s**-domain, 133
Improper rational function, 100, 110–111
Impulse functions, 16–19, 54, 107–108, 224–226, 352
Impulse invariance, IIR filters, 504–506
Impulse response, 31, 35–40, 54, 65–72, 107, 114, 286–287, 360–362
  cascaded LTI systems, 66
  comb filters, 286–287
  convolution with, 54
  critically damped, 67, 69–70
  discrete-time LTI systems, 360–362
  linear, constant-coefficient differential equations (LCCDE), 65–72
  linear time-invariant (LTI) systems, 31, 35–40, 54, 65–72, 114
  overdamped, 67–68
  static (memoryless) systems, 35–36
  underdamped, 67, 68–69
In-parallel LTI connections, 51–52
In-series LTI connections, 50–51, 66
Inductors in **s**-domain, 132
Infinite impulse response (IIR) filters, 492, 503–511
  BIBO stability of, 507
  bilinear transformation, 504, 506–507
  FIR filters compared with, 503, 510–511
  frequency warping, 507–508
  impulse invariance, 504–506
  lowpass, 509–511
  notch filter, 503–504
Infinite input resistance, op amps, 150
Initial-value theorem, 94–95, 374
Input derivatives, LCCDE with and without, 65–66
Input signals (excitations), 2
Instantaneous power, 21–22
Integrator circuit, op amps, 151–152
Interference, 254, 281–287, 290, 454–456, 477
Interpolation, 517–519, 523–525
Interrupted voltage source analysis, 134–136
Inverse systems, 86–88, 112–113, 164, 374–377, 440–443, 447–450

Inverted pendulum control, 178–182
Invertible systems, 111–113
Inverting amplifier, 150, 272–273
IRLS, 602–604
ISTA, 604–607, 613
Iterative reweighted least squares (IRLS) algorithm, 602–604
Iterative shrinking and thresholding algorithm (ISTA), 604–605, 606–607, 613

K

Kirchhoff's voltage law (KVL), 98

L

LabVIEW modules, 636
LandWeber algorithm, 604–605
Laplace transform pairs, 86, 88, 97, 106, 114
Laplace transform, 85–130, 131–191, 235–236
LASSO, 597–599, 602
Least absolute shrinkage and selection operator (LASSO), 597–599, 602
Line enhancement filter, 302
Line spectrum, 202–203, 206–207, 219–222, 238–241, 287–289, 309, 315–316, 323, 334, 391–392, 457–468, 500–553
Linear, constant coefficient, differential equations (LCCDE), 32–33, 63, 65–72, 86, 115, 117–122
Linear differential equations (LDE), 32
Linear-shift invariant (LSI) systems, 548
Linear time-invariant (LTI) systems, 30–84, 114–122, 356–362
Linearity property, 32–33, 91, 223, 359–360, 370
Linearization of systems, 178–179
Logarithmic spectrum, 550–551
Lowpass filters, 139–140, 254–255, 256–258, 259–260, 267–277, 289–297, 300–301, 305–306, 328–329, 422–424, 509–511, 597–560

M

M-periodic waveform, 209–212
Magnetic resonance imaging (MRI), 599
Magnitude (gain) frequency response, $M(\omega)$, 256–258, 278–281, 287–289
Mainlobe width, 477
Marginal stability, LTI systems, 60–61
Mathematical formulas, 625–627
Mathematical symbols, 619–620
MathScript installation, 628–629
MATLAB, 241–243, 628–635
    basic computations, 629–630
    discrete-time commands, 635
    frequency response, 634
    image reconstruction, 241–243
    partial fractions, 633–634
    plotting, 630–633
Mean-square convergence, 234
Mechanical systems, 141–143. *See also* Spring-mass-damper (SMD) systems
Median filtering, 567–568
Memory, LTI systems, 360
Minimax criterion, 493, 499–500
Minimum phase system, 112
Modulation, 254, 277, 303–319
    amplitude (AM), 312–313, 316–317
    bandwidth, 304–306
    Brick wall filter property, 277
    double-sideband (DSB), 303, 308–313, 316–317
    frequency division multiplexing (FDM), 304, 315–316
    frequency translation (mixing), 313–315
    index, 312
    multiplication of signals, 306–307
    signal fading, 311
    single-sideband (SSB), 318–319
    switching, 307–308
    tone, 313–314
Moment of inertia, motor shafts, 175
Motor systems, 174–177
Multiple system configuration, 154–156
Multirate signal processing, 512–526
Music applications, 287–291, 298–300, 430–432, 436–439, 454–456, 525–526. *See also* Noise
    audio signal processing, 525–526
    denoising signals, 298–300
    Fourier transform applications, 287–289, 298–300
    instrument (trumpet) filtering, 290–291, 302–303, 430–432, 436–439, 454–456
    separating simultaneously played notes, 288–289
    spectra of notes, 287–289

N

Natural response, 119
Natural/forced partition, 119
Negative feedback, *see* Feedback

Noise, 5, 110–111, 254, 290–291, 298–319, 430–432, 434–439, 443–444, 452–458, 526–533, 555, 557–560, 595–599
additive zero-mean white Gaussian, 298–299
autocorrelation, 527–528, 534–536
bandwidth, 304–306
comb filters, 291, 302–303, 434–439
correlation, 526–533
cross-correlation, 528–530
denoising, 298–301, 555, 557–560, 595–599
dereverberation, 443–444
discrete Fourier transforms (DFT), 452–458
discrete-time signals, 5, 430–432, 434–439, 443–444
Fourier transforms, 254, 298–319
images, 555, 557–560, 595–599
least absolute shrinkage and selection operator (LASSO), 597–599
lowpass filtering, 300–301, 597–560
modulation, 254, 303–319
notch filters, 290, 430–432
periodic interference removal, 454–456
resonator filters, 254, 300–303
reverberation, 443
shrinking, 596–597
signal classification, 531–533
signal filtering, 290–291, 298–303, 430–432, 434–439
signal-to-noise ratio (SNR), 110–111, 299–300, 557–559
spectrum computations, 457–458
thresholding, 452–454, 456, 595–597
trumpet signals, 298–301
Noncausal systems, 4, 58, 235
Non-minimum-phase systems, 449–452
Nonperiodic (signal) waveforms, 10–21, 219–221, 401, 462–468
Notch filters, 157, 254, 281–285, 290, 427–433, 503–504
Nyquist sampling criterion, 321, 323, 512

O

Octave, 525
OLHP, 202, 206–207
One-pole transfer functions, 152–154, 158–160
One-sided spectrum, 202, 206–207
Open left-hand plane (OLHP), 109
Open-loop systems, 163–165, 171–172, 175
Operational amplifiers (op amps), 132, 149–154, 158–160, 171–174, 296–297
Orthogonality property, 204, 390
Output-voltage saturation constraints, op amps, 150
Overdamped LTI system response, 67–68
Oversampling, 328–329, 512, 520–525
Overtones, 197, 288

P

Parallel realization, 160, 162
Parks-McClellan algorithm, 499–502
Parseval's theorem, 216–218, 230–232, 392–393
Partial fraction expansion, 99–106, 376–377
Partitions, *see* System response partitions
Passband filters, 258
Passive filters, 263–271
Pavement models, 73–77
Pendulums, *see* Inverted pendulum control
Periodic (signal) waveforms, 10–11, 15–16, 22, 199–212, 216–218, 232–233, 254, 285–287, 389–390, 401, 435–436, 457–462
Periodic interference, comb filters, 254, 285–287
Periodicity property, 10, 193
Phase/amplitude representation, 197, 201–204
Phase spectrum, 202–203, 206–207, 238–241
Phasor domain technique, 193–195
Physically realizable systems, 4
Pixels, 546
Point-spread function (PSF), 546, 548, 562
Pointwise convergence, 234
Pole factor, 101
Poles, 89–90, 100–106, 108–111, 114, 278–281, 282–283, 290–294, 421–426
Butterworth filters, 290–294
complex, 104–106
discrete-time filters, 421–426
distinct, 101–102, 104–105
filter design using, 278–281
notch filters, 282–283
parallel zeros, 282–283
placement of, 292–294
repeated, 102–106
s-plane positions, 89–90
single-pole transfer function, 279–280
strictly proper rational function, 100, 108–109
system stability, 108–111
zeros, 89–90, 100, 282–283, 421–426

Positive feedback, 160
Power, signals, 21–23
Proportional feedback, 164–166, 179–181
plus-derivative (PD), 166, 177, 180–181
plus-integral (PI), 178, 181
Pseudo-inverse solutions, 602
PSF, 546–548, 562
Pulse trains, 219–221, 321–323

Q

Quadrature-mirror filters (QMF), 570, 583
Quality factor, 265

R

Radix-2 Cooley–Tukey FFTs, 412–413
Ramp-function waveforms, 12–13
Ramp response, 39–40
Rational functions, 63, 89–90, 100–101, 108–111
RC circuits, 36–40, 42–46, 56–57, 62, 119–120, 213–216, 236–237, 259–261
Reaction time, 37–38
Rectangular function, 14–15
Rectangular pulses, 44–46, 53–54, 397–399
Rectangular window, 478–479
Recursive equations, 357
Region of convergence (ROC), 87, 367, 445–446, 449
Rejection bandwidth, 258
Repeated poles, 102–106
Residue method, 101
Resistors in **s**-domain, 132
Resonant frequency, 258–259, 264
Resonator filters, 254, 300–303
Reverberation, 443
Right-shifting property, **z**-transforms, 370–371
RL circuits, signal filtering, 261–263
RLC circuits, 98–99, 139–147
biomechanical model, 146–147
electromechanical analog of, 141–145
Laplace transforms, 98–99, 139–147
lowpass filter response, 139–140
SMD-RLC analysis procedure, 144
spring-mass-damper (SMD) systems compared with, 142–144, 146–147
ROC, 87, 367, 445–446, 449
Roll-off rate, 259

S

s-domain, 86, 89–90, 106–108, 132–140
s-plane, 89. *See also* **s**-domain
Sallen-Key op-amp filter, 296–297
Sampling (sifting) property, 18, 41–42
Sampling interval, 320, 512
Sampling rate, 320, 512
Sampling signals, 319–334, 458–460, 463–465, 547
aliasing, 325–328
antialiasing filter, 328–330
discrete Fourier transforms (DFT), 458–459, 463–465
discrete-time Fourier series (DTFS), 458–460
discretization length, 547
images, 547
Nyquist sampling rate, 321, 323
oversampling, 328–329
reconstruction, 325
sampling theorem, 319–320
Shannon's sampling theorem, 323–324
sinc interpolation formula, 324–325
undersampling, 325–327
Sawtooth waveforms, 15–16, 200–201
Scaling (homogeneity) property, 6–7, 18–19, 31–32, 41, 223–224, 372
Schmitt trigger, 160
Second-derivative property, 92–93
Semitone, 525
Shannon's sampling theorem, 323–324
Shep-Logan phantom, 592–593
Shrinking, image denoising by, 596–597
Sidelobe attenuation, 477–478
Signal flow (butterfly) graph, 408
Signal-to-noise ratio (SNR), 110–111, 299–300, 557–559
Signal transformations, 6–9, 192–193. *See also* Scaling property
Signum function, 226–227, 318–319
Sinc function, 219–221, 475
Sinc interpolation formula, 324–325
Sine/cosine representation, Fourier series, 197, 198–201
Single-pole, single-throw (SPST) switch, 98
Single sideband (SSB) modulation, 318–319
Singularity functions, 11–19, 87–88
Sinusoidal interference, notch filters, 282–283

Sinusoidal response, 61–65, 113–114, 193–195, 216–217, 327, 353–356, 369, 385–387
aliasing, 327
angular frequency, 385–387
bilateral Laplace transform, 113–114
differentiation of, 193–194
discrete-time signals, 353–356, 385–387
Fourier analysis, 193–195, 216–217
Fourier integrals, 216–217
frequency response function, 62–64
linear time-invariant (LTI) systems, 61–65
linear, constant-coefficient differential equations (LCCDE), 63
phasor domain technique, 193–195
time-varying function, 193–195
z-transforms, 369
SMD (springs, masses, dampers) system, 141–143
Smith-Barnwell condition, 570, 582
SNR, 110–111, 299–300, 557–559
Sobel edge detector, 562–563
Sparsity, image processing, 574, 584, 597, 601
Spatial Fourier transforms, 238–239
Spectral leakage, 460–462, 475. *See also* Data windows
Spectrograms, 485–492
chirp signals, 489–492
continuous-time, 486
discrete-time, 486
MATLAB/MathScript recipe, 486–487
signal varying-type spectra, 485–486
time and frequency resolution trade-off, 487–489
Spring constant (stiffness), 72
Spring force, 72–73, 142
Spring-mass-damper (SMD) systems, 72–73, 140–149
Square waveforms, Fourier series analysis, 199–200, 205
SSB, 318–319
Static (memoryless) systems, 35–36
Stator, 174
Steady-state response, 119
Stem plots, discrete-time signals, 349
Step functions, 11–12, 55, 226–227, 352
Step response, 35–40, 169–171, 174–177
Stiffness (spring constant), 72
Stopband filters, 258
Strictly proper rational function, 100, 108–109
Subband coding, wavelet transforms, 570–571
Summing amplifier, op amps, 152
Superposition, principle of, 32, 33–34, 196–197, 359
discrete-time LTI systems, 359
Fourier series analysis, 196–197
LCCDE application of, 33
LTI systems, 32, 33–34
Switching function, 514
Switching modulation, 307–308
System realization, *see* System synthesis
System response partitions, 86, 117–122, 379
System synthesis, 157–162
direct form I (DFI) realization, 157–160
direct form II (DFII) realization, 158–161
parallel realization, 160, 162
system realization, 157–161
transfer function, **H(s)**, 157–161

T

Temperature control, 162–163, 167–172
Thresholding, 452–454, 456, 595–597, 593–594
images, 595–597, 593–594
noise, 452–454, 456
Thumbnails, 555–556
Tikhonov regularization, 566–567
Time constant, 19–20, 36–37
Time derivative property, 229
Time modulation property, 229–230
Time-differentiation property, 92–93
Time-integration property, 93–94
Time-invariance property, 41, 360
Time-invariant systems, 34–35. *See also* Linear time-invariant (LTI) systems
Time-reversal transformation, 7, 351
Time-reversed step function, 12
Time-scaling transformation, 6–7, 18–19, 91, 351
Time-shift property, 6, 8, 52–54, 91–92, 350–351
Timelimited-bandlimited signals, 462–463
Tone modulation, 313–314
Total energy, 21–22
Transfer functions, 106–108, 114–116, 152–162, 163–166, 175–176, 235, 263–271, 278–281, 295–296, 380–384, 421–426
Transient response, 119
Transient signal component, 235
Transient/steady-state partition, 119–121
Trapezoidal pulse, 14–15
Triangle inequality, BIBO stability, 59–60
Triangle model for impulse functions, 17–18

Triangular pulses, LTI convolution of, 42–44, 56–57
Trigonometric relations, 625
Twiddle factors, 410
Two-dimensional spatial Fourier transforms, 238–239. *See also* Images
Two-sided spectrum, 202, 206–207

U

Ultrasound time delay, 536–537
Undamped natural frequency, 67
Underdamped LTI system response, 67, 68–69
Undersampling, 325–327
Unilateral Laplace transform, 86–89, 113. *See also* Laplace transform
Uniqueness property, 86
Unit impulse functions, 16–19, 107–108
Unit ramp functions, 12–13
Unit step functions, 11–12, 226–227
Unit time-shifted ramp function, 12–13
Unit time-shifted step function, 12
Units, abbreviations, 619–620
Universal property, 90–91
Upsampling (zero-padding), 351, 512, 516–519, 554–557

V

Valid deconvolution, 606, 609–611
Vertical velocity, electromechanical analogues, 140–141
Voltage restraints, op amps, 150

W

Wavelet transforms, 569–595
  basis functions, 569, 579, 583–584
  cyclic convolution, 571–572
  Daubechies scaling and wavelet functions, 584–590, 593–594
  DFT compared with, 570
  filter banks, 570, 574–576, 590–592
  frequency domain, 576
  Haar transform, 569, 572–577, 579–581, 592–593
  normalized functions, 576–577
  perfect reconstruction, 581–584
  recursion steps, 578
  subband coding, 570–571
  thresholding, 593–594
  two-dimensional transforms, 590–594
Wavenumber response, DSFT, 550
Weighted error, 499
Weighting coefficients, 407
White light, 555, 557
White noise, 298–299, 555
Width property, LTI convolution, 54
Wiener filter, 567–568
Windowing, 401–402, 493, 494–496
Windows, *see* Data windows

Z

z-domain transfer functions, 380–384
z-transforms, 366–379, 445–450
  bilateral, 445–450
  convolution property, 372–374
  discrete-time signals, 366–379
  finite duration signals, 368
  initial/final value theorems, 374
  inverse, 374–377, 447–450
  region of convergence (ROC), 367, 445–446
  right-shifting property, 370–371
  time delays, 370–372
  z-derivative property, 372
  z-scaling property, 372
Zero initial conditions, 42, 380
Zero input-current/input-voltage constraints, op amps, 150
Zero-input response (ZIR), 117–119, 379
Zero-mean noise, 298–299
Zero output resistance, op amps, 150
Zero-state response (ZSR), 117–119, 379
Zero-stuffing, 351, 512, 516–517
Zeros, 89–90, 100, 114, 278–281, 282–283, 421–426

Mark J.T. Smith was a professor at Georgia Tech when he and Thomas Barnwell derived the Smith-Barnwell condition for perfect reconstruction. Before that, he was a U.S. fencing champion in 1981 and 1983, and he competed for the U.S. in fencing in the 1984 Olympics in Los Angeles. At this writing, he is Dean of the Graduate School at Purdue University.

Baroness (in Belgium) Ingrid Daubechies contributed so much to the development of wavelets that she is informally known as ``mother wavelet.'' A MacArthur Fellow in 1992, she was the first woman president of the International Mathematics Union in 2011, and the first woman full professor of mathematics at Princeton University.

Jean-Baptiste Joseph Fourier made many contributions beyond Fourier series and Fourier transforms. He discovered the greenhouse effect, dimensional analysis, and Fourier's law of heat conduction. In politics, he was a scientific advisor to, and friend of, Napoleon Bonaparte, who appointed him to be a prefect in Grenoble, France.

CPSIA information can be obtained
at www.ICGtesting.com
Printed in the USA
BVHW021444031222
653270BV00001B/2